Common Formulas

Distance

$$d = rt$$

d = distance traveled
t = time
r = rate

Temperature

$$F = \frac{9}{5}C + 32$$

F = degrees Fahrenheit
C = degrees Celsius

Simple Interest

$$I = Prt$$

I = interest
P = principal
r = annual interest rate
t = time in years

Compound Interest

$$A = P\left(1 + \frac{r}{n}\right)^{nt}$$

A = balance
P = principal
r = annual interest rate
n = compoundings per year
t = time in years

Coordinate Plane: Midpoint Formula

Midpoint of line segment joining (x_1, y_1) and (x_2, y_2)

$$\left(\frac{x_1 + x_2}{2}, \frac{y_1 + y_2}{2}\right)$$

Coordinate Plane: Distance Formula

d = distance between points (x_1, y_1) and (x_2, y_2)

$$d = \sqrt{(x_2 - x_1)^2 + (y_2 - y_1)^2}$$

Quadratic Formula

Solutions of $ax^2 + bx + c = 0$

$$x = \frac{-b \pm \sqrt{b^2 - 4ac}}{2a}$$

Rules of Exponents

(Assume $a \neq 0$ and $b \neq 0$.)

$$a^0 = 1 \qquad a^m \cdot a^n = a^{m+n}$$

$$(ab)^m = a^m \cdot b^m \qquad (a^m)^n = a^{mn}$$

$$\frac{a^m}{a^n} = a^{m-n} \qquad \left(\frac{a}{b}\right)^m = \frac{a^m}{b^m}$$

$$a^{-n} = \frac{1}{a^n} \qquad \left(\frac{a}{b}\right)^{-n} = \frac{b^n}{a^n}$$

Basic Rules of Algebra

Commutative Property of Addition

$$a + b = b + a$$

Commutative Property of Multiplication

$$ab = ba$$

Associative Property of Addition

$$(a + b) + c = a + (b + c)$$

Associative Property of Multiplication

$$(ab)c = a(bc)$$

Left Distributive Property

$$a(b + c) = ab + ac$$

Right Distributive Property

$$(a + b)c = ac + bc$$

Additive Identity Property

$$a + 0 = 0 + a = a$$

Multiplicative Identity Property

$$a \cdot 1 = 1 \cdot a = a$$

Additive Inverse Property

$$a + (-a) = 0$$

Multiplicative Inverse Property

$$a \cdot \frac{1}{a} = 1, \quad a \neq 0$$

Properties of Equality

Addition Property of Equality

If $a = b$, then $a + c = b + c$.

Multiplication Property of Equality

If $a = b$, then $ac = bc$.

Cancellation Property of Addition

If $a + c = b + c$, then $a = b$.

Cancellation Property of Multiplication

If $ac = bc$, and $c \neq 0$, then $a = b$.

Zero Factor Property

If $ab = 0$, then $a = 0$ or $b = 0$.

All
Math 152

i2030330

3746 c.2

Getting Started Guide
for Students

HOUGHTON MIFFLIN

New Ways to Know®

What Is **Eduspace?**

Overview

Welcome to the *Eduspace Getting Started Guide for Students.* This guide will introduce you to the Eduspace environment—its features and functionality. Each section of this guide includes a general description of the area and step-by-step instructions on how to perform particular tasks.

If you have an Internet connection and a web browser, you can use Eduspace.

Eduspace is powered by the Blackboard online learning system and includes all of Blackboard's powerful features for teaching and learning. You will learn how to:

* Complete homework assignments and tests online.
* Communicate with your instructor and fellow students in a number of different ways.
* View your grades in the Gradebook.
* Access live, on-line tutoring from SMARTHINKING™.

Eduspace System Requirements

Minimum Requirements

* Microsoft Windows 2000 or Windows XP
  OR
  Macintosh OS 9.2, 10.1 or 10.2

* Microsoft Internet Explorer 5.5 (or above)
  OR
  Netscape Navigator 6.1 (or higher, 7.1 or higher for Windows XP or Macintosh OS 10.2)

* Acrobat Reader 4.0

* Display with 256 colors and 800x600 resolution or higher

* 56 K modem connection

Strongly Recommended Hardware/Software

* Microsoft 2000 with Microsoft Internet Explorer 6.0 or Netscape Navigator 6.1*
  OR
  Microsoft Windows XP with Internet Explorer 6.0 or Netscape Navigator 7.1*
  OR
  Macintosh OS 10.2 with Internet Explorer 5.2 or Netscape Navigator 7.0* (or higher)

* * Users of Math and Chemistry courses are strongly advised to use Internet Explorer to ensure that symbols in exercises and test questions appear properly.

* Adobe Acrobat 6.0

* Display with thousands of colors and 1024x768 resolution or higher

* T1 line, cable modem, DSL, or company Internet LAN

Note that Houghton Mifflin recommends using the Sun Java Run-time Environment (JRE) 1.4.x in your web browser. Use of the Microsoft JVM is not recommended.

Depending on your course content, you may need one or more of the following plug-ins: Flash Player 7 or higher, QuickTime 5.0 (6 recommended), RealPlayer 9.0 (10.0 recommended), Shockwave 8.5.1

What if I have Windows 98, Windows NT or AOL?

Although Houghton Mifflin has not performed certification of Eduspace on Windows 98, Windows NT, or AOL and therefore cannot say that we officially support those platforms, Houghton Mifflin is aware of users who are successfully using Eduspace in those environments. Note that when using any browser with Eduspace, cookies must be enabled.

Contacting Customer Support

If you encounter problems with using Eduspace or need help on using certain features, you may contact Houghton Mifflin Customer Support. Support is available 24 hours a day, seven days a week.

- Call Customer Support at **(800) 732-3223.**
- Click the Support tab in Eduspace to view support information and answers to frequently asked questions.

For Information Beyond This Guide

Eduspace contains a wealth of functionality beyond what is covered in this **Getting Started Guide.** To learn more:

- Click the How Do I? button on the My Eduspace home page.
- Click the Help button at the top of any screen for comprehensive information on all Eduspace features.
- Get personalized assistance by contacting Customer Support at **(800) 732-3223.**

Getting Started

To begin using Eduspace and to see some course material, complete the following steps:

- Obtain a registration code from the inside back cover of this guide.
- Obtain a course code from your instructor.
- Register for Eduspace.
- Log in to Eduspace.
- Enroll in a course.

You must register only once, but you must log in each time you want to use Eduspace.

Registering and Enrolling in a Course

1) Go to the Eduspace website: http://www.eduspace.com and choose to register.

2) The Eduspace registration page opens. Follow the on-screen directions to fill in the required fields and then click the Create Account button.

If you fail to fill in the required fields, you will receive an error screen highlighting the empty fields and allowing you to enter the required information.

3) Before you can enroll in a course, you must log in to Eduspace. Enter your username and password in the fields provided and click Login.

4) Click on Student and click the Continue button. The Eduspace Student Course Enrollment page opens. To enroll in your course, you must enter your registration code and your course code.

The registration code can be found on the inside back cover of this guide; the course code for your course will have been given to you by your instructor.

5) Click the Enroll button. Eduspace has now recorded your registration as a student.

How to Get Access if You Have Lost Your Username and/or Password

If you should forget the username you supplied to Eduspace during the registration process, you should call Customer Support at **(800) 732-3223.**

If you have forgotten your password, click on the "Forgot your password" link on the login page.

Entering Your Course

When you have registered and enrolled in your course, you'll automatically be logged into Eduspace. To log into Eduspace in the future:

1) Go to the Eduspace website: http://www.eduspace.com, and choose to login.

2) Enter the username and password you selected when you registered for Eduspace and login. You will be brought to the My Eduspace home page.

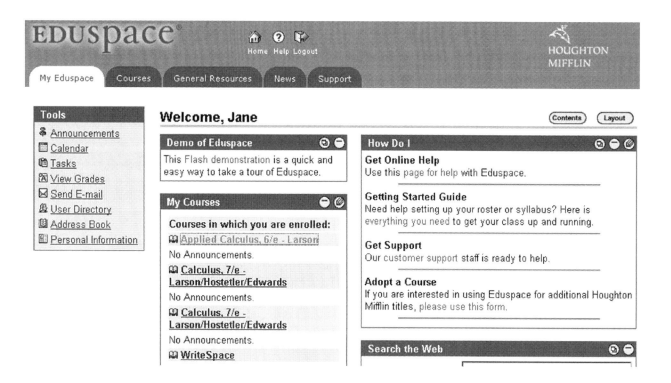

Look for the name of your course(s) in the My Courses module. Click the name of the course you wish to open. You will be brought to your course's home page.

Navigating in Eduspace

Once you have entered your course's home page, you can begin to explore the material in the Eduspace environment. This section describes how to view content and navigate through Eduspace.

Moving Around in Eduspace

Eduspace consists of five main pages that you access by clicking tabs. Pages can contain courses, modules, tools, and links to course-related material.

Tabs and Their Contents

This Tab	Contains This Information
My Eduspace	A series of modules—My Courses, My Announcements, How Do I? and Demo of Eduspace—give you access to Eduspace components. This page also displays a Tools module with links to common resources, such as your personal Tasks lists, e-mail, and Address Book. It also contains a link to where you can change your personal profile information, such as your email address or password.
Courses	This page lists all courses you are taking. Click on a course name to go a Course Home page.
General Resources	This page contains descriptions of and links to a variety of educational resource sites sponsored by Houghton Mifflin.
News	This page contains links to current news stories.
Support	This page contains links to a variety of Eduspace support resources including online help, customer support services, and an electronic version of this Getting Started Guide.

Course Home

Course Home is the first page you see when you enter your course:

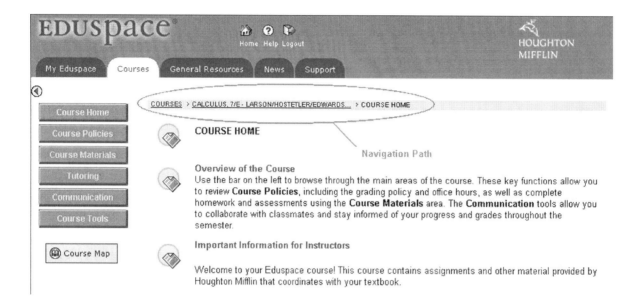

The left side of the page has buttons linking to the different areas of your course, as listed and explained in the table below:

Page Name	Description
Course Home	Clicking on this link will return you to the Course Home page.
Course Policies	This area contains basic information you need for class. Depending on what your instructor has made available, it may include: • A syllabus • Course objectives • Grading policies • Office hours • Staff information
Course Materials	The Course Materials area contains all of the content in your course, such as chapter information, homework assignments, and exercises, depending on the course. The Work Assigned by Your Instructor folder is the central place to view the assignments your instructor has given you.

Page Name	Description
Tutoring	The Tutoring area provides a direct link to SMARTHINKING™ online tutoring services.
Communication	The Communication area contains all the tools you need to use the discussion board, chat, e-mail, student and group pages, and the student roster. Your instructor may also use it to: • Contact students or organize group projects using tools such as the discussion board, virtual classroom, and e-mail. • Hold office hours online. • Post announcements and last-minute notices.
Course Tools	The Course Tools area provides features and functions that allow you to: • Manage your personal home page. • View your course calendar. • Check your grades. • View your personal task list. • Maintain a personal address book.

Clicking on any of these buttons will take you to the corresponding area of the course. Some areas may be secure or protected by a password to ensure that only registered students can enter and to keep grading information private between students and instructors.

You view course content by clicking on one of the content area buttons. These content area pages are visible to you once your instructor gives you access to your online course. The buttons appear on most pages of your course so you do not need to return to the Course Home page to move from one content area to another.

You can also navigate among different screens by clicking on the Navigation Path, indicated on the screen shot above. To return to or open a page, simply click on the page name in the Navigation Path.

Tutoring

If you are already logged into an Eduspace course that includes online tutoring through SMARTHINKING™, you can access SMARTHINKING™, an online tutoring service, from within your Eduspace course. Just click the Tutoring button and then read the on-screen instructions to link to SMARTHINKING™. You'll be taken directly to the SMARTHINKING™ system; you will not have to register or log in again.

Note: If the Tutoring button is not available in your course, either your instructor has opted not to include it or it is not available for this course.

INTERMEDIATE ALGEBRA

4

RON LARSON
Pennsylvania State University.
The Behrend College

ROBERT P. HOSTETLER
Pennsylvania State University
The Behrend College

with the assistance of
PATRICK M. KELLY
Mercyhurst College

HOUGHTON MIFFLIN COMPANY   BOSTON NEW YORK

Vice President and Publisher: Jack Shira
Associate Sponsoring Editor: Cathy Cantin
Development Manager: Maureen Ross
Associate Editor: Marika Hoe
Assistant Editor: James Cohen
Supervising Editor: Karen Carter
Senior Project Editor: Patty Bergin
Editorial Assistant: Allison Seymour
Production Technology Supervisor: Gary Crespo
Executive Marketing Manager: Michael Busnach
Senior Marketing Manager: Ben Rivera
Marketing Assistant: Lisa Lawler
Senior Manufacturing Coordinator: Priscilla Bailey
Composition and Art: Meridian Creative Group
Cover Design Manager: Diana Coe

Cover art © Dale Chihuly
Joslyn Ice, 2000
Ice blocks and halogen lights 150′ x 30′ x 14′ approx.
Joslyn Art Museum, Omaha Nebraska
Photo credit: Terry Rishel

Other information: Approximately 200 blocks were used, each weighing about
300 pounds. Each block was 44" long by 22" wide by 11" deep. The
installation was started 3 days before the exhibition's opening (Feb. 12)
and it lasted 4-5 days.

We have included examples and exercises that use real-life data as well as technology output
from a variety of software. This would not have been possible without the help of many people
and organizations. Our wholehearted thanks go to all their time and effort.

Trademark acknowledgement: TI is a registered trademark of Texas Instruments, Inc.

Custom Publishing Editor: Mary McGibbons
Custom Publishing Production Manager: Tina Kozik
Project Coordinator: Anisha Sandhu

This book contains select works from existing Houghton Mifflin Company resources and
was produced by Houghton Mifflin Custom Publishing for collegiate use. As such, those
adopting and/or contributing to this work are responsible for editorial content, accuracy,
continuity and completeness.

Printed in the United States of America.

ISBN: 0-618-66293-6
N-05118

1 2 3 4 5 6 7 8 9 – CM – 07 06 05

 **Houghton Mifflin**
Custom Publishing

222 Berkeley Street • Boston, MA 02116

Address all correspondence and order information to the above address.

Contents

A Word from the Authors (Preface)   vii
Features   ix
How to Study Algebra   xx

Motivating the Chapter   1

1 Fundamentals of Algebra   1

1.1   The Real Number System   2
1.2   Operations with Real Numbers   11
1.3   Properties of Real Numbers   23
   Mid-Chapter Quiz   31
1.4   Algebraic Expressions   32
1.5   Constructing Algebraic Expressions   41
   What Did You Learn? (Chapter Summary)   50
   Review Exercises   52   **Chapter Test**   55

Motivating the Chapter   56

2 Linear Equations and Inequalities   57

2.1   Linear Equations   58
2.2   Linear Equations and Problem Solving   69
2.3   Business and Scientific Problems   80
   Mid-Chapter Quiz   93
2.4   Linear Inequalities   94
2.5   Absolute Value Equations and Inequalities   107
   What Did You Learn? (Chapter Summary)   117
   Review Exercises   118   **Chapter Test**   123

Motivating the Chapter   124

3 Graphs and Functions   125

3.1   The Rectangular Coordinate System   126
3.2   Graphs of Equations   140
3.3   Slope and Graphs of Linear Equations   149
3.4   Equations of Lines   162
   Mid-Chapter Quiz   173
3.5   Graphs of Linear Inequalities   174
3.6   Relations and Functions   183
3.7   Graphs of Functions   197
What Did You Learn? (Chapter Summary)   208
Review Exercises   209   **Chapter Test**   215

Motivating the Chapter   216

4   Systems of Equations and Inequalities   217

4.1   Systems of Equations   218
4.2   Linear Systems in Two Variables   231
4.3   Linear Systems in Three Variables   241
       Mid-Chapter Quiz   253
4.4   Matrices and Linear Systems   254
4.5   Determinants and Linear Systems   267
4.6   Systems of Linear Inequalities   279
       What Did You Learn? (Chapter Summary)   289
       Review Exercises   290   **Chapter Test**   294
       Cumulative Test: Chapters 1–4   295

Motivating the Chapter   296

5   Polynomials and Factoring   297

5.1   Integer Exponents and Scientific Notation   298
5.2   Adding and Subtracting Polynomials   308
5.3   Multiplying Polynomials   318
       Mid-Chapter Quiz   327
5.4   Factoring by Grouping and Special Forms   328
5.5   Factoring Trinomials   338
5.6   Solving Polynomial Equations by Factoring   350
       What Did You Learn? (Chapter Summary)   360
       Review Exercises   361   **Chapter Test**   365

Motivating the Chapter   366

6   Rational Expressions, Equations, and Functions   367

6.1   Rational Expressions and Functions   368
6.2   Multiplying and Dividing Rational Expressions   380
6.3   Adding and Subtracting Rational Expressions   389
6.4   Complex Fractions   398
       Mid-Chapter Quiz   406
6.5   Dividing Polynomials and Synthetic Division   407
6.6   Solving Rational Equations   417
6.7   Applications and Variation   425
       What Did You Learn? (Chapter Summary)   438
       Review Exercises   439   **Chapter Test**   443

Motivating the Chapter   444

7 Radicals and Complex Numbers   445

7.1   **Radicals and Rational Exponents**   446

7.2   **Simplifying Radical Expressions**   457

7.3   **Adding and Subtracting Radical Expressions**   464

Mid-Chapter Quiz   470

7.4   **Multiplying and Dividing Radical Expressions**   471

7.5   **Radical Equations and Applications**   479

7.6   **Complex Numbers**   489

What Did You Learn? (Chapter Summary)   498

Review Exercises   499   **Chapter Test**   503

Cumulative Test: Chapters 5–7   504

Motivating the Chapter   506

8 Quadratic Equations, Functions, and Inequalities   507

8.1   **Solving Quadratic Equations: Factoring and Special Forms**   508

8.2   **Completing the Square**   517

8.3   **The Quadratic Formula**   525

Mid-Chapter Quiz   535

8.4   **Graphs of Quadratic Functions**   536

8.5   **Applications of Quadratic Equations**   546

8.6   **Quadratic and Rational Inequalities**   557

What Did You Learn? (Chapter Summary)   567

Review Exercises   568   **Chapter Test**   571

Motivating the Chapter   572

9 Exponential and Logarithmic Functions   573

9.1   **Exponential Functions**   574

9.2   **Composite and Inverse Functions**   587

9.3   **Logarithmic Functions**   601

Mid-Chapter Quiz   612

9.4   **Properties of Logarithms**   613

9.5   **Solving Exponential and Logarithmic Equations**   622

9.6   **Applications**   632

What Did You Learn? (Chapter Summary)   643

Review Exercises   644   **Chapter Test**   649

Motivating the Chapter   650

10 Conics   651

10.1 Circles and Parabolas   652

10.2 Ellipses   664

Mid-Chapter Quiz   674

10.3 Hyperbolas   675

10.4 Solving Nonlinear Systems of Equations   683

What Did You Learn? (Chapter Summary)   694

Review Exercises   695   **Chapter Test**   699

Cumulative Test: Chapters 8–10   700

Motivating the Chapter   702

11 Sequences, Series, and the Binomial Theorem   703

11.1 Sequences and Series   704

11.2 Arithmetic Sequences   715

Mid-Chapter Quiz   724

11.3 Geometric Sequences and Series   725

11.4 The Binomial Theorem   735

What Did You Learn? (Chapter Summary)   743

Review Exercises   744   **Chapter Test**   747

Appendices

Appendix A Introduction to Graphing Calculators   A1

Appendix B Further Concepts in Geometry   *Web

B.1   Exploring Congruence and Similarity

B.2   Angles

Appendix C Further Concepts in Statistics   *Web

Appendix D Introduction to Logic   *Web

D.1   Statements and Truth Tables

D.2   Implications, Quantifiers, and Venn Diagrams

D.3   Logical Arguments

Appendix E Counting Principles   *Web

Appendix F Probability   *Web

Answers to Odd-Numbered Exercises, Quizzes, and Tests   A9

Index of Applications   A99

Index   A103

Appendices B, C, D, E, and F are available on the textbook website. Go to math.college.hmco.com/students *and link to* **Intermediate Algebra**, *Fourth Edition.*

A Word from the Authors

Welcome to *Intermediate Algebra*, Fourth Edition. In this revision, we have continued to focus on developing students' proficiency and conceptual understanding of algebra. We hope you enjoy the Fourth Edition.

In response to suggestions from elementary and intermediate algebra instructors, we have revised and reorganized the coverage of topics for the Fourth Edition. Chapter P is now Chapter 1. All chapters, therefore, have been renumbered. The chapter on "Systems of Equations and Inequalities" (Chapter 4) has been moved to follow the chapter on "Graphs and Functions" (Chapter 3). A new section, 4.6 "Systems of Linear Equations," has been added to Chapter 4 "Systems of Equations and Inequalities." Section 5.1 "Integer Exponents and Scientific Notation" is now included in the "Polynomials and Factoring" chapter (Chapter 5) instead of the "Rational Expressions, Equations, and Functions" chapter. Two new sections, 6.4 "Complex Fractions" and 6.7 "Applications and Variation," have been added to Chapter 6 "Rational Expressions, Equations, and Functions." The sections from Chapter 7 "Linear Models and Graphs of Nonlinear Models" (of the Third Edition) have been integrated into different chapters throughout the Fourth Edition. "Graphs of Rational Functions" (Section 7.5 in the Third Edition) has been deleted from the Fourth Edition. A chapter on "Conics" (Chapter 10) has been added. Finally, Sections 10.5 "Counting Principles" and 10.6 "Probability" have been moved to the free website that supports the text.

In order to address the diverse needs and abilities of students, we offer a straightforward approach to the presentation of difficult concepts. In the Fourth Edition, the emphasis is on helping students learn a variety of techniques—symbolic, numeric, and visual—for solving problems. We are committed to providing students with a successful and meaningful course of study.

Our approach begins with *Motivating the Chapter,* a feature that introduces each chapter. These multipart problems are designed to show students the relevance of algebra to the world around them. Each *Motivating the Chapter* feature is a real-life application that requires students to apply the concepts of the chapter in order to solve each part of the problem. Problem-solving and critical thinking skills are emphasized here and throughout the text in applications that appear in the examples and exercise sets.

To improve the usefulness of the text as a study tool, we have added two new, paired features to the beginning of each section: *What You Should Learn* lists the main objectives that students will encounter throughout the section, and *Why You Should Learn It* provides a motivational explanation for learning the given objectives. To help keep students focused as they read the section, each objective presented in *What You Should Learn* is restated in the margin at the point where the concept is introduced.

In this edition, the *Study Tip, Technology: Tip,* and *Technology: Discovery* features have been revised. *Study Tip* features provide hints, cautionary notes, and words of advice for students as they learn the material. *Technology: Tip* features provide point-of-use instruction for using a graphing calculator, whereas *Technology: Discovery* features encourage students to explore mathematical concepts using their graphing or scientific calculators. All technology features are highlighted and can easily be omitted without loss of continuity in coverage of material.

The new chapter summary feature *What Did You Learn?* highlights important mathematical vocabulary (*Key Terms*) and primary concepts (*Key Concepts*) from the chapter. For easy reference, the *Key Terms* are correlated to the chapter by page number and the *Key Concepts* by section number.

As students proceed through each chapter, they have many opportunities to assess their understanding and practice skills. A set of *Exercises*, located at the end of each section, correlates to the *Examples* found within the section. *Mid-Chapter Quizzes* and *Chapter Tests* offer students self-assessment tools halfway through and at the conclusion of each chapter. *Review Exercises*, organized by section, restate the *What You Should Learn* objectives so that students may refer back to the appropriate topic discussion when working through the exercises. In addition, the *Review: Concepts, Skills, and Problem Solving* exercises that precede each exercise set, and the *Cumulative Tests* that follow Chapters 4, 7, and 10, give students more opportunities to revisit and review previously learned concepts.

To show students the practical uses of algebra, we highlight the connections between the mathematical concepts and the real world in the multitude of applications found throughout the text. We believe that students can overcome their difficulties in mathematics if they are encouraged and supported throughout the learning process. Too often, students become frustrated and lose interest in the material when they cannot follow the text. With this in mind, every effort has been made to write a readable text that can be understood by every student. We hope that your students find our approach engaging and effective.

Ron Larson

Robert P. Hostetler

Features

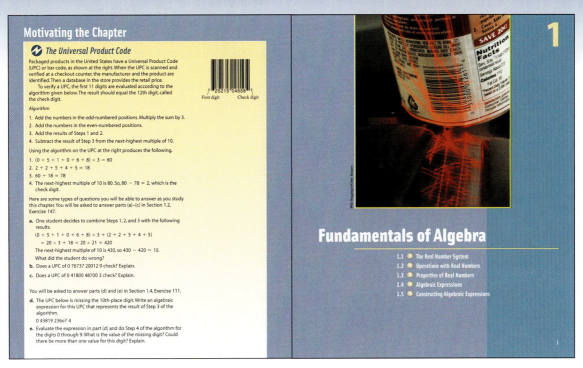

1

Fundamentals of Algebra

1.1 ● The Real Number System
1.2 ● Operations with Real Numbers
1.3 ● Properties of Real Numbers
1.4 ● Algebraic Expressions
1.5 ● Constructing Algebraic Expressions

Chapter Opener

Every chapter opens with *Motivating the Chapter.* These multipart problems use concepts discussed in the chapter and present them in the context of a single real-world application. *Motivating the Chapter* problems are correlated to specific sections and can be assigned as part of an exercise set or as an individual or group project. The icon ⊘ identifies an exercise that relates back to *Motivating the Chapter.*

Section Opener   *New*

Every section begins with a list of learning objectives called *What You Should Learn.* Each objective is restated in the margin at the point where it is covered. *Why You Should Learn It* provides a motivational explanation for learning the given objectives.

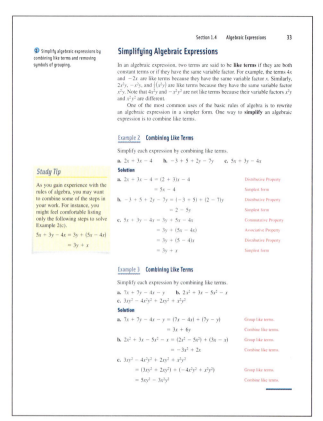

Examples

Each example has been carefully chosen to illustrate a particular mathematical concept or problem-solving technique. The examples cover a wide variety of problems and are titled for easy reference. Many examples include detailed, step-by-step solutions with side comments, which explain the key steps of the solution process.

Applications

A wide variety of real-life applications are integrated throughout the text in examples and exercises. These applications demonstrate the relevance of algebra in the real world. Many of the applications use current, real data. The icon 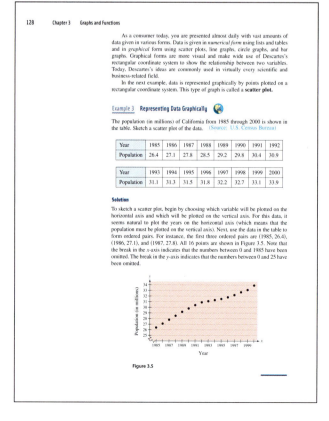 indicates an example involving a real-life application.

---

46   Chapter 1   Fundamentals of Algebra

Example 10 Constructing a Mathematical Model

Write an expression for the following phrase.

"The sum of two consecutive integers, the first of which is *n*"

Solution

The first integer is *n*. The next consecutive integer is $n + 1$. So the sum of two consecutive integers is

$$n + (n + 1) = 2n + 1.$$

Sometimes an expression may be written directly from a diagram using a common geometric formula, as shown in the next example.

Example 11 Constructing a Mathematical Model

Write expressions for the perimeter and area of the rectangle shown in Figure 1.16.

2*w* in.

$(w + 12)$ in.

Figure 1.16

Solution

For the perimeter of the rectangle, use the formula

Perimeter = 2(Length) + 2(Width).

Verbal Model: $2 \cdot$ **Length** $+ 2 \cdot$ **Width**

Labels: Length = $w + 12$ (inches)
        Width = $2w$ (inches)

Expression: $2(w + 12) + 2(2w) = 2w + 24 + 4w$
                      $= 6w + 24$ (inches)

For the area of the rectangle, use the formula

Area = (Length)(Width).

Verbal Model: **Length** $\cdot$ **Width**

Labels: Length = $w + 12$ (inches)
        Width = $2w$ (inches)

Expression: $(w + 12)(2w) = 2w^2 + 24w$ (square inches)

Problem Solving

This text provides many opportunities for students to sharpen their problem-solving skills. In both the examples and the exercises, students are asked to apply verbal, numerical, analytical, and graphical approaches to problem solving. In the spirit of the AMATYC and NCTM standards, students are taught a five-step strategy for solving applied problems, which begins with constructing a verbal model and ends with checking the answer.

Geometry

The Fourth Edition continues to provide coverage and integration of geometry in examples and exercises. The icon ▲ indicates an exercise involving geometry.

FEATURES

Section 2.3   Business and Scientific Problems   91

64. Work-Rate Problem You can complete a typing project in 5 hours, and your friend can complete it in 8 hours.
(a) What fractional part of the project can be accomplished by each person in 1 hour?
(b) How long will it take both of you to complete the project working together?

65. Work-Rate Problem You can mow a lawn in 3 hours, and your friend can mow it in 4 hours.
(a) What fractional part of the lawn can each of you mow in 1 hour?
(b) How long will it take both of you to mow the lawn working together?

66. Work-Rate Problem It takes 30 minutes for a pump to empty a water tank. A larger pump can empty the tank in half the time. How long would it take to empty the tank with both pumps operating?

In Exercises 67–76, solve for the specified variable. See Example 7.

67. Solve for *R*.
  Ohm's Law: $E = IR$

68. Solve for *r*.
  Simple Interest: $A = P + Prt$

69. Solve for *L*.
  Discount: $S = L - rL$

70. Solve for *C*.
  Markup: $S = C + rC$

71. Solve for *a*.
  Free-Falling Body: $h = 48t + \frac{1}{2}at^2$

72. Solve for *a*.
  Free-Falling Body: $h = 18t + \frac{1}{2}at^2$

73. Solve for *a*.
  Free-Falling Body: $h = -12t + \frac{1}{2}at^2$

74. Solve for *a*.
  Free-Falling Body: $h = 36t + \frac{1}{2}at^2 + 50$

75. Solve for *a*.
  Free-Falling Body: $h = -15t + \frac{1}{2}at^2 + 9.5$

76. Solve for *b*.
  Area of a Trapezoid: $A = \frac{1}{2}(a + b)h$

77. ▲ *Geometry* Find the volume of the circular cylinder shown in the figure.

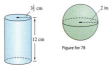

$3\frac{1}{2}$ cm

12 cm

Figure for 77

78. ▲ *Geometry* Find the volume of the sphere shown in the figure.

2 in.

Figure for 78

79. ▲ *Geometry* A rectangular picture frame has a perimeter of 3 feet. The width of the frame is 0.62 times its height. Find the height of the frame.

80. ▲ *Geometry* A rectangular stained glass window has a perimeter of 18 feet. The height of the window is 1.25 times its width. Find the width of the window.

81. ▲ *Geometry* A "Slow Moving Vehicle" sign has the shape of an equilateral triangle. The sign has a perimeter of 129 centimeters. Find the length of each side.

82. ▲ *Geometry* The length of a rectangle is three times its width. The perimeter of the rectangle is 64 inches. Find the dimensions of the rectangle.

83. *Meteorology* The average daily high temperature in Boston, Massachusetts is 59°F. What is Boston's average daily high temperature in degrees Celsius? (Source: U.S. National Oceanic and Atmospheric Administration.)

---

86   Chapter 2   Linear Equations and Inequalities

Example 7 Rewriting a Formula

In the perimeter formula $P = 2l + 2w$, solve for *w*.

Solution

$$P = 2l + 2w$$ Original formula

$$P - 2l = 2w$$ Subtract $2l$ from each side.

$$\frac{P - 2l}{2} = w$$ Divide each side by 2.

Example 8 Using a Geometric Formula

A local streets department plans to put sidewalks along the two streets that border your corner lot, which is 250 feet long on one side with an area of 30,000 square feet. Each lot owner is to pay $1.50 per foot of sidewalk bordering his or her lot.

a. Find the width of your lot.
b. How much will you have to pay for the sidewalks put on your lot?

Solution

Figure 2.3 shows a labeled diagram of your lot.

> **Study Tip**
>
> When solving problems such as the one in Example 8, you may find it helpful to draw and label a diagram.

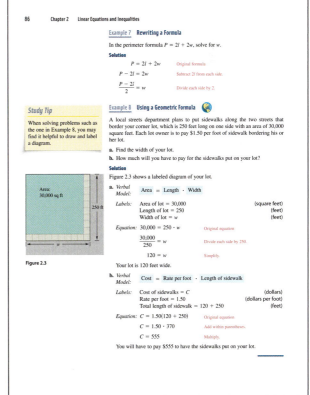

Area: 30,000 sq ft

250 ft

w

Figure 2.3

a. *Verbal Model:*   **Area** = **Length** $\cdot$ **Width**

Labels: Area of lot = 30,000   (square feet)
        Length of lot = 250   (feet)
        Width of lot = *w*   (feet)

Equation: $30,000 = 250 \cdot w$ Original equation

$$\frac{30,000}{250} = w$$ Divide each side by 250.

$$120 = w$$ Simplify.

Your lot is 120 feet wide.

b. *Verbal Model:*   **Cost** = **Rate per foot** $\cdot$ **Length of sidewalk**

Labels: Cost of sidewalks = *C*   (dollars)
        Rate per foot = 1.50   (dollars per foot)
        Total length of sidewalk = 120 + 250   (feet)

Equation: $C = 1.50(120 + 250)$ Original equation

$$C = 1.50 \cdot 370$$ Add within parentheses.

$$C = 555$$ Multiply.

You will have to pay $555 to have the sidewalks put on your lot.

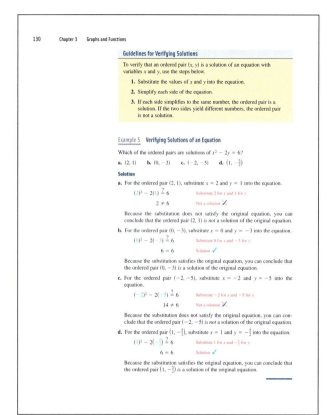

Definitions and Rules

All important definitions, rules, formulas, properties, and summaries of solution methods are highlighted for emphasis. Each of these features is also titled for easy reference.

Study Tips

Study Tips offer students specific point-of-use suggestions for studying algebra, as well as pointing out common errors and discussing alternative solution methods. They appear in the margins.

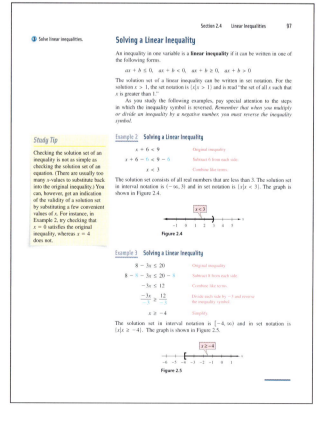

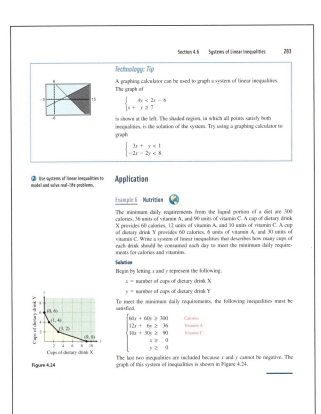

Graphics

Visualization is a critical problem-solving skill. To encourage the development of this skill, students are shown how to use graphs to reinforce algebraic and numeric solutions and to interpret data. The numerous figures in examples and exercises throughout the text were computer-generated for accuracy.

Technology: Tips

Point-of-use instructions for using graphing calculators appear in the margins. These features encourage the use of graphing technology as a tool for visualization of mathematical concepts, for verification of other solution methods, and for facilitation of computations. The *Technology: Tips* can easily be omitted without loss of continuity in coverage.

FEATURES

Technology: Discovery

Technology: Discovery features invite students to engage in active exploration of mathematical concepts and discovery of mathematical relationships through the use of scientific or graphing calculators. These activities encourage students to utilize their critical thinking skills and help them develop an intuitive understanding of theoretical concepts. *Technology: Discovery* features can easily be omitted without loss of continuity in coverage.

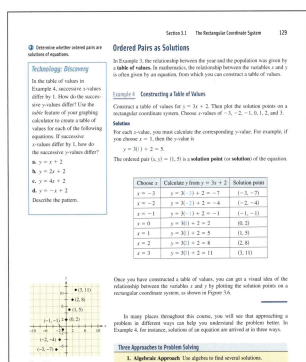

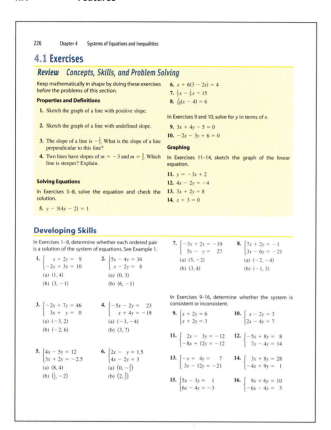

Review: Concepts, Skills, and Problem Solving

Each exercise set (except in Chapter 1) is preceded by these review exercises that are designed to help students keep up with concepts and skills learned in previous chapters. Answers to all *Review: Concepts, Skills, and Problem Solving* exercises are given in the back of the student text.

Exercises

The exercise sets are grouped into three categories: *Developing Skills, Solving Problems,* and *Explaining Concepts.* The exercise sets offer a diverse variety of computational, conceptual, and applied problems to accommodate many teaching and learning styles. Designed to build competence, skill, and understanding, each exercise set is graded in difficulty to allow students to gain confidence as they progress. Detailed solutions to all odd-numbered exercises are given in the *Student Solutions Guide,* and answers to all odd-numbered exercises are given in the back of the student text.

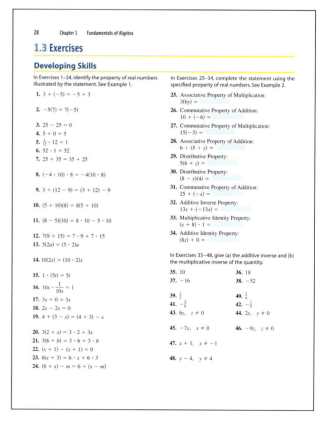

What Did You Learn?

Key Terms

system of equations, *p. 218*
solution of a system of
   equations, *p. 218*
points of intersection, *p. 219*
consistent system, *p. 219*
dependent system, *p. 219*
inconsistent system, *p. 219*
row-echelon form, *p. 241*

equivalent systems, *p. 242*
Gaussian elimination, *p. 242*
row operations, *p. 242*
matrix, *p. 254*
order (of a matrix), *p. 254*
square matrix, *p. 254*
augmented matrix, *p. 255*
coefficient matrix, *p. 255*

row-equivalent matrices, *p. 256*
minor (of an entry), *p. 268*
Cramer's Rule, *p. 270*
system of linear
   inequalities, *p. 279*
solution of a system of
   linear inequalities, *p. 279*
vertex, *p. 280*

Key Concepts

4.1 The method of substitution

1. Solve one of the equations for one variable in terms of the other.

2. Substitute the expression obtained in Step 1 in the other equation to obtain an equation in one variable.

3. Solve the equation obtained in Step 2.

4. Back-substitute the solution from Step 3 in the expression obtained in Step 1 to find the value of the other variable.

5. Check the solution to see that it satisfies both of the original equations.

4.2 The method of elimination

1. Obtain coefficients for x (or y) that differ only in sign by multiplying all terms of one or both equations by suitable constants.

2. Add the equations to eliminate one variable, and solve the resulting equation.

3. Back-substitute the value obtained in Step 2 in either of the original equations and solve for the other variable.

4. Check your solution in both of the original equations.

4.4 Elementary row operations

Two matrices are row-equivalent if one can be obtained from the other by a sequence of elementary row operations.

1. Interchange two rows.

2. Multiply a row by a nonzero constant.

3. Add a multiple of a row to another row.

4.4 Gaussian elimination with back-substitution

To use matrices and Gaussian elimination to solve a system of linear equations, use the following steps.

1. Write the augmented matrix of the system of linear equations.

2. Use elementary row operations to rewrite the augmented matrix in row-echelon form.

3. Write the system of linear equations corresponding to the matrix in row-echelon form, and use back-substitution to find the solution.

4.5 Determinant of a 2 x 2 matrix

$$\det(A) = |A| = \begin{vmatrix} a_1 & b_1 \\ a_2 & b_2 \end{vmatrix}$$
$$= a_1 b_2 - a_2 b_1$$

4.5 Expanding by minors

The determinant of a 3×3 matrix can be evaluated by expanding by minors.

$$\det(A) = \begin{vmatrix} a_1 & b_1 & c_1 \\ a_2 & b_2 & c_2 \\ a_3 & b_3 & c_3 \end{vmatrix}$$
$$= a_1 \begin{vmatrix} b_2 & c_2 \\ b_3 & c_3 \end{vmatrix} - b_1 \begin{vmatrix} a_2 & c_2 \\ a_3 & c_3 \end{vmatrix} + c_1 \begin{vmatrix} a_2 & b_2 \\ a_3 & b_3 \end{vmatrix}$$

4.6 Graphing a system of linear inequalities

1. Sketch the line that corresponds to each inequality.

2. Lightly shade the half-plane that is the graph of each linear inequality.

3. The graph of the system is the intersection of the half-planes.

What Did You Learn? (Chapter Summary)

Located at the end of every chapter, *What Did You Learn?* summarizes the *Key Terms* (referenced by page) and the *Key Concepts* (referenced by section) presented in the chapter. This effective study tool aids students as they review concepts and prepare for exams.

FEATURES

Review Exercises

The *Review Exercises* at the end of each chapter have been reorganized in the Fourth Edition. All skill-building and application exercises are first ordered by section, then grouped according to the objectives stated within *What You Should Learn.* This organization allows students to easily identify the appropriate sections and concepts for study and review.

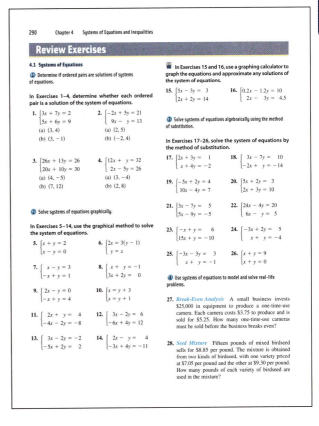

Review Exercises

4.1 Systems of Equations

1 Determine if ordered pairs are solutions of systems of equations.

In Exercises 1–4, determine whether each ordered pair is a solution of the system of equations.

1. $\begin{cases} 3x + 7y = 2 \\ 5x + 6y = 9 \end{cases}$
   (a) $(3, 4)$
   (b) $(3, -1)$

2. $\begin{cases} -2x + 5y = 21 \\ 9x - y = 13 \end{cases}$
   (a) $(2, 5)$
   (b) $(-2, 4)$

3. $\begin{cases} 26x + 13y = 26 \\ 20x + 10y = 30 \end{cases}$
   (a) $(4, -5)$
   (b) $(7, 12)$

4. $\begin{cases} 12x + y = 32 \\ 2x - 5y = 26 \end{cases}$
   (a) $(3, -4)$
   (b) $(2, 8)$

2 Solve systems of equations graphically.

In Exercises 5–14, use the graphical method to solve the system of equations.

5. $\begin{cases} x + y = 2 \\ x - y = 0 \end{cases}$

6. $\begin{cases} 2x = 3(y - 1) \\ y = x \end{cases}$

7. $\begin{cases} x - y = 3 \\ -x + y = 1 \end{cases}$

8. $\begin{cases} x + y = -1 \\ 3x + 2y = 0 \end{cases}$

9. $\begin{cases} 2x - y = 0 \\ -x + y = 4 \end{cases}$

10. $\begin{cases} x = y + 3 \\ x = y + 1 \end{cases}$

11. $\begin{cases} 2x + y = 4 \\ -4x - 2y = -8 \end{cases}$

12. $\begin{cases} 3x - 2y = 6 \\ -6x + 4y = 12 \end{cases}$

13. $\begin{cases} 3x - 2y = -2 \\ -5x + 2y = 2 \end{cases}$

14. $\begin{cases} 2x - y = 4 \\ -3x + 4y = -11 \end{cases}$

In Exercises 15 and 16, use a graphing calculator to graph the equations and approximate any solutions of the system of equations.

15. $\begin{cases} 5x - 3y = 3 \\ 2x + 2y = 14 \end{cases}$

16. $\begin{cases} 0.2x - 1.2y = 10 \\ 2x - 3y = 4.5 \end{cases}$

3 Solve systems of equations algebraically using the method of substitution.

In Exercises 17–26, solve the system of equations by the method of substitution.

17. $\begin{cases} 2x + 3y = 1 \\ x + 4y = -2 \end{cases}$

18. $\begin{cases} 3x - 7y = 10 \\ -2x + y = -14 \end{cases}$

19. $\begin{cases} -5x + 2y = 4 \\ 10x - 4y = 7 \end{cases}$

20. $\begin{cases} 5x + 2y = 3 \\ 2x + 3y = 10 \end{cases}$

21. $\begin{cases} 3x - 7y = 5 \\ 5x - 9y = -5 \end{cases}$

22. $\begin{cases} 24x - 4y = 20 \\ 6x - y = 5 \end{cases}$

23. $\begin{cases} -x + y = 6 \\ 15x + y = -10 \end{cases}$

24. $\begin{cases} -3x + 2y = 5 \\ x + y = -4 \end{cases}$

25. $\begin{cases} -3x - 3y = 3 \\ x + y = -1 \end{cases}$

26. $\begin{cases} x + y = 9 \\ x + y = 0 \end{cases}$

4 Use systems of equations to model and solve real-life problems.

27. *Break-Even Analysis* A small business invests $25,000 in equipment to produce a one-time-use camera. Each camera costs $3.75 to produce and is sold for $5.25. How many one-time-use cameras must be sold before the business breaks even?

28. *Seed Mixture* Fifteen pounds of mixed birdseed sells for $8.85 per pound. The mixture is obtained from two kinds of birdseed, with one variety priced at $7.05 per pound and the other at $9.30 per pound. How many pounds of each variety of birdseed are used in the mixture?

Mid-Chapter Quiz

Take this quiz as you would take a quiz in class. After you are done, check your work against the answers in the back of the book.

1. Determine the quadrants(s) in which the point $(x, 4)$ must be located if x is a real number. Explain your reasoning.

2. Determine whether each ordered pair is a solution point of the equation $4x - 3y = 10$.
   (a) $(2, 1)$ (b) $(1, -2)$ (c) $(2.5, 0)$ (d) $(2, -\frac{2}{3})$

In Exercises 3 and 4, plot the points on a rectangular coordinate system, find the distance between them, and determine the coordinates of the midpoint of the line segment joining the two points.

3. $(-1, 5), (3, 2)$ 4. $(-3, -2), (2, 10)$

In Exercises 5–7, sketch the graph of the equation and show the coordinates of three solution points (including x- and y-intercepts). (There are many correct answers.)

5. $3x + y - 6 = 0$ 6. $y = 6x - x^2$ 7. $y = |x - 2| - 3$

In Exercises 8–10, determine the slope of the line (if possible) through the two points. State whether the line rises, falls, is horizontal, or is vertical.

8. $(-3, 8), (7, 8)$ 9. $(3, 0), (6, 5)$ 10. $(-1, 4), (5, -6)$

In Exercises 11 and 12, write the equation of the line in slope-intercept form. Find the slope and y-intercept and use them to sketch the graph of the equation.

11. $3x + 6y = 6$ 12. $x - 2y = 4$

In Exercises 13 and 14, determine whether the lines are parallel, perpendicular, or neither.

13. $y = 3x + 2, y = -\frac{1}{3}x - 4$

14. L_1: $(4, 3), (-2, -9)$; L_2: $(0, -5), (5, 5)$

15. Write the general form of the equation of a line that passes through the point $(6, -1)$ and has a slope of $\frac{1}{2}$.

16. Your company purchases a new printing press for $85,000. For tax purposes, the printing press will be depreciated over a 10-year period. At the end of 10 years, the salvage value of the printing press is expected to be $4000. Find an equation that relates the depreciated value of the printing press to the number of years since it was purchased. Then sketch the graph of the equation.

Chapter Test

Take this test as you would take a test in class. After you are done, check your work against the answers in the back of the book.

1. Determine the quadrant in which the point (x, y) lies if $x > 0$ and $y < 0$.

2. Plot the points $(0, 5)$ and $(3, 1)$. Then find the distance between them and the coordinates of the midpoint of the line segment joining the two points.

3. Find the x- and y-intercepts of the graph of the equation $y = -3(x + 1)$.

4. Sketch the graph of the equation $y = |x - 2|$.

5. Find the slope (if possible) of the line passing through each pair of points.
   (a) $(-4, 7), (2, 3)$ (b) $(3, -2), (3, 6)$

6. Sketch the graph of the line passing through the point $(0, -6)$ with slope $m = \frac{4}{3}$.

7. Find the x- and y-intercepts of the graph of $2x + 5y - 10 = 0$. Use the results to sketch the graph.

8. Find an equation of the line through the points $(25, -15)$ and $(75, 10)$.

9. Find an equation of the vertical line through the point $(-2, 4)$.

10. Write equations of the lines that pass through the point $(-2, 3)$ and are (a) parallel and (b) perpendicular to the line $3x - 5y = 4$.

11. Sketch the graph of the inequality $2x - 3y \geq 9$.

12. The graph of $y^2(4 - x) = x^3$ is shown at the left. Does the graph represent y as a function of x? Explain your reasoning.

13. Determine whether the relation represents a function. Explain.
    (a) $\{(2, 4), (-6, 3), (3, 3), (1, -2)\}$ (b) $\{(0, 0), (1, 5), (-2, 1), (0, -4)\}$

14. Evaluate $g(x) = x/(x - 3)$ as indicated, and simplify.
    (a) $g(2)$ (b) $g(\frac{5}{2})$ (c) $g(x + 2)$

15. Find the domain of each function.
    (a) $h(t) = \sqrt{9 - t}$ (b) $f(x) = \dfrac{x + 1}{x - 4}$

Figure for 12

16. Sketch the graph of the function $g(x) = \sqrt{2 - x}$.

17. Describe the transformation of the graph of $f(x) = x^2$ that would produce the graph of $g(x) = -(x - 2)^2 + 1$.

18. After 4 years, the value of a $26,000 car will have depreciated to $10,000. Write the value V of the car as a linear function of t, the number of years since the car was purchased. When will the car be worth $16,000?

19. Use the graph of $f(x) = |x|$ to write a function that represents each graph.
    (a)   (b)   (c)

Cumulative Test: Chapters 1–4

Take this test as you would take a test in class. After you are done, check your work against the answers in the back of the book.

1. Place the correct symbol ($<$, $>$, or $=$) between the two real numbers.
   (a) -2 ☐ 5 (b) $\frac{1}{3}$ ☐ $\frac{1}{2}$ (c) $|2.3|$ ☐ $-|-4.5|$

2. Write an algebraic expression for the statement, "The number n is tripled and the product is decreased by 8."

In Exercises 3 and 4, perform the operations and simplify.

3. $t(3t - 1) - 2(t + 4)$

4. $3x(x^2 - 2) - 5(x^2 + 5)$

In Exercises 5–8, solve the equation or inequality.

5. $12 - 5(3 - x) = x + 3$ 6. $1 - \dfrac{x + 2}{4} = \dfrac{7}{8}$

7. $|x - 2| \geq 3$ 8. $-12 \leq 4x - 6 < 10$

9. Your annual automobile insurance premium is $1225. Because of a driving violation, your premium is increased by 15%. What is your new premium?

Figure for 10

10. The triangles at the left are similar. Solve for x by using the fact that corresponding sides of similar triangles are proportional.

11. The revenue R from selling x units of a product is $R = 12.90x$. The cost C of producing x units is $C = 8.50x + 450$. To obtain a profit, the revenue must be greater than the cost. For what values of x will this product produce a profit? Explain your reasoning.

12. Does the equation $x - y^3 = 0$ represent y as a function of x?

13. Find the domain of the function $f(x) = \sqrt{x - 2}$.

14. Given $f(x) = x^2 - 3x$, find (a) $f(4)$ and (b) $f(c + 3)$.

15. Find the slope of the line passing through $(-4, 0)$ and $(4, 6)$. Then find the distance between the points and the midpoint of the line segment joining the points.

16. Determine the equations of lines through the point $(-2, 1)$ (a) parallel to $2x - y = 1$ and (b) perpendicular to $3x + 2y = 5$.

In Exercises 17 and 18, graph the equation.

17. $4x + 3y - 12 = 0$ 18. $y = 1 - (x - 2)^2$

In Exercises 19–21, use the indicated method to solve the system.

19. Substitution:
$$\begin{cases} x + y = 6 \\ 2x - y = 3 \end{cases}$$

20. Elimination:
$$\begin{cases} 2x + y = 6 \\ 3x - 2y = 16 \end{cases}$$

21. Matrices:
$$\begin{cases} 2x + y - 2z = 1 \\ x \qquad - z = 1 \\ 3x + 3y + z = 12 \end{cases}$$

Mid-Chapter Quiz

Each chapter contains a *Mid-Chapter Quiz.* Answers to all questions in the *Mid-Chapter Quiz* are given in the back of the student text.

Chapter Test

Each chapter ends with a *Chapter Test.* Answers to all questions in the *Chapter Test* are given in the back of the student text.

Cumulative Test

The *Cumulative Tests* that follow Chapters 4, 7, and 10 provide a comprehensive self-assessment tool that helps students check their mastery of previously covered material. Answers to all questions in the *Cumulative Tests* are given in the back of the student text.

Intermediate Algebra, Fourth Edition, by Larson and Hostetler is accompanied by a comprehensive supplements package, which includes resources for both students and instructors. All items are keyed to the text.

Printed Resources

For Students

Student Solutions Guide by Gerry C. Fitch, Louisiana State University
(0-618-38829-X)

- Detailed, step-by-step solutions to all Review: Concepts, Skills, and Problem Solving exercises and to all odd-numbered exercises in the section exercise sets and in the review exercises
- Detailed, step-by-step solutions to all Mid-Chapter Quiz, Chapter Test, and Cumulative Test questions

For Instructors

Instructor's Annotated Edition
(0-618-38828-1)
Instructor's Resource Guide by Gerry C. Fitch and Ann Rutledge Kraus, The Pennsylvania State University, The Behrend College
(0-618-38830-3)

Technology Resources

For Students

HM mathSpace™ Student CD-ROM   (0-618-38835-4)

Website *(http://math.college.hmco.com/students)*

Houghton Mifflin Instructional Videos and DVDs by Dana Mosely
(Video ISBN: 0-618-38832-X; DVD ISBN: 0-618-38833-8)

SMARTHINKING™ Live, Online Tutoring   Houghton Mifflin has partnered with SMARTHINKING to provide an easy-to-use and effective online tutorial service. *Whiteboard Simulations* and *Practice Area* promote real-time visual interaction. Three levels of service are offered.

- **Text-Specific Tutoring** provides real-time, one-on-one instruction with a specially qualified "e-structor."
- *Questions Any Time* allows students to submit questions to the tutor outside the scheduled hours and receive a reply within 24 hours.
- *Independent Study Resources* connect students with around-the-clock access to additional educational services, including interactive websites, diagnostic tests, and Frequently Asked Questions posed to SMARTHINKING e-structors.

For Instructors

HMClassPrep™ with HM Testing   (0-618-38834-6)
Website *(http://math.college.hmco.com/instructors)*

Acknowledgments

We would like to thank the many people who have helped us revise the various editions of this text. Their encouragement, criticisms, and suggestions have been invaluable to us.

Reviewers

Mary Kay Best, Coastal Bend College; Patricia K. Bezona, Valdosta State University; Connie L. Buller, Metropolitan Community College; Mistye R. Canoy, Holmes Community College; Maggie W. Flint, Northeast State Technical Community College; William Hoard, Front Range Community College; Andrew J. Kaim, DePaul University; Jennifer L. Laveglia, Bellevue Community College; Aaron Montgomery, Purdue University North Central; William Naegele, South Suburban College; Jeanette O'Rourke, Middlesex County College; Judith Pranger, Binghamton University; Kent Sandefer, Mohave Community College; Robert L. Sartain, Howard Payne University; Jon W. Scott, Montgomery College; John Seims, Mesa Community College; Ralph Selensky, Eastern Arizona College; Charles I. Sherrill, Community College of Aurora; Kay Stroope, Phillips Community College of the University of Arkansas; Bettie Truitt, Black Hawk College; Betsey S. Whitman, Framingham State College; George J. Witt, Glendale Community College.

We would also like to thank the staff of Larson Texts, Inc. and the staff of Meridian Creative Group, who assisted in preparing the manuscript, rendering the art package, and typesetting and proofreading the pages and the supplements.

On a personal level, we are grateful to our wives, Deanna Gilbert Larson and Eloise Hostetler, for their love, patience, and support. Also, a special thanks goes to R. Scott O'Neil.

If you have suggestions for improving this text, please feel free to write to us. Over the past two decades we have received many useful comments from both instructors and students, and we value these comments very much.

Ron Larson
Robert P. Hostetler

How to Study Algebra

Your success in algebra depends on your active participation both in class and outside of class. Because the material you learn each day builds on the material you learned previously, it is important that you keep up with the course work every day and develop a clear plan of study. To help you learn how to study algebra, we have prepared a set of guidelines that highlight key study strategies.

Preparing for Class

The syllabus your instructor provides is an invaluable resource that outlines the major topics to be covered in the course. Use it to help you prepare. As a general rule, you should set aside two to four hours of study time for each hour spent in class. Being prepared is the first step toward success in algebra. Before class,

- Review your notes from the previous class.

- Read the portion of the text that will be covered in class.

- Use the *What You Should Learn* objectives listed at the beginning of each section to keep you focused on the main ideas of the section.

- Pay special attention to the definitions, rules, and concepts highlighted in boxes. Also, be sure you understand the meanings of mathematical symbols and of terms written in boldface type. Keep a vocabulary journal for easy reference.

- Read through the solved examples. Use the side comments that accompany the solution steps to help you follow the solution process. Also, read the *Study Tips* given in the margins.

- Make notes of anything you do not understand as you read through the text. If you still do not understand after your instructor covers the topic in question, ask questions before your instructor moves on to a new topic.

- If you are using technology in this course, read the *Technology: Tips* and try the *Technology: Discovery* exercises.

Keeping Up

Another important step toward success in algebra involves your ability to keep up with the work. It is very easy to fall behind, especially if you miss a class. To keep up with the course work, be sure to

- Attend every class. Bring your text, a notebook, and a pen or pencil. If you miss a class, get the notes from a classmate as soon as possible and review them carefully.

- Take notes in class. After class, read through your notes and add explanations so that your notes make sense to *you*.

- Reread the portion of the text that was covered in class. This time, work each example *before* reading through the solution.

- Do your homework as soon as possible, while concepts are still fresh in your mind.

Use your notes from class, the text discussion, the examples, and the *Study Tips* as you do your homework. Many exercises are keyed to specific examples in the text for easy reference.

Getting Extra Help

It can be very frustrating when you do not understand concepts and are unable to complete homework assignments. However, there are many resources available to help you with your study of algebra.

Your instructor may have office hours. If you are feeling overwhelmed and need help, make an appointment to discuss your difficulties with your instructor.

Find a study partner or a study group. Sometimes it helps to work through problems with another person.

Arrange to get regular assistance from a tutor. Many colleges have a math resource center available on campus as well.

Consult one of the many ancillaries available with this text: the *Student Solutions Guide,* HM mathSpace™ Student CD-ROM, videotapes, DVDs, and additional study resources available at our website at *http://math.college.hmco.com/students.*

Preparing for an Exam

The last step toward success in algebra lies in how you prepare for and complete exams. If you have followed the suggestions given above, then you are almost ready for exams. Do not assume that you can cram for the exam the night before—this seldom works. As a final preparation for the exam,

Read the *What Did You Learn?* chapter summary, which is keyed to each section, and review the concepts and terms.

Work through the *Review Exercises* if you need extra practice on material from a particular section.

Take the *Mid-Chapter Quiz* and the *Chapter Test* as if you were in class. You should set aside at least one hour per test. Check your answers against the answers given in the back of the book.

Review your notes and the portion of the text that will be covered on the exam.

Avoid studying up until the last minute. This will only make you anxious.

Once the exam begins, read through the directions and the entire exam before beginning. Work the problems that you know how to do first to avoid spending too much time on any one problem. Time management is extremely important when taking an exam.

If you finish early, use the remaining exam time to go over your work.

When you get an exam back, review it carefully and go over your errors. Rework the problems you answered incorrectly. Discovering the mistakes you made will help you improve your test-taking ability.

STUDY PLAN

Motivating the Chapter

⚡ The Universal Product Code

0 2 5 2 1 5 0 4 6 5 8 2
↗ First digit   Check digit ↖

Packaged products in the United States have a Universal Product Code (UPC) or bar code, as shown at the right. When the UPC is scanned and verified at a checkout counter, the manufacturer and the product are identified. Then a database in the store provides the retail price.

    To verify a UPC, the first 11 digits are evaluated according to the algorithm given below. The result should equal the 12th digit, called the check digit.

Algorithm

1. Add the numbers in the odd-numbered positions. Multiply the sum by 3.
2. Add the numbers in the even-numbered positions.
3. Add the results of Steps 1 and 2.
4. Subtract the result of Step 3 from the next-highest multiple of 10.

Using the algorithm on the UPC at the right produces the following.

1. $(0 + 5 + 1 + 0 + 6 + 8) \times 3 = 60$
2. $2 + 2 + 5 + 4 + 5 = 18$
3. $60 + 18 = 78$
4. The next-highest multiple of 10 is 80. So, $80 - 78 = 2$, which is the check digit.

Here are some types of questions you will be able to answer as you study this chapter. You will be asked to answer parts (a)–(c) in Section 1.2, Exercise 147.

a. One student decides to combine Steps 1, 2, and 3 with the following results.

    $(0 + 5 + 1 + 0 + 6 + 8) \times 3 + (2 + 2 + 5 + 4 + 5)$
       $= 20 \times 3 + 18 = 20 \times 21 = 420$

    The next-highest multiple of 10 is 430, so $430 - 420 = 10$.

    What did the student do wrong?

b. Does a UPC of 0 76737 20012 9 check? Explain.

c. Does a UPC of 0 41800 48700 3 check? Explain.

You will be asked to answer parts (d) and (e) in Section 1.4, Exercise 111.

d. The UPC below is missing the 10th-place digit. Write an algebraic expression for this UPC that represents the result of Step 3 of the algorithm.

    0 43819 236*a*7 4

e. Evaluate the expression in part (d) and do Step 4 of the algorithm for the digits 0 through 9. What is the value of the missing digit? Could there be more than one value for this digit? Explain.

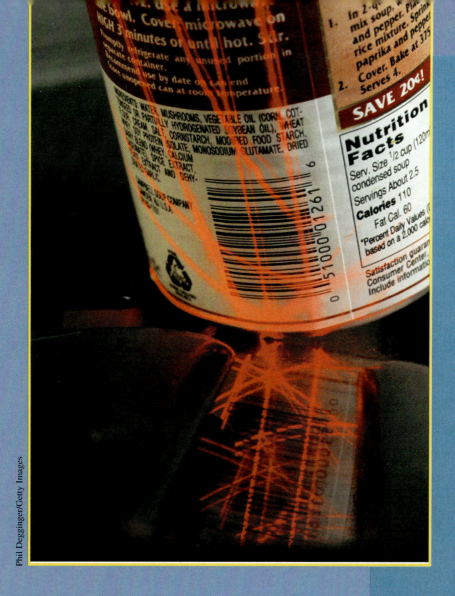

Phil Degginger/Getty Images

Fundamentals of Algebra

1.1 ● The Real Number System

1.2 ● Operations with Real Numbers

1.3 ● Properties of Real Numbers

1.4 ● Algebraic Expressions

1.5 ● Constructing Algebraic Expressions

1.1 The Real Number System

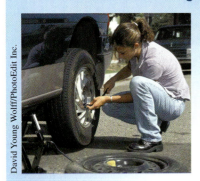

David Young Wolff/PhotoEdit Inc.

Why You Should Learn It

Inequality symbols can be used to represent many real-life situations, such as tire pressure (see Exercise 90 on page 10).

What You Should Learn

1 Understand the set of real numbers and the subsets of real numbers.

2 Use the real number line to order real numbers.

3 Use the real number line to find the distance between two real numbers.

4 Determine the absolute value of a real number.

Sets and Real Numbers

This chapter introduces the basic definitions, operations, and rules that form the fundamental concepts of algebra. Section 1.1 begins with real numbers and their representation on the real number line. Sections 1.2 and 1.3 discuss operations and properties of real numbers, and Sections 1.4 and 1.5 discuss algebraic expressions.

The formal term that is used in mathematics to refer to a collection of objects is the word **set.** For instance, the set

$$\{1, 2, 3\} \qquad \text{A set with three members}$$

contains the three numbers 1, 2, and 3. Note that the members of the set are enclosed in braces { }. Parentheses ( ) and brackets [ ] are used to represent other ideas.

The set of numbers that is used in arithmetic is called the set of **real numbers.** The term *real* distinguishes real numbers from *imaginary* or *complex* numbers—a type of number that you will study later in this text.

If all members of a set A are also members of a set B, then A is a **subset** of B. One of the most commonly used subsets of real numbers is the set of **natural numbers** or **positive integers.**

$$\{1, 2, 3, 4, \ldots\} \qquad \text{The set of positive integers}$$

Note that the three dots indicate that the pattern continues. For instance, the set also contains the numbers 5, 6, 7, and so on.

Positive integers can be used to describe many quantities in everyday life—for instance, you might be taking four classes this term, or you might be paying 240 dollars a month for rent. But even in everyday life, positive integers cannot describe some concepts accurately. For instance, you could have a zero balance in your checking account. To describe such a quantity, you need to expand the set of positive integers to include **zero,** forming the set of **whole numbers.** To describe a quantity such as $-5°$, you need to expand the set of whole numbers to include **negative integers.** This expanded set is called the set of **integers.** The set of integers is also a *subset* of the set of real numbers.

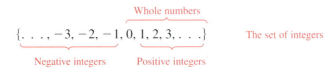

Whole numbers

$$\{\ldots, -3, -2, -1, 0, 1, 2, 3, \ldots\} \qquad \text{The set of integers}$$

Negative integers   Positive integers

1 Understand the set of real numbers and the subsets of real numbers.

Even with the set of integers, there are still many quantities in everyday life that you cannot describe accurately. The costs of many items are not in whole dollar amounts, but in parts of dollars, such as $1.19 and $39.98. You might work $8\frac{1}{2}$ hours, or you might miss the first *half* of a movie. To describe such quantities, you can expand the set of integers to include **fractions.** The expanded set is called the set of **rational numbers.** Formally, a real number is **rational** if it can be written as the ratio p/q of two integers, where $q \neq 0$ (the symbol $\neq$ means **does not equal**). Here are some examples of rational numbers.

$$2 = \frac{2}{1}, \quad \frac{1}{3} = 0.333 \ldots, \quad \frac{1}{8} = 0.125, \quad \text{and} \quad \frac{125}{111} = 1.126126 \ldots$$

The decimal representation of a rational number is either **terminating** or **repeating.** For instance, the decimal representation of $\frac{1}{4} = 0.25$ is terminating, and the decimal representation of

$$\frac{4}{11} = 0.363636 \ldots = 0.\overline{36}$$

is repeating. (The line over 36 indicates which digits repeat.) A real number that cannot be written as a ratio of two integers is **irrational.** For instance, the numbers

$$\sqrt{2} = 1.4142135 \ldots \quad \text{and} \quad \pi = 3.1415926 \ldots$$

are irrational.

The decimal representation of an irrational number neither terminates nor repeats. When you perform calculations using decimal representations of nonterminating, nonrepeating decimals, you usually use a decimal approximation that has been **rounded** to a certain number of decimal places. The rounding rule used in this text is to round up if the succeeding digit is 5 or more and round down if the succeeding digit is 4 or less. For example, to one decimal place, 7.35 would *round up* to 7.4. Similarly, to two decimal places, 2.364 would *round down* to 2.36. Rounded to four decimal places, the decimal approximations of the rational number $\frac{2}{3}$ and the irrational number π are

$$\frac{2}{3} \approx 0.6667 \quad \text{and} \quad \pi \approx 3.1416.$$

The symbol $\approx$ means **is approximately equal to.** Figure 1.1 shows several commonly used subsets of real numbers and their relationships to each other.

Example 1   Identifying Real Numbers

Which of the numbers in the set $\left\{-7, -\sqrt{3}, -1, -\frac{1}{5}, 0, \frac{3}{4}, \sqrt{2}, \pi, 5\right\}$ are (a) natural numbers, (b) integers, (c) rational numbers, and (d) irrational numbers?

Solution

a. Natural numbers: $\{5\}$

b. Integers: $\{-7, -1, 0, 5\}$

c. Rational numbers: $\left\{-7, -1, -\frac{1}{5}, 0, \frac{3}{4}, 5\right\}$

d. Irrational numbers: $\left\{-\sqrt{3}, \sqrt{2}, \pi\right\}$

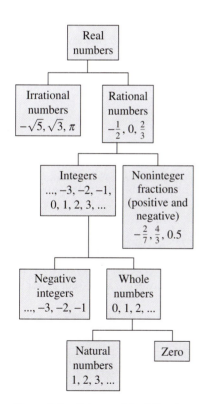

Figure 1.1   Subsets of Real Numbers

② Use the real number line to order real numbers.

The Real Number Line

The picture that represents the real numbers is called the **real number line.** It consists of a horizontal line with a point (the **origin**) labeled as 0. Numbers to the left of zero are **negative** and numbers to the right of 0 are **positive,** as shown in Figure 1.2.

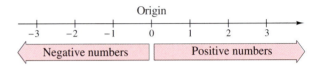

Figure 1.2   The Real Number Line

The real number zero is neither positive nor negative. So, to describe a real number that might be positive or zero, you can use the term **nonnegative real number.**

Each point on the real number line corresponds to exactly one real number, and each real number corresponds to exactly one point on the real number line, as shown in Figure 1.3. When you draw the point (on the real number line) that corresponds to a real number, you are **plotting** the real number.

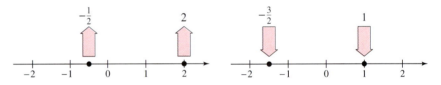

Each point on the real number line corresponds to a real number.

Each real number corresponds to a point on the real number line.

Figure 1.3

Example 2   Plotting Points on the Real Number Line

Plot the real numbers on the real number line.

a. $-\dfrac{5}{3}$ **b.** 2.3 **c.** $\dfrac{9}{4}$ **d.** -0.3

Solution

All four points are shown in Figure 1.4.

a. The point representing the real number $-\frac{5}{3} = -1.666\ldots$ lies between -2 and -1, but closer to -2, on the real number line.

b. The point representing the real number 2.3 lies between 2 and 3, but closer to 2, on the real number line.

c. The point representing the real number $\frac{9}{4} = 2.25$ lies between 2 and 3, but closer to 2, on the real number line. Note that the point representing $\frac{9}{4}$ lies slightly to the left of the point representing 2.3.

d. The point representing the real number -0.3 lies between -1 and 0, but closer to 0, on the real number line.

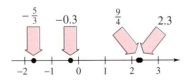

Figure 1.4

The real number line provides a way of comparing any two real numbers. For instance, if you choose any two (different) numbers on the real number line, one of the numbers must be to the left of the other. You can describe this by saying that the number to the left is **less than** the number to the right, or that the number to the right is **greater than** the number to the left, as shown in Figure 1.5.

Figure 1.5   *a* is to the left of *b*.

Order on the Real Number Line

If the real number a lies to the left of the real number b on the real number line, then a is **less than** b, which is written as

$$a < b.$$

This relationship can also be described by saying that b is **greater than** a and writing $b > a$. The expression $a \leq b$ means that a is **less than or equal to** b, and the expression $b \geq a$ means that b is **greater than or equal to** a. The symbols $<$, $>$, $\leq$, and $\geq$ are called **inequality symbols.**

Figure 1.6

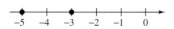

Figure 1.7

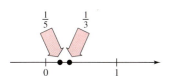

Figure 1.8

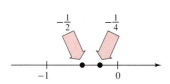

Figure 1.9

When asked to **order** two numbers, you are simply being asked to say which of the two numbers is greater.

Example 3   Ordering Real Numbers

Place the correct inequality symbol ($<$ or $>$) between each pair of numbers.

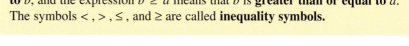

a. $-4 \quad\quad 0$ **b.** $-3 \quad\quad -5$ **c.** $\dfrac{1}{5} \quad\quad \dfrac{1}{3}$ **d.** $-\dfrac{1}{4} \quad\quad -\dfrac{1}{2}$

Solution

a. Because -4 lies to the left of 0 on the real number line, as shown in Figure 1.6, you can say that -4 is *less than* 0, and write $-4 < 0$.

b. Because -3 lies to the right of -5 on the real number line, as shown in Figure 1.7, you can say that -3 is *greater than* -5, and write $-3 > -5$.

c. Because $\frac{1}{5}$ lies to the left of $\frac{1}{3}$ on the real number line, as shown in Figure 1.8, you can say that $\frac{1}{5}$ is *less than* $\frac{1}{3}$, and write $\frac{1}{5} < \frac{1}{3}$.

d. Because $-\frac{1}{4}$ lies to the right of $-\frac{1}{2}$ on the real number line, as shown in Figure 1.9, you can say that $-\frac{1}{4}$ is *greater than* $-\frac{1}{2}$, and write $-\frac{1}{4} > -\frac{1}{2}$.

One effective way to order two fractions such as $\frac{5}{12}$ and $\frac{9}{23}$ is to compare their decimal equivalents. Because $\frac{5}{12} = 0.41\overline{6}$ and $\frac{9}{23} \approx 0.391$, you can write

$$\frac{5}{12} > \frac{9}{23}.$$

3 Use the real number line to find the distance between two real numbers.

Distance on the Real Number Line

Once you know how to represent real numbers as points on the real number line, it is natural to talk about the **distance between two real numbers.** Specifically, if a and b are two real numbers such that $a \le b$, then the distance between a and b is defined to be $b - a$.

Distance Between Two Real Numbers

If a and b are two real numbers such that $a \le b$, then the **distance between a and b** is given by

Distance between a and $b = b - a$.

Note from this definition that if $a = b$, the distance between a and b is zero. If $a \ne b$, the distance between a and b is positive.

Example 4   Finding the Distance Between Two Real Numbers

Find the distance between each pair of real numbers.

a. -2 and 3 **b.** 0 and 4 **c.** -4 and 0 **d.** 1 and $-\dfrac{1}{2}$

Solution

a. Because $-2 \le 3$, the distance between -2 and 3 is

$$3 - (-2) = 3 + 2 = 5. \qquad \text{See Figure 1.10.}$$

b. Because $0 \le 4$, the distance between 0 and 4 is

$$4 - 0 = 4. \qquad \text{See Figure 1.11.}$$

c. Because $-4 \le 0$, the distance between -4 and 0 is

$$0 - (-4) = 0 + 4 = 4. \qquad \text{See Figure 1.12.}$$

d. Because $-\frac{1}{2} \le 1$, let $a = -\frac{1}{2}$ and $b = 1$. So, the distance between 1 and $-\frac{1}{2}$ is

$$1 - \left(-\frac{1}{2}\right) = 1 + \frac{1}{2} = 1\frac{1}{2}. \qquad \text{See Figure 1.13.}$$

Study Tip

Recall that when you subtract a negative number, as in Example 4(a), you add the opposite of the second number to the first. Because the opposite of -2 is 2, you add 2 to 3.

Figure 1.10

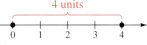

Figure 1.11

Figure 1.12

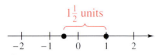

Figure 1.13

④ Determine the absolute value of a real number.

Absolute Value

Two real numbers are called **opposites** of each other if they lie the same distance from, but on opposite sides of, 0 on the real number line. For instance, -2 is the opposite of 2 (see Figure 1.14).

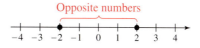

Figure1.14

The opposite of a negative number is called a **double negative** (see Figure 1.15).

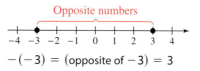

$-(-3) =$ (opposite of -3) $= 3$

Figure1.15

Opposite numbers are also referred to as **additive inverses** because their sum is zero. For instance, $3 + (-3) = 0$. In general, you have the following.

Opposites and Additive Inverses

Let a be a real number.

   1. $-a$ is the opposite of a.

   2. $-(-a) = a$ Double negative

   3. $a + (-a) = 0$ Additive inverse

The distance between a real number a and 0 (the origin) is called the **absolute value** of a. Absolute value is denoted by double vertical bars, $|\ \ |$. For example,

$$|5| = \text{“distance between 5 and 0”} = 5$$

and

$$|-8| = \text{“distance between } -8 \text{ and 0”} = 8.$$

Be sure you see from the following definition that the absolute value of a real number is never negative. For instance, if $a = -3$, then $|-3| = -(-3) = 3$. Moreover, the only real number whose absolute value is zero is 0. That is, $|0| = 0$.

Definition of Absolute Value

If a is a real number, then the **absolute value** of a is

$$|a| = \begin{cases} a, & \text{if } a \geq 0 \\ -a, & \text{if } a < 0. \end{cases}$$

Study Tip

Because *opposite* numbers lie the same distance from 0 on the real number line, they have the same absolute value. So, $|5| = 5$ and $|-5| = 5$.

Example 5   Finding Absolute Values

a. $|-10| = 10$ The absolute value of -10 is 10.

b. $\left|\dfrac{3}{4}\right| = \dfrac{3}{4}$ The absolute value of $\frac{3}{4}$ is $\frac{3}{4}$.

c. $|-3.2| = 3.2$ The absolute value of -3.2 is 3.2.

d. $-|-6| = -(6) = -6$ The opposite of $|-6|$ is -6.

Note that part (d) does not contradict the fact that the absolute value of a number cannot be negative. The expression $-|-6|$ calls for the *opposite* of an absolute value, and so it must be negative.

For any two real numbers a and b, exactly one of the following orders must be true: $a < b$, $a = b$, or $a > b$. This property of real numbers is called the **Law of Trichotomy.** In words, this property tells you that if a and b are any two real numbers, then a is less than b, a is equal to b, or a is greater than b.

Example 6   Comparing Real Numbers

Place the correct symbol ($<$, $>$, or $=$) between each pair of real numbers.

a. $|-2|$ ___ 1 **b.** $|-4|$ ___ $|4|$ **c.** $|12|$ ___ $|-15|$

d. $|-3|$ ___ -3 **e.** 2 ___ $-|-2|$ **f.** $-|-3|$ ___ -3

Solution

a. $|-2| > 1$, because $|-2| = 2$ and 2 is greater than 1.

b. $|-4| = |4|$, because $|-4| = 4$ and $|4| = 4$.

c. $|12| < |-15|$, because $|12| = 12$ and $|-15| = 15$, and 12 is less than 15.

d. $|-3| > -3$, because $|-3| = 3$ and 3 is greater than -3.

e. $2 > -|-2|$, because $-|-2| = -2$ and 2 is greater than -2.

f. $-|-3| = -3$, because $-|-3| = -3$ and -3 is equal to -3.

When the distance between the two real numbers a and b was defined to be $b - a$, the definition included the restriction $a \le b$. Using absolute value, you can generalize this definition. That is, if a and b are *any* two real numbers, then the distance between a and b is given by

Distance between a and $b = |b - a| = |a - b|$.

For instance, the distance between -2 and 1 is given by

$|-2 - 1| = |-3| = 3.$ Distance between -2 and 1

You could also find the distance between -2 and 1 as follows.

$|1 - (-2)| = |3| = 3$ Distance between -2 and 1

1.1 Exercises

Developing Skills

In Exercises 1–4, which of the real numbers in the set are (a) natural numbers, (b) integers, (c) rational numbers, and (d) irrational numbers? See Example 1.

1. $\left\{-10, -\sqrt{5}, -\frac{2}{3}, -\frac{1}{4}, 0, \frac{5}{8}, 1, \sqrt{3}, 4, 2\pi, 6\right\}$

2. $\left\{-\frac{7}{2}, -\sqrt{6}, -\frac{\pi}{2}, -\frac{3}{8}, 0, \sqrt{15}, \frac{10}{3}, 8, 245\right\}$

3. $\left\{-3.5, -\sqrt{4}, -\frac{1}{2}, -0.\overline{3}, 0, 3, \sqrt{5}, 3\pi, 25.2\right\}$

4. $\left\{-\sqrt{25}, -\sqrt{6}, -0.\overline{1}, -\frac{5}{3}, 0, 0.85, 3, 110\right\}$

In Exercises 5–8, list all members of the set.

5. The integers between -5.8 and 3.2

6. The even integers between -2.1 and 10.5

7. The odd integers between 0 and 3π

8. All prime numbers between 0 and 25

In Exercises 9 and 10, plot the real numbers on the real number line. See Example 2.

9. (a) 3 (b) $\frac{5}{2}$ (c) $-\frac{7}{2}$ (d) -5.2

10. (a) 8 (b) $\frac{4}{3}$ (c) -6.75 (d) $-\frac{9}{2}$

In Exercises 11–14, approximate the two numbers and order them.

11.

12.

13.

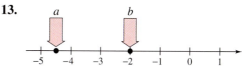

14.

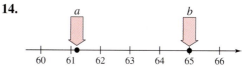

In Exercises 15–30, place the correct inequality symbol (< or >) between the pair of numbers. See Example 3.

15. 4 ⬜ 8   **16.** 8 ⬜ 3

17. $\frac{4}{5}$ ⬜ 1   **18.** 2 ⬜ $\frac{5}{3}$

19. -5 ⬜ 2   **20.** 9 ⬜ -1

21. -5 ⬜ -2 **22.** -8 ⬜ -3

23. -1 ⬜ -10 **24.** -2 ⬜ -3

25. $\frac{5}{8}$ ⬜ $\frac{1}{2}$ **26.** $\frac{3}{2}$ ⬜ $\frac{5}{2}$

27. $-\frac{2}{3}$ ⬜ $-\frac{10}{3}$ **28.** $-\frac{5}{3}$ ⬜ $-\frac{3}{2}$

29. 2.75 ⬜ π **30.** $-\pi$ ⬜ -3.1

In Exercises 31–42, find the distance between the pair of real numbers. See Example 4.

31. 4 and 10   **32.** 75 and 20

33. -12 and 7   **34.** -54 and 32

35. 18 and -32 **36.** 14 and -6

37. -8 and 0   **38.** 0 and 125

39. 0 and 35   **40.** -35 and 0

41. -6 and -9 **42.** -12 and -7

In Exercises 43–56, evaluate the expression. See Example 5.

43. $|10|$ **44.** $|62|$

45. $|-225|$ **46.** $|-14|$

47. $-|-85|$ **48.** $-|-36.5|$

49. $-|16|$ **50.** $-|-25|$

51. $-\left|-\frac{3}{4}\right|$

52. $-\left|\frac{3}{8}\right|$

53. $-|3.5|$

54. $|-1.4|$

55. $|-\pi|$

56. $-|\pi|$

In Exercises 57–64, place the correct symbol ($<$, $>$, or $=$) between the pair of real numbers. See Example 6.

57. $|-6|$ ___ $|2|$

58. $|-2|$ ___ $|2|$

59. $|47|$ ___ $|-27|$

60. $|150|$ ___ $|-310|$

61. $-|-1.8|$ ___ $-|1.8|$

62. $|12.5|$ ___ $-|-25|$

63. $\left|-\frac{3}{4}\right|$ ___ $-\left|\frac{4}{5}\right|$

64. $-\left|-\frac{7}{3}\right|$ ___ $-\left|\frac{1}{3}\right|$

In Exercises 65–74, find the opposite and the absolute value of the number.

65. 34

66. 225

67. -160

68. -52

69. $-\frac{3}{11}$

70. $\frac{7}{32}$

71. $\frac{5}{4}$

72. $\frac{4}{3}$

73. 4.7

74. -0.4

In Exercises 75–84, plot the number and its opposite on the real number line. Determine the distance of each from 0.

75. 7

76. -3

77. -5

78. 6

79. $-\frac{3}{5}$

80. $\frac{7}{4}$

81. $\frac{5}{3}$

82. $-\frac{3}{4}$

83. -4.25

84. 3.5

In Exercises 85–92, write the statement using inequality notation.

85. x is negative.

86. y is more than 25.

87. x is nonnegative.

88. u is at least 16.

89. z is greater than 2 and no more than 10.

90. The tire pressure p is at least 30 pounds per square inch and no more than 35 pounds per square inch.

91. The price p is less than \$225.

92. The average a will exceed 10,000.

Explaining Concepts

True or False? In Exercises 93–98, decide whether the statement is true or false. If the statement is false, give an example of a real number that makes the statement false.

93. Every integer is a rational number.

94. Every real number is either rational or irrational.

95. If x and y are real numbers, $|x + y| = |x| + |y|$.

96. $\frac{1}{6} = 0.17$

97. Every whole number is a natural number.

98. Every natural number is a whole number.

99. *Writing* Describe the difference between the set of natural numbers and the set of integers.

100. *Writing* Describe the difference between the rational numbers 0.15 and $0.\overline{15}$.

101. *Writing* Is there a difference between saying that a real number is positive and saying that a real number is nonnegative? Explain your answer.

102. *Writing* Which real number lies farther from -4: -8 or 6? Explain your answer.

103. *Writing* If you are given two real numbers a and b, explain how you can tell which is greater.

104. For each list of real numbers, arrange the entries in increasing order.

a. $3, -\sqrt{7}, \sqrt{2}, \pi, -3, 1, \frac{22}{7}, -\frac{15}{7}$

b. $\left|-\frac{3}{5}\right|, -\left|\frac{3}{5}\right|, \left|-\frac{5}{3}\right|, -\left|\frac{5}{3}\right|, \left|-\frac{3}{4}\right|, -\left|\frac{3}{4}\right|, \left|-\frac{4}{5}\right|, -\left|\frac{4}{5}\right|$

i 71.2 Operations with Real Numbers

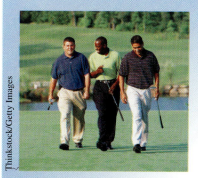

Thinkstock/Getty Images

What You Should Learn

1 Add, subtract, multiply, and divide real numbers.
2 Write repeated multiplication in exponential form and evaluate exponential expressions.
3 Use order of operations to evaluate expressions.
4 Evaluate expressions using a calculator and order of operations.

Why You Should Learn It

Real numbers can be used to represent many real-life quantities, such as the net profits for Calloway Golf Company (see Exercise 138 on page 21).

1 Add, subtract, multiply, and divide real numbers.

Operations with Real Numbers

There are four basic operations of arithmetic: addition, subtraction, multiplication, and division.

The result of adding two real numbers is the **sum** of the two numbers, and the two real numbers are the **terms** of the sum. The rules for adding real numbers are as follows.

Addition of Two Real Numbers

1. To **add** two real numbers with *like signs*, add their absolute values and attach the common sign to the result.

2. To **add** two real numbers with *unlike signs*, subtract the smaller absolute value from the greater absolute value and attach the sign of the number with the greater absolute value.

Example 1   Adding Integers

a. $-84 + 14 = -(84 - 14)$ Choose negative sign.

$= -70$ Subtract absolute values.

b. $-138 + (-62) = -(138 + 62)$ Use common sign.

$= -200$ Add absolute values.

Example 2   Adding Decimals

a. $-26.41 + (-0.53) = -(26.41 + 0.53)$ Use common sign.

$= -26.94$ Add absolute values.

b. $3.2 + (-0.4) = +(3.2 - 0.4)$ Choose positive sign.

$= 2.8$ Subtract absolute values.

The result of subtracting two real numbers is the **difference** of the two numbers. Subtraction of two real numbers is defined in terms of addition as follows.

Subtraction of Two Real Numbers

To **subtract** the real number b from the real number a, add the opposite of b to a. That is,

$$a - b = a + (-b).$$

Example 3 Subtracting Integers

Find each difference.

a. $9 - 21$ **b.** $-15 - 8$

Solution

a. $9 - 21 = 9 + (-21)$ Add opposite of 21.

$\qquad = -(21 - 9) = -12$ Choose negative sign and subtract absolute values.

b. $-15 - 8 = -15 + (-8)$ Add opposite of 8.

$\qquad = -(15 + 8) = -23$ Use common sign and add absolute values.

Example 4 Subtracting Decimals

Find each difference.

a. $-2.5 - (-2.7)$ **b.** $-7.02 - 13.8$

Solution

a. $-2.5 - (-2.7) = -2.5 + 2.7$ Add opposite of -2.7.

$\qquad = +(2.7 - 2.5) = 0.2$ Choose positive sign and subtract absolute values.

b. $-7.02 - 13.8 = -7.02 + (-13.8)$ Add opposite of 13.8.

$\qquad = -(7.02 + 13.8) = -20.82$ Use common sign and add absolute values.

Example 5 Evaluating an Expression

Evaluate the expression.

$$-13 - 7 + 11 - (-4)$$

Solution

$-13 - 7 + 11 - (-4) = -13 + (-7) + 11 + 4$ Add opposites.

$\qquad = -20 + 15$ Add two numbers at a time.

$\qquad = -5$ Add.

To add or subtract fractions, it is useful to recognize the equivalent forms of fractions, as illustrated below.

$$\frac{a}{b} = \frac{-a}{-b} = -\frac{-a}{b} = -\frac{a}{-b}$$ All are positive.

$$-\frac{a}{b} = \frac{-a}{b} = \frac{a}{-b} = -\frac{-a}{-b}$$ All are negative.

Study Tip

Here is an alternative method for adding and subtracting fractions with unlike denominators ($b \neq 0$ and $d \neq 0$).

$$\frac{a}{b} + \frac{c}{d} = \frac{ad + bc}{bd}$$

$$\frac{a}{b} - \frac{c}{d} = \frac{ad - bc}{bd}$$

For example,

$$\frac{1}{6} + \frac{3}{8} = \frac{1(8) + 6(3)}{6(8)}$$

$$= \frac{8 + 18}{48}$$

$$= \frac{26}{48}$$

$$= \frac{13}{24}.$$

Note that an additional step is needed to simplify the fraction after the numerators have been added.

Addition and Subtraction of Fractions

1. *Like Denominators:* The sum and difference of two fractions with like denominators ($c \neq 0$) are:

$$\frac{a}{c} + \frac{b}{c} = \frac{a + b}{c} \qquad \frac{a}{c} - \frac{b}{c} = \frac{a - b}{c}$$

2. *Unlike Denominators:* To add or subtract two fractions with unlike denominators, first rewrite the fractions so that they have the same denominator and apply the first rule.

To find the **least common denominator (LCD)** for two or more fractions, find the **least common multiple (LCM)** of their denominators. For instance, the LCM of 6 and 8 is 24. To see this, consider all multiples of 6 (6, 12, 18, 24, 30, 36, 42, 48, . . .) and all multiples of 8 (8, 16, 24, 32, 40, 48, . . .). The numbers 24 and 48 are common multiples, and the number 24 is the smallest of the common multiples. To add $\frac{1}{6}$ and $\frac{3}{8}$, proceed as follows.

$$\frac{1}{6} + \frac{3}{8} = \frac{1(4)}{6(4)} + \frac{3(3)}{8(3)} = \frac{4}{24} + \frac{9}{24} = \frac{4 + 9}{24} = \frac{13}{24}$$

Example 6   Adding and Subtracting Fractions

a. $\dfrac{5}{17} + \dfrac{9}{17} = \dfrac{5 + 9}{17}$ Add numerators.

$$= \frac{14}{17}$$ Simplify.

b. $\dfrac{3}{8} - \dfrac{5}{12} = \dfrac{3(3)}{8(3)} - \dfrac{5(2)}{12(2)}$ Least common denominator is 24.

$$= \frac{9}{24} - \frac{10}{24}$$ Simplify.

$$= \frac{9 - 10}{24}$$ Subtract numerators.

$$= -\frac{1}{24}$$ Simplify.

Example 7　Adding Mixed Numbers

Find the sum of $1\frac{4}{5}$ and $\frac{11}{7}$.

Solution

$$1\frac{4}{5} + \frac{11}{7} = \frac{9}{5} + \frac{11}{7} \qquad \text{Write } 1\frac{4}{5} \text{ as } \frac{9}{5}.$$

$$= \frac{9(7)}{5(7)} + \frac{11(5)}{7(5)} \qquad \text{Least common denominator is 35.}$$

$$= \frac{63}{35} + \frac{55}{35} \qquad \text{Simplify.}$$

$$= \frac{63 + 55}{35} = \frac{118}{35} \qquad \text{Add numerators and simplify.}$$

Multiplication of two real numbers can be described as *repeated addition*. For instance, 7×3 can be described as $3 + 3 + 3 + 3 + 3 + 3 + 3$. Multiplication is denoted in a variety of ways. For instance, 7×3, $7 \cdot 3$, $7(3)$, and $(7)(3)$ all denote the product "7 times 3." The result of multiplying two real numbers is their **product**, and each of the two numbers is a **factor** of the product.

Multiplication of Two Real Numbers

1. To multiply two real numbers with *like signs,* find the product of their absolute values. The product is *positive.*

2. To multiply two real numbers with *unlike signs,* find the product of their absolute values, and attach a minus sign. The product is *negative.*

3. The product of zero and any other real number is zero.

Example 8　Multiplying Integers

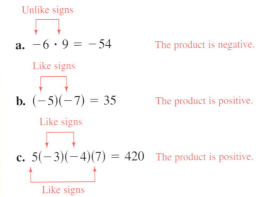

Unlike signs

a. $-6 \cdot 9 = -54$　　　The product is negative.

Like signs

b. $(-5)(-7) = 35$　　　The product is positive.

Like signs

c. $5(-3)(-4)(7) = 420$　　The product is positive.

Like signs

Study Tip

When operating with fractions, you should check to see whether your answers can be simplified by dividing out factors that are common to the numerator and denominator. For instance, the fraction $\frac{4}{6}$ can be written in simplified form as

$$\frac{4}{6} = \frac{\overset{1}{2 \cdot 2}}{\underset{1}{2 \cdot 3}} = \frac{2}{3}.$$

Note that dividing out a common factor is the division of a number by itself, and what remains is a factor of 1.

Multiplication of Two Fractions

The product of the two fractions a/b and c/d is given by

$$\frac{a}{b} \cdot \frac{c}{d} = \frac{ac}{bd}, \quad b \neq 0, \quad d \neq 0.$$

Example 9 Multiplying Fractions

Find the product.

$$\left(-\frac{3}{8}\right)\left(\frac{11}{6}\right)$$

Solution

$$\left(-\frac{3}{8}\right)\left(\frac{11}{6}\right) = -\frac{3(11)}{8(6)}$$ Multiply numerators and denominators.

$$= -\frac{\cancel{3}(11)}{8(2)(\cancel{3})}$$ Factor and divide out common factor.

$$= -\frac{11}{16}$$ Simplify.

The **reciprocal** of a nonzero real number a is defined as the number by which a must be multiplied to obtain 1. For instance, the reciprocal of 3 is $\frac{1}{3}$ because

$$3\left(\frac{1}{3}\right) = 1.$$

Similarly, the reciprocal of $-\frac{4}{5}$ is $-\frac{5}{4}$ because

$$-\frac{4}{5}\left(-\frac{5}{4}\right) = 1.$$

In general, the reciprocal of a/b is b/a. Note that the reciprocal of a positive number is positive, and the reciprocal of a negative number is negative.

Study Tip

Division by 0 is not defined because 0 has no reciprocal. If 0 had a reciprocal value b, then you would obtain the *false* result

$\frac{1}{0} = b$ The reciprocal of zero is b.

$1 = b \cdot 0$ Multiply each side by 0.

$1 = 0.$ False result, $1 \neq 0$

Division of Two Real Numbers

To divide the real number a by the nonzero real number b, multiply a by the reciprocal of b. That is,

$$a \div b = a \cdot \frac{1}{b}, \quad b \neq 0.$$

The result of dividing two real numbers is the **quotient** of the numbers. The number a is the **dividend** and the number b is the **divisor.** When the division is expressed as a/b or $\frac{a}{b}$, a is the **numerator** and b is the **denominator.**

Example 10   **Division of Real Numbers**

a. $-30 \div 5 = -30 \cdot \dfrac{1}{5}$ Invert divisor and multiply.

$\qquad\qquad = -\dfrac{30}{5}$ Multiply.

$\qquad\qquad = -\dfrac{6 \cdot 5}{5}$ Factor and divide out common factor.

$\qquad\qquad = -6$ Simplify.

b. $\dfrac{5}{16} \div 2\dfrac{3}{4} = \dfrac{5}{16} \div \dfrac{11}{4}$ Write $2\frac{3}{4}$ as $\frac{11}{4}$.

$\qquad\qquad = \dfrac{5}{16} \cdot \dfrac{4}{11}$ Invert divisor and multiply.

$\qquad\qquad = \dfrac{5(4)}{16(11)}$ Multiply.

$\qquad\qquad = \dfrac{5}{44}$ Simplify.

② Write repeated multiplication in exponential form and evaluate exponential expressions.

Positive Integer Exponents

Just as multiplication by a positive integer can be described as repeated addition, *repeated multiplication* can be written in what is called **exponential form.** Here is an example.

Repeated Multiplication		*Exponential Form*
$\underbrace{7 \cdot 7 \cdot 7 \cdot 7}_{\text{4 factors of 7}}$	$=$	7^4
$\underbrace{\left(-\dfrac{3}{4}\right)\left(-\dfrac{3}{4}\right)\left(-\dfrac{3}{4}\right)}_{\text{3 factors of } -\frac{3}{4}}$	$=$	$\left(-\dfrac{3}{4}\right)^3$

Exponential Notation

Let n be a positive integer and let a be a real number. Then the product of n factors of a is given by

$$a^n = \underbrace{a \cdot a \cdot a \cdots a}_{n \text{ factors}}.$$

In the exponential form a^n, a is the **base** and n is the **exponent.** Writing the exponential form a^n is called **"raising a to the nth power."**

When a number is raised to the *first* power, you usually do not write the exponent 1. For instance, you would usually write 5 rather than 5^1. Raising a number to the *second* power is called **squaring** the number. Raising a number to the *third* power is called **cubing** the number.

Study Tip

In parts (d) and (e) of Example 11, note that when a negative number is raised to an *odd* power, the result is *negative*, and when a negative number is raised to an *even* power, the result is *positive*.

Example 11   Evaluating Exponential Expressions

a. $(-3)^4 = (-3)(-3)(-3)(-3) = 81$ Negative sign is part of the base.

b. $-3^4 = -(3)(3)(3)(3) = -81$ Negative sign is not part of the base.

c. $\left(\dfrac{2}{5}\right)^3 = \left(\dfrac{2}{5}\right)\left(\dfrac{2}{5}\right)\left(\dfrac{2}{5}\right) = \dfrac{8}{125}$

d. $(-2)^5 = (-2)(-2)(-2)(-2)(-2) = -32$

e. $(-2)^6 = (-2)(-2)(-2)(-2)(-2)(-2) = 64$

In parts (a) and (b) of Example 11, be sure you see the distinction between the expressions $(-3)^4$ and -3^4. Here is a similar example.

$(-5)^2 = (-5)(-5) = 25$ Negative sign is part of the base.

$-5^2 = -(5)(5) = -25$ Negative sign is not part of the base.

3 Use order of operations to evaluate expressions.

Study Tip

The order of operations for multiplication applies when multiplication is written with the symbol $\times$ or $\cdot$. When multiplication is implied by parentheses, it has a higher priority than the Left-to-Right Rule. For instance,

$$8 \div 4(2) = 8 \div 8 = 1$$

but

$$8 \div 4 \cdot 2 = 2 \cdot 2 = 4.$$

Order of Operations

One of your goals in studying this book is to learn to communicate about algebra by reading and writing information about numbers. One way to help avoid confusion when communicating algebraic ideas is to establish an **order of operations.** This is done by giving priorities to different operations. First priority is given to exponents, second priority is given to multiplication and division, and third priority is given to addition and subtraction. To distinguish between operations with the same priority, use the *Left-to-Right Rule.*

Order of Operations

To evaluate an expression involving more than one operation, use the following order.

1. First do operations that occur within symbols of grouping.
2. Then evaluate powers.
3. Then do multiplications and divisions from left to right.
4. Finally, do additions and subtractions from left to right.

Example 12   Order of Operations Without Symbols of Grouping

a. $20 - 2 \cdot 3^2 = 20 - 2 \cdot 9$ Evaluate power.

$\qquad\qquad\quad = 20 - 18 = 2$ Multiply, then subtract.

b. $5 - 6 - 2 = (5 - 6) - 2$ Left-to-Right Rule

$\qquad\qquad\quad = -1 - 2 = -3$ Subtract.

c. $8 \div 2 \cdot 2 = (8 \div 2) \cdot 2$ Left-to-Right Rule

$\qquad\qquad = 4 \cdot 2 = 8$ Divide, then multiply.

When you want to change the established order of operations, you must use parentheses or other symbols of grouping. Part (d) in the next example shows that a fraction bar acts as a symbol of grouping.

Example 13   Order of Operations with Symbols of Grouping

a. $7 - 3(4 - 2) = 7 - 3(2)$ Subtract within symbols of grouping.

$\qquad = 7 - 6 = 1$ Multiply, then subtract.

b. $4 - 3(2)^3 = 4 - 3(8)$ Evaluate power.

$\qquad = 4 - 24 = -20$ Multiply, then subtract.

c. $1 - [4 - (5 - 3)] = 1 - (4 - 2)$ Subtract within symbols of grouping.

$\qquad = 1 - 2 = -1$ Subtract within symbols of grouping, then subtract.

d. $\dfrac{2 \cdot 5^2 - 10}{3^2 - 4} = (2 \cdot 5^2 - 10) \div (3^2 - 4)$ Rewrite using parentheses.

$\qquad = (50 - 10) \div (9 - 4)$ Evaluate powers and multiply within symbols of grouping.

$\qquad = 40 \div 5 = 8$ Subtract within symbols of grouping, then divide.

4 Evaluate expressions using a calculator and order of operations.

Calculators and Order of Operations

When using your own calculator, be sure that you are familiar with the use of each of the keys. For each of the calculator examples in the text, two possible keystroke sequences are given: one for a standard *scientific* calculator, and one for a *graphing* calculator.

Technology: Tip

Be sure you see the difference between the change sign key $\boxed{+/-}$ and the subtraction key $\boxed{-}$ on a scientific calculator. Also notice the difference between the minus key $\boxed{(-)}$ and the subtraction key $\boxed{-}$ on a graphing calculator.

Example 14   Evaluating Expressions on a Calculator

a. To evaluate the expression $7 - (5 \cdot 3)$, use the following keystrokes.

Keystrokes	Display	
7 $\boxed{-}$ $\boxed{(}$ 5 $\boxed{\times}$ 3 $\boxed{)}$ $\boxed{=}$	-8	Scientific
7 $\boxed{-}$ $\boxed{(}$ 5 $\boxed{\times}$ 3 $\boxed{)}$ $\boxed{\text{ENTER}}$	-8	Graphing

b. To evaluate the expression $(-3)^2 + 4$, use the following keystrokes.

Keystrokes	Display	
3 $\boxed{+/-}$ $\boxed{x^2}$ $\boxed{+}$ 4 $\boxed{=}$	13	Scientific
$\boxed{(}$ $\boxed{(-)}$ 3 $\boxed{)}$ $\boxed{x^2}$ $\boxed{+}$ 4 $\boxed{\text{ENTER}}$	13	Graphing

c. To evaluate the expression $5/(4 + 3 \cdot 2)$, use the following keystrokes.

Keystrokes	Display	
5 $\boxed{\div}$ $\boxed{(}$ 4 $\boxed{+}$ 3 $\boxed{\times}$ 2 $\boxed{)}$ $\boxed{=}$	0.5	Scientific
5 $\boxed{\div}$ $\boxed{(}$ 4 $\boxed{+}$ 3 $\boxed{\times}$ 2 $\boxed{)}$ $\boxed{\text{ENTER}}$	.5	Graphing

Technology: Discovery

To discover if your calculator performs the established order of operations, evaluate $7 + 5 \cdot 3 - 2^4 \div 4$ exactly as it appears. Does your calculator display 5 or 18? If your calculator performs the established order of operations, it will display 18.

1.2 Exercises

Developing Skills

In Exercises 1–38, evaluate the expression. See Examples 1–7.

1. $13 + 32$
2. $16 + 84$
3. $-8 + 12$
4. $-5 + 9$
5. $-6.4 + 3.7$
6. $-5.1 + 0.9$
7. $13 + (-6)$
8. $12 + (-10)$
9. $12.6 + (-38.5)$
10. $10.4 + (-43.5)$

11. $-7 - 15$
12. $-22 - 6$
13. $-13 + (-8)$
14. $-5 + (-52)$
15. $4 - 16 + (-8)$
16. $-15 + (-6) + 32$

17. $5.8 - 6.2 + 1.1$
18. $46.08 - 35.1 - 16.25$

19. $15 - 6 + 31 + (-18)$
20. $-21 + 8 - 13 - 51$

21. $\frac{3}{8} + \frac{7}{8}$
22. $\frac{5}{6} + \frac{7}{6}$
23. $\frac{3}{4} - \frac{1}{4}$
24. $\frac{5}{9} - \frac{1}{9}$
25. $\frac{3}{5} + \left(-\frac{1}{2}\right)$
26. $\frac{5}{6} - \frac{3}{4}$
27. $\frac{5}{8} + \frac{1}{4} - \frac{5}{6}$
28. $\frac{3}{10} - \frac{5}{2} + \frac{1}{5}$
29. $3\frac{1}{2} + 4\frac{3}{8}$
30. $10\frac{5}{8} - 6\frac{1}{4}$
31. $5\frac{3}{4} + 7\frac{3}{8}$
32. $8\frac{1}{2} - 4\frac{2}{3}$
33. $85 - |-25|$
34. $-36 + |-8|$
35. $-(-11.325) + |34.625|$
36. $|-16.25| - 54.78$
37. $-|-15.667| - 12.333$
38. $-\left|-15\frac{2}{3}\right| - 12\frac{1}{3}$

In Exercises 39–44, write the expression as a repeated addition problem.

39. $4(5)$
40. $5(2)$

41. $3(-4)$
42. $6(-2)$
43. $5\left(\frac{1}{2}\right)$
44. $4\left(\frac{3}{4}\right)$

In Exercises 45–50, write the expression as a multiplication problem.

45. $9 + 9 + 9 + 9$

46. $(-15) + (-15) + (-15) + (-15)$
47. $\frac{1}{4} + \frac{1}{4} + \frac{1}{4} + \frac{1}{4} + \frac{1}{4} + \frac{1}{4}$
48. $\frac{2}{3} + \frac{2}{3} + \frac{2}{3} + \frac{2}{3}$
49. $\left(-\frac{1}{5}\right) + \left(-\frac{1}{5}\right) + \left(-\frac{1}{5}\right) + \left(-\frac{1}{5}\right)$
50. $\left(-\frac{5}{22}\right) + \left(-\frac{5}{22}\right) + \left(-\frac{5}{22}\right)$

In Exercises 51–70, find the product. See Examples 8 and 9.

51. $5(-6)$
52. $-7(3)$
53. $(-8)(-6)$
54. $(-4)(-7)$
55. $-6(12)$
56. $7(10)$
57. $2(4)(-5)$
58. $3(-7)(10)$
59. $(-1)(12)(-3)$
60. $(-2)(-6)(4)$
61. $\left(-\frac{5}{8}\right)\left(-\frac{4}{5}\right)$
62. $\left(\frac{10}{13}\right)\left(-\frac{3}{5}\right)$
63. $-\frac{3}{2}\left(\frac{8}{5}\right)$
64. $\left(-\frac{4}{7}\right)\left(-\frac{4}{5}\right)$
65. $\frac{1}{2}\left(\frac{1}{6}\right)$
66. $\frac{1}{3}\left(\frac{2}{3}\right)$
67. $-\frac{9}{8}\left(\frac{16}{27}\right)\left(\frac{1}{2}\right)$
68. $\frac{2}{3}\left(-\frac{18}{5}\right)\left(-\frac{5}{6}\right)$
69. $\frac{1}{3}\left(-\frac{3}{4}\right)(2)$
70. $\frac{2}{5}(-3)\left(\frac{10}{9}\right)$

In Exercises 71–84, evaluate the expression. See Example 10.

71. $\frac{-18}{-3}$
72. $-\frac{30}{-15}$
73. $-48 \div 16$
74. $-27 \div (-9)$
75. $63 \div (-7)$
76. $-72 \div 12$
77. $-\frac{4}{5} \div \frac{8}{25}$
78. $\frac{8}{15} \div \frac{32}{5}$
79. $\left(-\frac{1}{3}\right) \div \left(-\frac{5}{6}\right)$
80. $-\frac{11}{12} \div \frac{5}{24}$
81. $5\frac{3}{4} \div 2\frac{1}{8}$
82. $-3\frac{5}{6} \div -2\frac{2}{3}$
83. $4\frac{1}{8} \div 3\frac{3}{2}$
84. $26\frac{2}{3} \div 10\frac{5}{6}$

In Exercises 85–90, write the expression as a repeated multiplication problem.

85. 4^3
86. $(-6)^5$
87. $\left(-\frac{3}{4}\right)^4$
88. $\left(\frac{2}{3}\right)^3$
89. $(-0.8)^6$
90. $(0.67)^4$

In Exercises 91–96, write the expression using exponential notation.

91. $(-7) \cdot (-7) \cdot (-7)$

92. $(-4)(-4)(-4)(-4)(-4)(-4)$

93. $(-5)(-5)(-5)(-5)$

94. $\left(\frac{5}{8}\right) \cdot \left(\frac{5}{8}\right) \cdot \left(\frac{5}{8}\right) \cdot \left(\frac{5}{8}\right)$

95. $-(7 \cdot 7 \cdot 7)$

96. $-(5 \cdot 5 \cdot 5 \cdot 5 \cdot 5 \cdot 5)$

In Exercises 97–106, evaluate the exponential expression. See Example 11.

97. $(-2)^4$

98. $(-3)^3$

99. -4^3

100. -3^4

101. $\left(\frac{4}{5}\right)^3$

102. $-\left(-\frac{1}{2}\right)^5$

103. $(0.3)^3$

104. $(0.2)^4$

105. $5(-0.4)^3$

106. $-3(0.8)^2$

In Exercises 107–126, evaluate the expression. See Examples 12 and 13.

107. $16 - 6 - 10$

108. $18 - 12 + 4$

109. $24 - 5 \cdot 2^2$

110. $18 + 3^2 - 12$

111. $28 \div 4 + 3 \cdot 5$

112. $6 \cdot 7 - 6^2 \div 4$

113. $14 - 2(8 - 4)$

114. $21 - 5(7 - 5)$

115. $45 + 3(16 \div 4)$

116. $72 - 8(6^2 \div 9)$

117. $5^2 - 2[9 - (18 - 8)]$

118. $8 \cdot 3^2 - 4(12 + 3)$

119. $5^3 + |-14 + 4|$

120. $|(-2)^5| - (25 + 7)$

121. $\dfrac{8 + 7}{12 - 15}$

122. $\dfrac{9 + 6(2)}{3 + 4}$

123. $\dfrac{4^2 - 5}{11} - 7$

124. $\dfrac{5^3 - 50}{-15} + 27$

125. $\dfrac{6 \cdot 2^2 - 12}{3^2 + 3}$

126. $\dfrac{7^2 - 2(11)}{5^2 + 8(-2)}$

In Exercises 127–132, evaluate the expression using a calculator. Round your answer to two decimal places. See Example 14.

127. $5.6[13 - 2.5(-6.3)]$

128. $35(1032 - 4650)$

129. $5^6 - 3(400)$

130. $300(1.09)^{10}$

131. $\dfrac{500}{(1.055)^{20}}$

132. $5(100 - 3.6^4) \div 4.1$

Solving Problems

Circle Graphs   In Exercises 133 and 134, find the unknown fractional part of the circle graph.

133.

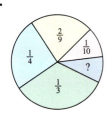

134.

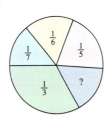

135. *Account Balance*   During one month, you made the following transactions in your non-interest-bearing checking account.

Initial Balance:	$2618.68
Deposit:	$1236.45
Withdrawal:	$25.62
Withdrawal:	$455.00
Withdrawal:	$125.00
Withdrawal:	$715.95

Find the balance at the end of the month.

136. *Profit*   The midyear financial statement of a clothing company showed a profit of $1,415,322.62. At the close of the year, the financial statement showed a profit for the year of $916,489.26. Find the profit (or loss) of the company for the second 6 months of the year.

137. *Stock Values*   On Monday you purchased $500 worth of stock. The value of the stock during the remainder of the week is shown in the bar graph.

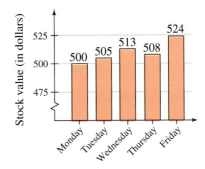

(a) Use the graph to complete the table.

Day	Daily gain or loss
Tuesday	
Wednesday	
Thursday	
Friday	

(b) Find the sum of the daily gains and losses. Interpret the result in the context of the problem. How could you determine this sum from the graph?

138. *Net Profit*   The net profits for Calloway Golf Company (in millions of dollars) for the years 1996 to 2001 are shown in the bar graph. Use the graph to create a table that shows the yearly gains or losses.   (Source: Calloway Golf Company)

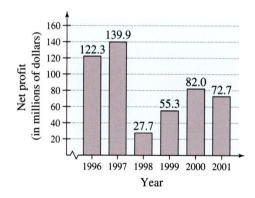

139. *Savings Plan*

(a) You save $50 per month for 18 years. How much money has been set aside during the 18 years?

(b) If the money in part (a) is deposited in a savings account earning 2% interest compounded monthly, the total amount in the account after 18 years will be

$$50\left[\left(1 + \frac{0.02}{12}\right)^{216} - 1\right]\left(1 + \frac{12}{0.02}\right).$$

Use a calculator to determine this amount.

(c) How much of the amount in part (b) is earnings from interest?

140. *Savings Plan*

(a) You save $75 per month for 30 years. How much money has been set aside during the 30 years?

(b) If the money in part (a) is deposited in a savings account earning 3% interest compounded monthly, the total amount in the account after 30 years will be

$$75\left[\left(1 + \frac{0.03}{12}\right)^{360} - 1\right]\left(1 + \frac{12}{0.03}\right).$$

Use a calculator to determine this amount.

(c) How much of the amount in part (b) is earnings from interest?

▲ *Geometry*   In Exercises 141–144, find the area of the figure. (The area A of a rectangle is given by $A = $ length · width, and the area A of a triangle is given by $A = \frac{1}{2}$ · base · height.)

141.   **142.**

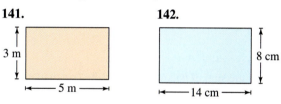

143.   **144.**

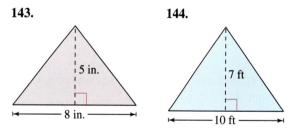

Volume   In Exercises 145 and 146, use the following information. A bale of hay is a rectangular solid weighing approximately 50 pounds. It has a length of 42 inches, a width of 18 inches, and a height of 14 inches. (The volume V of a rectangular solid is given by $V = $ length · width · height.)

145. Find the volume of a bale of hay in cubic feet if 1728 cubic inches equals 1 cubic foot.

146. Approximate the number of bales in a ton of hay. Then approximate the volume of a stack of baled hay that weighs 12 tons.

Explaining Concepts

147. ⚡ Answer parts (a)–(c) of Motivating the Chapter.

True or False? In Exercises 148–152, determine whether the statement is true or false. Justify your answer.

148. The reciprocal of every nonzero integer is an integer.

149. The reciprocal of every nonzero rational number is a rational number.

150. If a negative real number is raised to the 12th power, the result will be positive.

151. If a negative real number is raised to the 11th power, the result will be positive.

152. $a \div b = b \div a$

153. Can the sum of two real numbers be less than either number? If so, give an example.

154. Explain how to subtract one real number from another.

155. *Writing* In your own words, describe the rules for determining the sign of the product or the quotient of two real numbers.

156. If $a > 0$, state the values of n such that $(-a)^n = -a^n$.

157. *Writing* In your own words, describe the established order of operations for addition and subtraction. Without these priorities, explain why the expression $6 - 5 - 2$ would be ambiguous.

158. *Writing* Decide which expressions are equal to 27 when you follow the standard order of operations. For the expressions that are not equal to 27, see if you can discover a way to insert symbols of grouping (parentheses, brackets, or absolute value symbols) that make the expression equal to 27. Discuss the value of symbols of grouping in mathematical communication.

(a) $40 - 10 + 3$ (b) $5^2 + \frac{1}{2} \cdot 4$

(c) $8 \cdot 3 + 30 \div 2$ (d) $75 \div 2 + 1 + 2$

(e) $9 \cdot 4 - 18 \div 2$ (f) $7 \cdot 4 - 4 - 5$

Error Analysis In Exercises 159–162, describe the error.

159. $\dfrac{2}{3} + \dfrac{3}{2} = \dfrac{2 + 3}{3 + 2} = 1$

160. $\dfrac{5 + 12}{5} = \dfrac{5 + 12}{5} = 12$

161. $\dfrac{28}{83} = \dfrac{28}{83} = \dfrac{2}{3}$

162. $3 \cdot 4^2 = 12^2$

The symbol ⚡ indicates an exercise in which you are asked to answer parts of the Motivating the Chapter problem found on the Chapter Opener pages.

1.3 Properties of Real Numbers

What You Should Learn

1. Identify and use the properties of real numbers.
2. Develop additional properties of real numbers.

Why You Should Learn It

Understanding the properties of real numbers will help you to understand and use the properties of algebra.

1 Identify and use the properties of real numbers.

Basic Properties of Real Numbers

The following list summarizes the basic properties of addition and multiplication. Although the examples involve real numbers, you will learn later that these properties can also be applied to algebraic expressions.

Properties of Real Numbers

Let a, b, and c represent real numbers, variables, or algebraic expressions.

Property	Example
Commutative Property of Addition: $a + b = b + a$	$3 + 5 = 5 + 3$
Commutative Property of Multiplication: $ab = ba$	$2 \cdot 7 = 7 \cdot 2$
Associative Property of Addition: $(a + b) + c = a + (b + c)$	$(4 + 2) + 3 = 4 + (2 + 3)$
Associative Property of Multiplication: $(ab)c = a(bc)$	$(2 \cdot 5) \cdot 7 = 2 \cdot (5 \cdot 7)$
Distributive Property: $a(b + c) = ab + ac$	$4(7 + 3) = 4 \cdot 7 + 4 \cdot 3$
$(a + b)c = ac + bc$	$(2 + 5)3 = 2 \cdot 3 + 5 \cdot 3$
$a(b - c) = ab - ac$	$6(5 - 3) = 6 \cdot 5 - 6 \cdot 3$
$(a - b)c = ac - bc$	$(7 - 2)4 = 7 \cdot 4 - 2 \cdot 4$
Additive Identity Property: $a + 0 = 0 + a = a$	$9 + 0 = 0 + 9 = 9$
Multiplicative Identity Property: $a \cdot 1 = 1 \cdot a = a$	$-5 \cdot 1 = 1 \cdot (-5) = -5$
Additive Inverse Property: $a + (-a) = 0$	$3 + (-3) = 0$
Multiplicative Inverse Property: $a \cdot \dfrac{1}{a} = 1, \quad a \neq 0$	$8 \cdot \dfrac{1}{8} = 1$

Study Tip

The operations of subtraction and division are not listed at the right because they do not have many of the properties of real numbers. For instance, subtraction and division are not commutative or associative. To see this, consider the following.

$4 - 3 \neq 3 - 4$

$15 \div 5 \neq 5 \div 15$

$8 - (6 - 2) \neq (8 - 6) - 2$

$20 \div (4 \div 2) \neq (20 \div 4) \div 2$

Example 1   Identifying Properties of Real Numbers

Identify the property of real numbers illustrated by each statement.

a. $4(a + 3) = 4 \cdot a + 4 \cdot 3$ **b.** $6 \cdot \dfrac{1}{6} = 1$

c. $-3 + (2 + b) = (-3 + 2) + b$

d. $(b + 8) + 0 = b + 8$

Solution

a. This statement illustrates the Distributive Property.

b. This statement illustrates the Multiplicative Inverse Property.

c. This statement illustrates the Associative Property of Addition.

d. This statement illustrates the Additive Identity Property, where $(b + 8)$ is an algebraic expression.

The properties of real numbers make up the third component of what is called a **mathematical system.** These three components are a *set of numbers* (Section 1.1), *operations* with the set of numbers (Section 1.2), and *properties* of the operations with the numbers (Section 1.3).

Set of   Operations with   Properties of
Numbers   the Numbers   the Operations

Be sure you see that the properties of real numbers can be applied to variables and algebraic expressions as well as to real numbers.

Example 2   Using the Properties of Real Numbers

Complete each statement using the specified property of real numbers.

a. Multiplicative Identity Property: $(4a)1 =$

b. Associative Property of Addition: $(b + 8) + 3 =$

c. Additive Inverse Property: $0 = 5c +$

d. Distributive Property: $7 \cdot b + 7 \cdot 5 =$

Solution

a. By the Multiplicative Identity Property, $(4a)1 = 4a$.

b. By the Associative Property of Addition, $(b + 8) + 3 = b + (8 + 3)$.

c. By the Additive Inverse Property, $0 = 5c + (-5c)$.

d. By the Distributive Property, $7 \cdot b + 7 \cdot 5 = 7(b + 5)$.

To help you understand each property of real numbers, try stating the properties in your own words. For instance, the Associative Property of Addition can be stated as follows: *When three real numbers are added, it makes no difference which two are added first.*

② Develop additional properties of real numbers.

Additional Properties of Real Numbers

Once you have determined the basic properties (or *axioms*) of a mathematical system, you can go on to develop other properties. These additional properties are **theorems,** and the formal arguments that justify the theorems are **proofs.**

Additional Properties of Real Numbers

Let a, b, and c be real numbers, variables, or algebraic expressions.

Properties of Equality

Addition Property of Equality:	If $a = b$, then $a + c = b + c$.
Multiplication Property of Equality:	If $a = b$, then $ac = bc$.
Cancellation Property of Addition:	If $a + c = b + c$, then $a = b$.
Cancellation Property of Multiplication:	If $ac = bc$ and $c \neq 0$, then $a = b$.

Properties of Zero

Multiplication Property of Zero:	$0 \cdot a = 0$
Division Property of Zero:	$\dfrac{0}{a} = 0, \quad a \neq 0$
Division by Zero Is Undefined:	$\dfrac{a}{0}$ is undefined.

Properties of Negation

Multiplication by -1:	$(-1)(a) = -a,$ $(-1)(-a) = a$
Placement of Negative Signs:	$(-a)(b) = -(ab) = (a)(-b)$
Product of Two Opposites:	$(-a)(-b) = ab$

In Section 2.1, you will see that the properties of equality are useful for solving equations, as shown below. Note that the Addition and Multiplication Properties of Equality can be used to subtract the same nonzero quantity from each side of an equation or to divide each side of an equation by the same nonzero quantity.

$5x + 4 = -2x + 18$	Original equation
$5x + 4 - 4 = -2x + 18 - 4$	Subtract 4 from each side.
$5x = -2x + 14$	Simplify.
$5x + 2x = -2x + 2x + 14$	Add $2x$ to each side.
$7x = 14$	Simplify.
$\dfrac{7x}{7} = \dfrac{14}{7}$	Divide each side by 7.
$x = 2$	Simplify.

Study Tip

When the properties of real numbers are used in practice, the process is usually less formal than it would appear from the list of properties on this page. For instance, the steps shown at the right are less formal than those shown in Examples 5 and 6 on page 27. The importance of the properties is that they can be used to justify the steps of a solution. They do not always need to be listed for every step of the solution.

Each of the additional properties in the list on the preceding page can be proved using the basic properties of real numbers. Examples 3 and 4 illustrate such proofs.

Example 3   Proof of the Cancellation Property of Addition

Prove that if $a + c = b + c$, then $a = b$. (Use the Addition Property of Equality.)

Solution

Notice how each step is justified from the preceding step by means of a property of real numbers.

$a + c = b + c$	Write original equation.
$(a + c) + (-c) = (b + c) + (-c)$	Addition Property of Equality
$a + [c + (-c)] = b + [c + (-c)]$	Associative Property of Addition
$a + 0 = b + 0$	Additive Inverse Property
$a = b$	Additive Identity Property

Example 4   Proof of a Property of Negation

Prove that $(-1)a = -a$. (You may use any of the properties of equality and properties of zero.)

Solution

At first glance, it is a little difficult to see what you are being asked to prove. However, a good way to start is to consider carefully the definitions of the three numbers in the equation.

$$a = \text{given real number}$$

$$-1 = \text{the additive inverse of } 1$$

$$-a = \text{the additive inverse of } a$$

Now, by showing that $(-1)a$ has the same properties as the additive inverse of a, you will be showing that $(-1)a$ must be the additive inverse of a.

$(-1)a + a = (-1)a + (1)(a)$	Multiplicative Identity Property
$= (-1 + 1)a$	Distributive Property
$= (0)a$	Additive Inverse Property
$= 0$	Multiplication Property of Zero

Because $(-1)a + a = 0$, you can use the fact that $-a + a = 0$ to conclude that $(-1)a + a = -a + a$. From this, you can complete the proof as follows.

$(-1)a + a = -a + a$	Shown in first part of proof
$(-1)a = -a$	Cancellation Property of Addition

The list of additional properties of real numbers forms a very important part of algebra. Knowing the names of the properties is useful, but knowing how to use each property is extremely important. The next two examples show how several of the properties can be used to solve equations. (You will study these techniques in detail in Section 2.1.)

Example 5   Applying the Properties of Real Numbers

In the solution of the equation

$$b + 2 = 6$$

identify the property of real numbers that justifies each step.

Solution

$$b + 2 = 6 \qquad \text{Original equation}$$

Solution Step	Property
$(b + 2) + (-2) = 6 + (-2)$	Addition Property of Equality
$b + [2 + (-2)] = 6 - 2$	Associative Property of Addition
$b + 0 = 4$	Additive Inverse Property
$b = 4$	Additive Identity Property

Example 6   Applying the Properties of Real Numbers

In the solution of the equation

$$3x = 15$$

identify the property of real numbers that justifies each step.

Solution

$$3x = 15 \qquad \text{Original equation}$$

Solution Step	Property
$\left(\dfrac{1}{3}\right)3x = \left(\dfrac{1}{3}\right)15$	Multiplication Property of Equality
$\left(\dfrac{1}{3} \cdot 3\right)x = 5$	Associative Property of Multiplication
$(1)(x) = 5$	Multiplicative Inverse Property
$x = 5$	Multiplicative Identity Property

1.3 Exercises

Developing Skills

In Exercises 1–24, identify the property of real numbers illustrated by the statement. See Example 1.

1. $3 + (-5) = -5 + 3$

2. $-5(7) = 7(-5)$

3. $25 - 25 = 0$

4. $5 + 0 = 5$

5. $\frac{1}{12} \cdot 12 = 1$

6. $52 \cdot 1 = 52$

7. $25 + 35 = 35 + 25$

8. $(-4 \cdot 10) \cdot 8 = -4(10 \cdot 8)$

9. $3 + (12 - 9) = (3 + 12) - 9$

10. $(5 + 10)(8) = 8(5 + 10)$

11. $(8 - 5)(10) = 8 \cdot 10 - 5 \cdot 10$

12. $7(9 + 15) = 7 \cdot 9 + 7 \cdot 15$

13. $5(2a) = (5 \cdot 2)a$

14. $10(2x) = (10 \cdot 2)x$

15. $1 \cdot (5t) = 5t$

16. $10x \cdot \dfrac{1}{10x} = 1$

17. $3x + 0 = 3x$

18. $2x - 2x = 0$

19. $4 + (3 - x) = (4 + 3) - x$

20. $3(2 + x) = 3 \cdot 2 + 3x$

21. $3(6 + b) = 3 \cdot 6 + 3 \cdot b$

22. $(x + 1) - (x + 1) = 0$

23. $6(x + 3) = 6 \cdot x + 6 \cdot 3$

24. $(6 + x) - m = 6 + (x - m)$

In Exercises 25–34, complete the statement using the specified property of real numbers. See Example 2.

25. Associative Property of Multiplication:
$3(6y) =$

26. Commutative Property of Addition:
$10 + (-6) =$

27. Commutative Property of Multiplication:
$15(-3) =$

28. Associative Property of Addition:
$6 + (5 + y) =$

29. Distributive Property:
$5(6 + z) =$

30. Distributive Property:
$(8 - y)(4) =$

31. Commutative Property of Addition:
$25 + (-x) =$

32. Additive Inverse Property:
$13x + (-13x) =$

33. Multiplicative Identity Property:
$(x + 8) \cdot 1 =$

34. Additive Identity Property:
$(8x) + 0 =$

In Exercises 35–48, give (a) the additive inverse and (b) the multiplicative inverse of the quantity.

35. 10

36. 18

37. -16

38. -52

39. $\frac{1}{2}$

40. $\frac{3}{4}$

41. $-\frac{5}{8}$

42. $-\frac{1}{5}$

43. $6z, \quad z \neq 0$

44. $2y, \quad y \neq 0$

45. $-7x, \quad x \neq 0$

46. $-9z, \quad z \neq 0$

47. $x + 1, \quad x \neq -1$

48. $y - 4, \quad y \neq 4$

In Exercises 49–56, rewrite the expression using the Associative Property of Addition or the Associative Property of Multiplication.

49. $(x + 5) - 3$

50. $(z + 6) + 10$

51. $32 + (4 + y)$

52. $15 + (3 - x)$

53. $3(4 \cdot 5)$

54. $(1 \cdot 8)9$

55. $6(2y)$

56. $8(3x)$

In Exercises 57–64, rewrite the expression using the Distributive Property.

57. $20(2 + 5)$

58. $-3(4 - 8)$

59. $5(3x - 4)$

60. $6(2x + 5)$

61. $(x + 6)(-2)$

62. $(z - 10)(12)$

63. $-6(2y - 5)$

64. $-4(10 - b)$

In Exercises 65–68, the right side of the statement does not equal the left side. Change the right side so that it *does* equal the left side.

65. $3(x + 5) \neq 3x + 5$

66. $4(x + 2) \neq 4x + 2$

67. $-2(x + 8) \neq -2x + 16$

68. $-9(x + 4) \neq -9x + 36$

In Exercises 69 and 70, use the basic properties of real numbers to prove the statement. See Examples 3 and 4.

69. If $ac = bc$ and $c \neq 0$, then $a = b$.

70. $(a + b) + c = a + (b + c)$

In Exercises 71–74, identify the property of real numbers that justifies each step. See Examples 5 and 6.

71.
$$x + 5 = 3$$
$$(x + 5) + (-5) = 3 + (-5)$$
$$x + [5 + (-5)] = -2$$
$$x + 0 = -2$$
$$x = -2$$

72.
$$x - 8 = 20$$
$$(x - 8) + 8 = 20 + 8$$
$$x + (-8 + 8) = 28$$
$$x + 0 = 28$$
$$x = 28$$

73.
$$2x - 5 = 6$$
$$(2x - 5) + 5 = 6 + 5$$
$$2x + (-5 + 5) = 11$$
$$2x + 0 = 11$$
$$2x = 11$$
$$\tfrac{1}{2}(2x) = \tfrac{1}{2}(11)$$
$$\left(\tfrac{1}{2} \cdot 2\right)x = \tfrac{11}{2}$$
$$1 \cdot x = \tfrac{11}{2}$$
$$x = \tfrac{11}{2}$$

74.
$$3x + 4 = 10$$
$$(3x + 4) + (-4) = 10 + (-4)$$
$$3x + [4 + (-4)] = 6$$
$$3x + 0 = 6$$
$$3x = 6$$
$$\tfrac{1}{3}(3x) = \tfrac{1}{3}(6)$$
$$\left(\tfrac{1}{3} \cdot 3\right)x = 2$$
$$1 \cdot x = 2$$
$$x = 2$$

Mental Math   In Exercises 75–80, use the Distributive Property to perform the arithmetic mentally. For example, you work in an industry in which the wage is $14 per hour with "time and a half" for overtime. So, your hourly wage for overtime is

$$14(1.5) = 14\left(1 + \frac{1}{2}\right) = 14 + 7 = \$21.$$

75. $16(1.75) = 16\left(2 - \tfrac{1}{4}\right)$

76. $15\left(1\tfrac{2}{3}\right) = 15\left(2 - \tfrac{1}{3}\right)$

77. $7(62) = 7(60 + 2)$

78. $5(51) = 5(50 + 1)$

79. $9(6.98) = 9(7 - 0.02)$

80. $12(19.95) = 12(20 - 0.05)$

Solving Problems

81. ▲ *Geometry* The figure shows two adjoining rectangles. Demonstrate the Distributive Property by filling in the blanks to write the total area of the two rectangles in two ways.

$$(\quad + \quad) = \quad + \quad$$

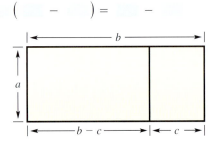

82. ▲ *Geometry* The figure shows two adjoining rectangles. Demonstrate the "subtraction version" of the Distributive Property by filling in the blanks to write the area of the left rectangle in two ways.

$$(\quad - \quad) = \quad - \quad$$

Dividends Per Share In Exercises 83–86, the dividends declared per share of common stock by the Gillette Company for the years 1996 through 2001 are approximated by the expression $0.063t - 0.0002$, where the dividends per share are measured in dollars and t represents the year, with $t = 6$ corresponding to 1996 (see figure). (Source: The Gillette Company 2001 Annual Report)

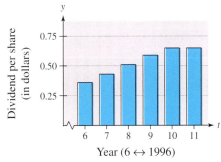

Year (6 ↔ 1996)

83. Use the graph to approximate the dividend paid in 1999.

84. Use the expression to approximate the annual increase in the dividend paid per share.

85. Use the expression to forecast the dividend per share in 2005.

86. In 2000, the actual dividend paid per share of common stock was $0.65. Compare this with the approximation given by the expression.

Explaining Concepts

87. *Writing* In your own words, give a verbal description of the Commutative Property of Addition.

88. What is the additive inverse of a real number? Give an example of the Additive Inverse Property.

89. What is the multiplicative inverse of a real number? Give an example of the Multiplicative Inverse Property.

90. *Writing* Does every real number have a multiplicative inverse? Explain.

91. State the Multiplication Property of Zero.

92. *Writing* Explain how the Addition Property of Equality can be used to allow you to subtract the same number from each side of an equation.

93. You define a new mathematical operation using the symbol ⊙. This operation is defined as $a \odot b = 2 \cdot a + b$. Give examples to show that this operation is neither commutative nor associative.

Mid-Chapter Quiz

Take this quiz as you would take a quiz in class. After you are done, check your work against the answers in the back of the book.

In Exercises 1 and 2, plot the two real numbers on the real number line and place the correct inequality symbol (< or >) between the two numbers.

1. -4.5 ___ -6

2. $\frac{3}{4}$ ___ $\frac{3}{2}$

In Exercises 3 and 4, find the distance between the two real numbers.

3. -15 and 7

4. -10.5 and -6.75

In Exercises 5 and 6, evaluate the expression.

5. $|-3.2|$

6. $-|9.8|$

In Exercises 7–16, evaluate the expression. Write fractions in simplest form.

7. $32 + (-18)$

8. $-10 - 12$

9. $\frac{3}{4} + \frac{7}{4}$

10. $\frac{2}{3} - \frac{1}{6}$

11. $(-3)(2)(-10)$

12. $\left(-\frac{4}{5}\right)\left(\frac{15}{32}\right)$

13. $\frac{7}{12} \div \frac{5}{6}$

14. $\left(-\frac{3}{2}\right)^3$

15. $3 - 2^2 + 25 \div 5$

16. $\dfrac{18 - 2(3 + 4)}{6^2 - (12 \cdot 2 + 10)}$

In Exercises 17 and 18, identify the property of real numbers illustrated by each statement.

17. (a) $8(u - 5) = 8 \cdot u - 8 \cdot 5$

   (b) $10x - 10x = 0$

18. (a) $(7 + y) - z = 7 + (y - z)$

   (b) $2x \cdot 1 = 2x$

19. During one month, you made the following transactions in your non-interest-bearing checking account.

Initial Balance:	$1522.76
Withdrawal:	$328.37
Withdrawal:	$65.99
Withdrawal:	$50.00
Deposit:	$413.88

Find the balance at the end of the month.

20. You deposit $30 in a retirement account twice each month. How much will you deposit in the account in 5 years?

21. Determine the unknown fractional part of the circle graph at the left. Explain how you were able to make this determination.

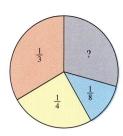

Figure for 21

1.4 Algebraic Expressions

David Young-Wolff/PhotoEdit

What You Should Learn

① Identify the terms and coefficients of algebraic expressions.

② Simplify algebraic expressions by combining like terms and removing symbols of grouping.

③ Evaluate algebraic expressions by substituting values for the variables.

Why You Should Learn It

Algebraic expressions can help you construct tables of values. For instance, in Example 9 on page 36, you can determine the percent of public schools in the United States with Internet access using an expression and a table of values.

Algebraic Expressions

One of the basic characteristics of algebra is the use of letters (or combinations of letters) to represent numbers. The letters used to represent the numbers are called **variables,** and combinations of letters and numbers are called **algebraic expressions.** Here are some examples.

$$3x, \quad x + 2, \quad \frac{x}{x^2 + 1}, \quad 2x - 3y, \quad 2x^3 - y^2$$

① Identify the terms and coefficients of algebraic expressions.

Algebraic Expression

A collection of letters (called **variables**) and real numbers (called **constants**) combined using the operations of addition, subtraction, multiplication, or division is called an **algebraic expression.**

Study Tip

It is important to understand the difference between a *term* and a *factor.* Terms are separated by addition whereas factors are separated by multiplication. For instance, the expression $4x(x + 2)$ has three factors: 4, x, and $(x + 2)$.

The **terms** of an algebraic expression are those parts that are separated by *addition.* For example, the algebraic expression $x^2 - 3x + 6$ has three terms: x^2, $-3x$, and 6. Note that $-3x$ is a term, rather than $3x$, because

$$x^2 - 3x + 6 = x^2 + (-3x) + 6. \qquad \text{Think of subtraction as a form of addition.}$$

The terms x^2 and $-3x$ are the **variable terms** of the expression, and 6 is the **constant term.** The numerical factor of a variable term is called the **coefficient.** For instance, the coefficient of the variable term $-3x$ is -3, and the coefficient of the variable term x^2 is 1. Example 1 identifies the terms and coefficients of three different algebraic expressions.

Example 1   Identifying Terms and Coefficients

Algebraic Expression	Terms	Coefficients
a. $5x - \dfrac{1}{3}$	$5x, \quad -\dfrac{1}{3}$	$5, \quad -\dfrac{1}{3}$
b. $4y + 6x - 9$	$4y, \quad 6x, \quad -9$	$4, \quad 6, \quad -9$
c. $\dfrac{2}{x} + 5x^4 - y$	$\dfrac{2}{x}, \quad 5x^4, \quad -y$	$2, \quad 5, \quad -1$

② Simplify algebraic expressions by combining like terms and removing symbols of grouping.

Simplifying Algebraic Expressions

In an algebraic expression, two terms are said to be **like terms** if they are both constant terms or if they have the same variable factor. For example, the terms $4x$ and $-2x$ are like terms because they have the same variable factor x. Similarly, $2x^2y$, $-x^2y$, and $\frac{1}{2}(x^2y)$ are like terms because they have the same variable factor x^2y. Note that $4x^2y$ and $-x^2y^2$ are not like terms because their variable factors x^2y and x^2y^2 are different.

One of the most common uses of the basic rules of algebra is to rewrite an algebraic expression in a simpler form. One way to **simplify** an algebraic expression is to combine like terms.

Example 2   Combining Like Terms

Simplify each expression by combining like terms.

a. $2x + 3x - 4$ **b.** $-3 + 5 + 2y - 7y$ **c.** $5x + 3y - 4x$

Solution

a. $2x + 3x - 4 = (2 + 3)x - 4$ Distributive Property

$\qquad\qquad\quad = 5x - 4$ Simplest form

b. $-3 + 5 + 2y - 7y = (-3 + 5) + (2 - 7)y$ Distributive Property

$\qquad\qquad\qquad\qquad = 2 - 5y$ Simplest form

c. $5x + 3y - 4x = 3y + 5x - 4x$ Commutative Property

$\qquad\qquad\quad = 3y + (5x - 4x)$ Associative Property

$\qquad\qquad\quad = 3y + (5 - 4)x$ Distributive Property

$\qquad\qquad\quad = 3y + x$ Simplest form

Example 3   Combining Like Terms

Simplify each expression by combining like terms.

a. $7x + 7y - 4x - y$ **b.** $2x^2 + 3x - 5x^2 - x$
c. $3xy^2 - 4x^2y^2 + 2xy^2 + x^2y^2$

Solution

a. $7x + 7y - 4x - y = (7x - 4x) + (7y - y)$ Group like terms.

$\qquad\qquad\qquad = 3x + 6y$ Combine like terms.

b. $2x^2 + 3x - 5x^2 - x = (2x^2 - 5x^2) + (3x - x)$ Group like terms.

$\qquad\qquad\qquad\qquad = -3x^2 + 2x$ Combine like terms.

c. $3xy^2 - 4x^2y^2 + 2xy^2 + x^2y^2$

$\qquad\quad = (3xy^2 + 2xy^2) + (-4x^2y^2 + x^2y^2)$ Group like terms.

$\qquad\quad = 5xy^2 - 3x^2y^2$ Combine like terms.

Study Tip

As you gain experience with the rules of algebra, you may want to combine some of the steps in your work. For instance, you might feel comfortable listing only the following steps to solve Example 2(c).

$5x + 3y - 4x = 3y + (5x - 4x)$

$\qquad\qquad\quad = 3y + x$

Another way to simplify an algebraic expression is to remove symbols of grouping. When removing symbols of grouping, remove the innermost symbols first and combine like terms. Then repeat the process for any remaining symbols of grouping.

A set of parentheses preceded by a *minus* sign can be removed by changing the sign of each term inside the parentheses. For instance,

$$3x - (2x - 7) = 3x - 2x + 7.$$

This is equivalent to using the Distributive Property with a multiplier of -1. That is,

$$3x - (2x - 7) = 3x + (-1)(2x - 7) = 3x - 2x + 7.$$

A set of parentheses preceded by a *plus* sign can be removed without changing the signs of the terms inside the parentheses. For instance,

$$3x + (2x - 7) = 3x + 2x - 7.$$

Study Tip

The exponential notation described in Section 1.2 can also be used when the base is a variable or algebraic expression. For instance, in Example 4(b), $x^2 \cdot x$ can be written as

$$x^2 \cdot x = x \cdot x \cdot x$$

$$= x^3. \qquad \text{3 factors of } x.$$

Example 4   Removing Symbols of Grouping

Simplify each expression.

a. $3(x - 5) - (2x - 7)$ **b.** $-4(x^2 + 4) + x^2(x + 4)$

Solution

a. $3(x - 5) - (2x - 7) = 3x - 15 - 2x + 7$ Distributive Property

$$= (3x - 2x) + (-15 + 7) \qquad \text{Group like terms.}$$

$$= x - 8 \qquad \text{Combine like terms.}$$

b. $-4(x^2 + 4) + x^2(x + 4) = -4x^2 - 16 + x^2 \cdot x + 4x^2$ Distributive Property

$$= -4x^2 - 16 + x^3 + 4x^2 \qquad \text{Exponential form}$$

$$= x^3 + (4x^2 - 4x^2) - 16 \qquad \text{Group like terms.}$$

$$= x^3 + 0 - 16 \qquad \text{Combine like terms.}$$

$$= x^3 - 16 \qquad \text{Simplest form}$$

Example 5   Removing Symbols of Grouping

a. $5x - 2x[3 + 2(x - 7)] = 5x - 2x(3 + 2x - 14)$ Distributive Property

$$= 5x - 2x(2x - 11) \qquad \text{Combine like terms.}$$

$$= 5x - 4x^2 + 22x \qquad \text{Distributive Property}$$

$$= -4x^2 + 27x \qquad \text{Combine like terms.}$$

b. $-3x(5x^4) + 2x^5 = -15x \cdot x^4 + 2x^5$ Multiply.

$$= -15x^5 + 2x^5 \qquad \text{Exponential form}$$

$$= -13x^5 \qquad \text{Combine like terms.}$$

③ Evaluate algebraic expressions by substituting values for the variables.

Evaluating Algebraic Expressions

To **evaluate** an algebraic expression, substitute numerical values for each of the variables in the expression. Here are some examples.

Expression	Value of Variable	Substitute	Value of Expression
$3x + 2$	$x = 2$	$3(2) + 2$	$6 + 2 = 8$
$4x^2 + 2x - 1$	$x = -1$	$4(-1)^2 + 2(-1) - 1$	$4 - 2 - 1 = 1$
$2x(x + 4)$	$x = -2$	$2(-2)(-2 + 4)$	$2(-2)(2) = -8$

Note that you must substitute the value for *each* occurrence of the variable.

Example 6 Evaluating Algebraic Expressions

Evaluate each algebraic expression when $x = -2$.

a. $5 + x^2$ **b.** $5 - x^2$

Solution

a. When $x = -2$, the expression $5 + x^2$ has a value of
$$5 + (-2)^2 = 5 + 4 = 9.$$

b. When $x = -2$, the expression $5 - x^2$ has a value of
$$5 - (-2)^2 = 5 - 4 = 1.$$

Example 7 Evaluating Algebraic Expressions

Evaluate each algebraic expression when $x = 2$ and $y = -1$.

a. $|y - x|$ **b.** $x^2 - 2xy + y^2$

Solution

a. When $x = 2$ and $y = -1$, the expression $|y - x|$ has a value of
$$|(-1) - (2)| = |-3| = 3.$$

b. When $x = 2$ and $y = -1$, the expression $x^2 - 2xy + y^2$ has a value of
$$2^2 - 2(2)(-1) + (-1)^2 = 4 + 4 + 1 = 9.$$

Example 8 Evaluating an Algebraic Expression

Evaluate $\dfrac{2xy}{x + 1}$ when $x = -4$ and $y = -3$.

Solution

When $x = -4$ and $y = -3$, the expression $(2xy)/(x + 1)$ has a value of
$$\frac{2(-4)(-3)}{-4 + 1} = \frac{24}{-3} = -8.$$

Algebraic expressions are often used to model real-life data, as shown in Example 9.

Example 9   Using a Mathematical Model

From 1994 to 2000, the percent of public schools in the United States with Internet access can be modeled by

$$\text{Percent} = -1.30t^2 + 29.0t - 61, \quad 4 \le t \le 10$$

where $t = 4$ represents 1994. Create a table that shows the percent for each of these years. (Source: National Center for Education Statistics)

Solution

To create a table of values that shows the percents of public schools with Internet access for the years 1994 to 2000, evaluate the expression $-1.30t^2 + 29.0t - 61$ for each integer value of t from $t = 4$ to $t = 10$.

Year	1994	1995	1996	1997	1998	1999	2000
t	4	5	6	7	8	9	10
Percent	34.2	51.5	66.2	78.3	87.8	94.7	99.0

Technology: Tip

Most graphing calculators can be used to evaluate an algebraic expression for several values of x and display the results in a table. For instance, to evaluate $2x^2 - 3x + 2$ when x is 0, 1, 2, 3, 4, 5, and 6, you can use the following steps.

1. Enter the expression into the graphing calculator.

2. Set the minimum value of the table to 0.

3. Set the table step or table increment to 1.

4. Display the table.

The results are shown below.

X	Y1
0	2
1	1
2	4
3	11
4	22
5	37
6	56

Consult the user's guide for your graphing calculator for specific instructions. Then complete a table for the expression $-4x^2 + 5x - 8$ when x is 0, 1, 2, 3, 4, 5, and 6.

1.4 Exercises

Developing Skills

In Exercises 1–18, identify the terms and coefficients of the algebraic expression. See Example 1.

1. $10x + 5$

2. $4 + 17y$

3. $12 - 6x^2$

4. $-16t^2 + 48$

5. $-3y^2 + 2y - 8$

6. $9t^2 + 2t + 10$

7. $1.2a - 4a^3$

8. $25z^3 - 4.8z^2$

9. $4x^2 - 3y^2 - 5x + 2y$

10. $7a^2 + 4a - b^2 + 19$

11. $xy - 5x^2y + 2y^2$

12. $14u^2 + 25uv - 3v^2$

13. $\frac{1}{4}x^2 - \frac{3}{8}x + 5$

14. $\frac{2}{3}y + 8z + \frac{5}{6}$

15. $\frac{x^2}{7} - \frac{2x}{5} + 3$

16. $\frac{2y^2}{3} - \frac{y}{9} - 11$

17. $x^2 - 2.5x - \frac{1}{x}$

18. $\frac{3}{t^2} - \frac{4}{t} + 6$

In Exercises 19–24, identify the property of algebra illustrated by the statement.

19. $4 - 3x = -3x + 4$

20. $(10 + x) - y = 10 + (x - y)$

21. $-5(2x) = (-5 \cdot 2)x$

22. $(x - 2)(3) = 3(x - 2)$

23. $(5 - 2)x = 5x - 2x$

24. $7y + 2y = (7 + 2)y$

In Exercises 25–28, use the property to rewrite the expression.

25. Distributive Property

$5(x + 6) = $

26. Commutative Property of Multiplication

$5(x + 6) = $

27. Distributive Property

$6x + 6 = $

28. Commutative Property of Addition

$6x + 6 = $

In Exercises 29–46, simplify the expression by combining like terms. See Examples 2 and 3.

29. $3x + 4x$

30. $18z + 14z$

31. $-2x^2 + 4x^2$

32. $20a^2 - 5a^2$

33. $7x - 11x$

34. $-23t + 11t$

35. $9y - 5y + 4y$

36. $8y + 7y - y$

37. $3x - 2y + 5x + 20y$

38. $-2a + 4b - 7a - b$

39. $7x^2 - 2x - x^2$

40. $9y + y^2 - 6y$

41. $-3z^4 + 6z - z + 8 + z^4$

42. $-5y^3 + 3y - 6y^2 + 8y^3 + y - 4$

43. $x^2 + 2xy - 2x^2 + xy + y$

44. $3a - 5ab + 9a^2 + 4ab - a$

45. $2x^2 + 7x^2y + xy^2 - 2x^2y + x^2$

46. $a^2b^2 - 3ab + a^2b + ab - 3a^2b^2$

In Exercises 47–58, use the Distributive Property to rewrite the expression.

47. $4(2x^2 + x - 3)$

48. $8(z^3 - 4z^2 + 2)$

49. $-3(6y^2 - y - 2)$

50. $-5(-x^2 + 2y + 1)$

51. $-(3x^2 - 2x + 4)$

52. $-(-5t^2 + 8t - 10)$

53. $x(5x + 2)$

54. $y(-y + 10)$

55. $3x(17 - 4x)$

56. $5y(2y - 1)$

57. $-5t(7 - 2t)$

58. $-6x(9x - 4)$

In Exercises 59–82, simplify the expression. See Examples 4 and 5.

59. $10(x - 3) + 2x - 5$

60. $3(x + 1) + x - 6$

61. $x - (5x + 9)$

62. $y - (3y - 1)$

63. $5a - (4a - 3)$

64. $7x - (2x + 5)$

65. $-3(3y - 1) + 2(y - 5)$

66. $5(a + 6) - 4(2a - 1)$

67. $-3(y^2 - 2) + y^2(y + 3)$

68. $x(x^2 - 5) - 4(4 - x)$

69. $x(x^2 + 3) - 3(x + 4)$

70. $5(x + 1) - x(2x + 6)$

71. $y^2(y + 1) + y(y^2 + 1)$

72. $3a(a^2 - 5) + a^2(a - 1)$

73. $9a - [7 - 5(7a - 3)]$

74. $12b - [9 - 7(5b - 6)]$

75. $3[2x - 4(x - 8)]$

76. $4[5 - 3(x^2 + 10)]$

77. $8x + 3x[10 - 4(3 - x)]$

78. $5y - y[9 + 6(y - 2)]$

79. $2[3(b - 5) - (b^2 + b + 3)]$

80. $5[3(z + 2) - (z^2 + z - 2)]$

81. $3[2(x + 7) + 3(x^2 + x - 5)]$

82. $4[3(2y - 1) + 5(2y^2 - y + 1)]$

In Exercises 83–100, evaluate the expression for the specified values of the variable(s). If not possible, state the reason. See Examples 6–8.

Expression	*Values*
83. $5 - 3x$	(a) $x = \frac{2}{3}$
	(b) $x = 5$
84. $\frac{3}{2}x - 2$	(a) $x = 6$
	(b) $x = -3$

Expression	*Values*
85. $10 - 4x^2$	(a) $x = -1$
	(b) $x = \frac{1}{2}$
86. $3y^2 + 10$	(a) $y = -2$
	(b) $y = \frac{1}{2}$
87. $y^2 - y + 5$	(a) $y = 2$
	(b) $y = -2$
88. $2x^2 + 5x - 3$	(a) $x = 2$
	(b) $x = -3$
89. $\dfrac{x}{x^2 + 1}$	(a) $x = 0$
	(b) $x = 3$
90. $5 - \dfrac{3}{x}$	(a) $x = 0$
	(b) $x = -6$
91. $3x + 2y$	(a) $x = 1, \quad y = 5$
	(b) $x = -6, \quad y = -9$
92. $6x - 5y$	(a) $x = -2, \quad y = -3$
	(b) $x = 1, \quad y = 1$
93. $x^2 - xy + y^2$	(a) $x = 2, \quad y = -1$
	(b) $x = -3, \quad y = -2$
94. $x^2y - xy^2$	(a) $x = 5, \quad y = 2$
	(b) $x = 3, \quad y = 3$
95. $\dfrac{2x + y}{x}$	(a) $x = 5, \quad y = 2$
	(b) $x = 3, \quad y = 3$
96. $\dfrac{x}{x - y}$	(a) $x = 0, \quad y = 10$
	(b) $x = 4, \quad y = 4$
97. $|y - x|$	(a) $x = 2, \quad y = 5$
	(b) $x = -2, \quad y = -2$
98. $|x^2 - y|$	(a) $x = 0, \quad y = -2$
	(b) $x = 3, \quad y = 15$
99. Distance traveled: rt	(a) $r = 40, \quad t = 5\frac{1}{4}$
	(b) $r = 35, \quad t = 4$
100. Simple interest: Prt	(a) $P = \$5000,$ $r = 0.085, \quad t = 10$
	(b) $P = \$750,$ $r = 0.07, \quad t = 3$

Solving Problems

▲ *Geometry*   In Exercises 101 and 102, write and simplify an expression for the area of the figure. Then evaluate the expression for the given value of the variable.

101. $b = 15$

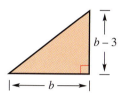

102. $h = 12$

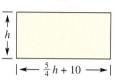

In Exercises 103 and 104, write an expression for the area of the figure. Then simplify the expression.

103.

104.

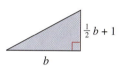

Using a Model   In Exercises 105 and 106, use the following model, which approximates the annual sales (in millions of dollars) of sporting goods in the United States from 1994 to 2000 (see figure), where $t = 4$ represents 1994.   (Source: National Sporting Goods Association)

Sales $= 3711.8t + 41,027$, $4 \le t \le 10$

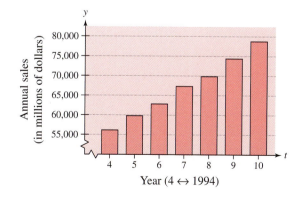

Year (4 ↔ 1994)

105. Graphically approximate the sales of sporting goods in 1998. Then use the model to confirm your estimate algebraically.

106. Use the model and a calculator to complete the table showing the sales from 1994 to 2000.

Year	1994	1995	1996	1997
Sales				

Year	1998	1999	2000
Sales			

Using a Model   In Exercises 107 and 108, use the following model, which approximates the median sale prices (in thousands of dollars) for homes in the United States from 1994 to 2000 (see figure), where $t = 4$ represents 1994.   (Source: U.S. Census Bureau, U.S. Department of Housing and Urban Development)

Sale price $= 6.6t + 102$, $4 \le t \le 10$

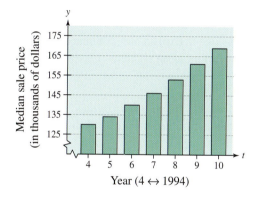

Year (4 ↔ 1994)

107. Graphically approximate the median sale price for homes in the United States in 1999. Then use the model to confirm your estimate algebraically.

108. Use the model and a calculator to complete the table showing the median sale prices for homes from 1994 to 2000.

Year	1994	1995	1996	1997
Price				

Year	1998	1999	2000
Price			

109. ▲ *Geometry*   The roof shown in the figure is made up of two trapezoids and two triangles. Find the total area of the roof.

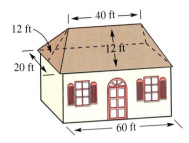

12 ft

40 ft

12 ft

20 ft

60 ft

110. *Exploration*

(a) A convex polygon with n sides has

$$\frac{n(n-3)}{2}, n \geq 4$$

diagonals. Verify the formula for a square, a pentagon, and a hexagon.

(b) Explain why the formula in part (a) will always yield a natural number.

Explaining Concepts

111. ⚡ Answer parts (d) and (e) of Motivating the Chapter.

112. *Writing*✎   Explain the difference between terms and factors in an algebraic expression.

113. *Writing*✎   In your own words, explain how to combine like terms. Give an example.

114. *Writing*✎   State the procedure for removing nested symbols of grouping. Nested symbols of grouping are symbols of grouping within symbols of grouping such as $5[1 + 4(3x + 2)]$.

115. *Writing*✎   Explain how the Distributive Property can be used to simplify the expression $5x + 3x$.

116. *Writing*✎   Explain how to rewrite $[x - (3 \cdot 4)] \div 5$ without using symbols of grouping.

117. *Writing*✎   Is it possible to evaluate the expression

$$\frac{x + 2}{y - 3}$$

when $x = 5$ and $y = 3$? Explain.

1.5 Constructing Algebraic Expressions

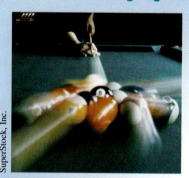

SuperStock, Inc.

What You Should Learn

1. Translate verbal phrases into algebraic expressions, and vice versa.
2. Construct algebraic expressions with hidden products.

Why You Should Learn It

Translating verbal phrases into algebraic expressions enables you to model real-life problems. For instance, in Exercise 73 on page 49, you will write an algebraic expression that models the area of the top of a billiard table.

Translating Phrases

In this section, you will study ways to *construct* algebraic expressions. When you translate a verbal sentence or phrase into an algebraic expression, watch for key words and phrases that indicate the four different operations of arithmetic.

1. Translate verbal phrases into algebraic expressions, and vice versa.

Translating Key Words and Phrases

Key Words and Phrases	Verbal Description	Algebraic Expression
Addition: Sum, plus, greater than, increased by, more than, exceeds, total of	The sum of 5 and x	$5 + x$
	Seven more than y	$y + 7$
Subtraction: Difference, minus, less than, decreased by, subtracted from, reduced by, the remainder	b is subtracted from 4.	$4 - b$
	Three less than z	$z - 3$
Multiplication: Product, multiplied by, twice, times, percent of	Two times x	$2x$
Division: Quotient, divided by, ratio, per	The ratio of x and 8	$\dfrac{x}{8}$

Example 1   Translating Verbal Phrases

Verbal Description	Algebraic Expression
a. Seven more than three times x	$3x + 7$
b. Four less than the product of 6 and n	$6n - 4$
c. The quotient of x and 3, decreased by 6	$\dfrac{x}{3} - 6$

Example 2 Translating Verbal Phrases

Verbal Description	*Algebraic Expression*
a. Eight added to the product of 2 and n	$2n + 8$
b. Four times the sum of y and 9	$4(y + 9)$
c. The difference of a and 7, all divided by 9	$\dfrac{a - 7}{9}$

In Examples 1 and 2, the verbal description specified the name of the variable. In most real-life situations, however, the variables are not specified and it is your task to assign variables to the *appropriate* quantities.

Example 3 Translating Verbal Phrases

Verbal Description	*Label*	*Algebraic Expression*
a. The sum of 7 and a number	The number $= x$	$7 + x$
b. Four decreased by the product of 2 and a number	The number $= n$	$4 - 2n$
c. Seven less than twice the sum of a number and 5	The number $= y$	$2(y + 5) - 7$

A good way to learn algebra is to do it forward and backward. For instance, the next example translates algebraic expressions into verbal form. Keep in mind that other key words could be used to describe the operations in each expression.

Example 4 Translating Expressions into Verbal Phrases

Without using a variable, write a verbal description for each expression.

a. $5x - 10$

b. $\dfrac{3 + x}{4}$

c. $2(3x + 4)$

d. $\dfrac{4}{x - 2}$

Solution

a. 10 less than the product of 5 and a number

b. The sum of 3 and a number, all divided by 4

c. Twice the sum of 3 times a number and 4

d. Four divided by a number reduced by 2

Study Tip

When you write a verbal model or construct an algebraic expression, watch out for statements that may be interpreted in more than one way. For instance, the statement

"The sum of x and 1 divided by 5"

is ambiguous because it could mean

$$\dfrac{x + 1}{5} \quad \text{or} \quad x + \dfrac{1}{5}.$$

Notice in Example 4(b) that the verbal description for

$$\dfrac{3 + x}{4}$$

contains the phrase "all divided by 4."

2 Construct algebraic expressions with hidden products.

Constructing Mathematical Models

Translating a verbal phrase into a mathematical model is critical in problem solving. The next four examples will demonstrate three steps for creating a mathematical model.

1. Construct a verbal model that represents the problem situation.

2. Assign labels to all quantities in the verbal model.

3. Construct a mathematical model (algebraic expression).

When verbal phrases are translated into algebraic expressions, products are often overlooked. Watch for hidden products in the next two examples.

Study Tip

Most real-life problems do not contain verbal expressions that clearly identify the arithmetic operations involved in the problem. You need to rely on past experience and the physical nature of the problem in order to identify the operations hidden in the problem statement.

Example 5   Constructing a Mathematical Model

A cash register contains x quarters. Write an algebraic expression for this amount of money in dollars.

Solution

Verbal Model:	Value of coin	$\cdot$	Number of coins	

Labels:	Value of coin $= 0.25$	(dollars per quarter)
	Number of coins $= x$	(quarters)

Expression: $0.25x$ (dollars)

Example 6   Constructing a Mathematical Model

A cash register contains n nickels and d dimes. Write an algebraic expression for this amount of money in cents.

Solution

Verbal Model:	Value of nickel	$\cdot$	Number of nickels	$+$	Value of dime	$\cdot$	Number of dimes

Labels:	Value of nickel $= 5$	(cents per nickel)
	Number of nickels $= n$	(nickels)
	Value of dime $= 10$	(cents per dime)
	Number of dimes $= d$	(dimes)

Expression: $5n + 10d$ (cents)

In Example 6, the final expression $5n + 10d$ is measured in cents. This makes "sense" as described below.

$$\frac{5 \text{ cents}}{\text{nickel}} \cdot n \text{ nickels} + \frac{10 \text{ cents}}{\text{dime}} \cdot d \text{ dimes}$$

Note that the nickels and dimes "divide out," leaving cents as the unit of measure for each term. This technique is called *unit analysis*, and it can be very helpful in determining the final unit of measure.

Example 7   Constructing a Mathematical Model

A person riding a bicycle travels at a constant rate of 12 miles per hour. Write an algebraic expression showing how far the person can ride in t hours.

Solution

For this problem, use the formula Distance = (Rate)(Time).

Verbal Model:	Rate · Time	

Labels:	Rate = 12	(miles per hour)
	Time = t	(hours)

Expression: 12t (miles)

Using unit analysis, you can see that the expression in Example 7 has *miles* as its unit of measure.

$$12\frac{\text{miles}}{\text{hour}} \cdot t \text{ hours}$$

When translating verbal phrases involving percents, be sure you write the percent *in decimal form*.

Percent	*Decimal Form*
4%	0.04
62%	0.62
140%	1.40
25%	0.25

Remember that when you find a percent of a number, you multiply. For instance, 25% of 78 is given by

$$0.25(78) = 19.5. \qquad \text{25\% of 78}$$

Example 8   Constructing a Mathematical Model

A person adds k liters of fluid containing 55% antifreeze to a car radiator. Write an algebraic expression that indicates how much antifreeze was added.

Solution

Verbal Model:	Percent antifreeze · Number of liters	

Labels:	Percent of antifreeze = 0.55	(in decimal form)
	Number of liters = k	(liters)

Expression: 0.55k (liters)

Note that the algebraic expression uses the decimal form of 55%. That is, you compute with 0.55 rather than 55%.

Hidden operations are often involved when labels are assigned to *two* unknown quantities. For example, two numbers add up to 18 and one of the numbers is assigned the variable *x*. What expression can you use to represent the second number? Let's try a specific case first, then apply it to a general case.

Specific case: If the first number is 7, the second number is $18 - 7 = 11$.

General case: If the first number is *x*, the second number is $18 - x$.

The strategy of using a *specific* case to help determine the general case is often helpful in applications. Observe the use of this strategy in the next example.

Example 9   Using Specific Cases to Model General Cases

a. A person's weekly salary is *d* dollars. Write an expression for the person's annual salary.

b. A person's annual salary is *y* dollars. Write an expression for the person's monthly salary.

Solution

a. *Specific Case:* If the weekly salary is $300, the annual salary is 52(300) dollars.

  General Case: If the weekly salary is *d* dollars, the annual salary is $52 \cdot d$ or $52d$ dollars.

b. *Specific Case:* If the annual salary is $24,000, the monthly salary is $24,000 \div 12$ dollars.

  General Case: If the annual salary is *y* dollars, the monthly salary is $y \div 12$ or $y/12$ dollars.

In mathematics it is useful to know how to represent certain types of integers algebraically. For instance, consider the set $\{2, 4, 6, 8, \ldots\}$ of *even* integers. Because every even integer has 2 as a factor,

$$2 = 2 \cdot 1, \quad 4 = 2 \cdot 2, \quad 6 = 2 \cdot 3, \quad 8 = 2 \cdot 4, \ldots$$

it follows that any integer *n* multiplied by 2 is sure to be the *even* number $2n$. Moreover, if $2n$ is even, then $2n - 1$ and $2n + 1$ are sure to be *odd* integers.

Two integers are called **consecutive integers** if they differ by 1. For any integer *n*, its next two larger consecutive integers are $n + 1$ and $(n + 1) + 1$ or $n + 2$. So, you can denote three consecutive integers by n, $n + 1$, and $n + 2$. These results are summarized below.

Labels for Integers

Let *n* represent an integer. Then even integers, odd integers, and consecutive integers can be represented as follows.

1. $2n$ denotes an *even* integer for $n = 1, 2, 3, \ldots$.

2. $2n - 1$ and $2n + 1$ denote *odd* integers for $n = 1, 2, 3, \ldots$.

3. $\{n, n + 1, n + 2, \ldots\}$ denotes a set of *consecutive* integers.

Example 10   Constructing a Mathematical Model

Write an expression for the following phrase.

"The sum of two consecutive integers, the first of which is n"

Solution

The first integer is n. The next consecutive integer is $n + 1$. So the sum of two consecutive integers is

$$n + (n + 1) = 2n + 1.$$

Sometimes an expression may be written directly from a diagram using a common geometric formula, as shown in the next example.

Example 11   Constructing a Mathematical Model

Write expressions for the perimeter and area of the rectangle shown in Figure 1.16.

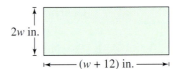

2w in.

←——— $(w + 12)$ in. ———→

Figure 1.16

Solution

For the perimeter of the rectangle, use the formula

Perimeter = 2(Length) + 2(Width).

Verbal
Model: $2 \cdot$ Length $+ 2 \cdot$ Width

Labels:	Length = $w + 12$	(inches)
	Width = $2w$	(inches)

Expression: $2(w + 12) + 2(2w) = 2w + 24 + 4w$
$\qquad\qquad\qquad\qquad\qquad\quad = 6w + 24$ (inches)

For the area of the rectangle, use the formula

Area = (Length)(Width).

Verbal
Model: Length $\cdot$ Width

Labels:	Length = $w + 12$	(inches)
	Width = $2w$	(inches)

Expression: $(w + 12)(2w) = 2w^2 + 24w$ (square inches)

1.5 Exercises

Developing Skills

In Exercises 1–24, translate the verbal phrase into an algebraic expression. See Examples 1–3.

1. The sum of 23 and a number n

2. Five more than a number n

3. The sum of 12 and twice a number n

4. The total of 25 and three times a number n

5. Six less than a number n

6. Fifteen decreased by three times a number n

7. Four times a number n minus 10

8. The product of a number y and 10 is decreased by 35.

9. Half of a number n

10. Seven-fifths of a number n

11. The quotient of a number x and 6

12. The ratio of y and 3

13. Eight times the ratio of N and 5

14. Twenty times the ratio of x and 9

15. The number c is quadrupled and the product is increased by 10.

16. The number u is tripled and the product is increased by 250.

17. Thirty percent of the list price L

18. Forty percent of the cost C

19. The sum of a number and 5, divided by 10

20. The sum of 7 and twice a number x, all divided by 8

21. The absolute value of the difference between a number and 8

22. The absolute value of the quotient of a number and 4

23. The product of 3 and the square of a number is decreased by 4.

24. The sum of 10 and one-fourth the square of a number

In Exercises 25–40, write a verbal description of the algebraic expression without using the variable. See Example 4.

25. $t - 2$

26. $5 - x$

27. $y + 50$

28. $2y + 3$

29. $2 - 3x$

30. $4x - 5$

31. $\dfrac{z}{2}$

32. $\dfrac{y}{8}$

33. $\dfrac{4}{5}x$

34. $\dfrac{2}{3}t$

35. $8(x - 5)$

36. $(x - 5)7$

37. $\dfrac{x + 10}{3}$

38. $\dfrac{x - 2}{3}$

39. $x(x + 7)$

40. $x^2 + 2$

In Exercises 41–64, write an algebraic expression that represents the specified quantity in the verbal statement, and simplify if possible. See Examples 5–10.

41. The amount of money (in dollars) represented by n quarters

42. The amount of money (in dollars) represented by x nickels

43. The amount of money (in dollars) represented by m dimes

44. The amount of money (in cents) represented by x nickels and y quarters

45. The amount of money (in cents) represented by m nickels and n dimes

46. The amount of money (in cents) represented by m dimes and n quarters

47. The distance traveled in t hours at an average speed of 65 miles per hour

48. The distance traveled in 5 hours at an average speed of r miles per hour

49. The time to travel 230 miles at an average speed of r miles per hour

50. The average rate of speed when traveling 360 miles in t hours

51. The amount of antifreeze in a cooling system containing y gallons of coolant that is 45% antifreeze

52. The amount of water in q quarts of a food product that is 65% water

53. The amount of wage tax due for a taxable income of I dollars that is taxed at the rate of 1.25%

54. The amount of sales tax on a purchase valued at L dollars if the tax rate is 6%

55. The sale price of a coat that has a list price of L dollars if it is a "20% off" sale

56. The total cost for a family to stay one night at a campground if the charge is $18 for the parents plus $3 for each of the n children

57. The total hourly wage for an employee when the base pay is $8.25 per hour plus 60 cents for each of q units produced per hour

58. The total hourly wage for an employee when the base pay is $11.65 per hour plus 80 cents for each of q units produced per hour

59. The sum of a number n and three times the number

60. The sum of three consecutive integers, the first of which is n

61. The sum of two consecutive odd integers, the first of which is $2n + 1$

62. The sum of two consecutive even integers, the first of which is $2n$

63. The product of two consecutive even integers, divided by 4

64. The absolute value of the difference of two consecutive integers, divided by 2

Solving Problems

▲ *Geometry*   In Exercises 65–68, write an expression for the area of the figure.

65.

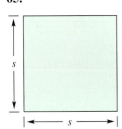

66.

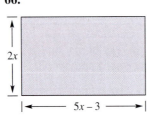

67.

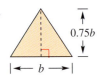

68.

▲ *Geometry*   In Exercises 69–72, write expressions for the perimeter and area of the region. Simplify the expressions. See Example 11.

69.

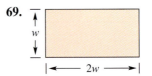

70.

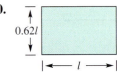

71.

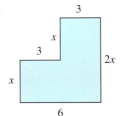

72.

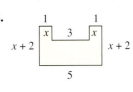

73. ▲ *Geometry*   Write an expression for the area of the top of the billiard table shown in the figure. What is the unit of measure for the area?

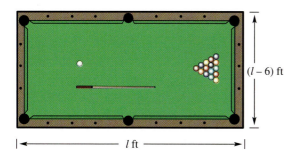

74. ▲ *Geometry*   Write an expression for the area of the advertising banner shown in the figure. What is the unit of measure for the area?

75. *Finding a Pattern*   Complete the table. The third row contains the differences between consecutive entries of the second row. Describe the pattern of the third row.

n	0	1	2	3	4	5
$5n - 3$						
Differences						

76. *Finding a Pattern*   Complete the table. The third row contains the differences between consecutive entries of the second row. Describe the pattern of the third row.

n	0	1	2	3	4	5
$3n + 1$						
Differences						

77. *Finding a Pattern*   Using the results of Exercises 75 and 76, guess the third-row difference that would result in a similar table if the algebraic expression were $an + b$.

78. *Think About It*   Find a and b such that the expression $an + b$ would yield the following table.

n	0	1	2	3	4	5
$an + b$	3	7	11	15	19	23

Explaining Concepts

79. The phrase *reduced by* implies what operation?

80. The word *ratio* indicates what operation?

81. Which are equivalent to $4x$?

  (a) x multiplied by 4

  (b) x increased by 4

  (c) the product of x and 4

  (d) the ratio of 4 and x

82. *Writing*   When each phrase is translated into an algebraic expression, is order important? Explain.

  (a) y multiplied by 5

  (b) 5 decreased by y

  (c) y divided by 5

  (d) the sum of 5 and y

83. *Writing*   When a statement is translated into an algebraic expression, explain why it may be helpful to use a specific case before writing the expression.

84. *Writing*   If n is an integer, how are the integers $2n - 1$ and $2n + 1$ related? Explain.

What Did You Learn?

Key Terms

set, *p. 2*
real numbers, *p. 2*
subset, *p. 2*
natural numbers, *p. 2*
whole numbers, *p. 2*
integers, *p. 2*
fractions, *p. 3*
rational numbers, *p. 3*
irrational numbers, *p. 3*
real number line, *p. 4*
origin, *p. 4*
nonnegative real number,   *p. 4*
inequality symbols, *p. 5*

opposites, *p. 7*
additive inverses, *p. 7*
absolute value, *p. 7*
sum, *p. 11*
difference, *p. 12*
least common
    denominator, *p. 13*
product, *p. 14*
factor, *p. 14*
reciprocal, *p. 15*
quotient, *p. 15*
dividend, *p. 15*
divisor, *p. 15*

numerator, *p. 15*
denominator, *p. 15*
exponential form, *p. 16*
base, *p. 16*
variables, *p. 32*
algebraic expressions, *p. 32*
variable terms, *p. 32*
constant term, *p. 32*
coefficient, *p. 32*
like terms, *p. 33*
simplify, *p. 33*
evaluate, *p. 35*
consecutive integers, *p. 45*

Key Concepts

1.1 ◯ Order on the real number line

If the real number a lies to the left of the real number b on the real number line, then a is less than b, which is written as $a < b$.

1.1 ◯ Distance between two real numbers

If a and b are two real numbers such that $a \leq b$, then the distance between a and b is $b - a$.

1.2 ◯ Addition of two real numbers

1. To add two real numbers with *like signs*, add their absolute values and attach the common sign to the result.

2. To add two real numbers with *unlike signs*, subtract the smaller absolute value from the greater absolute value and attach the sign of the number with the greater absolute value.

1.2 ◯ Subtraction of two real numbers

To subtract the real number b from the real number a, add the opposite of b to a: $a - b = a + (-b)$.

1.2 ◯ Addition and subtraction of fractions

1. *Like Denominators:* The sum and difference of two fractions with like denominators ($c \neq 0$) are:

$$\frac{a}{c} + \frac{b}{c} = \frac{a+b}{c} \qquad \frac{a}{c} - \frac{b}{c} = \frac{a-b}{c}$$

2. *Unlike Denominators*: To add or subtract two fractions with unlike denominators, first rewrite the fractions so that they have the same denominator and apply the first rule.

1.2 ◯ Multiplication of two real numbers

1. To multiply two real numbers with *like signs*, find the product of their absolute values. The product is *positive*.

2. To multiply two real numbers with *unlike signs*, find the product of their absolute values, and attach a minus sign. The product is *negative*.

3. The product of zero and any other real number is zero.

1.2 ◯ Multiplication of two fractions

The product of the two fractions a/b and c/d is given by

$$\frac{a}{b} \cdot \frac{c}{d} = \frac{ac}{bd}, \ b \neq 0, \ d \neq 0.$$

1.2 ◯ Division of two real numbers

To divide the real number a by the nonzero real number b, multiply a by the reciprocal of b. That is,

$$a \div b = a \cdot \frac{1}{b}, \ b \neq 0.$$

1.2 ◯ Exponential Notation

Let n be a positive integer and let a be a real number. Then the product of n factors of a is given by

$$a^n = a \cdot a \cdot a \cdots a$$

where a is the base and n is the exponent.

What Did You Learn?

Key Concepts   (continued)

1.2 ● Order of operations

To evaluate an expression involving more than one operation, use the following order.

1. First do operations that occur within symbols of grouping.
2. Then evaluate powers.
3. Then do multiplications and divisions from left to right.
4. Finally, do additions and subtractions from left to right.

1.3 ● Properties of real numbers

Let a, b, and c represent real numbers, variables, or algebraic expressions.

Commutative Property of Addition:

$a + b = b + a$

Commutative Property of Multiplication:

$ab = ba$

Associative Property of Addition:

$(a + b) + c = a + (b + c)$

Associative Property of Multiplication:

$(ab)c = a(bc)$

Distributive Property:

$a(b + c) = ab + ac \qquad a(b - c) = ab - ac$

$(a + b)c = ac + bc \qquad (a - b)c = ac - bc$

Additive Identity Property:

$a + 0 = 0 + a = a$

Multiplicative Identity Property:

$a \cdot 1 = 1 \cdot a = a$

Additive Inverse Property:

$a + (-a) = 0$

Multiplicative Inverse Property:

$a \cdot \dfrac{1}{a} = 1, \quad a \neq 0$

Addition Property of Equality:

If $a = b$, then $a + c = b + c$.

Multiplication Property of Equality:

If $a = b$, then $ac = bc$.

Cancellation Property of Addition:

If $a + c = b + c$, then $a = b$.

Cancellation Property of Multiplication:

If $ac = bc$ and $c \neq 0$, then $a = b$.

Multiplication Property of Zero: $0 \cdot a = 0$

Division Property of Zero: $\dfrac{0}{a} = 0, \quad a \neq 0$

Division by Zero Is Undefined: $\dfrac{a}{0}$ is undefined.

Multiplication by -1:

$(-1)(a) = -a, \quad (-1)(-a) = a$

Placement of Negative Signs:

$(-a)(b) = -(ab) = (a)(-b)$

Product of Two Opposites: $(-a)(-b) = ab$

1.5 ● Translating key words and phrases

Addition: sum, plus, greater than, increased by, more than, exceeds, total of

Subtraction: difference, minus, less than, decreased by, subtracted from, reduced by, the remainder

Multiplication: product, multiplied by, twice, times, percent of

Division: quotient, divided by, ratio, per

1.5 ● Labels for integers

Let n represent an integer. Then even integers, odd integers, and consecutive integers can be represented as follows.

1. $2n$ denotes an *even* integer for $n = 1, 2, 3, \ldots$.
2. $2n - 1$ and $2n + 1$ denote *odd* integers for $n = 1, 2, 3, \ldots$.
3. $\{n, n + 1, n + 2, \ldots\}$ denotes a set of *consecutive* integers.

Review Exercises

1.1 The Real Number System

1 Understand the set of real numbers and the subsets of real numbers.

In Exercises 1 and 2, which of the real numbers in the set are (a) natural numbers, (b) integers, (c) rational numbers, and (d) irrational numbers?

1. $\left\{\frac{3}{5}, -4, 0, \sqrt{2}, 52, -\frac{1}{8}, \sqrt{9}\right\}$

2. $\left\{98, -141, -\frac{7}{8}, 3.99, -\sqrt{12}, -\frac{54}{11}\right\}$

In Exercises 3 and 4, list all members of the set.

3. The natural numbers between -3.2 and 4.8

4. The even integers between -5.5 and 2.5

2 Use the real number line to order real numbers.

In Exercises 5–8, plot the real numbers on a real number line and place the correct inequality symbol ($<$ or $>$) between the numbers.

5. -5 3

6. -2 -8

7. $-\frac{8}{5}$ $-\frac{2}{5}$

8. 8.4 -3.2

3 Use the real number line to find the distance between two real numbers.

In Exercises 9–12, find the distance between the pair of real numbers.

9. 9 and -2

10. -7 and 4

11. -13.5 and -6.2

12. -8.4 and -0.3

4 Determine the absolute value of a real number.

In Exercises 13–16, evaluate the expression.

13. $|-5|$

14. $|6|$

15. $-|-7.2|$

16. $|-3.6|$

1.2 Operations with Real Numbers

1 Add, subtract, multiply, and divide real numbers.

In Exercises 17–38, evaluate the expression. If it is not possible, state the reason. Write all fractions in simplest form.

17. $15 + (-4)$

18. $-12 + 3$

19. $340 - 115 + 5$

20. $-154 + 86 - 240$

21. $-63.5 + 21.7$

22. $14.35 - 10.3$

23. $\frac{4}{21} + \frac{7}{21}$

24. $\frac{21}{16} - \frac{13}{16}$

25. $-\frac{5}{6} + 1$

26. $\frac{21}{32} + \frac{11}{24}$

27. $8\frac{3}{4} - 6\frac{5}{8}$

28. $-2\frac{9}{10} + 5\frac{3}{20}$

29. $-7 \cdot 4$

30. $(-8)(-3)$

31. $120(-5)(7)$

32. $(-16)(-15)(-4)$

33. $\frac{3}{8} \cdot \left(-\frac{2}{15}\right)$

34. $\frac{5}{21} \cdot \frac{21}{5}$

35. $\frac{-56}{-4}$

36. $\frac{85}{0}$

37. $-\frac{7}{15} \div \left(-\frac{7}{30}\right)$

38. $-\frac{2}{3} \div \frac{4}{15}$

2 Write repeated multiplication in exponential form and evaluate exponential expressions.

In Exercises 39 and 40, write the expression using exponential notation.

39. $6 \cdot 6 \cdot 6 \cdot 6 \cdot 6 \cdot 6 \cdot 6$

40. $\left(-\frac{1}{2}\right)\left(-\frac{1}{2}\right)\left(-\frac{1}{2}\right)$

In Exercises 41–46, evaluate the exponential expression.

41. $(-6)^4$

42. $-(-3)^4$

43. -4^2

44. 2^5

45. $-\left(-\frac{1}{2}\right)^3$

46. -5^2

3 Use order of operations to evaluate expressions.

In Exercises 47–50, evaluate the expression.

47. $120 - (5^2 \cdot 4)$

48. $45 - 45 \div 3^2$

49. $8 + 3[6^2 - 2(7 - 4)]$

50. $2^4 - [10 + 6(1 - 3)^2]$

4 Evaluate expressions using a calculator and order of operations.

In Exercises 51 and 52, evaluate the expression using a calculator. Round your answer to two decimal places.

51. $7(408.2^2 - 39.5 \div 0.3)$

52. $-59[4.6^3 + 5.8(-13.4)]$

53. *Total Charge* You purchased an entertainment system and made a down payment of $239 plus nine monthly payments of $45 each. What is the total amount you paid for the system?

54. *Savings Plan* You deposit $80 per month in a savings account for 10 years. The account earns 2% interest compounded monthly. The total amount in the account after 10 years will be

$$80\left[\left(1 + \frac{0.02}{12}\right)^{120} - 1\right]\left(1 + \frac{12}{0.02}\right).$$

Use a calculator to determine this amount.

1.3 Properties of Real Numbers

1 Identify and use the properties of real numbers.

In Exercises 55–64, identify the property of real numbers illustrated by the statement.

55. $13 - 13 = 0$

56. $7\left(\frac{1}{7}\right) = 1$

57. $7(9 + 3) = 7 \cdot 9 + 7 \cdot 3$

58. $15(4) = 4(15)$

59. $5 + (4 - y) = (5 + 4) - y$

60. $6(4z) = (6 \cdot 4)z$

61. $(u - v)(2) = 2(u - v)$

62. $(x + y) + 0 = x + y$

63. $8(x - y) = 8 \cdot x - 8 \cdot y$

64. $x(yz) = (xy)z$

In Exercises 65–68, rewrite the expression by using the Distributive Property.

65. $-(-u + 3v)$

66. $-5(2x - 4y)$

67. $-y(3y - 10)$

68. $x(3x + 4y)$

1.4 Algebraic Expressions

1 Identify the terms and coefficients of algebraic expressions.

In Exercises 69–72, identify the terms and coefficients of the algebraic expression.

69. $4y^3 - y^2 + \frac{17}{2}y$

70. $\frac{x}{3} + 2xy^2 + \frac{1}{5}$

71. $52 - 1.2x^3 + \frac{1}{x}$

72. $2ab^2 + a^2b^2 - \frac{1}{a}$

2 Simplify algebraic expressions by combining like terms and removing symbols of grouping.

In Exercises 73–82, simplify the expression.

73. $7x - 2x$ **74.** $25y + 32y$

75. $3u - 2v + 7v - 3u$

76. $7r - 4 - 9 + 3r$

77. $5(x - 4) + 10$

78. $15 - 7(z + 2)$

79. $3x - (y - 2x)$

80. $30x - (10x + 80)$

81. $3[b + 5(b - a)]$

82. $-2t[8 - (6 - t)] + 5t$

3 Evaluate algebraic expressions by substituting values for the variables.

In Exercises 83 and 84, evaluate the algebraic expression for the specified values of the variable(s). If not possible, state the reason.

Expression	Values
83. $x^2 - 2x - 3$	(a) $x = 3$
	(b) $x = 0$
84. $\dfrac{x}{y + 2}$	(a) $x = 0, \quad y = 3$
	(b) $x = 5, \quad y = -2$

1.5 Constructing Algebraic Expressions

1 Translate verbal phrases into algebraic expressions, and vice versa.

In Exercises 85–88, translate the verbal phrase into an algebraic expression.

85. Two hundred decreased by three times a number n

86. One hundred increased by the product of 15 and a number x

87. The sum of the square of a number y and 49

88. The absolute value of the sum of a number n and 10, all divided by 2

In Exercises 89–92, write a verbal description of the algebraic expression without using the variable.

89. $2y + 7$

90. $5u - 3$

91. $\dfrac{x - 5}{4}$

92. $4(a - 1)$

2 Construct algebraic expressions with hidden products.

In Exercises 93–96, write an algebraic expression that represents the quantity in the verbal statement, and simplify if possible.

93. The amount of income tax on a taxable income of I dollars when the tax rate is 18%

94. The distance traveled when you travel 8 hours at the average speed of r miles per hour

95. The area of a rectangle whose length is l units and whose width is five units less than the length

96. The sum of two consecutive integers, the first of which is n

Chapter Test

Take this test as you would take a test in class. After you are done, check your work against the answers in the back of the book.

1. Place the correct inequality symbol ($<$ or $>$) between the pair of numbers.
   (a) $\frac{5}{2}$ ___ $|-3|$ (b) $-\frac{2}{3}$ ___ $-\frac{3}{2}$

2. Find the distance between -6.2 and 5.7.

In Exercises 3–10, evaluate the expression.

3. $-14 + 9 - 15$

4. $\frac{2}{3} + \left(-\frac{7}{6}\right)$

5. $-2(225 - 150)$

6. $(-3)(4)(-5)$

7. $\left(-\frac{7}{16}\right)\left(-\frac{8}{21}\right)$

8. $\frac{5}{18} \div \frac{15}{8}$

9. $\left(-\frac{3}{5}\right)^3$

10. $\dfrac{4^2 - 6}{5} + 13$

11. Identify the property of real numbers illustrated by each statement.
    (a) $(-3 \cdot 5) \cdot 6 = -3(5 \cdot 6)$

    (b) $3y \cdot \dfrac{1}{3y} = 1$

12. Rewrite the expression $-6(2x - 1)$ using the Distributive Property.

In Exercises 13–16, simplify the expression.

13. $3x^2 - 2x - 5x^2 + 7x - 1$

14. $x(x + 2) - (x^2 + x - 13)$

15. $a(5a - 4) - 2(2a^2 - 2a)$

16. $4t - [3t - (10t + 7)]$

17. Explain the meaning of "evaluating an expression." Evaluate the expression $4 - (x + 1)^2$ for each value of x.

    (a) $x = -1$ (b) $x = 3$

18. It is necessary to cut a 144-foot rope into nine pieces of equal length. What is the length of each piece?

19. A *cord* of wood is a pile 4 feet high, 4 feet wide, and 8 feet long. The volume of a rectangular solid is its length times its width times its height. Find the number of cubic feet in 5 cords of wood.

20. Translate the phrase into an algebraic expression.

    "The product of a number n and 5, decreased by 8"

21. Write an algebraic expression for the sum of two consecutive even integers, the first of which is $2n$.

22. Write expressions for the perimeter and area of the rectangle shown at the left. Then simplify the expressions.

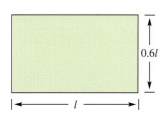

$0.6l$

l

Figure for 22

55

Motivating the Chapter

⚡ Cable Television and You

You are having cable television installed in your house. You need to decide whether you will purchase one or more premium movie channels or pay-per-view movies. You will not have both. Standard service is $35.20 per month and is required if you want a premium movie channel or pay-per-view movies. Each premium movie channel is $11.91 per month, and pay-per-view is $2.99 per month plus $3.95 per movie.

See Section 2.3, Exercise 91.

a. Write a verbal model that gives the monthly cost of cable television based on the number of premium movie channels that you order.

b. Write an algebraic equation for your verbal model from part (a). Create a table that shows the amounts paid per month for one, two, three, four, and five premium movie channels.

c. Write a verbal model that gives the monthly cost of cable television based on the number of pay-per-view movies you watch.

d. Write an algebraic equation for your verbal model from part (c). Create a table that shows the amounts paid per month for one, two, three, four, five, six, seven, and eight pay-per-view movies.

e. If you are paying for two premium movie channels, what percent of your bill goes to paying for these movie channels?

See Section 2.4, Exercise 117.

f. Your budget allows you to spend at most $50 per month on cable television. Use the algebraic model from part (b) to determine the number of premium movie channels you could purchase each month.

g. Your budget allows you to spend at most $50 per month on cable television. Use the algebraic model from part (d) to determine the number of pay-per-view movies that you could watch each month. Compare this with your answer to part (f). Which option would you choose, and why?

Syracuse Newspapers/The Image Works

Linear Equations and Inequalities

2.1 ● Linear Equations

2.2 ● Linear Equations and Problem Solving

2.3 ● Business and Scientific Problems

2.4 ● Linear Inequalities

2.5 ● Absolute Value Equations and Inequalities

2

2.1 Linear Equations

Hubert Stadler/Corbis

What You Should Learn

1 Check solutions of equations.

2 Solve linear equations in standard form.

3 Solve linear equations in nonstandard form.

Why You Should Learn It

Linear equations are used in many real-life applications. For instance, Example 9 on page 64 shows how a linear equation can model the number of trucks and buses sold in the United States.

1 Check solutions of equations.

Introduction

An **equation** is a statement that equates two algebraic expressions. Some examples are $x = 4$, $4x + 3 = 15$, $2x - 8 = 2(x - 4)$, and $x^2 - 16 = 0$.

   Solving an equation involving a variable means finding all values of the variable for which the equation is true. Such values are **solutions** and are said to **satisfy** the equation. For instance, $x = 3$ is a solution of $4x + 3 = 15$ because $4(3) + 3 = 15$ is a true statement.

   The **solution set** of an equation is the set of all solutions of the equation. Sometimes, an equation will have the set of all real numbers as its solution set. Such an equation is an **identity.** For instance, the equation

$$2x - 8 = 2(x - 4) \qquad \textcolor{red}{\text{Identity}}$$

is an identity because the equation is true for all real values of x. Try values such as 0, 1, -2, and 5 in this equation to see that each one is a solution.

   An equation whose solution set is not the entire set of real numbers is called a **conditional equation.** For instance, the equation

$$x^2 - 16 = 0 \qquad \textcolor{red}{\text{Conditional equation}}$$

is a conditional equation because it has only two solutions, $x = 4$ and $x = -4$. Example 1 shows how to **check** whether a given value is a solution.

Example 1 Checking a Solution of an Equation

Determine whether $x = -3$ is a solution of $-3x - 5 = 4x + 16$.

Solution

$$
\begin{aligned}
-3x - 5 &= 4x + 16 & &\textcolor{red}{\text{Write original equation.}} \\
-3(-3) - 5 &\overset{?}{=} 4(-3) + 16 & &\textcolor{red}{\text{Substitute } -3 \text{ for } x.} \\
9 - 5 &\overset{?}{=} -12 + 16 & &\textcolor{red}{\text{Simplify.}} \\
4 &= 4 & &\textcolor{red}{\text{Solution checks. } \checkmark}
\end{aligned}
$$

Because each side turns out to be the same number, you can conclude that $x = -3$ *is* a solution of the original equation. Try checking to see whether $x = -2$ is a solution.

Study Tip

When checking a solution, you should write a question mark over the equal sign to indicate that you are uncertain whether the "equation" is true for a given value of the variable.

It is helpful to think of an equation as having two sides that are in balance. Consequently, when you try to solve an equation, you must be careful to maintain that balance by performing the same operation(s) on each side.

Two equations that have the same set of solutions are **equivalent equations.** For instance, the equations $x = 3$ and $x - 3 = 0$ are equivalent equations because both have only one solution—the number 3. When any one of the four techniques in the following list is applied to an equation, the resulting equation is equivalent to the original equation.

Forming Equivalent Equations: Properties of Equality

An equation can be transformed into an *equivalent equation* using one or more of the following procedures.

	Original Equation	*Equivalent Equation*
1. *Simplify Either Side:* Remove symbols of grouping, combine like terms, or simplify fractions on one or both sides of the equation.	$4x - x = 8$	$3x = 8$
2. *Apply the Addition Property of Equality:* Add (or subtract) the same quantity to (from) *each* side of the equation.	$x - 3 = 5$	$x = 8$
3. *Apply the Multiplication Property of Equality:* Multiply (or divide) *each* side of the equation by the same *nonzero* quantity.	$3x = 12$	$x = 4$
4. *Interchange Sides:* Interchange the two sides of the equation.	$7 = x$	$x = 7$

When solving an equation, you can use any of the four techniques for forming equivalent equations to eliminate terms or factors in the equation. For example, to solve the equation

$$x + 4 = 2$$

you need to remove the term 4 from the left side. This is accomplished by subtracting 4 from each side.

$x + 4 = 2$	Write original equation.
$x + 4 - 4 = 2 - 4$	Subtract 4 from each side.
$x + 0 = -2$	Combine like terms.
$x = -2$	Simplify.

Although this solution method subtracted 4 from each side, you could just as easily have added -4 to each side. Both techniques are legitimate—which one you decide to use is a matter of personal preference.

2 Solve linear equations in standard form.

Solving Linear Equations in Standard Form

The most common type of equation in one variable is a **linear equation.**

> ### Definition of Linear Equation
>
> A **linear equation** in one variable x is an equation that can be written in the standard form
>
> $$ax + b = 0 \qquad \text{Standard form}$$
>
> where a and b are real numbers with $a \neq 0$.

A linear equation in one variable is also called a **first-degree equation** because its variable has an implied exponent of 1. Some examples of linear equations in the standard form $ax + b = 0$ are $3x + 2 = 0$ and $5x - 4 = 0$.

Remember that to *solve* an equation in x means to find the values of x that satisfy the equation. For a linear equation in the standard form

$$ax + b = 0$$

the goal is to **isolate** x by rewriting the standard equation in the form

$$x = \boxed{\text{a number}}.$$

Beginning with the original equation, you write a sequence of equivalent equations, each having the same solution as the original equation.

Example 2   Solving a Linear Equation in Standard Form

Solve $4x - 12 = 0$. Then check the solution.

Solution

$$
\begin{aligned}
4x - 12 &= 0 && \text{Write original equation.} \\
4x - 12 + 12 &= 0 + 12 && \text{Add 12 to each side.} \\
4x &= 12 && \text{Combine like terms.} \\
\frac{4x}{4} &= \frac{12}{4} && \text{Divide each side by 4.} \\
x &= 3 && \text{Simplify.}
\end{aligned}
$$

It appears that the solution is $x = 3$. You can check this as follows.

Check

$$
\begin{aligned}
4x - 12 &= 0 && \text{Write original equation.} \\
4(3) - 12 &\stackrel{?}{=} 0 && \text{Substitute 3 for } x. \\
12 - 12 &\stackrel{?}{=} 0 && \text{Simplify.} \\
0 &= 0 && \text{Solution checks. } \checkmark
\end{aligned}
$$

Study Tip

Be sure you see that solving an equation such as the one in Example 2 has two basic steps. The first step is to *find* the solution(s). The second step is to *check* that each solution you find actually satisfies the original equation. You can improve your accuracy in algebra by developing the habit of checking each solution.

You know that $x = 3$ is a solution of the equation in Example 2, but at this point you might be asking, "How can I be sure that the equation does not have other solutions?" The answer is that a linear equation in one variable always has *exactly one* solution. You can show this with the following steps.

$$ax + b = 0 \qquad \text{Original equation, with } a \neq 0$$

$$ax + b - b = 0 - b \qquad \text{Subtract } b \text{ from each side.}$$

$$ax = -b \qquad \text{Combine like terms.}$$

$$\frac{ax}{a} = \frac{-b}{a} \qquad \text{Divide each side by } a.$$

$$x = -\frac{b}{a} \qquad \text{Simplify.}$$

It is clear that the last equation has only one solution, $x = -b/a$. Because the last equation is equivalent to the original equation, you can conclude that every linear equation in one variable written in standard form has exactly one solution.

Example 3   Solving a Linear Equation in Standard Form

Solve $2x + 2 = 0$. Then check the solution.

Solution

$$2x + 2 = 0 \qquad \text{Write original equation.}$$

$$2x + 2 - 2 = 0 - 2 \qquad \text{Subtract 2 from each side.}$$

$$2x = -2 \qquad \text{Combine like terms.}$$

$$\frac{2x}{2} = \frac{-2}{2} \qquad \text{Divide each side by 2.}$$

$$x = -1 \qquad \text{Simplify.}$$

The solution is $x = -1$. You can check this as follows.

Check

$$2(-1) + 2 \overset{?}{=} 0 \qquad \text{Substitute } -1 \text{ for } x \text{ in original equation.}$$

$$-2 + 2 \overset{?}{=} 0 \qquad \text{Simplify.}$$

$$0 = 0 \qquad \text{Solution checks. } \checkmark$$

As you gain experience in solving linear equations, you will probably be able to perform some of the solution steps in your head. For instance, you might solve the equation given in Example 3 by performing two of the steps mentally and writing only three steps, as follows.

$$2x + 2 = 0 \qquad \text{Write original equation.}$$

$$2x = -2 \qquad \text{Subtract 2 from each side.}$$

$$x = -1 \qquad \text{Divide each side by 2.}$$

③ Solve linear equations in nonstandard form.

Solving Linear Equations in Nonstandard Form

Linear equations often occur in nonstandard forms that contain symbols of grouping or like terms that are not combined. Here are some examples.

$$x + 2 = 2x - 6, \quad 6(y - 1) = 2y - 3, \quad \frac{x}{18} + \frac{3x}{4} = 2$$

The next three examples show how to solve these linear equations.

Study Tip

Remember that the goal in solving any linear equation is to rewrite the original equation so that all the variable terms are on one side of the equal sign and all constant terms are on the other side.

Example 4   Solving a Linear Equation in Nonstandard Form

$x + 2 = 2x - 6$	Original equation
$-2x + x + 2 = -2x + 2x - 6$	Add $-2x$ to each side.
$-x + 2 = -6$	Combine like terms.
$-x + 2 - 2 = -6 - 2$	Subtract 2 from each side.
$-x = -8$	Combine like terms.
$(-1)(-x) = (-1)(-8)$	Multiply each side by -1.
$x = 8$	Simplify.

The solution is $x = 8$. Check this in the original equation.

In most cases, it helps to remove symbols of grouping as a first step in solving an equation. This is illustrated in Example 5.

Example 5   Solving a Linear Equation Containing Parentheses

$6(y - 1) = 2y - 3$	Original equation
$6y - 6 = 2y - 3$	Distributive Property
$6y - 2y - 6 = 2y - 2y - 3$	Subtract $2y$ from each side.
$4y - 6 = -3$	Combine like terms.
$4y - 6 + 6 = -3 + 6$	Add 6 to each side.
$4y = 3$	Combine like terms.
$\dfrac{4y}{4} = \dfrac{3}{4}$	Divide each side by 4.
$y = \dfrac{3}{4}$	Simplify.

The solution is $y = \frac{3}{4}$. Check this in the original equation.

If a linear equation contains fractions, you should first *clear the equation of fractions* by multiplying each side of the equation by the least common denominator (LCD) of the fractions.

Example 6   Solving a Linear Equation Containing Fractions

$$\frac{x}{18} + \frac{3x}{4} = 2$$ Original equation

$$36\left(\frac{x}{18} + \frac{3x}{4}\right) = 36(2)$$ Multiply each side by LCD of 36.

$$36 \cdot \frac{x}{18} + 36 \cdot \frac{3x}{4} = 36(2)$$ Distributive Property

$$2x + 27x = 72$$ Simplify.

$$29x = 72$$ Combine like terms.

$$\frac{29x}{29} = \frac{72}{29}$$ Divide each side by 29.

$$x = \frac{72}{29}$$ Simplify.

The solution is $x = \frac{72}{29}$. Check this in the original equation.

The next example shows how to solve a linear equation involving decimals. The procedure is basically the same, but the arithmetic can be messier.

Example 7   Solving a Linear Equation Involving Decimals

Solve $0.12x + 0.09(5000 - x) = 513$.

Solution

$$0.12x + 0.09(5000 - x) = 513$$ Write original equation.

$$0.12x + 450 - 0.09x = 513$$ Distributive Property

$$0.03x + 450 = 513$$ Combine like terms.

$$0.03x + 450 - 450 = 513 - 450$$ Subtract 450 from each side.

$$0.03x = 63$$ Combine like terms.

$$\frac{0.03x}{0.03} = \frac{63}{0.03}$$ Divide each side by 0.03.

$$x = 2100$$ Simplify.

The solution is $x = 2100$. Check this in the original equation.

Study Tip

A different approach to Example 7 would be to begin by multiplying each side of the equation by 100. This would clear the equation of decimals to produce

$$12x + 9(5000 - x) = 51,300.$$

Try solving this equation to see that you obtain the same solution.

Some equations in nonstandard form have *no solution* or *infinitely many solutions*. These cases are illustrated in Example 8.

Example 8   Solving Linear Equations: Special Cases

Solve each equation.

a. $2x - 4 = 2(x - 3)$ **b.** $3x + 2 + 2(x - 6) = 5(x - 2)$

Solution

a. $2x - 4 = 2(x - 3)$ Write original equation.

$2x - 4 = 2x - 6$ Distributive Property

$-4 \neq -6$ Subtract $2x$ from each side.

Because -4 does not equal -6, you can conclude that the original equation has no solution.

b. $3x + 2 + 2(x - 6) = 5(x - 2)$ Write original equation.

$3x + 2 + 2x - 12 = 5x - 10$ Distributive Property

$5x - 10 = 5x - 10$ Combine like terms.

$-10 = -10$ Subtract $5x$ from each side.

Because the last equation is true for any value of x, the equation is an identity, and you can conclude that the original equation has infinitely many solutions.

Example 9   Trucks and Buses

The bar graph in Figure 2.1 shows the numbers y of trucks and buses (in millions) sold in the United States for the years 1995 through 2000. A linear equation that models the data is given by

$$y = 0.27t + 4.3, \qquad 5 \leq t \leq 10$$

where t represents the year, with $t = 5$ corresponding to 1995. Determine when the number of trucks and buses sold reached 6.46 million.   (Source: Ward's Communications)

Solution

Let $y = 6.46$ and solve the resulting equation for t.

$y = 0.27t + 4.3$ Write original equation.

$6.46 = 0.27t + 4.3$ Substitute 6.46 for y.

$2.16 = 0.27t$ Subtract 4.3 from each side.

$8 = t$ Divide each side by 0.27.

Because $t = 5$ corresponds to 1995, it follows that $t = 8$ corresponds to 1998. So, the number of trucks and buses sold reached 6.46 million in 1998. From the bar graph in Figure 2.1, you can estimate the sales of trucks and buses to be about 6.5 million in 1998.

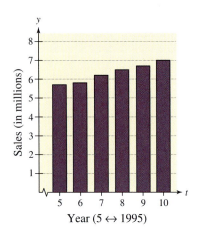

Year (5 ↔ 1995)

Figure 2.1

2.1 Exercises

Review   Concepts, Skills, and Problem Solving

Keep mathematically in shape by doing these exercises *before* the problems of this section.

Properties and Definitions

In Exercises 1–4, identify the property of real numbers illustrated by the statement.

1. $5 + x = x + 5$

2. $3x \cdot \dfrac{1}{3x} = 1$

3. $6(x - 2) = 6x - 6 \cdot 2$

4. $3 + (4 + x) = (3 + 4) + x$

Simplifying Expressions

In Exercises 5–10, evaluate the expression.

5. $4 - |-3|$

6. $-10 - (4 - 18)$

7. $\dfrac{3 - (5 - 20)}{4}$

8. $\dfrac{|3 - 18|}{3}$

9. $6\left(\dfrac{2}{15}\right)$

10. $\dfrac{5}{12} \div \left(-\dfrac{15}{16}\right)$

Problem Solving

11. *Savings Plan*   You plan to save $75 per month for 20 years. How much money will you set aside during the 20 years?

12. ▲ *Geometry*   It is necessary to cut a 135-foot rope into 15 pieces of equal length. What is the length of each piece?

Developing Skills

In Exercises 1–8, determine whether each value of the variable is a solution of the equation. See Example 1.

Equation	Values

1. $3x - 7 = 2$ (a) $x = 0$
       (b) $x = 3$

2. $5x + 9 = 4$ (a) $x = -1$
       (b) $x = 2$

3. $x + 8 = 3x$ (a) $x = 4$
       (b) $x = -4$

4. $10x - 3 = 7x$ (a) $x = 0$
       (b) $x = -1$

5. $3x + 3 = 2(x - 4)$ (a) $x = -11$
       (b) $x = 5$

6. $7x - 1 = 5(x + 5)$ (a) $x = 2$
       (b) $x = 13$

7. $\frac{1}{4}x = 3$ (a) $x = -4$
       (b) $x = 12$

8. $3(y + 2) = y - 5$ (a) $y = -\frac{3}{2}$
       (b) $y = -5.5$

In Exercises 9–12, identify the equation as a conditional equation, an identity, or an equation with no solution.

9. $3(x - 1) = 3x$

10. $2x + 8 = 6x$

11. $5(x + 3) = 2x + 3(x + 5)$

12. $\frac{2}{3}x + 4 = \frac{1}{3}x + 12$

In Exercises 13–16, determine whether the equation is linear. If not, state why.

13. $3x + 4 = 10$

14. $x^2 + 3 = 8$

15. $\dfrac{4}{x} - 3 = 5$

16. $3(x - 2) = 4x$

In Exercises 17–20, justify each step of the solution.

17.
$$3x + 15 = 0$$
$$3x + 15 - 15 = 0 - 15$$
$$3x = -15$$
$$\frac{3x}{3} = \frac{-15}{3}$$
$$x = -5$$

18.
$$7x - 21 = 0$$
$$7x - 21 + 21 = 0 + 21$$
$$7x = 21$$
$$\frac{7x}{7} = \frac{21}{7}$$
$$x = 3$$

19.
$$-2x + 5 = 12$$
$$-2x + 5 - 5 = 12 - 5$$
$$-2x = 7$$
$$\frac{-2x}{-2} = \frac{7}{-2}$$
$$x = -\frac{7}{2}$$

20.
$$25 - 3x = 10$$
$$25 - 3x - 25 = 10 - 25$$
$$-3x = -15$$
$$\frac{-3x}{-3} = \frac{-15}{-3}$$
$$x = 5$$

In Exercises 21–80, solve the equation and check the result. (If it has no solution, state the reason.) See Examples 2–8.

21. $x - 3 = 0$

22. $x + 8 = 0$

23. $3x - 12 = 0$

24. $-14x - 28 = 0$

25. $-6y - 4.2 = 0$

26. $0.5t - 7 = 0$

27. $6x + 4 = 0$

28. $8z - 10 = 0$

29. $-2u = 2$

30. $-5x = 15$

31. $4x - 7 = -11$

32. $5y + 9 = -6$

33. $23x - 4 = 42$

34. $15x - 18 = 27$

35. $3t + 8 = -2$

36. $10 - 6x = -5$

37. $4y - 3 = 4y$

38. $24 - 2x = x$

39. $-9y - 4 = -9y$

40. $6a + 2 = 6a$

41. $7 - 8x = 13x$

42. $2s - 16 = 34s$

43. $8 - 5t = 20 + t$

44. $3y + 14 = y + 20$

45. $4x - 5 = 2x - 1$

46. $8 - 7y = 5y - 4$

47. $3x - 1 = 2x + 14$

48. $9y + 4 = 12y - 2$

49. $8(x - 8) = 24$

50. $6(x + 2) = 30$

51. $-4(t + 2) = 0$

52. $8(z - 8) = 0$

53. $3(x - 4) = 7x + 6$

54. $-2(t + 3) = 9 - 5t$

55. $8x - 3(x - 2) = 12$

56. $12 = 6(y + 1) - 8y$

57. $5 - (2y - 4) = 15$

58. $26 - (3x - 10) = 6$

59. $12(x + 3) = 7(x + 3)$

60. $-5(x - 10) = 6(x - 10)$

61. $2(x + 7) - 9 = 5(x - 4)$

62. $4(2 - x) = -2(x + 7) - 2x$

63. $\dfrac{u}{5} = 10$

64. $-\dfrac{z}{2} = 7$

65. $t - \dfrac{2}{5} = \dfrac{3}{2}$

66. $z + \dfrac{1}{15} = -\dfrac{3}{10}$

67. $\dfrac{t}{5} - \dfrac{t}{2} = 1$

68. $\dfrac{t}{6} + \dfrac{t}{8} = 1$

69. $\dfrac{8x}{5} - \dfrac{x}{4} = -3$

70. $\dfrac{11x}{6} + \dfrac{1}{3} = 2x$

71. $\frac{1}{3}x + 1 = \frac{1}{12}x - 4$

72. $\frac{1}{9}x + \frac{1}{3} = \frac{11}{18}$

73. $\dfrac{25 - 4u}{3} = \dfrac{5u + 12}{4} + 6$

74. $\dfrac{8 - 3x}{4} - 4 = \dfrac{x}{6}$

75. $0.3x + 1.5 = 8.4$

76. $16.3 - 0.2x = 7.1$

77. $1.2(x - 3) = 10.8$

78. $6.5(1 - 2x) = 13$

79. $\frac{2}{3}(2x - 4) = \frac{1}{2}(x + 3) - 4$

80. $\frac{3}{4}(6 - x) = \frac{1}{3}(4x + 5) + 2$

Solving Problems

81. *Number Problem*   The sum of two consecutive integers is 251. Find the integers.

82. *Number Problem*   The sum of two consecutive integers is 137. Find the integers.

83. *Number Problem*   The sum of two consecutive even integers is 166. Find the integers.

84. *Number Problem*   The sum of two consecutive even integers is 626. Find the integers.

85. *Car Repair*   The bill for the repair of your car was $210. The cost for parts was $162. The cost for labor was $32 per hour. How many hours did the repair work take?

86. *Appliance Repair*   The bill for the repair of your refrigerator was $172. The cost for parts was $74. The cost for the service call and the first half hour of service was $50. The additional cost for labor was $16 per half hour. How many hours did the repair work take?

87. *Height*   Consider the fountain shown in the figure. The initial velocity of the stream of the water is 48 feet per second. The velocity v of the water at any time t (in seconds) is given by $v = 48 - 32t$. Find the time for a drop of water to travel from the base to the maximum height of the fountain.   (*Hint:* The maximum height is reached when $v = 0$.)

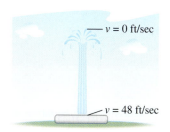

88. *Height*   The velocity v of an object projected vertically upward with an initial velocity of 64 feet per second is given by $v = 64 - 32t$, where t is time in seconds. When does the object reach its maximum height?

89. *Work Rate*   Two people can complete a task in t hours, where t must satisfy the equation

$$\frac{t}{10} + \frac{t}{15} = 1.$$

Find the required time t.

90. *Work Rate*   Two people can complete a task in t hours, where t must satisfy the equation

$$\frac{t}{12} + \frac{t}{20} = 1.$$

Find the required time t.

91. ▲ *Geometry*   The length of a rectangle is t times its width (see figure). So, the perimeter P is given by $P = 2w + 2(tw)$, where w is the width of the rectangle. The perimeter of the rectangle is 1200 meters.

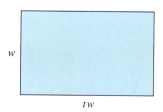

w

tw

(a) Complete the table of widths, lengths, and areas of the rectangle for the specified values of t.

t	1	1.5	2
Width			
Length			
Area			

t	3	4	5
Width			
Length			
Area			

(b) Use the table to write a short paragraph describing the relationship among the width, length, and area of a rectangle that has a *fixed* perimeter.

92. ▲ *Geometry*   The length of a rectangle is 10 meters greater than its width. The perimeter is 64 meters. Find the dimensions of the rectangle.

93. *Using a Model*   The average annual expenditures per student y (in dollars) for primary and secondary public schools in the United States from 1993 to 2000 can be approximated by the model $y = 233.6t + 4797$, $3 \le t \le 10$, where t represents the year, with $t = 3$ corresponding to 1993 (see figure). According to this model, during which year did the expenditures reach $5965? Explain how to answer the question graphically, numerically, and algebraically. (Source: National Education Association)

94. *Using a Model*   The average monthly rate y (in dollars) for basic cable television service in the United States from 1993 to 2000 can be approximated by the model $y = 1.519t + 15.35$, $3 \le t \le 10$, where t represents the year, with $t = 3$ corresponding to 1993 (see figure). According to this model, during which year was the average monthly rate $29? Explain how to answer the question graphically, numerically, and algebraically. (Source: Paul Kagan Associates, Inc.)

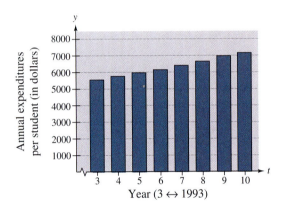

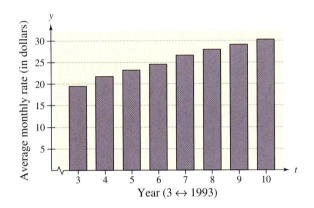

Explaining Concepts

95. *Writing*   Explain the difference between a conditional equation and an identity.

96. *Writing*   Explain the difference between evaluating an expression and solving an equation.

97. *Writing*   Give the standard form of a linear equation. Why is a linear equation sometimes called a first-degree equation?

98. *Writing*   What is meant by equivalent equations? Give an example of two equivalent equations.

99. *True or False?*   Multiplying each side of an equation by 0 yields an equivalent equation. Justify your answer.

100. *True or False?*   Subtracting 0 from each side of an equation yields an equivalent equation. Justify your answer.

101. *Writing*   Classify each equation as an identity, a conditional equation, or an equation with no solution. Discuss possible realistic situations in which the equations you classified as an identity and a conditional equation might apply. Write a brief description of these situations and explain how the equations could be used.

(a) $2x - 3 = -4 + 2x$

(b) $x + 0.05x = 37.75$

(c) $5x(3 + x) = 15x + 5x^2$

2.2 Linear Equations and Problem Solving

© Lester Lefkowitz/Corbis

Why You Should Learn It

Percents appear in many real-life situations. For instance, in Exercise 72 on page 77, percents are used to show the sources of energy consumption in the United States.

What You Should Learn

1 Use mathematical modeling to write algebraic equations representing real-life situations.

2 Solve percent problems using the percent equation.

3 Use ratios to compare unit prices for products.

4 Solve proportions.

Mathematical Modeling

In this section you will see how algebra can be used to solve problems that occur in real-life situations. This process is called **mathematical modeling,** and its basic steps are shown below.

Verbal description ⟹ Verbal model ⟹ Assign labels ⟹ Algebraic equation

1 Use mathematical modeling to write algebraic equations representing real-life situations.

Example 1   Mathematical Modeling

Write an algebraic equation that represents the following problem. Then solve the equation and answer the question.

You have accepted a job at an annual salary of $27,630. This salary includes a year-end bonus of $750. You are paid twice a month. What will your gross pay be for each paycheck?

Solution

Because there are 12 months in a year and you will be paid twice a month, it follows that you will receive 24 paychecks during the year. Construct an algebraic equation for this problem, as follows. Begin with a verbal model, then assign labels, and finally form an algebraic equation.

Verbal Model: $\dfrac{\text{Income}}{\text{for year}} = 24 \times \dfrac{\text{Amount of}}{\text{each paycheck}} + \text{Bonus}$

Labels: Income for year = 27,630   (dollars)
Amount of each paycheck = x (dollars)
Bonus = 750   (dollars)

Equation:

$27{,}630 = 24x + 750$ Original equation

$26{,}880 = 24x$ Subtract 750 from each side.

$\dfrac{26{,}880}{24} = \dfrac{24x}{24}$ Divide each side by 24.

$1120 = x$ Simplify.

Each paycheck will be $1120. Check this in the original statement of the problem.

Study Tip

You could solve the problem in Example 1 *without* algebra by simply subtracting the bonus of $750 from the annual salary of $27,630 and dividing the result by 24 pay periods. The reason for listing this example is to allow you to practice writing algebraic versions of the problem-solving skills *you already possess.* Your goals in this section are to practice formulating problems by logical reasoning *and* to use this reasoning to write algebraic versions of the problems. Later, you will encounter more complicated problems in which algebra is a necessary part of the solution.

2 Solve percent problems using the percent equation.

Percent Problems

Rates that describe increases, decreases, and discounts are often given as percents. **Percent** means *per hundred*, so 40% means 40 per hundred or, equivalently, $\frac{40}{100}$. The word "per" occurs in many other rates, such as price per ounce, miles per gallon, revolutions per minute, and cost per share. In applications involving percents, you need to convert the percent number to decimal or fractional form before performing any arithmetic operations. Some examples are listed below.

Percent	10%	$12\frac{1}{2}\%$	20%	25%	$33\frac{1}{3}\%$	50%	$66\frac{2}{3}\%$	75%
Decimal	0.1	0.125	0.2	0.25	0.33...	0.5	0.66...	0.75
Fraction	$\frac{1}{10}$	$\frac{1}{8}$	$\frac{1}{5}$	$\frac{1}{4}$	$\frac{1}{3}$	$\frac{1}{2}$	$\frac{2}{3}$	$\frac{3}{4}$

The primary use of percents is to compare two numbers. For example, you can compare 3 and 6 by saying that 3 is 50% of 6. In this statement, 6 is the **base number,** and 3 is the number being compared with the base number. The following model, which is called the **percent equation,** is helpful.

Verbal Model: $\dfrac{\text{Compared}}{\text{number}} = \dfrac{\text{Percent}}{\text{(decimal form)}} \cdot \dfrac{\text{Base}}{\text{number}}$

Labels: Compared number $= a$
Percent $= p$ (decimal form)
Base number $= b$

Equation: $a = p \cdot b$ Percent equation

Remember to convert p to a decimal value before multiplying by b.

Technology: Tip

Some scientific calculators and graphing calculators can convert percents to decimal and fractional forms. Consult the user's guide for your calculator for the proper keystrokes. Use your calculator to convert each percent to decimal and fractional forms.

a. 35%   **b.** 0.1%

c. 8%   **d.** 96%

e. 150%   **f.** 200%

Example 2   Solving a Percent Problem

The number 15.6 is 26% of what number?

Solution

Verbal Model: $\dfrac{\text{Compared}}{\text{number}} = \dfrac{\text{Percent}}{\text{(decimal form)}} \cdot \dfrac{\text{Base}}{\text{number}}$

Labels: Compared number $= 15.6$
Percent $= 0.26$ (decimal form)
Base number $= b$

Equation: $15.6 = 0.26b$ Original equation

$\dfrac{15.6}{0.26} = b$ Divide each side by 0.26.

$60 = b$ Simplify.

Check that 15.6 is 26% of 60 by multiplying 60 by 0.26 to get 15.6.

Example 3   Solving a Percent Problem

The number 28 is what percent of 80?

Solution

Verbal Model: $\boxed{\text{Compared number}} = \boxed{\begin{array}{c}\text{Percent}\\\text{(decimal form)}\end{array}} \cdot \boxed{\begin{array}{c}\text{Base}\\\text{number}\end{array}}$

Labels: Compared number $= 28$
Percent $= p$ (decimal form)
Base number $= 80$

Equation: $28 = p(80)$ Original equation

$\dfrac{28}{80} = p$ Divide each side by 80.

$0.35 = p$ Simplify.

So, 28 is 35% of 80. Check this solution by multiplying 80 by 0.35 to see that you obtain 28.

In most real-life applications, the base number b and the number a are much more disguised than in Examples 2 and 3. It sometimes helps to think of a as the "new" amount and b as the "original" amount.

Example 4   A Percent Application

A real estate agency receives a commission of $8092.50 for the sale of a $124,500 house. What percent commission is this?

Solution

A commission is a percent of the sale price paid to the agency for their services. To determine the percent commission, start with a verbal model.

Verbal Model: $\boxed{\text{Commission}} = \boxed{\begin{array}{c}\text{Percent}\\\text{(decimal form)}\end{array}} \cdot \boxed{\begin{array}{c}\text{Sale}\\\text{price}\end{array}}$

Labels: Commission $= 8092.50$ (dollars)
Percent $= p$ (decimal form)
Sale price $= 124,500$ (dollars)

Equation: $8092.50 = p(124,500)$ Original equation

$\dfrac{8092.50}{124,500} = p$ Divide each side by 124,500.

$0.065 = p$ Simplify.

The real estate agency receives a commission of 6.5%. Check this solution by multiplying 124,500 by 0.065 to see that you obtain 8092.50.

3 Use ratios to compare unit prices for products.

Ratios and Unit Prices

You know that a percent compares a number with 100. A **ratio** is a more generic rate form that compares one number with another. Specifically, if a and b have the same units of measure, then a/b is called the ratio of a to b. Note the *order* implied by a ratio. The ratio of a to b means a/b, whereas the ratio of b to a means b/a.

Study Tip

Conversions for common units of measure can be found on the inside back cover of the text.

Example 5   Using a Ratio

Find the ratio of 4 feet to 8 inches.

Solution

Because the units of feet and inches are not the same, you must first convert 4 feet into its equivalent in inches or convert 8 inches into its equivalent in feet. You can convert 4 feet to 48 inches (by multiplying 4 by 12) to obtain

$$\frac{4 \text{ feet}}{8 \text{ inches}} = \frac{48 \text{ inches}}{8 \text{ inches}} = \frac{48}{8} = \frac{6}{1}.$$

Or, you can convert 8 inches to $\frac{8}{12}$ feet (by dividing 8 by 12) to obtain

$$\frac{4 \text{ feet}}{8 \text{ inches}} = \frac{4 \text{ feet}}{\frac{8}{12} \text{ feet}} = 4 \div \frac{8}{12} = 4 \cdot \frac{12}{8} = \frac{6}{1}.$$

The **unit price** of an item is the quotient of the total price divided by the total units. That is,

$$\text{Unit price} = \frac{\text{Total price}}{\text{Total units}}.$$

To state unit prices, use the word "per." For instance, the unit price for a brand of coffee might be 4.79 dollars *per* pound.

Example 6   Comparing Unit Prices

Which is the better buy, a 12-ounce box of breakfast cereal for $2.79 or a 16-ounce box of the same cereal for $3.59?

Solution

The unit price for the 12-ounce box is

$$\text{Unit price} = \frac{\text{Total price}}{\text{Total units}} = \frac{\$2.79}{12 \text{ ounces}} = \$0.2325 \text{ per ounce.}$$

The unit price for the 16-ounce box is approximately

$$\text{Unit price} = \frac{\text{Total price}}{\text{Total units}} = \frac{\$3.59}{16 \text{ ounces}} \approx \$0.2244 \text{ per ounce.}$$

The 16-ounce box has a slightly lower unit price, and so it is the better buy.

4 Solve proportions.

Solving Proportions

A **proportion** is a statement that equates two ratios. For example, if the ratio of a to b is the same as the ratio of c to d, you can write the proportion as

$$\frac{a}{b} = \frac{c}{d}.$$

In typical problems, you know three of the values and need to find the fourth. The quantities a and d are called the **extremes** of the proportion, and the quantities b and c are called the **means** of the proportion. In a proportion, the product of the extremes is equal to the product of the means. This is done by **cross-multiplying.** That is, if

$$\frac{a}{b} = \frac{c}{d}$$

then $ad = bc$.

Proportions are often used in geometric applications involving similar triangles. Similar triangles have the same shape, but they may differ in size. The corresponding sides of similar triangles are proportional.

Example 7   Solving a Proportion in Geometry

The triangles shown in Figure 2.2 are similar triangles. Use this fact to find the length of the unknown side x of the larger triangle.

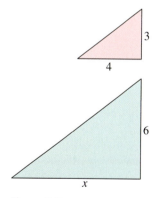

Figure 2.2

Study Tip

The proportion in Example 7 could also have been written as

$$\frac{6}{3} = \frac{x}{4}.$$

After cross-multiplying, you obtain the same equation

$$24 = 3x.$$

Solution

$$\frac{4}{3} = \frac{x}{6}$$ Set up proportion.

$$4 \cdot 6 = 3 \cdot x$$ Cross-multiply.

$$24 = 3x$$ Simplify.

$$8 = x$$ Divide each side by 3.

So, the length of the unknown side of the larger triangle is 8 units. Check this in the original statement of the problem.

Study Tip

You can write a proportion in several ways. Just be sure to put like quantities in similar positions on each side of the proportion.

Example 8   Gasoline Cost

You are driving from New York City to Phoenix, a trip of 2450 miles. You begin the trip with a full tank of gas and after traveling 424 miles, you refill the tank for $24.00. How much should you plan to spend on gasoline for the entire trip?

Solution

Verbal Model: $\dfrac{\text{Cost for trip}}{\text{Cost for tank}} = \dfrac{\text{Miles for trip}}{\text{Miles for tank}}$

Labels: Cost of gas for entire trip $= x$ (dollars)
Cost of gas for tank $= 24$ (dollars)
Miles for entire trip $= 2450$ (miles)
Miles for tank $= 424$ (miles)

Proportion: $\dfrac{x}{24} = \dfrac{2450}{424}$ Original proportion

$x \cdot 424 = 24 \cdot 2450$ Cross-multiply.

$424x = 58{,}800$ Simplify.

$x \approx 138.68$ Divide each side by 424.

You should plan to spend approximately $138.68 for gasoline on the trip. Check this in the original statement of the problem.

The following list summarizes a strategy for modeling and solving real-life problems.

Strategy for Solving Word Problems

1. Ask yourself what you need to know to solve the problem. Then *write a verbal model* that includes arithmetic operations to describe the problem.

2. *Assign labels* to each part of the verbal model—numbers to the known quantities and letters (or expressions) to the variable quantities.

3. Use the labels to *write an algebraic model* based on the verbal model.

4. *Solve* the resulting algebraic equation.

5. *Answer* the original question and check that your answer satisfies the original problem as stated.

In previous mathematics courses, you studied several other problem-solving strategies, such as *drawing a diagram, making a table, looking for a pattern,* and *solving a simpler problem.* Each of these strategies can also help you to solve problems in algebra.

2.2 Exercises

Review   Concepts, Skills, and Problem Solving

Keep mathematically in shape by doing these exercises *before* the problems of this section.

Properties and Definitions

1. In your own words, define an algebraic expression.

2. State the definition of the *terms* of an algebraic expression.

Algebraic Operations

In Exercises 3–10, perform the indicated operations.

3. $-360 + 120$

4. $-12 - (-46)$

5. $5(57 - 33)$

6. $-8(12 - 5)$

7. $-\frac{4}{15} \cdot \frac{15}{16}$

8. $\frac{3}{8} \div \frac{5}{16}$

9. $(12 - 15)^3$

10. $\left(-\frac{5}{8}\right)^2$

Problem Solving

🔺 *Geometry*   In Exercises 11 and 12, find and simplify an expression for the perimeter of the figure .

11.

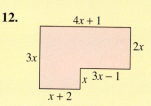

12.

Developing Skills

Mathematical Modeling   In Exercises 1–4, construct a verbal model and write an algebraic equation that represents the problem. Solve the equation. See Example 1.

1. Find a number such that the sum of the number and 30 is 82.

2. Find a number such that the difference of the number and 18 is 27.

3. You have accepted a job offer at an annual salary of $30,500. This salary includes a year-end bonus of $2300. You are paid every 2 weeks. What will your gross pay be for each paycheck?

4. You have a job on an assembly line for which you are paid $10 per hour plus $0.75 per unit assembled. Find the number of units produced in an eight-hour day if your earnings for the day are $146.

In Exercises 5–12, complete the table showing the equivalent forms of various percents.

Percent	Parts out of 100	Decimal	Fraction
5. 30%			
6. 75%			
7.		0.075	
8.		0.08	
9.			$\frac{2}{3}$
10.			$\frac{1}{8}$
11.	100		
12.	42		

In Exercises 13–34, solve using a percent equation. See Examples 2 and 3.

13. What is 35% of 250?

14. What is 68% of 800?

15. What is 8.5% of 816?

16. What is 70.2% of 980?

17. What is $12\frac{1}{2}\%$ of 1024?

18. What is $33\frac{1}{3}\%$ of 816?

19. What is 0.4% of 150,000?

20. What is 0.1% of 8925?

21. What is 250% of 32?

22. What is 300% of 16?

23. 84 is 24% of what number?

24. 416 is 65% of what number?

25. 42 is 120% of what number?

26. 168 is 350% of what number?

27. 96 is 0.8% of what number?

28. 18 is 2.4% of what number?

29. 496 is what percent of 800?

30. 1650 is what percent of 5000?

31. 2.4 is what percent of 480?

32. 3.8 is what percent of 190?

33. 2100 is what percent of 1200?

34. 900 is what percent of 500?

In Exercises 35–42, write the verbal expression as a ratio. Use the same units in both the numerator and denominator, and simplify. See Example 5.

35. 120 meters to 180 meters

36. 12 ounces to 20 ounces

37. 36 inches to 48 inches

38. 125 centimeters to 2 meters

39. 40 milliliters to 1 liter

40. 1 pint to 1 gallon

41. 5 pounds to 24 ounces

42. 45 minutes to 2 hours

In Exercises 43–52, solve the proportion. See Example 7.

43. $\dfrac{x}{6} = \dfrac{2}{3}$

44. $\dfrac{y}{36} = \dfrac{6}{7}$

45. $\dfrac{t}{4} = \dfrac{3}{2}$

46. $\dfrac{5}{16} = \dfrac{x}{4}$

47. $\dfrac{5}{4} = \dfrac{t}{6}$

48. $\dfrac{7}{8} = \dfrac{x}{2}$

49. $\dfrac{y}{6} = \dfrac{y-2}{4}$

50. $\dfrac{a}{5} = \dfrac{a+4}{8}$

51. $\dfrac{y+1}{10} = \dfrac{y-1}{6}$

52. $\dfrac{z-3}{3} = \dfrac{z+8}{12}$

Solving Problems

53. *College Enrollment* Approximately 15% of the students enrolled at Penn State University in the fall of 2002 were freshmen. The total enrollment for the school in the fall of 2002 was 83,038. How many freshmen were enrolled? (Source: Penn State University)

54. *Pension Fund* Your employer withholds $6\frac{1}{2}\%$ of your gross income of $3800 for your retirement. Determine the amount withheld each month.

55. *Passing Grade* There are 40 students in your class. On one test, 95% of the students received passing grades. How many students failed the test?

56. *Elections* In the 2000 presidential election, 105,586,274 votes were cast. This represented 67.5% of the registered voters in the United States. How many registered voters were in the United States in 2000? (Source: Federal Election Commission)

57. *Company Layoff* Because of slumping sales, a small company laid off 25 of its 160 employees. What percent of the work force was laid off?

58. *Monthly Rent* You spend $748 of your monthly income of $3400 for rent. What percent of your monthly income is your monthly rent payment?

59. *Tip Rate* A customer left $10 for a meal that cost $8.45. Determine the tip rate.

60. *Tip Rate* A customer left $40 for a meal that cost $34.73. Determine the tip rate.

61. *Tip Rate* A customer left $25 for a meal that cost $20.66. Determine the tip rate.

62. *Tip Rate* A customer left $60 for a meal that cost $47.24. Determine the tip rate.

63. *Tip Rate* A customer gave a taxi driver $9 for a ride that cost $8.20. Determine the tip rate.

64. *Tip Rate* A customer gave a taxi driver $21 for a ride that cost $18.80. Determine the tip rate.

65. *Real Estate Commission* A real estate agency receives a commission of $12,250 for the sale of a $175,000 house. What percent commission is this?

66. *Real Estate Commission*   A real estate agency receives a commission of $20,400 for the sale of a $240,000 house. What percent commission is this?

67. *Quality Control*   A quality control engineer reported that 1.5% of a sample of parts were defective. The engineer found three defective parts. How large was the sample?

68. *Price Inflation*   A new van costs $29,750, which is approximately 115% of what a comparable van cost 3 years ago. What did it cost 3 years ago?

69. *Floor Space*   You are planning to build a tool shed, but you are undecided about the size. The two sizes you are considering are 12 feet by 15 feet and 16 feet by 20 feet. The floor space of the larger is what percent of the floor space of the smaller? The floor space of the smaller is what percent of the floor space of the larger?

70. ▲ *Geometry*   The floor of a rectangular room that measures 10 feet by 12 feet is partially covered by a circular rug with a radius of 4 feet (see figure). What percent of the floor is covered by the rug? (*Hint:* The area of a circle is $A = \pi r^2$.)

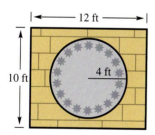

71. *Population*   The populations of six counties in Colorado in 2000 are shown in the circle graph. What percent of the total population is each county's population?   (Source: U.S. Census Bureau)

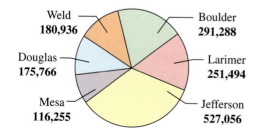

72. *Energy Use*   The circle graph shows the sources of the approximately 96 quadrillion British thermal units (Btu) of energy consumed in the United States in 2001. How many quadrillion Btu were obtained from coal?   (Source: Energy Information Administration)

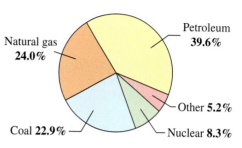

Graphical Estimation   In Exercises 73–76, use the bar graph to answer the questions. The graph shows the per capita food consumption of selected meats for 1990, 1995, and 1999.   (Source: U.S. Department of Agriculture)

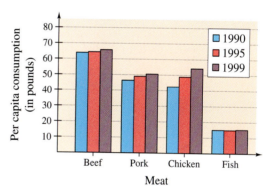

73. Approximate the increase in the per capita consumption of beef from 1990 to 1999. Use this estimate to approximate the percent increase.

74. Approximate the increase in the per capita consumption of chicken from 1990 to 1999. Use this estimate to approximate the percent increase.

75. Approximate the total number of pounds of pork consumed in 1990 if the population of the United States was approximately 250 million.

76. Of the four categories of meats shown in the table, about what percent of the meat diet was met by fish in 1995?

77. *Income Tax*   You have $12.50 of state tax withheld from your paycheck per week when your gross pay is $625. Find the ratio of tax to gross pay.

78. *Price-Earnings Ratio*   The **price-earnings ratio** is the ratio of the price of a stock to its earnings. Find the price-earnings ratio of a stock that sells for $56.25 per share and earns $6.25 per share.

79. *Compression Ratio*   The **compression ratio** of a cylinder is the ratio of its expanded volume to its compressed volume (see figure). The expanded volume of one cylinder of a small diesel engine is 425 cubic centimeters, and its compressed volume is 20 cubic centimeters. Find the compression ratio.

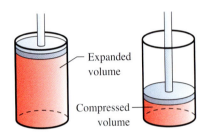

80. *Gear Ratio*   The **gear ratio** of two gears is the number of teeth in one gear to the number of teeth in the other gear. Two gears in a gearbox have 60 teeth and 40 teeth (see figure). Find the gear ratio.

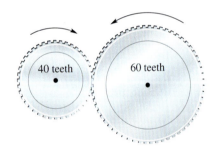

81. ▲ *Geometry*   Find the ratio of the areas of the two circles in the figure. (*Hint:* The area of a circle is $A = \pi r^2$.)

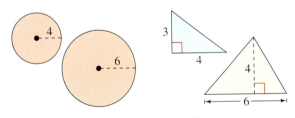

Figure for 81   Figure for 82

82. ▲ *Geometry*   Find the ratio of the areas of the two triangles in the figure.

Unit Prices   In Exercises 83–86, find the unit price (in dollars per ounce) of the product.

83. A 20-ounce can of pineapple for 95¢

84. A 64-ounce bottle of juice for $1.29

85. A one-pound, four-ounce loaf of bread for $1.69

86. An 18-ounce box of cereal for $4.29

Consumer Awareness   In Exercises 87–90, use unit prices to determine the better buy. See Example 6.

87. (a) A $14\frac{1}{2}$-ounce bag of chips for $2.32

(b) A $5\frac{1}{2}$-ounce bag of chips for $0.99

88. (a) A $10\frac{1}{2}$-ounce package of cookies for $1.79

(b) A 16-ounce package of cookies for $2.39

89. (a) A four-ounce tube of toothpaste for $1.69

(b) A six-ounce tube of toothpaste for $2.39

90. (a) A two-pound package of hamburger for $3.49

(b) A three-pound package of hamburger for $5.29

▲ *Geometry*   In Exercises 91–94, solve for the length x by using the fact that corresponding sides of similar triangles are proportional. See Example 7.

91.   **92.**

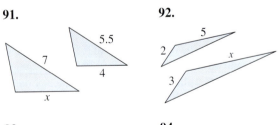

93.   **94.**

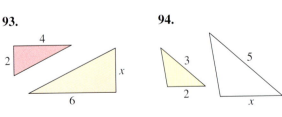

95. ▲ *Geometry*   A man who is 6 feet tall walks directly toward the tip of the shadow of a tree. When the man is 75 feet from the tree, he starts forming his own shadow beyond the shadow of the tree (see figure). The length of the shadow of the tree beyond this point is 11 feet. Find the height h of the tree.

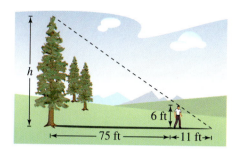

Figure for 95

96. ▲ *Geometry*   Find the length l of the shadow of a man who is 6 feet tall and is standing 15 feet from a streetlight that is 20 feet high (see figure).

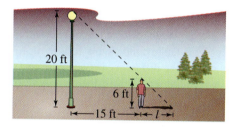

97. *Fuel Usage*   A tractor uses 5 gallons of diesel fuel to plow for 105 minutes. Assuming conditions remain the same, determine the number of gallons of fuel used in 6 hours.

98. *Spring Length*   A force of 32 pounds stretches a spring 6 inches. Determine the number of pounds of force required to stretch it 9 inches.

99. *Property Tax*   The tax on a property with an assessed value of $110,000 is $1650. Find the tax on a property with an assessed value of $160,000.

100. *Recipe*   Three cups of flour are required to make one batch of cookies. How many cups are required to make $3\frac{1}{2}$ batches?

101. *Quality Control*   A quality control engineer finds one defective unit in a sample of 75. At this rate, what is the expected number of defective units in a shipment of 200,000?

102. *Quality Control*   A quality control engineer finds 3 defective units in a sample of 120. At this rate, what is the expected number of defective units in a shipment of 5000?

103. *Quality Control*   A quality control engineer finds 6 defective units in a sample of 50. At this rate, what is the expected number of defective units in a shipment of 20,000?

104. *Quality Control*   A quality control inspector finds one color defect in a sample of 40 units. At this rate, what is the expected number of color defects in a shipment of 235?

105. *Quality Control*   A quality control inspector finds 4 scratch defects in a sample of 25 units. At this rate, what is the expected number of scratch defects in a shipment of 520?

106. *Public Opinion Poll*   In a public opinion poll, 870 people from a sample of 1500 indicate they will vote for the Republican candidate. Assuming this poll to be a correct indicator of the electorate, how many votes can the candidate expect to receive from 80,000 votes cast?

Explaining Concepts

107. *Writing*   Explain the meaning of the word *percent*.

108. *Writing*   Explain how to change percents to decimals and decimals to percents. Give examples.

109. *Writing*   Is it true that $\frac{1}{2}\% = 50\%$? Explain.

110. *Writing*   Define the term *ratio*. Give an example of a ratio.

111. *Writing*   During a year of financial difficulties, your company reduces your salary by 7%. What percent increase in this reduced salary is required to raise your salary to the amount it was prior to the reduction? Why isn't the percent increase the same as the percent of the reduction?

112. *Writing*   In your own words, describe the meaning of *mathematical modeling*. Give an example.

2.3 Business and Scientific Problems

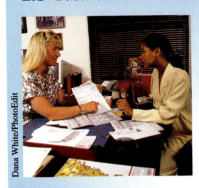

Dana White/PhotoEdit

What You Should Learn

1 Use mathematical models to solve business-related problems.

2 Use mathematical models to solve mixture problems.

3 Use mathematical models to solve classic rate problems.

4 Use formulas to solve application problems.

Why You Should Learn It

Mathematical models can be used to solve a wide variety of real-life problems. For instance, you can find the annual premium of an insurance policy using mathematical modeling. See Exercise 26 on page 89.

1 Use mathematical models to solve business-related problems.

Rates in Business Problems

Many business problems can be represented by mathematical models involving the sum of a fixed term and a variable term. The variable term is often a *hidden product* in which one of the factors is a percent or some other type of rate. Watch for these occurrences in the discussions and examples that follow.

The **markup** on a consumer item is the difference between the **cost** a retailer pays for an item and the **price** at which the retailer sells the item. A verbal model for this relationship is as follows.

Selling price = Cost + Markup   *Markup is a hidden product.*

The markup is the hidden product of the **markup rate** and the cost.

Markup = Markup rate · Cost

Example 1   Finding the Markup Rate

A clothing store sells a pair of jeans for $42. The cost of the jeans is $16.80. What is the markup rate?

Solution

Verbal Model: Selling price = Cost + Markup

Labels:
Selling price = 42   (dollars)
Cost = 16.80   (dollars)
Markup rate = p (percent in decimal form)
Markup = $p(16.80)$ (dollars)

Equation:

$42 = 16.80 + p(16.80)$ Original equation

$25.2 = p(16.80)$ Subtract 16.80 from each side.

$\dfrac{25.2}{16.80} = p$ Divide each side by 16.80.

$1.5 = p$ Simplify.

Because $p = 1.5$, it follows that the markup rate is 150%. Check this in the original statement of the problem.

The Granger Collection

In 1874, Levi Strauss designed the first pair of blue jeans. Today, billions of pairs of jeans are sold each year throughout the world.

Study Tip

Although markup and discount are similar, it is important to remember that markup is based on cost and discount is based on list price.

The model for a **discount** is similar to that for a markup.

Selling price = List price − Discount   *Discount is a hidden product.*

The discount is the hidden product of the **discount rate** and the list price.

Example 2   Finding the Discount and the Discount Rate

A DVD/VCR combination unit is marked down from its list price of \$310 to a sale price of \$217. What is the discount rate?

Solution

Verbal Model: Discount = $\dfrac{\text{Discount}}{\text{rate}}$ · $\dfrac{\text{List}}{\text{price}}$

Labels: Discount = 310 − 217 = 93   (dollars)
List price = 310   (dollars)
Discount rate = p (percent in decimal form)

Equation: $93 = p(310)$ *Original equation*

$$\frac{93}{310} = p$$ *Divide each side by 310.*

$0.30 = p$ *Simplify.*

The discount rate is 30%. Check this in the original statement of the problem.

Example 3   Finding the Hours of Labor

An auto repair bill of \$338 lists \$170 for parts and the rest for labor. It took 6 hours to repair the auto. What is the hourly rate for labor?

Solution

Verbal Model: $\dfrac{\text{Total}}{\text{bill}}$ = $\dfrac{\text{Price}}{\text{of parts}}$ + $\dfrac{\text{Price}}{\text{of labor}}$

Labels: Total bill = 338   (dollars)
Price of parts = 170   (dollars)
Hours of labor = 6   (hours)
Hourly rate for labor = x (dollars per hour)
Price of labor = 6x (dollars)

Equation: $338 = 170 + 6x$ *Original equation*
$168 = 6x$ *Subtract 170 from each side.*
$$\frac{168}{6} = x$$ *Divide each side by 6.*
$28 = x$ *Simplify.*

The hourly rate for labor is \$28 per hour. Check this in the original problem.

Notice in Example 3 that the price of labor is a hidden product.

2 Use mathematical models to solve mixture problems.

Rates in Mixture Problems

Many real-life problems involve combinations of two or more quantities that make up new or different quantities. Such problems are called **mixture problems.** They are usually composed of the sum of two or more "hidden products" that involve rate factors. Here is the generic form of the verbal model for mixture problems.

$$\begin{array}{c}\text{First}\\\text{rate}\end{array} \cdot \text{Amount} + \begin{array}{c}\text{Second}\\\text{rate}\end{array} \cdot \text{Amount} = \begin{array}{c}\text{Final}\\\text{rate}\end{array} \cdot \begin{array}{c}\text{Final}\\\text{amount}\end{array}$$

Study Tip

When you set up a verbal model, be sure to check that you are working with *the same type of units* in each part of the model. For instance, in Example 4 note that each of the three parts of the verbal model measures cost. (If two parts measured cost and the other part measured pounds, you would know that the model was incorrect.)

Example 4   A Mixture Problem

A nursery wants to mix two types of lawn seed. Type A sells for $10 per pound and type B sells for $15 per pound. To obtain 20 pounds of a mixture at $12 per pound, how many pounds of each type of seed are needed?

Solution

The rates are the unit prices for each type of seed.

Verbal | $\begin{array}{c}\text{Total cost}\\\text{of \$10 seed}\end{array} + \begin{array}{c}\text{Total cost}\\\text{of \$15 seed}\end{array} = \begin{array}{c}\text{Total cost}\\\text{of \$12 seed}\end{array}$
Model: |

Labels:
Unit price of type A = 10   (dollars per pound)
Pounds of $10 seed = x   (pounds)
Unit price of type B = 15   (dollars per pound)
Pounds of $15 seed = 20 − x   (pounds)
Unit price of mixture = 12   (dollars per pound)
Pounds of $12 seed = 20   (pounds)

Equation:
$10x + 15(20 - x) = 12(20)$ Original equation
$10x + 300 - 15x = 240$ Distributive Property
$300 - 5x = 240$ Combine like terms.
$-5x = -60$ Subtract 300 from each side.
$x = 12$ Divide each side by -5.

The mixture should contain 12 pounds of the $10 seed and $20 - x = 20 - 12 = 8$ pounds of the $15 seed.

Remember that when you have found a solution, you should always go back to the original statement of the problem and check to see that the solution makes sense—both algebraically and from a practical point of view. For instance, you can check the result of Example 4 as follows.

$$\overbrace{\left(\begin{array}{c}\$10 \text{ per}\\\text{pound}\end{array}\right)\left(\begin{array}{c}12\\\text{pounds}\end{array}\right)}^{\$10\text{ seed}} + \overbrace{\left(\begin{array}{c}\$15 \text{ per}\\\text{pound}\end{array}\right)\left(\begin{array}{c}8\\\text{pounds}\end{array}\right)}^{\$15\text{ seed}} \stackrel{?}{=} \overbrace{\left(\begin{array}{c}\$12 \text{ per}\\\text{pound}\end{array}\right)\left(\begin{array}{c}20\\\text{pounds}\end{array}\right)}^{\$12\text{ seed}}$$

$$\$120 + \$120 = \$240 \quad \text{Solution checks.} ✔$$

③ Use mathematical models to solve classic rate problems.

Classic Rate Problems

Time-dependent problems such as distance traveled at a given speed and work done at a specified rate are classic types of **rate problems.** The distance-rate-time problem fits the verbal model

$$\text{Distance} = \text{Rate} \cdot \text{Time} .$$

Study Tip

To convert minutes to hours, use the fact that there are 60 minutes in each hour. So, 45 minutes is equivalent to $\frac{45}{60}$ of 1 hour. In general, x minutes is $x/60$ of 1 hour.

For instance, if you travel at a constant (or average) rate of 55 miles per hour for 45 minutes, the total distance you travel is given by

$$\left(55\,\frac{\text{miles}}{\text{hour}}\right)\left(\frac{45}{60}\,\text{hour}\right) = 41.25 \text{ miles.}$$

As with all problems involving applications, be sure to check that the units in the verbal model make sense. For instance, in this problem the rate is given in *miles per hour*. So, in order for the solution to be given in *miles*, you must convert the time (from minutes) to *hours*. In the model, you can think of the hours as dividing out, as follows.

$$\left(55\,\frac{\text{miles}}{\text{hour}}\right)\left(\frac{45}{60}\,\text{hour}\right) = 41.25 \text{ miles}$$

Example 5   Distance-Rate Problem

Students are traveling in two cars to a football game 150 miles away. The first car leaves on time and travels at an average speed of 48 miles per hour. The second car starts $\frac{1}{2}$ hour later and travels at an average speed of 58 miles per hour. At these speeds, how long will it take the second car to catch up to the first car?

Solution

Verbal Model: Distance of first car = Distance of second car

Labels:

Time for first car $= t$	(hours)
Distance of first car $= 48t$	(miles)
Time for second car $= t - \frac{1}{2}$	(hours)
Distance of second car $= 58\left(t - \frac{1}{2}\right)$	(miles)

Equation:

$$48t = 58\left(t - \tfrac{1}{2}\right) \qquad \text{Original equation}$$

$$48t = 58t - 29 \qquad \text{Distributive Property}$$

$$48t - 58t = 58t - 58t - 29 \qquad \text{Subtract } 58t \text{ from each side.}$$

$$-10t = -29 \qquad \text{Combine like terms.}$$

$$\frac{-10t}{-10} = \frac{-29}{-10} \qquad \text{Divide each side by } -10.$$

$$t = 2.9 \qquad \text{Simplify.}$$

After the first car travels for 2.9 hours, the second car catches up to it. So, it takes the second car $t - 0.5 = 2.9 - 0.5 = 2.4$ hours to catch up to the first car.

In work-rate problems, the **rate of work** is the *reciprocal* of the time needed to do the entire job. For instance, if it takes 5 hours to complete a job, then the per hour work rate is $\frac{1}{5}$ job per hour. In general,

$$\text{Per hour work rate} = \frac{1}{\text{Total hours to complete a job}}.$$

The next example involves two rates of work and so fits the model for solving *mixture* problems.

Example 6   Work-Rate Problem

Consider two machines in a paper manufacturing plant. Machine 1 can complete one job (2000 pounds of paper) in 4 hours. Machine 2 is newer and can complete one job in $2\frac{1}{2}$ hours. How long will it take the two machines working together to complete one job?

Solution

Verbal Model:	$\dfrac{\text{Work}}{\text{done}} = \dfrac{\text{Portion done}}{\text{by machine 1}} + \dfrac{\text{Portion done}}{\text{by machine 2}}$

Labels:
Work done by both machines $= 1$ (job)

Time for each machine $= t$ (hours)

Per hour work rate for machine 1 $= \frac{1}{4}$ (job per hour)

Per hour work rate for machine 2 $= \frac{2}{5}$ (job per hour)

Equation:

$$1 = \left(\frac{1}{4}\right)(t) + \left(\frac{2}{5}\right)(t) \qquad \text{Rate} \cdot \text{time} + \text{rate} \cdot \text{time}$$

$$1 = \left(\frac{1}{4} + \frac{2}{5}\right)(t) \qquad \text{Distributive Property}$$

$$1 = \left(\frac{5}{20} + \frac{8}{20}\right)(t) \qquad \text{Least common denominator is 20.}$$

$$1 = \left(\frac{13}{20}\right)(t) \qquad \text{Simplify.}$$

$$1 \div \frac{13}{20} = t \qquad \text{Divide each side by } \tfrac{13}{20}.$$

$$\frac{20}{13} = t \qquad \text{Simplify.}$$

It will take $\frac{20}{13}$ hours (or about 1.54 hours) for both machines to complete the job. Check this solution in the original statement of the problem.

Note in Example 6 that the "2000 pounds of paper" was unnecessary information. The 2000 pounds is represented as one job. This type of unnecessary information in an applied problem is sometimes called a *red herring*. The 150 miles given in Example 5 is also a red herring.

Study Tip

Notice the hidden products, rate · time, in the portion of work done by each machine in Example 6. Watch for such products in the exercise set.

④ Use formulas to solve application problems.

Formulas

Many common types of geometric, scientific, and investment problems use ready-made equations called **formulas.** Knowing formulas such as those in the following lists will help you translate and solve a wide variety of real-life problems involving perimeter, circumference, area, volume, temperature, interest, and distance.

Common Formulas for Area, Perimeter, and Volume

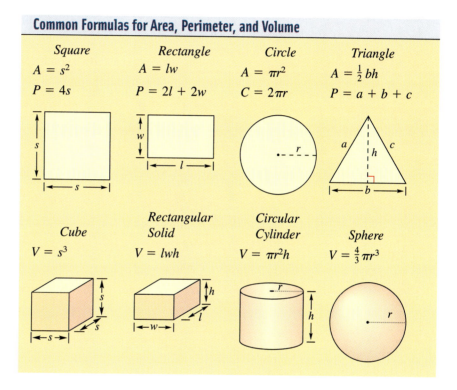

Square
$A = s^2$
$P = 4s$

Rectangle
$A = lw$
$P = 2l + 2w$

Circle
$A = \pi r^2$
$C = 2\pi r$

Triangle
$A = \frac{1}{2}bh$
$P = a + b + c$

Cube
$V = s^3$

Rectangular Solid
$V = lwh$

Circular Cylinder
$V = \pi r^2 h$

Sphere
$V = \frac{4}{3}\pi r^3$

Miscellaneous Common Formulas

Temperature: F = degrees Fahrenheit, C = degrees Celsius

$$F = \frac{9}{5}C + 32$$

Simple Interest: I = interest, P = principal, r = interest rate, t = time

$$I = Prt$$

Distance: d = distance traveled, r = rate, t = time

$$d = rt$$

When working with applied problems, you often need to rewrite one of the common formulas, as shown in the next example.

Example 7   Rewriting a Formula

In the perimeter formula $P = 2l + 2w$, solve for w.

Solution

$$P = 2l + 2w \qquad \text{Original formula}$$

$$P - 2l = 2w \qquad \text{Subtract } 2l \text{ from each side.}$$

$$\frac{P - 2l}{2} = w \qquad \text{Divide each side by 2.}$$

Example 8   Using a Geometric Formula

A local streets department plans to put sidewalks along the two streets that border your corner lot, which is 250 feet long on one side with an area of 30,000 square feet. Each lot owner is to pay $1.50 per foot of sidewalk bordering his or her lot.

a. Find the width of your lot.

b. How much will you have to pay for the sidewalks put on your lot?

Solution

Figure 2.3 shows a labeled diagram of your lot.

a. *Verbal Model:*   Area = Length · Width

> *Labels:*   Area of lot = 30,000   (square feet)
> Length of lot = 250   (feet)
> Width of lot = w (feet)

Equation: $30{,}000 = 250 \cdot w$ Original equation

$$\frac{30{,}000}{250} = w \qquad \text{Divide each side by 250.}$$

$$120 = w \qquad \text{Simplify.}$$

Your lot is 120 feet wide.

b. *Verbal Model:*   Cost = Rate per foot · Length of sidewalk

> *Labels:*   Cost of sidewalks = C (dollars)
> Rate per foot = 1.50   (dollars per foot)
> Total length of sidewalk = 120 + 250   (feet)

Equation: $C = 1.50(120 + 250)$ Original equation

$$C = 1.50 \cdot 370 \qquad \text{Add within parentheses.}$$

$$C = 555 \qquad \text{Multiply.}$$

You will have to pay $555 to have the sidewalks put on your lot.

Study Tip

When solving problems such as the one in Example 8, you may find it helpful to draw and label a diagram.

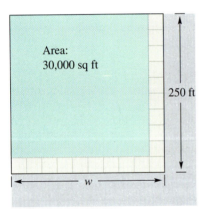

Area: 30,000 sq ft

250 ft

w

Figure 2.3

Example 9   Converting Temperature

The average daily temperature in July for Baltimore, Maryland is 77°F. What is Baltimore's average daily temperature in degrees Celsius? (Source: U.S. National Oceanic and Atmospheric Administration)

Solution

Verbal Model: $\dfrac{\text{Fahrenheit}}{\text{temperature}} = \dfrac{9}{5} \cdot \dfrac{\text{Celsius}}{\text{temperature}} + 32$

Labels: Fahrenheit temperature = 77   (degrees Fahrenheit)
            Celsius temperature = C (degrees Celsius)

Equation: $77 = \frac{9}{5}C + 32$ Original equation

              $45 = \frac{9}{5}C$ Subtract 32 from each side.

              $25 = C$ Multiply each side by $\frac{5}{9}$.

The average daily temperature in July for Baltimore is 25°C. Check this in the original statement of the problem.

Example 10   Simple Interest

A deposit of $8000 earned $300 in interest in 6 months.

a. What was the annual interest rate for this account?

b. At this rate, how long would it take to earn $800 in total interest?

Solution

a. *Verbal Model:* $\text{Interest} = \text{Principal} \cdot \text{Rate} \cdot \text{Time}$

Labels: Interest = 300   (dollars)
            Principal = 8000   (dollars)
            Annual interest rate = r (percent in decimal form)
            Time = $\frac{1}{2}$ (year)

Equation: $300 = 8000(r)\left(\dfrac{1}{2}\right)$ Original equation

              $\dfrac{300}{4000} = r \implies 0.075 = r$ Divide each side by 4000.

The annual interest rate is $r = 0.075$ or 7.5%.

b. Using the same verbal model as in part (a) with t representing time, you obtain the following equation.

$800 = 8000(0.075)(t)$ Original equation

$\dfrac{800}{600} = t \implies \dfrac{4}{3} = t$ Divide each side by 600.

So, it would take $\frac{4}{3}$ years, or $\frac{4}{3} \times 12 = 16$ months.

Technology: Tip

You can use a graphing calculator to solve simple interest problems by using the program found at our website *math.college.hmco.com/students.* Use the program and the guess, check, and revise method to find P when $I = \$3330$, $r = 6\%$, and $t = 3$ years.

2.3 Exercises

Review   *Concepts, Skills, and Problem Solving*

Keep mathematically in shape by doing these exercises *before* the problems of this section.

Properties and Definitions

1. What is the sign of the sum $(-7) + (-3)$? State the rule used.

2. What is the sign of the sum $-7 + 3$? State the rule used.

3. What is the sign of the product $(-6)(-2)$? State the rule used.

4. What is the sign of the product $6(-2)$? State the rule used.

Solving Equations

In Exercises 5–10, solve the equation.

5. $2x - 5 = x + 9$

6. $6x + 8 = 8 - 2x$

7. $2x + \dfrac{3}{2} = \dfrac{3}{2}$

8. $-\dfrac{x}{10} = 1000$

9. $-0.35x = 70$ **10.** $0.60x = 24$

Problem Solving

11. *Athletics*   The length of a relay race is 2.5 miles. The last change of runners occurs at the 1.8-mile marker. How far does the last person run?

12. *Agriculture*   During the months of January, February, and March, a farmer bought $34\frac{1}{3}$ tons, $18\frac{1}{5}$ tons, and $25\frac{5}{6}$ tons of soybeans, respectively. Find the total amount of soybeans purchased during the first quarter of the year.

Developing Skills

In Exercises 1–8, find the missing quantities. (Assume that the markup rate is a percent based on the cost.) See Example 1.

	Cost	Selling Price	Markup	Markup Rate
1.	$45.97	$64.33		
2.	$84.20	$113.67		
3.		$250.80	$98.80	
4.		$603.72	$184.47	
5.		$26,922.50	$4672.50	
6.		$16,440.50	$3890.50	
7.	$225.00			85.2%
8.	$732.00			$33\frac{1}{3}\%$

In Exercises 9–16, find the missing quantities. (Assume that the discount rate is a percent based on the list price.) See Example 2.

	List Price	Sale Price	Discount	Discount Rate
9.	$49.95	$25.74		
10.	$119.00	$79.73		
11.	$300.00		$189.00	
12.	$345.00		$134.55	
13.	$95.00			65%
14.		$19.90		20%
15.		$893.10	$251.90	
16.		$257.32	$202.18	

Solving Problems

17. *Markup*   The selling price of a jacket in a department store is $157.14. The cost of the jacket to the store is $130.95. What is the markup?

18. *Markup*   A shoe store sells a pair of shoes for $63.50. The cost of the shoes to the store is $43.50. What is the markup?

19. *Markup Rate* A jewelry store sells a pair of earrings for $84. The cost of the earrings to the store is $65.63. What is the markup rate?

20. *Markup Rate* A department store sells a sweater for $60. The cost of the sweater to the store is $35. What is the markup rate?

21. *Discount* A shoe store sells a pair of athletic shoes for $75. The shoes go on sale for $50. What is the discount?

22. *Discount* A bakery sells a dozen rolls for $1.75. You can buy a dozen day-old rolls for $0.75. What is the discount?

23. *Discount Rate* An auto store sells a pair of car mats for $20. On sale, the car mats sell for $16. What is the discount rate?

24. *Discount Rate* A department store sells a beach towel for $14. On sale, the beach towel sells for $10. What is the discount rate?

25. *Long-Distance Rate* The weekday rate for a telephone call is $0.75 for the first minute plus $0.55 for each additional minute. Determine the length of a call that cost $5.15. What would have been the cost of the call if it had been made during the weekend, when there is a 60% discount?

26. *Insurance Premium* The annual insurance premium for a policyholder is $862. The policyholder must pay a 20% surcharge because of an accident. Find the annual premium.

27. *Cost* An auto store gives the list price of a tire as $79.42. During a promotional sale, the store is selling four tires for the price of three. The store needs a markup on cost of 10% during the sale. What is the cost to the store of each tire?

28. *Price* The produce manager of a supermarket pays $22.60 for a 100-pound box of bananas. The manager estimates that 10% of the bananas will spoil before they are sold. At what price per pound should the bananas be sold to give the supermarket an average markup rate on cost of 30%?

29. *Amount Financed* A customer bought a lawn tractor for $4450 plus 6% sales tax. (a) Find the amount of the sales tax and the total bill. (b) A down payment of $1000 was made. Find the amount financed.

30. *Weekly Pay* The weekly salary of an employee is $375 plus a 6% commission on the employee's total sales. Find the weekly pay for a week in which the sales are $5500.

31. *Labor* An automobile repair bill of $216.37 lists $136.37 for parts and the rest for labor. The labor rate is $32 per hour. How many hours did it take to repair the automobile?

32. *Labor* An appliance repair store charges $50 for the first $\frac{1}{2}$ hour of a service call, and $18 for each additional $\frac{1}{2}$ hour of labor. Find the length of a service call for which the charge is $104.

33. *Labor* The bill for the repair of an automobile is $380. Included in this bill is a charge of $275 for parts, and the remainder of the bill is for labor. The charge for labor is $35 per hour. How many hours were spent in repairing the automobile?

34. *Labor* The bill for the repair of an automobile is $648. Included in this bill is a charge of $315 for parts, and the remainder of the bill is for labor. It took 9 hours to repair the automobile. What was the charge per hour for labor?

Mixture Problems In Exercises 35–38, determine the number of units of solutions 1 and 2 needed to obtain the desired amount and concentration of the final solution.

Concentration of Solution 1	Concentration of Solution 2	Concentration of Final Solution	Amount of Final Solution
35. 20%	60%	40%	100 gal
36. 50%	75%	60%	10 L
37. 15%	60%	45%	24 qt
38. 60%	80%	75%	55 gal

39. *Seed Mixture* A nursery wants to mix two types of lawn seed. One type sells for $12 per pound and the other type sells for $20 per pound. To obtain 100 pounds of a mixture at $14 per pound, how many pounds of each type of seed are needed?

40. *Nut Mixture* A grocer mixes two kinds of nuts costing $3.88 per pound and $4.88 per pound to make 100 pounds of a mixture costing $4.13 per pound. How many pounds of each kind of nut are in the mixture?

41. *Ticket Sales*   Ticket sales for a play total $2200. There are three times as many adult tickets sold as children's tickets. The prices of the tickets for adults and children are $6 and $4, respectively. Find the number of children's tickets sold.

42. *Ticket Sales*   Ticket sales for a spaghetti dinner total $1350. There are four times as many adult tickets sold as children's tickets. The prices of the tickets for adults and children are $6 and $3, respectively. Find the number of children's tickets sold.

43. *Antifreeze Mixture*   The cooling system on a truck contains 5 gallons of coolant that is 40% antifreeze. How much must be withdrawn and replaced with 100% antifreeze to bring the coolant in the system to 50% antifreeze?

44. *Fuel Mixture*   You mix gasoline and oil to obtain $2\frac{1}{2}$ gallons of mixture for an engine. The mixture is 40 parts gasoline and 1 part two-cycle oil. How much gasoline must be added to bring the mixture to 50 parts gasoline and 1 part oil?

Distance   In Exercises 45–50, determine the unknown distance, rate, or time. See Example 5.

	Distance, d	Rate, r	Time, t
45.		650 mi/hr	$3\frac{1}{2}$ hr
46.		45 ft/sec	10 sec
47.	1000 km	110 km/hr	
48.	250 ft	32 ft/sec	
49.	385 mi		7 hr
50.	1000 ft		$\frac{3}{2}$ sec

51. *Travel Time*   You ride your bike at an average speed of 12 miles per hour. How long will it take you to ride 30 miles?

52. *Travel Time*   You ride your bike at an average speed of 8 miles per hour. How long will it take you to ride 12 miles?

53. *Travel Time*   You ride your bike at an average speed of 16 miles per hour. How long will it take you to ride 20 miles?

54. *Travel Time*   You ride your bike at an average speed of 28 feet per second. How long will it take you to ride 252 feet?

55. *Travel Time*   You jog at a steady speed of 6 miles per hour. How long will it take you to jog 5 miles?

56. *Travel Time*   You jog at a steady speed of 150 meters per minute. How long will it take you to jog 5000 meters?

57. *Distance*   Two planes leave Chicago's O'Hare International Airport at approximately the same time and fly in opposite directions. How far apart are the planes after $1\frac{1}{3}$ hours if their average speeds are 480 miles per hour and 600 miles per hour?

58. *Distance*   Two trucks leave a depot at approximately the same time and travel the same route. How far apart are the trucks after $4\frac{1}{2}$ hours if their average speeds are 52 miles per hour and 56 miles per hour?

59. *Travel Time*   Determine the time for a space shuttle to travel a distance of 5000 miles in orbit when its average speed is 17,500 miles per hour.

60. *Speed of Light*   The distance between the sun and Earth is 93,000,000 miles and the speed of light is 186,282.397 miles per second. Determine the time for light to travel from the sun to Earth.

61. *Travel Time*   On the first part of a 317-mile trip, a sales representative averaged 58 miles per hour. The sales representative averaged only 52 miles per hour on the remainder of the trip because of an increased volume of traffic (see figure). The total time of the trip was 5 hours and 45 minutes. Find the amount of driving time at each speed.

62. *Travel Time*   Two cars start at the same location and travel in the same direction at average speeds of 30 miles per hour and 45 miles per hour. How much time must elapse before the two cars are 5 miles apart?

63. *Work-Rate Problems*   Determine the work rate for each task.

(a) A printer can print 8 pages per minute.

(b) A machine shop can produce 30 units in 8 hours.

64. *Work-Rate Problem*   You can complete a typing project in 5 hours, and your friend can complete it in 8 hours.

(a) What fractional part of the project can be accomplished by each person in 1 hour?

(b) How long will it take both of you to complete the project working together?

65. *Work-Rate Problem*   You can mow a lawn in 3 hours, and your friend can mow it in 4 hours.

(a) What fractional part of the lawn can each of you mow in 1 hour?

(b) How long will it take both of you to mow the lawn working together?

66. *Work-Rate Problem*   It takes 30 minutes for a pump to empty a water tank. A larger pump can empty the tank in half the time. How long would it take to empty the tank with both pumps operating?

In Exercises 67–76, solve for the specified variable. See Example 7.

67. Solve for R.

Ohm's Law: $E = IR$

68. Solve for r.

Simple Interest: $A = P + Prt$

69. Solve for L.

Discount: $S = L - rL$

70. Solve for C.

Markup: $S = C + rC$

71. Solve for a.

Free-Falling Body: $h = 48t + \dfrac{1}{2}at^2$

72. Solve for a.

Free-Falling Body: $h = 18t + \dfrac{1}{2}at^2$

73. Solve for a.

Free-Falling Body: $h = -12t + \dfrac{1}{2}at^2$

74. Solve for a.

Free-Falling Body: $h = 36t + \dfrac{1}{2}at^2 + 50$

75. Solve for a.

Free-Falling Body: $h = -15t + \dfrac{1}{2}at^2 + 9.5$

76. Solve for b.

Area of a Trapezoid: $A = \dfrac{1}{2}(a + b)h$

77. ▲ *Geometry*   Find the volume of the circular cylinder shown in the figure.

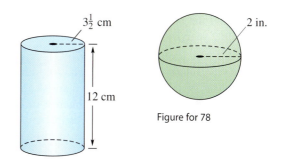

$3\frac{1}{2}$ cm

12 cm

2 in.

Figure for 78

Figure for 77

78. ▲ *Geometry*   Find the volume of the sphere shown in the figure.

79. ▲ *Geometry*   A rectangular picture frame has a perimeter of 3 feet. The width of the frame is 0.62 times its height. Find the height of the frame.

80. ▲ *Geometry*   A rectangular stained glass window has a perimeter of 18 feet. The height of the window is 1.25 times its width. Find the width of the window.

81. ▲ *Geometry*   A "Slow Moving Vehicle" sign has the shape of an equilateral triangle. The sign has a perimeter of 129 centimeters. Find the length of each side.

82. ▲ *Geometry*   The length of a rectangle is three times its width. The perimeter of the rectangle is 64 inches. Find the dimensions of the rectangle.

83. *Meteorology*   The average daily high temperature in Boston, Massachusetts is 59°F. What is Boston's average daily high temperature in degrees Celsius? (Source: U.S. National Oceanic and Atmospheric Administration.)

84. *Meteorology*   The average daily temperature in Jacksonville, Florida is 68°F. What is Jacksonville's average daily temperature in degrees Celsius? (Source: U.S. National Oceanic and Atmospheric Administration)

85. *Simple Interest*   Find the interest on a $5000 bond that pays an annual percentage rate of $6\frac{1}{2}\%$ for 6 years.

86. *Simple Interest*   Find the annual interest rate on a certificate of deposit that accumulated $400 interest in 2 years on a principal of $2500.

87. *Simple Interest*   The interest on a savings account is 7%. Find the principal required to earn $500 in interest in 2 years.

88. *Simple Interest*   An investment of $7000 is divided into two accounts earning 5% and 7% simple interest. (The 7% investment has a greater risk.) Your objective is to obtain a total annual interest income of $400 from the investments. What is the smallest amount you can invest at 7% in order to meet your objective?

89. *Average Wage*   The average hourly wage y (in dollars) for custodians at public schools in the United States from 1995 through 2000 can be approximated by the model $y = 0.264t + 8.75$, for $5 \le t \le 10$, where t represents the year, with $t = 5$ corresponding to 1995 (see figure). (Source: Educational Research Service)

(a) Use the graph to determine the year in which the average hourly wage was $10.35. Is the result the same when you use the model?

(b) What was the average annual hourly raise for custodians during this six-year period? Explain how you arrived at your answer.

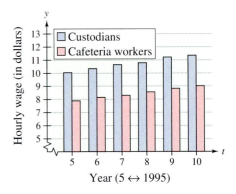

Figure for 89 and 90

90. *Average Wage*   The average hourly wage y (in dollars) for cafeteria workers at public schools in the United States from 1995 through 2000 can be approximated by the model $y = 0.226t + 6.76$, for $5 \le t \le 10$, where t represents the year, with $t = 5$ corresponding to 1995 (see figure). (Source: Educational Research Service)

(a) Use the graph to determine the year in which the average hourly wage was $8.56. Is the result the same when you use the model?

(b) What was the average annual hourly raise for cafeteria workers during this six-year period? Explain how you determined your answer.

Explaining Concepts

91. 🔷   Answer parts (a)–(e) of Motivating the Chapter on page 56.

92. *Writing*   Explain the difference between markup and markup rate.

93. *Writing*   Explain how to find the sale price of an item when you are given the list price and the discount rate.

94. It takes you t hours to complete a task. What portion of the task can you complete in 1 hour?

95. *Writing*   If the sides of a square are doubled, does the perimeter double? Explain.

96. *Writing*   If the sides of a square are doubled, does the area double? Explain.

97. *Writing*   If you forget the formula for the volume of a right circular cylinder, how can you derive it?

Mid-Chapter Quiz

Take this quiz as you would take a quiz in class. After you are done, check your work against the answers in the back of the book.

In Exercises 1–8, solve the equation and check the result. (If it is not possible, state the reason.)

1. $4x - 8 = 0$

2. $-3(z - 2) = 0$

3. $2(y + 3) = 18 - 4y$

4. $5t + 7 = 7(t + 1) - 2t$

5. $\dfrac{1}{4}x + 6 = \dfrac{3}{2}x - 1$

6. $\dfrac{u}{4} + \dfrac{u}{3} = 1$

7. $\dfrac{4 - x}{5} + 5 = \dfrac{5}{2}$

8. $3x + \frac{11}{12} = \frac{5}{16}$

In Exercises 9 and 10, solve the equation and round your answer to two decimal places.

9. $0.2x + 0.3 = 1.5$

10. $0.42x + 6 = 5.25x - 0.80$

11. Write the decimal 0.45 as a fraction and as a percent.

12. 500 is 250% of what number?

13. Find the unit price (in dollars per ounce) of a 12-ounce box of cereal that sells for $2.35.

14. A quality control engineer for a manufacturer finds one defective unit in a sample of 300. At this rate, what is the expected number of defective units in a shipment of 600,000?

15. A store is offering a discount of 25% on a computer with a list price of $1080. A mail-order catalog has the same computer for $799 plus $14.95 for shipping. Which is the better buy?

16. Last week you earned $616. Your regular hourly wage is $12.25 for the first 40 hours, and your overtime hourly wage is $18. How many hours of overtime did you work?

17. Fifty gallons of a 30% acid solution is obtained by combining solutions that are 25% acid and 50% acid. How much of each solution is required?

18. On the first part of a 300-mile trip, a sales representative averaged 62 miles per hour. The sales representative averaged 46 miles per hour on the remainder of the trip because of an increased volume of traffic. The total time of the trip was 6 hours. Find the amount of driving time at each speed.

19. You can paint a room in 6 hours, and your friend can paint it in 8 hours. How long will it take both of you to paint the room together?

20. The accompanying figure shows three squares. The perimeters of squares I and II are 20 inches and 32 inches, respectively. Find the area of square III.

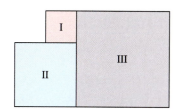

Figure for 20

2.4 Linear Inequalities

Royalty-Free/Corbis

What You Should Learn

① Sketch the graphs of inequalities.

② Identify the properties of inequalities that can be used to create equivalent inequalities.

③ Solve linear inequalities.

④ Solve compound inequalities.

⑤ Solve application problems involving inequalities.

Why You Should Learn It

Linear inequalities can be used to model and solve real-life problems. For instance, Exercises 115 and 116 on page 106 show how to use linear inequalities to analyze air pollutant emissions.

① Sketch the graphs of inequalities.

Intervals on the Real Number Line

In this section you will study **algebraic inequalities,** which are inequalities that contain one or more variable terms. Some examples are

$$x \leq 4, \quad x \geq -3, \quad x + 2 < 7, \quad \text{and} \quad 4x - 6 < 3x + 8.$$

As with an equation, you **solve** an inequality in the variable x by finding all values of x for which the inequality is true. Such values are called **solutions** and are said to **satisfy** the inequality. The set of all solutions of an inequality is the **solution set** of the inequality. The **graph** of an inequality is obtained by plotting its solution set on the real number line. Often, these graphs are intervals—either bounded or unbounded.

Bounded Intervals on the Real Number Line

Let a and b be real numbers such that $a < b$. The following intervals on the real number line are called **bounded intervals.** The numbers a and b are the **endpoints** of each interval. A bracket indicates that the endpoint is included in the interval, and a parenthesis indicates that the endpoint is excluded.

Notation	Interval Type	Inequality	Graph
$[a, b]$	Closed	$a \leq x \leq b$	
(a, b)	Open	$a < x < b$	
$[a, b)$		$a \leq x < b$	
$(a, b]$		$a < x \leq b$	

The **length** of the interval $[a, b]$ is the distance between its endpoints: $b - a$. The lengths of $[a, b]$, (a, b), $[a, b)$, and $(a, b]$ are the same. The reason that these four types of intervals are called "bounded" is that each has a finite length. An interval that *does not* have a finite length is **unbounded** (or **infinite**).

Unbounded Intervals on the Real Number Line

Let a and b be real numbers. The following intervals on the real number line are called **unbounded intervals**.

Notation	Interval Type	Inequality	Graph
$[a, \infty)$		$x \geq a$	
(a, ∞)	Open	$x > a$	
$(-\infty, b]$		$x \leq b$	
$(-\infty, b)$	Open	$x < b$	
$(-\infty, \infty)$	Entire real line		

The symbols ∞ (**positive infinity**) and $-\infty$ (**negative infinity**) do not represent real numbers. They are simply convenient symbols used to describe the unboundedness of an interval such as $(-5, \infty)$. This is read as the interval from -5 to infinity.

Example 1   Graphs of Inequalities

Sketch the graph of each inequality.

a. $-3 < x \leq 1$ **b.** $0 < x < 2$

c. $-3 < x$ **d.** $x \leq 2$

Solution

a. The graph of $-3 < x \leq 1$ is a bounded interval.

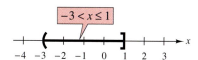

b. The graph of $0 < x < 2$ is a bounded interval.

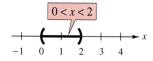

c. The graph of $-3 < x$ is an unbounded interval.

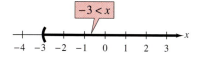

d. The graph of $x \leq 2$ is an unbounded interval.

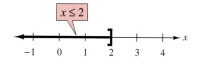

2 Identify the properties of inequalities that can be used to create equivalent inequalities.

Properties of Inequalities

Solving a linear inequality is much like solving a linear equation. You isolate the variable, using the **properties of inequalities.** These properties are similar to the properties of equality, but there are two important exceptions. *When each side of an inequality is multiplied or divided by a negative number, the direction of the inequality symbol must be reversed.* Here is an example.

$$-2 < 5 \qquad\qquad \text{Original inequality}$$

$$(-3)(-2) > (-3)(5) \qquad \text{Multiply each side by } -3 \text{ and reverse the inequality.}$$

$$6 > -15 \qquad\qquad \text{Simplify.}$$

Two inequalities that have the same solution set are **equivalent inequalities.** The following list of operations can be used to create equivalent inequalities.

Properties of Inequalities

1. *Addition and Subtraction Properties*

    Adding the same quantity to, or subtracting the same quantity from, each side of an inequality produces an equivalent inequality.

    If $a < b$, then $a + c < b + c$.

    If $a < b$, then $a - c < b - c$.

2. *Multiplication and Division Properties: Positive Quantities*

    Multiplying or dividing each side of an inequality by a positive quantity produces an equivalent inequality.

    If $a < b$ and c is positive, then $ac < bc$.

    If $a < b$ and c is positive, then $\dfrac{a}{c} < \dfrac{b}{c}$.

3. *Multiplication and Division Properties: Negative Quantities*

    Multiplying or dividing each side of an inequality by a negative quantity produces an equivalent inequality in which the inequality symbol is reversed.

    If $a < b$ and c is negative, then $ac > bc$.   Reverse inequality

    If $a < b$ and c is negative, then $\dfrac{a}{c} > \dfrac{b}{c}$.   Reverse inequality

4. *Transitive Property*

    Consider three quantities for which the first quantity is less than the second, and the second is less than the third. It follows that the first quantity must be less than the third quantity.

    If $a < b$ and $b < c$, then $a < c$.

These properties remain true if the symbols $<$ and $>$ are replaced by $\leq$ and $\geq$. Moreover, a, b, and c can represent real numbers, variables, or expressions. Note that you cannot multiply or divide each side of an inequality by zero.

③ Solve linear inequalities.

Solving a Linear Inequality

An inequality in one variable is a **linear inequality** if it can be written in one of the following forms.

$$ax + b \leq 0, \quad ax + b < 0, \quad ax + b \geq 0, \quad ax + b > 0$$

The solution set of a linear inequality can be written in set notation. For the solution $x > 1$, the set notation is $\{x \mid x > 1\}$ and is read "the set of all x such that x is greater than 1."

As you study the following examples, pay special attention to the steps in which the inequality symbol is reversed. *Remember that when you multiply or divide an inequality by a negative number, you must reverse the inequality symbol.*

Study Tip

Checking the solution set of an inequality is not as simple as checking the solution set of an equation. (There are usually too many x-values to substitute back into the original inequality.) You can, however, get an indication of the validity of a solution set by substituting a few convenient values of x. For instance, in Example 2, try checking that $x = 0$ satisfies the original inequality, whereas $x = 4$ does not.

Example 2   Solving a Linear Inequality

$x + 6 < 9$	Original inequality
$x + 6 - 6 < 9 - 6$	Subtract 6 from each side.
$x < 3$	Combine like terms.

The solution set consists of all real numbers that are less than 3. The solution set in interval notation is $(-\infty, 3)$ and in set notation is $\{x \mid x < 3\}$. The graph is shown in Figure 2.4.

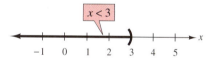

Figure 2.4

Example 3   Solving a Linear Inequality

$8 - 3x \leq 20$	Original inequality
$8 - 8 - 3x \leq 20 - 8$	Subtract 8 from each side.
$-3x \leq 12$	Combine like terms.
$\dfrac{-3x}{-3} \geq \dfrac{12}{-3}$	Divide each side by -3 and reverse the inequality symbol.
$x \geq -4$	Simplify.

The solution set in interval notation is $[-4, \infty)$ and in set notation is $\{x \mid x \geq -4\}$. The graph is shown in Figure 2.5.

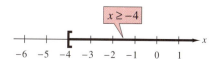

Figure 2.5

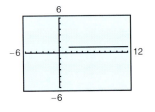
Example 4   Solving a Linear Inequality

$7x - 3 > 3(x + 1)$	Original inequality
$7x - 3 > 3x + 3$	Distributive Property
$7x - 3x - 3 > 3x - 3x + 3$	Subtract $3x$ from each side.
$4x - 3 > 3$	Combine like terms.
$4x - 3 + 3 > 3 + 3$	Add 3 to each side.
$4x > 6$	Combine like terms.
$\dfrac{4x}{4} > \dfrac{6}{4}$	Divide each side by 4.
$x > \dfrac{3}{2}$	Simplify.

The solution set consists of all real numbers that are greater than $\frac{3}{2}$. The solution set in interval notation is $\left(\frac{3}{2}, \infty\right)$ and in set notation is $\left\{x \mid x > \frac{3}{2}\right\}$. The graph is shown in Figure 2.6.

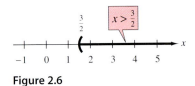

Figure 2.6

Example 5   Solving a Linear Inequality

$\dfrac{2x}{3} + 12 < \dfrac{x}{6} + 18$	Original inequality
$6 \cdot \left(\dfrac{2x}{3} + 12\right) < 6 \cdot \left(\dfrac{x}{6} + 18\right)$	Multiply each side by LCD of 6.
$4x + 72 < x + 108$	Distributive Property
$4x - x < 108 - 72$	Subtract x and 72 from each side.
$3x < 36$	Combine like terms.
$x < 12$	Divide each side by 3.

The solution set consists of all real numbers that are less than 12. The solution set in interval notation is $(-\infty, 12)$ and in set notation is $\{x \mid x < 12\}$. The graph is shown in Figure 2.7.

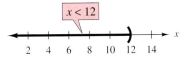

Figure 2.7

4 Solve compound inequalities.

Solving a Compound Inequality

Two inequalities joined by the word *and* or the word *or* constitute a **compound inequality.** When two inequalities are joined by the word *and*, the solution set consists of all real numbers that satisfy *both* inequalities. The solution set for the compound inequality $-4 \le 5x - 2$ *and* $5x - 2 < 7$ can be written more simply as the **double inequality**

$$-4 \le 5x - 2 < 7.$$

A compound inequality formed by the word *and* is called **conjunctive** and is the only kind that has the potential to form a double inequality. A compound inequality joined by the word *or* is called **disjunctive** and cannot be re-formed into a double inequality.

Example 6   Solving a Double Inequality

Solve the double inequality $-7 \le 5x - 2 < 8$.

Solution

$-7 \le 5x - 2 < 8$	Write original inequality.
$-7 + 2 \le 5x - 2 + 2 < 8 + 2$	Add 2 to all three parts.
$-5 \le 5x < 10$	Combine like terms.
$\dfrac{-5}{5} \le \dfrac{5x}{5} < \dfrac{10}{5}$	Divide each part by 5.
$-1 \le x < 2$	Simplify.

The solution set consists of all real numbers that are greater than or equal to -1 and less than 2. The solution set in interval notation is $[-1, 2)$ and in set notation is $\{x | -1 \le x < 2\}$. The graph is shown in Figure 2.8.

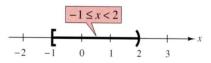

Figure 2.8

The double inequality in Example 6 could have been solved in two parts, as follows.

$$-7 \le 5x - 2 \qquad \text{and} \qquad 5x - 2 < 8$$
$$-5 \le 5x \qquad\qquad\qquad\quad 5x < 10$$
$$-1 \le x \qquad\qquad\qquad\quad\; x < 2$$

The solution set consists of all real numbers that satisfy both inequalities. In other words, the solution set is the set of all values of x for which $-1 \le x < 2$.

Example 7   Solving a Conjunctive Inequality

Solve the compound inequality $-1 \le 2x - 3$ and $2x - 3 < 5$.

Solution

Begin by writing the conjunctive inequality as a double inequality.

$-1 \le 2x - 3 < 5$	Write as double inequality.
$-1 + 3 \le 2x - 3 + 3 < 5 + 3$	Add 3 to all three parts.
$2 \le 2x < 8$	Combine like terms.
$\dfrac{2}{2} \le \dfrac{2x}{2} < \dfrac{8}{2}$	Divide each part by 2.
$1 \le x < 4$	Solution set

The solution set is $1 \le x < 4$ or, in set notation, $\{x \mid 1 \le x < 4\}$. The graph of the solution set is shown in Figure 2.9.

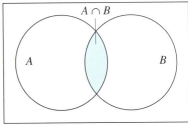

Figure 2.9

Example 8   Solving a Disjunctive Inequality

Solve the compound inequality $-3x + 6 \le 2$ or $-3x + 6 \ge 7$.

Solution

$-3x + 6 \le 2$	or	$-3x + 6 \ge 7$	Write original inequality.
$-3x + 6 - 6 \le 2 - 6$		$-3x + 6 - 6 \ge 7 - 6$	Subtract 6 from all parts.
$-3x \le -4$		$-3x \ge 1$	Combine like terms.
$\dfrac{-3x}{-3} \ge \dfrac{-4}{-3}$		$\dfrac{-3x}{-3} \le \dfrac{1}{-3}$	Divide all parts by -3 and reverse both inequality symbols.
$x \ge \dfrac{4}{3}$		$x \le -\dfrac{1}{3}$	Simplify.

The solution set is $x \le -\frac{1}{3}$ or $x \ge \frac{4}{3}$ or, in set notation, $\left\{ x \mid x \le -\frac{1}{3} \text{ or } x \ge \frac{4}{3} \right\}$. The graph of the solution set is shown in Figure 2.10.

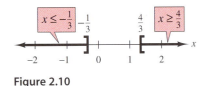

Figure 2.10

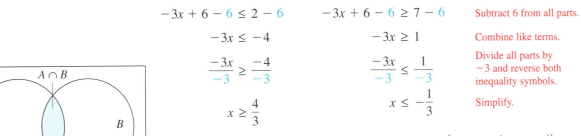

Compound inequalities can be written using *symbols*. For compound inequalities, the word *and* is represented by the symbol ∩, which is read as **intersection.** The word *or* is represented by the symbol ∪, which is read as **union.** Graphical representations are shown in Figure 2.11. If A and B are sets, then x is in $A \cap B$ if it is in both A and B. Similarly, x is in $A \cup B$ if it is in A or B, or possibly both.

Intersection of two sets

Union of two sets

Figure 2.11

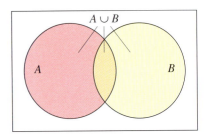

Figure 2.12

Example 9   Writing a Solution Set Using Union

A solution set is shown on the number line in Figure 2.12.

a. Write the solution set as a compound inequality.

b. Write the solution set using the union symbol.

Solution

a. As a compound inequality, you can write the solution set as $x \leq -1$ or $x > 2$.

b. Using set notation, you can write the left interval as $A = \{x \mid x \leq -1\}$ and the right interval as $B = \{x \mid x > 2\}$. So, using the union symbol, the entire solution set can be written as $A \cup B$.

Example 10   Writing a Solution Set Using Intersection

Write the compound inequality using the intersection symbol.

$$-3 \leq x \leq 4$$

Solution

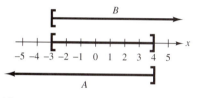

Figure 2.13

Consider the two sets $A = \{x \mid x \leq 4\}$ and $B = \{x \mid x \geq -3\}$. These two sets overlap, as shown on the number line in Figure 2.13. The compound inequality $-3 \leq x \leq 4$ consists of all numbers that are in $x \leq 4$ and $x \geq -3$, which means that it can be written as $A \cap B$.

⑤ Solve application problems involving inequalities.

Applications

Linear inequalities in real-life problems arise from statements that involve phrases such as "at least," "no more than," "minimum value," and so on. Study the meanings of the key phrases in the next example.

Example 11   Translating Verbal Statements

	Verbal Statement	*Inequality*	
a.	x is at most 3.	$x \leq 3$	"at most" means "less than or equal to."
b.	x is no more than 3.	$x \leq 3$	
c.	x is at least 3.	$x \geq 3$	"at least" means "greater than or equal to."
d.	x is no less than 3.	$x \geq 3$	
e.	x is more than 3.	$x > 3$	
f.	x is less than 3.	$x < 3$	
g.	x is a minimum of 3.	$x \geq 3$	
h.	x is at least 2, but less than 7.	$2 \leq x < 7$	
i.	x is greater than 2, but no more than 7.	$2 < x \leq 7$	

To solve real-life problems involving inequalities, you can use the same "verbal-model approach" you use with equations.

Example 12   Finding the Maximum Width of a Package

An overnight delivery service will not accept any package whose combined length and minimum girth (perimeter of a cross section) exceeds 132 inches. You are sending a rectangular package that has square cross sections. The length of the package is 68 inches. What is the maximum width of the sides of its square cross sections?

Solution

First make a sketch. In Figure 2.14, the length of the package is 68 inches, and each side is x inches wide because the package has a square cross section.

Figure 2.14

Verbal Model:	Length + Girth ≤ 132 inches

Labels:	Width of a side $= x$	(inches)
	Length $= 68$	(inches)
	Girth $= 4x$	(inches)

Inequality: $68 + 4x \le 132$

$$4x \le 64$$

$$x \le 16$$

The width of each side of the package must be less than or equal to 16 inches.

Example 13   Comparing Costs

A subcompact car can be rented from Company A for $240 per week with no extra charge for mileage. A similar car can be rented from Company B for $100 per week plus an additional 25 cents for each mile driven. How many miles must you drive in a week so that the rental fee for Company B is more than that for Company A?

Solution

Verbal Model:	Weekly cost for Company B	>	Weekly cost for Company A

Labels:	Number of miles driven in one week $= m$	(miles)
	Weekly cost for Company A $= 240$	(dollars)
	Weekly cost for Company B $= 100 + 0.25m$	(dollars)

Inequality: $100 + 0.25m > 240$

$$0.25m > 140$$

$$m > 560$$

Miles driven	Company A	Company B
520	$240.00	$230.00
530	$240.00	$232.50
540	$240.00	$235.00
550	$240.00	$237.50
560	$240.00	$240.00
570	$240.00	$242.50

So, the car from Company B is more expensive if you drive more than 560 miles in a week. The table shown at the left helps confirm this conclusion.

2.4 Exercises

Review   Concepts, Skills, and Problem Solving

Keep mathematically in shape by doing these exercises *before* the problems of this section.

Properties and Definitions

In Exercises 1–4, identify the property of real numbers illustrated by the statement.

1. $3yx = 3xy$

2. $3xy - 3xy = 0$

3. $6(x - 2) = 6x - 6 \cdot 2$

4. $3x + 0 = 3x$

Evaluating Expressions

In Exercises 5–10, evaluate the algebraic expression for the specified values of the variables. If not possible, state the reason.

5. $x^2 - y^2$

    $x = 4, \quad y = 3$

6. $4s + st$

    $s = 3, \quad t = -4$

7. $\dfrac{x}{x^2 + y^2}$

    $x = 0, \quad y = 3$

8. $\dfrac{z^2 + 2}{x^2 - 1}$

    $x = 2, \quad z = -1$

9. $\dfrac{a}{1 - r}$

    $a = 2, \quad r = \frac{1}{2}$

10. $2l + 2w$

    $l = 3, \quad w = 1.5$

Problem Solving

▲ *Geometry*   In Exercises 11 and 12, find the area of the trapezoid. The area of a trapezoid with parallel bases b_1 and b_2 and height h is $A = \frac{1}{2}(b_1 + b_2)h$.

11.

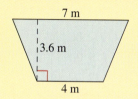

7 m

3.6 m

4 m

12.

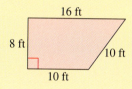

16 ft

8 ft

10 ft

10 ft

Developing Skills

In Exercises 1–4, determine whether each value of x satisfies the inequality.

    Inequality   *Values*

1. $7x - 10 > 0$ (a) $x = 3$ (b) $x = -2$

                    (c) $x = \frac{5}{2}$ (d) $x = \frac{1}{2}$

2. $3x + 2 < \dfrac{7x}{5}$ (a) $x = 0$ (b) $x = 4$

                    (c) $x = -4$ (d) $x = -1$

3. $0 < \dfrac{x + 5}{6} < 2$ (a) $x = 10$ (b) $x = 4$

                    (c) $x = 0$ (d) $x = -6$

4. $-2 < \dfrac{3 - x}{2} \leq 2$ (a) $x = 0$ (b) $x = 3$

                    (c) $x = 9$ (d) $x = -12$

In Exercises 5–10, match the inequality with its graph. [The graphs are labeled (a), (b), (c), (d), (e), and (f).]

(a)

(b)

(c)

(d)

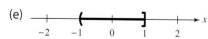

(e)

(f)

5. $x \geq -1$

6. $-1 < x \leq 1$

7. $x \leq -1$ or $x \geq 2$

8. $x < -1$ or $x \geq 1$

9. $-2 < x < 1$

10. $x < 2$

In Exercises 11–24, sketch the graph of the inequality. See Example 1.

11. $x \le 2$

12. $x > -6$

13. $x > 3.5$

14. $x \le -2.5$

15. $-5 < x \le 3$

16. $-1 < x \le 5$

17. $4 > x \ge 1$

18. $9 \ge x \ge 3$

19. $\frac{3}{2} \ge x > 0$

20. $-\frac{15}{4} < x < -\frac{5}{2}$

21. $x < -5$ or $x \ge -1$

22. $x \le -4$ or $x > 0$

23. $x \le 3$ or $x > 7$

24. $x \le -1$ or $x \ge 1$

25. Write an inequality equivalent to $5 - \frac{1}{3}x > 8$ by multiplying each side by -3.

26. Write an inequality equivalent to $5 - \frac{1}{3}x > 8$ by adding $\frac{1}{3}x$ to each side.

In Exercises 27–74, solve the inequality and sketch the solution on the real number line. See Examples 2–8.

27. $x - 4 \ge 0$

28. $x + 1 < 0$

29. $x + 7 \le 9$

30. $z - 4 > 0$

31. $2x < 8$

32. $3x \ge 12$

33. $-9x \ge 36$

34. $-6x \le 24$

35. $-\frac{3}{4}x < -6$

36. $-\frac{1}{5}x > -2$

37. $5 - x \le -2$

38. $1 - y \ge -5$

39. $2x - 5.3 > 9.8$

40. $1.6x + 4 \le 12.4$

41. $5 - 3x < 7$

42. $12 - 5x > 5$

43. $3x - 11 > -x + 7$

44. $21x - 11 \le 6x + 19$

45. $-3x + 7 < 8x - 13$

46. $6x - 1 > 3x - 11$

47. $\frac{x}{4} > 2 - \frac{x}{2}$

48. $\frac{x}{6} - 1 \le \frac{x}{4}$

49. $\frac{x - 4}{3} + 3 \le \frac{x}{8}$

50. $\frac{x + 3}{6} + \frac{x}{8} \ge 1$

51. $\frac{3x}{5} - 4 < \frac{2x}{3} - 3$

52. $\frac{4x}{7} + 1 > \frac{x}{2} + \frac{5}{7}$

53. $0 < 2x - 5 < 9$

54. $-6 \le 3x - 9 < 0$

55. $8 < 6 - 2x \le 12$

56. $-10 \le 4 - 7x < 10$

57. $-1 < -0.2x < 1$

58. $-2 < -0.5s \le 0$

59. $-3 < \frac{2x - 3}{2} < 3$

60. $0 \le \frac{x - 5}{2} < 4$

61. $1 > \frac{x - 4}{-3} > -2$

62. $-\frac{2}{3} < \frac{x - 4}{-6} \le \frac{1}{3}$

63. $2x - 4 \le 4$ and $2x + 8 > 6$

64. $7 + 4x < -5 + x$ and $2x + 10 \le -2$

65. $8 - 3x > 5$ and $x - 5 \ge -10$

66. $9 - x \le 3 + 2x$ and $3x - 7 \le -22$

67. $6.2 - 1.1x > 1$ or $1.2x - 4 > 2.7$

68. $0.4x - 3 \le 8.1$ or $4.2 - 1.6x \le 3$

69. $7x + 11 < 3 + 4x$ or $\frac{5}{2}x - 1 \ge 9 - \frac{3}{2}x$

70. $3x + 10 \le -x - 6$ or $\frac{x}{2} + 5 < \frac{5}{2}x - 4$

71. $-3(y + 10) \ge 4(y + 10)$

72. $2(4 - z) \ge 8(1 + z)$

73. $-4 \le 2 - 3(x + 2) < 11$

74. $16 < 4(y + 2) - 5(2 - y) \le 24$

In Exercises 75–80, write the solution set as a compound inequality. Then write the solution using set notation and the union or intersection symbol. See Example 9.

75.

76.

77.

78.

79.

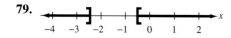

80.

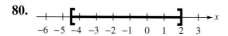

In Exercises 81–86, write the compound inequality using set notation and the union or intersection symbol. See Example 10.

81. $-7 \le x < 0$

82. $2 < x < 8$

83. $x < -5$ or $x > 3$

84. $x \ge -1$ or $x < -6$

85. $-\frac{9}{2} < x \le -\frac{3}{2}$

86. $x < 0$ or $x \ge \frac{2}{3}$

In Exercises 87–92, rewrite the statement using inequality notation. See Example 11.

87. x is nonnegative.

88. y is more than -2.

89. z is at least 8.

90. m is at least 4.

91. n is at least 10, but no more than 16.

92. x is at least 450, but no more than 500.

In Exercises 93–98, write a verbal description of the inequality.

93. $x \ge \frac{5}{2}$

94. $t < 4$

95. $3 \le y < 5$

96. $-4 \le t \le 4$

97. $0 < z \le \pi$

98. $-2 < x \le 5$

Solving Problems

99. *Budget*   A student group has $4500 budgeted for a field trip. The cost of transportation for the trip is $1900. To stay within the budget, all other costs C must be no more than what amount?

100. *Budget*   You have budgeted $1800 per month for your total expenses. The cost of rent per month is $600 and the cost of food is $350. To stay within your budget, all other costs C must be no more than what amount?

101. *Meteorology*   Miami's average temperature is greater than the average temperature in Washington, DC, and the average temperature in Washington, DC is greater than the average temperature in New York City. How does the average temperature in Miami compare with the average temperature in New York City?

102. *Elevation*   The elevation (above sea level) of San Francisco is less than the elevation of Dallas, and the elevation of Dallas is less than the elevation of Denver. How does the elevation of San Francisco compare with the elevation of Denver?

103. *Operating Costs*   A utility company has a fleet of vans. The annual operating cost per van is $C = 0.35m + 2900$, where m is the number of miles traveled by a van in a year. What is the maximum number of miles that will yield an annual operating cost that is less than $12,000?

104. *Operating Costs*   A fuel company has a fleet of trucks. The annual operating cost per truck is $C = 0.58m + 7800$, where m is the number of miles traveled by a truck in a year. What is the maximum number of miles that will yield an annual operating cost that is less than $25,000?

Cost, Revenue, and Profit   In Exercises 105 and 106, the revenue R from selling x units and the cost C of producing x units of a product are given. In order to obtain a profit, the revenue must be greater than the cost. For what values of x will this product produce a profit?

105. $R = 89.95x$

$C = 61x + 875$

106. $R = 105.45x$

$C = 78x + 25{,}850$

107. *Long-Distance Charges*   The cost of an international long-distance telephone call is $0.96 for the first minute and $0.75 for each additional minute. The total cost of the call cannot exceed $5. Find the interval of time that is available for the call.

108. *Long-Distance Charges*   The cost of an international long-distance telephone call is $1.45 for the first minute and $0.95 for each additional minute. The total cost of the call cannot exceed $15. Find the interval of time that is available for the call.

109. ▲ *Geometry*   The length of a rectangle is 16 centimeters. The perimeter of the rectangle must be at least 36 centimeters and not more than 64 centimeters. Find the interval for the width x.

110. ▲ *Geometry*   The width of a rectangle is 14 meters. The perimeter of the rectangle must be at least 100 meters and not more than 120 meters. Find the interval for the length x.

111. *Number Problem*   Four times a number n must be at least 12 and no more than 30. What interval contains this number?

112. *Number Problem*   Determine all real numbers n such that $\frac{1}{3} n$ must be more than 7.

113. *Hourly Wage*   Your company requires you to select one of two payment plans. One plan pays a straight $12.50 per hour. The second plan pays $8.00 per hour plus $0.75 per unit produced per hour. Write an inequality for the number of units that must be produced per hour so that the second option yields the greater hourly wage. Solve the inequality.

114. *Monthly Wage*   Your company requires you to select one of two payment plans. One plan pays a straight $3000 per month. The second plan pays $1000 per month plus a commission of 4% of your gross sales. Write an inequality for the gross sales per month for which the second option yields the greater monthly wage. Solve the inequality.

Environment   In Exercises 115 and 116, use the equation $y = -0.434t + 12.23$, for $4 \le t \le 10$, which models the air pollutant emissions y (in millions of metric tons) of methane caused by landfills in the continental United States from 1994 to 2000 (see figure). In this model, t represents the year, with $t = 4$ corresponding to 1994. (Source: U.S. Energy Information Administration)

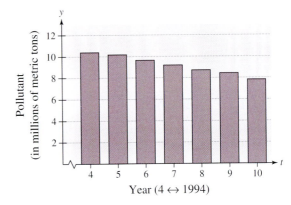

115. During which years was the air pollutant emission of methane caused by landfills greater than 10 million metric tons?

116. During which years was the air pollutant emission of methane caused by landfills less than 8.5 million metric tons?

Explaining Concepts

117. ⊘ Answer parts (f) and (g) of Motivating the Chapter on page 56.

118. *Writing*✐   Is adding -5 to each side of an inequality the same as subtracting 5 from each side? Explain.

119. *Writing*✐   Is dividing each side of an inequality by 5 the same as multiplying each side by $\frac{1}{5}$? Explain.

120. *Writing*✐   Describe any differences between properties of equalities and properties of inequalities.

121. Give an example of "reversing an inequality symbol."

122. If $-3 \le x \le 10$, then $-x$ must be in what interval?

2.5 Absolute Value Equations and Inequalities

Ronnie Kaufman/Corbis

What You Should Learn

1 Solve absolute value equations.

2 Solve inequalities involving absolute value.

Why You Should Learn It

Absolute value equations and inequalities can be used to model and solve real-life problems. For instance, in Exercise 125 on page 116, you will use an absolute value inequality to describe the normal body temperature range.

Solving Equations Involving Absolute Value

Consider the **absolute value equation**

$$|x| = 3.$$

The only solutions of this equation are $x = -3$ and $x = 3$, because these are the only two real numbers whose distance from zero is 3. (See Figure 2.15.) In other words, the absolute value equation $|x| = 3$ has exactly two solutions: $x = -3$ and $x = 3$.

1 Solve absolute value equations.

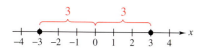

Figure 2.15

Solving an Absolute Value Equation

Let x be a variable or an algebraic expression and let a be a real number such that $a \geq 0$. The solutions of the equation $|x| = a$ are given by $x = -a$ and $x = a$. That is,

$$|x| = a \implies x = -a \quad \text{or} \quad x = a.$$

Study Tip

The strategy for solving absolute value equations is to *rewrite* the equation in *equivalent forms* that can be solved by previously learned methods. This is a common strategy in mathematics. That is, when you encounter a new type of problem, you try to rewrite the problem so that it can be solved by techniques you already know.

Example 1   Solving Absolute Value Equations

Solve each absolute value equation.

a. $|x| = 10$ **b.** $|x| = 0$ **c.** $|y| = -1$

Solution

a. This equation is equivalent to the two linear equations

$$x = -10 \quad \text{and} \quad x = 10. \quad \text{Equivalent linear equations}$$

So, the absolute value equation has two solutions: $x = -10$ and $x = 10$.

b. This equation is equivalent to the two linear equations

$$x = 0 \quad \text{and} \quad x = 0. \quad \text{Equivalent linear equations}$$

Because both equations are the same, you can conclude that the absolute value equation has only one solution: $x = 0$.

c. This absolute value equation has *no solution* because it is not possible for the absolute value of a real number to be negative.

Example 2   Solving Absolute Value Equations

Solve $|3x + 4| = 10$.

Solution

$$|3x + 4| = 10 \qquad \text{Write original equation.}$$

$$3x + 4 = -10 \quad \text{or} \quad 3x + 4 = 10 \qquad \text{Equivalent equations}$$

$$3x + 4 - 4 = -10 - 4 \qquad 3x + 4 - 4 = 10 - 4 \qquad \text{Subtract 4 from each side.}$$

$$3x = -14 \qquad\qquad 3x = 6 \qquad \text{Combine like terms.}$$

$$x = -\frac{14}{3} \qquad\qquad x = 2 \qquad \text{Divide each side by 3.}$$

Check

$$|3x + 4| = 10 \qquad\qquad |3x + 4| = 10$$

$$\left|3\left(-\tfrac{14}{3}\right) + 4\right| \stackrel{?}{=} 10 \qquad\qquad |3(2) + 4| \stackrel{?}{=} 10$$

$$|-14 + 4| \stackrel{?}{=} 10 \qquad\qquad |6 + 4| \stackrel{?}{=} 10$$

$$|-10| = 10 \checkmark \qquad\qquad |10| = 10 \checkmark$$

When solving absolute value equations, remember that it is possible that they have no solution. For instance, the equation $|3x + 4| = -10$ has no solution because the absolute value of a real number cannot be negative. Do not make the mistake of trying to solve such an equation by writing the "equivalent" linear equations as $3x + 4 = -10$ and $3x + 4 = 10$. These equations have solutions, but they are both extraneous.

The equation in the next example is not given in the **standard form**

$$|ax + b| = c, \quad c \geq 0.$$

Notice that the first step in solving such an equation is to write it in standard form.

Example 3   An Absolute Value Equation in Nonstandard Form

Solve $|2x - 1| + 3 = 8$.

Solution

$$|2x - 1| + 3 = 8 \qquad \text{Write original equation.}$$

$$|2x - 1| = 5 \qquad \text{Write in standard form.}$$

$$2x - 1 = -5 \quad \text{or} \quad 2x - 1 = 5 \qquad \text{Equivalent equations}$$

$$2x = -4 \qquad\qquad 2x = 6 \qquad \text{Add 1 to each side.}$$

$$x = -2 \qquad\qquad x = 3 \qquad \text{Divide each side by 2.}$$

The solutions are $x = -2$ and $x = 3$. Check these in the original equation.

If two algebraic expressions are equal in absolute value, they must either be equal to each other or be the *opposites* of each other. So, you can solve equations of the form

$$|ax + b| = |cx + d|$$

by forming the two linear equations

Expressions equal   Expressions opposite

$ax + b = cx + d$ and $ax + b = -(cx + d).$

Example 4 Solving an Equation Involving Two Absolute Values

Solve $|3x - 4| = |7x - 16|$.

Solution

$\|3x - 4\| = \|7x - 16\|$	Write original equation.
$3x - 4 = 7x - 16$ or $3x - 4 = -(7x - 16)$	Equivalent equations
$-4x - 4 = -16$ $3x - 4 = -7x + 16$	
$-4x = -12$ $10x = 20$	
$x = 3$ $x = 2$	Solutions

The solutions are $x = 3$ and $x = 2$. Check these in the original equation.

<table>
<tr><td>

Study Tip

When solving equations of the form

$$|ax + b| = |cx + d|$$

it is possible that one of the resulting equations will not have a solution. Note this occurrence in Example 5.

</td></tr>
</table>

Example 5 Solving an Equation Involving Two Absolute Values

Solve $|x + 5| = |x + 11|$.

Solution

By equating the expression $(x + 5)$ to the opposite of $(x + 11)$, you obtain

$x + 5 = -(x + 11)$	Equivalent equation
$x + 5 = -x - 11$	Distributive Property
$2x + 5 = -11$	Add x to each side.
$2x = -16$	Subtract 5 from each side.
$x = -8.$	Divide each side by 2.

However, by setting the two expressions equal to each other, you obtain

$x + 5 = x + 11$	Equivalent equation
$x = x + 6$	Subtract 5 from each side.
$0 = 6$	Subtract x from each side.

which is a false statement. So, the original equation has only one solution: $x = -8$. Check this solution in the original equation.

2 Solve inequalities involving absolute value.

Solving Inequalities Involving Absolute Value

To see how to solve inequalities involving absolute value, consider the following comparisons.

$\|x\| = 2$	$\|x\| < 2$	$\|x\| > 2$
$x = -2$ and $x = 2$	$-2 < x < 2$	$x < -2$ or $x > 2$

These comparisons suggest the following rules for solving inequalities involving absolute value.

Solving an Absolute Value Inequality

Let x be a variable or an algebraic expression and let a be a real number such that $a > 0$.

1. The solutions of $\|x\| < a$ are all values of x that lie between $-a$ and a. That is,

   $\|x\| < a$ if and only if $-a < x < a$.

2. The solutions of $\|x\| > a$ are all values of x that are less than $-a$ or greater than a. That is,

   $\|x\| > a$ if and only if $x < -a$ or $x > a$.

These rules are also valid if $<$ is replaced by $\leq$ and $>$ is replaced by $\geq$.

Example 6   Solving an Absolute Value Inequality

Solve $\|x - 5\| < 2$.

Solution

$\|x - 5\| < 2$	Write original inequality.
$-2 < x - 5 < 2$	Equivalent double inequality
$-2 + 5 < x - 5 + 5 < 2 + 5$	Add 5 to all three parts.
$3 < x < 7$	Combine like terms.

The solution set consists of all real numbers that are greater than 3 and less than 7. The solution set in interval notation is $(3, 7)$ and in set notation is $\{x | 3 < x < 7\}$. The graph of this solution set is shown in Figure 2.16.

Figure 2.16

To verify the solution of an absolute value inequality, you need to check values in the solution set and outside of the solution set. For instance, in Example 6 you can check that $x = 4$ is in the solution set and that $x = 2$ and $x = 8$ are not in the solution set.

Study Tip

In Example 6, note that an absolute value inequality of the form $|x| < a$ (or $|x| \le a$) can be solved with a double inequality. An inequality of the form $|x| > a$ (or $|x| \ge a$) cannot be solved with a double inequality. Instead, you must solve two separate inequalities, as demonstrated in Example 7.

Example 7   Solving an Absolute Value Inequality

Solve $|3x - 4| \ge 5$.

Solution

$|3x - 4| \ge 5$		Write original inequality.
$3x - 4 \le -5$ or $\quad 3x - 4 \ge 5$		Equivalent inequalities
$3x - 4 + 4 \le -5 + 4 \quad 3x - 4 + 4 \ge 5 + 4$		Add 4 to all parts.
$3x \le -1 \qquad\qquad 3x \ge 9$		Combine like terms.
$\dfrac{3x}{3} \le \dfrac{-1}{3} \qquad\qquad \dfrac{3x}{3} \ge \dfrac{9}{3}$		Divide each side by 3.
$x \le -\dfrac{1}{3} \qquad\qquad x \ge 3$		Simplify.

The solution set consists of all real numbers that are less than or equal to $-\frac{1}{3}$ or greater than or equal to 3. The solution set in interval notation is $\left(-\infty, -\frac{1}{3}\right] \cup [3, \infty)$ and in set notation is $\left\{ x \mid x \le -\frac{1}{3} \text{ or } x \ge 3 \right\}$. The graph is shown in Figure 2.17.

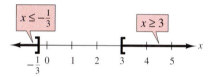

Figure 2.17

Example 8   Solving an Absolute Value Inequality

$\left| 2 - \dfrac{x}{3} \right| \le 0.01$	Original inequality
$-0.01 \le 2 - \dfrac{x}{3} \le 0.01$	Equivalent double inequality
$-2.01 \le -\dfrac{x}{3} \le -1.99$	Subtract 2 from all three parts.
$-2.01(-3) \ge -\dfrac{x}{3}(-3) \ge -1.99(-3)$	Multiply all three parts by -3 and reverse both inequality symbols.
$6.03 \ge x \ge 5.97$	Simplify.
$5.97 \le x \le 6.03$	Solution set in standard form

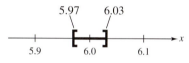

Figure 2.18

The solution set consists of all real numbers that are greater than or equal to 5.97 *and* less than or equal to 6.03. The solution set in interval notation is $[5.97, 6.03]$ and in set notation is $\{x \mid 5.97 \le x \le 6.03\}$. The graph is shown in Figure 2.18.

Technology: Tip

Most graphing calculators can graph absolute value inequalities. Consult your user's guide for specific instructions. The graph below shows the solution of the inequality $|x - 5| < 2$. Notice that the graph occurs as an interval above the x-axis.

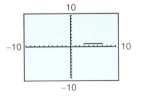

Example 9   Creating a Model

To test the accuracy of a rattlesnake's "pit-organ sensory system," a biologist blindfolded a rattlesnake and presented the snake with a warm "target." Of 36 strikes, the snake was on target 17 times. In fact, the snake was within 5 degrees of the target for 30 of the strikes. Let A represent the number of degrees by which the snake is off target. Then $A = 0$ represents a strike that is aimed directly at the target. Positive values of A represent strikes to the right of the target and negative values of A represent strikes to the left of the target. Use the diagram shown in Figure 2.19 to write an absolute value inequality that describes the interval in which the 36 strikes occurred.

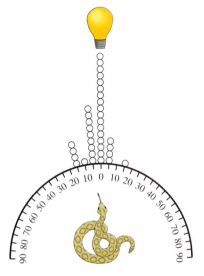

Figure 2.19

Solution

From the diagram, you can see that the snake was never off by more than 15 degrees in either direction. As a compound inequality, this can be represented by $-15 \leq A \leq 15$. As an absolute value inequality, the interval in which the strikes occurred can be represented by $|A| \leq 15$.

Example 10   Production

The estimated daily production at an oil refinery is given by the absolute value inequality $|x - 200{,}000| \leq 25{,}000$, where x is measured in barrels of oil. Solve the inequality to determine the maximum and minimum production levels.

Solution

$\lvert x - 200{,}000 \rvert \leq 25{,}000$	Write original inequality.
$-25{,}000 \leq x - 200{,}000 \leq 25{,}000$	Equivalent double inequality
$175{,}000 \leq x \leq 225{,}000$	Add 200,000 to all three parts.

So, the oil refinery can produce a maximum of 225,000 barrels of oil and a minimum of 175,000 barrels of oil per day.

2.5 Exercises

Review   Concepts, Skills, and Problem Solving

Keep mathematically in shape by doing these exercises *before* the problems of this section.

Properties and Definitions

1. *Writing* If n is an integer, how do the numbers $2n$ and $2n + 1$ differ? Explain.

2. *Writing* Are $-2x^4$ and $(-2x)^4$ equal? Explain.

3. *Writing* Show how to write $\frac{35}{14}$ in simplest form.

4. *Writing* Show how to divide $\frac{4}{5}$ by $\frac{z}{3}$.

Order of Real Numbers

In Exercises 5–10, place the correct inequality symbol ($<$ or $>$) between the two real numbers.

5. -5.6 ___ -4.8

6. $-|7.2|$ ___ $-|-13.1|$

7. $-\frac{3}{4}$ ___ -5

8. $-\frac{1}{5}$ ___ $-\frac{1}{3}$

9. π ___ -3

10. 6 ___ $\frac{13}{2}$

Problem Solving

Budget In Exercises 11 and 12, determine whether there is more or less than a $500 variance between the budgeted amount and the actual expense.

11. Wages

    Budgeted: $162,700

    Actual: $163,356

12. Taxes

    Budgeted: $42,640

    Actual: $42,335

Developing Skills

In Exercises 1–4, determine whether the value is a solution of the equation.

	Equation	Value
1.	$\|4x + 5\| = 10$	$x = -3$
2.	$\|2x - 16\| = 10$	$x = 3$
3.	$\|6 - 2w\| = 2$	$w = 4$
4.	$\left\|\frac{1}{2}t + 4\right\| = 8$	$t = 6$

In Exercises 5–8, transform the absolute value equation into two linear equations.

5. $|x - 10| = 17$

6. $|7 - 2t| = 5$

7. $|4x + 1| = \frac{1}{2}$

8. $|22k + 6| = 9$

In Exercises 9–52, solve the equation. (Some equations have no solution.) See Examples 1–5.

9. $|x| = 4$

10. $|x| = 3$

11. $|t| = -45$

12. $|s| = 16$

13. $|h| = 0$

14. $|x| = -82$

15. $|5x| = 15$

16. $\left|\frac{1}{3}x\right| = 2$

17. $|x + 1| = 5$

18. $|x + 5| = 7$

19. $\left|\frac{2s + 3}{5}\right| = 5$

20. $\left|\frac{7a + 6}{4}\right| = 2$

21. $|32 - 3y| = 16$

22. $|3 - 5x| = 13$

23. $|3x + 4| = -16$

24. $|20 - 5t| = 50$

25. $|4 - 3x| = 0$

26. $|3x - 2| = -5$

27. $\left|\frac{2}{3}x + 4\right| = 9$

28. $\left|3 - \frac{4}{5}x\right| = 1$

29. $|0.32x - 2| = 4$

30. $|2 - 1.5x| = 2$

31. $|5x - 3| + 8 = 22$

32. $|6x - 4| - 7 = 3$

33. $|3x + 9| - 12 = -8$

34. $|5 - 2x| + 10 = 6$

35. $\left|\dfrac{x - 3}{4}\right| + 4 = 4$

36. $\left|\dfrac{x - 3}{4}\right| - 3 = 2$

37. $\left|\dfrac{5x - 3}{2}\right| + 2 = 6$

38. $\left|\dfrac{3z + 5}{6}\right| - 3 = 6$

39. $-2|7 - 4x| = -16$

40. $4|5x + 1| = 24$

41. $3|2x - 5| + 4 = 7$

42. $2|4 - 3x| - 6 = -2$

43. $|x + 8| = |2x + 1|$

44. $|10 - 3x| = |x + 7|$

45. $|x + 2| = |3x - 1|$

46. $|x - 2| = |2x - 15|$

47. $|45 - 4x| = |32 - 3x|$

48. $|5x + 4| = |3x + 25|$

49. $\left|\frac{3}{4}x - 2\right| = \left|\frac{1}{2}x + 5\right|$

50. $\left|\frac{3}{2}r + 2\right| = \left|\frac{1}{2}r - 3\right|$

51. $|4x - 10| = 2|2x + 3|$

52. $3|2 - 3x| = |9x + 21|$

Think About It   In Exercises 53 and 54, write an absolute value equation that represents the verbal statement.

53. The distance between x and 5 is 3.

54. The distance between t and -2 is 6.

In Exercises 55–58, determine whether each x-value is a solution of the inequality.

Inequality		*Values*	
55. $|x| < 3$	(a) $x = 2$	(b) $x = -4$	
	(c) $x = 4$	(d) $x = -1$	
56. $|x| \le 5$	(a) $x = -7$	(b) $x = -4$	
	(c) $x = 4$	(d) $x = 9$	
57. $|x - 7| \ge 3$	(a) $x = 9$	(b) $x = -4$	
	(c) $x = 11$	(d) $x = 6$	
58. $|x - 3| > 5$	(a) $x = 16$	(b) $x = 3$	
	(c) $x = -2$	(d) $x = -3$	

In Exercises 59–62, transform the absolute value inequality into a double inequality or two separate inequalities.

59. $|y + 5| < 3$

60. $|6x + 7| \le 5$

61. $|7 - 2h| \ge 9$

62. $|8 - x| > 25$

In Exercises 63–66, sketch a graph that shows the real numbers that satisfy the statement.

63. All real numbers greater than -2 *and* less than 5

64. All real numbers greater than or equal to 3 *and* less than 10

65. All real numbers less than or equal to 4 *or* greater than 7

66. All real numbers less than -6 *or* greater than or equal to 6

In Exercises 67–104, solve the inequality. See Examples 6–8.

67. $|y| < 4$

68. $|x| < 6$

69. $|x| \ge 6$

70. $|y| \geq 4$

71. $|2x| < 14$

72. $|4z| \leq 9$

73. $\left|\dfrac{y}{3}\right| \leq 3$

74. $\left|\dfrac{t}{2}\right| < 4$

75. $|y - 2| \leq 4$

76. $|x - 3| \leq 6$

77. $|x + 6| > 10$

78. $|x - 4| \geq 3$

79. $|2x - 1| \leq 7$

80. $|3x + 4| < 2$

81. $|6t + 15| \geq 30$

82. $|3t + 1| > 5$

83. $|2 - 5x| > -8$

84. $|8 - 7x| < -6$

85. $|3x + 10| < -1$

86. $|4x - 5| > -3$

87. $\dfrac{|x + 2|}{10} \leq 8$

88. $\dfrac{|s - 3|}{5} > 4$

89. $\dfrac{|y - 16|}{4} < 30$

90. $\dfrac{|a + 6|}{2} \geq 16$

91. $\left|\dfrac{z}{10} - 3\right| > 8$

92. $\left|\dfrac{x}{8} + 1\right| < 0$

93. $\left|\dfrac{3x + 4}{5}\right| > \dfrac{7}{5}$

94. $\left|\dfrac{3 - 2x}{4}\right| \geq 5$

95. $|0.2x - 3| < 4$

96. $|1.5t - 8| \leq 16$

97. $\left|6 - \dfrac{3}{5}x\right| \leq 0.4$

98. $\left|3 - \dfrac{x}{4}\right| > 0.15$

99. $-2|3x + 6| < 4$

100. $-4|2x - 7| > -12$

101. $\left|9 - \dfrac{x}{2}\right| - 7 \leq 4$

102. $\left|8 - \dfrac{2}{3}x\right| + 6 \geq 10$

103. $\left|\dfrac{3x - 2}{4}\right| + 5 \geq 5$

104. $\left|\dfrac{2x - 4}{5}\right| - 9 \leq 3$

 In Exercises 105–110, use a graphing calculator to solve the inequality.

105. $|3x + 2| < 4$

106. $|2x - 1| \leq 3$

107. $|2x + 3| > 9$

108. $|7r - 3| > 11$

109. $|x - 5| + 3 \leq 5$

110. $|a + 1| - 4 < 0$

In Exercises 111–114, match the inequality with its graph. [The graphs are labeled (a), (b), (c), and (d).]

(a)

(b)

(c)

(d)

111. $|x - 4| \leq 4$

112. $|x - 4| < 1$

113. $\frac{1}{2}|x - 4| > 4$

114. $|2(x - 4)| \geq 4$

In Exercises 115–118, write an absolute value inequality that represents the interval.

115.

116.

117.

118.

The symbol  indicates an exercise in which you are instructed to use a graphing calculator.

In Exercises 119–122, write an absolute value inequality that represents the verbal statement.

119. The set of all real numbers x whose distance from 0 is less than 3.

120. The set of all real numbers x whose distance from 0 is more than 2.

121. The set of all real numbers x whose distance from 5 is more than 6.

122. The set of all real numbers x whose distance from 16 is less than 5.

Solving Problems

123. *Temperature*   The operating temperature of an electronic device must satisfy the inequality $|t - 72| \leq 10$, where t is given in degrees Fahrenheit. Sketch the graph of the solution of the inequality. What are the maximum and minimum temperatures?

124. *Time Study*   A time study was conducted to determine the length of time required to perform a task in a manufacturing process. The times required by approximately two-thirds of the workers in the study satisfied the inequality

$$\left| \frac{t - 15.6}{1.9} \right| \leq 1$$

where t is time in minutes. Sketch the graph of the solution of the inequality. What are the maximum and minimum times?

125. *Body Temperature*   Physicians consider an adult's body temperature x to be normal if it is between 97.6°F and 99.6°F. Write an absolute value inequality that describes this normal temperature range.

126. *Accuracy of Measurements*   In woodshop class, you must cut several pieces of wood to within $\frac{3}{16}$ inch of the teacher's specifications. Let $(s - x)$ represent the difference between the specification s and the measured length x of a cut piece.

(a) Write an absolute value inequality that describes the values of x that are within specifications.

(b) The length of one piece of wood is specified to be $s = 5\frac{1}{8}$ inches. Describe the acceptable lengths for this piece.

Explaining Concepts

127. Give a graphical description of the absolute value of a real number.

128. Give an example of an absolute value equation that has only one solution.

129. *Writing*   In your own words, explain how to solve an absolute value equation. Illustrate your explanation with an example.

130. The graph of the inequality $|x - 3| < 2$ can be described as *all real numbers that are within two units of 3*. Give a similar description of $|x - 4| < 1$.

131. Complete $|2x - 6| \leq$ _____ so that the solution is $0 \leq x \leq 6$.

132. When you buy a 16-ounce bag of chips, you probably expect to get *precisely* 16 ounces. Suppose the actual weight w (in ounces) of a "16-ounce" bag of chips is given by $|w - 16| \leq \frac{1}{2}$. You buy four 16-ounce bags. What is the greatest amount you can expect to get? What is the least? Explain.

133. *Writing*   You are teaching a class in algebra and one of your students hands in the following solution. What is wrong with this solution? What could you say to help your students avoid this type of error?

$$|3x - 4| \geq -5$$
$$3x - 4 \leq -5 \quad \text{or} \quad 3x - 4 \geq 5$$
$$3x \leq -1 \qquad\qquad 3x \geq 9$$
$$x \leq -\tfrac{1}{3} \qquad\qquad x \geq 3$$

What Did You Learn?

Key Terms

equation, *p.58*
solutions (equation),   *p.58*
solution set, *p.58*
identity, *p.58*
conditional equation, *p.58*
equivalent equations, *p.59*
linear equation, *p.60*
mathematical modeling,   *p.69*
percent, *p.70*
percent equation, *p.70*
ratio, *p.72*
unit price, *p.72*
proportion, *p.73*

cross-multiplying, *p.73*
markup, *p.80*
markup rate, *p.80*
discount, *p.81*
discount rate, *p.81*
rate of work, *p.84*
algebraic inequalities, *p.94*
solutions (inequality), *p.94*
solution set of an
   inequality, *p.94*
graph of an inequality,   *p.94*
bounded intervals, *p.94*
endpoints, *p.94*

length of [*a, b*], *p.94*
unbounded (infinite)
   intervals, *p.95*
positive infinity, *p.95*
negative infinity, *p.95*
equivalent inequalities, *p.96*
linear inequality, *p.97*
compound inequality, *p.99*
intersection, *p.100*
union, *p.100*
absolute value equation, *p.107*
standard form (absolute
   value equation), *p.108*

Key Concepts

2.1 ⚪ Forming equivalent equations: properties of equality

An equation can be transformed into an equivalent equation using one or more of the following:
1. Simplify either side.
2. Apply the Addition Property of Equality.
3. Apply the Multiplication Property of Equality.
4. Interchange sides.

2.2 ⚪ Strategy for solving word problems

1. Ask yourself what you need to know to solve the problem. Then write a verbal model that includes arithmetic operations to describe the problem.

2. Assign labels to each part of the verbal model—numbers to the known quantities and letters (or expressions) to the variable quantities.

3. Use the labels to write an algebraic model based on the verbal model.

4. Solve the resulting algebraic equation.

5. Answer the original question and check that your answer satisfies the original problem as stated.

2.3 ⚪ Common formulas

See the summary boxes on page 83.

2.4 ⚪ Properties of inequalities

1. Addition and subtraction properties
   If $a < b$, then $a + c < b + c$.
   If $a < b$, then $a - c < b - c$.

2. Multiplication and division properties: positive quantities
   If $a < b$ and c is positive, then $ac < bc$.
   If $a < b$ and c is positive, then $\dfrac{a}{c} < \dfrac{b}{c}$.

3. Multiplication and division properties: negative quantities
   If $a < b$ and c is negative, then $ac > bc$.
   If $a < b$ and c is negative, then $\dfrac{a}{c} > \dfrac{b}{c}$.

4. Transitive property
   If $a < b$ and $b < c$, then $a < c$.

2.5 ⚪ Solving an absolute value equation

Let x be a variable or an algebraic expression and let a be a real number such that $a \geq 0$. The solutions of the equation $|x| = a$ are given by $x = -a$ and $x = a$.

$$|x| = a \quad \Longrightarrow \quad x = -a \quad \text{or} \quad x = a.$$

2.5 ⚪ Solving an absolute value inequality

Let x be a variable or an algebraic expression and let a be a real number such that $a > 0$.
1. The solutions of $|x| < a$ are all values of x that lie between $-a$ and a. That is,

$$|x| < a \quad \text{if and only if} \quad -a < x < a.$$

2. The solutions of $|x| > a$ are all values of x that are less than $-a$ or greater than a. That is,

$$|x| > a \quad \text{if and only if} \quad x < -a \text{ or } x > a.$$

Review Exercises

2.1 Linear Equations

① Check solutions of equations.

In Exercises 1–4, determine whether each value of the variable is a solution of the equation.

	Equation		Values	

1. $45 - 7x = 3$ (a) $x = 3$ (b) $x = 6$

2. $3(3 - x) = -x$ (a) $x = \frac{9}{2}$ (b) $x = -\frac{2}{3}$

3. $\frac{x}{7} + \frac{x}{5} = 1$ (a) $x = \frac{35}{12}$ (b) $x = -\frac{2}{35}$

4. $\dfrac{t + 2}{6} = \dfrac{7}{2}$ (a) $t = -12$ (b) $t = 19$

② Solve linear equations in standard form.

In Exercises 5–8, solve the equation and check the result.

5. $3x + 21 = 0$ **6.** $-4x + 64 = 0$

7. $5x - 120 = 0$ **8.** $7x - 49 = 0$

③ Solve linear equations in nonstandard form.

In Exercises 9–30, solve the equation and check the result. (Some equations have no solution.)

9. $x + 2 = 5$

10. $x - 7 = 3$

11. $-3x = 36$

12. $11x = 44$

13. $-\frac{1}{8}x = 3$

14. $\frac{1}{10}x = 5$

15. $5x + 4 = 19$

16. $3 - 2x = 9$

17. $17 - 7x = 3$

18. $3 + 6x = 51$

19. $7x - 5 = 3x + 11$

20. $9 - 2x = 4x - 7$

21. $3(2y - 1) = 9 + 3y$

22. $-2(x + 4) = 2x - 7$

23. $4y - 6(y - 5) = 2$

24. $7x + 2(7 - x) = 8$

25. $4(3x - 5) = 6(2x + 3)$

26. $8(x - 2) = 3(x - 2)$

27. $\frac{4}{5}x - \frac{1}{10} = \frac{3}{2}$

28. $\frac{1}{4}s + \frac{3}{8} = \frac{5}{2}$

29. $1.4t + 2.1 = 0.9t$

30. $2.5x - 6.2 = 3.7x - 5.8$

31. *Number Problem*   Find two consecutive positive integers whose sum is 147.

32. *Number Problem*   Find two consecutive even integers whose sum is 74.

2.2 Linear Equations and Problem Solving

① Use mathematical modeling to write algebraic equations representing real-life situations.

Mathematical Modeling   In Exercises 33 and 34, construct a verbal model and write an algebraic equation that represents the problem. Solve the equation.

33. You have accepted a summer job that pays $200 per week plus a one-time bonus of $350. How many weeks would you need to work at this job to earn $2750?

34. You have a job as a salesperson for which you are paid $6 per hour plus $1.25 per sale made. Find the number of sales made in an eight-hour day if your earnings for the day are $88.

② Solve percent problems using the percent equation.

In Exercises 35–38, complete the table showing the equivalent forms of various percents.

	Percent	Parts out of 100	Decimal	Fraction
35.	87%			
36.			0.35	
37.		60		
38.				$\frac{1}{6}$

In Exercises 39–44, solve using a percent equation.

39. What is 130% of 50?

40. What is 0.4% of 7350?

41. 645 is $21\frac{1}{2}$% of what number?

42. 498 is 83% of what number?

43. 250 is what percent of 200?

44. 162.5 is what percent of 6500?

45. *Real Estate Commission*   A real estate agency receives a commission of $9000 for the sale of a $150,000 house. What percent commission is this?

46. *Quality Control*   A quality control engineer reported that 1.6% of a sample of parts were defective. The engineer found six defective parts. How large was the sample?

47. *Pension Fund*   Your employer withholds $216 of your gross income each month for your retirement. Determine the percent of your total monthly gross income of $3200 that is withheld for retirement.

48. *Sales Tax*   The state sales tax on an item you purchase is $7\frac{1}{4}$%. If the item costs $34, how much sales tax will you pay?

3 Use ratios to compare unit prices for products.

Consumer Awareness   **In Exercises 49 and 50, use unit prices to determine the better buy.**

49. (a) a 32-ounce package of candy for $12.16

  (b) an 18.5-ounce package of candy for $7.40

50. (a) a 3.5-pound bag of sugar for $3.08

  (b) a 2.2-pound bag of sugar for $1.87

51. *Income Tax*   You have $9.90 of state tax withheld from your paycheck per week when your gross pay is $396. Find the ratio of tax to gross pay.

52. *Price-Earnings Ratio*   The ratio of the price of a stock to its earnings is called the price-earnings ratio. Find the price-earnings ratio of a stock that sells for $46.75 per share and earns $5.50 per share.

4 Solve proportions.

In Exercises 53–56, solve the proportion.

53. $\dfrac{7}{8} = \dfrac{y}{4}$

54. $\dfrac{x}{16} = \dfrac{5}{12}$

55. $\dfrac{b}{6} = \dfrac{5+b}{15}$

56. $\dfrac{x+1}{3} = \dfrac{x-1}{2}$

▲ *Geometry*   **In Exercises 57 and 58, solve for the length x of the side of the triangle by using the fact that corresponding sides of similar triangles are proportional.**

57.

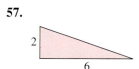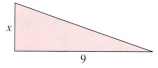

58.

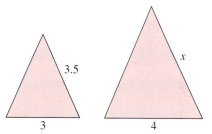

59. ▲ *Geometry*   You want to measure the height of a flagpole. To do this, you measure the flagpole's shadow and find that it is 30 feet long. You also measure the height of a five-foot lamp post and find its shadow to be 3 feet long (see figure). Find the height h of the flagpole.

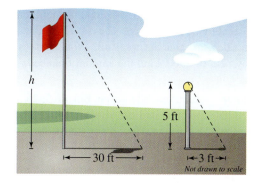

Not drawn to scale

60. ▲ *Geometry*   You want to measure the height of a silo. To do this, you measure the silo's shadow and find that it is 20 feet long. You are 6 feet tall and your shadow is $1\frac{1}{2}$ feet long. Find the height of the silo.

61. *Property Tax*   The tax on a property with an assessed value of $105,000 is $1680. Find the tax on a property with an assessed value of $125,000.

62. *Recipe*   One and one-half cups of milk are required to make one batch of pudding. How many cups are required to make $2\frac{1}{2}$ batches?

2.3 Business and Scientific Problems

❶ Use mathematical models to solve business-related problems.

In Exercises 63–66, find the missing quantities. (Assume that the markup rate is a percent based on the cost.)

	Cost	Selling Price	Markup	Markup Rate
63.	$99.95	$149.93		
64.	$23.50	$31.33		
65.		$125.85	$44.13	
66.		$895.00	$223.75	

In Exercises 67–70, find the missing quantities. (Assume that the discount rate is a percent based on the list price.)

	List Price	Sale Price	Discount	Discount Rate
67.	$71.95	$53.96		
68.	$559.95	$279.98		
69.	$1995.50		$598.65	
70.	$39.00		$15.60	

71. *Amount Financed*   You bought an entertainment system for $1250 plus 6% sales tax. Find the amount of sales tax and the total bill. You made a down payment of $300. Find the amount financed.

72. *Labor*   An automobile repair bill of $325.30 lists $200.30 for parts and the rest for labor. The labor rate is $25 per hour. How many hours did it take to repair the automobile?

❷ Use mathematical models to solve mixture problems.

73. *Mixture Problem*   Determine the number of liters of a 30% saline solution and the number of liters of a 60% saline solution that are required to obtain 10 liters of a 50% saline solution.

74. *Mixture Problem*   Determine the number of gallons of a 25% alcohol solution and the number of gallons of a 50% alcohol solution that are required to obtain 8 gallons of a 40% alcohol solution.

❸ Use mathematical models to solve classic rate problems.

75. *Distance*   Determine the distance an Air Force jet can travel in $2\frac{1}{3}$ hours when its average speed is 1500 miles per hour.

76. *Travel Time*   Determine the time for a bus to travel 330 miles when its average speed is 52 miles per hour.

77. *Speed*   A truck driver traveled at an average speed of 48 miles per hour on a 100-mile trip to pick up a load of freight. On the return trip with the truck fully loaded, the average speed was 40 miles per hour. Find the average speed for the round trip.

78. *Speed*   For 2 hours of a 400-mile trip, your average speed is only 40 miles per hour. Determine the average speed that must be maintained for the remainder of the trip if you want the average speed for the entire trip to be 50 miles per hour.

79. *Work-Rate Problem*   Find the time for two people working together to complete a task if it takes them 4.5 hours and 6 hours working individually.

80. *Work-Rate Problem*   Find the time for two people working together to complete half a task if it takes them 8 hours and 10 hours to complete the entire task working individually.

❹ Use formulas to solve application problems.

81. *Simple Interest*   Find the interest on a $1000 corporate bond that matures in 4 years and has an 8.5% interest rate.

82. *Simple Interest*   Find the annual interest rate on a certificate of deposit that pays $37.50 per year in interest on a principal of $500.

83. *Simple Interest*   The interest on a savings account is 9.5%. Find the principal required to earn $20,000 in interest in 4 years.

84. *Simple Interest*   A corporation borrows 3.25 million dollars for 2 years at an annual interest rate of 12% to modernize one of its manufacturing facilities. What is the total principal and interest that must be repaid?

85. *Simple Interest*   An inheritance of $50,000 is divided into two investments earning 8.5% and 10% simple interest. (The 10% investment has a greater risk.) Your objective is to obtain a total annual interest income of $4700 from the investments. What is the smallest amount you can invest at 10% in order to meet your objective?

86. *Simple Interest*   You invest $1000 in a certificate of deposit that has an annual interest rate of 7%. After 6 months, the interest is computed and added to the principal. During the second 6 months, the interest is computed using the original investment plus the interest earned during the first 6 months. What is the total interest earned during the first year of the investment?

87. ▲ *Geometry*   The perimeter of the rectangle shown in the figure is 110 feet. Find its dimensions.

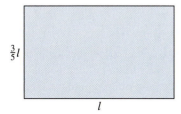

$\frac{3}{5}l$

l

88. ▲ *Geometry*   The area of the triangle shown in the figure is 90 square meters. Find the base of the triangle.

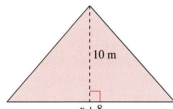

10 m

$x + 8$

89. *Meteorology*   The average daily temperature in Boise, Idaho is 10.5°C. What is Boise's average daily temperature in degrees Fahrenheit?   (Source: U.S. National Oceanic and Atmospheric Administration)

90. *Meteorology*   The average daily temperature in January in Atlanta, Georgia is 41°F. What is Atlanta's average daily temperature in January in degrees Celsius?   (Source: U.S. National Oceanic and Atmospheric Administration)

2.4 Linear Inequalities

① Sketch the graphs of inequalities.

In Exercises 91–94, sketch the graph of the inequality.

91. $-3 \leq x < 1$

92. $4 < x < 5.5$

93. $-7 < x$

94. $x \geq -2$

③ Solve linear inequalities.

In Exercises 95–106, solve the inequality and sketch the solution on the real number line.

95. $x - 5 \leq -1$

96. $x + 8 > 5$

97. $-5x < 30$

98. $-11x \geq 44$

99. $5x + 3 > 18$

100. $3x - 11 \leq 7$

101. $8x + 1 \geq 10x - 11$

102. $12 - 3x < 4x - 2$

103. $\frac{1}{3} - \frac{1}{2}y < 12$

104. $\frac{x}{4} - 2 < \frac{3x}{8} + 5$

105. $-4(3 - 2x) \leq 3(2x - 6)$

106. $3(2 - y) \geq 2(1 + y)$

4 Solve compound inequalities.

In Exercises 107–112, solve the compound inequality and sketch the solution on the real number line.

107. $-6 \le 2x + 8 < 4$

108. $-13 \le 3 - 4x < 13$

109. $5 > \dfrac{x + 1}{-3} > 0$

110. $12 \ge \dfrac{x - 3}{2} > 1$

111. $5x - 4 < 6$ and $3x + 1 > -8$

112. $6 - 2x \le 1$ or $10 - 4x > -6$

5 Solve application problems involving inequalities.

113. *Sales Goal*   The weekly salary of an employee is $150 plus a 6% commission on total sales. The employee needs a minimum salary of $650 per week. How much must be sold to produce this salary?

114. *Long-Distance Charges*   The cost of an international long-distance telephone call is $0.99 for the first minute and $0.49 for each additional minute. The total cost of the call cannot exceed $7.50. Find the interval of time that is available for the call.

2.5 Absolute Value Equations and Inequalities

1 Solve absolute value equations.

In Exercises 115–122, solve the equation.

115. $|x| = 6$

116. $|x| = -4$

117. $|4 - 3x| = 8$

118. $|2x + 3| = 7$

119. $|5x + 4| - 10 = -6$

120. $|x - 2| - 2 = 4$

121. $|3x - 4| = |x + 2|$

122. $|5x + 6| = |2x - 1|$

2 Solve inequalities involving absolute value.

In Exercises 123–130, solve the inequality.

123. $|x - 4| > 3$

124. $|t + 3| > 2$

125. $|3x| > 9$

126. $\left|\dfrac{t}{3}\right| < 1$

127. $|2x - 7| < 15$

128. $|5x - 1| < 9$

129. $|b + 2| - 6 > 1$

130. $|2y - 1| + 4 < -1$

In Exercises 131 and 132, use a graphing calculator to solve the inequality.

131. $|2x - 5| \ge 1$

132. $|5(1 - x)| \le 25$

In Exercises 133 and 134, write an absolute value inequality that represents the interval.

133.

134.

135. *Temperature*   The storage temperature of a computer must satisfy the inequality

$$|t - 78.3| \le 38.3$$

where t is given in degrees Fahrenheit. Sketch the graph of the solution of the inequality. What are the maximum and minimum temperatures?

136. *Temperature*   The operating temperature of a computer must satisfy the inequality

$$|t - 77| \le 27$$

where t is given in degrees Fahrenheit. Sketch the graph of the solution of the inequality. What are the maximum and minimum temperatures?

Chapter Test

Take this test as you would take a test in class. After you are done, check your work against the answers in the back of the book.

In Exercises 1–4, solve the equation.

1. $6x - 5 = 19$

2. $5x - 6 = 7x - 12$

3. $15 - 7(1 - x) = 3(x + 8)$

4. $\dfrac{2x}{3} = \dfrac{x}{2} + 4$

5. What is 125% of 3200?

6. 32 is what percent of 8000?

7. A store is offering a 20% discount on all items in its inventory. Find the list price on a tractor that has a sale price of $6400.

8. Which of the packages at the left is a better buy? Explain your reasoning.

$2.99

$2.49

12 oz   15 oz

Figure for 8

9. The tax on a property with an assessed value of $90,000 is $1200. Estimate the tax on a property with an assessed value of $110,000.

10. The bill (including parts and labor) for the repair of a home appliance was $165. The cost for parts was $85. The labor rate is $16 per half hour. How many hours were spent in repairing the appliance?

11. A pet store owner mixes two types of dog food costing $1.50 per pound and $3.05 per pound to make 25 pounds of a mixture costing $2.12 per pound. How many pounds of each kind of dog food are in the mixture.

12. Two cars start at the same location and travel in the same direction at average speeds of 40 miles per hour and 55 miles per hour. How much time must elapse before the two cars are 10 miles apart?

13. The interest on a savings account is 7.5%. Find the principal required to earn $300 in interest in 2 years.

14. Solve each equation.

(a) $|2x + 6| = 16$ (b) $|3x - 5| = |6x - 1|$ (c) $|9 - 4x| + 4 = 1$

15. Solve each inequality and sketch the solution on the real number line.

(a) $3x + 12 \geq -6$

(b) $1 + 2x > 7 - x$

(c) $0 \leq \dfrac{1 - x}{4} < 2$

(d) $-7 < 4(2 - 3x) \leq 20$

16. Rewrite the statement "t is at least 8" using inequality notation.

17. Solve each inequality.

(a) $|x - 3| \leq 2$ (b) $|5x - 3| > 12$ (c) $\left|\dfrac{x}{4} + 2\right| < 0.2$

18. A utility company has a fleet of vans. The annual operating cost per van is

$$C = 0.37m + 2700$$

where m is the number of miles traveled by a van in a year. What is the maximum number of miles that will yield an annual operating cost that is less than or equal to $11,950?

Motivating the Chapter

⚡ Automobile Depreciation

You are a sales representative for a pharmaceutical company. You have just paid $15,900 for a new car that you will use for traveling between clients. After 3 years, the car is expected to have a value of $10,200. Because you will use the car for business purposes, its depreciation is deductible from your taxable income. Let x represent the time, with $x = 0$ corresponding to the time of purchase and $x = 3$ to the time 3 years later. Let y represent the value of the car. Using the straight-line method of depreciation, answer the following.

See Section 3.3, Exercise 89.

a. Plot the value of the car when new and the value 3 years later on a rectangular coordinate system. Connect the points with a straight line.

b. Calculate the slope of the line.

c. Interpret the meaning of the slope in part (b) in the context of the problem.

See Section 3.4, Exercise 83.

d. Create a table of values showing the expected value of the car over the first 4 years ($x = 0, 1, 2, 3$, and 4) of its use.

e. Write a linear equation that models this depreciation problem.

f. Does a fifth-year value ($x = 5$) of $7000 fit this model? Explain.

See Section 3.6, Exercise 93.

g. Write the linear equation from part (e) as a function of x. What is the value of the car after 7 years?

h. Why might straight-line depreciation not be a fair model for automobile depreciation? Include a sketch of a graph that might better represent automobile depreciation.

i. From a practical perspective, describe the *domain* and the *range* of the linear function in part (e).

Spencer Grant/PhotoEdit

3

Graphs and Functions

3.1 ● The Rectangular Coordinate System

3.2 ● Graphs of Equations

3.3 ● Slope and Graphs of Linear Equations

3.4 ● Equations of Lines

3.5 ● Graphs of Linear Inequalities

3.6 ● Relations and Functions

3.7 ● Graphs of Functions

3.1 The Rectangular Coordinate System

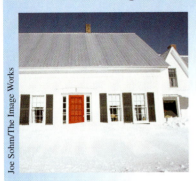

Joe Sohm/The Image Works

What You Should Learn

1. Plot points on a rectangular coordinate system.
2. Determine whether ordered pairs are solutions of equations.
3. Use the Distance Formula to find the distance between two points.
4. Use the Midpoint Formula to find the midpoints of line segments.

Why You Should Learn It

A rectangular coordinate system can be used to represent relationships between two variables. For instance, Exercise 98 on page 139 shows the relationship between temperatures and the corresponding amounts of natural gas used to heat a home.

1. Plot points on a rectangular coordinate system.

The Rectangular Coordinate System

Just as you can represent real numbers by points on the real number line, you can represent ordered pairs of real numbers by points in a plane. This plane is called a **rectangular coordinate system** or the **Cartesian plane,** after the French mathematician René Descartes.

A rectangular coordinate system is formed by two real number lines intersecting at a right angle, as shown in Figure 3.1. The horizontal number line is usually called the **x-axis,** and the vertical number line is usually called the **y-axis.** (The plural of axis is *axes*.) The point of intersection of the two axes is called the **origin,** and the axes separate the plane into four regions called **quadrants.**

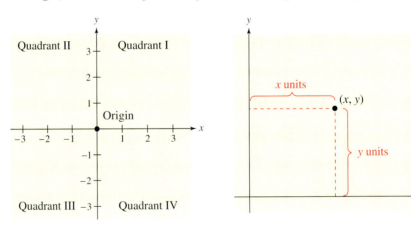

Figure 3.1

Figure 3.2

Each point in the plane corresponds to an **ordered pair** (x, y) of real numbers x and y, called the **coordinates** of the point. The first number (or **x-coordinate**) tells how far to the left or right the point is from the vertical axis, and the second number (or **y-coordinate**) tells how far up or down the point is from the horizontal axis, as shown in Figure 3.2.

A positive x-coordinate implies that the point lies to the *right* of the vertical axis; a negative x-coordinate implies that the point lies to the *left* of the vertical axis; and an x-coordinate of zero implies that the point lies *on* the vertical axis. Similarly, a positive y-coordinate implies that the point lies *above* the horizontal axis; a negative y-coordinate implies that the point lies *below* the horizontal axis; and a y-coordinate of zero implies that the point lies *on* the horizontal axis.

Study Tip

The signs of the coordinates tell you in which quadrant the point lies. For instance, if x and y are positive, the point (x, y) lies in Quadrant I.

Locating a given point in a plane is called **plotting** the point. Example 1 shows how this is done.

Example 1 Plotting Points on a Rectangular Coordinate System

Plot the points $(-2, 1)$, $(4, 0)$, $(3, -1)$, $(4, 3)$, $(0, 0)$, and $(-1, -3)$ on a rectangular coordinate system.

Solution

The point $(-2, 1)$ is two units to the *left* of the vertical axis and one unit *above* the horizontal axis.

Similarly, the point $(4, 0)$ is four units to the *right* of the vertical axis and *on* the horizontal axis. (It is on the horizontal axis because its y-coordinate is 0.) The other four points can be plotted in a similar way, as shown in Figure 3.3.

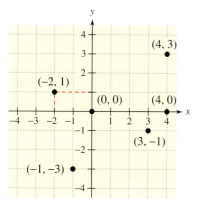

Figure 3.3

In Example 1, you were given the coordinates of several points and asked to plot the points on a rectangular coordinate system. Example 2 looks at the reverse problem. That is, you are given points on a rectangular coordinate system and are asked to determine their coordinates.

Example 2 Finding Coordinates of Points

Determine the coordinates of each of the points shown in Figure 3.4.

Solution

Point A lies two units to the *right* of the vertical axis and one unit *below* the horizontal axis. So, point A must be given by the ordered pair $(2, -1)$. The coordinates of the other four points can be determined in a similar way. The results are summarized as follows.

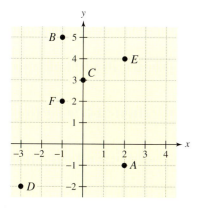

Figure 3.4

Point	Position	Coordinates
A	Two units *right*, one unit *down*	$(2, -1)$
B	One unit *left*, five units *up*	$(-1, 5)$
C	Zero units *right* (or *left*), three units *up*	$(0, 3)$
D	Three units *left*, two units *down*	$(-3, -2)$
E	Two units *right*, four units *up*	$(2, 4)$
F	One unit *left*, two units *up*	$(-1, 2)$

In Example 2, note that point $A(2, -1)$ and point $F(-1, 2)$ are different points. The order in which the numbers appear in an ordered pair is important.

As a consumer today, you are presented almost daily with vast amounts of data given in various forms. Data is given in *numerical form* using lists and tables and in *graphical* form using scatter plots, line graphs, circle graphs, and bar graphs. Graphical forms are more visual and make wide use of Descartes's rectangular coordinate system to show the relationship between two variables. Today, Descartes's ideas are commonly used in virtually every scientific and business-related field.

In the next example, data is represented graphically by points plotted on a rectangular coordinate system. This type of graph is called a **scatter plot.**

Example 3   Representing Data Graphically

The population (in millions) of California from 1985 through 2000 is shown in the table. Sketch a scatter plot of the data.   (Source: U.S. Census Bureau)

Year	1985	1986	1987	1988	1989	1990	1991	1992
Population	26.4	27.1	27.8	28.5	29.2	29.8	30.4	30.9

Year	1993	1994	1995	1996	1997	1998	1999	2000
Population	31.1	31.3	31.5	31.8	32.2	32.7	33.1	33.9

Solution

To sketch a scatter plot, begin by choosing which variable will be plotted on the horizontal axis and which will be plotted on the vertical axis. For this data, it seems natural to plot the years on the horizontal axis (which means that the population must be plotted on the vertical axis). Next, use the data in the table to form ordered pairs. For instance, the first three ordered pairs are (1985, 26.4), (1986, 27.1), and (1987, 27.8). All 16 points are shown in Figure 3.5. Note that the break in the x-axis indicates that the numbers between 0 and 1985 have been omitted. The break in the y-axis indicates that the numbers between 0 and 25 have been omitted.

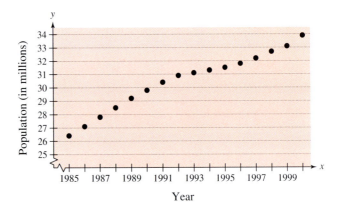

Figure 3.5

② Determine whether ordered pairs are solutions of equations.

Technology: Discovery

In the table of values in Example 4, successive *x*-values differ by 1. How do the successive *y*-values differ? Use the *table* feature of your graphing calculator to create a table of values for each of the following equations. If successive *x*-values differ by 1, how do the successive *y*-values differ?

a. $y = x + 2$

b. $y = 2x + 2$

c. $y = 4x + 2$

d. $y = -x + 2$

Describe the pattern.

Ordered Pairs as Solutions

In Example 3, the relationship between the year and the population was given by a **table of values.** In mathematics, the relationship between the variables *x* and *y* is often given by an equation, from which you can construct a table of values.

Example 4   Constructing a Table of Values

Construct a table of values for $y = 3x + 2$. Then plot the solution points on a rectangular coordinate system. Choose *x*-values of $-3, -2, -1, 0, 1, 2,$ and 3.

Solution

For each *x*-value, you must calculate the corresponding *y*-value. For example, if you choose $x = 1$, then the *y*-value is

$$y = 3(1) + 2 = 5.$$

The ordered pair $(x, y) = (1, 5)$ is a **solution point** (or **solution**) of the equation.

Choose x	Calculate y from $y = 3x + 2$	Solution point
$x = -3$	$y = 3(-3) + 2 = -7$	$(-3, -7)$
$x = -2$	$y = 3(-2) + 2 = -4$	$(-2, -4)$
$x = -1$	$y = 3(-1) + 2 = -1$	$(-1, -1)$
$x = 0$	$y = 3(0) + 2 = 2$	$(0, 2)$
$x = 1$	$y = 3(1) + 2 = 5$	$(1, 5)$
$x = 2$	$y = 3(2) + 2 = 8$	$(2, 8)$
$x = 3$	$y = 3(3) + 2 = 11$	$(3, 11)$

Once you have constructed a table of values, you can get a visual idea of the relationship between the variables *x* and *y* by plotting the solution points on a rectangular coordinate system, as shown in Figure 3.6.

In many places throughout this course, you will see that approaching a problem in different ways can help you understand the problem better. In Example 4, for instance, solutions of an equation are arrived at in three ways.

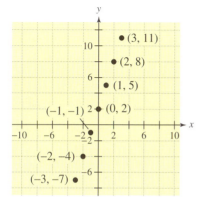

Figure 3.6

Three Approaches to Problem Solving

1. **Algebraic Approach** Use algebra to find several solutions.

2. **Numerical Approach** Construct a table that shows several solutions.

3. **Graphical Approach** Draw a graph that shows several solutions.

Guidelines for Verifying Solutions

To verify that an ordered pair (x, y) is a solution of an equation with variables x and y, use the steps below.

1. Substitute the values of x and y into the equation.

2. Simplify each side of the equation.

3. If each side simplifies to the same number, the ordered pair is a solution. If the two sides yield different numbers, the ordered pair is not a solution.

Example 5   Verifying Solutions of an Equation

Which of the ordered pairs are solutions of $x^2 - 2y = 6$?

a. $(2, 1)$ **b.** $(0, -3)$ **c.** $(-2, -5)$ **d.** $\left(1, -\frac{5}{2}\right)$

Solution

a. For the ordered pair $(2, 1)$, substitute $x = 2$ and $y = 1$ into the equation.

$$(2)^2 - 2(1) \overset{?}{=} 6 \qquad \text{Substitute 2 for } x \text{ and 1 for } y.$$

$$2 \neq 6 \qquad \text{Not a solution } \times$$

Because the substitution does not satisfy the original equation, you can conclude that the ordered pair $(2, 1)$ *is not* a solution of the original equation.

b. For the ordered pair $(0, -3)$, substitute $x = 0$ and $y = -3$ into the equation.

$$(0)^2 - 2(-3) \overset{?}{=} 6 \qquad \text{Substitute 0 for } x \text{ and } -3 \text{ for } y.$$

$$6 = 6 \qquad \text{Solution } \checkmark$$

Because the substitution satisfies the original equation, you can conclude that the ordered pair $(0, -3)$ *is* a solution of the original equation.

c. For the ordered pair $(-2, -5)$, substitute $x = -2$ and $y = -5$ into the equation.

$$(-2)^2 - 2(-5) \overset{?}{=} 6 \qquad \text{Substitute } -2 \text{ for } x \text{ and } -5 \text{ for } y.$$

$$14 \neq 6 \qquad \text{Not a solution } \times$$

Because the substitution does not satisfy the original equation, you can conclude that the ordered pair $(-2, -5)$ *is not* a solution of the original equation.

d. For the ordered pair $\left(1, -\frac{5}{2}\right)$, substitute $x = 1$ and $y = -\frac{5}{2}$ into the equation.

$$(1)^2 - 2\left(-\frac{5}{2}\right) \overset{?}{=} 6 \qquad \text{Substitute 1 for } x \text{ and } -\frac{5}{2} \text{ for } y.$$

$$6 = 6 \qquad \text{Solution } \checkmark$$

Because the substitution satisfies the original equation, you can conclude that the ordered pair $\left(1, -\frac{5}{2}\right)$ *is* a solution of the original equation.

③ Use the Distance Formula to find the distance between two points.

The Distance Formula

You know from Section 1.1 that the distance d between two points a and b on the real number line is simply

$$d = |b - a|.$$

The same "absolute value rule" is used to find the distance between two points that lie on the same *vertical or horizontal line* in the coordinate plane, as shown in Example 6.

Example 6   Finding Horizontal and Vertical Distances

a. Find the distance between the points $(2, -2)$ and $(2, 4)$.

b. Find the distance between the points $(-3, -2)$ and $(2, -2)$.

Solution

a. Because the x-coordinates are equal, you can visualize a vertical line through the points $(2, -2)$ and $(2, 4)$, as shown in Figure 3.7. The distance between these two points is the absolute value of the difference of their y-coordinates.

$$\text{Vertical distance} = |4 - (-2)| \qquad \text{Subtract } y\text{-coordinates.}$$
$$= |6| \qquad \text{Simplify.}$$
$$= 6 \qquad \text{Evaluate absolute value.}$$

b. Because the y-coordinates are equal, you can visualize a horizontal line through the points $(-3, -2)$ and $(2, -2)$, as shown in Figure 3.7. The distance between these two points is the absolute value of the difference of their x-coordinates.

$$\text{Horizontal distance} = |2 - (-3)| \qquad \text{Subtract } x\text{-coordinates.}$$
$$= |5| \qquad \text{Simplify.}$$
$$= 5 \qquad \text{Evaluate absolute value.}$$

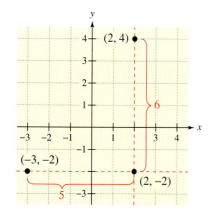

Figure 3.7

In Figure 3.7, note that the horizontal distance between the points $(-3, -2)$ and $(2, -2)$ is the absolute value of the difference of the x-coordinates, and the vertical distance between the points $(2, -2)$ and $(2, 4)$ is the absolute value of the difference of the y-coordinates.

The technique applied in Example 6 can be used to develop a general formula for finding the distance between two points in the plane. This general formula will work for any two points, even if they do not lie on the same vertical or horizontal line. To develop the formula, you use the **Pythagorean Theorem,** which states that for a right triangle, the hypotenuse c and sides a and b are related by the formula

$$a^2 + b^2 = c^2 \qquad \text{Pythagorean Theorem}$$

as shown in Figure 3.8. (The converse is also true. That is, if $a^2 + b^2 = c^2$, then the triangle is a right triangle.)

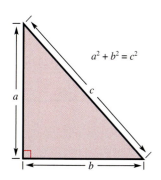

Figure 3.8

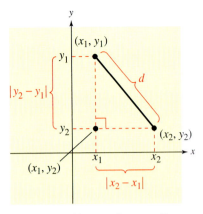

Figure 3.9 Distance Between Two Points

To develop a general formula for the distance between two points, let (x_1, y_1) and (x_2, y_2) represent two points in the plane (that do not lie on the same horizontal or vertical line). With these two points, a right triangle can be formed, as shown in Figure 3.9. Note that the third vertex of the triangle is (x_1, y_2). Because (x_1, y_1) and (x_1, y_2) lie on the same vertical line, the length of the vertical side of the triangle is $|y_2 - y_1|$. Similarly, the length of the horizontal side is $|x_2 - x_1|$. By the Pythagorean Theorem, the square of the distance between (x_1, y_1) and (x_2, y_2) is

$$d^2 = |x_2 - x_1|^2 + |y_2 - y_1|^2.$$

Because the distance d must be positive, you can choose the positive square root and write

$$d = \sqrt{|x_2 - x_1|^2 + |y_2 - y_1|^2}.$$

Finally, replacing $|x_2 - x_1|^2$ and $|y_2 - y_1|^2$ by the equivalent expressions $(x_2 - x_1)^2$ and $(y_2 - y_1)^2$ yields the **Distance Formula.**

The Distance Formula

The distance d between two points (x_1, y_1) and (x_2, y_2) is

$$d = \sqrt{(x_2 - x_1)^2 + (y_2 - y_1)^2}.$$

Note that for the special case in which the two points lie on the same vertical or horizontal line, the Distance Formula still works. For instance, applying the Distance Formula to the points $(2, -2)$ and $(2, 4)$ produces

$$d = \sqrt{(2 - 2)^2 + [4 - (-2)]^2} = \sqrt{6^2} = 6$$

which is the same result obtained in Example 6(a).

Example 7 Finding the Distance Between Two Points

Find the distance between the points $(-1, 2)$ and $(2, 4)$, as shown in Figure 3.10.

Solution

Let $(x_1, y_1) = (-1, 2)$ and let $(x_2, y_2) = (2, 4)$, and apply the Distance Formula.

$$d = \sqrt{(x_2 - x_1)^2 + (y_2 - y_1)^2} \qquad \text{Distance Formula}$$
$$d = \sqrt{[2 - (-1)]^2 + (4 - 2)^2} \qquad \text{Substitute coordinates of points.}$$
$$= \sqrt{3^2 + 2^2} \qquad \text{Simplify.}$$
$$= \sqrt{13} \approx 3.61 \qquad \text{Simplify and use calculator.}$$

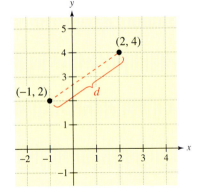

Figure 3.10

When the Distance Formula is used, it does not matter which point is considered (x_1, y_1) and which is (x_2, y_2), because the result will be the same. For instance, in Example 7, let $(x_1, y_1) = (2, 4)$ and let $(x_2, y_2) = (-1, 2)$. Then

$$d = \sqrt{[(-1) - 2]^2 + (2 - 4)^2} = \sqrt{(-3)^2 + (-2)^2} = \sqrt{13} \approx 3.61.$$

The Distance Formula has many applications in mathematics. In the next example, you can use the Distance Formula and the converse of the Pythagorean Theorem to verify that three points form the vertices of a right triangle.

Example 8   Verifying a Right Triangle

Show that the points $(1, 2)$, $(3, 1)$, and $(4, 3)$ are vertices of a right triangle.

Solution

The three points are plotted in Figure 3.11. Using the Distance Formula, you can find the lengths of the three sides of the triangle.

$$d_1 = \sqrt{(3 - 1)^2 + (1 - 2)^2} = \sqrt{4 + 1} = \sqrt{5}$$

$$d_2 = \sqrt{(4 - 3)^2 + (3 - 1)^2} = \sqrt{1 + 4} = \sqrt{5}$$

$$d_3 = \sqrt{(4 - 1)^2 + (3 - 2)^2} = \sqrt{9 + 1} = \sqrt{10}$$

Because

$$d_1^2 + d_2^2 = \left(\sqrt{5}\right)^2 + \left(\sqrt{5}\right)^2 = 5 + 5 = 10$$

and

$$d_3^2 = \left(\sqrt{10}\right)^2 = 10$$

you can conclude from the converse of the Pythagorean Theorem that the triangle is a right triangle.

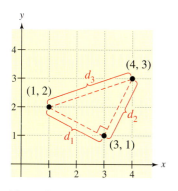

Figure 3.11

Three or more points are **collinear** if they all lie on the same line. You can use the Distance Formula to determine whether points are collinear as shown in Example 9.

Example 9   Collinear Points

Determine whether the set of points is collinear.

$$\{A(2, 6), B(5, 2), C(8, -2)\}$$

Solution

Let d_1 equal the distance between A and B, let d_2 equal the distance between B and C, and let d_3 equal the distance between A and C. You can find these distances as follows.

$$d_1 = \sqrt{(5 - 2)^2 + (2 - 6)^2} = \sqrt{9 + 16} = \sqrt{25} = 5$$

$$d_2 = \sqrt{(8 - 5)^2 + (-2 - 2)^2} = \sqrt{9 + 16} = \sqrt{25} = 5$$

$$d_3 = \sqrt{(8 - 2)^2 + (-2 - 6)^2} = \sqrt{36 + 64} = \sqrt{100} = 10$$

So, the set of points is collinear because $d_1 + d_2 = 5 + 5 = 10 = d_3$. These three points are plotted in Figure 3.12. From the figure, the points appear to lie on the same line.

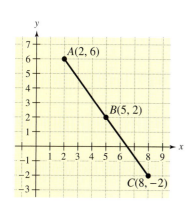

Figure 3.12

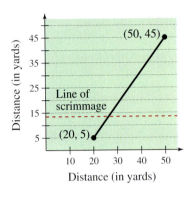

Figure 3.13

Example 10 Finding the Length of a Pass

A football quarterback throws a pass from the five-yard line, 20 yards from the sideline. The pass is caught by a wide receiver on the 45-yard line, 50 yards from the same sideline, as shown in Figure 3.13. How long is the pass?

Solution

You can find the length of the pass by finding the distance between points $(20, 5)$ and $(50, 45)$.

$$d = \sqrt{(50 - 20)^2 + (45 - 5)^2}$$ Substitute coordinates of points into Distance Formula.

$$= \sqrt{900 + 1600} = 50$$ Simplify.

So, the pass is 50 yards long.

In Example 10, the scale along the goal line does not normally appear on a football field. However, when you use coordinate geometry to solve real-life problems, you are free to place the coordinate system in any way that is convenient to the solution of the problem.

4 Use the Midpoint Formula to find the midpoints of line segments.

The Midpoint Formula

The **midpoint** of a line segment that joins two points is the point that divides the segment into two equal parts. To find the midpoint of the line segment that joins two points in a coordinate plane, you can simply find the average values of the respective coordinates of the two endpoints using the **Midpoint Formula.**

The Midpoint Formula

The midpoint of the line segment joining the points (x_1, y_1) and (x_2, y_2) is given by the Midpoint Formula

$$\text{Midpoint} = \left(\frac{x_1 + x_2}{2}, \frac{y_1 + y_2}{2} \right).$$

Example 11 Finding the Midpoint of a Line Segment

Find the midpoint of the line segment joining the points $(-5, -3)$ and $(9, 3)$.

Solution

Let $(x_1, y_1) = (-5, -3)$ and let $(x_2, y_2) = (9, 3)$.

$$\text{Midpoint} = \left(\frac{x_1 + x_2}{2}, \frac{y_1 + y_2}{2} \right)$$ Midpoint Formula

$$= \left(\frac{-5 + 9}{2}, \frac{-3 + 3}{2} \right)$$ Substitute for x_1, y_1, x_2, and y_2.

$$= (2, 0)$$ Simplify.

The line segment and the midpoint are shown in Figure 3.14.

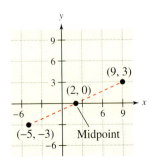

Figure 3.14

3.1 Exercises

Review   Concepts, Skills, and Problem Solving

Keep mathematically in shape by doing these exercises *before* the problems of this section.

Properties and Definitions

1. *Writing* ✐   (a) Is $3x = 7$ a linear equation? Explain.
(b) Is $x^2 + 3x = 2$ a linear equation? Explain.

2. *Writing* ✐   Explain how to check if $x = 3$ is a solution of the equation $5x - 4 = 11$.

Simplifying Expressions

In Exercises 3–10, simplify the expression.

3. $5x - 7 + x + 1 - 2x$

4. $18 - y + 2y^2 - 6 + 4y$

5. $3t + 7 - 2t + 4$

6. $-6m^2 + 8m - 10 + 5m^2$

7. $4 - 3(2x + 1)$

8. $5(x + 2) - 4(2x - 3)$

9. $24\left(\dfrac{y}{3} + \dfrac{y}{6}\right)$

10. $0.12x + 0.05(2000 - 2x)$

Problem Solving

11. *Work Rate*   You can mow a lawn in 4 hours, and your friend can mow it in 5 hours. What fractional part of the lawn can each of you mow in 1 hour? Working together, how long will it take to mow the lawn?

12. *Average Speed*   A truck driver traveled at an average speed of 50 miles per hour on a 200-mile trip. On the return trip with the truck fully loaded, the average speed was 42 miles per hour. Find the average speed for the round trip.

Developing Skills

In Exercises 1–8, plot the points on a rectangular coordinate system. See Example 1.

1. $(4, 3), (-5, 3), (3, -5)$ **2.** $(-2, 5), (-2, -5), (3, 5)$

3. $(-8, -2), (6, -2), (6, 5)$

4. $(0, 4), (0, 0), (3, 0)$ **5.** $\left(\frac{5}{2}, -2\right), \left(-2, \frac{1}{4}\right), \left(\frac{3}{2}, -\frac{7}{2}\right)$

6. $\left(-\frac{2}{3}, 3\right), \left(\frac{1}{4}, -\frac{5}{4}\right), \left(-5, -\frac{7}{4}\right)$

7. $\left(\frac{3}{2}, 1\right), (4, -3), \left(-\frac{4}{3}, \frac{7}{3}\right)$ **8.** $(-3, -5), \left(\frac{9}{4}, \frac{3}{4}\right), \left(\frac{5}{2}, -2\right)$

In Exercises 9–12, determine the coordinates of the points. See Example 2.

9.

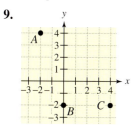

10.

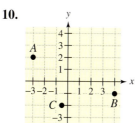

11.   **12.**

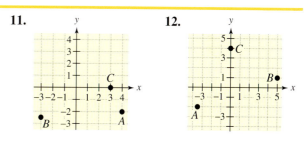

In Exercises 13–20, plot the points and connect them with line segments to form the figure. (*Note:* A *rhombus* is a parallelogram whose sides are all of the same length.)

13. *Square:* $(0, 6), (3, 3), (0, 0), (-3, 3)$

14. *Rectangle:* $(7, 0), (9, 1), (4, 6), (6, 7)$

15. *Triangle:* $(-1, 2), (2, 0), (3, 5)$

16. *Triangle:* $(1, 4), (0, -1), (5, 9)$

17. *Parallelogram:* $(4, 0), (6, -2), (0, -4), (-2, -2)$

18. *Parallelogram:* $(-1, 1), (0, 4), (4, -2), (5, 1)$

19. *Rhombus:* $(0, 0), (3, 2), (2, 3), (5, 5)$

20. *Rhombus:* $(-3, -3), (-2, -1), (-1, -2), (0, 0)$

In Exercises 21–28, find the coordinates of the point.

21. The point is located five units to the left of the *y*-axis and two units above the *x*-axis.

22. The point is located 10 units to the right of the *y*-axis and four units below the *x*-axis.

23. The point is located three units to the right of the *y*-axis and two units below the *x*-axis.

24. The point is located four units to the right of the *y*-axis and six units above the *x*-axis.

25. The coordinates of the point are equal, and the point is located in the third quadrant eight units to the left of the *y*-axis.

26. The coordinates of the point are equal in magnitude and opposite in sign, and the point is located seven units to the right of the *y*-axis.

27. The point is on the positive *x*-axis 10 units from the origin.

28. The point is on the negative *y*-axis five units from the origin.

In Exercises 29–42, determine the quadrant in which the point is located without plotting it. (*x* and *y* are real numbers.)

29. $(-3, -5)$

30. $(4, -2)$

31. $\left(-15, \frac{7}{8}\right)$

32. $\left(\frac{5}{11}, \frac{3}{8}\right)$

33. $(-200, -1365.6)$

34. $(-6.2, 8.05)$

35. (x, y), $x > 0, y < 0$

36. (x, y), $x > 0, y > 0$

37. $(x, 4)$

38. $(10, y)$

39. $(-3, y)$

40. $(x, -6)$

41. (x, y), $xy > 0$

42. (x, y), $xy < 0$

In Exercises 43–46, sketch a scatter plot of the points whose coordinates are shown in the table. See Example 3.

43. *Exam Scores*   The table shows the times *x* in hours invested in studying for five different algebra exams and the resulting scores *y*.

x	5	2	3	6.5	4
y	81	71	88	92	86

44. *Net Sales*   The net sales *y* (in billions of dollars) of Wal-Mart for the years 1996 through 2002 are shown in the table. The time in years is given by *x*. (Source: Wal-Mart 2002 Annual Report)

x	1996	1997	1998	1999	2000	2001	2002
y	93.6	104.9	118.0	137.6	165.0	191.3	217.8

45. *Meteorology*   The table shows the monthly normal temperatures *y* (in degrees Fahrenheit) for Duluth, Minnesota for each month of the year, with $x = 1$ representing January. (Source: National Climatic Data Center)

x	1	2	3	4	5	6
y	7	12	24	39	51	60

x	7	8	9	10	11	12
y	66	64	54	44	28	13

46. *Fuel Efficiency*   The table shows various speeds *x* of a car in miles per hour and the corresponding approximate fuel efficiencies *y* in miles per gallon.

x	50	55	60	65	70
y	28	26.4	24.8	23.4	22

Shifting a Graph   In Exercises 47 and 48, the figure is shifted to a new location in the plane. Find the coordinates of the vertices of the figure in its new location.

47.

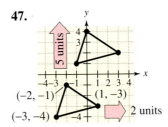

48.

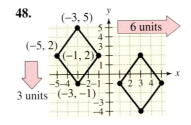

In Exercises 49–52, complete the table of values. Then plot the solution points on a rectangular coordinate system. See Example 4.

49.

x	-2	0	2	4	6
$y = \frac{3}{4}x + 2$					

50.

x	-6	-3	0	$\frac{3}{4}$	10
$y = \frac{4}{3}x - \frac{1}{3}$					

51.

x	-2	0	2	4	6
$y = \lvert 3x - 4 \rvert + 1$					

52.

x	-2	0	2	4	6
$y = 5x - 1$					

In Exercises 53 and 54, use the *table* feature of a graphing calculator to complete the table of values.

53.

x	-4	$\frac{2}{5}$	4	8	12
$y = -\frac{5}{2}x + 4$					

54.

x	-2	0	2	4	6
$y = 4x^2 + x - 2$					

In Exercises 55–60, determine whether each ordered pair is a solution of the equation. See Example 5.

55. $x^2 + 3y = -5$
(a) $(3, -2)$
(b) $(-2, -3)$
(c) $(3, -5)$
(d) $(4, -7)$

56. $y^2 - 4x = 8$
(a) $(0, 6)$
(b) $(-4, 2)$
(c) $(-1, 3)$
(d) $(7, 6)$

57. $4y - 2x + 1 = 0$
(a) $(0, 0)$
(b) $\left(\frac{1}{2}, 0\right)$
(c) $\left(-3, -\frac{7}{4}\right)$
(d) $\left(1, -\frac{3}{4}\right)$

58. $5x - 2y + 50 = 0$
(a) $(-10, 0)$
(b) $(-5, 5)$
(c) $(0, 25)$
(d) $(20, -2)$

59. $y = \frac{7}{8}x + 3$
(a) $\left(\frac{8}{7}, 4\right)$
(b) $(8, 10)$
(c) $(0, 0)$
(d) $(-16, 14)$

60. $y = \frac{5}{8}x - 2$
(a) $(0, 0)$
(b) $(8, 3)$
(c) $(16, -7)$
(d) $\left(-\frac{8}{5}, 3\right)$

In Exercises 61–68, plot the points and find the distance between them. State whether the points lie on a horizontal or a vertical line. See Example 6.

61. $(3, -2), (3, 5)$

62. $(-2, 8), (-2, 1)$

63. $(3, 2), (10, 2)$

64. $(-120, -2), (130, -2)$

65. $\left(-3, \frac{3}{2}\right), \left(-3, \frac{9}{4}\right)$

66. $\left(\frac{3}{4}, 1\right), \left(\frac{3}{4}, -10\right)$

67. $\left(-4, \frac{1}{3}\right), \left(\frac{5}{2}, \frac{1}{3}\right)$

68. $\left(\frac{1}{2}, \frac{7}{8}\right), \left(\frac{11}{2}, \frac{7}{8}\right)$

In Exercises 69–78, find the distance between the points. See Example 7.

69. $(3, 7), (4, 5)$ **70.** $(5, 2), (8, 3)$

71. $(1, 3), (5, 6)$ **72.** $(3, 10), (15, 5)$

73. $(-3, 0), (4, -3)$ **74.** $(0, -5), (2, -8)$

75. $(-2, -3), (4, 2)$ **76.** $(-5, 4), (10, -3)$

77. $(1, 3), (3, -2)$ **78.** $\left(\frac{1}{2}, 1\right), \left(\frac{3}{2}, 2\right)$

▲ *Geometry*   In Exercises 79–82, show that the points are vertices of a right triangle. See Example 8.

79.

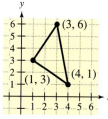

80.

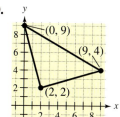

81.

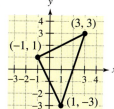

82.

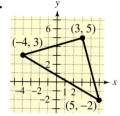

In Exercises 83–86, use the Distance Formula to determine whether the three points are collinear. See Example 9.

83. $(2, 3), (2, 6), (6, 3)$

84. $(2, 4), (-1, 6), (-3, 1)$

85. $(8, 3), (5, 2), (2, 1)$

86. $(2, 4), (1, 1), (0, -2)$

▲ *Geometry* In Exercises 87 and 88, find the perimeter of the triangle with the given vertices.

87. $(-2, 0), (0, 5), (1, 0)$

88. $(-5, -2), (-1, 4), (3, -1)$

In Exercises 89–92, find the midpoint of the line segment joining the points, and then plot the points and the midpoint. See Example 11.

89. $(-2, 0), (4, 8)$

90. $(-3, -2), (7, 2)$

91. $(1, 6), (6, 3)$

92. $(2, 7), (9, -1)$

Solving Problems

93. *Numerical Interpretation* For a handyman to install x windows in your home, the cost y is given by $y = 200x + 450$. Use x-values of 1, 2, 3, 4, and 5 to construct a table of values for y. Then use the table to help describe the relationship between the number of windows x and the cost of installation y.

94. *Numerical Interpretation* When an employee works x hours of overtime in a week, the employee's weekly pay y is given by $y = 15x + 600$. Use x-values of 0, 1, 2, 3, and 4 to construct a table of values for y. Then use the table to help describe the relationship between the number of overtime hours x and the weekly pay y.

95. *Football Pass* A football quarterback throws a pass from the 10-yard line, 10 yards from the sideline. The pass is caught by a wide receiver on the 40-yard line, 35 yards from the same sideline, as shown in the figure. How long is the pass? See Example 10.

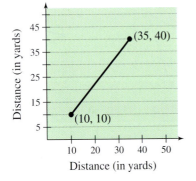

96. *Soccer Pass* A soccer player passes the ball from a point that is 18 yards from the endline and 12 yards from the sideline. The pass is received by a teammate who is 42 yards from the same endline and 50 yards from the same sideline, as shown in the figure. How long is the pass?

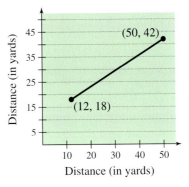

97. *Fuel Efficiency* The scatter plot shows the speed of a car in kilometers per hour and the amount of fuel, in liters used per 100 kilometers traveled, that the car needs to maintain that speed. From the graph, how would you describe the relationship between the speed of the car and the fuel used?

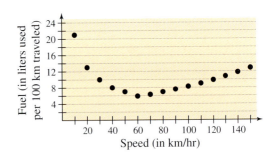

98. *Heating a Home* A family wished to study the relationship between the temperature outside and the amount of natural gas used to heat their house. For 16 months, the family observed the average temperature in degrees Fahrenheit for the month and the average amount of natural gas in hundreds of cubic feet used in that month. The scatter plot shows this data for the 16 months. Does the graph suggest that there is a strong relationship between the temperature outside and the amount of natural gas used? If a month were to have an average temperature of 45°F, about how much natural gas would you expect this family to use on average to heat their house for that month?

99. *Net Sales* PetsMart, Inc. had net sales of $2110 million in 1999 and $2501 million in 2001. Use the Midpoint Formula to estimate the net sales in 2000. (Source: PetsMart, Inc. 2001 Annual Report)

100. *Revenue* Best Buy Co. had annual revenues of $12,494 million in 2000 and $19,597 million in 2002. Use the Midpoint Formula to estimate the revenues in 2001. (Source: Best Buy Co., Inc. 2002 Annual Report)

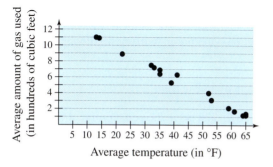

Explaining Concepts

101. *Writing* Discuss the significance of the word *ordered* when referring to an ordered pair (x, y).

102. *Writing* When plotting the point (x, y), what does the x-coordinate measure? What does the y-coordinate measure?

103. What is the x-coordinate of any point on the y-axis? What is the y-coordinate of any point on the x-axis?

104. *Writing* Explain why the ordered pair $(-3, 4)$ is not a solution point of the equation $y = 4x + 15$.

105. *Writing* When points are plotted on the rectangular coordinate system, is it true that the scales on the x- and y-axes must be the same? Explain.

106. State the Pythagorean Theorem and give examples of its use.

107. *Conjecture* Plot the points $(2, 1), (-3, 5)$, and $(7, -3)$ on a rectangular coordinate system. Then change the sign of the x-coordinate of each point and plot the three new points on the same rectangular coordinate system. What conjecture can you make about the location of a point when the sign of the x-coordinate is changed?

108. *Conjecture* Plot the points $(2, 1), (-3, 5)$, and $(7, -3)$ on a rectangular coordinate system. Then change the sign of the y-coordinate of each point and plot the three new points on the same rectangular coordinate system. What conjecture can you make about the location of a point when the sign of the y-coordinate is changed?

3.2 Graphs of Equations

Peter Grumann/Getty Images

What You Should Learn

1. Sketch graphs of equations using the point-plotting method.
2. Find and use *x*- and *y*-intercepts as aids to sketching graphs.
3. Use a pattern to write an equation for an application problem, and sketch its graph.

Why You Should Learn It

The graph of an equation can help you see relationships between real-life quantities. For instance, in Exercise 84 on page 147, a graph can be used to relate the change in U.S. farm assets to time.

1 Sketch graphs of equations using the point-plotting method.

The Graph of an Equation

In Section 3.1, you saw that the solutions of an equation in *x* and *y* can be represented by points on a rectangular coordinate system. The set of *all* solutions of an equation is called its **graph.** In this section, you will study a basic technique for sketching the graph of an equation—the **point-plotting method.**

Example 1   Sketching the Graph of an Equation

To sketch the graph of $3x - y = 2$, first solve the equation for *y*.

$$3x - y = 2 \qquad \text{Write original equation.}$$
$$-y = -3x + 2 \qquad \text{Subtract } 3x \text{ from each side.}$$
$$y = 3x - 2 \qquad \text{Divide each side by } -1.$$

Next, create a table of values. The choice of *x*-values to use in the table is somewhat arbitrary. However, the more *x*-values you choose, the easier it will be to recognize a pattern.

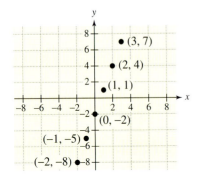

Figure 3.15

x	-2	-1	0	1	2	3
$y = 3x - 2$	-8	-5	-2	1	4	7
Solution point	$(-2, -8)$	$(-1, -5)$	$(0, -2)$	$(1, 1)$	$(2, 4)$	$(3, 7)$

Now, plot the solution points, as shown in Figure 3.15. It appears that all six points lie on a line, so complete the sketch by drawing a line through the points, as shown in Figure 3.16.

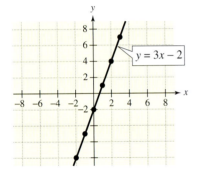

Figure 3.16

The equation in Example 1 is an example of a **linear equation** in two variables—the variables are raised to the first power and the graph of the equation is a line. By drawing a line through the plotted points, you are implying that every point on this line is a solution point of the given equation.

Technology: Discovery

Use a graphing calculator to graph each equation, and then answer the questions.

i. $y = 3x + 2$

ii. $y = 4 - x$

iii. $y = x^2 + 3x$

iv. $y = x^2 - 5$

v. $y = |x - 4|$

vi. $y = |x + 1|$

a. Which of the graphs are lines?

b. Which of the graphs are U-shaped?

c. Which of the graphs are V-shaped?

d. Describe the graph of the equation $y = x^2 + 7$ before you graph it. Use a graphing calculator to confirm your answer.

The Point-Plotting Method of Sketching a Graph

1. If possible, rewrite the equation by isolating one of the variables.
2. Make a table of values showing several solution points.
3. Plot these points on a rectangular coordinate system.
4. Connect the points with a smooth curve or line.

Example 2   Sketching the Graph of a Nonlinear Equation

Sketch the graph of $-x^2 + 2x + y = 0$.

Solution

Begin by solving the equation for y.

$$-x^2 + 2x + y = 0 \qquad \text{Write original equation.}$$

$$2x + y = x^2 \qquad \text{Add } x^2 \text{ to each side.}$$

$$y = x^2 - 2x \qquad \text{Subtract } 2x \text{ from each side.}$$

Next, create a table of values.

x	-2	-1	0	1	2	3	4
$y = x^2 - 2x$	8	3	0	-1	0	3	8
Solution point	$(-2, 8)$	$(-1, 3)$	$(0, 0)$	$(1, -1)$	$(2, 0)$	$(3, 3)$	$(4, 8)$

Study Tip

Example 2 shows three common ways to represent the relationship between two variables. The equation $y = x^2 - 2x$ is the *analytical* or *algebraic* representation, the table of values is the *numerical* representation, and the graph in Figure 3.18 is the *graphical* representation. You will see and use analytical, numerical, and graphical representations throughout this course.

Now, plot the seven solution points, as shown in Figure 3.17. Finally, connect the points with a smooth curve, as shown in Figure 3.18.

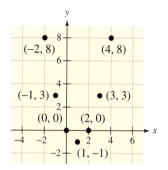

Figure 3.17

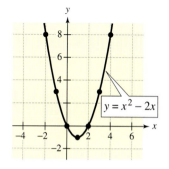

Figure 3.18

The graph of the equation given in Example 2 is called a **parabola.** You will study this type of graph in detail in a later chapter.

Example 3 examines the graph of an equation that involves an absolute value. Remember that the absolute value of a number is its distance from zero on the real number line. For instance, $|-5| = 5$, $|2| = 2$, and $|0| = 0$.

Example 3   The Graph of an Absolute Value Equation

Sketch the graph of $y = |x - 2|$.

Solution

This equation is already written in a form with y isolated on the left. So begin by creating a table of values. Be sure that you understand how the absolute value is evaluated. For instance, when $x = -2$, the value of y is

$$y = |-2 - 2| \qquad \text{Substitute } -2 \text{ for } x.$$
$$= |-4| \qquad \text{Simplify.}$$
$$= 4 \qquad \text{Simplify.}$$

and when $x = 3$, the value of y is

$$y = |3 - 2| \qquad \text{Substitute } 3 \text{ for } x.$$
$$= |1| \qquad \text{Simplify.}$$
$$= 1. \qquad \text{Simplify.}$$

x	-2	-1	0	1	2	3	4	5
$y = |x - 2|$	4	3	2	1	0	1	2	3
Solution point	$(-2, 4)$	$(-1, 3)$	$(0, 2)$	$(1, 1)$	$(2, 0)$	$(3, 1)$	$(4, 2)$	$(5, 3)$

Next, plot the solution points, as shown in Figure 3.19. It appears that the points lie in a "V-shaped" pattern, with the point $(2, 0)$ lying at the bottom of the "V." Connect the points to form the graph shown in Figure 3.20.

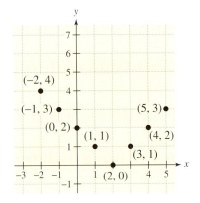

Figure 3.19

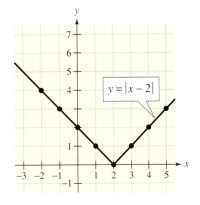

Figure 3.20

② Find and use x- and y-intercepts as aids to sketching graphs.

Intercepts: Aids to Sketching Graphs

Two types of solution points that are especially useful are those having zero as the *y*-coordinate and those having zero as the *x*-coordinate. Such points are called **intercepts** because they are the points at which the graph intersects, respectively, the *x*- and *y*-axes.

Study Tip

When you create a table of values for a graph, include any intercepts you have found. You should also include points to the left and right of the intercepts of the graph. This helps to give a more complete view of the graph.

Definition of Intercepts

The point $(a, 0)$ is called an **x-intercept** of the graph of an equation if it is a solution point of the equation. To find the *x*-intercept(s), let $y = 0$ and solve the equation for *x*.

The point $(0, b)$ is called a **y-intercept** of the graph of an equation if it is a solution point of the equation. To find the *y*-intercept(s), let $x = 0$ and solve the equation for *y*.

Example 4   Finding the Intercepts of a Graph

Find the intercepts and sketch the graph of $y = 2x - 3$.

Solution

Find the *x*-intercept by letting $y = 0$ and solving for *x*.

$$y = 2x - 3 \qquad \text{Write original equation.}$$

$$0 = 2x - 3 \qquad \text{Substitute 0 for } y.$$

$$3 = 2x \qquad \text{Add 3 to each side.}$$

$$\tfrac{3}{2} = x \qquad \text{Solve for } x.$$

Find the *y*-intercept by letting $x = 0$ and solving for *y*.

$$y = 2x - 3 \qquad \text{Write original equation.}$$

$$y = 2(0) - 3 \qquad \text{Substitute 0 for } x.$$

$$y = -3 \qquad \text{Solve for } y.$$

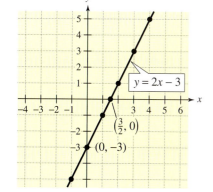

Figure 3.21

So, the graph has one *x*-intercept, which occurs at the point $\left(\tfrac{3}{2}, 0\right)$, and one *y*-intercept, which occurs at the point $(0, -3)$. To sketch the graph of the equation, create a table of values. (Include the intercepts in the table.) Finally, using the solution points given in the table, sketch the graph of the equation, as shown in Figure 3.21.

x	-1	0	1	$\tfrac{3}{2}$	2	3	4
$y = 2x - 3$	-5	-3	-1	0	1	3	5
Solution point	$(-1, -5)$	$(0, -3)$	$(1, -1)$	$\left(\tfrac{3}{2}, 0\right)$	$(2, 1)$	$(3, 3)$	$(4, 5)$

3 Use a pattern to write an equation for an application problem, and sketch its graph.

Real-Life Application of Graphs

Newspapers and news magazines frequently use graphs to show real-life relationships between variables. Example 5 shows how such a graph can help you visualize the concept of **straight-line depreciation.**

Example 5   Straight-Line Depreciation: Finding the Pattern

Your small business buys a new printing press for $65,000. For income tax purposes, you decide to depreciate the printing press over a 10-year period. At the end of the 10 years, the salvage value of the printing press is expected to be $5000.

a. Find an equation that relates the depreciated value of the printing press to the number of years since it was purchased.

b. Sketch the graph of the equation.

c. What is the y-intercept of the graph and what does it represent?

Solution

a. The total depreciation over the 10-year period is

$$\$65,000 - 5000 = \$60,000.$$

Because the same amount is depreciated each year, it follows that the annual depreciation is

$$\frac{\$60,000}{10} = \$6000.$$

So, after 1 year, the value of the printing press is

Value after 1 year $= \$65,000 - (1)6000 = \$59,000.$

By similar reasoning, you can see that the values after 2, 3, and 4 years are

Value after 2 years $= \$65,000 - (2)6000 = \$53,000$

Value after 3 years $= \$65,000 - (3)6000 = \$47,000$

Value after 4 years $= \$65,000 - (4)6000 = \$41,000.$

Let y represent the value of the printing press after t years and follow the pattern determined for the first 4 years to obtain

$$y = 65,000 - 6000t.$$

b. A sketch of the graph of the depreciation equation is shown in Figure 3.22.

c. To find the y-intercept of the graph, let $t = 0$ and solve the equation for y.

$y = 65,000 - 6000t$	Write original equation.
$y = 65,000 - 6000(0)$	Substitute 0 for t.
$y = 65,000$	Simplify.

So, the y-intercept is $(0, 65,000)$ and it corresponds to the initial value of the printing press.

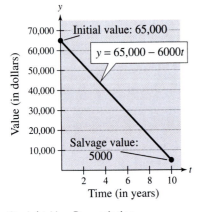

Straight-Line Depreciation

Figure 3.22

3.2 Exercises

Review Concepts, Skills, and Problem Solving

Keep mathematically in shape by doing these exercises *before* the problems of this section.

Properties and Definitions

1. If $t - 3 > 7$ and c is an algebraic expression, then what is the relationship between $t - 3 + c$ and $7 + c$?

2. If $t - 3 < 7$ and $c < 0$, then what is the relationship between $(t - 3)c$ and $7c$?

3. Complete the Multiplicative Inverse Property: $y(1/y) = $ ____

4. Identify the property illustrated by $u + v = v + u$.

Simplifying Expressions

In Exercises 5–10, solve the inequality.

5. $3x + 9 \geq 12$

6. $5 - 3x > 14$

7. $-4 < 10x + 1 < 6$

8. $-2 \leq 1 - 2x \leq 2$

9. $-3 \leq -\dfrac{x}{2} \leq 3$

10. $-5 < x - 25 < 5$

Problem Solving

11. *Inflation* The price of a new van is approximately 112% of what it was 3 years ago. What was the approximate price 3 years ago if the current price is $32,500?

12. *Gross Income* Your employer withholds $3\frac{1}{2}\%$ of your gross income for medical insurance coverage. Determine the amount withheld each month if your gross monthly income is $3100.

Developing Skills

In Exercises 1–6, match the equation with its graph. [The graphs are labeled (a), (b), (c), (d), (e), and (f).]

(a)
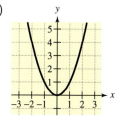

(b)
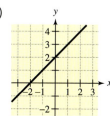

(c)
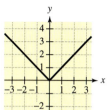

(d)
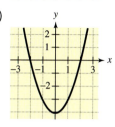

(e)
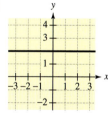

(f)
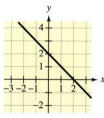

1. $y = 2$

2. $y = 2 + x$

3. $y = 2 - x$

4. $y = x^2$

5. $y = x^2 - 4$

6. $y = |x|$

In Exercises 7–30, sketch the graph of the equation. See Examples 1–3.

7. $y = 3x$

8. $y = -2x$

9. $y = 4 - x$

10. $y = x - 7$

11. $2x - y = 3$

12. $3x - y = -2$

13. $3x + 2y = 2$

14. $2y + 5x = 6$

15. $y = -x^2$

16. $y = x^2$

17. $y = x^2 - 3$

18. $y = 4 - x^2$

19. $-x^2 - 3x + y = 0$

20. $-x^2 + x + y = 0$

21. $x^2 - 2x - y = 1$

22. $x^2 + 3x - y = 4$

23. $y = |x|$

24. $y = -|x|$

25. $y = |x| + 3$

26. $y = |x| - 1$

27. $y = |x + 3|$

28. $y = |x - 1|$

29. $y = x^3$

30. $y = -x^3$

In Exercises 31–44, find the x- and y-intercepts (if any) of the graph of the equation. See Example 4.

31. $y = 6x - 3$

32. $y = 4 - 3x$

33. $y = 12 - \frac{2}{5}x$

34. $y = \frac{3}{4}x + 15$

35. $x + 2y = 10$

36. $3x - 2y = 12$

37. $4x - y + 3 = 0$

38. $2x + 3y - 8 = 0$

39. $y = |x| - 1$

40. $y = |x| + 4$

41. $y = -|x + 5|$

42. $y = |x - 4|$

43. $y = |x - 1| - 3$

44. $y = |x + 3| - 1$

In Exercises 45–50, graphically estimate the x- and y-intercepts (if any) of the graph.

45. $2x + 3y = 6$

46. $3x - y + 9 = 0$

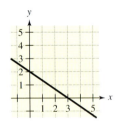

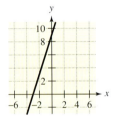

47. $y = x^2 + 3$

48. $y = -x^2 + 4x$

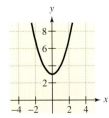

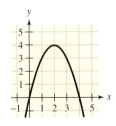

49. $y = -2$

50. $x = 3$

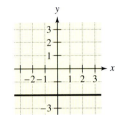

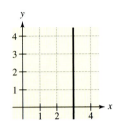

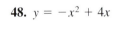 In Exercises 51–56, use a graphing calculator to graph the equation. Approximate the x- and y-intercepts (if any).

51. $y = 8 - 4x$

52. $y = 5x - 20$

53. $y = (x - 1)(x - 6)$

54. $y = (x + 2)(x - 3)$

55. $y = |4x + 6| - 2$

56. $y = |2x - 4| + 1$

In Exercises 57–78, sketch the graph of the equation and show the coordinates of three solution points (including x- and y-intercepts).

57. $y = 3 - x$

58. $y = x - 3$

59. $y = 2x - 3$

60. $y = -4x + 8$

61. $4x + y = 3$

62. $y - 2x = -4$

63. $2x - 3y = 6$

64. $3x + 4y = 12$

65. $x + 5y = 10$

66. $5x - y = 10$

67. $y = x^2 - 9$

68. $y = 9 - x^2$

69. $y = 1 - x^2$

70. $y = x^2 - 4$

71. $y = x(x - 2)$

72. $y = -x(x + 4)$

73. $y = |x| - 3$

74. $y = |x| + 2$

75. $y = |x + 2|$

76. $y = |x - 3|$

77. $y = -|x| + |x + 1|$

78. $y = |x| + |x - 2|$

Solving Problems

79. *Straight-Line Depreciation* A manufacturing plant purchases a new molding machine for $225,000. The depreciated value y after t years is given by

$$y = 225,000 - 20,000t, \quad 0 \le t \le 8.$$

Sketch a graph of this model.

80. *Straight-Line Depreciation* A manufacturing plant purchases a new computer system for $20,000. The depreciated value y after t years is given by

$$y = 20,000 - 3000t, \quad 0 \le t \le 6.$$

Sketch a graph of this model.

81. *Straight-Line Depreciation* Your company purchases a new delivery van for $40,000. For tax purposes, the van will be depreciated over a seven-year period. At the end of 7 years, the value of the van is expected to be $5000.

(a) Find an equation that relates the depreciated value of the van to the number of years since it was purchased.

(b) Sketch the graph of the equation.

(c) What is the y-intercept of the graph and what does it represent?

82. *Straight-Line Depreciation* Your company purchases a new limousine for $55,000. For tax purposes, the limousine will be depreciated over a 10-year period. At the end of 10 years, the value of the limousine is expected to be $10,000.

(a) Find an equation that relates the depreciated value of the limousine to the number of years since it was purchased.

(b) Sketch the graph of the equation.

(c) What is the y-intercept of the graph and what does it represent?

83. *Hooke's Law* The force F (in pounds) required to stretch a spring x inches from its natural length is given by

$$F = \frac{4}{3}x, \quad 0 \le x \le 12.$$

(a) Use the model to complete the table.

x	0	3	6	9	12
F					

(b) Sketch a graph of the model.

(c) Determine the required change in F if x is doubled. Explain your reasoning.

84. *Agriculture* The total assets A (in billions of dollars) for the farming sector of the United States for the years 1993 through 2000 are shown in the table. (Source: U.S. Department of Agriculture)

Year	1993	1994	1995	1996
A	910	936	968	1005

Year	1997	1998	1999	2000
A	1053	1086	1141	1188

A model for this data is

$$A = 40.2t + 775$$

where t is time in years, with $t = 3$ corresponding to 1993.

(a) Sketch a graph of the model and plot the data on the same set of coordinate axes.

(b) How well does the model represent the data? Explain your reasoning.

(c) Use the model to predict the total assets in the farming sector in the year 2005.

85. *Misleading Graphs* Graphs can help you visualize relationships between two variables, but they can also be misused to imply results that are not correct. The two graphs below represent the same data points. Why do the graphs appear different? Identify ways in which this could be misleading.

86. *Exploration* Graph the equations $y = x^2 + 1$ and $y = -(x^2 + 1)$ on the same set of coordinate axes. Explain how the graph of an equation changes when the expression for y is multiplied by -1. Justify your answer by giving additional examples.

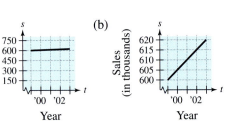

Explaining Concepts

87. *Writing* Define the graph of an equation.

88. *Writing* How many solution points make up the graph of $y = 2x - 1$? Explain.

89. An equation gives the relationship between profit y and time t. Profit has been decreasing at a lower rate than in the past. Is it possible to sketch the graph of such an equation? If so, sketch a representative graph.

90. *Writing* Explain how to find the intercepts of a graph. Give examples.

91. *Writing* The graph shown represents the distance d in miles that a person drives during a 10-minute trip from home to work.

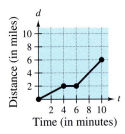

(a) How far is the person's home from the person's place of work? Explain.

(b) Describe the trip for time $4 < t < 6$. Explain.

(c) During what time interval is the person's speed greatest? Explain.

3.3 Slope and Graphs of Linear Equations

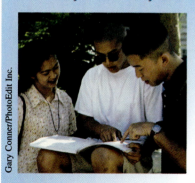

Gary Conner/PhotoEdit Inc.

Why You Should Learn It

Slopes of lines can be used to describe rates of change. For instance, in Exercise 87 on page 161, you will use slope to describe the average rate of change in the tuition and fees paid by college students.

What You Should Learn

1 Determine the slope of a line through two points.

2 Write linear equations in slope-intercept form and graph the equations.

3 Use slopes to determine whether two lines are parallel, perpendicular, or neither.

4 Use slopes to describe rates of change in real-life problems.

The Slope of a Line

1 Determine the slope of a line through two points.

The **slope** of a nonvertical line is the number of units the line rises or falls vertically for each unit of horizontal change from left to right. For example, the line in Figure 3.23 rises two units for each unit of horizontal change from left to right, and so this line has a slope of $m = 2$.

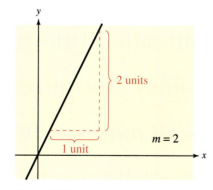

Figure 3.23

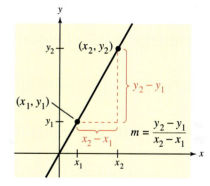

Figure 3.24

Definition of the Slope of a Line

The **slope** m of the nonvertical line passing through the points (x_1, y_1) and (x_2, y_2) is

$$m = \frac{y_2 - y_1}{x_2 - x_1} = \frac{\text{Change in } y}{\text{Change in } x} = \frac{\text{Rise}}{\text{Run}}$$

where $x_1 \neq x_2$ (see Figure 3.24).

When the formula for slope is used, the *order of subtraction* is important. Given two points on a line, you are free to label either of them (x_1, y_1) and the other (x_2, y_2). However, once this has been done, you must form the numerator and denominator using the same order of subtraction.

$$m = \frac{y_2 - y_1}{x_2 - x_1} \qquad m = \frac{y_1 - y_2}{x_1 - x_2} \qquad m = \frac{y_2 - y_1}{x_1 - x_2}$$

　　Correct　　　　　　Correct　　　　　　Incorrect

For instance, the slope of the line passing through the points $(-1, 3)$ and $(4, -2)$ is

$$m = \frac{y_2 - y_1}{x_2 - x_1} = \frac{-2 - 3}{4 - (-1)} = \frac{-5}{5} = -1$$

or

$$m = \frac{y_1 - y_2}{x_1 - x_2} = \frac{3 - (-2)}{-1 - 4} = \frac{5}{-5} = -1.$$

Example 1 Finding the Slope of a Line Through Two Points

Find the slope of the line passing through each pair of points.

a. $(1, 2)$ and $(4, 5)$ **b.** $(-1, 4)$ and $(2, 1)$

Solution

a. Let $(x_1, y_1) = (1, 2)$ and $(x_2, y_2) = (4, 5)$. The slope of the line through these points is

$$m = \frac{y_2 - y_1}{x_2 - x_1} \qquad \text{Difference in } y\text{-values}$$
$$ \text{Difference in } x\text{-values}$$

$$= \frac{5 - 2}{4 - 1}$$

$$= 1.$$

The graph of the line is shown in Figure 3.25.

b. The slope of the line through $(-1, 4)$ and $(2, 1)$ is

$$m = \frac{1 - 4}{2 - (-1)}$$

$$= \frac{-3}{3}$$

$$= -1.$$

The graph of the line is shown in Figure 3.26.

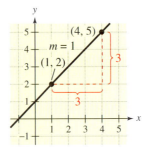

Positive Slope
Figure 3.25

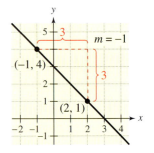

Negative Slope
Figure 3.26

Example 2   Finding the Slope of a Line Through Two Points

Find the slope of the line passing through each pair of points.

a. $(1, 4)$ and $(3, 4)$ **b.** $(3, 1)$ and $(3, 3)$

Solution

a. Let $(x_1, y_1) = (1, 4)$ and $(x_2, y_2) = (3, 4)$. The slope of the line through these points is

$$m = \frac{y_2 - y_1}{x_2 - x_1} \qquad \text{Difference in } y\text{-values}$$
$$\text{Difference in } x\text{-values}$$

$$= \frac{4 - 4}{3 - 1}$$

$$= \frac{0}{2} = 0.$$

The graph of the line is shown in Figure 3.27.

b. The slope of the line through $(3, 1)$ and $(3, 3)$ is undefined. Applying the formula for slope, you have

$$\frac{3 - 1}{3 - 3} = \frac{2}{0}. \qquad \text{Division by 0 is undefined.}$$

Because division by zero is not defined, the slope of a vertical line is not defined. The graph of the line is shown in Figure 3.28.

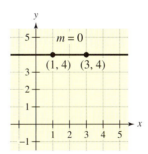

Zero Slope
Figure 3.27

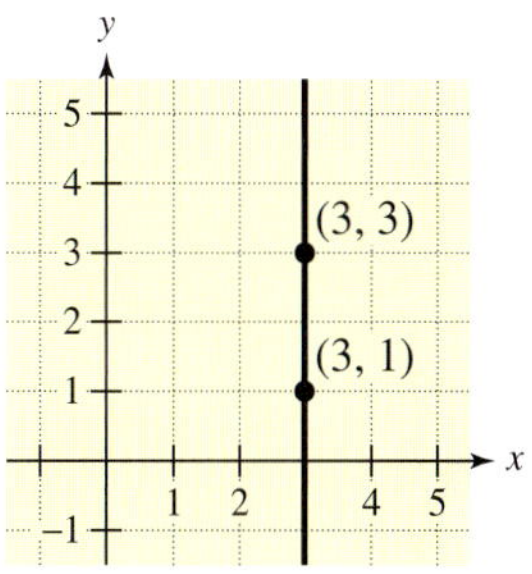

Slope is undefined.
Figure 3.28

From the slopes of the lines shown in Examples 1 and 2, you can make the following generalizations about the slope of a line.

Slope of a Line

1. A line with positive slope $(m > 0)$ *rises* from left to right.

2. A line with negative slope $(m < 0)$ *falls* from left to right.

3. A line with zero slope $(m = 0)$ is *horizontal.*

4. A line with undefined slope is *vertical.*

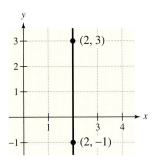

Vertical line
Undefined slope
Figure 3.29

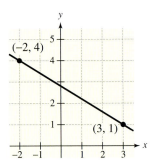

Line falls
Negative slope
Figure 3.30

Example 3   **Using Slope to Describe Lines**

Describe the line through each pair of points.

a. $(2, -1)$ and $(2, 3)$ **b.** $(-2, 4)$ and $(3, 1)$

Solution

a. Let $(x_1, y_1) = (2, -1)$ and $(x_2, y_2) = (2, 3)$.

$$m = \frac{3 - (-1)}{2 - 2} = \frac{4}{0}$$ Undefined slope (See Figure 3.29.)

Because the slope is undefined, the line is vertical.

b. Let $(x_1, y_1) = (-2, 4)$ and $(x_2, y_2) = (3, 1)$.

$$m = \frac{1 - 4}{3 - (-2)} = -\frac{3}{5} < 0$$ Negative slope (See Figure 3.30.)

Because the slope is negative, the line falls from left to right.

Example 4   **Using Slope to Describe Lines**

Describe the line through each pair of points.

a. $(1, 3)$ and $(4, 3)$ **b.** $(-1, 1)$ and $(2, 5)$

Solution

a. Let $(x_1, y_1) = (1, 3)$ and $(x_2, y_2) = (4, 3)$.

$$m = \frac{3 - 3}{4 - 1} = \frac{0}{3} = 0$$ Zero slope (See Figure 3.31.)

Because the slope is zero, the line is horizontal.

b. Let $(x_1, y_1) = (-1, 1)$ and $(x_2, y_2) = (2, 5)$.

$$m = \frac{5 - 1}{2 - (-1)} = \frac{4}{3} > 0$$ Positive slope (See Figure 3.32.)

Because the slope is positive, the line rises from the left to the right.

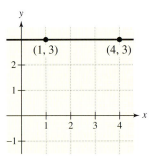

Horizontal line
Zero slope
Figure 3.31

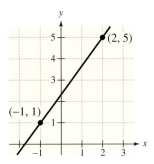

Line rises
Positive slope
Figure 3.32

Any two points on a nonvertical line can be used to calculate its slope. This is demonstrated in the next example.

Example 5   Finding the Slope of a Line

Sketch the graph of the line given by $2x + 3y = 6$. Then find the slope of the line. (Choose two different pairs of points on the line and show that the same slope is obtained from either pair.)

Solution

Begin by solving the equation for y.

$$2x + 3y = 6 \qquad \text{Write original equation.}$$

$$3y = -2x + 6 \qquad \text{Subtract } 2x \text{ from each side.}$$

$$y = \frac{-2x + 6}{3} \qquad \text{Divide each side by 3.}$$

$$y = -\frac{2}{3}x + 2 \qquad \text{Simplify.}$$

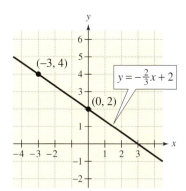

(a)

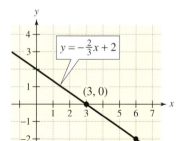

(b)

Figure 3.33

Then construct a table of values, as shown below.

x	-3	0	3	6
$y = -\frac{2}{3}x + 2$	4	2	0	-2
Solution point	$(-3, 4)$	$(0, 2)$	$(3, 0)$	$(6, -2)$

From the solution points shown in the table, sketch the graph of the line (see Figure 3.33). To calculate the slope of the line using two different sets of points, first use the points $(-3, 4)$ and $(0, 2)$, as shown in Figure 3.33(a), and obtain

$$m = \frac{2 - 4}{0 - (-3)} = -\frac{2}{3}.$$

Next, use the points $(3, 0)$ and $(6, -2)$, as shown in Figure 3.33(b), and obtain

$$m = \frac{-2 - 0}{6 - 3} = -\frac{2}{3}.$$

Try some other pairs of points on the line to see that you obtain a slope of $m = -\frac{2}{3}$ regardless of which two points you use.

Technology: Tip

Setting the viewing window on a graphing calculator affects the appearance of a line's slope. When you are using a graphing calculator, you cannot judge whether a slope is steep or shallow unless you use a *square setting*. See Appendix A for more information on setting a viewing window.

2 Write linear equations in slope-intercept form and graph the equations.

Slope as a Graphing Aid

You have seen that, before creating a table of values for an equation, you should first solve the equation for y. When you do this for a linear equation, you obtain some very useful information. Consider the results of Example 5.

$2x + 3y = 6$	Write original equation.
$3y = -2x + 6$	Subtract $2x$ from each side.
$y = -\dfrac{2}{3}x + 2$	Divide each side by 3.

Observe that the coefficient of x is the slope of the graph of this equation (see Example 5). Moreover, the constant term, 2, gives the y-intercept of the graph.

$$y = -\frac{2}{3}x + 2$$

Slope y-Intercept $(0, 2)$

This form is called the **slope-intercept form** of the equation of the line.

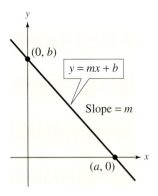

Figure 3.34

Slope-Intercept Form of the Equation of a Line

The graph of the equation

$$y = mx + b$$

is a line whose slope is m and whose y-intercept is $(0, b)$. (See Figure 3.34.)

Example 6   Slope and y-Intercept of a Line

Find the slope and y-intercept of the graph of the equation

$$4x - 5y = 15.$$

Solution

Begin by writing the equation in slope-intercept form, as follows.

$4x - 5y = 15$	Write original equation.
$-4x + 4x - 5y = -4x + 15$	Add $-4x$ to each side.
$-5y = -4x + 15$	Combine like terms.
$y = \dfrac{-4x + 15}{-5}$	Divide each side by -5.
$y = \dfrac{4}{5}x - 3$	Slope-intercept form

From the slope-intercept form, you can see that $m = \frac{4}{5}$ and $b = -3$. So, the slope of the graph of the equation is $\frac{4}{5}$ and the y-intercept is $(0, -3)$.

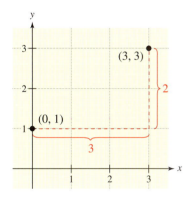

Figure 3.35

So far, you have been plotting several points in order to sketch the equation of a line. However, now that you can recognize equations of lines (linear equations), you don't have to plot as many points—two points are enough. (You might remember from geometry that *two points are all that are necessary to determine a line.*)

Example 7   Using the Slope and y-Intercept to Sketch a Line

Use the slope and y-intercept to sketch the graph of

$$y = \frac{2}{3}x + 1.$$

Solution

The equation is already in slope-intercept form.

$$y = mx + b$$

$$y = \frac{2}{3}x + 1 \qquad\qquad \text{Slope-intercept form}$$

So, the slope of the line is $m = \frac{2}{3}$ and the y-intercept is $(0, b) = (0, 1)$. Now you can sketch the graph of the line as follows. First, plot the y-intercept. Then, using a slope of $\frac{2}{3}$

$$m = \frac{2}{3} = \frac{\text{Change in } y}{\text{Change in } x}$$

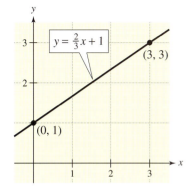

Figure 3.36

locate a second point on the line by moving three units to the right and two units upward (or two units upward and three units to the right), as shown in Figure 3.35. Finally, obtain the graph by drawing a line through the two points, as shown in Figure 3.36.

Example 8   Using the Slope and y-Intercept to Sketch a Line

Use the slope and y-intercept to sketch the graph of $12x + 3y = 6$.

Solution

Begin by writing the equation in slope-intercept form.

$$12x + 3y = 6 \qquad\qquad \text{Write original equation.}$$

$$3y = -12x + 6 \qquad\qquad \text{Subtract } 12x \text{ from each side.}$$

$$y = \frac{-12x + 6}{3} \qquad\qquad \text{Divide each side by 3.}$$

$$y = -4x + 2 \qquad\qquad \text{Slope-intercept form}$$

So, the slope of the line is $m = -4$ and the y-intercept is $(0, b) = (0, 2)$. Now you can sketch the graph of the line as follows. First, plot the y-intercept. Then, using a slope of -4, locate a second point on the line by moving one unit to the right and four units downward (or four units downward and one unit to the right). Finally, obtain the graph by drawing a line through the two points, as shown in Figure 3.37.

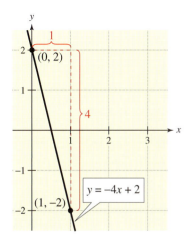

Figure 3.37

3 Use slopes to determine whether two lines are parallel, perpendicular, or neither.

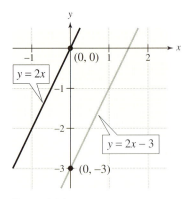

Figure 3.38

Parallel and Perpendicular Lines

You know from geometry that two lines in a plane are *parallel* if they do not intersect. What this means in terms of their slopes is suggested by Example 9.

Example 9   Lines That Have the Same Slope

On the same set of coordinate axes, sketch the lines given by

$$y = 2x \quad \text{and} \quad y = 2x - 3.$$

Solution

For the line given by $y = 2x$, the slope is $m = 2$ and the y-intercept is $(0, 0)$. For the line given by $y = 2x - 3$, the slope is also $m = 2$ and the y-intercept is $(0, -3)$. The graphs of these two lines are shown in Figure 3.38.

In Example 9, notice that the two lines have the same slope *and* appear to be parallel. The following rule states that this is always the case.

Parallel Lines

Two distinct nonvertical lines are parallel if and only if they have the same slope.

The phrase "if and only if" in this rule is used in mathematics as a way to write two statements in one. The first statement says that *if two distinct nonvertical lines have the same slope, they must be parallel.* The second statement says that *if two distinct nonvertical lines are parallel, they must have the same slope.*

Another rule from geometry is that two lines in a plane are perpendicular if they intersect at right angles. In terms of their slopes, this means that two nonvertical lines are perpendicular if their slopes are negative reciprocals of each other. For instance, the negative reciprocal of 5 is $-\frac{1}{5}$, so the lines

$$y = 5x + 2 \quad \text{and} \quad y = -\frac{1}{5}x - 4$$

are perpendicular to each other, as shown in Figure 3.39.

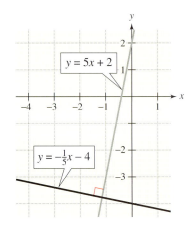

Figure 3.39

Perpendicular Lines

Consider two nonvertical lines whose slopes are m_1 and m_2. The two lines are perpendicular if and only if their slopes are *negative reciprocals* of each other. That is,

$$m_1 = -\frac{1}{m_2}, \quad \text{or equivalently,} \quad m_1 \cdot m_2 = -1.$$

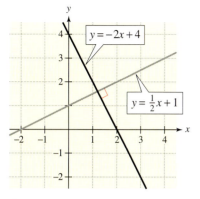

Figure 3.40

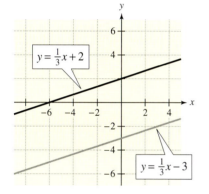

Figure 3.41

④ Use slopes to describe rates of change in real-life problems.

Example 10 Parallel or Perpendicular?

Are the pairs of lines parallel, perpendicular, or neither?

a. $y = -2x + 4$, $y = \frac{1}{2}x + 1$ **b.** $y = \frac{1}{3}x + 2$, $y = \frac{1}{3}x - 3$

Solution

a. The first line has a slope of $m_1 = -2$, and the second line has a slope of $m_2 = \frac{1}{2}$. Because these slopes are negative reciprocals of each other, the two lines must be perpendicular, as shown in Figure 3.40.

b. Each of these two lines has a slope of $m = \frac{1}{3}$. So, the two lines must be parallel, as shown in Figure 3.41.

Slope as a Rate of Change

In real-life problems, slope is often used to describe a **constant rate of change** or an **average rate of change.** In such cases, units of measure are assigned, such as miles per hour or dollars per year.

Example 11 Slope as a Rate of Change

In 1991, the average cost of a hotel room was $58.08. By 2000, the average cost had risen to $85.89. Find the average rate of change in the cost of a hotel room from 1991 to 2000.   (Source: American Hotel & Lodging Association)

Solution

Let c represent the cost of a hotel room and let t represent the year. The two given data points are represented by (t_1, c_1) and (t_2, c_2).

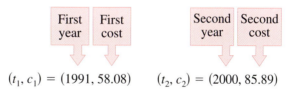

$$(t_1, c_1) = (1991, 58.08) \qquad (t_2, c_2) = (2000, 85.89)$$

Now use the formula for slope to find the average rate of change.

$$\text{Rate of change} = \frac{c_2 - c_1}{t_2 - t_1} \qquad \text{Slope formula}$$

$$= \frac{85.89 - 58.08}{2000 - 1991} \qquad \text{Substitute values.}$$

$$= \frac{27.81}{9} = 3.09 \qquad \text{Simplify.}$$

From 1991 through 2000, the average rate of change in the cost of a hotel room was $3.09 per year. The exact change in cost varied from one year to the next, as shown in the scatter plot in Figure 3.42.

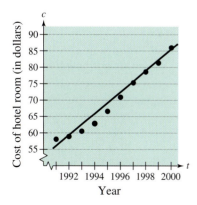

Figure 3.42

3.3 Exercises

Review   Concepts, Skills, and Problem Solving

Keep mathematically in shape by doing these exercises *before* the problems of this section.

Properties and Definitions

1. When two equations such as $2x - 3 = 5$ and $2x = 8$ have the same set of solutions, the equations are called _____.

2. Use the Addition Property of Equality to fill in the blank.

   $12x - 5 = 13$

   $12x = 13 +$ ▯

Solving Equations

In Exercises 3–10, solve the equation.

3. $x + \dfrac{x}{2} = 4$

4. $\dfrac{1}{3}x + 1 = 10$

5. $-4(x - 5) = 0$

6. $\dfrac{3}{8}x + \dfrac{3}{4} = 2$

7. $8(x - 14) = 32$

8. $12(3 - x) = 5 - 7(2x + 1)$

9. $-(2x + 8) + \dfrac{1}{3}(6x + 5) = 0$

10. $(1 + r)500 = 550$

Problem Solving

11. *Telephone Charges*   The cost of an international telephone call is $1.10 for the first minute and $0.45 for each additional minute. The total cost of the call cannot exceed $11. Find the interval of time that is available for the call.

12. *Operating Cost*   A fuel company has a fleet of trucks. The annual operating cost per truck is $C = 0.65m + 4500$, where m is the number of miles traveled by a truck in a year. What is the maximum number of miles that will yield an annual operating cost that is less than $20,000?

Developing Skills

In Exercises 1–6, estimate the slope of the line from its graph.

1.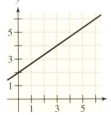

2.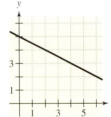

3.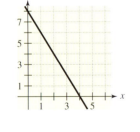

4.

5.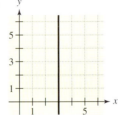

6.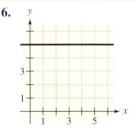

In Exercises 7 and 8, identify the line that has each slope m.

7. (a) $m = \dfrac{3}{4}$

   (b) $m = 0$

   (c) $m = -3$

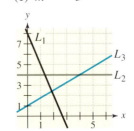

8. (a) $m = -\dfrac{5}{2}$

   (b) m is undefined.

   (c) $m = 2$

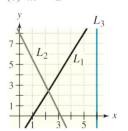

In Exercises 9–24, plot the points and find the slope (if possible) of the line passing through them. State whether the line rises, falls, is horizontal, or is vertical. See Examples 1–4.

9. $(0, 0), (7, 5)$ **10.** $(0, 0), (-3, -4)$

11. $(0, 0), (5, -4)$ **12.** $(0, 0), (-2, 1)$

13. $(-4, 3), (-2, 5)$ **14.** $(7, 1), (4, -5)$

15. $(-5, -3), (-5, 4)$ **16.** $(9, 2), (-9, 2)$

17. $(2, -5), (7, -5)$ **18.** $(-3, 4), (-3, 8)$

19. $\left(\frac{3}{4}, 2\right), \left(5, -\frac{5}{2}\right)$ **20.** $\left(\frac{1}{2}, -1\right), \left(3, \frac{2}{3}\right)$

21. $\left(\frac{3}{4}, \frac{1}{4}\right), \left(-\frac{3}{2}, \frac{1}{8}\right)$ **22.** $\left(-\frac{3}{2}, -\frac{1}{2}\right), \left(\frac{5}{8}, \frac{1}{2}\right)$

23. $(4.2, -1), (-4.2, 6)$ **24.** $(3.4, 0), (3.4, 1)$

In Exercises 25–30, sketch the graph of the line. Then find the slope of the line. See Example 5.

25. $y = 2x - 1$

26. $y = 3x + 2$

27. $y = -\frac{1}{2}x + 4$

28. $y = \frac{3}{4}x - 5$

29. $4x + 5y = 10$

30. $3x - 2y = 8$

In Exercises 31 and 32, solve for x so that the line through the points has the given slope.

31. $(4, 5), (x, 7)$

$m = -\frac{2}{3}$

32. $(x, -2), (5, 0)$

$m = \frac{3}{4}$

In Exercises 33 and 34, solve for y so that the line through the points has the given slope.

33. $(-3, y), (9, 3)$

$m = \frac{3}{2}$

34. $(-3, 20), (2, y)$

$m = -6$

In Exercises 35–42, a point on a line and the slope of the line are given. Find two additional points on the line. (There are many correct answers.)

35. $(5, 2)$ **36.** $(-4, 3)$

$m = 0$ m is undefined.

37. $(3, -4)$ **38.** $(-1, -5)$

$m = 3$ $m = 2$

39. $(0, 3)$ **40.** $(-2, 6)$

$m = -1$ $m = -3$

41. $(-5, 0)$ **42.** $(-1, 1)$

$m = \frac{4}{3}$ $m = -\frac{3}{4}$

In Exercises 43–50, write the equation of the line in slope-intercept form.

43. $6x - 3y = 9$ **44.** $2x + 4y = 16$

45. $4y - x = -4$ **46.** $3x - 2y = -10$

47. $2x + 5y - 3 = 0$ **48.** $8x - 6y + 1 = 0$

49. $x = 2y - 4$ **50.** $x = -\frac{3}{2}y + \frac{2}{3}$

In Exercises 51–56, find the slope and y-intercept of the line. See Example 6.

51. $y = 3x - 2$

52. $y = 4 - 2x$

53. $3y - 2x = 3$

54. $4x + 8y = -1$

55. $5x + 3y - 2 = 0$

56. $6y - 5x + 18 = 0$

In Exercises 57–66, write the equation of the line in slope-intercept form, and then use the slope and y-intercept to sketch the line. See Examples 7 and 8.

57. $3x - y - 2 = 0$ **58.** $x - y - 5 = 0$

59. $x + y = 0$ **60.** $x - y = 0$

61. $3x + 2y - 2 = 0$ **62.** $x - 2y - 2 = 0$

63. $x - 4y + 2 = 0$ **64.** $8x + 6y - 3 = 0$

65. $x - 0.2y - 1 = 0$ **66.** $0.5x + 0.6y - 3 = 0$

In Exercises 67–70, sketch the graph of a line through the point $(3, 2)$ having the given slope.

67. $m = -\frac{1}{3}$ **68.** $m = \frac{3}{2}$

69. m is undefined.   **70.** $m = 0$

In Exercises 71–74, plot the x- and y-intercepts and sketch the line.

71. $2x - y + 4 = 0$

72. $3x + 5y + 15 = 0$

73. $-5x + 2y - 20 = 0$

74. $3x - 5y - 15 = 0$

In Exercises 75–78, determine whether the two lines are parallel, perpendicular, or neither. See Examples 9 and 10.

75. L_1: $y = \frac{1}{2}x - 2$ **76.** L_1: $y = 3x - 2$
L_2: $y = \frac{1}{2}x + 3$ L_2: $y = 3x + 1$

77. L_1: $y = \frac{3}{4}x - 3$ **78.** L_1: $y = -\frac{2}{3}x - 5$
L_2: $y = -\frac{4}{3}x + 1$ L_2: $y = \frac{3}{2}x + 1$

In Exercises 79–82, determine whether the lines L_1 and L_2 passing through the pair of points are parallel, perpendicular, or neither.

79. L_1: $(0, 4), (2, 8)$ **80.** L_1: $(3, 4), (-2, 3)$
L_2: $(0, -1), (3, 5)$ L_2: $(0, -3), (2, -1)$

81. L_1: $(0, 2), (6, -2)$ **82.** L_1: $(3, 2), (-1, -2)$
L_2: $(4, 0), (6, 3)$ L_2: $(2, 0), (3, -1)$

Solving Problems

83. *Road Grade*   When driving down a mountain road, you notice warning signs indicating a "12% grade." This means that the slope of the road is $-\frac{12}{100}$. Over a stretch of the road, your elevation drops by 2000 feet (see figure). What is the horizontal change in your position?

2000 ft

x

Not drawn to scale

84. *Ramp*   A loading dock ramp rises 4 feet above the ground. The ramp has a slope of $\frac{1}{10}$. What is the length of the ramp?

85. *Roof Pitch*   The slope, or pitch, of a roof (see figure) is such that it rises (or falls) 3 feet for every 4 feet of horizontal distance. Determine the maximum height of the attic of the house for a 30-foot wide house.

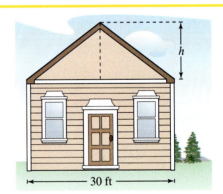

h

30 ft

Figure for 85 and 86

86. *Roof Pitch*   The slope, or pitch, of a roof (see figure) is such that it rises (or falls) 4 feet for every 5 feet of horizontal distance. Determine the maximum height of the attic of the house for a 30-foot wide house.

87. *Education* The average annual amount of tuition and fees y paid by an in-state student attending a private college in the United States from 1994 to 2000 can be approximated by the model

$$y = 580.7t + 8276, \quad 4 \le t \le 10$$

where t represents the year, with $t = 4$ corresponding to 1994. (Source: U.S. National Center for Education Statistics)

(a) Use the model to complete the table.

t	4	5	6	7
y				

t	8	9	10
y			

(b) Graph the model on a rectangular coordinate system.

(c) Find the average rate of change of tuition and fees from 1994 to 2000.

(d) If this increase continued at the current rate, predict the amount of tuition and fees that would be paid in the year 2010.

88. *Simple Interest* An inheritance of $8000 is invested in an account that pays $7\frac{1}{2}\%$ simple interest. The amount of money in the account after t years is given by the model

$$y = 8000 + 600t, \quad t \ge 0.$$

(a) Use the model to complete the table.

t	0	1	2
y			

t	3	4	5
y			

(b) Graph the equation on a rectangular coordinate system.

(c) Find the average rate of change of the amount in the account over 5 years.

Explaining Concepts

89. ⬥ Answer parts (a)–(c) of Motivating the Chapter on page 124.

90. *Writing* Can any pair of points on a line be used to calculate the slope of the line? Explain.

91. *Writing* In your own words, give interpretations of a negative slope, a zero slope, and a positive slope.

92. *Writing* The slopes of two lines are -3 and $\frac{3}{2}$. Which is steeper? Explain.

93. In the form $y = mx + b$, what does m represent? What does b represent?

94. *Writing* What is the relationship between the x-intercept of the graph of the line $y = mx + b$ and the solution to the equation $mx + b = 0$? Explain.

95. *Writing* Is it possible for two lines with positive slopes to be perpendicular to each other? Explain.

3.4 Equations of Lines

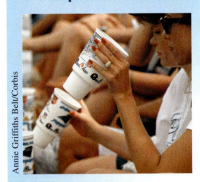

Annie Griffiths Belt/Corbis

What You Should Learn

1. Write equations of lines using point-slope form.
2. Write equations of horizontal, vertical, parallel, and perpendicular lines.
3. Use linear models to solve application problems.

Why You Should Learn It

Linear equations can be used to model and solve real-life problems. For instance, in Exercise 78 on page 171, a linear equation is used to model the relationship between the price of soft drinks and the demand for that product.

① Write equations of lines using point-slope form.

The Point-Slope Form of the Equation of a Line

In Sections 3.1 through 3.3, you have been studying analytic (or coordinate) geometry. Analytic geometry uses a coordinate plane to give visual representations of algebraic concepts, such as equations or functions.

There are two basic types of problems in analytic geometry.

1. Given an equation, sketch its graph: Algebra ⟹ Geometry

2. Given a graph, write its equation: Geometry ⟹ Algebra

In Section 3.3, you worked primarily with the first type of problem. In this section, you will study the second type. Specifically, you will learn how to write the equation of a line when you are given its slope and a point on the line. Before giving a general formula for doing this, consider the following example.

Example 1   Writing an Equation of a Line

Write an equation of the line with slope $\frac{4}{3}$ that passes through the point $(-2, 1)$.

Solution

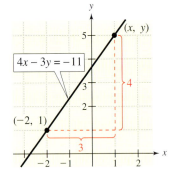

Figure 3.43

Begin by sketching the line, as shown in Figure 3.43. You know that the slope of a line is the same through any two points on the line. So, to find an equation of the line, let (x, y) represent any point on the line. Using the representative point (x, y) and the point $(-2, 1)$, it follows that the slope of the line is

$$m = \frac{y - 1}{x - (-2)}.$$ Difference in y-values
      Difference in x-values

Because the slope of the line is $m = \frac{4}{3}$, this equation can be rewritten as follows.

$$\frac{4}{3} = \frac{y - 1}{x + 2}$$ Slope formula

$$4(x + 2) = 3(y - 1)$$ Cross-multiply.

$$4x + 8 = 3y - 3$$ Distributive Property

$$4x - 3y = -11$$ Subtract 8 and $3y$ from each side.

An equation of the line is $4x - 3y = -11$.

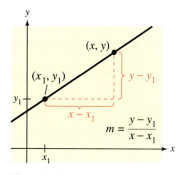

Figure 3.44

The procedure in Example 1 can be used to derive a *formula* for the equation of a line, given its slope and a point on the line. In Figure 3.44, let (x_1, y_1) be a given point on the line whose slope is m. If (x, y) is any *other* point on the line, it follows that

$$\frac{y - y_1}{x - x_1} = m.$$

This equation in variables x and y can be rewritten in the form

$$y - y_1 = m(x - x_1)$$

which is called the **point-slope form** of the equation of a line.

Point-Slope Form of the Equation of a Line

The **point-slope form** of the equation of the line that passes through the point (x_1, y_1) and has a slope of m is

$$y - y_1 = m(x - x_1).$$

Example 2   The Point-Slope Form of the Equation of a Line

Write an equation of the line that passes through the point $(2, -3)$ and has slope $m = -2$.

Solution

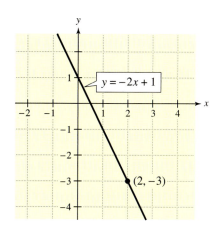

Figure 3.45

Use the point-slope form with $(x_1, y_1) = (2, -3)$ and $m = -2$.

$$y - y_1 = m(x - x_1) \qquad \text{Point-slope form}$$
$$y - (-3) = -2(x - 2) \qquad \text{Substitute } y_1 = -3,\ x_1 = 2,\ \text{and } m = -2.$$
$$y + 3 = -2x + 4 \qquad \text{Simplify.}$$
$$y = -2x + 1 \qquad \text{Subtract 3 from each side.}$$

So, an equation of the line is $y = -2x + 1$. Note that this is the slope-intercept form of the equation. The graph of the line is shown in Figure 3.45.

In Example 2, note that it was stated that $y = -2x + 1$ is "an" equation of the line rather than "the" equation of the line. The reason for this is that every equation can be written in many equivalent forms. For instance,

$$y = -2x + 1, \quad 2x + y = 1, \quad \text{and} \quad 2x + y - 1 = 0$$

are all equations of the line in Example 2. The first of these equations $(y = -2x + 1)$ is the slope-intercept form

$$y = mx + b \qquad \text{Slope-intercept form}$$

and it provides the most information about the line. The last of these equations $(2x + y - 1 = 0)$ is the **general form** of the equation of a line.

$$ax + by + c = 0 \qquad \text{General form}$$

Technology: Tip

A program for several models of graphing calculators that uses the two-point form to find the equation of a line is available at our website *math.college.hmco.com/students.*

The program prompts for the coordinates of the two points and then outputs the slope and the *y*-intercept of the line that passes through the two points. Verify Example 3 using this program.

The point-slope form can be used to find the equation of a line passing through two points (x_1, y_1) and (x_2, y_2). First, use the formula for the slope of a line passing through two points.

$$m = \frac{y_2 - y_1}{x_2 - x_1}$$

Then, substitute this value for m into the point-slope form to obtain the equation

$$y - y_1 = \frac{y_2 - y_1}{x_2 - x_1}(x - x_1). \qquad \text{Two-point form}$$

This is sometimes called the **two-point form** of the equation of a line.

Example 3   An Equation of a Line Passing Through Two Points

Write the general form of the equation of the line that passes through the points $(4, 2)$ and $(-2, 3)$.

Solution

Let $(x_1, y_1) = (4, 2)$ and $(x_2, y_2) = (-2, 3)$. Then apply the formula for the slope of a line passing through two points, as follows.

$$m = \frac{y_2 - y_1}{x_2 - x_1}$$

$$= \frac{3 - 2}{-2 - 4}$$

$$= -\frac{1}{6}$$

Now, using the point-slope form, you can find the equation of the line.

$$y - y_1 = m(x - x_1) \qquad \text{Point-slope form}$$

$$y - 2 = -\frac{1}{6}(x - 4) \qquad \text{Substitute } y_1 = 2, \ x_1 = 4, \text{ and } m = -\frac{1}{6}.$$

$$6(y - 2) = -(x - 4) \qquad \text{Multiply each side by 6.}$$

$$6y - 12 = -x + 4 \qquad \text{Distributive Property}$$

$$x + 6y - 12 = 4 \qquad \text{Add } x \text{ to each side.}$$

$$x + 6y - 16 = 0 \qquad \text{Subtract 4 from each side.}$$

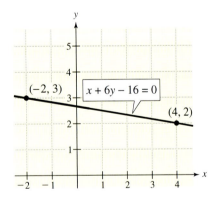

Figure 3.46

The general form of the equation of the line is $x + 6y - 16 = 0$. The graph of this line is shown in Figure 3.46.

In Example 3, it does not matter which of the two points is labeled (x_1, y_1) and which is labeled (x_2, y_2). Try switching these labels to

$$(x_1, y_1) = (-2, 3) \quad \text{and} \quad (x_2, y_2) = (4, 2)$$

and reworking the problem to see that you obtain the same equation.

2 Write equations of horizontal, vertical, parallel, and perpendicular lines.

Other Equations of Lines

Recall from Section 3.3 that a horizonal line has a slope of zero. From the slope-intercept form of the equation of a line, you can see that a horizontal line has an equation of the form

$$y = (0)x + b \quad \text{or} \quad y = b. \qquad \text{Horizontal line}$$

This is consistent with the fact that each point on a horizontal line through $(0, b)$ has a y-coordinate of b, as shown in Figure 3.47. Similarly, each point on a vertical line through $(a, 0)$ has an x-coordinate of a, as shown in Figure 3.48. Because you know that a vertical line has an undefined slope, you know that it has an equation of the form

$$x = a. \qquad \text{Vertical line}$$

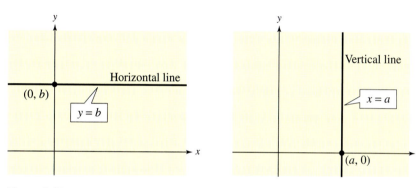

Figure 3.47   **Figure 3.48**

Example 4   Writing Equations of Horizontal and Vertical Lines

Write an equation for each line.

a. Vertical line through $(-2, 4)$

b. Horizontal line through $(0, 6)$

c. Line passing through $(-2, 3)$ and $(3, 3)$

d. Line passing through $(-1, 2)$ and $(-1, 3)$

Solution

a. Because the line is vertical and passes through the point $(-2, 4)$, you know that every point on the line has an x-coordinate of -2. So, the equation is $x = -2$.

b. Because the line is horizontal and passes through the point $(0, 6)$, you know that every point on the line has a y-coordinate of 6. So, the equation of the line is $y = 6$.

c. Because both points have the same y-coordinate, the line through $(-2, 3)$ and $(3, 3)$ is horizontal. So, its equation is $y = 3$.

d. Because both points have the same x-coordinate, the line through $(-1, 2)$ and $(-1, 3)$ is vertical. So, its equation is $x = -1$.

In Section 3.3, you learned that parallel lines have the same slope and perpendicular lines have slopes that are negative reciprocals of each other. You can use these facts to write an equation of a line parallel or perpendicular to a given line.

Example 5   Parallel and Perpendicular Lines

Write equations of the lines that pass through the point $(3, -2)$ and are (a) parallel and (b) perpendicular to the line $x - 4y = 6$, as shown in Figure 3.49.

Solution

By writing the given line in slope-intercept form, $y = \frac{1}{4}x - \frac{3}{2}$, you can see that it has a slope of $\frac{1}{4}$. So, a line parallel to it must also have a slope of $\frac{1}{4}$ and a line perpendicular to it must have a slope of -4.

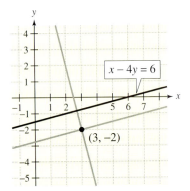

Figure 3.49

a. $y - y_1 = m(x - x_1)$ Point-slope form

$y - (-2) = \frac{1}{4}(x - 3)$ Substitute $y_1 = -2$, $x_1 = 3$, and $m = \frac{1}{4}$.

$y = \frac{1}{4}x - \frac{11}{4}$ Equation of parallel line

b. $y - y_1 = m(x - x_1)$ Point-slope form

$y - (-2) = -4(x - 3)$ Substitute $y_1 = -2$, $x_1 = 3$, and $m = -4$.

$y = -4x + 10$ Equation of perpendicular line

The equation of a vertical line cannot be written in slope-intercept form because the slope of a vertical line is undefined. However, *every* line has an equation that can be written in the general form $ax + by + c = 0$ where a and b are not both zero.

Study Tip

The slope-intercept form of the equation of a line is better suited for *sketching a line*. On the other hand, the point-slope form of the equation of a line is better suited for *creating the equation of a line*, given its slope and a point on the line.

Summary of Equations of Lines

1. Slope of a line through (x_1, y_1) and (x_2, y_2): $m = \dfrac{y_2 - y_1}{x_2 - x_1}$

2. General form of equation of line: $ax + by + c = 0$

3. Equation of vertical line: $x = a$

4. Equation of horizontal line: $y = b$

5. Slope-intercept form of equation of line: $y = mx + b$

6. Point-slope form of equation of line: $y - y_1 = m(x - x_1)$

7. Parallel lines (equal slopes): $m_1 = m_2$

8. Perpendicular lines (negative reciprocal slopes): $m_1 = -\dfrac{1}{m_2}$

3 Use linear models to solve application problems.

Application

Example 6 Total Sales

Home Depot, Inc. had total sales of $38.4 billion in 1999 and $45.7 billion in 2000. (Source: Home Depot, Inc.)

a. Using only this information, write a linear equation that models the sales (in billions of dollars) terms of the year.

b. Interpret the meaning of the slope in the context of the problem.

c. Predict the sales for 2001.

Solution

a. Let $t = 9$ represent 1999. Then the two given values are represented by the data points $(9, 38.4)$ and $(10, 45.7)$. The slope of the line through these points is

$$m = \frac{y_2 - y_1}{t_2 - t_1}$$

$$= \frac{45.7 - 38.4}{10 - 9}$$

$$= 7.3.$$

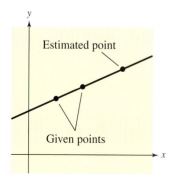

Figure 3.50

Using the point-slope form, you can find the equation that relates the sales y and the year t to be

$y - y_1 = m(t - t_1)$	Point-slope form
$y - 38.4 = 7.3(t - 9)$	Substitute for y_1, m, and t_1.
$y - 38.4 = 7.3t - 65.7$	Distributive Property
$y = 7.3t - 27.3.$	Write in slope-intercept form.

b. The slope of the equation in part (a) indicates that the total sales for Home Depot, Inc. increased by $7.3 billion each year.

c. Using the equation from part (a), you can predict the sales for 2001 ($t = 11$) to be

$y = 7.3t - 27.3$	Equation from part (a)
$y = 7.3(11) - 27.3$	Substitute 11 for t.
$y = 53.0.$	Simplify.

So, the predicted sales for 2001 are $53.0 billion. In this case, the prediction is quite good—the actual sales in 2001 were $53.6 billion. The graph of this equation is shown in Figure 3.50.

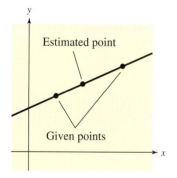

(a) Linear Extrapolation

(b) Linear Interpolation

Figure 3.51

The estimation method illustrated in Example 6 is called **linear extrapolation.** Note in Figure 3.51(a) that for linear extrapolation, the estimated point lies to the right of the given points. When the estimated point lies *between* two given points, as in Figure 3.51(b), the procedure is called **linear interpolation.**

3.4 Exercises

Review Concepts, Skills, and Problem Solving

Keep mathematically in shape by doing these exercises *before* the problems of this section.

Properties and Definitions

1. State the definition of the ratio of the real number a to the real number b.

2. $\dfrac{4}{5} = \dfrac{12}{u}$ is a statement of equality of two ratios. What is this statement of equality called?

Solving Percent Problems

In Exercises 3–10, solve the percent problem.

3. What is $7\frac{1}{2}\%$ of 25?

4. What is 150% of 6000?

5. 225 is what percent of 150?

6. 93 is what percent of 600?

7. What percent of 240 is 160?

8. What percent of 350 is 450?

9. 0.5% of what number is 400?

10. 48% of what number is 132?

Problem Solving

11. *Construction* The ratio of cement to sand in a 90-pound bag of dry mix is 1 to 4. Find the number of pounds of sand in the bag.

12. *Velocity* The velocity v of an object projected vertically upward with an initial velocity of 96 feet per second is given by $v = 96 - 32t$, where t is time in seconds and air resistance is neglected. Find the time at which the maximum height ($v = 0$) of the object is attained.

Developing Skills

In Exercises 1–4, match the equation with its graph. [The graphs are labeled (a), (b), (c), and (d).]

(a)

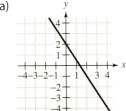

(b)

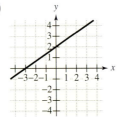

(c)

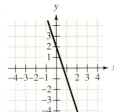

(d)

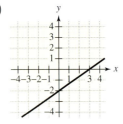

1. $y = \frac{2}{3}x + 2$

2. $y = \frac{2}{3}x - 2$

3. $y = -\frac{3}{2}x + 2$

4. $y = -3x + 2$

In Exercises 5–10, write an equation of the line that passes through the point and has the specified slope. See Example 1.

5. $(2, -3)$
   $m = 4$

6. $(-1, 5)$
   $m = -4$

7. $(-3, 1)$
   $m = -\frac{1}{2}$

8. $(6, 9)$
   $m = \frac{2}{3}$

9. $\left(\frac{3}{4}, -1\right)$
   $m = \frac{4}{5}$

10. $\left(-2, \frac{3}{2}\right)$
    $m = -\frac{1}{6}$

In Exercises 11–26, use the point-slope form of the equation of a line to write an equation of the line that passes through the point and has the specified slope. When possible, write the equation in slope-intercept form. See Example 2.

11. $(0, 0)$
    $m = -\frac{1}{2}$

12. $(0, 0)$
    $m = \frac{1}{5}$

13. $(0, -4)$
    $m = 3$

14. $(0, 5)$
    $m = 2$

15. $(0, 6)$

$m = -\frac{3}{4}$

16. $(0, -8)$

$m = \frac{2}{3}$

17. $(-2, 8)$

$m = -2$

18. $(4, -1)$

$m = 3$

19. $(-4, -7)$

$m = \frac{5}{4}$

20. $(6, -8)$

$m = -\frac{2}{3}$

21. $\left(-2, \frac{7}{2}\right)$

$m = -4$

22. $\left(1, -\frac{3}{2}\right)$

$m = 1$

23. $\left(\frac{3}{4}, \frac{5}{2}\right)$

$m = \frac{4}{3}$

24. $\left(-\frac{5}{2}, \frac{1}{2}\right)$

$m = -\frac{2}{5}$

25. $(2, -1)$

$m = 0$

26. $(-8, 5)$

$m = 0$

In Exercises 27–42, write the general form of the equation of the line that passes through the two points. See Example 3.

27. $(0, 0), (2, 3)$

28. $(0, 0), (3, -5)$

29. $(0, 4), (4, 0)$

30. $(0, -2), (2, 0)$

31. $(1, 4), (5, 6)$

32. $(2, 6), (4, 1)$

33. $(-5, 2), (5, -2)$

34. $(-2, -3), (-4, -6)$

35. $\left(\frac{3}{2}, 3\right), \left(\frac{9}{2}, 4\right)$

36. $\left(4, \frac{7}{3}\right), \left(-1, \frac{1}{3}\right)$

37. $\left(10, \frac{1}{2}\right), \left(\frac{3}{2}, \frac{7}{4}\right)$

38. $\left(-4, \frac{3}{5}\right), \left(\frac{3}{4}, -\frac{2}{5}\right)$

39. $(5, 9), (8, -1.4)$

40. $(2, -8), (6, 2.3)$

41. $(2, 0.6), (8, -4.2)$

42. $(-5, 0.6), (3, -3.4)$

In Exercises 43–46, write the slope-intercept form of the equation of the line that passes through the two points.

43. $(-2, 2), (4, 5)$

44. $(0, 10), (5, 0)$

45. $(-2, 3), (4, 3)$

46. $(-6, -3), (4, 3)$

In Exercises 47–52, write an equation of the line. See Example 4.

47. Vertical line through $(-1, 5)$

48. Vertical line through $(2, -3)$

49. Horizontal line through $(0, -5)$

50. Horizontal line through $(-4, 6)$

51. Line through $(-7, 2)$ and $(-7, -1)$

52. Line through $(6, 4)$ and $(-9, 4)$

In Exercises 53–62, write equations of the lines that pass through the point and are (a) parallel and (b) perpendicular to the given line. See Example 5.

53. $(2, 1)$

$6x - 2y = 3$

54. $(-3, 4)$

$x + 6y = 12$

55. $(-5, 4)$

$5x + 4y = 24$

56. $(6, -4)$

$3x + 10y = 24$

57. $(3, 7)$

$4x - y - 3 = 0$

58. $(-5, -10)$

$2x + 5y - 12 = 0$

59. $\left(\frac{2}{3}, \frac{4}{3}\right)$

$x - 5 = 0$

60. $\left(\frac{5}{8}, \frac{9}{4}\right)$

$-5x + 4y = 0$

61. $(-1, 2)$

$y + 5 = 0$

62. $(3, -4)$

$x - 10 = 0$

In Exercises 63–66, write the **intercept form** of the equation of the line with intercepts $(a, 0)$ and $(0, b)$. The equation is given by

$$\frac{x}{a} + \frac{y}{b} = 1, \ a \neq 0, \ b \neq 0.$$

63. x-intercept: $(3, 0)$
     y-intercept: $(0, 2)$

64. x-intercept: $(-6, 0)$
     y-intercept: $(0, 2)$

65. x-intercept: $\left(-\frac{5}{6}, 0\right)$
     y-intercept: $\left(0, -\frac{7}{3}\right)$

66. x-intercept: $\left(-\frac{8}{3}, 0\right)$
     y-intercept: $(0, -4)$

Solving Problems

67. *Cost*   The cost C (in dollars) of producing x units of a product is shown in the table. Find a linear model to represent the data. Estimate the cost of producing 400 units.

x	0	50	100	150	200
C	5000	6000	7000	8000	9000

68. *Temperature Conversion*   The relationship between the Fahrenheit F and Celsius C temperature scales is shown in the table. Find a linear model to represent the data. Estimate the Celsius temperature when the Fahrenheit temperature is 72 degrees.

F	41	50	59	68	77
C	5	10	15	20	25

69. *Sales*   The total sales for a new camera equipment store were $200,000 for the second year and $500,000 for the fifth year. Find a linear model to represent the data. Estimate the total sales for the sixth year.

70. *Sales*   The total sales for a new sportswear store were $150,000 for the third year and $250,000 for the fifth year. Find a linear model to represent the data. Estimate the total sales for the sixth year.

71. *Sales Commission*   The salary for a sales representative is $1500 per month plus a commission of total monthly sales. The table shows the relationship between the salary S and total monthly sales M. Write an equation of the line giving the salary S in terms of the monthly sales M. What is the commission rate?

M	0	1000	2000	3000	4000
S	1500	1530	1560	1590	1620

72. *Reimbursed Expenses*   A sales representative is reimbursed $125 per day for lodging and meals plus an amount per mile driven. The table below shows the relationship between the daily cost C to the company and the number of miles driven x. Write an equation giving the daily cost C to the company in terms of x, the number of miles driven. How much is the sales representative reimbursed per mile?

x	50	100	150	200	250
C	142	159	176	193	210

73. *Discount*   A store is offering a 30% discount on all items in its inventory.

(a) Write an equation of the line giving the sale price S for an item in terms of its list price L.

(b) Use the equation in part (a) to find the sale price of an item that has a list price of $135.

74. *Reimbursed Expenses*   A sales representative is reimbursed $150 per day for lodging and meals plus $0.35 per mile driven.

(a) Write an equation of the line giving the daily cost C to the company in terms of x, the number of miles driven.

(b) 🖩 Use a graphing calculator to graph the line in part (a) and graphically estimate the daily cost to the company when the representative drives 230 miles. Confirm your estimate algebraically.

(c) 🖩 Use the graph in part (b) to estimate graphically the number of miles driven when the daily cost to the company is $200. Confirm your estimate algebraically.

75. *Straight-Line Depreciation*   A small business purchases a photocopier for $7400. After 4 years, its depreciated value will be $1500.

(a) Assuming straight-line depreciation, write an equation of the line giving the value V of the copier in terms of time t in years.

(b) Use the equation in part (a) to find the value of the copier after 2 years.

76. *Straight-Line Depreciation*   A business purchases a van for $27,500. After 5 years, its depreciated value will be $12,000.

(a) Assuming straight-line depreciation, write an equation of the line giving the value V of the van in terms of time t in years.

(b) Use the equation in part (a) to find the value of the van after 2 years.

77. *Education*   A small college had an enrollment of 1500 students in 1995. During the next 10 years, the enrollment increased by approximately 60 students per year.

(a) Write an equation of the line giving the enrollment N in terms of the year t. (Let $t = 5$ correspond to the year 1995.)

(b) *Linear Extrapolation*   Use the equation in part (a) to predict the enrollment in the year 2010.

(c) *Linear Interpolation*   Use the equation in part (a) to estimate the enrollment in 2000.

78. *Demand*   When soft drinks sold for $0.80 per can at football games, approximately 6000 cans were sold. When the price was raised to $1.00 per can, the demand dropped to 4000. Assume that the relationship between the price p and demand x is linear.

(a) Write an equation of the line giving the demand x in terms of the price p.

(b) *Linear Extrapolation*   Use the equation in part (a) to predict the number of cans of soft drinks sold if the price is raised to $1.10.

(c) *Linear Interpolation*   Use the equation in part (a) to estimate the number of cans of soft drinks sold if the price is $0.90.

79. *Data Analysis*   The table shows the expected number of additional years of life E for a person of age A. (Source: U.S. National Center for Health Statistics)

A	Birth	10	20	40	60	80
E	76.7	67.4	57.7	38.8	21.5	8.6

(a) Sketch a scatter plot of the data.

(b) Use a straightedge to sketch the "best-fitting" line through the points.

(c) Find an equation of the line you sketched in part (b).

(d) Use the equation in part (c) to estimate the expected number of additional years of life of a person who is 30 years old.

80. *Data Analysis*   An instructor gives 20-point quizzes and 100-point exams in a mathematics course. The average quiz and test scores for six students are given as ordered pairs (x, y), where x is the average quiz score and y is the average test score. The ordered pairs are (18, 87), (10, 55), (19, 96), (16, 79), (13, 76), and (15, 82).

(a) Sketch a scatter plot of the data.

(b) Use a straightedge to sketch the "best-fitting" line through the points.

(c) Find an equation of the line you sketched in part (b).

(d) Use the equation in part (c) to estimate the average test score for a person with an average quiz score of 17.

81. *Depth Markers* A swimming pool is 40 feet long, 20 feet wide, 4 feet deep at the shallow end, and 9 feet deep at the deep end. Position the side of the pool on a rectangular coordinate system as shown in the figure and find an equation of the line representing the edge of the inclined bottom of the pool. Use this equation to determine the distances from the deep end at which markers must be placed to indicate each one-foot change in the depth of the pool.

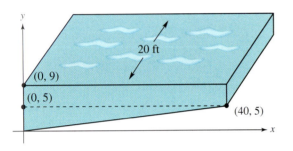

82. *Carpentry* A carpenter uses a wedge-shaped block of wood to support heavy objects. The block of wood is 12 inches long, 2 inches wide, 6 inches high at the tall end, and 2 inches high at the short end. Position the side of the block on a rectangular coordinate system as shown in the figure and find an equation of the line representing the edge of the slanted top of the block. Use this equation to determine the distances from the tall end at which marks must be made to indicate each one-inch change in the height of the block.

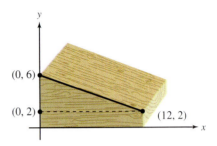

Explaining Concepts

83. ⚡ Answer parts (d)–(f) of Motivating the Chapter on page 124.

84. *Writing* ✏ Can any pair of points on a line be used to determine the equation of the line? Explain.

85. Write, from memory, the point-slope form, the slope-intercept form, and the general form of an equation of a line.

86. In the equation $y = 3x + 5$, what does the 3 represent? What does the 5 represent?

87. *Writing* ✏ In the equation of a vertical line, the variable y is missing. Explain why.

Mid-Chapter Quiz

Take this quiz as you would take a quiz in class. After you are done, check your work against the answers in the back of the book.

1. Determine the quadrants(s) in which the point $(x, 4)$ must be located if x is a real number. Explain your reasoning.

2. Determine whether each ordered pair is a solution point of the equation $4x - 3y = 10$.

    (a) $(2, 1)$ (b) $(1, -2)$ (c) $(2.5, 0)$ (d) $\left(2, -\frac{2}{3}\right)$

In Exercises 3 and 4, plot the points on a rectangular coordinate system, find the distance between them, and determine the coordinates of the midpoint of the line segment joining the two points.

3. $(-1, 5), (3, 2)$ 4. $(-3, -2), (2, 10)$

In Exercises 5–7, sketch the graph of the equation and show the coordinates of three solution points (including *x*- and *y*-intercepts). (There are many correct answers.)

5. $3x + y - 6 = 0$ 6. $y = 6x - x^2$ 7. $y = |x - 2| - 3$

In Exercises 8–10, determine the slope of the line (if possible) through the two points. State whether the line rises, falls, is horizontal, or is vertical.

8. $(-3, 8), (7, 8)$ 9. $(3, 0), (6, 5)$ 10. $(-1, 4), (5, -6)$

In Exercises 11 and 12, write the equation of the line in slope-intercept form. Find the slope and *y*-intercept and use them to sketch the graph of the equation.

11. $3x + 6y = 6$ 12. $x - 2y = 4$

In Exercises 13 and 14, determine whether the lines are parallel, perpendicular, or neither.

13. $y = 3x + 2, y = -\frac{1}{3}x - 4$

14. L_1: $(4, 3), (-2, -9)$; L_2: $(0, -5), (5, 5)$

15. Write the general form of the equation of a line that passes through the point $(6, -1)$ and has a slope of $\frac{1}{2}$.

16. Your company purchases a new printing press for $85,000. For tax purposes, the printing press will be depreciated over a 10-year period. At the end of 10 years, the salvage value of the printing press is expected to be $4000. Find an equation that relates the depreciated value of the printing press to the number of years since it was purchased. Then sketch the graph of the equation.

3.5 Graphs of Linear Inequalities

Helen King/Corbis

What You Should Learn

1 Verify solutions of linear inequalities in two variables.

2 Sketch graphs of linear inequalities in two variables.

Why You Should Learn It

Linear inequalities in two variables can be used to model and solve real-life problems. For instance, in Exercise 64 on page 182, a linear inequality is used to model the target heart rate for an adult.

1 Verify solutions of linear inequalities in two variables.

Linear Inequalities in Two Variables

A **linear inequality in two variables,** x and y, is an inequality that can be written in one of the forms below (where a and b are not both zero).

$$ax + by < c, \quad ax + by > c, \quad ax + by \leq c, \quad ax + by \geq c$$

Here are some examples.

$$4x - 3y < 7, \quad x - y > -3, \quad x \leq 2, \quad y \geq -4$$

An ordered pair (x_1, y_1) is a **solution** of a linear inequality in x and y if the inequality is true when x_1 and y_1 are substituted for x and y, respectively. For instance, the ordered pair $(3, 2)$ is a solution of the inequality $x - y > 0$ because $3 - 2 > 0$ is a true statement.

Example 1 Verifying Solutions of Linear Inequalities

Determine whether each point is a solution of $2x - 3y \geq -2$.

a. $(0, 0)$ **b.** $(0, 1)$

Solution

To determine whether a point (x_1, y_1) is a solution of the inequality, substitute the coordinates of the point into the inequality.

a. $2x - 3y \geq -2$ Write original inequality.

      $2(0) - 3(0) \overset{?}{\geq} -2$ Substitute 0 for x and 0 for y.

              $0 \geq -2$ Inequality is satisfied. ✔

Because the inequality is satisfied, the point $(0, 0)$ *is* a solution.

b. $2x - 3y \geq -2$ Write original inequality.

      $2(0) - 3(1) \overset{?}{\geq} -2$ Substitute 0 for x and 1 for y.

              $-3 \ngeq -2$ Inequality is not satisfied. ✗

Because the inequality is not satisfied, the point $(0, 1)$ *is not* a solution.

② Sketch graphs of linear inequalities in two variables.

The Graph of a Linear Inequality in Two Variables

The **graph** of a linear inequality is the collection of all solution points of the inequality. To sketch the graph of a linear inequality such as

$$4x - 3y < 12 \qquad \text{Original linear inequality}$$

begin by sketching the graph of the *corresponding linear equation*

$$4x - 3y = 12. \qquad \text{Corresponding linear equation}$$

Use *dashed* lines for the inequalities < and > and *solid* lines for the inequalities ≤ and ≥. The graph of the equation separates the plane into two regions, called **half-planes.** In each half-plane, one of the following *must* be true.

1. All points in the half-plane are solutions of the inequality.

2. No point in the half-plane is a solution of the inequality.

So, you can determine whether the points in an entire half-plane satisfy the inequality by simply testing *one* point in the region. This graphing procedure is summarized as follows.

Sketching the Graph of a Linear Inequality in Two Variables

1. Replace the inequality sign by an equal sign, and sketch the graph of the resulting equation. (Use a dashed line for < or >, and a solid line for ≤ or ≥.)

2. Test one point in one of the half-planes formed by the graph in Step 1.

   a. If the point satisfies the inequality, shade the entire half-plane to denote that every point in the region satisfies the inequality.

   b. If the point does not satisfy the inequality, then shade the other half-plane.

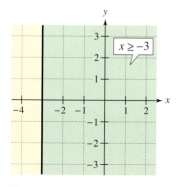

Figure 3.52

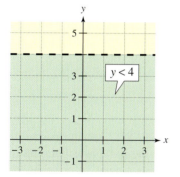

Figure 3.53

Example 2   Sketching the Graphs of Linear Inequalities

Sketch the graph of each linear inequality.

a. $x \geq -3$ **b.** $y < 4$

Solution

a. The graph of the corresponding equation $x = -3$ is a vertical line. The points that satisfy the inequality $x \geq -3$ are those lying on or to the right of this line, and the line is solid because the inequality sign is ≥. The graph is shown in Figure 3.52.

b. The graph of the corresponding equation $y = 4$ is a horizontal line. The points that satisfy the inequality $y < 4$ are those lying below this line, and the line is dashed because the inequality sign is <. The graph is shown in Figure 3.53.

Example 3   Sketching the Graphs of a Linear Inequality

Sketch the graph of the linear inequality $x + y > 3$.

Solution

The graph of the corresponding equation $x + y = 3$ is a line. To begin, find the x-intercept by letting $y = 0$ and solving for x.

$$x + 0 = 3 \qquad\Longrightarrow\qquad x = 3 \qquad\text{Substitute 0 for } y \text{ and solve for } x.$$

Find the y-intercept by letting $x = 0$ and solving for y.

$$0 + y = 3 \qquad\Longrightarrow\qquad y = 3 \qquad\text{Substitute 0 for } x \text{ and solve for } y.$$

So, the graph has an x-intercept at the point $(3, 0)$ and a y-intercept at the point $(0, 3)$. Plot these points and connect them with a dashed line. Because the origin $(0, 0)$ does not satisfy the inequality, the graph consists of the half-plane lying above the line, as shown in Figure 3.54. (Try checking a point above the line. Regardless of which point you choose, you will see that it is a solution.)

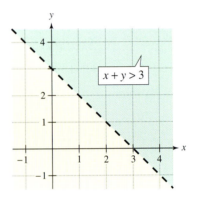

Figure 3.54

Example 4   Sketching the Graph of a Linear Inequality

Sketch the graph of the linear inequality $2x + y \leq 2$.

Solution

The graph of the corresponding equation $2x + y = 2$ is a line. To begin, find the x-intercept by letting $y = 0$ and solving for x.

$$2x + 0 = 2 \qquad\Longrightarrow\qquad x = 1 \qquad\text{Substitute 0 for } y \text{ and solve for } x.$$

Find the y-intercept by letting $x = 0$ and solving for y.

$$2(0) + y = 2 \qquad\Longrightarrow\qquad y = 2 \qquad\text{Substitute 0 for } x \text{ and solve for } y.$$

So, the graph has an x-intercept at the point $(1, 0)$ and a y-intercept at the point $(0, 2)$. Plot these points and connect them with a solid line. Because the origin $(0, 0)$ satisfies the inequality, the graph consists of the half-plane lying on or below the line, as shown in Figure 3.55. (Try checking a point on or below the line. Regardless of which point you choose, you will see that it is a solution.)

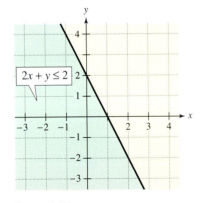

Figure 3.55

For a linear inequality in two variables, you can sometimes simplify the graphing procedure by writing the inequality in *slope-intercept form*. For instance, by writing $x + y > 1$ in the form

$$y > -x + 1 \qquad \text{Slope-intercept form}$$

you can see that the solution points lie *above* the line $y = -x + 1$, as shown in Figure 3.56. So, when $y > mx + b$ (or $y \geq mx + b$), shade above the line $y = mx + b$. Similarly, by writing the inequality $4x - 3y > 12$ in the form

$$y < \frac{4}{3}x - 4 \qquad \text{Slope-intercept form}$$

you can see that the solutions lie *below* the line $y = \frac{4}{3}x - 4$, as shown in Figure 3.57. So, when $y < mx + b$ (or $y \leq mx + b$), shade below the line $y = mx + b$.

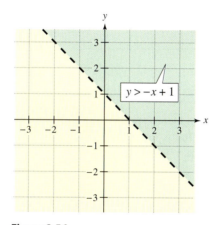

Figure 3.56   **Figure 3.57**

Example 5 Sketching the Graph of a Linear Inequality

Use the slope-intercept form of a linear equation as an aid in sketching the graph of the inequality $2x - 3y \leq 15$.

Solution

To begin, rewrite the inequality in slope-intercept form.

$$2x - 3y \leq 15 \qquad \text{Write original inequality.}$$

$$-3y \leq -2x + 15 \qquad \text{Subtract } 2x \text{ from each side.}$$

$$y \geq \frac{2}{3}x - 5 \qquad \text{Divide each side by } -3 \text{ and reverse the inequality symbol.}$$

From this form, you can conclude that the solution is the half-plane lying *on or above* the line

$$y = \frac{2}{3}x - 5.$$

The graph is shown in Figure 3.58. To verify the solution, test any point in the shaded region.

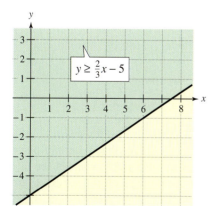

Figure 3.58

Technology: Tip

Most graphing calculators can graph inequalities in two variables. Consult the user's guide of your graphing calculator for specific instructions. The graph of $y \leq \frac{1}{2}x - 3$ is shown below.

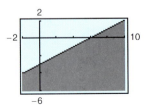

Try using a graphing calculator to graph each inequality.

a. $2x + 3y \geq 3$ **b.** $3x - 2y \leq 5$

Example 6   An Application: Working to Meet a Budget

Your budget requires you to earn *at least* $160 per week. You work two part-time jobs. One is tutoring, which pays $10 per hour, and the other is at a fast-food restaurant, which pays $8 per hour. Let x represent the number of hours tutoring and let y represent the number of hours worked at the fast-food restaurant.

a. Write an inequality that represents the number of hours worked at each job in order to meet your budget requirements.

b. Graph the inequality and identify at least two ordered pairs (x, y) that represent the number of hours you must work at each job in order to meet your budget requirements.

Solution

a. *Verbal Model:* $10 \cdot \boxed{\begin{array}{c}\text{Number of} \\ \text{hours tutoring}\end{array}} + 8 \cdot \boxed{\begin{array}{c}\text{Number of hours at} \\ \text{fast-food restaurant}\end{array}} \geq 160$

Labels: Number of hours tutoring $= x$ (hours)
            Number of hours at fast-food restaurant $= y$ (hours)

Inequality: $10x + 8y \geq 160$

b. Rewrite the inequality in slope-intercept form.

$10x + 8y \geq 160$ Write original inequality.

$8y \geq -10x + 160$ Subtract $10x$ from each side.

$y \geq -1.25x + 20$ Divide each side by 8.

Graph the corresponding equation $y = -1.25x + 20$ and shade the half-plane lying above the line, as shown in Figure 3.59. From the graph, you can see that two solutions that will yield the desired weekly earnings of at least $160 are $(8, 10)$ and $(12, 15)$. There are many other solutions.

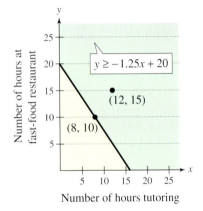

Figure 3.59

3.5 Exercises

Review   Concepts, Skills, and Problem Solving

Keep mathematically in shape by doing these exercises *before* the problems of this section.

Properties and Definitions

1. *Writing* ✎   Explain what the slope-intercept form of a line is, and explain how it can be used to help sketch the line.

2. *Writing* ✎   Explain what the point-slope form of a line is, and explain how it can be used to help determine an equation of the line.

Solving Inequalities

In Exercises 3–8, solve the inequality.

3. $7 - 3x > 4 - x$

4. $2(x + 6) - 20 < 2$

5. $\dfrac{x}{6} + \dfrac{x}{4} < 1$

6. $\dfrac{5 - x}{2} \geq 8$

7. $|x - 3| < 2$

8. $|x - 5| > 3$

Graphing

In Exercises 9–12, graph the equation.

9. $3x - y = 4$

10. $2x + 2y = 6$

11. $y = 3x^2 + 1$

12. $2x^2 + y - 4 = 0$

Developing Skills

In Exercises 1–8, determine whether each point is a solution of the inequality. See Example 1.

Inequality	Points

1. $x - 2y < 4$ (a) $(0, 0)$ (b) $(2, -1)$
(c) $(3, 4)$ (d) $(5, 1)$

2. $x + y < 3$ (a) $(0, 6)$ (b) $(4, 0)$
(c) $(0, -2)$ (d) $(1, 1)$

3. $3x + y \geq 10$ (a) $(1, 3)$ (b) $(-3, 1)$
(c) $(3, 1)$ (d) $(2, 15)$

4. $-3x + 5y \geq 6$ (a) $(2, 8)$ (b) $(-10, -3)$
(c) $(0, 0)$ (d) $(3, 3)$

Inequality	Points

5. $y > 0.2x - 1$ (a) $(0, 2)$ (b) $(6, 0)$
(c) $(4, -1)$ (d) $(-2, 7)$

6. $y < -3.5x + 7$ (a) $(1, 5)$ (b) $(5, -1)$
(c) $(-1, 4)$ (d) $\left(0, \frac{4}{3}\right)$

7. $y \leq 3 - |x|$ (a) $(-1, 4)$ (b) $(2, -2)$
(c) $(6, 0)$ (d) $(5, -2)$

8. $y \geq |x - 3|$ (a) $(0, 0)$ (b) $(1, 2)$
(c) $(4, 10)$ (d) $(5, -1)$

In Exercises 9–14, match the inequality with its graph. [The graphs are labeled (a), (b), (c), (d), (e), and (f).]

(a)
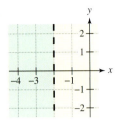

(b)
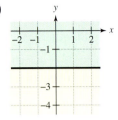

(c)
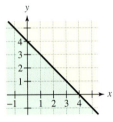

(d)
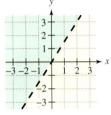

(e)
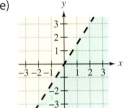

(f)
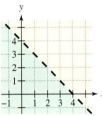

9. $y \geq -2$

10. $x < -2$

11. $3x - 2y < 0$

12. $3x - 2y > 0$

13. $x + y < 4$

14. $x + y \leq 4$

In Exercises 15–38, sketch the graph of the solution of the linear inequality. See Examples 2–5.

15. $x \geq 6$

16. $x < -3$

17. $y < 5$

18. $y > 2$

19. $y > \frac{1}{2}x$

20. $y \leq 2x$

21. $y \geq 3 - x$

22. $y > x + 6$

23. $y \leq x + 2$

24. $y \leq 1 - x$

25. $x + y \geq 4$

26. $x + y \leq 5$

27. $x - 2y \geq 6$

28. $3x + y \leq 9$

29. $3x + 2y \geq 2$

30. $3x + 5y \leq 15$

31. $3x - 2 \leq 5x + y$

32. $2x - 2y \geq 8 + 2y$

33. $0.2x + 0.3y < 2$

34. $0.25x - 0.75y > 6$

35. $y - 1 > -\frac{1}{2}(x - 2)$

36. $y - 2 < -\frac{2}{3}(x - 3)$

37. $\frac{x}{3} + \frac{y}{4} \leq 1$

38. $\frac{x}{2} + \frac{y}{6} \geq 1$

In Exercises 39–46, use a graphing calculator to graph the solution of the inequality.

39. $y \geq \frac{3}{4}x - 1$

40. $y \leq 9 - \frac{3}{2}x$

41. $y \leq -\frac{2}{3}x + 6$

42. $y \geq \frac{1}{4}x + 3$

43. $x - 2y - 4 \geq 0$

44. $2x + 4y - 3 \leq 0$

45. $2x + 3y - 12 \leq 0$

46. $x - 3y + 9 \geq 0$

In Exercises 47–52, write an inequality for the shaded region shown in the figure.

47.
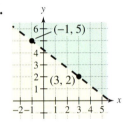

48.
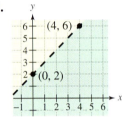

49.
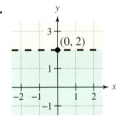

50.
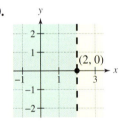

51.
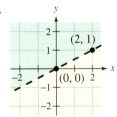

52.
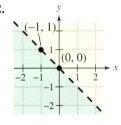

Solving Problems

53. ▲ *Geometry* The perimeter of a rectangle of length x and width y cannot exceed 500 inches.

(a) Write a linear inequality for this constraint.

(b) ▦ Use a graphing calculator to graph the solution of the inequality.

54. ▲ *Geometry* The perimeter of a rectangle of length x and width y must be at least 100 centimeters.

(a) Write a linear inequality for this constraint.

(b) ▦ Use a graphing utility to graph the solution of the inequality.

55. *Storage Space* A warehouse for storing chairs and tables has 1000 square feet of floor space. Each chair requires 10 square feet of floor space and each table requires 15 square feet.

(a) Write a linear inequality for this space constraint where x is the number of chairs and y is the number of tables stored.

(b) Sketch a graph of the solution of the inequality.

56. *Storage Space* A warehouse for storing desks and filing cabinets has 2000 square feet of floor space. Each desk requires 15 square feet of floor space and each filing cabinet requires 6 square feet.

(a) Write a linear inequality for this space constraint where x is the number of desks and y is the number of filing cabinets stored.

(b) Sketch a graph of the solution of the inequality.

57. *Consumerism* You and some friends go out for pizza. Together you have $26. You want to order two large pizzas with cheese at $8 each. Each additional topping costs $0.40, and each small soft drink costs $0.80.

(a) Write an inequality that represents the number of toppings x and drinks y that your group can afford. (Assume there is no sales tax.)

(b) Sketch a graph of the solution of the inequality.

(c) What are the coordinates for an order of six soft drinks and two large pizzas with cheese, each with three additional toppings? Is this a solution of the inequality?

58. *Consumerism* You and some friends go out for pizza. Together you have $48. You want to order three large pizzas with cheese at $9 each. Each additional topping costs $1, and each soft drink costs $1.50.

(a) Write an inequality that represents the number of toppings x and drinks y that your group can afford. (Assume there is no sales tax.)

(b) Sketch a graph of the solution of the inequality.

(c) What are the coordinates for an order of eight soft drinks and three large pizzas with cheese, each with two additional toppings? Is this a solution of the inequality?

59. *Nutrition* A dietitian is asked to design a special diet supplement using two foods. Each ounce of food X contains 12 units of protein and each ounce of food Y contains 16 units of protein. The minimum daily requirement in the diet is 250 units of protein.

(a) Write an inequality that represents the different numbers of units of food X and food Y required.

(b) Sketch a graph of the solution of the inequality. From the graph, find several ordered pairs with positive integer coordinates that are solutions of the inequality.

60. *Nutrition* A dietitian is asked to design a special diet supplement using two foods. Each ounce of food X contains 30 units of calcium and each ounce of food Y contains 20 units of calcium. The minimum daily requirement in the diet is 300 units of calcium.

(a) Write an inequality that represents the different numbers of units of food X and food Y required.

(b) Sketch a graph of the solution of the inequality. From the graph, find several ordered pairs with positive integer coordinates that are solutions of the inequality.

61. *Weekly Pay*   You have two part-time jobs. One is at a grocery store, which pays $9 per hour, and the other is mowing lawns, which pays $6 per hour. Between the two jobs, you want to earn at least $150 a week.

(a) Write an inequality that shows the different numbers of hours you can work at each job.

(b) Sketch the graph of the solution of the inequality. From the graph, find several ordered pairs with positive integer coordinates that are solutions of the inequality.

62. *Weekly Pay*   You have two part-time jobs. One is at a candy store, which pays $6 per hour, and the other is providing childcare, which pays $5 per hour. Between the two jobs, you want to earn at least $120 a week.

(a) Write an inequality that shows the different numbers of hours you can work at each job.

(b) Sketch the graph of the solution of the inequality. From the graph, find several ordered pairs with positive integer coordinates that are solutions of the inequality.

63. *Sports*   Your hockey team needs at least 70 points for the season in order to advance to the playoffs. Your team finishes with w wins, each worth 2 points, and t ties, each worth 1 point.

(a) Write a linear inequality that shows the different numbers of wins and ties your team must record in order to advance to the playoffs.

(b) Sketch the graph of the solution of the inequality. From the graph, find several ordered pairs with positive integer coordinates that are solutions of the inequality.

64. *Exercise*   The maximum heart rate r (in beats per minute) of a person in normal health is related to the person's age A (in years). The relationship between r and A is given by $r \le 220 - A$. (Source: American Heart Association)

(a) Sketch a graph of the solution of the inequality.

(b) Physicians recommend that during a workout a person strive to increase his or her heart rate to 75% of the maximum rate for the person's age. Sketch the graph of $r = 0.75(220 - A)$ on the same set of coordinate axes used in part (a).

Explaining Concepts

65. List the four forms of a linear inequality in variables x and y.

66. *Writing* ✐   What is meant by saying that (x_1, y_1) is a solution of a linear inequality in x and y?

67. *Writing* ✐   Explain the meaning of the term *half-plane*. Give an example of an inequality whose graph is a half-plane.

68. How does the solution of $x - y > 1$ differ from the solution of $x - y \ge 1$?

69. *Writing* ✐   After graphing the corresponding equation, how do you decide which half-plane is the solution of a linear inequality?

70. *Writing* ✐   Explain the difference between graphing the solution of the inequality $x \le 3$ on the real number line and graphing it on a rectangular coordinate system.

3.6 Relations and Functions

Charles Gupton/Corbis

What You Should Learn

1. Identify the domains and ranges of relations.
2. Determine if relations are functions by inspection.
3. Use function notation and evaluate functions.
4. Identify the domains and ranges of functions.

Why You Should Learn It

Functions can be used to model and solve real-life problems. For instance, in Exercise 79 on page 195, a function is used to find the cost of producing a new video game.

Relations

Many everyday occurrences involve two quantities that are paired or matched with each other by some rule of correspondence. The mathematical term for such a correspondence is **relation.**

① Identify the domains and ranges of relations.

Definition of Relation

A **relation** is any set of ordered pairs. The set of first components in the ordered pairs is the **domain** of the relation. The set of second components is the **range** of the relation.

Example 1   Analyzing a Relation

Find the domain and range of the relation $\{(0, 1), (1, 3), (2, 5), (3, 5), (0, 3)\}$.

Solution

The domain is the set of all first components of the relation, and the range is the set of all second components.

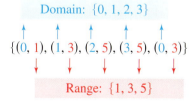

A graphical representation of this relation is shown in Figure 3.60.

Study Tip

It is not necessary to list repeated components of the domain and range of a relation.

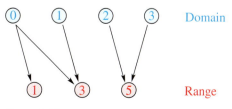

Figure 3.60

② Determine if relations are functions by inspection.

Functions

In modeling real-life situations, you will work with a special type of relation called a function. A **function** is a relation in which no two ordered pairs have the same first component and different second components. For instance, (2, 3) and (2, 4) could not be ordered pairs of a function.

Definition of a Function

A **function** f from a set A to a set B is a rule of correspondence that assigns to each element x in the set A exactly one element y in the set B. The set A is called the **domain** (or set of inputs) of the function f, and the set B is called the **range** (or set of outputs) of the function.

The rule of correspondence for a function establishes a set of "input-output" ordered pairs of the form (x, y), where x is an input and y is the corresponding output. In some cases, the rule may generate only a finite set of ordered pairs, whereas in other cases the rule may generate an infinite set of ordered pairs.

Example 2   Input-Output Ordered Pairs for Functions

Write a set of ordered pairs that represents the rule of correspondence.

a. Winner of the Super Bowl in 2000, 2001, 2002, and 2003

b. The squares of all real numbers

Solution

a. For the function that pairs the year from 2000 to 2003 with the winner of the Super Bowl, each ordered pair is of the form (year, winner).

$\{$(2000, Rams), (2001, Ravens), (2002, Patriots), (2003, Buccaneers)$\}$

b. For the function that pairs each real number with its square, each ordered pair is of the form (x, x^2).

$\{$All points (x, x^2), where x is a real number$\}$

> **Study Tip**
>
> In Example 2, the set in part (a) has only finite numbers of ordered pairs, whereas the set in part (b) has an infinite number of ordered pairs.

Functions are commonly represented in four ways.

1. *Verbally* by a sentence that describes how the input variable is related to the output variable.

2. *Numerically* by a table or a list of ordered pairs that matches input values with output values.

3. *Graphically* by points on a graph in a coordinate plane in which the input values are represented by the horizontal axis and the output values are represented by the vertical axis.

4. *Algebraically* by an equation in two variables.

A function has certain characteristics that distinguish it from a relation. To determine whether or not a relation is a function, use the following list of characteristics of a function.

Characteristics of a Function

1. Each element in the domain A must be matched with an element in the range, which is contained in the set B.

2. Some elements in set B may not be matched with any element in the domain A.

3. Two or more elements of the domain may be matched with the same element in the range.

4. No element of the domain is matched with two different elements in the range.

Example 3   Test for Functions

Decide whether the description represents a function.

a. The input value x is any of the 50 states in the United States and the output value y is the number of governors of that state.

b. $\{(1, 1), (2, 4), (3, 9), (4, 16), (5, 25), (6, 36)\}$

c.

Input, x	Output, y
-1	7
3	2
4	0
3	4

d. Let $A = \{a, b, c\}$ and let $B = \{1, 2, 3, 4, 5\}$.

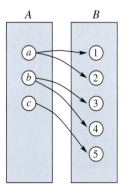

Solution

a. This set of ordered pairs *does* represent a function. Regardless of the input value x, the output value is always 1.

b. This set of ordered pairs *does* represent a function. No input value is matched with *two* output values.

c. This table *does not* represent a function. The input value 3 is matched with *two* different output values, 2 and 4.

d. This diagram *does not* represent a function. The element a in set A is matched with *two* elements in set B. This is also true of b.

Representing functions by sets of ordered pairs is a common practice in the study of *discrete mathematics*, which deals mainly with finite sets of data or with finite subsets of the set of real numbers. In algebra, however, it is more common to represent functions by equations or formulas involving two variables. For instance, the equation

$$y = x^2 \qquad \text{Squaring function}$$

represents the variable y as a function of the variable x. The variable x is the **independent variable** and the variable y is the **dependent variable.** In this context, the domain of the function is the set of all *allowable* real values for the independent variable x, and the range of the function is the *resulting* set of all values taken on by the dependent variable y.

Example 4   Testing for Functions Represented by Equations

Which of the equations represent y as a function of x?

a. $y = x^2 + 1$ **b.** $x - y^2 = 2$ **c.** $-2x + 3y = 4$

Solution

a. For the equation

$$y = x^2 + 1$$

just one value of y corresponds to each value of x. For instance, when $x = 1$, the value of y is

$$y = 1^2 + 1 = 2.$$

So, y *is* a function of x.

b. By writing the equation $x - y^2 = 2$ in the form

$$y^2 = x - 2$$

you can see that two values of y correspond to some values of x. For instance, when $x = 3$,

$$y^2 = 3 - 2$$
$$y^2 = 1$$
$$y = 1 \quad \text{or} \quad y = -1.$$

So, the solution points $(3, 1)$ and $(3, -1)$ show that y *is not* a function of x.

c. By writing the equation $-2x + 3y = 4$ in the form

$$y = \frac{2}{3}x + \frac{4}{3}$$

you can see that just one value of y corresponds to each value of x. For instance, when $x = 2$, the value of y is $\frac{4}{3} + \frac{4}{3} = \frac{8}{3}$. So, y *is* a function of x.

An equation that defines y as a function of x may or may not also define x as a function of y. For instance, the equation in part (a) of Example 4 does not define x as a function of y, but the equation in part (c) does.

③ Use function notation and evaluate functions.

Function Notation

When an equation is used to represent a function, it is convenient to name the function so that it can be easily referenced. For example, the function $y = x^2 + 1$ in Example 4(a) can be given the name "f" and written in **function notation** as

$$f(x) = x^2 + 1.$$

Function Notation

In the notation $f(x)$:

f is the **name** of the function.

x is the **domain** (or input) value.

$f(x)$ is the **range** (or output) value y for a given x.

The symbol $f(x)$ is read as *the value of f at x* or simply *f of x*.

The process of finding the value of $f(x)$ for a given value of x is called **evaluating a function.** This is accomplished by substituting a given x-value (input) into the equation to obtain the value of $f(x)$ (output). Here is an example.

Function	*x-Value*	*Function Value*
$f(x) = 3 - 4x$	$x = -1$	$f(-1) = 3 - 4(-1)$
		$= 3 + 4$
		$= 7$

Although f is often used as a convenient function name and x as the independent variable, you can use other letters. For instance, the equations

$$f(x) = 2x^2 + 5, \quad f(t) = 2t^2 + 5, \quad \text{and} \quad g(s) = 2s^2 + 5$$

all define the same function. In fact, the letters used are simply "placeholders," and this same function is well described by the form

$$f(\quad) = 2(\quad)^2 + 5$$

where the parentheses are used in place of a letter. To evaluate $f(-2)$, simply place -2 in each set of parentheses, as follows.

$$f(-2) = 2(-2)^2 + 5$$
$$= 2(4) + 5$$
$$= 8 + 5$$
$$= 13$$

When evaluating a function, you are not restricted to substituting only numerical values into the parentheses. For instance, the value of $f(3x)$ is

$$f(3x) = 2(3x)^2 + 5$$
$$= 2(9x^2) + 5$$
$$= 18x^2 + 5.$$

Example 5   Evaluating a Function

Let $g(x) = 3x - 4$. Find each value of the function.

a. $g(1)$ **b.** $g(-2)$ **c.** $g(y)$ **d.** $g(x + 1)$ **e.** $g(x) + g(1)$

Solution

a. Replacing x with 1 produces $g(1) = 3(1) - 4 = 3 - 4 = -1$.

b. Replacing x with -2 produces $g(-2) = 3(-2) - 4 = -6 - 4 = -10$.

c. Replacing x with y produces $g(y) = 3(y) - 4 = 3y - 4$.

d. Replacing x with $(x + 1)$ produces

$$g(x + 1) = 3(x + 1) - 4 = 3x + 3 - 4 = 3x - 1.$$

e. Using the result of part (a) for $g(1)$, you have

$$g(x) + g(1) = (3x - 4) + (-1) = 3x - 4 - 1 = 3x - 5.$$

Study Tip

Note that

$$g(x + 1) \neq g(x) + g(1).$$

In general, $g(a + b)$ is not equal to $g(a) + g(b)$.

Sometimes a function is defined by more than one equation, each of which is given a portion of the domain. Such a function is called a **piecewise-defined function.** To evaluate a piecewise-defined function f for a given value of x, first determine the portion of the domain in which the x-value lies and then use the corresponding equation to evaluate f. This is illustrated in Example 6.

Example 6   A Piecewise-Defined Function

Let $f(x) = \begin{cases} x^2 + 1, & \text{if } x < 0 \\ x - 2, & \text{if } x \geq 0 \end{cases}$. Find each value of the function.

a. $f(-1)$ **b.** $f(0)$ **c.** $f(-2)$ **d.** $f(-3) + f(4)$

Solution

a. Because $x = -1 < 0$, use $f(x) = x^2 + 1$ to obtain

$$f(-1) = (-1)^2 + 1 = 1 + 1 = 2.$$

b. Because $x = 0 \geq 0$, use $f(x) = x - 2$ to obtain

$$f(0) = 0 - 2 = -2.$$

c. Because $x = -2 < 0$, use $f(x) = x^2 + 1$ to obtain

$$f(-2) = (-2)^2 + 1 = 4 + 1 = 5.$$

d. Because $x = -3 < 0$, use $f(x) = x^2 + 1$ to obtain

$$f(-3) = (-3)^2 + 1 = 9 + 1 = 10.$$

Because $x = 4 \geq 0$, use $f(x) = x - 2$ to obtain

$$f(4) = 4 - 2 = 2.$$

So, $f(-3) + f(4) = 10 + 2 = 12$.

④ Identify the domains and ranges of functions.

Finding the Domain and Range of a Function

The domain of a function may be explicitly described along with the function, or it may be *implied* by the expression used to define the function. The **implied domain** is the set of all real numbers (inputs) that yield real number values for the function. For instance, the function given by

$$f(x) = \frac{1}{x - 3} \qquad \text{Domain: all } x \neq 3$$

has an implied domain that consists of all real values of x other than $x = 3$. The value $x = 3$ is excluded from the domain because division by zero is undefined. Another common type of implied domain is that used to avoid even roots of negative numbers. For instance, the function given by

$$f(x) = \sqrt{x} \qquad \text{Domain: all } x \geq 0$$

is defined only for $x \geq 0$. So, its implied domain is the set of all real numbers x such that $x \geq 0$. More will be said about the domains of square root functions in Chapter 7.

Example 7   Finding the Domain of a Function

Find the domain of each function.

a. $f(x) = \sqrt{2x - 6}$ **b.** $g(x) = \dfrac{4x}{(x - 1)(x + 5)}$

Solution

a. The domain of f consists of all real numbers x such that $2x - 6 \geq 0$. Solving this inequality yields

$2x - 6 \geq 0$	Original inequality
$2x \geq 6$	Add 6 to each side.
$x \geq 3.$	Divide each side by 2.

So, the domain consists of all real numbers x such that $x \geq 3$.

b. The domain of g consists of all real numbers x such that the denominator is not equal to zero. The denominator will be equal to zero when either factor of the denominator is zero.

First Factor

$x - 1 = 0$	Set the first factor equal to zero.
$x = 1$	Add 1 to each side.

Second Factor

$x + 5 = 0$	Set the second factor equal to zero.
$x = -5$	Subtract 5 from each side.

So, the domain consists of all real numbers x such that $x \neq 1$ and $x \neq -5$.

Example 8   Finding the Domain and Range of a Function

Find the domain and range of each function.

a. $f: \{(-3, 0), (-1, 2), (0, 4), (2, 4), (4, -1)\}$

b. Area of a circle: $A = \pi r^2$

Solution

a. The domain of f consists of all first coordinates in the set of ordered pairs. The range consists of all second coordinates in the set of ordered pairs. So, the domain and range are as follows.

$$\text{Domain} = \{-3, -1, 0, 2, 4\} \qquad \text{Range} = \{0, 2, 4, -1\}$$

b. For the area of a circle, you must choose positive values for the radius r. So, the domain is the set of all real numbers r such that $r > 0$. The range is therefore the set of all real numbers A such that $A > 0$.

Note in Example 8(b) that the domain of a function can be implied by a physical context. For instance, from the equation $A = \pi r^2$, you would have no strictly mathematical reason to restrict r to positive values. However, because you know that this function represents the area of a circle, you can conclude that the radius must be positive.

Example 9   Geometry: The Dimensions of a Container

You work in the marketing department of a soft-drink company and are experimenting with a new can for iced tea that is slightly narrower and taller that a standard can. For your experimental can, the ratio of the height to the radius is 4, as shown in Figure 3.61.

a. Write the volume of the can as a function of the radius r. Find the domain of the function.

b. Write the volume of the can as a function of the height h. Find the domain of the function.

Solution

The volume of a circular cylinder is given by the formula $V = \pi r^2 h$.

a. Because the ratio of the height of the can to the radius is 4, $h = 4r$. Substitute this value of h into the formula to obtain

$$V(r) = \pi r^2 h = \pi r^2 (4r) = 4\pi r^3. \qquad \text{\color{red} Write } V \text{ as a function of } r.$$

The domain is the set of all real numbers r such that $r > 0$.

b. You know that $h = 4r$ from part (a), so you can determine that $r = h/4$. Substitute this value of r into the formula to obtain

$$V(h) = \pi r^2 h = \pi \left(\frac{h}{4}\right)^2 h = \frac{\pi h^3}{16}. \qquad \text{\color{red} Write } V \text{ as a function of } h.$$

The domain is the set of all real numbers h such that $h > 0$.

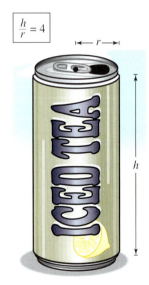

$\dfrac{h}{r} = 4$

Figure 3.61

3.6 Exercises

Review   Concepts, Skills, and Problem Solving

Keep mathematically in shape by doing these exercises *before* the problems of this section.

Properties and Definitions

1. If $a < b$ and $b < c$, then what is the relationship between a and c? Identify this property of inequalities.

2. Demonstrate the Multiplication Property of Equality for the equation $9x = 36$.

3. Use inequality notation to write the statement, "y is no more than 45."

4. Use inequality notation to write the statement, "x is at least 15."

Simplifying Expressions

In Exercises 5–10, simplify the expression.

5. $6y - 3x + 3x - 10y$

6. $8(x - 2) - 3(x - 2)$

7. $\frac{2}{3}t - \frac{5}{8} + \frac{5}{6}t$

8. $\frac{3}{8}x - \frac{1}{12}x + 8$

9. $3x^2 - 5x + 3 + 28x - 33x^2$

10. $4x^3 - 3x^2y + 4xy^2 + 15x^2y + y^3$

Problem Solving

11. *Cooking*   Two and one-half cups of flour are required to make one batch of cookies. How many cups are required to make $3\frac{1}{2}$ batches?

12. *Fuel*   The gasoline-to-oil ratio of a two-cycle engine is 32 to 1. Determine the amount of gasoline required to produce a mixture that has $\frac{1}{2}$ pint of oil.

Developing Skills

In Exercises 1–4, find the domain and the range of the relation. Then draw a graphical representation of the relation. See Example 1.

1. $\{(-2, 0), (0, 1), (1, 4), (0, -1)\}$

2. $\{(3, 10), (4, 5), (6, -2), (8, 3)\}$

3. $\{(0, 0), (4, -3), (2, 8), (5, 5), (6, 5)\}$

4. $\{(-3, 6), (-3, 2), (-3, 5)\}$

In Exercises 5–10, write a set of ordered pairs that represents the rule of correspondence. See Example 2.

5. In a week, a salesperson travels a distance d in t hours at an average speed of 50 miles per hour. The travel times for each day are 3 hours, 2 hours, 8 hours, 6 hours, and $\frac{1}{2}$ hour.

6. A court stenographer translates and types a court record of w words for t minutes at a rate of 60 words per minute. The amounts of time spent for each page of the court record are 8 minutes, 10 minutes, 7.5 minutes, and 4 minutes.

7. The cubes of all positive integers less than 8

8. The cubes of all integers greater than -2 and less than 5

9. The winners of the World Series from 1999 to 2002

10. The men inaugurated as president of the United States in 1981, 1985, 1989, 1993, 1997, and 2001.

In Exercises 11–22, determine whether the relation is a function. See Example 3.

11. Domain   Range
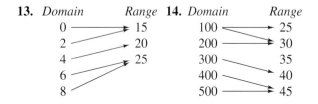

12. Domain   Range

13. Domain   Range **14.** Domain   Range
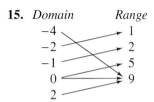

15. Domain   Range
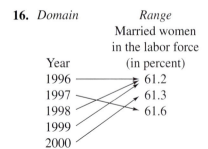

16. Domain   Range
Married women
in the labor force
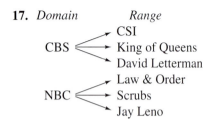

Year	(in percent)
1996	61.2
1997	61.3
1998	61.6
1999	
2000	

(Source: U.S. Bureau of Labor Statistics)

17. Domain   Range

CBS — CSI
      King of Queens
      David Letterman

NBC — Law & Order
      Scrubs
      Jay Leno

18.   Domain   Range

CSI
King of Queens — CBS
David Letterman
Law & Order
Scrubs — NBC
Jay Leno

19.

Input value, x	Output value, y
0	0
1	1
2	4
3	9
4	16

20.

Input value, x	Output value, y
0	1
1	8
2	12
1	15
0	20

21.

Input value, x	Output value, y
4	2
7	4
9	6
7	8
4	10

22.

Input value, x	Output value, y
0	5
2	5
4	5
6	5
8	5

In Exercises 23 and 24, determine which sets of ordered pairs represent functions from A to B. See Example 3.

23. $A = \{0, 1, 2, 3\}$ and $B = \{-2, -1, 0, 1, 2\}$

(a) $\{(0, 1), (1, -2), (2, 0), (3, 2)\}$

(b) $\{(0, -1), (2, 2), (1, -2), (3, 0), (1, 1)\}$

(c) $\{(0, 0), (1, 0), (2, 0), (3, 0)\}$

(d) $\{(0, 2), (3, 0), (1, 1)\}$

24. $A = \{1, 2, 3\}$ and $B = \{9, 10, 11, 12\}$

(a) $\{(1, 10), (3, 11), (3, 12), (2, 12)\}$

(b) $\{(1, 10), (2, 11), (3, 12)\}$

(c) $\{(1, 10), (1, 9), (3, 11), (2, 12)\}$

(d) $\{(3, 9), (2, 9), (1, 12)\}$

In Exercises 25–28, show that both ordered pairs are solutions of the equation and explain why this implies that y is not a function of x.

25. $x^2 + y^2 = 25$; $(0, 5)$, $(0, -5)$

26. $x^2 + 4y^2 = 16$; $(0, 2)$, $(0, -2)$

27. $|y| = x + 2$; $(1, 3)$, $(1, -3)$

28. $|y - 2| = x$; $(2, 4)$, $(2, 0)$

In Exercises 29–34, explain why the equation represents y as a function of x. See Example 4.

29. $y = 10x + 12$

30. $y = 3 - 8x$

31. $3x + 7y - 2 = 0$

32. $x - 9y + 3 = 0$

33. $y = x(x - 10)$

34. $y = (x + 2)^2 + 3$

In Exercises 35–40, fill in each blank and simplify.

35. $f(x) = 3x + 5$
  (a) $f(2) = 3(\ \ \ \ \ \ \ \ ) + 5$
  (b) $f(-2) = 3(\ \ \ \ \ \ \ \ ) + 5$
  (c) $f(k) = 3(\ \ \ \ \ \ \ \ ) + 5$
  (d) $f(k + 1) = 3(\ \ \ \ \ \ \ \ ) + 5$

36. $f(x) = 6 - 2x$
  (a) $f(3) = 6 - 2(\ \ \ \ \ \ \ \ )$
  (b) $f(-4) = 6 - 2(\ \ \ \ \ \ \ \ )$
  (c) $f(n) = 6 - 2(\ \ \ \ \ \ \ \ )$
  (d) $f(n - 2) = 6 - 2(\ \ \ \ \ \ \ \ )$

37. $f(x) = 3 - x^2$
  (a) $f(0) = 3 - (\ \ \ \ \ \ \ \ )^2$
  (b) $f(-3) = 3 - (\ \ \ \ \ \ \ \ )^2$
  (c) $f(m) = 3 - (\ \ \ \ \ \ \ \ )^2$
  (d) $f(2t) = 3 - (\ \ \ \ \ \ \ \ )^2$

38. $f(x) = \sqrt{x + 8}$
  (a) $f(1) = \sqrt{(\ \ \ \ \ \ \ \ ) + 8}$
  (b) $f(-4) = \sqrt{(\ \ \ \ \ \ \ \ ) + 8}$
  (c) $f(h) = \sqrt{(\ \ \ \ \ \ \ \ ) + 8}$
  (d) $f(h - 8) = \sqrt{(\ \ \ \ \ \ \ \ ) + 8}$

39. $f(x) = \dfrac{x}{x + 2}$
  (a) $f(3) = \dfrac{(\ \ \ \ \ \ \ \ )}{(\ \ \ \ \ \ \ \ ) + 2}$
  (b) $f(-4) = \dfrac{(\ \ \ \ \ \ \ \ )}{(\ \ \ \ \ \ \ \ ) + 2}$
  (c) $f(s) = \dfrac{(\ \ \ \ \ \ \ \ )}{(\ \ \ \ \ \ \ \ ) + 2}$
  (d) $f(s - 2) = \dfrac{(\ \ \ \ \ \ \ \ )}{(\ \ \ \ \ \ \ \ ) + 2}$

40. $f(x) = \dfrac{2x}{x - 7}$
  (a) $f(2) = \dfrac{2(\ \ \ \ \ \ \ \ )}{(\ \ \ \ \ \ \ \ ) - 7}$
  (b) $f(-3) = \dfrac{2(\ \ \ \ \ \ \ \ )}{(\ \ \ \ \ \ \ \ ) - 7}$
  (c) $f(t) = \dfrac{2(\ \ \ \ \ \ \ \ )}{(\ \ \ \ \ \ \ \ ) - 7}$
  (d) $f(t + 5) = \dfrac{2(\ \ \ \ \ \ \ \ )}{(\ \ \ \ \ \ \ \ ) - 7}$

In Exercises 41–56, evaluate the function as indicated, and simplify. See Examples 5 and 6.

41. $f(x) = 12x - 7$
  (a) $f(3)$ (b) $f\left(\frac{3}{2}\right)$
  (c) $f(a) + f(1)$ (d) $f(a + 1)$

42. $f(x) = 3 - 7x$
  (a) $f(-1)$ (b) $f\left(\frac{1}{2}\right)$
  (c) $f(t) + f(-2)$ (d) $f(2t - 3)$

43. $g(x) = 2 - 4x + x^2$
  (a) $g(4)$ (b) $g(0)$
  (c) $g(2y)$ (d) $g(4) + g(6)$

44. $h(x) = x^2 - 2x$
  (a) $h(2)$ (b) $h(0)$
  (c) $h(1) - h(-4)$ (d) $h(4t)$

45. $f(x) = \sqrt{x + 5}$

   (a) $f(-1)$ (b) $f(4)$

   (c) $f(z - 5)$ (d) $f(5z)$

46. $h(x) = \sqrt{2x - 3}$

   (a) $h(4)$ (b) $h(2)$

   (c) $h(4n)$ (d) $h(n + 2)$

47. $g(x) = 8 - |x - 4|$

   (a) $g(0)$ (b) $g(8)$

   (c) $g(16) - g(-1)$ (d) $g(x - 2)$

48. $g(x) = 2|x + 1| - 2$

   (a) $g(2)$ (b) $g(-1)$

   (c) $g(-4)$ (d) $g(3) + g(-5)$

49. $f(x) = \dfrac{3x}{x - 5}$

   (a) $f(0)$ (b) $f\left(\frac{5}{3}\right)$

   (c) $f(2) - f(-1)$ (d) $f(x + 4)$

50. $f(x) = \dfrac{x + 2}{x - 3}$

   (a) $f(-3)$ (b) $f\left(-\frac{3}{2}\right)$

   (c) $f(4) + f(8)$ (d) $f(x - 5)$

51. $f(x) = \begin{cases} x + 8, & \text{if } x < 0 \\ 10 - 2x, & \text{if } x \geq 0 \end{cases}$

   (a) $f(4)$ (b) $f(-10)$

   (c) $f(0)$ (d) $f(6) - f(-2)$

52. $f(x) = \begin{cases} -x, & \text{if } x \leq 0 \\ 6 - 3x, & \text{if } x > 0 \end{cases}$

   (a) $f(0)$ (b) $f\left(-\frac{3}{2}\right)$

   (c) $f(4)$ (d) $f(-2) + f(25)$

53. $h(x) = \begin{cases} 4 - x^2, & \text{if } x \leq 2 \\ x - 2, & \text{if } x > 2 \end{cases}$

   (a) $h(2)$ (b) $h\left(-\frac{3}{2}\right)$

   (c) $h(5)$ (d) $h(-3) + h(7)$

54. $f(x) = \begin{cases} x^2, & \text{if } x < 1 \\ x^2 - 3x + 2, & \text{if } x \geq 1 \end{cases}$

   (a) $f(1)$ (b) $f(-1)$

   (c) $f(2)$ (d) $f(-3) + f(3)$

55. $f(x) = 2x + 5$

   (a) $\dfrac{f(x + 2) - f(2)}{x}$ (b) $\dfrac{f(x - 3) - f(3)}{x}$

56. $f(x) = 3x + 4$

   (a) $\dfrac{f(x + 1) - f(1)}{x}$ (b) $\dfrac{f(x - 5) - f(5)}{x}$

In Exercises 57–66, find the domain of the function. See Example 7.

57. $f(x) = x^2 + x - 2$

58. $h(x) = 3x^2 - x$

59. $f(t) = \dfrac{t + 3}{t(t + 2)}$

60. $g(s) = \dfrac{s - 2}{(s - 6)(s - 10)}$

61. $g(x) = \sqrt{x + 4}$

62. $f(x) = \sqrt{2 - x}$

63. $f(x) = \sqrt{2x - 1}$

64. $G(x) = \sqrt{8 - 3x}$

65. $f(t) = |t - 4|$

66. $f(x) = |x + 3|$

In Exercises 67–74, find the domain and range of the function. See Example 8.

67. f: $\{(0, 0), (2, 1), (4, 8), (6, 27)\}$

68. f: $\{(-3, 4), (-1, 3), (2, 0), (5, 3)\}$

69. f: $\left\{\left(-3, -\frac{17}{2}\right), \left(-1, -\frac{5}{2}\right), (4, 2), (10, 2)\right\}$

70. f: $\left\{\left(\frac{1}{2}, 4\right), \left(\frac{3}{4}, 5\right), (1, 6), \left(\frac{5}{4}, 7\right)\right\}$;

71. Circumference of a circle: $C = 2\pi r$

72. Area of a square with side s: $A = s^2$

73. Area of a circle with radius r: $A = \pi r^2$

74. Volume of a sphere with radius r: $V = \frac{4}{3}\pi r^3$

Solving Problems

75. ▲ *Geometry*   Write the perimeter P of a square as a function of the length x of one of its sides.

76. ▲ *Geometry*   Write the surface area S of a cube as a function of the length x of one of its edges.

77. ▲ *Geometry*   Write the volume V of a cube as a function of the length x of one of its edges.

78. ▲ *Geometry*   Write the length L of the diagonal of a square as a function of the length x of one of its sides.

79. *Cost*   The inventor of a new video game believes that the variable cost for producing the game is \$1.95 per unit and the fixed costs are \$8000. Write the total cost C as a function of x, the number of video games produced.

80. *Distance*   An airplane is flying at a speed of 230 miles per hour. Write the distance d traveled by the airplane as a function of time t in hours.

81. *Distance*   An airplane is flying at a speed of 120 miles per hour. Write the distance d traveled by the airplane as a function of time t in hours.

82. *Distance*   A car travels for 4 hours on a highway at a steady speed. Write the distance d traveled by the car as a function of its speed s in miles per hour.

83. *Distance*   A train travels at a speed of 65 miles per hour. Write the distance d traveled by the train as a function of time t in hours. Then find d when the value of t is 4.

84. *Distance*   A migrating bird flies at a steady speed for 8 hours. Write the distance d traveled by the bird as a function of its speed s in miles per hour. Then find d when the value of s is 35.

85. ▲ *Geometry*   An open box is to be made from a square piece of material 24 inches on a side by cutting equal squares from the corners and turning up the sides (see figure). Write the volume V of the box as a function of x.

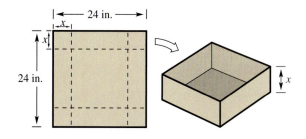

86. ▲ *Geometry*   Strips of width x are cut from the four sides of a square that is 32 inches on a side (see figure). Write the area A of the remaining square as a function of x.

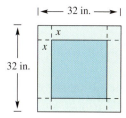

87. ▲ *Geometry*   Strips of width x are cut from two adjacent sides of a square that is 32 inches on a side (see figure). Write the area A of the remaining square as a function of x.

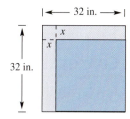

88. *Profit*   The marketing department of a business has determined that the profit from selling x units of a product is approximated by the model

$$P(x) = 50\sqrt{x} - 0.5x - 500.$$

Find (a) $P(1600)$ and (b) $P(2500)$.

89. *Safe Load*   A solid rectangular beam has a height of 6 inches and a width of 4 inches. The safe load S of the beam with the load at the center is a function of its length L and is approximated by the model

$$S(L) = \frac{128{,}160}{L}$$

where S is measured in pounds and L is measured in feet. Find (a) $S(12)$ and (b) $S(16)$.

90. *Wages*   A wage earner is paid $12.00 per hour for regular time and time-and-a-half for overtime. The weekly wage function is

$$W(h) = \begin{cases} 12h, & 0 < h \le 40 \\ 18(h - 40) + 480, & h > 40, \end{cases}$$

where h represents the number of hours worked in a week.

(a) Evaluate $W(30)$, $W(40)$, $W(45)$, and $W(50)$.

(b) Could you use values of h for which $h < 0$ in this model? Why or why not?

Data Analysis   In Exercises 91 and 92, use the graph, which shows the numbers of students (in millions) enrolled at all levels in public and private schools in the United States. (Source: U.S. National Center for Education Statistics)

91. Is the public school enrollment a function of the year? Is the private school enrollment a function of the year? Explain.

92. Let $f(x)$ represent the number of public school students in year x. Approximate $f(1999)$.

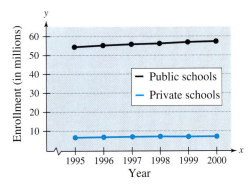

Figure for 91 and 92

Explaining Concepts

93. ⚡ Answer parts (g)–(i) of Motivating the Chapter on page 124.

94. *Writing* Explain the difference between a relation and a function.

95. *Writing* Is every relation a function? Explain.

96. *Writing* In your own words, explain the meanings of *domain* and *range*.

97. *Writing* Compile a list of statements describing relationships in everyday life. For each statement, identify the dependent and independent variables and discuss whether the statement *is* a function or *is not* a function and why.

98. *Writing* Describe an advantage of function notation.

In Exercises 99 and 100, determine whether the statements use the word *function* in ways that are *mathematically* correct.

99. (a) The sales tax on a purchased item is a function of the selling price.

(b) Your score on the next algebra exam is a function of the number of hours you study the night before the exam.

100. (a) The amount in your savings account is a function of your salary.

(b) The speed at which a free-falling baseball strikes the ground is a function of the height from which it was dropped.

3.7 Graphs of Functions

Robert Brenner/PhotoEdit Inc.

What You Should Learn

① Sketch graphs of functions on rectangular coordinate systems.

② Identify the graphs of basic functions.

③ Use the Vertical Line Test to determine if graphs represent functions.

④ Use vertical and horizontal shifts and reflections to sketch graphs of functions.

Why You Should Learn It

Graphs of functions can help you visualize relationships between variables in real-life situations. For instance, in Exercise 81 on page 207, the graph of a function visually represents the resident population of the United States.

The Graph of a Function

① Sketch graphs of functions on rectangular coordinate systems.

Consider a function f whose domain and range are the set of real numbers. The **graph** of f is the set of ordered pairs $(x, f(x))$, where x is in the domain of f.

$$x = x\text{-coordinate of the ordered pair}$$

$$f(x) = y\text{-coordinate of the ordered pair}$$

Figure 3.62 shows a typical graph of such a function.

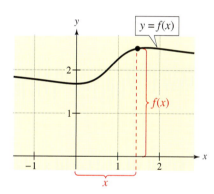

Figure 3.62

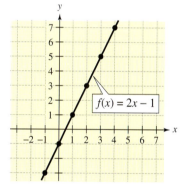

Figure 3.63

Study Tip

In Example 1, the (implied) domain of the function is the set of all real numbers. When writing the equation of a function, you may choose to restrict its domain by writing a condition to the right of the equation. For instance, the domain of the function

$$f(x) = 4x + 5, \quad x \geq 0$$

is the set of all nonnegative real numbers (all $x \geq 0$).

Example 1   Sketching the Graph of a Function

Sketch the graph of $f(x) = 2x - 1$.

Solution

One way to sketch the graph is to begin by making a table of values.

x	-1	0	1	2	3	4
$f(x)$	-3	-1	1	3	5	7

Next, plot the six points shown in the table. Finally, connect the points with a line, as shown in Figure 3.63.

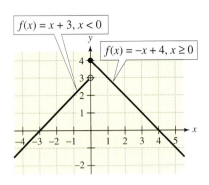

$f(x) = x + 3, x < 0$

$f(x) = -x + 4, x \geq 0$

Figure 3.64

Example 2   Sketching the Graph of a Piecewise-Defined Function

Sketch the graph of $f(x) = \begin{cases} x + 3, & x < 0 \\ -x + 4, & x \geq 0 \end{cases}$.

Solution

Begin by graphing $f(x) = x + 3$ for $x < 0$, as shown in Figure 3.64. You will recognize that this is the graph of the line $y = x + 3$ with the restriction that the x-values are negative. Because $x = 0$ is not in the domain, the right endpoint of the line is an open dot. Then graph $f(x) = -x + 4$ for $x \geq 0$ on the same set of coordinate axes, as shown in Figure 3.64. This is the graph of the line $y = -x + 4$ with the restriction that the x-values are nonnegative. Because $x = 0$ is in the domain, the left endpoint of the line is a solid dot.

2 Identify the graphs of basic functions.

Graphs of Basic Functions

To become good at sketching the graphs of functions, it helps to be familiar with the graphs of some basic functions. The functions shown in Figure 3.65, and variations of them, occur frequently in applications.

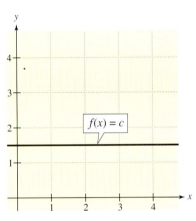

$f(x) = c$

(a) Constant function

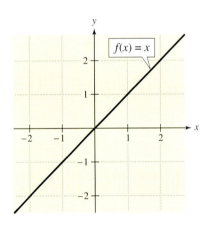

$f(x) = x$

(b) Identity function

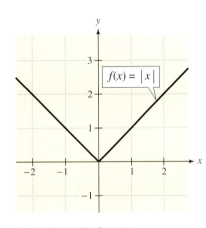

$f(x) = |x|$

(c) Absolute value function

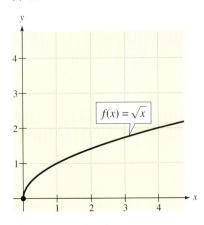

$f(x) = \sqrt{x}$

(d) Square root function

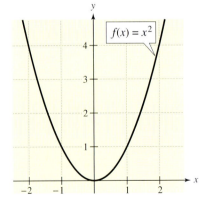

$f(x) = x^2$

(e) Squaring function

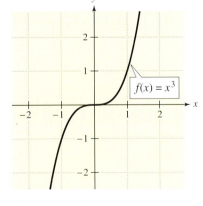

$f(x) = x^3$

(f) Cubing function

Figure 3.65

③ Use the Vertical Line Test to determine if graphs represent functions.

The Vertical Line Test

By the definition of a function, at most one y-value corresponds to a given x-value. This implies that any vertical line can intersect the graph of a function at most once.

Study Tip

The **Vertical Line Test** provides you with an easy way to determine whether an equation represents y as a function of x. If the graph of an equation has the property that no vertical line intersects the graph at two (or more) points, then the equation represents y as a function of x. On the other hand, if you can find a vertical line that intersects the graph at two (or more) points, then the equation does not represent y as a function of x, because there are two (or more) values of y that correspond to certain values of x.

Vertical Line Test for Functions

A set of points on a rectangular coordinate system is the graph of y as a function of x if and only if no vertical line intersects the graph at more than one point.

Example 3   Using the Vertical Line Test

Determine whether each equation represents y as a function of x.

a. $y = x^2 - 3x + \frac{1}{4}$

b. $x = y^2 - 1$

c. $x = y^3$

Solution

a. From the graph of the equation in Figure 3.66, you can see that every vertical line intersects the graph at most once. So, by the Vertical Line Test, the equation *does* represent y as a function of x.

b. From the graph of the equation in Figure 3.67, you can see that a vertical line intersects the graph twice. So, by the Vertical Line Test, the equation *does not* represent y as a function of x.

c. From the graph of the equation in Figure 3.68, you can see that every vertical line intersects the graph at most once. So, by the Vertical Line Test, the equation *does* represent y as a function of x.

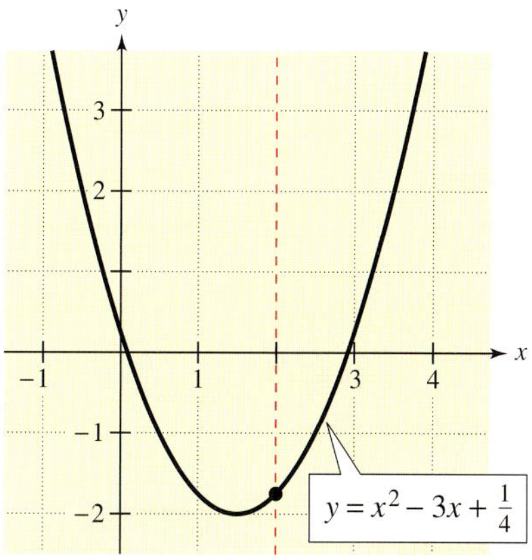

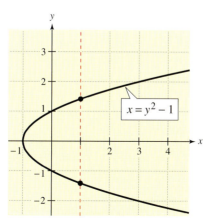

Graph of a function of x.

Vertical line intersects once.

Figure 3.66

Not a graph of a function of x.

Vertical line intersects twice.

Figure 3.67

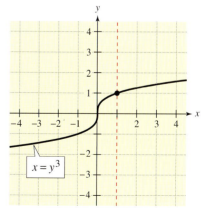

Graph of a function of x.

Vertical line intersects once.

Figure 3.68

④ Use vertical and horizontal shifts and reflections to sketch graphs of functions.

Transformations of Graphs of Functions

Many functions have graphs that are simple transformations of the basic graphs shown in Figure 3.65. The following list summarizes the various types of **vertical** and **horizontal shifts** of the graphs of functions.

Technology: Discovery

Use a graphing calculator to display the graphs of $y = x^2 + c$ where c is equal to $-2, 0, 2$, and 4. What conclusions can you make?

Use a graphing calculator to display the graphs of $y = (x + c)^2$ where c is equal to $-3, -1, 0, 1$, and 3. What conclusions can you make?

Vertical and Horizontal Shifts

Let c be a positive real number. **Vertical** and **horizontal shifts** of the graph of the function $y = f(x)$ are represented as follows.

1. Vertical shift c units *upward:* $h(x) = f(x) + c$

2. Vertical shift c units *downward:* $h(x) = f(x) - c$

3. Horizontal shift c units to the *right:* $h(x) = f(x - c)$

4. Horizontal shift c units to the *left:* $h(x) = f(x + c)$

Note that for a vertical transformation the addition of a positive number c yields a shift upward (in the positive direction) and the subtraction of a positive number c yields a shift downward (in the negative direction). For a horizontal transformation, replacing x with $x + c$ yields a shift to the left (in the negative direction) and replacing x with $x - c$ yields a shift to the right (in the positive direction).

Example 4   Shifts of the Graphs of Functions

Use the graph of $f(x) = x^2$ to sketch the graph of each function.

a. $g(x) = x^2 - 2$ **b.** $h(x) = (x + 3)^2$

Solution

a. Relative to the graph of $f(x) = x^2$, the graph of $g(x) = x^2 - 2$ represents a shift of two units *downward*, as shown in Figure 3.69.

b. Relative to the graph of $f(x) = x^2$, the graph of $h(x) = (x + 3)^2$ represents a shift of three units to the *left*, as shown in Figure 3.70.

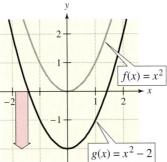

Vertical Shift: Two Units Downward
Figure 3.69

Horizontal Shift: Three Units Left
Figure 3.70

Some graphs can be obtained from *combinations* of vertical and horizontal shifts, as shown in part (b) of the next example.

Example 5   Shifts of the Graphs of Functions

Use the graph of $f(x) = x^3$ to sketch the graph of each function.

a. $g(x) = x^3 + 2$

b. $h(x) = (x - 1)^3 + 2$

Solution

a. Relative to the graph of $f(x) = x^3$, the graph of $g(x) = x^3 + 2$ represents a shift of two units *upward*, as shown in Figure 3.71.

b. Relative to the graph of $f(x) = x^3$, the graph of $h(x) = (x - 1)^3 + 2$ represents a shift of one unit to the *right*, followed by a shift of two units *upward*, as shown in Figure 3.72.

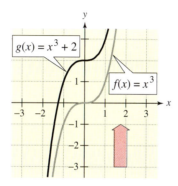

Vertical Shift: Two Units Upward
Figure 3.71

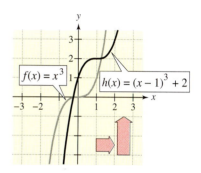

Horizontal Shift: One Unit Right
Vertical Shift: Two Units Upward
Figure 3.72

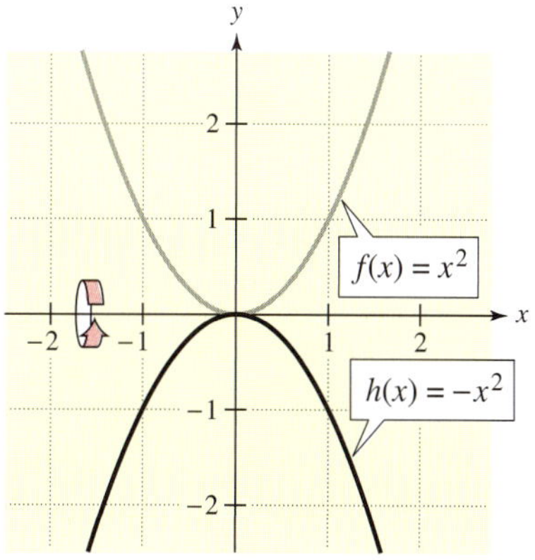

The second basic type of transformation is a **reflection.** For instance, if you imagine that the x-axis represents a mirror, then the graph of

$$h(x) = -x^2$$

is the mirror image (or reflection) of the graph of

$$f(x) = x^2$$

as shown in Figure 3.73.

Figure 3.73   Reflection

Reflections in the Coordinate Axes

Reflections of the graph of $y = f(x)$ are represented as follows.

1. Reflection in the x-axis: $h(x) = -f(x)$

2. Reflection in the y-axis: $h(x) = f(-x)$

Technology: Tip

A program called *parabola* can be found at our website *math.college.hmco.com/students.* This program will give you practice in working with reflections, horizontal shifts, and vertical shifts for a variety of graphing calculator models. The program will sketch the function

$$y = R(x + H)^2 + V$$

where R is ± 1, H is an integer between -6 and 6, and V is an integer between -3 and 3. After you determine the values for R, H, and V, the program will confirm your values by listing R, H, and V.

Example 6   Reflections of the Graphs of Functions

Use the graph of $f(x) = \sqrt{x}$ to sketch the graph of each function.

a. $g(x) = -\sqrt{x}$ **b.** $h(x) = \sqrt{-x}$

Solution

a. Relative to the graph of $f(x) = \sqrt{x}$, the graph of $g(x) = -\sqrt{x} = -f(x)$ represents a *reflection in the x-axis,* as shown in Figure 3.74.

b. Relative to the graph of $f(x) = \sqrt{x}$, the graph of $h(x) = \sqrt{-x} = f(-x)$ represents a *reflection in the y-axis,* as shown in Figure 3.75.

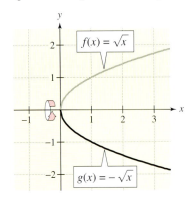

Reflection in *x*-Axis

Figure 3.74

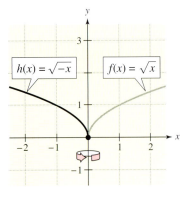

Reflection in *y*-Axis

Figure 3.75

Example 7   Graphical Reasoning

Identify the basic function, and any transformation shown in Figure 3.76. Write the equation for the graphed function.

Solution

In Figure 3.76, the basic function is $f(x) = |x|$. The transformation is a reflection in the *x*-axis and a vertical shift one unit downward. The equation of the graphed function is $f(x) = -|x| - 1$.

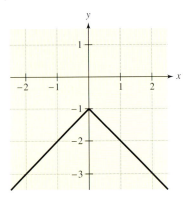

Figure 3.76

3.7 Exercises

Review   Concepts, Skills, and Problem Solving

Keep mathematically in shape by doing these exercises *before* the problems of this section.

Properties and Definitions

In Exercises 1–4, identify the property of real numbers illustrated by the statement.

1. $8x \cdot \dfrac{1}{8x} = 1$

2. $3x + 0 = 3x$

3. $-4(x + 10) = -4 \cdot x + (-4)(10)$

4. $5 + (-3 + x) = (5 - 3) + x$

Simplifying Expressions

In Exercises 5–10, simplify the expression.

5. $-(x + 2) - x$

6. $5y + 2 - (2y - 1)$

7. $7x + 3(8x - 1)$

8. $4y - 2(y + 9)$

9. $-1.2t + 0.9t^2 - 7.5t + 2.1t^2$

10. $4.8m^2 - 3.7m + 6.5m^2 + 10.8m$

Problem Solving

11. *Consumer Awareness*   A department store is offering a discount of 20% on a sewing machine with a list price of $239.95. A mail-order catalog has the same machine for $188.95 plus $4.32 for shipping. Which is the better bargain?

12. *Insurance Premium*   The annual automobile insurance premium for a policyholder is normally $739. However, after having an accident, the policyholder was charged an additional 30%. What is the new annual premium?

Developing Skills

In Exercises 1–28, sketch the graph of the function. Then determine its domain and range. See Examples 1 and 2.

1. $f(x) = 2x - 7$

2. $f(x) = 3 - 2x$

3. $g(x) = \frac{1}{2}x^2$

4. $h(x) = \frac{1}{4}x^2 - 1$

5. $f(x) = -(x - 1)^2$

6. $g(x) = (x + 2)^2 + 3$

7. $h(x) = x^2 - 6x + 8$

8. $f(x) = -x^2 - 2x + 1$

9. $C(x) = \sqrt{x} - 1$

10. $Q(x) = 4 - \sqrt{x}$

11. $f(t) = \sqrt{t - 2}$

12. $h(x) = \sqrt{4 - x}$

13. $G(x) = 8$

14. $H(x) = -4$

15. $g(s) = s^3 + 1$

16. $f(x) = x^3 - 4$

17. $f(x) = |x + 3|$

18. $g(x) = |x - 1|$

19. $K(s) = |s - 4| + 1$

20. $Q(t) = 1 - |t + 1|$

21. $f(x) = 6 - 3x, \quad 0 \le x \le 2$

22. $f(x) = \frac{1}{3}x - 2, \quad 6 \le x \le 12$

23. $h(x) = x^3, \quad -2 \le x \le 2$

24. $h(x) = 6x - x^2, \quad 0 \le x \le 6$

25. $h(x) = \begin{cases} 2x + 3, & x < 0 \\ 3 - x, & x \geq 0 \end{cases}$

26. $f(x) = \begin{cases} x + 6, & x < 0 \\ 6 - 2x, & x \geq 0 \end{cases}$

27. $f(x) = \begin{cases} 3 - x, & x < -3 \\ x^2 + x, & x \geq -3 \end{cases}$

28. $h(x) = \begin{cases} 4 - x^2 & x \leq 2 \\ x - 2, & x > 2 \end{cases}$

In Exercises 29–32, use a graphing calculator to graph the function and find its domain and range.

29. $g(x) = 1 - x^2$

30. $f(x) = 3x^3 - 4$

31. $f(x) = \sqrt{x - 2}$

32. $h(x) = \sqrt{4 - x^2}$

In Exercises 33–38, use the Vertical Line Test to determine whether y is a function of x. See Example 3.

33. $y = \frac{1}{3}x^3$

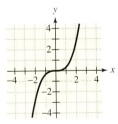

34. $y = x^2 - 2x$

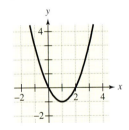

35. $y = (x + 2)^2$

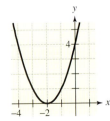

36. $x - 2y^2 = 0$

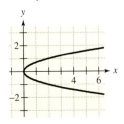

37. $x^2 + y^2 = 16$

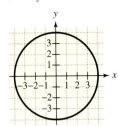

38. $|y| = x$

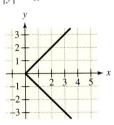

In Exercises 39–42, sketch a graph of the equation. Use the Vertical Line Test to determine whether y is a function of x.

39. $3x - 5y = 15$

40. $y = x^2 + 2$

41. $y^2 = x + 1$

42. $x = y^4$

In Exercises 43–46, match the function with its graph. [The graphs are labeled (a), (b), (c), and (d).]

(a)
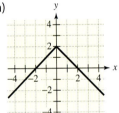

(b)
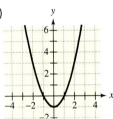

(c)
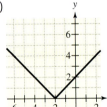

(d)
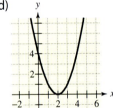

43. $f(x) = x^2 - 1$

44. $f(x) = (x - 2)^2$

45. $f(x) = 2 - |x|$

46. $f(x) = |x + 2|$

In Exercises 47 and 48, identify the transformation of f and sketch a graph of the function h. See Examples 4–6.

47. $f(x) = x^2$

(a) $h(x) = x^2 + 2$

(b) $h(x) = x^2 - 4$

(c) $h(x) = (x + 2)^2$

(d) $h(x) = (x - 4)^2$

(e) $h(x) = (x - 3)^2 + 1$

(f) $h(x) = -x^2 + 4$

48. $f(x) = x^3$

    (a) $h(x) = x^3 + 3$ (b) $h(x) = x^3 - 5$

    (c) $h(x) = (x - 3)^3$ (d) $h(x) = (x + 2)^3$

    (e) $h(x) = 2 - (x - 1)^3$ (f) $h(x) = -x^3$

In Exercises 49–54, identify the transformation of the graph of $f(x) = |x|$ and sketch the graph of h.

49. $h(x) = |x - 5|$

50. $h(x) = |x + 3|$

51. $h(x) = |x| - 5$

52. $h(x) = |-x|$

53. $h(x) = -|x|$

54. $h(x) = 5 - |x|$

In Exercises 55–60, use the graph of $f(x) = x^2$ to write a function that represents the graph.

55.
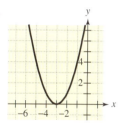

56.
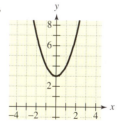

57.
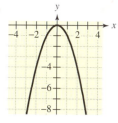

58.
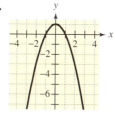

59.
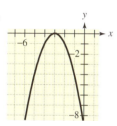

60.
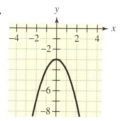

In Exercises 61–66, use the graph of $f(x) = \sqrt{x}$ to write a function that represents the graph.

61.
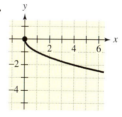

62.
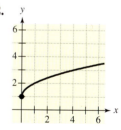

63.
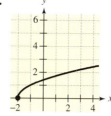

64.
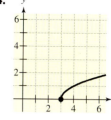

65.
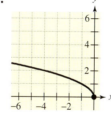

66.
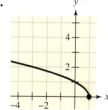

In Exercises 67 and 68, use a graphing calculator to graph f. Decide how to alter the function to produce each of the transformation descriptions. Graph each transformation in the same viewing window with f; confirm that the transformation moved f as described.

a. The graph of f shifted to the left three units

b. The graph of f shifted downward five units

c. The graph of f shifted upward one unit

d. The graph of f shifted to the right two units

67. $f(x) = x^2 + 2$ **68.** $f(x) = \sqrt{x + 3}$

In Exercises 69–74, identify the basic function, and any transformation shown in the graph. Write the equation for the graphed function. See Example 7.

69.

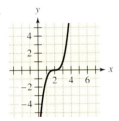

70.

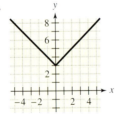

73.

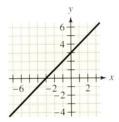

74.

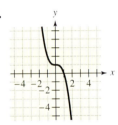

75. Use the graph of f to sketch each graph.

(a) $y = f(x) + 2$

(b) $y = -f(x)$

(c) $y = f(x - 2)$

(d) $y = f(x + 2)$

(e) $y = f(x) - 1$

(f) $y = f(-x)$

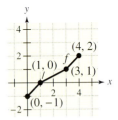

71.

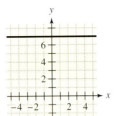

72.

76. Use the graph of f to sketch each graph.

(a) $y = f(x) - 1$

(b) $y = f(x + 1)$

(c) $y = f(x - 1)$

(d) $y = -f(x - 2)$

(e) $y = f(-x)$

(f) $y = f(x) + 2$

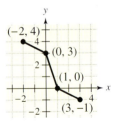

Solving Problems

77. 📷 *Automobile Engines* The percent x of antifreeze that prevents an automobile's engine coolant from freezing down to P degrees Fahrenheit is modeled by

$P = 26.0 - 0.0242x^2, \ 20 \le x \le 60.$

(Source: Standard Handbook for Mechanical Engineers)

(a) Use a graphing calculator to graph the model over the specified domain.

(b) Use the graph to approximate the value of x that yields protection against freezing at $-25°$ F.

78. *Profit* The profit P when x units of a product are sold is given by

$P(x) = 0.47x - 100, \ 0 \le x \le 1000.$

(a) 📷 Use a graphing calculator to graph the profit function over the specified domain.

(b) Approximately how many units must be sold for the company to break even $(P = 0)$?

(c) Approximately how many units must be sold for the company to make a profit of $300?

79. ▲ *Geometry*   The perimeter of a rectangle is 200 meters.

 (a) Show algebraically that the area of the rectangle is given by $A = l(100 - l)$, where l is its length.

 (b) ▦ Use a graphing calculator to graph the area function.

 (c) ▦ Use the graph to determine the value of l that yields the largest value of A. Interpret the result.

80. ▲ *Geometry*   The length and width of a rectangular flower garden are 40 feet and 30 feet, respectively. A walkway of uniform width x surrounds the garden.

 (a) Write the outside perimeter y of the walkway as a function of x.

 (b) ▦ Use a graphing calculator to graph the function for the perimeter.

 (c) ▦ Determine the slope of the graph in part (b). For each additional one-foot increase in the width of the walkway, determine the increase in its outside perimeter.

81. *Population*   For the years 1950 through 2000, the resident population P (in millions) of the United States can be modeled by

$$P(t) = 0.003t^2 + 2.37t + 153.1, \quad 0 \le t \le 50$$

where $t = 0$ represents 1950. (Source: U.S. Census Bureau)

 (a) ▦ Use a graphing calculator to graph the function over the appropriate domain.

 (b) In the transformation of the population function $P_1(t) = 0.003(t + 30)^2 + 2.37(t + 30) + 153.1$, $t = 0$ corresponds to what calendar year? Explain.

 (c) ▦ Use a graphing calculator to graph P_1 over the appropriate domain.

82. *Graphical Reasoning*   An electronically controlled thermostat in a home is programmed to lower the temperature automatically during the night. The temperature T, in degrees Fahrenheit, is given in terms of t, the time on a 24-hour clock (see figure).

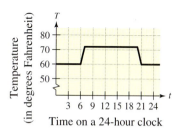

 (a) Explain why T is a function of t.

 (b) Find $T(4)$ and $T(15)$.

 (c) The thermostat is reprogrammed to produce a temperature H where $H(t) = T(t - 1)$. Explain how this changes the temperature in the house.

 (d) The thermostat is reprogrammed to produce a temperature H where $H(t) = T(t) - 1$. Explain how this changes the temperature in the house.

Explaining Concepts

83. *Writing*✎   Explain the change in the range of the function $f(x) = 2x$ if the domain is changed from $0 \le x \le 2$ to $0 \le x \le 4$.

84. *Writing*✎   In your own words, explain how to use the Vertical Line Test.

85. *Writing*✎   Describe the four types of shifts of the graph of a function.

86. *Writing*✎   Describe the relationship between the graphs of $f(x)$ and $g(x) = -f(x)$.

87. *Writing*✎   Describe the relationship between the graphs of $f(x)$ and $g(x) = f(-x)$.

88. *Writing*✎   Describe the relationship between the graphs of $f(x)$ and $g(x) = f(x - 2)$.

What Did You Learn?

Key Terms

rectangular coordinate system, *p. 126*
x-axis, *p. 126*
y-axis, *p. 126*
origin, *p. 126*
quadrants, *p. 126*
ordered pair, *p. 126*
x-coordinate, *p. 126*
y-coordinate, *p. 126*
Pythagorean Theorem, *p. 131*
Distance Formula, *p. 132*
Midpoint Formula, *p. 134*
graph (of an equation), *p. 140*

linear equation, *p. 140*
x-intercept, *p. 143*
y-intercept, *p. 143*
slope, *p. 149*
slope-intercept form, *p. 154*
point-slope form, *p. 163*
general form, *p. 163*
two-point form, *p. 164*
linear extrapolation, *p. 167*
linear interpolation, *p. 167*
linear inequality, *p. 174*
graph (of a linear inequality), *p. 175*

relation, *p. 183*
domain, *pp. 183, 184, 187*
range, *pp. 183, 184, 187*
function, *p. 184*
independent variable, *p. 186*
dependent variable, *p. 186*
function notation, *p. 187*
piecewise-defined function, *p. 188*
implied domain, *p. 189*
graph (of a function), *p. 197*
Vertical Line Test, *p. 199*

Key Concepts

3.1 ◯ Guidelines for verifying solutions

To verify that an ordered pair (x, y) is a solution of an equation with variables x and y, use the steps below.

1. Substitute the values of x and y into the equation.

2. Simplify each side of the equation.

3. If each side simplifies to the same number, the ordered pair is a solution. If the two sides yield different numbers, the ordered pair is not a solution.

3.2 ◯ The point-plotting method of sketching a graph

1. If possible, rewrite the equation by isolating one of the variables.

2. Make a table of values showing several solution points.

3. Plot these points on a rectangular coordinate system.

4. Connect the points with a smooth curve or line.

3.4 ◯ Summary of equations of lines

1. Slope of a line through (x_1, y_1) and (x_2, y_2):
   $$m = (y_2 - y_1)/(x_2 - x_1)$$

2. General form: $ax + by + c = 0$

3. Equation of vertical line: $x = a$

4. Equation of horizontal line: $y = b$

5. Slope-intercept form: $y = mx + b$

6. Point-slope form: $y - y_1 = m(x - x_1)$

7. Parallel lines (equal slopes): $m_1 = m_2$

8. Perpendicular lines (negative reciprocal slopes):
   $$m_1 = -1/m_2$$

3.6 ◯ Characteristics of a function

1. Each element in the domain A must be matched with an element in the range, which is contained in the set B.

2. Some elements in set B may not be matched with any element in the domain A.

3. Two or more elements of the domain may be matched with the same element in the range.

4. No element of the domain is matched with two different elements in the range.

3.7 ◯ Vertical and horizontal shifts

Let c be a positive real number. Vertical and horizontal shifts of the graph of the function $y = f(x)$ are represented as follows.

1. Vertical shift c units upward: $\quad h(x) = f(x) + c$

2. Vertical shift c units downward: $\quad h(x) = f(x) - c$

3. Horizontal shift c units to the right: $h(x) = f(x - c)$

4. Horizontal shift c units to the left: $h(x) = f(x + c)$

3.7 ◯ Reflections in the coordinate axes

Reflections of the graph of $y = f(x)$ are represented as:

1. Reflection in the x-axis: $\quad h(x) = -f(x)$

2. Reflection in the y-axis: $\quad h(x) = f(-x)$

Review Exercises

3.1 The Rectangular Coordinate System

1 Plot points on a rectangular coordinate system.

In Exercises 1 and 2, plot the points on a rectangular coordinate system.

1. $(0, -3)$, $\left(\frac{5}{2}, 5\right)$, $(-2, -4)$
2. $\left(1, -\frac{3}{2}\right)$, $\left(-2, 2\frac{3}{4}\right)$, $(5, 10)$

In Exercises 3 and 4, plot the points and connect them with line segments to form the figure.

3. *Right triangle:* $(1, 1)$, $(12, 9)$, $(4, 20)$
4. *Parallelogram:* $(0, 0)$, $(7, 1)$, $(8, 4)$, $(1, 3)$

In Exercises 5–8, determine the quadrant in which the point is located without plotting it. (x and y are real numbers.)

5. $(2, -6)$

6. $(-4.8, -2)$

7. $(4, y)$

8. (x, y), $xy > 0$

2 Determine whether ordered pairs are solutions of equations.

In Exercises 9 and 10, determine whether each ordered pair is a solution of the equation.

9. $y = 4 - \frac{1}{2}x$
   - (a) $(4, 2)$
   - (b) $(-1, 5)$
   - (c) $(-4, 0)$
   - (d) $(8, 0)$
10. $3x - 2y + 18 = 0$
    - (a) $(3, 10)$
    - (b) $(0, 9)$
    - (c) $(-4, 3)$
    - (d) $(-8, 0)$

3 Use the Distance Formula to find the distance between two points.

In Exercises 11–14, plot the points and find the distance between them.

11. $(4, 3)$, $(4, 8)$
12. $(2, -5)$, $(6, -5)$
13. $(-5, -1)$, $(1, 2)$
14. $(-2, 10)$, $(3, -2)$

In Exercises 15 and 16, use the Distance Formula to determine whether the three points are collinear.

15. $(-3, -2)$, $(1, 0)$, $(5, 2)$
16. $(-5, 7)$, $(-1, 2)$, $(3, -4)$

4 Use the Midpoint Formula to find the midpoints of line segments.

In Exercises 17–20, find the midpoint of the line segment joining the points, and then plot the points and the midpoint.

17. $(1, 4)$, $(7, 2)$
18. $(-1, 3)$, $(5, 5)$
19. $(-4, 0)$, $(-2, -4)$
20. $(1, 6)$, $(6, 1)$

3.2 Graphs of Equations

1 Sketch graphs of equations using the point-plotting method.

In Exercises 21–28, sketch the graph of the equation.

21. $y = 6 - \frac{1}{3}x$

22. $y = \frac{3}{4}x - 2$

23. $3y - 2x - 3 = 0$

24. $3x + 4y + 12 = 0$

25. $y = x^2 - 1$

26. $y = (x - 2)^2$

27. $y = |x| - 2$

28. $y = |x - 3|$

2 Find and use x- and y-intercepts as aids to sketching graphs.

In Exercises 29–36, find the x- and y-intercepts of the graph of the equation. Then sketch the graph of the equation and show the coordinates of three solution points (including x- and y-intercepts).

29. $y = 4x - 6$

30. $y = 3x + 9$

31. $7x - 2y = -14$

32. $5x + 4y = 10$

33. $y = |x - 5|$

34. $y = |x| + 4$

35. $y = |2x + 1| - 5$

36. $y = |3 - 6x| - 15$

In Exercises 37–42, use a graphing calculator to graph the equation. Approximate the *x*- and *y*-intercepts (if any).

37. $y = (x - 3)^2 - 3$

38. $y = \frac{1}{4}(x - 2)^3$

39. $y = -|x - 4| - 7$

40. $y = 3 - |x - 3|$

41. $y = \sqrt{3 - x}$

42. $y = x - 2\sqrt{x}$

3 Use a pattern to write an equation for an application problem, and sketch its graph.

43. *Straight-Line Depreciation*   Your family purchases a new SUV for $28,000. For financing purposes, the SUV will be depreciated over a five-year period. At the end of 5 years, the value of the SUV is expected to be $13,000.

    (a) Find an equation that relates the depreciated value of the SUV to the number of years since it was purchased.

    (b) Sketch the graph of the equation.

    (c) What is the *y*-intercept of the graph and what does it represent?

44. *Straight-Line Depreciation*   A company purchases a new computer system for $20,000. For tax purposes, the computer system will be depreciated over a six-year period. At the end of the 6 years, the value of the system is expected to be $2000.

    (a) Find an equation that relates the depreciated value of the computer system to the number of years since it was purchased.

    (b) Sketch the graph of the equation.

    (c) What is the *y*-intercept of the graph and what does it represent?

3.3 Slope and Graphs of Linear Equations

1 Determine the slope of a line through two points.

In Exercises 45–50, find the slope of the line through the points.

45. $(-1, 1), (6, 3)$

46. $(-2, 5), (3, -8)$

47. $(-1, 3), (4, 3)$

48. $(7, 2), (7, 8)$

49. $(0, 6), (8, 0)$

50. $(0, 0), \left(\frac{7}{2}, 6\right)$

In Exercises 51–56, a point on a line and the slope of the line are given. Find two additional points on the line. (There are many correct answers.)

51. $(2, -4)$

    $m = -3$

52. $\left(-4, \frac{1}{2}\right)$

    $m = 2$

53. $(3, 1)$

    $m = \frac{5}{4}$

54. $\left(-3, -\frac{3}{2}\right)$

    $m = -\frac{1}{3}$

55. $(3, 7)$

    m is undefined.

56. $(7, -2)$

    $m = 0$

57. *Ramp*   A loading dock ramp rises 3 feet above the ground. The ramp has a slope of $\frac{1}{12}$. What is the length of the ramp?

58. *Road Grade*   When driving down a mountain road, you notice a warning sign indicating a "9% grade." This means that the slope of the road is $-\frac{9}{100}$. Over a stretch of road, your elevation drops by 1500 feet. What is the horizontal change in your position?

2 Write linear equations in slope-intercept form and graph the equations.

In Exercises 59–62, write the equation of the line in slope-intercept form and use the slope and *y*-intercept to sketch the line.

59. $5x - 2y - 4 = 0$

60. $x - 3y - 6 = 0$

61. $x + 2y - 2 = 0$

62. $y - 6 = 0$

3 Use slopes to determine whether two lines are parallel, perpendicular, or neither.

In Exercises 63–68, determine whether the lines are parallel, perpendicular, or neither.

63. L_1: $y = \frac{3}{2}x + 1$

    L_2: $y = \frac{2}{3}x - 1$

64. L_1: $y = 2x - 5$

    L_2: $y = 2x + 3$

65. L_1: $y = \frac{3}{2}x - 2$

L_2: $y = -\frac{2}{3}x + 1$

66. L_1: $y = -0.3x - 2$

L_2: $y = 0.3x + 1$

67. L_1: $2x - 3y - 5 = 0$

L_2: $x + 2y - 6 = 0$

68. L_1: $4x + 3y - 6 = 0$

L_2: $3x - 4y - 8 = 0$

4 Use slopes to describe rates of change in real-life problems.

69. *Consumer Awareness* In 1993, the average cost of regular unleaded gasoline was $1.11 per gallon. By 2000, the average cost had risen to $1.51 per gallon. Find the average rate of change in the cost of a gallon of regular unleaded gasoline from 1993 to 2000. (Source: U.S. Energy Information Administration)

70. *Consumer Awareness* In 1990, the average cost of a T-bone steak was $5.45 per pound. By 2000, the average cost was $6.82 per pound. Find the average rate of change in the cost of a pound of T-bone steak from 1990 to 2000 . (Source: U.S. Bureau of Labor Statistics)

3.4 Equations of Lines

1 Write equations of lines using point-slope form.

In Exercises 71–78, write an equation of the line that passes through the point and has the specified slope.

71. $(1, -4)$

$m = 2$

72. $(-5, -5)$

$m = 3$

73. $(-1, 4)$

$m = -4$

74. $(5, -2)$

$m = -2$

75. $\left(\frac{5}{2}, 4\right)$

$m = -\frac{2}{3}$

76. $\left(-2, -\frac{4}{3}\right)$

$m = \frac{3}{2}$

77. $(-6, 5)$

$m = \frac{1}{4}$

78. $(7, 8)$

$m = -\frac{3}{5}$

In Exercises 79–84, write the slope-intercept form of the equation of the line that passes through the two points.

79. $(-6, 0), (0, -3)$

80. $(0, 10), (6, 8)$

81. $(-2, -3), (4, 6)$

82. $(-10, 2), (4, -7)$

83. $\left(\frac{4}{3}, \frac{1}{6}\right), \left(4, \frac{7}{6}\right)$

84. $\left(\frac{1}{2}, 0\right), \left(\frac{5}{2}, 5\right)$

2 Write equations of horizontal, vertical, parallel, and perpendicular lines.

In Exercises 85–88, write an equation of the line.

85. Horizontal line through $(-4, 7)$

86. Vertical line through $(-2, 13)$

87. Line through $(-5, -2)$ and $(-5, 3)$

88. Line through $(1, 8)$ and $(15, 8)$

In Exercises 89–92, write equations of the lines that pass through the point and are (a) parallel and (b) perpendicular to the given line.

89. $\left(\frac{3}{5}, -\frac{4}{5}\right)$

$3x + y = 2$

90. $(-1, 5)$

$2x + 4y = 1$

91. $(12, 1)$

$5x = 3$

92. $\left(\frac{3}{8}, 3\right)$

$4x - 3y = 12$

3 Use linear models to solve application problems.

93. *Annual Salary* Your annual salary in 1998 was $28,500. During the next 5 years, your annual salary increased by approximately $1100 per year.

(a) Write an equation of the line giving the annual salary S in terms of the year t. (Let $t = 8$ correspond to the year 1998.)

(b) *Linear Extrapolation* Use the equation in part (a) to predict your annual salary in the year 2006.

(c) *Linear Interpolation* Use the equation in part (a) to estimate your annual salary in 2002.

94. *Rent*   The rent for a two-bedroom apartment was $525 per month in 1998. During the next 5 years, the rent increased by approximately $55 per year.

    (a) Write an equation of the line giving the rent R in terms of the year t. (Let $t = 8$ correspond to the year 1998.)

    (b) *Linear Extrapolation*   Use the equation in part (a) to predict the rent in the year 2007.

    (c) *Linear Interpolation*   Use the equation in part (a) to estimate the rent in 2001.

3.5 Graphs of Linear Inequalities

1 Verify solutions of linear inequalities in two variables.

In Exercises 95 and 96, determine whether each point is a solution of the inequality.

95. $5x - 8y \geq 12$

    (a) $(-1, 2)$ (b) $(3, -1)$

    (c) $(4, 0)$ (d) $(0, 3)$

96. $2x + 4y - 14 < 0$

    (a) $(0, 0)$ (b) $(3, 2)$

    (c) $(1, 3)$ (d) $(-4, 1)$

2 Sketch graphs of linear inequalities in two variables.

In Exercises 97–104, sketch the graph of the solution of the linear inequality.

97. $y > 4$ **98.** $x \leq 5$

99. $x - 2 \geq 0$ **100.** $y + 3 < 0$

101. $2x + y < 1$ **102.** $3x - 4y > 2$

103. $-(x - 1) \leq 4y - 2$ **104.** $(y - 3) \geq 2(x - 5)$

In Exercises 105–108, use a graphing calculator to graph the solution of the inequality.

105. $y \leq 12 - \frac{3}{2}x$ **106.** $y \leq \frac{1}{3}x + 1$

107. $x + y \geq 0$ **108.** $4x - 3y \geq 2$

109. ▲ *Geometry*   The perimeter of a rectangle of length x and width y cannot exceed 800 feet.

    (a) Write a linear inequality for this constraint.

    (b) ▦ Use a graphing calculator to graph the solution of the inequality.

110. *Weekly Pay*   You have two part-time jobs. One is at a grocery store, which pays $7 per hour, and the other is mowing lawns, which pays $10 per hour. Between the two jobs, you want to earn at least $200 a week.

    (a) Write an inequality that shows the different numbers of hours you can work at each job.

    (b) Sketch the graph of the inequality. From the graph, find several ordered pairs with positive integer coordinates that are solutions of the inequality.

3.6 Relations and Functions

1 Identify the domains and ranges of relations.

In Exercises 111 and 112, find the domain and the range of the relation. Then draw a graphical representation of the relation.

111. $\{(-3, 4), (-1, 0), (0, 1), (1, 4), (-3, 5)\}$

112. $\{(-2, 4), (-1, 1), (0, 0), (1, 1), (2, 4)\}$

2 Determine if relations are functions by inspection.

In Exercises 113–116, determine whether the relation is a function.

113.

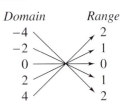

Domain   *Range*
6   0
7   1
8   2
9

114.

Domain   *Range*
−4   2
−2   1
0   0
2   1
4   2

115.

Input value, x	Output value, y
1	10
2	10
3	10
4	10
5	10

116.

Input value, x	Output value, y
3	0
6	3
9	8
6	12
0	2

③ Use function notation and evaluate functions.

In Exercises 117–124, evaluate the function as indicated, and simplify.

117. $f(x) = 4 - \frac{5}{2}x$

(a) $f(-10)$

(b) $f\left(\frac{2}{5}\right)$

(c) $f(t) + f(-4)$

(d) $f(x + h)$

118. $h(x) = x(x - 8)$

(a) $h(8)$

(b) $h(10)$

(c) $h(-3) + h(4)$

(d) $h(4t)$

119. $f(t) = \sqrt{5 - t}$

(a) $f(-4)$

(b) $f(5)$

(c) $f(3)$

(d) $f(5z)$

120. $g(x) = |x + 4|$

(a) $g(0)$

(b) $g(-8)$

(c) $g(2) - g(-5)$

(d) $g(x - 2)$

121. $f(x) = \begin{cases} -3x, & x \le 0 \\ 1 - x^2, & x > 0 \end{cases}$

(a) $f(2)$

(b) $f\left(-\frac{2}{3}\right)$

(c) $f(1)$

(d) $f(4) - f(3)$

122. $h(x) = \begin{cases} x^3, & x \le 1 \\ (x - 1)^2 + 1, & x > 1 \end{cases}$

(a) $h(2)$

(b) $h\left(-\frac{1}{2}\right)$

(c) $h(0)$

(d) $h(4) - h(3)$

123. $f(x) = 3 - 2x$

(a) $\dfrac{f(x + 2) - f(2)}{x}$

(b) $\dfrac{f(x - 3) - f(3)}{x}$

124. $f(x) = 7x + 10$

(a) $\dfrac{f(x + 1) - f(1)}{x}$

(b) $\dfrac{f(x - 5) - f(5)}{x}$

④ Identify the domains and ranges of functions.

In Exercises 125–128, find the domain of the function.

125. $h(x) = 4x^2 - 7$

126. $g(s) = \dfrac{s + 1}{(s - 1)(s + 5)}$

127. $f(x) = \sqrt{5 - 2x}$

128. $f(x) = |x - 6| + 10$

129. ▲ *Geometry* A wire 150 inches long is to be cut into four pieces to form a rectangle whose shortest side has a length of x. Write the area A of the rectangle as a function of x. What is the domain of the function?

130. ▲ *Geometry* A wire 100 inches long is to be cut into four pieces to form a rectangle whose shortest side has a length of x. Write the area A of the rectangle as a function of x. What is the domain of the function?

3.7 Graphs of Functions

① Sketch graphs of functions on rectangular coordinate systems.

In Exercises 131–140, sketch the graph of the function. Then determine its domain and range.

131. $y = 4 - (x - 3)^2$

132. $h(x) = 9 - (x - 2)^2$

133. $y = \sqrt{x} + 2$

134. $f(t) = \sqrt{t - 2}$

135. $y = 8 - |x|$

136. $f(x) = |x + 1| - 2$

137. $g(x) = 6 - 3x, \quad -2 \le x < 4$

138. $h(x) = x(4 - x), \quad 0 \le x \le 4$

139. $f(x) = \begin{cases} 2 - (x - 1)^2, & x < 1 \\ 2 + (x - 1)^2, & x \ge 1 \end{cases}$

140. $f(x) = \begin{cases} 2x, & x \le 0 \\ x^2 + 1, & x > 0 \end{cases}$

2 Identify the graphs of basic functions.

In Exercises 141–146, match the function with its graph. [The graphs are labeled (a), (b), (c), (d), (e), and (f).]

(a)
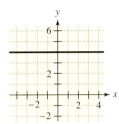

(b)
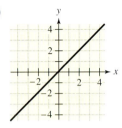

(c)
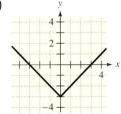

(d)
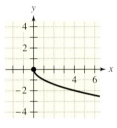

(e)
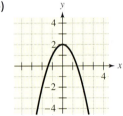

(f)
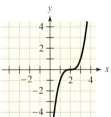

141. $f(x) = -x^2 + 2$ **142.** $f(x) = |x| - 3$

143. $f(x) = 4$ **144.** $f(x) = (x - 2)^3$

145. $f(x) = -\sqrt{x}$ **146.** $f(x) = x$

3 Use the Vertical Line Test to determine if graphs represent functions.

In Exercises 147–150, use the Vertical Line Test to determine if the graph represents y as a function of x.

147. $9y^2 = 4x^3$ **148.** $y = x^4 - 1$

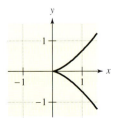
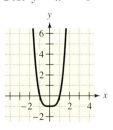

149. $y = x^3 - 3x^2$ **150.** $x^2 + y^2 = 9$

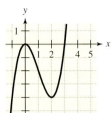
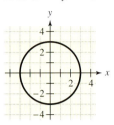

4 Use vertical and horizontal shifts and reflections to sketch graphs of functions.

In Exercises 151–154, identify the transformation of the graph of $f(x) = \sqrt{x}$ and sketch the graph of h.

151. $h(x) = -\sqrt{x}$

152. $h(x) = \sqrt{x} + 3$

153. $h(x) = \sqrt{x - 1}$

154. $h(x) = 1 - \sqrt{x + 4}$

In Exercises 155–158, use the graph of $f(x) = x^2$ to write a function that represents the graph.

155.
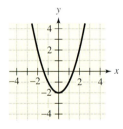

156.
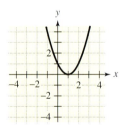

157.
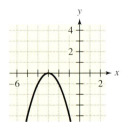

158.
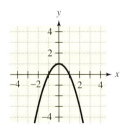

Chapter Test

Take this test as you would take a test in class. After you are done, check your work against the answers in the back of the book.

1. Determine the quadrant in which the point (x, y) lies if $x > 0$ and $y < 0$.

2. Plot the points $(0, 5)$ and $(3, 1)$. Then find the distance between them and the coordinates of the midpoint of the line segment joining the two points.

3. Find the x- and y-intercepts of the graph of the equation $y = -3(x + 1)$.

4. Sketch the graph of the equation $y = |x - 2|$.

5. Find the slope (if possible) of the line passing through each pair of points.
   (a) $(-4, 7), (2, 3)$ (b) $(3, -2), (3, 6)$

6. Sketch the graph of the line passing through the point $(0, -6)$ with slope $m = \frac{4}{3}$.

7. Find the x- and y-intercepts of the graph of $2x + 5y - 10 = 0$. Use the results to sketch the graph.

8. Find an equation of the line through the points $(25, -15)$ and $(75, 10)$.

9. Find an equation of the vertical line through the point $(-2, 4)$.

10. Write equations of the lines that pass through the point $(-2, 3)$ and are (a) parallel and (b) perpendicular to the line $3x - 5y = 4$.

11. Sketch the graph of the inequality $2x - 3y \geq 9$.

12. The graph of $y^2(4 - x) = x^3$ is shown at the left. Does the graph represent y as a function of x? Explain your reasoning.

13. Determine whether the relation represents a function. Explain.
    (a) $\{(2, 4), (-6, 3), (3, 3), (1, -2)\}$ (b) $\{(0, 0), (1, 5), (-2, 1), (0, -4)\}$

14. Evaluate $g(x) = x/(x - 3)$ as indicated, and simplify.
    (a) $g(2)$ (b) $g\left(\frac{7}{2}\right)$ (c) $g(x + 2)$

15. Find the domain of each function.
    (a) $h(t) = \sqrt{9 - t}$ (b) $f(x) = \dfrac{x + 1}{x - 4}$

16. Sketch the graph of the function $g(x) = \sqrt{2 - x}$.

17. Describe the transformation of the graph of $f(x) = x^2$ that would produce the graph of $g(x) = -(x - 2)^2 + 1$.

18. After 4 years, the value of a \$26,000 car will have depreciated to \$10,000. Write the value V of the car as a linear function of t, the number of years since the car was purchased. When will the car be worth \$16,000?

19. Use the graph of $f(x) = |x|$ to write a function that represents each graph.
    (a) 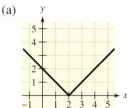   (b) 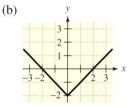   (c)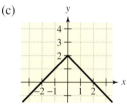

Figure for 12

215

Motivating the Chapter

 ## Soccer Club Fundraiser

A collegiate soccer club has a fundraising dinner. Student tickets sell for $8 and nonstudent tickets sell for $15. There are 115 tickets sold and the total revenue is $1445.

See Section 4.2, Exercise 83.

a. Set up a system of linear equations that can be used to determine how many tickets of each type were sold.

b. Solve the system in part (a) by the method of substitution.

c. Solve the system in part (a) by the method of elimination.

The soccer club decides to set goals for the next fundraising dinner. To meet these goals, a "major contributor" category is added. A person donating $100 is considered a major contributor to the soccer club and receives a "free" ticket to the dinner. The club's goals are to have 200 people in attendance, with the number of major contributors being one-fourth the number of students, and to raise $4995.

See Section 4.3, Exercise 55.

d. Set up a system of linear equations to determine how many of each kind of ticket would need to be sold for the second fundraising dinner.

e. Solve the system in part (d) by Gaussian elimination.

f. Would it be possible for the soccer club to meet its goals if only 18 people donated $100? Explain.

See Section 4.5, Exercise 93.

g. Solve the system in part (d) using matrices.

h. Solve the system in part (d) using determinants.

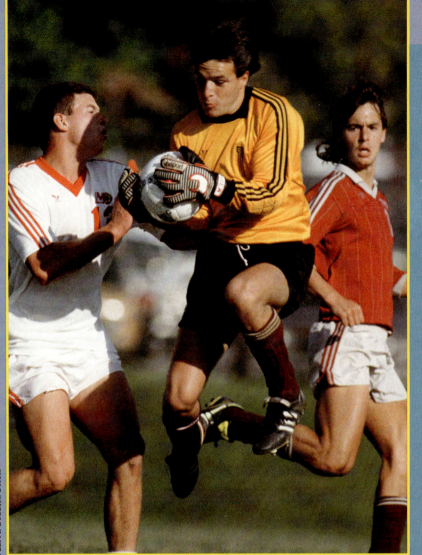

Paul A. Souders/Corbis

Systems of Equations and Inequalities

4.1 ● Systems of Equations

4.2 ● Linear Systems in Two Variables

4.3 ● Linear Systems in Three Variables

4.4 ● Matrices and Linear Systems

4.5 ● Determinants and Linear Systems

4.6 ● Systems of Linear Inequalities

4.1 Systems of Equations

David Lassman/The Image Works

Why You Should Learn It

Systems of equations can be used to model and solve real-life problems. For instance, in Exercise 104 on page 230, a system of equations is used to model the numbers of morning and evening newspapers in the United States.

What You Should Learn

1. Determine if ordered pairs are solutions of systems of equations.
2. Solve systems of equations graphically.
3. Solve systems of equations algebraically using the method of substitution.
4. Use systems of equations to model and solve real-life problems.

Systems of Equations

Many problems in business and science involve **systems of equations.** These systems consist of two or more equations, each containing two or more variables.

$$\begin{cases} ax + by = c & \text{Equation 1} \\ dx + ey = f & \text{Equation 2} \end{cases}$$

A **solution** of such a system is an ordered pair (x, y) of real numbers that satisfies *each* equation in the system. When you find the set of all solutions of the system of equations, you are **solving the system of equations.**

---

1. Determine if ordered pairs are solutions of systems of equations.

Example 1   Checking Solutions of a System of Equations

Check whether each ordered pair is a solution of the system of equations.

$$\begin{cases} x + y = 6 & \text{Equation 1} \\ 2x - 5y = -2 & \text{Equation 2} \end{cases}$$

a. $(3, 3)$ **b.** $(4, 2)$

Solution

a. To determine whether the ordered pair $(3, 3)$ is a solution of the system of equations, substitute 3 for x and 3 for y in *each* of the original equations.

$3 + 3 = 6.$ ✓   Substitute 3 for x and 3 for y in Equation 1.

$2(3) - 5(3) \neq -2.$ ✗   Substitute 3 for x and 3 for y in Equation 2.

Because the ordered pair $(3, 3)$ fails to check in *both* equations, you can conclude that it *is not* a solution of the original system of equations.

b. By substituting 4 for x and 2 for y in each of the original equations, you can determine that the ordered pair $(4, 2)$ is a solution of *both* equations.

$4 + 2 = 6$ ✓   Substitute 4 for x and 2 for y in Equation 1.

$2(4) - 5(2) = -2$ ✓   Substitute 4 for x and 2 for y in Equation 2.

So, $(4, 2)$ *is* a solution of the original system of equations.

② Solve systems of equations graphically.

Solving Systems of Equations by Graphing

You can gain insight about the location and number of solutions of a system of equations by sketching the graph of each equation in the same coordinate plane. The solutions of the system correspond to the **points of intersection** of the graphs.

A system of linear equations can have exactly one solution, infinitely many solutions, or no solution. To see why this is true, consider the graphical interpretations of three systems of two linear equations shown below.

Graphs			
Graphical Interpretation	The two lines intersect.	The two lines coincide (are identical).	The two lines are parallel.
Intersection	Single point of intersection	Infinitely many points of intersection	No point of intersection
Slopes of Lines	Slopes are not equal.	Slopes are equal.	Slopes are equal.
Number of Solutions	Exactly one solution	Infinitely many solutions	No solution
Type of System	**Consistent system**	**Dependent (consistent) system**	**Inconsistent system**

Technology: Discovery

Rewrite each system of equations in slope-intercept form and graph the equations using a graphing calculator. What is the relationship between the slopes of the two lines and the number of points of intersection?

a. $\begin{cases} 2x + 4y = 8 \\ 4x - 3y = -6 \end{cases}$

b. $\begin{cases} -x + 5y = 15 \\ 2x - 10y = -7 \end{cases}$

c. $\begin{cases} x - y = 9 \\ 2x - 2y = 18 \end{cases}$

Note that the word *consistent* is used to mean that the system of linear equations has at least one solution, whereas the word *inconsistent* is used to mean that the system of linear equations has no solution.

You can see from the graphs above that a comparison of the slopes of two lines gives useful information about the number of solutions of the corresponding system of equations. For instance:

Consistent systems have lines with different slopes.
Dependent (consistent) systems have lines with equal slopes and equal *y*-intercepts.
Inconsistent systems have lines with equal slopes, but different *y*-intercepts.

So, when solving a system of equations graphically, it is helpful to know the slopes of the lines directly. Writing these linear equations in the slope-intercept form

$$y = mx + b \qquad \text{Slope-intercept form}$$

allows you to identify the slopes quickly.

Example 2   The Graphical Method of Solving a System

Use the graphical method to solve the system of equations.

$$\begin{cases} 2x + 3y = 7 & \text{Equation 1} \\ 2x - 5y = -1 & \text{Equation 2} \end{cases}$$

Solution

Because both equations in the system are linear, you know that they have graphs that are straight lines. To sketch these lines, first write each equation in slope-intercept form, as follows.

$$y = -\frac{2}{3}x + \frac{7}{3} \qquad \text{Equation 1}$$

$$y = \frac{2}{5}x + \frac{1}{5} \qquad \text{Equation 2}$$

Because their slopes are not equal, you can conclude that the graphs will intersect at a single point. The lines corresponding to these two equations are shown in Figure 4.1. From this figure, it appears that the two lines intersect at the point $(2, 1)$. You can check these coordinates as follows.

Check

Substitute in 1st Equation

$$2x + 3y = 7$$
$$2(2) + 3(1) \stackrel{?}{=} 7$$
$$4 + 3 \stackrel{?}{=} 7$$
$$7 = 7 \quad \checkmark$$

Substitute in 2nd Equation

$$2x - 5y = -1$$
$$2(2) - 5(1) \stackrel{?}{=} -1$$
$$4 - 5 \stackrel{?}{=} -1$$
$$-1 = -1$$

Because both equations in the system are satisfied, the point $(2, 1)$ is a solution of the system.

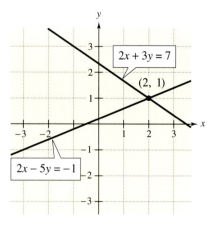

Figure 4.1

③ Solve systems of equations algebraically using the method of substitution.

Solving Systems of Equations by Substitution

Solving a system of equations graphically is limited by the ability to sketch an accurate graph. It is difficult to obtain an accurate solution if one or both coordinates of a solution point are fractional or irrational. One analytic way to determine an exact solution of a system of two equations in two variables is to convert the system to *one* equation in *one* variable by an appropriate substitution.

Example 3   The Method of Substitution: One-Solution Case

Solve the system of equations.

$$\begin{cases} -x + y = 3 & \text{Equation 1} \\ 3x + y = -1 & \text{Equation 2} \end{cases}$$

Solution

Begin by solving for y in Equation 1.

$$y = x + 3 \qquad \text{Revised Equation 1}$$

Next, substitute this expression for y in Equation 2.

$$3x + y = -1 \qquad \text{Equation 2}$$
$$3x + (x + 3) = -1 \qquad \text{Substitute } x + 3 \text{ for } y.$$
$$4x + 3 = -1 \qquad \text{Combine like terms.}$$
$$4x = -4 \qquad \text{Subtract 3 from each side.}$$
$$x = -1 \qquad \text{Divide each side by 4.}$$

At this point, you know that the x-coordinate of the solution is -1. To find the y-coordinate, *back-substitute* the x-value into the revised Equation 1.

$$y = x + 3 \qquad \text{Revised Equation 1}$$
$$y = -1 + 3 \qquad \text{Substitute } -1 \text{ for } x.$$
$$y = 2 \qquad \text{Simplify.}$$

The solution is $(-1, 2)$. Check this in the original system of equations.

Study Tip

The term **back-substitute** implies that you work backwards. After finding a value for one of the variables, substitute that value back into one of the equations in the original (or revised) system to find the value of the other variable.

When you use substitution, it does not matter which variable you solve for first. You will obtain the same solution regardless. When making your choice, you should choose the variable that is easier to work with. For instance, in the system

$$\begin{cases} 3x - 2y = 1 & \text{Equation 1} \\ x + 4y = 3 & \text{Equation 2} \end{cases}$$

it is easier to begin by solving for x in the second equation. But in the system

$$\begin{cases} 2x + y = 5 & \text{Equation 1} \\ 3x - 2y = 11 & \text{Equation 2} \end{cases}$$

it is easier to begin by solving for y in the first equation.

The steps for using the method of substitution to solve a system of two equations involving two variables are summarized as follows.

The Method of Substitution

1. Solve one of the equations for one variable in terms of the other.

2. Substitute the expression obtained in Step 1 in the other equation to obtain an equation in one variable.

3. Solve the equation obtained in Step 2.

4. Back-substitute the solution from Step 3 in the expression obtained in Step 1 to find the value of the other variable.

5. Check the solution to see that it satisfies *both* of the original equations.

Example 4   The Method of Substitution: No-Solution Case

To solve the system of equations

$$\begin{cases} 2x - 2y = 0 & \text{Equation 1} \\ x - y = 1 & \text{Equation 2} \end{cases}$$

begin by solving for y in Equation 2.

$$y = x - 1 \qquad \text{Revised Equation 2}$$

Next, substitute this expression for y in Equation 1.

$$2x - 2(x - 1) = 0 \qquad \text{Substitute } x - 1 \text{ for } y \text{ in Equation 1.}$$

$$2x - 2x + 2 = 0 \qquad \text{Distributive Property}$$

$$2 = 0 \qquad \text{False statement}$$

Because the substitution process produced a false statement $(2 = 0)$, you can conclude that the original system of equations is inconsistent and has no solution. You can check your solution graphically, as shown in Figure 4.2.

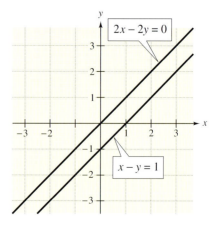

Figure 4.2

4 Use systems of equations to model and solve real-life problems.

Applications

To model a real-life situation with a system of equations, you can use the same basic problem-solving strategy that has been used throughout the text.

Write a verbal model. ⟹ Assign labels. ⟹ Write an algebraic model. ⟹ Solve the algebraic model. ⟹ Answer the question.

After answering the question, remember to check the answer in the original statement of the problem.

Example 5   A Mixture Problem

A roofing contractor bought 30 bundles of shingles and four rolls of roofing paper for $528. A second purchase (at the same prices) cost $140 for eight bundles of shingles and one roll of roofing paper. Find the price per bundle of shingles and the price per roll of roofing paper.

Solution

Verbal Model:

$$30\left(\begin{array}{c}\text{Price of}\\\text{a bundle}\end{array}\right) + 4\left(\begin{array}{c}\text{Price of}\\\text{a roll}\end{array}\right) = 528$$

$$8\left(\begin{array}{c}\text{Price of}\\\text{a bundle}\end{array}\right) + 1\left(\begin{array}{c}\text{Price of}\\\text{a roll}\end{array}\right) = 140$$

Labels: Price of a bundle of shingles $= x$ (dollars)
           Price of a roll of roofing paper $= y$ (dollars)

System: $\begin{cases} 30x + 4y = 528 & \quad\text{Equation 1} \\ 8x + y = 140 & \quad\text{Equation 2} \end{cases}$

Solving the second equation for y produces $y = 140 - 8x$, and substituting this expression for y in the first equation produces the following.

$$30x + 4(140 - 8x) = 528 \qquad \text{Substitute } 140 - 8x \text{ for } y.$$
$$30x + 560 - 32x = 528 \qquad \text{Distributive Property}$$
$$-2x = -32 \qquad \text{Combine like terms.}$$
$$x = 16 \qquad \text{Divide each side by } -2.$$

Back-substituting 16 for x in revised Equation 2 produces

$$y = 140 - 8(16) = 12.$$

So, you can conclude that the price of shingles is $16 per bundle and the price of roofing paper is $12 per roll. Check this in the original statement of the problem.

Check

1st Equation	*2nd Equation*	
$30(16) + 4(12) \overset{?}{=} 528$	$8(16) + 12 \overset{?}{=} 140$	Substitute 16 for x and 12 for y.
$480 + 48 = 528$	$128 + 12 = 140$	Solution checks. ✓

The total cost C of producing x units of a product usually has two components—the initial cost and the cost per unit. When enough units have been sold so that the total revenue R equals the total cost, the sales are said to have reached the **break-even point.** You can find this break-even point by setting C equal to R and solving for x. In other words, the break-even point corresponds to the point of intersection of the cost and revenue graphs.

Example 6   Break-Even Analysis

A small business invests \$14,000 to produce a new energy bar. Each bar costs \$0.80 to produce and is sold for \$1.50. How many bars must be sold before the business breaks even?

Solution

Verbal Model: $\dfrac{\text{Total}}{\text{cost}} = \dfrac{\text{Cost}}{\text{per bar}} \cdot \dfrac{\text{Number}}{\text{of bars}} + \dfrac{\text{Initial}}{\text{cost}}$

$\dfrac{\text{Total}}{\text{revenue}} = \dfrac{\text{Price}}{\text{per bar}} \cdot \dfrac{\text{Number}}{\text{of bars}}$

Labels: Total cost $= C$ (dollars)
Cost per bar $= 0.80$ (dollars per bar)
Number of bars $= x$ (bars)
Initial cost $= 14{,}000$ (dollars)
Total revenue $= R$ (dollars)
Price per bar $= 1.50$ (dollars per bar)

System: $\begin{cases} C = 0.80x + 14{,}000 & \text{Equation 1} \\ R = 1.50x & \text{Equation 2} \end{cases}$

Because the break-even point occurs when $R = C$, you have

$1.50x = 0.80x + 14{,}000$ $R = C$

$0.7x = 14{,}000$ Subtract $0.80x$ from each side.

$x = 20{,}000.$ Divide each side by 0.7.

So, it follows that the business must sell 20,000 bars before it breaks even. Profit P (or loss) for the business can be determined by the equation $P = R - C$. Note in Figure 4.3 that sales less than the break-even point correspond to a loss for the business, whereas sales greater than the break-even point correspond to a profit for the business. The following table helps confirm this conclusion.

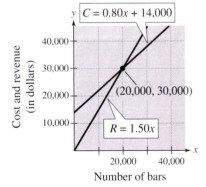

Figure 4.3

Units, x	0	5000	10,000	15,000	20,000	25,000
Revenue, R	\$0	\$7500	\$15,000	\$22,500	\$30,000	\$37,500
Cost, C	\$14,000	\$18,000	\$22,000	\$26,000	\$30,000	\$34,000
Profit, P	$-\$14{,}000$	$-\$10{,}500$	$-\$7000$	$-\$3500$	\$0	\$3500

Example 7   Investment

A total of $12,000 was invested in two funds paying 6% and 8% simple interest. The interest for 1 year is $880. How much of the $12,000 was invested in each fund?

Solution

Verbal Model:

$$\boxed{\text{Amount in 6\% fund}} + \boxed{\text{Amount in 8\% fund}} = 12{,}000$$

$$6\% \cdot \boxed{\text{Amount in 6\% fund}} + 8\% \cdot \boxed{\text{Amount in 8\% fund}} = 880$$

Labels: Amount in 6% fund $= x$ (dollars)
Amount in 8% fund $= y$ (dollars)

System:

$$\begin{cases} x + y = 12{,}000 & \text{Equation 1} \\ 0.06x + 0.08y = 880 & \text{Equation 2} \end{cases}$$

Begin by solving for x in the first equation.

$$x + y = 12{,}000 \qquad \text{Equation 1}$$

$$x = 12{,}000 - y \qquad \text{Revised Equation 1}$$

Substituting this expression for x in the second equation produces the following.

$$0.06x + 0.08y = 880 \qquad \text{Equation 2}$$

$$0.06(12{,}000 - y) + 0.08y = 880 \qquad \text{Substitute } 12{,}000 - y \text{ for } x.$$

$$720 - 0.06y + 0.08y = 880 \qquad \text{Distributive Property}$$

$$720 + 0.02y = 880 \qquad \text{Simplify.}$$

$$0.02y = 160 \qquad \text{Combine like terms.}$$

$$y = 8000 \qquad \text{Divide each side by 0.02.}$$

Back-substitute this value for y in revised Equation 1.

$$x = 12{,}000 - y \qquad \text{Revised Equation 1}$$

$$x = 12{,}000 - 8000 \qquad \text{Substitute 8000 for } y.$$

$$x = 4000 \qquad \text{Simplify.}$$

So, $4000 was invested in the fund paying 6% and $8000 was invested in the fund paying 8%. Check this in the original statement of the problem, as follows.

Check

Substitute in 1st Equation	*Substitute in 2nd Equation*
$x + y = 12{,}000$	$0.06x + 0.08y = 880$
$4000 + 8000 \overset{?}{=} 12{,}000$	$0.06(4000) + 0.08(8000) \overset{?}{=} 880$
$12{,}000 = 12{,}000$ ✓	$880 = 880$ ✓

4.1 Exercises

Review Concepts, Skills, and Problem Solving

Keep mathematically in shape by doing these exercises *before* the problems of this section.

Properties and Definitions

1. Sketch the graph of a line with positive slope.

2. Sketch the graph of a line with undefined slope.

3. The slope of a line is $-\frac{2}{3}$. What is the slope of a line perpendicular to this line?

4. Two lines have slopes of $m = -3$ and $m = \frac{3}{2}$. Which line is steeper? Explain.

Solving Equations

In Exercises 5–8, solve the equation and check the solution.

5. $y - 3(4y - 2) = 1$

6. $x + 6(3 - 2x) = 4$

7. $\frac{1}{2}x - \frac{1}{5}x = 15$

8. $\frac{1}{10}(x - 4) = 6$

In Exercises 9 and 10, solve for y in terms of x.

9. $3x + 4y - 5 = 0$

10. $-2x - 3y + 6 = 0$

Graphing

In Exercises 11–14, sketch the graph of the linear equation.

11. $y = -3x + 2$

12. $4x - 2y = -4$

13. $3x + 2y = 8$

14. $x + 3 = 0$

Developing Skills

In Exercises 1–8, determine whether each ordered pair is a solution of the system of equations. See Example 1.

1. $\begin{cases} x + 2y = 9 \\ -2x + 3y = 10 \end{cases}$

(a) $(1, 4)$

(b) $(3, -1)$

2. $\begin{cases} 5x - 4y = 34 \\ x - 2y = 8 \end{cases}$

(a) $(0, 3)$

(b) $(6, -1)$

3. $\begin{cases} -2x + 7y = 46 \\ 3x + y = 0 \end{cases}$

(a) $(-3, 2)$

(b) $(-2, 6)$

4. $\begin{cases} -5x - 2y = 23 \\ x + 4y = -19 \end{cases}$

(a) $(-3, -4)$

(b) $(3, 7)$

5. $\begin{cases} 4x - 5y = 12 \\ 3x + 2y = -2.5 \end{cases}$

(a) $(8, 4)$

(b) $\left(\frac{1}{2}, -2\right)$

6. $\begin{cases} 2x - y = 1.5 \\ 4x - 2y = 3 \end{cases}$

(a) $\left(0, -\frac{3}{2}\right)$

(b) $\left(2, \frac{5}{2}\right)$

7. $\begin{cases} -3x + 2y = -19 \\ 5x - y = 27 \end{cases}$

(a) $(5, -2)$

(b) $(3, 4)$

8. $\begin{cases} 7x + 2y = -1 \\ 3x - 6y = -21 \end{cases}$

(a) $(-2, -4)$

(b) $(-1, 3)$

In Exercises 9–16, determine whether the system is consistent or inconsistent.

9. $\begin{cases} x + 2y = 6 \\ x + 2y = 3 \end{cases}$

10. $\begin{cases} x - 2y = 3 \\ 2x - 4y = 7 \end{cases}$

11. $\begin{cases} 2x - 3y = -12 \\ -8x + 12y = -12 \end{cases}$

12. $\begin{cases} -5x + 8y = 8 \\ 7x - 4y = 14 \end{cases}$

13. $\begin{cases} -x + 4y = 7 \\ 3x - 12y = -21 \end{cases}$

14. $\begin{cases} 3x + 8y = 28 \\ -4x + 9y = 1 \end{cases}$

15. $\begin{cases} 5x - 3y = 1 \\ 6x - 4y = -3 \end{cases}$

16. $\begin{cases} 9x + 6y = 10 \\ -6x - 4y = 3 \end{cases}$

⊞ In Exercises 17–20, use a graphing calculator to graph the equations in the system. Use the graphs to determine whether the system is consistent or inconsistent. If the system is consistent, determine the number of solutions.

17. $\begin{cases} \frac{1}{3}x - \frac{1}{2}y = 1 \\ -2x + 3y = 6 \end{cases}$ **18.** $\begin{cases} x + y = 5 \\ x - y = 5 \end{cases}$

19. $\begin{cases} -2x + 3y = 6 \\ x - y = -1 \end{cases}$ **20.** $\begin{cases} 2x - 4y = 9 \\ x - 2y = 4.5 \end{cases}$

In Exercises 21–28, use the graphs of the equations to determine whether the system has any solutions. Find any solutions that exist.

21. $\begin{cases} x + y = 4 \\ x + y = -1 \end{cases}$ **22.** $\begin{cases} -x + y = 5 \\ x + 2y = 4 \end{cases}$

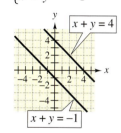

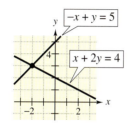

23. $\begin{cases} 5x - 3y = 4 \\ 2x + 3y = 3 \end{cases}$ **24.** $\begin{cases} 2x - y = 4 \\ -4x + 2y = -12 \end{cases}$

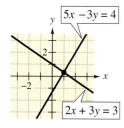

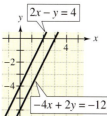

25. $\begin{cases} -2x + 3y = 6 \\ 8x - 12y = -24 \end{cases}$ **26.** $\begin{cases} 2x - 3y = 6 \\ 4x + 3y = 12 \end{cases}$

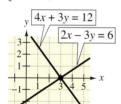

27. $\begin{cases} 0.5x + 0.5y = 1.5 \\ -x + y = 1 \end{cases}$ **28.** $\begin{cases} 0.25x - 0.5y = -1 \\ -0.5x + y = 2 \end{cases}$

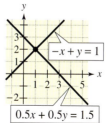

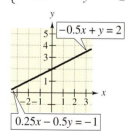

In Exercises 29–42, use the graphical method to solve the system of equations. See Example 2.

29. $\begin{cases} y = -x + 3 \\ y = x + 1 \end{cases}$ **30.** $\begin{cases} y = 2x - 4 \\ y = -\frac{1}{2}x + 1 \end{cases}$

31. $\begin{cases} x - y = 2 \\ x + y = 2 \end{cases}$ **32.** $\begin{cases} x - y = 0 \\ x + y = 4 \end{cases}$

33. $\begin{cases} 3x - 4y = 5 \\ x = 3 \end{cases}$ **34.** $\begin{cases} 5x + 2y = 18 \\ y = 2 \end{cases}$

35. $\begin{cases} 4x + 5y = 20 \\ \frac{4}{5}x + y = 4 \end{cases}$ **36.** $\begin{cases} -x + 3y = 7 \\ 2x - 6y = 6 \end{cases}$

37. $\begin{cases} 2x - 5y = 20 \\ 4x - 5y = 40 \end{cases}$ **38.** $\begin{cases} 5x + 3y = 24 \\ x - 2y = 10 \end{cases}$

39. $\begin{cases} x + y = 2 \\ 3x + 3y = 6 \end{cases}$ **40.** $\begin{cases} 4x - 3y = -3 \\ 8x - 6y = -6 \end{cases}$

41. $\begin{cases} 4x + 5y = 7 \\ 2x - 3y = 9 \end{cases}$ **42.** $\begin{cases} 7x + 4y = 6 \\ 5x - 3y = -25 \end{cases}$

⊞ In Exercises 43–46, use a graphing calculator to graph the equations and approximate any solutions of the system of equations.

43. $\begin{cases} 2x - y = 7.7 \\ -x - 3y = 1.4 \end{cases}$ **44.** $\begin{cases} 3x + 2y = 7.8 \\ -x + 3y = 15 \end{cases}$

45. $\begin{cases} 3.4x - 5.6y = 10.2 \\ 5.8x + 1.4y = -33.6 \end{cases}$

46. $\begin{cases} -2.3x + 7.9y = 88.3 \\ -5.3x - 2.7y = -16.5 \end{cases}$

In Exercises 47–70, solve the system of equations by the method of substitution. See Examples 3 and 4.

47. $\begin{cases} x - 2y = 0 \\ 3x + 2y = 8 \end{cases}$

48. $\begin{cases} x - y = 0 \\ 5x - 2y = 6 \end{cases}$

49. $\begin{cases} x = 4 \\ x - 2y = -2 \end{cases}$

50. $\begin{cases} y = 2 \\ x - 6y = -6 \end{cases}$

51. $\begin{cases} x + y = 3 \\ 2x - y = 0 \end{cases}$

52. $\begin{cases} -x + y = 5 \\ x - 4y = 0 \end{cases}$

53. $\begin{cases} x + y = 2 \\ x - 4y = 12 \end{cases}$

54. $\begin{cases} x - 2y = -1 \\ x - 5y = 2 \end{cases}$

55. $\begin{cases} x + 6y = 19 \\ x - 7y = -7 \end{cases}$

56. $\begin{cases} x - 5y = -6 \\ 4x - 3y = 10 \end{cases}$

57. $\begin{cases} 8x + 5y = 100 \\ 9x - 10y = 50 \end{cases}$

58. $\begin{cases} x + 4y = 300 \\ x - 2y = 0 \end{cases}$

59. $\begin{cases} -13x + 16y = 10 \\ 5x + 16y = -26 \end{cases}$

60. $\begin{cases} 2x + 5y = 29 \\ 5x + 2y = 13 \end{cases}$

61. $\begin{cases} 4x - 14y = -15 \\ 18x - 12y = 9 \end{cases}$

62. $\begin{cases} 5x - 24y = -12 \\ 17x - 24y = 36 \end{cases}$

63. $\begin{cases} x + 2y = 9 \\ x - 3y = -3 \end{cases}$

64. $\begin{cases} 2x + 2y = 2 \\ 3x - y = 1 \end{cases}$

65. $\begin{cases} 3x - y = 7 \\ 2x + 2y = 5 \end{cases}$

66. $\begin{cases} 5x + 3y = 4 \\ x - 4y = 3 \end{cases}$

67. $\begin{cases} \frac{1}{5}x + \frac{1}{2}y = 8 \\ x + y = 20 \end{cases}$

68. $\begin{cases} \frac{1}{2}x + \frac{3}{4}y = 10 \\ \frac{3}{2}x - y = 4 \end{cases}$

69. $\begin{cases} \frac{1}{8}x + \frac{1}{2}y = 1 \\ \frac{3}{5}x + y = \frac{3}{5} \end{cases}$

70. $\begin{cases} \frac{1}{8}x - \frac{1}{4}y = \frac{3}{4} \\ -\frac{1}{4}x + \frac{3}{4}y = -1 \end{cases}$

In Exercises 71–74, use a graphing calculator to graph the equations in the system. The graphs appear to be parallel. Yet, from the slope-intercept forms of the lines, you can find that the slopes are not equal and the graphs intersect. Find the point of intersection of the two lines.

71. $\begin{cases} x - 100y = -200 \\ 3x - 275y = 198 \end{cases}$

72. $\begin{cases} 35x - 33y = 0 \\ 12x - 11y = 92 \end{cases}$

73. $\begin{cases} 3x - 25y = 50 \\ 9x - 100y = 50 \end{cases}$

74. $\begin{cases} x + 40y = 80 \\ 2x + 150y = 195 \end{cases}$

Think About It   In Exercises 75–82, write a system of equations having the given solution. (There are many correct answers.)

75. $(4, 5)$

76. No solution

77. $(-1, -2)$

78. $\left(\frac{1}{2}, 3\right)$

79. Infinitely many solutions, including $\left(4, -\frac{1}{2}\right)$ and $(-1, 1)$

80. $(-2, 6)$

81. No solution

82. Infinitely many solutions, including $(-3, -1)$ and $(0, 3)$

Solving Problems

83. *Hay Mixture*   A farmer wants to mix two types of hay. The first type sells for $125 per ton and the second type sells for $75 per ton. The farmer wants a total of 100 tons of hay at a cost of $90 per ton. How many tons of each type of hay should be used in the mixture?

84. *Seed Mixture*   Ten pounds of mixed birdseed sells for $6.97 per pound. The mixture is obtained from two kinds of birdseed, with one variety priced at $5.65 per pound and the other at $8.95 per pound. How many pounds of each variety of birdseed are used in the mixture?

85. *Break-Even Analysis*   A small business invests $8000 in equipment to produce a new candy bar. Each bar costs $1.20 to produce and is sold for $2.00. How many candy bars must be sold before the business breaks even?

86. *Break-Even Analysis*   A business invests $50,000 in equipment to produce a new hand-held video game. Each game costs $19.25 to produce and is sold for $35.95. How many hand-held video games must be sold before the business breaks even?

87. *Break-Even Analysis*   You are setting up a small business and have invested $10,000 to produce hand cream that will sell for $3.25 a bottle. Each bottle can be produced for $1.65. How many bottles of hand cream must you sell to break even?

88. *Break-Even Analysis*   You are setting up a small business and have made an initial investment of $30,000. The unit cost of the guitar you are producing is $26.50, and the selling price is $76.50. How many guitars must you sell to break even?

89. *Investment*   A total of $12,000 is invested in two bonds that pay 8.5% and 10% simple interest. The annual interest is $1140. How much is invested in each bond?

90. *Investment*   A total of $25,000 is invested in two funds paying 8% and 8.5% simple interest. The annual interest is $2060. How much is invested in each fund?

Number Problems   In Exercises 91–98, find two positive integers that satisfy the given requirements.

91. The sum of the two numbers is 80 and their difference is 18.

92. The sum of the two numbers is 93 and their difference is 31.

93. The sum of the larger number and 3 times the smaller number is 51 and their difference is 3.

94. The sum of the larger number and twice the smaller number is 61 and their difference is 7.

95. The sum of the two numbers is 52 and the larger number is 8 less than twice the smaller number.

96. The sum of the two numbers is 160 and the larger number is 3 times the smaller number.

97. The difference of twice the smaller number and the larger number is 13 and the sum of the smaller number and twice the larger number is 114.

98. The difference of the numbers is 86 and the larger number is 3 times the smaller number.

▲ *Geometry*   In Exercises 99–102, find the dimensions of the rectangle meeting the specified conditions.

	Perimeter	*Condition*
99.	50 feet	The length is 5 feet greater than the width.
100.	320 inches	The width is 20 inches less than the length.
101.	68 yards	The width is $\frac{7}{10}$ of the length.
102.	90 meters	The length is $1\frac{1}{2}$ times the width.

103. ⊞ *Graphical Analysis*   From 1991 through 2000, the population of Alabama grew at a lower rate than the population of Arizona. Models that represent the populations of the two states are

$A_1 = 35.8t + 4073$ Alabama

$A_2 = 141.8t + 3584$ Arizona

where A_1 and A_2 are the populations in thousands and t is the calendar year, with $t = 1$ corresponding to 1991. Use a graphing calculator to determine the year during which the population of Arizona first exceeded the population of Alabama. (Source: U.S. Census Bureau)

104. ⊞ *Graphical Analysis*   From 1994 to 2001, the number of daily morning newspapers in the United States increased, while the number of daily evening newspapers decreased. Models that represent the circulations of the two types of daily papers are

$M = 20.3t + 558$ Morning

$E = -32.5t + 1051$ Evening

where M and E are the numbers of newspapers and t is the year, with $t = 4$ corresponding to 1994. Use a graphing calculator to determine the year during which the number of morning papers first exceeded the number of evening papers. (Source: Editor & Publisher Co.)

Explaining Concepts

105. *Writing*   What is meant by a solution of a system of equations in two variables?

106. *Writing*   List and explain the basic steps in solving a system of equations by substitution.

107. *Writing*   When solving a system of equations by substitution, how do you recognize that the system has no solution?

108. *Writing*   What does it mean to *back-substitute* when solving a system of equations?

109. *Writing*   Give a geometric description of the solution of a system of equations in two variables.

110. *Writing*   Describe any advantages of the method of substitution over the graphical method of solving a system of equations.

111. *Writing*   Is it possible for a consistent system of linear equations to have exactly two solutions? Explain.

112. You want to create several systems of equations with relatively simple solutions that students can use for practice. Discuss how to create a system of equations that has a given solution. Illustrate your method by creating a system of linear equations that has one of the following solutions: $(1, 4), (-2, 5), (-3, 1),$ or $(4, 2)$.

4.2 Linear Systems in Two Variables

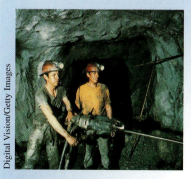

Digital Vision/Getty Images

What You Should Learn

1 Solve systems of linear equations algebraically using the method of elimination.

2 Use systems of linear equations to model and solve real-life problems.

Why You Should Learn It

Systems of linear equations can be used to find the best-fitting line that models a data set. For instance, in Exercise 81 on page 240, a system of linear equations is used to find the best-fitting line to model the hourly wages for miners in the United States.

1 Solve systems of linear equations algebraically using the method of elimination.

The Method of Elimination

In Section 4.1, you studied two ways of solving a system of equations—substitution and graphing. Now you will study a third way—the **method of elimination.**

The key step in the method of elimination is to obtain, for one of the variables, coefficients that differ only in sign, so that by *adding* the two equations this variable will be eliminated. Notice how this is accomplished in Example 1—when the two equations are added, the y-terms are eliminated.

Example 1   The Method of Elimination

Solve the system of linear equations.

$$\begin{cases} 3x + 2y = 4 & \text{Equation 1} \\ 5x - 2y = 8 & \text{Equation 2} \end{cases}$$

Solution

Begin by noting that the coefficients of y differ only in sign. By adding the two equations, you can eliminate y.

$$
\begin{array}{ll}
3x + 2y = 4 & \text{Equation 1} \\
\underline{5x - 2y = 8} & \text{Equation 2} \\
8x = 12 & \text{Add equations.}
\end{array}
$$

So, $x = \frac{3}{2}$. By back-substituting this value in the first equation, you can solve for y, as follows.

$$
\begin{array}{ll}
3x + 2y = 4 & \text{Equation 1} \\
3\left(\dfrac{3}{2}\right) + 2y = 4 & \text{Substitute } \frac{3}{2} \text{ for } x. \\
2y = -\dfrac{1}{2} & \text{Subtract } \frac{9}{2} \text{ from each side.} \\
y = -\dfrac{1}{4} & \text{Divide each side by 2.}
\end{array}
$$

The solution is $\left(\frac{3}{2}, -\frac{1}{4}\right)$. Check this solution in the original system.

The steps for solving a system of linear equations by the method of elimination are summarized as follows.

The Method of Elimination

1. Obtain coefficients for x (or y) that differ only in sign by multiplying all terms of one or both equations by suitable constants.

2. Add the equations to eliminate one variable, and solve the resulting equation.

3. Back-substitute the value obtained in Step 2 in either of the original equations and solve for the other variable.

4. Check your solution in *both* of the original equations.

Example 2 shows how Step 1 above is used in the method of elimination.

Example 2   The Method of Elimination

Solve the system of linear equations.

$$\begin{cases} 4x - 5y = 13 & \text{Equation 1} \\ 3x - \ y = \ 7 & \text{Equation 2} \end{cases}$$

Solution

To obtain coefficients of y that differ only in sign, multiply Equation 2 by -5.

$$\begin{cases} 4x - 5y = 13 \\ 3x - \ y = \ 7 \end{cases} \implies \begin{array}{ll} 4x - 5y = \ \ 13 & \text{Equation 1} \\ \underline{-15x + 5y = -35} & \text{Multiply Equation 2 by } -5. \\ -11x \ \ \ \ \ \ \ = -22 & \text{Add equations.} \end{array}$$

So, $x = 2$. Back-substitute this value in Equation 2 and solve for y.

$$3x - y = 7 \qquad\qquad\qquad \text{Equation 2}$$

$$3(2) - y = 7 \qquad\qquad\qquad \text{Substitute 2 for } x.$$

$$y = -1 \qquad\qquad\qquad \text{Solve for } y.$$

The solution is $(2, -1)$. Check this in the original system of equations, as follows.

Substitute in 1st Equation

$$4x - 5y = 13 \qquad\qquad\qquad \text{Equation 1}$$

$$4(2) - 5(-1) \stackrel{?}{=} 13 \qquad\qquad\qquad \text{Substitute 2 for } x \text{ and } -1 \text{ for } y.$$

$$8 + 5 = 13 \qquad\qquad\qquad \text{Solution checks.} \checkmark$$

Substitute in 2nd Equation

$$3x - y = 7 \qquad\qquad\qquad \text{Equation 2}$$

$$3(2) - (-1) \stackrel{?}{=} 7 \qquad\qquad\qquad \text{Substitute 2 for } x \text{ and } -1 \text{ for } y.$$

$$6 + 1 = 7 \qquad\qquad\qquad \text{Solution checks.} \checkmark$$

Example 3 shows how the method of elimination can be used to determine that a system of linear equations has no solution, while Example 4 shows how the method of elimination works with a system that has infinitely many solutions.

Example 3   The Method of Elimination: No-Solution Case

To solve the system of linear equations

$$\begin{cases} 3x + 9y = 8 & \text{Equation 1} \\ 2x + 6y = 7 & \text{Equation 2} \end{cases}$$

obtain coefficients of x that differ only in sign by multiplying Equation 1 by 2 and Equation 2 by -3.

$$\begin{cases} 3x + 9y = 8 \\ 2x + 6y = 7 \end{cases} \implies \begin{array}{rcr} 6x + 18y = & 16 & \text{Multiply Equation 1 by 2.} \\ -6x - 18y = & -21 & \text{Multiply Equation 2 by } -3. \\ \hline 0 = & -5 & \text{False statement} \end{array}$$

Because $0 = -5$ is a false statement, you can conclude that the system is inconsistent and has no solution. You can confirm this by graphing the two lines, as shown in Figure 4.4. Because the lines are parallel, the system is inconsistent.

Example 4   The Method of Elimination: Many-Solutions Case

To solve the system of linear equations

$$\begin{cases} -2x + 6y = & 3 & \text{Equation 1} \\ 4x - 12y = & -6 & \text{Equation 2} \end{cases}$$

obtain coefficients of x that differ only in sign by multiplying Equation 1 by 2.

$$\begin{cases} -2x + 6y = & 3 \\ 4x - 12y = & -6 \end{cases} \implies \begin{array}{rcr} -4x + 12y = & 6 & \text{Multiply Equation 1 by 2.} \\ 4x - 12y = & -6 & \text{Equation 2} \\ \hline 0 = & 0 & \text{Add equations.} \end{array}$$

Because $0 = 0$ is a true statement, the system has infinitely many solutions. You can confirm this by graphing the two lines, as shown in Figure 4.5. So, the solution set consists of all points (x, y) lying on the line $-2x + 6y = 3$.

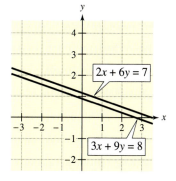

Figure 4.4

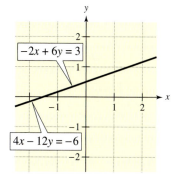

Figure 4.5

2 Use systems of linear equations to model and solve real-life problems.

Applications

To determine whether a real-life problem can be solved using a system of linear equations, consider these questions:

1. Does the problem involve more than one unknown quantity?

2. Are there two (or more) equations or conditions to be satisfied?

If one or both of these conditions occur, the appropriate mathematical model for the problem may be a system of linear equations.

Technology: Tip

The general solution of the linear system

$$\begin{cases} ax + by = c \\ dx + ey = f \end{cases}$$

is $x = (ce - bf)/(ae - bd)$ and $y = (af - cd)/(ae - bd)$. If $ae - bd = 0$, the system does not have a unique solution. Graphing calculator programs for solving such a system can be found at our website *math.college.hmco.com/students*. Try using this program to solve the systems in Examples 5, 6, and 7.

Example 5   A Mixture Problem

A company with two stores buys six large delivery vans and five small delivery vans. The first store receives four of the large vans and two of the small vans for a total cost of $160,000. The second store receives two of the large vans and three of the small vans for a total cost of $128,000. What is the cost of each type of van?

Solution

The two unknowns in this problem are the costs of the two types of vans.

Verbal Model: $4 \left(\begin{array}{c} \text{Cost of} \\ \text{large van} \end{array} \right) + 2 \left(\begin{array}{c} \text{Cost of} \\ \text{small van} \end{array} \right) = \$160,000$

$2 \left(\begin{array}{c} \text{Cost of} \\ \text{large van} \end{array} \right) + 3 \left(\begin{array}{c} \text{Cost of} \\ \text{small van} \end{array} \right) = \$128,000$

Labels: Cost of large van = x (dollars)
       Cost of small van = y (dollars)

System: $\begin{cases} 4x + 2y = 160,000 & \text{Equation 1} \\ 2x + 3y = 128,000 & \text{Equation 2} \end{cases}$

To solve this system of linear equations, use the method of elimination. To obtain coefficients of x that differ only in sign, multiply Equation 2 by -2.

$\begin{cases} 4x + 2y = 160,000 \\ 2x + 3y = 128,000 \end{cases}$ ⟹

$\begin{array}{rll} 4x + 2y = & 160,000 & \text{Equation 1} \\ -4x - 6y = & -256,000 & \text{Multiply Equation 2} \\ & & \text{by } -2. \\ \hline -4y = & -96,000 & \text{Add equations.} \\ y = & 24,000 & \text{Divide each side by } -4. \end{array}$

So, the cost of each small van is $y = \$24,000$. Back-substitute this value in Equation 1 to find the cost of each large van.

$$4x + 2y = 160,000 \qquad\qquad \text{Equation 1}$$

$$4x + 2(24,000) = 160,000 \qquad\qquad \text{Substitute 24,000 for } y.$$

$$4x = 112,000 \qquad\qquad \text{Simplify.}$$

$$x = 28,000 \qquad\qquad \text{Divide each side by 4.}$$

The cost of each large van is $x = \$28,000$. Check this solution in the original statement of the problem.

Study Tip

When solving application problems, make sure your answers make sense. For instance, a negative result for the x- or y-value in Example 5 would not make sense.

Example 6   An Application Involving Two Speeds

You take a motorboat trip on a river (18 miles upstream and 18 miles downstream). You run the motor at the same speed going up and down the river, but because of the river's current, the trip upstream takes $1\frac{1}{2}$ hours and the trip downstream takes only 1 hour. Determine the speed of the current.

Solution

Verbal Model:

$$\text{Boat speed} - \text{Current speed} = \text{Upstream speed}$$

$$\text{Boat speed} + \text{Current speed} = \text{Downstream speed}$$

Labels: Boat speed (in still water) $= x$ (miles per hour)
           Current speed $= y$ (miles per hour)
           Upstream speed $= 18/1.5 = 12$ (miles per hour)
           Downstream speed $= 18/1 = 18$ (miles per hour)

System: $\begin{cases} x - y = 12 & \text{Equation 1} \\ x + y = 18 & \text{Equation 2} \end{cases}$

To solve this system of linear equations, use the method of elimination.

$$
\begin{array}{ll}
x - y = 12 & \text{Equation 1} \\
\underline{x + y = 18} & \text{Equation 2} \\
2x \quad\ = 30 & \text{Add equations.}
\end{array}
$$

So, the speed of the boat in still water is 15 miles per hour. Back-substitution yields $y = 3$. So, the speed of the current is 3 miles per hour. Check this solution in the original statement of the problem.

Example 7   Data Analysis: Best-Fitting Line

The slope and y-intercept of the line $y = mx + b$ that best fits the three noncollinear points $(1, 0)$, $(2, 1)$, and $(3, 4)$ are given by the solution of the system of linear equations below.

$$\begin{cases} 3b + 6m = 5 & \text{Equation 1} \\ 6b + 14m = 14 & \text{Equation 2} \end{cases}$$

Solve this system. Then find the equation of the best-fitting line.

Solution

To solve this system of linear equations, use the method of elimination.

$$
\begin{array}{lll}
\begin{cases} 3b + 6m = 5 \\ 6b + 14m = 14 \end{cases}
&
\begin{array}{l}
-6b - 12m = -10 \quad \text{Multiply Equation 1 by } -2. \\
\underline{6b + 14m = 14} \quad \text{Equation 2} \\
2m = 4 \quad \text{Add equations.}
\end{array}
\end{array}
$$

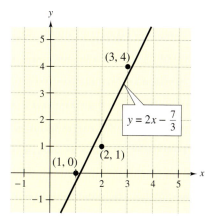

Figure 4.6

So, the slope is 2. Back-substitution of this value in Equation 2 yields a y-intercept of $-\frac{7}{3}$. So, the equation of the best-fitting line is $y = 2x - \frac{7}{3}$. The points and the best-fitting line are shown in Figure 4.6.

4.2 Exercises

Review   Concepts, Skills, and Problem Solving

Keep mathematically in shape by doing these exercises *before* the problems of this section.

Properties and Definitions

1. Identify the property of real numbers illustrated by $2(x + y) = 2x + 2y$.

2. What property of equality is demonstrated in solving the following equation?

$$x - 4 = 7$$
$$x - 4 + 4 = 7 + 4$$
$$x = 11$$

Solving Inequalities

In Exercises 3–8, solve the inequality.

3. $1 < 2x + 5 < 9$

4. $0 \le \dfrac{x - 4}{2} < 6$

5. $|6x| > 12$

6. $|1 - 2x| < 5$

7. $4x - 12 < 0$

8. $4x + 4 \ge 9$

Problem Solving

9. *Operating Cost*   The annual operating cost of a truck is $C = 0.45m + 6200$, where m is the number of miles traveled by the truck in a year. What number of miles will yield an annual operating cost that is less than \$15,000?

10. *Monthly Wage*   You must select one of two plans of payment when working for a company. One plan pays \$2500 per month. The second pays \$1500 per month plus a commission of 4% of your gross sales x. Write an inequality for which the second option yields the greater monthly wage. Solve the inequality.

Developing Skills

In Exercises 1–12, solve the system of linear equations by the method of elimination. Identify and label each line with its equation and label the point of intersection (if any). See Examples 1–4.

1. $\begin{cases} 2x + y = 4 \\ x - y = 2 \end{cases}$

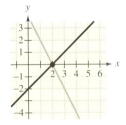

2. $\begin{cases} x + 3y = 2 \\ -x + 2y = 3 \end{cases}$

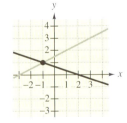

3. $\begin{cases} -x + 2y = 1 \\ x - y = 2 \end{cases}$

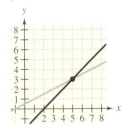

4. $\begin{cases} x + y = 0 \\ -3x - 3y = 0 \end{cases}$

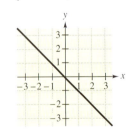

5. $\begin{cases} 3x + y = 3 \\ 2x - y = 7 \end{cases}$

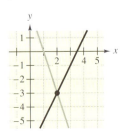

6. $\begin{cases} -x + 2y = 2 \\ 3x + y = 15 \end{cases}$

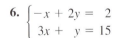
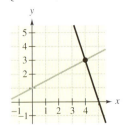

7. $\begin{cases} x - y = 1 \\ -3x + 3y = 8 \end{cases}$

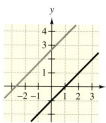

8. $\begin{cases} 3x + 4y = 2 \\ 0.6x + 0.8y = 1.6 \end{cases}$

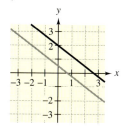

9. $\begin{cases} x - 3y = 5 \\ -2x + 6y = -10 \end{cases}$

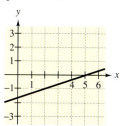

10. $\begin{cases} x - 4y = 5 \\ 5x + 4y = 7 \end{cases}$

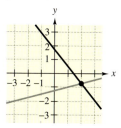

11. $\begin{cases} 2x - 8y = -11 \\ 5x + 3y = 7 \end{cases}$

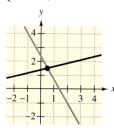

12. $\begin{cases} 3x + 4y = 0 \\ 9x - 5y = 17 \end{cases}$

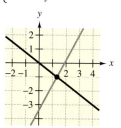

In Exercises 13–40, solve the system of linear equations by the method of elimination. See Examples 1–4.

13. $\begin{cases} 6x - 6y = 25 \\ 3y = 11 \end{cases}$

14. $\begin{cases} x + 3y = 4 \\ 2x = 2 \end{cases}$

15. $\begin{cases} x + y = 0 \\ x - y = 4 \end{cases}$

16. $\begin{cases} x - y = 12 \\ x + y = 2 \end{cases}$

17. $\begin{cases} 3x - 5y = 1 \\ 2x + 5y = 9 \end{cases}$

18. $\begin{cases} -x + 2y = 12 \\ x + 6y = 20 \end{cases}$

19. $\begin{cases} 5x + 2y = 7 \\ 3x - y = 13 \end{cases}$

20. $\begin{cases} 4x + 3y = 8 \\ x - 2y = 13 \end{cases}$

21. $\begin{cases} x - 3y = 2 \\ 3x - 7y = 4 \end{cases}$

22. $\begin{cases} 2s - t = 9 \\ 3s + 4t = -14 \end{cases}$

23. $\begin{cases} 2x + y = 9 \\ 3x - y = 16 \end{cases}$

24. $\begin{cases} 7r - s = -25 \\ 2r + 5s = 14 \end{cases}$

25. $\begin{cases} 2u + 3v = 8 \\ 3u + 4v = 13 \end{cases}$

26. $\begin{cases} 4x - 3y = 25 \\ -3x + 8y = 10 \end{cases}$

27. $\begin{cases} 12x - 5y = 2 \\ -24x + 10y = 6 \end{cases}$

28. $\begin{cases} -2x + 3y = 9 \\ 6x - 9y = -27 \end{cases}$

29. $\begin{cases} \frac{2}{3}r - s = 0 \\ 10r + 4s = 19 \end{cases}$

30. $\begin{cases} x - y = -\frac{1}{2} \\ 4x - 48y = -35 \end{cases}$

31. $\begin{cases} 0.05x - 0.03y = 0.21 \\ x + y = 9 \end{cases}$

32. $\begin{cases} 0.02x - 0.05y = -0.19 \\ 0.03x + 0.04y = 0.52 \end{cases}$

33. $\begin{cases} 0.7u - v = -0.4 \\ 0.3u - 0.8v = 0.2 \end{cases}$

34. $\begin{cases} 0.15x - 0.35y = -0.5 \\ -0.12x + 0.25y = 0.1 \end{cases}$

35. $\begin{cases} 5x + 7y = 25 \\ x + 1.4y = 5 \end{cases}$

36. $\begin{cases} 6b - \frac{13}{2}m = 1 \\ -b + \frac{13}{12}m = -\frac{1}{12} \end{cases}$

37. $\begin{cases} \frac{1}{2}x - \frac{1}{3}y = 1 \\ \frac{1}{4}x - \frac{1}{9}y = \frac{2}{3} \end{cases}$

38. $\begin{cases} \frac{1}{3}x = 4 - \frac{1}{4}y \\ \frac{3}{5}x - \frac{4}{5}y = 0 \end{cases}$

39. $\begin{cases} \frac{1}{5}x + \frac{1}{5}y = 4 \\ \frac{2}{3}x - y = \frac{8}{3} \end{cases}$

40. $\begin{cases} \frac{2}{3}x - 4 = \frac{1}{2}y \\ x - 3y = \frac{1}{3} \end{cases}$

In Exercises 41–48, solve the system of linear equations by any convenient method.

41. $\begin{cases} 3x + 2y = 5 \\ y = 2x + 13 \end{cases}$

42. $\begin{cases} 4x + y = -2 \\ -6x + y = 18 \end{cases}$

43. $\begin{cases} y = 5x - 3 \\ y = -2x + 11 \end{cases}$

44. $\begin{cases} 3y = 2x + 21 \\ x = 50 - 4y \end{cases}$

45. $\begin{cases} 2x - y = 20 \\ -x + y = -5 \end{cases}$

46. $\begin{cases} 3x - 2y = -20 \\ 5x + 6y = 32 \end{cases}$

47. $\begin{cases} \frac{3}{2}x + 2y = 12 \\ \frac{1}{4}x + y = 4 \end{cases}$

48. $\begin{cases} x + 2y = 4 \\ \frac{1}{2}x + \frac{1}{3}y = 1 \end{cases}$

In Exercises 49–54, decide whether the system is consistent or inconsistent.

49. $\begin{cases} 4x - 5y = 3 \\ -8x + 10y = -6 \end{cases}$

50. $\begin{cases} 4x - 5y = 3 \\ -8x + 10y = 14 \end{cases}$

51. $\begin{cases} -2x + 5y = 3 \\ 5x + 2y = 8 \end{cases}$

52. $\begin{cases} x + 10y = 12 \\ -2x + 5y = 2 \end{cases}$

53. $\begin{cases} -10x + 15y = 25 \\ 2x - 3y = -24 \end{cases}$

54. $\begin{cases} 4x - 5y = 28 \\ -2x + 2.5y = -14 \end{cases}$

In Exercises 55 and 56, determine the value of k such that the system of linear equations is inconsistent.

55. $\begin{cases} 5x - 10y = 40 \\ -2x + ky = 30 \end{cases}$

56. $\begin{cases} 12x - 18y = 5 \\ -18x + ky = 10 \end{cases}$

In Exercises 57 and 58, find a system of linear equations that has the given solution. (There are many correct answers.)

57. $\left(3, -\frac{3}{2}\right)$

58. $(-8, 12)$

Solving Problems

59. *Break-Even Analysis* To open a small business, you need an initial investment of $85,000. Each week your costs will be about $7400. Your projected weekly revenue is $8100. How many weeks will it take to break even?

60. *Break-Even Analysis* To open a small business, you need an initial investment of $250,000. Each week your costs will be about $8650. Your projected weekly revenue is $9950. How many weeks will it take to break even?

61. *Investment* A total of $20,000 is invested in two bonds that pay 8% and 9.5% simple interest. The annual interest is $1675. How much is invested in each bond?

62. *Investment* A total of $4500 is invested in two funds paying 4% and 5% simple interest. The annual interest is $210. How much is invested in each fund?

63. *Average Speed* A van travels for 2 hours at an average speed of 40 miles per hour. How much longer must the van travel at an average speed of 55 miles per hour so that the average speed for the total trip will be 50 miles per hour?

64. *Average Speed* A truck travels for 4 hours at an average speed of 42 miles per hour. How much longer must the truck travel at an average speed of 55 miles per hour so that the average speed for the total trip will be 50 miles per hour?

65. *Air Speed* An airplane flying into a headwind travels 1800 miles in 3 hours and 36 minutes. On the return flight, the same distance is traveled in 3 hours. Find the speed of the plane in still air and the speed of the wind, assuming that both remain constant throughout the round trip.

66. *Air Speed* An airplane flying into a headwind travels 3000 miles in 6 hours and 15 minutes. On the return flight, the same distance is traveled in 5 hours. Find the speed of the plane in still air and the speed of the wind, assuming that both remain constant throughout the round trip.

67. *Ticket Sales*   Five hundred tickets were sold for a fundraising dinner. The receipts totaled $3312.50. Adult tickets were $7.50 each and children's tickets were $4.00 each. How many tickets of each type were sold?

68. *Ticket Sales*   A fundraising dinner was held on two consecutive nights. On the first night, 100 adult tickets and 175 children's tickets were sold, for a total of $937.50. On the second night, 200 adult tickets and 316 children's tickets were sold, for a total of $1790.00. Find the price of each type of ticket.

69. *Gasoline Mixture*   Twelve gallons of regular unleaded gasoline plus 8 gallons of premium unleaded gasoline cost $23.08. Premium unleaded gasoline costs $0.11 more per gallon than regular unleaded. Find the price per gallon for each grade of gasoline.

70. *Gasoline Mixture*   The total cost of 8 gallons of regular unleaded gasoline and 12 gallons of premium unleaded gasoline is $27.84. Premium unleaded gasoline costs $0.17 more per gallon than regular unleaded. Find the price per gallon for each grade of gasoline.

71. *Alcohol Mixture*   How many liters of a 40% alcohol solution must be mixed with a 65% solution to obtain 20 liters of a 50% solution?

72. *Alcohol Mixture*   How many liters of a 20% alcohol solution must be mixed with a 60% solution to obtain 40 liters of a 35% solution?

73. *Alcohol Mixture*   How many fluid ounces of a 50% alcohol solution must be mixed with a 90% solution to obtain 32 fluid ounces of a 75% solution?

74. *Acid Mixture*   Five gallons of a 45% acid solution is obtained by mixing a 90% solution with a 30% solution. How many gallons of each solution must be used to obtain the desired mixture?

75. *Acid Mixture*   Thirty liters of a 46% acid solution is obtained by mixing a 40% solution with a 70% solution. How many liters of each solution must be used to obtain the desired mixture?

76. *Acid Mixture*   Fifty gallons of a 70% acid solution is obtained by mixing an 80% solution with a 50% solution. How many gallons of each solution must be used to obtain the desired mixture?

77. *Nut Mixture*   Ten pounds of mixed nuts sells for $6.87 per pound. The mixture is obtained from two kinds of nuts, peanuts priced at $5.70 per pound and cashews at $8.70 per pound. How many pounds of each variety of nut are used in the mixture?

78. *Nut Mixture*   Thirty pounds of mixed nuts sells for $6.30 per pound. The mixture is obtained from two kinds of nuts, walnuts priced at $6.50 per pound and peanuts at $5.90 per pound. How many pounds of each variety of nut are used in the mixture?

79. *Best-Fitting Line*   The slope and y-intercept of the line $y = mx + b$ that best fits the three noncollinear points $(0, 0)$, $(1, 1)$, and $(2, 3)$ are given by the solution of the following system of linear equations.
$$\begin{cases} 5m + 3b = 7 \\ 3m + 3b = 4 \end{cases}$$
(a) Solve the system and find the equation of the best-fitting line.

(b) Plot the three points and sketch the graph of the best-fitting line.

80. *Best-Fitting Line*   The slope and y-intercept of the line $y = mx + b$ that best fits the three noncollinear points $(0, 4)$, $(1, 2)$, and $(2, 1)$ are given by the solution of the following system of linear equations.
$$\begin{cases} 3b + 3m = 7 \\ 3b + 5m = 4 \end{cases}$$
(a) Solve the system and find the equation of the best-fitting line.

(b) Plot the three points and sketch the graph of the best-fitting line.

81. *Hourly Wages*   The average hourly wages for those employed in the mining industry in the United States for selected years from 1990 through 2000 are shown in the table.   (Source: U.S. Bureau of Labor Statistics)

Year	1990	1995	2000
Wage, y	$13.68	$15.30	$17.14

(a) Plot the data shown in the table. Let x represent the year, with $x = 0$ corresponding to 1990.

(b) The line $y = mx + b$ that best fits the data is given by the solution of the following system.

$$\begin{cases} 3b + 15m = 46.12 \\ 15b + 125m = 247.9 \end{cases}$$

Solve the system and find the equation of the best-fitting line. Sketch the graph of the line on the same set of coordinate axes used in part (a).

(c) Interpret the meaning of the slope of the line in the context of this problem.

82. *Weekly Earnings*   The average weekly earnings for those employed in the manufacturing industry in the United States for selected years from 1990 through 2000 are shown in the table.   (Source: U.S. Bureau of Labor Statistics)

Year	1990	1995	2000
Weekly earnings, y	$442	$515	$597

(a) Plot the data shown in the table. Let x represent the year, with $x = 0$ corresponding to 1990.

(b) The line $y = mx + b$ that best fits the data is given by the solution of the following system.

$$\begin{cases} 3b + 15m = 1554 \\ 15b + 125m = 8545 \end{cases}$$

Solve the system and find the equation of the best-fitting line. Sketch the graph of the line on the same set of coordinate axes used in part (a).

(c) Interpret the meaning of the slope of the line in the context of this problem.

Explaining Concepts

83.   Answer parts (a)–(c) of Motivating the Chapter on page 216.

84. *Writing*   When solving a system by elimination, how do you recognize that it has infinitely many solutions?

85. *Writing*   Explain what is meant by an *inconsistent* system of linear equations.

86. *Writing*   In your own words, explain how to solve a system of linear equations by elimination.

87. *Writing*   How can you recognize that a system of linear equations has no solution? Give an example.

88. *Writing*   Under what conditions might substitution be better than elimination for solving a system of linear equations?

4.3 Linear Systems in Three Variables

Frank Whitney/Getty Images

What You Should Learn

1 Solve systems of linear equations using row-echelon form with back-substitution.

2 Solve systems of linear equations using the method of Gaussian elimination.

3 Solve application problems using elimination with back-substitution.

Why You Should Learn It

Systems of linear equations in three variables can be used to model and solve real-life problems. For instance, in Exercise 47 on page 251, a system of linear equations can be used to determine a chemical mixture for a pesticide.

1 Solve systems of linear equations using row-echelon form with back-substitution.

Row-Echelon Form

The method of elimination can be applied to a system of linear equations in more than two variables. In fact, this method easily adapts to computer use for solving systems of linear equations with dozens of variables.

When the method of elimination is used to solve a system of linear equations, the goal is to rewrite the system in a form to which back-substitution can be applied. For instance, consider the following two systems of linear equations.

$$\begin{cases} x - 2y + 2z = 9 \\ -x + 3y = -4 \\ 2x - 5y + z = 10 \end{cases} \qquad \begin{cases} x - 2y + 2z = 9 \\ y + 2z = 5 \\ z = 3 \end{cases}$$

Which of these two systems do you think is easier to solve? After comparing the two systems, it should be clear that it is easier to solve the system on the right because the value of z is already shown and back-substitution will readily yield the values of x and y. The system on the right is said to be in **row-echelon form,** which means that it has a "stair-step" pattern with leading coefficients of 1.

Example 1 Using Back-Substitution

In the following system of linear equations, you know the value of z from Equation 3.

$$\begin{cases} x - 2y + 2z = 9 & \text{Equation 1} \\ y + 2z = 5 & \text{Equation 2} \\ z = 3 & \text{Equation 3} \end{cases}$$

To solve for y, substitute $z = 3$ in Equation 2 to obtain

$$y + 2(3) = 5 \quad \Longrightarrow \quad y = -1. \qquad \text{Substitute 3 for } z.$$

Finally, substitute $y = -1$ and $z = 3$ in Equation 1 to obtain

$$x - 2(-1) + 2(3) = 9 \quad \Longrightarrow \quad x = 1. \qquad \text{Substitute } -1 \text{ for } y \text{ and 3 for } z.$$

The solution is $x = 1$, $y = -1$, and $z = 3$, which can also be written as the **ordered triple** $(1, -1, 3)$. Check this in the original system of equations.

Study Tip

When checking a solution, remember that the solution must satisfy each equation in the original system.

② Solve systems of linear equations using the method of Gaussian elimination.

The Method of Gaussian Elimination

Two systems of equations are **equivalent systems** if they have the same solution set. To solve a system that is not in row-echelon form, first convert it to an *equivalent* system that is in row-echelon form. To see how this is done, let's take another look at the method of elimination, as applied to a system of two linear equations.

Example 2   The Method of Elimination

Solve the system of linear equations.

$$\begin{cases} 3x - 2y = -1 & \text{Equation 1} \\ x - y = 0 & \text{Equation 2} \end{cases}$$

Solution

$$\begin{cases} x - y = 0 \\ 3x - 2y = -1 \end{cases} \qquad \text{Interchange the two equations in the system.}$$

$$\begin{aligned} -3x + 3y &= 0 \qquad \text{Multiply new Equation 1 by } -3 \text{ and add it to new} \\ \underline{3x - 2y} &= -1 \qquad \text{Equation 2.} \\ y &= -1 \end{aligned}$$

$$\begin{cases} x - y = 0 \\ y = -1 \end{cases} \qquad \text{New system in row-echelon form}$$

Using back-substitution, you can determine that the solution is $(-1, -1)$. Check the solution in each equation in the original system, as follows.

$$\begin{array}{cc} \text{Equation 1} & \text{Equation 2} \\ 3x - 2y \overset{?}{=} -1 & x - y \overset{?}{=} 0 \\ 3(-1) - 2(-1) = -1 \ \checkmark & (-1) - (-1) = 0 \ \checkmark \end{array}$$

Rewriting a system of linear equations in row-echelon form usually involves a chain of equivalent systems, each of which is obtained by using one of the three basic row operations. This process is called **Gaussian elimination.**

Operations That Produce Equivalent Systems

Each of the following **row operations** on a system of linear equations produces an *equivalent* system of linear equations.

1. Interchange two equations.

2. Multiply one of the equations by a nonzero constant.

3. Add a multiple of one of the equations to another equation to replace the latter equation.

Example 3   Using Gaussian Elimination to Solve a System

Solve the system of linear equations.

$$\begin{cases} x - 2y + 2z = 9 & \text{Equation 1} \\ -x + 3y = -4 & \text{Equation 2} \\ 2x - 5y + z = 10 & \text{Equation 3} \end{cases}$$

Solution

Because the leading coefficient of the first equation is 1, you can begin by saving the x in the upper left position and eliminating the other x terms from the first column, as follows.

$$\begin{cases} x - 2y + 2z = 9 \\ y + 2z = 5 \\ 2x - 5y + z = 10 \end{cases}$$

> Adding the first equation to the second equation produces a new second equation.

$$\begin{cases} x - 2y + 2z = 9 \\ y + 2z = 5 \\ -y - 3z = -8 \end{cases}$$

> Adding -2 times the first equation to the third equation produces a new third equation.

Now that all but the first x have been eliminated from the first column, go to work on the second column. (You need to eliminate y from the third equation.)

$$\begin{cases} x - 2y + 2z = 9 \\ y + 2z = 5 \\ -z = -3 \end{cases}$$

> Adding the second equation to the third equation produces a new third equation.

Finally, you need a coefficient of 1 for z in the third equation.

$$\begin{cases} x - 2y + 2z = 9 \\ y + 2z = 5 \\ z = 3 \end{cases}$$

> Multiplying the third equation by -1 produces a new third equation.

This is the same system that was solved in Example 1, and, as in that example, you can conclude by back-substitution that the solution is

$$x = 1, \quad y = -1, \quad \text{and} \quad z = 3.$$

The solution can be written as the ordered triple

$$(1, -1, 3).$$

You can check the solution by substituting 1 for x, -1 for y, and 3 for z in each equation of the original system, as follows.

Check

$$\text{Equation 1:} \quad x - 2y + 2z \stackrel{?}{=} 9$$
$$1 - 2(-1) + 2(3) = 9 \checkmark$$
$$\text{Equation 2:} \quad -x + 3y \stackrel{?}{=} -4$$
$$-(1) + 3(-1) = -4 \checkmark$$
$$\text{Equation 3:} \quad 2x - 5y + z \stackrel{?}{=} 10$$
$$2(1) - 5(-1) + 3 = 10 \checkmark$$

Example 4   Using Gaussian Elimination to Solve a System

Solve the system of linear equations.

$$\begin{cases} 4x + y - 3z = 11 & \text{Equation 1} \\ 2x - 3y + 2z = 9 & \text{Equation 2} \\ x + y + z = -3 & \text{Equation 3} \end{cases}$$

Solution

$$\begin{cases} x + y + z = -3 \\ 2x - 3y + 2z = 9 \\ 4x + y - 3z = 11 \end{cases}$$

Interchange the first and third equations.

$$\begin{cases} x + y + z = -3 \\ -5y = 15 \\ 4x + y - 3z = 11 \end{cases}$$

Adding -2 times the first equation to the second equation produces a new second equation.

$$\begin{cases} x + y + z = -3 \\ -5y = 15 \\ -3y - 7z = 23 \end{cases}$$

Adding -4 times the first equation to the third equation produces a new third equation.

$$\begin{cases} x + y + z = -3 \\ y = -3 \\ -3y - 7z = 23 \end{cases}$$

Multiplying the second equation by $-\frac{1}{5}$ produces a new second equation.

$$\begin{cases} x + y + z = -3 \\ y = -3 \\ -7z = 14 \end{cases}$$

Adding 3 times the second equation to the third equation produces a new third equation.

$$\begin{cases} x + y + z = -3 \\ y = -3 \\ z = -2 \end{cases}$$

Multiplying the third equation by $-\frac{1}{7}$ produces a new third equation.

Now you can see that $z = -2$ and $y = -3$. Moreover, by back-substituting these values in Equation 1, you can determine that $x = 2$. So, the solution is

$$x = 2, \quad y = -3, \quad \text{and} \quad z = -2$$

which can be written as the ordered triple $(2, -3, -2)$. You can check this solution as follows.

Check

Equation 1: $4x + y - 3z \stackrel{?}{=} 11$
$$4(2) + (-3) - 3(-2) \stackrel{?}{=} 11$$
$$11 = 11 \checkmark$$

Equation 2: $2x - 3y + 2z \stackrel{?}{=} 9$
$$2(2) - 3(-3) + 2(-2) \stackrel{?}{=} 9$$
$$9 = 9 \checkmark$$

Equation 3: $x + y + z \stackrel{?}{=} -3$
$$(2) + (-3) + (-2) \stackrel{?}{=} -3$$
$$-3 = -3 \checkmark$$

The next example involves an inconsistent system—one that has no solution. The key to recognizing an inconsistent system is that at some stage in the elimination process, you obtain a false statement such as $0 = 6$. Watch for such statements as you do the exercises for this section.

Example 5   An Inconsistent System

Solve the system of linear equations.

$$\begin{cases} x - 3y + z = 1 & \text{Equation 1} \\ 2x - y - 2z = 2 & \text{Equation 2} \\ x + 2y - 3z = -1 & \text{Equation 3} \end{cases}$$

Solution

$$\begin{cases} x - 3y + z = 1 \\ 5y - 4z = 0 \\ x + 2y - 3z = -1 \end{cases}$$

Adding -2 times the first equation to the second equation produces a new second equation.

$$\begin{cases} x - 3y + z = 1 \\ 5y - 4z = 0 \\ 5y - 4z = -2 \end{cases}$$

Adding -1 times the first equation to the third equation produces a new third equation.

$$\begin{cases} x - 3y + z = 1 \\ 5y - 4z = 0 \\ 0 = -2 \end{cases}$$

Adding -1 times the second equation to the third equation produces a new third equation.

Because the third "equation" is a false statement, you can conclude that this system is inconsistent and so has no solution. Moreover, because this system is equivalent to the original system, you can conclude that the original system also has no solution.

As with a system of linear equations in two variables, the solution(s) of a system of linear equations in more than two variables must fall into one of three categories.

The Number of Solutions of a Linear System

For a system of linear equations, exactly one of the following is true.

1. There is exactly one solution.

2. There are infinitely many solutions.

3. There is no solution.

The graph of a system of three linear equations in three variables consists of *three planes*. When these planes intersect in a single point, the system has exactly one solution. (See Figure 4.7.) When the three planes intersect in a line or a plane, the system has infinitely many solutions. (See Figures 4.8 and 4.9.) When the three planes have no point in common, the system has no solution. (See Figures 4.10 and 4.11.)

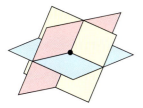

Solution: one point

Figure 4.7

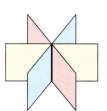

Solution: one line

Figure 4.8

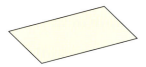

Solution: one plane

Figure 4.9

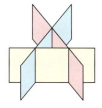

Solution: none

Figure 4.10

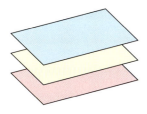

Solution: none

Figure 4.11

Example 6   A System with Infinitely Many Solutions

Solve the system of linear equations.

$$\begin{cases} x + y - 3z = -1 & \text{Equation 1} \\ \quad\quad y - z = 0 & \text{Equation 2} \\ -x + 2y \quad\quad = 1 & \text{Equation 3} \end{cases}$$

Solution

Begin by rewriting the system in row-echelon form.

$$\begin{cases} x + y - 3z = -1 \\ \quad\quad y - z = 0 \\ \quad\quad 3y - 3z = 0 \end{cases}$$

> Adding the first equation to the third equation produces a new third equation.

$$\begin{cases} x + y - 3z = -1 \\ \quad\quad y - z = 0 \\ \quad\quad\quad\quad 0 = 0 \end{cases}$$

> Adding -3 times the second equation to the third equation produces a new third equation.

This means that Equation 3 depends on Equations 1 and 2 in the sense that it gives us no additional information about the variables. So, the original system is equivalent to the system

$$\begin{cases} x + y - 3z = -1 \\ \quad\quad y - z = 0. \end{cases}$$

In this last equation, solve for y in terms of z to obtain $y = z$. Back-substituting for y in the previous equation produces $x = 2z - 1$. Finally, letting $z = a$, where a is any real number, you can see that solutions to the original system are all of the form

$$x = 2a - 1, \ y = a, \text{ and } z = a.$$

So, every ordered triple of the form

$$(2a - 1, a, a), \quad\quad a \text{ is a real number}$$

is a solution of the system.

In Example 6, there are other ways to write the same infinite set of solutions. For instance, letting $x = b$, the solutions could have been written as

$$\left(b, \frac{1}{2}(b + 1), \frac{1}{2}(b + 1) \right), \quad\quad b \text{ is a real number.}$$

To convince yourself that this description produces the same set of solutions, consider the comparison shown below.

Substitution	Solution	
$a = 0$	$(2(0)-1, 0, 0) = (-1, 0, 0)$	
$b = -1$	$\left(-1, \frac{1}{2}(-1 + 1), \frac{1}{2}(-1 + 1)\right) = (-1, 0, 0)$	Same solution
$a = 1$	$(2(1)-1, 1, 1) = (1, 1, 1)$	
$b = 1$	$\left(1, \frac{1}{2}(1 + 1), \frac{1}{2}(1 + 1)\right) = (1, 1, 1)$	Same solution

3 Solve application problems using elimination with back-substitution.

Applications

Example 7   Vertical Motion

The height at time t of an object that is moving in a (vertical) line with constant acceleration a is given by the **position equation**

$$s = \frac{1}{2}at^2 + v_0 t + s_0.$$

The height s is measured in feet, the acceleration a is measured in feet per second squared, the time t is measured in seconds, v_0 is the initial velocity (at time $t = 0$), and s_0 is the initial height. Find the values of a, v_0, and s_0, if $s = 164$ feet at 1 second, $s = 180$ feet at 2 seconds, and $s = 164$ feet at 3 seconds.

Solution

By substituting the three values of t and s into the position equation, you obtain three linear equations in a, v_0, and s_0.

When $t = 1$, $s = 164$: $\quad \frac{1}{2}a(1)^2 + v_0(1) + s_0 = 164$

When $t = 2$, $s = 180$: $\quad \frac{1}{2}a(2)^2 + v_0(2) + s_0 = 180$

When $t = 3$, $s = 164$: $\quad \frac{1}{2}a(3)^2 + v_0(3) + s_0 = 164$

By multiplying the first and third equations by 2, this system can be rewritten as

$$\begin{cases} a + 2v_0 + 2s_0 = 328 & \text{Equation 1} \\ 2a + 2v_0 + s_0 = 180 & \text{Equation 2} \\ 9a + 6v_0 + 2s_0 = 328 & \text{Equation 3} \end{cases}$$

and you can apply Gaussian elimination to obtain

$$\begin{cases} a + 2v_0 + 2s_0 = 328 & \text{Equation 1} \\ -2v_0 - 3s_0 = -476 & \text{Equation 2} \\ 2s_0 = 232. & \text{Equation 3} \end{cases}$$

From the third equation, $s_0 = 116$, so back-substitution in Equation 2 yields

$$-2v_0 - 3(116) = -476$$

$$-2v_0 = -128$$

$$v_0 = 64.$$

Finally, back-substituting $s_0 = 116$ and $v_0 = 64$ in Equation 1 yields

$$a + 2(64) + 2(116) = 328$$

$$a = -32.$$

So, the position equation for this object is $s = -16t^2 + 64t + 116$.

Example 8   A Geometry Application

The sum of the measures of two angles of a triangle is twice the measure of the third angle. The measure of the first angle is $18°$ more than the measure of the third angle. Find the measures of the three angles.

Solution

Let x, y, and z represent the measures of the first, second, and third angles, respectively. The sum of the measures of the three angles of a triangle is $180°$. From the given information, you can write the system of equations as follows.

$$\begin{cases} x + y + z = 180 & \text{Equation 1} \\ x + y = 2z & \text{Equation 2} \\ x = z + 18 & \text{Equation 3} \end{cases}$$

By rewriting this system in the standard form you obtain

$$\begin{cases} x + y + z = 180 & \text{Equation 1} \\ x + y - 2z = 0 & \text{Equation 2} \\ x - z = 18. & \text{Equation 3} \end{cases}$$

Using Gaussian elimination to solve this system yields $x = 78$, $y = 42$, and $z = 60$. So, the measures of the three angles are $78°$, $42°$, and $60°$, respectively. You can check these solutions as follows.

Check

Equation 1: $78 + 42 + 60 = 180$ ✓

Equation 2: $78 + 42 - 2(60) = 0$ ✓

Equation 3: $78 - 60 = 18$ ✓

Example 9   Grades of Paper

A paper manufacturer sells a 50-pound package that consists of three grades of computer paper. Grade A costs $6.00 per pound, grade B costs $4.50 per pound, and grade C costs $3.50 per pound. Half of the 50-pound package consists of the two cheaper grades. The cost of the 50-pound package is $252.50. How many pounds of each grade of paper are there in the 50-pound package?

Solution

Let A represent grade A paper, B represent grade B paper, and C represent grade C paper. From the given information you can write the system of equations as follows.

$$\begin{cases} A + B + C = 50 & \text{Equation 1} \\ 6A + 4.50B + 3.50C = 252.50 & \text{Equation 2} \\ B + C = 25 & \text{Equation 3} \end{cases}$$

Using Gaussian elimination to solve this system yields $A = 25$, $B = 15$, and $C = 10$. So, there are 25 pounds of grade A paper, 15 pounds of grade B paper, and 10 pounds of grade C paper in the 50-pound package. Check this solution in the original statement of the problem.

4.3 Exercises

Review   Concepts, Skills, and Problem Solving

Keep mathematically in shape by doing these exercises *before* the problems of this section.

Properties and Definitions

1. A linear equation of the form $2x + 8 = 7$ has how many solutions?

2. What is the usual first step in solving an equation such as

$$\frac{t}{6} + \frac{5}{8} = \frac{7}{4}?$$

Solving Equations

In Exercises 3–8, solve the equation.

3. $\dfrac{x}{6} + \dfrac{5}{8} = \dfrac{7}{4}$

4. $0.25x + 1.75 = 4.5$

5. $|2x - 4| = 6$

6. $2|7 - x| = 10$

7. $6x - (x + 1) = 5x - 1$

8. $\frac{1}{4}(5 - 2x) = 9x - 7x$

Models and Graphs

9. The length of each edge of a cube is s inches. Write the volume V of the cube as a function of s.

10. Write the area A of a circle as a function of its circumference C.

11. The speed of a ship is 15 knots. Write the distance d the ship travels as a function of time t. Graph the model.

12. Your weekly pay is $180 plus $1.25 per sale. Write your weekly pay P as a function of the number of sales n. Graph the model.

Developing Skills

In Exercises 1 and 2, determine whether each ordered triple is a solution of the system of linear equations.

1. $\begin{cases} x + 3y + 2z = 1 \\ 5x - y + 3z = 16 \\ -3x + 7y + z = -14 \end{cases}$

   (a) $(0, 3, -2)$ (b) $(12, 5, -13)$
   (c) $(1, -2, 3)$ (d) $(-2, 5, -3)$

2. $\begin{cases} 3x - y + 4z = -10 \\ -x + y + 2z = 6 \\ 2x - y + z = -8 \end{cases}$

   (a) $(-2, 4, 0)$ (b) $(0, -3, 10)$
   (c) $(1, -1, 5)$ (d) $(7, 19, -3)$

In Exercises 3–6, use back-substitution to solve the system of linear equations. See Example 1.

3. $\begin{cases} x - 2y + 4z = 4 \\ 3y - z = 2 \\ z = -5 \end{cases}$

4. $\begin{cases} 5x + 4y - z = 0 \\ 10y - 3z = 11 \\ z = 3 \end{cases}$

5. $\begin{cases} x - 2y + 4z = 4 \\ y = 3 \\ y + z = 2 \end{cases}$

6. $\begin{cases} x = 10 \\ 3x + 2y = 2 \\ x + y + 2z = 0 \end{cases}$

In Exercises 7 and 8, determine whether the two systems of linear equations are equivalent. Give reasons for your answer.

7. $\begin{cases} x + 3y - z = 6 \\ 2x - y + 2z = 1 \\ 3x + 2y - z = 2 \end{cases}$ $\begin{cases} x + 3y - z = 6 \\ -7y + 4z = 1 \\ -7y - 4z = -16 \end{cases}$

8. $\begin{cases} x - 2y + 3z = 9 \\ -x + 3y = -4 \\ 2x - 5y + 5z = 17 \end{cases}$ $\begin{cases} x - 2y + 3z = 9 \\ y + 3z = 5 \\ -y - z = -1 \end{cases}$

In Exercises 9 and 10, perform the row operation and write the equivalent system of linear equations. See Example 2.

9. Add Equation 1 to Equation 2.

$$\begin{cases} x - 2y = 8 & \text{Equation 1} \\ -x + 3y = 6 & \text{Equation 2} \end{cases}$$

What did this operation accomplish?

10. Add -2 times Equation 1 to Equation 3.

$$\begin{cases} x - 2y + 3z = 5 & \text{Equation 1} \\ -x + y + 5z = 4 & \text{Equation 2} \\ 2x - 3z = 0 & \text{Equation 3} \end{cases}$$

What did this operation accomplish?

In Exercises 11–34, solve the system of linear equations. See Examples 3–6.

11. $\begin{cases} x + z = 4 \\ y = 2 \\ 4x + z = 7 \end{cases}$

12. $\begin{cases} x = 3 \\ -x + 3y = 3 \\ y + 2z = 4 \end{cases}$

13. $\begin{cases} x + y + z = 6 \\ 2x - y + z = 3 \\ 3x - z = 0 \end{cases}$

14. $\begin{cases} x + y + z = 2 \\ -x + 3y + 2z = 8 \\ 4x + y = 4 \end{cases}$

15. $\begin{cases} x + y + z = -3 \\ 4x + y - 3z = 11 \\ 2x - 3y + 2z = 9 \end{cases}$

16. $\begin{cases} x - y + 2z = -4 \\ 3x + y - 4z = -6 \\ 2x + 3y - 4z = 4 \end{cases}$

17. $\begin{cases} x + 2y + 6z = 5 \\ -x + y - 2z = 3 \\ x - 4y - 2z = 1 \end{cases}$

18. $\begin{cases} x + 6y + 2z = 9 \\ 3x - 2y + 3z = -1 \\ 5x - 5y + 2z = 7 \end{cases}$

19. $\begin{cases} 2x + 2z = 2 \\ 5x + 3y = 4 \\ 3y - 4z = 4 \end{cases}$

20. $\begin{cases} x + y + 8z = 3 \\ 2x + y + 11z = 4 \\ x + 3z = 0 \end{cases}$

21. $\begin{cases} 6y + 4z = -12 \\ 3x + 3y = 9 \\ 2x - 3z = 10 \end{cases}$

22. $\begin{cases} 2x - 4y + z = 0 \\ 3x + 2z = -1 \\ -6x + 3y + 2z = -10 \end{cases}$

23. $\begin{cases} 2x + y + 3z = 1 \\ 2x + 6y + 8z = 3 \\ 6x + 8y + 18z = 5 \end{cases}$

24. $\begin{cases} 3x - y - 2z = 5 \\ 2x + y + 3z = 6 \\ 6x - y - 4z = 9 \end{cases}$

25. $\begin{cases} y + z = 5 \\ 2x + 4z = 4 \\ 2x - 3y = -14 \end{cases}$

26. $\begin{cases} 5x + 2y = -8 \\ z = 5 \\ 3x - y + z = 9 \end{cases}$

27. $\begin{cases} 2x + z = 1 \\ 5y - 3z = 2 \\ 6x + 20y - 9z = 11 \end{cases}$

28. $\begin{cases} 2x + y - z = 4 \\ y + 3z = 2 \\ 3x + 2y = 4 \end{cases}$

29. $\begin{cases} 3x + y + z = 2 \\ 4x + 2z = 1 \\ 5x - y + 3z = 0 \end{cases}$

30. $\begin{cases} 2x + 3z = 4 \\ 5x + y + z = 2 \\ 11x + 3y - 3z = 0 \end{cases}$

31. $\begin{cases} 0.2x + 1.3y + 0.6z = 0.1 \\ 0.1x + 0.3z = 0.7 \\ 2x + 10y + 8z = 8 \end{cases}$

32. $\begin{cases} 0.3x - 0.1y + 0.2z = 0.35 \\ 2x + y - 2z = -1 \\ 2x + 4y + 3z = 10.5 \end{cases}$

33. $\begin{cases} x + 4y - 2z = 2 \\ -3x + y + z = -2 \\ 5x + 7y - 5z = 6 \end{cases}$

34. $\begin{cases} x - 2y - z = 3 \\ 2x + y - 3z = 1 \\ x + 8y - 3z = -7 \end{cases}$

In Exercises 35 and 36, find a system of linear equations in three variables with integer coefficients that has the given point as a solution. (There are many correct answers.)

35. $(4, -3, 2)$ **36.** $(5, 7, -10)$

Solving Problems

Vertical Motion   In Exercises 37–40, find the position equation $s = \frac{1}{2}at^2 + v_0t + s_0$ for an object that has the indicated heights at the specified times. See Example 7.

37. $s = 128$ feet at $t = 1$ second

$s = 80$ feet at $t = 2$ seconds

$s = 0$ feet at $t = 3$ seconds

38. $s = 48$ feet at $t = 1$ second

$s = 64$ feet at $t = 2$ seconds

$s = 48$ feet at $t = 3$ seconds

39. $s = 32$ feet at $t = 1$ second

$s = 32$ feet at $t = 2$ seconds

$s = 0$ feet at $t = 3$ seconds

40. $s = 10$ feet at $t = 0$ seconds

$s = 54$ feet at $t = 1$ second

$s = 46$ feet at $t = 3$ seconds

41. ▲ *Geometry*   The sum of the measures of two angles of a triangle is twice the measure of the third angle. The measure of the second angle is 28° less than the measure of the third angle. Find the measures of the three angles.

42. ▲ *Geometry*   The measure of the second angle of a triangle is one-half the measure of the first angle. The measure of the third angle is 70° less than 2 times the measure of the second angle. Find the measures of the three angles.

43. *Investment*   An inheritance of $80,000 is divided among three investments yielding a total of $8850 in interest per year. The interest rates for the three investments are 6%, 10%, and 15%. The amount invested at 10% is $750 more than the amount invested at 15%. Find the amount invested at each rate.

44. *Investment*   An inheritance of $16,000 is divided among three investments yielding a total of $940 in interest per year. The interest rates for the three investments are 5%, 6%, and 7%. The amount invested at 6% is $3000 less than the amount invested at 5%. Find the amount invested at each rate.

45. *Investment*   You receive a total of $708 a year in interest from three investments. The interest rates for the three investments are 6%, 8%, and 9%. The 8% investment is half of the 6% investment, and the 9% investment is $1000 less than the 6% investment. What is the amount of each investment?

46. *Investment*   You receive a total of $1520 a year in interest from three investments. The interest rates for the three investments are 5%, 7%, and 8%. The 5% investment is half of the 7% investment, and the 7% investment is $1500 less than the 8% investment. What is the amount of each investment?

47. *Chemical Mixture*   A mixture of 12 gallons of chemical A, 16 gallons of chemical B, and 26 gallons of chemical C is required to kill a destructive crop insect. Commercial spray X contains one, two, and two parts of these chemicals. Spray Y contains only chemical C. Spray Z contains only chemicals A and B in equal amounts. How much of each type of commercial spray is needed to obtain the desired mixture?

48. *Fertilizer Mixture*   A mixture of 5 pounds of fertilizer A, 13 pounds of fertilizer B, and 4 pounds of fertilizer C provides the optimal nutrients for a plant. Commercial brand X contains equal parts of fertilizer B and fertilizer C. Brand Y contains one part of fertilizer A and two parts of fertilizer B. Brand Z contains two parts of fertilizer A, five parts of fertilizer B, and two parts of fertilizer C. How much of each fertilizer brand is needed to obtain the desired mixture?

49. *Floral Arrangements*   A florist sells three types of floral arrangements for $40, $30, and $20 per arrangement. In one year the total revenue for the arrangements was $25,000, which corresponds to the sale of 850 arrangements. The florist sold 4 times as many of the $20 arrangements as the $30 arrangements. How many arrangements of each type were sold?

50. *Coffee*   A coffee manufacturer sells a 10-pound package of coffee that consists of three flavors of coffee. Vanilla flavored coffee costs $2 per pound, Hazelnut flavored coffee costs $2.50 per pound, and French Roast flavored coffee costs $3 per pound. The package contains the same amount of Hazelnut coffee as French Roast coffee. The cost of the 10-pound package is $26. How many pounds of each type of coffee are in the package?

51. *Mixture Problem*   A chemist needs 12 gallons of a 20% acid solution. It is mixed from three solutions whose concentrations are 10%, 15%, and 25%. How many gallons of each solution will satisfy each condition?

(a) Use 4 gallons of the 25% solution.

(b) Use as little as possible of the 25% solution.

(c) Use as much as possible of the 25% solution.

52. *Mixture Problem*   A chemist needs 10 liters of a 25% acid solution. It is mixed from three solutions whose concentrations are 10%, 20%, and 50%. How many liters of each solution will satisfy each condition?

(a) Use 2 liters of the 50% solution.

(b) Use as little as possible of the 50% solution.

(c) Use as much as possible of the 50% solution.

53. *School Orchestra*   The table shows the percents of each section of the North High School orchestra that were chosen to participate in the city orchestra, the county orchestra, and the state orchestra. Thirty members of the city orchestra, 17 members of the county orchestra, and 10 members of the state orchestra are from North High. How many members are in each section of North High's orchestra?

Orchestra	String	Wind	Percussion
City orchestra	40%	30%	50%
County orchestra	20%	25%	25%
State orchestra	10%	15%	25%

54. *Sports*   The table shows the percents of each unit of the North High School football team that were chosen for academic honors, as city all-stars, and as county all-stars. Of all the players on the football team, 5 were awarded with academic honors, 13 were named city all-stars, and 4 were named county all-stars. How many members of each unit are there on the football team?

	Defense	Offense	Special teams
Academic honors	0%	10%	20%
City all-stars	10%	20%	50%
County all-stars	10%	0%	20%

Explaining Concepts

55. ⊘ Answer parts (d)–(f) of Motivating the Chapter on page 216.

56. Give an example of a system of linear equations that is in row-echelon form.

57. Show how to use back-substitution to solve the system you found in Exercise 56.

58. *Writing* Describe the row operations that are performed on a system of linear equations to produce an equivalent system of equations.

59. Write a system of four linear equations in four unknowns, and solve it by elimination.

Mid-Chapter Quiz

Take this quiz as you would take a quiz in class. After you are done, check your work against the answers in the back of the book.

1. Determine whether each ordered pair is a solution of the system of linear equations: (a) $(1, -2)$ (b) $(10, 4)$

$$\begin{cases} 5x - 12y = 2 \\ 2x + 1.5y = 26 \end{cases}$$

In Exercises 2–4, graph the equations in the system. Use the graphs to determine the number of solutions of the system.

2. $\begin{cases} -6x + 9y = 9 \\ 2x - 3y = 6 \end{cases}$

3. $\begin{cases} x - 2y = -4 \\ 3x - 2y = 4 \end{cases}$

4. $\begin{cases} 0.5x - 1.5y = 7 \\ -2x + 6y = -28 \end{cases}$

In Exercises 5–7, use the graphical method to solve the system of equations.

5. $\begin{cases} x = 4 \\ 2x - y = 6 \end{cases}$

6. $\begin{cases} 2x + 7y = 16 \\ 3x + 2y = 24 \end{cases}$

7. $\begin{cases} 4x - y = 9 \\ x - 3y = 16 \end{cases}$

In Exercises 8–10, solve the system of equations by the method of substitution.

8. $\begin{cases} 2x - 3y = 4 \\ y = 2 \end{cases}$

9. $\begin{cases} 6x - 2y = 2 \\ 9x - 3y = 1 \end{cases}$

10. $\begin{cases} 5x - y = 32 \\ 6x - 9y = 18 \end{cases}$

In Exercises 11–14, use elimination or Gaussian elimination to solve the linear system.

11. $\begin{cases} x + 10y = 18 \\ 5x + 2y = 42 \end{cases}$

12. $\begin{cases} 3x + 11y = 38 \\ 7x - 5y = -34 \end{cases}$

13. $\begin{cases} a + b + c = 1 \\ 4a + 2b + c = 2 \\ 9a + 3b + c = 4 \end{cases}$

14. $\begin{cases} x + 4z = 17 \\ -3x + 2y - z = -20 \\ x - 5y + 3z = 19 \end{cases}$

In Exercises 15 and 16, write a system of linear equations having the given solution. (There are many correct answers.)

15. $(10, -12)$

16. $(1, 3, -7)$

17. Twenty gallons of a 30% brine solution is obtained by mixing a 20% solution with a 50% solution. How many gallons of each solution are required?

18. The measure of the second angle of a triangle is 10° less than twice the measure of the first angle. The measure of the third angle is 10° greater than the measure of the first angle. Find the measures of the three angles.

4.4 Matrices and Linear Systems

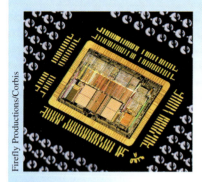

Firefly Productions/Corbis

What You Should Learn

(1) Determine the order of matrices.

(2) Form coefficient and augmented matrices and form linear systems from augmented matrices.

(3) Perform elementary row operations to solve systems of linear equations.

(4) Use matrices and Gaussian elimination with back-substitution to solve systems of linear equations.

Why You Should Learn It

Systems of linear equations that model real-life situations can be solved using matrices. For instance, in Exercise 81 on page 266, the numbers of computer parts a company produces can be found using a matrix.

Matrices

In this section, you will study a streamlined technique for solving systems of linear equations. This technique involves the use of a rectangular array of real numbers called a **matrix.** (The plural of matrix is matrices.) Here is an example of a matrix.

$$
\begin{array}{cccc}
\text{Column} & \text{Column} & \text{Column} & \text{Column} \\
1 & 2 & 3 & 4
\end{array}
$$

$$
\begin{array}{c}
\text{Row 1} \\
\text{Row 2} \\
\text{Row 3}
\end{array}
\begin{bmatrix}
3 & -2 & 4 & 1 \\
0 & 1 & -1 & 2 \\
2 & 0 & -3 & 0
\end{bmatrix}
$$

This matrix has three rows and four columns, which means that its **order** is 3×4, which is read as "3 by 4." Each number in the matrix is an **entry** of the matrix.

(1) Determine the order of matrices.

Example 1   Order of Matrices

Determine the order of each matrix.

a. $\begin{bmatrix} 1 & -2 & 4 \\ 0 & 1 & -2 \end{bmatrix}$ **b.** $\begin{bmatrix} 0 & 0 \\ 0 & 0 \end{bmatrix}$ **c.** $\begin{bmatrix} 1 & -3 \\ -2 & 0 \\ 4 & -2 \end{bmatrix}$

Solution

a. This matrix has two rows and three columns, so the order is 2×3.

b. This matrix has two rows and two columns, so the order is 2×2.

c. This matrix has three rows and two columns, so the order is 3×2.

Study Tip

The order of a matrix is always given as *row by column*. A matrix with the same number of rows as columns is called a **square matrix.** For instance, the 2×2 matrix in Example 1(b) is square.

2 Form coefficient and augmented matrices and form linear systems from augmented matrices.

Augmented and Coefficient Matrices

A matrix derived from a system of linear equations (each written in standard form) is the **augmented matrix** of the system. Moreover, the matrix derived from the coefficients of the system (but that does not include the constant terms) is the **coefficient matrix** of the system. Here is an example.

Study Tip

Note the use of 0 for the missing y-variable in the third equation, and also note the fourth column of constant terms in the augmented matrix.

$$
\begin{array}{cccc}
\textit{System} & \textit{Coefficient Matrix} & \textit{Augmented Matrix}
\end{array}
$$

$$
\begin{cases}
x - 4y + 3z = 5 \\
-x + 3y - z = -3 \\
2x - 4z = 6
\end{cases}
\qquad
\begin{bmatrix}
1 & -4 & 3 \\
-1 & 3 & -1 \\
2 & 0 & -4
\end{bmatrix}
\qquad
\left[\begin{array}{ccc:c}
1 & -4 & 3 & 5 \\
-1 & 3 & -1 & -3 \\
2 & 0 & -4 & 6
\end{array}\right]
$$

When forming either the coefficient matrix or the augmented matrix of a system, you should begin by vertically aligning the variables in the equations.

$$
\begin{array}{ccc}
\textit{Given System} & \textit{Align Variables} & \textit{Form Augmented Matrix}
\end{array}
$$

$$
\begin{cases}
x + 3y = 9 \\
-y + 4z = -2 \\
x - 5z = 0
\end{cases}
\qquad
\begin{cases}
x + 3y = 9 \\
-y + 4z = -2 \\
x - 5z = 0
\end{cases}
\qquad
\left[\begin{array}{ccc:c}
1 & 3 & 0 & 9 \\
0 & -1 & 4 & -2 \\
1 & 0 & -5 & 0
\end{array}\right]
$$

Example 2 Forming Coefficient and Augmented Matrices

Form the coefficient matrix and the augmented matrix for each system.

a. $\begin{cases} -x + 5y = 2 \\ 7x - 2y = -6 \end{cases}$
b. $\begin{cases} 3x + 2y - z = 1 \\ x + 2z = -3 \\ -2x - y = 4 \end{cases}$

Solution

$$
\begin{array}{cccc}
\textit{System} & \textit{Coefficient Matrix} & \textit{Augmented Matrix}
\end{array}
$$

a. $\begin{cases} -x + 5y = 2 \\ 7x - 2y = -6 \end{cases}$
$\begin{bmatrix} -1 & 5 \\ 7 & -2 \end{bmatrix}$
$\left[\begin{array}{cc:c} -1 & 5 & 2 \\ 7 & -2 & -6 \end{array}\right]$

b. $\begin{cases} 3x + 2y - z = 1 \\ x + 2z = -3 \\ -2x - y = 4 \end{cases}$
$\begin{bmatrix} 3 & 2 & -1 \\ 1 & 0 & 2 \\ -2 & -1 & 0 \end{bmatrix}$
$\left[\begin{array}{ccc:c} 3 & 2 & -1 & 1 \\ 1 & 0 & 2 & -3 \\ -2 & -1 & 0 & 4 \end{array}\right]$

Example 3 Forming Linear Systems from Their Matrices

Write the system of linear equations that is represented by each matrix.

a. $\left[\begin{array}{cc:c} 3 & -5 & 4 \\ -1 & 2 & 0 \end{array}\right]$
b. $\left[\begin{array}{cc:c} 1 & 3 & 2 \\ 0 & 1 & -3 \end{array}\right]$
c. $\left[\begin{array}{ccc:c} 2 & 0 & -8 & 1 \\ -1 & 1 & 1 & 2 \\ 5 & -1 & 7 & 3 \end{array}\right]$

Solution

a. $\begin{cases} 3x - 5y = 4 \\ -x + 2y = 0 \end{cases}$
b. $\begin{cases} x + 3y = 2 \\ y = -3 \end{cases}$
c. $\begin{cases} 2x - 8z = 1 \\ -x + y + z = 2 \\ 5x - y + 7z = 3 \end{cases}$

3 Perform elementary row operations to solve systems of linear equations.

Elementary Row Operations

In Section 4.3, you studied three operations that can be used on a system of linear equations to produce an equivalent system: (1) interchange two equations, (2) multiply an equation by a nonzero constant, and (3) add a multiple of an equation to another equation. In matrix terminology, these three operations correspond to **elementary row operations.**

Elementary Row Operations

Any of the following **elementary row operations** performed on an augmented matrix will produce a matrix that is row-equivalent to the original matrix. Two matrices are **row-equivalent** if one can be obtained from the other by a sequence of elementary row operations.

1. Interchange two rows.

2. Multiply a row by a nonzero constant.

3. Add a multiple of a row to another row.

Example 4   Elementary Row Operations

a. Interchange the first and second rows.

Original Matrix

$$\begin{bmatrix} 0 & 1 & 3 & 4 \\ -1 & 2 & 0 & 3 \\ 2 & -3 & 4 & 1 \end{bmatrix}$$

New Row-Equivalent Matrix

$$\begin{matrix} R_2 \\ R_1 \end{matrix} \begin{bmatrix} -1 & 2 & 0 & 3 \\ 0 & 1 & 3 & 4 \\ 2 & -3 & 4 & 1 \end{bmatrix}$$

b. Multiply the first row by $\frac{1}{2}$.

Original Matrix

$$\begin{bmatrix} 2 & -4 & 6 & -2 \\ 1 & 3 & -3 & 0 \\ 5 & -2 & 1 & 2 \end{bmatrix}$$

New Row-Equivalent Matrix

$$\frac{1}{2}R_1 \rightarrow \begin{bmatrix} 1 & -2 & 3 & -1 \\ 1 & 3 & -3 & 0 \\ 5 & -2 & 1 & 2 \end{bmatrix}$$

c. Add -2 times the first row to the third row.

Original Matrix

$$\begin{bmatrix} 1 & 2 & -4 & 3 \\ 0 & 3 & -2 & -1 \\ 2 & 1 & 5 & -2 \end{bmatrix}$$

New Row-Equivalent Matrix

$$\begin{bmatrix} 1 & 2 & -4 & 3 \\ 0 & 3 & -2 & -1 \\ 0 & -3 & 13 & -8 \end{bmatrix}$$

$-2R_1 + R_3 \rightarrow$

d. Add 6 times the first row to the second row.

Original Matrix

$$\begin{bmatrix} 1 & 2 & 2 & -4 \\ -6 & -11 & 3 & 18 \\ 0 & 0 & 4 & 7 \end{bmatrix}$$

New Row-Equivalent Matrix

$$6R_1 + R_2 \rightarrow \begin{bmatrix} 1 & 2 & 2 & -4 \\ 0 & 1 & 15 & -6 \\ 0 & 0 & 4 & 7 \end{bmatrix}$$

Example 7   Gaussian Elimination with Back-Substitution

Solve the system of linear equations.

$$\begin{cases} 3x + 3y & = & 9 \\ 2x & - 3z = & 10 \\ & 6y + 4z = & -12 \end{cases}$$

Solution

$$\begin{bmatrix} 3 & 3 & 0 & \vdots & 9 \\ 2 & 0 & -3 & \vdots & 10 \\ 0 & 6 & 4 & \vdots & -12 \end{bmatrix}$$

Augmented matrix for system of linear equations

$$\frac{1}{3}R_1 \rightarrow \begin{bmatrix} 1 & 1 & 0 & \vdots & 3 \\ 2 & 0 & -3 & \vdots & 10 \\ 0 & 6 & 4 & \vdots & -12 \end{bmatrix}$$

First column has leading 1 in upper left corner.

$$-2R_1 + R_2 \rightarrow \begin{bmatrix} 1 & 1 & 0 & \vdots & 3 \\ 0 & -2 & -3 & \vdots & 4 \\ 0 & 6 & 4 & \vdots & -12 \end{bmatrix}$$

First column has zeros under its leading 1.

$$-\frac{1}{2}R_2 \rightarrow \begin{bmatrix} 1 & 1 & 0 & \vdots & 3 \\ 0 & 1 & \frac{3}{2} & \vdots & -2 \\ 0 & 6 & 4 & \vdots & -12 \end{bmatrix}$$

Second column has leading 1 in second row.

$$-6R_2 + R_3 \rightarrow \begin{bmatrix} 1 & 1 & 0 & \vdots & 3 \\ 0 & 1 & \frac{3}{2} & \vdots & -2 \\ 0 & 0 & -5 & \vdots & 0 \end{bmatrix}$$

Second column has zero under its leading 1.

$$-\frac{1}{5}R_3 \rightarrow \begin{bmatrix} 1 & 1 & 0 & \vdots & 3 \\ 0 & 1 & \frac{3}{2} & \vdots & -2 \\ 0 & 0 & 1 & \vdots & 0 \end{bmatrix}$$

Third column has leading 1 in third row.

The system of linear equations that corresponds to this (row-echelon) matrix is

$$\begin{cases} x + y & = & 3 \\ y + \frac{3}{2}z & = & -2 \\ z & = & 0. \end{cases}$$

Using back-substitution, you can find that the solution is

$$x = 5, \quad y = -2, \quad \text{and} \quad z = 0$$

which can be written as the ordered triple $(5, -2, 0)$. Check this in the original system, as follows.

Check

Equation 1: $3(5) + 3(-2)$ = 9 ✓

Equation 2: $2(5)$ $- 3(0)$ = 10 ✓

Equation 3: $6(-2) + 4(0)$ = -12 ✓

Example 8   A System with No Solution

Solve the system of linear equations.

$$\begin{cases} 6x - 10y = -4 \\ 9x - 15y = 5 \end{cases}$$

Solution

$$\begin{bmatrix} 6 & -10 & \vdots & -4 \\ 9 & -15 & \vdots & 5 \end{bmatrix}$$

Augmented matrix for system of linear equations

$$\tfrac{1}{6}R_1 \rightarrow \begin{bmatrix} 1 & -\tfrac{5}{3} & \vdots & -\tfrac{2}{3} \\ 9 & -15 & \vdots & 5 \end{bmatrix}$$

First column has leading 1 in upper left corner.

$$-9R_1 + R_2 \rightarrow \begin{bmatrix} 1 & -\tfrac{5}{3} & \vdots & -\tfrac{2}{3} \\ 0 & 0 & \vdots & 11 \end{bmatrix}$$

First column has a zero under its leading 1.

The "equation" that corresponds to the second row of this matrix is $0 = 11$. Because this is a false statement, the system of equations has no solution.

Example 9   A System with Infinitely Many Solutions

Solve the system of linear equations.

$$\begin{cases} 12x - 6y = -3 \\ -8x + 4y = 2 \end{cases}$$

Solution

$$\begin{bmatrix} 12 & -6 & \vdots & -3 \\ -8 & 4 & \vdots & 2 \end{bmatrix}$$

Augmented matrix for system of linear equations

$$\tfrac{1}{12}R_1 \rightarrow \begin{bmatrix} 1 & -\tfrac{1}{2} & \vdots & -\tfrac{1}{4} \\ -8 & 4 & \vdots & 2 \end{bmatrix}$$

First column has leading 1 in upper left corner.

$$8R_1 + R_2 \rightarrow \begin{bmatrix} 1 & -\tfrac{1}{2} & \vdots & -\tfrac{1}{4} \\ 0 & 0 & \vdots & 0 \end{bmatrix}$$

First column has a zero under its leading 1.

Because the second row of the matrix is all zeros, you can conclude that the system of equations has an infinite number of solutions, represented by all points (x, y) on the line

$$x - \frac{1}{2}y = -\frac{1}{4}.$$

Because this line can be written as

$$x = \frac{1}{2}y - \frac{1}{4}$$

you can write the solution set as

$$\left(\frac{1}{2}a - \frac{1}{4}, a \right), \quad \text{where } a \text{ is any real number.}$$

Example 10   Investment Portfolio

You have a portfolio totaling $219,000 and want to invest in municipal bonds, blue-chip stocks, and growth or speculative stocks. The municipal bonds pay 6% annually. Over a five-year period, you expect blue-chip stocks to return 10% annually and growth stocks to return 15% annually. You want a combined annual return of 8%, and you also want to have only one-fourth of the portfolio invested in stocks. How much should be allocated to each type of investment?

Solution

Let M represent municipal bonds, B represent blue-chip stocks, and G represent growth stocks. These three equations make up the following system.

$$\begin{cases} M + B + G = 219,000 & \text{Equation 1: total investment is \$219,000.} \\ 0.06M + 0.10B + 0.15G = 17,520 & \text{Equation 2: combined annual return is 8\%.} \\ B + G = 54,750 & \text{Equation 3: } \tfrac{1}{4} \text{ of investment is allocated to stocks.} \end{cases}$$

Form the augmented matrix for this system of equations, and then use elementary row operations to obtain the row-echelon form of the matrix.

$$\begin{bmatrix} 1 & 1 & 1 & \vdots & 219,000 \\ 0.06 & 0.10 & 0.15 & \vdots & 17,520 \\ 0 & 1 & 1 & \vdots & 54,750 \end{bmatrix}$$

Augmented matrix for system of linear equations

$$-0.06R_1 + R_2 \rightarrow \begin{bmatrix} 1 & 1 & 1 & \vdots & 219,000 \\ 0 & 0.04 & 0.09 & \vdots & 4,380 \\ 0 & 1 & 1 & \vdots & 54,750 \end{bmatrix}$$

First column has zeros under its leading 1.

$$25R_2 \rightarrow \begin{bmatrix} 1 & 1 & 1 & \vdots & 219,000 \\ 0 & 1 & 2.25 & \vdots & 109,500 \\ 0 & 1 & 1 & \vdots & 54,750 \end{bmatrix}$$

Second column has leading 1 in second row.

$$-R_2 + R_3 \rightarrow \begin{bmatrix} 1 & 1 & 1 & \vdots & 219,000 \\ 0 & 1 & 2.25 & \vdots & 109,500 \\ 0 & 0 & -1.25 & \vdots & -54,750 \end{bmatrix}$$

Second column has zero under its leading 1.

$$-0.8R_3 \rightarrow \begin{bmatrix} 1 & 1 & 1 & \vdots & 219,000 \\ 0 & 1 & 2.25 & \vdots & 109,500 \\ 0 & 0 & 1 & \vdots & 43,800 \end{bmatrix}$$

Third column has leading 1 in third row and matrix is in row-echelon form.

From the row-echelon form, you can see that $G = 43,800$. By back-substituting G into the revised second equation, you can determine the value of B.

$$B + 2.25(43,800) = 109,500 \implies B = 10,950$$

By back-substituting B and G into Equation 1, you can solve for M.

$$M + 10,950 + 43,800 = 219,000 \implies M = 164,250$$

So, you should invest $164,250 in municipal bonds, $10,950 in blue-chip stocks, and $43,800 in growth or speculative stocks. Check this solution by substituting these values into the original system of equations.

4.4 Exercises

Review   *Concepts, Skills, and Problem Solving*

Keep mathematically in shape by doing these exercises *before* the problems of this section.

Properties and Definitions

In Exercises 1–4, identify the property of real numbers illustrated by the statement.

1. $2ab - 2ab = 0$

2. $8t \cdot 1 = 8t$

3. $b + 3a = 3a + b$

4. $3(2x) = (3 \cdot 2)x$

Algebraic Operations

In Exercises 5–8, plot the points on the rectangular coordinate system. Find the slope of the line passing through the points. If not possible, state why.

5. $(0, -6), (8, 0)$

6. $\left(\frac{5}{2}, \frac{7}{2}\right), \left(\frac{5}{2}, 4\right)$

7. $\left(-\frac{5}{8}, -\frac{3}{4}\right), \left(1, -\frac{9}{2}\right)$

8. $(3, 1.2), (-3, 2.1)$

In Exercises 9 and 10, find the distance between the two points and the midpoint of the line segment joining the two points.

9. $(12, 8), (6, 8)$

10. $\left(-3, 2\right), \left(-\frac{3}{2}, -2\right)$

Problem Solving

11. *Membership Drive*   Through a membership drive, the membership of a public television station increased by 10%. The current number of members is 8415. How many members did the station have before the membership drive?

12. *Consumer Awareness*   A sales representative indicates that if a customer waits another month for a new car that currently costs $23,500, the price will increase by 4%. The customer has a certificate of deposit that comes due in 1 month and will pay a penalty for early withdrawal if the money is withdrawn before the due date. Determine the penalty for early withdrawal that would equal the cost increase of waiting to buy the car.

Developing Skills

In Exercises 1–10, determine the order of the matrix. See Example 1.

1. $\begin{bmatrix} 3 & -2 \\ -4 & 0 \\ 2 & -7 \\ -1 & -3 \end{bmatrix}$

2. $\begin{bmatrix} 4 & 0 & -5 \\ -1 & 8 & 9 \\ 0 & -3 & 4 \end{bmatrix}$

3. $\begin{bmatrix} -2 & 5 \\ 0 & -1 \end{bmatrix}$

4. $\begin{bmatrix} 5 & -8 & 32 \\ 7 & 15 & 28 \end{bmatrix}$

5. $\begin{bmatrix} 4 \\ -2 \\ 0 \\ 1 \end{bmatrix}$

6. $\begin{bmatrix} 1 & -1 & 2 & 3 \end{bmatrix}$

7. $\begin{bmatrix} 5 \end{bmatrix}$

8. $\begin{bmatrix} 3 & 4 \\ 2 & -1 \\ 8 & 10 \\ -6 & -6 \\ 12 & 50 \end{bmatrix}$

9. $\begin{bmatrix} 13 & 12 & -9 & 0 \end{bmatrix}$

10. $\begin{bmatrix} 6 \\ -13 \\ 22 \end{bmatrix}$

In Exercises 11–16, form (a) the coefficient matrix and (b) the augmented matrix for the system of linear equations. See Example 2.

11. $\begin{cases} 4x - 5y = -2 \\ -x + 8y = 10 \end{cases}$

12. $\begin{cases} 8x + 3y = 25 \\ 3x - 9y = 12 \end{cases}$

13. $\begin{cases} x + 10y - 3z = 2 \\ 5x - 3y + 4z = 0 \\ 2x + 4y = 6 \end{cases}$

14. $\begin{cases} 9x - 3y + z = 13 \\ 12x - 8z = 5 \\ 3x + 4y - z = 6 \end{cases}$

15. $\begin{cases} 5x + y - 3z = 7 \\ 2y + 4z = 12 \end{cases}$

16. $\begin{cases} 10x + 6y - 8z = -4 \\ -4x - 7y = 9 \end{cases}$

22. $\begin{bmatrix} 0 & 1 & -5 & 8 & \vdots & 10 \\ 2 & 4 & -1 & 0 & \vdots & 15 \\ 1 & 1 & 7 & 9 & \vdots & -8 \end{bmatrix}$

23. $\begin{bmatrix} 13 & 1 & 4 & -2 & \vdots & -4 \\ 5 & 4 & 0 & -1 & \vdots & 0 \\ 1 & 2 & 6 & 8 & \vdots & 5 \\ -10 & 12 & 3 & 1 & \vdots & -2 \end{bmatrix}$

24. $\begin{bmatrix} 7 & 3 & -2 & 4 & \vdots & 2 \\ -1 & 0 & 4 & -1 & \vdots & 6 \\ 8 & 3 & 0 & 0 & \vdots & -4 \\ 0 & 2 & -4 & 3 & \vdots & 12 \end{bmatrix}$

In Exercises 17–24, write the system of linear equations represented by the augmented matrix. (Use variables x, y, z, and w.) See Example 3.

17. $\begin{bmatrix} 4 & 3 & \vdots & 8 \\ 1 & -2 & \vdots & 3 \end{bmatrix}$

18. $\begin{bmatrix} 9 & -4 & \vdots & 0 \\ 6 & 1 & \vdots & -4 \end{bmatrix}$

19. $\begin{bmatrix} 1 & 0 & 2 & \vdots & -10 \\ 0 & 3 & -1 & \vdots & 5 \\ 4 & 2 & 0 & \vdots & 3 \end{bmatrix}$

20. $\begin{bmatrix} 4 & -1 & 3 & \vdots & 5 \\ 2 & 0 & -2 & \vdots & -1 \\ -1 & 6 & 0 & \vdots & 3 \end{bmatrix}$

21. $\begin{bmatrix} 5 & 8 & 2 & 0 & \vdots & -1 \\ -2 & 15 & 5 & 1 & \vdots & 9 \\ 1 & 6 & -7 & 0 & \vdots & -3 \end{bmatrix}$

In Exercises 25–30, fill in the blank(s) by using elementary row operations to form a row-equivalent matrix. See Examples 4 and 5.

25. $\begin{bmatrix} 1 & 4 & 3 \\ 2 & 10 & 5 \end{bmatrix}$

$\begin{bmatrix} 1 & 4 & 3 \\ 0 & & -1 \end{bmatrix}$

26. $\begin{bmatrix} 3 & 6 & 8 \\ 4 & -3 & 6 \end{bmatrix}$

$\begin{bmatrix} 3 & 6 & 8 \\ 1 & -9 & \end{bmatrix}$

27. $\begin{bmatrix} 9 & -18 & 6 \\ 2 & 8 & 15 \end{bmatrix}$

$\begin{bmatrix} 1 & & \\ 2 & 8 & 15 \end{bmatrix}$

28. $\begin{bmatrix} 2 & 3 & -5 & 6 \\ 5 & -7 & 12 & 9 \\ -4 & 6 & 9 & 5 \end{bmatrix}$

$\begin{bmatrix} 2 & 3 & -5 & 6 \\ 5 & -7 & 12 & 9 \\ 0 & 12 & & \end{bmatrix}$

29. $\begin{bmatrix} 1 & 1 & 4 & -1 \\ 3 & 8 & 10 & 3 \\ -2 & 1 & 12 & 6 \end{bmatrix}$

$\begin{bmatrix} 1 & 1 & 4 & -1 \\ 0 & 5 & & \\ 0 & 3 & & \end{bmatrix}$

$\begin{bmatrix} 1 & 1 & 4 & -1 \\ 0 & 1 & -\frac{2}{5} & \frac{6}{5} \\ 0 & 3 & & \end{bmatrix}$

30. $\begin{bmatrix} 2 & 4 & 8 & 3 \\ 1 & -1 & -3 & 2 \\ 2 & 6 & 4 & 9 \end{bmatrix}$

$\begin{bmatrix} 1 & & & \\ 1 & -1 & -3 & 2 \\ 2 & 6 & 4 & 9 \end{bmatrix}$

$\begin{bmatrix} 1 & 2 & 4 & \frac{3}{2} \\ 0 & & -7 & \frac{1}{2} \\ 0 & 2 & & \end{bmatrix}$

In Exercises 31–36, convert the matrix to row-echelon form. (There are many correct answers.)

31. $\begin{bmatrix} 1 & 2 & 3 \\ 2 & -1 & -4 \end{bmatrix}$ **32.** $\begin{bmatrix} 1 & 3 & 6 \\ -4 & -9 & 3 \end{bmatrix}$

33. $\begin{bmatrix} 4 & 6 & 1 \\ -2 & 2 & 5 \end{bmatrix}$ **34.** $\begin{bmatrix} 3 & 2 & 6 \\ 2 & 3 & -3 \end{bmatrix}$

35. $\begin{bmatrix} 1 & 1 & 0 & 5 \\ -2 & -1 & 2 & -10 \\ 3 & 6 & 7 & 14 \end{bmatrix}$ **36.** $\begin{bmatrix} 1 & 2 & -1 & 3 \\ 3 & 7 & -5 & 14 \\ -2 & -1 & -3 & 8 \end{bmatrix}$

▦ In Exercises 37–40, use the matrix capabilities of a graphing calculator to write the matrix in row-echelon form. (There are many correct answers.)

37. $\begin{bmatrix} 1 & -1 & -1 & 1 \\ 4 & -4 & 1 & 8 \\ -6 & 8 & 18 & 0 \end{bmatrix}$

38. $\begin{bmatrix} 1 & -3 & 0 & -7 \\ -3 & 10 & 1 & 23 \\ 4 & -10 & 2 & -24 \end{bmatrix}$

39. $\begin{bmatrix} 1 & 1 & -1 & 3 \\ 2 & 1 & 2 & 5 \\ 3 & 2 & 1 & 8 \end{bmatrix}$

40. $\begin{bmatrix} 1 & -3 & -2 & -8 \\ 1 & 3 & -2 & 17 \\ 1 & 2 & -2 & -5 \end{bmatrix}$

In Exercises 41–46, write the system of linear equations represented by the augmented matrix. Then use back-substitution to find the solution. (Use variables x, y, and z.)

41. $\begin{bmatrix} 1 & -2 & \vdots & 4 \\ 0 & 1 & \vdots & -3 \end{bmatrix}$ **42.** $\begin{bmatrix} 1 & 5 & \vdots & 0 \\ 0 & 1 & \vdots & -1 \end{bmatrix}$

43. $\begin{bmatrix} 1 & 5 & \vdots & 3 \\ 0 & 1 & \vdots & -2 \end{bmatrix}$ **44.** $\begin{bmatrix} 1 & 5 & -3 & \vdots & 0 \\ 0 & 1 & 0 & \vdots & 6 \\ 0 & 0 & 1 & \vdots & -5 \end{bmatrix}$

45. $\begin{bmatrix} 1 & -1 & 2 & \vdots & 4 \\ 0 & 1 & -1 & \vdots & 2 \\ 0 & 0 & 1 & \vdots & -2 \end{bmatrix}$

46. $\begin{bmatrix} 1 & 2 & -2 & \vdots & -1 \\ 0 & 1 & 1 & \vdots & 9 \\ 0 & 0 & 1 & \vdots & -3 \end{bmatrix}$

In Exercises 47–72, use matrices to solve the system of linear equations. See Examples 5–9.

47. $\begin{cases} x + 2y = 7 \\ 3x + y = 8 \end{cases}$ **48.** $\begin{cases} 2x + 6y = 16 \\ 2x + 3y = 7 \end{cases}$

49. $\begin{cases} 6x - 4y = 2 \\ 5x + 2y = 7 \end{cases}$ **50.** $\begin{cases} x - 3y = 5 \\ -2x + 6y = -10 \end{cases}$

51. $\begin{cases} 12x + 10y = -14 \\ 4x - 3y = -11 \end{cases}$ **52.** $\begin{cases} -x - 5y = -10 \\ 2x - 3y = 7 \end{cases}$

53. $\begin{cases} -x + 2y = 1.5 \\ 2x - 4y = 3 \end{cases}$ **54.** $\begin{cases} 2x - y = -0.1 \\ 3x + 2y = 1.6 \end{cases}$

55. $\begin{cases} x - 2y - z = 6 \\ y + 4z = 5 \\ 4x + 2y + 3z = 8 \end{cases}$ **56.** $\begin{cases} x - 3z = -2 \\ 3x + y - 2z = 5 \\ 2x + 2y + z = 4 \end{cases}$

57. $\begin{cases} x + y - 5z = 3 \\ x - 2z = 1 \\ 2x - y - z = 0 \end{cases}$ **58.** $\begin{cases} 2y + z = 3 \\ -4y - 2z = 0 \\ x + y + z = 2 \end{cases}$

59. $\begin{cases} 2x + 4y = 10 \\ 2x + 2y + 3z = 3 \\ -3x + y + 2z = -3 \end{cases}$

60. $\begin{cases} 2x - y + 3z = 24 \\ 2y - z = 14 \\ 7x - 5y = 6 \end{cases}$ **61.** $\begin{cases} x - 3y + 2z = 8 \\ 2y - z = -4 \\ x + z = 3 \end{cases}$

62. $\begin{cases} 2x + 3z = 3 \\ 4x - 3y + 7z = 5 \\ 8x - 9y + 15z = 9 \end{cases}$

63. $\begin{cases} -2x - 2y - 15z = 0 \\ x + 2y + 2z = 18 \\ 3x + 3y + 22z = 2 \end{cases}$

64. $\begin{cases} 2x + 4y + 5z = 5 \\ x + 3y + 3z = 2 \\ 2x + 4y + 4z = 2 \end{cases}$ **65.** $\begin{cases} 2x + 4z = 1 \\ x + y + 3z = 0 \\ x + 3y + 5z = 0 \end{cases}$

66. $\begin{cases} 3x + y - 2z = 2 \\ 6x + 2y - 4z = 1 \\ -3x - y + 2z = 1 \end{cases}$ **67.** $\begin{cases} x + 3y = 2 \\ 2x + 6y = 4 \\ 2x + 5y + 4z = 3 \end{cases}$

68. $\begin{cases} 4x + 3y = 10 \\ 2x - y = 10 \\ -2x + z = -9 \end{cases}$

69. $\begin{cases} 4x - y + z = 4 \\ -6x + 3y - 2z = -5 \\ 2x + 5y - z = 7 \end{cases}$

70. $\begin{cases} 2x + 2y + z = 8 \\ 2x + 3y + z = 7 \\ 6x + 8y + 3z = 22 \end{cases}$

71. $\begin{cases} 2x + y - 2z = 4 \\ 3x - 2y + 4z = 6 \\ -4x + y + 6z = 12 \end{cases}$

72. $\begin{cases} 3x + 3y + z = 4 \\ 2x + 6y + z = 5 \\ -x - 3y + 2z = -5 \end{cases}$

Solving Problems

73. *Investment* A corporation borrowed $1,500,000 to expand its line of clothing. Some of the money was borrowed at 8%, some at 9%, and the remainder at 12%. The annual interest payment to the lenders was $133,000. The amount borrowed at 8% was 4 times the amount borrowed at 12%. How much was borrowed at each rate?

74. *Investment* An inheritance of $16,000 was divided among three investments yielding a total of $990 in simple interest per year. The interest rates for the three investments were 5%, 6%, and 7%. The 5% and 6% investments were $3000 and $2000 less than the 7% investment, respectively. Find the amount placed in each investment.

Investment Portfolio In Exercises 75 and 76, consider an investor with a portfolio totaling $500,000 that is to be allocated among the following types of investments: certificates of deposit, municipal bonds, blue-chip stocks, and growth or speculative stocks. How much should be allocated to each type of investment?

75. The certificates of deposit pay 10% annually, and the municipal bonds pay 8% annually. Over a five-year period, the investor expects the blue-chip stocks to return 12% annually and the growth stocks to return 13% annually. The investor wants a combined annual return of 10% and also wants to have only one-fourth of the portfolio invested in stocks.

76. The certificates of deposit pay 9% annually, and the municipal bonds pay 5% annually. Over a five-year period, the investor expects the blue-chip stocks to return 12% annually and the growth stocks to return 14% annually. The investor wants a combined annual return of 10% and also wants to have only one-fourth of the portfolio invested in stocks.

77. *Nut Mixture*   A grocer wishes to mix three kinds of nuts to obtain 50 pounds of a mixture priced at $4.95 per pound. Peanuts cost $3.50 per pound, almonds cost $4.50 per pound, and pistachios cost $6.00 per pound. Half of the mixture is composed of peanuts and almonds. How many pounds of each variety should the grocer use?

78. *Nut Mixture*   A grocer wishes to mix three kinds of nuts to obtain 50 pounds of a mixture priced at $4.10 per pound. Peanuts cost $3.00 per pound, pecans cost $4.00 per pound, and cashews cost $6.00 per pound. Three-quarters of the mixture is composed of peanuts and pecans. How many pounds of each variety should the grocer use?

79. *Number Problem*   The sum of three positive numbers is 33. The second number is 3 greater than the first, and the third is 4 times the first. Find the three numbers.

80. *Number Problem*   The sum of three positive numbers is 24. The second number is 4 greater than the first, and the third is 3 times the first. Find the three numbers.

81. *Production*   A company produces computer chips, resistors, and transistors. Each computer chip requires 2 units of copper, 2 units of zinc, and 1 unit of glass. Each resistor requires 1 unit of copper, 3 units of zinc, and 2 units of glass. Each transistor requires 3 units of copper, 2 units of zinc, and 2 units of glass. There are 70 units of copper, 80 units of zinc, and 55 units of glass available for use. Find the number of computer chips, resistors, and transistors the company can produce.

82. *Production*   A gourmet baked goods company specializes in chocolate muffins, chocolate cookies, and chocolate brownies. Each muffin requires 2 units of chocolate, 3 units of flour, and 2 units of sugar. Each cookie requires 1 unit of chocolate, 1 unit of flour, and 1 unit of sugar. Each brownie requires 2 units of chocolate, 1 unit of flour, and 1.5 units of sugar. There are 550 units of chocolate, 525 units of flour, and 500 units of sugar available for use. Find the number of chocolate muffins, chocolate cookies, and chocolate brownies the company can produce.

Explaining Concepts

83. *Writing*   Describe the three elementary row operations that can be performed on an augmented matrix.

84. *Writing*   What is the relationship between the three elementary row operations on an augmented matrix and the row operations on a system of linear equations?

85. *Writing*   What is meant by saying that two augmented matrices are *row-equivalent?*

86. Give an example of a matrix in *row-echelon form.* There are many correct answers.

87. *Writing*   Describe the row-echelon form of an augmented matrix that corresponds to a system of linear equations that is inconsistent.

88. *Writing*   Describe the row-echelon form of an augmented matrix that corresponds to a system of linear equations that has an infinite number of solutions.

4.5 Determinants and Linear Systems

Kevin R. Morris/Corbis

What You Should Learn

1. Find determinants of 2 × 2 matrices and 3 × 3 matrices.
2. Use determinants and Cramer's Rule to solve systems of linear equations.
3. Use determinants to find areas of triangles, to test for collinear points, and to find equations of lines.

The Determinant of a Matrix

Associated with each square matrix is a real number called its **determinant.** The use of determinants arose from special number patterns that occur during the solution of systems of linear equations. For instance, the system

$$\begin{cases} a_1 x + b_1 y = c_1 \\ a_2 x + b_2 y = c_2 \end{cases}$$

1 Find determinants of 2 × 2 matrices and 3 × 3 matrices.

has a solution given by

$$x = \frac{b_2 c_1 - b_1 c_2}{a_1 b_2 - a_2 b_1} \qquad \text{and} \qquad y = \frac{a_1 c_2 - a_2 c_1}{a_1 b_2 - a_2 b_1}$$

provided that $a_1 b_2 - a_2 b_1 \neq 0$. Note that the denominator of each fraction is the same. This denominator is called the **determinant** of the coefficient matrix of the system.

$$\begin{array}{cc} \textit{Coefficient Matrix} & \textit{Determinant} \\ A = \begin{bmatrix} a_1 & b_1 \\ a_2 & b_2 \end{bmatrix} & \det(A) = a_1 b_2 - a_2 b_1 \end{array}$$

The determinant of the matrix A can also be denoted by vertical bars on both sides of the matrix, as indicated in the following definition.

Study Tip

Note that $\det(A)$ and $|A|$ are used interchangeably to represent the determinant of A. Although vertical bars are also used to denote the absolute value of a real number, the context will show which use is intended.

Definition of the Determinant of a 2 × 2 Matrix

$$\det(A) = |A| = \begin{vmatrix} a_1 & b_1 \\ a_2 & b_2 \end{vmatrix} = a_1 b_2 - a_2 b_1$$

A convenient method for remembering the formula for the determinant of a 2 × 2 matrix is shown in the diagram below.

$$\det(A) = \begin{vmatrix} a_1 & b_1 \\ a_2 & b_2 \end{vmatrix} = a_1 b_2 - a_2 b_1$$

Note that the determinant is given by the difference of the products of the two diagonals of the matrix.

Example 1   The Determinant of a 2×2 Matrix

Find the determinant of each matrix.

a. $A = \begin{bmatrix} 2 & -3 \\ 1 & 4 \end{bmatrix}$ **b.** $B = \begin{bmatrix} -1 & 2 \\ 2 & -4 \end{bmatrix}$ **c.** $C = \begin{bmatrix} 1 & 3 \\ 2 & 5 \end{bmatrix}$

Solution

a. $\det(A) = \begin{vmatrix} 2 & -3 \\ 1 & 4 \end{vmatrix} = 2(4) - 1(-3) = 8 + 3 = 11$

b. $\det(B) = \begin{vmatrix} -1 & 2 \\ 2 & -4 \end{vmatrix} = (-1)(-4) - 2(2) = 4 - 4 = 0$

c. $\det(C) = \begin{vmatrix} 1 & 3 \\ 2 & 5 \end{vmatrix} = 1(5) - 2(3) = 5 - 6 = -1$

Notice in Example 1 that the determinant of a matrix can be positive, zero, or negative.

One way to evaluate the determinant of a 3×3 matrix, called **expanding by minors,** allows you to write the determinant of a 3×3 matrix in terms of three 2×2 determinants. The **minor** of an entry in a 3×3 matrix is the determinant of the 2×2 matrix that remains after deletion of the row and column in which the entry occurs. Here are three examples.

Determinant	Entry	Minor of Entry	Value of Minor
$\begin{vmatrix} 1 & -1 & 3 \\ 0 & 2 & 5 \\ -2 & 4 & -7 \end{vmatrix}$	1	$\begin{vmatrix} 2 & 5 \\ 4 & -7 \end{vmatrix}$	$2(-7) - 4(5) = -34$
$\begin{vmatrix} 1 & -1 & 3 \\ 0 & 2 & 5 \\ -2 & 4 & -7 \end{vmatrix}$	-1	$\begin{vmatrix} 0 & 5 \\ -2 & -7 \end{vmatrix}$	$0(-7) - (-2)(5) = 10$
$\begin{vmatrix} 1 & -1 & 3 \\ 0 & 2 & 5 \\ -2 & 4 & -7 \end{vmatrix}$	3	$\begin{vmatrix} 0 & 2 \\ -2 & 4 \end{vmatrix}$	$0(4) - (-2)(2) = 4$

Technology: Tip

A graphing calculator with matrix capabilities can be used to evaluate the determinant of a square matrix. Consult the user's guide of your graphing calculator to learn how to evaluate a determinant. Use the graphing calculator to check the result in Example 1(a). Then try to evaluate the determinant of the 3×3 matrix at the right using a graphing calculator. Finish the evaluation of the determinant by expanding by minors to check the result.

Expanding by Minors

$$\det(A) = \begin{vmatrix} a_1 & b_1 & c_1 \\ a_2 & b_2 & c_2 \\ a_3 & b_3 & c_3 \end{vmatrix}$$

$$= a_1(\text{minor of } a_1) - b_1(\text{minor of } b_1) + c_1(\text{minor of } c_1)$$

$$= a_1 \begin{vmatrix} b_2 & c_2 \\ b_3 & c_3 \end{vmatrix} - b_1 \begin{vmatrix} a_2 & c_2 \\ a_3 & c_3 \end{vmatrix} + c_1 \begin{vmatrix} a_2 & b_2 \\ a_3 & b_3 \end{vmatrix}$$

This pattern is called **expanding by minors** along the first row. A similar pattern can be used to expand by minors along any row or column.

$$\begin{bmatrix} + & - & + \\ - & + & - \\ + & - & + \end{bmatrix}$$

Figure 4.12   Sign Pattern for a 3×3 Matrix

The *signs* of the terms used in expanding by minors follow the alternating pattern shown in Figure 4.12. For instance, the signs used to expand by minors along the second row are $-, +, -$, as shown below.

$$\det(A) = \begin{vmatrix} a_1 & b_1 & c_1 \\ a_2 & b_2 & c_2 \\ a_3 & b_3 & c_3 \end{vmatrix}$$

$$= -a_2(\text{minor of } a_2) + b_2(\text{minor of } b_2) - c_2(\text{minor of } c_2)$$

Example 2   Finding the Determinant of a 3×3 Matrix

Find the determinant of $A = \begin{bmatrix} -1 & 1 & 2 \\ 0 & 2 & 3 \\ 3 & 4 & 2 \end{bmatrix}$.

Solution

By expanding by minors along the *first column*, you obtain

$$\det(A) = \begin{vmatrix} -1 & 1 & 2 \\ 0 & 2 & 3 \\ 3 & 4 & 2 \end{vmatrix}$$

$$= (-1)\begin{vmatrix} 2 & 3 \\ 4 & 2 \end{vmatrix} - (0)\begin{vmatrix} 1 & 2 \\ 4 & 2 \end{vmatrix} + (3)\begin{vmatrix} 1 & 2 \\ 2 & 3 \end{vmatrix}$$

$$= (-1)(4 - 12) - (0)(2 - 8) + (3)(3 - 4)$$

$$= 8 - 0 - 3 = 5$$

Example 3   Finding the Determinant of a 3×3 Matrix

Find the determinant of $A = \begin{bmatrix} 1 & 2 & 1 \\ 3 & 0 & 2 \\ 4 & 0 & -1 \end{bmatrix}$.

Solution

By expanding by minors along the *second column*, you obtain

$$\det(A) = \begin{vmatrix} 1 & 2 & 1 \\ 3 & 0 & 2 \\ 4 & 0 & -1 \end{vmatrix}$$

$$= -(2)\begin{vmatrix} 3 & 2 \\ 4 & -1 \end{vmatrix} + (0)\begin{vmatrix} 1 & 1 \\ 4 & -1 \end{vmatrix} - (0)\begin{vmatrix} 1 & 1 \\ 3 & 2 \end{vmatrix}$$

$$= -(2)(-3 - 8) + 0 - 0 = 22$$

Note in the expansions in Examples 2 and 3 that a zero entry will always yield a zero term when expanding by minors. So, when you are evaluating the determinant of a matrix, you should choose to expand along the row or column that has the most zero entries.

② Use determinants and Cramer's Rule to solve systems of linear equations.

Cramer's Rule

So far in this chapter, you have studied four methods for solving a system of linear equations: graphing, substitution, elimination (with equations), and elimination (with matrices). You will now learn one more method, called *Cramer's Rule,* which is named after Gabriel Cramer (1704–1752). This rule uses determinants to write the solution of a system of linear equations.

In Cramer's Rule, the value of a variable is expressed as the quotient of two determinants of the coefficient matrix of the system. The numerator is the determinant of the matrix formed by using the column of constants as replacements for the coefficients of the variable. In the definition below, note the notation for the different determinants.

Study Tip

Cramer's Rule is not as general as the elimination method because Cramer's Rule requires that the coefficient matrix of the system be square *and* that the system have exactly one solution.

Cramer's Rule

1. For the system of linear equations

$$\begin{cases} a_1 x + b_1 y = c_1 \\ a_2 x + b_2 y = c_2 \end{cases}$$

the solution is given by

$$x = \frac{D_x}{D} = \frac{\begin{vmatrix} c_1 & b_1 \\ c_2 & b_2 \end{vmatrix}}{\begin{vmatrix} a_1 & b_1 \\ a_2 & b_2 \end{vmatrix}}, \qquad y = \frac{D_y}{D} = \frac{\begin{vmatrix} a_1 & c_1 \\ a_2 & c_2 \end{vmatrix}}{\begin{vmatrix} a_1 & b_1 \\ a_2 & b_2 \end{vmatrix}}$$

provided that $D \neq 0$.

2. For the system of linear equations

$$\begin{cases} a_1 x + b_1 y + c_1 z = d_1 \\ a_2 x + b_2 y + c_2 z = d_2 \\ a_3 x + b_3 y + c_3 z = d_3 \end{cases}$$

the solution is given by

$$x = \frac{D_x}{D} = \frac{\begin{vmatrix} d_1 & b_1 & c_1 \\ d_2 & b_2 & c_2 \\ d_3 & b_3 & c_3 \end{vmatrix}}{\begin{vmatrix} a_1 & b_1 & c_1 \\ a_2 & b_2 & c_2 \\ a_3 & b_3 & c_3 \end{vmatrix}}, \qquad y = \frac{D_y}{D} = \frac{\begin{vmatrix} a_1 & d_1 & c_1 \\ a_2 & d_2 & c_2 \\ a_3 & d_3 & c_3 \end{vmatrix}}{\begin{vmatrix} a_1 & b_1 & c_1 \\ a_2 & b_2 & c_2 \\ a_3 & b_3 & c_3 \end{vmatrix}},$$

$$z = \frac{D_z}{D} = \frac{\begin{vmatrix} a_1 & b_1 & d_1 \\ a_2 & b_2 & d_2 \\ a_3 & b_3 & d_3 \end{vmatrix}}{\begin{vmatrix} a_1 & b_1 & c_1 \\ a_2 & b_2 & c_2 \\ a_3 & b_3 & c_3 \end{vmatrix}}, D \neq 0$$

Example 4 Using Cramer's Rule for a 2 × 2 System

Use Cramer's Rule to solve the system of linear equations.

$$\begin{cases} 4x - 2y = 10 \\ 3x - 5y = 11 \end{cases}$$

Solution

Begin by finding the determinant of the coefficient matrix.

$$D = \begin{vmatrix} 4 & -2 \\ 3 & -5 \end{vmatrix} = -20 - (-6) = -14$$

Then, use the formulas for x and y given by Cramer's Rule.

$$x = \frac{D_x}{D} = \frac{\begin{vmatrix} 10 & -2 \\ 11 & -5 \end{vmatrix}}{-14} = \frac{(-50) - (-22)}{-14} = \frac{-28}{-14} = 2$$

$$y = \frac{D_y}{D} = \frac{\begin{vmatrix} 4 & 10 \\ 3 & 11 \end{vmatrix}}{-14} = \frac{44 - 30}{-14} = \frac{14}{-14} = -1$$

The solution is $(2, -1)$. Check this in the original system of equations.

Example 5 Using Cramer's Rule for a 3 × 3 System

Use Cramer's Rule to solve the system of linear equations.

$$\begin{cases} -x + 2y - 3z = 1 \\ 2x \qquad\;\; + \;\; z = 0 \\ 3x - 4y + 4z = 2 \end{cases}$$

Solution

The determinant of the coefficient matrix is $D = 10$.

$$x = \frac{D_x}{D} = \frac{\begin{vmatrix} 1 & 2 & -3 \\ 0 & 0 & 1 \\ 2 & -4 & 4 \end{vmatrix}}{10} = \frac{8}{10} = \frac{4}{5}$$

$$y = \frac{D_y}{D} = \frac{\begin{vmatrix} -1 & 1 & -3 \\ 2 & 0 & 1 \\ 3 & 2 & 4 \end{vmatrix}}{10} = \frac{-15}{10} = -\frac{3}{2}$$

$$z = \frac{D_z}{D} = \frac{\begin{vmatrix} -1 & 2 & 1 \\ 2 & 0 & 0 \\ 3 & -4 & 2 \end{vmatrix}}{10} = \frac{-16}{10} = -\frac{8}{5}$$

The solution is $\left(\frac{4}{5}, -\frac{3}{2}, -\frac{8}{5}\right)$. Check this in the original system of equations.

Study Tip

When using Cramer's Rule, remember that the method *does not* apply if the determinant of the coefficient matrix is zero.

③ Use determinants to find areas of triangles, to test for collinear points, and to find equations of lines.

Applications of Determinants

In addition to Cramer's Rule, determinants have many other practical applications. For instance, you can use a determinant to find the area of a triangle whose vertices are given by three points on a rectangular coordinate system.

Area of a Triangle

The area of a triangle with vertices (x_1, y_1), (x_2, y_2), and (x_3, y_3) is

$$\text{Area} = \pm\frac{1}{2}\begin{vmatrix} x_1 & y_1 & 1 \\ x_2 & y_2 & 1 \\ x_3 & y_3 & 1 \end{vmatrix}$$

where the symbol $(\pm)$ indicates that the appropriate sign should be chosen to yield a positive area.

Example 6   Finding the Area of a Triangle

Find the area of the triangle whose vertices are $(2, 0)$, $(1, 3)$, and $(3, 2)$, as shown in Figure 4.13.

Solution

Choose $(x_1, y_1) = (2, 0)$, $(x_2, y_2) = (1, 3)$, and $(x_3, y_3) = (3, 2)$. To find the area of the triangle, evaluate the determinant by expanding by minors along the first row.

$$\begin{vmatrix} x_1 & y_1 & 1 \\ x_2 & y_2 & 1 \\ x_3 & y_3 & 1 \end{vmatrix} = \begin{vmatrix} 2 & 0 & 1 \\ 1 & 3 & 1 \\ 3 & 2 & 1 \end{vmatrix}$$

$$= 2\begin{vmatrix} 3 & 1 \\ 2 & 1 \end{vmatrix} - 0\begin{vmatrix} 1 & 1 \\ 3 & 1 \end{vmatrix} + 1\begin{vmatrix} 1 & 3 \\ 3 & 2 \end{vmatrix}$$

$$= 2(1) - 0 + 1(-7)$$

$$= -5$$

Using this value, you can conclude that the area of the triangle is

$$\text{Area} = -\frac{1}{2}\begin{vmatrix} 2 & 0 & 1 \\ 1 & 3 & 1 \\ 3 & 2 & 1 \end{vmatrix}$$

$$= -\frac{1}{2}(-5) = \frac{5}{2}.$$

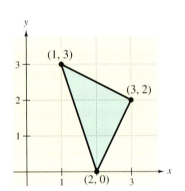

Figure 4.13

To see the benefit of the "determinant formula," try finding the area of the triangle in Example 6 using the standard formula:

$$\text{Area} = \frac{1}{2}(\text{Base})(\text{Height}).$$

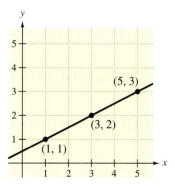

Figure 4.14

Suppose the three points in Example 6 had been on the same line. What would have happened had the area formula been applied to three such points? The answer is that the determinant would have been zero. Consider, for instance, the three collinear points $(1, 1)$, $(3, 2)$, and $(5, 3)$, as shown in Figure 4.14. The area of the "triangle" that has these three points as vertices is

$$\frac{1}{2}\begin{vmatrix} 1 & 1 & 1 \\ 3 & 2 & 1 \\ 5 & 3 & 1 \end{vmatrix} = \frac{1}{2}\left(1\begin{vmatrix} 2 & 1 \\ 3 & 1 \end{vmatrix} - 1\begin{vmatrix} 3 & 1 \\ 5 & 1 \end{vmatrix} + 1\begin{vmatrix} 3 & 2 \\ 5 & 3 \end{vmatrix} \right)$$

$$= \frac{1}{2}[-1 - (-2) + (-1)]$$

$$= 0.$$

This result is generalized as follows.

Test for Collinear Points

Three points (x_1, y_1), (x_2, y_2), and (x_3, y_3) are collinear (lie on the same line) if and only if

$$\begin{vmatrix} x_1 & y_1 & 1 \\ x_2 & y_2 & 1 \\ x_3 & y_3 & 1 \end{vmatrix} = 0$$

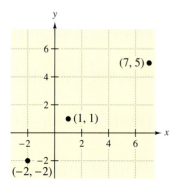

Figure 4.15

Example 7   Testing for Collinear Points

Determine whether the points $(-2, -2)$, $(1, 1)$, and $(7, 5)$ are collinear. (See Figure 4.15.)

Solution

Letting $(x_1, y_1) = (-2, -2)$, $(x_2, y_2) = (1, 1)$, and $(x_3, y_3) = (7, 5)$, you have

$$\begin{vmatrix} x_1 & y_1 & 1 \\ x_2 & y_2 & 1 \\ x_3 & y_3 & 1 \end{vmatrix} = \begin{vmatrix} -2 & -2 & 1 \\ 1 & 1 & 1 \\ 7 & 5 & 1 \end{vmatrix}$$

$$= -2\begin{vmatrix} 1 & 1 \\ 5 & 1 \end{vmatrix} - (-2)\begin{vmatrix} 1 & 1 \\ 7 & 1 \end{vmatrix} + 1\begin{vmatrix} 1 & 1 \\ 7 & 5 \end{vmatrix}$$

$$= -2(-4) - (-2)(-6) + 1(-2)$$

$$= -6.$$

Because the value of this determinant *is not* zero, you can conclude that the three points *do not* lie on the same line and are not collinear.

As a good review, look at how the slope can be used to verify the result in Example 7. Label the points $A(-2, -2)$, $B(1, 1)$, and $C(7, 5)$. Because the slopes from A to B and from A to C are different, the points are not collinear.

You can also use determinants to find the equation of a line through two points. In this case, the first row consists of the variables x and y and the number 1. By expanding by minors along the first row, the resulting 2×2 determinants are the coefficients of the variables x and y and the constant of the linear equation, as shown in Example 8.

Two-Point Form of the Equation of a Line

An equation of the line passing through the distinct points (x_1, y_1) and (x_2, y_2) is given by

$$\begin{vmatrix} x & y & 1 \\ x_1 & y_1 & 1 \\ x_2 & y_2 & 1 \end{vmatrix} = 0.$$

Example 8   Finding an Equation of a Line

Find an equation of the line passing through $(-2, 1)$ and $(3, -2)$.

Solution

Applying the determinant formula for the equation of a line produces

$$\begin{vmatrix} x & y & 1 \\ -2 & 1 & 1 \\ 3 & -2 & 1 \end{vmatrix} = 0.$$

To evaluate this determinant, you can expand by minors along the first row to obtain the following.

$$x \begin{vmatrix} 1 & 1 \\ -2 & 1 \end{vmatrix} - y \begin{vmatrix} -2 & 1 \\ 3 & 1 \end{vmatrix} + 1 \begin{vmatrix} -2 & 1 \\ 3 & -2 \end{vmatrix} = 0$$

$$3x + 5y + 1 = 0$$

So, an equation of the line is $3x + 5y + 1 = 0$.

Note that this method of finding the equation of a line works for all lines, including horizontal and vertical lines, as shown below.

Vertical Line Through $(2, 0)$ *and* $(2, 2)$:

$$\begin{vmatrix} x & y & 1 \\ 2 & 0 & 1 \\ 2 & 2 & 1 \end{vmatrix} = 0$$

$$-2x - 0y + 4 = 0$$

$$-2x = -4$$

$$x = 2$$

Horizontal Line Through $(-3, 4)$ *and* $(2, 4)$:

$$\begin{vmatrix} x & y & 1 \\ -3 & 4 & 1 \\ 2 & 4 & 1 \end{vmatrix} = 0$$

$$0x + 5y - 20 = 0$$

$$5y = 20$$

$$y = 4$$

4.5 Exercises

Review   Concepts, Skills, and Problem Solving

Keep mathematically in shape by doing these exercises *before* the problems of this section.

Properties and Definitions

1. *Writing* Explain what is meant by the *domain* of a function.

2. *Writing* Explain what is meant by the *range* of a function.

3. *Writing* What distinguishes a *function* from a relation?

4. *Writing* In your own words, explain what the notation $f(4)$ means when $f(x) = x^2 - x + 2$.

Functions and Graphs

In Exercises 5–8, evaluate the function as indicated.

5. For $f(x) = 3x + 2$, find $f(-2)$.

6. For $f(x) = -x^2 - x + 5$, find $f(-1)$.

7. For $f(x) = |10 - 3x|$, find $f(5)$.

8. For $f(x) = \begin{cases} -2x, & x < 0 \\ x^2 + 4, & x \geq 0 \end{cases}$, find $f(0)$.

In Exercises 9 and 10, sketch the graph of the function, and then determine its domain and range.

9. $f(x) = 2 - \sqrt{x}$

10. $f(x) = x^3 - 3$

Models

In Exercises 11 and 12, translate the phrase into an algebraic expression.

11. The time to travel 320 miles if the average speed is r miles per hour

12. The perimeter of a triangle if the sides are $x + 1$, $\frac{1}{2}x + 5$, and $3x + 1$

Developing Skills

In Exercises 1–12, find the determinant of the matrix. See Example 1.

1. $\begin{bmatrix} 2 & 1 \\ 3 & 4 \end{bmatrix}$

2. $\begin{bmatrix} -3 & 1 \\ 5 & 2 \end{bmatrix}$

3. $\begin{bmatrix} 5 & 2 \\ -6 & 3 \end{bmatrix}$

4. $\begin{bmatrix} 2 & -2 \\ 4 & 3 \end{bmatrix}$

5. $\begin{bmatrix} 5 & -4 \\ -10 & 8 \end{bmatrix}$

6. $\begin{bmatrix} 4 & -3 \\ 0 & 0 \end{bmatrix}$

7. $\begin{bmatrix} 2 & 6 \\ 0 & 3 \end{bmatrix}$

8. $\begin{bmatrix} -2 & 3 \\ 6 & -9 \end{bmatrix}$

9. $\begin{bmatrix} -7 & 6 \\ \frac{1}{2} & 3 \end{bmatrix}$

10. $\begin{bmatrix} \frac{2}{3} & \frac{5}{6} \\ 14 & -2 \end{bmatrix}$

11. $\begin{bmatrix} 0.3 & 0.5 \\ 0.5 & 0.3 \end{bmatrix}$

12. $\begin{bmatrix} -1.2 & 4.5 \\ 0.4 & -0.9 \end{bmatrix}$

In Exercises 13–30, evaluate the determinant of the matrix. Expand by minors along the row or column that appears to make the computation easiest. See Examples 2 and 3.

13. $\begin{vmatrix} 2 & 3 & -1 \\ 6 & 0 & 0 \\ 4 & 1 & 1 \end{vmatrix}$

14. $\begin{bmatrix} 10 & 2 & -4 \\ 8 & 0 & -2 \\ 4 & 0 & 2 \end{bmatrix}$

15. $\begin{vmatrix} 1 & 1 & 2 \\ 3 & 1 & 0 \\ -2 & 0 & 3 \end{vmatrix}$

16. $\begin{bmatrix} 2 & 1 & 3 \\ 1 & 4 & 4 \\ 1 & 0 & 2 \end{bmatrix}$

17. $\begin{vmatrix} 2 & 4 & 6 \\ 0 & 3 & 1 \\ 0 & 0 & -5 \end{vmatrix}$

18. $\begin{vmatrix} 2 & 3 & 1 \\ 0 & 5 & -2 \\ 0 & 0 & -2 \end{vmatrix}$

19. $\begin{vmatrix} -2 & 2 & 3 \\ 1 & -1 & 0 \\ 0 & 1 & 4 \end{vmatrix}$

20. $\begin{bmatrix} 3 & 2 & 2 \\ 2 & 2 & 2 \\ -4 & 4 & 3 \end{bmatrix}$

21. $\begin{bmatrix} 1 & 4 & -2 \\ 3 & 6 & -6 \\ -2 & 1 & 4 \end{bmatrix}$ **22.** $\begin{bmatrix} 2 & -1 & 0 \\ 4 & 2 & 1 \\ 4 & 2 & 1 \end{bmatrix}$

23. $\begin{bmatrix} 1 & 4 & -2 \\ 3 & 2 & 0 \\ -1 & 4 & 3 \end{bmatrix}$ **24.** $\begin{bmatrix} 6 & 8 & -7 \\ 0 & 0 & 0 \\ 4 & -6 & 22 \end{bmatrix}$

25. $\begin{bmatrix} 2 & -5 & 0 \\ 4 & 7 & 0 \\ -7 & 25 & 3 \end{bmatrix}$ **26.** $\begin{bmatrix} 8 & 7 & 6 \\ -4 & 0 & 0 \\ 5 & 1 & 4 \end{bmatrix}$

27. $\begin{bmatrix} 0.1 & 0.2 & 0.3 \\ -0.3 & 0.2 & 0.2 \\ 5 & 4 & 4 \end{bmatrix}$ **28.** $\begin{bmatrix} -0.4 & 0.4 & 0.3 \\ 0.2 & 0.2 & 0.2 \\ 0.3 & 0.2 & 0.2 \end{bmatrix}$

29. $\begin{bmatrix} x & y & 1 \\ 3 & 1 & 1 \\ -2 & 0 & 1 \end{bmatrix}$ **30.** $\begin{bmatrix} x & y & 1 \\ -2 & -2 & 1 \\ 1 & 5 & 1 \end{bmatrix}$

In Exercises 31–36, use a graphing calculator to evaluate the determinant of the matrix.

31. $\begin{vmatrix} 5 & -3 & 2 \\ 7 & 5 & -7 \\ 0 & 6 & -1 \end{vmatrix}$ **32.** $\begin{vmatrix} 3 & -1 & 2 \\ 1 & -1 & 2 \\ -2 & 3 & 10 \end{vmatrix}$

33. $\begin{vmatrix} -\frac{1}{2} & -1 & 6 \\ 8 & -\frac{1}{4} & -4 \\ 1 & 2 & 1 \end{vmatrix}$ **34.** $\begin{vmatrix} \frac{1}{2} & \frac{3}{2} & \frac{1}{2} \\ 4 & 8 & 10 \\ -2 & -6 & 12 \end{vmatrix}$

35. $\begin{bmatrix} 0.2 & 0.8 & -0.3 \\ 0.1 & 0.8 & 0.6 \\ -10 & -5 & 1 \end{bmatrix}$

36. $\begin{bmatrix} 0.4 & 0.3 & 0.3 \\ -0.2 & 0.6 & 0.6 \\ 3 & 1 & 1 \end{bmatrix}$

In Exercises 37–52, use Cramer's Rule to solve the system of linear equations. (If not possible, state the reason.) See Examples 4 and 5.

37. $\begin{cases} x + 2y = 5 \\ -x + y = 1 \end{cases}$ **38.** $\begin{cases} 2x - y = -10 \\ 3x + 2y = -1 \end{cases}$

39. $\begin{cases} 3x + 4y = -2 \\ 5x + 3y = 4 \end{cases}$ **40.** $\begin{cases} 18x + 12y = 13 \\ 30x + 24y = 23 \end{cases}$

41. $\begin{cases} 20x + 8y = 11 \\ 12x - 24y = 21 \end{cases}$ **42.** $\begin{cases} 13x - 6y = 17 \\ 26x - 12y = 8 \end{cases}$

43. $\begin{cases} -0.4x + 0.8y = 1.6 \\ 2x - 4y = 5 \end{cases}$ **44.** $\begin{cases} -0.4x + 0.8y = 1.6 \\ 0.2x + 0.3y = 2.2 \end{cases}$

45. $\begin{cases} 3u + 6v = 5 \\ 6u + 14v = 11 \end{cases}$ **46.** $\begin{cases} 3x_1 + 2x_2 = 1 \\ 2x_1 + 10x_2 = 6 \end{cases}$

47. $\begin{cases} 4x - y + z = -5 \\ 2x + 2y + 3z = 10 \\ 5x - 2y + 6z = 1 \end{cases}$

48. $\begin{cases} 4x - 2y + 3z = -2 \\ 2x + 2y + 5z = 16 \\ 8x - 5y - 2z = 4 \end{cases}$

49. $\begin{cases} 3a + 3b + 4c = 1 \\ 3a + 5b + 9c = 2 \\ 5a + 9b + 17c = 4 \end{cases}$ **50.** $\begin{cases} 2x + 3y + 5z = 4 \\ 3x + 5y + 9z = 7 \\ 5x + 9y + 17z = 13 \end{cases}$

51. $\begin{cases} 5x - 3y + 2z = 2 \\ 2x + 2y - 3z = 3 \\ x - 7y + 8z = -4 \end{cases}$ **52.** $\begin{cases} 3x + 2y + 5z = 4 \\ 4x - 3y - 4z = 1 \\ -8x + 2y + 3z = 0 \end{cases}$

In Exercises 53–56, solve the system of linear equations using a graphing calculator and Cramer's Rule. See Examples 4 and 5.

53. $\begin{cases} -3x + 10y = 22 \\ 9x - 3y = 0 \end{cases}$ **54.** $\begin{cases} 3x + 7y = 3 \\ 7x + 25y = 11 \end{cases}$

55. $\begin{cases} 3x - 2y + 3z = 8 \\ x + 3y + 6z = -3 \\ x + 2y + 9z = -5 \end{cases}$

56. $\begin{cases} 6x + 4y - 8z = -22 \\ -2x + 2y + 3z = 13 \\ -2x + 2y - z = 5 \end{cases}$

In Exercises 57 and 58, solve the equation.

57. $\begin{vmatrix} 5 - x & 4 \\ 1 & 2 - x \end{vmatrix} = 0$

58. $\begin{vmatrix} 4 - x & -2 \\ 1 & 1 - x \end{vmatrix} = 0$

Solving Problems

Area of a Triangle   In Exercises 59–66, use a determinant to find the area of the triangle with the given vertices. See Example 6.

59. $(0, 3), (4, 0), (8, 5)$

60. $(2, 0), (0, 5), (6, 3)$

61. $(0, 0), (3, 1), (1, 5)$

62. $(-2, -3), (2, -3), (0, 4)$

63. $(-2, 1), (3, -1), (1, 6)$

64. $(-4, 2), (1, 5), (4, -4)$

65. $\left(0, \frac{1}{2}\right), \left(\frac{5}{2}, 0\right), (4, 3)$

66. $\left(\frac{1}{4}, 0\right), \left(0, \frac{3}{4}\right), (8, -2)$

Area of a Region   In Exercises 67–70, find the area of the shaded region of the figure.

67.

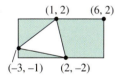

68.

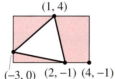

69.

70.

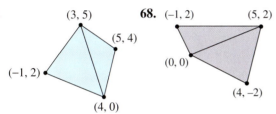

71. *Area of a Region*   A large region of forest has been infested with gypsy moths. The region is roughly triangular, as shown in the figure. Approximate the number of square miles in this region.

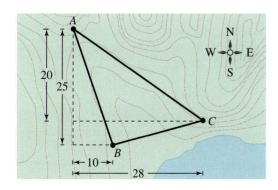

72. *Area of a Region*   You have purchased a triangular tract of land, as shown in the figure. How many square feet are there in the tract of land?

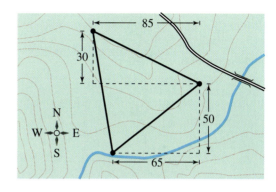

Collinear Points   In Exercises 73–78, determine whether the points are collinear. See Example 7.

73. $(-1, 11), (0, 8), (2, 2)$

74. $(-1, -1), (1, 9), (2, 13)$

75. $(-1, -5), (1, -1), (4, 5)$

76. $(-1, 8), (1, 2), (2, 0)$

77. $\left(-2, \frac{1}{3}\right), (2, 1), \left(3, \frac{1}{5}\right)$

78. $\left(0, \frac{1}{2}\right), \left(1, \frac{7}{6}\right), \left(9, \frac{13}{2}\right)$

Equation of a Line   In Exercises 79–86, use a determinant to find the equation of the line through the points. See Example 8.

79. $(0, 0), (5, 3)$

80. $(-4, 3), (2, 1)$

81. $(10, 7), (-2, -7)$

82. $(-8, 3), (4, 6)$

83. $\left(-2, \frac{3}{2}\right), (3, -3)$

84. $\left(-\frac{1}{2}, 3\right), \left(\frac{5}{2}, 1\right)$

85. $(2, 3.6), (8, 10)$

86. $(3, 1.6), (5, -2.2)$

87. *Electrical Networks*   Laws that deal with electrical currents are known as *Kirchhoff's Laws*. When Kirchhoff's Laws are applied to the electrical network shown in the figure, the currents I_1, I_2, and I_3 are the solution of the system

$$\begin{cases} I_1 - I_2 + I_3 = 0 \\ 3I_1 + 2I_2 \quad\;\; = 7 \\ \quad\;\; 2I_2 + 4I_3 = 8. \end{cases}$$

Find the currents.

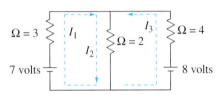

88. *Electrical Networks*   When Kirchhoff's Laws are applied to the electrical network shown in the figure, the currents I_1, I_2, and I_3 are the solution of the system

$$\begin{cases} I_1 + I_2 - I_3 = 0 \\ I_1 \quad\;\; + 2I_3 = 12 \\ I_1 - 2I_2 \quad\;\; = -4. \end{cases}$$

Find the currents.

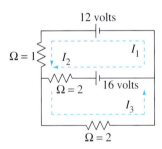

89. *Electrical Networks*   When Kirchhoff's Laws are applied to the electrical network shown in the figure, the currents I_1, I_2, and I_3 are the solution of the system

$$\begin{cases} I_1 - I_2 + I_3 = 0 \\ \quad\;\; I_2 + 4I_3 = 8 \\ 4I_1 + I_2 \quad\;\; = 16. \end{cases}$$

Find the currents.

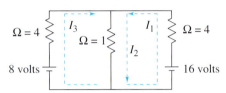

90. *Force*   When three forces are applied to a beam, Newton's Laws suggests that the forces F_1, F_2, and F_3 are the solution of the system

$$\begin{cases} 3F_1 + F_2 - F_3 = 2 \\ F_1 - 2F_2 + F_3 = 0 \\ 4F_1 - F_2 + F_3 = 0. \end{cases}$$

Find the forces.

91. (a) Use Cramer's Rule to solve the system of linear equations.

$$\begin{cases} kx + 3ky = 2 \\ (2 + k)x + ky = 5 \end{cases}$$

(b) For what values of k can Cramer's Rule not be used?

92. (a) Use Cramer's Rule to solve the system of linear equations.

$$\begin{cases} kx + (1 - k)y = 1 \\ (1 - k)x + ky = 3 \end{cases}$$

(b) For what value(s) of k will the system be inconsistent?

Explaining Concepts

93. ✪   Answer parts (g) and (h) of Motivating the Chapter on page 216.

94. *Writing*✎   Explain the difference between a square matrix and its determinant.

95. *Writing*✎   Is it possible to find the determinant of a 2×3 matrix? Explain.

96. *Writing*✎   What is meant by the minor of an entry of a square matrix?

97. *Writing*✎   What conditions must be met in order to use Cramer's Rule to solve a system of linear equations?

4.6 Systems of Linear Inequalities

Bonnie Kamin/PhotoEdit

What You Should Learn

1. Solve systems of linear inequalities in two variables.
2. Use systems of linear inequalities to model and solve real-life problems.

Why You Should Learn It

Systems of linear inequalities can be used to model and solve real-life problems. For instance, in Exercise 65 on page 288, a system of linear inequalities can be used to analyze the compositions of dietary supplements.

1. Solve systems of linear inequalities in two variables.

Systems of Linear Inequalities in Two Variables

You have already graphed linear inequalities in two variables. However, many practical problems in business, science, and engineering involve **systems of linear inequalities.** This type of system arises in problems that have *constraint* statements that contain phrases such as "more than," "less than," "at least," "no more than," "a minimum of," and "a maximum of." A **solution** of a system of linear inequalities in x and y is a point (x, y) that satisfies each inequality in the system.

To sketch the graph of a system of inequalities in two variables, first sketch (on the same coordinate system) the graph of each individual inequality. The **solution set** is the region that is *common* to every graph in the system.

Example 1   Graphing a System of Linear Inequalities

Sketch the graph of the system of linear inequalities: $\begin{cases} 2x - y \leq 5 \\ x + 2y \geq 2. \end{cases}$

Solution

Begin by rewriting each inequality in slope-intercept form. Then sketch the line for each corresponding equation of each inequality. See Figures 4.16–4.18.

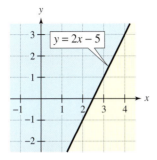

Graph of $2x - y \leq 5$ is all points on and above $y = 2x - 5$.
Figure 4.16

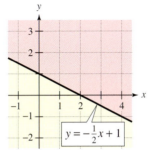

Graph of $x + 2y \geq 2$ is all points on and above $y = -\frac{1}{2}x + 1$.
Figure 4.17

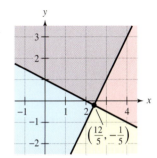

Graph of system is the purple wedge-shaped region.
Figure 4.18

In Figure 4.18, note that the two borderlines of the region

$$y = 2x - 5 \quad \text{and} \quad y = -\frac{1}{2}x + 1$$

intersect at the point $\left(\frac{12}{5}, -\frac{1}{5}\right)$. Such a point is called a **vertex** of the region. The region shown in the figure has only one vertex. Some regions, however, have several vertices. When you are sketching the graph of a system of linear inequalities, it is helpful to find and label any vertices of the region.

Graphing a System of Linear Inequalities

1. Sketch the line that corresponds to each inequality. (Use dashed lines for inequalities with $<$ or $>$ and solid lines for inequalities with $\leq$ or $\geq$.)

2. Lightly shade the half-plane that is the graph of each linear inequality. (Colored pencils may help distinguish different half-planes.)

3. The graph of the system is the intersection of the half-planes. (If you use colored pencils, it is the region that is selected with *every* color.)

Example 2   Graphing a System of Linear Inequalities

Sketch the graph of the system of linear inequalities: $\begin{cases} y < 4 \\ y > 1 \end{cases}$.

Solution

The graph of the first inequality is the half-plane below the horizontal line

$$y = 4. \qquad \text{Upper boundary}$$

The graph of the second inequality is the half-plane above the horizontal line

$$y = 1. \qquad \text{Lower boundary}$$

The graph of the system is the horizontal band that lies *between* the two horizontal lines (where $y < 4$ *and* $y > 1$), as shown in Figure 4.19.

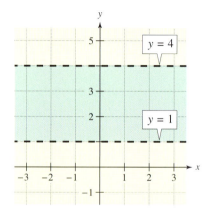

Figure 4.19

Example 3   Graphing a System of Linear Inequalities

Sketch the graph of the system of linear inequalities, and label the vertices.

$$\begin{cases} x - y < & 2 \\ x & > -2 \\ & y \le & 3 \end{cases}$$

Solution

Begin by sketching the half-planes represented by the three linear inequalities. The graph of

$$x - y < 2$$

is the half-plane lying above the line $y = x - 2$, the graph of

$$x > -2$$

is the half-plane lying to the right of the line $x = -2$, and the graph of

$$y \le 3$$

is the half-plane lying on or below the line $y = 3$. As shown in Figure 4.20, the region that is common to all three of these half-planes is a triangle. The vertices of the triangle are found as follows.

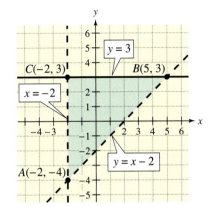

Figure 4.20

Vertex A: $(-2, -4)$	*Vertex B:* $(5, 3)$	*Vertex C:* $(-2, 3)$
Solution of the system	Solution of the system	Solution of the system
$\begin{cases} x - y = & 2 \\ x & = -2 \end{cases}$	$\begin{cases} x - y = 2 \\ y = 3 \end{cases}$	$\begin{cases} x = -2 \\ y = & 3 \end{cases}$

For the triangular region shown in Figure 4.20, each point of intersection of a pair of boundary lines corresponds to a vertex. With more complicated regions, two border lines can sometimes intersect at a point that is not a vertex of the region, as shown in Figure 4.21. To keep track of which points of intersection are actually vertices of the region, you should sketch the region and refer to your sketch as you find each point of intersection.

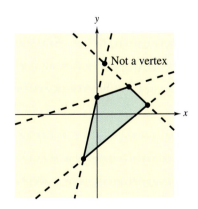

Figure 4.21

Example 4   Graphing a System of Linear Inequalities

Sketch the graph of the system of linear inequalities, and label the vertices.

$$\begin{cases} x + y \le 5 \\ 3x + 2y \le 12 \\ x \ge 0 \\ y \ge 0 \end{cases}$$

Solution

Begin by sketching the half-planes represented by the four linear inequalities. The graph of

$$x + y \le 5$$

is the half-plane lying on and below the line $y = -x + 5$. The graph of

$$3x + 2y \le 12$$

is the half-plane lying on and below the line $y = -\frac{3}{2}x + 6$. The graph of $x \ge 0$ is the half-plane lying on and to the right of the y-axis, and the graph of $y \ge 0$ is the half-plane lying on and above the x-axis. As shown in Figure 4.22, the region that is common to all four of these half-planes is a four-sided polygon. The vertices of the region are found as follows.

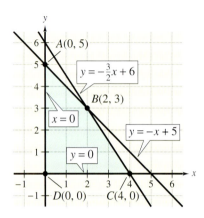

Figure 4.22

Vertex A: $(0, 5)$	Vertex B: $(2, 3)$	Vertex C: $(4, 0)$	Vertex D: $(0, 0)$
Solution of the system	Solution of the system	Solution of the system	Solution of the system
$\begin{cases} x + y = 5 \\ x = 0 \end{cases}$	$\begin{cases} x + y = 5 \\ 3x + 2y = 12 \end{cases}$	$\begin{cases} 3x + 2y = 12 \\ y = 0 \end{cases}$	$\begin{cases} x = 0 \\ y = 0 \end{cases}$

Example 5   Finding the Boundaries of a Region

Find a system of inequalities that defines the region shown in Figure 4.23.

Solution

Three of the boundaries of the region are horizontal or vertical—they are easy to find. To find the diagonal boundary line, use the techniques in Section 3.4 to find the equation of the line passing through the points $(4, 4)$ and $(6, 0)$. You can use the formula for slope to find $m = -2$, and then use the point-slope form with point $(6, 0)$ and $m = -2$ to obtain

$$y - 0 = -2(x - 6).$$

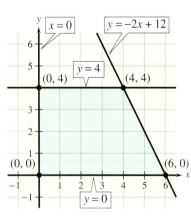

Figure 4.23

So, the equation is $y = -2x + 12$. The system of linear inequalities that describes the region is as follows.

$$\begin{cases} y \le 4 & \text{Region lies on and below line } y = 4. \\ y \ge 0 & \text{Region lies on and above } x\text{-axis.} \\ x \ge 0 & \text{Region lies on and to the right of } y\text{-axis.} \\ y \le -2x + 12 & \text{Region lies on and below line } y = -2x + 12. \end{cases}$$

Technology: Tip

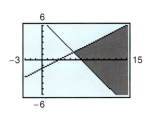

A graphing calculator can be used to graph a system of linear inequalities. The graph of

$$\begin{cases} 4y < 2x - 6 \\ x + y \geq 7 \end{cases}$$

is shown at the left. The shaded region, in which all points satisfy both inequalities, is the solution of the system. Try using a graphing calculator to graph

$$\begin{cases} 3x + y < 1 \\ -2x - 2y < 8. \end{cases}$$

2 Use systems of linear inequalities to model and solve real-life problems.

Application

Example 6   Nutrition

The minimum daily requirements from the liquid portion of a diet are 300 calories, 36 units of vitamin A, and 90 units of vitamin C. A cup of dietary drink X provides 60 calories, 12 units of vitamin A, and 10 units of vitamin C. A cup of dietary drink Y provides 60 calories, 6 units of vitamin A, and 30 units of vitamin C. Write a system of linear inequalities that describes how many cups of each drink should be consumed each day to meet the minimum daily requirements for calories and vitamins.

Solution

Begin by letting x and y represent the following.

$$x = \text{number of cups of dietary drink X}$$

$$y = \text{number of cups of dietary drink Y}$$

To meet the minimum daily requirements, the following inequalities must be satisfied.

$$\begin{cases} 60x + 60y \geq 300 & \text{Calories} \\ 12x + 6y \geq 36 & \text{Vitamin A} \\ 10x + 30y \geq 90 & \text{Vitamin C} \\ x \geq 0 \\ y \geq 0 \end{cases}$$

The last two inequalities are included because x and y cannot be negative. The graph of this system of inequalities is shown in Figure 4.24.

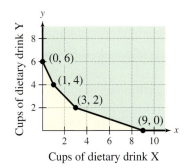

Figure 4.24

4.6 Exercises

Review   Concepts, Skills, and Problem Solving

Keep mathematically in shape by doing these exercises *before* the problems of this section.

Properties and Definitions

1. *Writing* Given a function $f(x)$, describe how the graph of $h(x) = f(x) + c$ compares with the graph of $f(x)$ for a positive real number c.

2. *Writing* Given a function $f(x)$, describe how the graph of $h(x) = f(x) - c$ compares with the graph of $f(x)$ for a positive real number c.

3. *Writing* Given a function $f(x)$, describe how the graph of $h(x) = f(x - c)$ compares with the graph of $f(x)$ for a positive real number c.

4. *Writing* Given a function $f(x)$, describe how the graph of $h(x) = f(x + c)$ compares with the graph of $f(x)$ for a positive real number c.

5. *Writing* Given a function $f(x)$, describe how the graph of $h(x) = -f(x)$ compares with the graph of $f(x)$.

6. *Writing* Given a function $f(x)$, describe how the graph of $h(x) = f(-x)$ compares with the graph of $f(x)$.

Graphing

In Exercises 7 and 8, use the graph of $f(x) = x^2$ to write a function that represents the graph.

7.
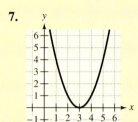

8.
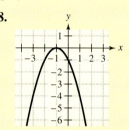

In Exercises 9 and 10, use the graph of f to sketch the graph.

9. $f(x) - 3$

10. $f(x - 1) + 2$

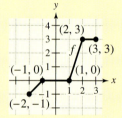

Problem Solving

11. *Retail Price*   A video game system that costs a retailer $169.50 is marked up by 36%. Find the price to the consumer.

12. *Sale Price*   A sweater is listed at $44. Find the sale price if there is a 20%-off sale.

Developing Skills

In Exercises 1–6, match the system of linear inequalities with its graph. [The graphs are labeled (a), (b), (c), (d), (e), and (f).]

(a)

(b)

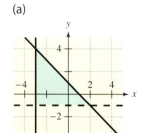

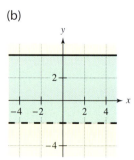

(c)   (d)

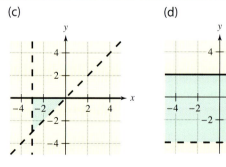

(e)

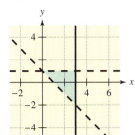

(f)

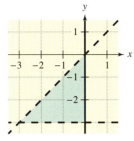

1. $\begin{cases} y > & x \\ x > -3 \\ y \le & 0 \end{cases}$

2. $\begin{cases} y \le & 4 \\ y > -2 \end{cases}$

3. $\begin{cases} y < & x \\ y > -3 \\ x \le & 0 \end{cases}$

4. $\begin{cases} x \le & 3 \\ y < & 1 \\ y > -x + 1 \end{cases}$

5. $\begin{cases} y > & -1 \\ x \ge & -3 \\ y \le -x + & 1 \end{cases}$

6. $\begin{cases} y > -4 \\ y \le & 2 \end{cases}$

In Exercises 7–44, sketch a graph of the solution of the system of linear inequalities. See Examples 1–4.

7. $\begin{cases} x < & 3 \\ x > -2 \end{cases}$

8. $\begin{cases} y > -1 \\ y \le & 2 \end{cases}$

9. $\begin{cases} x + y \le 3 \\ x - 1 \le 1 \end{cases}$

10. $\begin{cases} x + y \ge 2 \\ x - y \le 2 \end{cases}$

11. $\begin{cases} 2x - 4y \le 6 \\ x + & y \ge 2 \end{cases}$

12. $\begin{cases} 4x + 10y \le 5 \\ x - & y \le 4 \end{cases}$

13. $\begin{cases} x + 2y \le 6 \\ x - 2y \le 0 \end{cases}$

14. $\begin{cases} 2x + y \le 0 \\ x - y \le 8 \end{cases}$

15. $\begin{cases} x - 2y > 4 \\ 2x + & y > 6 \end{cases}$

16. $\begin{cases} 3x + & y < 6 \\ x + 2y > 2 \end{cases}$

17. $\begin{cases} x + y > -1 \\ x + y < & 3 \end{cases}$

18. $\begin{cases} x - y > & 2 \\ x - y < -4 \end{cases}$

19. $\begin{cases} y \ge \frac{4}{3}x + 1 \\ y \le 5x - 2 \end{cases}$

20. $\begin{cases} y \ge \frac{1}{2}x + \frac{1}{2} \\ y \le 4x - \frac{1}{2} \end{cases}$

21. $\begin{cases} y > & x - 2 \\ y > -\frac{1}{3}x + 5 \end{cases}$

22. $\begin{cases} y > & x - 4 \\ y > \frac{2}{3}x + \frac{1}{3} \end{cases}$

23. $\begin{cases} y \ge 3x - 3 \\ y \le -x + 1 \end{cases}$

24. $\begin{cases} y \ge 2x - 3 \\ y \le 3x + 1 \end{cases}$

25. $\begin{cases} y > & 2x \\ y > -x + 4 \end{cases}$

26. $\begin{cases} y \le -x \\ y \le & x + 1 \end{cases}$

27. $\begin{cases} x + 2y \le -4 \\ y \ge x + 5 \end{cases}$

28. $\begin{cases} x + y \le -3 \\ y \ge 3x - 4 \end{cases}$

29. $\begin{cases} x + y \le 4 \\ x \ge 0 \\ y \ge 0 \end{cases}$

30. $\begin{cases} 2x + y \le 6 \\ x \ge 0 \\ y \ge 0 \end{cases}$

31. $\begin{cases} 4x - 2y > 8 \\ x \ge 0 \\ y \le 0 \end{cases}$

32. $\begin{cases} 2x - 6y > 6 \\ x \le 0 \\ y \le 0 \end{cases}$

33. $\begin{cases} y > -5 \\ x \le & 2 \\ y \le & x + 2 \end{cases}$

34. $\begin{cases} y \ge -1 \\ x \le & 2 \\ y \le & x + 2 \end{cases}$

35. $\begin{cases} x + y \le 1 \\ -x + y \le 1 \\ y \ge 0 \end{cases}$

36. $\begin{cases} 3x + 2y < 6 \\ x - 3y \ge 1 \\ y \ge 0 \end{cases}$

37. $\begin{cases} x + & y \le 5 \\ x - 2y \ge 2 \\ y \ge 3 \end{cases}$

38. $\begin{cases} 2x + & y \ge 2 \\ x - 3y \le 2 \\ y \le 1 \end{cases}$

39. $\begin{cases} -3x + 2y < & 6 \\ x - 4y > -2 \\ 2x + & y < & 3 \end{cases}$

40. $\begin{cases} x - 7y > -36 \\ 5x + 2y > & 5 \\ 6x + 5y > & 6 \end{cases}$

41. $\begin{cases} 2x + & y < 2 \\ 6x + 3y > 2 \end{cases}$

42. $\begin{cases} x - 2y < -6 \\ 5x - 3y > -9 \end{cases}$

43. $\begin{cases} x \ge 1 \\ x - 2y \le 3 \\ 3x + 2y \ge 9 \\ x + & y \le 6 \end{cases}$

44. $\begin{cases} x + y \le & 4 \\ x + y \ge -1 \\ x - y \ge -2 \\ x - y \le & 2 \end{cases}$

In Exercises 45–50, use a graphing calculator to graph the solution of the system of linear inequalities.

45. $\begin{cases} 2x - 3y \le 6 \\ y \le 4 \end{cases}$ **46.** $\begin{cases} 6x + 3y \ge 12 \\ y \le 4 \end{cases}$

47. $\begin{cases} 2x - 2y \le 5 \\ y \le 6 \end{cases}$ **48.** $\begin{cases} 2x + 3y \ge 12 \\ y \ge 2 \end{cases}$

49. $\begin{cases} 2x + y \le 2 \\ y \ge -4 \end{cases}$ **50.** $\begin{cases} x - 2y \ge -6 \\ y \le 6 \end{cases}$

In Exercises 51–56, write a system of linear inequalities that describes the shaded region. See Example 5.

53.
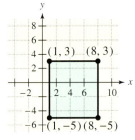

54.
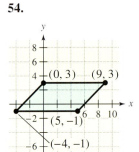

51.
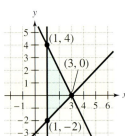

52.
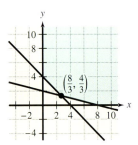

55.
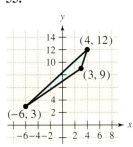

56.
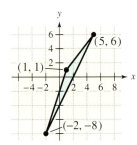

Solving Problems

57. *Production*   A furniture company can sell all the tables and chairs it produces. Each table requires 1 hour in the assembly center and $1\frac{1}{3}$ hours in the finishing center. Each chair requires $1\frac{1}{2}$ hours in the assembly center and $1\frac{1}{2}$ hours in the finishing center. The company's assembly center is available 12 hours per day, and its finishing center is available 15 hours per day. Write a system of linear inequalities describing the different production levels. Graph the system.

58. *Production*   An electronics company can sell all the VCRs and DVD players it produces. Each VCR requires 2 hours on the assembly line and $1\frac{1}{2}$ hours on the testing line. Each DVD player requires $2\frac{1}{2}$ hours on the assembly line and 3 hours on the testing line. The company's assembly line is available 18 hours per day, and its testing line is available 16 hours per day. Write a system of linear inequalities describing the different production levels. Graph the system.

59. *Investment* A person plans to invest up to $20,000 in two different interest-bearing accounts, account X and account Y. Account X is to contain at least $5000. Moreover, account Y should have at least twice the amount in account X. Write a system of linear inequalities describing the various amounts that can be deposited in each account. Graph the system.

60. *Investment* A person plans to invest up to $10,000 in two different interest-bearing accounts, account X and account Y. Account Y is to contain at least $3000. Moreover, account X should have at least three times the amount in account Y. Write a system of linear inequalities describing the various amounts that can be deposited in each account. Graph the system.

61. *Ticket Sales* Two types of tickets are to be sold for a concert. General admission tickets cost $15 per ticket and stadium seat tickets cost $25 per ticket. The promoter of the concert must sell at least 15,000 tickets, including at least 8000 general admission tickets and at least 4000 stadium seat tickets. Moreover, the gross receipts must total at least $275,000 in order for the concert to be held. Write a system of linear inequalities describing the different numbers of tickets that can be sold. Use a graphing calculator to graph the system.

62. *Ticket Sales* For a concert event, there are $30 reserved seat tickets and $20 general admission tickets. There are 2000 reserved seats available, and fire regulations limit the number of paid ticket holders to 3000. The promoter must take in at least $75,000 in ticket sales. Write a system of linear inequalities describing the different numbers of tickets that can be sold. Use a graphing calculator to graph the system.

63. *Geometry* The figure shows a cross section of a roped-off swimming area at a beach. Write a system of linear inequalities describing the cross section. (Each unit in the coordinate system represents 1 foot.)

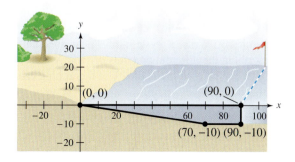

64. *Geometry* The figure shows the chorus platform on a stage. Write a system of linear inequalities describing the part of the audience that can see the full chorus. (Each unit in the coordinate system represents 1 meter.)

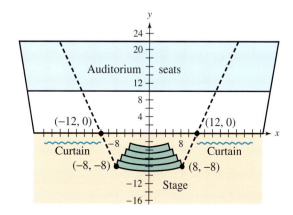

65. ▦ *Nutrition* A dietitian is asked to design a special diet supplement using two different foods. Each ounce of food X contains 20 units of calcium, 15 units of iron, and 10 units of vitamin B. Each ounce of food Y contains 10 units of calcium, 10 units of iron, and 20 units of vitamin B. The minimum daily requirements in the diet are 280 units of calcium, 160 units of iron, and 180 units of vitamin B. Write a system of linear inequalities describing the different amounts of food X and food Y that can be used in the diet. Use a graphing calculator to graph the system.

66. ▦ *Nutrition* A veterinarian is asked to design a special canine dietary supplement using two different dog foods. Each ounce of food X contains 12 units of calcium, 8 units of iron, and 6 units of protein. Each ounce of food Y contains 10 units of calcium, 10 units of iron, and 8 units of protein. The minimum daily requirements of the diet are 200 units of calcium, 100 units of iron, and 120 units of protein. Write a system of linear inequalities describing the different amounts of dog food X and dog food Y that can be used. Use a graphing calculator to graph the system.

Explaining Concepts

67. *Writing* Explain the meaning of the term *half-plane*. Give an example of an inequality whose graph is a half-plane.

68. *Writing* Explain how you can check any single point (x_1, y_1) to determine whether the point is a solution of a system of linear inequalities.

69. *Writing* Explain how to determine the vertices of the solution region for a system of linear inequalities.

70. *Writing* Describe the difference between the solution set of a system of linear equations and the solution set of a system of linear inequalities.

What Did You Learn?

Key Terms

system of equations, *p. 218*

solution of a system of
    equations, *p. 218*

points of intersection, *p. 219*

consistent system, *p. 219*

dependent system, *p. 219*

inconsistent system, *p. 219*

row-echelon form, *p. 241*

equivalent systems, *p. 242*

Gaussian elimination, *p. 242*

row operations, *p. 242*

matrix, *p. 254*

order (of a matrix), *p. 254*

square matrix, *p. 254*

augmented matrix, *p. 255*

coefficient matrix, *p. 255*

row-equivalent matrices, *p. 256*

minor (of an entry), *p. 268*

Cramer's Rule, *p. 270*

system of linear
    inequalities, *p. 279*

solution of a system of
    linear inequalities, *p. 279*

vertex, *p. 280*

Key Concepts

4.1 ⦿ The method of substitution

1. Solve one of the equations for one variable in terms of the other.

2. Substitute the expression obtained in Step 1 in the other equation to obtain an equation in one variable.

3. Solve the equation obtained in Step 2.

4. Back-substitute the solution from Step 3 in the expression obtained in Step 1 to find the value of the other variable.

5. Check the solution to see that it satisfies both of the original equations.

4.2 ⦿ The method of elimination

1. Obtain coefficients for x (or y) that differ only in sign by multiplying all terms of one or both equations by suitable constants.

2. Add the equations to eliminate one variable, and solve the resulting equation.

3. Back-substitute the value obtained in Step 2 in either of the original equations and solve for the other variable.

4. Check your solution in both of the original equations.

4.4 ⦿ Elementary row operations

Two matrices are row-equivalent if one can be obtained from the other by a sequence of elementary row operations.

1. Interchange two rows.

2. Multiply a row by a nonzero constant.

3. Add a multiple of a row to another row.

4.4 ⦿ Gaussian elimination with back-substitution

To use matrices and Gaussian elimination to solve a system of linear equations, use the following steps.

1. Write the augmented matrix of the system of linear equations.

2. Use elementary row operations to rewrite the augmented matrix in row-echelon form.

3. Write the system of linear equations corresponding to the matrix in row-echelon form, and use back-substitution to find the solution.

4.5 ⦿ Determinant of a 2 x 2 matrix

$$\det(A) = |A| = \begin{vmatrix} a_1 & b_1 \\ a_2 & b_2 \end{vmatrix}$$
$$= a_1 b_2 - a_2 b_1$$

4.5 ⦿ Expanding by minors

The determinant of a 3×3 matrix can be evaluated by expanding by minors.

$$\det(A) = \begin{vmatrix} a_1 & b_1 & c_1 \\ a_2 & b_2 & c_2 \\ a_3 & b_3 & c_3 \end{vmatrix}$$
$$= a_1 \begin{vmatrix} b_2 & c_2 \\ b_3 & c_3 \end{vmatrix} - b_1 \begin{vmatrix} a_2 & c_2 \\ a_3 & c_3 \end{vmatrix} + c_1 \begin{vmatrix} a_2 & b_2 \\ a_3 & b_3 \end{vmatrix}$$

4.6 ⦿ Graphing a system of linear inequalities

1. Sketch the line that corresponds to each inequality.

2. Lightly shade the half-plane that is the graph of each linear inequality.

3. The graph of the system is the intersection of the half-planes.

Review Exercises

4.1 Systems of Equations

1 Determine if ordered pairs are solutions of systems of equations.

In Exercises 1–4, determine whether each ordered pair is a solution of the system of equations.

1. $\begin{cases} 3x + 7y = 2 \\ 5x + 6y = 9 \end{cases}$

(a) $(3, 4)$

(b) $(3, -1)$

2. $\begin{cases} -2x + 5y = 21 \\ 9x - y = 13 \end{cases}$

(a) $(2, 5)$

(b) $(-2, 4)$

3. $\begin{cases} 26x + 13y = 26 \\ 20x + 10y = 30 \end{cases}$

(a) $(4, -5)$

(b) $(7, 12)$

4. $\begin{cases} 12x + y = 32 \\ 2x - 5y = 26 \end{cases}$

(a) $(3, -4)$

(b) $(2, 8)$

2 Solve systems of equations graphically.

In Exercises 5–14, use the graphical method to solve the system of equations.

5. $\begin{cases} x + y = 2 \\ x - y = 0 \end{cases}$

6. $\begin{cases} 2x = 3(y - 1) \\ y = x \end{cases}$

7. $\begin{cases} x - y = 3 \\ -x + y = 1 \end{cases}$

8. $\begin{cases} x + y = -1 \\ 3x + 2y = 0 \end{cases}$

9. $\begin{cases} 2x - y = 0 \\ -x + y = 4 \end{cases}$

10. $\begin{cases} x = y + 3 \\ x = y + 1 \end{cases}$

11. $\begin{cases} 2x + y = 4 \\ -4x - 2y = -8 \end{cases}$

12. $\begin{cases} 3x - 2y = 6 \\ -6x + 4y = 12 \end{cases}$

13. $\begin{cases} 3x - 2y = -2 \\ -5x + 2y = 2 \end{cases}$

14. $\begin{cases} 2x - y = 4 \\ -3x + 4y = -11 \end{cases}$

In Exercises 15 and 16, use a graphing calculator to graph the equations and approximate any solutions of the system of equations.

15. $\begin{cases} 5x - 3y = 3 \\ 2x + 2y = 14 \end{cases}$

16. $\begin{cases} 0.2x - 1.2y = 10 \\ 2x - 3y = 4.5 \end{cases}$

3 Solve systems of equations algebraically using the method of substitution.

In Exercises 17–26, solve the system of equations by the method of substitution.

17. $\begin{cases} 2x + 3y = 1 \\ x + 4y = -2 \end{cases}$

18. $\begin{cases} 3x - 7y = 10 \\ -2x + y = -14 \end{cases}$

19. $\begin{cases} -5x + 2y = 4 \\ 10x - 4y = 7 \end{cases}$

20. $\begin{cases} 5x + 2y = 3 \\ 2x + 3y = 10 \end{cases}$

21. $\begin{cases} 3x - 7y = 5 \\ 5x - 9y = -5 \end{cases}$

22. $\begin{cases} 24x - 4y = 20 \\ 6x - y = 5 \end{cases}$

23. $\begin{cases} -x + y = 6 \\ 15x + y = -10 \end{cases}$

24. $\begin{cases} -3x + 2y = 5 \\ x + y = -4 \end{cases}$

25. $\begin{cases} -3x - 3y = 3 \\ x + y = -1 \end{cases}$

26. $\begin{cases} x + y = 9 \\ x + y = 0 \end{cases}$

4 Use systems of equations to model and solve real-life problems.

27. *Break-Even Analysis* A small business invests $25,000 in equipment to produce a one-time-use camera. Each camera costs $3.75 to produce and is sold for $5.25. How many one-time-use cameras must be sold before the business breaks even?

28. *Seed Mixture* Fifteen pounds of mixed birdseed sells for $8.85 per pound. The mixture is obtained from two kinds of birdseed, with one variety priced at $7.05 per pound and the other at $9.30 per pound. How many pounds of each variety of birdseed are used in the mixture?

4.2 Linear Systems in Two Variables

① Solve systems of linear equations algebraically using the method of elimination.

In Exercises 29–36, solve the system of linear equations by the method of elimination.

29. $\begin{cases} x + y = 0 \\ 2x + y = 0 \end{cases}$ 30. $\begin{cases} 4x + y = 1 \\ x - y = 4 \end{cases}$

31. $\begin{cases} 2x - y = 2 \\ 6x + 8y = 39 \end{cases}$ 32. $\begin{cases} 3x + 2y = 11 \\ x - 3y = -11 \end{cases}$

33. $\begin{cases} 4x + y = -3 \\ -4x + 3y = 23 \end{cases}$ 34. $\begin{cases} -3x + 5y = -23 \\ 2x - 5y = 22 \end{cases}$

35. $\begin{cases} 0.2x + 0.3y = 0.14 \\ 0.4x + 0.5y = 0.20 \end{cases}$

36. $\begin{cases} 0.1x + 0.5y = -0.17 \\ -0.3x - 0.2y = -0.01 \end{cases}$

② Use systems of linear equations to model and solve real-life problems.

37. *Acid Mixture*   One hundred gallons of a 60% acid solution is obtained by mixing a 75% solution with a 50% solution. How many gallons of each solution must be used to obtain the desired mixture?

38. *Alcohol Mixture*   Fifty gallons of a 90% alcohol solution is obtained by mixing a 100% solution with a 75% solution. How many gallons of each solution must be used to obtain the desired mixture?

39. *Average Speed*   A bus travels for 3 hours at an average speed of 50 miles per hour. How much longer must the bus travel at an average speed of 60 miles per hour so that the average speed for the total trip will be 52 miles per hour?

40. *Employment*   The numbers of people (in thousands) employed in the cellular telephone industry in the United States from 1998 to 2001 are shown in the table. (Source: Cellular Telecommunications Industry Association)

Year	1998	1999	2000	2001
Employees, y	135	156	188	204

(a) Plot the data shown in the table. Let x represent the year, with $x = 8$ corresponding to 1998.

(b) The line $y = mx + b$ that best fits the data is given by the solution of the system below. Solve the system and find the equation of the best-fitting line. Sketch the graph of the line on the same set of coordinate axes used in part (a).

$\begin{cases} 4b + 38m = 683 \\ 38b + 366m = 6608 \end{cases}$

(c) Interpret the meaning of the slope of the line in the context of this problem.

4.3 Linear Systems in Three Variables

① Solve systems of linear equations using row-echelon form with back-substitution.

In Exercises 41–44, use back-substitution to solve the system of linear equations.

41. $\begin{cases} x = 3 \\ x + 2y = 7 \\ -3x - y + 4z = 9 \end{cases}$ 42. $\begin{cases} 2x + 3y = 9 \\ 4x - 6z = 12 \\ y = 5 \end{cases}$

43. $\begin{cases} x + 2y = 6 \\ 3y = 9 \\ x + 2z = 12 \end{cases}$ 44. $\begin{cases} 5x - 6z = -17 \\ 3x - 4y + 5z = -1 \\ 2z = -6 \end{cases}$

② Solve systems of linear equations using the method of Gaussian elimination.

In Exercises 45–48, solve the system of linear equations.

45. $\begin{cases} -x + y + 2z = 1 \\ 2x + 3y + z = -2 \\ 5x + 4y + 2z = 4 \end{cases}$

46. $\begin{cases} 2x + 3y + z = 10 \\ 2x - 3y - 3z = 22 \\ 4x - 2y + 3z = -2 \end{cases}$

47. $\begin{cases} x - y - z = 1 \\ -2x + y + 3z = -5 \\ 3x + 4y - z = 6 \end{cases}$

48. $\begin{cases} -3x + y + 2z = -13 \\ -x - y + z = 0 \\ 2x + 2y - 3z = -1 \end{cases}$

③ Solve application problems using elimination with back-substitution.

49. *Investment* An inheritance of $20,000 is divided among three investments yielding a total of $1780 in interest per year. The interest rates for the three investments are 7%, 9%, and 11%. The amounts invested at 9% and 11% are $3000 and $1000 less than the amount invested at 7%, respectively. Find the amount invested at each rate.

50. *Vertical Motion* Find the position equation

$$s = \frac{1}{2}at^2 + v_0t + s_0$$

for an object that has the indicated heights at the specified times.

$s = 192$ feet at $t = 1$ second

$s = 128$ feet at $t = 2$ seconds

$s = 80$ feet at $t = 3$ seconds

4.4 Matrices and Linear Systems

① Determine the order of matrices.

In Exercises 51–54, determine the order of the matrix.

51. $\begin{bmatrix} 4 & -5 \end{bmatrix}$

52. $\begin{bmatrix} 1 & 5 \\ 3 & -4 \end{bmatrix}$

53. $\begin{bmatrix} 5 & 7 & 9 \\ 11 & -12 & 0 \end{bmatrix}$

54. $\begin{bmatrix} 15 \\ 13 \\ -9 \end{bmatrix}$

② Form coefficient and augmented matrices and form linear systems from augmented matrices.

In Exercises 55 and 56, form (a) the coefficient matrix and (b) the augmented matrix for the system of linear equations.

55. $\begin{cases} 3x - 2y = 12 \\ -x + y = -2 \end{cases}$

56. $\begin{cases} x + 2y + z = 4 \\ 3x - z = 2 \\ -x + 5y - 2z = -6 \end{cases}$

In Exercises 57 and 58, write the system of linear equations represented by the matrix. (Use variables x, y, and z.)

57. $\begin{bmatrix} 4 & -1 & 0 & \vdots & 2 \\ 6 & 3 & 2 & \vdots & 1 \\ 0 & 1 & 4 & \vdots & 0 \end{bmatrix}$

58. $\begin{bmatrix} -15 & 2 & \vdots & -7 \\ 3 & 7 & \vdots & 8 \end{bmatrix}$

③ Perform elementary row operations to solve systems of linear equations.

In Exercises 59–62, use matrices and elementary row operations to solve the system.

59. $\begin{cases} 5x + 4y = 2 \\ -x + y = -22 \end{cases}$

60. $\begin{cases} 2x - 5y = 2 \\ 3x - 7y = 1 \end{cases}$

61. $\begin{cases} 0.2x - 0.1y = 0.07 \\ 0.4x - 0.5y = -0.01 \end{cases}$

62. $\begin{cases} 2x + y = 0.3 \\ 3x - y = -1.3 \end{cases}$

④ Use matrices and Gaussian elimination with back-substitution to solve systems of linear equations.

In Exercises 63–68, use matrices to solve the system of linear equations.

63. $\begin{cases} x + 2y + 6z = 4 \\ -3x + 2y - z = -4 \\ 4x + 2z = 16 \end{cases}$

64. $\begin{cases} -x + 3y - z = -4 \\ 2x + 6z = 14 \\ -3x - y + z = 10 \end{cases}$

65. $\begin{cases} 2x_1 + 3x_2 + 3x_3 = 3 \\ 6x_1 + 6x_2 + 12x_3 = 13 \\ 12x_1 + 9x_2 - x_3 = 2 \end{cases}$

66. $\begin{cases} -x_1 + 2x_2 + 3x_3 = 4 \\ 2x_1 - 4x_2 - x_3 = -13 \\ 3x_1 + 2x_2 - 4x_3 = -1 \end{cases}$

67. $\begin{cases} x - 4z = 17 \\ -2x + 4y + 3z = -14 \\ 5x - y + 2z = -3 \end{cases}$

68. $\begin{cases} 2x + 3y - 5z = 3 \\ -x + 2y = 3 \\ 3x + 5y + 2z = 15 \end{cases}$

4.5 Determinants and Linear Systems

1 Find determinants of 2 × 2 matrices and 3 × 3 matrices.

In Exercises 69–74, find the determinant of the matrix using any appropriate method.

69. $\begin{bmatrix} 7 & 10 \\ 10 & 15 \end{bmatrix}$ **70.** $\begin{bmatrix} -3.4 & 1.2 \\ -5 & 2.5 \end{bmatrix}$

71. $\begin{bmatrix} 8 & 6 & 3 \\ 6 & 3 & 0 \\ 3 & 0 & 2 \end{bmatrix}$ **72.** $\begin{bmatrix} 7 & -1 & 10 \\ -3 & 0 & -2 \\ 12 & 1 & 1 \end{bmatrix}$

73. $\begin{bmatrix} 8 & 3 & 2 \\ 1 & -2 & 4 \\ 6 & 0 & 5 \end{bmatrix}$ **74.** $\begin{bmatrix} 4 & 0 & 10 \\ 0 & 10 & 0 \\ 10 & 0 & 34 \end{bmatrix}$

2 Use determinants and Cramer's Rule to solve systems of linear equations.

In Exercises 75–78, use Cramer's Rule to solve the system of linear equations. (If not possible, state the reason.)

75. $\begin{cases} 7x + 12y = 63 \\ 2x + 3y = 15 \end{cases}$ **76.** $\begin{cases} 12x + 42y = -17 \\ 30x - 18y = 19 \end{cases}$

77. $\begin{cases} -x + y + 2z = 1 \\ 2x + 3y + z = -2 \\ 5x + 4y + 2z = 4 \end{cases}$ **78.** $\begin{cases} 2x_1 + x_2 + 2x_3 = 4 \\ 2x_1 + 2x_2 = 5 \\ 2x_1 - x_2 + 6x_3 = 2 \end{cases}$

3 Use determinants to find areas of triangles, to test for collinear points, and to find equations of lines.

Area of a Triangle   In Exercises 79–82, use a determinant to find the area of the triangle with the given vertices.

79. $(1, 0), (5, 0), (5, 8)$

80. $(-4, 0), (4, 0), (0, 6)$

81. $(1, 2), (4, -5), (3, 2)$

82. $\left(\frac{3}{2}, 1\right), \left(4, -\frac{1}{2}\right), (4, 2)$

Collinear Points   In Exercises 83 and 84, determine whether the points are collinear.

83. $(1, 2), (5, 0), (10, -2)$

84. $(-3, 7), (1, 3), (5, -1)$

Equation of a Line   In Exercises 85 and 86, use a determinant to find the equation of the line through the points.

85. $(-4, 0), (4, 4)$ **86.** $\left(-\frac{5}{2}, 3\right), \left(\frac{7}{2}, 1\right)$

4.6 Systems of Linear Inequalities

1 Solve systems of linear inequalities in two variables.

In Exercises 87–90, sketch a graph of the solution of the system of linear inequalities.

87. $\begin{cases} x + y < 5 \\ x > 2 \\ y \geq 0 \end{cases}$ **88.** $\begin{cases} 2x + y > 2 \\ x < 2 \\ y < 1 \end{cases}$

89. $\begin{cases} x + 2y \leq 160 \\ 3x + y \leq 180 \\ x \geq 0 \\ y \geq 0 \end{cases}$ **90.** $\begin{cases} 2x + 3y \leq 24 \\ 2x + y \leq 16 \\ x \geq 0 \\ y \geq 0 \end{cases}$

2 Use systems of linear inequalities to model and solve real-life problems.

91. *Fruit Distribution*   A Pennsylvania fruit grower has up to 1500 bushels of apples that are to be divided between markets in Harrisburg and Philadelphia. These two markets need at least 400 bushels and 600 bushels, respectively. Write a system of linear inequalities describing the various ways the fruit can be divided between the cities. Graph the system.

92. *Inventory Costs*   A warehouse operator has up to 24,000 square feet of floor space in which to store two products. Each unit of product I requires 20 square feet of floor space and costs $12 per day to store. Each unit of product II requires 30 square feet of floor space and costs $8 per day to store. The total storage cost per day cannot exceed $12,400. Write a system of linear inequalities describing the various ways the two products can be stored. Graph the system.

Chapter Test

Take this test as you would take a test in class. After you are done, check your work against the answers in the back of the book.

1. Determine whether each ordered pair is a solution of the system at the left.

    (a) $(3, -4)$
    (b) $\left(1, \frac{1}{2}\right)$

$$\begin{cases} 2x - 2y = 1 \\ -x + 2y = 0 \end{cases}$$

System for 1

In Exercises 2–11, use the indicated method to solve the system.

2. *Graphical:* $\begin{cases} x - 2y = -1 \\ 2x + 3y = 12 \end{cases}$

3. *Substitution:* $\begin{cases} 5x - y = 6 \\ 4x - 3y = -4 \end{cases}$

4. *Substitution:* $\begin{cases} 2x - 2y = -2 \\ 3x + y = 9 \end{cases}$

5. *Elimination:* $\begin{cases} 3x - 4y = -14 \\ -3x + y = 8 \end{cases}$

6. *Elimination:* $\begin{cases} x + 2y - 4z = 0 \\ 3x + y - 2z = 5 \\ 3x - y + 2z = 7 \end{cases}$

7. *Matrices:* $\begin{cases} x \qquad - 3z = -10 \\ -2y + 2z = 0 \\ x - 2y \qquad = -7 \end{cases}$

8. *Matrices:* $\begin{cases} x - 3y + z = -3 \\ 3x + 2y - 5z = 18 \\ y + z = -1 \end{cases}$

9. *Cramer's Rule:* $\begin{cases} 2x - 7y = 7 \\ 3x + 7y = 13 \end{cases}$

10. *Any Method:* $\begin{cases} 3x - 2y + z = 12 \\ x - 3y \qquad = 2 \\ -3x \qquad - 9z = -6 \end{cases}$

11. *Any Method:* $\begin{cases} 4x + y + 2z = -4 \\ 3y + z = 8 \\ -3x + y - 3z = 5 \end{cases}$

$$\begin{bmatrix} 3 & -2 & 0 \\ -1 & 5 & 3 \\ 2 & 7 & 1 \end{bmatrix}$$

Matrix for 12

12. Evaluate the determinant of the matrix shown at the left.

13. Use a determinant to find the area of the triangle with vertices $(0, 0)$, $(5, 4)$, and $(6, 0)$.

14. Graph the solution of the system of linear inequalities.

$$\begin{cases} x - 2y > -3 \\ 2x + 3y \le 22 \\ y \ge 0 \end{cases}$$

15. The perimeter of a rectangle is 68 feet and its width is $\frac{8}{9}$ times its length. Find the dimensions of the rectangle.

16. An inheritance of $25,000 is divided among three investments yielding a total of $1275 in interest per year. The interest rates for the three investments are 4.5%, 5%, and 8%. The amounts invested at 5% and 8% are $4000 and $10,000 less than the amount invested at 4.5%, respectively. Find the amount invested at each rate.

17. Two types of tickets are sold for a concert. Reserved seat tickets cost $20 per ticket and floor seat tickets cost $30 per ticket. The promoter of the concert must sell at least 16,000 tickets, including at least 9000 reserved seat tickets and at least 4000 floor seat tickets. Gross receipts must total at least $400,000 in order for the concert to be held. Write a system of linear inequalities describing the different numbers of tickets that can be sold. Graph the system.

Cumulative Test: Chapters 1–4

Take this test as you would take a test in class. After you are done, check your work against the answers in the back of the book.

1. Place the correct symbol ($<$, $>$, or $=$) between the two real numbers.

(a) -2 ▭ 5 (b) $\frac{1}{3}$ ▭ $\frac{1}{2}$ (c) $|2.3|$ ▭ $-|-4.5|$

2. Write an algebraic expression for the statement, "The number n is tripled and the product is decreased by 8."

In Exercises 3 and 4, perform the operations and simplify.

3. $t(3t - 1) - 2(t + 4)$

4. $3x(x^2 - 2) - 5(x^2 + 5)$

In Exercises 5–8, solve the equation or inequality.

5. $12 - 5(3 - x) = x + 3$

6. $1 - \dfrac{x + 2}{4} = \dfrac{7}{8}$

7. $|x - 2| \geq 3$

8. $-12 \leq 4x - 6 < 10$

9. Your annual automobile insurance premium is $1225. Because of a driving violation, your premium is increased by 15%. What is your new premium?

10. The triangles at the left are similar. Solve for x by using the fact that corresponding sides of similar triangles are proportional.

11. The revenue R from selling x units of a product is $R = 12.90x$. The cost C of producing x units is $C = 8.50x + 450$. To obtain a profit, the revenue must be greater than the cost. For what values of x will this product produce a profit? Explain your reasoning.

12. Does the equation $x - y^3 = 0$ represent y as a function of x?

13. Find the domain of the function $f(x) = \sqrt{x - 2}$.

14. Given $f(x) = x^2 - 3x$, find (a) $f(4)$ and (b) $f(c + 3)$.

15. Find the slope of the line passing through $(-4, 0)$ and $(4, 6)$. Then find the distance between the points and the midpoint of the line segment joining the points.

16. Determine the equations of lines through the point $(-2, 1)$ (a) parallel to $2x - y = 1$ and (b) perpendicular to $3x + 2y = 5$.

In Exercises 17 and 18, graph the equation.

17. $4x + 3y - 12 = 0$

18. $y = 1 - (x - 2)^2$

In Exercises 19–21, use the indicated method to solve the system.

19. *Substitution:*

$$\begin{cases} x + y = 6 \\ 2x - y = 3 \end{cases}$$

20. *Elimination:*

$$\begin{cases} 2x + y = 6 \\ 3x - 2y = 16 \end{cases}$$

21. *Matrices:*

$$\begin{cases} 2x + y - 2z = 1 \\ x \quad\quad - z = 1 \\ 3x + 3y + z = 12 \end{cases}$$

9

4.5

13

x

Figure for 10

295

Motivating the Chapter

A Storage Bin for Drying Grain

A rectangular grain bin has a width that is 5 feet greater than its height and a length that is 2 feet less than three times its height. A screen in the shape of a pyramid is centered at the base of the bin (see figure). Air is pumped into the bin through the screen pyramid in order to dry the grain. The screen pyramid has a height that is 3 feet less than the height of the bin, a width that is 6 feet less than the width of the bin, and a length that is twice the height of the pyramid.

See Section 5.3, Exercise 115.

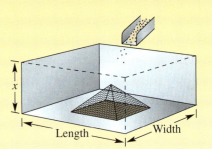

a. Write the dimensions, in feet, of the bin in terms of its height x. Write a polynomial function $V_B(x)$ that represents the volume of the rectangular grain bin before the screen pyramid is inserted.

b. Write the dimensions, in feet, of the screen pyramid in terms of x. Write a polynomial function $V_p(x)$ that represents the volume of the screen pyramid. [The formula for the volume of a pyramid is $V = \frac{1}{3}$ (area of base)(height).]

c. Write a polynomial function $V_s(x)$ that represents the volume of grain that can be stored in the bin when the screen pyramid is in place.

See Section 5.6, Exercise 113.

d. The opening for airflow (the base of the screen pyramid) must be 30 square feet. Find the dimensions of the pyramid.

e. Under the conditions of part (d), calculate the maximum volume of grain that can be dried in this bin.

f. Under the conditions of part (d), what is the domain of the volume function $V_s(x)$ from part (c)?

Richard A. Cooke/Corbis

Polynomials and Factoring

5.1 ● Integer Exponents and Scientific Notation
5.2 ● Adding and Subtracting Polynomials
5.3 ● Multiplying Polynomials
5.4 ● Factoring by Grouping and Special Forms
5.5 ● Factoring Trinomials
5.6 ● Solving Polynomial Equations by Factoring

5.1 Integer Exponents and Scientific Notation

What You Should Learn

1. Use the rules of exponents to simplify expressions.
2. Rewrite exponential expressions involving negative and zero exponents.
3. Write very large and very small numbers in scientific notation.

Why You Should Learn It

Scientific notation can be used to represent very large real-life quantities. For instance, in Exercise 140 on page 307, you will use scientific notation to represent the average amount of poultry produced per person.

1. Use the rules of exponents to simplify expressions.

Rules of Exponents

Recall from Section 1.2 that *repeated multiplication* can be written in what is called **exponential form.** Let n be a positive integer and let a be a real number. Then the product of n factors of a is given by

$$a^n = \underbrace{a \cdot a \cdot a \cdot \cdots a}_{n \text{ factors}}. \qquad a \text{ is the base and } n \text{ is the exponent.}$$

When multiplying two exponential expressions that have the *same base*, you add exponents. To see why this is true, consider the product $a^3 \cdot a^2$. Because the first expression represents three factors of a and the second represents two factors of a, the product of the two expressions represents five factors of a, as follows.

$$a^3 \cdot a^2 = \underbrace{(a \cdot a \cdot a)}_{3 \text{ factors}} \cdot \underbrace{(a \cdot a)}_{2 \text{ factors}} = \underbrace{(a \cdot a \cdot a \cdot a \cdot a)}_{5 \text{ factors}} = a^{3+2} = a^5$$

Rules of Exponents

Let m and n be positive integers, and let a and b represent real numbers, variables, or algebraic expressions.

Rule	Example
1. Product: $a^m \cdot a^n = a^{m+n}$	$x^5(x^4) = x^{5+4} = x^9$
2. Product-to-Power: $(ab)^m = a^m \cdot b^m$	$(2x)^3 = 2^3(x^3) = 8x^3$
3. Power-to-Power: $(a^m)^n = a^{mn}$	$(x^2)^3 = x^{2 \cdot 3} = x^6$
4. Quotient: $\dfrac{a^m}{a^n} = a^{m-n}, m > n, a \neq 0$	$\dfrac{x^5}{x^3} = x^{5-3} = x^2, x \neq 0$
5. Quotient-to-Power: $\left(\dfrac{a}{b}\right)^m = \dfrac{a^m}{b^m}, b \neq 0$	$\left(\dfrac{x}{4}\right)^2 = \dfrac{x^2}{4^2} = \dfrac{x^2}{16}$

The product rule and the product-to-power rule can be extended to three or more factors. For example,

$$a^m \cdot a^n \cdot a^k = a^{m+n+k} \quad \text{and} \quad (abc)^m = a^m b^m c^m.$$

Example 1 Using Rules of Exponents

Simplify: **a.** $(x^2y^4)(3x)$ **b.** $-2(y^2)^3$ **c.** $(-2y^2)^3$ **d.** $(3x^2)(-5x)^3$

Solution

a. $(x^2y^4)(3x) = 3(x^2 \cdot x)(y^4) = 3(x^{2+1})(y^4) = 3x^3y^4$

b. $-2(y^2)^3 = (-2)(y^{2 \cdot 3}) = -2y^6$

c. $(-2y^2)^3 = (-2)^3(y^2)^3 = -8(y^{2 \cdot 3}) = -8y^6$

d. $(3x^2)(-5x)^3 = 3(-5)^3(x^2 \cdot x^3) = 3(-125)(x^{2+3}) = -375x^5$

Example 2 Using Rules of Exponents

Simplify: **a.** $\dfrac{14a^5b^3}{7a^2b^2}$ **b.** $\left(\dfrac{x^2}{2y}\right)^3$ **c.** $\dfrac{x^n y^{3n}}{x^2 y^4}$ **d.** $\dfrac{(2a^2b^3)^2}{a^3b^2}$

Solution

a. $\dfrac{14a^5b^3}{7a^2b^2} = 2(a^{5-2})(b^{3-2}) = 2a^3b$

b. $\left(\dfrac{x^2}{2y}\right)^3 = \dfrac{(x^2)^3}{(2y)^3} = \dfrac{x^{2 \cdot 3}}{2^3 y^3} = \dfrac{x^6}{8y^3}$

c. $\dfrac{x^n y^{3n}}{x^2 y^4} = x^{n-2} y^{3n-4}$

d. $\dfrac{(2a^2b^3)^2}{a^3b^2} = \dfrac{2^2(a^{2 \cdot 2})(b^{3 \cdot 2})}{a^3b^2} = \dfrac{4a^4b^6}{a^3b^2} = 4(a^{4-3})(b^{6-2}) = 4ab^4$

Integer Exponents

2 Rewrite exponential expressions involving negative and zero exponents.

The definition of an exponent can be extended to include zero and negative integers. If a is a real number such that $a \neq 0$, then a^0 is defined as 1. Moreover, if m is an integer, then a^{-m} is defined as the reciprocal of a^m.

Definitions of Zero Exponents and Negative Exponents

Let a and b be real numbers such that $a \neq 0$ and $b \neq 0$, and let m be an integer.

1. $a^0 = 1$ **2.** $a^{-m} = \dfrac{1}{a^m}$ **3.** $\left(\dfrac{a}{b}\right)^{-m} = \left(\dfrac{b}{a}\right)^m$

These definitions are consistent with the rules of exponents given on page 298. For instance, consider the following.

$$x^0 \cdot x^m = x^{0+m} = x^m = 1 \cdot x^m$$

(x^0 is the same as 1)

Example 3   Zero Exponents and Negative Exponents

Rewrite each expression without using zero exponents or negative exponents.

a. 3^0 **b.** 3^{-2} **c.** $\left(\frac{3}{4}\right)^{-1}$

Solution

a. $3^0 = 1$ Definition of zero exponents

b. $3^{-2} = \dfrac{1}{3^2} = \dfrac{1}{9}$ Definition of negative exponents

c. $\left(\dfrac{3}{4}\right)^{-1} = \left(\dfrac{4}{3}\right)^1 = \dfrac{4}{3}$ Definition of negative exponents

> **Study Tip**
>
> Because the expression a^0 is equal to 1 for any real number *a such that* $a \neq 0$, zero cannot have a zero exponent. So, 0^0 is undefined.

The following rules are valid for all integer exponents, including integer exponents that are zero or negative. (The first five rules were listed on page 298.)

Summary of Rules of Exponents

Let m and n be integers, and let a and b represent real numbers, variables, or algebraic expressions. (All denominators and bases are nonzero.)

Product and Quotient Rules | *Example*

1. $a^m \cdot a^n = a^{m+n}$ $x^4(x^3) = x^{4+3} = x^7$

2. $\dfrac{a^m}{a^n} = a^{m-n}$ $\dfrac{x^3}{x} = x^{3-1} = x^2$

Power Rules

3. $(ab)^m = a^m \cdot b^m$ $(3x)^2 = 3^2(x^2) = 9x^2$

4. $(a^m)^n = a^{mn}$ $(x^3)^3 = x^{3 \cdot 3} = x^9$

5. $\left(\dfrac{a}{b}\right)^m = \dfrac{a^m}{b^m}$ $\left(\dfrac{x}{3}\right)^2 = \dfrac{x^2}{3^2} = \dfrac{x^2}{9}$

Zero and Negative Exponent Rules

6. $a^0 = 1$ $(x^2 + 1)^0 = 1$

7. $a^{-m} = \dfrac{1}{a^m}$ $x^{-2} = \dfrac{1}{x^2}$

8. $\left(\dfrac{a}{b}\right)^{-m} = \left(\dfrac{b}{a}\right)^m$ $\left(\dfrac{x}{3}\right)^{-2} = \left(\dfrac{3}{x}\right)^2 = \dfrac{3^2}{x^2} = \dfrac{9}{x^2}$

Example 4   Using Rules of Exponents

a. $2x^{-1} = 2(x^{-1}) = 2\left(\dfrac{1}{x}\right) = \dfrac{2}{x}$ Use negative exponent rule and simplify.

b. $(2x)^{-1} = \dfrac{1}{(2x)^1} = \dfrac{1}{2x}$ Use negative exponent rule and simplify.

Example 5   Using Rules of Exponents

Rewrite each expression using only positive exponents. (For each expression, assume that $x \neq 0$).

a. $\dfrac{3}{x^{-2}} = \dfrac{3}{\left(\dfrac{1}{x^2}\right)}$ Negative exponent rule

$\qquad\quad = 3\left(\dfrac{x^2}{1}\right)$ Invert divisor and multiply.

$\qquad\quad = 3x^2$ Simplify.

b. $\dfrac{1}{(3x)^{-2}} = \dfrac{1}{\left[\dfrac{1}{(3x)^2}\right]}$ Use negative exponent rule.

$\qquad\quad = \dfrac{1}{\left(\dfrac{1}{(9x^2)}\right)}$ Use product-to-power rule and simplify.

$\qquad\quad = (1)\left(\dfrac{9x^2}{1}\right)$ Invert divisor and multiply.

$\qquad\quad = 9x^2$ Simplify.

Example 6   Using Rules of Exponents

Rewrite each expression using only positive exponents. (For each expression, assume that $x \neq 0$ and $y \neq 0$.)

a. $(-5x^{-3})^2 = (-5)^2(x^{-3})^2$ Product-to-power rule

$\qquad\qquad\quad = 25x^{-6}$ Power-to-product rule

$\qquad\qquad\quad = \dfrac{25}{x^6}$ Negative exponent rule

b. $-\left(\dfrac{7x}{y^2}\right)^{-2} = -\left(\dfrac{y^2}{7x}\right)^2$ Negative exponent rule

$\qquad\qquad\quad = -\dfrac{(y^2)^2}{(7x)^2}$ Quotient-to-power rule

$\qquad\qquad\quad = -\dfrac{y^4}{49x^2}$ Power-to-power and product-to-power rules

c. $\dfrac{12x^2y^{-4}}{6x^{-1}y^2} = 2(x^{2-(-1)})(y^{-4-2})$ Quotient rule

$\qquad\qquad\quad = 2x^3y^{-6}$ Simplify.

$\qquad\qquad\quad = \dfrac{2x^3}{y^6}$ Negative exponent rule

Example 7   Using Rules of Exponents

Rewrite each expression using only positive exponents. (For each expression, assume that $x \neq 0$ and $y \neq 0$.)

a. $\left(\dfrac{8x^{-1}y^4}{4x^3y^2}\right)^{-3} = \left(\dfrac{2y^2}{x^4}\right)^{-3}$ Simplify.

$= \left(\dfrac{x^4}{2y^2}\right)^{3}$ Negative exponent rule

$= \dfrac{x^{12}}{2^3y^6}$ Quotient-to-power rule

$= \dfrac{x^{12}}{8y^6}$ Simplify.

b. $\dfrac{3xy^0}{x^2(5y)^0} = \dfrac{3x(1)}{x^2(1)} = \dfrac{3}{x}$ Zero exponent rule

③ Write very large and very small numbers in scientific notation.

Scientific Notation

Exponents provide an efficient way of writing and computing with very large and very small numbers. For instance, a drop of water contains more than 33 billion billion molecules—that is, 33 followed by 18 zeros. It is convenient to write such numbers in **scientific notation.** This notation has the form $c \times 10^n$, where $1 \leq c < 10$ and n is an integer. So, the number of molecules in a drop of water can be written in scientific notation as follows.

$$33{,}000{,}000{,}000{,}000{,}000{,}000 = 3.3 \times 10^{19}$$

19 places

The *positive* exponent 19 indicates that the number being written in scientific notation is *large* (10 or more) and that the decimal point has been moved 19 places. A *negative* exponent in scientific notation indicates that the number is *small* (less than 1).

Example 8   Writing Scientific Notation

Write each number in scientific notation.

a. 0.0000684   **b.** 937,200,000

Solution

a. $0.0000684 = 6.84 \times 10^{-5}$ Small number ⟹ negative exponent

Five places

b. $937{,}200{,}000.0 = 9.372 \times 10^8$ Large number ⟹ positive exponent

Eight places

Example 9   Writing Decimal Notation

Write each number in decimal notation.

a. 2.486×10^2 **b.** 1.81×10^{-6}

Solution

a. $2.486 \times 10^2 = 248.6$ Positive exponent ⟹ large number

Two places

b. $1.81 \times 10^{-6} = 0.00000181$ Negative exponent ⟹ small number

Six places

Example 10   Using Scientific Notation

Rewrite the factors in scientific notation and then evaluate

$$\frac{(2,400,000,000)(0.0000045)}{(0.00003)(1500)}.$$

Solution

$$\frac{(2,400,000,000)(0.0000045)}{(0.00003)(1500)} = \frac{(2.4 \times 10^9)(4.5 \times 10^{-6})}{(3.0 \times 10^{-5})(1.5 \times 10^3)}$$

$$= \frac{(2.4)(4.5)(10^3)}{(4.5)(10^{-2})}$$

$$= (2.4)(10^5)$$

$$= 240,000$$

Technology: Tip

Most scientific and graphing calculators automatically switch to scientific notation when they are showing large or small numbers that exceed the display range.

To *enter* numbers in scientific notation, your calculator should have an exponential entry key labeled [EE] or [EXP]. Consult the user's guide of your calculator for instructions on keystrokes and how numbers in scientific notation are displayed.

Example 11   Using Scientific Notation with a Calculator

Use a calculator to evaluate each expression.

a. $65,000 \times 3,400,000,000$ **b.** $0.000000348 \div 870$

Solution

a. 6.5 [EXP] 4 [×] 3.4 [EXP] 9 [=]   Scientific

6.5 [EE] 4 [×] 3.4 [EE] 9 [ENTER]   Graphing

The calculator display should read [2.21E 14], which implies that

$$(6.5 \times 10^4)(3.4 \times 10^9) = 2.21 \times 10^{14} = 221,000,000,000,000.$$

b. 3.48 [EXP] 7 [+/−] [÷] 8.7 [EXP] 2 [=]   Scientific

3.48 [EE] [(−)] 7 [÷] 8.7 [EE] 2 [ENTER]   Graphing

The calculator display should read [4E −10], which implies that

$$\frac{3.48 \times 10^{-7}}{8.7 \times 10^2} = 4.0 \times 10^{-10} = 0.0000000004.$$

5.1 Exercises

Review   Concepts, Skills, and Problem Solving

Keep mathematically in shape by doing these exercises *before* the problems of this section.

Properties and Definitions

1. *Writing* In your own words, describe the graph of an equation.

2. *Writing* Describe the point-plotting method of graphing an equation.

3. Find the coordinates of two points on the graph of $g(x) = \sqrt{x - 2}$.

4. *Writing* Describe the procedure for finding the x- and y-intercepts of the graph of an equation.

Evaluating Functions

In Exercises 5–8, evaluate the function as indicated, and simplify.

5. $f(x) = 3x - 9$
   (a) $f(-2)$
   (b) $f\left(\frac{1}{2}\right)$

6. $f(x) = x^2 + x$
   (a) $f(4)$
   (b) $f(-2)$

7. $f(x) = 6x - x^2$
   (a) $f(0)$ (b) $f(t + 1)$

8. $f(x) = \dfrac{x - 2}{x + 2}$
   (a) $f(10)$ (b) $f(4 - z)$

Graphing Equations

In Exercises 9–12, use a graphing calculator to graph the function. Identify any intercepts.

9. $f(x) = 5 - 2x$

10. $h(x) = \frac{1}{2}x + |x|$

11. $g(x) = x^2 - 4x$

12. $f(x) = 2\sqrt{x + 1}$

Developing Skills

In Exercises 1–20, use the rules of exponents to simplify the expression (if possible). See Examples 1 and 2.

1. (a) $-3x^3 \cdot x^5$ (b) $(-3x)^2 \cdot x^5$

2. (a) $5^2 y^4 \cdot y^2$ (b) $(5y)^2 \cdot y^4$

3. (a) $(-5z^2)^3$ (b) $(-5z^4)^2$

4. (a) $(-5z^3)^2$ (b) $(-5z)^4$

5. (a) $(u^3 v)(2v^2)$ (b) $(-4u^4)(u^5 v)$

6. (a) $(6xy^7)(-x)$ (b) $(x^5 y^3)(2y^3)$

7. (a) $5u^2 \cdot (-3u^6)$ (b) $(2u)^4(4u)$

8. (a) $(3y)^3(2y^2)$ (b) $3y^3 \cdot 2y^2$

9. (a) $-(m^5 n)^3(-m^2 n^2)^2$ (b) $(-m^5 n)(m^2 n^2)$

10. (a) $-(m^3 n^2)(mn^3)$ (b) $-(m^3 n^2)^2(-mn^3)$

11. (a) $\dfrac{27m^5 n^6}{9mn^3}$ (b) $\dfrac{-18m^3 n^6}{-6mn^3}$

12. (a) $\dfrac{28x^2 y^3}{2xy^2}$ (b) $\dfrac{24xy^2}{8y}$

13. (a) $\left(\dfrac{3x}{4y}\right)^2$ (b) $\left(\dfrac{5u}{3v}\right)^3$

14. (a) $\left(\dfrac{2a}{3y}\right)^5$ (b) $-\left(\dfrac{2a}{3y}\right)^2$

15. (a) $-\dfrac{(-2x^2 y)^3}{9x^2 y^2}$ (b) $-\dfrac{(-2xy^3)^2}{6y^2}$

16. (a) $\dfrac{(-4xy)^3}{8xy^2}$ (b) $\dfrac{(-xy)^4}{-3(xy)^2}$

17. (a) $\left[\dfrac{(-5u^3 v)^2}{10u^2 v}\right]^2$ (b) $\left[\dfrac{-5(u^3 v)^2}{10u^2 v}\right]^2$

18. (a) $\left[\dfrac{(3x^2)(2x)^2}{(-2x)(6x)}\right]^2$ (b) $\left[\dfrac{(3x^2)(2x)^4}{(-2x)^2(6x)}\right]^2$

19. (a) $\dfrac{x^{2n+4}\,y^{4n}}{x^5\,y^{2n+1}}$ (b) $\dfrac{x^{6n}\,y^{n-7}}{x^{4n+2}\,y^5}$

20. (a) $\dfrac{x^{3n}\,y^{2n-1}}{x^n\,y^{n+3}}$ (b) $\dfrac{x^{4n-6}\,y^{n+10}}{x^{2n-5}\,y^{n-2}}$

In Exercises 21–50, evaluate the expression. See Example 3.

21. 5^{-2} **22.** 2^{-4}

23. -10^{-3} **24.** -20^{-2}

25. $(-3)^0$ **26.** 25^0

27. $\dfrac{1}{4^{-3}}$ **28.** $\dfrac{1}{-8^{-2}}$

29. $\dfrac{1}{(-2)^{-5}}$ **30.** $-\dfrac{1}{6^{-2}}$

31. $\left(\frac{2}{3}\right)^{-1}$ **32.** $\left(\frac{4}{5}\right)^{-3}$

33. $\left(\frac{3}{16}\right)^0$ **34.** $\left(-\frac{5}{8}\right)^{-2}$

35. $27 \cdot 3^{-3}$ **36.** $4^2 \cdot 4^{-3}$

37. $\dfrac{3^4}{3^{-2}}$ **38.** $\dfrac{5^{-1}}{5^2}$

39. $\dfrac{10^3}{10^{-2}}$ **40.** $\dfrac{10^{-5}}{10^{-6}}$

41. $(4^2 \cdot 4^{-1})^{-2}$

42. $(5^3 \cdot 5^{-4})^{-3}$

43. $(2^{-3})^2$

44. $(-4^{-1})^{-2}$

45. $2^{-3} + 2^{-4}$

46. $4 - 3^{-2}$

47. $\left(\frac{3}{4} + \frac{5}{8}\right)^{-2}$

48. $\left(\frac{1}{2} - \frac{2}{3}\right)^{-1}$

49. $(5^0 - 4^{-2})^{-1}$

50. $(32 + 4^{-3})^0$

In Exercises 51–90, rewrite the expression using only positive exponents, and simplify. (Assume that any variables in the expression are nonzero.) See Examples 4–7.

51. $y^4 \cdot y^{-2}$ **52.** $x^{-2} \cdot x^{-5}$

53. $z^5 \cdot z^{-3}$ **54.** $t^{-1} \cdot t^{-6}$

55. $7x^{-4}$ **56.** $3y^{-3}$

57. $(4x)^{-3}$

58. $(5u)^{-2}$

59. $\dfrac{1}{x^{-6}}$

60. $\dfrac{4}{y^{-1}}$

61. $\dfrac{8a^{-6}}{6a^{-7}}$

62. $\dfrac{6u^{-2}}{15u^{-1}}$

63. $\dfrac{(4t)^0}{t^{-2}}$

64. $\dfrac{(5u)^{-4}}{(5u)^0}$

65. $(2x^2)^{-2}$

66. $(4a^{-2})^{-3}$

67. $(-3x^{-3}y^2)(4x^2y^{-5})$

68. $(5s^5t^{-5})(-6s^{-2}t^4)$

69. $(3x^2y^{-2})^{-2}$

70. $(-4y^{-3}z)^{-3}$

71. $\left(\dfrac{x}{10}\right)^{-1}$

72. $\left(\dfrac{4}{z}\right)^{-2}$

73. $\dfrac{6x^3y^{-3}}{12x^{-2}y}$

74. $\dfrac{2y^{-1}z^{-3}}{4yz^{-3}}$

75. $\left(\dfrac{3u^2v^{-1}}{3^3u^{-1}v^3}\right)^{-2}$

76. $\left(\dfrac{5^2x^3y^{-3}}{125xy}\right)^{-1}$

77. $\left(\dfrac{a^{-2}}{b^{-2}}\right)\left(\dfrac{b}{a}\right)^3$

78. $\left(\dfrac{a^{-3}}{b^{-3}}\right)\left(\dfrac{b}{a}\right)^3$

79. $(2x^3y^{-1})^{-3}(4xy^{-6})$

80. $(ab)^{-2}(a^2b^2)^{-1}$

81. $u^4(6u^{-3}v^0)(7v)^0$

82. $x^5(3x^0y^4)(7y)^0$

83. $[(x^{-4}y^{-6})^{-1}]^2$

84. $[(2x^{-3}y^{-2})^2]^{-2}$

85. $\dfrac{(2a^{-2}b^4)^3b}{(10a^3b)^2}$

86. $\dfrac{(5x^2y^{-5})^{-1}}{2x^{-5}y^4}$

87. $(u + v^{-2})^{-1}$

88. $x^{-2}(x^2 + y^2)$

89. $\dfrac{a + b}{ba^{-1} - ab^{-1}}$

90. $\dfrac{u^{-1} - v^{-1}}{u^{-1} + v^{-1}}$

In Exercises 91–104, write the number in scientific notation. See Example 8.

91. 3,600,000

92. 98,100,000

93. 47,620,000

94. 956,300,000

95. 0.00031

96. 0.00625

97. 0.0000000381

98. 0.0007384

99. *Land Area of Earth*: 57,300,000 square miles

100. *Water Area of Earth*: 139,500,000 square miles

101. *Light Year*: 9,460,800,000,000 kilometers

102. *Thickness of Soap Bubble*: 0.0000001 meter

103. *Relative Density of Hydrogen*: 0.0899 grams per milliliter.

104. *One Micron (Millionth of Meter)*: 0.00003937 inch

In Exercises 105–114, write the number in decimal notation. See Example 9.

105. 6×10^7

106. 5.05×10^{12}

107. 1.359×10^{-7}

108. 8.6×10^{-9}

109. *2001 Merrill Lynch Revenues:* $\$3.8757 \times 10^{10}$
(Source: 2001 Merrill Lynch Annual Report)

110. *Number of Air Sacs in Lungs*: 3.5×10^8

111. *Interior Temperature of Sun*: 1.5×10^7 degrees Celsius

112. *Width of Air Molecule*: 9.0×10^{-9} meter

113. *Charge of Electron*: 4.8×10^{-10} electrostatic unit

114. *Width of Human Hair*: 9.0×10^{-4} meter

In Exercises 115–124, evaluate the expression without a calculator. See Example 10.

115. $(2 \times 10^9)(3.4 \times 10^{-4})$

116. $(6.5 \times 10^6)(2 \times 10^4)$

117. $(5 \times 10^4)^2$

118. $(4 \times 10^6)^3$

119. $\dfrac{3.6 \times 10^{12}}{6 \times 10^5}$

120. $\dfrac{2.5 \times 10^{-3}}{5 \times 10^2}$

121. $(4,500,000)(2,000,000,000)$

122. $(62,000,000)(0.0002)$

123. $\dfrac{64,000,000}{0.00004}$

124. $\dfrac{72,000,000,000}{0.00012}$

In Exercises 125–132, evaluate with a calculator. Write the answer in scientific notation, $c \times 10^n$, with c rounded to two decimal places. See Example 11.

125. $\dfrac{(0.0000565)(2,850,000,000,000)}{0.00465}$

126. $\dfrac{(3,450,000,000)(0.000125)}{(52,000,000)(0.000003)}$

127. $\dfrac{1.357 \times 10^{12}}{(4.2 \times 10^2)(6.87 \times 10^{-3})}$

128. $\dfrac{(3.82 \times 10^5)^2}{(8.5 \times 10^4)(5.2 \times 10^{-3})}$

129. $72{,}400 \times 2{,}300{,}000{,}000$

130. $(8.67 \times 10^4)^7$

131. $\dfrac{(5{,}000{,}000)^3(0.000037)^2}{(0.005)^4}$

132. $\dfrac{(6{,}200{,}000)(0.005)^3}{(0.00035)^5}$

Solving Problems

133. *Distance*   The distance from Earth to the sun is approximately 93 million miles. Write this distance in scientific notation.

134. *Electrons*   A cube of copper with an edge of 1 centimeter has approximately 8.483×10^{22} free electrons. Write this real number in decimal notation.

135. *Light Year*   One light year (the distance light can travel in 1 year) is approximately 9.46×10^{15} meters. Approximate the time (in minutes) for light to travel from the sun to Earth if that distance is approximately 1.50×10^{11} meters.

136. *Distance*   The star Alpha Andromeda is approximately 95 light years from Earth. Determine this distance in meters. (See Exercise 135 for the definition of a light year.)

137. *Masses of Earth and Sun*   The masses of Earth and the sun are approximately 5.98×10^{24} kilograms and 1.99×10^{30} kilograms, respectively. The mass of the sun is approximately how many times that of Earth?

138. *Metal Expansion*   When the temperature of an iron steam pipe 200 feet long is increased by 75°C, the length of the pipe will increase by an amount $75(200)(1.1 \times 10^{-5})$. Find this amount of increase in length.

139. *Federal Debt*   In July 2000, the estimated population of the United States was 275 million people, and the estimated federal debt was 5629 billion dollars. Use these two numbers to determine the amount each person would have to pay to eliminate the debt.   (Source: U.S. Census Bureau and U.S. Office of Management and Budget)

140. *Poultry Production*   In 2000, the estimated population of the world was 6 billion people, and the world-wide production of poultry meat was 58 million metric tons. Use these two numbers to determine the average amount of poultry produced per person in 2000.   (Source: U.S. Census Bureau and U.S. Department of Agriculture)

Explaining Concepts

141. In $(3x)^4$, what is $3x$ called? What is 4 called?

142. *Writing*   Discuss any differences between $(-2x)^{-4}$ and $-2x^{-4}$.

143. *Writing*   In your own words, describe how you can "move" a factor from the numerator to the denominator or vice versa.

144. *Writing*   Is the number 32.5×10^5 written in scientific notation? Explain.

145. *Writing*   When is scientific notation an efficient way of writing and computing real numbers?

5.2 Adding and Subtracting Polynomials

Herbert Hartmann/Getty Images

Why You Should Learn It

Polynomials can be used to model many aspects of the physical world. For instance, in Exercise 103 on page 316, a polynomial is used to model the height of an object thrown from the top of the Eiffel Tower.

What You Should Learn

1️⃣ Identify leading coefficients and degrees of polynomials.

2️⃣ Add and subtract polynomials using a horizontal format and a vertical format.

3️⃣ Use polynomials to model and solve real-life problems.

Basic Definitions

1️⃣ Identify leading coefficients and degrees of polynomials.

A **polynomial in x** is an algebraic expression whose terms are all of the form ax^k, where a is any real number and k is a nonnegative integer. The following *are not* polynomials for the reasons stated.

- The expression $2x^{-1} + 5$ is not a polynomial because the exponent in $2x^{-1}$ is negative.
- The expression $x^3 + 3x^{1/2}$ is not a polynomial because the exponent in $3x^{1/2}$ is not an integer.

Definition of a Polynomial in x

Let $a_n, a_{n-1}, \ldots, a_2, a_1, a_0$ be real numbers and let n be a *nonnegative integer*. A **polynomial in x** is an expression of the form

$$a_n x^n + a_{n-1} x^{n-1} + \cdots + a_2 x^2 + a_1 x + a_0$$

where $a_n \neq 0$. The polynomial is of **degree n,** and the number a_n is called the **leading coefficient.** The number a_0 is called the **constant term.**

In the term ax^k, a is the **coefficient** and k is the degree of the term. Note that the degree of the term ax is 1, and the degree of a constant term is 0. Because a polynomial is an algebraic sum, the coefficients take on the signs between the terms. For instance, the polynomial

$$x^3 - 4x^2 + 3 = (1)x^3 + (-4)x^2 + (0)x + 3$$

has coefficients 1, $-4, 0$, and 3. A polynomial that is written in order of descending powers of the variable is said to be in **standard form.** A polynomial with only one term is a **monomial.** Polynomials with two *unlike* terms are called **binomials,** and those with three *unlike* terms are called **trinomials.** Some examples of polynomials are shown below.

Polynomial	Standard Form	Degree	Leading Coefficient
$5x^2 - 2x^7 + 4 - 2x$	$-2x^7 + 5x^2 - 2x + 4$	7	-2
$16 + x^3$	$x^3 + 16$	3	1
10	10	0	10

② Add and subtract polynomials using a horizontal format and a vertical format.

Adding and Subtracting Polynomials

To add two polynomials, simply combine like terms. This can be done in either a horizontal or a vertical format, as shown in Examples 1 and 2.

Example 1   Adding Polynomials Horizontally

a. $(2x^3 + x^2 - 5) + (x^2 + x + 6)$ Original polynomials

$\quad = (2x^3) + (x^2 + x^2) + (x) + (-5 + 6)$ Group like terms.

$\quad = 2x^3 + 2x^2 + x + 1$ Combine like terms.

b. $(3x^2 + 2x + 4) + (3x^2 - 6x + 3) + (-x^2 + 2x - 4)$

$\quad = (3x^2 + 3x^2 - x^2) + (2x - 6x + 2x) + (4 + 3 - 4)$

$\quad = 5x^2 - 2x + 3$

Example 2   Using a Vertical Format to Add Polynomials

Use a vertical format to find the sum.

$\quad (5x^3 + 2x^2 - x + 7) + (3x^2 - 4x + 7) + (-x^3 + 4x^2 - 8)$

Solution

To use a vertical format, align the terms of the polynomials by their degrees.

$$
\begin{array}{r}
5x^3 + 2x^2 - x + 7 \\
3x^2 - 4x + 7 \\
-\ x^3 + 4x^2 - 8 \\
\hline
4x^3 + 9x^2 - 5x + 6
\end{array}
$$

To subtract one polynomial from another, *add the opposite*. You can do this by changing the sign of each term of the polynomial that is being subtracted and then adding the resulting like terms.

Example 3   Subtracting Polynomials Horizontally

Use a horizontal format to subtract $x^3 + 2x^2 - x - 4$ from $3x^3 - 5x^2 + 3$.

Solution

$\quad (3x^3 - 5x^2 + 3) - (x^3 + 2x^2 - x - 4)$ Write original polynomials.

$\quad = (3x^3 - 5x^2 + 3) + (-x^3 - 2x^2 + x + 4)$ Add the opposite.

$\quad = (3x^3 - x^3) + (-5x^2 - 2x^2) + (x) + (3 + 4)$ Group like terms.

$\quad = 2x^3 - 7x^2 + x + 7$ Combine like terms.

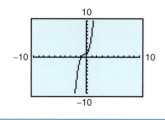

Be especially careful to use the correct signs when subtracting one polynomial from another. One of the most common mistakes in algebra is to forget to change signs correctly when subtracting one expression from another. Here is an example.

Wrong sign

$(x^2 - 2x + 3) - (x^2 + 2x - 2) \ne x^2 - 2x + 3 - x^2 + 2x - 2$ Common error

Wrong sign

The error illustrated above is forgetting to change two of the signs in the polynomial that is being subtracted. Be sure to add the opposite of *every* term of the subtracted polynomial.

Example 4   Using a Vertical Format to Subtract Polynomials

Use a vertical format to find the difference.

$$(4x^4 - 2x^3 + 5x^2 - x + 8) - (3x^4 - 2x^3 + 3x - 4)$$

Solution

$$
\begin{array}{l}
(4x^4 - 2x^3 + 5x^2 - x + 8) \\
-(3x^4 - 2x^3 \qquad\quad + 3x - 4)
\end{array}
\Longrightarrow
\begin{array}{l}
4x^4 - 2x^3 + 5x^2 - x + 8 \\
-3x^4 + 2x^3 \qquad\quad - 3x + 4 \\
\hline
x^4 \qquad\qquad + 5x^2 - 4x + 12
\end{array}
$$

Example 5   Combining Polynomials

Use a horizontal format to perform the indicated operations and simplify.

a. $(2x^2 - 7x + 2) - (4x^2 + 5x - 1) + (-x^2 + 4x + 4)$

b. $(-x^2 + 4x - 3) - [(4x^2 - 3x + 8) - (-x^2 + x + 7)]$

Solution

a. $(2x^2 - 7x + 2) - (4x^2 + 5x - 1) + (-x^2 + 4x + 4)$

$= 2x^2 - 7x + 2 - 4x^2 - 5x + 1 - x^2 + 4x + 4$

$= (2x^2 - 4x^2 - x^2) + (-7x - 5x + 4x) + (2 + 1 + 4)$

$= -3x^2 - 8x + 7$

b. $(-x^2 + 4x - 3) - [(4x^2 - 3x + 8) - (-x^2 + x + 7)]$

$= (-x^2 + 4x - 3) - (4x^2 - 3x + 8 + x^2 - x - 7)$

$= (-x^2 + 4x - 3) - [(4x^2 + x^2) + (-3x - x) + (8 - 7)]$

$= (-x^2 + 4x - 3) - (5x^2 - 4x + 1)$

$= -x^2 + 4x - 3 - 5x^2 + 4x - 1$

$= (-x^2 - 5x^2) + (4x + 4x) + (-3 - 1)$

$= -6x^2 + 8x - 4$

③ Use polynomials to model and solve real-life problems.

Applications

Function notation can be used to represent polynomials in a single variable. Such notation is useful for evaluating polynomials and is used in applications involving polynomials, as shown in Example 6.

There are many applications that involve polynomials. One commonly used second-degree polynomial is called a **position function.** This polynomial is a function of time and has the form

$$h(t) = -16t^2 + v_0t + s_0 \qquad \text{Position function}$$

where the height h is measured in feet and the time t is measured in seconds.

This position function gives the height (above ground) of a free-falling object. The coefficient of t, v_0, is called the **initial velocity** of the object, and the constant term, s_0, is called the **initial height** of the object. If the initial velocity is positive, the object was projected upward (at $t = 0$), if the initial velocity is negative, the object was projected downward, and if the initial velocity is zero, the object was dropped.

$t = 0$
$t = 1$
$t = 4$
1050 ft
$t = 7$

Figure 5.1

Example 6   Free-Falling Object

An object is thrown downward from the 86th floor observatory at the Empire State Building, which is 1050 feet high. The initial velocity is -15 feet per second. Use the position function

$$h(t) = -16t^2 - 15t + 1050$$

to find the height of the object when $t = 1$, $t = 4$, and $t = 7$. (See Figure 5.1.)

Solution

When $t = 1$, the height of the object is

$$h(1) = -16(1)^2 - 15(1) + 1050 \qquad \text{Substitute 1 for } t.$$
$$= -16 - 15 + 1050 \qquad \text{Simplify.}$$
$$= 1019 \text{ feet.}$$

When $t = 4$, the height of the object is

$$h(4) = -16(4)^2 - 15(4) + 1050 \qquad \text{Substitute 4 for } t.$$
$$= -256 - 60 + 1050 \qquad \text{Simplify.}$$
$$= 734 \text{ feet.}$$

When $t = 7$, the height of the object is

$$h(7) = -16(7)^2 - 15(7) + 1050 \qquad \text{Substitute 7 for } t.$$
$$= -784 - 105 + 1050 \qquad \text{Simplify.}$$
$$= 161 \text{ feet.}$$

Use your calculator to determine the height of the object in Example 6 when $t = 7.6457$. What can you conclude?

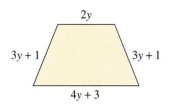

Figure 5.2

Example 7   Geometry: Perimeter

Write an expression for the perimeter of the trapezoid shown in Figure 5.2. Then find the perimeter when $y = 5$.

Solution

To write an expression for the perimeter P of the trapezoid, find the sum of the lengths of the sides.

$$P = (2y) + (3y + 1) + (3y + 1) + (4y + 3)$$
$$= (2y + 3y + 3y + 4y) + (1 + 1 + 3)$$
$$= 12y + 5$$

To find the perimeter when $y = 5$, substitute 5 for y in the expression for the perimeter.

$$P = 12y + 5$$
$$= 12(5) + 5$$
$$= 60 + 5 = 65 \text{ units}$$

Example 8   Using Polynomial Models

The numbers of endangered plant species E and of threatened plant species T in the United States from 1991 to 2001 can be modeled by

$$E = -3.50t^2 + 79.9t + 146, \quad 1 \le t \le 11 \qquad \text{Endangered plant species}$$
$$T = -0.01t^2 + 9.0t + 54, \quad 1 \le t \le 11 \qquad \text{Threatened plant species}$$

where t represents the year, with $t = 1$ corresponding to 1991. Find a polynomial that models the total number P of endangered *and* threatened plant species from 1991 to 2001. Estimate the total number P of endangered and threatened plant species in 2000.   (Source: U.S. Fish and Wildlife Service)

Solution

The sum of the two polynomial models is as follows.

$$E + T = (-3.50t^2 + 79.9t + 146) + (-0.01t^2 + 9.0t + 54)$$
$$= -3.51t^2 + 88.9t + 200$$

So, the polynomial that models the total number of endangered and threatened plant species is

$$P = E + T$$
$$= -3.51t^2 + 88.9t + 200.$$

Using this model, and substituting $t = 10$, you can estimate the total number of endangered and threatened plant species in 2000 to be

$$P = -3.51(10)^2 + 88.9(10) + 200$$
$$= 738 \text{ plant species.}$$

5.2 Exercises

Review   Concepts, Skills, and Problem Solving

Keep mathematically in shape by doing these exercises *before* the problems of this section.

Properties and Definitions

1. The product of two real numbers is -96 and one of the factors is 12. What is the sign of the other factor?

2. Determine the sum of the digits of 576. Because this sum is divisible by 9, the number 576 is divisible by what numbers?

3. *True or False?* -6^2 is positive.

4. *True or False?* $(-6)^2$ is positive.

Solving Inequalities

In Exercises 5–10, solve the inequality.

5. $2x - 12 \geq 0$

6. $7 - 3x < 4 - x$

7. $-2 < 4 - 2x < 10$

8. $4 \leq x + 5 < 8$

9. $|x - 3| < 2$

10. $|x - 5| > 3$

Problem Solving

11. *Property Tax*   The tax on a property with an assessed value of \$145,000 is \$2400. Find the tax on a property with an assessed value of \$90,000.

12. *Fuel Efficiency*   A car uses 7 gallons of gasoline for a trip of 200 miles. How many gallons would be used on a trip of 325 miles? (Assume there is no change in the fuel efficiency.)

Developing Skills

In Exercises 1–12, write the polynomial in standard form, and find its degree and leading coefficient.

1. $3x - 10x^2 + 5 - 42x^3$

2. $4 - 14t^4 + t^5 - 20t$

3. $4y + 16$

4. $50 - x$

5. $5x^2 + x - 7$

6. $12 + 4y - y^2$

7. $-3x^3 - 2x^2 - 3$

8. $6t + 4t^5 - t^2 + 3$

9. -4

10. 28

11. $v_0 t - 16t^2$ (v_0 is constant.)

12. $48 - \frac{1}{2}at^2$ (a is constant.)

In Exercises 13–18, determine whether the polynomial is a monomial, a binomial, or a trinomial.

13. $12 - 5y^2$

14. $-6y + 3 + y^3$

15. $x^3 + 2x^2 - 4$

16. t^3

17. 5

18. $25 - 2u^2$

In Exercises 19–22, give an example of a polynomial in x that satisfies the conditions. (There are many correct answers.)

19. A monomial of degree 3

20. A trinomial of degree 4 and leading coefficient -2

21. A binomial of degree 2 and leading coefficient 8

22. A monomial of degree 0

In Exercises 23–26, state why the expression is not a polynomial.

23. $y^{-3} - 2$

24. $x^3 - 4x^{1/3}$

25. $\dfrac{8}{x}$

26. $\dfrac{2}{x - 4}$

In Exercises 27–42, use a horizontal format to find the sum. See Example 1.

27. $5 + (2 + 3x)$

28. $(6 - 2x) + 4x$

29. $(2x^2 - 3) + (5x^2 + 6)$

30. $(3x + 1) + (6x - 1)$

31. $(5y + 6) + (4y^2 - 6y - 3)$

32. $(3x^3 - 2x + 8) + (3x - 5)$

33. $(2 - 8y) + (-2y^4 + 3y + 2)$

34. $(z^3 + 6z - 2) + (3z^2 - 6z)$

35. $(8 - t^4) + (5 + t^4)$

36. $(y^5 - 4y) + (3y - y^5) + (y^5 - 5)$

37. $(x^2 - 3x + 8) + (2x^2 - 4x) + 3x^2$

38. $(3a^2 + 5a) + (7 - a^2 - 5a) + (2a^2 + 8)$

39. $\left(\frac{2}{3}x^3 - 4x + 1\right) + \left(-\frac{3}{5} + 7x - \frac{1}{2}x^3\right)$

40. $\left(2 - \frac{1}{4}y^2 + y^4\right) + \left(\frac{1}{3}y^4 - \frac{3}{2}y^2 - 3\right)$

41. $(6.32t - 4.51t^2) + (7.2t^2 + 1.03t - 4.2)$

42. $(0.13x^4 - 2.25x - 1.63) + (5.3x^4 + 1.76x^2 + 1.29x)$

In Exercises 43–50, use a vertical format to find the sum. See Example 2.

43. $\begin{array}{r} 5x^2 - 3x + 4 \\ -3x^2 \quad\;\; - 4 \\ \hline \end{array}$

44. $\begin{array}{r} 3x^4 - 2x^2 - 9 \\ -5x^4 + \;\; x^2 \\ \hline \end{array}$

45. $(4x^3 - 2x^2 + 8x) + (4x^2 + x - 6)$

46. $(4x^3 + 8x^2 - 5x + 3) + (x^3 - 3x^2 - 7)$

47. $(5p^2 - 4p + 2) + (-3p^2 + 2p - 7)$

48. $(16 - 32t) + (64 + 48t - 16t^2)$

49. $(2.5b - 3.6b^2) + (7.1 - 3.1b - 2.4b^2) + 6.6b^2$

50. $(1.7y^3 - 6.2y^2 + 5.9) + (2.2y + 6.7y^2 - 3.5y^3)$

In Exercises 51–62, use a horizontal format to find the difference. See Example 3.

51. $(4 - y^3) - (4 + y^3)$

52. $(5y^4 - 2) - (3y^4 + 2)$

53. $(3x^2 - 2x + 1) - (2x^2 + x - 1)$

54. $(5q^2 - 3q + 5) - (4q^2 - 3q - 10)$

55. $(6t^3 - 12) - (-t^3 + t - 2)$

56. $(-10s^2 - 5) - (2s^2 + 6s)$

57. $\left(\frac{1}{4}y^2 - 5y\right) - \left(12 + 4y - \frac{3}{2}y^2\right)$

58. $\left(12 - \frac{2}{3}x + \frac{1}{2}x^2\right) - \left(x^3 + 3x^2 - \frac{1}{6}x\right)$

59. $(10.4t^4 - 0.23t^5 + 1.3t^2) - (2.6 - 7.35t + 6.7t^2 - 9.6t^5)$

60. $(u^3 - 9.75u^2 + 0.12u - 3) - (0.7u^3 - 6.9u^2 - 4.83)$

61. Subtract $3x^3 - (x^2 + 5x)$ from $x^3 - 3x$.

62. Subtract $y^4 - (y^2 - 8y)$ from $y^2 + 3y^4$.

In Exercises 63–68, use a vertical format to find the difference. See Example 4.

63. $\begin{array}{r} x^2 - \;\; x + 3 \\ - \quad (x - 2) \\ \hline \end{array}$ **64.** $\begin{array}{r} 3t^4 - 5t^2 \\ -(-t^4 + 2t^2 - 14) \\ \hline \end{array}$

65. $(25 - 15x - 2x^3) - (12 - 13x + 2x^3)$

66. $(4x^2 + 5x - 6) - (2x^2 - 4x + 5)$

67. $(6x^4 - 3x^7 + 4) - (8x^7 + 10x^5 - 2x^4 - 12)$

68. $(13x^3 - 9x^2 + 4x - 5) - (5x^3 + 7x + 3)$

In Exercises 69–84, perform the indicated operations and simplify. See Example 5.

69. $-(2x^3 - 3) + (4x^3 - 2x)$

70. $(2x^2 + 1) - (x^2 - 2x + 1)$

71. $(4x^5 - 10x^3 + 6x) - (8x^5 - 3x^3 + 11) + (4x^5 + 5x^3 - x^2)$

72. $(15 - 2y + y^2) + (3y^2 - 6y + 1) - (4y^2 - 8y + 16)$

73. $(5y^2 - 2y) - [(y^2 + y) - (3y^2 - 6y + 2)]$

74. $(p^3 + 4) - [(p^2 + 4) + (3p - 9)]$

75. $(8x^3 - 4x^2 + 3x) -$
$[(x^3 - 4x^2 + 5) + (x - 5)]$

76. $(5x^4 - 3x^2 + 9) -$
$[(2x^4 + x^3 - 7x^2) - (x^2 + 6)]$

77. $3(4x^2 - 1) + (3x^3 - 7x^2 + 5)$

78. $(x^3 - 2x^2 - x) - 5(2x^3 + x^2 - 4x)$

79. $2(t^2 + 12) - 5(t^2 + 5) + 6(t^2 + 5)$

80. $-10(v + 2) + 8(v - 1) - 3(v - 9)$

81. $15v - 3(3v - v^2) + 9(8v + 3)$

82. $9(7x^2 - 3x + 3) - 4(15x + 2) - (3x^2 - 7x)$

83. $5s - [6s - (30s + 8)]$

84. $3x^2 - 2[3x + (9 - x^2)]$

In Exercises 85–90, perform the indicated operations and simplify. (Assume that all exponents represent positive integers.)

85. $(2x^{2r} - 6x^r - 3) + (3x^{2r} - 2x^r + 6)$

86. $(6x^{2r} - 5x^r + 4) + (2x^{2r} + 2x^r + 3)$

87. $(3x^{2m} + 2x^m - 8) - (x^{2m} - 4x^m + 3)$

88. $(x^{2m} - 6x^m + 4) - (2x^{2m} - 4x^m - 3)$

89. $(7x^{4n} - 3x^{2n} - 1) - (4x^{4n} + x^{3n} - 6x^{2n})$

90. $(-4x^{3n} + 5x^{2n} + x^n) - (x^{2n} + 9x^n - 14)$

■ *Graphical Reasoning*   In Exercises 91 and 92, use a graphing calculator to graph the expressions for y_1 and y_2 in the same viewing window. What conclusion can you make?

91. $y_1 = (x^3 - 3x^2 - 2) - (x^2 + 1)$
$y_2 = x^3 - 4x^2 - 3$

92. $y_1 = \left(\frac{1}{2}x^3 + 2x\right) + (x^3 - x^2 - x + 1)$
$y_2 = \frac{3}{2}x^3 - x^2 + x + 1$

In Exercises 93 and 94, $f(x) = 4x^3 - 3x^2 + 7$ and $g(x) = 9 - x - x^2 - 5x^3$. Find $h(x)$.

93. $h(x) = f(x) + g(x)$

94. $h(x) = f(x) - g(x)$

Solving Problems

Free-Falling Object   In Exercises 95–98, find the height in feet of a free-falling object at the specified times using the position function. Then write a paragraph that describes the vertical path of the object. See Example 6.

95. $h(t) = -16t^2 + 64$
(a) $t = 0$ (b) $t = \frac{1}{2}$
(c) $t = 1$ (d) $t = 2$

96. $h(t) = -16t^2 + 256$
(a) $t = 0$ (b) $t = 1$
(c) $t = \frac{5}{2}$ (d) $t = 4$

97. $h(t) = -16t^2 + 80t + 50$
(a) $t = 0$ (b) $t = 2$
(c) $t = 4$ (d) $t = 5$

98. $h(t) = -16t^2 + 96t$
(a) $t = 0$ (b) $t = 2$
(c) $t = 3$ (d) $t = 6$

Free-Falling Object   In Exercises 99–102, use the position function to determine whether the free-falling object was dropped, was thrown upward, or was thrown downward. Also determine the height in feet of the object at time $t = 0$.

99. $h(t) = -16t^2 + 100$

100. $h(t) = -16t^2 + 50t$

101. $h(t) = -16t^2 - 24t + 50$

102. $h(t) = -16t^2 + 32t + 300$

103. *Free-Falling Object* An object is thrown upward from the top of the Eiffel Tower, which is 984 feet tall. The initial velocity is 40 feet per second. Use the position function

$h(t) = -16t^2 + 40t + 984$

to find the height of the object when $t = 1$, $t = 5$, and $t = 9$.

104. *Free-Falling Object* An object is dropped from a hot-air balloon that is 200 feet above the ground (see figure). Use the position function

$h(t) = -16t^2 + 200$

to find the height of the object when $t = 1$, $t = 2$, and $t = 3$.

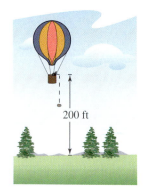

200 ft

105. *Cost, Revenue, and Profit* A manufacturer can produce and sell x radios per week. The total cost C (in dollars) for producing the radios is given by

$C = 8x + 15,000$

and the total revenue R is given by

$R = 14x$.

Find the profit P obtained by selling 5000 radios per week. (*Note: $P = R - C$*)

106. *Cost, Revenue, and Profit* A manufacturer can produce and sell x golf clubs per week. The total cost C (in dollars) for producing the golf clubs is given by $C = 12x + 8000$ and the total revenue R is given by $R = 17x$. Find the profit P obtained by selling 10,000 golf clubs per week. (*Note: $P = R - C$*)

▲ *Geometry* In Exercises 107 and 108, write and simplify an expression for the perimeter of the figure. See Example 7.

107.

$2x + 4$ $2x + 4$

x x
x x

$3x$ $3x$

108.

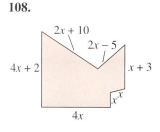

$2x + 10$

$2x - 5$

$4x + 2$ $x + 3$

x x

$4x$

▲ *Geometry* In Exercises 109 and 110, find an expression that represents the area of the entire region.

109. |←— 6 —→|

$\frac{3}{2}x$

$\frac{9}{2}x$

110. |←— 5 —→|

x

x

$3x$

▲ *Geometry* In Exercises 111 and 112, find an expression for the area of the shaded region of the figure.

111.

x $x + 6$

7

12

112.

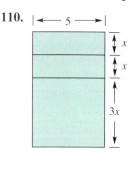

$4x + 7$

4

5

$2x$

113. *Stopping Distance*   The total stopping distance of an automobile is the distance traveled during the driver's reaction time plus the distance traveled after the brakes are applied. In an experiment, these distances were measured (in feet) when an automobile was traveling at a speed of x miles per hour on dry, level pavement. The distance traveled during the reaction time was $R = 1.1x$, and the braking distance was $B = 0.0475x^2 - 0.001x + 0.23$.

(a) Determine the polynomial that represents the total stopping distance T.

(b) 🖩 Use a graphing calculator to graph R, B, and T in the same viewing window.

(c) 🖩 Use the graph to estimate the total stopping distance when $x = 30$ and $x = 60$ miles per hour.

114. *Beverage Consumption*   The per capita consumptions (average consumptions per person) of all beverage milks and coffee in the United States from 1996 to 2000 can be approximated by the two polynomial models

$y = -0.33t + 25.9$ Beverage milks

$y = 0.043t^2 + 0.33t + 18.5$ Coffee

In these models, y represents the average consumption per person in gallons and t represents the year, with $t = 6$ corresponding to 1996.   (Source: U.S. Department of Agriculture)

(a) Find a polynomial model that represents the per capita consumption of both beverage milks and coffee during the time period.

(b) 🖩 During the given period, the per capita consumption of beverage milks was decreasing and the per capita consumption of coffee was increasing (see figure). Use a graphing calculator to graph the model from part (a). Was the total per capita consumption of beverage milks and coffee increasing or decreasing over this period?   Explain.

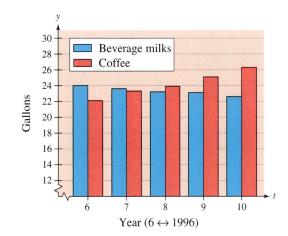

Year (6 ↔ 1996)

Explaining Concepts

115. *Writing* 🖉 Explain the difference between the degree of a term of a polynomial in x and the degree of a polynomial.

116. What algebraic operation separates the terms of a polynomial? What operation separates the factors of a term?

117. Give an example of combining like terms.

118. Can two third-degree polynomials be added to produce a second-degree polynomial? If so, give an example.

119. *Writing* 🖉 Is every trinomial a second-degree polynomial? If not, give an example of a trinomial that is not a second-degree polynomial.

120. *Writing* 🖉 Describe the method for subtracting polynomials.

5.3 Multiplying Polynomials

Paul Poplis/Getty Images

What You Should Learn

1 Use the Distributive Property and the FOIL Method to multiply polynomials.

2 Use special product formulas to multiply two binomials.

3 Use multiplication of polynomials in application problems.

Why You Should Learn It

Multiplication of polynomials can be used to model many aspects of the business world. For instance, in Exercise 108 on page 326, you will use a polynomial to model the total revenue from selling apple pies.

Multiplying Polynomials

The simplest type of polynomial multiplication involves a monomial multiplier. The product is obtained by direct application of the Distributive Property. For instance, to multiply the monomial $3x$ by the polynomial $(2x^2 - 5x + 3)$, multiply *each* term of the polynomial by $3x$.

$$(3x)(2x^2 - 5x + 3) = (3x)(2x^2) - (3x)(5x) + (3x)(3)$$

$$= 6x^3 - 15x^2 + 9x$$

1 Use the Distributive Property and the FOIL Method to multiply polynomials.

Example 1   Finding Products with Monomial Multipliers

Multiply the polynomial by the monomial.

a. $(2x - 7)(3x)$ **b.** $4x^2(3x - 2x^3 + 1)$

Solution

a. $(2x - 7)(3x) = 2x(3x) - 7(3x)$ Distributive Property

$$= 6x^2 - 21x$$ Rules of exponents

b. $4x^2(3x - 2x^3 + 1)$

$$= 4x^2(3x) - 4x^2(2x^3) + 4x^2(1)$$ Distributive Property

$$= 12x^3 - 8x^5 + 4x^2$$ Rules of exponents

$$= -8x^5 + 12x^3 + 4x^2$$ Standard form

Example 2   Finding Products with Negative Monomial Multipliers

a. $(-x)(5x^2 - x) = (-x)(5x^2) - (-x)(x)$ Distributive Property

$$= -5x^3 + x^2$$ Rules of exponents

b. $(-2x^2)(x^2 - 8x + 4)$

$$= (-2x^2)(x^2) - (-2x^2)(8x) + (-2x^2)(4)$$ Distributive Property

$$= -2x^4 + 16x^3 - 8x^2$$ Rules of exponents

To multiply two *binomials*, you can use both (left and right) forms of the Distributive Property. For example, if you treat the binomial $(2x + 7)$ as a single quantity, you can multiply $(3x - 2)$ by $(2x + 7)$ as follows.

$$(3x - 2)(2x + 7) = 3x(2x + 7) - 2(2x + 7)$$

$$= (3x)(2x) + (3x)(7) - (2)(2x) - (2)(7)$$

$$= 6x^2 + 21x - 4x - 14$$

Product of First terms	Product of Outer terms	Product of Inner terms	Product of Last terms

$$= 6x^2 + 17x - 14$$

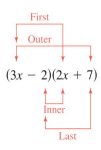

First

Outer

$(3x - 2)(2x + 7)$

Inner

Last

The FOIL Method

With practice, you should be able to multiply two binomials without writing out all of the steps shown above. In fact, the four products in the boxes above suggest that you can write the product of two binomials in just one step. This is called the **FOIL Method.** Note that the words *First*, *Outer*, *Inner*, and *Last* refer to the positions of the terms in the original product, as shown at the left.

Example 3 Multiplying Binomials (FOIL Method)

Use the FOIL Method to find the product.

$$(x - 3)(x + 3)$$

Solution

$$(x - 3)(x + 3) = \overset{\text{F}}{x^2} + \overset{\text{O}}{3x} - \overset{\text{I}}{3x} - \overset{\text{L}}{9}$$

$$= x^2 - 9 \qquad \text{Combine like terms.}$$

Example 4 Multiplying Binomials (FOIL Method)

Use the FOIL Method to find the product.

$$(3x + 4)(2x + 1)$$

Solution

$$(3x + 4)(2x + 1) = \overset{\text{F}}{6x^2} + \overset{\text{O}}{3x} + \overset{\text{I}}{8x} + \overset{\text{L}}{4}$$

$$= 6x^2 + 11x + 4 \qquad \text{Combine like terms.}$$

To multiply two polynomials that have three or more terms, you can use the same basic principle as for multiplying monomials and binomials. That is, *each term of one polynomial must be multiplied by each term of the other polynomial.* This can be done using either a horizontal or a vertical format.

Example 5   Multiplying Polynomials (Horizontal Format)

Use a horizontal format to find the product.

$$(4x^2 - 3x - 1)(2x - 5)$$

Solution

$(4x^2 - 3x - 1)(2x - 5)$

$\quad = (4x^2 - 3x - 1)(2x) - (4x^2 - 3x - 1)(5)$ Distributive Property

$\quad = 8x^3 - 6x^2 - 2x - (20x^2 - 15x - 5)$ Distributive Property

$\quad = 8x^3 - 6x^2 - 2x - 20x^2 + 15x + 5$ Subtract (change signs).

$\quad = 8x^3 - 26x^2 + 13x + 5$ Combine like terms.

Example 6   Multiplying Polynomials (Vertical Format)

Write the polynomials in standard form and use a vertical format to find the product.

$$(4x^2 + x - 2)(5 + 3x - x^2)$$

Solution

$$
\begin{array}{r}
4x^2 + \ x - \ 2 \\
\times \quad\quad -x^2 + 3x + \ 5 \\
\hline
20x^2 + 5x - 10 \\
12x^3 + \ 3x^2 - 6x \\
-4x^4 - \ \ x^3 + \ 2x^2 \\
\hline
-4x^4 + 11x^3 + 25x^2 - \ \ x - 10
\end{array}
$$

Standard form
Standard form
$5(4x^2 + x - 2)$
$3x(4x^2 + x - 2)$
$-x^2(4x^2 + x - 2)$
Combine like terms.

Example 7   An Area Model for Multiplying Polynomials

Show that $(2x + 1)(x + 2) = 2x^2 + 5x + 2$.

Solution

An appropriate area model for demonstrating the multiplication of two binomials would be $A = lw$, the area formula for a rectangle. Think of a rectangle whose sides are $x + 2$ and $2x + 1$. The area of this rectangle is

$$(2x + 1)(x + 2). \quad \text{Area} = (\text{length})(\text{width})$$

Another way to find the area is to add the areas of the rectangular parts, as shown in Figure 5.3. There are two squares whose sides are x, five rectangles whose sides are x and 1, and two squares whose sides are 1. The total area of these nine rectangles is

$$2x^2 + 5x + 2. \quad \text{Area} = \text{sum of rectangular areas}$$

Because each method must produce the same area, you can conclude that

$$(2x + 1)(x + 2) = 2x^2 + 5x + 2.$$

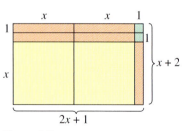

Figure 5.3

2 Use special product formulas to multiply two binomials.

Special Products

Some binomial products, such as that in Example 3, have special forms that occur frequently in algebra.

Special Products

Let u and v be real numbers, variables, or algebraic expressions. Then the following formulas are true.

Special Product	*Example*
Sum and Difference of Two Terms	
$(u + v)(u - v) = u^2 - v^2$	$(3x - 4)(3x + 4) = 9x^2 - 16$
Square of a Binomial	
$(u + v)^2 = u^2 + 2uv + v^2$	$(2x + 5)^2 = 4x^2 + 2(2x)(5) + 25$
	$= 4x^2 + 20x + 25$
$(u - v)^2 = u^2 - 2uv + v^2$	$(x - 6)^2 = x^2 - 2(x)(6) + 36$
	$= x^2 - 12x + 36$

When squaring a binomial, note that the resulting middle term, $\pm 2uv$, is always *twice* the product of the two terms.

Study Tip

A frequent error in calculating special products is to forget the middle term when squaring a binomial. Use $x = 3$ and $y = 2$ to verify the following statements.

1. $(x + y)^2 = x^2 + 2xy + y^2$
2. $(x + y)^2 \neq x^2 + y^2$
3. $(x - y)^2 = x^2 - 2xy + y^2$
4. $(x - y)^2 \neq x^2 - y^2$

Example 8   Product of the Sum and Difference of Two Terms

a. $(3x - 2)(3x + 2) = (3x)^2 - 2^2$ $(u + v)(u - v) = u^2 - v^2$

$= 9x^2 - 4$ Simplify.

b. $(6 + 5x)(6 - 5x) = 6^2 - (5x)^2 = 36 - 25x^2$

Example 9   Squaring a Binomial

$(2x - 7)^2 = (2x)^2 - 2(2x)(7) + 7^2$ Square of a binomial

$= 4x^2 - 28x + 49$ Simplify.

Example 10   Cubing a Binomial

$(x + 4)^3 = (x + 4)^2(x + 4)$ Rules of exponents

$= (x^2 + 8x + 16)(x + 4)$ Square of a binomial

$= x^2(x + 4) + 8x(x + 4) + 16(x + 4)$ Distributive Property

$= x^3 + 4x^2 + 8x^2 + 32x + 16x + 64$ Distributive Property

$= x^3 + 12x^2 + 48x + 64$ Combine like terms.

3 Use multiplication of polynomials in application problems.

Application

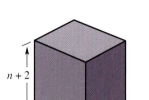

Figure 5.4

Example 11   Geometry: Area and Volume

The closed rectangular box shown in Figure 5.4 has sides whose lengths (in inches) are consecutive integers.

a. Write a polynomial function $V(n)$ that represents the volume of the box.

b. What is the volume if the length of the shortest side is 4 inches?

c. Write a polynomial function $A(n)$ for the area of the base of the box.

d. Write a polynomial function for the area of the base if the length and width increase by 3. That is, find $A(n + 3)$.

Solution

a. The volume of a rectangular box is equal to the product of its length, width, and height. So, the volume can be represented by the following function.

$$V(n) = n(n + 1)(n + 2) = n(n^2 + 3n + 2) = n^3 + 3n^2 + 2n$$

b. If $n = 4$, the volume of the box is

$$V(4) = (4)^3 + 3(4)^2 + 2(4) = 64 + 48 + 8 = 120 \text{ cubic inches.}$$

c. The length of the base is n and the width is $n + 1$. Find the area as follows.

$$A(n) = (\text{Length})(\text{Width}) = n(n + 1) = n^2 + n$$

d. $A(n + 3) = (n + 3)^2 + (n + 3) = n^2 + 6n + 9 + n + 3 = n^2 + 7n + 12$

Example 12   Revenue

A software manufacturer has determined that the demand for its new video game is given by the equation $p = 50 - 0.001x$, where p is the price of the game (in dollars) and x is the number of units sold. The total revenue R from selling x units of a product is given by the equation $R = xp$. Find the revenue equation for the video game. Then find the revenue when 3000 units are sold.

Solution

$R = xp$	Revenue equation
$\quad = x(50 - 0.001x)$	Substitute for p.
$\quad = 50x - 0.001x^2$	Distributive Property

So, the revenue equation for the video game is $R = 50x - 0.001x^2$. To find the revenue when 3000 units are sold, substitute 3000 for x in the revenue equation.

$R = 50x - 0.001x^2$	Revenue equation
$\quad = 50(3000) - 0.001(3000)^2$	Substitute 3000 for x.
$\quad = 141,000$	Simplify.

So, the revenue when 3000 units are sold is $141,000.

5.3 Exercises

Review   Concepts, Skills, and Problem Solving

Keep mathematically in shape by doing these exercises *before* the problems of this section.

Properties and Definitions

1. *Writing* ✎   Relative to the *x*- and *y*-axes, explain the meaning of each coordinate of the point $(-2, 3)$.

2. A point lies three units from the *x*-axis and four units from the *y*-axis. Give the ordered pair for such a point in each quadrant.

Evaluating Expressions

In Exercises 3–6, find the missing coordinate of the solution point.

3. $y = \frac{3}{5}x + 4$
   $(15, \quad)$

4. $y = 3 - \frac{5}{9}x$
   $(12, \quad)$

5. $y = 5.46 - 0.95x$
   $(\quad, -1)$

6. $y = 3 + 0.2x$
   $(\quad, 4.4)$

In Exercises 7–10, evaluate the function.

7. $f(x) = x + 1$
   (a) $f(6) - f(1)$
   (b) $f(x - 2) - f(5)$

8. $f(x) = 3 - 2x$
   (a) $f(5) - f(0)$
   (b) $f(x + 3) - f(3)$

9. $g(x) = \dfrac{x}{x + 10}$
   (a) $g(5) - g(-2)$
   (b) $g(c - 6)$

10. $h(x) = \sqrt{x - 4}$
    (a) $h(8) - h(20)$
    (b) $h(t + 3)$

Graphing

In Exercises 11 and 12, graph the function.

11. $g(x) = 7 - \frac{3}{2}x$

12. $h(x) = |3 - x|$

Developing Skills

In Exercises 1–14, perform the indicated multiplication(s). See Examples 1 and 2.

1. $(-2a^2)(-8a)$
2. $(-6n)(3n^2)$
3. $2y(5 - y)$
4. $5z(2z - 7)$
5. $4x(2x^2 - 3x + 5)$
6. $3y(-3y^2 + 7y - 3)$
7. $-2x^2(5 + 3x^2 - 7x^3)$
8. $-3a^2(8 - 2a - a^2)$
9. $-x^3(x^4 - 2x^3 + 5x - 6)$
10. $-y^4(7y^3 - 4y^2 + y - 4)$
11. $-3x(-5x)(5x + 2)$
12. $4t(-3t)(t^2 - 1)$
13. $u^2v(3u^4 - 5u^2v + 6uv^3)$
14. $ab^3(2a - 9a^2b + 3b)$

In Exercises 15–32, multiply using the FOIL Method. See Examples 3 and 4.

15. $(x + 2)(x + 4)$
16. $(x - 5)(x - 3)$
17. $(x - 6)(x + 5)$
18. $(x + 7)(x - 1)$
19. $(x - 3)(x - 3)$
20. $(x - 6)(x + 6)$
21. $(2x - 3)(x + 5)$
22. $(3x + 1)(x - 4)$
23. $(5x - 2)(2x - 6)$
24. $(4x + 7)(3x + 7)$
25. $(8 - 3x^2)(4x + 1)$
26. $(6x^2 + 2)(9 - 2x)$
27. $\left(4y - \frac{1}{3}\right)(12y + 9)$
28. $\left(5t - \frac{3}{4}\right)(2t - 16)$
29. $(2x + y)(3x + 2y)$
30. $(2x - y)(3x - 2y)$

31. $(2t - 1)(t + 1) + (2t - 5)(t - 1)$

32. $(s - 3t)(s + t) - (s - 3t)(s - t)$

In Exercises 33–44, use a horizontal format to find the product. See Example 5.

33. $(x - 1)(x^2 - 4x + 6)$ **34.** $(z + 2)(z^2 - 4z + 4)$

35. $(3a + 2)(a^2 + 3a + 1)$ **36.** $(2t + 3)(t^2 - 5t + 1)$

37. $(2u^2 + 3u - 4)(4u + 5)$

38. $(2x^2 - 5x + 1)(3x - 4)$

39. $(x^3 - 3x + 2)(x - 2)$

40. $(x^2 + 4)(x^2 - 2x - 4)$

41. $(5x^2 + 2)(x^2 + 4x - 1)$

42. $(2x^2 - 3)(2x^2 - 2x + 3)$

43. $(t^2 + t - 2)(t^2 - t + 2)$

44. $(y^2 + 3y + 5)(2y^2 - 3y - 1)$

In Exercises 45–52, use a vertical format to find the product. See Example 6.

45. $7x^2 - 14x + 9$
$$\underline{\times \qquad 4x^3 + 3}$$

46. $4x^4 - 6x^2 + 9$
$$\underline{\times \qquad 2x^2 + 3}$$

47. $(u - 2)(2u^2 + 5u + 3)$

48. $(z - 2)(z^2 + z + 1)$

49. $(-x^2 + 2x - 1)(2x + 1)$

50. $(2s^2 - 5s + 6)(3s - 4)$

51. $(t^2 + t - 2)(t^2 - t + 2)$

52. $(y^2 + 3y + 5)(2y^2 - 3y - 1)$

In Exercises 53–82, use a special product formula to find the product. See Examples 8 and 9.

53. $(x + 2)(x - 2)$ **54.** $(x - 5)(x + 5)$

55. $(x - 8)(x + 8)$ **56.** $(x + 1)(x - 1)$

57. $(2 + 7y)(2 - 7y)$ **58.** $(4 + 3z)(4 - 3z)$

59. $(6 - 4x)(6 + 4x)$ **60.** $(8 - 3x)(8 + 3x)$

61. $(2a + 5b)(2a - 5b)$

62. $(5u + 12v)(5u - 12v)$ **63.** $(6x - 9y)(6x + 9y)$

64. $(8x - 5y)(8x + 5y)$ **65.** $\left(2x - \frac{1}{4}\right)\left(2x + \frac{1}{4}\right)$

66. $\left(\frac{2}{3}x + 7\right)\left(\frac{2}{3}x - 7\right)$

67. $(0.2t + 0.5)(0.2t - 0.5)$

68. $(4a - 0.1b)(4a + 0.1b)$

69. $(x + 5)^2$ **70.** $(x + 2)^2$

71. $(x - 10)^2$ **72.** $(u - 7)^2$

73. $(2x + 5)^2$ **74.** $(3x + 8)^2$

75. $(6x - 1)^2$ **76.** $(5 - 3z)^2$

77. $(2x - 7y)^2$ **78.** $(3m + 4n)^2$

79. $[(x + 2) + y]^2$

80. $[(x - 4) - y]^2$

81. $[u - (v - 3)][u + (v - 3)]$

82. $[z + (y + 1)][z - (y + 1)]$

In Exercises 83–86, simplify the expression. See Example 10.

83. $(x + 3)^3$ **84.** $(y - 2)^3$

85. $(u + v)^3$ **86.** $(u - v)^3$

In Exercises 87–92, simplify the expression. (Assume that all variables represent positive integers.)

87. $3x^r(5x^{2r} + 4x^{3r-1})$ **88.** $5x^r(4x^{r+2} - 3x^r)$

89. $(6x^m - 5)(2x^{2m} - 3)$

90. $(x^{3m} - x^{2m})(x^{2m} + 2x^{4m})$

91. $(x^{m-n})^{m+n}$ **92.** $(y^{m+n})^{m+n}$

In Exercises 93–96, use a graphing calculator to graph the expressions for y_1 and y_2 in the same viewing window. What can you conclude? Verify the conclusion algebraically.

93. $y_1 = (x + 1)(x^2 - x + 2)$

$y_2 = x^3 + x + 2$

94. $y_1 = (x - 3)^2$

$y_2 = x^2 - 6x + 9$

95. $y_1 = (2x - 3)(x + 2)$

$y_2 = 2x^2 + x - 6$

96. $y_1 = \left(x + \frac{1}{2}\right)\left(x - \frac{1}{2}\right)$

$y_2 = x^2 - \frac{1}{4}$

97. For the function $f(x) = x^2 - 2x$, find and simplify each of the following.

(a) $f(t - 3)$

(b) $f(2 + h) - f(2)$

98. For the function $f(x) = 2x^2 - 5x + 4$, find and simplify each of the following.

(a) $f(y + 2)$

(b) $f(1 + h) - f(1)$

Solving Problems

99. ▲ *Geometry* A closed rectangular box has sides of lengths n, $n + 2$, and $n + 4$ inches. (See figure.)

(a) Write a polynomial function $V(n)$ that represents the volume of the box.

(b) What is the volume if the length of the shortest side is 2 inches?

(c) Write a polynomial function $A(n)$ that represents the area of the base of the box.

(d) Write a polynomial function for the area of the base if the length and width increase by 4. Show that the polynomial function is $A(n + 4)$.

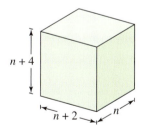

Figure for 99

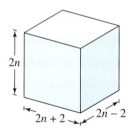

Figure for 100

100. ▲ *Geometry* A closed rectangular box has sides of lengths $2n - 2$, $2n + 2$, and $2n$ inches. (See figure.)

(a) Write a polynomial function $V(n)$ that represents the volume of the box.

(b) What is the volume if the length of the shortest side is 6 inches?

(c) Write a polynomial function $A(n)$ that represents the area of the base of the box.

(d) Write a polynomial function for the area of the base if the length and width increase by 2. Show that the area of the base is not $A(n + 4)$.

▲ *Geometry* In Exercises 101–104, write an expression for the area of the shaded region of the figure. Then simplify the expression.

101.

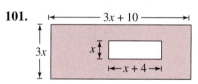

102.

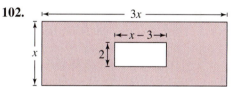

103.

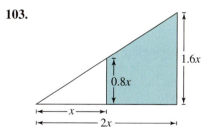

104.

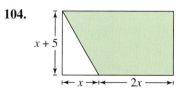

105. ▲ *Geometry* The length of a rectangle is $1\frac{1}{2}$ times its width w. Find expressions for (a) the perimeter and (b) the area of the rectangle.

106. ▲ *Geometry* The base of a triangle is $3x$ and its height is $x + 5$. Find an expression for the area A of the triangle.

107. *Revenue* A shop owner has determined that the demand for his daily newspapers is given by the equation $p = 125 - 0.02x$, where p is the price of the newspaper (in cents) and x is the number of papers sold. The total revenue R from selling x units of a product is given by the equation $R = xp$. Find the revenue equation for the shop owner's daily newspapers. Then find the revenue when 1000 newspapers are sold.

108. *Revenue* A supermarket has determined that the demand for its apple pies is given by the equation $p = 20 - 0.015x$, where p is the price of the apple pie (in dollars) and x is the number of pies sold. The total revenue R from selling x units of a product is given by the equation $R = xp$. Find the revenue equation for the supermarket's apple pies. Then find the revenue when 50 apple pies are sold.

109. *Compound Interest* After 2 years, an investment of $1000 compounded annually at interest rate r will yield an amount $1000(1 + r)^2$. Find this product.

110. *Compound Interest* After 2 years, an investment of $1000 compounded annually at an interest rate of 9.5% will yield an amount $1000(1 + 0.095)^2$. Find this product.

▲ *Geometric Modeling* In Exercises 111 and 112, use the area model to write two different expressions for the total area. Then equate the two expressions and name the algebraic property that is illustrated. See Example 7.

111.

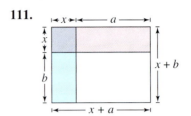

112.

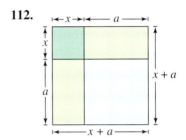

113. *Finding a Pattern* Perform the multiplications.

(a) $(x - 1)(x + 1)$

(b) $(x - 1)(x^2 + x + 1)$

(c) $(x - 1)(x^3 + x^2 + x + 1)$

From the pattern formed by these products, can you predict the result of $(x - 1)(x^4 + x^3 + x^2 + x + 1)$?

114. Use the FOIL Method to verify each of the following.

(a) $(x + y)^2 = x^2 + 2xy + y^2$

(b) $(x - y)^2 = x^2 - 2xy + y^2$

(c) $(x - y)(x + y) = x^2 - y^2$

Explaining Concepts

115. ⟳ Answer parts (a)–(c) of Motivating the Chapter on page 296.

116. Give an example of how to use the Distributive Property to multiply two binomials.

117. *Writing* Explain the meaning of each letter of FOIL as it relates to multiplying two binomials.

118. What is the degree of the product of two polynomials of degrees m and n?

119. *True or False?* Determine whether the statement is true or false. Justify your answer.

(a) The product of two monomials is a monomial.

(b) The product of two binomials is a binomial.

Mid-Chapter Quiz

Take this quiz as you would take a quiz in class. After you are done, check your work against the answers in the back of the book.

1. Determine the degree and leading coefficient of the polynomial
   $3 - 2x + 4x^3 - 2x^4$.

2. Explain why $2x - 3x^{1/2} + 5$ is not a polynomial.

In Exercises 3–22, perform the indicated operations and simplify (use only positive exponents).

3. $(5y^2)(-y^4)(2y^3)$

4. $(-6x)(-3x^2)^2$

5. $(-5n^2)(-2n^3)$

6. $(-2x^2)^3(x^4)$

7. $\dfrac{6x^{-7}}{(-2x^2)^{-3}}$

8. $\left(\dfrac{4y^2}{5x}\right)^{-2}$

9. $\left(\dfrac{3a^{-2}b^5}{9a^{-4}b^0}\right)^{-2}$

10. $\left(\dfrac{5x^0y^{-7}}{2x^{-2}y^4}\right)^{-3}$

11. Add $2t^3 + 3t^2 - 2$ to $t^3 + 9$.

12. $(3 - 7y) + (7y^2 + 2y - 3)$

13. $(7x^3 - 3x^2 + 1) - (x^2 - 2x^3)$

14. $(5 - u) - 2[3 - (u^2 + 1)]$

15. $7y(4 - 3y)$

16. $(x - 7)(x + 3)$

17. $(4x - y)(6x - 5y)$

18. $2z(z + 5) - 7(z + 5)$

19. $(6r + 5)(6r - 5)$

20. $(2x - 3)^2$

21. $(x + 1)(x^2 - x + 1)$

22. $(x^2 - 3x + 2)(x^2 + 5x - 10)$

23. Find the area of the shaded region of the figure.

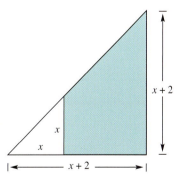

24. An object is thrown upward from the top of a 100-foot building with an initial velocity of 32 feet per second. Use the position function $h(t) = -16t^2 + 32t + 100$ to find the height of the object when $t = 1$ and $t = 2$.

25. A manufacturer can produce and sell x T-shirts per week. The total cost C (in dollars) of producing the T-shirts is given by $C = 5x + 2000$, and the total revenue R is given by $R = 19x$. Find the profit P obtained by selling 1000 T-shirts per week.

5.4 Factoring by Grouping and Special Forms

What You Should Learn

1. Factor greatest common monomial factors from polynomials.
2. Factor polynomials by grouping terms.
3. Factor the difference of two squares and factor the sum or difference of two cubes.
4. Factor polynomials completely by repeated factoring.

Why You Should Learn It

In some cases, factoring a polynomial enables you to determine unknown quantities. For instance, in Exercise 126 on page 336, you will factor the expression for the area of a rectangle to determine the length of the rectangle.

1. Factor greatest common monomial factors from polynomials.

Common Monomial Factors

In Section 5.3, you studied ways of multiplying polynomials. In this section and the next two sections, you will study the reverse process—**factoring polynomials.**

Use Distributive Property to multiply.   *Use Distributive Property to factor.*

$$3x(4 - 5x) = 12x - 15x^2 \qquad\qquad 12x - 15x^2 = 3x(4 - 5x)$$

Notice that factoring changes a *sum of terms* into a *product of factors.*

To be efficient in factoring expressions, you need to understand the concept of the *greatest common factor* of two (or more) integers or terms. Recall from arithmetic that every integer can be factored into a product of prime numbers. The **greatest common factor** (or **GCF**) of two or more integers is the greatest integer that is a factor of each number.

Example 1   Finding the Greatest Common Factor

Find the greatest common factor of $6x^5$, $30x^4$, and $12x^3$.

Solution

From the factorizations

$$6x^5 = 2 \cdot 3 \cdot x \cdot x \cdot x \cdot x \cdot x = (6x^3)(x^2)$$

$$30x^4 = 2 \cdot 3 \cdot 5 \cdot x \cdot x \cdot x \cdot x = (6x^3)(5x)$$

$$12x^3 = 2 \cdot 2 \cdot 3 \cdot x \cdot x \cdot x = (6x^3)(2)$$

you can conclude that the greatest common factor is $6x^3$.

Consider the three terms given in Example 1 as terms of the polynomial $6x^5 + 30x^4 + 12x^3$. The greatest common factor, $6x^3$, of these terms is the **greatest common monomial factor** of the polynomial. When you use the Distributive Property to remove this factor from each term of the polynomial, you are **factoring out** the greatest common monomial factor.

$$6x^5 + 30x^4 + 12x^3 = 6x^3(x^2) + 6x^3(5x) + 6x^3(2) \qquad \text{Factor each term.}$$

$$= 6x^3(x^2 + 5x + 2) \qquad\qquad \text{Factor out common monomial factor.}$$

Greatest Common Monomial Factor

If a polynomial in x with integer coefficients has a greatest common monomial factor of the form ax^n, the following statements must be true.

1. The coefficient a must be the greatest integer that *divides* each of the coefficients in the polynomial.

2. The variable factor x^n is the highest-powered variable factor that *is common* to all terms of the polynomial.

Example 2   Factoring Out a Greatest Common Monomial Factor

Factor out the greatest common monomial factor from $24x^3 - 32x^2$.

Solution

For the terms $24x^3$ and $32x^2$, 8 is the greatest integer factor of 24 and 32 and x^2 is the highest-powered variable factor common to x^3 and x^2. So, the greatest common monomial factor of $24x^3$ and $32x^2$ is $8x^2$. You can factor the polynomial as follows.

$$24x^3 - 32x^2 = (8x^2)(3x) - (8x^2)(4)$$
$$= 8x^2(3x - 4)$$

The greatest common monomial factor of a polynomial is usually considered to have a positive coefficient. However, sometimes it is convenient to factor a negative number out of a polynomial. You can see how this is done in the next example.

Study Tip

When factoring a polynomial, remember that you can check your results by multiplying. That is, if you multiply the factors, you should obtain the original polynomial.

Example 3   A Negative Common Monomial Factor

Factor the polynomial $-3x^2 + 12x - 18$ in two ways.

a. Factor out a 3.   **b.** Factor out a -3.

Solution

a. By factoring out the common monomial factor of 3, you obtain

$$-3x^2 + 12x - 18 = 3(-x^2) + 3(4x) + 3(-6)$$
$$= 3(-x^2 + 4x - 6).$$

b. By factoring out the common monomial factor of -3, you obtain

$$-3x^2 + 12x - 18 = -3(x^2) + (-3)(-4x) + (-3)(6)$$
$$= -3(x^2 - 4x + 6).$$

Check these results by multiplying.

2 Factor polynomials by grouping terms.

Factoring by Grouping

Some expressions have common factors that are not simple monomials. For instance, the expression $x^2(2x - 3) + 4(2x - 3)$ has the common *binomial* factor $(2x - 3)$. Factoring out this common factor produces

$$x^2(2x - 3) + 4(2x - 3) = (2x - 3)(x^2 + 4).$$

This type of factoring is part of a more general procedure called **factoring by grouping.**

Example 4   Common Binomial Factors

Factor the expression $5x^2(6x - 5) - 2(6x - 5)$.

Solution

Each of the terms of this expression has a binomial factor of $(6x - 5)$. Factoring this binomial out of each term produces the following.

$$5x^2(6x - 5) - 2(6x - 5) = (6x - 5)(5x^2 - 2)$$

In Example 4, the original expression was already grouped so that it was easy to determine the common binomial factor. In practice, you will have to do the grouping as well as the factoring.

Example 5   Factoring By Grouping

Factor each polynomial by grouping.

a. $x^3 - 5x^2 + x - 5$ **b.** $4x^3 + 3x - 8x^2 - 6$

Solution

a. $x^3 - 5x^2 + x - 5 = (x^3 - 5x^2) + (x - 5)$ Group terms.

$$= x^2(x - 5) + 1(x - 5)$$ Factor grouped terms.

$$= (x - 5)(x^2 + 1)$$ Common binomial factor

b. $4x^3 + 3x - 8x^2 - 6 = 4x^3 - 8x^2 + 3x - 6$ Write in standard form.

$$= (4x^3 - 8x^2) + (3x - 6)$$ Group terms.

$$= 4x^2(x - 2) + 3(x - 2)$$ Factor grouped terms.

$$= (x - 2)(4x^2 + 3)$$ Common binomial factor

> **Study Tip**
>
> You should write a polynomial in standard form before trying to factor by grouping. Then group and remove a common monomial factor from the first two terms and the last two terms. Finally, if possible, factor out the common binomial factor.

Note that in Example 5(a) the polynomial is factored by grouping the first and second terms and the third and fourth terms. You could just as easily have grouped the first and third terms and the second and fourth terms, as follows.

$$x^3 - 5x^2 + x - 5 = (x^3 + x) - (5x^2 + 5)$$

$$= x(x^2 + 1) - 5(x^2 + 1) = (x^2 + 1)(x - 5)$$

③ Factor the difference of two squares and factor the sum or difference of two cubes.

Factoring Special Products

Some polynomials have special forms that you should learn to recognize so that you can factor them easily. One of the easiest special polynomial forms to recognize and to factor is the form $u^2 - v^2$, called a **difference of two squares.** This form arises from the special product $(u + v)(u - v)$ in Section 5.3.

Difference of Two Squares

Let u and v be real numbers, variables, or algebraic expressions. Then the expression $u^2 - v^2$ can be factored as follows.

$$u^2 - v^2 = (u + v)(u - v)$$

Difference   Opposite signs

To recognize perfect squares, look for coefficients that are squares of integers and for variables raised to *even* powers.

Example 6   Factoring the Difference of Two Squares

Factor each polynomial.

a. $x^2 - 64$ **b.** $49x^2 - 81y^2$

Solution

a. $x^2 - 64 = x^2 - 8^2$ Write as difference of two squares.

$\qquad\qquad = (x + 8)(x - 8)$ Factored form

b. $49x^2 - 81y^2 = (7x)^2 - (9y)^2$ Write as difference of two squares.

$\qquad\qquad\qquad = (7x + 9y)(7x - 9y)$ Factored form

Remember that the rule $u^2 - v^2 = (u + v)(u - v)$ applies to polynomials or expressions in which u and v are themselves expressions.

Example 7   Factoring the Difference of Two Squares

Factor the expression $(x + 2)^2 - 9$.

Solution

$(x + 2)^2 - 9 = (x + 2)^2 - 3^2$ Write as difference of two squares.

$\qquad\qquad = [(x + 2) + 3][(x + 2) - 3]$ Factored form

$\qquad\qquad = (x + 5)(x - 1)$ Simplify.

To check this result, write the original polynomial in standard form. Then multiply the factored form to see that you obtain the same standard form.

Sum or Difference of Two Cubes

Let u and v be real numbers, variables, or algebraic expressions. Then the expressions $u^3 + v^3$ and $u^3 - v^3$ can be factored as follows.

Like signs

1. $u^3 + v^3 = (u + v)(u^2 - uv + v^2)$

Unlike signs

Like signs

2. $u^3 - v^3 = (u - v)(u^2 + uv + v^2)$

Unlike signs

Example 8   Factoring Sums and Differences of Cubes

Factor each polynomial.

a. $x^3 - 125$ **b.** $8y^3 + 1$ **c.** $y^3 - 27x^3$

Solution

a. This polynomial is the difference of two cubes because x^3 is the cube of x and 125 is the cube of 5.

$$x^3 - 125 = x^3 - 5^3 \qquad \text{Write as difference of two cubes.}$$
$$= (x - 5)(x^2 + 5x + 5^2) \qquad \text{Factored form}$$
$$= (x - 5)(x^2 + 5x + 25) \qquad \text{Simplify.}$$

b. This polynomial is the sum of two cubes because $8y^3$ is the cube of $2y$ and 1 is the cube of 1.

$$8y^3 + 1 = (2y)^3 + 1^3 \qquad \text{Write as sum of two cubes.}$$
$$= (2y + 1)[(2y)^2 - (2y)(1) + 1^2] \qquad \text{Factored form}$$
$$= (2y + 1)(4y^2 - 2y + 1) \qquad \text{Simplify.}$$

c. $y^3 - 27x^3 = y^3 - (3x)^3 \qquad \text{Write as difference of two cubes.}$

$$= (y - 3x)[y^2 + 3xy + (3x)^2] \qquad \text{Factored form}$$
$$= (y - 3x)(y^2 + 3xy + 9x^2) \qquad \text{Simplify.}$$

You can check the results of Example 8(a) by multiplying, as follows.

$$(x - 5)(x^2 + 5x + 25) = x(x^2 + 5x + 25) - 5(x^2 + 5x + 25)$$
$$= x(x^2) + x(5x) + x(25) - 5(x^2) - 5(5x) - 5(25)$$
$$= x^3 + 5x^2 + 25x - 5x^2 - 25x - 125$$
$$= x^3 - 125$$

④ Factor polynomials completely by repeated factoring.

Factoring Completely

Sometimes the difference of two squares can be hidden by the presence of a common monomial factor. Remember that with *all* factoring techniques, you should first factor out any common monomial factors.

Example 9   Factoring Completely

Factor the polynomial $125x^2 - 80$ completely.

Solution

Because both terms have a common factor of 5, begin by factoring 5 from the expression.

$$125x^2 - 80 = 5(25x^2 - 16) \qquad \text{Factor out common monomial factor.}$$
$$= 5[(5x)^2 - 4^2] \qquad \text{Write as difference of two squares.}$$
$$= 5(5x + 4)(5x - 4) \qquad \text{Factored form}$$

The polynomial in Example 9 is said to be **completely factored** because none of its factors can be further factored using integer coefficients.

Example 10   Factoring Completely

Factor each polynomial completely: **a.** $x^4 - y^4$ **b.** $81m^4 - 1$

Solution

a. $x^4 - y^4 = (x^2 + y^2)(x^2 - y^2)$ \qquad Factor as difference of two squares.
$$= (x^2 + y^2)(x^2 - y^2) \qquad \text{Find second difference of two squares.}$$
$$= (x^2 + y^2)(x + y)(x - y) \qquad \text{Factored completely}$$

b. $81m^4 - 1 = (9m^2 + 1)(9m^2 - 1)$ \qquad Factor as difference of two squares.
$$= (9m^2 + 1)(9m^2 - 1) \qquad \text{Find second difference of two squares.}$$
$$= (9m^2 + 1)(3m + 1)(3m - 1) \qquad \text{Factored completely}$$

> **Study Tip**
>
> The sum of two squares, such as $9m^2 + 1$ in Example 10(b), cannot be factored further using integer coefficients. Such polynomials are called **prime** with respect to the integers. Some other prime polynomials are $x^2 + 4$ and $4x^2 + 9$.

Example 11   Geometry: Area of a Rectangle

The area of a rectangle of width $4x$ is given by the polynomial $12x^2 + 32x$, as shown in Figure 5.5. Factor this expression to determine the length of the rectangle.

Solution

The polynomial for the area of the rectangle factors as follows.

$$12x^2 + 32x = 4x(3x + 8) \qquad \text{Factor out common monomial factor.}$$

So, the length of the rectangle is $3x + 8$.

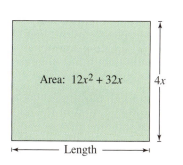

Area: $12x^2 + 32x$ � $4x$

Length

Figure 5.5

5.4 Exercises

Developing Skills

In Exercises 1–10, find the greatest common factor of the expressions. See Example 1.

1. 48, 90, 96

2. 36, 150, 100

3. $3x^2, 12x$

4. $27x^4, 18x^3$

5. $30z^2, -12z^3$

6. $-45y, 150y^3$

7. $28b^2, 14b^3, 42b^5$

8. $16x^2y, 84xy^2, 36x^2y^2$

9. $42(x + 8)^2, 63(x + 8)^3$

10. $66(3 - y), 44(3 - y)^2$

In Exercises 11–30, factor out the greatest common monomial factor. (Some of the polynomials have no common monomial factor.) See Example 2.

11. $4x + 4$

12. $7y - 7$

13. $6y - 20$

14. $9x + 30$

15. $24t^2 - 36$

16. $54x^2 - 36$

17. $x^2 + 9x$

18. $y^2 - 5y$

19. $2t^2 + 8t$

20. $3x^2 + 6x$

21. $11u^2 + 9$

22. $16 - 3y^3$

23. $28x^2 + 16x - 8$

24. $9 - 27y - 15y^2$

25. $3x^2y^2 - 15y$

26. $4uv + 6u^2v^2$

27. $15xy^2 - 3x^2y + 9xy$

28. $4x^2 - 2xy + 3y^2$

29. $14x^4y^3 + 21x^3y^2 + 9x^2$

30. $17x^5y^3 - xy^2 + 34y^2$

In Exercises 31–38, factor a negative real number from the polynomial and then write the polynomial factor in standard form. See Example 3.

31. $7 - y^2$

32. $4 - x^3$

33. $7 - 14x$

34. $15 - 5x$

35. $16 + 4x - 6x^2$

36. $12x - 6x^2 - 18$

37. $y - 3y^3 - 2y^2$

38. $-2t^3 + 4t^2 + 7$

In Exercises 39–42, fill in the missing factor.

39. $2y - \frac{3}{5} = \frac{1}{5}\left(\right)$

40. $3z + \frac{3}{8} = \frac{1}{8}\left(\right)$

41. $\frac{3}{2}x + \frac{5}{4} = \frac{1}{4}\left(\right)$

42. $\frac{1}{3}x - \frac{5}{6} = \frac{1}{6}\left(\right)$

In Exercises 43–52, factor the expression by factoring out the common binomial factor. See Example 4.

43. $2y(y - 4) + 5(y - 4)$

44. $7t(s + 9) - 6(s + 9)$

45. $5x(3x + 2) - 3(3x + 2)$

46. $6(4t - 3) - 5t(4t - 3)$

47. $2(7a + 6) - 3a^2(7a + 6)$

48. $4(5y - 12) + 3y^2(5y - 12)$

49. $8t^3(4t - 1)^2 + 3(4t - 1)^2$

50. $2y^2(y^2 + 6)^3 + 7(y^2 + 6)^3$

51. $(x - 5)(4x + 9) - (3x + 4)(4x + 9)$

52. $(3x + 7)(2x - 1) + (x - 6)(2x - 1)$

In Exercises 53–64, factor the polynomial by grouping. See Example 5.

53. $x^2 + 25x + x + 25$

54. $x^2 - 9x + x - 9$

55. $y^2 - 6y + 2y - 12$

56. $y^2 + 3y + 4y + 12$

57. $x^3 + 2x^2 + x + 2$

58. $t^3 - 11t^2 + t - 11$

59. $3a^3 - 12a^2 - 2a + 8$

60. $3s^3 + 6s^2 + 5s + 10$

61. $z^4 - 2z + 3z^3 - 6$

62. $4u^4 - 6u - 2u^3 + 3$

63. $5x^3 - 10x^2y + 7xy^2 - 14y^3$

64. $10u^4 - 8u^2v^3 - 12v^4 + 15u^2v$

In Exercises 65–86, factor the difference of two squares. See Examples 6 and 7.

65. $x^2 - 25$

66. $y^2 - 144$

67. $1 - a^2$

68. $16 - b^2$

69. $16y^2 - 9$

70. $9z^2 - 36$

71. $81 - 4x^2$

72. $49 - 64x^2$

73. $4z^2 - y^2$

74. $9u^2 - v^2$

75. $36x^2 - 25y^2$

76. $100a^2 - 49b^2$

77. $u^2 - \frac{1}{16}$

78. $v^2 - \frac{9}{25}$

79. $\frac{4}{9}x^2 - \frac{16}{25}y^2$

80. $\frac{1}{4}x^2 - \frac{36}{49}y^2$

81. $(x - 1)^2 - 16$

82. $(x - 3)^2 - 4$

83. $81 - (z + 5)^2$

84. $36 - (y - 6)^2$

85. $(2x + 5)^2 - (x - 4)^2$

86. $(3y - 1)^2 - (x + 6)^2$

In Exercises 87–98, factor the sum or difference of cubes. See Example 8.

87. $x^3 - 8$

88. $t^3 - 1$

89. $y^3 + 64$

90. $z^3 + 125$

91. $8t^3 - 27$

92. $27s^3 + 64$

93. $27u^3 + 1$

94. $64v^3 - 125$

95. $64a^3 + b^3$

96. $m^3 - 8n^3$

97. $x^3 + 27y^3$

98. $u^3 + 125v^3$

In Exercises 99–108, factor the polynomial completely. See Examples 9 and 10.

99. $8 - 50x^2$

100. $8y^2 - 18$

101. $8x^3 + 64$

102. $a^3 - 16a$

103. $y^4 - 81$

104. $u^4 - 16$

105. $3x^4 - 300x^2$

106. $6x^5 + 30x^3$

107. $6x^6 - 48y^6$

108. $2u^6 + 54v^6$

In Exercises 109–114, factor the expression. (Assume that all exponents represent positive integers.)

109. $4x^{2n} - 25$

110. $81 - 16y^{4n}$

111. $2x^{3r} + 8x^r + 4x^{2r}$

112. $3x^{n+1} + 6x^n - 15x^{n+2}$

113. $4y^{m+n} + 7y^{2m+n} - y^{m+2n}$

114. $x^{2r+s} - 5x^{r+3s} + 10x^{2r+2s}$

Graphical Reasoning   In Exercises 115–118, use a graphing calculator to graph y_1 and y_2 in the same viewing window. What can you conclude?

115. $y_1 = 3x - 6$
$y_2 = 3(x - 2)$

116. $y_1 = x^3 - 2x^2$
$y_2 = x^2(x - 2)$

117. $y_1 = x^2 - 4$
$y_2 = (x + 2)(x - 2)$

118. $y_1 = x(x + 1) - 4(x + 1)$
$y_2 = (x + 1)(x - 4)$

Think About It   In Exercises 119 and 120, show all the different groupings that can be used to factor the polynomial completely. Carry out the various factorizations to show that they yield the same result.

119. $3x^3 + 4x^2 - 3x - 4$

120. $6x^3 - 8x^2 + 9x - 12$

Solving Problems

Revenue   The revenue from selling x units of a product at a price of p dollars per unit is given by $R = xp$. In Exercises 121 and 122, factor the expression for revenue and determine an expression that gives the price in terms of x.

121. $R = 800x - 0.25x^2$

122. $R = 1000x - 0.4x^2$

123. *Simple Interest*   The total amount of money accrued from a principal of P dollars invested at $r\%$ simple interest for t years is given by $P + Prt$. Factor this expression.

124. *Chemical Reaction*   The rate of change of a chemical reaction is given by $kQx - kx^2$, where Q is the amount of the original substance, x is the amount of substance formed, and k is a constant of proportionality. Factor this expression.

125. ▲ *Geometry*   The area of a rectangle of length l is given by the polynomial $45l - l^2$. Factor this expression to determine the width of the rectangle.

126. ▲ *Geometry*   The area of a rectangle of width w is given by the polynomial $32w - w^2$. Factor this expression to determine the length of the rectangle.

127. ▲ *Geometry*   The surface area of a rectangular solid of height h and square base with edge of length x is given by $2x^2 + 4xh$. Factor this expression.

128. ▲ *Geometry*   The surface area of a right circular cylinder is given by $S = \pi r^2 + 2\pi rh$ (see figure). Factor this expression.

129. *Product Design*   A washer on the drive train of a car has an inside radius of r centimeters and an outside radius of R centimeters (see figure). Find the area of one of the flat surfaces of the washer and write the area in factored form.

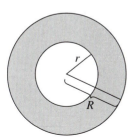

130. ▲ *Geometry*   The cube shown in the figure is formed by solids I, II, III, and IV.

(a) Explain how you could determine each expression for volume.

	Volume
Entire cube	a^3
Solid I	$a^2(a - b)$
Solid II	$ab(a - b)$
Solid III	$b^2(a - b)$
Solid IV	b^3

(b) Add the volumes of solids I, II, and III. Factor the result to show that the total volume can be expressed as $(a - b)(a^2 + ab + b^2)$.

(c) Explain why the total volume of solids I, II, and III can also be expressed as $a^3 - b^3$. Then explain how the figure can be used as a geometric model for the *difference of two cubes* factoring pattern.

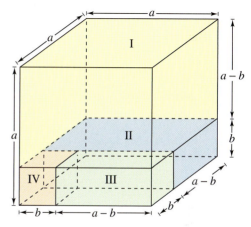

Figure for 130

Explaining Concepts

131. *Writing* ✏   Explain what is meant by saying that a polynomial is in factored form.

132. *Writing* ✏   How do you check your result after factoring a polynomial?

133. *Writing* ✏   Describe the method of finding the greatest common factor of two (or more) integers.

134. *Writing* ✏   Explain how the word *factor* can be used as a noun or as a verb.

135. Give an example of using the Distributive Property to factor a polynomial.

136. Give an example of a polynomial that is prime with respect to the integers.

5.5 Factoring Trinomials

What You Should Learn

1. Recognize and factor perfect square trinomials.
2. Factor trinomials of the forms $x^2 + bx + c$ and $ax^2 + bx + c$.
3. Factor trinomials of the form $ax^2 + bx + c$ by grouping.
4. Factor polynomials using the guidelines for factoring.

Why You Should Learn It

The technique for factoring trinomials will help you in solving quadratic equations in Section 5.6.

Perfect Square Trinomials

A **perfect square trinomial** is the square of a binomial. For instance,

$$x^2 + 6x + 9 = (x + 3)(x + 3) = (x + 3)^2$$

is the square of the binomial $(x + 3)$. Perfect square trinomials come in one of two forms: the middle term is either positive or negative.

1. Recognize and factor perfect square trinomials.

Perfect Square Trinomials

Let u and v represent real numbers, variables, or algebraic expressions.

1. $u^2 + 2uv + v^2 = (u + v)^2$ **2.** $u^2 - 2uv + v^2 = (u - v)^2$

                Same sign   Same sign

To recognize a perfect square trinomial, remember that the first and last terms must be perfect squares and positive, and the middle term must be *twice* the product of u and v. (The middle term can be positive or negative.)

Example 1   Factoring Perfect Square Trinomials

a. $x^2 - 4x + 4 = x^2 - 2(x)(2) + 2^2 = (x - 2)^2$

b. $16y^2 + 24y + 9 = (4y)^2 + 2(4y)(3) + 3^2 = (4y + 3)^2$

c. $9x^2 - 30xy + 25y^2 = (3x)^2 - 2(3x)(5y) + (5y)^2 = (3x - 5y)^2$

Example 2   Factoring Out a Common Monomial Factor First

a. $3x^2 - 30x + 75 = 3(x^2 - 10x + 25)$ Factor out common monomial factor.

                   $= 3(x - 5)^2$ Factor as perfect square trinomial.

b. $16y^3 + 80y^2 + 100y = 4y(4y^2 + 20y + 25)$ Factor out common monomial factor.

                   $= 4y(2y + 5)^2$ Factor as perfect square trinomial.

2 Factor trinomials of the forms $x^2 + bx + c$ and $ax^2 + bx + c$.

Factoring Trinomials

To factor a trinomial of the form $x^2 + bx + c$, consider the following.

$$(x + m)(x + n) = x^2 + nx + mx + mn$$

$$= x^2 + \underbrace{(m + n)}x + \underbrace{mn}$$

Sum of   Product
terms   of terms

$$= x^2 + \quad b \quad x + \quad c$$

From this, you can see that to factor a trinomial $x^2 + bx + c$ into a product of two binomials, you must find *two factors of c whose sum is b*. There are many different techniques for factoring trinomials. The most common technique is to use *guess, check, and revise* with mental math.

Example 3   Factoring a Trinomial of the Form $x^2 + bx + c$

Factor the trinomial $x^2 + 3x - 4$.

Solution

You need to find two numbers whose product is -4 and whose sum is 3. Using mental math, you can determine that the numbers are 4 and -1.

The product of 4 and -1 is -4.

$$x^2 + 3x - 4 = (x + 4)(x - 1)$$

The sum of 4 and -1 is 3.

Example 4   Factoring Trinomials of the Form $x^2 + bx + c$

Factor each trinomial.

a. $x^2 - 2x - 8$

b. $x^2 - 5x + 6$

Solution

a. You need to find two numbers whose product is -8 and whose sum is -2.

The product of -4 and 2 is -8.

$$x^2 - 2x - 8 = (x - 4)(x + 2)$$

The sum of -4 and 2 is -2.

b. You need to find two numbers whose product is 6 and whose sum is -5.

The product of -3 and -2 is 6.

$$x^2 - 5x + 6 = (x - 3)(x - 2)$$

The sum of -3 and -2 is -5.

Study Tip

Use a list to help you find the two numbers with the required product and sum. For Example 4(a):

Factors of -8	Sum
$1, -8$	-7
$-1, 8$	7
$2, -4$	-2
$-2, 4$	2

Because -2 is the required sum, the correct factorization is

$$x^2 - 2x - 8 = (x - 4)(x + 2).$$

When factoring a trinomial of the form $x^2 + bx + c$, if you have trouble finding two factors of c whose sum is b, it may be helpful to list all of the distinct pairs of factors, and then choose the appropriate pair from the list. For instance, consider the trinomial

$$x^2 - 2x - 24.$$

For this trinomial, $c = -24$ and $b = -2$. So, you need two factors of -24 whose sum is -2. Here is the complete list.

Factors of -24	*Sum of Factors*
$1, -24$	$1 - 24 = -23$
$-1, 24$	$-1 + 24 = 23$
$2, -12$	$2 - 12 = -10$
$-2, 12$	$-2 + 12 = 10$
$3, -8$	$3 - 8 = -5$
$-3, 8$	$-3 + 8 = 5$
$4, -6$	$4 - 6 = -2$ ⬅ Correct choice
$-4, 6$	$-4 + 6 = 2$

With experience, you will be able to narrow this list down *mentally* to only two or three possibilities whose sums can then be tested to determine the correct factorization. Here are some suggestions for narrowing down the list.

Guidelines for Factoring $x^2 + bx + c$

1. If c is *positive*, its factors have like signs that match the sign of b.

2. If c is *negative*, its factors have unlike signs.

3. If $|b|$ is small relative to $|c|$, first try those factors of c that are closest to each other in absolute value.

4. If $|b|$ is near $|c|$, first try those factors of c that are farthest from each other in absolute value.

Study Tip

With *any* factoring problem, remember that you can check your result by multiplying. For instance, in Example 5, you can check the result by multiplying $(x - 18)$ by $(x + 1)$ to see that you obtain $x^2 - 17x - 18$.

Remember that not all trinomials are factorable using integers. For instance, $x^2 - 2x - 4$ is not factorable using integers because there is no pair of factors of -4 whose sum is -2.

Example 5   Factoring a Trinomial of the Form $x^2 + bx + c$

Factor $x^2 - 17x - 18$.

Solution

You need to find two numbers whose product is -18 and whose sum is -17. Because $|b| = |-17| = 17$ and $|c| = |-18| = 18$ are close in value, choose factors of -18 that are farthest from each other.

The product of -18 and 1 is -18.

$$x^2 - 17x - 18 = (x - 18)(x + 1)$$

The sum of -18 and 1 is -17.

To factor a trinomial whose leading coefficient is not 1, use the following pattern.

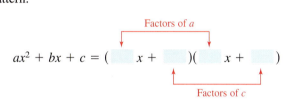

Factors of a

$$ax^2 + bx + c = (\quad x + \quad)(\quad x + \quad)$$

Factors of c

The goal is to find a combination of factors of a and c such that the outer and inner products add up to the middle term bx.

Example 6   Factoring a Trinomial of the Form $ax^2 + bx + c$

Factor the trinomial $4x^2 + 5x - 6$.

Solution

First, observe that $4x^2 + 5x - 6$ has no common monomial factor. For this trinomial, $a = 4$, which factors as $(1)(4)$ or $(2)(2)$, and $c = -6$, which factors as $(-1)(6)$, $(1)(-6)$, $(-2)(3)$, or $(2)(-3)$. A test of the many possibilities is shown below.

Factors	$O + I$	
$(x + 1)(4x - 6)$	$-6x + 4x = -2x$	$-2x$ does not equal $5x$.
$(x - 1)(4x + 6)$	$6x - 4x = 2x$	$2x$ does not equal $5x$.
$(x + 6)(4x - 1)$	$-x + 24x = 23x$	$23x$ does not equal $5x$.
$(x - 6)(4x + 1)$	$x - 24x = -23x$	$-23x$ does not equal $5x$.
$(x - 2)(4x + 3)$	$3x - 8x = -5x$	$-5x$ does not equal $5x$.
$(x + 2)(4x - 3)$	$-3x + 8x = 5x$	$5x$ equals $5x$. ✔
$(2x + 1)(2x - 6)$	$-12x + 2x = -10x$	$-10x$ does not equal $5x$.
$(2x - 1)(2x + 6)$	$12x - 2x = 10x$	$10x$ does not equal $5x$.
$(2x + 2)(2x - 3)$	$-6x + 4x = -2x$	$-2x$ does not equal $5x$.
$(2x - 2)(2x + 3)$	$6x - 4x = 2x$	$2x$ does not equal $5x$.
$(x + 3)(4x - 2)$	$-2x + 12x = 10x$	$10x$ does not equal $5x$.
$(x - 3)(4x + 2)$	$2x - 12x = -10x$	$-10x$ does not equal $5x$.

So, you can conclude that the correct factorization is

$$4x^2 + 5x - 6 = (x + 2)(4x - 3).$$

Check this result by multiplying $(x + 2)$ by $(4x - 3)$.

Study Tip

If the original trinomial has no common monomial factor, its binomial factors cannot have common monomial factors. So, in Example 6, you do not have to test factors, such as $(4x - 6)$, that have a common monomial factor of 2. Which of the other factors in Example 6 did not need to be tested?

The guidelines on the following page can help shorten the list of possible factorizations of a trinomial.

Guidelines for Factoring $ax^2 + bx + c$

1. If the trinomial has a common monomial factor, you should factor out the common factor before trying to find binomial factors.

2. Because the resulting trinomial has no common monomial factors, you do not have to test any binomial factors that have a common monomial factor.

3. Do not switch the signs of the factors of c unless the middle term $(O + I)$ is correct except in sign.

Example 7   Factoring a Trinomial of the Form $ax^2 + bx + c$

Factor the trinomial $2x^2 - x - 21$.

Solution

First observe that $2x^2 - x - 21$ has no common monomial factor. For this trinomial, $a = 2$, which factors as $(1)(2)$, and $c = -21$, which factors as $(1)(-21)$, $(-1)(21)$, $(3)(-7)$, or $(-3)(7)$. Because b is small, avoid the large factors of -21, and test the smaller ones.

Factors	$O + I$	
$(2x + 3)(x - 7)$	$-14x + 3x = -11x$	$-11x$ does not equal $-x$.
$(2x + 7)(x - 3)$	$-6x + 7x = x$	x does not equal $-x$.

Because $(2x + 7)(x - 3)$ results in a middle term that is correct except in sign, you need only switch the signs of the factors of c to obtain the correct factorization.

$$2x^2 - x - 21 = (2x - 7)(x + 3) \qquad \text{Correct factorization}$$

Check this result by multiplying $(2x - 7)$ by $(x + 3)$.

Study Tip

Notice in Example 8 that a factorization such as $(2x + 2)(3x + 5)$ was not considered because $(2x + 2)$ has a common monomial factor of 2.

Example 8   Factoring a Trinomial of the Form $ax^2 + bx + c$

Factor the trinomial $6x^2 + 19x + 10$.

Solution

First observe that $6x^2 + 19x + 10$ has no common monomial factor. For this trinomial, $a = 6$, which factors as $(1)(6)$ or $(2)(3)$, and $c = 10$, which factors as $(1)(10)$ or $(2)(5)$. You can test the potential factors as follows.

Factors	$O + I$	
$(x + 10)(6x + 1)$	$x + 60x = 61x$	$61x$ does not equal $19x$.
$(x + 2)(6x + 5)$	$5x + 12x = 17x$	$17x$ does not equal $19x$.
$(2x + 1)(3x + 10)$	$20x + 3x = 23x$	$23x$ does not equal $19x$.
$(2x + 5)(3x + 2)$	$4x + 15x = 19x$	$19x$ equals $19x$. ✓

So, the correct factorization is $6x^2 + 19x + 10 = (2x + 5)(3x + 2)$.

Example 9   Factoring Completely

Factor the trinomial $8x^2y - 60xy + 28y$ completely.

Solution

Begin by factoring out the common monomial factor $4y$.

$$8x^2y - 60xy + 28y = 4y(2x^2 - 15x + 7)$$

Now, for the new trinomial $2x^2 - 15x + 7$, $a = 2$ and $c = 7$. The possible factorizations of this trinomial are as follows.

Factors	$O + I$	
$(2x - 7)(x - 1)$	$-2x - 7x = -9x$	$-9x$ does not equal $-15x$.
$(2x - 1)(x - 7)$	$-14x - x = -15x$	$-15x$ equals $-15x$. ✓

So, the complete factorization of the original trinomial is

$$8x^2y - 60xy + 28y = 4y(2x^2 - 15x + 7) = 4y(2x - 1)(x - 7).$$

Check this result by multiplying.

When factoring a trinomial with a negative leading coefficient, first factor -1 out of the trinomial, as demonstrated in Example 10.

Example 10   A Trinomial with a Negative Leading Coefficient

Factor the trinomial $-3x^2 + 16x + 35$.

Solution

Begin by factoring (-1) out of the trinomial.

$$-3x^2 + 16x + 35 = (-1)(3x^2 - 16x - 35)$$

For the new trinomial $3x^2 - 16x - 35$, you have $a = 3$ and $c = -35$. Some possible factorizations of this trinomial are as follows.

Factors	$O + I$	
$(3x - 1)(x + 35)$	$105x - x = 104x$	$104x$ does not equal $-16x$.
$(3x - 35)(x + 1)$	$3x - 35x = -32x$	$-32x$ does not equal $-16x$.
$(3x - 7)(x + 5)$	$15x - 7x = 8x$	$8x$ does not equal $-16x$.
$(3x - 5)(x + 7)$	$21x - 5x = 16x$	$16x$ does not equal $-16x$.

Because $(3x - 5)(x + 7)$ results in a middle term that is correct except in sign, you need only switch the signs of the factors of c to obtain the correct factorization.

$(3x + 5)(x - 7)$	$-21x + 5x = -16x$	$-16x$ equals $-16x$. ✓

So, the correct factorization is

$$-3x^2 + 16x + 35 = (-1)(3x + 5)(x - 7) = (3x + 5)(-x + 7).$$

③ Factor trinomials of the form $ax^2 + bx + c$ by grouping.

Factoring Trinomials by Grouping (Optional)

So far in this section, you have been using *guess, check, and revise* to factor trinomials. An alternative technique is to use *factoring by grouping* to factor a trinomial. For instance, suppose you rewrite the trinomial $2x^2 + 7x - 15$ as

$$2x^2 + 7x - 15 = 2x^2 + 10x - 3x - 15.$$

Then, by grouping the first two terms and the third and fourth terms, you can factor the polynomial as follows.

$$2x^2 + 7x - 15 = 2x^2 + (10x - 3x) - 15 \qquad \text{Rewrite middle term.}$$
$$= (2x^2 + 10x) - (3x + 15) \qquad \text{Group terms.}$$
$$= 2x(x + 5) - 3(x + 5) \qquad \text{Factor out common monomial factor in each group.}$$
$$= (x + 5)(2x - 3) \qquad \text{Distributive Property}$$

Guidelines for Factoring $ax^2 + bx + c$ by Grouping

1. If necessary, write the trinomial in standard form.

2. Choose factors of the product ac that add up to b.

3. Use these factors to rewrite the middle term as a sum or difference.

4. Group and remove a common monomial factor from the first two terms and the last two terms.

5. If possible, factor out the common binomial factor.

Example 11   Factoring a Trinomial by Grouping

Use factoring by grouping to factor the trinomial $3x^2 + 5x - 2$.

Solution

For the trinomial $3x^2 + 5x - 2$, $a = 3$ and $c = -2$, which implies that the product ac is -6. Now, because -6 factors as $(6)(-1)$, and $6 - 1 = 5 = b$, you can rewrite the middle term as $5x = 6x - x$. This produces the following result.

$$3x^2 + 5x - 2 = 3x^2 + (6x - x) - 2 \qquad \text{Rewrite middle term.}$$
$$= (3x^2 + 6x) - (x + 2) \qquad \text{Group terms.}$$
$$= 3x(x + 2) - (x + 2) \qquad \text{Factor out common monomial factor in first group.}$$
$$= (x + 2)(3x - 1) \qquad \text{Distributive Property}$$

So, the trinomial factors as $3x^2 + 5x - 2 = (x + 2)(3x - 1)$. Check this result as follows.

$$(x + 2)(3x - 1) = 3x^2 - x + 6x - 2 \qquad \text{FOIL Method}$$
$$= 3x^2 + 5x - 2 \qquad \text{Combine like terms.}$$

Study Tip

Factoring by grouping can be more efficient than the *guess, check, and revise* method, especially when the coefficients a and c have many factors.

4 Factor polynomials using the guidelines for factoring.

Summary of Factoring

Although the basic factoring techniques have been discussed one at a time, from this point on you must decide which technique to apply to any given problem situation. The guidelines below should assist you in this selection process.

Guidelines for Factoring Polynomials

1. Factor out any common factors.

2. Factor according to one of the special polynomial forms: difference of two squares, sum or difference of two cubes, or perfect square trinomials.

3. Factor trinomials, $ax^2 + bx + c$, using the methods for $a = 1$ and $a \neq 1$.

4. For polynomials with four terms, factor by grouping.

5. Check to see whether the factors themselves can be factored.

6. Check the results by multiplying the factors.

Example 12   Factoring Polynomials

Factor each polynomial completely.

a. $3x^2 - 108$

b. $4x^3 - 32x^2 + 64x$

c. $6x^3 + 27x^2 - 15x$

d. $x^3 - 3x^2 - 4x + 12$

Solution

a. $3x^2 - 108 = 3(x^2 - 36)$ Factor out common factor.

$= 3(x + 6)(x - 6)$ Difference of two squares

b. $4x^3 - 32x^2 + 64x = 4x(x^2 - 8x + 16)$ Factor out common factor.

$= 4x(x - 4)^2$ Factor as perfect square trinomial.

c. $6x^3 + 27x^2 - 15x = 3x(2x^2 + 9x - 5)$ Factor out common factor.

$= 3x(2x - 1)(x + 5)$ Factor.

d. $x^3 - 3x^2 - 4x + 12 = (x^3 - 3x^2) + (-4x + 12)$ Group terms.

$= x^2(x - 3) - 4(x - 3)$ Factor out common factors.

$= (x - 3)(x^2 - 4)$ Distributive Property

$= (x - 3)(x + 2)(x - 2)$ Difference of two squares

5.5 Exercises

Review   Concepts, Skills, and Problem Solving

Keep mathematically in shape by doing these exercises *before* the problems of this section.

Properties and Definitions

1. *Writing* Explain why a function of x cannot have two y-intercepts.

2. What is the leading coefficient of the polynomial $5t - 3t^2 + 6t^3 - 4$?

In Exercises 3 and 4, write an inequality involving absolute values that represents the verbal statement.

3. The set of all real numbers x whose distance from 0 is less than 5.

4. The set of all real numbers x whose distance from 6 is more than 3.

Slope

In Exercises 5–10, plot the points on a rectangular coordinate system and find the slope (if possible) of the line passing through the points.

5. $(-3, 2), (5, -4)$

6. $(2, 8), (7, -3)$

7. $\left(\frac{5}{2}, \frac{7}{2}\right), \left(\frac{7}{3}, -2\right)$

8. $\left(-\frac{9}{4}, -\frac{1}{4}\right), \left(-3, \frac{9}{2}\right)$

9. $(6, 4), (6, -3)$

10. $(-4, 5), (7, 5)$

Problem Solving

11. *Simple Interest*   You borrow $12,000 for 6 months. You agree to pay back the principal and interest in one lump sum. The simple interest rate for the loan is 12%. What will be the amount of the payment?

12. *Average Speed*   A truck driver traveled at an average speed of 54 miles per hour on a 100-mile trip. On the return trip with the truck fully loaded, the average speed was 45 miles per hour. Find the average speed for the round trip.

Developing Skills

In Exercises 1–20, factor the perfect square trinomial. See Examples 1 and 2.

1. $x^2 + 4x + 4$

2. $z^2 + 6z + 9$

3. $a^2 - 12a + 36$

4. $y^2 - 14y + 49$

5. $25y^2 - 10y + 1$

6. $4z^2 + 28z + 49$

7. $9b^2 + 12b + 4$

8. $4x^2 - 4x + 1$

9. $u^2 + 8uv + 16v^2$

10. $x^2 - 14xy + 49y^2$

11. $36x^2 - 60xy + 25y^2$

12. $4y^2 + 20yz + 25z^2$

13. $5x^2 + 30x + 45$

14. $4x^2 - 32x + 64$

15. $2x^3 + 24x^2 + 72x$

16. $3u^3 - 48u^2 + 192u$

17. $20v^4 - 60v^3 + 45v^2$

18. $-18y^3 - 12y^2 - 2y$

19. $\frac{1}{4}x^2 - \frac{2}{3}x + \frac{4}{9}$

20. $\frac{1}{9}x^2 + \frac{8}{15}x + \frac{16}{25}$

In Exercises 21–24, find two real numbers b such that the expression is a perfect square trinomial.

21. $x^2 + bx + 81$

22. $x^2 + bx + \frac{9}{16}$

23. $4x^2 + bx + 9$

24. $16x^2 + bxy + 25y^2$

In Exercises 25–28, find a real number c such that the expression is a perfect square trinomial.

25. $x^2 + 8x + c$

26. $x^2 + 12x + c$

27. $y^2 - 6y + c$

28. $z^2 - 20z + c$

In Exercises 29–36, fill in the missing factor.

29. $x^2 + 5x + 4 = (x + 4)()$

30. $a^2 + 2a - 8 = (a + 4)()$

31. $y^2 - y - 20 = (y + 4)()$

32. $y^2 + 6y + 8 = (y + 4)()$

33. $x^2 - 2x - 24 = (x + 4)()$

34. $x^2 + 7x + 12 = (x + 4)()$

35. $z^2 - 6z + 8 = (z - 4)()$

36. $z^2 + 2z - 24 = (z - 4)()$

In Exercises 37–50, factor the trinomial. See Examples 3–5.

37. $x^2 + 4x + 3$ **38.** $x^2 + 7x + 10$

39. $x^2 - 5x + 6$ **40.** $x^2 - 10x + 24$

41. $y^2 + 7y - 30$ **42.** $m^2 - 3m - 10$

43. $t^2 - 4t - 21$ **44.** $x^2 + 4x - 12$

45. $x^2 - 20x + 96$ **46.** $y^2 - 35y + 300$

47. $x^2 - 2xy - 35y^2$ **48.** $u^2 + 5uv + 6v^2$

49. $x^2 + 30xy + 216y^2$ **50.** $a^2 - 21ab + 110b^2$

In Exercises 51–56, find all integers b such that the trinomial can be factored.

51. $x^2 + bx + 18$ **52.** $x^2 + bx + 14$

53. $x^2 + bx - 21$ **54.** $x^2 + bx - 7$

55. $x^2 + bx + 35$ **56.** $x^2 + bx - 38$

In Exercises 57–60, find two integers c such that the trinomial can be factored. (There are many correct answers.)

57. $x^2 + 6x + c$ **58.** $x^2 + 9x + c$

59. $x^2 - 3x + c$ **60.** $x^2 - 12x + c$

In Exercises 61–66, fill in the missing factor.

61. $5x^2 + 18x + 9 = (x + 3)()$

62. $5x^2 + 19x + 12 = (x + 3)()$

63. $5a^2 + 12a - 9 = (a + 3)()$

64. $5c^2 + 11c - 12 = (c + 3)()$

65. $2y^2 - 3y - 27 = (y + 3)()$

66. $3y^2 - y - 30 = (y + 3)()$

In Exercises 67–92, factor the trinomial, if possible. (*Note:* Some of the trinomials may be prime.) See Examples 6–10.

67. $6x^2 - 5x - 25$ **68.** $3x^2 - 16x - 12$

69. $10y^2 - 7y - 12$ **70.** $6x^2 - x - 15$

71. $12x^2 - 7x + 1$ **72.** $3y^2 - 10y + 8$

73. $5z^2 + 2z - 3$ **74.** $15x^2 + 4x - 3$

75. $2t^2 - 7t - 4$ **76.** $3z^2 - z - 4$

77. $6b^2 + 19b - 7$ **78.** $10x^2 - 24x - 18$

79. $18y^2 + 35y + 12$ **80.** $20x^2 + x - 12$

81. $-2x^2 - x + 6$ **82.** $-6x^2 + 5x - 6$

83. $1 - 11x - 60x^2$ **84.** $2 + 5x - 12x^2$

85. $6x^2 - 3x - 84$ **86.** $12x^2 + 32x - 12$

87. $60y^3 + 35y^2 - 50y$ **88.** $12x^2 + 42x^3 - 54x^4$

89. $10a^2 + 23ab + 6b^2$ **90.** $6u^2 - 5uv - 4v^2$

91. $24x^2 - 14xy - 3y^2$ **92.** $10x^2 + 9xy - 9y^2$

In Exercises 93–98, factor the trinomial by grouping. See Example 11.

93. $3x^2 + 10x + 8$ **94.** $2x^2 + 9x + 9$

95. $6x^2 + x - 2$ **96.** $6x^2 - x - 15$

97. $15x^2 - 11x + 2$ **98.** $12x^2 - 28x + 15$

In Exercises 99–114, factor the expression completely.
See Example 12.

99. $3x^4 - 12x^3$ **100.** $20y^2 - 45$

101. $10t^3 + 2t^2 - 36t$ **102.** $16z^3 - 56z^2 + 49z$

103. $54x^3 - 2$ **104.** $3t^3 - 24$

105. $27a^3b^4 - 9a^2b^3 - 18ab^2$
106. $8m^3n + 20m^2n^2 - 48mn^3$
107. $x^3 + 2x^2 - 16x - 32$ **108.** $x^3 - 7x^2 - 4x + 28$

109. $36 - (z + 3)^2$
110. $(x + 7y)^2 - 4a^2$
111. $x^2 - 10x + 25 - y^2$
112. $a^2 - 2ab + b^2 - 16$
113. $x^8 - 1$
114. $x^4 - 16y^4$

In Exercises 115–120, factor the trinomial. (Assume that
n represents a positive integer.)

115. $x^{2n} - 5x^n - 24$ **116.** $y^{2n} + y^n - 2$

117. $x^{2n} + 3x^n - 10$ **118.** $x^{2n} + 4x^n - 12$

119. $4y^{2n} - 4y^n - 3$ **120.** $3x^{2n} - 16x^n - 12$

Graphical Reasoning   In Exercises 121–124, use a
graphing calculator to graph the two equations in the
same viewing window. What can you conclude?

121. $y_1 = x^2 + 6x + 9$
    $y_2 = (x + 3)^2$
122. $y_1 = 4x^2 - 4x + 1$
    $y_2 = (2x - 1)^2$
123. $y_1 = x^2 + 2x - 3$
    $y_2 = (x - 1)(x + 3)$
124. $y_1 = 3x^2 - 8x - 16$
    $y_2 = (3x + 4)(x - 4)$

Solving Problems

Geometric Modeling   In Exercises 125–128,
match the geometric factoring model with the correct
factoring formula. [The models are labeled (a), (b), (c),
and (d).]

(a)

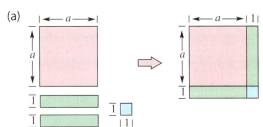

(b)

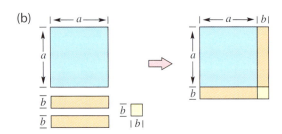

(c)

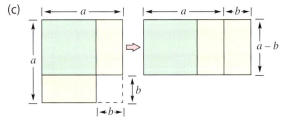

(d)

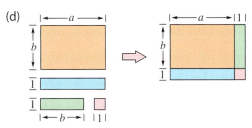

125. $a^2 - b^2 = (a + b)(a - b)$
126. $a^2 + 2a + 1 = (a + 1)^2$
127. $a^2 + 2ab + b^2 = (a + b)^2$
128. $ab + a + b + 1 = (a + 1)(b + 1)$

▲ *Geometry*   In Exercises 129 and 130, write, in factored form, an expression for the area of the shaded region of the figure.

129.

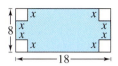

130.

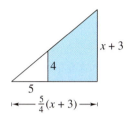

131. *Number Problem*   Let n be an integer.

(a) Factor $8n^3 - 8n$ so as to verify that it represents three consecutive even integers.   (*Hint:* Show that each factor has a common factor of 2.)

(b) If $n = 10$, what are the three integers?

132. *Number Problem*   Let n be an integer.

(a) Factor $8n^3 + 12n^2 - 2n - 3$ so as to verify that it represents three consecutive odd integers.

(b) If $n = 15$, what are the three integers?

Explaining Concepts

133. *Writing*✐   In your own words, explain how you would factor $x^2 - 5x + 6$.

134. Give an example of a prime trinomial.

135. *Writing*✐   Explain how you can check the factors of a trinomial. Give an example.

136. *Error Analysis*   Describe the error.

$$9x^2 - 9x - 54 = (3x + 6)(3x - 9)$$
$$= 3(x + 2)(x - 3)$$

137. Is $x(x + 2) - 2(x + 2)$ completely factored? If not, show the complete factorization.

138. Is $(2x - 4)(x + 1)$ completely factored? If not, show the complete factorization.

139. *Writing*✐   Create five factoring problems that you think represent a fair test of a person's factoring skills. Discuss how it is possible to *create* polynomials that are factorable.

5.6 Solving Polynomial Equations by Factoring

Carol Havens/Corbis

What You Should Learn

1 Use the Zero-Factor Property to solve equations.

2 Solve quadratic equations by factoring.

3 Solve higher-degree polynomial equations by factoring.

4 Solve application problems by factoring.

Why You Should Learn It

Quadratic equations can be used to model and solve real-life problems. For instance, Exercise 103 on page 358 shows how a quadratic equation can be used to model the time it takes an object thrown from the Royal Gorge Bridge to reach the ground.

1 Use the Zero-Factor Property to solve equations.

Study Tip

The Zero-Factor Property is just another way of saying that the only way the product of two or more factors can be zero is if one (or more) of the factors is zero.

The Zero-Factor Property

You have spent the first five sections of this chapter developing skills for *rewriting* (simplifying and factoring) polynomials. In this section you will use these skills, together with the **Zero-Factor Property,** to solve polynomial equations.

Zero-Factor Property

Let a and b be real numbers, variables, or algebraic expressions. If a and b are factors such that

$$ab = 0$$

then $a = 0$ or $b = 0$. This property also applies to three or more factors.

The Zero-Factor Property is the primary property for solving equations in algebra. For instance, to solve the equation

$$(x - 1)(x + 2) = 0 \qquad \text{Original equation}$$

you can use the Zero-Factor Property to conclude that either $(x - 1)$ or $(x + 2)$ must be zero. Setting the first factor equal to zero implies that $x = 1$ is a solution.

$$x - 1 = 0 \quad \Longrightarrow \quad x = 1 \qquad \text{First solution}$$

Similarly, setting the second factor equal to zero implies that $x = -2$ is a solution.

$$x + 2 = 0 \quad \Longrightarrow \quad x = -2 \qquad \text{Second solution}$$

So, the equation $(x - 1)(x + 2) = 0$ has exactly two solutions: $x = 1$ and $x = -2$. Check these solutions by substituting them in the original equation.

$$(x - 1)(x + 2) = 0 \qquad \text{Write original equation.}$$

$$(1 - 1)(1 + 2) \overset{?}{=} 0 \qquad \text{Substitute 1 for } x.$$

$$(0)(3) = 0 \qquad \text{First solution checks.} \ \checkmark$$

$$(-2 - 1)(-2 + 2) \overset{?}{=} 0 \qquad \text{Substitute } -2 \text{ for } x.$$

$$(-3)(0) = 0 \qquad \text{Second solution checks.} \ \checkmark$$

2 Solve quadratic equations by factoring.

Solving Quadratic Equations by Factoring

Definition of Quadratic Equation

A **quadratic equation** is an equation that can be written in the general form

$$ax^2 + bx + c = 0 \qquad \text{Quadratic equation}$$

where a, b, and c are real numbers with $a \neq 0$.

Here are some examples of quadratic equations.

$$x^2 - 2x - 3 = 0, \quad 2x^2 + x - 1 = 0, \quad x^2 - 5x = 0$$

In the next four examples, note how you can combine your factoring skills with the Zero-Factor Property to solve quadratic equations.

Example 1   Using Factoring to Solve a Quadratic Equation

Solve $x^2 - x - 6 = 0$.

Solution

First, make sure that the right side of the equation is zero. Next, factor the left side of the equation. Finally, apply the Zero-Factor Property to find the solutions.

$x^2 - x - 6 = 0$	Write original equation.
$(x + 2)(x - 3) = 0$	Factor left side of equation.
$x + 2 = 0 \implies x = -2$	Set 1st factor equal to 0 and solve for x.
$x - 3 = 0 \implies x = 3$	Set 2nd factor equal to 0 and solve for x.

The equation has two solutions: $x = -2$ and $x = 3$.

Check

$(-2)^2 - (-2) - 6 \stackrel{?}{=} 0$	Substitute -2 for x in original equation.
$4 + 2 - 6 \stackrel{?}{=} 0$	Simplify.
$0 = 0$	Solution checks. ✔
$(3)^2 - 3 - 6 \stackrel{?}{=} 0$	Substitute 3 for x in original equation.
$9 - 3 - 6 \stackrel{?}{=} 0$	Simplify.
$0 = 0$	Solution checks. ✔

Study Tip

In Section 2.1, you learned that the general strategy for solving a linear equation is to *isolate the variable.* Notice in Example 1 that the general strategy for solving a quadratic equation is to factor the equation into linear factors.

Factoring and the Zero-Factor Property allow you to solve a quadratic equation by converting it into two *linear* equations, which you already know how to solve. This is a common strategy of algebra—to break down a given problem into simpler parts, each of which can be solved by previously learned methods.

In order for the Zero-Factor Property to be used, a polynomial equation *must* be written in **general form.** That is, the polynomial must be on one side of the equation and zero must be the only term on the other side of the equation. To write $x^2 - 3x = 10$ in general form, subtract 10 from each side of the equation.

$x^2 - 3x = 10$	Write original equation.
$x^2 - 3x - 10 = 10 - 10$	Subtract 10 from each side.
$x^2 - 3x - 10 = 0$	General form

To solve this equation, factor the left side as $(x - 5)(x + 2)$, then form the linear equations $x - 5 = 0$ and $x + 2 = 0$. The solutions of these two linear equations are $x = 5$ and $x = -2$, respectively. Be sure you see that the Zero-Factor Property can be applied only to a product that is equal to *zero*. For instance, you cannot factor the left side as $x(x - 3) = 10$ and assume that $x = 10$ and $x - 3 = 10$ yield solutions. For instance, if you substitute $x = 10$ into the original equation you obtain the false statement $70 = 10$. Similarly, when $x = 13$ is substituted into the original equation you obtain another false statement, $130 = 10$. The general strategy for solving a quadratic equation by factoring is summarized in the following guidelines.

Guidelines for Solving Quadratic Equations

1. Write the quadratic equation in general form.

2. Factor the left side of the equation.

3. Set each factor with a variable equal to zero.

4. Solve each linear equation.

5. Check each solution in the original equation.

Example 2   Solving a Quadratic Equation by Factoring

Solve $2x^2 + 5x = 12$.

Solution

$2x^2 + 5x = 12$	Write original equation.
$2x^2 + 5x - 12 = 0$	Write in general form.
$(2x - 3)(x + 4) = 0$	Factor left side of equation.
$2x - 3 = 0$	Set 1st factor equal to 0.
$x = \frac{3}{2}$	Solve for x.
$x + 4 = 0$	Set 2nd factor equal to 0.
$x = -4$	Solve for x.

The solutions are $x = \frac{3}{2}$ and $x = -4$. Check these solutions in the original equation.

In Examples 1 and 2, the original equations each involved a second-degree (quadratic) polynomial and each had *two different* solutions. You will sometimes encounter second-degree polynomial equations that have only one (repeated) solution. This occurs when the left side of the general form of the equation is a perfect square trinomial, as shown in Example 3.

Example 3   A Quadratic Equation with a Repeated Solution

Solve $x^2 - 2x + 16 = 6x$.

Solution

$x^2 - 2x + 16 = 6x$	Write original equation.
$x^2 - 8x + 16 = 0$	Write in general form.
$(x - 4)^2 = 0$	Factor.
$x - 4 = 0$ or $x - 4 = 0$	Set factors equal to 0.
$x = 4$	Solve for x.

Note that even though the left side of this equation has two factors, the factors are the same. So, the only solution of the equation is $x = 4$. This solution is called a **repeated solution.**

Check

$x^2 - 2x + 16 = 6x$	Write original equation.
$(4)^2 - 2(4) + 16 \stackrel{?}{=} 6(4)$	Substitute 4 for x.
$16 - 8 + 16 \stackrel{?}{=} 24$	Simplify.
$24 = 24$	Solution checks. ✔

Example 4   Solving a Quadratic Equation by Factoring

Solve $(x + 3)(x + 6) = 4$.

Solution

Begin by multiplying the factors on the left side.

$(x + 3)(x + 6) = 4$	Write original equation.
$x^2 + 9x + 18 = 4$	Multiply factors.
$x^2 + 9x + 14 = 0$	Write in general form.
$(x + 2)(x + 7) = 0$	Factor.
$x + 2 = 0$ ⟹ $x = -2$	Set 1st factor equal to 0 and solve for x.
$x + 7 = 0$ ⟹ $x = -7$	Set 2nd factor equal to 0 and solve for x.

The equation has two solutions: $x = -2$ and $x = -7$. Check these in the original equation.

Technology: Discovery

Write the equation in Example 3 in general form. Graph this equation on your graphing calculator.

$$y = x^2 - 8x + 16$$

What are the x-intercepts of the graph of the equation?

Write the equation in Example 4 in general form. Graph this equation on your graphing calculator.

$$y = x^2 + 9x + 14$$

What are the x-intercepts of the graph of the equation?

How do the x-intercepts relate to the solutions of the equations? What can you conclude about the solutions to the equations and the x-intercepts of the graphs of the equations?

③ Solve higher-degree polynomial equations by factoring.

Solving Higher-Degree Equations by Factoring

Example 5   Solving a Polynomial Equation with Three Factors

Solve $3x^3 = 15x^2 + 18x$.

Solution

$$3x^3 = 15x^2 + 18x \qquad \text{Write original equation.}$$

$$3x^3 - 15x^2 - 18x = 0 \qquad \text{Write in general form.}$$

$$3x(x^2 - 5x - 6) = 0 \qquad \text{Factor out common factor.}$$

$$3x(x - 6)(x + 1) = 0 \qquad \text{Factor.}$$

$$3x = 0 \implies x = 0 \qquad \text{Set 1st factor equal to 0.}$$

$$x - 6 = 0 \implies x = 6 \qquad \text{Set 2nd factor equal to 0.}$$

$$x + 1 = 0 \implies x = -1 \qquad \text{Set 3rd factor equal to 0.}$$

So, $x = 0$, $x = 6$, and $x = -1$. Check these three solutions.

> **Technology: Discovery**
>
> Use a graphing calculator to graph
>
> $$y = x^2 + 3x - 40.$$
>
> From the graph, determine the number of solutions of the equation. Explain how to use a graphing calculator to solve
>
> $$2x^3 - 3x^2 - 5x + 1 = 0.$$
>
> How many solutions does the equation have? How does the number of solutions relate to the degree of the equation?

Notice that the equation in Example 5 is a third-degree equation and has three solutions. This is not a coincidence. In general, a polynomial equation can have *at most* as many solutions as its degree. For instance, a second-degree equation can have zero, one, or two solutions. Notice that the equation in Example 6 is a fourth-degree equation and has four solutions.

Example 6   Solving a Polynomial Equation with Four Factors

Solve $x^4 + x^3 - 4x^2 - 4x = 0$.

Solution

$$x^4 + x^3 - 4x^2 - 4x = 0 \qquad \text{Write original equation.}$$

$$x(x^3 + x^2 - 4x - 4) = 0 \qquad \text{Factor out common factor.}$$

$$x[(x^3 + x^2) + (-4x - 4)] = 0 \qquad \text{Group terms.}$$

$$x[x^2(x + 1) - 4(x + 1)] = 0 \qquad \text{Factor grouped terms.}$$

$$x[(x + 1)(x^2 - 4)] = 0 \qquad \text{Distributive Property}$$

$$x(x + 1)(x + 2)(x - 2) = 0 \qquad \text{Difference of two squares}$$

$$x = 0 \implies x = 0$$

$$x + 1 = 0 \implies x = -1$$

$$x + 2 = 0 \implies x = -2$$

$$x - 2 = 0 \implies x = 2$$

So, $x = 0$, $x = -1$, $x = -2$, and $x = 2$. Check these four solutions.

④ Solve application problems by factoring.

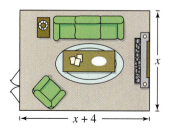

Figure 5.6

Applications

Example 7   Geometry: Dimensions of a Room

A rectangular room has an area of 192 square feet. The length of the room is 4 feet more than its width, as shown in Figure 5.6. Find the dimensions of the room.

Solution

Verbal Model: | Length · Width = Area |

Labels: Length $= x + 4$ (feet)
            Width $= x$ (feet)
            Area $= 192$ (square feet)

Equation: $(x + 4)x = 192$

$$x^2 + 4x - 192 = 0$$

$$(x + 16)(x - 12) = 0$$

$$x = -16 \quad \text{or} \quad x = 12$$

Because the negative solution does not make sense, choose the positive solution $x = 12$. When the width of the room is 12 feet, the length of the room is

Length $= x + 4 = 12 + 4 = 16$ feet.

So, the dimensions of the room are 12 feet by 16 feet. Check this solution in the original statement of the problem.

Example 8   Free-Falling Object

The height of a rock dropped into a well that is 64 feet deep above the water level is given by the position function $h(t) = -16t^2 + 64$, where the height is measured in feet and the time t is measured in seconds. (See Figure 5.7.) How long will it take the rock to hit the water at the bottom of the well?

Solution

In Figure 5.7, note that the water level of the well corresponds to a height of 0 feet. So, substitute a height of 0 for $h(t)$ in the equation and solve for t.

$$0 = -16t^2 + 64 \qquad \text{Substitute 0 for } h(t).$$

$$16t^2 - 64 = 0 \qquad \text{Write in general form.}$$

$$16(t^2 - 4) = 0 \qquad \text{Factor out common factor.}$$

$$16(t + 2)(t - 2) = 0 \qquad \text{Difference of two squares}$$

$$t = -2 \quad \text{or} \quad t = 2 \qquad \text{Solutions using Zero-Factor Property}$$

Because a time of -2 seconds does not make sense in this problem, choose the positive solution $t = 2$, and conclude that the rock hits the water 2 seconds after it is dropped. Check this solution in the original statement of the problem.

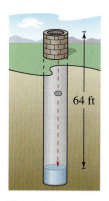

Figure 5.7

64 ft

5.6 Exercises

Review   Concepts, Skills, and Problem Solving

Keep mathematically in shape by doing these exercises *before* the problems of this section.

Properties and Definitions

In Exercises 1–4, identify the property of real numbers illustrated by the statement.

1. $3uv - 3uv = 0$

2. $5z \cdot 1 = 5z$

3. $2s(1 - s) = 2s - 2s^2$

4. $(3x)y = 3(xy)$

Solving Equations

In Exercises 5–10, solve the equation.

5. $4 - \frac{1}{2}x = 6$

6. $500 - 0.75x = 235$

7. $4(x - 3) - (4x + 5) = 0$

8. $12(3 - x) = 5 - 7(2x + 1)$

9. $\dfrac{12 + x}{4} = 13$

10. $8(t - 24) = 0$

Problem Solving

11. *Cost, Revenue, and Profit*   The cost C of producing x units of a product is $C = 12 + 8x$. The revenue R from selling x units of the product is $R = 16x - \frac{1}{4}x^2$, where $0 \le x \le 20$. The profit P is $P = R - C$.

    (a) Perform the subtraction required to find the polynomial representing profit.

    (b) 📱 Use a graphing calculator to graph the polynomial representing profit.

    (c) Determine the profit when $x = 16$ units are produced and sold.

12. *Sales*   A clothing retailer had annual sales of $1118.7 million in 2002 and $1371.4 million in 2004. Use the Midpoint Formula to estimate the sales in 2003.

Developing Skills

In Exercises 1–12, use the Zero-Factor Property to solve the equation.

1. $2x(x - 8) = 0$

2. $z(z + 6) = 0$

3. $(y - 3)(y + 10) = 0$

4. $(s - 16)(s + 15) = 0$

5. $25(a + 4)(a - 2) = 0$

6. $17(t - 3)(t + 8) = 0$

7. $(2t + 5)(3t + 1) = 0$

8. $(5x - 3)(x - 8) = 0$

9. $4x(2x - 3)(2x + 25) = 0$

10. $\frac{1}{5}x(x - 2)(3x + 4) = 0$

11. $(x - 3)(2x + 1)(x + 4) = 0$

12. $(y - 39)(2y + 7)(y + 12) = 0$

In Exercises 13–78, solve the equation by factoring. See Examples 1–6.

13. $5y - y^2 = 0$

14. $3x^2 + 9x = 0$

15. $9x^2 + 15x = 0$

16. $4x^2 - 6x = 0$

17. $2x^2 = 32x$

18. $8x^2 = 5x$

19. $5y^2 = 15y$

20. $3x^2 = 7x$

21. $x^2 - 25 = 0$

22. $x^2 - 121 = 0$

23. $3y^2 - 48 = 0$

24. $25z^2 - 100 = 0$

25. $x^2 - 3x - 10 = 0$

26. $x^2 - x - 12 = 0$

27. $x^2 - 10x + 24 = 0$

28. $20 - 9x + x^2 = 0$

29. $4x^2 + 15x = 25$

30. $14x^2 + 9x = -1$

31. $7 + 13x - 2x^2 = 0$

32. $11 + 32y - 3y^2 = 0$

33. $3y^2 - 2 = -y$

34. $-2x - 15 = -x^2$

35. $-13x + 36 = -x^2$

36. $x^2 - 15 = -2x$

37. $m^2 - 8m + 18 = 2$

38. $a^2 + 4a + 10 = 6$

39. $x^2 + 16x + 57 = -7$

40. $x^2 - 12x + 21 = -15$

41. $4z^2 - 12z + 15 = 6$

42. $16t^2 + 48t + 40 = 4$

43. $x(x + 2) - 10(x + 2) = 0$

44. $x(x - 15) + 3(x - 15) = 0$

45. $u(u - 3) + 3(u - 3) = 0$

46. $x(x + 10) - 2(x + 10) = 0$

47. $x(x - 5) = 36$ **48.** $s(s + 4) = 96$

49. $y(y + 6) = 72$ **50.** $x(x - 4) = 12$

51. $t(2t - 3) = 35$ **52.** $3u(3u + 1) = 20$

53. $(a + 2)(a + 5) = 10$ **54.** $(x - 8)(x - 7) = 20$

55. $(x - 4)(x + 5) = 10$

56. $(u - 6)(u + 4) = -21$

57. $(t - 2)^2 = 16$ **58.** $(s + 4)^2 = 49$

59. $9 = (x + 2)^2$ **60.** $1 = (y + 3)^2$

61. $(x - 3)^2 - 25 = 0$ **62.** $1 - (x + 1)^2 = 0$

63. $81 - (x + 4)^2 = 0$ **64.** $(s + 5)^2 - 49 = 0$

65. $x^3 - 19x^2 + 84x = 0$ **66.** $x^3 + 18x^2 + 45x = 0$

67. $6t^3 = t^2 + t$ **68.** $3u^3 = 5u^2 + 2u$

69. $z^2(z + 2) - 4(z + 2) = 0$

70. $16(3 - u) - u^2(3 - u) = 0$

71. $a^3 + 2a^2 - 9a - 18 = 0$

72. $x^3 - 2x^2 - 4x + 8 = 0$

73. $c^3 - 3c^2 - 9c + 27 = 0$

74. $v^3 + 4v^2 - 4v - 16 = 0$

75. $x^4 - 3x^3 - x^2 + 3x = 0$

76. $x^4 + 2x^3 - 9x^2 - 18x = 0$

77. $8x^4 + 12x^3 - 32x^2 - 48x = 0$

78. $9x^4 - 15x^3 - 9x^2 + 15x = 0$

Graphical Reasoning   In Exercises 79–82, determine the *x*-intercepts of the graph and explain how the *x*-intercepts correspond to the solutions of the polynomial equation when $y = 0$.

79. $y = x^2 - 9$ **80.** $y = x^2 - 4x + 4$

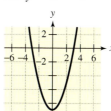 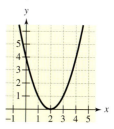

81. $y = x^3 - 6x^2 + 9x$ **82.** $y = x^3 - 3x^2 - x + 3$

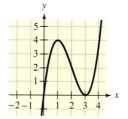 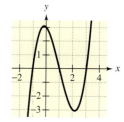

In Exercises 83–90, use a graphing calculator to graph the equation and find any *x*-intercepts of the graph. Verify algebraically that any *x*-intercepts are solutions of the polynomial equation when $y = 0$.

83. $y = x^2 - 6x$ **84.** $y = x^2 - 11x + 28$

85. $y = x^2 - 8x + 12$ **86.** $y = (x - 2)^2 - 9$

87. $y = 2x^2 + 5x - 12$ **88.** $y = x^3 - 4x$

89. $y = 2x^3 - 5x^2 - 12x$ **90.** $y = 2 + x - 2x^2 - x^3$

91. Let a and b be real numbers such that $a \neq 0$. Find the solutions of $ax^2 + bx = 0$.

92. Let a be a nonzero real number. Find the solutions of $ax^2 - ax = 0$.

Solving Problems

Think About It   In Exercises 93 and 94, find a quadratic equation with the given solutions.

93. $x = -3$, $x = 5$

94. $x = 1$, $x = 6$

95. *Number Problem*   The sum of a positive number and its square is 240. Find the number.

96. *Number Problem*   Find two consecutive positive integers whose product is 132.

97. ▲ *Geometry*   The rectangular floor of a storage shed has an area of 330 square feet. The length of the floor is 7 feet more than its width (see figure). Find the dimensions of the floor.

Figure for 97

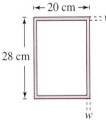

Figure for 98

98. ▲ *Geometry*   The outside dimensions of a picture frame are 28 centimeters and 20 centimeters (see figure). The area of the exposed part of the picture is 468 square centimeters. Find the width w of the frame.

99. ▲ *Geometry*   A triangle has an area of 48 square inches. The height of the triangle is $1\frac{1}{2}$ times its base. Find the base and height of the triangle.

100. ▲ *Geometry*   The height of a triangle is 4 inches less than its base. The area of the triangle is 70 square inches. Find the base and height of the triangle.

101. ▲ *Geometry*   An open box is to be made from a rectangular piece of material that is 5 meters long and 4 meters wide. The box is made by cutting squares of dimension x from the corners and turning up the sides, as shown in the figure. The volume V of a rectangular solid is the product of its length, width, and height.

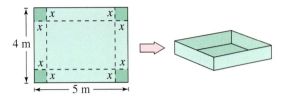

Figure for 101

(a) Show algebraically that the volume of the box is given by $V = (5 - 2x)(4 - 2x)x$.

(b) Determine the values of x for which $V = 0$. Determine an appropriate domain for the function V in the context of this problem.

(c) Complete the table.

x	0.25	0.50	0.75	1.00	1.25	1.50	1.75
V							

(d) Use the table to determine x when $V = 3$. Verify the result algebraically.

(e) ▦ Use a graphing calculator to graph the volume function. Use the graph to approximate the value of x that yields the box of greatest volume.

102. ▲ *Geometry*   An open box with a square base is to be constructed from 880 square inches of material. The height of the box is 6 inches. What are the dimensions of the base?   (*Hint:* The surface area is given by $S = x^2 + 4xh$.)

103. *Free-Falling Object*   An object is thrown upward from the Royal Gorge Bridge in Colorado, 1053 feet above the Arkansas River, with an initial velocity of 48 feet per second. The height h (in feet) of the object is modeled by the position equation $h = -16t^2 + 48t + 1053$ where t is the time measured in seconds. How long will it take for the object to reach the ground?

104. *Free-Falling Object*   A hammer is dropped from a construction project 576 feet above the ground. The height h (in feet) of the hammer is modeled by the position equation $h = -16t^2 + 576$ where t is the time in seconds. How long will it take for the hammer to reach the ground?

105. *Free-Falling Object*   A penny is dropped from the roof of a building 256 feet above the ground. The height h (in feet) of the penny after t seconds is modeled by the equation $h = -16t^2 + 256$. How long will it take for the penny to reach the ground?

106. *Free-Falling Object*   An object is thrown upward from a height of 32 feet with an initial velocity of 16 feet per second. The height h (in feet) of the object after t seconds is modeled by the equation $h = -16t^2 + 16t + 32$. How long will it take for the object to reach the ground?

107. *Free-Falling Object*   An object falls from the roof of a building 194 feet above the ground toward a balcony 50 feet above the ground. The height h (in feet) of the object after t seconds is modeled by the equation $h = -16t^2 + 194$. How long will it take for the object to reach the balcony?

108. *Free-Falling Object*   Your friend stands 96 feet above you on a cliff. You throw an object upward with an initial velocity of 80 feet per second. The height h (in feet) of the object after t seconds is modeled by the equation $h = -16t^2 + 80t$. How long will it take for the object to reach your friend on the way up? On the way down?

109. *Break-Even Analysis*   The revenue R from the sale of x VCRs is given by $R = 90x - x^2$. The cost of producing x VCRs is given by $C = 200 + 60x$. How many VCRs must be produced and sold in order to break even?

110. *Break-Even Analysis*   The revenue R from the sale of x cameras is given by $R = 60x - x^2$. The cost of producing x cameras is given by $C = 75 + 40x$. How many cameras must be produced and sold in order to break even?

111. *Investigation*   Solve the equation $2(x + 3)^2 + (x + 3) - 15 = 0$ in the following two ways.

(a) Let $u = x + 3$, and solve the resulting equation for u. Then find the corresponding values of x that are solutions of the original equation.

(b) Expand and collect like terms in the original equation, and solve the resulting equation for x.

(c) Which method is easier? Explain.

112. *Investigation*   Solve each equation using both methods described in Exercise 107.

(a) $3(x + 6)^2 - 10(x + 6) - 8 = 0$

(b) $8(x + 2)^2 - 18(x + 2) + 9 = 0$

Explaining Concepts

113. ⟳   Answer parts (d)–(f) of Motivating the Chapter on page 296.

114. Give an example of how the Zero-Factor Property can be used to solve a quadratic equation.

115. *True or False?*   If $(2x - 5)(x + 4) = 1$, then $2x - 5 = 1$ or $x + 4 = 1$. Justify your answer.

116. *Writing* ✐   Is it possible for a quadratic equation to have only one solution? Explain.

117. What is the maximum number of solutions of an nth-degree polynomial equation? Give an example of a third-degree equation that has only one real number solution.

118. ▦   The polynomial equation $x^3 - x - 3 = 0$ *cannot* be solved algebraically using any of the techniques described in this book. It does, however, have one solution that is a real number.

(a) *Graphical Solution:* Use a graphing calculator to graph the equation and estimate the solution.

(b) *Numerical Solution:* Use the *table* feature of a graphing calculator to create a table and estimate the solution.

What Did You Learn?

Key Terms

exponential form, *p. 298*
scientific notation, *p. 302*
polynomial in *x*, *p. 308*
degree *n*, *p. 308*
leading coefficient, *p. 308*
constant term, *p. 308*

standard form, *p. 308*
monomial, *p. 308*
binomial, *p. 308*
trinomial , *p. 308*
FOIL Method, *p. 319*
factoring polynomials, *p. 328*

greatest common
    monomial factor, *p. 328*
factoring by grouping, *p. 330*
completely factored, *p. 333*
perfect square trinomial, *p. 338*
quadratic equation, *p. 351*

Key Concepts

5.1 ◯ Summary of rules of exponents

Let m and n be integers, and let a and b represent real numbers, variables, or algebraic expressions. (All denominators and bases are nonzero.)

1. Product Rule: $a^m \cdot a^n = a^{m+n}$

2. Quotient Rule: $\dfrac{a^m}{a^n} = a^{m-n}$

3. Product-to-Power Rule: $(ab)^m = a^m \cdot b^m$

4. Power-to-Power Rule: $(a^m)^n = a^{mn}$

5. Quotient-to-Power Rule: $\left(\dfrac{a}{b}\right)^m = \dfrac{a^m}{b^m}$

6. Zero Exponent Rule: $a^0 = 1$

7. Negative Exponent Rule: $a^{-m} = \dfrac{1}{a^m}$

8. Negative Exponent Rule: $\left(\dfrac{a}{b}\right)^{-m} = \left(\dfrac{b}{a}\right)^m$

5.3 ◯ Special products

Let u and v be real numbers, variables, or algebraic expressions. Then the following formulas are true.

1. Sum and Difference of Two Terms:
   $(u + v)(u - v) = u^2 - v^2$

2. Square of a Binomial: $(u \pm v)^2 = u^2 \pm 2uv + v^2$

5.4 ◯ Difference of two squares

Let u and v be real numbers, variables, or algebraic expressions. Then the expression $u^2 - v^2$ can be factored as follows: $u^2 - v^2 = (u + v)(u - v)$.

5.4 ◯ Sum or difference of two cubes

Let u and v be real numbers, variables, or algebraic expressions. Then the expressions $u^3 \pm v^3$ can be factored as follows: $u^3 \pm v^3 = (u \pm v)(u^2 \mp uv + v^2)$.

5.5 ◯ Perfect square trinomials

Let u and v be real numbers, variables, or algebraic expressions: $u^2 \pm 2uv + v^2 = (u \pm v)^2$

5.5 ◯ Guidelines for factoring $x^2 + bx + c$

See page 340 for factoring guidelines.

5.5 ◯ Guidelines for factoring $ax^2 + bx + c$

See page 342 for factoring guidelines.

5.5 ◯ Guidelines for factoring $ax^2 + bx + c$ by grouping

See page 344 for factoring guidelines.

5.5 ◯ Guidelines for factoring polynomials

1. Factor out any common factors.

2. Factor according to one of the special polynomial forms: difference of two squares, sum or difference of two cubes, or perfect square trinomials.

3. Factor trinomials, $ax^2 + bx + c$, using the methods for $a = 1$ and $a \neq 1$.

4. For polynomials with four terms, factor by grouping.

5. Check to see whether the factors themselves can be factored.

6. Check the results by multiplying the factors.

5.6 ◯ Zero-Factor Property

Let a and b be real numbers, variables, or algebraic expressions. If a and b are factors such that $ab = 0$, then $a = 0$ or $b = 0$. This property also applies to three or more factors.

5.6 ◯ Guidelines for solving quadratic equations

1. Write the quadratic equation in general form.

2. Factor the left side of the equation.

3. Set each factor with a variable equal to zero.

4. Solve each linear equation.

5. Check each solution in the original equation.

Review Exercises

5.1 Integer Exponents and Scientific Notation

① Use the rules of exponents to simplify expressions.

In Exercises 1–14, use the rules of exponents to simplify the expression (if possible).

1. $x^2 \cdot x^3$

2. $-3y^2 \cdot y^4$

3. $(u^2)^3$

4. $(v^4)^2$

5. $(-2z)^3$

6. $(-3y)^2(2)$

7. $-(u^2v)^2(-4u^3v)$

8. $(12x^2y)(3x^2y^4)^2$

9. $\dfrac{12z^5}{6z^2}$

10. $\dfrac{15m^3}{25m}$

11. $\dfrac{120u^5v^3}{15u^3v}$

12. $-\dfrac{(-2x^2y^3)^2}{-3xy^2}$

13. $\left(\dfrac{72x^4}{6x^2}\right)^2$

14. $\left(-\dfrac{y^2}{2}\right)^3$

② Rewrite exponential expressions involving negative and zero exponents.

In Exercises 15–18, evaluate the expression.

15. $(2^3 \cdot 3^2)^{-1}$

16. $(2^{-2} \cdot 5^2)^{-2}$

17. $\left(\dfrac{2}{5}\right)^{-3}$

18. $\left(\dfrac{1}{3^{-2}}\right)^2$

In Exercises 19–30, rewrite the expression using only positive exponents, and simplify. (Assume that any variables in the expression are nonzero.)

19. $(6y^4)(2y^{-3})$

20. $4(-3x)^{-3}$

21. $\dfrac{4x^{-2}}{2x}$

22. $\dfrac{15t^5}{24t^{-3}}$

23. $(x^3y^{-4})^0$

24. $(5x^{-2}y^4)^{-2}$

25. $\dfrac{2a^{-3}b^4}{4a^5b^{-5}}$

26. $\dfrac{2u^0v^{-2}}{10u^{-1}v^{-3}}$

27. $\left(\dfrac{3x^{-1}y^2}{12x^5y^{-3}}\right)^{-1}$

28. $\left(\dfrac{4x^{-3}z^{-1}}{8x^4z}\right)^{-2}$

29. $u^3(5u^0v^{-1})(9u)^2$

30. $a^4(2a^{-1}b^2)(ab)^0$

③ Write very large and very small numbers in scientific notation.

In Exercises 31 and 32, write the number in scientific notation.

31. 0.0000538

32. 30,296,000,000

In Exercises 33 and 34, write the number in decimal form.

33. 4.833×10^8

34. 2.74×10^{-4}

In Exercises 35–38, evaluate the expression without a calculator.

35. $(6 \times 10^3)^2$

36. $(3 \times 10^{-3})(8 \times 10^7)$

37. $\dfrac{3.5 \times 10^7}{7 \times 10^4}$

38. $\dfrac{1}{(6 \times 10^{-3})^2}$

5.2 Adding and Subtracting Polynomials

① Identify leading coefficients and degrees of polynomials.

In Exercises 39–42, write the polynomial in standard form, and find its degree and leading coefficient.

39. $6x^3 - 4x + 5x^2 - x^4$

40. $2x^6 - 5x^3 + x^5 - 7$

41. $14 - 6x + 3x^2 - 7x^3$

42. $9x - 2x^3 + x^5 - 8x^7$

In Exercises 43 and 44, give an example of a polynomial in x that satisfies the conditions. (There are many correct answers.)

43. A trinomial of degree 5 and leading coefficient -6

44. A binomial of degree 2 and leading coefficient 7

2 Add and subtract polynomials using a horizontal format and a vertical format.

In Exercises 45–56, use a horizontal format to find the sum or difference.

45. $(10x + 8) + (x^2 + 3x)$

46. $(-7x + 3) + (x^2 - 18)$

47. $(5x^3 - 6x + 11) + (5 + 6x - x^2 - 8x^3)$

48. $(7 - 12x^2 + 8x^3) + (x^4 - 6x^3 + 7x^2 - 5)$

49. $(3y - 4) - (2y^2 + 1)$

50. $(x^2 - 5) - (3 - 6x)$

51. $(-x^3 - 3x) - 4(2x^3 - 3x + 1)$

52. $(7z^2 + 6z) - 3(5z^2 + 2z)$

53. $3y^2 - [2y + 3(y^2 + 5)]$

54. $(16a^3 + 5a) - 5[a + (2a^3 - 1)]$

55. $(3x^5 + 4x^2 - 8x + 12) - (2x^5 + x) +$
$(3x^2 - 4x^3 - 9)$

56. $(7x^4 - 10x^2 + 4x) + (x^3 - 3x) - (3x^4 - 5x^2 + 1)$

In Exercises 57–60, use a vertical format to find the sum or difference.

57. $\quad 3x^2 + 5x$
$\quad \underline{-4x^2 - \quad x + 6}$

58. $\quad\quad 6x + 1$
$\quad \underline{x^2 - 4x}$

59. $\quad\quad 3t - 5$
$\quad \underline{-(t^2 - t - 5)}$

60. $10y^2 \quad\quad + 3$
$\quad \underline{-(y^2 + 4y - 9)}$

3 Use polynomials to model and solve real-life problems.

▲ *Geometry* **In Exercises 61 and 62, write and simplify an expression for the perimeter of the figure.**

61.

62.

63. ***Cost, Revenue, and Profit*** A manufacturer can produce and sell x backpacks per week. The total cost C (in dollars) for producing the backpacks is given by $C = 12x + 3000$, and the total revenue R is given by $R = 20x$. Find the profit P obtained by selling 1200 backpacks per week.

64. ***Cost, Revenue, and Profit*** A manufacturer can produce and sell x notepads per week. The total cost C (in dollars) for producing the notepads is given by $C = 0.5x + 1000$, and the total revenue R is given by $R = 1.1x$. Find the profit P obtained by selling 5000 notepads per week.

5.3 Multiplying Polynomials

1 Use the Distributive Property and the FOIL Method to multiply polynomials.

In Exercises 65–78, perform the multiplication and simplify.

65. $(-2x)^3(x + 4)$

66. $(-4y)^2(y - 2)$

67. $3x(2x^2 - 5x + 3)$

68. $-2y(5y^2 - y - 4)$

69. $(x - 2)(x + 7)$

70. $(x + 6)(x - 9)$

71. $(5x + 3)(3x - 4)$

72. $(4x - 1)(2x - 5)$

73. $(4x^2 + 3)(6x^2 + 1)$

74. $(3y^2 + 2)(4y^2 - 5)$

75. $(2x^2 - 3x + 2)(2x + 3)$

76. $(5s^3 + 4s - 3)(4s - 5)$

77. $2u(u - 7) - (u + 1)(u - 7)$

78. $(3v + 2)(-5v) + 5v(3v + 2)$

2 Use special product formulas to multiply two binomials.

In Exercises 79–84, use a special product formula to find the product.

79. $(4x - 7)^2$

80. $(2x + 3y)^2$

81. $(5u - 8)(5u + 8)$

82. $(5x - 2y)(5x + 2y)$

83. $[(u - 3) + v][(u - 3) - v]$

84. $[(m - 5) + n]^2$

③ Use multiplication of polynomials in application problems.

▲ *Geometry*   In Exercises 85 and 86, write an expression for the area of the shaded region of the figure. Then simplify the expression.

85.

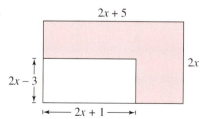

86.

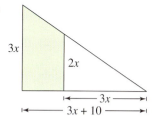

87. **▲** *Geometry*   The length of a rectangle is five times its width w. Find expressions for (a) the perimeter and (b) the area of the rectangle.

88. *Compound Interest*   After 2 years, an investment of $750 compounded annually at an interest rate r will yield an amount $750(1 + r)^2$. Find this product.

5.4 Factoring by Grouping and Special Forms

① Factor greatest common monomial factors from polynomials.

In Exercises 89–94, factor out the greatest common monomial factor.

89. $24x^2 - 18$

90. $14z^3 + 21$

91. $2x^2 + x$

92. $-a^3 - 4a$

93. $6x^2 + 15x^3 - 3x$

94. $8y - 12y^2 + 24y^3$

② Factor polynomials by grouping terms.

In Exercises 95–100, factor the polynomial by grouping.

95. $28(x + 5) - 70(x + 5)$

96. $(u - 9v)(u - v) + v(u - 9v)$

97. $v^3 - 2v^2 - v + 2$

98. $y^3 + 4y^2 - y - 4$

99. $t^3 + 3t^2 + 3t + 9$

100. $x^3 + 7x^2 + 3x + 21$

③ Factor the difference of two squares and factor the sum or difference of two cubes.

In Exercises 101–104, factor the difference of two squares.

101. $x^2 - 36$

102. $16y^2 - 49$

103. $(u + 6)^2 - 81$

104. $(y - 3)^2 - 16$

In Exercises 105–108, factor the sum or difference of two cubes.

105. $u^3 - 1$

106. $t^3 - 125$

107. $8x^3 + 27$

108. $64y^3 + 8$

④ Factor polynomials completely by repeated factoring.

In Exercises 109–112, factor the polynomial completely.

109. $x^3 - x$

110. $y^4 - 4y^2$

111. $24 + 3u^3$

112. $54 - 2x^3$

5.5 Factoring Trinomials

① Recognize and factor perfect square trinomials.

In Exercises 113–116, factor the perfect square trinomial.

113. $x^2 - 18x + 81$

114. $y^2 + 16y + 64$

115. $4s^2 + 40st + 100t^2$

116. $u^2 - 10uv + 25v^2$

② Factor trinomials of the forms $x^2 + bx + c$ and $ax^2 + bx + c$

In Exercises 117–122, factor the trinomial.

117. $x^2 + 2x - 35$

118. $x^2 - 12x + 32$

119. $2x^2 - 7x + 6$ **120.** $5x^2 + 11x - 12$

121. $18x^2 + 27x + 10$ **122.** $12x^2 - 13x - 14$

3 Factor trinomials of the form $ax^2 + bx + c$ by grouping.

In Exercises 123–128, factor the trinomial by grouping.

123. $4x^2 - 3x - 1$ **124.** $12x^2 - 7x + 1$

125. $5x^2 - 12x + 7$ **126.** $3u^2 + 7u - 6$

127. $2s^2 - 13s + 21$ **128.** $3x^2 - 13x - 10$

4 Factor polynomials using the guidelines for factoring.

In Exercises 129–136, factor the expression completely.

129. $4a - 64a^3$ **130.** $3b + 27b^3$

131. $8x(2x - 3) - 4(2x - 3)$
132. $x^3 + 3x^2 - 4x - 12$
133. $\frac{1}{4}x^2 + xy + y^2$ **134.** $x^2 - \frac{2}{3}x + \frac{1}{9}$

135. $x^2 - 10x + 25 - y^2$
136. $u^6 - 8v^6$

5.6 Solving Polynomial Equations by Factoring

1 Use the Zero-Factor Property to solve equations.

In Exercises 137–142, use the Zero-Factor Property to solve the equation.

137. $4x(x - 2) = 0$ **138.** $-7x(2x + 5) = 0$

139. $(2x + 1)(x - 3) = 0$
140. $(x - 7)(3x - 8) = 0$
141. $(x + 10)(4x - 1)(5x + 9) = 0$
142. $3x(x + 8)(2x - 7) = 0$

2 Solve quadratic equations by factoring.

In Exercises 143–150, solve the quadratic equation by factoring.

143. $3s^2 - 2s - 8 = 0$ **144.** $x^2 - 25x = -150$

145. $10x(x - 3) = 0$ **146.** $3x(4x + 7) = 0$

147. $z(5 - z) + 36 = 0$ **148.** $(x + 3)^2 - 25 = 0$

149. $v^2 - 100 = 0$ **150.** $x^2 - 121 = 0$

3 Solve higher-degree polynomial equations by factoring.

In Exercises 151–158, solve the polynomial equation by factoring.

151. $2y^4 + 2y^3 - 24y^2 = 0$
152. $9x^4 - 15x^3 - 6x^2 = 0$
153. $x^3 - 11x^2 + 18x = 0$
154. $x^3 + 20x^2 + 36x = 0$
155. $b^3 - 6b^2 - b + 6 = 0$
156. $x^3 + 3x^2 - 5x - 15 = 0$
157. $x^4 - 5x^3 - 9x^2 + 45x = 0$
158. $2x^4 + 6x^3 - 50x^2 - 150x = 0$

4 Solve application problems by factoring.

159. *Number Problem*   Find two consecutive positive odd integers whose product is 195.

160. *Number Problem*   Find two consecutive positive even integers whose product is 224.

161. ▲ *Geometry*   A rectangle has an area of 900 square inches. The length of the rectangle is $2\frac{1}{4}$ times its width. Find the dimensions of the rectangle.

162. ▲ *Geometry*   A rectangle has an area of 432 square inches. The width of the rectangle is $\frac{3}{4}$ times its length. Find the dimensions of the rectangle.

163. *Free-Falling Object*   An object is dropped from a weather balloon 6400 feet above the ground. The height h (in feet) of the object is modeled by the position equation $h = -16t^2 + 6400$, where t is the time (in seconds). How long will it take the object to reach the ground?

164. *Free-Falling Object*   An object is thrown upward from the Trump Tower in New York City, which is 664 feet tall, with an initial velocity of 45 feet per second. The height h (in feet) of the object is modeled by the position equation $h = -16t^2 + 45t + 664$, where t is the time (in seconds). How long will it take the object to reach the ground?

Chapter Test

Take this test as you would take a test in class. After you are done, check your work against the answers in the back of the book.

1. Determine the degree and leading coefficient of $-5.2x^3 + 3x^2 - 8$.

2. Explain why the following expression is not a polynomial.

$$\frac{4}{x^2 + 2}$$

In Exercises 3 and 4, rewrite each expression using only positive exponents, and simplify. (Assume that any variables in the expression are nonzero.)

3. (a) $\dfrac{4x^{-2}y^3}{5^{-1}x^3y^{-2}}$
   (b) $\left(\dfrac{-2x^2y}{z^{-3}}\right)^{-2}$

4. (a) $\left(-\dfrac{2u^2}{v^{-1}}\right)^3\left(\dfrac{3v^2}{u^{-3}}\right)$
   (b) $\dfrac{(-3x^2y^{-1})^4}{6x^2y^0}$

In Exercises 5–9, perform the indicated operations and simplify.

5. (a) $(5a^2 - 3a + 4) + (a^2 - 4)$
   (b) $(16 - y^2) - (16 + 2y + y^2)$

6. (a) $-2(2x^4 - 5) + 4x(x^3 + 2x - 1)$
   (b) $4t - [3t - (10t + 7)]$

7. (a) $-3x(x - 4)$
   (b) $(2x - 3y)(x + 5y)$

8. (a) $(x - 1)[2x + (x - 3)]$
   (b) $(2s - 3)(3s^2 - 4s + 7)$

9. (a) $(4x - 3)^2$
   (b) $[4 - (a + b)][4 + (a + b)]$

In Exercises 10–15, factor the expression completely.

10. $18y^2 - 12y$

11. $v^2 - \dfrac{16}{9}$

12. $x^3 - 3x^2 - 4x + 12$

13. $9u^2 - 6u + 1$

14. $6x^2 - 26x - 20$

15. $x^3 + 27$

In Exercises 16–19, solve the equation.

16. $(x + 5)(x - 2) = 60$

17. $(y + 2)^2 - 9 = 0$

18. $12 + 5y - 3y^2 = 0$

19. $2x^3 + 10x^2 + 8x = 0$

20. Write an expression for the area of the shaded region in the figure. Then simplify the expression.

21. The area of a rectangle is 54 square centimeters. The length of the rectangle is $1\frac{1}{2}$ times its width. Find the dimensions of the rectangle.

22. The area of a triangle is 35 square feet. The height of the triangle is 4 feet more than twice its base. Find the base and height of the triangle.

23. The revenue R from the sale of x computer desks is given by $R = x^2 - 35x$. The cost C of producing x computer desks is given by $C = 150 + 12x$. How many computer desks must be produced and sold in order to break even?

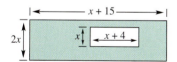

Figure for 20

Motivating the Chapter

⚡ A Canoe Trip

You and a friend are planning a canoe trip on a river. You want to travel 10 miles upstream and 10 miles back downstream during daylight hours. You know that in still water you are able to paddle the canoe at an average speed of 5 miles per hour. While traveling upstream your average speed will be 5 miles per hour minus the speed of the current, and while traveling downstream your average speed will be 5 miles per hour plus the speed of the current.

See Section 6.3, Exercise 85.

a. Write an expression that represents the time it will take to travel upstream in terms of the speed x (in miles per hour) of the current. Write an expression that represents the time it will take to travel downstream in terms of the speed of the current.

b. Write a function f for the entire time (in hours) of the trip in terms of x.

c. Write the rational function f as a single fraction.

See Section 6.6, Exercise 87.

d. The time for the entire trip is $6\frac{1}{4}$ hours. What is the speed of the current? Explain.

e. The speed of the current is 4 miles per hour. Can you and your friend make the trip during 12 hours of daylight? Explain.

Larry Prosor/Superstock, Inc.

6

Rational Expressions, Equations, and Functions

6.1 ● Rational Expressions and Functions

6.2 ● Multiplying and Dividing Rational Expressions

6.3 ● Adding and Subtracting Rational Expressions

6.4 ● Complex Fractions

6.5 ● Dividing Polynomials and Synthetic Division

6.6 ● Solving Rational Equations

6.7 ● Applications and Variation

367

6.1 Rational Expressions and Functions

Paul Barton/Corbis

What You Should Learn

1 Find the domain of a rational function.

2 Simplify rational expressions.

Why You Should Learn It

Rational expression can be used to solve real-life problems. For instance, in Exercise 91 on page 379, you will find a rational expression that models the average cable television revenue per subscriber.

1 Find the domain of a rational function.

The Domain of a Rational Function

A fraction whose numerator and denominator are polynomials is called a **rational expression.** Some examples are

$$\frac{3}{x+4}, \quad \frac{2x}{x^2-4x+4}, \quad \text{and} \quad \frac{x^2-5x}{x^2+2x-3}.$$

In Section 3.6, you learned that because division by zero is undefined, the denominator of a rational expression cannot be zero. So, in your work with rational expressions, you must assume that all real number values of the variable that make the denominator zero are excluded. For the three fractions above, $x = -4$ is excluded from the first fraction, $x = 2$ from the second, and both $x = 1$ and $x = -3$ from the third. The set of *usable* values of the variable is called the **domain** of the rational expression.

Definition of a Rational Expression

Let u and v be polynomials. The algebraic expression

$$\frac{u}{v}$$

is a **rational expression.** The **domain** of this rational expression is the set of all real numbers for which $v \neq 0$.

Like polynomials, rational expressions can be used to describe functions. Such functions are called **rational functions.**

Study Tip

Every polynomial is also a rational expression because you can consider the denominator to be 1. The domain of every polynomial is the set of all real numbers.

Definition of a Rational Function

Let $u(x)$ and $v(x)$ be polynomial functions. The function

$$f(x) = \frac{u(x)}{v(x)}$$

is a **rational function.** The **domain** of f is the set of all real numbers for which $v(x) \neq 0$.

Example 1   Finding the Domains of Rational Functions

Find the domain of each rational function.

a. $f(x) = \dfrac{4}{x - 2}$ **b.** $g(x) = \dfrac{2x + 5}{8}$

Solution

a. The denominator is zero when $x - 2 = 0$ or $x = 2$. So, the domain is all real values of x such that $x \neq 2$. In interval notation, you can write the domain as

$$\text{Domain} = (-\infty, 2) \cup (2, \infty).$$

b. The denominator, 8, is never zero, so the domain is the set of *all* real numbers. In interval notation, you can write the domain as

$$\text{Domain} = (-\infty, \infty).$$

Technology: Discovery

Use a graphing calculator to graph the equation that corresponds to part (a) of Example 1. Then use the *trace* or *table* feature of the calculator to determine the behavior of the graph near $x = 2$. Graph the equation that corresponds to part (b) of Example 1. How does this graph differ from the graph in part (a)?

Example 2   Finding the Domains of Rational Functions

Find the domain of each rational function.

a. $f(x) = \dfrac{5x}{x^2 - 16}$ **b.** $h(x) = \dfrac{3x - 1}{x^2 - 2x - 3}$

Solution

a. The denominator is zero when $x^2 - 16 = 0$. Solving this equation by factoring, you find that the denominator is zero when $x = -4$ or $x = 4$. So, the domain is all real values of x such that $x \neq -4$ and $x \neq 4$. In interval notation, you can write the domain as

$$\text{Domain} = (-\infty, -4) \cup (-4, 4) \cup (4, \infty).$$

b. The denominator is zero when $x^2 - 2x - 3 = 0$. Solving this equation by factoring, you find that the denominator is zero when $x = 3$ or when $x = -1$. So, the domain is all real values of x such that $x \neq 3$ and $x \neq -1$. In interval notation, you can write the domain as

$$\text{Domain} = (-\infty, -1) \cup (-1, 3) \cup (3, \infty).$$

Study Tip

Remember that when interval notation is used, the symbol $\cup$ means *union* and the symbol $\cap$ means *intersection*.

Study Tip

When a rational function is written, the domain is usually not listed with the function. It is *implied* that the real numbers that make the denominator zero are excluded from the function. For instance, you know to exclude $x = 2$ and $x = -2$ from the function

$$f(x) = \frac{3x + 2}{x^2 - 4}$$

without having to list this information with the function.

In applications involving rational functions, it is often necessary to restrict the domain further. To indicate such a restriction, you should write the domain to the right of the fraction. For instance, the domain of the rational function

$$f(x) = \frac{x^2 + 20}{x + 4}, \qquad x > 0$$

is the set of *positive* real numbers, as indicated by the inequality $x > 0$. Note that the normal domain of this function would be all real values of x such that $x \neq -4$. However, because "$x > 0$" is listed to the right of the function, the domain is further restricted by this inequality.

Example 3   An Application Involving a Restricted Domain

You have started a small business that manufactures lamps. The initial investment for the business is $120,000. The cost of each lamp that you manufacture is $15. So, your total cost of producing x lamps is

$$C = 15x + 120{,}000. \qquad \text{Cost function}$$

Your average cost per lamp depends on the number of lamps produced. For instance, the average cost per lamp $\overline{C}$ for producing 100 lamps is

$$\overline{C} = \frac{15(100) + 120{,}000}{100} \qquad \text{Substitute 100 for } x.$$

$$= \$1215. \qquad \text{Average cost per lamp for 100 lamps}$$

The average cost per lamp decreases as the number of lamps increases. For instance, the average cost per lamp $\overline{C}$ for producing 1000 lamps is

$$\overline{C} = \frac{15(1000) + 120{,}000}{1000} \qquad \text{Substitute 1000 for } x.$$

$$= \$135. \qquad \text{Average cost per lamp for 1000 lamps}$$

In general, the average cost of producing x lamps is

$$\overline{C} = \frac{15x + 120{,}000}{x}. \qquad \text{Average cost per lamp for } x \text{ lamps}$$

What is the domain of this rational function?

Solution

If you were considering this function from only a mathematical point of view, you would say that the domain is all real values of x such that $x \neq 0$. However, because this function is a mathematical model representing a real-life situation, you must decide which values of x make sense in real life. For this model, the variable x represents the number of lamps that you produce. Assuming that you cannot produce a fractional number of lamps, you can conclude that the domain is the set of positive integers—that is,

$$\text{Domain} = \{1, 2, 3, 4, \ldots \}.$$

② Simplify rational expressions.

Simplifying Rational Expressions

As with numerical fractions, a rational expression is said to be in **simplified** (or **reduced**) **form** if its numerator and denominator have no common factors (other than ± 1). To simplify rational expressions, you can apply the rule below.

Simplifying Rational Expressions

Let u, v, and w represent real numbers, variables, or algebraic expressions such that $v \neq 0$ and $w \neq 0$. Then the following is valid.

$$\frac{uw}{vw} = \frac{u\cancel{w}}{v\cancel{w}} = \frac{u}{v}$$

Be sure you divide out only *factors*, not *terms*. For instance, consider the expressions below.

$$\frac{2 \cdot 2}{2(x + 5)} \qquad \text{You } can \text{ divide out the common factor 2.}$$

$$\frac{3 + x}{3 + 2x} \qquad \text{You } cannot \text{ divide out the common term 3.}$$

Simplifying a rational expression requires two steps: (1) completely factor the numerator and denominator and (2) divide out any *factors* that are common to both the numerator and denominator. So, your success in simplifying rational expressions actually lies in your ability to *factor completely* the polynomials in both the numerator and denominator.

Example 4   Simplifying a Rational Expression

Simplify the rational expression $\dfrac{2x^3 - 6x}{6x^2}$.

Solution

First note that the domain of the rational expression is all real values of x such that $x \neq 0$. Then, completely factor both the numerator and denominator.

$$\frac{2x^3 - 6x}{6x^2} = \frac{2x(x^2 - 3)}{2x(3x)} \qquad \text{Factor numerator and denominator.}$$

$$= \frac{2\cancel{x}(x^2 - 3)}{2\cancel{x}(3x)} \qquad \text{Divide out common factor } 2x.$$

$$= \frac{x^2 - 3}{3x} \qquad \text{Simplified form}$$

In simplified form, the domain of the rational expression is the same as that of the original expression—all real values of x such that $x \neq 0$.

Technology: Tip

Use the *table* feature of a graphing calculator to compare the two functions in Example 5.

$$y_1 = \frac{x^2 + 2x - 15}{3x - 9}$$

$$y_2 = \frac{x + 5}{3}$$

Set the increment value of the table to 1 and compare the values at $x = 0, 1, 2, 3, 4,$ and 5. Next set the increment value to 0.1 and compare the values at $x = 2.8, 2.9, 3.0, 3.1,$ and 3.2. From the table you can see that the functions differ only at $x = 3$. This shows why $x \neq 3$ must be written as part of the simplified form of the original expression.

Example 5 Simplifying a Rational Expression

Simplify the rational expression $\dfrac{x^2 + 2x - 15}{3x - 9}$.

Solution

The domain of the rational expression is all real values of x such that $x \neq 3$.

$$\frac{x^2 + 2x - 15}{3x - 9} = \frac{(x + 5)(x - 3)}{3(x - 3)} \qquad \text{Factor numerator and denominator.}$$

$$= \frac{(x + 5)(x - 3)}{3(x - 3)} \qquad \text{Divide out common factor } (x - 3).$$

$$= \frac{x + 5}{3}, \; x \neq 3 \qquad \text{Simplified form}$$

Dividing out common factors from the numerator and denominator of a rational expression can change its domain. For instance, in Example 5 the domain of the original expression is all real values of x such that $x \neq 3$. So, the original expression is equal to the simplified expression for all real numbers *except* 3.

Example 6 Simplifying a Rational Expression

Simplify the rational expression $\dfrac{x^3 - 16x}{x^2 - 2x - 8}$.

Solution

The domain of the rational expression is all real values of x such that $x \neq -2$ and $x \neq 4$.

$$\frac{x^3 - 16x}{x^2 - 2x - 8} = \frac{x(x^2 - 16)}{(x + 2)(x - 4)} \qquad \text{Partially factor.}$$

$$= \frac{x(x + 4)(x - 4)}{(x + 2)(x - 4)} \qquad \text{Factor completely.}$$

$$= \frac{x(x + 4)(x - 4)}{(x + 2)(x - 4)} \qquad \text{Divide out common factor } (x - 4).$$

$$= \frac{x(x + 4)}{x + 2}, \; x \neq 4 \qquad \text{Simplified form}$$

When simplifying a rational expression, be aware of the domain. If the domain in the original expression is no longer the same as the domain in the simplified expression, it is important to list the domain next to the simplified expression so that both the original and simplified expressions are equal. For instance, in Example 6 the restriction $x \neq 4$ is listed so that the domains agree for the original and simplified expressions. The example does not list $x \neq -2$ because it is apparent by looking at either expression.

Study Tip

Be sure to factor *completely* the numerator and denominator of a rational expression before concluding that there is no common factor. This may involve a change in signs. Remember that the Distributive Property allows you to write $(b - a)$ as $-(a - b)$. Watch for this in Example 7.

Example 7   Simplification Involving a Change in Sign

Simplify the rational expression $\dfrac{2x^2 - 9x + 4}{12 + x - x^2}$.

Solution

The domain of the rational expression is all real values of x such that $x \neq -3$ and $x \neq 4$.

$$\frac{2x^2 - 9x + 4}{12 + x - x^2} = \frac{(2x - 1)(x - 4)}{(4 - x)(3 + x)} \qquad \text{Factor numerator and denominator.}$$

$$= \frac{(2x - 1)(x - 4)}{-(x - 4)(3 + x)} \qquad (4 - x) = -(x - 4)$$

$$= \frac{(2x - 1)\cancel{(x - 4)}}{-\cancel{(x - 4)}(3 + x)} \qquad \text{Divide out common factor } (x - 4).$$

$$= -\frac{2x - 1}{3 + x}, \quad x \neq 4 \qquad \text{Simplified form}$$

The simplified form is equivalent to the original expression for all values of x such that $x \neq 4$. Note that $x = -3$ is excluded from the domains of both the original and simplified expressions.

In Example 7, be sure you see that when dividing the numerator and denominator by the common factor of $(x - 4)$, you keep the minus sign. In the simplified form of the fraction, this text uses the convention of moving the minus sign out in front of the fraction. However, this is a personal preference. All of the following forms are equivalent.

$$-\frac{2x - 1}{3 + x} = \frac{-(2x - 1)}{3 + x} = \frac{-2x + 1}{3 + x} = \frac{2x - 1}{-3 - x} = \frac{2x - 1}{-(3 + x)}$$

In the next three examples, rational expressions that involve more than one variable are simplified.

Example 8   A Rational Expression Involving Two Variables

Simplify the rational expression $\dfrac{3xy + y^2}{2y}$.

Solution

The domain of the rational expression is all real values of y such that $y \neq 0$.

$$\frac{3xy + y^2}{2y} = \frac{y(3x + y)}{2y} \qquad \text{Factor numerator and denominator.}$$

$$= \frac{\cancel{y}(3x + y)}{2\cancel{y}} \qquad \text{Divide out common factor } y.$$

$$= \frac{3x + y}{2}, \quad y \neq 0 \qquad \text{Simplified form}$$

Example 9   A Rational Expression Involving Two Variables

$$\frac{2x^2 + 2xy - 4y^2}{5x^3 - 5xy^2} = \frac{2(x - y)(x + 2y)}{5x(x - y)(x + y)} \qquad \text{Factor numerator and denominator.}$$

$$= \frac{2(x - y)(x + 2y)}{5x(x - y)(x + y)} \qquad \text{Divide out common factor } (x - y).$$

$$= \frac{2(x + 2y)}{5x(x + y)}, \quad x \neq y \qquad \text{Simplified form}$$

The domain of the original rational expression is all real numbers such that $x \neq 0$ and $x \neq \pm y$.

Example 10   A Rational Expression Involving Two Variables

$$\frac{4x^2y - y^3}{2x^2y - xy^2} = \frac{(2x - y)(2x + y)y}{(2x - y)xy} \qquad \text{Factor numerator and denominator.}$$

$$= \frac{(2x - y)(2x + y)y}{(2x - y)xy} \qquad \text{Divide out common factors } (2x - y) \text{ and } y.$$

$$= \frac{2x + y}{x}, \quad y \neq 0, \ y \neq 2x \qquad \text{Simplified form}$$

The domain of the original rational expression is all real numbers such that $x \neq 0$, $y \neq 0$, and $y \neq 2x$.

Example 11   Geometry: Area

Find the ratio of the area of the shaded portion of the triangle to the total area of the triangle. (See Figure 6.1.)

Solution

The area of the shaded portion of the triangle is given by

$$\text{Area} = \tfrac{1}{2}(4x)(x + 2) = \tfrac{1}{2}(4x^2 + 8x) = 2x^2 + 4x.$$

The total area of the triangle is given by

$$\text{Area} = \tfrac{1}{2}(4x + 4x)(x + 4) = \tfrac{1}{2}(8x)(x + 4) = \tfrac{1}{2}(8x^2 + 32x) = 4x^2 + 16x.$$

So, the ratio of the area of the shaded portion of the triangle to the total area of the triangle is

$$\frac{2x^2 + 4x}{4x^2 + 16x} = \frac{2x(x + 2)}{4x(x + 4)} = \frac{x + 2}{2(x + 4)}, \quad x \neq 0.$$

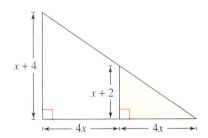

Figure 6.1

As you study the examples and work the exercises in this section and the next four sections, keep in mind that you are *rewriting expressions in simpler forms*. You are not solving equations. Equal signs are used in the steps of the simplification process only to indicate that the new form of the expression is *equivalent* to the original form.

6.1 Exercises

Review   Concepts, Skills, and Problem Solving

Keep mathematically in shape by doing these exercises *before* the problems of this section.

Properties and Definitions

1. *Writing*   Define the slope of the line through the points (x_1, y_1) and (x_2, y_2).

2. *Writing*   Make a statement about the slope m of the line for each condition.

   (a) The line rises from left to right.

   (b) The line falls from left to right.

   (c) The line is horizontal.

   (d) The line is vertical.

Simplifying Expressions

In Exercises 3–8, simplify the expression.

3. $2(x + 5) - 3 - (2x - 3)$

4. $3(y + 4) + 5 - (3y + 5)$

5. $4 - 2[3 + 4(x + 1)]$ **6.** $5x + x[3 - 2(x - 3)]$

7. $\left(\dfrac{5}{x^2}\right)^2$ **8.** $-\dfrac{(2u^2v)^2}{-3uv^2}$

Problem Solving

9. *Mixture Problem*   Determine the numbers of gallons of a 30% solution and a 60% solution that must be mixed to obtain 20 gallons of a 40% solution.

10. *Original Price*   A suit sells for $375 during a 25% off storewide clearance sale. What was the original price of the suit?

Developing Skills

In Exercises 1–20, find the domain of the rational function. See Examples 1 and 2.

1. $f(x) = \dfrac{x^2 + 9}{4}$ **2.** $f(y) = \dfrac{y^2 - 3}{7}$

3. $f(x) = \dfrac{7}{x - 5}$ **4.** $g(x) = \dfrac{-3}{x - 9}$

5. $f(x) = \dfrac{12x}{4 - x}$ **6.** $h(y) = \dfrac{2y}{6 - y}$

7. $g(x) = \dfrac{2x}{x + 10}$ **8.** $f(x) = \dfrac{4x}{x + 1}$

9. $h(x) = \dfrac{x}{x^2 + 4}$ **10.** $h(x) = \dfrac{4x}{x^2 + 16}$

11. $f(y) = \dfrac{y - 4}{y(y + 3)}$

12. $f(z) = \dfrac{z + 2}{z(z - 4)}$

13. $f(t) = \dfrac{5t}{t^2 - 16}$

14. $f(x) = \dfrac{x}{x^2 - 4}$

15. $g(y) = \dfrac{y + 5}{y^2 - 3y}$

16. $g(t) = \dfrac{t - 6}{t^2 + 5t}$

17. $g(x) = \dfrac{x + 1}{x^2 - 5x + 6}$

18. $h(t) = \dfrac{3t^2}{t^2 - 2t - 3}$

19. $f(u) = \dfrac{u^2}{3u^2 - 2u - 5}$

20. $g(y) = \dfrac{y + 5}{4y^2 - 5y - 6}$

In Exercises 21–26, evaluate the rational function as indicated and simplify. If not possible, state the reason.

21. $f(x) = \dfrac{4x}{x + 3}$

   (a) $f(1)$ (b) $f(-2)$

   (c) $f(-3)$ (d) $f(0)$

22. $f(x) = \dfrac{x - 10}{4x}$

   (a) $f(10)$ (b) $f(0)$

   (c) $f(-2)$ (d) $f(12)$

23. $g(x) = \dfrac{x^2 - 4x}{x^2 - 9}$

   (a) $g(0)$ (b) $g(4)$

   (c) $g(3)$ (d) $g(-3)$

24. $g(t) = \dfrac{t - 2}{2t - 5}$

   (a) $g(2)$ (b) $g\left(\tfrac{5}{2}\right)$

   (c) $g(-2)$ (d) $g(0)$

25. $h(s) = \dfrac{s^2}{s^2 - s - 2}$

   (a) $h(10)$ (b) $h(0)$

   (c) $h(-1)$ (d) $h(2)$

26. $f(x) = \dfrac{x^3 + 1}{x^2 - 6x + 9}$

   (a) $f(-1)$ (b) $f(3)$

   (c) $f(-2)$ (d) $f(2)$

In Exercises 27–32, describe the domain. See Example 3.

27. ▲ *Geometry*   A rectangle of length x inches has an area of 500 square inches. The perimeter P of the rectangle is given by

$$P = 2\left(x + \frac{500}{x}\right).$$

28. *Cost*   The cost C in millions of dollars for the government to seize $p\%$ of an illegal drug as it enters the country is given by

$$C = \frac{528p}{100 - p}.$$

29. *Inventory Cost*   The inventory cost I when x units of a product are ordered from a supplier is given by

$$I = \frac{0.25x + 2000}{x}.$$

30. *Average Cost*   The average cost $\overline{C}$ for a manufacturer to produce x units of a product is given by

$$\overline{C} = \frac{1.35x + 4570}{x}.$$

31. *Pollution Removal*   The cost C in dollars of removing $p\%$ of the air pollutants in the stack emission of a utility company is given by the rational function

$$C = \frac{80,000p}{100 - p}.$$

32. *Consumer Awareness*   The average cost of a movie video rental $\overline{M}$ when you consider the cost of purchasing a video cassette recorder and renting x movie videos at $3.49 per movie is

$$\overline{M} = \frac{75 + 3.49x}{x}.$$

In Exercises 33–40, fill in the missing factor.

33. $\dfrac{5(\qquad\qquad)}{6(x + 3)} = \dfrac{5}{6}, \quad x \neq -3$

34. $\dfrac{7(\qquad\qquad)}{15(x - 10)} = \dfrac{7}{15}, \quad x \neq 10$

35. $\dfrac{3x(x + 16)^2}{2(\qquad\qquad)} = \dfrac{x}{2}, \quad x \neq -16$

36. $\dfrac{25x^2(x - 10)}{12(\qquad\qquad)} = \dfrac{5x}{12}, \quad x \neq 10, \quad x \neq 0$

37. $\dfrac{(x + 5)(\qquad\qquad)}{3x^2(x - 2)} = \dfrac{x + 5}{3x}, \quad x \neq 2$

38. $\dfrac{(3y - 7)(\qquad\qquad)}{y^2 - 4} = \dfrac{3y - 7}{y + 2}, \quad y \neq 2$

39. $\dfrac{8x(\qquad\qquad)}{x^2 - 3x - 10} = \dfrac{8x}{x - 5}, \quad x \neq -2$

40. $\dfrac{(3 - z)(\qquad\qquad)}{z^3 + 2z^2} = \dfrac{3 - z}{z^2}, \quad z \neq -2$

In Exercises 41–78, simplify the rational expression. See Examples 4–10.

41. $\dfrac{5x}{25}$

42. $\dfrac{32y}{24}$

43. $\dfrac{12y^2}{2y}$

44. $\dfrac{15z^3}{15z^3}$

45. $\dfrac{18x^2y}{15xy^4}$

46. $\dfrac{16y^2z^2}{60y^5z}$

47. $\dfrac{3x^2 - 9x}{12x^2}$

48. $\dfrac{8x^3 + 4x^2}{20x}$

49. $\dfrac{x^2(x - 8)}{x(x - 8)}$

50. $\dfrac{a^2b(b - 3)}{b^3(b - 3)^2}$

51. $\dfrac{2x - 3}{4x - 6}$

52. $\dfrac{y^2 - 81}{2y - 18}$

53. $\dfrac{5 - x}{3x - 15}$

54. $\dfrac{x^2 - 36}{6 - x}$

55. $\dfrac{a + 3}{a^2 + 6a + 9}$

56. $\dfrac{u^2 - 12u + 36}{u - 6}$

57. $\dfrac{x^2 - 7x}{x^2 - 14x + 49}$

58. $\dfrac{z^2 + 22z + 121}{3z + 33}$

59. $\dfrac{y^3 - 4y}{y^2 + 4y - 12}$

60. $\dfrac{x^2 - 7x}{x^2 - 4x - 21}$

61. $\dfrac{x^3 - 4x}{x^2 - 5x + 6}$

62. $\dfrac{x^4 - 25x^2}{x^2 + 2x - 15}$

63. $\dfrac{3x^2 - 7x - 20}{12 + x - x^2}$

64. $\dfrac{2x^2 + 3x - 5}{7 - 6x - x^2}$

65. $\dfrac{2x^2 + 19x + 24}{2x^2 - 3x - 9}$

66. $\dfrac{2y^2 + 13y + 20}{2y^2 + 17y + 30}$

67. $\dfrac{15x^2 + 7x - 4}{25x^2 - 16}$

68. $\dfrac{56z^2 - 3z - 20}{49z^2 - 16}$

69. $\dfrac{3xy^2}{xy^2 + x}$

70. $\dfrac{x + 3x^2y}{3xy + 1}$

71. $\dfrac{y^2 - 64x^2}{5(3y + 24x)}$

72. $\dfrac{x^2 - 25z^2}{x + 5z}$

73. $\dfrac{5xy + 3x^2y^2}{xy^3}$

74. $\dfrac{4u^2v - 12uv^2}{18uv}$

75. $\dfrac{u^2 - 4v^2}{u^2 + uv - 2v^2}$

76. $\dfrac{x^2 + 4xy}{x^2 - 16y^2}$

77. $\dfrac{3m^2 - 12n^2}{m^2 + 4mn + 4n^2}$

78. $\dfrac{x^2 + xy - 2y^2}{x^2 + 3xy + 2y^2}$

In Exercises 79 and 80, complete the table. What can you conclude?

79.

x	-2	-1	0	1	2	3	4
$\dfrac{x^2 - x - 2}{x - 2}$							
$x + 1$							

80.

x	-2	-1	0	1	2	3	4
$\dfrac{x^2 + 5x}{x}$							
$x + 5$							

Solving Problems

▲ *Geometry* In Exercises 81–84, find the ratio of the area of the shaded portion to the total area of the figure. See Example 11.

81.
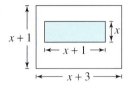

82.
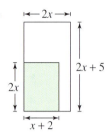

83.
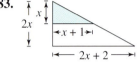

84.
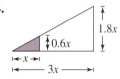

85. *Average Cost* A machine shop has a setup cost of $2500 for the production of a new product. The cost of labor and material for producing each unit is $9.25.

(a) Write the total cost C as a function of x, the number of units produced.

(b) Write the average cost per unit $\overline{C} = C/x$ as a function of x, the number of units produced.

(c) Determine the domain of the function in part (b).

(d) Find the value of $\overline{C}(100)$.

86. *Average Cost* A greeting card company has an initial investment of $60,000. The cost of producing one dozen cards is $6.50.

(a) Write the total cost C as a function of x, the number of cards in dozens produced.

(b) Write the average cost per dozen $\overline{C} = C/x$ as a function of x, the number of cards in dozens produced.

(c) Determine the domain of the function in part (b).

(d) Find the value of $\overline{C}(11,000)$.

87. *Distance Traveled* A van starts on a trip and travels at an average speed of 45 miles per hour. Three hours later, a car starts on the same trip and travels at an average speed of 60 miles per hour.

(a) Find the distance each vehicle has traveled when the car has been on the road for t hours.

(b) Use the result of part (a) to write the distance between the van and the car as a function of t.

(c) Write the ratio of the distance the car has traveled to the distance the van has traveled as a function of t.

88. *Distance Traveled* A car starts on a trip and travels at an average speed of 55 miles per hour. Two hours later, a second car starts on the same trip and travels at an average speed of 65 miles per hour.

(a) Find the distance each vehicle has traveled when the second car has been on the road for t hours.

(b) Use the result of part (a) to write the distance between the first car and the second car as a function of t.

(c) Write the ratio of the distance the second car has traveled to the distance the first car has traveled as a function of t.

89. ▲ *Geometry* One swimming pool is circular and another is rectangular. The rectangular pool's width is three times its depth. Its length is 6 feet more than its width. The circular pool has a diameter that is twice the width of the rectangular pool, and it is 2 feet deeper. Find the ratio of the circular pool's volume to the rectangular pool's volume.

90. ▲ *Geometry* A circular pool has a radius five times its depth. A rectangular pool has the same depth as the circular pool. Its width is 4 feet more than three times its depth and its length is 2 feet less than six times its depth. Find the ratio of the rectangular pool's volume to the circular pool's volume.

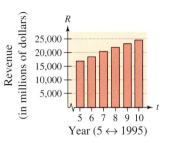
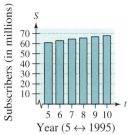

Figures for 91 and 92

Cable TV Revenue In Exercises 91 and 92, use the following polynomial models, which give the total basic cable television revenue R (in millions of dollars) and the number of basic cable subscribers S (in millions) from 1995 through 2000 (see figures).

$R = 1531.1t + 9358, \quad 5 \leq t \leq 10$

$S = 1.33t + 54.6, \quad 5 \leq t \leq 10$

In these models, t represents the year, with $t = 5$ corresponding to 1995. (Source: Paul Kagen Associates, Inc.)

91. Find a rational model that represents the average basic cable television revenue per subscriber during the years 1995 to 2000.

92. Use the model found in Exercise 91 to complete the table, which shows the average basic cable television revenue per subscriber.

Year	1995	1996	1997	1998	1999	2000
Average revenue						

Explaining Concepts

93. *Writing* Define the term *rational expression*.

94. Give an example of a rational function whose domain is the set of all real numbers.

95. *Writing* How do you determine whether a rational expression is in simplified form?

96. *Writing* Can you divide out common terms from the numerator and denominator of a rational expression? Explain.

97. *Error Analysis* Describe the error.

$$\frac{2x^2}{x^2+4} = \frac{2x^2}{x^2+4} = \frac{2}{1+4} = \frac{2}{5}$$

98. *Writing* Is the following statement true? Explain.

$$\frac{6x-5}{5-6x} = -1$$

99. You are the instructor of an algebra course. One of your students turns in the following incorrect solutions. Find the errors, discuss the student's misconceptions, and construct correct solutions.

a. $\dfrac{3x^2+5x-4}{x} = 3x+5-4 = 3x+1$

b. $\dfrac{x^2+7x}{x+7} = \dfrac{x^2}{x}+\dfrac{7x}{7} = x+x = 2x$

6.2 Multiplying and Dividing Rational Expressions

What You Should Learn

1️⃣ Multiply rational expressions and simplify.

2️⃣ Divide rational expressions and simplify.

Why You Should Learn It

Multiplication and division of rational expressions can be used to solve real-life applications. For instance, Example 9 on page 384 shows how a rational expression is used to model the amount Americans spent per person on books and maps from 1995 to 2000.

1️⃣ Multiply rational expressions and simplify.

Multiplying Rational Expressions

The rule for multiplying rational expressions is the same as the rule for multiplying numerical fractions. That is, you *multiply numerators, multiply denominators, and write the new fraction in simplified form.*

$$\frac{3}{4} \cdot \frac{7}{6} = \frac{21}{24} = \frac{\cancel{3} \cdot 7}{\cancel{3} \cdot 8} = \frac{7}{8}$$

Multiplying Rational Expressions

Let u, v, w, and z represent real numbers, variables, or algebraic expressions such that $v \neq 0$ and $z \neq 0$. Then the product of u/v and w/z is

$$\frac{u}{v} \cdot \frac{w}{z} = \frac{uw}{vz}.$$

In order to recognize common factors in the product, write the numerators and denominators in completely factored form, as demonstrated in Example 1.

Example 1   Multiplying Rational Expressions

Multiply the rational expressions $\dfrac{4x^3y}{3xy^4} \cdot \dfrac{-6x^2y^2}{10x^4}$.

Solution

$$\frac{4x^3y}{3xy^4} \cdot \frac{-6x^2y^2}{10x^4} = \frac{(4x^3y) \cdot (-6x^2y^2)}{(3xy^4) \cdot (10x^4)} \qquad \text{Multiply numerators and denominators.}$$

$$= \frac{-24x^5y^3}{30x^5y^4} \qquad \text{Simplify.}$$

$$= \frac{-4\cancel{(6)}\cancel{(x^5)}\cancel{(y^3)}}{5\cancel{(6)}\cancel{(x^5)}\cancel{(y^3)}(y)} \qquad \text{Factor and divide out common factors.}$$

$$= -\frac{4}{5y}, \quad x \neq 0 \qquad \text{Simplified form}$$

Example 2   Multiplying Rational Expressions

Multiply the rational expressions.

$$\frac{x}{5x^2 - 20x} \cdot \frac{x - 4}{2x^2 + x - 3}$$

Solution

$$\frac{x}{5x^2 - 20x} \cdot \frac{x - 4}{2x^2 + x - 3}$$

$$= \frac{x \cdot (x - 4)}{(5x^2 - 20x) \cdot (2x^2 + x - 3)} \qquad \text{Multiply numerators and denominators.}$$

$$= \frac{x(x - 4)}{5x(x - 4)(x - 1)(2x + 3)} \qquad \text{Factor.}$$

$$= \frac{\cancel{x(x - 4)}}{5\cancel{x(x - 4)}(x - 1)(2x + 3)} \qquad \text{Divide out common factors.}$$

$$= \frac{1}{5(x - 1)(2x + 3)}, \ x \neq 0, \ x \neq 4 \qquad \text{Simplified form}$$

Example 3   Multiplying Rational Expressions

Multiply the rational expressions.

$$\frac{4x^2 - 4x}{x^2 + 2x - 3} \cdot \frac{x^2 + x - 6}{4x}$$

Solution

$$\frac{4x^2 - 4x}{x^2 + 2x - 3} \cdot \frac{x^2 + x - 6}{4x}$$

$$= \frac{4x(x - 1)(x + 3)(x - 2)}{(x - 1)(x + 3)(4x)} \qquad \text{Multiply and factor.}$$

$$= \frac{4\cancel{x(x - 1)(x + 3)}(x - 2)}{\cancel{(x - 1)(x + 3)(4x)}} \qquad \text{Divide out common factors.}$$

$$= x - 2, \ x \neq 0, \ x \neq 1, \ x \neq -3 \qquad \text{Simplified form}$$

The rule for multiplying rational expressions can be extended to cover products involving expressions that are not in fractional form. To do this, rewrite the (nonfractional) expression as a fraction whose denominator is 1. Here is a simple example.

$$\frac{x + 3}{x - 2} \cdot (5x) = \frac{x + 3}{x - 2} \cdot \frac{5x}{1}$$

$$= \frac{(x + 3)(5x)}{x - 2}$$

$$= \frac{5x(x + 3)}{x - 2}$$

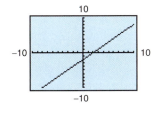

In the next example, note how to divide out a factor that differs only in sign. The Distributive Property is used in the step in which $(y - x)$ is rewritten as $(-1)(x - y)$.

Example 4   Multiplying Rational Expressions

Multiply the rational expressions.

$$\frac{x - y}{y^2 - x^2} \cdot \frac{x^2 - xy - 2y^2}{3x - 6y}$$

Solution

$$\frac{x - y}{y^2 - x^2} \cdot \frac{x^2 - xy - 2y^2}{3x - 6y}$$

$$= \frac{(x - y)(x - 2y)(x + y)}{(y + x)(y - x)(3)(x - 2y)} \qquad \text{Multiply and factor.}$$

$$= \frac{(x - y)(x - 2y)(x + y)}{(y + x)(-1)(x - y)(3)(x - 2y)} \qquad (y - x) = -1(x - y)$$

$$= \frac{\cancel{(x - y)}\cancel{(x - 2y)}\cancel{(x + y)}}{\cancel{(x + y)}(-1)\cancel{(x - y)}(3)\cancel{(x - 2y)}} \qquad \text{Divide out common factors.}$$

$$= -\frac{1}{3}, \quad x \neq y, \ x \neq -y, \ x \neq 2y \qquad \text{Simplified form}$$

The rule for multiplying rational expressions can be extended to cover products of three or more expressions, as shown in Example 5.

Example 5   Multiplying Three Rational Expressions

Multiply the rational expressions.

$$\frac{x^2 - 3x + 2}{x + 2} \cdot \frac{3x}{x - 2} \cdot \frac{2x + 4}{x^2 - 5x}$$

Solution

$$\frac{x^2 - 3x + 2}{x + 2} \cdot \frac{3x}{x - 2} \cdot \frac{2x + 4}{x^2 - 5x}$$

$$= \frac{(x - 1)(x - 2)(3)(x)(2)(x + 2)}{(x + 2)(x - 2)(x)(x - 5)} \qquad \text{Multiply and factor.}$$

$$= \frac{(x - 1)\cancel{(x - 2)}(3)\cancel{(x)}(2)\cancel{(x + 2)}}{\cancel{(x + 2)}\cancel{(x - 2)}\cancel{(x)}(x - 5)} \qquad \text{Divide out common factors.}$$

$$= \frac{6(x - 1)}{x - 5}, \quad x \neq 0, \ x \neq 2, \ x \neq -2 \qquad \text{Simplified form}$$

② Divide rational expressions and simplify.

Dividing Rational Expressions

To divide two rational expressions, multiply the first expression by the *reciprocal* of the second. That is, *invert the divisor and multiply*.

Dividing Rational Expressions

Let u, v, w, and z represent real numbers, variables, or algebraic expressions such that $v \neq 0$, $w \neq 0$, and $z \neq 0$. Then the quotient of u/v and w/z is

$$\frac{u}{v} \div \frac{w}{z} = \frac{u}{v} \cdot \frac{z}{w} = \frac{uz}{vw}.$$

Example 6   Dividing Rational Expressions

Divide the rational expressions.

$$\frac{x}{x+3} \div \frac{4}{x-1}$$

Solution

$$\frac{x}{x+3} \div \frac{4}{x-1} = \frac{x}{x+3} \cdot \frac{x-1}{4} \qquad \text{Invert divisor and multiply.}$$

$$= \frac{x(x-1)}{(x+3)(4)} \qquad \text{Multiply numerators and denominators.}$$

$$= \frac{x(x-1)}{4(x+3)}, \quad x \neq 1 \qquad \text{Simplify.}$$

Example 7   Dividing Rational Expressions

$$\frac{2x}{3x-12} \div \frac{x^2-2x}{x^2-6x+8} \qquad \text{Original expressions}$$

$$= \frac{2x}{3x-12} \cdot \frac{x^2-6x+8}{x^2-2x} \qquad \text{Invert divisor and multiply.}$$

$$= \frac{(2)(x)(x-2)(x-4)}{(3)(x-4)(x)(x-2)} \qquad \text{Factor.}$$

$$= \frac{(2)(x)(x-2)(x-4)}{(3)(x-4)(x)(x-2)} \qquad \text{Divide out common factors.}$$

$$= \frac{2}{3}, \quad x \neq 0, \ x \neq 2, \ x \neq 4 \qquad \text{Simplified form}$$

Remember that the original expression is equivalent to $\frac{2}{3}$ except for $x = 0$, $x = 2$, and $x = 4$.

Example 8   Dividing Rational Expressions

Divide the rational expressions.

$$\frac{x^2 - y^2}{2x + 2y} \div \frac{2x^2 - 3xy + y^2}{6x + 2y}$$

Solution

$$\frac{x^2 - y^2}{2x + 2y} \div \frac{2x^2 - 3xy + y^2}{6x + 2y}$$

$$= \frac{x^2 - y^2}{2x + 2y} \cdot \frac{6x + 2y}{2x^2 - 3xy + y^2} \qquad \text{Invert divisor and multiply.}$$

$$= \frac{(x + y)(x - y)(2)(3x + y)}{(2)(x + y)(2x - y)(x - y)} \qquad \text{Factor.}$$

$$= \frac{\cancel{(x + y)}\cancel{(x - y)}(2)(3x + y)}{\cancel{(2)}\cancel{(x + y)}(2x - y)\cancel{(x - y)}} \qquad \text{Divide out common factors.}$$

$$= \frac{3x + y}{2x - y}, \; x \neq y, x \neq -y \qquad \text{Simplified form}$$

Example 9   Amount Spent on Books and Maps

The amount A (in millions of dollars) Americans spent on books and maps and the population P (in millions) of the United States for the period 1995 through 2000 can be modeled by

$$A = \frac{-24.86t + 17,862.7}{-0.05t + 1.0}, \quad 5 \leq t \leq 10$$

and

$$P = 2.46t + 250.8, \quad 5 \leq t \leq 10$$

where t represents the year, with $t = 5$ corresponding to 1995. Find a model T for the amount Americans spent *per person* on books and maps.   (Source: U.S. Bureau of Economic Analysis and U.S. Census Bureau)

Solution

To find a model T for the amount Americans spent per person on books and maps, divide the total amount by the population.

$$T = \frac{-24.86t + 17,862.7}{-0.05t + 1.0} \div 2.46t + 250.8 \qquad \text{Divide amount spent by population.}$$

$$= \frac{-24.86t + 17,862.7}{-0.05t + 1.0} \cdot \frac{1}{2.46t + 250.8} \qquad \text{Invert divisor and multiply.}$$

$$= \frac{-24.86t + 17,862.7}{(-0.05t + 1.0)(2.46t + 250.8)}, \; 5 \leq t \leq 10 \qquad \text{Model}$$

6.2 Exercises

Review   Concepts, Skills, and Problem Solving

Keep mathematically in shape by doing these exercises *before* the problems of this section.

Properties and Definitions

1. *Writing* ✎   Explain how to factor the difference of two squares $9t^2 - 4$.

2. *Writing* ✎   Explain how to factor the perfect square trinomial $4x^2 - 12x + 9$.

3. *Writing* ✎   Explain how to factor the sum of two cubes $8x^3 + 64$.

4. *Writing* ✎   Factor $3x^2 + 13x - 10$, and explain how you can check your answer.

Algebraic Operations

In Exercises 5–10, factor the expression completely.

5. $5x - 20x^2$

6. $64 - (x - 6)^2$

7. $15x^2 - 16x - 15$

8. $16t^2 + 8t + 1$

9. $y^3 - 64$

10. $8x^3 + 1$

Graphs

In Exercises 11 and 12, sketch the line through the point with each indicated slope on the same set of coordinate axes.

	Point		Slopes
11.	$(2, -3)$	(a) 0	(b) Undefined
		(c) 2	(d) $-\frac{1}{3}$
12.	$(-1, 4)$	(a) 2	(b) -1
		(c) $\frac{1}{2}$	(d) Undefined

Developing Skills

In Exercises 1–8, fill in the missing factor.

1. $\dfrac{7x^2}{3y(\quad\quad)} = \dfrac{7}{3y}, \quad x \neq 0$

2. $\dfrac{14x(x - 3)^2}{(x - 3)(\quad\quad)} = \dfrac{2x}{x - 3}, \quad x \neq 3$

3. $\dfrac{3x(x + 2)^2}{(x - 4)(\quad\quad)} = \dfrac{3x}{x - 4}, \quad x \neq -2$

4. $\dfrac{(x + 1)^3}{x(\quad\quad)} = \dfrac{x + 1}{x}, \quad x \neq -1$

5. $\dfrac{3u(\quad\quad)}{7v(u + 1)} = \dfrac{3u}{7v}, \quad u \neq -1$

6. $\dfrac{(3t + 5)(\quad\quad)}{5t^2(3t - 5)} = \dfrac{3t + 5}{t}, \quad t \neq \dfrac{5}{3}$

7. $\dfrac{13x(\quad\quad)}{4 - x^2} = \dfrac{13x}{x - 2}, \quad x \neq -2$

8. $\dfrac{x^2(\quad\quad)}{x^2 - 10x} = \dfrac{x^2}{10 - x}, \quad x \neq 0$

In Exercises 9–36, multiply and simplify. See Examples 1–5.

9. $7x \cdot \dfrac{9}{14x}$

10. $\dfrac{6}{5a} \cdot (25a)$

11. $\dfrac{8s^3}{9s} \cdot \dfrac{6s^2}{32s}$

12. $\dfrac{3x^4}{7x} \cdot \dfrac{8x^2}{9}$

13. $16u^4 \cdot \dfrac{12}{8u^2}$

14. $25x^3 \cdot \dfrac{8}{35x}$

15. $\dfrac{8}{3 + 4x} \cdot (9 + 12x)$

16. $(6 - 4x) \cdot \dfrac{10}{3 - 2x}$

17. $\dfrac{8u^2v}{3u + v} \cdot \dfrac{u + v}{12u}$

18. $\dfrac{1 - 3xy}{4x^2y} \cdot \dfrac{46x^4y^2}{15 - 45xy}$

19. $\dfrac{12 - r}{3} \cdot \dfrac{3}{r - 12}$ **20.** $\dfrac{8 - z}{8 + z} \cdot \dfrac{z + 8}{z - 8}$

21. $\dfrac{(2x - 3)(x + 8)}{x^3} \cdot \dfrac{x}{3 - 2x}$

22. $\dfrac{x + 14}{x^3(10 - x)} \cdot \dfrac{x(x - 10)}{5}$

23. $\dfrac{4r - 12}{r - 2} \cdot \dfrac{r^2 - 4}{r - 3}$

24. $\dfrac{5y - 20}{5y + 15} \cdot \dfrac{2y + 6}{y - 4}$

25. $\dfrac{2t^2 - t - 15}{t + 2} \cdot \dfrac{t^2 - t - 6}{t^2 - 6t + 9}$

26. $\dfrac{y^2 - 16}{y^2 + 8y + 16} \cdot \dfrac{3y^2 - 5y - 2}{y^2 - 6y + 8}$

27. $(x^2 - 4y^2) \cdot \dfrac{xy}{(x - 2y)^2}$

28. $(u - 2v)^2 \cdot \dfrac{u + 2v}{u - 2v}$

29. $\dfrac{x^2 + 2xy - 3y^2}{(x + y)^2} \cdot \dfrac{x^2 - y^2}{x + 3y}$

30. $\dfrac{(x - 2y)^2}{x + 2y} \cdot \dfrac{x^2 + 7xy + 10y^2}{x^2 - 4y^2}$

31. $\dfrac{x + 5}{x - 5} \cdot \dfrac{2x^2 - 9x - 5}{3x^2 + x - 2} \cdot \dfrac{x^2 - 1}{x^2 + 7x + 10}$

32. $\dfrac{t^2 + 4t + 3}{2t^2 - t - 10} \cdot \dfrac{t}{t^2 + 3t + 2} \cdot \dfrac{2t^2 + 4t^3}{t^2 + 3t}$

33. $\dfrac{9 - x^2}{2x + 3} \cdot \dfrac{4x^2 + 8x - 5}{4x^2 - 8x + 3} \cdot \dfrac{6x^4 - 2x^3}{8x^2 + 4x}$

34. $\dfrac{16x^2 - 1}{4x^2 + 9x + 5} \cdot \dfrac{5x^2 - 9x - 18}{x^2 - 12x + 36} \cdot \dfrac{12 + 4x - x^2}{4x^2 - 13x + 3}$

35. $\dfrac{x^3 + 3x^2 - 4x - 12}{x^3 - 3x^2 - 4x + 12} \cdot \dfrac{x^2 - 9}{x}$

36. $\dfrac{xu - yu + xv - yv}{xu + yu - xv - yv} \cdot \dfrac{xu + yu + xv + yv}{xu - yu - xv + yv}$

In Exercises 37–50, divide and simplify. See Examples 6–8.

37. $x^2 \div \dfrac{3x}{4}$ **38.** $\dfrac{u}{10} \div u^2$

39. $\dfrac{2x}{5} \div \dfrac{x^2}{15}$ **40.** $\dfrac{3y^2}{20} \div \dfrac{y}{15}$

41. $\dfrac{7xy^2}{10u^2v} \div \dfrac{21x^3}{45uv}$ **42.** $\dfrac{25x^2y}{60x^3y^2} \div \dfrac{5x^4y^3}{16x^2y}$

43. $\dfrac{3(a + b)}{4} \div \dfrac{(a + b)^2}{2}$

44. $\dfrac{x^2 + 9}{5(x + 2)} \div \dfrac{x + 3}{5(x^2 - 4)}$

45. $\dfrac{(x^3y)^2}{(x + 2y)^2} \div \dfrac{x^2y}{(x + 2y)^3}$

46. $\dfrac{x^2 - y^2}{2x^2 - 8x} \div \dfrac{(x - y)^2}{2xy}$

47. $\dfrac{y^2 - 2y - 15}{y^2 - 9} \div \dfrac{12 - 4y}{y^2 - 6y + 9}$

48. $\dfrac{x + 3}{x^2 + 7x + 10} \div \dfrac{x^2 + 6x + 9}{x^2 + 5x + 6}$

49. $\dfrac{x^2 + 2x - 15}{x^2 + 11x + 30} \div \dfrac{x^2 - 8x + 15}{x^2 + 2x - 24}$

50. $\dfrac{y^2 + 5y - 14}{y^2 + 10y + 21} \div \dfrac{y^2 + 5y + 6}{y^2 + 7y + 12}$

In Exercises 51–58, perform the operations and simplify. (In Exercises 57 and 58, n is a positive integer.)

51. $\left[\dfrac{x^2}{9} \cdot \dfrac{3(x+4)}{x^2 + 2x}\right] \div \dfrac{x}{x+2}$

52. $\left(\dfrac{x^2 + 6x + 9}{x^2} \cdot \dfrac{2x + 1}{x^2 - 9}\right) \div \dfrac{4x^2 + 4x + 1}{x^2 - 3x}$

53. $\left[\dfrac{xy + y}{4x} \div (3x + 3)\right] \div \dfrac{y}{3x}$

54. $\dfrac{3u^2 - u - 4}{u^2} \div \dfrac{3u^2 + 12u + 4}{u^4 - 3u^3}$

55. $\dfrac{2x^2 + 5x - 25}{3x^2 + 5x + 2} \cdot \dfrac{3x^2 + 2x}{x + 5} \div \left(\dfrac{x}{x + 1}\right)^2$

56. $\dfrac{t^2 - 100}{4t^2} \cdot \dfrac{t^3 - 5t^2 - 50t}{t^4 + 10t^3} \div \dfrac{(t - 10)^2}{5t}$

57. $x^3 \cdot \dfrac{x^{2n} - 9}{x^{2n} + 4x^n + 3} \div \dfrac{x^{2n} - 2x^n - 3}{x}$

58. $\dfrac{x^{n+1} - 8x}{x^{2n} + 2x^n + 1} \cdot \dfrac{x^{2n} - 4x^n - 5}{x} \div x^n$

In Exercises 59 and 60, use a graphing calculator to graph the two equations in the same viewing window. Use the graphs and a table of values to verify that the expressions are equivalent. Verify the results algebraically.

59. $y_1 = \dfrac{x^2 - 10x + 25}{x^2 - 25} \cdot \dfrac{x + 5}{2}$

$y_2 = \dfrac{x - 5}{2}, \quad x \neq \pm 5$

60. $y_1 = \dfrac{3x + 15}{x^4} \div \dfrac{x + 5}{x^2}$

$y_2 = \dfrac{3}{x^2}, \quad x \neq -5$

Solving Problems

▲ *Geometry*   In Exercises 61 and 62, write and simplify an expression for the area of the shaded region.

61.

| $\dfrac{2w + 3}{3}$ | $\dfrac{2w + 3}{3}$ | $\dfrac{2w + 3}{3}$ |

$\dfrac{w}{2}$

$\dfrac{w}{2}$

62.

| $\dfrac{2w - 1}{2}$ | $\dfrac{2w - 1}{2}$ |

$\dfrac{w}{3}$

$\dfrac{w}{3}$

$\dfrac{w}{3}$

Probability   In Exercises 63–66, consider an experiment in which a marble is tossed into a rectangular box with dimensions $2x$ centimeters by $4x + 2$ centimeters.

The probability that the marble will come to rest in the unshaded portion of the box is equal to the ratio of the unshaded area to the total area of the figure. Find the probability in simplified form.

63.

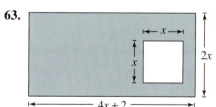

64.

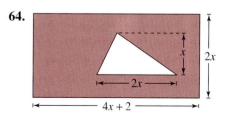

65.

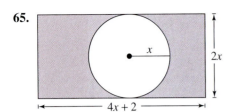

66.

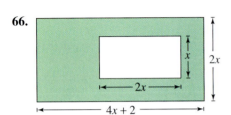

67. *Photocopy Rate* A photocopier produces copies at a rate of 20 pages per minute.

(a) Determine the time required to copy 1 page.

(b) Determine the time required to copy x pages.

(c) Determine the time required to copy 35 pages.

68. *Pumping Rate* The rate for a pump is 15 gallons per minute.

(a) Determine the time required to pump 1 gallon.

(b) Determine the time required to pump x gallons.

(c) Determine the time required to pump 130 gallons.

Explaining Concepts

69. *Writing* In your own words, explain how to divide rational expressions.

70. *Writing* Explain how to divide a rational expression by a polynomial.

71. *Error Analysis* Describe the error.

$$\frac{x^2 - 4}{5x} \div \frac{x + 2}{x - 2} = \frac{5x}{x^2 - 4} \cdot \frac{x + 2}{x - 2}$$

$$= \frac{5x}{(x + 2)(x - 2)} \cdot \frac{x + 2}{x - 2}$$

$$= \frac{5x}{(x - 2)^2}$$

72. *Writing* Complete the table for the given values of x.

x	60	100	1000
$\dfrac{x - 10}{x + 10}$			
$\dfrac{x + 50}{x - 50}$			
$\dfrac{x - 10}{x + 10} \cdot \dfrac{x + 50}{x - 50}$			

x	10,000	100,000	1,000,000
$\dfrac{x - 10}{x + 10}$			
$\dfrac{x + 50}{x - 50}$			
$\dfrac{x - 10}{x + 10} \cdot \dfrac{x + 50}{x - 50}$			

What kind of pattern do you see? Try to explain what is going on. Can you see why?

6.3 Adding and Subtracting Rational Expressions

Tom Carter/PhotoEdit

What You Should Learn

1 Add or subtract rational expressions with like denominators and simplify.

2 Add or subtract rational expressions with unlike denominators and simplify.

Why You Should Learn It

Addition and subtraction of rational expressions can be used to solve real-life applications. For instance, in Exercise 83 on page 397, you will find a rational expression that models the number of participants in high school athletic programs.

1 Add or subtract rational expressions with like denominators and simplify.

Adding or Subtracting with Like Denominators

As with numerical fractions, the procedure used to add or subtract two rational expressions depends on whether the expressions have *like* or *unlike* denominators. To add or subtract two rational expressions with *like* denominators, simply combine their numerators and place the result over the common denominator.

Adding or Subtracting with Like Denominators

If u, v, and w are real numbers, variables, or algebraic expressions, and $w \neq 0$, the following rules are valid.

1. $\dfrac{u}{w} + \dfrac{v}{w} = \dfrac{u + v}{w}$ Add fractions with like denominators.

2. $\dfrac{u}{w} - \dfrac{v}{w} = \dfrac{u - v}{w}$ Subtract fractions with like denominators.

Example 1   Adding and Subtracting with Like Denominators

a. $\dfrac{x}{4} + \dfrac{5 - x}{4} = \dfrac{x + (5 - x)}{4} = \dfrac{5}{4}$ Add numerators.

b. $\dfrac{7}{2x - 3} - \dfrac{3x}{2x - 3} = \dfrac{7 - 3x}{2x - 3}$ Subtract numerators.

Example 2   Subtracting Rational Expressions and Simplifying

$$\frac{x}{x^2 - 2x - 3} - \frac{3}{x^2 - 2x - 3} = \frac{x - 3}{x^2 - 2x - 3}$$ Subtract numerators.

$$= \frac{(1)(x - 3)}{(x - 3)(x + 1)}$$ Factor.

$$= \frac{1}{x + 1}, \quad x \neq 3$$ Simplified form

Study Tip

After adding or subtracting two (or more) rational expressions, check the resulting fraction to see if it can be simplified, as illustrated in Example 2.

The rules for adding and subtracting rational expressions with like denominators can be extended to cover sums and differences involving three or more rational expressions, as illustrated in Example 3.

Example 3 Combining Three Rational Expressions

$$\frac{x^2 - 26}{x - 5} - \frac{2x + 4}{x - 5} + \frac{10 + x}{x - 5} \qquad \text{Original expressions}$$

$$= \frac{(x^2 - 26) - (2x + 4) + (10 + x)}{x - 5} \qquad \text{Write numerator over common denominator.}$$

$$= \frac{x^2 - 26 - 2x - 4 + 10 + x}{x - 5} \qquad \text{Distributive Property}$$

$$= \frac{x^2 - x - 20}{x - 5} \qquad \text{Combine like terms.}$$

$$= \frac{(x - 5)(x + 4)}{x - 5} \qquad \text{Factor and divide out common factor.}$$

$$= x + 4, \quad x \ne 5 \qquad \text{Simplified form}$$

2 Add or subtract rational expressions with unlike denominators and simplify.

Adding or Subtracting with Unlike Denominators

To add or subtract rational expressions with *unlike* denominators, you must first rewrite each expression using the **least common multiple (LCM)** of the denominators of the individual expressions. The least common multiple of two (or more) polynomials is the simplest polynomial that is a multiple of each of the original polynomials. This means that the LCM must contain all the *different* factors in the polynomials and each of these factors must be repeated the maximum number of times it occurs in any one of the polynomials.

Example 4 Finding Least Common Multiples

a. The least common multiple of

$$6x = 2 \cdot 3 \cdot x, \quad 2x^2 = 2 \cdot x \cdot x, \quad \text{and} \quad 9x^3 = 3 \cdot 3 \cdot x \cdot x \cdot x$$

is $2 \cdot 3 \cdot 3 \cdot x \cdot x \cdot x = 18x^3$.

b. The least common multiple of

$$x^2 - x = x(x - 1) \quad \text{and} \quad 2x - 2 = 2(x - 1)$$

is $2x(x - 1)$.

c. The least common multiple of

$$3x^2 + 6x = 3x(x + 2) \quad \text{and} \quad x^2 + 4x + 4 = (x + 2)^2$$

is $3x(x + 2)^2$.

To add or subtract rational expressions with *unlike* denominators, you must first rewrite the rational expressions so that they have *like* denominators. The like denominator that you use is the least common multiple of the original denominators and is called the **least common denominator (LCD)** of the original rational expressions. Once the rational expressions have been written with like denominators, you can simply add or subtract these rational expressions using the rules given at the beginning of this section.

Technology: Tip

You can use a graphing calculator to check your results when adding or subtracting rational expressions. For instance, in Example 5, try graphing the equations

$$y_1 = \frac{7}{6x} + \frac{5}{8x}$$

and

$$y_2 = \frac{43}{24x}$$

in the same viewing window. If the two graphs coincide, as shown below, you can conclude that the solution checks.

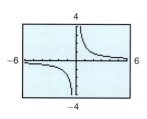

Example 5   Adding with Unlike Denominators

Add the rational expressions: $\dfrac{7}{6x} + \dfrac{5}{8x}$.

Solution

By factoring the denominators, $6x = 2 \cdot 3 \cdot x$ and $8x = 2^3 \cdot x$, you can conclude that the least common denominator is $2^3 \cdot 3 \cdot x = 24x$.

$$\frac{7}{6x} + \frac{5}{8x} = \frac{7(4)}{6x(4)} + \frac{5(3)}{8x(3)}$$ Rewrite expressions using LCD of $24x$.

$$= \frac{28}{24x} + \frac{15}{24x}$$ Like denominators

$$= \frac{28 + 15}{24x} = \frac{43}{24x}$$ Add fractions and simplify.

Example 6   Subtracting with Unlike Denominators

Subtract the rational expressions: $\dfrac{3}{x - 3} - \dfrac{5}{x + 2}$.

Solution

The only factors of the denominators are $x - 3$ and $x + 2$. So, the least common denominator is $(x - 3)(x + 2)$.

$$\frac{3}{x - 3} - \frac{5}{x + 2}$$ Write original expressions.

$$= \frac{3(x + 2)}{(x - 3)(x + 2)} - \frac{5(x - 3)}{(x - 3)(x + 2)}$$ Rewrite expressions using LCD of $(x - 3)(x + 2)$.

$$= \frac{3x + 6}{(x - 3)(x + 2)} - \frac{5x - 15}{(x - 3)(x + 2)}$$ Distributive Property

$$= \frac{(3x + 6) - (5x - 15)}{(x - 3)(x + 2)}$$ Subtract fractions.

$$= \frac{3x + 6 - 5x + 15}{(x - 3)(x + 2)}$$ Distributive Property

$$= \frac{-2x + 21}{(x - 3)(x + 2)}$$ Simplified form

Study Tip

In Example 7, notice that the denominator $2 - x$ is rewritten as $(-1)(x - 2)$ and then the problem is changed from addition to subtraction.

Example 7   Adding with Unlike Denominators

$$\frac{6x}{x^2 - 4} + \frac{3}{2 - x}$$
Original expressions

$$= \frac{6x}{(x + 2)(x - 2)} + \frac{3}{(-1)(x - 2)}$$
Factor denominators.

$$= \frac{6x}{(x + 2)(x - 2)} - \frac{3(x + 2)}{(x + 2)(x - 2)}$$
Rewrite expressions using LCD of $(x + 2)(x - 2)$.

$$= \frac{6x}{(x + 2)(x - 2)} - \frac{3x + 6}{(x + 2)(x - 2)}$$
Distributive Property

$$= \frac{6x - (3x + 6)}{(x + 2)(x - 2)}$$
Subtract.

$$= \frac{6x - 3x - 6}{(x + 2)(x - 2)}$$
Distributive Property

$$= \frac{3x - 6}{(x + 2)(x - 2)}$$
Simplify.

$$= \frac{3(x - 2)}{(x + 2)(x - 2)}$$
Factor and divide out common factor.

$$= \frac{3}{x + 2}, \quad x \neq 2$$
Simplified form

Example 8   Subtracting with Unlike Denominators

$$\frac{x}{x^2 - 5x + 6} - \frac{1}{x^2 - x - 2}$$
Original expressions

$$= \frac{x}{(x - 3)(x - 2)} - \frac{1}{(x - 2)(x + 1)}$$
Factor denominators.

$$= \frac{x(x + 1)}{(x - 3)(x - 2)(x + 1)} - \frac{1(x - 3)}{(x - 3)(x - 2)(x + 1)}$$
Rewrite expressions using LCD of $(x - 3)(x - 2)(x + 1)$.

$$= \frac{x^2 + x}{(x - 3)(x - 2)(x + 1)} - \frac{x - 3}{(x - 3)(x - 2)(x + 1)}$$
Distributive Property

$$= \frac{(x^2 + x) - (x - 3)}{(x - 3)(x - 2)(x + 1)}$$
Subtract fractions.

$$= \frac{x^2 + x - x + 3}{(x - 3)(x - 2)(x + 1)}$$
Distributive Property

$$= \frac{x^2 + 3}{(x - 3)(x - 2)(x + 1)}$$
Simplified form

Example 9   Combining Three Rational Expressions

$$\frac{4x}{x^2 - 16} + \frac{x}{x + 4} - \frac{2}{x} = \frac{4x}{(x + 4)(x - 4)} + \frac{x}{x + 4} - \frac{2}{x}$$

$$= \frac{4x(x)}{x(x + 4)(x - 4)} + \frac{x(x)(x - 4)}{x(x + 4)(x - 4)} - \frac{2(x + 4)(x - 4)}{x(x + 4)(x - 4)}$$

$$= \frac{4x^2 + x^2(x - 4) - 2(x^2 - 16)}{x(x + 4)(x - 4)}$$

$$= \frac{4x^2 + x^3 - 4x^2 - 2x^2 + 32}{x(x + 4)(x - 4)}$$

$$= \frac{x^3 - 2x^2 + 32}{x(x + 4)(x - 4)}$$

To add or subtract two rational expressions, you can use the LCD method or the basic definition

$$\frac{a}{b} \pm \frac{c}{d} = \frac{ad \pm bc}{bd}, \quad b \neq 0, d \neq 0. \qquad \textcolor{red}{\text{Basic definition}}$$

This definition provides an efficient way of adding or subtracting two rational expressions that have no common factors in their denominators.

Example 10   Dog Registrations

For the years 1997 through 2001, the number of rottweilers R (in thousands) and the number of collies C (in thousands) registered with the American Kennel Club can be modeled by

$$R = \frac{-1.9t^2 + 3726}{t^2} \quad \text{and} \quad C = \frac{-1.04t + 5}{-0.18t + 1}, \quad 7 \leq t \leq 11$$

where t represents the year, with $t = 7$ corresponding to 1997. Find a rational model T for the number of rottweilers and collies registered with the American Kennel Club.   (Source: American Kennel Club)

Solution

To find a model for T, find the sum of R and C.

$$T = \frac{-1.9t^2 + 3726}{t^2} + \frac{(-1.04t + 5)}{(-0.18t + 1)} \qquad \textcolor{red}{\text{Sum of } R \text{ and } C.}$$

$$= \frac{(-1.9t^2 + 3726)(-0.18t + 1) + t^2(-1.04t + 5)}{t^2(-0.18t + 1)} \qquad \textcolor{red}{\text{Basic definition}}$$

$$= \frac{0.342t^3 - 1.9t^2 - 670.68t + 3726 - 1.04t^3 + 5t^2}{t^2(-0.18t + 1)} \qquad \textcolor{red}{\text{FOIL Method and Distributive Property}}$$

$$= \frac{-0.698t^3 + 3.1t^2 - 670.68t + 3726}{t^2(-0.18t + 1)} \qquad \textcolor{red}{\text{Combine like terms.}}$$

6.3 Exercises

Review   *Concepts, Skills, and Problem Solving*

Keep mathematically in shape by doing these exercises *before* the problems of this section.

Properties and Definitions

1. Write the equation $5y - 3x - 4 = 0$ in the following forms.

   (a) Slope-intercept form

   (b) Point-slope form (many correct answers)

2. *Writing* ✎ Explain how you can visually determine the sign of the slope of a line by observing its graph.

Simplifying Expressions

In Exercises 3–10, perform the multiplication and simplify.

3. $-6x(10 - 7x)$

4. $(2 - y)(3 + 2y)$

5. $(11 - x)(11 + x)$

6. $(4 - 5z)(4 + 5z)$

7. $(x + 1)^2$

8. $t(t^2 + 1) - t(t^2 - 1)$

9. $(x - 2)(x^2 + 2x + 4)$

10. $t(t - 4)(2t + 3)$

Creating Expressions

▲ *Geometry*   In Exercises 11 and 12, write and simplify expressions for the perimeter and area of the figure.

11.

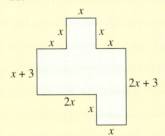

12.

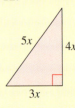

Developing Skills

In Exercises 1–16, combine and simplify. See Examples 1–3.

1. $\dfrac{5x}{8} - \dfrac{7x}{8}$

2. $\dfrac{7y}{12} + \dfrac{9y}{12}$

3. $\dfrac{2}{3a} - \dfrac{11}{3a}$

4. $\dfrac{6}{19x} - \dfrac{7}{19x}$

5. $\dfrac{x}{9} - \dfrac{x + 2}{9}$

6. $\dfrac{4 - y}{4} + \dfrac{3y}{4}$

7. $\dfrac{z^2}{3} + \dfrac{z^2 - 2}{3}$

8. $\dfrac{10x^2 + 1}{3} - \dfrac{10x^2}{3}$

9. $\dfrac{2x + 5}{3x} + \dfrac{1 - x}{3x}$

10. $\dfrac{16 + z}{5z} - \dfrac{11 - z}{5z}$

11. $\dfrac{3y}{3} - \dfrac{3y - 3}{3} - \dfrac{7}{3}$

12. $\dfrac{-16u}{9} - \dfrac{27 - 16u}{9} + \dfrac{2}{9}$

13. $\dfrac{3y - 22}{y - 6} - \dfrac{2y - 16}{y - 6}$

14. $\dfrac{5x - 1}{x + 4} + \dfrac{5 - 4x}{x + 4}$

15. $\dfrac{2x - 1}{x(x - 3)} + \dfrac{1 - x}{x(x - 3)}$

16. $\dfrac{7s - 5}{2s + 5} + \dfrac{3(s + 10)}{2s + 5}$

In Exercises 17–28, find the least common multiple of the expressions. See Example 4.

17. $5x^2, 20x^3$

18. $14t^2, 42t^5$

19. $9y^3, 12y$

20. $44m^2, 10m$

21. $15x^2, 3(x + 5)$

22. $6x^2, 15x(x - 1)$

23. $63z^2(z + 1), 14(z + 1)^4$

24. $18y^3, 27y(y - 3)^2$

25. $8t(t + 2), 14(t^2 - 4)$

26. $2y^2 + y - 1, 4y^2 - 2y$

27. $6(x^2 - 4), 2x(x + 2)$

28. $t^3 + 3t^2 + 9t, 2t^2(t^2 - 9)$

41. $\dfrac{x - 8}{x^2 - 25}, \dfrac{9x}{x^2 - 10x + 25}$

42. $\dfrac{3y}{y^2 - y - 12}, \dfrac{y - 4}{y^2 + 3y}$

In Exercises 29–34, fill in the missing factor.

29. $\dfrac{7x^2}{4a(\quad\quad)} = \dfrac{7}{4a}, \quad x \neq 0$

30. $\dfrac{3y(x - 3)^2}{(x - 3)(\quad\quad)} = \dfrac{21y}{x - 3}$

31. $\dfrac{5r(\quad\quad)}{3v(u + 1)} = \dfrac{5r}{3v}, \quad u \neq -1$

32. $\dfrac{(3t + 5)(\quad\quad)}{10t^2(3t - 5)} = \dfrac{3t + 5}{2t}, \quad t \neq \dfrac{5}{3}$

33. $\dfrac{7y(\quad\quad)}{4 - x^2} = \dfrac{7y}{x - 2}, \quad x \neq -2$

34. $\dfrac{4x^2(\quad\quad)}{x^2 - 10x} = \dfrac{4x^2}{10 - x}, \quad x \neq 0$

In Exercises 35–42, find the least common denominator of the two fractions and rewrite each fraction using the least common denominator.

35. $\dfrac{n + 8}{3n - 12}, \dfrac{10}{6n^2}$

36. $\dfrac{8s}{(s + 2)^2}, \dfrac{3}{s^3 + s^2 - 2s}$

37. $\dfrac{2}{x^2(x - 3)}, \dfrac{5}{x(x + 3)}$

38. $\dfrac{5t}{2t(t - 3)^2}, \dfrac{4}{t(t - 3)}$

39. $\dfrac{v}{2v^2 + 2v}, \dfrac{4}{3v^2}$

40. $\dfrac{4x}{(x + 5)^2}, \dfrac{x - 2}{x^2 - 25}$

In Exercises 43–76, combine and simplify. See Examples 5–9.

43. $\dfrac{5}{4x} - \dfrac{3}{5}$

44. $\dfrac{10}{b} + \dfrac{1}{10b}$

45. $\dfrac{7}{a} + \dfrac{14}{a^2}$

46. $\dfrac{1}{6u^2} - \dfrac{2}{9u}$

47. $\dfrac{20}{x - 4} + \dfrac{20}{4 - x}$

48. $\dfrac{15}{2 - t} - \dfrac{7}{t - 2}$

49. $\dfrac{3x}{x - 8} - \dfrac{6}{8 - x}$

50. $\dfrac{1}{y - 6} + \dfrac{y}{6 - y}$

51. $25 + \dfrac{10}{x + 4}$

52. $\dfrac{100}{x - 10} - 8$

53. $\dfrac{3x}{3x - 2} + \dfrac{2}{2 - 3x}$

54. $\dfrac{y}{5y - 3} - \dfrac{3}{3 - 5y}$

55. $\dfrac{7}{2x} + \dfrac{1}{x - 2}$

56. $\dfrac{3}{y - 1} + \dfrac{5}{4y}$

57. $\dfrac{x}{x + 3} - \dfrac{5}{x - 2}$

58. $\dfrac{1}{x + 4} - \dfrac{1}{x + 2}$

59. $\dfrac{12}{x^2 - 9} - \dfrac{2}{x - 3}$

60. $\dfrac{12}{x^2 - 4} - \dfrac{3}{x + 2}$

61. $\dfrac{3}{x - 5} + \dfrac{2}{x + 5}$

62. $\dfrac{7}{2x - 3} + \dfrac{3}{2x + 3}$

63. $\dfrac{4}{x^2} - \dfrac{4}{x^2 + 1}$

64. $\dfrac{2}{y^2 + 2} + \dfrac{1}{2y^2}$

65. $\dfrac{x}{x^2 - 9} + \dfrac{3}{x^2 - 5x + 6}$

66. $\dfrac{x}{x^2 - x - 30} - \dfrac{1}{x + 5}$

67. $\dfrac{4}{x - 4} + \dfrac{16}{(x - 4)^2}$

68. $\dfrac{3}{x - 2} - \dfrac{1}{(x - 2)^2}$

69. $\dfrac{y}{x^2 + xy} - \dfrac{x}{xy + y^2}$

70. $\dfrac{5}{x + y} + \dfrac{5}{x^2 - y^2}$

71. $\dfrac{4}{x} - \dfrac{2}{x^2} + \dfrac{4}{x + 3}$

72. $\dfrac{5}{2} - \dfrac{1}{2x} - \dfrac{3}{x + 1}$

73. $\dfrac{3u}{u^2 - 2uv + v^2} + \dfrac{2}{u - v} - \dfrac{u}{u - v}$

74. $\dfrac{1}{x - y} - \dfrac{3}{x + y} + \dfrac{3x - y}{x^2 - y^2}$

75. $\dfrac{x + 2}{x - 1} - \dfrac{2}{x + 6} - \dfrac{14}{x^2 + 5x - 6}$

76. $\dfrac{x}{x^2 + 15x + 50} + \dfrac{7}{x + 10} - \dfrac{x - 1}{x + 5}$

In Exercises 77 and 78, use a graphing calculator to graph the two equations in the same viewing window. Use the graphs to verify that the expressions are equivalent. Verify the results algebraically.

77. $y_1 = \dfrac{2}{x} + \dfrac{4}{x - 2}, \, y_2 = \dfrac{6x - 4}{x(x - 2)}$

78. $y_1 = 3 - \dfrac{1}{x - 1}, \, y_2 = \dfrac{3x - 4}{x - 1}$

Solving Problems

79. *Work Rate*   After working together for t hours on a common task, two workers have completed fractional parts of the job equal to $t/4$ and $t/6$. What fractional part of the task has been completed?

80. *Work Rate*   After working together for t hours on a common task, two workers have completed fractional parts of the job equal to $t/3$ and $t/5$. What fractional part of the task has been completed?

81. *Rewriting a Fraction*   The fraction $4/(x^3 - x)$ can be rewritten as a sum of three fractions, as follows.

$$\frac{4}{x^3 - x} = \frac{A}{x} + \frac{B}{x + 1} + \frac{C}{x - 1}$$

The numbers A, B, and C are the solutions of the system

$$\begin{cases} A + B + C = 0 \\ \quad\,\, - B + C = 0 \\ -A \qquad\qquad = 4. \end{cases}$$

Solve the system and verify that the sum of the three resulting fractions is the original fraction.

82. *Rewriting a Fraction*   The fraction

$$\frac{x + 1}{x^3 - x^2}$$

can be rewritten as a sum of three fractions, as follows.

$$\frac{x + 1}{x^3 - x^2} = \frac{A}{x} + \frac{B}{x^2} + \frac{C}{x - 1}$$

The numbers A, B, and C are the solutions of the system

$$\begin{cases} A \qquad\quad + C = 0 \\ -A + B \qquad = 1 \\ \quad\,\, - B \qquad = 1. \end{cases}$$

Solve the system and verify that the sum of the three resulting fractions is the original fraction.

Sports In Exercises 83 and 84, use the following models, which give the number of males M (in thousands) and the number of females F (in thousands) participating in high school athletic programs from 1995 through 2001.

$$M = \frac{463.76t + 2911.4}{0.09t + 1.0} \quad \text{and} \quad F = \frac{3183.41t - 4827.2}{t}$$

In these models, t represents the year, with $t = 5$ corresponding to 1995. (Source: 2001 High School Participation Survey)

83. Find a rational model that represents the total number of participants in high school athletic programs.

84. Use the model you found in Exercise 83 to complete the table showing the total number of participants in high school athletic programs.

Year	1995	1996	1997	1998
Participants				

Year	1999	2000	2001
Participants			

Explaining Concepts

85. ⚡ Answer parts (a)–(c) of Motivating the Chapter on page 366.

86. *Writing* In your own words, describe how to add or subtract rational expressions with like denominators.

87. *Writing* In your own words, describe how to add or subtract rational expressions with unlike denominators.

88. *Writing* Is it possible for the least common denominator of two fractions to be the same as one of the fraction's denominators? If so, give an example.

89. *Error Analysis* Describe the error.

$$\frac{x - 1}{x + 4} - \frac{4x - 11}{x + 4} = \frac{x - 1 - 4x - 11}{x + 4}$$
$$= \frac{-3x - 12}{x + 4} = \frac{-3(x + 4)}{x + 4}$$
$$= -3$$

90. *Error Analysis* Describe the error.

$$\frac{2}{x} - \frac{3}{x + 1} + \frac{x + 1}{x^2}$$
$$= \frac{2x(x + 1) - 3x^2 + (x + 1)^2}{x^2(x + 1)}$$
$$= \frac{2x^2 + x - 3x^2 + x^2 + 1}{x^2(x + 1)}$$
$$= \frac{x + 1}{x^2(x + 1)} = \frac{1}{x^2}$$

91. *Writing* Evaluate each expression at the given value of the variable in two different ways: (1) combine and simplify the rational expressions first and then evaluate the simplified expression at the given variable value, and (2) substitute the given value of the variable first and then simplify the resulting expression. Do you get the same result with each method? Discuss which method you prefer and why. List any advantages and/or disadvantages of each method.

(a) $\dfrac{1}{m - 4} - \dfrac{1}{m + 4} + \dfrac{3m}{m^2 - 16}$, $m = 2$

(b) $\dfrac{x - 2}{x^2 - 9} + \dfrac{3x + 2}{x^2 - 5x + 6}$, $x = 4$

(c) $\dfrac{3y^2 + 16y - 8}{y^2 + 2y - 8} - \dfrac{y - 1}{y - 2} + \dfrac{y}{y + 4}$, $y = 3$

6.4 Complex Fractions

David Forbert/SuperStock, Inc.

Why You Should Learn It

Complex fractions can be used to model real-life situations. For instance, in Exercise 66 on page 405, a complex fraction is used to model the annual percent rate for a home-improvement loan.

What You Should Learn

1. Simplify complex fractions using rules for dividing rational expressions.
2. Simplify complex fractions having a sum or difference in the numerator and/or denominator.

Complex Fractions

Problems involving the division of two rational expressions are sometimes written as **complex fractions.** A complex fraction is a fraction that has a fraction in its numerator or denominator, or both. The rules for dividing rational expressions still apply. For instance, consider the following complex fraction.

$$\dfrac{\left(\dfrac{x+2}{3}\right)}{\left(\dfrac{x-2}{x}\right)} \longrightarrow$$

Numerator fraction

Main fraction line

Denominator fraction

1. Simplify complex fractions using rules for dividing rational expressions.

To perform the division implied by this complex fraction, invert the denominator fraction and multiply, as follows.

$$\dfrac{\left(\dfrac{x+2}{3}\right)}{\left(\dfrac{x-2}{x}\right)} = \dfrac{x+2}{3} \cdot \dfrac{x}{x-2}$$

$$= \dfrac{x(x+2)}{3(x-2)}, \quad x \neq 0$$

Note that for complex fractions you make the main fraction line slightly longer than the fraction lines in the numerator and denominator.

Example 1   Simplifying a Complex Fraction

$$\dfrac{\left(\dfrac{5}{14}\right)}{\left(\dfrac{25}{8}\right)} = \dfrac{5}{14} \cdot \dfrac{8}{25}$$

Invert divisor and multiply.

$$= \dfrac{5 \cdot 2 \cdot 2 \cdot 2}{2 \cdot 7 \cdot 5 \cdot 5}$$

Multiply, factor, and divide out common factors.

$$= \dfrac{4}{35}$$

Simplified form

Example 2   Simplifying a Complex Fraction

Simplify the complex fraction.

$$\dfrac{\left(\dfrac{4y^3}{(5x)^2}\right)}{\left(\dfrac{(2y)^2}{10x^3}\right)}$$

Solution

$$\dfrac{\left(\dfrac{4y^3}{(5x)^2}\right)}{\left(\dfrac{(2y)^2}{10x^3}\right)} = \dfrac{4y^3}{25x^2} \cdot \dfrac{10x^3}{4y^2}$$

Invert divisor and multiply.

$$= \dfrac{4y^2 \cdot y \cdot 2 \cdot 5x^2 \cdot x}{5 \cdot 5x^2 \cdot 4y^2}$$

Multiply and factor.

$$= \dfrac{\cancel{4y^2} \cdot y \cdot 2 \cdot 5\cancel{x^2} \cdot x}{5 \cdot 5\cancel{x^2} \cdot \cancel{4y^2}}$$

Divide out common factors.

$$= \dfrac{2xy}{5}, \quad x \neq 0, \ y \neq 0$$

Simplified form

Example 3   Simplifying a Complex Fraction

Simplify the complex fraction.

$$\dfrac{\left(\dfrac{x+1}{x+2}\right)}{\left(\dfrac{x+1}{x+5}\right)}$$

Solution

$$\dfrac{\left(\dfrac{x+1}{x+2}\right)}{\left(\dfrac{x+1}{x+5}\right)} = \dfrac{x+1}{x+2} \cdot \dfrac{x+5}{x+1}$$

Invert divisor and multiply.

$$= \dfrac{(x+1)(x+5)}{(x+2)(x+1)}$$

Multiply numerators and denominators.

$$= \dfrac{\cancel{(x+1)}(x+5)}{(x+2)\cancel{(x+1)}}$$

Divide out common factors.

$$= \dfrac{x+5}{x+2}, \quad x \neq -1, \ x \neq -5$$

Simplified form

In Example 3, the domain of the complex fraction is restricted by the denominators in the original expression and by the denominators in the original expression after the divisor has been inverted. So, the domain of the original expression is all real values of x such that $x \neq -2, x \neq -5$, and $x \neq -1$.

Example 4   Simplifying a Complex Fraction

Simplify the complex fraction.

$$\frac{\left(\dfrac{x^2 + 4x + 3}{x - 2}\right)}{2x + 6}$$

Solution

Begin by writing the denominator in fractional form.

$$\frac{\left(\dfrac{x^2 + 4x + 3}{x - 2}\right)}{2x + 6} = \frac{\left(\dfrac{x^2 + 4x + 3}{x - 2}\right)}{\left(\dfrac{2x + 6}{1}\right)} \qquad \text{\textcolor{red}{Rewrite denominator.}}$$

$$= \frac{x^2 + 4x + 3}{x - 2} \cdot \frac{1}{2x + 6} \qquad \text{\textcolor{red}{Invert divisor and multiply.}}$$

$$= \frac{(x + 1)(x + 3)}{(x - 2)(2)(x + 3)} \qquad \text{\textcolor{red}{Multiply and factor.}}$$

$$= \frac{(x + 1)\cancel{(x + 3)}}{(x - 2)(2)\cancel{(x + 3)}} \qquad \text{\textcolor{red}{Divide out common factor.}}$$

2 Simplify complex fractions having a sum or difference in the numerator and/or denominator.

$$= \frac{x + 1}{2(x - 2)}, \quad x \ne -3 \qquad \text{\textcolor{red}{Simplified form}}$$

Study Tip

Another way of simplifying the complex fraction in Example 5 is to multiply the numerator and denominator by the least common denominator for all fractions in the numerator and denominator. For this fraction, when you multiply the numerator and denominator by $3x$, you obtain the same result.

$$\frac{\left(\dfrac{x}{3} + \dfrac{2}{3}\right)}{\left(1 - \dfrac{2}{x}\right)} = \frac{\left(\dfrac{x}{3} + \dfrac{2}{3}\right)}{\left(1 - \dfrac{2}{x}\right)} \cdot \frac{3x}{3x}$$

$$= \frac{\dfrac{x}{3}(3x) + \dfrac{2}{3}(3x)}{(1)(3x) - \dfrac{2}{x}(3x)}$$

$$= \frac{x^2 + 2x}{3x - 6}$$

$$= \frac{x(x + 2)}{3(x - 2)}, \ x \ne 0$$

Complex Fractions with Sums or Differences

Complex fractions can have numerators and/or denominators that are sums or differences of fractions. To simplify a complex fraction, combine its numerator and its denominator into single fractions. Then divide by inverting the denominator and multiplying.

Example 5   Simplifying a Complex Fraction

$$\frac{\left(\dfrac{x}{3} + \dfrac{2}{3}\right)}{\left(1 - \dfrac{2}{x}\right)} = \frac{\left(\dfrac{x}{3} + \dfrac{2}{3}\right)}{\left(\dfrac{x}{x} - \dfrac{2}{x}\right)} \qquad \text{\textcolor{red}{Rewrite with least common denominators.}}$$

$$= \frac{\left(\dfrac{x + 2}{3}\right)}{\left(\dfrac{x - 2}{x}\right)} \qquad \text{\textcolor{red}{Add fractions.}}$$

$$= \frac{x + 2}{3} \cdot \frac{x}{x - 2} \qquad \text{\textcolor{red}{Invert divisor and multiply.}}$$

$$= \frac{x(x + 2)}{3(x - 2)}, \quad x \ne 0 \qquad \text{\textcolor{red}{Simplified form}}$$

Study Tip

In Example 6, the domain of the complex fraction is restricted by every denominator in the expression: $x + 2$, x, and $\left(\dfrac{3}{x + 2} + \dfrac{2}{x}\right)$.

Example 6 Simplifying a Complex Fraction

$$\frac{\left(\dfrac{2}{x + 2}\right)}{\left(\dfrac{3}{x + 2} + \dfrac{2}{x}\right)} = \frac{\left(\dfrac{2}{x + 2}\right)(x)(x + 2)}{\left(\dfrac{3}{x + 2}\right)(x)(x + 2) + \left(\dfrac{2}{x}\right)(x)(x + 2)}$$

$x(x + 2)$ is the least common denominator.

$$= \frac{2x}{3x + 2(x + 2)}$$

Multiply and simplify.

$$= \frac{2x}{3x + 2x + 4}$$

Distributive Property

$$= \frac{2x}{5x + 4}, \quad x \neq -2, \quad x \neq 0$$

Simplify.

Notice that the numerator and denominator of the complex fraction were multiplied by $(x)(x + 2)$, which is the least common denominator of the fractions in the original complex fraction.

When simplifying a rational expression containing negative exponents, first rewrite the expression with positive exponents and then proceed with simplifying the expression. This is demonstrated in Example 7.

Example 7 Simplifying a Complex Fraction

$$\frac{5 + x^{-2}}{8x^{-1} + x} = \frac{\left(5 + \dfrac{1}{x^2}\right)}{\left(\dfrac{8}{x} + x\right)}$$

Rewrite with positive exponents.

$$= \frac{\left(\dfrac{5x^2}{x^2} + \dfrac{1}{x^2}\right)}{\left(\dfrac{8}{x} + \dfrac{x^2}{x}\right)}$$

Rewrite with least common denominators.

$$= \frac{\left(\dfrac{5x^2 + 1}{x^2}\right)}{\left(\dfrac{x^2 + 8}{x}\right)}$$

Add fractions.

$$= \frac{5x^2 + 1}{x^2} \cdot \frac{x}{x^2 + 8}$$

Invert divisor and multiply.

$$= \frac{\cancel{x}(5x^2 + 1)}{\cancel{x}(x)(x^2 + 8)}$$

Divide out common factor.

$$= \frac{5x^2 + 1}{x(x^2 + 8)}$$

Simplified form

6.4 Exercises

Review   Concepts, Skills, and Problem Solving

Keep mathematically in shape by doing these exercises *before* the problems of this section.

Properties and Definitions

1. *Writing* ✏️   In your own words, explain how to use the zero and negative exponent rules.

2. *Writing* ✏️   Explain how to determine the exponent when writing 0.00000237 in scientific notation.

Simplifying Expressions

In Exercises 3–6, simplify the expression. (Assume that any variables in the expression are nonzero.)

3. $(7x^2)(2x^{-3})$

4. $(y^0 z^3)(z^2)^{-4}$

5. $\dfrac{a^4 b^{-2}}{a^{-1} b^5}$

6. $(x + 2)^4 \div (x + 2)^3$

Graphing Functions

In Exercises 7–10, identify the transformation of the graph of $f(x) = x^2$ and sketch the graph of g.

7. $g(x) = x^2 - 2$

8. $g(x) = (x - 2)^2$

9. $g(x) = -x^2$

10. $g(x) = (-x)^2$

Problem Solving

11. *Number Problem*   The sum of three positive numbers is 33. The second number is three greater than the first, and the third is four times the first. Find the three numbers.

12. *Nut Mixture*   A grocer wishes to mix peanuts that cost \$3 per pound, almonds that cost \$4 per pound, and pistachios that cost \$6 per pound to make a 50-pound mixture priced at \$4.10 per pound. Three-quarters of the mixture should be composed of peanuts and almonds. How many pounds of each type of nut should the grocer use in the mixture?

Developing Skills

In Exercises 1–26, simplify the complex fraction. See Examples 1–4.

1. $\dfrac{\left(\dfrac{x^3}{4}\right)}{\left(\dfrac{x}{8}\right)}$

2. $\dfrac{\left(\dfrac{y^4}{12}\right)}{\left(\dfrac{y}{16}\right)}$

3. $\dfrac{\left(\dfrac{x^2}{12}\right)}{\left(\dfrac{5x}{18}\right)}$

4. $\dfrac{\left(\dfrac{3u^2}{6v^3}\right)}{\left(\dfrac{u}{3v}\right)}$

5. $\dfrac{\left(\dfrac{8x^2 y}{3z^2}\right)}{\left(\dfrac{4xy}{9z^5}\right)}$

6. $\dfrac{\left(\dfrac{36x^4}{5y^4 z^5}\right)}{\left(\dfrac{9xy^2}{20z^5}\right)}$

7. $\dfrac{\left(\dfrac{6x^3}{(5y)^2}\right)}{\left(\dfrac{(3x)^2}{15y^4}\right)}$

8. $\dfrac{\left(\dfrac{(3r)^3}{10r^4}\right)}{\left(\dfrac{9r}{(2t)^2}\right)}$

9. $\dfrac{\left(\dfrac{y}{3 - y}\right)}{\left(\dfrac{y^2}{y - 3}\right)}$

10. $\dfrac{\left(\dfrac{x}{x - 4}\right)}{\left(\dfrac{x}{4 - x}\right)}$

11. $\dfrac{\left(\dfrac{25x^2}{x-5}\right)}{\left(\dfrac{10x}{5+4x-x^2}\right)}$

12. $\dfrac{\left(\dfrac{5x}{x+7}\right)}{\left(\dfrac{10}{x^2+8x+7}\right)}$

13. $\dfrac{\left(\dfrac{2x-10}{x+1}\right)}{\left(\dfrac{x-5}{x+1}\right)}$

14. $\dfrac{\left(\dfrac{a+5}{6a-15}\right)}{\left(\dfrac{a+5}{2a-5}\right)}$

15. $\dfrac{\left(\dfrac{x^2+3x-10}{x+4}\right)}{3x-6}$

16. $\dfrac{\left(\dfrac{x^2-2x-8}{x-1}\right)}{5x-20}$

17. $\dfrac{2x-14}{\left(\dfrac{x^2-9x+14}{x+3}\right)}$

18. $\dfrac{4x+16}{\left(\dfrac{x^2+9x+20}{x-1}\right)}$

19. $\dfrac{\left(\dfrac{6x^2-17x+5}{3x^2+3x}\right)}{\left(\dfrac{3x-1}{3x+1}\right)}$

20. $\dfrac{\left(\dfrac{6x^2-13x-5}{5x^2+5x}\right)}{\left(\dfrac{2x-5}{5x+1}\right)}$

21. $\dfrac{16x^2+8x+1}{3x^2+8x-3} \div \dfrac{4x^2-3x-1}{x^2+6x+9}$

22. $\dfrac{9x^2-24x+16}{x^2+10x+25} \div \dfrac{6x^2-5x-4}{2x^2+3x-35}$

23. $\dfrac{x^2+3x-2x-6}{x^2-4} \div \dfrac{x+3}{x^2+4x+4}$

24. $\dfrac{t^3+t^2-9t-9}{t^2-5t+6} \div \dfrac{t^2+6t+9}{t-2}$

25. $\dfrac{\left(\dfrac{x^2-3x-10}{x^2-4x+4}\right)}{\left(\dfrac{21+4x-x^2}{x^2-5x-14}\right)}$

26. $\dfrac{\left(\dfrac{x^2+5x+6}{4x^2-20x+25}\right)}{\left(\dfrac{x^2-5x-24}{4x^2-25}\right)}$

In Exercises 27–46, simplify the complex fraction. See Examples 5 and 6.

27. $\dfrac{\left(1+\dfrac{3}{y}\right)}{y}$

28. $\dfrac{x}{\left(\dfrac{5}{x}+2\right)}$

29. $\dfrac{\left(\dfrac{x}{2}\right)}{\left(2+\dfrac{3}{x}\right)}$

30. $\dfrac{\left(1-\dfrac{2}{x}\right)}{\left(\dfrac{x}{2}\right)}$

31. $\dfrac{\left(\dfrac{4}{x}+3\right)}{\left(\dfrac{4}{x}-3\right)}$

32. $\dfrac{\left(\dfrac{1}{t}-1\right)}{\left(\dfrac{1}{t}+1\right)}$

33. $\dfrac{\left(3+\dfrac{9}{x-3}\right)}{\left(4+\dfrac{12}{x-3}\right)}$

34. $\dfrac{\left(x+\dfrac{2}{x-3}\right)}{\left(x+\dfrac{6}{x-3}\right)}$

35. $\dfrac{\left(\dfrac{3}{x^2} + \dfrac{1}{x}\right)}{\left(2 - \dfrac{4}{5x}\right)}$

36. $\dfrac{\left(16 - \dfrac{1}{x^2}\right)}{\left(\dfrac{1}{4x^2} - 4\right)}$

37. $\dfrac{\left(\dfrac{y}{x} - \dfrac{x}{y}\right)}{\left(\dfrac{x + y}{xy}\right)}$

38. $\dfrac{\left(x - \dfrac{2y^2}{x - y}\right)}{x - 2y}$

39. $\dfrac{\left(1 - \dfrac{1}{y}\right)}{\left(\dfrac{1 - 4y}{y - 3}\right)}$

40. $\dfrac{\left(\dfrac{x + 1}{x + 2} - \dfrac{1}{x}\right)}{\left(\dfrac{2}{x + 2}\right)}$

41. $\dfrac{\left(\dfrac{10}{x + 1}\right)}{\left(\dfrac{1}{2x + 2} + \dfrac{3}{x + 1}\right)}$

42. $\dfrac{\left(\dfrac{2}{x + 5}\right)}{\left(\dfrac{2}{x + 5} + \dfrac{1}{4x + 20}\right)}$

43. $\dfrac{\left(\dfrac{1}{x} - \dfrac{1}{x + 1}\right)}{\left(\dfrac{1}{x + 1}\right)}$

44. $\dfrac{\left(\dfrac{5}{y} - \dfrac{6}{2y + 1}\right)}{\left(\dfrac{5}{2y + 1}\right)}$

45. $\dfrac{\left(\dfrac{x}{x - 3} - \dfrac{2}{3}\right)}{\left(\dfrac{10}{3x} + \dfrac{x^2}{x - 3}\right)}$

46. $\dfrac{\left(\dfrac{1}{2x} - \dfrac{6}{x + 5}\right)}{\left(\dfrac{x}{x - 5} + \dfrac{1}{x}\right)}$

In Exercises 47–54, simplify the expression. See Example 7.

47. $\dfrac{2y - y^{-1}}{10 - y^{-2}}$

48. $\dfrac{9x - x^{-1}}{3 + x^{-1}}$

49. $\dfrac{7x^2 + 2x^{-1}}{5x^{-3} + x}$

50. $\dfrac{3x^{-2} - x}{4x^{-1} + 6x}$

51. $\dfrac{x^{-1} + y^{-1}}{x^{-1} - y^{-1}}$

52. $\dfrac{x^{-1} - y^{-1}}{x^{-2} - y^{-2}}$

53. $\dfrac{x^{-2} - y^{-2}}{(x + y)^2}$

54. $\dfrac{x - y}{x^{-2} - y^{-2}}$

In Exercises 55 and 56, use the function to find and simplify the expression for

$$\dfrac{f(2 + h) - f(2)}{h}.$$

55. $f(x) = \dfrac{1}{x}$

56. $f(x) = \dfrac{x}{x - 1}$

Solving Problems

57. *Average of Two Numbers* Determine the average of the two real numbers $x/5$ and $x/6$.

58. *Average of Two Numbers* Determine the average of the two real numbers $2x/3$ and $3x/5$.

59. *Average of Two Numbers* Determine the average of the two real numbers $2x/3$ and $x/4$.

60. *Average of Two Numbers* Determine the average of the two real numbers $4/a^2$ and $2/a$.

61. *Average of Two Numbers* Determine the average of the two real numbers $(b + 5)/4$ and $2/b$.

62. *Average of Two Numbers* Determine the average of the two real numbers $5/2s$ and $(s + 1)/5$.

63. *Number Problem* Find three real numbers that divide the real number line between $x/9$ and $x/6$ into four equal parts (see figure).

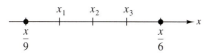

64. *Number Problem* Find two real numbers that divide the real number line between $x/3$ and $5x/4$ into three equal parts (see figure).

65. *Electrical Resistance* When two resistors of resistance R_1 and R_2 are connected in parallel, the total resistance is modeled by

$$\frac{1}{\left(\dfrac{1}{R_1} + \dfrac{1}{R_2}\right)}.$$

Simplify this complex fraction.

66. *Monthly Payment* The approximate annual percent rate r of a monthly installment loan is

$$r = \frac{\left[\dfrac{24(MN - P)}{N}\right]}{\left(P + \dfrac{MN}{12}\right)}$$

where N is the total number of payments, M is the monthly payment, and P is the amount financed.

(a) Simplify the expression.

(b) Approximate the annual percent rate for a four-year home-improvement loan of $15,000 with monthly payments of $350.

In Exercises 67 and 68, use the following models, which give the number N (in thousands) of cellular telephone subscribers and the annual service revenue R (in millions of dollars) generated by subscribers in the United States from 1994 through 2000.

$$N = \frac{4568.33t + 1042.7}{-0.06t + 1.0} \quad \text{and} \quad R = \frac{1382.16t + 5847.9}{-0.06t + 1.0}$$

In these models, t represents the year, with $t = 4$ corresponding to 1994. (Source: Cellular Telecommunications and Internet Association)

67. (a) 📊 Use a graphing calculator to graph the two models in the same viewing window.

(b) Find a model for the average monthly bill per subscriber. (*Note:* Modify the revenue model from years to months.)

68. (a) Use the model in Exercise 67 (b) to complete the table.

Year, t	4	6	8	10
Monthly bill				

(b) The number of subscribers and the revenue were increasing over the last few years, and yet the average monthly bill was decreasing. Explain how this is possible.

Explaining Concepts

69. *Writing* Define the term *complex fraction*. Give an example and show how to simplify the fraction.

70. What are the numerator and denominator of each complex fraction?

(a) $\dfrac{5}{\left(\dfrac{3}{x^2 + 5x + 6}\right)}$ (b) $\dfrac{\left(\dfrac{5}{3}\right)}{x^2 + 5x + 6}$

71. What are the numerator and denominator of each complex fraction?

(a) $\dfrac{\left(\dfrac{x - 1}{5}\right)}{\left(\dfrac{2}{x^2 + 2x - 35}\right)}$ (b) $\dfrac{\left(\dfrac{1}{2y} + x\right)}{\left(\dfrac{3}{y} + x\right)}$

72. *Writing* Of the two methods discussed in this section for simplifying complex fractions, select the method you prefer and explain the method in your own words.

Mid-Chapter Quiz

Take this quiz as you would take a quiz in class. After you are done, check your work against the answers in the back of the book.

1. Determine the domain of $f(y) = \dfrac{y + 2}{y(y - 4)}$.

2. Evaluate $h(x) = (x^2 - 9)/(x^2 - x - 2)$ for the indicated values of x and simplify. If it is not possible, state the reason.

    (a) $h(-3)$ (b) $h(0)$ (c) $h(-1)$ (d) $h(5)$

In Exercises 3–8, simplify the rational expression.

3. $\dfrac{9y^2}{6y}$

4. $\dfrac{8u^3v^2}{36uv^3}$

5. $\dfrac{4x^2 - 1}{x - 2x^2}$

6. $\dfrac{(z + 3)^2}{2z^2 + 5z - 3}$

7. $\dfrac{7ab + 3a^2b^2}{a^2b}$

8. $\dfrac{2mn^2 - n^3}{2m^2 + mn - n^2}$

In Exercises 9–20, perform the indicated operations and simplify.

9. $\dfrac{11t^2}{6} \cdot \dfrac{9}{33t}$

10. $(x^2 + 2x) \cdot \dfrac{5}{x^2 - 4}$

11. $\dfrac{4}{3(x - 1)} \cdot \dfrac{12x}{6(x^2 + 2x - 3)}$

12. $\dfrac{80z^4}{49x^5y^7} \div \dfrac{25z^5}{14x^{12}y^5}$

13. $\dfrac{a - b}{9a + 9b} \div \dfrac{a^2 - b^2}{a^2 + 2a + 1}$

14. $\dfrac{5u}{3(u + v)} \cdot \dfrac{2(u^2 - v^2)}{3v} \div \dfrac{25u^2}{18(u - v)}$

15. $\dfrac{5x - 6}{x - 2} + \dfrac{2x - 5}{x - 2}$

16. $\dfrac{x}{x^2 - 9} - \dfrac{4(x - 3)}{x + 3}$

17. $\dfrac{x^2 + 2}{x^2 - x - 2} + \dfrac{1}{x + 1} - \dfrac{x}{x - 2}$

18. $\dfrac{\left(\dfrac{9t^2}{3 - t} \right)}{\left(\dfrac{6t}{t - 3} \right)}$

19. $\dfrac{\left(\dfrac{10}{x^2 + 2x} \right)}{\left(\dfrac{15}{x^2 + 3x + 2} \right)}$

20. $\dfrac{3x^{-1} - y^{-1}}{(x - y)^{-1}}$

21. You open a floral shop with a setup cost of \$25,000. The cost of creating one dozen floral arrangements is \$144.

    (a) Write the total cost C as a function of x, the number of floral arrangements (in dozens) created.

    (b) Write the average cost per dozen $\overline{C} = C/x$ as a function of x, the number of floral arrangements (in dozens) created.

    (c) Find the value of $\overline{C}(500)$.

22. Determine the average of the three real numbers x, $x/2$, and $x/3$.

6.5 Dividing Polynomials and Synthetic Division

What You Should Learn

1. Divide polynomials by monomials and write in simplest form.
2. Use long division to divide polynomials by polynomials.
3. Use synthetic division to divide polynomials by polynomials of the form $x - k$.
4. Use synthetic division to factor polynomials.

Why You Should Learn It

Division of polynomials is useful in higher-level mathematics when factoring and finding zeros of polynomials.

Dividing a Polynomial by a Monomial

To divide a polynomial by a monomial, *reverse* the procedure used to add or subtract two rational expressions. Here is an example.

$$2 + \frac{1}{x} = \frac{2x}{x} + \frac{1}{x} = \frac{2x + 1}{x} \qquad \text{Add fractions.}$$

$$\frac{2x + 1}{x} = \frac{2x}{x} + \frac{1}{x} = 2 + \frac{1}{x} \qquad \text{Divide by monomial.}$$

1 Divide polynomials by monomials and write in simplest form.

Dividing a Polynomial by a Monomial

Let u, v, and w represent real numbers, variables, or algebraic expressions such that $w \neq 0$.

$$\textbf{1.} \ \frac{u + v}{w} = \frac{u}{w} + \frac{v}{w} \qquad \textbf{2.} \ \frac{u - v}{w} = \frac{u}{w} - \frac{v}{w}$$

When dividing a polynomial by a monomial, remember to write the resulting expressions in simplest form, as illustrated in Example 1.

Example 1   Dividing a Polynomial by a Monomial

Perform the division and simplify.

$$\frac{12x^2 - 20x + 8}{4x}$$

Solution

$$\frac{12x^2 - 20x + 8}{4x} = \frac{12x^2}{4x} - \frac{20x}{4x} + \frac{8}{4x} \qquad \text{Divide each term in the numerator by } 4x.$$

$$= \frac{3(4x)(x)}{4x} - \frac{5(4x)}{4x} + \frac{2(4)}{4x} \qquad \text{Divide out common factors.}$$

$$= 3x - 5 + \frac{2}{x} \qquad \text{Simplified form}$$

② Use long division to divide polynomials by polynomials.

Long Division

In Section 6.1, you learned how to divide one polynomial by another by factoring and dividing out common factors. For instance, you can divide $x^2 - 2x - 3$ by $x - 3$ as follows.

$$(x^2 - 2x - 3) \div (x - 3) = \frac{x^2 - 2x - 3}{x - 3} \qquad \text{Write as fraction.}$$

$$= \frac{(x + 1)(x - 3)}{x - 3} \qquad \text{Factor numerator.}$$

$$= \frac{(x + 1)(x - 3)}{x - 3} \qquad \text{Divide out common factor.}$$

$$= x + 1, \quad x \neq 3 \qquad \text{Simplified form}$$

This procedure works well for polynomials that factor easily. For those that do not, you can use a more general procedure that follows a "long division algorithm" similar to the algorithm used for dividing positive integers, which is reviewed in Example 2.

Example 2   Long Division Algorithm for Positive Integers

Use the long division algorithm to divide 6584 by 28.

Solution

Think $\frac{65}{28} \approx 2$.

Think $\frac{98}{28} \approx 3$.

Think $\frac{144}{28} \approx 5$.

$$
\begin{array}{r}
235 \\
28 \overline{)\ 6584} \\
\underline{56} \\
98 \\
\underline{84} \\
144 \\
\underline{140} \\
4
\end{array}
$$

Multiply 2 by 28.

Subtract and bring down 8.

Multiply 3 by 28.

Subtract and bring down 4.

Multiply 5 by 28.

Remainder

So, you have

$$6584 \div 28 = 235 + \frac{4}{28}$$

$$= 235 + \frac{1}{7}.$$

In Example 2, the numerator 6584 is the **dividend**, 28 is the **divisor**, 235 is the **quotient**, and 4 is the **remainder**.

In the next several examples, you will see how the long division algorithm can be extended to cover the division of one polynomial by another.

Along with the long division algorithm, follow the steps below when performing long division of polynomials.

Long Division of Polynomials

1. Write the dividend and divisor in descending powers of the variable.

2. Insert placeholders with zero coefficients for missing powers of the variable. (See Example 5.)

3. Perform the long division of the polynomials as you would with integers.

4. Continue the process until the degree of the remainder is less than that of the divisor.

Example 3   Long Division Algorithm for Polynomials

Think $x^2/x = x$.

Think $3x/x = 3$.

$$
\begin{array}{r}
x + 3 \\
x - 1 \overline{\smash{)}\, x^2 + 2x + 4} \\
\underline{x^2 - x} \\
3x + 4 \\
\underline{3x - 3} \\
7
\end{array}
$$

Multiply x by $(x - 1)$.
Subtract and bring down 4.
Multiply 3 by $(x - 1)$.
Subtract.

The remainder is a fractional part of the divisor, so you can write

Dividend   Quotient   Remainder

$$
\frac{x^2 + 2x + 4}{x - 1} = x + 3 + \frac{7}{x - 1}.
$$

Divisor   Divisor

Study Tip

Note that in Example 3, the division process requires $3x - 3$ to be subtracted from $3x + 4$. Therefore, the difference

$$
\begin{array}{r}
3x + 4 \\
-(3x - 3)
\end{array}
$$

is implied and written simply as

$$
\begin{array}{r}
3x + 4 \\
\underline{3x - 3} \\
7
\end{array}.
$$

You can check a long division problem by multiplying by the divisor. For instance, you can check the result of Example 3 as follows.

$$
\frac{x^2 + 2x + 4}{x - 1} \overset{?}{=} x + 3 + \frac{7}{x - 1}
$$

$$
(x - 1)\left(\frac{x^2 + 2x + 4}{x - 1}\right) \overset{?}{=} (x - 1)\left(x + 3 + \frac{7}{x - 1}\right)
$$

$$
x^2 + 2x + 4 \overset{?}{=} (x + 3)(x - 1) + 7
$$

$$
x^2 + 2x + 4 \overset{?}{=} (x^2 + 2x - 3) + 7
$$

$$
x^2 + 2x + 4 = x^2 + 2x + 4 \quad ✔
$$

Technology: Tip

You can check the result of a division problem *graphically* with a graphing calculator by comparing the graphs of the original quotient and the simplified form. The graphical check for Example 4 is shown below. Because the graphs coincide, you can conclude that the solution checks.

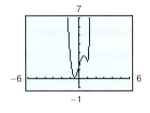

Example 4   Writing in Standard Form Before Dividing

Divide $-13x^3 + 10x^4 + 8x - 7x^2 + 4$ by $3 - 2x$.

Solution

First write the divisor and dividend in standard polynomial form.

$$
\begin{array}{r}
-5x^3 - x^2 + 2x - 1 \\
-2x + 3 \overline{\smash{)}\ 10x^4 - 13x^3 - 7x^2 + 8x + 4} \\
\underline{10x^4 - 15x^3} \\
2x^3 - 7x^2 \\
\underline{2x^3 - 3x^2} \\
-4x^2 + 8x \\
\underline{-4x^2 + 6x} \\
2x + 4 \\
\underline{2x - 3} \\
7
\end{array}
$$

Multiply $-5x^3$ by $(-2x + 3)$.
Subtract and bring down $-7x^2$.
Multiply $-x^2$ by $(-2x + 3)$.
Subtract and bring down $8x$.
Multiply $2x$ by $(-2x + 3)$.
Subtract and bring down 4.
Multiply -1 by $(-2x + 3)$.

This shows that

$$
\underbrace{\frac{\overbrace{10x^4 - 13x^3 - 7x^2 + 8x + 4}^{\text{Dividend}}}{\underbrace{-2x + 3}_{\text{Divisor}}}}_{} = \overbrace{-5x^3 - x^2 + 2x - 1}^{\text{Quotient}} + \frac{\overset{\text{Remainder}}{7}}{\underbrace{-2x + 3}_{\text{Divisor}}}.
$$

When the dividend is missing one or more powers of x, the long division algorithm requires that you account for the missing powers, as shown in Example 5.

Example 5   Accounting for Missing Powers of x

Divide $x^3 - 2$ by $x - 1$.

Solution

To account for the missing x^2- and x-terms, insert $0x^2$ and $0x$.

$$
\begin{array}{r}
x^2 + x + 1 \\
x - 1 \overline{\smash{)}\ x^3 + 0x^2 + 0x - 2} \\
\underline{x^3 - x^2} \\
x^2 + 0x \\
\underline{x^2 - x} \\
x - 2 \\
\underline{x - 1} \\
-1
\end{array}
$$

Insert $0x^2$ and $0x$.
Multiply x^2 by $(x - 1)$.
Subtract and bring down $0x$.
Multiply x by $(x - 1)$.
Subtract and bring down -2.
Multiply 1 by $(x - 1)$.
Subtract.

So, you have

$$
\frac{x^3 - 2}{x - 1} = x^2 + x + 1 - \frac{1}{x - 1}.
$$

In each of the long division examples presented so far, the divisor has been a first-degree polynomial. The long division algorithm works just as well with polynomial divisors of degree two or more, as shown in Example 6.

Example 6   A Second-Degree Divisor

Divide $x^4 + 6x^3 + 6x^2 - 10x - 3$ by $x^2 + 2x - 3$.

Solution

$$
\begin{array}{r}
x^2 + 4x + 1 \\
x^2 + 2x - 3 \overline{)\, x^4 + 6x^3 + 6x^2 - 10x - 3} \\
\underline{x^4 + 2x^3 - 3x^2} \\
4x^3 + 9x^2 - 10x \\
\underline{4x^3 + 8x^2 - 12x} \\
x^2 + 2x - 3 \\
\underline{x^2 + 2x - 3} \\
0
\end{array}
$$

Multiply x^2 by $(x^2 + 2x - 3)$.
Subtract and bring down $-10x$.
Multiply $4x$ by $(x^2 + 2x - 3)$.
Subtract and bring down -3.
Multiply 1 by $(x^2 + 2x - 3)$.
Subtract.

Study Tip

If the remainder of a division problem is zero, the divisor is said to **divide evenly** into the dividend.

So, $x^2 + 2x - 3$ divides evenly into $x^4 + 6x^3 + 6x^2 - 10x - 3$. That is,

$$
\frac{x^4 + 6x^3 + 6x^2 - 10x - 3}{x^2 + 2x - 3} = x^2 + 4x + 1, \; x \neq -3, x \neq 1.
$$

3 Use synthetic division to divide polynomials by polynomials of the form $x - k$.

Synthetic Division

There is a nice shortcut for division by polynomials of the form $x - k$. It is called **synthetic division** and is outlined for a third-degree polynomial as follows.

Synthetic Division of a Third-Degree Polynomial

Use synthetic division to divide $ax^3 + bx^2 + cx + d$ by $x - k$, as follows.

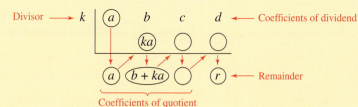

Vertical Pattern: Add terms.
Diagonal Pattern: Multiply by k.

Keep in mind that synthetic division works *only* for divisors of the form $x - k$. Remember that $x + k = x - (-k)$. Moreover, the degree of the quotient is always one less than the degree of the dividend.

Example 7 Using Synthetic Division

Use synthetic division to divide $x^3 + 3x^2 - 4x - 10$ by $x - 2$.

Solution

The coefficients of the dividend form the top row of the synthetic division array. Because you are dividing by $x - 2$, write 2 at the top left of the array. To begin the algorithm, bring down the first coefficient. Then multiply this coefficient by 2, write the result in the second row, and add the two numbers in the second column. By continuing this pattern, you obtain the following.

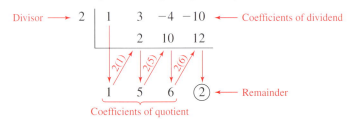

The bottom row shows the coefficients of the quotient. So, the quotient is

$$1x^2 + 5x + 6$$

and the remainder is 2. So, the result of the division problem is

$$\frac{x^3 + 3x^2 - 4x - 10}{x - 2} = x^2 + 5x + 6 + \frac{2}{x - 2}.$$

④ Use synthetic division to factor polynomials.

Factoring and Division

Synthetic division (or long division) can be used to factor polynomials. If the remainder in a synthetic division problem is zero, you know that the divisor divides *evenly* into the dividend. So, the original polynomial can be factored as the product of two polynomials of lesser degrees, as in Example 8.

Example 8 Factoring a Polynomial

The polynomial $x^3 - 7x + 6$ can be factored completely using synthetic division. Because $x - 1$ is a factor of the polynomial, you can divide as follows.

$$
\begin{array}{r|rrrr}
1 & 1 & 0 & -7 & 6 \\
  &   & 1 & 1 & -6 \\
\hline
  & 1 & 1 & -6 & \boxed{0} \leftarrow \text{Remainder}
\end{array}
$$

Because the remainder is zero, the divisor divides evenly into the dividend:

$$\frac{x^3 - 7x + 6}{x - 1} = x^2 + x - 6.$$

From this result, you can factor the original polynomial as follows.

$$x^3 - 7x + 6 = (x - 1)(x^2 + x - 6) = (x - 1)(x + 3)(x - 2)$$

6.5 Exercises

Review   Concepts, Skills, and Problem Solving

Keep mathematically in shape by doing these exercises *before* the problems of this section.

Properties and Definitions

1. *Writing*✏️   Show how to write the fraction $120y/90$ in simplified form.

2. Write an algebraic expression that represents the product of two consecutive odd integers, the first of which is $2n + 1$.

3. Write an algebraic expression that represents the sum of two consecutive odd integers, the first of which is $2n + 1$.

4. Write an algebraic expression that represents the product of two consecutive even integers, the first of which is $2n$.

Solving Equations

In Exercises 5–10, solve the equation.

5. $3(2 - x) = 5x$

6. $125 - 50x = 0$

7. $8y^2 - 50 = 0$

8. $t^2 - 8t = 0$

9. $x^2 + x - 42 = 0$

10. $x(10 - x) = 25$

Models and Graphs

11. *Monthly Wages*   You receive a monthly salary of $1500 plus a commission of 12% of sales. Find a model for the monthly wages y as a function of sales x. Graph the model.

12. *Education*   In the year 2003, a college had an enrollment of 3680 students. Enrollment was projected to increase by 60 students per year. Find a model for the enrollment N as a function of time t in years. (Let $t = 3$ represent the year 2003.) Graph the function for the years 2003 through 2013.

Developing Skills

In Exercises 1–14, perform the division. See Example 1.

1. $(7x^3 - 2x^2) \div x$

2. $(6a^2 + 7a) \div a$

3. $(4x^2 - 2x) \div (-x)$

4. $(5y^3 + 6y^2 - 3y) \div (-y)$

5. $(m^4 + 2m^2 - 7) \div m$

6. $(x^3 + x - 2) \div x$

7. $\dfrac{50z^3 + 30z}{-5z}$

8. $\dfrac{18c^4 - 24c^2}{-6c}$

9. $\dfrac{8z^3 + 3z^2 - 2z}{2z}$

10. $\dfrac{6x^4 + 8x^3 - 18x^2}{3x^2}$

11. $\dfrac{4x^5 - 6x^4 + 12x^3 - 8x^2}{4x^2}$

12. $\dfrac{15x^{12} - 5x^9 + 30x^6}{5x^6}$

13. $(5x^2y - 8xy + 7xy^2) \div 2xy$

14. $(-14s^4t^2 + 7s^2t^2 - 18t) \div 2s^2t$

In Exercises 15–52, perform the division. See Examples 2–6.

15. $\dfrac{x^2 - 8x + 15}{x - 3}$

16. $\dfrac{t^2 - 18t + 72}{t - 6}$

17. $(x^2 + 15x + 50) \div (x + 5)$

18. $(y^2 - 6y - 16) \div (y + 2)$

19. Divide $x^2 - 5x + 8$ by $x - 2$.

20. Divide $x^2 + 10x - 9$ by $x - 3$.

21. Divide $21 - 4x - x^2$ by $3 - x$.

22. Divide $5 + 4x - x^2$ by $1 + x$.

23. $\dfrac{5x^2 + 2x + 3}{x + 2}$

24. $\dfrac{2x^2 + 13x + 15}{x + 5}$

25. $\dfrac{12x^2 + 17x - 5}{3x + 2}$

26. $\dfrac{8x^2 + 2x + 3}{4x - 1}$

27. $(12t^2 - 40t + 25) \div (2t - 5)$

28. $(15 - 14u - 8u^2) \div (5 + 2u)$

29. Divide $2y^2 + 7y + 3$ by $2y + 1$.

30. Divide $10t^2 - 7t - 12$ by $2t - 3$.

31. $\dfrac{x^3 - 2x^2 + 4x - 8}{x - 2}$ **32.** $\dfrac{x^3 + 4x^2 + 7x + 28}{x + 4}$

33. $\dfrac{9x^3 - 3x^2 - 3x + 4}{3x + 2}$ **34.** $\dfrac{4y^3 + 12y^2 + 7y - 3}{2y + 3}$

35. $(2x + 9) \div (x + 2)$ **36.** $(12x - 5) \div (2x + 3)$

37. $\dfrac{x^2 + 16}{x + 4}$ **38.** $\dfrac{y^2 + 8}{y + 2}$

39. $\dfrac{6z^2 + 7z}{5z - 1}$ **40.** $\dfrac{8y^2 - 2y}{3y + 5}$

41. $\dfrac{16x^2 - 1}{4x + 1}$ **42.** $\dfrac{81y^2 - 25}{9y - 5}$

43. $\dfrac{x^3 + 125}{x + 5}$ **44.** $\dfrac{x^3 - 27}{x - 3}$

45. $(x^3 + 4x^2 + 7x + 6) \div (x^2 + 2x + 3)$

46. $(2x^3 + 2x^2 - 2x - 15) \div (2x^2 + 4x + 5)$

47. $(4x^4 - 3x^2 + x - 5) \div (x^2 - 3x + 2)$

48. $(8x^5 + 6x^4 - x^3 + 1) \div (2x^3 - x^2 - 3)$

49. Divide $x^6 - 1$ by $x - 1$.

50. Divide x^3 by $x - 1$.

51. $x^5 \div (x^2 + 1)$

52. $x^4 \div (x - 2)$

In Exercises 53–56, simplify the expression.

53. $\dfrac{4x^4}{x^3} - 2x$ **54.** $\dfrac{15x^3y}{10x^2} + \dfrac{3xy^2}{2y}$

55. $\dfrac{8u^2v}{2u} + \dfrac{3(uv)^2}{uv}$ **56.** $\dfrac{x^2 + 2x - 3}{x - 1} - (3x - 4)$

In Exercises 57–68, use synthetic division to divide. See Example 7.

57. $(x^2 + x - 6) \div (x - 2)$

58. $(x^2 + 5x - 6) \div (x + 6)$

59. $\dfrac{x^3 + 3x^2 - 1}{x + 4}$

60. $\dfrac{x^4 - 4x^2 + 6}{x - 4}$

61. $\dfrac{x^4 - 4x^3 + x + 10}{x - 2}$

62. $\dfrac{2x^5 - 3x^3 + x}{x - 3}$

63. $\dfrac{5x^3 - 6x^2 + 8}{x - 4}$

64. $\dfrac{5x^3 + 6x + 8}{x + 2}$

65. $\dfrac{10x^4 - 50x^3 - 800}{x - 6}$

66. $\dfrac{x^5 - 13x^4 - 120x + 80}{x + 3}$

67. $\dfrac{0.1x^2 + 0.8x + 1}{x - 0.2}$

68. $\dfrac{x^3 - 0.8x + 2.4}{x + 0.1}$

In Exercises 69–76, completely factor the polynomial given one of its factors. See Example 8.

	Polynomial	Factor
69.	$x^3 - 13x + 12$	$x - 3$
70.	$x^3 + x^2 - 32x - 60$	$x + 5$
71.	$6x^3 - 13x^2 + 9x - 2$	$x - 1$
72.	$9x^3 - 3x^2 - 56x - 48$	$x - 3$
73.	$x^4 + 7x^3 + 3x^2 - 63x - 108$	$x + 3$
74.	$x^4 - 6x^3 - 8x^2 + 96x - 128$	$x - 4$
75.	$15x^2 - 2x - 8$	$x - \frac{4}{5}$
76.	$18x^2 - 9x - 20$	$x + \frac{5}{6}$

In Exercises 77 and 78, find the constant c such that the denominator divides evenly into the numerator.

77. $\dfrac{x^3 + 2x^2 - 4x + c}{x - 2}$

78. $\dfrac{x^4 - 3x^2 + c}{x + 6}$

In Exercises 79 and 80, use a graphing calculator to graph the two equations in the same viewing window. Use the graphs to verify that the expressions are equivalent. Verify the results algebraically.

79. $y_1 = \dfrac{x + 4}{2x}$

$y_2 = \dfrac{1}{2} + \dfrac{2}{x}$

80. $y_1 = \dfrac{x^2 + 2}{x + 1}$

$y_2 = x - 1 + \dfrac{3}{x + 1}$

In Exercises 81 and 82, perform the division assuming that n is a positive integer.

81. $\dfrac{x^{3n} + 3x^{2n} + 6x^n + 8}{x^n + 2}$ **82.** $\dfrac{x^{3n} - x^{2n} + 5x^n - 5}{x^n - 1}$

Think About It   In Exercises 83 and 84, the divisor, quotient, and remainder are given. Find the dividend.

	Divisor	Quotient	Remainder
83.	$x - 6$	$x^2 + x + 1$	-4
84.	$x + 3$	$x^3 + x^2 - 4$	8

Finding a Pattern   In Exercises 85 and 86, complete the table for the function. The first row is completed for Exercise 85. What conclusion can you draw as you compare the values of $f(k)$ with the remainders? (Use synthetic division to find the remainders.)

85. $f(x) = x^3 - x^2 - 2x$

86. $f(x) = 2x^3 - x^2 - 2x + 1$

k	$f(k)$	Divisor $(x - k)$	Remainder
-2	-8	$x + 2$	-8
-1			
0			
$\frac{1}{2}$			
1			
2			

Solving Problems

87. ▲ *Geometry* The area of a rectangle is $2x^3 + 3x^2 - 6x - 9$ and its length is $2x + 3$. Find the width of the rectangle.

88. ▲ *Geometry* A rectangular house has a volume of $x^3 + 55x^2 + 650x + 2000$ cubic feet (the space in the attic is not included). The height of the house is $x + 5$ feet (see figure). Find the number of square feet of floor space *on the first floor* of the house.

▲ *Geometry* In Exercises 89 and 90, you are given the expression for the volume of the solid shown. Find the expression for the missing dimension.

89. $V = x^3 + 18x^2 + 80x + 96$

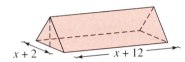

90. $V = h^4 + 3h^3 + 2h^2$

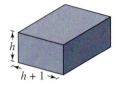

Explaining Concepts

91. *Error Analysis* Describe the error.

$$\frac{6x + 5y}{x} = \frac{\cancel{6x} + 5y}{\cancel{x}} = 6 + 5y$$

92. Create a polynomial division problem and identify the dividend, divisor, quotient, and remainder.

93. *Writing* Explain what it means for a divisor to divide *evenly* into a dividend.

94. *Writing* Explain how you can check polynomial division.

95. *True or False?* If the divisor divides evenly into the dividend, the divisor and quotient are factors of the dividend. Justify your answer.

96. *Writing* For synthetic division, what form must the divisor have?

97. ⊞ *Writing* Use a graphing calculator to graph each polynomial in the same viewing window using the standard setting. Use the *zero* or *root* feature to find the *x*-intercepts. What can you conclude about the polynomials? Verify your conclusion algebraically.

(a) $y = (x - 4)(x - 2)(x + 1)$

(b) $y = (x^2 - 6x + 8)(x + 1)$

(c) $y = x^3 - 5x^2 + 2x + 8$

98. ⊞ *Writing* Use a graphing calculator to graph the function

$$f(x) = \frac{x^3 - 5x^2 + 2x + 8}{x - 2}.$$

Use the *zero* or *root* feature to find the *x*-intercepts. Why does this function have only two *x*-intercepts? To what other function does the graph of $f(x)$ appear to be equivalent? What is the difference between the two graphs?

6.6 Solving Rational Equations

Thinkstock/Getty Images

What You Should Learn

1 Solve rational equations containing constant denominators.

2 Solve rational equations containing variable denominators.

Why You Should Learn It

Rational equations can be used to model and solve real-life applications. For instance, in Exercise 85 on page 424, you will use a rational equation to determine the speeds of two runners.

Equations Containing Constant Denominators

In Section 2.1, you studied a strategy for solving equations that contain fractions with *constant* denominators. That procedure is reviewed here because it is the basis for solving more general equations involving fractions. Recall from Section 2.1 that you can "clear an equation of fractions" by multiplying each side of the equation by the least common denominator (LCD) of the fractions in the equation. Note how this is done in the next three examples.

1 Solve rational equations containing constant denominators.

Example 1   An Equation Containing Constant Denominators

Solve $\dfrac{3}{5} = \dfrac{x}{2} + 1$.

Solution

The least common denominator of the fractions is 10, so begin by multiplying each side of the equation by 10.

$$\frac{3}{5} = \frac{x}{2} + 1 \qquad \text{Write original equation.}$$

$$10\left(\frac{3}{5}\right) = 10\left(\frac{x}{2} + 1\right) \qquad \text{Multiply each side by LCD of 10.}$$

$$6 = 5x + 10 \qquad \text{Distribute and simplify.}$$

$$-4 = 5x \quad \Longrightarrow \quad -\frac{4}{5} = x \qquad \text{Subtract 10 from each side, then divide each side by 5.}$$

The solution is $x = -\frac{4}{5}$. You can check this in the original equation as follows.

Check

$$\frac{3}{5} \stackrel{?}{=} \frac{-4/5}{2} + 1 \qquad \text{Substitute } -\tfrac{4}{5} \text{ for } x \text{ in the original equation.}$$

$$\frac{3}{5} \stackrel{?}{=} -\frac{4}{5} \cdot \frac{1}{2} + 1 \qquad \text{Invert divisor and multiply.}$$

$$\frac{3}{5} = -\frac{2}{5} + 1 \qquad \text{Solution checks.} \checkmark$$

Example 2   An Equation Containing Constant Denominators

Solve $\dfrac{x-3}{6} = 7 - \dfrac{x}{12}$.

Solution

The least common denominator of the fractions is 12, so begin by multiplying each side of the equation by 12.

$$\frac{x-3}{6} = 7 - \frac{x}{12} \qquad \text{Write original equation.}$$

$$12\left(\frac{x-3}{6}\right) = 12\left(7 - \frac{x}{12}\right) \qquad \text{Multiply each side by LCD of 12.}$$

$$2x - 6 = 84 - x \qquad \text{Distribute and simplify.}$$

$$3x - 6 = 84 \qquad \text{Add } x \text{ to each side.}$$

$$3x = 90 \implies x = 30 \qquad \text{Add 6 to each side, then divide each side by 3.}$$

Check

$$\frac{30-3}{6} \stackrel{?}{=} 7 - \frac{30}{12} \qquad \text{Substitute 30 for } x \text{ in the original equation.}$$

$$\frac{27}{6} = \frac{42}{6} - \frac{15}{6} \qquad \text{Solution checks. } \checkmark$$

Example 3   An Equation That Has Two Solutions

Solve $\dfrac{x^2}{3} + \dfrac{x}{2} = \dfrac{5}{6}$.

Solution

The least common denominator of the fractions is 6, so begin by multiplying each side of the equation by 6.

$$\frac{x^2}{3} + \frac{x}{2} = \frac{5}{6} \qquad \text{Write original equation.}$$

$$6\left(\frac{x^2}{3} + \frac{x}{2}\right) = 6\left(\frac{5}{6}\right) \qquad \text{Multiply each side by LCD of 6.}$$

$$\frac{6x^2}{3} + \frac{6x}{2} = \frac{30}{6} \qquad \text{Distributive Property}$$

$$2x^2 + 3x = 5 \qquad \text{Simplify.}$$

$$2x^2 + 3x - 5 = 0 \qquad \text{Subtract 5 from each side.}$$

$$(2x + 5)(x - 1) = 0 \qquad \text{Factor.}$$

$$2x + 5 = 0 \implies x = -\frac{5}{2} \qquad \text{Set 1st factor equal to 0.}$$

$$x - 1 = 0 \implies x = 1 \qquad \text{Set 2nd factor equal to 0.}$$

The solutions are $x = -\frac{5}{2}$ and $x = 1$. Check these in the original equation.

② Solve rational equations containing variable denominators.

Equations Containing Variable Denominators

Remember that you always *exclude* those values of a variable that make the denominator of a rational expression equal to zero. This is especially critical in solving equations that contain variable denominators. You will see why in the examples that follow.

Example 4   An Equation Containing Variable Denominators

Solve the equation.

$$\frac{7}{x} - \frac{1}{3x} = \frac{8}{3}$$

Solution

The least common denominator of the fractions is $3x$, so begin by multiplying each side of the equation by $3x$.

$$\frac{7}{x} - \frac{1}{3x} = \frac{8}{3}$$ Write original equation.

$$3x\left(\frac{7}{x} - \frac{1}{3x}\right) = 3x\left(\frac{8}{3}\right)$$ Multiply each side by LCD of $3x$.

$$\frac{21x}{x} - \frac{3x}{3x} = \frac{24x}{3}$$ Distributive Property

$$21 - 1 = 8x$$ Simplify.

$$\frac{20}{8} = x$$ Combine like terms and divide each side by 8.

$$\frac{5}{2} = x$$ Simplify.

The solution is $x = \frac{5}{2}$. You can check this in the original equation as follows.

Check

$$\frac{7}{x} - \frac{1}{3x} = \frac{8}{3}$$ Write original equation.

$$\frac{7}{5/2} - \frac{1}{3(5/2)} \stackrel{?}{=} \frac{8}{3}$$ Substitute $\frac{5}{2}$ for x.

$$7\left(\frac{2}{5}\right) - \left(\frac{1}{3}\right)\left(\frac{2}{5}\right) \stackrel{?}{=} \frac{8}{3}$$ Invert divisors and multiply.

$$\frac{14}{5} - \frac{2}{15} \stackrel{?}{=} \frac{8}{3}$$ Simplify.

$$\frac{40}{15} \stackrel{?}{=} \frac{8}{3}$$ Combine like terms.

$$\frac{8}{3} = \frac{8}{3}$$ Solution checks. ✔

Technology: Tip

You can use a graphing calculator to approximate the solution of the equation in Example 4. To do this, graph the left side of the equation and the right side of the equation in the same viewing window.

$$y_1 = \frac{7}{x} - \frac{1}{3x} \text{ and } y_2 = \frac{8}{3}$$

The solution of the equation is the x-coordinate of the point at which the two graphs intersect, as shown below. You can use the *intersect* feature of the graphing calculator to approximate the point of intersection to be $\left(\frac{5}{2}, \frac{8}{3}\right)$. So, the solution is $x = \frac{5}{2}$, which is the same solution obtained in Example 4.

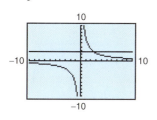

Throughout the text, the importance of checking solutions is emphasized. Up to this point, the main reason for checking has been to make sure that you did not make arithmetic errors in the solution process. In the next example, you will see that there is another reason for checking solutions in the *original* equation. That is, even with no mistakes in the solution process, it can happen that a "trial solution" does not satisfy the original equation. This type of solution is called an **extraneous solution.** An extraneous solution of an equation does not, by definition, satisfy the original equation, and so *must not* be listed as an actual solution.

Example 5   An Equation with No Solution

Solve $\dfrac{5x}{x-2} = 7 + \dfrac{10}{x-2}$.

Solution

The least common denominator of the fractions is $x - 2$, so begin by multiplying each side of the equation by $x - 2$.

$$\dfrac{5x}{x-2} = 7 + \dfrac{10}{x-2} \qquad \text{Write original equation.}$$

$$(x-2)\left(\dfrac{5x}{x-2}\right) = (x-2)\left(7 + \dfrac{10}{x-2}\right) \qquad \text{Multiply each side by } x - 2.$$

$$5x = 7(x-2) + 10, \quad x \neq 2 \qquad \text{Distribute and simplify.}$$

$$5x = 7x - 14 + 10 \qquad \text{Distributive Property}$$

$$5x = 7x - 4 \qquad \text{Combine like terms.}$$

$$-2x = -4 \qquad \text{Subtract } 7x \text{ from each side.}$$

$$x = 2 \qquad \text{Divide each side by } -2.$$

At this point, the solution appears to be $x = 2$. However, by performing a check, you can see that this "trial solution" is extraneous.

Check

$$\dfrac{5x}{x-2} = 7 + \dfrac{10}{x-2} \qquad \text{Write original equation.}$$

$$\dfrac{5(2)}{2-2} \stackrel{?}{=} 7 + \dfrac{10}{2-2} \qquad \text{Substitute 2 for } x.$$

$$\dfrac{10}{0} \stackrel{?}{=} 7 + \dfrac{10}{0} \qquad \text{Solution does not check.} \;✗$$

Because the check results in *division by zero*, you can conclude that 2 is extraneous. So, the original equation has no solution.

Notice that $x = 2$ is excluded from the domains of the two fractions in the original equation in Example 5. You may find it helpful when solving these types of equations to list the domain restrictions *before* beginning the solution process.

Study Tip

Although cross-multiplication can be a little quicker than multiplying by the least common denominator, remember that *it can be used only with equations that have a single fraction on each side* of the equation.

An equation with a single fraction on each side can be cleared of fractions by **cross-multiplying,** which is equivalent to multiplying by the LCD and then dividing out. To do this, multiply the left numerator by the right denominator and multiply the right numerator by the left denominator, as shown below.

$$\frac{a}{b} = \frac{c}{d} \quad \Longrightarrow \quad ad = bc, \, b \neq 0, \, d \neq 0$$

Example 6   Cross-Multiplying

Solve $\dfrac{2x}{x + 4} = \dfrac{3}{x - 1}$.

Solution

The domain is all real values of x such that $x \neq -4$ and $x \neq 1$. You can use cross-multiplication to solve this equation.

$\dfrac{2x}{x + 4} = \dfrac{3}{x - 1}$	Write original equation.
$2x(x - 1) = 3(x + 4), \, x \neq -4, \, x \neq 1$	Cross-multiply.
$2x^2 - 2x = 3x + 12$	Distributive Property
$2x^2 - 5x - 12 = 0$	Subtract $3x$ and 12 from each side.
$(2x + 3)(x - 4) = 0$	Factor.
$2x + 3 = 0 \quad \Longrightarrow \quad x = -\frac{3}{2}$	Set 1st factor equal to 0.
$x - 4 = 0 \quad \Longrightarrow \quad x = 4$	Set 2nd factor equal to 0.

The solutions are $x = -\frac{3}{2}$ and $x = 4$. Check these in the original equation.

Example 7   An Equation That Has Two Solutions

Technology: Discovery

Use a graphing calculator to graph the equation

$$y = \frac{3x}{x + 1} - \frac{12}{x^2 - 1} - 2.$$

Then use the *zero* or *root* feature of the calculator to determine the x-intercepts. How do the x-intercepts compare with the solutions to Example 7? What can you conclude?

Solve $\dfrac{3x}{x + 1} = \dfrac{12}{x^2 - 1} + 2$.

The domain is all real values of x such that $x \neq 1$ and $x \neq -1$. The least common denominator is $(x + 1)(x - 1) = x^2 - 1$.

$(x^2 - 1)\left(\dfrac{3x}{x + 1}\right) = (x^2 - 1)\left(\dfrac{12}{x^2 - 1} + 2\right)$	Multiply each side of original equation by LCD of $x^2 - 1$.
$(x - 1)(3x) = 12 + 2(x^2 - 1), \quad x \neq \pm 1$	Simplify.
$3x^2 - 3x = 12 + 2x^2 - 2$	Distributive Property
$x^2 - 3x - 10 = 0$	Subtract $2x^2$ and 10 from each side.
$(x + 2)(x - 5) = 0$	Factor.
$x + 2 = 0 \quad \Longrightarrow \quad x = -2$	Set 1st factor equal to 0.
$x - 5 = 0 \quad \Longrightarrow \quad x = 5$	Set 2nd factor equal to 0.

The solutions are $x = -2$ and $x = 5$. Check these in the original equation.

6.6 Exercises

Review   *Concepts, Skills, and Problem Solving*

Keep mathematically in shape by doing these exercises *before* the problems of this section.

Properties and Definitions

In Exercises 1 and 2, determine the quadrants in which the point must be located.

1. $(-2, y)$, y is a real number.

2. $(x, 3)$, x is a real number.

3. Give the positions of points whose y-coordinates are 0.

4. Find the coordinates of the point nine units to the right of the y-axis and six units below the x-axis.

Solving Inequalities

In Exercises 5–10, solve the inequality.

5. $7 - 3x > 4 - x$

6. $2(x + 6) - 20 < 2$

7. $|x - 3| < 2$

8. $|x - 5| > 3$

9. $|\frac{1}{4}x - 1| \geq 3$

10. $|2 - \frac{1}{3}x| \leq 10$

Problem Solving

11. *Distance*   A jogger leaves a point on a fitness trail running at a rate of 6 miles per hour. Five minutes later, a second jogger leaves from the same location running at 8 miles per hour. How long will it take the second runner to overtake the first, and how far will each have run at that point?

12. *Investment*   An inheritance of $24,000 is invested in two bonds that pay 7.5% and 9% simple interest. The annual interest is $1935. How much is invested in each bond?

Developing Skills

In Exercises 1–4, determine whether each value of x is a solution to the equation.

Equation	*Values*
1. $\dfrac{x}{3} - \dfrac{x}{5} = \dfrac{4}{3}$	(a) $x = 0$ (b) $x = -1$
	(c) $x = \frac{1}{8}$ (d) $x = 10$
2. $x = 4 + \dfrac{21}{x}$	(a) $x = 0$ (b) $x = -3$
	(c) $x = 7$ (d) $x = -1$
3. $\dfrac{x}{4} + \dfrac{3}{4x} = 1$	(a) $x = -1$ (b) $x = 1$
	(c) $x = 3$ (d) $x = -3$
4. $5 - \dfrac{1}{x - 3} = 2$	(a) $x = \frac{10}{3}$ (b) $x = -\frac{1}{3}$
	(c) $x = 0$ (d) $x = 1$

In Exercises 5–22, solve the equation. See Examples 1–3.

5. $\dfrac{x}{6} - 1 = \dfrac{2}{3}$

6. $\dfrac{y}{8} + 7 = -\dfrac{1}{2}$

7. $\dfrac{1}{4} = \dfrac{z + 1}{8}$

8. $\dfrac{a}{5} = \dfrac{a - 3}{2}$

9. $\dfrac{x}{3} - \dfrac{3x}{4} = \dfrac{5x}{12}$

10. $\dfrac{x}{4} - \dfrac{x}{6} = \dfrac{1}{4}$

11. $\dfrac{z + 2}{3} = 4 - \dfrac{z}{12}$

12. $\dfrac{x - 5}{5} + 3 = -\dfrac{x}{4}$

13. $\dfrac{2y - 9}{6} = 3y - \dfrac{3}{4}$

14. $\dfrac{4x - 2}{7} - \dfrac{5}{14} = 2x$

15. $\dfrac{t}{2} = 12 - \dfrac{3t^2}{2}$

16. $\dfrac{x^2}{2} - \dfrac{3x}{5} = -\dfrac{1}{10}$

17. $\dfrac{5y - 1}{12} + \dfrac{y}{3} = -\dfrac{1}{4}$

18. $\dfrac{z - 4}{9} - \dfrac{3z + 1}{18} = \dfrac{3}{2}$

19. $\dfrac{h + 2}{5} - \dfrac{h - 1}{9} = \dfrac{2}{3}$ **20.** $\dfrac{u - 2}{6} + \dfrac{2u + 5}{15} = 3$

21. $\dfrac{x + 5}{4} - \dfrac{3x - 8}{3} = \dfrac{4 - x}{12}$

22. $\dfrac{2x - 7}{10} - \dfrac{3x + 1}{5} = \dfrac{6 - x}{5}$

In Exercises 23–66, solve the equation. (Check for extraneous solutions.) See Examples 4–7.

23. $\dfrac{9}{25 - y} = -\dfrac{1}{4}$ **24.** $\dfrac{2}{u + 4} = \dfrac{5}{8}$

25. $5 - \dfrac{12}{a} = \dfrac{5}{3}$ **26.** $\dfrac{6}{b} + 22 = 24$

27. $\dfrac{4}{x} - \dfrac{7}{5x} = -\dfrac{1}{2}$ **28.** $\dfrac{5}{3} = \dfrac{6}{7x} + \dfrac{2}{x}$

29. $\dfrac{12}{y + 5} + \dfrac{1}{2} = 2$ **30.** $\dfrac{7}{8} - \dfrac{16}{t - 2} = \dfrac{3}{4}$

31. $\dfrac{5}{x} = \dfrac{25}{3(x + 2)}$ **32.** $\dfrac{10}{x + 4} = \dfrac{15}{4(x + 1)}$

33. $\dfrac{8}{3x + 5} = \dfrac{1}{x + 2}$ **34.** $\dfrac{500}{3x + 5} = \dfrac{50}{x - 3}$

35. $\dfrac{3}{x + 2} - \dfrac{1}{x} = \dfrac{1}{5x}$ **36.** $\dfrac{12}{x + 5} + \dfrac{5}{x} = \dfrac{20}{x}$

37. $\dfrac{1}{2} = \dfrac{18}{x^2}$ **38.** $\dfrac{1}{4} = \dfrac{16}{z^2}$

39. $\dfrac{32}{t} = 2t$ **40.** $\dfrac{20}{u} = \dfrac{u}{5}$

41. $x + 1 = \dfrac{72}{x}$ **42.** $\dfrac{48}{x} = x - 2$

43. $1 = \dfrac{16}{y} - \dfrac{39}{y^2}$ **44.** $x - \dfrac{24}{x} = 5$

45. $\dfrac{2x}{3x - 10} - \dfrac{5}{x} = 0$ **46.** $\dfrac{x + 42}{x} = x$

47. $\dfrac{2x}{5} = \dfrac{x^2 - 5x}{5x}$ **48.** $\dfrac{3x}{4} = \dfrac{x^2 + 3x}{8x}$

49. $\dfrac{y + 1}{y + 10} = \dfrac{y - 2}{y + 4}$ **50.** $\dfrac{x - 3}{x + 1} = \dfrac{x - 6}{x + 5}$

51. $\dfrac{15}{x} + \dfrac{9x - 7}{x + 2} = 9$

52. $\dfrac{3z - 2}{z + 1} = 4 - \dfrac{z + 2}{z - 1}$

53. $\dfrac{2}{6q + 5} - \dfrac{3}{4(6q + 5)} = \dfrac{1}{28}$

54. $\dfrac{10}{x(x - 2)} + \dfrac{4}{x} = \dfrac{5}{x - 2}$

55. $\dfrac{4}{2x + 3} + \dfrac{17}{5x - 3} = 3$

56. $\dfrac{1}{x - 1} + \dfrac{3}{x + 1} = 2$

57. $\dfrac{2}{x - 10} - \dfrac{3}{x - 2} = \dfrac{6}{x^2 - 12x + 20}$

58. $\dfrac{5}{x + 2} + \dfrac{2}{x^2 - 6x - 16} = \dfrac{-4}{x - 8}$

59. $\dfrac{x + 3}{x^2 - 9} + \dfrac{4}{3 - x} - 2 = 0$

60. $1 - \dfrac{6}{4 - x} = \dfrac{x + 2}{x^2 - 16}$

61. $\dfrac{x}{x - 2} + \dfrac{3x}{x - 4} = \dfrac{-2(x - 6)}{x^2 - 6x + 8}$

62. $\dfrac{2(x + 1)}{x^2 - 4x + 3} + \dfrac{6x}{x - 3} = \dfrac{3x}{x - 1}$

63. $\dfrac{5}{x^2 + 4x + 3} + \dfrac{2}{x^2 + x - 6} = \dfrac{3}{x^2 - x - 2}$

64. $\dfrac{2}{x^2 + 2x - 8} - \dfrac{1}{x^2 + 9x + 20} = \dfrac{4}{x^2 + 3x - 10}$

65. $\dfrac{x}{2} = \dfrac{2 - \dfrac{3}{x}}{1 - \dfrac{1}{x}}$ **66.** $\dfrac{2x}{3} = \dfrac{1 + \dfrac{2}{x}}{1 + \dfrac{1}{x}}$

In Exercises 67–70, (a) use the graph to determine any x-intercepts of the graph and (b) set $y = 0$ and solve the resulting rational equation to confirm the result of part (a).

67. $y = \dfrac{x + 2}{x - 2}$ **68.** $y = \dfrac{2x}{x + 4}$

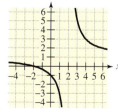

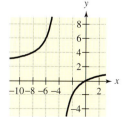

69. $y = x - \dfrac{1}{x}$

70. $y = x - \dfrac{2}{x} - 1$

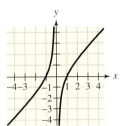

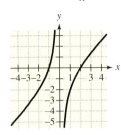

73. $y = \dfrac{1}{x} + \dfrac{4}{x - 5}$

74. $y = 20\left(\dfrac{2}{x} - \dfrac{3}{x - 1}\right)$

75. $y = (x + 1) - \dfrac{6}{x}$

76. $y = \dfrac{x^2 - 4}{x}$

In Exercises 71–76, (a) use a graphing calculator to graph the equation and determine any x-intercepts of the graph and (b) set $y = 0$ and solve the resulting rational equation to confirm the result of part (a).

71. $y = \dfrac{x - 4}{x + 5}$

72. $y = \dfrac{1}{x} - \dfrac{3}{x + 4}$

Think About It In Exercises 77–80, if the exercise is an equation, solve it; if it is an expression, simplify it.

77. $\dfrac{16}{x^2 - 16} + \dfrac{x}{2x - 8} = \dfrac{1}{2}$

78. $\dfrac{5}{x + 3} + \dfrac{5}{3} + 3$

79. $\dfrac{16}{x^2 - 16} + \dfrac{x}{2x - 8} + \dfrac{1}{2}$

80. $\dfrac{5}{x + 3} + \dfrac{5}{3} = 3$

Solving Problems

81. *Number Problem* Find a number such that the sum of the number and its reciprocal is $\frac{65}{8}$.

82. *Number Problem* Find a number such that the sum of two times the number and three times its reciprocal is $\frac{97}{4}$.

83. *Wind Speed* A plane has a speed of 300 miles per hour in still air. The plane travels a distance of 680 miles with a tail wind in the same time it takes to travel 520 miles into a head wind. Find the speed of the wind.

84. *Average Speed* During the first part of a six-hour trip, you travel 240 miles at an average speed of r miles per hour. For the next 72 miles, you increase your speed by 10 miles per hour. What are your two average speeds?

85. *Speed* One person runs 2 miles per hour faster than a second person. The first person runs 5 miles in the same time the second person runs 4 miles. Find the speed of each person.

86. *Speed* The speed of a commuter plane is 150 miles per hour lower than that of a passenger jet. The commuter plane travels 450 miles in the same time the jet travels 1150 miles. Find the speed of each plane.

Explaining Concepts

87. ✪ Answer parts (d) and (e) of Motivating the Chapter on page 366.

88. *Writing* Describe how to solve a rational equation.

89. *Writing* Define the term *extraneous solution*. How do you identify an extraneous solution?

90. *Writing* Explain how you can use a graphing calculator to estimate the solution of a rational equation.

91. *Writing* When can you use cross-multiplication to solve a rational equation? Explain.

6.7 Applications and Variation

NASA

What You Should Learn

1️⃣ Solve application problems involving rational equations.

2️⃣ Solve application problems involving direct variation.

3️⃣ Solve application problems involving inverse variation.

4️⃣ Solve application problems involving joint variation.

Why You Should Learn It

You can use mathematical models in a wide variety of applications including variation. For instance, in Exercise 56 on page 436, you will use direct variation to model the weight of a person on the moon.

1️⃣ Solve application problems involving rational equations.

Rational Equation Applications

The examples that follow are types of application problems that you have seen earlier in the text. The difference now is that the variable appears in the denominator of a rational expression.

Example 1   Average Speeds

You and your friend travel to separate colleges in the same amount of time. You drive 380 miles and your friend drives 400 miles. Your friend's average speed is 3 miles per hour faster than your average speed. What is your average speed and what is your friend's average speed?

Solution

Begin by setting your time equal to your friend's time. Then use an alternative version for the formula for distance that gives the time in terms of the distance and the rate.

Verbal Model:

| Your time = Your friend's time |

$$\frac{\text{Your distance}}{\text{Your rate}} = \frac{\text{Friend's distance}}{\text{Friend's rate}}$$

Labels: Your distance = 380   (miles)
Your rate = r (miles per hour)
Friend's distance = 400   (miles)
Friend's rate = $r + 3$ (miles per hour)

Equation:

$$\frac{380}{r} = \frac{400}{r + 3} \qquad \text{Original equation.}$$

$$380(r + 3) = 400(r), \quad r \neq 0, r \neq -3 \qquad \text{Cross-multiply.}$$

$$380r + 1140 = 400r \qquad \text{Distributive Property}$$

$$1140 = 20r \quad \Longrightarrow \quad 57 = r \qquad \text{Simplify.}$$

Your average speed is 57 miles per hour and your friend's average speed is $57 + 3 = 60$ miles per hour. Check this in the original statement of the problem.

Example 2   A Work-Rate Problem

With the cold water valve open, it takes 8 minutes to fill a washing machine tub. With both the hot and cold water valves open, it takes 5 minutes to fill the tub. How long will it take to fill the tub with only the hot water valve open?

Solution

Verbal Model:
$$\text{Rate for cold water} + \text{Rate for hot water} = \text{Rate for warm water}$$

Labels:
$$\text{Rate for cold water} = \frac{1}{8} \qquad \text{(tub per minute)}$$

$$\text{Rate for hot water} = \frac{1}{t} \qquad \text{(tub per minute)}$$

$$\text{Rate for warm water} = \frac{1}{5} \qquad \text{(tub per minute)}$$

Equation:
$$\frac{1}{8} + \frac{1}{t} = \frac{1}{5} \qquad \text{Original equation}$$

$$5t + 40 = 8t \qquad \text{Multiply each side by LCD of } 40t \text{ and simplify.}$$

$$40 = 3t \implies \frac{40}{3} = t \qquad \text{Simplify.}$$

So, it takes $13\frac{1}{3}$ minutes to fill the tub with hot water. Check this solution.

Example 3   Cost-Benefit Model

A utility company burns coal to generate electricity. The cost C (in dollars) of removing $p\%$ of the pollutants from smokestack emissions is modeled by

$$C = \frac{80,000p}{100 - p}, \quad 0 \le p < 100.$$

Determine the percent of air pollutants in the stack emissions that can be removed for $420,000.

Solution

To determine the percent of air pollutants in the stack emissions that can be removed for $420,000, substitute 420,000 for C in the model.

$$420,000 = \frac{80,000p}{100 - p} \qquad \text{Substitute 420,000 for } C.$$

$$420,000(100 - p) = 80,000p \qquad \text{Cross-multiply.}$$

$$42,000,000 - 420,000p = 80,000p \qquad \text{Distributive Property}$$

$$42,000,000 = 500,000p \qquad \text{Add } 420,000p \text{ to each side.}$$

$$84 = p \qquad \text{Divide each side by 500,000.}$$

So, 84% of air pollutants in the stack emissions can be removed for $420,000. Check this in the original statement of the problem.

② Solve application problems involving direct variation.

Direct Variation

In the mathematical model for **direct variation,** y is a *linear* function of x. Specifically,

$$y = kx.$$

To use this mathematical model in applications involving direct variation, you need to use the given values of x and y to find the value of the constant k.

Direct Variation

The following statements are equivalent.

1. y **varies directly** as x.

2. y is **directly proportional** to x.

3. $y = kx$ for some constant k.

The number k is called the **constant of proportionality.**

Example 4   Direct Variation

The total revenue R (in dollars) obtained from selling x ice show tickets is directly proportional to the number of tickets sold x. When 10,000 tickets are sold, the total revenue is $142,500.

a. Find a mathematical model that relates the total revenue R to the number of tickets sold x.

b. Find the total revenue obtained from selling 12,000 tickets.

Solution

a. Because the total revenue is directly proportional to the number of tickets sold, the linear model is $R = kx$. To find the value of the constant k, use the fact that $R = 142,500$ when $x = 10,000$. Substituting those values into the model produces

$$142,500 = k(10,000) \qquad \text{Substitute for } R \text{ and } x.$$

which implies that

$$k = \frac{142,500}{10,000} = 14.25.$$

So, the equation relating the total revenue to the total number of tickets sold is

$$R = 14.25x. \qquad \text{Direct variation model}$$

The graph of this equation is shown in Figure 6.2.

b. When $x = 12,000$, the total revenue is

$$R = 14.25(12,000) = \$171,000.$$

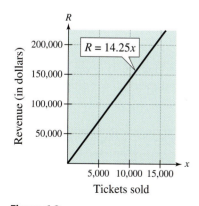

Figure 6.2

Example 5   Direct Variation

Hooke's Law for springs states that the distance a spring is stretched (or compressed) is proportional to the force on the spring. A force of 20 pounds stretches a spring 5 inches.

a. Find a mathematical model that relates the distance the spring is stretched to the force applied to the spring.

b. How far will a force of 30 pounds stretch the spring?

Solution

a. For this problem, let d represent the distance (in inches) that the spring is stretched and let F represent the force (in pounds) that is applied to the spring. Because the distance d is proportional to the force F, the model is

$$d = kF.$$

To find the value of the constant k, use the fact that $d = 5$ when $F = 20$. Substituting these values into the model produces

$$5 = k(20) \qquad \text{Substitute 5 for } d \text{ and 20 for } F.$$

$$\frac{5}{20} = k \qquad \text{Divide each side by 20.}$$

$$\frac{1}{4} = k. \qquad \text{Simplify.}$$

So, the equation relating distance and force is

$$d = \frac{1}{4}F. \qquad \text{Direct variation model}$$

b. When $F = 30$, the distance is

$$d = \frac{1}{4}(30) = 7.5 \text{ inches.} \qquad \text{See Figure 6.3.}$$

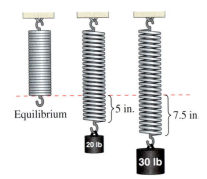

Equilibrium   5 in.   7.5 in.

20 lb

30 lb

Figure 6.3

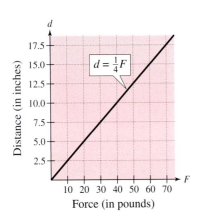

$d = \frac{1}{4}F$

Distance (in inches)

17.5
15.0
12.5
10.0
7.5
5.0
2.5

10 20 30 40 50 60 70
Force (in pounds)

Figure 6.4

In Example 5, you can get a clearer understanding of Hooke's Law by using the model $d = \frac{1}{4}F$ to create a table or a graph (see Figure 6.4). From the table or from the graph, you can see what it means for the distance to be "proportional to the force."

Force, F	10 lb	20 lb	30 lb	40 lb	50 lb	60 lb
Distance, d	2.5 in.	5.0 in.	7.5 in.	10.0 in.	12.5 in.	15.0 in.

In Examples 4 and 5, the direct variations are such that an *increase* in one variable corresponds to an *increase* in the other variable. There are, however, other applications of direct variation in which an increase in one variable corresponds to a *decrease* in the other variable. For instance, in the model $y = -2x$, an increase in x will yield a decrease in y.

Another type of direct variation relates one variable to a power of another.

Direct Variation as *n*th Power

The following statements are equivalent.

1. *y* **varies directly as the *n*th power** of *x*.

2. *y* is **directly proportional to the *n*th power** of *x*.

3. $y = kx^n$ for some constant *k*.

Example 6   Direct Variation as a Power

The distance a ball rolls down an inclined plane is directly proportional to the square of the time it rolls. During the first second, a ball rolls down a plane a distance of 6 feet.

a. Find a mathematical model that relates the distance traveled to the time.

b. How far will the ball roll during the first 2 seconds?

Solution

a. Letting *d* be the distance (in feet) that the ball rolls and letting *t* be the time (in seconds), you obtain the model

$$d = kt^2.$$

Because $d = 6$ when $t = 1$, you obtain

$d = kt^2$ Write original equation.

$6 = k(1)^2$ ⟹ $6 = k.$ Substitute 6 for *d* and 1 for *t*.

So, the equation relating distance to time is

$d = 6t^2.$ Direct variation as 2nd power model

The graph of this equation is shown in Figure 6.5.

b. When $t = 2$, the distance traveled is

$$d = 6(2)^2 = 6(4) = 24 \text{ feet.} \qquad \text{See Figure 6.6.}$$

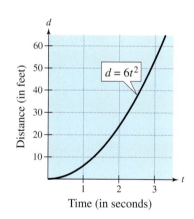

Figure 6.5

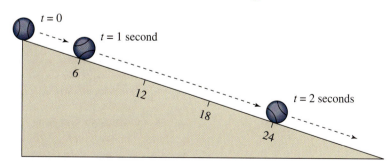

Figure 6.6

③ Solve application problems involving inverse variation.

Inverse Variation

A second type of variation is called **inverse variation.** With this type of variation, one of the variables is said to be inversely proportional to the other variable.

Inverse Variation

1. The following three statements are equivalent.

   a. y **varies inversely** as x.

   b. y is **inversely proportional** to x.

   c. $y = \dfrac{k}{x}$ for some constant k.

2. If $y = \dfrac{k}{x^n}$, then y is inversely proportional to the nth power of x.

Example 7   Inverse Variation

The marketing department of a large company has found that the demand for one of its hand tools varies inversely as the price of the product. (When the price is low, more people are willing to buy the product than when the price is high.) When the price of the tool is $7.50, the monthly demand is 50,000 tools. Approximate the monthly demand if the price is reduced to $6.00.

Solution

Let x represent the number of tools that are sold each month (the demand), and let p represent the price per tool (in dollars). Because the demand is inversely proportional to the price, the model is

$$x = \frac{k}{p}.$$

By substituting $x = 50,000$ when $p = 7.50$, you obtain

$$50,000 = \frac{k}{7.50} \qquad \text{Substitute 50,000 for } x \text{ and 7.50 for } p.$$

$$375,000 = k. \qquad \text{Multiply each side by 7.50.}$$

So, the inverse variation model is $x = \dfrac{375,000}{p}$.

The graph of this equation is shown in Figure 6.7. To find the demand that corresponds to a price of $6.00, substitute 6 for p in the equation and obtain

$$x = \frac{375,000}{6} = 62,500 \text{ tools.}$$

So, if the price is lowered from $7.50 per tool to $6.00 per tool, you can expect the monthly demand to increase from 50,000 tools to 62,500 tools.

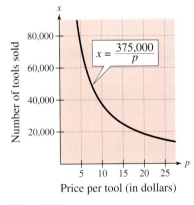

$$x = \frac{375,000}{p}$$

Number of tools sold

Price per tool (in dollars)

Figure 6.7

Some applications of variation involve problems with *both* direct and inverse variation in the same model. These types of models are said to have **combined variation.**

Example 8   Direct and Inverse Variation

An electronics manufacturer determines that the demand for its portable radio is directly proportional to the amount spent on advertising and inversely proportional to the price of the radio. When $40,000 is spent on advertising and the price per radio is $20, the monthly demand is 10,000 radios.

a. If the amount of advertising were increased to $50,000, how much could the price be increased to maintain a monthly demand of 10,000 radios?

b. If you were in charge of the advertising department, would you recommend this increased expense in advertising?

Solution

a. Let x represent the number of radios that are sold each month (the demand), let a represent the amount spent on advertising (in dollars), and let p represent the price per radio (in dollars). Because the demand is directly proportional to the advertising expense and inversely proportional to the price, the model is

$$x = \frac{ka}{p}.$$

By substituting 10,000 for x when $a = 40,000$ and $p = 20$, you obtain

$$10,000 = \frac{k(40,000)}{20} \qquad \text{Substitute 10,000 for } x, \text{ 40,000 for } a, \text{ and 20 for } p.$$

$$200,000 = 40,000k \qquad \text{Multiply each side by 20.}$$

$$5 = k. \qquad \text{Divide each side by 40,000.}$$

So, the model is

$$x = \frac{5a}{p}. \qquad \text{Direct and inverse variation model}$$

To find the price that corresponds to a demand of 10,000 and an advertising expense of $50,000, substitute 10,000 for x and 50,000 for a into the model and solve for p.

$$10,000 = \frac{5(50,000)}{p} \qquad p = \frac{5(50,000)}{10,000} = \$25$$

So, the price increase would be $25 - $20 = $5.

b. The total revenue for selling 10,000 radios at $20 each is $200,000, and the revenue for selling 10,000 radios at $25 each is $250,000. So, increasing the advertising expense from $40,000 to $50,000 would increase the revenue by $50,000. This implies that you should recommend the increased expense in advertising.

④ Solve application problems involving joint variation.

Joint Variation

The model used in Example 8 involved both direct and inverse variation, and the word "and" was used to couple the two types of variation together. To describe two different *direct* variations in the same statement, the word "jointly" is used. For instance, the model $z = kxy$ can be described by saying that z is *jointly* proportional to x and y.

Joint Variation

1. The following three statements are equivalent.

   a. z **varies jointly** as x and y.

   b. z is **jointly proportional** to x and y.

   c. $z = kxy$ for some constant k.

2. If $z = kx^n y^m$, then z is jointly proportional to the nth power of x and the mth power of y.

Example 9   Joint Variation

The *simple interest* for a savings account is jointly proportional to the time and the principal. After one quarter (3 months), the interest for a principal of $6000 is $120. How much interest would a principal of $7500 earn in 5 months?

Solution

To begin, let I represent the interest earned (in dollars), let P represent the principal (in dollars), and let t represent the time (in years). Because the interest is jointly proportional to the time and the principal, the model is

$$I = ktP.$$

Because $I = 120$ when $P = 6000$ and $t = \frac{1}{4}$, you have

$$120 = k\left(\frac{1}{4}\right)(6000) \qquad \text{Substitute 120 for } I, \tfrac{1}{4} \text{ for } t, \text{ and 6000 for } P.$$

$$120 = 1500\,k \qquad \text{Simplify.}$$

$$0.08 = k. \qquad \text{Divide each side by 1500.}$$

So, the model that relates interest to time and principal is

$$I = 0.08tP. \qquad \text{Joint variation model}$$

To find the interest earned on a principal of $7500 over a five-month period of time, substitute $P = 7500$ and $t = \frac{5}{12}$ into the model and obtain an interest of

$$I = 0.08\left(\frac{5}{12}\right)(7500) = \$250.$$

6.7 Exercises

Review   Concepts, Skills, and Problem Solving

Keep mathematically in shape by doing these exercises *before* the problems of this section.

Properties and Definitions

1. Determine the domain of $f(x) = x^2 - 4x + 9$.

2. Determine the domain of $h(x) = \dfrac{x - 1}{x^2(x^2 + 1)}$.

Functions

In Exercises 3–6, consider the function

$$f(x) = 2x^3 - 3x^2 - 18x + 27$$
$$= (2x - 3)(x + 3)(x - 3).$$

3. ▦ Use a graphing calculator to graph both expressions for the function. Are the graphs the same?

4. Verify the factorization by multiplying the polynomials in the factored form of f.

5. Verify the factorization by performing the long division

$$\frac{2x^3 - 3x^2 - 18x + 27}{2x - 3}$$

and then factoring the quotient.

6. Verify the factorization by performing the long division

$$\frac{2x^3 - 3x^2 - 18x + 27}{x^2 - 9}.$$

In Exercises 7 and 8, use the function to find and simplify the expression for

$$\frac{f(2 + h) - f(2)}{h}.$$

7. $f(x) = x^2 - 3$

8. $f(x) = \dfrac{3}{x + 5}$

Modeling

9. *Cost*   The inventor of a new game believes that the variable cost for producing the game is \$5.75 per unit and the fixed costs are \$12,000. Write the total cost C as a function of x, the number of games produced.

10. ▲ *Geometry*   The length of a rectangle is one and one-half times its width. Write the perimeter P of the rectangle as a function of the rectangle's width w.

Developing Skills

In Exercises 1–14, write a model for the statement.

1. I varies directly as V.

2. C varies directly as r.

3. V is directly proportional to t.

4. s varies directly as the cube of t.

5. u is directly proportional to the square of v.

6. V varies directly as the cube root of x.

7. p varies inversely as d.

8. S is inversely proportional to the square of v.

9. A varies inversely as the fourth power of t.

10. P is inversely proportional to the square root of $1 + r$.

11. A varies jointly as l and w.

12. V varies jointly as h and the square of r.

13. *Boyle's Law*   If the temperature of a gas is not allowed to change, its absolute pressure P is inversely proportional to its volume V.

14. *Newton's Law of Universal Gravitation*   The gravitational attraction F between two particles of masses m_1 and m_2 is directly proportional to the product of the masses and inversely proportional to the square of the distance r between the particles.

In Exercises 15–20, write a verbal sentence using variation terminology to describe the formula.

15. *Area of a Triangle:* $A = \frac{1}{2}bh$

16. *Area of a Rectangle:* $A = lw$

17. *Volume of a Right Circular Cylinder:* $V = \pi r^2 h$

18. *Volume of a Sphere:* $V = \frac{4}{3}\pi r^3$

19. *Average Speed:* $r = \dfrac{d}{t}$

20. *Height of a Cylinder:* $h = \dfrac{V}{\pi r^2}$

In Exercises 21–32, find the constant of proportionality and write an equation that relates the variables.

21. s varies directly as t, and $s = 20$ when $t = 4$.

22. h is directly proportional to r, and $h = 28$ when $r = 12$.

23. F is directly proportional to the square of x, and $F = 500$ when $x = 40$.

24. M varies directly as the cube of n, and $M = 0.012$ when $n = 0.2$.

25. n varies inversely as m, and $n = 32$ when $m = 1.5$.

26. q is inversely proportional to p, and $q = \frac{3}{2}$ when $p = 50$.

27. g varies inversely as the square root of z, and $g = \frac{4}{5}$ when $z = 25$.

28. u varies inversely as the square of v, and $u = 40$ when $v = \frac{1}{2}$.

29. F varies jointly as x and y, and $F = 500$ when $x = 15$ and $y = 8$.

30. V varies jointly as h and the square of b, and $V = 288$ when $h = 6$ and $b = 12$.

31. d varies directly as the square of x and inversely with r, and $d = 3000$ when $x = 10$ and $r = 4$.

32. z is directly proportional to x and inversely proportional to the square root of y, and $z = 720$ when $x = 48$ and $y = 81$.

Solving Problems

33. *Current Speed* A boat travels at a speed of 20 miles per hour in still water. It travels 48 miles upstream and then returns to the starting point in a total of 5 hours. Find the speed of the current.

34. *Average Speeds* You and your college roommate travel to your respective hometowns in the same amount of time. You drive 210 miles and your friend drives 190 miles. Your friend's average speed is 6 miles per hour lower than your average speed. What are your average speed and your friend's average speed?

35. *Partnership Costs* A group plans to start a new business that will require $240,000 for start-up capital. The individuals in the group share the cost equally. If two additional people join the group, the cost per person will decrease by $4000. How many people are presently in the group?

36. *Partnership Costs* A group of people share equally the cost of a $150,000 endowment. If they could find four more people to join the group, each person's share of the cost would decrease by $6250. How many people are presently in the group?

37. *Population Growth* A biologist introduces 100 insects into a culture. The population P of the culture is approximated by the model below, where t is the time in hours. Find the time required for the population to increase to 1000 insects.

$$P = \frac{500(1 + 3t)}{5 + t}$$

38. *Pollution Removal* The cost C in dollars of removing $p\%$ of the air pollutants in the stack emissions of a utility company is modeled by the equation below. Determine the percent of air pollutants in the stack emissions that can be removed for $680,000.

$$C = \frac{120{,}000p}{100 - p}.$$

39. *Work Rate*   One landscaper works $1\frac{1}{2}$ times as fast as another landscaper. It takes them 9 hours working together to complete a job. Find the time it takes each landscaper to complete the job working alone.

40. *Flow Rate*   The flow rate for one pipe is $1\frac{1}{4}$ times that of another pipe. A swimming pool can be filled in 5 hours using both pipes. Find the time required to fill the pool using only the pipe with the lower flow rate.

41. *Nail Sizes*   The unit for determining the size of a nail is a *penny*. For example, 8d represents an 8-penny nail. The number N of finishing nails per pound can be modeled by

$$N = -139.1 + \frac{2921}{x}$$

where x is the size of the nail.

(a) What is the domain of the function?

(b) 🖩 Use a graphing calculator to graph the function.

(c) Use the graph to determine the size of the finishing nail if there are 153 nails per pound.

(d) Verify the result of part (c) algebraically.

42. *Learning Curve*   A psychologist observed that a four-year-old child could memorize N lines of a poem, where N depended on the number x of short sessions that the psychologist worked with the child. The number of lines N memorized can be easily modeled by

$$N = \frac{20x}{x + 1}, \qquad x \geq 0.$$

(a) 🖩 Use a graphing calculator to graph the function.

(b) Use the graph to determine the number of sessions if the child can memorize 18 lines of the poem.

(c) Verify the result of part (b) algebraically.

43. *Revenue*   The total revenue R is directly proportional to the number of units sold x. When 500 units are sold, the revenue is $3875. Find the revenue when 635 units are sold. Then interpret the constant of proportionality.

44. *Revenue*   The total revenue R is directly proportional to the number of units sold x. When 25 units are sold, the revenue is $300. Find the revenue when 42 units are sold. Then interpret the constant of proportionality.

45. *Hooke's Law*   A force of 50 pounds stretches a spring 5 inches.

(a) How far will a force of 20 pounds stretch the spring?

(b) What force is required to stretch the spring 1.5 inches?

46. *Hooke's Law*   A force of 50 pounds stretches a spring 3 inches.

(a) How far will a force of 20 pounds stretch the spring?

(b) What force is required to stretch the spring 1.5 inches?

47. *Hooke's Law*   A baby weighing $10\frac{1}{2}$ pounds compresses the spring of a baby scale 7 millimeters. Determine the weight of a baby that compresses the spring 12 millimeters.

48. *Hooke's Law*   A force of 50 pounds stretches the spring of a scale 1.5 inches.

(a) Write the force F as a function of the distance x the spring is stretched.

(b) Graph the function in part (a) where $0 \leq x \leq 5$. Identify the graph.

49. *Free-Falling Object*   The velocity v of a free-falling object is proportional to the time that the object has fallen. The constant of proportionality is the acceleration due to gravity. The velocity of a falling object is 96 feet per second after the object has fallen for 3 seconds. Find the acceleration due to gravity.

50. *Free-Falling Object*   Neglecting air resistance, the distance d that an object falls varies directly as the square of the time t it has fallen. An object falls 64 feet in 2 seconds. Determine the distance it will fall in 6 seconds.

51. *Stopping Distance* The stopping distance d of an automobile is directly proportional to the square of its speed s. On a road surface, a car requires 75 feet to stop when its speed is 30 miles per hour. The brakes are applied when the car is traveling at 50 miles per hour under similar road conditions. Estimate the stopping distance.

52. *Frictional Force* The frictional force F between the tires and the road that is required to keep a car on a curved section of a highway is directly proportional to the square of the speed s of the car. If the speed of the car is doubled, the force will change by what factor?

53. *Power Generation* The power P generated by a wind turbine varies directly as the cube of the wind speed w. The turbine generates 750 watts of power in a 25-mile-per-hour wind. Find the power it generates in a 40-mile-per-hour wind.

54. *Demand* A company has found that the daily demand x for its boxes of chocolates is inversely proportional to the price p. When the price is \$5, the demand is 800 boxes. Approximate the demand when the price is increased to \$6.

55. *Predator-Prey* The number N of prey t months after a natural predator is introduced into a test area is inversely proportional to $t + 1$. If $N = 500$ when $t = 0$, find N when $t = 4$.

56. *Weight of an Astronaut* A person's weight on the moon varies directly with his or her weight on Earth. An astronaut weighs 360 pounds on Earth, including heavy equipment. On the moon the astronaut weighs only 60 pounds with the equipment. If the first woman in space, Valentina Tereshkova, had landed on the moon and weighed 54 pounds with equipment, how much would she have weighed on Earth with her equipment?

57. *Pressure* When a person walks, the pressure P on each sole varies inversely with the area A of the sole. A person is trudging through deep snow, wearing boots that have a sole area of 29 square inches each. The sole pressure is 4 pounds per square inch. If the person was wearing snowshoes, each with an area 11 times that of their boot soles, what would be the pressure on each snowshoe? The constant of variation in this problem is the weight of the person. How much does the person weigh?

58. *Environment* The graph shows the percent p of oil that remained in Chedabucto Bay, Nova Scotia, after an oil spill. The cleaning of the spill was left primarily to natural actions such as wave motion, evaporation, photochemical decomposition, and bacterial decomposition. After about a year, the percent that remained varied inversely as time. Find a model that relates p and t, where t is the number of years since the spill. Then use it to find the percent of oil that remained $6\frac{1}{2}$ years after the spill, and compare the result with the graph.

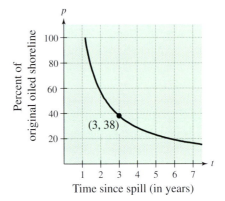

59. *Meteorology* The graph shows the temperature of the water in the north central Pacific Ocean. At depths greater than 900 meters, the water temperature varies inversely with the water depth. Find a model that relates the temperature T to the depth d. Then use it to find the water temperature at a depth of 4385 meters, and compare the result with the graph.

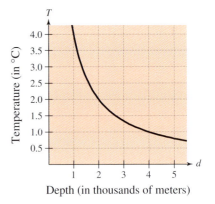

60. *Engineering* The load P that can be safely supported by a horizontal beam varies jointly as the product of the width W of the beam and the square of the depth D and inversely as the length L. (See figure).

(a) Write a model for the statement.

(b) How does P change when the width and length of the beam are both doubled?

(c) How does P change when the width and depth of the beam are doubled?

(d) How does P change when all three of the dimensions are doubled?

(e) How does P change when the depth of the beam is cut in half?

(f) A beam with width 3 inches, depth 8 inches, and length 10 feet can safely support 2000 pounds. Determine the safe load of a beam made from the same material if its depth is increased to 10 inches.

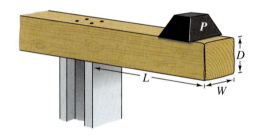

In Exercises 61–64, complete the table and plot the resulting points.

x		2	4	6	8	10
$y = kx^2$						

61. $k = 1$ **62.** $k = 2$

63. $k = \frac{1}{2}$ **64.** $k = \frac{1}{4}$

In Exercises 65–68, complete the table and plot the resulting points.

x		2	4	6	8	10
$y = \dfrac{k}{x^2}$						

65. $k = 2$ **66.** $k = 5$

67. $k = 10$ **68.** $k = 20$

In Exercises 69 and 70, determine whether the variation model is of the form $y = kx$ or $y = k/x$, and find k.

69.

x	10	20	30	40	50
y	$\frac{2}{5}$	$\frac{1}{5}$	$\frac{2}{15}$	$\frac{1}{10}$	$\frac{2}{25}$

70.

x	10	20	30	40	50
y	-3	-6	-9	-12	-15

Explaining Concepts

71. *Writing* Suppose the constant of proportionality is positive and y varies directly as x. If one of the variables increases, how will the other change? Explain.

72. *Writing* Suppose the constant of proportionality is positive and y varies inversely as x. If one of the variables increases, how will the other change? Explain.

73. *Writing* If y varies directly as the square of x and x is doubled, how will y change? Use the rules of exponents to explain your answer.

74. *Writing* If y varies inversely as the square of x and x is doubled, how will y change? Use the rules of exponents to explain your answer.

75. *Writing* Describe a real life problem for each type of variation (direct, inverse, and joint).

What Did You Learn?

Key Terms

rational expression, *p. 368*
rational function, *p. 368*
domain (of a rational
   function), *p. 368*
simplified form, *p. 371*
least common multiple,
   p. 390

least common
   denominator, *p. 391*
complex fraction, *p. 398*
dividend, *p. 408*
divisor, *p. 408*
quotient, *p. 408*
remainder, *p. 408*
synthetic division, *p. 411*

extraneous solution, *p. 420*
cross-multiplying, *p. 421*
direct variation, *p. 427*
constant of proportionality,
   p. 427
inverse variation, *p. 430*
combined variation, *p. 431*

Key Concepts

6.1 ◯ Simplifying rational expressions

Let u, v, and w represent real numbers, variables, or algebraic expressions such that $v \neq 0$ and $w \neq 0$. Then the following is valid.

$$\frac{uw}{vw} = \frac{u\cancel{w}}{v\cancel{w}} = \frac{u}{v}$$

6.2 ◯ Multiplying rational expressions

Let u, v, w, and z represent real numbers, variables, or algebraic expressions such that $v \neq 0$ and $z \neq 0$. Then the product of u/v and w/z is

$$\frac{u}{v} \cdot \frac{w}{z} = \frac{uw}{vz}.$$

6.2 ◯ Dividing rational expressions

Let u, v, w, and z represent real numbers, variables, or algebraic expressions such that $v \neq 0$, $w \neq 0$, and $z \neq 0$. Then the quotient of u/v and w/z is

$$\frac{u}{v} \div \frac{w}{z} = \frac{u}{v} \cdot \frac{z}{w} = \frac{uz}{vw}.$$

6.3 ◯ Adding or subtracting with like denominators

If u, v, and w are real numbers, variables, or algebraic expressions, and $w \neq 0$, the following rules are valid.

1. $\dfrac{u}{w} + \dfrac{v}{w} = \dfrac{u+v}{w}$ 2. $\dfrac{u}{w} - \dfrac{v}{w} = \dfrac{u-v}{w}$

6.3 ◯ Adding or subtracting with unlike denominators

Rewrite the rational expressions with like denominators by finding the least common denominator. Then add or subtract as with like denominators.

6.5 ◯ Dividing a polynomial by a monomial

Let u, v, and w represent real numbers, variables, or algebraic expressions such that $w \neq 0$.

1. $\dfrac{u+v}{w} = \dfrac{u}{w} + \dfrac{v}{w}$ 2. $\dfrac{u-v}{w} = \dfrac{u}{w} - \dfrac{v}{w}$

6.5 ◯ Synthetic division of a third-degree polynomial

Use synthetic division to divide $ax^3 + bx^2 + cx + d$ by $x - k$, as follows.

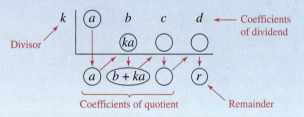

Vertical Pattern: Add terms
Diagonal Pattern: Multiply by k.

6.6 ◯ Solving rational equations

1. Determine the domain of each of the fractions in the equation.

2. Obtain an equivalent equation by multiplying each side of the equation by the least common denominator of all the fractions in the equation.

3. Solve the resulting equation.

4. Check your solution(s) in the original equation.

6.7 ◯ Variation models

In the following, k is a constant.

1. *Direction variation:* $y = kx$

2. *Direction variation as nth power:* $y = kx^n$

3. *Inverse variation:* $y = k/x$

4. *Inverse variation as nth power:* $y = k/x^n$

5. *Joint variation:* $z = kxy$

6. *Joint variation as nth and mth powers:* $z = kx^n y^m$

Review Exercises

6.1 Rational Expressions and Functions

1 Find the domain of a rational function.

In Exercises 1–4, find the domain of the rational function.

1. $f(y) = \dfrac{3y}{y - 8}$

2. $g(t) = \dfrac{t + 4}{t + 12}$

3. $g(u) = \dfrac{u}{u^2 - 7u + 6}$

4. $f(x) = \dfrac{x - 12}{x(x^2 - 16)}$

5. ▲ *Geometry* A rectangle with a width of w inches has an area of 36 square inches. The perimeter P of the rectangle is given by

$$P = 2\left(w + \frac{36}{w}\right).$$

Describe the domain of the function.

6. *Average Cost* The average cost $\overline{C}$ for a manufacturer to produce x units of a product is given by

$$\overline{C} = \frac{15,000 + 0.75x}{x}.$$

Describe the domain of the function.

2 Simplify rational expressions.

In Exercises 7-14, simplify the rational expression.

7. $\dfrac{6x^4y^2}{15xy^2}$

8. $\dfrac{2(y^3z)^2}{28(yz^2)^2}$

9. $\dfrac{5b - 15}{30b - 120}$

10. $\dfrac{4a}{10a^2 + 26a}$

11. $\dfrac{9x - 9y}{y - x}$

12. $\dfrac{x + 3}{x^2 - x - 12}$

13. $\dfrac{x^2 - 5x}{2x^2 - 50}$

14. $\dfrac{x^2 + 3x + 9}{x^3 - 27}$

6.2 Multiplying and Dividing Rational Expressions

1 Multiply rational expressions and simplify.

In Exercises 15–22, multiply and simplify.

15. $3x(x^2y)^2$

16. $2b(-3b)^3$

17. $\dfrac{7}{8} \cdot \dfrac{2x}{y} \cdot \dfrac{y^2}{14x^2}$

18. $\dfrac{15(x^2y)^3}{3y^3} \cdot \dfrac{12y}{x}$

19. $\dfrac{60z}{z + 6} \cdot \dfrac{z^2 - 36}{5}$

20. $\dfrac{x^2 - 16}{6} \cdot \dfrac{3}{x^2 - 8x + 16}$

21. $\dfrac{u}{u - 3} \cdot \dfrac{3u - u^2}{4u^2}$

22. $x^2 \cdot \dfrac{x + 1}{x^2 - x} \cdot \dfrac{(5x - 5)^2}{x^2 + 6x + 5}$

2 Divide rational expressions and simplify.

In Exercises 23–30, divide and simplify.

23. $\dfrac{24x^4}{15x}$

24. $\dfrac{8u^2v}{6v}$

25. $25y^2 \div \dfrac{xy}{5}$

26. $\dfrac{6}{z^2} \div 4z^2$

27. $\dfrac{x^2 + 3x + 2}{3x^2 + x - 2} \div (x + 2)$

28. $\dfrac{x^2 - 14x + 48}{x^2 - 6x} \div (3x - 24)$

29. $\dfrac{x^2 - 7x}{x + 1} \div \dfrac{x^2 - 14x + 49}{x^2 - 1}$

30. $\dfrac{x^2 - x}{x + 1} \div \dfrac{5x - 5}{x^2 + 6x + 5}$

6.3 Adding and Subtracting Rational Expressions

① Add or subtract rational expressions with like denominators and simplify.

In Exercises 31–38, combine and simplify.

31. $\dfrac{4x}{5} + \dfrac{11x}{5}$

32. $\dfrac{7y}{12} - \dfrac{4y}{12}$

33. $\dfrac{15}{3x} - \dfrac{3}{3x}$

34. $\dfrac{4}{5x} + \dfrac{1}{5x}$

35. $\dfrac{2(3y + 4)}{2y + 1} + \dfrac{3 - y}{2y + 1}$

36. $\dfrac{4x - 2}{3x + 1} - \dfrac{x + 1}{3x + 1}$

37. $\dfrac{4x}{x + 2} + \dfrac{3x - 7}{x + 2} - \dfrac{9}{x + 2}$

38. $\dfrac{3}{2y - 3} - \dfrac{y - 10}{2y - 3} + \dfrac{5y}{2y - 3}$

② Add or subtract rational expressions with unlike denominators and simplify.

In Exercises 39–46, combine and simplify.

39. $\dfrac{1}{x + 5} + \dfrac{3}{x - 12}$

40. $\dfrac{2}{x - 10} + \dfrac{3}{4 - x}$

41. $5x + \dfrac{2}{x - 3} - \dfrac{3}{x + 2}$

42. $4 - \dfrac{4x}{x + 6} + \dfrac{7}{x - 5}$

43. $\dfrac{6}{x - 5} - \dfrac{4x + 7}{x^2 - x - 20}$

44. $\dfrac{5}{x + 2} + \dfrac{25 - x}{x^2 - 3x - 10}$

45. $\dfrac{5}{x + 3} - \dfrac{4x}{(x + 3)^2} - \dfrac{1}{x - 3}$

46. $\dfrac{8}{y} - \dfrac{3}{y + 5} + \dfrac{4}{y - 2}$

In Exercises 47 and 48, use a graphing calculator to graph the two equations in the same viewing window. Use the graphs to verify that the expressions are equivalent. Verify the results algebraically.

47. $y_1 = \dfrac{1}{x} - \dfrac{3}{x + 3}$

$y_2 = \dfrac{3 - 2x}{x(x + 3)}$

48. $y_1 = \dfrac{5x}{x - 5} + \dfrac{7}{x + 1}$

$y_2 = \dfrac{5x^2 + 12x - 35}{x^2 - 4x - 5}$

6.4 Complex Fractions

① Simplify complex fractions using rules for dividing rational expressions.

In Exercises 49–52, simplify the complex fraction.

49. $\dfrac{\left(\dfrac{6}{x}\right)}{\left(\dfrac{2}{x^3}\right)}$

50. $\dfrac{xy}{\left(\dfrac{5x^2}{2y}\right)}$

51. $\dfrac{\left(\dfrac{6x^2}{x^2 + 2x - 35}\right)}{\left(\dfrac{x^3}{x^2 - 25}\right)}$

52. $\dfrac{\left[\dfrac{24 - 18x}{(2 - x)^2}\right]}{\left(\dfrac{60 - 45x}{x^2 - 4x + 4}\right)}$

② Simplify complex fractions having a sum or difference in the numerator and/or denominator.

In Exercises 53–58, simplify the complex fraction.

53. $\dfrac{3t}{\left(5 - \dfrac{2}{t}\right)}$

54. $\dfrac{\left(\dfrac{1}{x} - \dfrac{1}{2}\right)}{2x}$

55. $\dfrac{\left(x - 3 + \dfrac{2}{x}\right)}{\left(1 - \dfrac{2}{x}\right)}$

56. $\dfrac{3x - 1}{\left(\dfrac{2}{x^2} + \dfrac{5}{x}\right)}$

57. $\dfrac{\left(\dfrac{1}{a^2 - 16} - \dfrac{1}{a}\right)}{\left(\dfrac{1}{a^2 + 4a} + 4\right)}$

58. $\dfrac{\left(\dfrac{1}{x^2} - \dfrac{1}{y^2}\right)}{\left(\dfrac{1}{x} + \dfrac{1}{y}\right)}$

6.5 Dividing Polynomials and Synthetic Division

1 Divide polynomials by monomials and write in simplest form.

In Exercises 59–62, perform the division.

59. $(4x^3 - x) \div (2x)$ **60.** $(10x + 15) \div (5x)$

61. $\dfrac{3x^3y^2 - x^2y^2 + x^2y}{x^2y}$

62. $\dfrac{6a^3b^3 + 2a^2b - 4ab^2}{2ab}$

2 Use long division to divide polynomials by polynomials.

In Exercises 63–68, perform the division.

63. $\dfrac{6x^3 + 2x^2 - 4x + 2}{3x - 1}$

64. $\dfrac{4x^4 - x^3 - 7x^2 + 18x}{x - 2}$

65. $\dfrac{x^4 - 3x^2 + 2}{x^2 - 1}$

66. $\dfrac{x^4 - 4x^3 + 3x}{x^2 - 1}$

67. $\dfrac{x^5 - 3x^4 + x^2 + 6}{x^3 - 2x^2 + x - 1}$

68. $\dfrac{x^6 + 4x^5 - 3x^2 + 5x}{x^3 + x^2 - 4x + 3}$

3 Use synthetic division to divide polynomials by polynomials of the form $x - k$.

In Exercises 69–72, use synthetic division to divide.

69. $\dfrac{x^3 + 7x^2 + 3x - 14}{x + 2}$

70. $\dfrac{x^4 - 2x^3 - 15x^2 - 2x + 10}{x - 5}$

71. $(x^4 - 3x^2 - 25) \div (x - 3)$

72. $(2x^3 + 5x - 2) \div \left(x + \dfrac{1}{2}\right)$

4 Use synthetic division to factor polynomials.

In Exercises 73 and 74, completely factor the polynomial given one of its factors.

Polynomial	*Factor*
73. $x^3 + 2x^2 - 5x - 6$	$x - 2$
74. $2x^3 + x^2 - 2x - 1$	$x + 1$

6.6 Solving Rational Equations

1 Solve rational equations containing constant denominators.

In Exercises 75 and 76, solve the equation.

75. $\dfrac{3x}{8} = -15 + \dfrac{x}{4}$

76. $\dfrac{t + 1}{6} = \dfrac{1}{2} - 2t$

2 Solve rational equations containing variable denominators.

In Exercises 77–90, solve the equation.

77. $8 - \dfrac{12}{t} = \dfrac{1}{3}$ **78.** $5 + \dfrac{2}{x} = \dfrac{1}{4}$

79. $\dfrac{2}{y} - \dfrac{1}{3y} = \dfrac{1}{3}$ **80.** $\dfrac{7}{4x} - \dfrac{6}{8x} = 1$

81. $r = 2 + \dfrac{24}{r}$ **82.** $\dfrac{2}{x} - \dfrac{x}{6} = \dfrac{2}{3}$

83. $8\left(\dfrac{6}{x} - \dfrac{1}{x + 5}\right) = 15$ **84.** $\dfrac{3}{y + 1} - \dfrac{8}{y} = 1$

85. $\dfrac{4x}{x-5} + \dfrac{2}{x} = -\dfrac{4}{x-5}$ **86.** $\dfrac{2x}{x-3} - \dfrac{3}{x} = 0$

87. $\dfrac{12}{x^2+x-12} - \dfrac{1}{x-3} = -1$

88. $\dfrac{3}{x-1} + \dfrac{6}{x^2-3x+2} = 2$

89. $\dfrac{5}{x^2-4} - \dfrac{6}{x-2} = -5$

90. $\dfrac{3}{x^2-9} + \dfrac{4}{x+3} = 1$

6.7 Applications and Variation

① Solve application problems involving rational equations.

91. *Average Speed*   You drive 56 miles on a service call for your company. On the return trip, which takes 10 minutes less than the original trip, your average speed is 8 miles per hour greater. What is your average speed on the return trip?

92. *Average Speed*   You drive 220 miles to see a friend. On the return trip, which takes 20 minutes less than the original trip, your average speed is 5 miles per hour faster. What is your average speed on the return trip?

93. *Partnership Costs*   A group of people starting a business agree to share equally in the cost of a $60,000 piece of machinery. If they could find two more people to join the group, each person's share of the cost would decrease by $5000. How many people are presently in the group?

94. *Work Rate*   One painter works $1\frac{1}{2}$ times as fast as another painter. It takes them 4 hours working together to paint a room. Find the time it takes each painter to paint the room working alone.

95. *Population Growth*   The Parks and Wildlife Commission introduces 80,000 fish into a large lake. The population P (in thousands) of the fish is approximated by the model

$$P = \dfrac{20(4+3t)}{1+0.05t}$$

where t is the time in years. Find the time required for the population to increase to 400,000 fish.

96. *Average Cost*   The average cost $\overline{C}$ for producing x units of a product is given by

$$\overline{C} = 1.5 + \dfrac{4200}{x}.$$

Determine the number of units that must be produced to obtain an average cost of $2.90 per unit.

② Solve application problems involving direct variation.

97. *Hooke's Law*   A force of 100 pounds stretches a spring 4 inches. Find the force required to stretch the spring 6 inches.

98. *Stopping Distance*   The stopping distance d of an automobile is directly proportional to the square of its speed s. How will the stopping distance be changed by doubling the speed of the car?

③ Solve application problems involving inverse variation.

99. *Travel Time*   The travel time between two cities is inversely proportional to the average speed. A train travels between the cities in 3 hours at an average speed of 65 miles per hour. How long would it take to travel between the cities at an average speed of 80 miles per hour?

100. *Demand*   A company has found that the daily demand x for its cordless telephones is inversely proportional to the price p. When the price is $25, the demand is 1000 telephones. Approximate the demand when the price is increased to $28.

④ Solve application problems involving joint variation.

101. *Simple Interest*   The simple interest for a savings account is jointly proportional to the time and the principal. After three quarters (9 months), the interest for a principal of $12,000 is $675. How much interest would a principal of $8200 earn in 18 months?

102. *Cost*   The cost of constructing a wooden box with a square base varies jointly as the height of the box and the square of the width of the box. A box of height 16 inches and of width 6 inches costs $28.80. How much would a box of height 14 inches and of width 8 inches cost?

Chapter Test

Take this test as you would take a test in class. After you are done, check your work against the answers in the back of the book.

1. Find the domain of $f(x) = \dfrac{2x}{x^2 - 5x + 6}$.

In Exercises 2 and 3, simplify the rational expression.

2. $\dfrac{2 - x}{3x - 6}$

3. $\dfrac{2a^2 - 5a - 12}{5a - 20}$

4. Find the least common multiple of x^2, $3x^3$, and $(x + 4)^2$.

In Exercises 5–18, perform the operation and simplify.

5. $\dfrac{4z^3}{5} \cdot \dfrac{25}{12z^2}$

6. $\dfrac{y^2 + 8y + 16}{2(y - 2)} \cdot \dfrac{8y - 16}{(y + 4)^3}$

7. $(4x^2 - 9) \div \dfrac{2x + 3}{2x^2 - x - 3}$

8. $\dfrac{(2xy^2)^3}{15} \div \dfrac{12x^3}{21}$

9. $2x + \dfrac{1 - 4x^2}{x + 1}$

10. $\dfrac{5x}{x + 2} - \dfrac{2}{x^2 - x - 6}$

11. $\dfrac{3}{x} - \dfrac{5}{x^2} + \dfrac{2x}{x^2 + 2x + 1}$

12. $\dfrac{4}{x + 1} + \dfrac{4x}{x + 1}$

13. $\dfrac{\left(\dfrac{3x}{x + 2}\right)}{\left(\dfrac{12}{x^3 + 2x^2}\right)}$

14. $\dfrac{\left(9x - \dfrac{1}{x}\right)}{\left(\dfrac{1}{x} - 3\right)}$

15. $\dfrac{3x^{-2} + y^{-1}}{(x + y)^{-1}}$

16. $\dfrac{4x^2 + 2x - 7}{2x}$

17. $\dfrac{t^4 + t^2 - 6t}{t^2 - 2}$

18. $\dfrac{2x^4 - 15x^2 - 7}{x - 3}$

In Exercises 19–21, solve the equation.

19. $\dfrac{3}{h + 2} = \dfrac{1}{8}$

20. $\dfrac{2}{x + 5} - \dfrac{3}{x + 3} = \dfrac{1}{x}$

21. $\dfrac{1}{x + 1} + \dfrac{1}{x - 1} = \dfrac{2}{x^2 - 1}$

22. Find a mathematical model that relates u and v if v varies directly as the square root of u, and $v = \frac{3}{2}$ when $u = 36$.

23. If the temperature of a gas is not allowed to change, the absolute pressure P of the gas is inversely proportional to its volume V, according to Boyle's Law. A large balloon is filled with 180 cubic meters of helium at atmospheric pressure (1 atm) at sea level. What is the volume of the helium if the balloon rises to an altitude at which the atmospheric pressure is 0.75 atm? (Assume that the temperature does not change.)

443

Motivating the Chapter

 ## Building a Greenhouse

You are building a greenhouse in the form of a half cylinder. The volume of the greenhouse is to be approximately 35,350 cubic feet.

See Section 7.1, Exercise 153.

$l = 100$ ft

a. The formula for the radius r (in feet) of a half cylinder is

$$r = \sqrt{\frac{2V}{\pi l}}$$

where V is the volume (in cubic feet) and l is the length (in feet). Find the radius of the greenhouse shown and round your result to the nearest whole number. Use this value of r in parts (b)–(d).

b. Beams for holding a sprinkler system are to be placed across the building. The formula for the height h at which the beams are to be placed is

$$h = \sqrt{r^2 - \left(\frac{a}{2}\right)^2}$$

where a is the length of the beam. Rewrite h as a function of a.

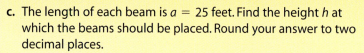

Cross Section of Greenhouse

c. The length of each beam is $a = 25$ feet. Find the height h at which the beams should be placed. Round your answer to two decimal places.

d. The equation from part (b) can be rewritten as

$$a = 2\sqrt{r^2 - h^2}.$$

The height is $h = 8$ feet. What is the length a of each beam? Round your answer to two decimal places.

See Section 7.5, Exercise 103.

e. The cost of building the greenhouse is estimated to be $25,000. The money to pay for the greenhouse was invested in an interest-bearing account 10 years ago at an annual percent rate of 7%. The amount of money earned can be found using the formula

$$r = \left(\frac{A}{P}\right)^{1/n} - 1$$

where r is the annual percent rate (in decimal form), A is the amount in the account after 10 years, P is the initial deposit, and n is the number of years. What initial deposit P would have generated enough money to cover the building cost of $25,000?

John A. Rizzo/Photodisc/Getty Images

Radicals and Complex Numbers

7.1 ● Radicals and Rational Exponents

7.2 ● Simplifying Radical Expressions

7.3 ● Adding and Subtracting Radical Expressions

7.4 ● Multiplying and Dividing Radical Expressions

7.5 ● Radical Equations and Applications

7.6 ● Complex Numbers

7.1 Radicals and Rational Exponents

What You Should Learn

1 Determine the *n*th roots of numbers and evaluate radical expressions.

2 Use the rules of exponents to evaluate or simplify expressions with rational exponents.

3 Use a calculator to evaluate radical expressions.

4 Evaluate radical functions and find the domains of radical functions.

Why You Should Learn It

Algebraic equations often involve rational exponents. For instance, in Exercise 147 on page 455, you will use an equation involving a rational exponent to find the depreciation rate for a truck.

1 Determine the *n*th roots of numbers and evaluate radical expressions.

Roots and Radicals

A **square root** of a number is defined as one of its two equal factors. For example, 5 is a square root of 25 because 5 is one of the two equal factors of 25. In a similar way, a **cube root** of a number is one of its three equal factors.

Number	Equal Factors	Root	Type
$9 = 3^2$	$3 \cdot 3$	3	Square root
$25 = (-5)^2$	$(-5)(-5)$	-5	Square root
$-27 = (-3)^3$	$(-3)(-3)(-3)$	-3	Cube root
$64 = 4^3$	$4 \cdot 4 \cdot 4$	4	Cube root
$16 = 2^4$	$2 \cdot 2 \cdot 2 \cdot 2$	2	Fourth root

Definition of *n*th Root of a Number

Let a and b be real numbers and let n be an integer such that $n \geq 2$. If

$$a = b^n$$

then b is an ***n*th root of *a*.** If $n = 2$, the root is a **square root.** If $n = 3$, the root is a **cube root.**

Some numbers have more than one *n*th root. For example, both 5 and -5 are square roots of 25. To avoid ambiguity about which root you are referring to, the **principal *n*th root** of a number is defined in terms of a **radical symbol** $\sqrt[n]{\ }$. So the *principal square root* of 25, written as $\sqrt{25}$, is the positive root, 5.

Principal *n*th Root of a Number

Let a be a real number that has at least one (real number) *n*th root. The **principal *n*th root of *a*** is the *n*th root that has the same sign as a, and it is denoted by the **radical symbol**

$$\sqrt[n]{a}. \qquad \text{Principal } n\text{th root}$$

The positive integer n is the **index** of the radical, and the number a is the **radicand.** If $n = 2$, omit the index and write $\sqrt{a}$ rather than $\sqrt[2]{a}$.

Study Tip

In the definition at the right, "the *n*th root that has the same sign as a" means that the principal *n*th root of a is positive if a is positive and negative if a is negative. For example, $\sqrt{4} = 2$ and $\sqrt[3]{-8} = -2$. When a negative root is needed, you must use the negative sign with the square root sign. For example, $-\sqrt{4} = -2$.

Example 1   Finding Roots of Numbers

Find each root.

a. $\sqrt{36}$ b. $-\sqrt{36}$ c. $\sqrt{-4}$ d. $\sqrt[3]{8}$ e. $\sqrt[3]{-8}$

Solution

a. $\sqrt{36} = 6$ because $6 \cdot 6 = 6^2 = 36$.

b. $-\sqrt{36} = -6$ because $6 \cdot 6 = 6^2 = 36$. So, $(-1)(\sqrt{36}) = (-1)(6) = -6$.

c. $\sqrt{-4}$ is not real because there is no real number that when multiplied by itself yields -4.

d. $\sqrt[3]{8} = 2$ because $2 \cdot 2 \cdot 2 = 2^3 = 8$.

e. $\sqrt[3]{-8} = -2$ because $(-2)(-2)(-2) = (-2)^3 = -8$.

Properties of nth Roots

Property	Example
1. If a is a positive real number and n is even, then a has exactly two (real) nth roots, which are denoted by $\sqrt[n]{a}$ and $-\sqrt[n]{a}$.	The two real square roots of 81 are $\sqrt{81} = 9$ and $-\sqrt{81} = -9$.
2. If a is any real number and n is odd, then a has only one (real) nth root, which is denoted by $\sqrt[n]{a}$.	$\sqrt[3]{27} = 3$ $\sqrt[3]{-64} = -4$
3. If a is a negative real number and n is even, then a has no (real) nth root.	$\sqrt{-64}$ is not a real number.

Integers such as 1, 4, 9, 16, 49, and 81 are called **perfect squares** because they have integer square roots. Similarly, integers such as 1, 8, 27, 64, and 125 are called **perfect cubes** because they have integer cube roots.

Example 2   Classifying Perfect nth Powers

State whether each number is a perfect square, a perfect cube, both, or neither.

a. 81   b. -125 c. 64   d. 32

Solution

a. 81 is a perfect square because $9^2 = 81$. It is not a perfect cube.

b. -125 is a perfect cube because $(-5)^3 = -125$. It is not a perfect square.

c. 64 is a perfect square because $8^2 = 64$, and it is also a perfect cube because $4^3 = 64$.

d. 32 is not a perfect square or a perfect cube. (It is, however, a perfect fifth power, because $2^5 = 32$.)

Study Tip

The square roots of perfect squares are rational numbers, so $\sqrt{25}$, $\sqrt{49}$, and $\sqrt{100}$ are examples of rational numbers. However, square roots such as $\sqrt{5}$, $\sqrt{19}$, and $\sqrt{34}$ are irrational numbers. Similarly, $\sqrt[3]{27}$ and $\sqrt[4]{16}$ are rational numbers, whereas $\sqrt[3]{6}$ and $\sqrt[4]{21}$ are irrational numbers.

Raising a number to the nth power and taking the principal nth root of a number can be thought of as *inverse* operations. Here are some examples.

$$\left(\sqrt{4}\right)^2 = (2)^2 = 4 \quad \text{and} \quad \sqrt{4} = \sqrt{2^2} = 2$$

$$\left(\sqrt[3]{27}\right)^3 = (3)^3 = 27 \quad \text{and} \quad \sqrt[3]{27} = \sqrt[3]{3^3} = 3$$

$$\left(\sqrt[4]{16}\right)^4 = (2)^4 = 16 \quad \text{and} \quad \sqrt[4]{16} = \sqrt[4]{2^4} = 2$$

$$\left(\sqrt[5]{-243}\right)^5 = (-3)^5 = -243 \quad \text{and} \quad \sqrt[5]{-243} = \sqrt[5]{(-3)^5} = -3$$

Inverse Properties of nth Powers and nth Roots

Let a be a real number, and let n be an integer such that $n \geq 2$.

Property	*Example*
1. If a has a principal nth root, then $$\left(\sqrt[n]{a}\right)^n = a.$$	$\left(\sqrt{5}\right)^2 = 5$
2. If n is odd, then $$\sqrt[n]{a^n} = a.$$	$\sqrt[3]{5^3} = 5$
If n is even, then $$\sqrt[n]{a^n} = |a|.$$	$\sqrt{(-5)^2} = |-5| = 5$

Example 3   Evaluating Radical Expressions

Evaluate each radical expression.

a. $\sqrt[3]{4^3}$ **b.** $\sqrt[3]{(-2)^3}$ **c.** $\left(\sqrt{7}\right)^2$

d. $\sqrt{(-3)^2}$ **e.** $\sqrt{-3^2}$

Solution

a. Because the index of the radical is odd, you can write

$$\sqrt[3]{4^3} = 4.$$

b. Because the index of the radical is odd, you can write

$$\sqrt[3]{(-2)^3} = -2.$$

c. Using the inverse property of powers and roots, you can write

$$\left(\sqrt{7}\right)^2 = 7.$$

d. Because the index of the radical is even, you must include absolute value signs, and write

$$\sqrt{(-3)^2} = |-3| = 3.$$

e. Because $\sqrt{-3^2} = \sqrt{-9}$ is an even root of a negative number, its value is not a real number.

Study Tip

In parts (d) and (e) of Example 3, notice that the two expressions inside the radical are different. In part (d), the negative sign is part of the base. In part (e), the negative sign is not part of the base.

② Use the rules of exponents to evaluate or simplify expressions with rational exponents.

Rational Exponents

So far in the text you have worked with algebraic expressions involving only integer exponents. Next you will see that algebraic expressions may also contain **rational exponents.**

Definition of Rational Exponents

Let a be a real number, and let n be an integer such that $n \geq 2$. If the principal nth root of a exists, then $a^{1/n}$ is defined as

$$a^{1/n} = \sqrt[n]{a}.$$

If m is a positive integer that has no common factor with n, then

$$a^{m/n} = (a^{1/n})^m = \left(\sqrt[n]{a}\right)^m \quad \text{and} \quad a^{m/n} = (a^m)^{1/n} = \sqrt[n]{a^m}.$$

It does not matter in which order the two operations are performed, provided the nth root exists. Here is an example.

$$8^{2/3} = \left(\sqrt[3]{8}\right)^2 = 2^2 = 4 \qquad \text{Cube root, then second power}$$
$$8^{2/3} = \sqrt[3]{8^2} = \sqrt[3]{64} = 4 \qquad \text{Second power, then cube root}$$

The rules of exponents that were listed in Section 5.1 also apply to rational exponents (provided the roots indicated by the denominators exist). These rules are listed below, with different examples.

Summary of Rules of Exponents

Let r and s be rational numbers, and let a and b be real numbers, variables, or algebraic expressions. (All denominators and bases are nonzero.)

Product and Quotient Rules — *Example*

1. $a^r \cdot a^s = a^{r+s}$ $\qquad$ $4^{1/2}(4^{1/3}) = 4^{5/6}$

2. $\dfrac{a^r}{a^s} = a^{r-s}$ $\qquad$ $\dfrac{x^2}{x^{1/2}} = x^{2-(1/2)} = x^{3/2}$

Power Rules

3. $(ab)^r = a^r \cdot b^r$ $\qquad$ $(2x)^{1/2} = 2^{1/2}(x^{1/2})$

4. $(a^r)^s = a^{rs}$ $\qquad$ $(x^3)^{1/2} = x^{3/2}$

5. $\left(\dfrac{a}{b}\right)^r = \dfrac{a^r}{b^r}$ $\qquad$ $\left(\dfrac{x}{3}\right)^{2/3} = \dfrac{x^{2/3}}{3^{2/3}}$

Zero and Negative Exponent Rules

6. $a^0 = 1$ $\qquad$ $(3x)^0 = 1$

7. $a^{-r} = \dfrac{1}{a^r}$ $\qquad$ $4^{-3/2} = \dfrac{1}{4^{3/2}} = \dfrac{1}{(2)^3} = \dfrac{1}{8}$

8. $\left(\dfrac{a}{b}\right)^{-r} = \left(\dfrac{b}{a}\right)^r$ $\qquad$ $\left(\dfrac{x}{4}\right)^{-1/2} = \left(\dfrac{4}{x}\right)^{1/2} = \dfrac{2}{x^{1/2}}$

Study Tip

The numerator of a rational exponent denotes the *power* to which the base is raised, and the denominator denotes the *root* to be taken.

Technology: Discovery

Use a calculator to evaluate the expressions below.

$$\dfrac{3.4^{4.6}}{3.4^{3.1}} \quad \text{and} \quad 3.4^{1.5}$$

How are these two expressions related? Use your calculator to verify some of the other rules of exponents.

Example 4   Evaluating Expressions with Rational Exponents

Evaluate each expression.

a. $8^{4/3}$ **b.** $(4^2)^{3/2}$ **c.** $25^{-3/2}$

d. $\left(\dfrac{64}{125}\right)^{2/3}$ **e.** $-16^{1/2}$ **f.** $(-16)^{1/2}$

Solution

a. $8^{4/3} = (8^{1/3})^4 = \left(\sqrt[3]{8}\right)^4 = 2^4 = 16$ Root is 3. Power is 4.

b. $(4^2)^{3/2} = 4^{2\cdot(3/2)} = 4^{6/2} = 4^3 = 64$ Root is 2. Power is 3.

c. $25^{-3/2} = \dfrac{1}{25^{3/2}} = \dfrac{1}{\left(\sqrt{25}\right)^3} = \dfrac{1}{5^3} = \dfrac{1}{125}$ Root is 2. Power is 3.

d. $\left(\dfrac{64}{125}\right)^{2/3} = \dfrac{64^{2/3}}{125^{2/3}} = \dfrac{\left(\sqrt[3]{64}\right)^2}{\left(\sqrt[3]{125}\right)^2} = \dfrac{4^2}{5^2} = \dfrac{16}{25}$ Root is 3. Power is 2.

e. $-16^{1/2} = -\sqrt{16} = -(4) = -4$ Root is 2. Power is 1.

f. $(-16)^{1/2} = \sqrt{-16}$ is not a real number.   Root is 2. Power is 1.

In parts (e) and (f) of Example 4, be sure that you see the distinction between the expressions $-16^{1/2}$ and $(-16)^{1/2}$.

Example 5   Using Rules of Exponents

Rewrite each expression using rational exponents.

a. $x\sqrt[4]{x^3}$ **b.** $\dfrac{\sqrt[3]{x^2}}{\sqrt{x^3}}$ **c.** $\sqrt[3]{x^2 y}$

Solution

a. $x\sqrt[4]{x^3} = x(x^{3/4}) = x^{1+(3/4)} = x^{7/4}$

b. $\dfrac{\sqrt[3]{x^2}}{\sqrt{x^3}} = \dfrac{x^{2/3}}{x^{3/2}} = x^{(2/3)-(3/2)} = x^{-5/6} = \dfrac{1}{x^{5/6}}$

c. $\sqrt[3]{x^2 y} = (x^2 y)^{1/3} = (x^2)^{1/3} y^{1/3} = x^{2/3} y^{1/3}$

Example 6   Using Rules of Exponents

Use rules of exponents to simplify each expression.

a. $\sqrt{\sqrt[3]{x}}$ **b.** $\dfrac{(2x-1)^{4/3}}{\sqrt[3]{2x-1}}$

Solution

a. $\sqrt{\sqrt[3]{x}} = \sqrt{x^{1/3}} = (x^{1/3})^{1/2} = x^{(1/3)(1/2)} = x^{1/6}$

b. $\dfrac{(2x-1)^{4/3}}{\sqrt[3]{2x-1}} = \dfrac{(2x-1)^{4/3}}{(2x-1)^{1/3}} = (2x-1)^{(4/3)-(1/3)} = (2x-1)^{3/3} = 2x-1$

③ Use a calculator to evaluate radical expressions.

Radicals and Calculators

There are two methods of evaluating radicals on most calculators. For square roots, you can use the *square root key* $\boxed{\sqrt{\ }}$ or $\boxed{\sqrt{x}}$. For other roots, you should first convert the radical to exponential form and then use the *exponential key* $\boxed{y^x}$ or $\boxed{\wedge}$.

Technology: Tip

Some calculators have cube root functions $\sqrt[3]{\ }$ or $\sqrt[3]{x}$ and xth root functions $\sqrt[x]{\ }$ or $\sqrt[x]{y}$ that can be used to evaluate roots other than square roots. Consult the user's guide of your calculator for specific keystrokes.

Example 7   Evaluating Roots with a Calculator

Evaluate each expression. Round the result to three decimal places.

a. $\sqrt{5}$ **b.** $\sqrt[5]{25}$ **c.** $\sqrt[3]{-4}$ **d.** $(-8)^{3/2}$

Solution

a. 5 $\boxed{\sqrt{x}}$ Scientific

   $\boxed{\sqrt{\ }}$ 5 $\boxed{\text{ENTER}}$ Graphing

   The display is 2.236067977. Rounded to three decimal places, $\sqrt{5} \approx 2.236$.

b. First rewrite the expression as $\sqrt[5]{25} = 25^{1/5}$. Then use one of the following keystroke sequences.

   25 $\boxed{y^x}$ $\boxed{(}$ 1 $\boxed{\div}$ 5 $\boxed{)}$ $\boxed{=}$ Scientific

   25 $\boxed{\wedge}$ $\boxed{(}$ 1 $\boxed{\div}$ 5 $\boxed{)}$ $\boxed{\text{ENTER}}$ Graphing

   The display is 1.903653939. Rounded to three decimal places, $\sqrt[5]{25} \approx 1.904$.

c. If your calculator does not have a cube root key, use the fact that

$$\sqrt[3]{-4} = \sqrt[3]{(-1)(4)} = \sqrt[3]{-1}\sqrt[3]{4} = -\sqrt[3]{4} = -4^{1/3}$$

   and attach the negative sign of the radical as the last keystroke.

   4 $\boxed{+/-}$ $\boxed{y^x}$ $\boxed{(}$ 1 $\boxed{\div}$ 3 $\boxed{)}$ $\boxed{=}$ Scientific

   $\boxed{\sqrt[3]{\ }}$ $\boxed{(-)}$ 4 $\boxed{)}$ $\boxed{\text{ENTER}}$ Graphing

   The display is −1.587401052. Rounded to three decimal places, $\sqrt[3]{-4} \approx -1.587$.

d. 8 $\boxed{+/-}$ $\boxed{y^x}$ $\boxed{(}$ 3 $\boxed{\div}$ 2 $\boxed{)}$ $\boxed{=}$ Scientific

   $\boxed{(}$ $\boxed{(-)}$ 8 $\boxed{)}$ $\boxed{\wedge}$ $\boxed{(}$ 3 $\boxed{\div}$ 2 $\boxed{)}$ $\boxed{\text{ENTER}}$ Graphing

   The display should indicate an error because an even root of a negative number is not real.

---

④ Evaluate radical functions and find the domains of radical functions.

Radical Functions

A **radical function** is a function that contains a radical such as

$$f(x) = \sqrt{x} \quad \text{or} \quad g(x) = \sqrt[3]{x}.$$

When evaluating a radical function, note that the radical symbol is a symbol of grouping.

Technology: Discovery

Consider the function
$f(x) = x^{2/3}$.

a. What is the domain of the function?

b. Use your graphing calculator to graph each equation, in order.

$y_1 = x^{(2 \div 3)}$

$y_2 = (x^2)^{1/3}$ Power, then root

$y_3 = (x^{1/3})^2$ Root, then power

c. Are the graphs all the same? Are their domains all the same?

d. On your graphing calculator, which of the forms properly represent the function $f(x) = x^{m/n}$?

$y_1 = x^{(m \div n)}$

$y_2 = (x^m)^{1/n}$

$y_3 = (x^{1/n})^m$

e. Explain how the domains of $f(x) = x^{2/3}$ and $g(x) = x^{-2/3}$ differ.

Example 8   Evaluating Radical Functions

Evaluate each radical function when $x = 4$.

a. $f(x) = \sqrt[3]{x - 31}$ **b.** $g(x) = \sqrt{16 - 3x}$

Solution

a. $f(4) = \sqrt[3]{4 - 31} = \sqrt[3]{-27} = -3$

b. $g(4) = \sqrt{16 - 3(4)} = \sqrt{16 - 12} = \sqrt{4} = 2$

The **domain** of the radical function $f(x) = \sqrt[n]{x}$ is the set of all real numbers such that x has a principal nth root.

Domain of a Radical Function

Let n be an integer that is greater than or equal to 2.

1. If n is odd, the domain of $f(x) = \sqrt[n]{x}$ is the set of all real numbers.

2. If n is even, the domain of $f(x) = \sqrt[n]{x}$ is the set of all nonnegative real numbers.

Example 9   Finding the Domains of Radical Functions

Describe the domain of each function.

a. $f(x) = \sqrt[3]{x}$ **b.** $f(x) = \sqrt{x^3}$

Solution

a. The domain of $f(x) = \sqrt[3]{x}$ is the set of all real numbers because for any real number x, the expression $\sqrt[3]{x}$ is a real number.

b. The domain of $f(x) = \sqrt{x^3}$ is the set of all nonnegative real numbers. For instance, 1 is in the domain but -1 is not because $\sqrt{(-1)^3} = \sqrt{-1}$ is not a real number.

Example 10   Finding the Domain of a Radical Function

Find the domain of $f(x) = \sqrt{2x - 1}$.

Solution

The domain of f consists of all x such that $2x - 1 \geq 0$. Using the methods described in Section 2.4, you can solve this inequality as follows.

$2x - 1 \geq 0$ Write original inequality.

$2x \geq 1$ Add 1 to each side.

$x \geq \frac{1}{2}$ Divide each side by 2.

So, the domain is the set of all real numbers x such that $x \geq \frac{1}{2}$.

Study Tip

In general, the domain of a radical function where the index n is even includes all real values for which the expression under the radical is greater than or equal to zero.

7.1 Exercises

Properties and Definitions

In Exercises 1–4, complete the rule of exponents.

1. $a^m \cdot a^n =$

2. $(ab)^m =$

3. $(a^m)^n =$

4. $\dfrac{a^m}{a^n} =$ if $a \neq 0$

Solving Equations

In Exercises 5–8, solve for y.

5. $3x + y = 4$

6. $2x + 3y = 2$

7. $x - 5y = 2y + 7$

8. $\dfrac{3 + 2y}{4} = 6x$

In Exercises 9–12, solve for x.

9. $x(3x + 5) = 0$

10. $2x^2(x - 10) = 0$

11. $x^2 + 6x + 8 = 0$

12. $x^2 - x = 42$

Developing Skills

In Exercises 1–8, find the root if it exists. See Example 1.

1. $\sqrt{64}$

2. $-\sqrt{100}$

3. $-\sqrt{49}$

4. $\sqrt{-25}$

5. $\sqrt[3]{-27}$

6. $\sqrt[3]{-64}$

7. $\sqrt{-1}$

8. $-\sqrt[3]{1}$

In Exercises 9–14, state whether the number is a perfect square, a perfect cube, or neither. See Example 2.

9. 49

10. -27

11. 1728

12. 964

13. 96

14. 225

In Exercises 15–18, determine whether the square root is a rational or irrational number.

15. $\sqrt{6}$

16. $\sqrt{\dfrac{9}{16}}$

17. $\sqrt{900}$

18. $\sqrt{72}$

In Exercises 19–48, evaluate the radical expression without using a calculator. If not possible, state the reason. See Example 3.

19. $\sqrt{8^2}$

20. $-\sqrt{10^2}$

21. $\sqrt{(-10)^2}$

22. $\sqrt{(-12)^2}$

23. $\sqrt{-9^2}$

24. $\sqrt{-12^2}$

25. $-\sqrt{\left(\frac{2}{3}\right)^2}$

26. $\sqrt{\left(\frac{3}{4}\right)^2}$

27. $\sqrt{-\left(\frac{3}{10}\right)^2}$

28. $\sqrt{\left(-\frac{3}{5}\right)^2}$

29. $\left(\sqrt{5}\right)^2$

30. $-\left(\sqrt{10}\right)^2$

31. $-\left(\sqrt{23}\right)^2$

32. $\left(-\sqrt{18}\right)^2$

33. $\sqrt[3]{5^3}$

34. $\sqrt[3]{(-7)^3}$

35. $\sqrt[3]{10^3}$

36. $\sqrt[3]{4^3}$

37. $-\sqrt[3]{(-6)^3}$

38. $-\sqrt[3]{9^3}$

39. $\sqrt[3]{\left(-\frac{1}{4}\right)^3}$

40. $-\sqrt[3]{\left(\frac{1}{5}\right)^3}$

41. $\left(\sqrt[3]{11}\right)^3$

42. $\left(\sqrt[3]{-6}\right)^3$

43. $\left(-\sqrt[3]{24}\right)^3$

44. $\left(\sqrt[3]{21}\right)^3$

45. $\sqrt[4]{3^4}$

46. $\sqrt[5]{(-2)^5}$

47. $-\sqrt[4]{-5^4}$

48. $-\sqrt[4]{2^4}$

In Exercises 49–52, fill in the missing description.

Radical Form	Rational Exponent Form
49. $\sqrt{16} = 4$	
50. $\sqrt[3]{27^2} = 9$	
51.	$125^{1/3} = 5$
52.	$256^{3/4} = 64$

In Exercises 53–68, evaluate without using a calculator. See Example 4.

53. $25^{1/2}$ **54.** $49^{1/2}$

55. $-36^{1/2}$ **56.** $-121^{1/2}$

57. $32^{-2/5}$ **58.** $81^{-3/4}$

59. $(-27)^{-2/3}$ **60.** $(-243)^{-3/5}$

61. $\left(\frac{8}{27}\right)^{2/3}$ **62.** $\left(\frac{256}{625}\right)^{1/4}$

63. $\left(\frac{121}{9}\right)^{-1/2}$ **64.** $\left(\frac{27}{1000}\right)^{-4/3}$

65. $(3^3)^{2/3}$ **66.** $(8^2)^{3/2}$

67. $-(4^4)^{3/4}$ **68.** $(-2^3)^{5/3}$

In Exercises 69–86, rewrite the expression using rational exponents. See Example 5.

69. $\sqrt{t}$ **70.** $\sqrt[3]{x}$

71. $x\sqrt[3]{x^6}$ **72.** $t\sqrt[5]{t^2}$

73. $u^2\sqrt[3]{u}$ **74.** $y\sqrt[4]{y^2}$

75. $\dfrac{\sqrt{x}}{\sqrt{x^3}}$ **76.** $\dfrac{\sqrt[3]{x^2}}{\sqrt[3]{x^4}}$

77. $\dfrac{\sqrt[4]{t}}{\sqrt{t^5}}$ **78.** $\dfrac{\sqrt[3]{x^4}}{\sqrt{x^3}}$

79. $\sqrt[3]{x^2} \cdot \sqrt[3]{x^7}$ **80.** $\sqrt[5]{z^3} \cdot \sqrt[5]{z^2}$

81. $\sqrt[4]{y^3} \cdot \sqrt[3]{y}$ **82.** $\sqrt[6]{x^5} \cdot \sqrt[3]{x^4}$

83. $\sqrt[4]{x^3y}$ **84.** $\sqrt[3]{u^4v^2}$

85. $z^2\sqrt{y^5z^4}$ **86.** $x^2\sqrt[3]{xy^4}$

In Exercises 87–108, simplify the expression. See Example 6.

87. $3^{1/4} \cdot 3^{3/4}$

88. $2^{2/5} \cdot 2^{3/5}$

89. $(2^{1/2})^{2/3}$

90. $(4^{1/3})^{9/4}$

91. $\dfrac{2^{1/5}}{2^{6/5}}$

92. $\dfrac{5^{-3/4}}{5}$

93. $(c^{3/2})^{1/3}$

94. $(k^{-1/3})^{3/2}$

95. $\dfrac{18y^{4/3}z^{-1/3}}{24y^{-2/3}z}$

96. $\dfrac{a^{3/4} \cdot a^{1/2}}{a^{5/2}}$

97. $(3x^{-1/3}y^{3/4})^2$

98. $(-2u^{3/5}v^{-1/5})^3$

99. $\left(\dfrac{x^{1/4}}{x^{1/6}}\right)^3$

100. $\left(\dfrac{3m^{1/6}n^{1/3}}{4n^{-2/3}}\right)^2$

101. $\sqrt{\sqrt[4]{y}}$

102. $\sqrt[3]{\sqrt{2x}}$

103. $\sqrt[4]{\sqrt{x^3}}$

104. $\sqrt[5]{\sqrt[3]{y^4}}$

105. $\dfrac{(x+y)^{3/4}}{\sqrt[4]{x+y}}$

106. $\dfrac{(a-b)^{1/3}}{\sqrt[3]{a-b}}$

107. $\dfrac{(3u-2v)^{2/3}}{\sqrt{(3u-2v)^3}}$

108. $\dfrac{\sqrt[4]{2x+y}}{(2x+y)^{3/2}}$

In Exercises 109–122, use a calculator to evaluate the expression. Round your answer to four decimal places. If not possible, state the reason. See Example 7.

109. $\sqrt{45}$

110. $\sqrt{-23}$

111. $315^{2/5}$

112. $962^{2/3}$

113. $1698^{-3/4}$

114. $382.5^{-3/2}$

115. $\sqrt[4]{212}$

116. $\sqrt[3]{-411}$

117. $\sqrt[3]{545^2}$

118. $\sqrt[5]{-35^3}$

119. $\dfrac{8 - \sqrt{35}}{2}$

120. $\dfrac{-5 + \sqrt{3215}}{10}$

121. $\dfrac{3 + \sqrt{17}}{9}$

122. $\dfrac{7 - \sqrt{241}}{12}$

In Exercises 123–128, evaluate the function as indicated, if possible, and simplify. See Example 8.

123. $f(x) = \sqrt{2x + 9}$

    (a) $f(0)$ (b) $f(8)$ (c) $f(-6)$ (d) $f(36)$

124. $g(x) = \sqrt{5x - 6}$

    (a) $g(0)$ (b) $g(2)$ (c) $g(30)$ (d) $g\left(\frac{7}{5}\right)$

125. $g(x) = \sqrt[3]{x + 1}$

    (a) $g(7)$ (b) $g(26)$ (c) $g(-9)$ (d) $g(-65)$

126. $f(x) = \sqrt[3]{2x - 1}$

    (a) $f(0)$ (b) $f(-62)$ (c) $f(-13)$ (d) $f(63)$

127. $f(x) = \sqrt[4]{x - 3}$

    (a) $f(19)$ (b) $f(1)$ (c) $f(84)$ (d) $f(4)$

128. $g(x) = \sqrt[4]{x + 1}$

    (a) $g(0)$ (b) $g(15)$ (c) $g(-82)$ (d) $g(80)$

In Exercises 129–138, describe the domain of the function. See Examples 9 and 10.

129. $f(x) = 3\sqrt{x}$

130. $h(x) = \sqrt[4]{x}$

131. $g(x) = \dfrac{2}{\sqrt[4]{x}}$

132. $g(x) = \dfrac{10}{\sqrt[3]{x}}$

133. $f(x) = \sqrt[3]{x^4}$

134. $f(x) = \sqrt{-x}$

135. $h(x) = \sqrt{3x + 7}$

136. $f(x) = \sqrt{8x - 1}$

137. $g(x) = \sqrt{4 - 9x}$

138. $g(x) = \sqrt{10 - 2x}$

In Exercises 139–142, describe the domain of the function algebraically. Use a graphing calculator to graph the function. Did the graphing calculator omit part of the domain? If so, complete the graph by hand.

139. $y = \dfrac{5}{\sqrt[4]{x^3}}$

140. $y = 4\sqrt[3]{x}$

141. $g(x) = 2x^{3/5}$

142. $h(x) = 5x^{2/3}$

In Exercises 143–146, perform the multiplication. Use a graphing calculator to confirm your result.

143. $x^{1/2}(2x - 3)$

144. $x^{4/3}(3x^2 - 4x + 5)$

145. $y^{-1/3}(y^{1/3} + 5y^{4/3})$

146. $(x^{1/2} - 3)(x^{1/2} + 3)$

Solving Problems

Mathematical Modeling   In Exercises 147 and 148, use the formula for the *declining balances method*

$$r = 1 - \left(\frac{S}{C}\right)^{1/n}$$

to find the depreciation rate r. In the formula, n is the useful life of the item (in years), S is the salvage value (in dollars), and C is the original cost (in dollars).

147. A \$75,000 truck depreciates over an eight-year period, as shown in the graph. Find r.

Figure for 147

148. A \$125,000 printing press depreciates over a 10-year period, as shown in the graph. Find r.

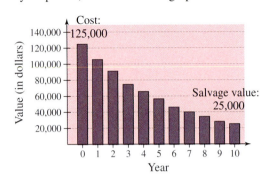

149. ▲ *Geometry* Find the dimensions of a piece of carpet for a classroom with 529 square feet of floor space, assuming the floor is square.

150. ▲ *Geometry* Find the dimensions of a square mirror with an area of 1024 square inches.

151. ▲ *Geometry* The length D of a diagonal of a rectangular solid of length l, width w, and height h is represented by $D = \sqrt{l^2 + w^2 + h^2}$. Approximate to two decimal places the length of D of the solid shown in the figure.

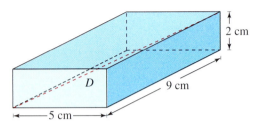

152. *Velocity* A stream of water moving at a rate of v feet per second can carry particles of size $0.03\sqrt{v}$ inches.

(a) Find the particle size that can be carried by a stream flowing at the rate of $\frac{3}{4}$ foot per second. Round your answer to three decimal places.

(b) Find the particle size that can be carried by a stream flowing at the rate of $\frac{3}{16}$ foot per second. Round your answer to three decimal places.

Explaining Concepts

153. ✹ Answer parts (a)–(d) of Motivating the Chapter on page 444.

154. *Writing* In your own words, define the nth root of a number.

155. *Writing* Define the *radicand* and the *index* of a radical.

156. *Writing* If n is even, what must be true about the radicand for the nth root to be a real number? Explain.

157. *Writing* Is it true that $\sqrt{2} = 1.414$? Explain.

158. Given a real number x, state the conditions on n for each of the following.

(a) $\sqrt[n]{x^n} = x$

(b) $\sqrt[n]{x^n} = |x|$

159. *Investigation* Find all possible "last digits" of perfect squares. (For instance, the last digit of 81 is 1 and the last digit of 64 is 4.) Is it possible that 4,322,788,986 is a perfect square?

7.2 Simplifying Radical Expressions

Fundamental Photographs

What You Should Learn

1 Use the Product and Quotient Rules for Radicals to simplify radical expressions.

2 Use rationalization techniques to simplify radical expressions.

3 Use the Pythagorean Theorem in application problems.

Why You Should Learn It

Algebraic equations often involve radicals. For instance, in Exercise 74 on page 463, you will use a radical equation to find the period of a pendulum.

Simplifying Radicals

In this section, you will study ways to simplify radicals. For instance, the expression $\sqrt{12}$ can be simplified as

$$\sqrt{12} = \sqrt{4 \cdot 3} = \sqrt{4}\sqrt{3} = 2\sqrt{3}.$$

This rewritten form is based on the following rules for multiplying and dividing radicals.

1 Use the Product and Quotient Rules for Radicals to simplify radical expressions.

Product and Quotient Rules for Radicals

Let u and v be real numbers, variables, or algebraic expressions. If the nth roots of u and v are real, the following rules are true.

1. $\sqrt[n]{uv} = \sqrt[n]{u}\sqrt[n]{v}$ Product Rule for Radicals

2. $\sqrt[n]{\dfrac{u}{v}} = \dfrac{\sqrt[n]{u}}{\sqrt[n]{v}}, \quad v \neq 0$ Quotient Rule for Radicals

Study Tip

The Product and Quotient Rules for Radicals can be shown to be true by converting the radicals to exponential form and using the rules of exponents on page 449.

Using Rule 3

$$\sqrt[n]{uv} = (uv)^{1/n}$$
$$= u^{1/n}v^{1/n}$$
$$= \sqrt[n]{u}\sqrt[n]{v}$$

Using Rule 5

$$\sqrt[n]{\frac{u}{v}} = \left(\frac{u}{v}\right)^{1/n}$$
$$= \frac{u^{1/n}}{v^{1/n}} = \frac{\sqrt[n]{u}}{\sqrt[n]{v}}$$

You can use the Product Rule for Radicals to *simplify* square root expressions by finding the largest perfect square factor and removing it from the radical, as follows.

$$\sqrt{48} = \sqrt{16 \cdot 3} = \sqrt{16}\sqrt{3} = 4\sqrt{3}$$

This process is called **removing perfect square factors from the radical.**

Example 1   Removing Constant Factors from Radicals

Simplify each radical by removing as many factors as possible.

a. $\sqrt{75}$ **b.** $\sqrt{72}$ **c.** $\sqrt{162}$

Solution

a. $\sqrt{75} = \sqrt{25 \cdot 3} = \sqrt{25}\sqrt{3} = 5\sqrt{3}$ 25 is a perfect square factor of 75.

b. $\sqrt{72} = \sqrt{36 \cdot 2} = \sqrt{36}\sqrt{2} = 6\sqrt{2}$ 36 is a perfect square factor of 72.

c. $\sqrt{162} = \sqrt{81 \cdot 2} = \sqrt{81}\sqrt{2} = 9\sqrt{2}$ 81 is a perfect square factor of 162.

When removing *variable* factors from a square root radical, remember that it is not valid to write $\sqrt{x^2} = x$ *unless* you happen to know that x is nonnegative. Without knowing anything about x, the only way you can simplify $\sqrt{x^2}$ is to include absolute value signs when you remove x from the radical.

$$\sqrt{x^2} = |x| \qquad \text{Restricted by absolute value signs}$$

When simplifying the expression $\sqrt{x^3}$, it is not necessary to include absolute value signs because the domain does not include negative numbers.

$$\sqrt{x^3} = \sqrt{x^2(x)} = x\sqrt{x} \qquad \text{Restricted by domain of radical}$$

Example 2 Removing Variable Factors from Radicals

Simplify each radical expression.

a. $\sqrt{25x^2}$ **b.** $\sqrt{12x^3}, \quad x \geq 0$ **c.** $\sqrt{144x^4}$ **d.** $\sqrt{72x^3y^2}$

Solution

a. $\sqrt{25x^2} = \sqrt{5^2x^2} = \sqrt{5^2}\sqrt{x^2}$ — Product Rule for Radicals

$\qquad = 5|x|$ — $\sqrt{x^2} = |x|$

b. $\sqrt{12x^3} = \sqrt{2^2x^2(3x)} = \sqrt{2^2}\sqrt{x^2}\sqrt{3x}$ — Product Rule for Radicals

$\qquad = 2x\sqrt{3x}$ — $\sqrt{2^2}\sqrt{x^2} = 2x, \quad x \geq 0$

c. $\sqrt{144x^4} = \sqrt{12^2(x^2)^2} = \sqrt{12^2}\sqrt{(x^2)^2}$ — Product Rule for Radicals

$\qquad = 12x^2$ — $\sqrt{12^2}\sqrt{(x^2)^2} = 12|x^2| = 12x^2$

d. $\sqrt{72x^3y^2} = \sqrt{6^2x^2y^2} \cdot \sqrt{2x}$ — Product Rule for Radicals

$\qquad = \sqrt{6^2}\sqrt{x^2}\sqrt{y^2} \cdot \sqrt{2x}$ — Product Rule for Radicals

$\qquad = 6x|y|\sqrt{2x}$ — $\sqrt{6^2}\sqrt{x^2}\sqrt{y^2} = 6x|y|$

In the same way that perfect squares can be removed from square root radicals, perfect nth powers can be removed from nth root radicals.

Example 3 Removing Factors from Radicals

Simplify each radical expression.

a. $\sqrt[3]{40}$ **b.** $\sqrt[4]{x^5}, \quad x \geq 0$

Solution

a. $\sqrt[3]{40} = \sqrt[3]{8(5)} = \sqrt[3]{2^3} \cdot \sqrt[3]{5}$ — Product Rule for Radicals

$\qquad = 2\sqrt[3]{5}$ — $\sqrt[3]{2^3} = 2$

b. $\sqrt[4]{x^5} = \sqrt[4]{x^4(x)} = \sqrt[4]{x^4}\sqrt[4]{x}$ — Product Rule for Radicals

$\qquad = x\sqrt[4]{x}$ — $\sqrt[4]{x^4} = x, \quad x \geq 0$

Study Tip

To find the perfect nth root factor of 486 in Example 4(a), you can write the prime factorization of 486.

$$486 = 2 \cdot 3 \cdot 3 \cdot 3 \cdot 3 \cdot 3$$
$$= 2 \cdot 3^5$$

From its prime factorization, you can see that 3^5 is a fifth root factor of 486.

$$\sqrt[5]{486} = \sqrt[5]{2 \cdot 3^5}$$
$$= \sqrt[5]{3^5}\,\sqrt[5]{2}$$
$$= 3\sqrt[5]{2}$$

Example 4   Removing Factors from Radicals

Simplify each radical expression.

a. $\sqrt[5]{486x^7}$ **b.** $\sqrt[3]{128x^3y^5}$

Solution

a. $\sqrt[5]{486x^7} = \sqrt[5]{243x^5(2x^2)}$

$\qquad\qquad = \sqrt[5]{3^5x^5} \cdot \sqrt[5]{2x^2}$ Product Rule for Radicals

$\qquad\qquad = 3x\,\sqrt[5]{2x^2}$ $\sqrt[5]{3^5}\,\sqrt[5]{x^5} = 3x$

b. $\sqrt[3]{128x^3y^5} = \sqrt[3]{64x^3y^3(2y^2)}$

$\qquad\qquad = \sqrt[3]{4^3x^3y^3} \cdot \sqrt[3]{2y^2}$ Product Rule for Radicals

$\qquad\qquad = 4xy\,\sqrt[3]{2y^2}$ $\sqrt[3]{4^3}\,\sqrt[3]{x^3}\,\sqrt[3]{y^3} = 4xy$

Example 5   Removing Factors from Radicals

Simplify each radical expression.

a. $\sqrt{\dfrac{81}{25}}$ **b.** $\dfrac{\sqrt{56x^2}}{\sqrt{8}}$

Solution

a. $\sqrt{\dfrac{81}{25}} = \dfrac{\sqrt{81}}{\sqrt{25}} = \dfrac{9}{5}$ Quotient Rule for Radicals

b. $\dfrac{\sqrt{56x^2}}{\sqrt{8}} = \sqrt{\dfrac{56x^2}{8}}$ Quotient Rule for Radicals

$\qquad\quad = \sqrt{7x^2}$ Simplify.

$\qquad\quad = \sqrt{7} \cdot \sqrt{x^2}$ Product Rule for Radicals

$\qquad\quad = \sqrt{7}\,|x|$ $\sqrt{x^2} = |x|$

Example 6   Removing Factors from Radicals

Simplify the radical expression.

$$-\sqrt[3]{\dfrac{y^5}{27x^3}}$$

Solution

$$-\sqrt[3]{\dfrac{y^5}{27x^3}} = -\dfrac{\sqrt[3]{y^3y^2}}{\sqrt[3]{27x^3}}$$ Quotient Rule for Radicals

$$= -\dfrac{\sqrt[3]{y^3} \cdot \sqrt[3]{y^2}}{\sqrt[3]{27} \cdot \sqrt[3]{x^3}}$$ Product Rule for Radicals

$$= -\dfrac{y\,\sqrt[3]{y^2}}{3x}$$ Simplify.

② Use rationalization techniques to simplify radical expressions.

Rationalization Techniques

Removing factors from radicals is only one of two techniques used to simplify radicals. Three conditions must be met in order for a radical expression to be in simplest form. These three conditions are summarized as follows.

Simplifying Radical Expressions

A radical expression is said to be in *simplest form* if all three of the statements below are true.

1. All possible nth powered factors have been removed from each radical.

2. No radical contains a fraction.

3. No denominator of a fraction contains a radical.

To meet the last two conditions, you can use a second technique for simplifying radical expressions called **rationalizing the denominator.** This involves multiplying both the numerator and denominator by a *rationalizing factor* that creates a perfect nth power in the denominator.

Study Tip

When rationalizing a denominator, remember that for square roots you want a perfect square in the denominator, for cube roots you want a perfect cube, and so on. For instance, to find the rationalizing factor needed to create a perfect square in the denominator of Example 7(c) you can write the prime factorization of 18.

$$18 = 2 \cdot 3 \cdot 3$$

$$= 2 \cdot 3^2$$

From its prime factorization, you can see that 3^2 is a square root factor of 18. You need one more factor of 2 to create a perfect square in the denominator:

$$2 \cdot (2 \cdot 3^2) = 2 \cdot 2 \cdot 3^2$$

$$= 2^2 \cdot 3^2$$

$$= 4 \cdot 9 = 36.$$

Example 7 Rationalizing the Denominator

Rationalize the denominator in each expression.

a. $\sqrt{\dfrac{3}{5}}$ **b.** $\dfrac{4}{\sqrt[3]{9}}$ **c.** $\dfrac{8}{3\sqrt{18}}$

Solution

a. $\sqrt{\dfrac{3}{5}} = \dfrac{\sqrt{3}}{\sqrt{5}} = \dfrac{\sqrt{3}}{\sqrt{5}} \cdot \dfrac{\sqrt{5}}{\sqrt{5}} = \dfrac{\sqrt{15}}{\sqrt{5^2}} = \dfrac{\sqrt{15}}{5}$ Multiply by $\sqrt{5}/\sqrt{5}$ to create a perfect square in the denominator.

b. $\dfrac{4}{\sqrt[3]{9}} = \dfrac{4}{\sqrt[3]{9}} \cdot \dfrac{\sqrt[3]{3}}{\sqrt[3]{3}} = \dfrac{4\sqrt[3]{3}}{\sqrt[3]{27}} = \dfrac{4\sqrt[3]{3}}{3}$ Multiply by $\sqrt[3]{3}/\sqrt[3]{3}$ to create a perfect cube in the denominator.

c. $\dfrac{8}{3\sqrt{18}} = \dfrac{8}{3\sqrt{18}} \cdot \dfrac{\sqrt{2}}{\sqrt{2}} = \dfrac{8\sqrt{2}}{3\sqrt{36}} = \dfrac{8\sqrt{2}}{3\sqrt{6^2}}$ Multiply by $\sqrt{2}/\sqrt{2}$ to create a perfect square in the denominator.

$$= \dfrac{8\sqrt{2}}{3(6)} = \dfrac{4\sqrt{2}}{9}$$

Example 8 Rationalizing the Denominator

a. $\sqrt{\dfrac{8x}{12y^5}} = \sqrt{\dfrac{(4)(2)x}{(4)(3)y^5}} = \sqrt{\dfrac{2x}{3y^5}} = \dfrac{\sqrt{2x}}{\sqrt{3y^5}} \cdot \dfrac{\sqrt{3y}}{\sqrt{3y}} = \dfrac{\sqrt{6xy}}{\sqrt{3^2y^6}} = \dfrac{\sqrt{6xy}}{3|y^3|}$

b. $\sqrt[3]{\dfrac{54x^6y^3}{5z^2}} = \dfrac{\sqrt[3]{(3^3)(2)(x^6)(y^3)}}{\sqrt[3]{5z^2}} \cdot \dfrac{\sqrt[3]{25z}}{\sqrt[3]{25z}} = \dfrac{3x^2y\sqrt[3]{50z}}{\sqrt[3]{5^3z^3}} = \dfrac{3x^2y\sqrt[3]{50z}}{5z}$

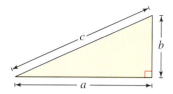

③ Use the Pythagorean Theorem in application problems.

Figure 7.1

Applications of Radicals

Radicals commonly occur in applications involving right triangles. Recall that a right triangle is one that contains a right (or 90°) angle, as shown in Figure 7.1. The relationship among the three sides of a right triangle is described by the **Pythagorean Theorem,** which states that if a and b are the lengths of the legs and c is the length of the hypotenuse, then

$$c = \sqrt{a^2 + b^2} \text{ and } a = \sqrt{c^2 - b^2}.$$

Pythagorean Theorem: $a^2 + b^2 = c^2$

Example 9   The Pythagorean Theorem

Find the length of the hypotenuse of the right triangle shown in Figure 7.2.

Solution

Because you know that $a = 6$ and $b = 9$, you can use the Pythagorean Theorem to find c as follows.

$$
\begin{aligned}
c &= \sqrt{a^2 + b^2} && \text{Pythagorean Theorem} \\
&= \sqrt{6^2 + 9^2} && \text{Substitute 6 for } a \text{ and 9 for } b. \\
&= \sqrt{117} && \text{Simplify.} \\
&= \sqrt{9}\sqrt{13} && \text{Product Rule for Radicals} \\
&= 3\sqrt{13} && \text{Simplify.}
\end{aligned}
$$

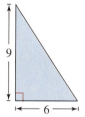

Figure 7.2

Example 10   An Application of the Pythagorean Theorem

A softball diamond has the shape of a square with 60-foot sides, as shown in Figure 7.3. The catcher is 5 feet behind home plate. How far does the catcher have to throw to reach second base?

Solution

In Figure 7.3, let x be the hypotenuse of a right triangle with 60-foot sides. So, by the Pythagorean Theorem, you have the following.

$$
\begin{aligned}
x &= \sqrt{60^2 + 60^2} && \text{Pythagorean Theorem} \\
&= \sqrt{7200} && \text{Simplify.} \\
&= \sqrt{3600}\sqrt{2} && \text{Product Rule for Radicals} \\
&= 60\sqrt{2} && \text{Simplify.} \\
&\approx 84.9 \text{ feet} && \text{Use a calculator.}
\end{aligned}
$$

So, the distance from home plate to second base is approximately 84.9 feet. Because the catcher is 5 feet behind home plate, the catcher must make a throw of

$$x + 5 \approx 84.9 + 5 = 89.9 \text{ feet.}$$

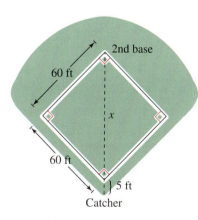

Figure 7.3

7.2 Exercises

Review   Concepts, Skills, and Problem Solving

Keep mathematically in shape by doing these exercises *before* the problems of this section.

Properties and Definitions

1. *Writing* ✎   Explain how to determine the half-plane satisfying $x - y > -3$.

2. *Writing* ✎   Describe the difference between the graphs of $3x + 4y \leq 4$ and $3x + 4y < 4$.

Factoring

In Exercises 3–8, factor the expression completely.

3. $-x^3 + 3x^2 - x + 3$ 4. $4t^2 - 169$

5. $x^2 - 3x + 2$

6. $2x^2 + 5x - 7$

7. $11x^2 + 6x - 5$

8. $4x^2 - 28x + 49$

Problem Solving

9. *Ticket Sales*   Twelve hundred tickets were sold for a theater production, and the receipts for the performance totaled $21,120. The tickets for adults and students sold for $20 and $12.50, respectively. How many of each kind of ticket were sold?

10. *Quality Control*   A quality control engineer for a buyer found two defective units in a sample of 75. At that rate, what is the expected number of defective units in a shipment of 10,000 units?

Developing Skills

In Exercises 1–18, simplify the radical. (Do not use a calculator.) See Example 1.

1. $\sqrt{20}$ 2. $\sqrt{27}$

3. $\sqrt{50}$ 4. $\sqrt{125}$

5. $\sqrt{96}$ 6. $\sqrt{84}$

7. $\sqrt{216}$ 8. $\sqrt{147}$

9. $\sqrt{1183}$ 10. $\sqrt{1176}$

11. $\sqrt{0.04}$ 12. $\sqrt{0.25}$

13. $\sqrt{0.0072}$ 14. $\sqrt{0.0027}$

15. $\sqrt{2.42}$ 16. $\sqrt{9.8}$

17. $\sqrt{\frac{13}{25}}$ 18. $\sqrt{\frac{15}{36}}$

In Exercises 19–52, simplify the radical expression. See Examples 2–6.

19. $\sqrt{9x^5}$ 20. $\sqrt{64x^3}$

21. $\sqrt{48y^4}$ 22. $\sqrt{32x}$

23. $\sqrt{117y^6}$ 24. $\sqrt{160x^8}$

25. $\sqrt{120x^2y^3}$ 26. $\sqrt{125u^4v^6}$

27. $\sqrt{192a^5b^7}$ 28. $\sqrt{363x^{10}y^9}$

29. $\sqrt[3]{48}$ 30. $\sqrt[3]{81}$

31. $\sqrt[3]{112}$ 32. $\sqrt[4]{112}$

33. $\sqrt[3]{40x^5}$ 34. $\sqrt[3]{54z^7}$

35. $\sqrt[4]{324y^6}$ 36. $\sqrt[5]{160x^8}$

37. $\sqrt[3]{x^4y^3}$ 38. $\sqrt[3]{a^5b^6}$

39. $\sqrt[4]{3x^4y^2}$ 40. $\sqrt[4]{128u^4v^7}$

41. $\sqrt[5]{32x^5y^6}$ 42. $\sqrt[3]{16x^4y^5}$

43. $\sqrt[3]{\frac{35}{64}}$ 44. $\sqrt[4]{\frac{5}{16}}$

45. $\sqrt[5]{\frac{32x^2}{y^5}}$ 46. $\sqrt[3]{\frac{16z^3}{y^6}}$

47. $\sqrt[3]{\frac{54a^4}{b^9}}$ 48. $\sqrt[4]{\frac{3u^2}{16v^8}}$

49. $\sqrt{\frac{32a^4}{b^2}}$ 50. $\sqrt{\frac{18x^2}{z^6}}$

51. $\sqrt[4]{(3x^2)^4}$ 52. $\sqrt[5]{96x^5}$

In Exercises 53–70, rationalize the denominator and simplify further, if possible. See Examples 7 and 8.

53. $\sqrt{\frac{1}{3}}$ 54. $\sqrt{\frac{1}{5}}$

55. $\frac{1}{\sqrt{7}}$ 56. $\frac{12}{\sqrt{3}}$

57. $\sqrt[4]{\dfrac{5}{4}}$ **58.** $\sqrt[3]{\dfrac{9}{25}}$

59. $\dfrac{6}{\sqrt[3]{32}}$ **60.** $\dfrac{10}{\sqrt[5]{16}}$

61. $\dfrac{1}{\sqrt{y}}$ **62.** $\sqrt{\dfrac{5}{c}}$

63. $\sqrt{\dfrac{4}{x}}$ **64.** $\sqrt{\dfrac{4}{x^3}}$

65. $\dfrac{1}{\sqrt{2x}}$ **66.** $\dfrac{5}{\sqrt{8x^5}}$

67. $\dfrac{6}{\sqrt{3b^3}}$ **68.** $\dfrac{1}{\sqrt{xy}}$

69. $\sqrt[3]{\dfrac{2x}{3y}}$ **70.** $\sqrt[3]{\dfrac{20x^2}{9y^2}}$

 Geometry   In Exercises 71 and 72, find the length of the hypotenuse of the right triangle. See Example 9.

71.   **72.**

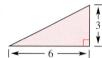 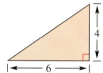

Solving Problems

73. *Frequency*   The frequency f in cycles per second of a vibrating string is given by

$$f = \dfrac{1}{100}\sqrt{\dfrac{400 \times 10^6}{5}}.$$

Use a calculator to approximate this number. (Round the result to two decimal places.)

74. *Period of a Pendulum*   The time t (in seconds) for a pendulum of length L (in feet) to go through one complete cycle (its period) is given by

$$t = 2\pi\sqrt{\dfrac{L}{32}}.$$

Find the period of a pendulum whose length is 4 feet. (Round your answer to two decimal places.)

75. *Geometry*   A ladder is to reach a window that is 26 feet high. The ladder is placed 10 feet from the base of the wall (see figure). How long must the ladder be?

Figure for 75

Figure for 76

76. *Geometry*   A string is attached to opposite corners of a piece of wood that is 6 inches wide and 14 inches long (see figure). How long must the string be?

77. *Investigation*   Enter any positive real number into your calculator and find its square root. Then repeatedly take the square root of the result.

$$\sqrt{x},\ \sqrt{\sqrt{x}},\ \sqrt{\sqrt{\sqrt{x}}},\ \dots$$

What real number does the display appear to be approaching?

Explaining Concepts

78. Give an example of multiplying two radicals.

79. *Writing*   Describe the three conditions that characterize a simplified radical expression.

80. *Writing*   Describe how you would simplify $1/\sqrt{3}$.

81. *Writing*   When is $\sqrt{x^2} \neq x$? Explain.

82. Square the real number $5/\sqrt{3}$ and note that the radical is eliminated from the denominator. Is this equivalent to rationalizing the denominator? Why or why not?

7.3 Adding and Subtracting Radical Expressions

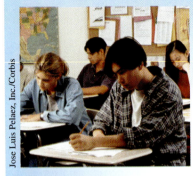

What You Should Learn

1. Use the Distributive Property to add and subtract like radicals.
2. Use radical expressions in application problems.

Why You Should Learn It

Radical expressions can be used to model and solve real-life problems. For instance, Example 6 on page 466 shows how to find a radical expression that models the total number of SAT and ACT tests taken.

Adding and Subtracting Radical Expressions

Two or more radical expressions are called **like radicals** if they have the same index and the same radicand. For instance, the expressions $\sqrt{2}$ and $3\sqrt{2}$ are like radicals, whereas the expressions $\sqrt{3}$ and $\sqrt[3]{3}$ are not. Two radical expressions that are like radicals can be added or subtracted by adding or subtracting their coefficients.

1. Use the Distributive Property to add and subtract like radicals.

Example 1   Combining Radical Expressions

Simplify each expression by combining like terms.

a. $\sqrt{7} + 5\sqrt{7} - 2\sqrt{7}$
b. $6\sqrt{x} - \sqrt[3]{4} - 5\sqrt{x} + 2\sqrt[3]{4}$
c. $3\sqrt[3]{x} + 2\sqrt[3]{x} + \sqrt{x} - 8\sqrt{x}$

Solution

a. $\sqrt{7} + 5\sqrt{7} - 2\sqrt{7} = (1 + 5 - 2)\sqrt{7}$ Distributive Property

$= 4\sqrt{7}$ Simplify.

b. $6\sqrt{x} - \sqrt[3]{4} - 5\sqrt{x} + 2\sqrt[3]{4}$

$= \left(6\sqrt{x} - 5\sqrt{x}\right) + \left(-\sqrt[3]{4} + 2\sqrt[3]{4}\right)$ Group like terms.

$= (6 - 5)\sqrt{x} + (-1 + 2)\sqrt[3]{4}$ Distributive Property

$= \sqrt{x} + \sqrt[3]{4}$ Simplify.

c. $3\sqrt[3]{x} + 2\sqrt[3]{x} + \sqrt{x} - 8\sqrt{x}$

$= (3 + 2)\sqrt[3]{x} + (1 - 8)\sqrt{x}$ Distributive Property

$= 5\sqrt[3]{x} - 7\sqrt{x}$ Simplify.

Study Tip

It is important to realize that the expression $\sqrt{a} + \sqrt{b}$ is not equal to $\sqrt{a + b}$. For instance, you may be tempted to add $\sqrt{6} + \sqrt{3}$ and get $\sqrt{9} = 3$. But remember, you cannot add unlike radicals. So, $\sqrt{6} + \sqrt{3}$ cannot be simplified further.

Before concluding that two radicals cannot be combined, you should first rewrite them in simplest form. This is illustrated in Examples 2 and 3.

Example 2   Simplifying Before Combining Radical Expressions

Simplify each expression by combining like terms.

a. $\sqrt{45x} + 3\sqrt{20x}$

b. $5\sqrt{x^3} - x\sqrt{4x}$

Solution

a. $\sqrt{45x} + 3\sqrt{20x} = 3\sqrt{5x} + 6\sqrt{5x}$ Simplify radicals.

$\qquad\qquad\qquad\qquad = 9\sqrt{5x}$ Combine like radicals.

b. $5\sqrt{x^3} - x\sqrt{4x} = 5x\sqrt{x} - 2x\sqrt{x}$ Simplify radicals.

$\qquad\qquad\qquad\qquad = 3x\sqrt{x}$ Combine like radicals.

Example 3   Simplifying Before Combining Radical Expressions

Simplify each expression by combining like terms.

a. $\sqrt[3]{54y^5} + 4\sqrt[3]{2y^2}$

b. $\sqrt[3]{6x^4} + \sqrt[3]{48x} - \sqrt[3]{162x^4}$

Solution

a. $\sqrt[3]{54y^5} + 4\sqrt[3]{2y^2} = 3y\sqrt[3]{2y^2} + 4\sqrt[3]{2y^2}$ Simplify radicals.

$\qquad\qquad\qquad\qquad = (3y + 4)\sqrt[3]{2y^2}$ Combine like radicals.

b. $\sqrt[3]{6x^4} + \sqrt[3]{48x} - \sqrt[3]{162x^4}$ Write original expression.

$\qquad = x\sqrt[3]{6x} + 2\sqrt[3]{6x} - 3x\sqrt[3]{6x}$ Simplify radicals.

$\qquad = (x + 2 - 3x)\sqrt[3]{6x}$ Distributive Property

$\qquad = (2 - 2x)\sqrt[3]{6x}$ Combine like terms.

In some instances, it may be necessary to rationalize denominators before combining radicals.

Example 4   Rationalizing Denominators Before Simplifying

$$\sqrt{7} - \frac{5}{\sqrt{7}} = \sqrt{7} - \left(\frac{5}{\sqrt{7}} \cdot \frac{\sqrt{7}}{\sqrt{7}}\right)$$ Multiply by $\sqrt{7}/\sqrt{7}$ to create a perfect square in the denominator.

$$= \sqrt{7} - \frac{5\sqrt{7}}{7}$$ Simplify.

$$= \left(1 - \frac{5}{7}\right)\sqrt{7}$$ Distributive Property

$$= \frac{2}{7}\sqrt{7}$$ Simplify.

2 Use radical expressions in application problems.

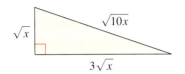

Figure 7.4

Applications

Example 5   Geometry: Perimeter of a Triangle

Write and simplify an expression for the perimeter of the triangle shown in Figure 7.4.

Solution

$$P = a + b + c \qquad \text{Formula for perimeter of a triangle}$$

$$= \sqrt{x} + 3\sqrt{x} + \sqrt{10x} \qquad \text{Substitute.}$$

$$= (1 + 3)\sqrt{x} + \sqrt{10x} \qquad \text{Distributive Property}$$

$$= 4\sqrt{x} + \sqrt{10x} \qquad \text{Simplify.}$$

Example 6   SAT and ACT Participants

The number S (in thousands) of SAT tests taken and the number A (in thousands) of ACT tests taken from 1996 to 2001 can be modeled by the equations

$$S = 496 + 239.7\sqrt{t}, \quad 6 \leq t \leq 11 \qquad \text{SAT tests}$$

$$A = 494 + 176.5\sqrt{t}, \quad 6 \leq t \leq 11 \qquad \text{ACT tests}$$

where t represents the year, with $t = 6$ corresponding to 1996. Find a radical expression that models the total number T of SAT and ACT tests taken from 1996 to 2001. Estimate the total number T of SAT and ACT tests taken in 2000. (Source: College Entrance Examination Board and ACT, Inc.)

Solution

The sum of the two models is as follows.

$$S + A = 496 + 239.7\sqrt{t} + 494 + 176.5\sqrt{t}$$

$$= \left(239.7\sqrt{t} + 176.5\sqrt{t}\right) + (496 + 494)$$

$$= 416.2\sqrt{t} + 990$$

So, the radical expression that models the total number of SAT and ACT tests taken is

$$T = S + A$$

$$= 416.2\sqrt{t} + 990.$$

Using this model, substitute $t = 10$ to estimate the total number of SAT and ACT tests taken in 2000.

$$T = 416.2\sqrt{10} + 990$$

$$\approx 2306$$

7.3 Exercises

Review *Concepts, Skills, and Problem Solving*

Keep mathematically in shape by doing these exercises *before* the problems of this section.

Properties and Definitions

1. *Writing* 🖉   Explain what is meant by a solution to a system of linear equations.

2. *Writing* 🖉   Is it possible for a system of linear equations to have no solution? Explain.

3. *Writing* 🖉   Is it possible for a system of linear equations to have infinitely many solutions? Explain.

4. *Writing* 🖉   Is it possible for a system of linear equations to have exactly two solutions? Explain.

Solving Systems of Equations

In Exercises 5 and 6, sketch the graphs of the equations and approximate any solutions of the system of linear equations.

5. $\begin{cases} 3x + 2y = -4 \\ y = 3x + 7 \end{cases}$

6. $\begin{cases} 2x + 3y = 12 \\ 4x - y = 10 \end{cases}$

In Exercises 7 and 8, solve the system by the method of substitution.

7. $\begin{cases} x - 3y = -2 \\ 7y - 4x = 6 \end{cases}$

8. $\begin{cases} y = x + 2 \\ y - x = 8 \end{cases}$

In Exercises 9 and 10, solve the system by the method of elimination.

9. $\begin{cases} 1.5x - 3 = -2y \\ 3x + 4y = 6 \end{cases}$

10. $\begin{cases} x + 4y + 3z = 2 \\ 2x + y + z = 10 \\ -x + y + 2z = 8 \end{cases}$

Problem Solving

11. *Cost*   Two DVDs and one videocassette tape cost $72. One DVD and two videocassette tapes cost $57. What is the price of each item?

12. *Money*   A collection of $20, $5, and $1 bills totals $159. There are as many $20 bills as there are $5 and $1 bills combined. There are 14 bills in total. How many of each type of bill are there?

Developing Skills

In Exercises 1–46, combine the radical expressions, if possible. See Examples 1–3.

1. $3\sqrt{2} - \sqrt{2}$

2. $6\sqrt{5} - 2\sqrt{5}$

3. $4\sqrt{32} + 7\sqrt{32}$

4. $3\sqrt{7} + 2\sqrt{7}$

5. $8\sqrt{5} + 9\sqrt[3]{5}$

6. $12\sqrt{8} - 3\sqrt[3]{8}$

7. $9\sqrt[3]{5} - 6\sqrt[3]{5}$

8. $14\sqrt[5]{2} - 6\sqrt[5]{2}$

9. $4\sqrt[3]{y} + 9\sqrt[3]{y}$

10. $13\sqrt{x} + \sqrt{x}$

11. $15\sqrt[4]{s} - \sqrt[4]{s}$

12. $9\sqrt[4]{t} - 3\sqrt[4]{t}$

13. $8\sqrt{2} + 6\sqrt{2} - 5\sqrt{2}$

14. $2\sqrt{6} + 8\sqrt{6} - 3\sqrt{6}$

15. $\sqrt[4]{3} - 5\sqrt[4]{7} - 12\sqrt[4]{3}$

16. $9\sqrt[3]{17} + 7\sqrt[3]{2} - 4\sqrt[3]{17} + \sqrt[3]{2}$

17. $9\sqrt[3]{7} - \sqrt{3} + 4\sqrt[3]{7} + 2\sqrt{3}$

18. $5\sqrt{7} - 8\sqrt[4]{11} + \sqrt{7} + 9\sqrt[4]{11}$

19. $8\sqrt{27} - 3\sqrt{3}$

20. $9\sqrt{50} - 4\sqrt{2}$

21. $3\sqrt{45} + 7\sqrt{20}$

22. $5\sqrt{12} + 16\sqrt{27}$

23. $2\sqrt[3]{54} + 12\sqrt[3]{16}$

24. $4\sqrt[4]{48} - \sqrt[4]{243}$

25. $5\sqrt{9x} - 3\sqrt{x}$

26. $4\sqrt{y} + 2\sqrt{16y}$

27. $3\sqrt{x+1} + 10\sqrt{x+1}$

28. $4\sqrt{a-1} + \sqrt{a-1}$

29. $\sqrt{25y} + \sqrt{64y}$

30. $\sqrt[3]{16t^4} - \sqrt[3]{54t^4}$

31. $10\sqrt[3]{z} - \sqrt[3]{z^4}$

32. $5\sqrt[3]{24u^2} + 2\sqrt[3]{81u^5}$

33. $\sqrt{5a} + 2\sqrt{45a^3}$

34. $4\sqrt{3x^3} - \sqrt{12x}$

35. $\sqrt[3]{6x^4} + \sqrt[3]{48x}$

36. $\sqrt[3]{54x} - \sqrt[3]{2x^4}$

37. $\sqrt{9x-9} + \sqrt{x-1}$

38. $\sqrt{4y+12} + \sqrt{y+3}$

39. $\sqrt{x^3-x^2} + \sqrt{4x-4}$

40. $\sqrt{9x-9} - \sqrt{x^3-x^2}$

41. $2\sqrt[3]{a^4b^2} + 3a\sqrt[3]{ab^2}$

42. $3|y|\sqrt[4]{48x^5} - x\sqrt[4]{3x^5y^4}$

43. $\sqrt{4r^7s^5} + 3r^2\sqrt{r^3s^5} - 2rs\sqrt{r^5s^3}$

44. $x\sqrt[3]{27x^5y^2} - x^2\sqrt[3]{x^2y^2} + z\sqrt[3]{x^8y^2}$

45. $\sqrt[3]{128x^9y^{10}} - 2x^2y\sqrt[3]{16x^3y^7}$

46. $5\sqrt[3]{320x^5y^8} + 2x\sqrt[3]{135x^2y^8}$

In Exercises 47–56, perform the addition or subtraction and simplify your answer. See Example 4.

47. $\sqrt{5} - \dfrac{3}{\sqrt{5}}$

48. $\sqrt{10} + \dfrac{5}{\sqrt{10}}$

49. $\sqrt{20} - \sqrt{\dfrac{1}{5}}$

50. $\sqrt{\dfrac{1}{3}} + \sqrt{48}$

51. $\sqrt{12y} - \dfrac{y}{\sqrt{3y}}$

52. $\dfrac{x}{\sqrt{3x}} + \sqrt{27x}$

53. $\sqrt{2x} - \dfrac{3}{\sqrt{2x}}$

54. $\dfrac{8}{\sqrt{5x}} + \sqrt{5x}$

55. $\sqrt{7y^3} - \sqrt{\dfrac{9}{7y^3}}$

56. $\sqrt{\dfrac{4}{3x^3}} + \sqrt{3x^3}$

In Exercises 57–60, place the correct symbol ($<$, $>$, or $=$) between the numbers.

57. $\sqrt{7} + \sqrt{18}$ ____ $\sqrt{7+18}$

58. $\sqrt{10} - \sqrt{6}$ ____ $\sqrt{10-6}$

59. 5 ____ $\sqrt{3^2 + 2^2}$

60. 5 ____ $\sqrt{3^2 + 4^2}$

Solving Problems

▲ *Geometry*　In Exercises 61–64, write and simplify an expression for the perimeter of the figure.

61.

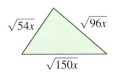

62.

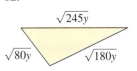

63.

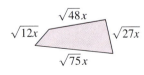

64.

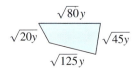

65. ▲ *Geometry* The foundation of a house is 40 feet long and 30 feet wide. The height of the attic is 5 feet (see figure).

(a) Use the Pythagorean Theorem to find the length of the hypotenuse of each of the two right triangles formed by the roof line. (Assume no overhang.)

(b) Use the result of part (a) to determine the total area of the roof.

66. ▲ *Geometry* The four corners are cut from a four-foot-by-eight-foot sheet of plywood, as shown in the figure. Find the perimeter of the remaining piece of plywood.

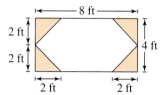

Explaining Concepts

67. *Writing*✐ Is $\sqrt{2} + \sqrt{18}$ in simplest form? Explain.

68. *Writing*✐ Explain what it means for two radical expressions to be like radicals.

69. Will the sum of two radicals always be a radical? Give an example to support your answer.

70. Will the difference of two radicals always be a radical? Give an example to support your answer.

71. You are an algebra instructor, and one of your students hands in the following work. Find and correct the errors, and discuss how you can help your student avoid such errors in the future.

(a) $7\sqrt{3} + 4\sqrt{2} = 11\sqrt{5}$

(b) $3\sqrt[3]{k} - 6\sqrt{k} = -3\sqrt{k}$

Mid-Chapter Quiz

Take this quiz as you would take a quiz in class. After you are done, check your work against the answers in the back of the book.

In Exercises 1–4, evaluate the expression.

1. $\sqrt{225}$

2. $\sqrt[4]{\frac{81}{16}}$

3. $64^{1/2}$

4. $(-27)^{2/3}$

In Exercises 5 and 6, evaluate the function as indicated, if possible, and simplify.

5. $f(x) = \sqrt{3x - 5}$

    (a) $f(0)$ (b) $f(2)$ (c) $f(10)$

6. $g(x) = \sqrt{9 - x}$

    (a) $g(0)$ (b) $g(5)$ (c) $g(10)$

In Exercises 7 and 8, describe the domain of the function.

7. $g(x) = \dfrac{12}{\sqrt[3]{x}}$

8. $h(x) = \sqrt{4x - 5}$

In Exercises 9–14, simplify the expression.

9. $\sqrt{27x^2}$

10. $\sqrt[4]{81x^6}$

11. $\sqrt{\dfrac{4u^3}{9}}$

12. $\sqrt[3]{\dfrac{16}{u^6}}$

13. $\sqrt{125x^3y^2z^4}$

14. $2a\sqrt[3]{16a^3b^5}$

In Exercises 15 and 16, rationalize the denominator and simplify further, if possible.

15. $\dfrac{24}{\sqrt{12}}$

16. $\dfrac{10}{\sqrt{5x}}$

In Exercises 17–22, combine the radical expressions, if possible.

17. $2\sqrt{3} - 4\sqrt{7} + \sqrt{3}$

18. $\sqrt{200y} - 3\sqrt{8y}$

19. $5\sqrt{12} + 2\sqrt{3} - \sqrt{75}$

20. $\sqrt{25x + 50} - \sqrt{x + 2}$

21. $6x\sqrt[3]{5x^2} + 2\sqrt[3]{40x^4}$

22. $3\sqrt{x^3y^4z^5} + 2xy^2\sqrt{xz^5} - xz^2\sqrt{xy^4z}$

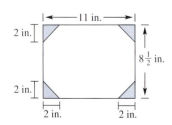

Figure for 23

23. The four corners are cut from an $8\frac{1}{2}$-inch-by-11-inch sheet of paper, as shown in the figure at the left. Find the perimeter of the remaining piece of paper.

7.4 Multiplying and Dividing Radical Expressions

Paul A. Souders/Corbis

Why You Should Learn It

Multiplication of radicals is often used in real-life applications. For instance, in Exercise 107 on page 477, you will multiply two radical expressions to find the area of the cross section of a wooden beam.

What You Should Learn

1 Use the Distributive Property or the FOIL Method to multiply radical expressions.

2 Determine the products of conjugates.

3 Simplify quotients involving radicals by rationalizing the denominators.

Multiplying Radical Expressions

You can multiply radical expressions by using the Distributive Property or the FOIL Method. In both procedures, you also make use of the Product Rule for Radicals from Section 7.2, which is given by $\sqrt[n]{uv} = \sqrt[n]{u}\,\sqrt[n]{v}$, where u and v are real numbers whose nth roots are also real numbers.

1 Use the Distributive Property or the FOIL Method to multiply radical expressions.

Example 1   Multiplying Radical Expressions

Find each product and simplify.

a. $\sqrt{6} \cdot \sqrt{3}$ **b.** $\sqrt[3]{5} \cdot \sqrt[3]{16}$

Solution

a. $\sqrt{6} \cdot \sqrt{3} = \sqrt{6 \cdot 3} = \sqrt{18} = \sqrt{9 \cdot 2} = 3\sqrt{2}$

b. $\sqrt[3]{5} \cdot \sqrt[3]{16} = \sqrt[3]{5 \cdot 16} = \sqrt[3]{80} = \sqrt[3]{8 \cdot 10} = 2\sqrt[3]{10}$

Example 2   Multiplying Radical Expressions

Find each product and simplify.

a. $\sqrt{3}\left(2 + \sqrt{5}\right)$ **b.** $\sqrt{2}\left(4 - \sqrt{8}\right)$ **c.** $\sqrt{6}\left(\sqrt{12} - \sqrt{3}\right)$

Solution

a. $\sqrt{3}\left(2 + \sqrt{5}\right) = 2\sqrt{3} + \sqrt{3}\sqrt{5}$ Distributive Property

$\qquad\qquad\quad\; = 2\sqrt{3} + \sqrt{15}$ Product Rule for Radicals

b. $\sqrt{2}\left(4 - \sqrt{8}\right) = 4\sqrt{2} - \sqrt{2}\sqrt{8}$ Distributive Property

$\qquad\qquad\quad\; = 4\sqrt{2} - \sqrt{16} = 4\sqrt{2} - 4$ Product Rule for Radicals

c. $\sqrt{6}\left(\sqrt{12} - \sqrt{3}\right) = \sqrt{6}\sqrt{12} - \sqrt{6}\sqrt{3}$ Distributive Property

$\qquad\qquad\qquad\quad = \sqrt{72} - \sqrt{18}$ Product Rule for Radicals

$\qquad\qquad\qquad\quad = 6\sqrt{2} - 3\sqrt{2} = 3\sqrt{2}$ Find perfect square factors.

In Example 2, the Distributive Property was used to multiply radical expressions. In Example 3, note how the FOIL Method can be used to multiply binomial radical expressions.

Example 3 Using the FOIL Method

$$
\overset{\text{F} \qquad \text{0} \qquad \text{I} \qquad \text{L}}{}
$$

a. $\left(2\sqrt{7} - 4\right)\left(\sqrt{7} + 1\right) = 2\left(\sqrt{7}\right)^2 + 2\sqrt{7} - 4\sqrt{7} - 4$ FOIL Method

$$= 2(7) + (2 - 4)\sqrt{7} - 4$$ Combine like radicals.

$$= 10 - 2\sqrt{7}$$ Combine like terms.

b. $\left(3 - \sqrt{x}\right)\left(1 + \sqrt{x}\right) = 3 + 3\sqrt{x} - \sqrt{x} - \left(\sqrt{x}\right)^2$ FOIL Method

$$= 3 + 2\sqrt{x} - x, \quad x \geq 0$$ Combine like radicals.

---

② Determine the products of conjugates.

Conjugates

The expressions $3 + \sqrt{6}$ and $3 - \sqrt{6}$ are called **conjugates** of each other. Notice that they differ only in the sign between the terms. The product of two conjugates is the difference of two squares, which is given by the special product formula $(a + b)(a - b) = a^2 - b^2$. Here are some other examples.

Expression	Conjugate	Product
$1 - \sqrt{3}$	$1 + \sqrt{3}$	$(1)^2 - \left(\sqrt{3}\right)^2 = 1 - 3 = -2$
$\sqrt{5} + \sqrt{2}$	$\sqrt{5} - \sqrt{2}$	$\left(\sqrt{5}\right)^2 - \left(\sqrt{2}\right)^2 = 5 - 2 = 3$
$\sqrt{10} - 3$	$\sqrt{10} + 3$	$\left(\sqrt{10}\right)^2 - (3)^2 = 10 - 9 = 1$
$\sqrt{x} + 2$	$\sqrt{x} - 2$	$\left(\sqrt{x}\right)^2 - (2)^2 = x - 4, x \geq 0$

Example 4 Multiplying Conjugates

Find the conjugate of the expression and multiply the expression by its conjugate.

a. $2 - \sqrt{5}$ **b.** $\sqrt{3} + \sqrt{x}$

Solution

a. The conjugate of $2 - \sqrt{5}$ is $2 + \sqrt{5}$.

$$\left(2 - \sqrt{5}\right)\left(2 + \sqrt{5}\right) = 2^2 - \left(\sqrt{5}\right)^2$$ Special product formula

$$= 4 - 5 = -1$$ Simplify.

b. The conjugate of $\sqrt{3} + \sqrt{x}$ is $\sqrt{3} - \sqrt{x}$.

$$\left(\sqrt{3} + \sqrt{x}\right)\left(\sqrt{3} - \sqrt{x}\right) = \left(\sqrt{3}\right)^2 - \left(\sqrt{x}\right)^2$$ Special product formula

$$= 3 - x, \quad x \geq 0$$ Simplify.

③ Simplify quotients involving radicals by rationalizing the denominators.

Dividing Radical Expressions

To simplify a *quotient* involving radicals, you rationalize the denominator. For single-term denominators, you can use the rationalization process described in Section 7.2. To rationalize a denominator involving two terms, multiply both the numerator and denominator by the *conjugate of the denominator.*

Example 5   Simplifying Quotients Involving Radicals

Simplify each expression.

a. $\dfrac{\sqrt{3}}{1 - \sqrt{5}}$ b. $\dfrac{4}{2 - \sqrt{3}}$

Solution

a. $\dfrac{\sqrt{3}}{1 - \sqrt{5}} = \dfrac{\sqrt{3}}{1 - \sqrt{5}} \cdot \dfrac{1 + \sqrt{5}}{1 + \sqrt{5}}$ Multiply numerator and denominator by conjugate of denominator.

$= \dfrac{\sqrt{3}(1 + \sqrt{5})}{1^2 - (\sqrt{5})^2}$ Special product formula

$= \dfrac{\sqrt{3} + \sqrt{15}}{1 - 5}$ Simplify.

$= -\dfrac{\sqrt{3} + \sqrt{15}}{4}$ Simplify.

b. $\dfrac{4}{2 - \sqrt{3}} = \dfrac{4}{2 - \sqrt{3}} \cdot \dfrac{2 + \sqrt{3}}{2 + \sqrt{3}}$ Multiply numerator and denominator by conjugate of denominator.

$= \dfrac{4(2 + \sqrt{3})}{2^2 - (\sqrt{3})^2}$ Special product formula

$= \dfrac{8 + 4\sqrt{3}}{4 - 3}$ Simplify.

$= 8 + 4\sqrt{3}$ Simplify.

Example 6   Simplifying a Quotient Involving Radicals

$\dfrac{5\sqrt{2}}{\sqrt{7} + \sqrt{2}} = \dfrac{5\sqrt{2}}{\sqrt{7} + \sqrt{2}} \cdot \dfrac{\sqrt{7} - \sqrt{2}}{\sqrt{7} - \sqrt{2}}$ Multiply numerator and denominator by conjugate of denominator.

$= \dfrac{5(\sqrt{14} - \sqrt{4})}{(\sqrt{7})^2 - (\sqrt{2})^2}$ Special product formula

$= \dfrac{5(\sqrt{14} - 2)}{7 - 2}$ Simplify.

$= \dfrac{\cancel{5}(\sqrt{14} - 2)}{\cancel{5}}$ Divide out common factor.

$= \sqrt{14} - 2$ Simplest form

Example 7 Dividing Radical Expressions

Perform each division and simplify.

a. $\dfrac{6}{\sqrt{x} - 2}$

b. $\dfrac{2 - \sqrt{3}}{\sqrt{6} + \sqrt{2}}$

Solution

a. $\dfrac{6}{\sqrt{x} - 2} = \dfrac{6}{\sqrt{x} - 2} \cdot \dfrac{\sqrt{x} + 2}{\sqrt{x} + 2}$ Multiply numerator and denominator by conjugate of denominator.

$= \dfrac{6\left(\sqrt{x} + 2\right)}{\left(\sqrt{x}\right)^2 - 2^2}$ Special product formula

$= \dfrac{6\sqrt{x} + 12}{x - 4}$ Simplify.

b. $\dfrac{2 - \sqrt{3}}{\sqrt{6} + \sqrt{2}} = \dfrac{2 - \sqrt{3}}{\sqrt{6} + \sqrt{2}} \cdot \dfrac{\sqrt{6} - \sqrt{2}}{\sqrt{6} - \sqrt{2}}$ Multiply numerator and denominator by conjugate of denominator.

$= \dfrac{2\sqrt{6} - 2\sqrt{2} - \sqrt{18} + \sqrt{6}}{\left(\sqrt{6}\right)^2 - \left(\sqrt{2}\right)^2}$ FOIL Method and special product formula

$= \dfrac{3\sqrt{6} - 2\sqrt{2} - 3\sqrt{2}}{6 - 2}$ Simplify.

$= \dfrac{3\sqrt{6} - 5\sqrt{2}}{4}$ Simplify.

Example 8 Dividing Radical Expressions

Perform the division and simplify.

$$\dfrac{1}{\sqrt{x} - \sqrt{x + 1}}$$

Solution

$\dfrac{1}{\sqrt{x} - \sqrt{x + 1}} = \dfrac{1}{\sqrt{x} - \sqrt{x + 1}} \cdot \dfrac{\sqrt{x} + \sqrt{x + 1}}{\sqrt{x} + \sqrt{x + 1}}$ Multiply numerator and denominator by conjugate of denominator.

$= \dfrac{\sqrt{x} + \sqrt{x + 1}}{\left(\sqrt{x}\right)^2 - \left(\sqrt{x + 1}\right)^2}$ Special product formula

$= \dfrac{\sqrt{x} + \sqrt{x + 1}}{x - (x + 1)}$ Simplify.

$= \dfrac{\sqrt{x} + \sqrt{x + 1}}{-1}$ Combine like terms.

$= -\sqrt{x} - \sqrt{x + 1}$ Simplify.

7.4 Exercises

Review   Concepts, Skills, and Problem Solving

Keep mathematically in shape by doing these exercises *before* the problems of this section.

Properties and Definitions

In Exercises 1–4, use $x^2 + bx + c = (x + m)(x + n)$.

1. $mn =$ ☐

2. If $c > 0$, then what must be true about the signs of m and n?

3. If $c < 0$, then what must be true about the signs of m and n?

4. If m and n have like signs, then $m + n =$ ☐.

Equations of Lines

In Exercises 5–10, find an equation of the line through the two points.

5. $(-1, -2), (3, 6)$

6. $(1, 5), (6, 0)$

7. $(6, 3), (10, 3)$

8. $(4, -2), (4, 5)$

9. $\left(\frac{4}{3}, 8\right), (5, 6)$

10. $(7, 4), (10, 1)$

Models

In Exercises 11 and 12, translate the phrase into an algebraic expression.

11. The time to travel 360 miles if the average speed is r miles per hour

12. The perimeter of a rectangle of length L and width $L/3$

Developing Skills

In Exercises 1–46, multiply and simplify. See Examples 1–3.

1. $\sqrt{2} \cdot \sqrt{8}$

2. $\sqrt{6} \cdot \sqrt{18}$

3. $\sqrt{3} \cdot \sqrt{6}$

4. $\sqrt{5} \cdot \sqrt{10}$

5. $\sqrt[3]{12} \cdot \sqrt[3]{6}$

6. $\sqrt[3]{9} \cdot \sqrt[3]{9}$

7. $\sqrt[4]{8} \cdot \sqrt[4]{6}$

8. $\sqrt[4]{54} \cdot \sqrt[4]{3}$

9. $\sqrt{7}(3 - \sqrt{7})$

10. $\sqrt{3}(4 + \sqrt{3})$

11. $\sqrt{2}(\sqrt{20} + 8)$

12. $\sqrt{7}(\sqrt{14} + 3)$

13. $\sqrt{6}(\sqrt{12} - \sqrt{3})$

14. $\sqrt{10}(\sqrt{5} + \sqrt{6})$

15. $4\sqrt{2}(\sqrt{3} - \sqrt{5})$

16. $3\sqrt{5}(\sqrt{5} - \sqrt{2})$

17. $\sqrt{y}(\sqrt{y} + 4)$

18. $\sqrt{x}(5 - \sqrt{x})$

19. $\sqrt{a}(4 - \sqrt{a})$

20. $\sqrt{z}(\sqrt{z} + 5)$

21. $\sqrt[3]{4}(\sqrt[3]{2} - 7)$

22. $\sqrt[3]{9}(\sqrt[3]{3} + 2)$

23. $(\sqrt{3} + 2)(\sqrt{3} - 2)$

24. $(3 - \sqrt{5})(3 + \sqrt{5})$

25. $(\sqrt{5} + 3)(\sqrt{3} - 5)$

26. $(\sqrt{7} + 6)(\sqrt{2} + 6)$

27. $(\sqrt{20} + 2)^2$

28. $(4 - \sqrt{20})^2$

29. $(\sqrt[3]{6} - 3)(\sqrt[3]{4} + 3)$

30. $(\sqrt[3]{9} + 5)(\sqrt[3]{5} - 5)$

31. $(10 + \sqrt{2x})^2$

32. $(5 - \sqrt{3v})^2$

33. $(9\sqrt{x} + 2)(5\sqrt{x} - 3)$

34. $(16\sqrt{u} - 3)(\sqrt{u} - 1)$

35. $(3\sqrt{x} - 5)(3\sqrt{x} + 5)$

36. $(7 - 3\sqrt{3t})(7 + 3\sqrt{3t})$

37. $(\sqrt[3]{2x} + 5)^2$

38. $(\sqrt[3]{3x} - 4)^2$

39. $(\sqrt[3]{y} + 2)(\sqrt[3]{y^2} - 5)$

40. $(\sqrt[3]{2y} + 10)(\sqrt[3]{4y^2} - 10)$

41. $\left(\sqrt[3]{t} + 1\right)\left(\sqrt[3]{t^2} + 4\sqrt[3]{t} - 3\right)$

42. $\left(\sqrt{x} - 2\right)\left(\sqrt{x^3} - 2\sqrt{x^2} + 1\right)$

43. $\sqrt{x^3 y^4}\left(2\sqrt{xy^2} - \sqrt{x^3 y}\right)$

44. $3\sqrt{xy^3}\left(\sqrt{x^3 y} + 2\sqrt{xy^2}\right)$

45. $2\sqrt[3]{x^4 y^5}\left(\sqrt[3]{8x^{12} y^4} + \sqrt[3]{16xy^9}\right)$

46. $\sqrt[4]{8x^3 y^5}\left(\sqrt[4]{4x^5 y^7} - \sqrt[4]{3x^7 y^6}\right)$

In Exercises 47–52, complete the statement.

47. $5x\sqrt{3} + 15\sqrt{3} = 5\sqrt{3}\left(\right)$

48. $x\sqrt{7} - x^2\sqrt{7} = x\sqrt{7}\left(\right)$

49. $4\sqrt{12} - 2x\sqrt{27} = 2\sqrt{3}\left(\right)$

50. $5\sqrt{50} + 10y\sqrt{8} = 5\sqrt{2}\left(\right)$

51. $6u^2 + \sqrt{18u^3} = 3u\left(\right)$

52. $12s^3 - \sqrt{32s^4} = 4s^2\left(\right)$

In Exercises 53–66, find the conjugate of the expression. Then multiply the expression by its conjugate and simplify. See Example 4.

53. $2 + \sqrt{5}$

54. $\sqrt{2} - 9$

55. $\sqrt{11} - \sqrt{3}$

56. $\sqrt{10} + \sqrt{7}$

57. $\sqrt{15} + 3$

58. $\sqrt{11} + 3$

59. $\sqrt{x} - 3$

60. $\sqrt{t} + 7$

61. $\sqrt{2u} - \sqrt{3}$

62. $\sqrt{5a} + \sqrt{2}$

63. $2\sqrt{2} + \sqrt{4}$

64. $4\sqrt{3} + \sqrt{2}$

65. $\sqrt{x} + \sqrt{y}$

66. $3\sqrt{u} + \sqrt{3v}$

In Exercises 67–70, evaluate the function as indicated and simplify.

67. $f(x) = x^2 - 6x + 1$

 (a) $f\left(2 - \sqrt{3}\right)$ (b) $f\left(3 - 2\sqrt{2}\right)$

68. $g(x) = x^2 + 8x + 11$

 (a) $g\left(-4 + \sqrt{5}\right)$ (b) $g\left(-4\sqrt{2}\right)$

69. $f(x) = x^2 - 2x - 1$

 (a) $f\left(1 + \sqrt{2}\right)$ (b) $f\left(\sqrt{4}\right)$

70. $g(x) = x^2 - 4x + 1$

 (a) $g\left(1 + \sqrt{5}\right)$ (b) $g\left(2 - \sqrt{3}\right)$

In Exercises 71–94, rationalize the denominator of the expression and simplify. See Examples 5–8.

71. $\dfrac{6}{\sqrt{11} - 2}$

72. $\dfrac{8}{\sqrt{7} + 3}$

73. $\dfrac{7}{\sqrt{3} + 5}$

74. $\dfrac{5}{9 - \sqrt{6}}$

75. $\dfrac{3}{2\sqrt{10} - 5}$

76. $\dfrac{4}{3\sqrt{5} - 1}$

77. $\dfrac{2}{\sqrt{6} + \sqrt{2}}$

78. $\dfrac{10}{\sqrt{9} + \sqrt{5}}$

79. $\dfrac{9}{\sqrt{3} - \sqrt{7}}$

80. $\dfrac{12}{\sqrt{5} + \sqrt{8}}$

81. $\left(\sqrt{7} + 2\right) \div \left(\sqrt{7} - 2\right)$

82. $\left(5 - \sqrt{3}\right) \div \left(3 + \sqrt{3}\right)$

83. $\left(\sqrt{x} - 5\right) \div \left(2\sqrt{x} - 1\right)$

84. $\left(2\sqrt{t} + 1\right) \div \left(2\sqrt{t} - 1\right)$

85. $\dfrac{3x}{\sqrt{15} - \sqrt{3}}$

86. $\dfrac{5y}{\sqrt{12} + \sqrt{10}}$

87. $\dfrac{2t^2}{\sqrt{5} - \sqrt{t}}$

88. $\dfrac{5x}{\sqrt{x} - \sqrt{2}}$

89. $\dfrac{8a}{\sqrt{3a} + \sqrt{a}}$

90. $\dfrac{7z}{\sqrt{5z} - \sqrt{z}}$

91. $\dfrac{3(x - 4)}{x^2 - \sqrt{x}}$

92. $\dfrac{6(y + 1)}{y^2 + \sqrt{y}}$

93. $\dfrac{\sqrt{u + v}}{\sqrt{u - v} - \sqrt{u}}$

94. $\dfrac{z}{\sqrt{u + z} - \sqrt{u}}$

In Exercises 95–98, use a graphing calculator to graph the functions in the same viewing window. Use the graphs to verify that the expressions are equivalent. Verify your results algebraically.

95. $y_1 = \dfrac{10}{\sqrt{x} + 1}$

$y_2 = \dfrac{10(\sqrt{x} - 1)}{x - 1}$

96. $y_1 = \dfrac{4x}{\sqrt{x} + 4}$

$y_2 = \dfrac{4x(\sqrt{x} - 4)}{x - 16}$

97. $y_1 = \dfrac{2\sqrt{x}}{2 - \sqrt{x}}$

$y_2 = \dfrac{2(2\sqrt{x} + x)}{4 - x}$

98. $y_1 = \dfrac{\sqrt{2x} + 6}{\sqrt{2x} - 2}$

$y_2 = \dfrac{x + 6 + 4\sqrt{2x}}{x - 2}$

Rationalizing Numerators   In the study of calculus, students sometimes rewrite an expression by rationalizing the numerator. In Exercises 99–106, rationalize the numerator. (*Note:* The results will not be in simplest radical form.)

99. $\dfrac{\sqrt{2}}{7}$

100. $\dfrac{\sqrt{10}}{\sqrt{3x}}$

101. $\dfrac{\sqrt{5}}{\sqrt{7x}}$

102. $\dfrac{\sqrt{10}}{5}$

103. $\dfrac{\sqrt{7} + \sqrt{3}}{5}$

104. $\dfrac{\sqrt{2} - \sqrt{5}}{4}$

105. $\dfrac{\sqrt{y} - 5}{\sqrt{3}}$

106. $\dfrac{\sqrt{x} + 6}{\sqrt{2}}$

Solving Problems

107. *Geometry*   The rectangular cross section of a wooden beam cut from a log of diameter 24 inches (see figure) will have maximum strength if its width w and height h are given by

$$w = 8\sqrt{3} \quad \text{and} \quad h = \sqrt{24^2 - (8\sqrt{3})^2}.$$

Find the area of the rectangular cross section and write the area in simplest form.

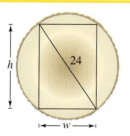

Figure for 107

108. ▲ *Geometry*   The areas of the circles in the figure are 15 square centimeters and 20 square centimeters. Find the ratio of the radius of the small circle to the radius of the large circle.

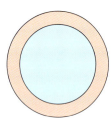

109. *Force*   The force required to slide a steel block weighing 500 pounds across a milling machine is

$$\frac{500k}{\dfrac{1}{\sqrt{k^2+1}} + \dfrac{k^2}{\sqrt{k^2+1}}}$$

where k is the friction constant (see figure). Simplify this expression.

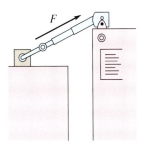

110. The ratio of the width of the Temple of Hephaestus to its height (see figure) is approximately

$$\frac{w}{h} \approx \frac{2}{\sqrt{5}-1}.$$

This number is called the **golden section.** Early Greeks believed that the most aesthetically pleasing rectangles were those whose sides had this ratio.

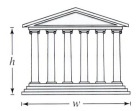

(a) Rationalize the denominator for this expression. Approximate your answer, rounded to two decimal places.

(b) Use the Pythagorean Theorem, a straightedge, and a compass to construct a rectangle whose sides have the golden section as their ratio.

Explaining Concepts

111. Multiply $\sqrt{3}(1 - \sqrt{6})$. State an algebraic property to justify each step.

112. *Writing*   Describe the differences and similarities of using the FOIL Method with polynomial expressions and with radical expressions.

113. *Writing*   Multiply $3 - \sqrt{2}$ by its conjugate. Explain why the result has no radicals.

114. *Writing*   Is the number $3/(1 + \sqrt{5})$ in simplest form? If not, explain the steps for writing it in simplest form.

7.5 Radical Equations and Applications

What You Should Learn

1. Solve a radical equation by raising each side to the nth power.
2. Solve application problems involving radical equations.

Why You Should Learn It

Radical equations can be used to model and solve real-life applications. For instance, in Exercise 100 on page 488, a radical equation is used to model the total monthly cost of daily flights between Chicago and Denver.

Jeff Greenberg/The Image Works

1. Solve a radical equation by raising each side to the nth power.

Solving Radical Equations

Solving equations involving radicals is somewhat like solving equations that contain fractions—first try to eliminate the radicals and obtain a polynomial equation. Then, solve the polynomial equation using the standard procedures. The following property plays a key role.

Raising Each Side of an Equation to the nth Power

Let u and v be real numbers, variables, or algebraic expressions, and let n be a positive integer. If $u = v$, then it follows that

$$u^n = v^n.$$

This is called **raising each side of an equation to the nth power.**

To use this property to solve a radical equation, first try to isolate one of the radicals on one side of the equation. When using this property to solve radical equations, it is critical that you check your solutions in the original equation.

Technology: Tip

To use a graphing calculator to check the solution in Example 1, graph

$$y = \sqrt{x} - 8$$

as shown below. Notice that the graph crosses the x-axis at $x = 64$, which confirms the solution that was obtained algebraically.

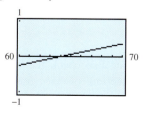

Example 1   Solving an Equation Having One Radical

Solve $\sqrt{x} - 8 = 0$.

Solution

$\sqrt{x} - 8 = 0$	Write original equation.
$\sqrt{x} = 8$	Isolate radical.
$\left(\sqrt{x}\right)^2 = 8^2$	Square each side.
$x = 64$	Simplify.

Check

$\sqrt{64} - 8 \stackrel{?}{=} 0$	Substitute 64 for x in original equation.
$8 - 8 = 0$	Solution checks. ✔

So, the equation has one solution: $x = 64$.

Checking solutions of a radical equation is especially important because raising each side of an equation to the *n*th power to remove the radical(s) often introduces *extraneous* solutions.

Example 2   Solving an Equation Having One Radical

Solve $\sqrt{3x} + 6 = 0$.

Solution

$$\sqrt{3x} + 6 = 0 \qquad \text{Write original equation.}$$

$$\sqrt{3x} = -6 \qquad \text{Isolate radical.}$$

$$\left(\sqrt{3x}\right)^2 = (-6)^2 \qquad \text{Square each side.}$$

$$3x = 36 \qquad \text{Simplify.}$$

$$x = 12 \qquad \text{Divide each side by 3.}$$

Check

$$\sqrt{3(12)} + 6 \stackrel{?}{=} 0 \qquad \text{Substitute 12 for } x \text{ in original equation.}$$

$$6 + 6 \neq 0 \qquad \text{Solution does not check. } \times$$

The solution $x = 12$ is an extraneous solution. So, the original equation has no solution. You can also check this graphically, as shown in Figure 7.5. Notice that the graph does not cross the x-axis and so has no x-intercept.

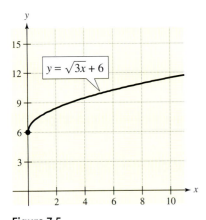

$y = \sqrt{3x} + 6$

Figure 7.5

Example 3   Solving an Equation Having One Radical

Solve $\sqrt[3]{2x + 1} - 2 = 3$.

Solution

$$\sqrt[3]{2x + 1} - 2 = 3 \qquad \text{Write original equation.}$$

$$\sqrt[3]{2x + 1} = 5 \qquad \text{Isolate radical.}$$

$$\left(\sqrt[3]{2x + 1}\right)^3 = 5^3 \qquad \text{Cube each side.}$$

$$2x + 1 = 125 \qquad \text{Simplify.}$$

$$2x = 124 \qquad \text{Subtract 1 from each side.}$$

$$x = 62 \qquad \text{Divide each side by 2.}$$

Check

$$\sqrt[3]{2(62) + 1} - 2 \stackrel{?}{=} 3 \qquad \text{Substitute 62 for } x \text{ in original equation.}$$

$$\sqrt[3]{125} - 2 \stackrel{?}{=} 3 \qquad \text{Simplify.}$$

$$5 - 2 = 3 \qquad \text{Solution checks. } \checkmark$$

So, the equation has one solution: $x = 62$. You can also check the solution graphically by determining the point of intersection of the graphs of $y = \sqrt[3]{2x + 1} - 2$ (left side of equation) and $y = 3$ (right side of equation), as shown in Figure 7.6.

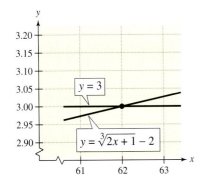

$y = 3$

$y = \sqrt[3]{2x + 1} - 2$

Figure 7.6

Technology: Tip

In Example 4, you can graphically check the solution of the equation by graphing the left side and right side in the same viewing window. That is, graph the equations

$$y_1 = \sqrt{5x + 3}$$

and

$$y_2 = \sqrt{x + 11}$$

in the same viewing window, as shown below. Using the *intersect* feature of the graphing calculator will enable you to approximate the point(s) at which the graphs intersect. From the figure, you can see that the two graphs intersect at $x = 2$, which is the solution obtained in Example 4.

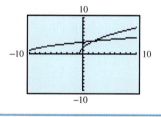

Example 4   Solving an Equation Having Two Radicals

Solve $\sqrt{5x + 3} = \sqrt{x + 11}$.

Solution

$$\sqrt{5x + 3} = \sqrt{x + 11} \qquad \text{Write original equation.}$$

$$\left(\sqrt{5x + 3}\right)^2 = \left(\sqrt{x + 11}\right)^2 \qquad \text{Square each side.}$$

$$5x + 3 = x + 11 \qquad \text{Simplify.}$$

$$4x + 3 = 11 \qquad \text{Subtract } x \text{ from each side.}$$

$$4x = 8 \qquad \text{Subtract 3 from each side.}$$

$$x = 2 \qquad \text{Divide each side by 4.}$$

Check

$$\sqrt{5x + 3} = \sqrt{x + 11} \qquad \text{Write original equation.}$$

$$\sqrt{5(2) + 3} \stackrel{?}{=} \sqrt{2 + 11} \qquad \text{Substitute 2 for } x.$$

$$\sqrt{13} = \sqrt{13} \qquad \text{Solution checks. } ✓$$

So, the equation has one solution: $x = 2$.

Example 5   Solving an Equation Having Two Radicals

Solve $\sqrt[4]{3x} + \sqrt[4]{2x - 5} = 0$.

Solution

$$\sqrt[4]{3x} + \sqrt[4]{2x - 5} = 0 \qquad \text{Write original equation.}$$

$$\sqrt[4]{3x} = -\sqrt[4]{2x - 5} \qquad \text{Isolate radicals.}$$

$$\left(\sqrt[4]{3x}\right)^4 = \left(-\sqrt[4]{2x - 5}\right)^4 \qquad \text{Raise each side to fourth power.}$$

$$3x = 2x - 5 \qquad \text{Simplify.}$$

$$x = -5 \qquad \text{Subtract } 2x \text{ from each side.}$$

Check

$$\sqrt[4]{3x} + \sqrt[4]{2x - 5} = 0 \qquad \text{Write original equation.}$$

$$\sqrt[4]{3(-5)} + \sqrt[4]{2(-5) - 5} \stackrel{?}{=} 0 \qquad \text{Substitute } -5 \text{ for } x.$$

$$\sqrt[4]{-15} + \sqrt[4]{-15} \neq 0 \qquad \text{Solution does not check. } ✗$$

The solution does not check because it yields fourth roots of negative radicands. So, this equation has no solution. Try checking this graphically. If you graph both sides of the equation, you will discover that the graphs do not intersect.

In the next example you will see that squaring each side of the equation results in a quadratic equation. Remember that you must check the solutions in the *original* radical equation.

Example 6   An Equation That Converts to a Quadratic Equation

Solve $\sqrt{x} + 2 = x$.

Solution

$\sqrt{x} + 2 = x$	Write original equation.
$\sqrt{x} = x - 2$	Isolate radical.
$\left(\sqrt{x}\right)^2 = (x - 2)^2$	Square each side.
$x = x^2 - 4x + 4$	Simplify.
$-x^2 + 5x - 4 = 0$	Write in general form.
$(-1)(x - 4)(x - 1) = 0$	Factor.
$x - 4 = 0 \implies x = 4$	Set 1st factor equal to 0.
$x - 1 = 0 \implies x = 1$	Set 2nd factor equal to 0.

Check

First Solution	Second Solution
$\sqrt{4} + 2 \stackrel{?}{=} 4$	$\sqrt{1} + 2 \stackrel{?}{=} 1$
$2 + 2 = 4$	$1 + 2 \neq 1$

From the check you can see that $x = 1$ is an extraneous solution. So, the only solution is $x = 4$.

When an equation contains two radicals, it may not be possible to isolate both. In such cases, you may have to raise each side of the equation to a power at *two* different stages in the solution.

Example 7   Repeatedly Squaring Each Side of an Equation

Solve $\sqrt{3t + 1} = 2 - \sqrt{3t}$.

Solution

$\sqrt{3t + 1} = 2 - \sqrt{3t}$	Write original equation.
$\left(\sqrt{3t + 1}\right)^2 = \left(2 - \sqrt{3t}\right)^2$	Square each side (1st time).
$3t + 1 = 4 - 4\sqrt{3t} + 3t$	Simplify.
$-3 = -4\sqrt{3t}$	Isolate radical.
$(-3)^2 = \left(-4\sqrt{3t}\right)^2$	Square each side (2nd time).
$9 = 16(3t)$	Simplify.
$\dfrac{3}{16} = t$	Divide each side by 48 and simplify.

The solution is $t = \frac{3}{16}$. Check this in the original equation.

② Solve application problems involving radical equations.

Applications

Example 8   Electricity

The amount of power consumed by an electrical appliance is given by

$$I = \sqrt{\frac{P}{R}}$$

where I is the current measured in amps, R is the resistance measured in ohms, and P is the power measured in watts. Find the power used by an electric heater for which $I = 10$ amps and $R = 16$ ohms.

Solution

$$10 = \sqrt{\frac{P}{16}}$$ Substitute 10 for I and 16 for R in original equation.

$$10^2 = \left(\sqrt{\frac{P}{16}}\right)^2$$ Square each side.

$$100 = \frac{P}{16} \implies 1600 = P$$ Simplify and multiply each side by 16.

So, the solution is $P = 1600$ watts. Check this in the original equation.

Study Tip

An alternative way to solve the problem in Example 8 would be first to solve the equation for P.

$$I = \sqrt{\frac{P}{R}}$$

$$I^2 = \left(\sqrt{\frac{P}{R}}\right)^2$$

$$I^2 = \frac{P}{R}$$

$$I^2 R = P$$

At this stage, you can substitute the known values of I and R to obtain

$$P = (10)^2 16 = 1600.$$

Example 9   An Application of the Pythagorean Theorem

The distance between a house on shore and a playground on shore is 40 meters. The distance between the playground and a house on an island is 50 meters (see Figure 7.7). What is the distance between the two houses?

Solution

From Figure 7.7, you can see that the distances form a right triangle. So, you can use the Pythagorean Theorem to find the distance between the two houses.

$$c = \sqrt{a^2 + b^2}$$ Pythagorean Theorem

$$50 = \sqrt{40^2 + b^2}$$ Substitute 40 for a and 50 for c.

$$50 = \sqrt{1600 + b^2}$$ Simplify.

$$50^2 = (\sqrt{1600 + b^2})^2$$ Square each side.

$$2500 = 1600 + b^2$$ Simplify.

$$0 = b^2 - 900$$ Write in general form.

$$0 = (b + 30)(b - 30)$$ Factor.

$$b + 30 = 0 \implies b = -30$$ Set 1st factor equal to 0.

$$b - 30 = 0 \implies b = 30$$ Set 2nd factor equal to 0.

Choose the positive solution to obtain a distance of 30 meters. Check this solution in the original equation.

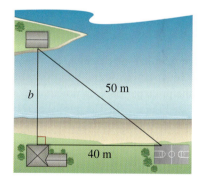

Figure 7.7

Example 10   Velocity of a Falling Object

The velocity of a free-falling object can be determined from the equation $v = \sqrt{2gh}$, where v is the velocity measured in feet per second, $g = 32$ feet per second per second, and h is the distance (in feet) the object has fallen. Find the height from which a rock has been dropped when it strikes the ground with a velocity of 50 feet per second.

Solution

$v = \sqrt{2gh}$	Write original equation.
$50 = \sqrt{2(32)h}$	Substitute 50 for v and 32 for g.
$50^2 = \left(\sqrt{64h}\right)^2$	Square each side.
$2500 = 64h$	Simplify.
$39 \approx h$	Divide each side by 64.

Check

Because the value of h was rounded in the solution, the check will not result in an equality. If the solution is valid, the expressions on each side of the equal sign will be *approximately* equal to each other.

$v = \sqrt{2gh}$	Write original equation.
$50 \overset{?}{\approx} \sqrt{2(32)(39)}$	Substitute 50 for v, 32 for g, and 39 for h.
$50 \overset{?}{\approx} \sqrt{2496}$	Simplify.
$50 \approx 49.96$	Solution checks. ✔

So, the height from which the rock has been dropped is approximately 39 feet.

Example 11   Market Research

The marketing department at a publisher determines that the demand for a book depends on the price of the book in accordance with the formula $p = 40 - \sqrt{0.0001x + 1}$, $x \geq 0$, where p is the price per book in dollars and x is the number of books sold at the given price (see Figure 7.8). The publisher sets the price at $12.95. How many copies can the publisher expect to sell?

Solution

$p = 40 - \sqrt{0.0001x + 1}$	Write original equation.
$12.95 = 40 - \sqrt{0.0001x + 1}$	Substitute 12.95 for p.
$\sqrt{0.0001x + 1} = 27.05$	Isolate radical.
$0.0001x + 1 = 731.7025$	Square each side.
$0.0001x = 730.7025$	Subtract 1 from each side.
$x = 7{,}307{,}025$	Divide each side by 0.0001.

So, by setting the book's price at $12.95, the publisher can expect to sell about 7.3 million copies. Check this in the original equation.

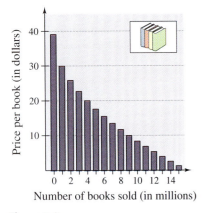

Figure 7.8

7.5 Exercises

Review   Concepts, Skills, and Problem Solving

Keep mathematically in shape by doing these exercises *before* the problems of this section.

Properties and Definitions

1. *Writing*✏️   Explain how to determine the domain of the function

$$f(x) = \frac{4}{(x+2)(x-3)}.$$

2. *Writing*✏️   Explain the excluded value $(x \neq -3)$ in the following.

$$\frac{2x^2 + 5x - 3}{x^2 - 9} = \frac{2x - 1}{x - 3}, \quad x \neq 3, \quad x \neq -3$$

Simplifying Expressions

In Exercises 3–6, simplify the expression. (Assume that any variables in the expression are nonzero.)

3. $(-3x^2y^3)^2 \cdot (4xy^2)$

4. $(x^2 - 3xy)^0$

5. $\dfrac{64r^2s^4}{16rs^2}$

6. $\left(\dfrac{3x}{4y^3}\right)^2$

In Exercises 7–10, perform the indicated operation and simplify.

7. $\dfrac{x+13}{x^3(3-x)} \cdot \dfrac{x(x-3)}{5}$

8. $\dfrac{x+2}{5x+15} \cdot \dfrac{x-2}{5(x-3)}$

9. $\dfrac{2x}{x-5} - \dfrac{5}{5-x}$

10. $\dfrac{3}{x-1} - 5$

Graphs

In Exercises 11 and 12, graph the function and identify any intercepts.

11. $f(x) = 2x - 3$

12. $f(x) = -\frac{3}{4}x + 2$

Developing Skills

In Exercises 1–4, determine whether each value of *x* is a solution of the equation.

	Equation	*Values of x*

1. $\sqrt{x} - 10 = 0$ (a) $x = -4$ (b) $x = -100$
   (c) $x = \sqrt{10}$ (d) $x = 100$

2. $\sqrt{3x} - 6 = 0$ (a) $x = \frac{2}{3}$ (b) $x = 2$
   (c) $x = 12$ (d) $x = -\frac{1}{3}\sqrt{6}$

3. $\sqrt[3]{x-4} = 4$ (a) $x = -60$ (b) $x = 68$
   (c) $x = 20$ (d) $x = 0$

4. $\sqrt[4]{2x} + 2 = 6$ (a) $x = 128$ (b) $x = 2$
   (c) $x = -2$ (d) $x = 0$

In Exercises 5–54, solve the equation and check your solution(s). (Some of the equations have no solution.) See Examples 1–7.

5. $\sqrt{x} = 12$

6. $\sqrt{x} = 5$

7. $\sqrt{y} = 7$

8. $\sqrt{t} = 4$

9. $\sqrt[3]{z} = 3$

10. $\sqrt[4]{x} = 2$

11. $\sqrt{y} - 7 = 0$

12. $\sqrt{t} - 13 = 0$

13. $\sqrt{u} + 13 = 0$

14. $\sqrt{y} + 15 = 0$

15. $\sqrt{x} - 8 = 0$

16. $\sqrt{x} - 10 = 0$

17. $\sqrt{10x} = 30$

18. $\sqrt{8x} = 6$

19. $\sqrt{-3x} = 9$

20. $\sqrt{-4y} = 4$

21. $\sqrt{5t} - 2 = 0$

22. $10 - \sqrt{6x} = 0$

23. $\sqrt{3y+1} = 4$

24. $\sqrt{3-2x} = 2$

25. $\sqrt{4-5x} = -3$

26. $\sqrt{2t-7} = -5$

27. $\sqrt{3y + 5} - 3 = 4$ **28.** $\sqrt{a - 3} + 5 = 6$

29. $5\sqrt{x + 2} = 8$ **30.** $2\sqrt{x + 4} = 7$

31. $\sqrt{x + 3} = \sqrt{2x - 1}$ **32.** $\sqrt{3t + 1} = \sqrt{t + 15}$

33. $\sqrt{3y - 5} - 3\sqrt{y} = 0$

34. $\sqrt{2u + 10} - 2\sqrt{u} = 0$

35. $\sqrt[3]{3x - 4} = \sqrt[3]{x + 10}$

36. $2\sqrt[3]{10 - 3x} = \sqrt[3]{2 - x}$

37. $\sqrt[3]{2x + 15} - \sqrt[3]{x} = 0$

38. $\sqrt[4]{2x} + \sqrt[4]{x + 3} = 0$

39. $\sqrt{x^2 - 2} = x + 4$ **40.** $\sqrt{x^2 - 4} = x - 2$

41. $\sqrt{2x} = x - 4$ **42.** $\sqrt{x} = x - 6$

43. $\sqrt{8x + 1} = x + 2$ **44.** $\sqrt{3x + 7} = x + 3$

45. $\sqrt{3x + 4} = \sqrt{4x + 3}$

46. $\sqrt{2x - 7} = \sqrt{3x - 12}$

47. $\sqrt{z + 2} = 1 + \sqrt{z}$

48. $\sqrt{2x + 5} = 7 - \sqrt{2x}$

49. $\sqrt{2t + 3} = 3 - \sqrt{2t}$

50. $\sqrt{x} + \sqrt{x + 2} = 2$

51. $\sqrt{x + 5} - \sqrt{x} = 1$

52. $\sqrt{x + 1} = 2 - \sqrt{x}$

53. $\sqrt{x - 6} + 3 = \sqrt{x + 9}$

54. $\sqrt{x + 3} - \sqrt{x - 1} = 1$

In Exercises 55–62, solve the equation and check your solution(s).

55. $t^{3/2} = 8$ **56.** $v^{2/3} = 25$

57. $3y^{1/3} = 18$ **58.** $2x^{3/4} = 54$

59. $(x + 4)^{2/3} = 4$ **60.** $(u - 2)^{4/3} = 81$

61. $(2x + 5)^{1/3} + 3 = 0$ **62.** $(x - 6)^{3/2} - 27 = 0$

In Exercises 63–72, use a graphing calculator to graph each side of the equation in the same viewing window. Use the graphs to approximate the solution(s). Verify your answer algebraically.

63. $\sqrt{x} = 2(2 - x)$ **64.** $\sqrt{2x + 3} = 4x - 3$

65. $\sqrt{x^2 + 1} = 5 - 2x$ **66.** $\sqrt{8 - 3x} = x$

67. $\sqrt{x + 3} = 5 - \sqrt{x}$ **68.** $\sqrt[3]{5x - 8} = 4 - \sqrt[3]{x}$

69. $4\sqrt[3]{x} = 7 - x$ **70.** $\sqrt[3]{x + 4} = \sqrt{6 - x}$

71. $\sqrt{15 - 4x} = 2x$ **72.** $\dfrac{4}{\sqrt{x}} = 3\sqrt{x} - 4$

In Exercises 73–76, use the given function to find the indicated value of x.

73. For $f(x) = \sqrt{x} - \sqrt{x - 9}$,
find x such that $f(x) = 1$.

74. For $g(x) = \sqrt{x} + \sqrt{x - 5}$,
find x such that $g(x) = 5$.

75. For $h(x) = \sqrt{x - 2} - \sqrt{4x + 1}$,
find x such that $h(x) = -3$.

76. For $f(x) = \sqrt{2x + 7} - \sqrt{x + 15}$,
find x such that $f(x) = -1$.

Solving Problems

▲ *Geometry*   In Exercises 77–80, find the length x of the unknown side of the right triangle. (Round your answer to two decimal places.)

77.

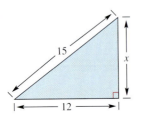

78.

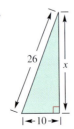

79.

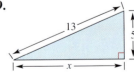

80.

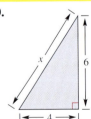

81. ▲ *Geometry*   The screen of a computer monitor has a diagonal of 13.75 inches and a width of 8.25 inches. Draw a diagram of the computer monitor and find the length of the screen.

82. ▲ *Geometry*   A basketball court is 50 feet wide and 94 feet long. Draw a diagram of the basketball court and find the length of a diagonal of the court.

83. ▲ *Geometry*   An extension ladder is placed against the side of a house such that the base of the ladder is 2 meters from the base of the house and the ladder reaches 6 meters up the side of the house. How far is the ladder extended?

84. ▲ *Geometry*   A guy wire on a 100-foot radio tower is attached to the top of the tower and to an anchor 50 feet from the base of the tower. Find the length of the guy wire.

85. ▲ *Geometry*   A ladder is 17 feet long, and the bottom of the ladder is 8 feet from the side of a house. How far does the ladder reach up the side of the house?

86. ▲ *Geometry*   A 10-foot plank is used to brace a basement wall during construction of a home. The plank is nailed to the wall 6 feet above the floor. Find the slope of the plank.

87. ▲ *Geometry*   Determine the length and width of a rectangle with a perimeter of 92 inches and a diagonal of 34 inches.

88. ▲ *Geometry*   Determine the length and width of a rectangle with a perimeter of 68 inches and a diagonal of 26 inches.

89. ▲ *Geometry*   The lateral surface area of a cone (see figure) is given by $S = \pi r \sqrt{r^2 + h^2}$. Solve the equation for h. Then find the height of a cone with a lateral surface area of $364\pi\sqrt{2}$ square centimeters and a radius of 14 centimeters.

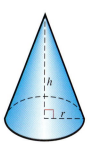

90. ▦ ▲ *Geometry*   Write a function that gives the radius r of a circle in terms of the circle's area A. Use a graphing calculator to graph this function.

Height   In Exercises 91 and 92, use the formula $t = \sqrt{d/16}$, which gives the time t in seconds for a free-falling object to fall d feet.

91. A construction worker drops a nail from a building and observes it strike a water puddle after approximately 2 seconds. Estimate the height from which the nail was dropped.

92. A construction worker drops a nail from a building and observes it strike a water puddle after approximately 3 seconds. Estimate the height from which the nail was dropped.

Free-Falling Object   In Exercises 93–96, use the equation for the velocity of a free-falling object, $v = \sqrt{2gh}$, as described in Example 10.

93. An object is dropped from a height of 50 feet. Estimate the velocity of the object when it strikes the ground.

94. An object is dropped from a height of 200 feet. Estimate the velocity of the object when it strikes the ground.

95. An object strikes the ground with a velocity of 60 feet per second. Estimate the height from which the object was dropped.

96. An object strikes the ground with a velocity of 120 feet per second. Estimate the height from which the object was dropped.

Period of a Pendulum   In Exercises 97 and 98, the time t (in seconds) for a pendulum of length L (in feet) to go through one complete cycle (its period) is given by $t = 2\pi\sqrt{L/32}$.

97. How long is the pendulum of a grandfather clock with a period of 1.5 seconds?

98. How long is the pendulum of a mantel clock with a period of 0.75 second?

99. *Demand*   The demand equation for a sweater is

$$p = 50 - \sqrt{0.8(x - 1)}$$

where x is the number of units demanded per day and p is the price per sweater. Find the demand when the price is set at $30.02.

100. *Airline Passengers*   An airline offers daily flights between Chicago and Denver. The total monthly cost C (in millions of dollars) of these flights is

$$C = \sqrt{0.2x + 1}, \quad x \geq 0$$

where x is measured in thousands of passengers (see figure). The total cost of the flights for June is 2.5 million dollars. Approximately how many passengers flew in June?

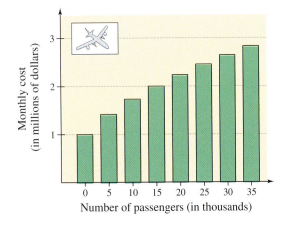

Monthly cost (in millions of dollars)

Number of passengers (in thousands)

101. *Consumer Spending*   The total amount m (in dollars) consumers spent per person on movies in theaters in the United States for the years 1995 through 2000 can be modeled by

$$m = 1.63 + 10.463\sqrt{t}, \quad 5 \leq t \leq 10$$

where t represents the year, with $t = 5$ corresponding to 1995. (Source: Veronis, Suhler & Associates Inc.)

(a) 🖩 Use a graphing calculator to graph the model.

(b) In what year did the amount consumers spent per person on movies in theaters reach $34?

102. *Falling Object*   Without using a stopwatch, you can find the length of time an object has been falling by using the equation from physics

$$t = \sqrt{\frac{h}{384}}$$

where t is the time (in seconds) and h is the distance (in inches) the object has fallen. How far does an object fall in 0.25 second? in 0.10 second?

Explaining Concepts

103. 🔷   Answer part (e) of Motivating the Chapter on page 444.

104. *Writing* ✏️   In your own words, describe the steps that can be used to solve a radical equation.

105. *Writing* ✏️   Does raising each side of an equation to the nth power always yield an equivalent equation? Explain.

106. *Writing* ✏️   One reason for checking a solution in the original equation is to discover errors that were made when solving the equation. Describe another reason.

107. *Error Analysis*   Describe the error.

$$\sqrt{x} + \sqrt{6} = 8$$
$$\left(\sqrt{x}\right)^2 + \left(\sqrt{6}\right)^2 = 8^2$$
$$x + 6 = 64$$
$$x = 58$$

108. *Exploration*   The solution of the equation $x + \sqrt{x - a} = b$ is $x = 20$. Discuss how to find a and b. (There are many correct values for a and b.)

7.6 Complex Numbers

What You Should Learn

① Write square roots of negative numbers in i-form and perform operations on numbers in i-form.
② Determine the equality of two complex numbers.
③ Add, subtract, and multiply complex numbers.
④ Use complex conjugates to write the quotient of two complex numbers in standard form.

Why You Should Learn It

Understanding complex numbers can help you in Section 8.3 to identify quadratic equations that have no real solutions.

① Write square roots of negative numbers in i-form and perform operations on numbers in i-form.

The Imaginary Unit i

In Section 7.1, you learned that a negative number has no *real* square root. For instance, $\sqrt{-1}$ is not real because there is no real number x such that $x^2 = -1$. So, as long as you are dealing only with real numbers, the equation $x^2 = -1$ has no solution. To overcome this deficiency, mathematicians have expanded the set of numbers by including the **imaginary unit i,** defined as

$$i = \sqrt{-1}. \qquad \text{Imaginary unit}$$

This number has the property that $i^2 = -1$. So, the imaginary unit i is a solution of the equation $x^2 = -1$.

The Square Root of a Negative Number

Let c be a positive real number. Then the square root of $-c$ is given by

$$\sqrt{-c} = \sqrt{c(-1)} = \sqrt{c}\sqrt{-1} = \sqrt{c}\,i.$$

When writing $\sqrt{-c}$ in the **i-form,** $\sqrt{c}\,i$, note that i is outside the radical.

Technology: Discovery

Use a calculator to evaluate each radical. Does one result in an error message? Explain why.

a. $\sqrt{121}$
b. $\sqrt{-121}$
c. $-\sqrt{121}$

Example 1   Writing Numbers in i-Form

Write each number in i-form.

a. $\sqrt{-36}$ b. $\sqrt{-\dfrac{16}{25}}$ c. $\sqrt{-54}$ d. $\dfrac{\sqrt{-48}}{\sqrt{-3}}$

Solution

a. $\sqrt{-36} = \sqrt{36(-1)} = \sqrt{36}\sqrt{-1} = 6i$

b. $\sqrt{-\dfrac{16}{25}} = \sqrt{\dfrac{16}{25}(-1)} = \sqrt{\dfrac{16}{25}}\sqrt{-1} = \dfrac{4}{5}i$

c. $\sqrt{-54} = \sqrt{54(-1)} = \sqrt{54}\sqrt{-1} = 3\sqrt{6}\,i$

d. $\dfrac{\sqrt{-48}}{\sqrt{-3}} = \dfrac{\sqrt{48}\sqrt{-1}}{\sqrt{3}\sqrt{-1}} = \dfrac{\sqrt{48}\,i}{\sqrt{3}\,i} = \sqrt{\dfrac{48}{3}} = \sqrt{16} = 4$

To perform operations with square roots of negative numbers, you must *first* write the numbers in *i*-form. Once the numbers have been written in *i*-form, you add, subtract, and multiply as follows.

$ai + bi = (a + b)i$	Addition
$ai - bi = (a - b)i$	Subtraction
$(ai)(bi) = ab(i^2) = ab(-1) = -ab$	Multiplication

Study Tip

When performing operations with numbers in *i*-form, you sometimes need to be able to evaluate powers of the imaginary unit *i*. The first several powers of *i* are as follows.

$$i^1 = i$$

$$i^2 = -1$$

$$i^3 = i(i^2) = i(-1) = -i$$

$$i^4 = (i^2)(i^2) = (-1)(-1) = 1$$

$$i^5 = i(i^4) = i(1) = i$$

$$i^6 = (i^2)(i^4) = (-1)(1) = -1$$

$$i^7 = (i^3)(i^4) = (-i)(1) = -i$$

$$i^8 = (i^4)(i^4) = (1)(1) = 1$$

Note how the pattern of values $i, -1, -i,$ and 1 repeats itself for powers greater than 4.

Example 2   Operations with Square Roots of Negative Numbers

Perform each operation.

a. $\sqrt{-9} + \sqrt{-49}$ **b.** $\sqrt{-32} - 2\sqrt{-2}$

Solution

a. $\sqrt{-9} + \sqrt{-49} = \sqrt{9}\sqrt{-1} + \sqrt{49}\sqrt{-1}$ Product Rule for Radicals

$$= 3i + 7i$$ Write in *i*-form.

$$= 10i$$ Simplify.

b. $\sqrt{-32} - 2\sqrt{-2} = \sqrt{32}\sqrt{-1} - 2\sqrt{2}\sqrt{-1}$ Product Rule for Radicals

$$= 4\sqrt{2}i - 2\sqrt{2}i$$ Write in *i*-form.

$$= 2\sqrt{2}i$$ Simplify.

Example 3   Multiplying Square Roots of Negative Numbers

Find each product.

a. $\sqrt{-15}\sqrt{-15}$ **b.** $\sqrt{-5}(\sqrt{-45} - \sqrt{-4})$

Solution

a. $\sqrt{-15}\sqrt{-15} = (\sqrt{15}i)(\sqrt{15}i)$ Write in *i*-form.

$$= (\sqrt{15})^2 i^2$$ Multiply.

$$= 15(-1)$$ $i^2 = -1$

$$= -15$$ Simplify.

b. $\sqrt{-5}(\sqrt{-45} - \sqrt{-4}) = \sqrt{5}i(3\sqrt{5}i - 2i)$ Write in *i*-form.

$$= (\sqrt{5}i)(3\sqrt{5}i) - (\sqrt{5}i)(2i)$$ Distributive Property

$$= 3(5)(-1) - 2\sqrt{5}(-1)$$ Multiply.

$$= -15 + 2\sqrt{5}$$ Simplify.

When multiplying square roots of negative numbers, be sure to write them in *i*-form *before multiplying*. If you do not do this, you can obtain incorrect answers. For instance, in Example 3(a) be sure you see that

$$\sqrt{-15}\sqrt{-15} \neq \sqrt{(-15)(-15)} = \sqrt{225} = 15.$$

② Determine the equality of two complex numbers.

Complex Numbers

A number of the form $a + bi$, where a and b are real numbers, is called a **complex number.** The real number a is called the **real part** of the complex number $a + bi$, and the number bi is called the **imaginary part.**

Definition of Complex Number

If a and b are real numbers, the number $a + bi$ is a **complex number,** and it is said to be written in **standard form.** If $b = 0$, the number $a + bi = a$ is a real number. If $b \neq 0$, the number $a + bi$ is called an **imaginary number.** A number of the form bi, where $b \neq 0$, is called a **pure imaginary number.**

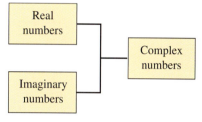

Figure 7.9

A number cannot be both real and imaginary. For instance, the numbers -2, $0, 1, \frac{1}{2}$, and $\sqrt{2}$ are real numbers (but they are *not* imaginary numbers), and the numbers $-3i, 2 + 4i$, and $-1 + i$ are imaginary numbers (but they are *not* real numbers). The diagram shown in Figure 7.9 further illustrates the relationship among real, complex, and imaginary numbers.

Two complex numbers $a + bi$ and $c + di$, in standard form, are equal if and only if $a = c$ and $b = d$.

Example 4   Equality of Two Complex Numbers

To determine whether the complex numbers $\sqrt{9} + \sqrt{-48}$ and $3 - 4\sqrt{3}i$ are equal, begin by writing the first number in standard form.

$$\sqrt{9} + \sqrt{-48} = \sqrt{3^2} + \sqrt{4^2(3)(-1)} = 3 + 4\sqrt{3}i$$

From this form, you can see that the two numbers are not equal because they have imaginary parts that differ in sign.

Example 5   Equality of Two Complex Numbers

Find values of x and y that satisfy the equation $3x - \sqrt{-25} = -6 + 3yi$.

Solution

Begin by writing the left side of the equation in standard form.

$$3x - 5i = -6 + 3yi \qquad \text{\textcolor{red}{Each side is in standard form.}}$$

For these two numbers to be equal, their real parts must be equal to each other and their imaginary parts must be equal to each other.

Real Parts	*Imaginary Parts*
$3x = -6$	$3yi = -5i$
$x = -2$	$3y = -5$
	$y = -\frac{5}{3}$

So, $x = -2$ and $y = -\frac{5}{3}$.

3 Add, subtract, and multiply complex numbers.

Operations with Complex Numbers

To add or subtract two complex numbers, you add (or subtract) the real and imaginary parts separately. This is similar to combining like terms of a polynomial.

$$(a + bi) + (c + di) = (a + c) + (b + d)i \qquad \text{Addition of complex numbers}$$

$$(a + bi) - (c + di) = (a - c) + (b - d)i \qquad \text{Subtraction of complex numbers}$$

Example 6   Adding and Subtracting Complex Numbers

a. $(3 - i) + (-2 + 4i) = (3 - 2) + (-1 + 4)i = 1 + 3i$

b. $3i + (5 - 3i) = 5 + (3 - 3)i = 5$

c. $4 - (-1 + 5i) + (7 + 2i) = [4 - (-1) + 7] + (-5 + 2)i = 12 - 3i$

d. $(6 + 3i) + \left(2 - \sqrt{-8}\right) - \sqrt{-4} = (6 + 3i) + \left(2 - 2\sqrt{2}i\right) - 2i$

$$= (6 + 2) + \left(3 - 2\sqrt{2} - 2\right)i$$

$$= 8 + \left(1 - 2\sqrt{2}\right)i$$

> **Study Tip**
>
> Note in part (b) of Example 6 that the sum of two complex numbers can be a real number.

The Commutative, Associative, and Distributive Properties of real numbers are also valid for complex numbers, as is the FOIL Method.

Example 7   Multiplying Complex Numbers

Perform each operation and write the result in standard form.

a. $(7i)(-3i)$ **b.** $(1 - i)\left(\sqrt{-9}\right)$

c. $(2 - i)(4 + 3i)$ **d.** $(3 + 2i)(3 - 2i)$

Solution

a. $(7i)(-3i) = -21i^2$ Multiply.

$$= -21(-1) = 21 \qquad i^2 = -1$$

b. $(1 - i)\left(\sqrt{-9}\right) = (1 - i)(3i)$ Write in i-form.

$$= 3i - 3(i^2) \qquad \text{Distributive Property}$$

$$= 3i - 3(-1) = 3 + 3i \qquad i^2 = -1$$

c. $(2 - i)(4 + 3i) = 8 + 6i - 4i - 3i^2$ FOIL Method

$$= 8 + 6i - 4i - 3(-1) \qquad i^2 = -1$$

$$= 11 + 2i \qquad \text{Combine like terms.}$$

d. $(3 + 2i)(3 - 2i) = 3^2 - (2i)^2$ Special product formula

$$= 9 - 4i^2 \qquad \text{Simplify.}$$

$$= 9 - 4(-1) = 13 \qquad i^2 = -1$$

④ Use complex conjugates to write the quotient of two complex numbers in standard form.

Complex Conjugates

In Example 7(d), note that the product of two complex numbers can be a real number. This occurs with pairs of complex numbers of the form $a + bi$ and $a - bi$, called **complex conjugates.** In general, the product of complex conjugates has the following form.

$$(a + bi)(a - bi) = a^2 - (bi)^2 = a^2 - b^2i^2 = a^2 - b^2(-1) = a^2 + b^2$$

Here are some examples.

Complex Number	Complex Conjugate	Product
$4 - 5i$	$4 + 5i$	$4^2 + 5^2 = 41$
$3 + 2i$	$3 - 2i$	$3^2 + 2^2 = 13$
$-2 = -2 + 0i$	$-2 = -2 - 0i$	$(-2)^2 + 0^2 = 4$
$i = 0 + i$	$-i = 0 - i$	$0^2 + 1^2 = 1$

To write the quotient of $a + bi$ and $c + di$ in standard form, where c and d are not both zero, multiply the numerator and denominator by the *complex conjugate of the denominator*, as shown in Example 8.

Example 8   Writing Quotients of Complex Numbers in Standard Form

a. $\dfrac{2 - i}{4i} = \dfrac{2 - i}{4i} \cdot \dfrac{(-4i)}{(-4i)}$ Multiply numerator and denominator by complex conjugate of denominator.

$= \dfrac{-8i + 4i^2}{-16i^2}$ Multiply fractions.

$= \dfrac{-8i + 4(-1)}{-16(-1)}$ $i^2 = -1$

$= \dfrac{-8i - 4}{16}$ Simplify.

$= -\dfrac{1}{4} - \dfrac{1}{2}i$ Write in standard form.

b. $\dfrac{5}{3 - 2i} = \dfrac{5}{3 - 2i} \cdot \dfrac{3 + 2i}{3 + 2i}$ Multiply numerator and denominator by complex conjugate of denominator.

$= \dfrac{5(3 + 2i)}{(3 - 2i)(3 + 2i)}$ Multiply fractions.

$= \dfrac{5(3 + 2i)}{3^2 + 2^2}$ Product of complex conjugates

$= \dfrac{15 + 10i}{13}$ Simplify.

$= \dfrac{15}{13} + \dfrac{10}{13}i$ Write in standard form.

Example 9   Writing a Quotient of Complex Numbers in Standard Form

$$\frac{8 - i}{8 + i} = \frac{8 - i}{8 + i} \cdot \frac{8 - i}{8 - i}$$ Multiply numerator and denominator by complex conjugate of denominator.

$$= \frac{64 - 16i + i^2}{8^2 + 1^2}$$ Multiply fractions.

$$= \frac{64 - 16i + (-1)}{8^2 + 1^2}$$ $i^2 = -1$

$$= \frac{63 - 16i}{65}$$ Simplify.

$$= \frac{63}{65} - \frac{16}{65}i$$ Write in standard form.

Example 10   Writing a Quotient of Complex Numbers in Standard Form

$$\frac{2 + 3i}{4 - 2i} = \frac{2 + 3i}{4 - 2i} \cdot \frac{4 + 2i}{4 + 2i}$$ Multiply numerator and denominator by complex conjugate of denominator.

$$= \frac{8 + 16i + 6i^2}{4^2 + 2^2}$$ Multiply fractions.

$$= \frac{8 + 16i + 6(-1)}{4^2 + 2^2}$$ $i^2 = -1$

$$= \frac{2 + 16i}{20}$$ Simplify.

$$= \frac{1}{10} + \frac{4}{5}i$$ Write in standard form.

Example 11   Verifying a Complex Solution of an Equation

Show that $x = 2 + i$ is a solution of the equation

$$x^2 - 4x + 5 = 0.$$

Solution

$$x^2 - 4x + 5 = 0$$ Write original equation.

$$(2 + i)^2 - 4(2 + i) + 5 \overset{?}{=} 0$$ Substitute $2 + i$ for x.

$$4 + 4i + i^2 - 8 - 4i + 5 \overset{?}{=} 0$$ Expand.

$$i^2 + 1 \overset{?}{=} 0$$ Combine like terms.

$$(-1) + 1 \overset{?}{=} 0$$ $i^2 = -1$

$$0 = 0$$ Solution checks. ✓

So, $x = 2 + i$ is a solution of the original equation.

7.6 Exercises

Review Concepts, Skills, and Problem Solving

Keep mathematically in shape by doing these exercises *before* the problems of this section.

Properties and Definitions

1. *Writing* In your own words, describe how to multiply $\dfrac{3t}{5} \cdot \dfrac{8t^2}{15}$.

2. *Writing* In your own words, describe how to divide $\dfrac{3t}{5} \div \dfrac{8t^2}{15}$.

3. *Writing* In your own words, describe how to add $\dfrac{3t}{5} + \dfrac{8t^2}{15}$.

4. *Writing* What is the value of $\dfrac{t-5}{5-t}$? Explain.

Simplifying Expressions

In Exercises 5–10, simplify the expression.

5. $\dfrac{x^2}{2x+3} \div \dfrac{5x}{2x+3}$

6. $\dfrac{x-y}{5x} \div \dfrac{x^2-y^2}{x^2}$

7. $\dfrac{\dfrac{9}{x}}{\left(\dfrac{6}{x}+2\right)}$

8. $\dfrac{\left(1+\dfrac{2}{x}\right)}{\left(x-\dfrac{4}{x}\right)}$

9. $\dfrac{\left(\dfrac{4}{x^2-9}+\dfrac{2}{x-2}\right)}{\left(\dfrac{1}{x+3}+\dfrac{1}{x-3}\right)}$

10. $\dfrac{\left(\dfrac{1}{x+1}+\dfrac{1}{2}\right)}{\left(\dfrac{3}{2x^2+4x+2}\right)}$

Problem Solving

11. *Number Problem* Find two real numbers that divide the real number line between $x/2$ and $4x/3$ into three equal parts.

12. *Capacitance* When two capacitors with capacitances C_1 and C_2 are connected in series, the equivalent capacitance is given by

$$\dfrac{1}{\left(\dfrac{1}{C_1}+\dfrac{1}{C_2}\right)}.$$

Simplify this complex fraction.

Developing Skills

In Exercises 1–16, write the number in *i*-form. See Example 1.

1. $\sqrt{-4}$

2. $\sqrt{-9}$

3. $-\sqrt{-144}$

4. $\sqrt{-49}$

5. $\sqrt{-\frac{4}{25}}$

6. $-\sqrt{-\frac{36}{121}}$

7. $\sqrt{-0.09}$

8. $\sqrt{-0.0004}$

9. $\sqrt{-8}$

10. $\sqrt{-75}$

11. $\sqrt{-7}$

12. $\sqrt{-15}$

13. $\dfrac{\sqrt{-12}}{\sqrt{-3}}$

14. $\dfrac{\sqrt{-45}}{\sqrt{-5}}$

15. $\sqrt{-\frac{18}{64}}$

16. $\sqrt{-\frac{8}{25}}$

In Exercises 17–38, perform the operation(s) and write the result in standard form. See Examples 2 and 3.

17. $\sqrt{-16}+\sqrt{-36}$

18. $\sqrt{-25}-\sqrt{-9}$

19. $\sqrt{-50}-\sqrt{-8}$

20. $\sqrt{-500}+\sqrt{-45}$

21. $\sqrt{-48}+\sqrt{-12}-\sqrt{-27}$

22. $\sqrt{-32}-\sqrt{-18}+\sqrt{-50}$

23. $\sqrt{-8}\sqrt{-2}$

24. $\sqrt{-25}\sqrt{-6}$

25. $\sqrt{-18}\sqrt{-3}$

26. $\sqrt{-7}\sqrt{-7}$

27. $\sqrt{-0.16}\sqrt{-1.21}$

28. $\sqrt{-0.49}\sqrt{-1.44}$

29. $\sqrt{-3}\left(\sqrt{-3}+\sqrt{-4}\right)$

30. $\sqrt{-12}\left(\sqrt{-3} - \sqrt{-12}\right)$

31. $\sqrt{-5}\left(\sqrt{-16} - \sqrt{-10}\right)$

32. $\sqrt{-24}\left(\sqrt{-9} + \sqrt{-4}\right)$

33. $\sqrt{-2}\left(3 - \sqrt{-8}\right)$ **34.** $\sqrt{-9}\left(1 + \sqrt{-16}\right)$

35. $\left(\sqrt{-16}\right)^2$ **36.** $\left(\sqrt{-2}\right)^2$

37. $\left(\sqrt{-4}\right)^3$ **38.** $\left(\sqrt{-5}\right)^3$

In Exercises 39–46, determine the values of a and b that satisfy the equation. See Examples 4 and 5.

39. $3 - 4i = a + bi$

40. $-8 + 6i = a + bi$

41. $5 - 4i = (a + 3) + (b - 1)i$

42. $-10 + 12i = 2a + (5b - 3)i$

43. $-4 - \sqrt{-8} = a + bi$

44. $\sqrt{-36} - 3 = a + bi$

45. $(a + 5) + (b - 1)i = 7 - 3i$

46. $(2a + 1) + (2b + 3)i = 5 + 12i$

In Exercises 47–60, perform the operation(s) and write the result in standard form. See Example 6.

47. $(4 - 3i) + (6 + 7i)$

48. $(-10 + 2i) + (4 - 7i)$

49. $(-4 - 7i) + (-10 - 33i)$

50. $(15 + 10i) - (2 + 10i)$

51. $13i - (14 - 7i)$

52. $(-21 - 50i) + (21 - 20i)$

53. $(30 - i) - (18 + 6i) + 3i^2$

54. $(4 + 6i) + (15 + 24i) - (1 - i)$

55. $6 - (3 - 4i) + 2i$

56. $22 + (-5 + 8i) + 10i$

57. $\left(\frac{4}{3} + \frac{1}{3}i\right) + \left(\frac{5}{6} + \frac{7}{6}i\right)$

58. $(0.05 + 2.50i) - (6.2 + 11.8i)$

59. $15i - (3 - 25i) + \sqrt{-81}$

60. $(-1 + i) - \sqrt{2} - \sqrt{-2}$

In Exercises 61–88, perform the operation and write the result in standard form. See Example 7.

61. $(3i)(12i)$ **62.** $(-5i)(4i)$

63. $(3i)(-8i)$ **64.** $(-2i)(-10i)$

65. $(-6i)(-i)(6i)$ **66.** $(10i)(12i)(-3i)$

67. $(-3i)^3$ **68.** $(8i)^2$

69. $(-3i)^2$ **70.** $(2i)^4$

71. $-5(13 + 2i)$ **72.** $10(8 - 6i)$

73. $4i(-3 - 5i)$ **74.** $-3i(10 - 15i)$

75. $(9 - 2i)\left(\sqrt{-4}\right)$ **76.** $(11 + 3i)\left(\sqrt{-25}\right)$

77. $(4 + 3i)(-7 + 4i)$ **78.** $(3 + 5i)(2 + 15i)$

79. $(-7 + 7i)(4 - 2i)$ **80.** $(3 + 5i)(2 - 15i)$

81. $\left(-2 + \sqrt{-5}\right)\left(-2 - \sqrt{-5}\right)$

82. $\left(-3 - \sqrt{-12}\right)\left(4 - \sqrt{-12}\right)$

83. $(3 - 4i)^2$ **84.** $(7 + i)^2$

85. $(2 + 5i)^2$ **86.** $(8 - 3i)^2$

87. $(2 + i)^3$ **88.** $(3 - 2i)^3$

In Exercises 89–98, simplify the expression.

89. i^7 **90.** i^{11}

91. i^{24} **92.** i^{35}

93. i^{42} **94.** i^{64}

95. i^9 **96.** i^{71}

97. $(-i)^6$ **98.** $(-i)^4$

In Exercises 99–110, multiply the number by its complex conjugate and simplify.

99. $2 + i$ **100.** $3 + 2i$

101. $-2 - 8i$ **102.** $10 - 3i$

103. $5 - \sqrt{6}i$ **104.** $-4 + \sqrt{2}i$

105. $10i$ **106.** 20

107. $1 + \sqrt{-3}$ **108.** $-3 - \sqrt{-5}$

109. $1.5 + \sqrt{-0.25}$ **110.** $3.2 - \sqrt{-0.04}$

In Exercises 111–124, write the quotient in standard form. See Examples 8–10.

111. $\dfrac{20}{2i}$ **112.** $\dfrac{-5}{-3i}$

113. $\dfrac{2 + i}{-5i}$ **114.** $\dfrac{1 + i}{3i}$

115. $\dfrac{4}{1-i}$

116. $\dfrac{20}{3+i}$

117. $\dfrac{7i+14}{7i}$

118. $\dfrac{6i+3}{3i}$

119. $\dfrac{-12}{2+7i}$

120. $\dfrac{15}{2(1-i)}$

121. $\dfrac{3i}{5+2i}$

122. $\dfrac{4i}{5-3i}$

123. $\dfrac{4+5i}{3-7i}$

124. $\dfrac{5+3i}{7-4i}$

In Exercises 125–130, perform the operation and write the result in standard form.

125. $\dfrac{5}{3+i}+\dfrac{1}{3-i}$

126. $\dfrac{1}{1-2i}+\dfrac{4}{1+2i}$

127. $\dfrac{3i}{1+i}+\dfrac{2}{2+3i}$

128. $\dfrac{i}{4-3i}-\dfrac{5}{2+i}$

129. $\dfrac{1+i}{i}-\dfrac{3}{5-2i}$

130. $\dfrac{3-2i}{i}-\dfrac{1}{7+i}$

In Exercises 131–134, determine whether each number is a solution of the equation. See Example 11.

131. $x^2+2x+5=0$

    (a) $x=-1+2i$ (b) $x=-1-2i$

132. $x^2-4x+13=0$

    (a) $x=2-3i$ (b) $x=2+3i$

133. $x^3+4x^2+9x+36=0$

    (a) $x=-4$ (b) $x=-3i$

134. $x^3-8x^2+25x-26=0$

    (a) $x=2$ (b) $x=3-2i$

135. *Cube Roots* The principal cube root of 125, $\sqrt[3]{125}$, is 5. Evaluate the expression x^3 for each value of x.

    (a) $x=\dfrac{-5+5\sqrt{3}i}{2}$

    (b) $x=\dfrac{-5-5\sqrt{3}i}{2}$

136. *Cube Roots* The principal cube root of 27, $\sqrt[3]{27}$, is 3. Evaluate the expression x^3 for each value of x.

    (a) $x=\dfrac{-3+3\sqrt{3}i}{2}$

    (b) $x=\dfrac{-3-3\sqrt{3}i}{2}$

137. *Pattern Recognition* Compare the results of Exercises 135 and 136. Use the results to list possible cube roots of (a) 1, (b) 8, and (c) 64. Verify your results algebraically.

138. *Algebraic Properties* Consider the complex number $1+5i$.

    (a) Find the additive inverse of the number.

    (b) Find the multiplicative inverse of the number.

In Exercises 139–142, perform the operations.

139. $(a+bi)+(a-bi)$

140. $(a+bi)(a-bi)$

141. $(a+bi)-(a-bi)$

142. $(a+bi)^2+(a-bi)^2$

Explaining Concepts

143. *Writing* Define the imaginary unit i.

144. *Writing* Explain why the equation $x^2=-1$ does not have real number solutions.

145. *Writing* Describe the error.

$$\sqrt{-3}\sqrt{-3}=\sqrt{(-3)(-3)}=\sqrt{9}=3$$

146. *True or False?* Some numbers are both real and imaginary. Justify your answer.

147. The polynomial x^2+1 is prime *with respect to the integers*. It is not, however, prime *with respect to the complex numbers*. Show how x^2+1 can be factored using complex numbers.

What Did You Learn?

Key Terms

square root, *p. 446*
cube root, *p. 446*
nth root of a, *p. 446*
principal nth root of a, *p. 446*
radical symbol, *p. 446*
index, *p. 446*
radicand, *p. 446*
perfect square, *p. 447*

perfect cube, *p. 447*
rational exponent, *p. 449*
radical function, *p. 451*
rationalizing the
    denominator, *p. 460*
Pythagorean Theorem, *p. 461*
like radicals, *p. 464*
conjugates, *p. 472*

imaginary unit *i*, *p. 489*
i-form, *p. 489*
complex number, *p. 491*
real part, *p. 491*
imaginary part, *p. 491*
imaginary number, *p. 491*
complex conjugates,
    p. 493

Key Concepts

7.1 ◯ Properties of *n*th roots

1. If a is a positive real number and n is even, then a has exactly two (real) *n*th roots, which are denoted by $\sqrt[n]{a}$ and $-\sqrt[n]{a}$.

2. If a is any real number and n is odd, then a has only one (real) *n*th root, which is denoted by $\sqrt[n]{a}$.

3. If a is a negative real number and n is even, then a has no (real) *n*th root.

7.1 ◯ Inverse properties of *n*th powers and *n*th roots

Let a be a real number, and let n be an integer such that $n \geq 2$.

1. If a has a principal *n*th root, then $\left(\sqrt[n]{a}\right)^n = a$.

2. If n is odd, then $\sqrt[n]{a^n} = a$. If n is even, then $\sqrt[n]{a^n} = |a|$.

7.1 ◯ Rules of exponents

Let r and s be rational numbers, and let a and b be real numbers, variables, or algebraic expressions. (All denominators and bases are nonzero.)

1. $a^r \cdot a^s = a^{r+s}$ 2. $\dfrac{a^r}{a^s} = a^{r-s}$

3. $(ab)^r = a^r \cdot b^r$ 4. $(a^r)^s = a^{rs}$

5. $\left(\dfrac{a}{b}\right)^r = \dfrac{a^r}{b^r}$ 6. $a^0 = 1$

7. $a^{-r} = \dfrac{1}{a^r}$ 8. $\left(\dfrac{a}{b}\right)^{-r} = \left(\dfrac{b}{a}\right)^r$

7.1 ◯ Domain of a radical function

Let n be an integer that is greater than or equal to 2.

1. If n is odd, the domain of $f(x) = \sqrt[n]{x}$ is the set of all real numbers.

2. If n is even, the domain of $f(x) = \sqrt[n]{x}$ is the set of all nonnegative real numbers.

7.2 ◯ Product and Quotient Rules for Radicals

Let u and v be real numbers, variables, or algebraic expressions. If the *n*th roots of u and v are real, the following rules are true.

1. $\sqrt[n]{uv} = \sqrt[n]{u}\,\sqrt[n]{v}$ 2. $\sqrt[n]{\dfrac{u}{v}} = \dfrac{\sqrt[n]{u}}{\sqrt[n]{v}}, \quad v \neq 0$

7.2 ◯ Simplifying radical expressions

A radical expression is said to be in simplest form if all three of the statements below are true.

1. All possible *n*th powered factors have been removed from each radical.

2. No radical contains a fraction.

3. No denominator of a fraction contains a radical.

7.5 ◯ Raising each side of an equation to the *n*th power

Let u and v be real numbers, variables, or algebraic expressions, and let n be a positive integer. If $u = v$, then it follows that $u^n = v^n$.

7.6 ◯ The square root of a negative number

Let c be a positive real number. Then the square root of $-c$ is given by

$$\sqrt{-c} = \sqrt{c(-1)} = \sqrt{c}\,\sqrt{-1} = \sqrt{c}\,i.$$

When writing $\sqrt{-c}$ in the *i*-form, $\sqrt{c}\,i$, note that i is outside the radical.

Review Exercises

7.1 Radicals and Rational Exponents

① Determine the *n*th roots of numbers and evaluate radical expressions.

In Exercises 1–14, evaluate the radical expression without using a calculator. If not possible, state the reason.

1. $\sqrt{49}$

2. $\sqrt{25}$

3. $-\sqrt{81}$

4. $\sqrt{-16}$

5. $\sqrt[3]{-8}$

6. $\sqrt[3]{-1}$

7. $-\sqrt[3]{64}$

8. $-\sqrt[3]{125}$

9. $\sqrt{\left(\frac{5}{6}\right)^2}$

10. $\sqrt{\left(\frac{8}{15}\right)^2}$

11. $\sqrt[3]{-\left(\frac{1}{5}\right)^3}$

12. $-\sqrt[3]{\left(-\frac{27}{64}\right)^3}$

13. $\sqrt{-2^2}$

14. $\sqrt{-4^2}$

② Use the rules of exponents to evaluate or simplify expressions with rational exponents.

In Exercises 15–18, fill in the missing description.

Radical Form	Rational Exponent Form
15. $\sqrt{49} = 7$	
16. $\sqrt[3]{0.125} = 0.5$	
17.	$216^{1/3} = 6$
18.	$16^{1/4} = 2$

In Exercises 19–24, evaluate without using a calculator.

19. $27^{4/3}$

20. $16^{3/4}$

21. $-(5^2)^{3/2}$

22. $(-9)^{5/2}$

23. $8^{-4/3}$

24. $243^{-2/5}$

In Exercises 25–36, rewrite the expression using rational exponents.

25. $x^{3/4} \cdot x^{-1/6}$

26. $a^{2/3} \cdot a^{3/5}$

27. $z\sqrt[3]{z^2}$

28. $x^2\sqrt[4]{x^3}$

29. $\dfrac{\sqrt[4]{x^3}}{\sqrt{x^4}}$

30. $\dfrac{\sqrt{x^3}}{\sqrt[3]{x^2}}$

31. $\sqrt[3]{a^3b^2}$

32. $\sqrt[5]{x^6y^2}$

33. $\sqrt[4]{\sqrt{x}}$

34. $\sqrt{\sqrt[3]{x^4}}$

35. $\dfrac{(3x+2)^{2/3}}{\sqrt[3]{3x+2}}$

36. $\dfrac{\sqrt[5]{3x+6}}{(3x+6)^{4/5}}$

③ Use a calculator to evaluate radical expressions.

In Exercises 37–40, use a calculator to evaluate the expression. Round the answer to four decimal places.

37. $75^{-3/4}$

38. $510^{5/3}$

39. $\sqrt{13^2 - 4(2)(7)}$

40. $\dfrac{-3.7 + \sqrt{15.8}}{2(2.3)}$

④ Evaluate radical functions and find the domains of radical functions.

In Exercises 41–44, evaluate the function as indicated, if possible, and simplify.

41. $f(x) = \sqrt{x - 2}$
    (a) $f(11)$ (b) $f(83)$

42. $f(x) = \sqrt{6x - 5}$
    (a) $f(5)$ (b) $f(-1)$

43. $g(x) = \sqrt[3]{2x - 1}$
    (a) $g(0)$ (b) $g(14)$

44. $g(x) = \sqrt[4]{x + 5}$
    (a) $g(11)$ (b) $g(-10)$

In Exercises 45 and 46, describe the domain of the function.

45. $f(x) = \sqrt{9 - 2x}$

46. $g(x) = \sqrt[3]{x + 2}$

7.2 Simplifying Radical Expressions

① Use the Product and Quotient Rules for Radicals to simplify radical expressions.

In Exercises 47–54, simplify the radical expression.

47. $\sqrt{75u^5v^4}$

48. $\sqrt{24x^3y^4}$

49. $\sqrt{0.25x^4y}$

50. $\sqrt{0.16s^6t^3}$

51. $\sqrt[4]{64a^2b^5}$

52. $\sqrt{36x^3y^2}$

53. $\sqrt[3]{48a^3b^4}$

54. $\sqrt[4]{32u^4v^5}$

2 Use rationalization techniques to simplify radical expressions.

In Exercises 55–60, rationalize the denominator and simplify further, if possible.

55. $\sqrt{\dfrac{5}{6}}$

56. $\sqrt{\dfrac{3}{20}}$

57. $\dfrac{3}{\sqrt{12x}}$

58. $\dfrac{4y}{\sqrt{10z}}$

59. $\dfrac{2}{\sqrt[3]{2x}}$

60. $\sqrt[3]{\dfrac{16t}{s^2}}$

3 Use the Pythagorean Theorem in application problems.

 Geometry   In Exercises 61 and 62, find the length of the hypotenuse of the right triangle.

61.

62.

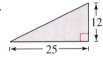

7.3 Adding and Subtracting Radical Expressions

1 Use the Distributive Property to add and subtract like radicals.

In Exercises 63–74, combine the radical expressions, if possible.

63. $2\sqrt{7} - 5\sqrt{7} + 4\sqrt{7}$

64. $3\sqrt{5} - 7\sqrt{5} + 2\sqrt{5}$

65. $3\sqrt{40} - 10\sqrt{90}$

66. $9\sqrt{50} - 5\sqrt{8} + \sqrt{48}$

67. $5\sqrt{x} - \sqrt[3]{x} + 9\sqrt{x} - 8\sqrt[3]{x}$

68. $\sqrt{3x} - \sqrt[4]{6x^2} + 2\sqrt[4]{6x^2} - 4\sqrt{3x}$

69. $10\sqrt[4]{y+3} - 3\sqrt[4]{y+3}$

70. $5\sqrt[3]{x-3} + 4\sqrt[3]{x-3}$

71. $2x\sqrt[3]{24x^2y} - \sqrt[3]{3x^5y}$

72. $4xy^2\sqrt[4]{243x} + 2y^2\sqrt[4]{48x^5}$

73. $x\sqrt{9x^4y^5} - 2x^3\sqrt{8y^5} + 4xy^2\sqrt{4x^4y}$

74. $2x^2y^2\sqrt{75xy} + 5xy\sqrt{3x^3y^3} - xy^2\sqrt{300x^3y}$

2 Use radical expressions in application problems.

75. 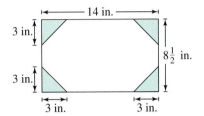 *Geometry*   The four corners are cut from an $8\frac{1}{2}$-inch-by-14-inch sheet of paper (see figure). Find the perimeter of the remaining piece of paper.

76. *Geometry*   Write and simplify an expression for the perimeter of the triangle.

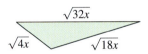

7.4 Multiplying and Dividing Radical Expressions

1 Use the Distributive Property or the FOIL Method to multiply radical expressions.

In Exercises 77–86, multiply and simplify.

77. $\sqrt{15} \cdot \sqrt{20}$

78. $\sqrt{42} \cdot \sqrt{21}$

79. $\sqrt{5}\left(\sqrt{10} + 3\right)$

80. $\sqrt{6}\left(\sqrt{24} - 8\right)$

81. $\sqrt{10}\left(\sqrt{2} + \sqrt{5}\right)$

82. $\sqrt{12}\left(\sqrt{6} - \sqrt{8}\right)$

83. $\left(\sqrt{5} + 6\right)^2$

84. $\left(4 - 3\sqrt{2}\right)^2$

85. $\left(\sqrt{3} - \sqrt{x}\right)\left(\sqrt{3} + \sqrt{x}\right)$

86. $\left(2 + 3\sqrt{5}\right)\left(2 - 3\sqrt{5}\right)$

2 Determine the products of conjugates.

In Exercises 87–90, find the conjugate of the expression. Then multiply the expression by its conjugate and simplify.

87. $3 - \sqrt{7}$

88. $\sqrt{5} + 10$

89. $\sqrt{x} + 20$

90. $9 - \sqrt{2y}$

3 Simplify quotients involving radicals by rationalizing the denominators.

In Exercises 91–98, rationalize the denominator of the expression and simplify.

91. $\dfrac{3}{1 - \sqrt{2}}$

92. $\dfrac{\sqrt{5}}{\sqrt{10} + 3}$

93. $\dfrac{3\sqrt{8}}{2\sqrt{2} + \sqrt{3}}$

94. $\dfrac{7\sqrt{6}}{\sqrt{3} - 4\sqrt{2}}$

95. $\dfrac{\sqrt{2} - 1}{\sqrt{3} - 4}$

96. $\dfrac{3 + \sqrt{3}}{5 - \sqrt{3}}$

97. $\left(\sqrt{x} + 10\right) \div \left(\sqrt{x} - 10\right)$

98. $\left(3\sqrt{s} + 4\right) \div \left(\sqrt{s} + 2\right)$

7.5 Radical Equations and Applications

1 Solve a radical equation by raising each side to the nth power.

In Exercises 99–114, solve the equation and check your solution(s).

99. $\sqrt{y} = 15$

100. $\sqrt{x} - 3 = 0$

101. $\sqrt{3x} + 9 = 0$

102. $\sqrt{4x} + 6 = 9$

103. $\sqrt{2(a - 7)} = 14$

104. $\sqrt{5(4 - 3x)} = 10$

105. $\sqrt[3]{5x - 7} - 3 = -1$

106. $\sqrt[4]{2x + 3} + 4 = 5$

107. $\sqrt[3]{5x + 2} - \sqrt[3]{7x - 8} = 0$

108. $\sqrt[4]{9x - 2} - \sqrt[4]{8x} = 0$

109. $\sqrt{2(x + 5)} = x + 5$

110. $y - 2 = \sqrt{y + 4}$

111. $\sqrt{v - 6} = 6 - v$

112. $\sqrt{5t} = 1 + \sqrt{5(t - 1)}$

113. $\sqrt{1 + 6x} = 2 - \sqrt{6x}$

114. $\sqrt{2 + 9b} + 1 = 3\sqrt{b}$

2 Solve application problems involving radical equations.

115. ▲ *Geometry* Determine the length and width of a rectangle with a perimeter of 34 inches and a diagonal of 13 inches.

116. ▲ *Geometry* Determine the length and width of a rectangle with a perimeter of 84 inches and a diagonal of 30 inches.

117. ▲ *Geometry* A ladder is 18 feet long, and the bottom of the ladder is 9 feet from the side of a house. How far does the ladder reach up the side of the house?

118. *Period of a Pendulum* The time t (in seconds) for a pendulum of length L (in feet) to go through one complete cycle (its period) is given by

$$t = 2\pi \sqrt{\frac{L}{32}}.$$

How long is the pendulum of a grandfather clock with a period of 1.3 seconds?

119. *Height* The time t (in seconds) for a free-falling object to fall d feet is given by

$$t = \sqrt{\frac{d}{16}}.$$

A child drops a pebble from a bridge and observes it strike the water after approximately 6 seconds. Estimate the height from which the pebble was dropped.

120. *Free-Falling Object*   The velocity of a free-falling object can be determined from the equation

$$v = \sqrt{2gh}$$

where v is the velocity (in feet per second), $g = 32$ feet per second per second, and h is the distance (in feet) the object has fallen. Find the height from which a rock has been dropped when it strikes the ground with a velocity of 25 feet per second.

7.6 Complex Numbers

① Write square roots of negative numbers in i-form and perform operations on numbers in i-form.

In Exercises 121–126, write the number in i-form.

121. $\sqrt{-48}$

122. $\sqrt{-0.16}$

123. $10 - 3\sqrt{-27}$

124. $3 + 2\sqrt{-500}$

125. $\frac{3}{4} - 5\sqrt{-\frac{3}{25}}$

126. $-0.5 + 3\sqrt{-1.21}$

In Exercises 127–134, perform the operation(s) and write the result in standard form.

127. $\sqrt{-81} + \sqrt{-36}$

128. $\sqrt{-49} + \sqrt{-1}$

129. $\sqrt{-121} - \sqrt{-84}$

130. $\sqrt{-169} - \sqrt{-4}$

131. $\sqrt{-5}\sqrt{-5}$

132. $\sqrt{-24}\sqrt{-6}$

133. $\sqrt{-10}\left(\sqrt{-4} - \sqrt{-7}\right)$

134. $\sqrt{-5}\left(\sqrt{-10} + \sqrt{-15}\right)$

② Determine the equality of two complex numbers.

In Exercises 135–138, determine the values of a and b that satisfy the equation.

135. $12 - 5i = (a + 2) + (b - 1)i$

136. $-48 + 9i = (a - 5) + (b + 10)i$

137. $\sqrt{-49} + 4 = a + bi$

138. $-3 - \sqrt{-4} = a + bi$

③ Add, subtract, and multiply complex numbers.

In Exercises 139–146, perform the operation and write the result in standard form.

139. $(-4 + 5i) - (-12 + 8i)$

140. $(-8 + 3i) - (6 + 7i)$

141. $(3 - 8i) + (5 + 12i)$

142. $(-6 + 3i) + (-1 + i)$

143. $(4 - 3i)(4 + 3i)$

144. $(12 - 5i)(2 + 7i)$

145. $(6 - 5i)^2$

146. $(2 - 9i)^2$

④ Use complex conjugates to write the quotient of two complex numbers in standard form.

In Exercises 147–152, write the quotient in standard form.

147. $\dfrac{7}{3i}$

148. $\dfrac{4}{5i}$

149. $\dfrac{4i}{2 - 8i}$

150. $\dfrac{5i}{2 + 9i}$

151. $\dfrac{3 - 5i}{6 + i}$

152. $\dfrac{2 + i}{1 - 9i}$

Take this test as you would take a test in class. After you are done, check your work against the answers in the back of the book.

In Exercises 1 and 2, evaluate each expression without using a calculator.

1. (a) $16^{3/2}$

(b) $\sqrt{5}\sqrt{20}$

2. (a) $27^{-2/3}$

(b) $\sqrt{2}\sqrt{18}$

3. For $f(x) = \sqrt{9 - 5x}$, find $f(-8)$ and $f(0)$.

4. Find the domain of $g(x) = \sqrt{7x - 3}$.

In Exercises 5–7, simplify each expression.

5. (a) $\left(\dfrac{x^{1/2}}{x^{1/3}}\right)^2$

(b) $5^{1/4} \cdot 5^{7/4}$

6. (a) $\sqrt{\dfrac{32}{9}}$

(b) $\sqrt[3]{24}$

7. (a) $\sqrt{24x^3}$

(b) $\sqrt[4]{16x^5y^8}$

In Exercises 8 and 9, rationalize the denominator of the expression and simplify.

8. $\dfrac{2}{\sqrt[3]{9y}}$

9. $\dfrac{10}{\sqrt{6} - \sqrt{2}}$

10. Subtract: $5\sqrt{3x} - 3\sqrt{75x}$

11. Multiply and simplify: $\sqrt{5}\left(\sqrt{15x} + 3\right)$

12. Expand: $\left(4 - \sqrt{2x}\right)^2$

13. Factor: $7\sqrt{27} + 14y\sqrt{12} = 7\sqrt{3}\left(\right)$

In Exercises 14–16, solve the equation.

14. $\sqrt{3y} - 6 = 3$

15. $\sqrt{x^2 - 1} = x - 2$

16. $\sqrt{x} - x + 6 = 0$

In Exercises 17–20, perform the operation(s) and simplify.

17. $(2 + 3i) - \sqrt{-25}$

18. $(2 - 3i)^2$

19. $\sqrt{-16}\left(1 + \sqrt{-4}\right)$

20. $(3 - 2i)(1 + 5i)$

21. Write $\dfrac{5 - 2i}{3 + i}$ in standard form.

22. The velocity v (in feet per second) of an object is given by $v = \sqrt{2gh}$, where $g = 32$ feet per second per second and h is the distance (in feet) the object has fallen. Find the height from which a rock has been dropped when it strikes the ground with a velocity of 80 feet per second.

Take this test as you would take a test in class. After you are done, check your work against the answers in the back of the book.

In Exercises 1–3, simplify the expression.

1. $(-2x^5y^{-2}z^0)^{-1}$

2. $\dfrac{12s^5t^{-2}}{20s^{-2}t^{-1}}$

3. $\left(\dfrac{2x^{-4}y^3}{3x^5y^{-3}z^0}\right)^{-2}$

4. Evaluate $(4 \times 10^3)^2$ without using a calculator.

In Exercises 5–8, perform the operation(s) and simplify.

5. $(x^5 + 2x^3 + x^2 - 10x) - (2x^3 - x^2 + x - 4)$

6. $-3(3x^3 - 4x^2 + x) + 3x(2x^2 + x - 1)$

7. $(2x + 1)(x - 5)$

8. $(3x + 2)(3x^2 - x + 1)$

In Exercises 9–12, factor the polynomial completely.

9. $3x^2 - 8x - 35$

10. $9x^2 - 144$

11. $y^3 - 3y^2 - 9y + 27$

12. $8t^3 - 40t^2 + 50t$

In Exercises 13 and 14, solve the polynomial equation.

13. $3x^2 + x - 24 = 0$

14. $7x^3 - 63x = 0$

In Exercises 15–20, perform the operation(s) and simplify.

15. $\dfrac{x^2 + 8x + 16}{18x^2} \cdot \dfrac{2x^4 + 4x^3}{x^2 - 16}$

16. $\dfrac{x^2 + 4x}{2x^2 - 7x + 3} \div \dfrac{x^2 - 16}{x - 3}$

17. $\dfrac{5x}{x + 2} - \dfrac{2}{x^2 - x - 6}$

18. $\dfrac{2}{x} - \dfrac{x}{x^3 + 3x^2} + \dfrac{1}{x + 3}$

19. $\dfrac{\left(\dfrac{3x}{x + 2}\right)}{\left(\dfrac{12}{x^3 + 2x^2}\right)}$

20. $\dfrac{\left(\dfrac{x}{y} - \dfrac{y}{x}\right)}{\left(\dfrac{x - y}{xy}\right)}$

21. Use synthetic division to divide $2x^3 + 7x^2 - 5$ by $x + 3$.

22. Use long division to divide $4x^4 - 6x^3 + x - 4$ by $2x - 1$.

In Exercises 23 and 24, solve the rational equation.

23. $\dfrac{1}{x} + \dfrac{4}{10 - x} = 1$

24. $\dfrac{x - 3}{x} + 1 = \dfrac{x - 4}{x - 6}$

In Exercises 25–28, simplify the expression.

25. $\sqrt{24x^2y^3}$

26. $\sqrt[3]{80a^{15}b^8}$

27. $(12a^{-4}b^6)^{1/2}$

28. $\left(\dfrac{t^{1/2}}{t^{1/4}}\right)^2$

29. Add: $10\sqrt{20x} + 3\sqrt{125x}$.

30. Expand: $\left(\sqrt{2x} - 3\right)^2$.

31. Rationalize the denominator and simplify the result: $\dfrac{3}{\sqrt{10} - \sqrt{x}}$

In Exercises 32–35, solve the radical equation.

32. $\sqrt{x - 5} - 6 = 0$

33. $\sqrt{3 - x} + 10 = 11$

34. $\sqrt{x + 5} - \sqrt{x - 7} = 2$

35. $\sqrt{x - 4} = \sqrt{x + 7} - 1$

In Exercises 36–38, perform the operation and simplify.

36. $\sqrt{-2}\left(\sqrt{-8} + 3\right)$

37. $(-4 + 11i) - (3 - 5i)$

38. $(3 - 4i)^2$

39. Write $\dfrac{1 - 2i}{4 + i}$ in standard form.

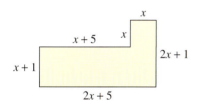

Figure for 40

40. Write and simplify an expression for the perimeter of the figure shown at the left.

41. You can mow a lawn in 2 hours and your friend can mow it in 3.5 hours. What fractional part of the lawn can each of you mow in 1 hour? How long will it take both of you to mow the lawn together?

42. The game commission introduces 50 deer into newly acquired state game lands. The population P of the herd is approximated by the model

$$P = \dfrac{10(5 + 3t)}{1 + 0.004t}$$

where t is the time in years. Find the time required for the population to increase to 250 deer.

43. A guy wire on a 120-foot radio tower is attached to the top of the tower and to an anchor 60 feet from the base of the tower. Find the length of the guy wire.

44. The time t (in seconds) for a free-falling object to fall d feet is given by

$$t = \sqrt{\dfrac{d}{16}}.$$

A construction worker drops a nail from a building and observes it strike a water puddle after approximately 4 seconds. Estimate the height from which the nail was dropped.

Motivating the Chapter

⊘ *Height of a Falling Object*

You drop or throw a rock from the Tacoma Narrows Bridge 192 feet above Puget Sound. The height h (in feet) of the rock at any time t (in seconds) is

$$h = -16t^2 + v_0 t + h_0$$

where v_0 is the initial velocity (in feet per second) of the rock and h_0 is the initial height.

See Section 8.1, Exercise 143.

a. You drop the rock ($v_0 = 0$ ft/sec). How long will it take to hit the water? What method did you use to solve the quadratic equation? Explain why you used that method.

b. You throw the rock straight upward with an initial velocity of 32 feet per second. Find the time(s) when h is 192 feet. What method did you use to solve this quadratic equation? Explain why you used this method.

See Section 8.3, Exercise 121.

c. You throw the rock straight upward with an initial velocity of 32 feet per second. Find the time when h is 100 feet. What method did you use to solve this quadratic equation? Explain why you used this method.

d. You move to a lookout point that is 128 feet above the water. You throw the rock straight upward at the same rate as when you were 192 feet above the water. Would you expect it to reach the water in less time? Verify your conclusion algebraically.

See Section 8.6, Exercise 117.

e. You throw a rock straight upward with an initial velocity of 32 feet per second from a height of 192 feet. During what interval of time is the height greater than 144 feet?

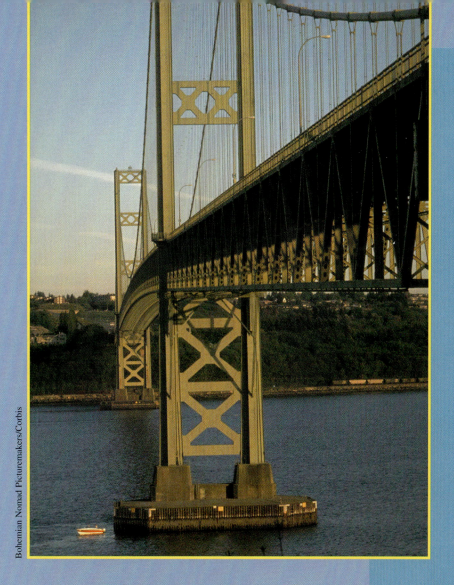

Bohemian Nomad Picturemakers/Corbis

Quadratic Equations, Functions, and Inequalities

8.1 ● Solving Quadratic Equations: Factoring and Special Forms

8.2 ● Completing the Square

8.3 ● The Quadratic Formula

8.4 ● Graphs of Quadratic Functions

8.5 ● Applications of Quadratic Equations

8.6 ● Quadratic and Rational Inequalities

8.1 Solving Quadratic Equations: Factoring and Special Forms

Chris Whitehead/Getty Images

What You Should Learn

1. Solve quadratic equations by factoring.
2. Solve quadratic equations by the Square Root Property.
3. Solve quadratic equations with complex solutions by the Square Root Property.
4. Use substitution to solve equations of quadratic form.

Why You Should Learn It

Quadratic equations can be used to model and solve real-life problems. For instance, in Exercises 141 and 142 on page 516, you will use a quadratic equation to determine national health care expenditures in the United States.

Solving Quadratic Equations by Factoring

In this chapter, you will study methods for solving quadratic equations and equations of quadratic form. To begin, let's review the method of factoring that you studied in Section 5.6.

Remember that the first step in solving a quadratic equation by factoring is to write the equation in general form. Next, factor the left side. Finally, set each factor equal to zero and solve for x. Check each solution in the original equation.

1. Solve quadratic equations by factoring.

Example 1   Solving Quadratic Equations by Factoring

a.

$x^2 + 5x = 24$	Original equation
$x^2 + 5x - 24 = 0$	Write in general form.
$(x + 8)(x - 3) = 0$	Factor.
$x + 8 = 0 \implies x = -8$	Set 1st factor equal to 0.
$x - 3 = 0 \implies x = 3$	Set 2nd factor equal to 0.

b.

$3x^2 = 4 - 11x$	Original equation
$3x^2 + 11x - 4 = 0$	Write in general form.
$(3x - 1)(x + 4) = 0$	Factor.
$3x - 1 = 0 \implies x = \dfrac{1}{3}$	Set 1st factor equal to 0.
$x + 4 = 0 \implies x = -4$	Set 2nd factor equal to 0.

c.

$9x^2 + 12 = 3 + 12x + 5x^2$	Original equation
$4x^2 - 12x + 9 = 0$	Write in general form.
$(2x - 3)(2x - 3) = 0$	Factor.
$2x - 3 = 0 \implies x = \dfrac{3}{2}$	Set factor equal to 0.

Check each solution in its original equation.

Study Tip

In Example 1(c), the quadratic equation produces two identical solutions. This is called a **double** or **repeated solution.**

② Solve quadratic equations by the Square Root Property.

The Square Root Property

Consider the following equation, where $d > 0$ and u is an algebraic expression.

$$u^2 = d \qquad \text{Original equation}$$

$$u^2 - d = 0 \qquad \text{Write in general form.}$$

$$\left(u + \sqrt{d}\right)\left(u - \sqrt{d}\right) = 0 \qquad \text{Factor.}$$

$$u + \sqrt{d} = 0 \implies u = -\sqrt{d} \qquad \text{Set 1st factor equal to 0.}$$

$$u - \sqrt{d} = 0 \implies u = \sqrt{d} \qquad \text{Set 2nd factor equal to 0.}$$

Because the solutions differ only in sign, they can be written together using a "plus or minus sign": $u = \pm\sqrt{d}$. This form of the solution is read as "u is equal to plus or minus the square root of d." When you are solving an equation of the form $u^2 = d$ *without* going through the steps of factoring, you are using the Square Root Property.

Technology: Tip

To check graphically the solutions of an equation written in general form, graph the left side of the equation and locate its x-intercepts. For instance, in Example 2(b), write the equation as

$$(x - 2)^2 - 10 = 0$$

and then use a graphing calculator to graph

$$y = (x - 2)^2 - 10$$

as shown below. You can use the *zoom* and *trace* features or the *zero* or *root* feature to approximate the x-intercepts of the graph to be $x \approx 5.16$ and $x \approx -1.16$.

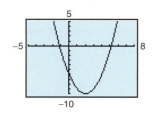

> ### Square Root Property
>
> The equation $u^2 = d$, where $d > 0$, has exactly two solutions:
>
> $$u = \sqrt{d} \quad \text{and} \quad u = -\sqrt{d}.$$
>
> These solutions can also be written as $u = \pm\sqrt{d}$. This solution process is also called **extracting square roots.**

Example 2   Square Root Property

a. $3x^2 = 15 \qquad\qquad\qquad\qquad$ Original equation

$\qquad x^2 = 5 \qquad\qquad\qquad\qquad$ Divide each side by 3.

$\qquad x = \pm\sqrt{5} \qquad\qquad\qquad$ Square Root Property

The solutions are $x = \sqrt{5}$ and $x = -\sqrt{5}$. Check these in the original equation.

b. $(x - 2)^2 = 10 \qquad\qquad\qquad$ Original equation

$\qquad x - 2 = \pm\sqrt{10} \qquad\qquad$ Square Root Property

$\qquad x = 2 \pm\sqrt{10} \qquad\qquad$ Add 2 to each side.

The solutions are $x = 2 + \sqrt{10} \approx 5.16$ and $x = 2 - \sqrt{10} \approx -1.16$.

c. $(3x - 6)^2 - 8 = 0 \qquad\qquad$ Original equation

$\qquad (3x - 6)^2 = 8 \qquad\qquad$ Add 8 to each side.

$\qquad 3x - 6 = \pm 2\sqrt{2} \qquad\qquad$ Square Root Property and rewrite $\sqrt{8}$ as $2\sqrt{2}$.

$\qquad 3x = 6 \pm 2\sqrt{2} \qquad\qquad$ Add 6 to each side.

$\qquad x = \dfrac{6 \pm 2\sqrt{2}}{3} \qquad\qquad$ Divide each side by 3.

The solutions are $x = \left(6 + 2\sqrt{2}\right)/3 \approx 2.94$ and $x = \left(6 - 2\sqrt{2}\right)/3 \approx 1.06$.

3 Solve quadratic equations with complex solutions by the Square Root Property.

Quadratic Equations with Complex Solutions

Prior to Section 7.6, the only solutions to find were real numbers. But now that you have studied complex numbers, it makes sense to look for other types of solutions. For instance, although the quadratic equation $x^2 + 1 = 0$ has no solutions that are real numbers, it does have two solutions that are complex numbers: i and $-i$. To check this, substitute i and $-i$ for x.

$$(i)^2 + 1 = -1 + 1 = 0 \qquad \text{Solution checks. } \checkmark$$
$$(-i)^2 + 1 = -1 + 1 = 0 \qquad \text{Solution checks. } \checkmark$$

One way to find complex solutions of a quadratic equation is to extend the Square Root Property to cover the case in which d is a negative number.

Square Root Property (Complex Square Root)

The equation $u^2 = d$, where $d < 0$, has exactly two solutions:

$$u = \sqrt{|d|}\,i \quad \text{and} \quad u = -\sqrt{|d|}\,i.$$

These solutions can also be written as $u = \pm\sqrt{|d|}\,i$.

Technology: Discovery

Solve each quadratic equation below algebraically. Then use a graphing calculator to check the solutions. Which equations have real solutions and which have complex solutions? Which graphs have x-intercepts and which have no x-intercepts? Compare the type of solution(s) of each quadratic equation with the x-intercept(s) of the graph of the equation.

a. $y = 2x^2 + 3x - 5$

b. $y = 2x^2 + 4x + 2$

c. $y = x^2 + 4$

d. $y = (x + 7)^2 + 2$

Example 3   Square Root Property

a. $x^2 + 8 = 0$ Original equation

$\qquad x^2 = -8$ Subtract 8 from each side.

$\qquad x = \pm\sqrt{8}\,i = \pm 2\sqrt{2}\,i$ Square Root Property

The solutions are $x = 2\sqrt{2}\,i$ and $x = -2\sqrt{2}\,i$. Check these in the original equation.

b. $(x - 4)^2 = -3$ Original equation

$\qquad x - 4 = \pm\sqrt{3}\,i$ Square Root Property

$\qquad x = 4 \pm \sqrt{3}\,i$ Add 4 to each side.

The solutions are $x = 4 + \sqrt{3}\,i$ and $x = 4 - \sqrt{3}\,i$. Check these in the original equation.

c. $2(3x - 5)^2 + 32 = 0$ Original equation

$\qquad 2(3x - 5)^2 = -32$ Subtract 32 from each side.

$\qquad (3x - 5)^2 = -16$ Divide each side by 2.

$\qquad 3x - 5 = \pm 4i$ Square Root Property

$\qquad 3x = 5 \pm 4i$ Add 5 to each side.

$\qquad x = \dfrac{5 \pm 4i}{3}$ Divide each side by 3.

The solutions are $x = (5 + 4i)/3$ and $x = (5 - 4i)/3$. Check these in the original equation.

4 Use substitution to solve equations of quadratic form.

Equations of Quadratic Form

Both the factoring method and the Square Root Property can be applied to nonquadratic equations that are of **quadratic form.** An equation is said to be of quadratic form if it has the form

$$au^2 + bu + c = 0$$

where u is an algebraic expression. Here are some examples.

Equation	Written in Quadratic Form
$x^4 + 5x^2 + 4 = 0$	$(x^2)^2 + 5(x^2) + 4 = 0$
$x - 5\sqrt{x} + 6 = 0$	$(\sqrt{x})^2 - 5(\sqrt{x}) + 6 = 0$
$2x^{2/3} + 5x^{1/3} - 3 = 0$	$2(x^{1/3})^2 + 5(x^{1/3}) - 3 = 0$
$18 + 2x^2 + (x^2 + 9)^2 = 8$	$(x^2 + 9)^2 + 2(x^2 + 9) - 8 = 0$

To solve an equation of quadratic form, it helps to make a substitution and rewrite the equation in terms of u, as demonstrated in Examples 4 and 5.

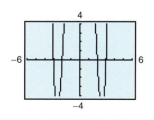
Example 4   Solving an Equation of Quadratic Form

Solve $x^4 - 13x^2 + 36 = 0$.

Solution

Begin by writing the original equation in quadratic form, as follows.

$$x^4 - 13x^2 + 36 = 0 \qquad \text{Write original equation.}$$
$$(x^2)^2 - 13(x^2) + 36 = 0 \qquad \text{Write in quadratic form.}$$

Next, let $u = x^2$ and substitute u into the equation written in quadratic form. Then, factor and solve the equation.

$$u^2 - 13u + 36 = 0 \qquad \text{Substitute } u \text{ for } x^2.$$
$$(u - 4)(u - 9) = 0 \qquad \text{Factor.}$$
$$u - 4 = 0 \implies u = 4 \qquad \text{Set 1st factor equal to 0.}$$
$$u - 9 = 0 \implies u = 9 \qquad \text{Set 2nd factor equal to 0.}$$

At this point you have found the "u-solutions." To find the "x-solutions," replace u with x^2 and solve for x.

$$u = 4 \implies x^2 = 4 \implies x = \pm 2$$
$$u = 9 \implies x^2 = 9 \implies x = \pm 3$$

The solutions are $x = 2$, $x = -2$, $x = 3$, and $x = -3$. Check these in the original equation.

Be sure you see in Example 4 that the u-solutions of 4 and 9 represent only a temporary step. They are not solutions of the original equation and cannot be substituted into the original equation.

Example 5 **Solving an Equation of Quadratic Form**

a. $x - 5\sqrt{x} + 6 = 0$ Original equation

This equation is of quadratic form with $u = \sqrt{x}$.

$$\left(\sqrt{x}\right)^2 - 5\left(\sqrt{x}\right) + 6 = 0$$ Write in quadratic form.

$$u^2 - 5u + 6 = 0$$ Substitute u for $\sqrt{x}$.

$$(u - 2)(u - 3) = 0$$ Factor.

$$u - 2 = 0 \implies u = 2$$ Set 1st factor equal to 0.

$$u - 3 = 0 \implies u = 3$$ Set 2nd factor equal to 0.

Now, using the u-solutions of 2 and 3, you obtain the x-solutions as follows.

$$u = 2 \implies \sqrt{x} = 2 \implies x = 4$$

$$u = 3 \implies \sqrt{x} = 3 \implies x = 9$$

b. $x^{2/3} - x^{1/3} - 6 = 0$ Original equation

This equation is of quadratic form with $u = x^{1/3}$.

$$(x^{1/3})^2 - (x^{1/3}) - 6 = 0$$ Write in quadratic form.

$$u^2 - u - 6 = 0$$ Substitute u for $x^{1/3}$.

$$(u + 2)(u - 3) = 0$$ Factor.

$$u + 2 = 0 \implies u = -2$$ Set 1st factor equal to 0.

$$u - 3 = 0 \implies u = 3$$ Set 2nd factor equal to 0.

Now, using the u-solutions of -2 and 3, you obtain the x-solutions as follows.

$$u = -2 \implies x^{1/3} = -2 \implies x = -8$$

$$u = 3 \implies x^{1/3} = 3 \implies x = 27$$

Example 6 **Surface Area of a Softball**

The surface area of a sphere of radius r is given by $S = 4\pi r^2$. The surface area of a softball is $144/\pi$ square inches. Find the diameter d of the softball.

Solution

$$\frac{144}{\pi} = 4\pi r^2$$ Substitute $144/\pi$ for S.

$$\frac{36}{\pi^2} = r^2 \implies \pm\sqrt{\frac{36}{\pi^2}} = r$$ Divide each side by 4π and use Square Root Property.

Choosing the positive root, you obtain $r = 6/\pi$, and so the diameter of the softball is

$$d = 2r = 2\left(\frac{6}{\pi}\right) = \frac{12}{\pi} \approx 3.82 \text{ inches.}$$

8.1 Exercises

Review　Concepts, Skills, and Problem Solving

Keep mathematically in shape by doing these exercises *before* the problems of this section.

Properties and Definitions

1. *Writing* ✐　Identify the leading coefficient in $5t - 3t^3 + 7t^2$. Explain.

2. *Writing* ✐　State the degree of the product $(y^2 - 2)(y^3 + 7)$. Explain.

3. *Writing* ✐　Sketch a graph for which y is not a function of x. Explain why it is not a function.

4. *Writing* ✐　Sketch a graph for which y is a function of x. Explain why it is a function.

Simplifying Expressions

In Exercises 5–10, simplify the expression.

5. $(x^3 \cdot x^{-2})^{-3}$

6. $(5x^{-4}y^5)(-3x^2y^{-1})$

7. $\left(\dfrac{2x}{3y}\right)^{-2}$

8. $\left(\dfrac{7u^{-4}}{3v^{-2}}\right)\left(\dfrac{14u}{6v^2}\right)^{-1}$

9. $\dfrac{6u^2v^{-3}}{27uv^3}$

10. $\dfrac{-14r^4s^2}{-98rs^2}$

Problem Solving

11. *Predator-Prey*　The number N of prey t months after a natural predator is introduced into the test area is inversely proportional to the square root of $t + 1$. If $N = 300$ when $t = 0$, find N when $t = 8$.

12. *Travel Time*　The travel time between two cities is inversely proportional to the average speed. A train travels between two cities in 2 hours at an average speed of 58 miles per hour. How long would the trip take at an average speed of 72 miles per hour? What does the constant of proportionality measure in this problem?

Developing Skills

In Exercises 1–20, solve the equation by factoring. See Example 1.

1. $x^2 - 12x + 35 = 0$

2. $x^2 + 15x + 44 = 0$

3. $x^2 - x - 30 = 0$

4. $x^2 + 2x - 63 = 0$

5. $x^2 + 4x = 45$

6. $x^2 - 7x = 18$

7. $x^2 - 12x + 36 = 0$

8. $x^2 + 60x + 900 = 0$

9. $9x^2 + 24x + 16 = 0$

10. $8x^2 - 10x + 3 = 0$

11. $4x^2 - 12x = 0$

12. $25y^2 - 75y = 0$

13. $u(u - 9) - 12(u - 9) = 0$

14. $16x(x - 8) - 12(x - 8) = 0$

15. $3x(x - 6) - 5(x - 6) = 0$

16. $3(4 - x) - 2x(4 - x) = 0$

17. $(y - 4)(y - 3) = 6$

18. $(6 + u)(1 - u) = 10$

19. $2x(3x + 2) = 5 - 6x^2$

20. $(2z + 1)(2z - 1) = -4z^2 - 5z + 2$

In Exercises 21–42, solve the equation by using the Square Root Property. See Example 2.

21. $x^2 = 49$

22. $z^2 = 144$

23. $6x^2 = 54$

24. $5t^2 = 5$

25. $25x^2 = 16$

26. $9z^2 = 121$

27. $\dfrac{y^2}{2} = 32$

28. $\dfrac{x^2}{6} = 24$

29. $4x^2 - 25 = 0$

30. $16y^2 - 121 = 0$

31. $4u^2 - 225 = 0$

32. $16x^2 - 1 = 0$

33. $(x + 4)^2 = 64$

34. $(y - 20)^2 = 25$

35. $(x - 3)^2 = 0.25$

36. $(x + 2)^2 = 0.81$

37. $(x - 2)^2 = 7$

38. $(x + 8)^2 = 28$

39. $(2x + 1)^2 = 50$

40. $(3x - 5)^2 = 48$

41. $(4x - 3)^2 - 98 = 0$

42. $(5x + 11)^2 - 300 = 0$

In Exercises 43–64, solve the equation by using the Square Root Property. See Example 3.

43. $z^2 = -36$

44. $x^2 = -16$

45. $x^2 + 4 = 0$

46. $y^2 + 16 = 0$

47. $9u^2 + 17 = 0$

48. $4v^2 + 9 = 0$

49. $(t - 3)^2 = -25$

50. $(x + 5)^2 = -81$

51. $(3z + 4)^2 + 144 = 0$

52. $(2y - 3)^2 + 25 = 0$

53. $(2x + 3)^2 = -54$

54. $(6y - 5)^2 = -8$

55. $9(x + 6)^2 = -121$

56. $4(x - 4)^2 = -169$

57. $(x - 1)^2 = -27$

58. $(2x + 3)^2 = -54$

59. $(x + 1)^2 + 0.04 = 0$

60. $(x - 3)^2 + 2.25 = 0$

61. $\left(c - \frac{2}{3}\right)^2 + \frac{1}{9} = 0$

62. $\left(u + \frac{5}{8}\right)^2 + \frac{49}{16} = 0$

63. $\left(x + \frac{7}{3}\right)^2 = -\frac{38}{9}$

64. $\left(y - \frac{5}{6}\right)^2 = -\frac{4}{5}$

In Exercises 65–80, find all real and complex solutions of the quadratic equation.

65. $2x^2 - 5x = 0$

66. $3t^2 + 6t = 0$

67. $2x^2 + 5x - 12 = 0$

68. $3x^2 + 8x - 16 = 0$

69. $x^2 - 900 = 0$

70. $y^2 - 225 = 0$

71. $x^2 + 900 = 0$

72. $y^2 + 225 = 0$

73. $\frac{2}{3}x^2 = 6$

74. $\frac{1}{3}x^2 = 4$

75. $(x - 5)^2 - 100 = 0$

76. $(y + 12)^2 - 400 = 0$

77. $(x - 5)^2 + 100 = 0$

78. $(y + 12)^2 + 400 = 0$

79. $(x + 2)^2 + 18 = 0$

80. $(x + 2)^2 - 18 = 0$

In Exercises 81–90, use a graphing calculator to graph the function. Use the graph to approximate any x-intercepts. Set $y = 0$ and solve the resulting equation. Compare the result with the x-intercepts of the graph.

81. $y = x^2 - 9$

82. $y = 5x - x^2$

83. $y = x^2 - 2x - 15$

84. $y = 9 - 4(x - 3)^2$

85. $y = 4 - (x - 3)^2$

86. $y = 4(x + 1)^2 - 9$

87. $y = 2x^2 - x - 6$

88. $y = 4x^2 - x - 14$

89. $y = 3x^2 - 8x - 16$

90. $y = 5x^2 + 9x - 18$

In Exercises 91–96, use a graphing calculator to graph the function and observe that the graph has no x-intercepts. Set $y = 0$ and solve the resulting equation. Identify the type of solutions of the equation.

91. $y = x^2 + 7$

92. $y = x^2 + 5$

93. $y = (x - 1)^2 + 1$

94. $y = (x + 2)^2 + 3$

95. $y = (x + 3)^2 + 5$

96. $y = (x - 2)^2 + 3$

In Exercises 97–100, solve for y in terms of x. Let f and g be functions representing, respectively, the positive square root and the negative square root. Use a graphing calculator to graph f and g in the same viewing window.

97. $x^2 + y^2 = 4$

98. $x^2 - y^2 = 4$

99. $x^2 + 4y^2 = 4$

100. $x - y^2 = 0$

In Exercises 101–130, solve the equation of quadratic form. (Find all real *and* complex solutions.) See Examples 4–5.

101. $x^4 - 5x^2 + 4 = 0$

102. $x^4 - 10x^2 + 25 = 0$

103. $x^4 - 5x^2 + 6 = 0$

104. $x^4 - 11x^2 + 30 = 0$

105. $(x^2 - 4)^2 + 2(x^2 - 4) - 3 = 0$

106. $(x^2 - 1)^2 + (x^2 - 1) - 6 = 0$

107. $x - 3\sqrt{x} - 4 = 0$

108. $x - \sqrt{x} - 6 = 0$

109. $x - 7\sqrt{x} + 10 = 0$

110. $x - 11\sqrt{x} + 24 = 0$

111. $x^{2/3} - x^{1/3} - 6 = 0$

112. $x^{2/3} + 3x^{1/3} - 10 = 0$

113. $2x^{2/3} - 7x^{1/3} + 5 = 0$

114. $3x^{2/3} + 8x^{1/3} + 5 = 0$

115. $x^{2/5} - 3x^{1/5} + 2 = 0$

116. $x^{2/5} + 5x^{1/5} + 6 = 0$

117. $2x^{2/5} - 7x^{1/5} + 3 = 0$

118. $2x^{2/5} + 3x^{1/5} + 1 = 0$

119. $x^{1/3} - x^{1/6} - 6 = 0$

120. $x^{1/3} + 2x^{1/6} - 3 = 0$

121. $x^{1/2} - 3x^{1/4} + 2 = 0$

122. $x^{1/2} - 5x^{1/4} + 6 = 0$

123. $\dfrac{1}{x^2} - \dfrac{3}{x} + 2 = 0$

124. $\dfrac{1}{x^2} - \dfrac{1}{x} - 6 = 0$

125. $4x^{-2} - x^{-1} - 5 = 0$

126. $2x^{-2} - x^{-1} - 1 = 0$

127. $(x^2 - 3x)^2 - 2(x^2 - 3x) - 24 = 0$

128. $(x^2 - 6x)^2 - 2(x^2 - 6x) - 35 = 0$

129. $16\left(\dfrac{x - 1}{x - 8}\right)^2 + 8\left(\dfrac{x - 1}{x - 8}\right) + 1 = 0$

130. $9\left(\dfrac{x + 2}{x + 3}\right)^2 - 6\left(\dfrac{x + 2}{x + 3}\right) + 1 = 0$

Solving Problems

131. ▲ *Geometry*   The surface area S of a spherical float for a parade is 289π square feet. Find the diameter d of the float.

132. ▲ *Geometry*   The surface area S of a basketball is $900/\pi$ square inches. Find the radius r of the basketball.

Free-Falling Object   In Exercises 133–136, find the time required for an object to reach the ground when it is dropped from a height of s_0 feet. The height h (in feet) is given by

$$h = -16t^2 + s_0$$

where t measures the time (in seconds) after the object is released.

133. $s_0 = 256$ **134.** $s_0 = 48$

135. $s_0 = 128$ **136.** $s_0 = 500$

137. *Free-Falling Object*   The height h (in feet) of an object thrown vertically upward from a tower 144 feet tall is given by $h = 144 + 128t - 16t^2$, where t measures the time in seconds from the time when the object is released. How long does it take for the object to reach the ground?

138. *Revenue*   The revenue R (in dollars) from selling x televisions is given by $R = x\left(120 - \frac{1}{2}x\right)$. Find the number of televisions that must be sold to produce a revenue of $7000.

Compound Interest   The amount A after 2 years when a principal of P dollars is invested at annual interest rate r compounded annually is given by $A = P(1 + r)^2$. In Exercises 139 and 140, find r.

139. $P = \$1500$, $A = \$1685.40$

140. $P = \$5000$, $A = \$5724.50$

National Health Expenditures   In Exercises 141 and 142, the national expenditures for health care in the United States from 1995 through 2001 is given by

$$y = 4.43t^2 + 872, \quad 5 \le t \le 11.$$

In this model, y represents the expenditures (in billions of dollars) and t represents the year, with $t = 5$ corresponding to 1995 (see figure).   (Source: U.S. Centers for Medicare & Medicaid Services)

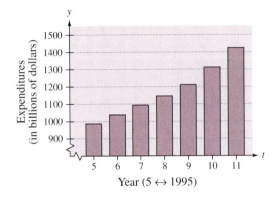

Figure for 141 and 142

141. Algebraically determine the year when expenditures were approximately $1100 billion. Graphically confirm the result.

142. Algebraically determine the year when expenditures were approximately $1200 billion. Graphically confirm the result.

Explaining Concepts

143. Answer parts (a) and (b) of Motivating the Chapter on page 506.

144. *Writing*   For a quadratic equation $ax^2 + bx + c = 0$, where a, b, and c are real numbers with $a \ne 0$, explain why b and c can equal 0, but a cannot.

145. *Writing*   Explain the Zero-Factor Property and how it can be used to solve a quadratic equation.

146. Is it possible for a quadratic equation to have only one solution? If so, give an example.

147. *True or False?*   The only solution of the equation $x^2 = 25$ is $x = 5$. Justify your answer.

148. *Writing*   Describe the steps in solving a quadratic equation by using the Square Root Property.

149. *Writing*   Describe the procedure for solving an equation of quadratic form. Give an example.

8.2 Completing the Square

Lon C. Diehl/PhotoEdit, Inc.

What You Should Learn

1. Rewrite quadratic expressions in completed square form.
2. Solve quadratic equations by completing the square.

Why You Should Learn It

You can use techniques such as completing the square to solve quadratic equations that model real-life situations. For instance, Example 7 on page 520 shows how to find the dimensions of a cereal box by completing the square.

1 Rewrite quadratic expressions in completed square form.

Constructing Perfect Square Trinomials

Consider the quadratic equation

$$(x - 2)^2 = 10. \qquad \text{\textcolor{red}{Completed square form}}$$

You know from Example 2(b) in the preceding section that this equation has two solutions: $x = 2 + \sqrt{10}$ and $x = 2 - \sqrt{10}$. Suppose you were given the equation in its general form

$$x^2 - 4x - 6 = 0. \qquad \text{\textcolor{red}{General form}}$$

How could you solve this form of the quadratic equation? You could try factoring, but after attempting to do so you would find that the left side of the equation is not factorable using integer coefficients.

In this section, you will study a technique for rewriting an equation in a completed square form. This technique is called **completing the square.** Note that prior to completing the square, the coefficient of the second-degree term must be 1.

Completing the Square

To **complete the square** for the expression $x^2 + bx$, add $(b/2)^2$, which is the square of half the coefficient of x. Consequently,

$$x^2 + bx + \left(\frac{b}{2}\right)^2 = \left(x + \frac{b}{2}\right)^2.$$

$$\underbrace{\qquad}_{\text{(half)}^2}$$

Example 1   Constructing a Perfect Square Trinomial

What term should be added to $x^2 - 8x$ so that it becomes a perfect square trinomial? To find this term, notice that the coefficient of the x-term is -8. Take half of this coefficient and square the result to get $(-4)^2 = 16$. Add this term to the expression to make it a perfect square trinomial.

$$x^2 - 8x + (-4)^2 = x^2 - 8x + 16 \qquad \text{\textcolor{red}{Add $(-4)^2 = 16$ to the expression.}}$$

You can then rewrite the expression as the square of a binomial, $(x - 4)^2$.

2 Solve quadratic equations by completing the square.

Solving Equations by Completing the Square

Completing the square can be used to solve quadratic equations. When using this procedure, remember to *preserve the equality* by adding the same constant to each side of the equation.

Example 2 Completing the Square: Leading Coefficient Is 1

Solve $x^2 + 12x = 0$ by completing the square.

Solution

$x^2 + 12x = 0$	Write original equation.
$x^2 + 12x + 6^2 = 36$	Add $6^2 = 36$ to each side.

$\underbrace{\qquad}_{\text{(half)}^2}$

$(x + 6)^2 = 36$	Completed square form
$x + 6 = \pm\sqrt{36}$	Square Root Property
$x = -6 \pm 6$	Subtract 6 from each side.
$x = -6 + 6$ or $x = -6 - 6$	Separate solutions.
$x = 0 \qquad\qquad x = -12$	Solutions

The solutions are $x = 0$ and $x = -12$. Check these in the original equation.

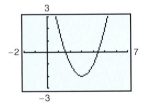
Example 3 Completing the Square: Leading Coefficient Is 1

Solve $x^2 - 6x + 7 = 0$ by completing the square.

Solution

$x^2 - 6x + 7 = 0$	Write original equation.
$x^2 - 6x = -7$	Subtract 7 from each side.
$x^2 - 6x + (-3)^2 = -7 + 9$	Add $(-3)^2 = 9$ to each side.

$\underbrace{\qquad}_{\text{(half)}^2}$

$(x - 3)^2 = 2$	Completed square form
$x - 3 = \pm\sqrt{2}$	Square Root Property
$x = 3 \pm \sqrt{2}$	Add 3 to each side.
$x = 3 + \sqrt{2}$ or $x = 3 - \sqrt{2}$	Solutions

The solutions are $x = 3 + \sqrt{2} \approx 4.41$ and $x = 3 - \sqrt{2} \approx 1.59$. Check these in the original equation.

If the leading coefficient of a quadratic equation is not 1, you must divide each side of the equation by this coefficient *before* completing the square. This process is demonstrated in Examples 4 and 5.

Example 4   Completing the Square: Leading Coefficient Is Not 1

$$2x^2 - x - 2 = 0 \qquad \text{Original equation}$$

$$2x^2 - x = 2 \qquad \text{Add 2 to each side.}$$

$$x^2 - \frac{1}{2}x = 1 \qquad \text{Divide each side by 2.}$$

$$x^2 - \frac{1}{2}x + \left(-\frac{1}{4}\right)^2 = 1 + \frac{1}{16} \qquad \text{Add } \left(-\frac{1}{4}\right)^2 = \frac{1}{16} \text{ to each side.}$$

$$\left(x - \frac{1}{4}\right)^2 = \frac{17}{16} \qquad \text{Completed square form}$$

$$x - \frac{1}{4} = \pm \frac{\sqrt{17}}{4} \qquad \text{Square Root Property}$$

$$x = \frac{1}{4} \pm \frac{\sqrt{17}}{4} \qquad \text{Add } \tfrac{1}{4} \text{ to each side.}$$

The solutions are $x = \frac{1}{4}\left(1 + \sqrt{17}\right)$ and $x = \frac{1}{4}\left(1 - \sqrt{17}\right)$. Check these in the original equation.

Example 5   Completing the Square: Leading Coefficient Is Not 1

$$3x^2 - 6x + 1 = 0 \qquad \text{Original equation}$$

$$3x^2 - 6x = -1 \qquad \text{Subtract 1 from each side.}$$

$$x^2 - 2x = -\frac{1}{3} \qquad \text{Divide each side by 3.}$$

$$x^2 - 2x + (-1)^2 = -\frac{1}{3} + 1 \qquad \text{Add } (-1)^2 = 1 \text{ to each side.}$$

$$(x - 1)^2 = \frac{2}{3} \qquad \text{Completed square form}$$

$$x - 1 = \pm \sqrt{\frac{2}{3}} \qquad \text{Square Root Property}$$

$$x - 1 = \pm \frac{\sqrt{6}}{3} \qquad \text{Rationalize the denominator.}$$

$$x = 1 \pm \frac{\sqrt{6}}{3} \qquad \text{Add 1 to each side.}$$

The solutions are $x = 1 + \sqrt{6}/3$ and $x = 1 - \sqrt{6}/3$. Check these in the original equation.

Example 6   A Quadratic Equation with Complex Solutions

Solve $x^2 - 4x + 8 = 0$ by completing the square.

Solution

$$x^2 - 4x + 8 = 0 \qquad \text{Write original equation.}$$

$$x^2 - 4x = -8 \qquad \text{Subtract 8 from each side.}$$

$$x^2 - 4x + (-2)^2 = -8 + 4 \qquad \text{Add } (-2)^2 = 4 \text{ to each side.}$$

$$(x - 2)^2 = -4 \qquad \text{Completed square form}$$

$$x - 2 = \pm 2i \qquad \text{Square Root Property}$$

$$x = 2 \pm 2i \qquad \text{Add 2 to each side.}$$

The solutions are $x = 2 + 2i$ and $x = 2 - 2i$. Check these in the original equation.

Example 7   Dimensions of a Cereal Box

A cereal box has a volume of 441 cubic inches. Its height is 12 inches and its base has the dimensions x by $x + 7$ (see Figure 8.1). Find the dimensions of the base in inches.

12 in.

$x + 7$ x

Figure 8.1

Solution

$$lwh = V \qquad \text{Formula for volume of a rectangular box}$$

$$(x + 7)(x)(12) = 441 \qquad \text{Substitute 441 for } V, x + 7 \text{ for } l, x \text{ for } w, \text{ and 12 for } h.$$

$$12x^2 + 84x = 441 \qquad \text{Multiply factors.}$$

$$x^2 + 7x = \frac{441}{12} \qquad \text{Divide each side by 12.}$$

$$x^2 + 7x + \left(\frac{7}{2}\right)^2 = \frac{147}{4} + \frac{49}{4} \qquad \text{Add } \left(\frac{7}{2}\right)^2 = \frac{49}{4} \text{ to each side.}$$

$$\left(x + \frac{7}{2}\right)^2 = \frac{196}{4} \qquad \text{Completed square form}$$

$$x + \frac{7}{2} = \pm \sqrt{49} \qquad \text{Square Root Property}$$

$$x = -\frac{7}{2} \pm 7 \qquad \text{Subtract } \tfrac{7}{2} \text{ from each side.}$$

Choosing the positive root, you obtain

$$x = -\frac{7}{2} + 7 = \frac{7}{2} = 3.5 \text{ inches} \qquad \text{Width of base}$$

and

$$x + 7 = 3.5 + 7 = 10.5 \text{ inches.} \qquad \text{Length of base}$$

8.2 Exercises

Review   Concepts, Skills, and Problem Solving

Keep mathematically in shape by doing these exercises *before* the problems of this section.

Properties and Definitions

In Exercises 1–4, complete the rule of exponents and/or simplify.

1. $(ab)^4 = \rule{1cm}{0.4pt}$ **2.** $(a^r)^s = \rule{1cm}{0.4pt}$

3. $\left(\dfrac{a}{b}\right)^{-r} = \rule{1cm}{0.4pt}$, $a \neq 0, b \neq 0$

4. $a^{-r} = \rule{1cm}{0.4pt}$, $a \neq 0$

Solving Equations

In Exercises 5–8, solve the equation.

5. $\dfrac{4}{x} - \dfrac{2}{3} = 0$

6. $2x - 3[1 + (4 - x)] = 0$

7. $3x^2 - 13x - 10 = 0$

8. $x(x - 3) = 40$

Graphing

In Exercises 9–12, graph the function.

9. $g(x) = \frac{2}{3}x - 5$

10. $h(x) = 5 - \sqrt{x}$

11. $f(x) = \dfrac{4}{x + 2}$

12. $f(x) = 2x + |x - 1|$

Developing Skills

In Exercises 1–16, add a term to the expression so that it becomes a perfect square trinomial. See Example 1.

1. $x^2 + 8x + \rule{1cm}{0.4pt}$ **2.** $x^2 + 12x + \rule{1cm}{0.4pt}$

3. $y^2 - 20y + \rule{1cm}{0.4pt}$ **4.** $y^2 - 2y + \rule{1cm}{0.4pt}$

5. $x^2 - 16x + \rule{1cm}{0.4pt}$ **6.** $x^2 + 18x + \rule{1cm}{0.4pt}$

7. $t^2 + 5t + \rule{1cm}{0.4pt}$ **8.** $u^2 + 7u + \rule{1cm}{0.4pt}$

9. $x^2 - 9x + \rule{1cm}{0.4pt}$ **10.** $y^2 - 11y + \rule{1cm}{0.4pt}$

11. $a^2 - \frac{1}{3}a + \rule{1cm}{0.4pt}$ **12.** $y^2 + \frac{4}{3}y + \rule{1cm}{0.4pt}$

13. $y^2 - \frac{3}{5}y + \rule{1cm}{0.4pt}$ **14.** $x^2 - \frac{6}{5}x + \rule{1cm}{0.4pt}$

15. $r^2 - 0.4r + \rule{1cm}{0.4pt}$ **16.** $s^2 + 4.6s + \rule{1cm}{0.4pt}$

In Exercises 17–32, solve the equation first by completing the square and then by factoring. See Examples 2–5.

17. $x^2 - 20x = 0$ **18.** $x^2 + 32x = 0$

19. $x^2 + 6x = 0$ **20.** $t^2 - 10t = 0$

21. $y^2 - 5y = 0$ **22.** $t^2 - 9t = 0$

23. $t^2 - 8t + 7 = 0$ **24.** $y^2 - 8y + 12 = 0$

25. $x^2 + 7x + 12 = 0$ **26.** $z^2 + 3z - 10 = 0$

27. $x^2 - 3x - 18 = 0$ **28.** $t^2 - 5t - 36 = 0$

29. $2x^2 - 14x + 12 = 0$ **30.** $3x^2 - 3x - 6 = 0$

31. $4x^2 + 4x - 15 = 0$ **32.** $3x^2 - 13x + 12 = 0$

In Exercises 33–72, solve the equation by completing the square. Give the solutions in exact form and in decimal form rounded to two decimal places. (The solutions may be complex numbers.) See Examples 2–6.

33. $x^2 - 4x - 3 = 0$ **34.** $x^2 - 6x + 7 = 0$

35. $x^2 + 4x - 3 = 0$ **36.** $x^2 + 6x + 7 = 0$

37. $x^2 + 6x = 7$ **38.** $x^2 + 8x = 9$

39. $x^2 - 10x = 22$ **40.** $x^2 - 4x = -9$

41. $x^2 + 8x + 7 = 0$ **42.** $x^2 + 10x + 9 = 0$

43. $x^2 - 10x + 21 = 0$ **44.** $x^2 - 10x + 24 = 0$ **61.** $3x^2 + 9x + 5 = 0$ **62.** $5x^2 - 15x + 7 = 0$

45. $y^2 + 5y + 3 = 0$

63. $4y^2 + 4y - 9 = 0$

46. $y^2 + 6y + 7 = 0$ **47.** $x^2 + 10 = 6x$

48. $x^2 + 23 = 10x$ **49.** $z^2 + 4z + 13 = 0$ **64.** $4z^2 - 3z + 2 = 0$

50. $z^2 + 12z + 25 = 0$

65. $5x^2 - 3x + 10 = 0$

51. $-x^2 + x - 1 = 0$ **52.** $1 - x - x^2 = 0$

66. $7x^2 + 4x + 3 = 0$

53. $x^2 - 7x + 12 = 0$
54. $y^2 + 5y + 9 = 0$

67. $x(x - 7) = 2$ **68.** $2x\left(x + \dfrac{4}{3}\right) = 5$

55. $x^2 - \frac{2}{3}x - 3 = 0$ **56.** $x^2 + \frac{4}{5}x - 1 = 0$

69. $0.5t^2 + t + 2 = 0$

57. $v^2 + \frac{3}{4}v - 2 = 0$

70. $0.1x^2 + 0.5x = -0.2$

58. $u^2 - \frac{2}{3}u + 5 = 0$

71. $0.1x^2 + 0.2x + 0.5 = 0$
72. $0.02x^2 + 0.10x - 0.05 = 0$

59. $2x^2 + 8x + 3 = 0$ **60.** $3x^2 - 24x - 5 = 0$

In Exercises 73–78, find the real solutions.

73. $\dfrac{x}{2} - \dfrac{1}{x} = 1$

74. $\dfrac{x}{2} + \dfrac{5}{x} = 4$

75. $\dfrac{x^2}{4} = \dfrac{x+1}{2}$

76. $\dfrac{x^2+2}{24} = \dfrac{x-1}{3}$

77. $\sqrt{2x+1} = x - 3$

78. $\sqrt{3x-2} = x - 2$

In Exercises 79–86, use a graphing calculator to graph the function. Use the graph to approximate any x-intercepts of the graph. Set $y = 0$ and solve the resulting equation. Compare the result with the x-intercepts of the graph.

79. $y = x^2 + 4x - 1$

80. $y = x^2 + 6x - 4$

81. $y = x^2 - 2x - 5$

82. $y = 2x^2 - 6x - 5$

83. $y = \frac{1}{3}x^2 + 2x - 6$

84. $y = \frac{1}{2}x^2 - 3x + 1$

85. $y = -x^2 - x + 3$

86. $y = \sqrt{x} - x + 2$

Solving Problems

87. ▲ *Geometric Modeling*

(a) Find the area of the two adjoining rectangles and large square in the figure.

(b) Find the area of the small square in the lower right-hand corner of the figure and add it to the area found in part (a).

(c) Find the dimensions and the area of the entire figure after adjoining the small square in the lower right-hand corner of the figure. Note that you have shown geometrically the technique of completing the square.

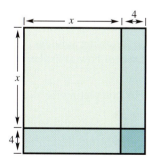

Figure for 87

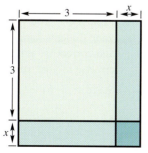

Figure for 88

88. ▲ *Geometric Modeling*   Repeat Exercise 87 for the model shown above.

89. ▲ *Geometry*   The area of the triangle in the figure is 12 square centimeters. Find the base and height.

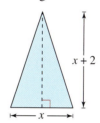

Figure for 89

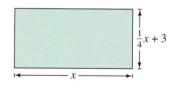

Figure for 90

90. ▲ *Geometry*   The area of the rectangle in the figure is 160 square feet. Find the rectangle's dimensions.

91. ▲ *Geometry*   You have 200 meters of fencing to enclose two adjacent rectangular corrals (see figure). The total area of the enclosed region is 1400 square meters. What are the dimensions of each corral? (The corrals are the same size.)

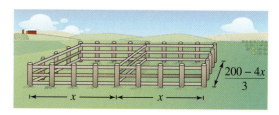

92. ▲ *Geometry* An open box with a rectangular base of x inches by $x + 4$ inches has a height of 6 inches (see figure). The volume of the box is 840 cubic inches. Find the dimensions of the box.

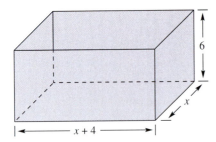

93. *Revenue* The revenue R (in dollars) from selling x pairs of running shoes is given by

$$R = x\left(50 - \frac{1}{2}x\right).$$

Find the number of pairs of running shoes that must be sold to produce a revenue of $1218.

94. *Revenue* The revenue R (in dollars) from selling x golf clubs is given by

$$R = x\left(100 - \frac{1}{10}x\right).$$

Find the number of golf clubs that must be sold to produce a revenue of $11,967.90.

Explaining Concepts

95. *Writing* What is a perfect square trinomial?

96. *Writing* What term must be added to $x^2 + 5x$ to complete the square? Explain how you found the term.

97. *Writing* Explain the use of the Square Root Property when solving a quadratic equation by the method of completing the square.

98. Is it possible for a quadratic equation to have no real number solution? If so, give an example.

99. *Writing* When using the method of completing the square to solve a quadratic equation, what is the first step if the leading coefficient is not 1? Is the resulting equation equivalent to the original equation? Explain.

100. *True or False?* If you solve a quadratic equation by completing the square and obtain solutions that are rational numbers, then you could have solved the equation by factoring. Justify your answer.

101. *Writing* Consider the quadratic equation $(x - 1)^2 = d$.

(a) What value(s) of d will produce a quadratic equation that has exactly one (repeated) solution?

(b) Describe the value(s) of d that will produce two different solutions, both of which are *rational* numbers.

(c) Describe the value(s) of d that will produce two different solutions, both of which are *irrational* numbers.

(d) Describe the value(s) of d that will produce two different solutions, both of which are *complex* numbers.

102. *Writing* You teach an algebra class and one of your students hands in the following solution. Find and correct the error(s). Discuss how to explain the error(s) to your student.

Solve $x^2 + 6x - 13 = 0$ by completing the square.

$$x^2 + 6x = 13$$
$$x^2 + 6x + \left(\frac{6}{2}\right)^2 = 13$$
$$(x + 3)^2 = 13$$
$$x + 3 = \pm\sqrt{13}$$
$$x = -3 \pm \sqrt{13}$$

8.3 The Quadratic Formula

Mug Shots/Corbis

What You Should Learn

1. Derive the Quadratic Formula by completing the square for a general quadratic equation.
2. Use the Quadratic Formula to solve quadratic equations.
3. Determine the types of solutions of quadratic equations using the discriminant.
4. Write quadratic equations from solutions of the equations.

Why You Should Learn It

Knowing the Quadratic Formula can be helpful in solving quadratic equations that model real-life situations. For instance, in Exercise 117 on page 533, you will solve a quadratic equation that models the number of people employed in the construction industry.

1 Derive the Quadratic Formula by completing the square for a general quadratic equation.

The Quadratic Formula

A fourth technique for solving a quadratic equation involves the **Quadratic Formula.** This formula is derived by completing the square for a general quadratic equation.

$$ax^2 + bx + c = 0 \qquad \text{General form, } a \neq 0$$

$$ax^2 + bx = -c \qquad \text{Subtract } c \text{ from each side.}$$

$$x^2 + \frac{b}{a}x = -\frac{c}{a} \qquad \text{Divide each side by } a.$$

$$x^2 + \frac{b}{a}x + \left(\frac{b}{2a}\right)^2 = -\frac{c}{a} + \left(\frac{b}{2a}\right)^2 \qquad \text{Add } \left(\frac{b}{2a}\right)^2 \text{ to each side.}$$

$$\left(x + \frac{b}{2a}\right)^2 = \frac{b^2 - 4ac}{4a^2} \qquad \text{Simplify.}$$

$$x + \frac{b}{2a} = \pm\sqrt{\frac{b^2 - 4ac}{4a^2}} \qquad \text{Square Root Property}$$

$$x = -\frac{b}{2a} \pm \frac{\sqrt{b^2 - 4ac}}{2|a|} \qquad \text{Subtract } \frac{b}{2a} \text{ from each side.}$$

$$x = \frac{-b \pm \sqrt{b^2 - 4ac}}{2a} \qquad \text{Simplify.}$$

Study Tip

The Quadratic Formula is one of the most important formulas in algebra, and you should memorize it. It helps to try to memorize a verbal statement of the rule. For instance, you might try to remember the following verbal statement of the Quadratic Formula: "The opposite of b, plus or minus the square root of b squared minus $4ac$, all divided by $2a$."

The Quadratic Formula

The solutions of $ax^2 + bx + c = 0$, $a \neq 0$, are given by the **Quadratic Formula**

$$x = \frac{-b \pm \sqrt{b^2 - 4ac}}{2a}.$$

The expression inside the radical, $b^2 - 4ac$, is called the **discriminant.**

1. If $b^2 - 4ac > 0$, the equation has *two* real solutions.

2. If $b^2 - 4ac = 0$, the equation has *one* (repeated) real solution.

3. If $b^2 - 4ac < 0$, the equation has *no* real solutions.

② Use the Quadratic Formula to solve quadratic equations.

Solving Equations by the Quadratic Formula

When using the Quadratic Formula, remember that *before* the formula can be applied, you must first write the quadratic equation in general form in order to determine the values of a, b, and c.

Example 1   The Quadratic Formula: Two Distinct Solutions

$$x^2 + 6x = 16 \qquad \text{Original equation}$$

$$x^2 + 6x - 16 = 0 \qquad \text{Write in general form.}$$

$$x = \frac{-b \pm \sqrt{b^2 - 4ac}}{2a} \qquad \text{Quadratic Formula}$$

$$x = \frac{-6 \pm \sqrt{6^2 - 4(1)(-16)}}{2(1)} \qquad \text{Substitute 1 for } a, \text{ 6 for } b, \text{ and } -16 \text{ for } c.$$

$$x = \frac{-6 \pm \sqrt{100}}{2} \qquad \text{Simplify.}$$

$$x = \frac{-6 \pm 10}{2} \qquad \text{Simplify.}$$

$$x = 2 \quad \text{or} \quad x = -8 \qquad \text{Solutions}$$

The solutions are $x = 2$ and $x = -8$. Check these in the original equation.

> **Study Tip**
>
> In Example 1, the solutions are rational numbers, which means that the equation could have been solved by factoring. Try solving the equation by factoring.

Example 2   The Quadratic Formula: Two Distinct Solutions

$$-x^2 - 4x + 8 = 0 \qquad \text{Leading coefficient is negative.}$$

$$x^2 + 4x - 8 = 0 \qquad \text{Multiply each side by } -1.$$

$$x = \frac{-b \pm \sqrt{b^2 - 4ac}}{2a} \qquad \text{Quadratic Formula}$$

$$x = \frac{-4 \pm \sqrt{4^2 - 4(1)(-8)}}{2(1)} \qquad \text{Substitute 1 for } a, \text{ 4 for } b, \text{ and } -8 \text{ for } c.$$

$$x = \frac{-4 \pm \sqrt{48}}{2} \qquad \text{Simplify.}$$

$$x = \frac{-4 \pm 4\sqrt{3}}{2} \qquad \text{Simplify.}$$

$$x = \frac{2(-2 \pm 2\sqrt{3})}{2} \qquad \text{Factor numerator.}$$

$$x = \frac{2(-2 \pm 2\sqrt{3})}{2} \qquad \text{Divide out common factor.}$$

$$x = -2 \pm 2\sqrt{3} \qquad \text{Solutions}$$

> **Study Tip**
>
> If the leading coefficient of a quadratic equation is negative, you should begin by multiplying each side of the equation by -1, as shown in Example 2. This will produce a positive leading coefficient, which is less cumbersome to work with.

The solutions are $x = -2 + 2\sqrt{3}$ and $x = -2 - 2\sqrt{3}$. Check these in the original equation.

Example 3   The Quadratic Formula: One Repeated Solution

$18x^2 - 24x + 8 = 0$	Original equation
$9x^2 - 12x + 4 = 0$	Divide each side by 2.
$x = \dfrac{-b \pm \sqrt{b^2 - 4ac}}{2a}$	Quadratic Formula
$x = \dfrac{-(-12) \pm \sqrt{(-12)^2 - 4(9)(4)}}{2(9)}$	Substitute 9 for a, -12 for b, and 4 for c.
$x = \dfrac{12 \pm \sqrt{144 - 144}}{18}$	Simplify.
$x = \dfrac{12 \pm \sqrt{0}}{18}$	Simplify.
$x = \dfrac{2}{3}$	Solution

The only solution is $x = \frac{2}{3}$. Check this in the original equation.

Note in the next example how the Quadratic Formula can be used to solve a quadratic equation that has complex solutions.

Example 4   The Quadratic Formula: Complex Solutions

$2x^2 - 4x + 5 = 0$	Original equation
$x = \dfrac{-b \pm \sqrt{b^2 - 4ac}}{2a}$	Quadratic Formula
$x = \dfrac{-(-4) \pm \sqrt{(-4)^2 - 4(2)(5)}}{2(2)}$	Substitute 2 for a, -4 for b, and 5 for c.
$x = \dfrac{4 \pm \sqrt{-24}}{4}$	Simplify.
$x = \dfrac{4 \pm 2\sqrt{6}\,i}{4}$	Write in i-form.
$x = \dfrac{2(2 \pm \sqrt{6}\,i)}{2 \cdot 2}$	Factor numerator and denominator.
$x = \dfrac{2(2 \pm \sqrt{6}\,i)}{2 \cdot 2}$	Divide out common factor.
$x = \dfrac{2 \pm \sqrt{6}\,i}{2}$	Solutions

The solutions are $x = \frac{1}{2}(2 + \sqrt{6}\,i)$ and $x = \frac{1}{2}(2 - \sqrt{6}\,i)$. Check these in the original equation.

③ Determine the types of solutions of quadratic equations using the discriminant.

The Discriminant

The radicand in the Quadratic Formula, $b^2 - 4ac$, is called the discriminant because it allows you to "discriminate" among different types of solutions.

Using the Discriminant

Let a, b, and c be rational numbers such that $a \neq 0$. The discriminant of the quadratic equation $ax^2 + bx + c = 0$ is given by $b^2 - 4ac$, and can be used to classify the solutions of the equation as follows.

Discriminant	Solution Type
1. Perfect square	Two distinct rational solutions (Example 1)
2. Positive nonperfect square	Two distinct irrational solutions (Example 2)
3. Zero	One repeated rational solution (Example 3)
4. Negative number	Two distinct complex solutions (Example 4)

Example 5   Using the Discriminant

Determine the type of solution(s) for each quadratic equation.

a. $x^2 - x + 2 = 0$
b. $2x^2 - 3x - 2 = 0$
c. $x^2 - 2x + 1 = 0$
d. $x^2 - 2x - 1 = 9$

Solution

Equation	Discriminant	Solution Type
a. $x^2 - x + 2 = 0$	$b^2 - 4ac = (-1)^2 - 4(1)(2)$ $= 1 - 8 = -7$	Two distinct complex solutions
b. $2x^2 - 3x - 2 = 0$	$b^2 - 4ac = (-3)^2 - 4(2)(-2)$ $= 9 + 16 = 25$	Two distinct rational solutions
c. $x^2 - 2x + 1 = 0$	$b^2 - 4ac = (-2)^2 - 4(1)(1)$ $= 4 - 4 = 0$	One repeated rational solution
d. $x^2 - 2x - 1 = 9$	$b^2 - 4ac = (-2)^2 - 4(1)(-10)$ $= 4 + 40 = 44$	Two distinct irrational solutions

Technology: Discovery

Use a graphing calculator to graph each equation.

a. $y = x^2 - x + 2$
b. $y = 2x^2 - 3x - 2$
c. $y = x^2 - 2x + 1$
d. $y = x^2 - 2x - 10$

Describe the solution type of each equation and check your results with those shown in Example 5. Why do you think the discriminant is used to determine solution types?

Summary of Methods for Solving Quadratic Equations

Method	Example

1. Factoring

$$3x^2 + x = 0$$

$$x(3x + 1) = 0 \implies x = 0 \quad \text{and} \quad x = -\frac{1}{3}$$

2. Square Root Property

$$(x + 2)^2 = 7$$

$$x + 2 = \pm\sqrt{7} \implies x = -2 + \sqrt{7} \quad \text{and} \quad x = -2 - \sqrt{7}$$

3. Completing the square

$$x^2 + 6x = 2$$

$$x^2 + 6x + 3^2 = 2 + 9$$

$$(x + 3)^2 = 11 \implies x = -3 + \sqrt{11} \quad \text{and} \quad x = -3 - \sqrt{11}$$

4. Quadratic Formula

$$3x^2 - 2x + 2 = 0 \implies x = \frac{-(-2) \pm \sqrt{(-2)^2 - 4(3)(2)}}{2(3)} = \frac{1}{3} \pm \frac{\sqrt{5}}{3}i$$

---

4 Write quadratic equations from solutions of the equations.

Writing Quadratic Equations from Solutions

Using the Zero-Factor Property, you know that the equation $(x + 5)(x - 2) = 0$ has two solutions, $x = -5$ and $x = 2$. You can use the Zero-Factor Property in reverse to find a quadratic equation given its solutions. This process is demonstrated in Example 6.

Reverse of Zero-Factor Property

Let a and b be real numbers, variables, or algebraic expressions. If $a = 0$ or $b = 0$, then a and b are factors such that $ab = 0$.

Technology: Tip

A program for several models of graphing calculators that uses the Quadratic Formula to solve quadratic equations can be found at our website, *math.college.hmco.com/students*. The program will display real solutions to quadratic equations.

Example 6   Writing a Quadratic Equation from Its Solutions

Write a quadratic equation that has the solutions $x = 4$ and $x = -7$. Using the solutions $x = 4$ and $x = -7$, you can write the following.

$$x = 4 \qquad \text{and} \qquad x = -7 \quad \text{Solutions}$$

$$x - 4 = 0 \qquad\qquad x + 7 = 0 \quad \text{Obtain zero on one side of each equation.}$$

$$(x - 4)(x + 7) = 0 \quad \text{Reverse of Zero-Factor Property}$$

$$x^2 + 3x - 28 = 0 \quad \text{Foil Method}$$

So, a quadratic equation that has the solutions $x = 4$ and $x = -7$ is

$$x^2 + 3x - 28 = 0.$$

This is not the only quadratic equation with the solutions $x = 4$ and $x = -7$. You can obtain other quadratic equations with these solutions by multiplying $x^2 + 3x - 28 = 0$ by any nonzero real number.

8.3 Exercises

Developing Skills

In Exercises 1–4, write the quadratic equation in general form.

1. $2x^2 = 7 - 2x$

2. $7x^2 + 15x = 5$

3. $x(10 - x) = 5$

4. $x(3x + 8) = 15$

In Exercises 5–16, solve the equation first by using the Quadratic Formula and then by factoring. See Examples 1–4.

5. $x^2 - 11x + 28 = 0$

6. $x^2 - 12x + 27 = 0$

7. $x^2 + 6x + 8 = 0$

8. $x^2 + 9x + 14 = 0$

9. $4x^2 + 4x + 1 = 0$

10. $9x^2 + 12x + 4 = 0$

11. $4x^2 + 12x + 9 = 0$

12. $9x^2 - 30x + 25 = 0$

13. $6x^2 - x - 2 = 0$

14. $10x^2 - 11x + 3 = 0$

15. $x^2 - 5x - 300 = 0$

16. $x^2 + 20x - 300 = 0$

In Exercises 17–46, solve the equation by using the Quadratic Formula. (Find all real *and* complex solutions.) See Examples 1–4.

17. $x^2 - 2x - 4 = 0$

18. $x^2 - 2x - 6 = 0$

19. $t^2 + 4t + 1 = 0$

20. $y^2 + 6y + 4 = 0$

21. $x^2 + 6x - 3 = 0$

22. $x^2 + 8x - 4 = 0$

23. $x^2 - 10x + 23 = 0$

24. $u^2 - 12u + 29 = 0$

25. $2x^2 + 3x + 3 = 0$

26. $2x^2 - x + 1 = 0$

27. $3v^2 - 2v - 1 = 0$ **28.** $4x^2 + 6x + 1 = 0$

29. $2x^2 + 4x - 3 = 0$ **30.** $x^2 - 8x + 19 = 0$

31. $9z^2 + 6z - 4 = 0$ **32.** $8y^2 - 8y - 1 = 0$

33. $-4x^2 - 6x + 3 = 0$

34. $-5x^2 - 15x + 10 = 0$

35. $8x^2 - 6x + 2 = 0$

36. $6x^2 + 3x - 9 = 0$

37. $-4x^2 + 10x + 12 = 0$

38. $-15x^2 - 10x + 25 = 0$

39. $9x^2 = 1 + 9x$ **40.** $7x^2 = 3 - 5x$

41. $3x - 2x^2 = 4 - 5x^2$

42. $x - x^2 = 1 - 6x^2$ **43.** $x^2 - 0.4x - 0.16 = 0$

44. $x^2 + 0.6x - 0.41 = 0$

45. $2.5x^2 + x - 0.9 = 0$

46. $0.09x^2 - 0.12x - 0.26 = 0$

In Exercises 47–56, use the discriminant to determine the type of solutions of the quadratic equation. See Example 5.

47. $x^2 + x + 1 = 0$

48. $x^2 + x - 1 = 0$

49. $2x^2 - 5x - 4 = 0$

50. $10x^2 + 5x + 1 = 0$

51. $5x^2 + 7x + 3 = 0$

52. $3x^2 - 2x - 5 = 0$

53. $4x^2 - 12x + 9 = 0$

54. $2x^2 + 10x + 6 = 0$

55. $3x^2 - x + 2 = 0$

56. $9x^2 - 24x + 16 = 0$

In Exercises 57–74, solve the quadratic equation by using the most convenient method. (Find all real *and* complex solutions.)

57. $z^2 - 169 = 0$ **58.** $t^2 = 144$

59. $5y^2 + 15y = 0$ **60.** $7u^2 + 49u = 0$

61. $25(x - 3)^2 - 36 = 0$

62. $9(x + 4)^2 + 16 = 0$

63. $2y(y - 18) + 3(y - 18) = 0$

64. $4y(y + 7) - 5(y + 7) = 0$

65. $x^2 + 8x + 25 = 0$ **66.** $x^2 - 3x - 4 = 0$

67. $x^2 - 24x + 128 = 0$ **68.** $y^2 + 21y + 108 = 0$

69. $3x^2 - 13x + 169 = 0$

70. $2x^2 - 15x + 225 = 0$

71. $18x^2 + 15x - 50 = 0$

72. $14x^2 + 11x - 40 = 0$

73. $7x(x + 2) + 5 = 3x(x + 1)$

74. $5x(x - 1) - 7 = 4x(x - 2)$

In Exercises 75–84, write a quadratic equation having the given solutions. See Example 6.

75. $5, -2$ **76.** $-2, 3$

77. $1, 7$ **78.** $3, 9$

79. $1 + \sqrt{2}, 1 - \sqrt{2}$

80. $-3 + \sqrt{5}, -3 - \sqrt{5}$

81. $5i, -5i$ **82.** $2i, -2i$

83. 12 **84.** -4

In Exercises 85–92, use a graphing calculator to graph the function. Use the graph to approximate any x-intercepts of the graph. Set $y = 0$ and solve the resulting equation. Compare the result with the x-intercepts of the graph.

85. $y = 3x^2 - 6x + 1$ **86.** $y = x^2 + x + 1$

87. $y = -(4x^2 - 20x + 25)$

88. $y = x^2 - 4x + 3$ **89.** $y = 5x^2 - 18x + 6$

90. $y = 15x^2 + 3x - 105$

91. $y = -0.04x^2 + 4x - 0.8$

92. $y = 3.7x^2 - 10.2x + 3.2$

In Exercises 93–96, use a graphing calculator to determine the number of real solutions of the quadratic equation. Verify your answer algebraically.

93. $2x^2 - 5x + 5 = 0$ **94.** $2x^2 - x - 1 = 0$

95. $\frac{1}{5}x^2 + \frac{6}{5}x - 8 = 0$ **96.** $\frac{1}{3}x^2 - 5x + 25 = 0$

In Exercises 97–100, determine all real values of x for which the function has the indicated value.

97. $f(x) = 2x^2 - 7x + 1, \ f(x) = -3$

98. $f(x) = 2x^2 - 7x + 5, \ f(x) = 0$

99. $g(x) = 2x^2 - 3x + 16, \ g(x) = 14$

100. $h(x) = 6x^2 + x + 10, \ h(x) = -2$

In Exercises 101–104, solve the equation.

101. $\dfrac{2x^2}{5} - \dfrac{x}{2} = 1$ **102.** $\dfrac{x^2 - 9x}{6} = \dfrac{x - 1}{2}$

103. $\sqrt{x + 3} = x - 1$ **104.** $\sqrt{2x - 3} = x - 2$

Think About It In Exercises 105–108, describe the values of c such that the equation has (a) two real number solutions, (b) one real number solution, and (c) two complex number solutions.

105. $x^2 - 6x + c = 0$

106. $x^2 - 12x + c = 0$

107. $x^2 + 8x + c = 0$

108. $x^2 + 2x + c = 0$

Solving Problems

109. ▲ *Geometry* A rectangle has a width of x inches, a length of $x + 6.3$ inches, and an area of 58.14 square inches. Find its dimensions.

110. ▲ *Geometry* A rectangle has a length of $x + 1.5$ inches, a width of x inches, and an area of 18.36 square inches. Find its dimensions.

111. *Free-Falling Object* A stone is thrown vertically upward at a velocity of 40 feet per second from a bridge that is 50 feet above the level of the water (see figure). The height h (in feet) of the stone at time t (in seconds) after it is thrown is

$h = -16t^2 + 40t + 50.$

(a) Find the time when the stone is again 50 feet above the water.

(b) Find the time when the stone strikes the water.

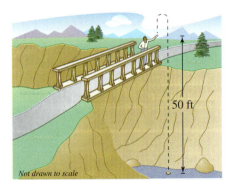

Not drawn to scale

50 ft

Figure for 111

112. *Free-Falling Object* A stone is thrown vertically upward at a velocity of 20 feet per second from a bridge that is 40 feet above the level of the water. The height h (in feet) of the stone at time t (in seconds) after it is thrown is

$$h = -16t^2 + 20t + 40.$$

(a) Find the time when the stone is again 40 feet above the water.

(b) Find the time when the stone strikes the water.

113. *Free-Falling Object* You stand on a bridge and throw a stone upward from 25 feet above a lake with an initial velocity of 20 feet per second. The height h (in feet) of the stone at time t (in seconds) after it is thrown is modeled by

$$h = -16t^2 + 20t + 25.$$

(a) Find the time when the stone is again 25 feet above the lake.

(b) Find the time when the stone strikes the water.

114. *Free-Falling Object* You stand on a bridge and throw a stone upward from 61 feet above a lake with an initial velocity of 36 feet per second. The height h (in feet) of the stone at time t (in seconds) after it is thrown is modeled by

$$h = -16t^2 + 36t + 61.$$

(a) Find the time when the stone is again 61 feet above the lake.

(b) Find the time when the stone strikes the water.

115. *Free-Falling Object* From the roof of a building, 100 feet above the ground, you toss a coin upward with an initial velocity of 5 feet per second. The height h (in feet) of the coin at time t (in seconds) after it is tossed is modeled by

$$h = -16t^2 + 5t + 100.$$

(a) Find the time when the coin is again 100 feet above the ground.

(b) Find the time when the coin hits the ground.

116. *Free-Falling Object* You throw an apple upward from 42 feet above the ground in an apple tree, with an initial velocity of 30 feet per second. The height h (in feet) of the apple at time t (in seconds) after it is thrown is modeled by

$$h = -16t^2 + 30t + 42.$$

(a) Find the time when the apple is again 42 feet above the ground.

(b) Find the time when the apple hits the ground.

117. *Employment* The number y (in thousands) of people employed in the construction industry in the United States from 1994 through 2001 can be modeled by

$$y = 10.29t^2 + 164.5t + 6624, \; 4 \le t \le 11$$

where t represents the year, with $t = 4$ corresponding to 1994. (Source: U.S. Bureau of Labor Statistics)

(a) ▦ Use a graphing calculator to graph the model.

(b) ▦ Use the graph in part (a) to find the year in which there were approximately 9,000,000 employed in the construction industry in the United States. Verify your answer algebraically.

(c) Use the model to estimate the number employed in the construction industry in 2002.

118. ▦ *Cellular Phone Subscribers* The number s (in thousands) of cellular phone subscribers in the United States for the years 1994 through 2001 can be modeled by

$$s = 1178.29t^2 - 2816.5t + 17,457, \; 4 \le t \le 11$$

where $t = 4$ corresponds to 1994.

(Source: Cellular Telecommunications & Internet Association)

(a) Use a graphing calculator to graph the model.

(b) Use the graph in part (a) to determine the year in which there were 44 million cellular phone subscribers. Verify your answer algebraically.

119. *Exploration* Determine the two solutions, x_1 and x_2, of each quadratic equation. Use the values of x_1 and x_2 to fill in the boxes.

Equation	x_1, x_2	$x_1 + x_2$	$x_1 x_2$

(a) $x^2 - x - 6 = 0$

(b) $2x^2 + 5x - 3 = 0$

(c) $4x^2 - 9 = 0$

(d) $x^2 - 10x + 34 = 0$

120. *Think About It* Consider a general quadratic equation $ax^2 + bx + c = 0$ whose solutions are x_1 and x_2. Use the results of Exercise 119 to determine a relationship among the coefficients a, b, and c, and the sum $(x_1 + x_2)$ and product $(x_1 x_2)$ of the solutions.

Explaining Concepts

121. ⊘ Answer parts (c) and (d) of Motivating the Chapter on page 506.

122. *Writing* State the Quadratic Formula *in words*.

123. *Writing* What is the discriminant of $ax^2 + bx + c = 0$? How is the discriminant related to the number and type of solutions of the equation?

124. *Writing* Explain how completing the square can be used to develop the Quadratic Formula.

125. *Writing* List the four methods for solving a quadratic equation.

Mid-Chapter Quiz

Take this quiz as you would take a quiz in class. After you are done, check your work against the answers in the back of the book.

In Exercises 1–8, solve the quadratic equation by the specified method.

1. Factoring:

   $2x^2 - 72 = 0$

2. Factoring:

   $2x^2 + 3x - 20 = 0$

3. Square Root Property:

   $3x^2 = 36$

4. Square Root Property:

   $(u - 3)^2 - 16 = 0$

5. Completing the square:

   $s^2 + 10s + 1 = 0$

6. Completing the square:

   $2y^2 + 6y - 5 = 0$

7. Quadratic Formula:

   $x^2 + 4x - 6 = 0$

8. Quadratic Formula:

   $6v^2 - 3v - 4 = 0$

In Exercises 9–16, solve the equation by using the most convenient method. (Find all real *and* complex solutions.)

9. $x^2 + 5x + 7 = 0$

10. $36 - (t - 4)^2 = 0$

11. $x(x - 10) + 3(x - 10) = 0$

12. $x(x - 3) = 10$

13. $4b^2 - 12b + 9 = 0$

14. $3m^2 + 10m + 5 = 0$

15. $x - 2\sqrt{x} - 24 = 0$

16. $x^4 + 7x^2 + 12 = 0$

In Exercises 17 and 18, solve the equation of quadratic form. (Find all real *and* complex solutions.)

17. $x - 4\sqrt{x} - 1 = 0$

18. $x^4 - 12x^2 + 27 = 0$

In Exercises 19 and 20, use a graphing calculator to graph the function. Use the graph to approximate any *x*-intercepts of the graph. Set $y = 0$ and solve the resulting equation. Compare the results with the *x*-intercepts of the graph.

19. $y = \frac{1}{2}x^2 - 3x - 1$

20. $y = x^2 + 0.45x - 4$

21. The revenue R from selling x alarm clocks is given by

    $R = x(20 - 0.2x)$.

    Find the number of alarm clocks that must be sold to produce a revenue of $500.

22. A rectangle has a length of x meters, a width of $100 - x$ meters, and an area of 2275 square meters. Find its dimensions.

8.4 Graphs of Quadratic Functions

Joaquin Palting/Photodisc/Getty Images

What You Should Learn

① Determine the vertices of parabolas by completing the square.

② Sketch parabolas.

③ Write the equation of a parabola given the vertex and a point on the graph.

④ Use parabolas to solve application problems.

Why You Should Learn It

Real-life situations can be modeled by graphs of quadratic functions. For instance, in Exercise 101 on page 544, a quadratic equation is used to model the maximum height of a diver.

Graphs of Quadratic Functions

In this section, you will study graphs of quadratic functions of the form

$$f(x) = ax^2 + bx + c. \qquad \text{Quadratic function}$$

Figure 8.2 shows the graph of a simple quadratic function, $f(x) = x^2$.

① Determine the vertices of parabolas by completing the square.

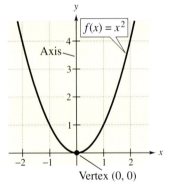

Figure 8.2

Graphs of Quadratic Functions

The graph of $f(x) = ax^2 + bx + c$, $a \neq 0$, is a **parabola.** The completed square form

$$f(x) = a(x - h)^2 + k \qquad \text{Standard form}$$

is the **standard form** of the function. The **vertex** of the parabola occurs at the point (h, k), and the vertical line passing through the vertex is the **axis** of the parabola.

Every parabola is *symmetric* about its axis, which means that if it were folded along its axis, the two parts would match.

If a is positive, the graph of $f(x) = ax^2 + bx + c$ opens upward, and if a is negative, the graph opens downward, as shown in Figure 8.3. Observe in Figure 8.3 that the y-coordinate of the vertex identifies the minimum function value if $a > 0$ and the maximum function value if $a < 0$.

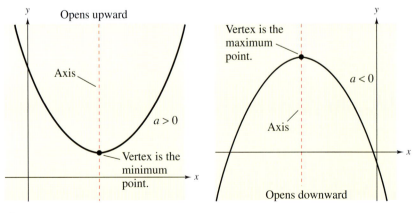

Figure 8.3

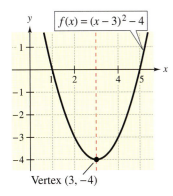

$$f(x) = (x - 3)^2 - 4$$

Vertex $(3, -4)$

Figure 8.4

Example 1   Finding the Vertex by Completing the Square

Find the vertex of the parabola given by $f(x) = x^2 - 6x + 5$.

Solution

Begin by writing the function in standard form.

$f(x) = x^2 - 6x + 5$	Original function
$f(x) = x^2 - 6x + (-3)^2 - (-3)^2 + 5$	Complete the square.
$f(x) = (x^2 - 6x + 9) - 9 + 5$	Regroup terms.
$f(x) = (x - 3)^2 - 4$	Standard form

From the standard form, you can see that the vertex of the parabola occurs at the point $(3, -4)$, as shown in Figure 8.4. The minimum value of the function is $f(3) = -4$.

In Example 1, the vertex of the graph was found by *completing the square*. Another approach to finding the vertex is to complete the square once for a general function and then use the resulting formula for the vertex.

$f(x) = ax^2 + bx + c$	Quadratic function
$= a\left(x^2 + \dfrac{b}{a}x\right) + c$	Factor a out of first two terms.
$= a\left[x^2 + \dfrac{b}{a}x + \left(\dfrac{b}{2a}\right)^2\right] + c - \dfrac{b}{4a}$	Complete the square.
$= a\left(x + \dfrac{b}{2a}\right)^2 + c - \dfrac{b^2}{4a}$	Standard form

From this form you can see that the vertex occurs when $x = -b/(2a)$.

Example 2   Finding the Vertex with a Formula

Find the vertex of the parabola given by $f(x) = 3x^2 - 9x$.

Solution

From the original function, it follows that $a = 3$ and $b = -9$. So, the x-coordinate of the vertex is

$$x = \frac{-b}{2a} = \frac{-(-9)}{2(3)} = \frac{3}{2}.$$

Substitute $\frac{3}{2}$ for x into the original equation to find the y-coordinate.

$$f\left(-\frac{b}{2a}\right) = f\left(\frac{3}{2}\right) = 3\left(\frac{3}{2}\right)^2 - 9\left(\frac{3}{2}\right) = -\frac{27}{4}$$

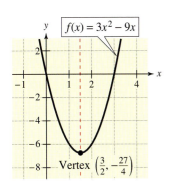

$$f(x) = 3x^2 - 9x$$

Vertex $\left(\frac{3}{2}, -\frac{27}{4}\right)$

Figure 8.5

So, the vertex of the parabola is $\left(\frac{3}{2}, -\frac{27}{4}\right)$, the minimum value of the function is $f\left(\frac{3}{2}\right) = -\frac{27}{4}$, and the parabola opens upward, as shown in Figure 8.5.

② Sketch parabolas.

Sketching a Parabola

To obtain an accurate sketch of a parabola, the following guidelines are useful.

Sketching a Parabola

1. Determine the vertex and axis of the parabola by completing the square or by using the formula $x = -b/(2a)$.

2. Plot the vertex, axis, x- and y-intercepts, and a few additional points on the parabola. (Using the symmetry about the axis can reduce the number of points you need to plot.)

3. Use the fact that the parabola opens *upward* if $a > 0$ and opens *downward* if $a < 0$ to complete the sketch.

Study Tip

The x- and y-intercepts are useful points to plot. Another convenient fact is that the x-coordinate of the vertex lies halfway between the x-intercepts. Keep this in mind as you study the examples and do the exercises in this section.

Example 3   Sketching a Parabola

To sketch the parabola given by $y = x^2 + 6x + 8$, begin by writing the equation in standard form.

$$y = x^2 + 6x + 8 \qquad \text{Write original equation.}$$

$$y = (x^2 + 6x + 3^2 - 3^2) + 8 \qquad \text{Complete the square.}$$

$$\text{(half of 6)}^2$$

$$y = (x^2 + 6x + 9) - 9 + 8 \qquad \text{Regroup terms.}$$

$$y = (x + 3)^2 - 1 \qquad \text{Standard form}$$

The vertex occurs at the point $(-3, -1)$ and the axis is the line $x = -3$. After plotting this information, calculate a few additional points on the parabola, as shown in the table. Note that the y-intercept is $(0, 8)$ and the x-intercepts are solutions to the equation

$$x^2 + 6x + 8 = (x + 4)(x + 2) = 0.$$

x	-5	-4	-3	-2	-1
$y = (x + 3)^2 - 1$	3	0	-1	0	3
Solution point	$(-5, 3)$	$(-4, 0)$	$(-3, -1)$	$(-2, 0)$	$(-1, 3)$

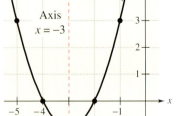

Figure 8.6

The graph of the parabola is shown in Figure 8.6. Note that the parabola opens upward because the leading coefficient (in general form) is positive.

The graph of the parabola in Example 3 can also be obtained by shifting the graph of $y = x^2$ to the left three units and downward one unit, as discussed in Section 3.7.

③ Write the equation of a parabola given the vertex and a point on the graph.

Writing the Equation of a Parabola

To write the equation of a parabola with a vertical axis, use the fact that its standard equation has the form $y = a(x - h)^2 + k$, where (h, k) is the vertex.

Example 4   Writing the Equation of a Parabola

Write the equation of the parabola with vertex $(-2, 1)$ and y-intercept $(0, -3)$, as shown in Figure 8.7.

Solution

Because the vertex occurs at $(h, k) = (-2, 1)$, the equation has the form

$$y = a(x - h)^2 + k \qquad \text{Standard form}$$

$$y = a[x - (-2)]^2 + 1 \qquad \text{Substitute } -2 \text{ for } h \text{ and } 1 \text{ for } k.$$

$$y = a(x + 2)^2 + 1. \qquad \text{Simplify.}$$

To find the value of a, use the fact that the y-intercept is $(0, -3)$.

$$y = a(x + 2)^2 + 1 \qquad \text{Write standard form.}$$

$$-3 = a(0 + 2)^2 + 1 \qquad \text{Substitute } 0 \text{ for } x \text{ and } -3 \text{ for } y.$$

$$-1 = a \qquad \text{Simplify.}$$

So, the standard form of the equation of the parabola is

$$y = -(x + 2)^2 + 1.$$

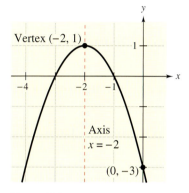

Vertex $(-2, 1)$

Axis $x = -2$

$(0, -3)$

Figure 8.7

Example 5   Writing the Equation of a Parabola

Write the equation of the parabola with vertex $(3, -4)$ and that passes through the point $(5, -2)$, as shown in Figure 8.8.

Solution

Because the vertex occurs at $(h, k) = (3, -4)$, the equation has the form

$$y = a(x - h)^2 + k \qquad \text{Standard form}$$

$$y = a(x - 3)^2 + (-4) \qquad \text{Substitute } 3 \text{ for } h \text{ and } -4 \text{ for } k.$$

$$y = a(x - 3)^2 - 4. \qquad \text{Simplify.}$$

To find the value of a, use the fact that the parabola passes through the point $(5, -2)$.

$$y = a(x - 3)^2 - 4 \qquad \text{Write standard form.}$$

$$-2 = a(5 - 3)^2 - 4 \qquad \text{Substitute } 5 \text{ for } x \text{ and } -2 \text{ for } y.$$

$$\frac{1}{2} = a \qquad \text{Simplify.}$$

So, the standard form of the equation of the parabola is $y = \frac{1}{2}(x - 3)^2 - 4$.

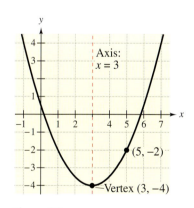

Axis: $x = 3$

$(5, -2)$

Vertex $(3, -4)$

Figure 8.8

4 Use parabolas to solve application problems.

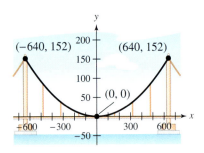

Figure 8.9

Application

Example 6   Golden Gate Bridge

Each cable of the Golden Gate Bridge is suspended (in the shape of a parabola) between two towers that are 1280 meters apart. The top of each tower is 152 meters above the roadway. The cables touch the roadway at the midpoint between the towers (see Figure 8.9).

a. Write an equation that models the cables of the bridge.

b. Find the height of the suspension cables over the roadway at a distance of 320 meters from the center of the bridge.

Solution

a. From Figure 8.9, you can see that the vertex of the parabola occurs at (0, 0). So, the equation has the form

$$y = a(x - h)^2 + k \qquad \text{Standard form}$$

$$y = a(x - 0)^2 + 0 \qquad \text{Substitute 0 for } h \text{ and 0 for } k.$$

$$y = ax^2. \qquad \text{Simplify.}$$

To find the value of a, use the fact that the parabola passes through the point (640, 152).

$$y = ax^2 \qquad \text{Write standard form.}$$

$$152 = a(640)^2 \qquad \text{Substitute 640 for } x \text{ and 152 for } y.$$

$$\frac{19}{51,200} = a \qquad \text{Simplify.}$$

So, an equation that models the cables of the bridge is

$$y = \frac{19}{51,200}x^2.$$

b. To find the height of the suspension cables over the roadway at a distance of 320 meters from the center of the bridge, evaluate the equation from part (a) when $x = 320$.

$$y = \frac{19}{51,200}x^2 \qquad \text{Write original equation.}$$

$$y = \frac{19}{51,200}(320)^2 \qquad \text{Substitute 320 for } x.$$

$$y = 38 \qquad \text{Simplify.}$$

So, the height of the suspension cables over the roadway is 38 meters.

8.4 Exercises

Review   Concepts, Skills, and Problem Solving

Keep mathematically in shape by doing these exercises *before* the problems of this section.

Properties and Definitions

1. Fill in the blanks: $(x + b)^2 = x^2 + \boxed{} \, x + \boxed{}$.

2. *Writing* Fill in the blank so that the expression is a perfect square trinomial. Explain how the constant is determined.

$x^2 + 5x + \boxed{}$

Simplifying Expressions

In Exercises 3–10, simplify the expression.

3. $(4x + 3y) - 3(5x + y)$

4. $(-15u + 4v) + 5(3u - 9v)$

5. $2x^2 + (2x - 3)^2 + 12x$

6. $y^2 - (y + 2)^2 + 4y$

7. $\sqrt{24x^2y^3}$

8. $\sqrt[3]{9} \cdot \sqrt[3]{15}$

9. $(12a^{-4}b^6)^{1/2}$

10. $(16^{1/3})^{3/4}$

Problem Solving

11. *Alcohol Mixture* How many liters of an 18% alcohol solution must be mixed with a 45% solution to obtain 12 liters of a 36% solution?

12. *Television* During a television show there were 12 commercials. Some of the commercials were 30-seconds long and some were 60-seconds long. The total amount of time for the 30-second commercials was 6 minutes less than the total amount of time for all the commercials during the show. How many 30-second commercials and how many 60-second commercials were there?

Developing Skills

In Exercises 1–6, match the equation with its graph. [The graphs are labeled (a), (b), (c), (d), (e), and (f).]

1. $y = (x + 1)^2 - 3$

2. $y = -(x + 1)^2$

3. $y = x^2 - 3$

4. $y = -x^2 + 3$

5. $y = (x - 2)^2$

6. $y = 2 - (x - 2)^2$

(a)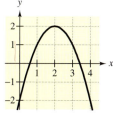

(b)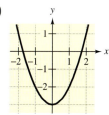

(c)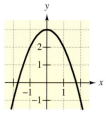

(d)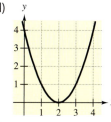

(e)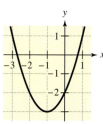

(f)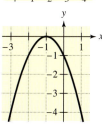

In Exercises 7–18, write the equation of the parabola in standard form and find the vertex of its graph. See Example 1.

7. $y = x^2 + 2$

8. $y = x^2 + 2x$

9. $y = x^2 - 4x + 7$

10. $y = x^2 + 6x - 5$

11. $y = x^2 + 6x + 5$

12. $y = x^2 - 4x + 5$

13. $y = -x^2 + 6x - 10$

14. $y = 4 - 8x - x^2$

15. $y = -x^2 + 2x - 7$

16. $y = -x^2 - 10x + 10$

17. $y = 2x^2 + 6x + 2$

18. $y = 3x^2 - 3x - 9$

In Exercises 19–24, find the vertex of the graph of the function by using the formula $x = -b/(2a)$. See Example 2.

19. $f(x) = x^2 - 8x + 15$ **20.** $f(x) = x^2 + 4x + 1$

21. $g(x) = -x^2 - 2x + 1$

22. $h(x) = -x^2 + 14x - 14$

23. $y = 4x^2 + 4x + 4$ **24.** $y = 9x^2 - 12x$

In Exercises 25–34, state whether the graph opens upward or downward and find the vertex.

25. $y = 2(x - 0)^2 + 2$ **26.** $y = -3(x + 5)^2 - 3$

27. $y = 4 - (x - 10)^2$ **28.** $y = 2(x - 12)^2 + 3$

29. $y = x^2 - 6$ **30.** $y = -(x + 1)^2$

31. $y = -(x - 3)^2$ **32.** $y = x^2 - 6x$

33. $y = -x^2 + 6x$ **34.** $y = -x^2 - 5$

In Exercises 35–46, find the x- and y-intercepts of the graph.

35. $y = 25 - x^2$ **36.** $y = x^2 - 49$

37. $y = x^2 - 9x$ **38.** $y = x^2 + 4x$

39. $y = x^2 + 2x - 3$ **40.** $y = -x^2 + 4x - 5$

41. $y = 4x^2 - 12x + 9$ **42.** $y = 10 - x - 2x^2$

43. $y = x^2 - 3x + 3$ **44.** $y = x^2 - 3x - 10$

45. $y = -2x^2 - 6x + 5$

46. $y = -4x^2 + 6x - 9$

In Exercises 47–70, sketch the parabola. Identify the vertex and any x-intercepts. Use a graphing calculator to verify your results. See Example 3.

47. $g(x) = x^2 - 4$

48. $h(x) = x^2 - 9$

49. $f(x) = -x^2 + 4$

50. $f(x) = -x^2 + 9$

51. $f(x) = x^2 - 3x$

52. $g(x) = x^2 - 4x$

53. $y = -x^2 + 3x$

54. $y = -x^2 + 4x$

55. $y = (x - 4)^2$

56. $y = -(x + 4)^2$

57. $y = x^2 - 8x + 15$

58. $y = x^2 + 4x + 2$

59. $y = -(x^2 + 6x + 5)$

60. $y = -x^2 + 2x + 8$

61. $q(x) = -x^2 + 6x - 7$

62. $g(x) = x^2 + 4x + 7$

63. $y = 2(x^2 + 6x + 8)$

64. $y = 3x^2 - 6x + 4$

65. $y = \frac{1}{2}(x^2 - 2x - 3)$

66. $y = -\frac{1}{2}(x^2 - 6x + 7)$

67. $y = \frac{1}{5}(3x^2 - 24x + 38)$

68. $y = \frac{1}{5}(2x^2 - 4x + 7)$

69. $f(x) = 5 - \frac{1}{3}x^2$

70. $f(x) = \frac{1}{3}x^2 - 2$

In Exercises 71–78, identify the transformation of the graph of $f(x) = x^2$ and sketch a graph of h.

71. $h(x) = x^2 + 2$

72. $h(x) = x^2 - 4$

73. $h(x) = (x + 2)^2$

74. $h(x) = (x - 4)^2$

75. $h(x) = -(x + 5)^2$

76. $h(x) = -x^2 - 6$

77. $h(x) = -(x + 1)^2 - 1$

78. $h(x) = -(x - 3)^2 + 2$

🖩 In Exercises 79–82, use a graphing calculator to approximate the vertex of the graph. Verify the result algebraically.

79. $y = \frac{1}{6}(2x^2 - 8x + 11)$

80. $y = -\frac{1}{4}(4x^2 - 20x + 13)$

81. $y = -0.7x^2 - 2.7x + 2.3$

82. $y = 0.75x^2 - 7.50x + 23.00$

In Exercises 83–86, write an equation of the parabola. See Example 4.

83.

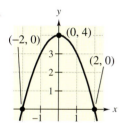

84.

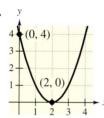

85.

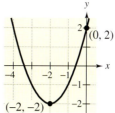

86.

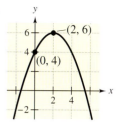

In Exercises 87–94, write an equation of the parabola $y = a(x - h)^2 + k$ that satisfies the conditions. See Example 5.

87. Vertex: $(2, 1)$; $a = 1$

88. Vertex: $(-3, -3)$; $a = 1$

89. Vertex: $(2, -4)$; Point on the graph: $(0, 0)$

90. Vertex: $(-2, -4)$; Point on the graph: $(0, 0)$

91. Vertex: $(3, 2)$; Point on the graph: $(1, 4)$

92. Vertex: $(-1, -1)$; Point on the graph: $(0, 4)$

93. Vertex: $(-1, 5)$; Point on the graph: $(0, 1)$

94. Vertex: $(5, 2)$; Point on the graph: $(10, 3)$

Solving Problems

95. *Path of a Ball*　The height y (in feet) of a ball thrown by a child is given by

$$y = -\frac{1}{12}x^2 + 2x + 4$$

where x is the horizontal distance (in feet) from where the ball is thrown.

　(a) How high is the ball when it leaves the child's hand?

　(b) How high is the ball when it reaches its maximum height?

　(c) How far from the child does the ball strike the ground?

96. *Path of a Ball*　Repeat Exercise 95 if the path of the ball is modeled by

$$y = -\frac{1}{16}x^2 + 2x + 5.$$

97. *Path of an Object*　A child launches a toy rocket from a table. The height y (in feet) of the rocket is given by

$$y = -\frac{1}{5}x^2 + 6x + 3$$

where x is the horizontal distance (in feet) from where the rocket is launched.

　(a) Determine the height from which the rocket is launched.

　(b) How high is the rocket at its maximum height?

　(c) How far away does the rocket land from where it is launched?

98. *Path of an Object*   You use a fishing rod to cast a lure into the water. The height y (in feet) of the lure is given by

$$y = -\tfrac{1}{90}x^2 + \tfrac{1}{5}x + 9$$

where x is the horizontal distance (in feet) from the point where the lure is released.

(a) Determine the height from which the lure is released.

(b) How high is the lure at its maximum height?

(c) How far away does the lure land from where it is released?

99. *Path of a Ball*   The height y (in feet) of a ball that you throw is given by

$$y = -\tfrac{1}{200}x^2 + x + 6$$

where x is the horizontal distance (in feet) from where you release the ball.

(a) How high is the ball when you release it?

(b) How high is the ball when it reaches its maximum height?

(c) How far away does the ball strike the ground from where you released it?

100. *Path of a Ball*   The height y (in feet) of a softball that you hit is given by

$$y = -\tfrac{1}{70}x^2 + 2x + 2$$

where x is the horizontal distance (in feet) from where you hit the ball.

(a) How high is the ball when you hit it?

(b) How high is the ball at its maximum height?

(c) How far from where you hit the ball does it strike the ground?

101. *Path of a Diver*   The path of a diver is given by

$$y = -\tfrac{4}{9}x^2 + \tfrac{24}{9}x + 10$$

where y is the height in feet and x is the horizontal distance from the end of the diving board in feet. What is the maximum height of the diver?

102. *Path of a Diver*   Repeat Exercise 101 if the path of the diver is modeled by

$$y = -\tfrac{4}{3}x^2 + \tfrac{10}{3}x + 10.$$

103. ⊞ *Cost*   The cost C of producing x units of a product is given by

$$C = 800 - 10x + \tfrac{1}{4}x^2, \quad 0 < x < 40.$$

Use a graphing calculator to graph this function and approximate the value of x when C is minimum.

104. ⊞ ▲ *Geometry*   The area A of a rectangle is given by the function

$$A = \frac{2}{\pi}(100x - x^2), \quad 0 < x < 100$$

where x is the length of the base of the rectangle in feet. Use a graphing calculator to graph the function and to approximate the value of x when A is maximum.

105. ⊞ *Graphical Estimation*   The number N (in thousands) of military reserve personnel in the United States for the years 1992 through 2000 is approximated by the model

$$N = 4.64t^2 - 85.5t + 1263, \quad 2 \le t \le 10$$

where t is the time in years, with $t = 2$ corresponding to 1992.   (Source: U.S. Department of Defense)

(a) Use a graphing calculator to graph the model.

(b) Use your graph from part (a) to determine the year when the number of military reserves was greatest. Approximate the number for that year.

106. ⊞ *Graphical Estimation*   The profit P (in thousands of dollars) for a landscaping company is given by

$$P = 230 + 20s - \tfrac{1}{2}s^2$$

where s is the amount (in hundreds of dollars) spent on advertising. Use a graphing calculator to graph the profit function and approximate the amount of advertising that yields a maximum profit. Verify the maximum profit algebraically.

107. *Bridge Design*   A bridge is to be constructed over a gorge with the main supporting arch being a parabola (see figure). The equation of the parabola is $y = 4[100 - (x^2/2500)]$, where x and y are measured in feet.

(a) Find the length of the road across the gorge.

(b) Find the height of the parabolic arch at the center of the span.

(c) Find the lengths of the vertical girders at intervals of 100 feet from the center of the bridge.

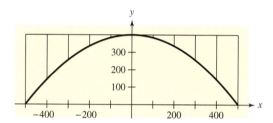

108. *Highway Design*   A highway department engineer must design a parabolic arc to create a turn in a freeway around a city. The vertex of the parabola is placed at the origin, and the parabola must connect with roads represented by the equations

$$y = -0.4x - 100, \quad x < -500$$

and

$$y = 0.4x - 100, \quad x > 500$$

(see figure). Find an equation of the parabolic arc.

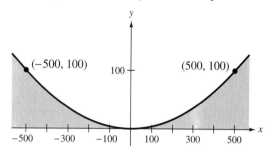

Explaining Concepts

109. *Writing*   In your own words, describe the graph of the quadratic function $f(x) = ax^2 + bx + c$.

110. *Writing*   Explain how to find the vertex of the graph of a quadratic function.

111. *Writing*   Explain how to find any x- or y-intercepts of the graph of a quadratic function.

112. *Writing*   Explain how to determine whether the graph of a quadratic function opens upward or downward.

113. *Writing*   How is the discriminant related to the graph of a quadratic function?

114. *Writing*   Is it possible for the graph of a quadratic function to have two y-intercepts? Explain.

115. *Writing*   Explain how to determine the maximum (or minimum) value of a quadratic function.

8.5 Applications of Quadratic Equations

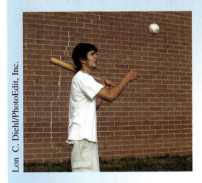

Lon C. Diehl/PhotoEdit, Inc.

Why You Should Learn It

Quadratic equations are used in a wide variety of real-life problems. For instance, in Exercise 46 on page 555, a quadratic equation is used to model the height of a baseball after you hit the ball.

1 Use quadratic equations to solve application problems.

What You Should Learn

1 Use quadratic equations to solve application problems.

Applications of Quadratic Equations

Example 1 **An Investment Problem**

A car dealer bought a fleet of cars from a car rental agency for a total of $120,000. By the time the dealer had sold all but four of the cars, at an average profit of $2500 each, the original investment of $120,000 had been regained. How many cars did the dealer sell, and what was the average price per car?

Solution

Although this problem is stated in terms of average price and average profit per car, you can use a model that assumes that each car sold for the same price.

Verbal Model: $\dfrac{\text{Selling price}}{\text{per car}} = \dfrac{\text{Cost}}{\text{per car}} + \dfrac{\text{Profit}}{\text{per car}}$

Labels:

Number of cars sold $= x$	(cars)
Number of cars bought $= x + 4$	(cars)
Selling price per car $= 120{,}000/x$	(dollars per car)
Cost per car $= 120{,}000/(x + 4)$	(dollars per car)
Profit per car $= 2500$	(dollars per car)

Equation:

$$\frac{120{,}000}{x} = \frac{120{,}000}{x + 4} + 2500$$

$$120{,}000(x + 4) = 120{,}000x + 2500x(x + 4), \quad x \neq 0, \ x \neq -4$$

$$120{,}000x + 480{,}000 = 120{,}000x + 2500x^2 + 10{,}000x$$

$$0 = 2500x^2 + 10{,}000x - 480{,}000$$

$$0 = x^2 + 4x - 192$$

$$0 = (x - 12)(x + 16)$$

$$x - 12 = 0 \quad \Longrightarrow \quad x = 12$$

$$x + 16 = 0 \quad \Longrightarrow \quad x = -16$$

Choosing the positive value, it follows that the dealer sold 12 cars at an average price of $120{,}000/12 = 10{,}000$ per car. Check this in the original statement.

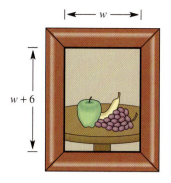

Figure 8.10

Example 2   Geometry: Dimensions of a Picture

A picture is 6 inches taller than it is wide and has an area of 216 square inches. What are the dimensions of the picture?

Solution

Begin by drawing a diagram, as shown in Figure 8.10.

Verbal Model:.   Area of picture = Width · Height

Labels: Picture width = w (inches)
Picture height = $w + 6$ (inches)
Area = 216   (square inches)

Equation: $216 = w(w + 6)$

$$0 = w^2 + 6w - 216$$

$$0 = (w + 18)(w - 12)$$

$w + 18 = 0 \quad \Longrightarrow \quad w = -18$

$w - 12 = 0 \quad \Longrightarrow \quad w = 12$

Of the two possible solutions, choose the positive value of w and conclude that the picture is $w = 12$ inches wide and $w + 6 = 12 + 6 = 18$ inches tall. Check these dimensions in the original statement of the problem.

Example 3   An Interest Problem

The formula $A = P(1 + r)^2$ represents the amount of money A in an account in which P dollars is deposited for 2 years at an annual interest rate of r (in decimal form). Find the interest rate if a deposit of $6000 increases to $6933.75 over a two-year period.

Solution

$A = P(1 + r)^2$ Write given formula.

$6933.75 = 6000(1 + r)^2$ Substitute 6933.75 for A and 6000 for P.

$1.155625 = (1 + r)^2$ Divide each side by 6000.

$\pm 1.075 = 1 + r$ Square Root Property

$0.075 = r$ Choose positive solution.

The annual interest rate is $r = 0.075 = 7.5\%$.

Check

$A = P(1 + r)^2$ Write given formula.

$6933.75 \overset{?}{=} 6000(1 + 0.075)^2$ Substitute 6933.75 for A, 6000 for P, and 0.075 for r.

$6933.75 \overset{?}{=} 6000(1.155625)$ Simplify.

$6933.75 = 6933.75$ Solution checks. ✓

Example 4 Reduced Rates

A ski club chartered a bus for a ski trip at a cost of $720. In an attempt to lower the bus fare per skier, the club invited nonmembers to go along. When four nonmembers agreed to go on the trip, the fare per skier decreased by $6. How many club members are going on the trip?

Solution

Verbal Model: $\boxed{\text{Cost per skier}} \cdot \boxed{\text{Number of skiers}} = \$720$

Labels:

Number of ski club members $= x$ (people)

Number of skiers $= x + 4$ (people)

Original cost per skier $= \dfrac{720}{x}$ (dollars per person)

New cost per skier $= \dfrac{720}{x} - 6$ (dollars per person)

Equation:

$$\left(\frac{720}{x} - 6\right)(x + 4) = 720 \qquad \text{Original equation}$$

$$\left(\frac{720 - 6x}{x}\right)(x + 4) = 720 \qquad \text{Rewrite 1st factor.}$$

$$(720 - 6x)(x + 4) = 720x, \ \ x \neq 0 \qquad \text{Multiply each side by } x.$$

$$720x + 2880 - 6x^2 - 24x = 720x \qquad \text{Multiply factors.}$$

$$-6x^2 - 24x + 2880 = 0 \qquad \text{Subtract } 720x \text{ from each side.}$$

$$x^2 + 4x - 480 = 0 \qquad \text{Divide each side by } -6.$$

$$(x + 24)(x - 20) = 0 \qquad \text{Factor left side of equation.}$$

$$x + 24 = 0 \implies x = -24 \qquad \text{Set 1st factor equal to 0.}$$

$$x - 20 = 0 \implies x = 20 \qquad \text{Set 2nd factor equal to 0.}$$

Choosing the positive value of x, you can conclude that 20 ski club members are going on the trip. Check this solution in the original statement of the problem, as follows.

Check

Original cost for 20 ski club members:

$$\frac{720}{x} = \frac{720}{20} = \$36 \qquad \text{Substitute 20 for } x.$$

New cost with 4 nonmembers:

$$\frac{720}{x + 4} = \frac{720}{24} = \$30$$

Decrease in fare with 4 nonmembers:

$$36 - 30 = \$6 \qquad \text{Solution checks. } \checkmark$$

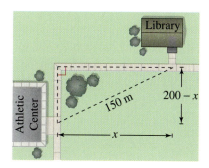

Figure 8.11

Example 5   An Application Involving the Pythagorean Theorem

An L-shaped sidewalk from the athletic center to the library on a college campus is 200 meters long, as shown in Figure 8.11. By cutting diagonally across the grass, students shorten the walking distance to 150 meters. What are the lengths of the two legs of the sidewalk?

Solution

Common Formula: $a^2 + b^2 = c^2$ Pythagorean Theorem

Labels: Length of one leg $= x$ (meters)
Length of other leg $= 200 - x$ (meters)
Length of diagonal $= 150$ (meters)

Equation:
$$x^2 + (200 - x)^2 = 150^2$$
$$x^2 + 40{,}000 - 400x + x^2 = 22{,}500$$
$$2x^2 - 400x + 40{,}000 = 22{,}500$$
$$2x^2 - 400x + 17{,}500 = 0$$
$$x^2 - 200x + 8750 = 0$$

By the Quadratic Formula, you can find the solutions as follows.

$$x = \frac{-(-200) \pm \sqrt{(-200)^2 - 4(1)(8750)}}{2(1)}$$ Substitute 1 for a, -200 for b, and 8750 for c.

$$= \frac{200 \pm \sqrt{5000}}{2}$$

$$= \frac{200 \pm 50\sqrt{2}}{2}$$

$$= \frac{2\left(100 \pm 25\sqrt{2}\right)}{2}$$

$$= 100 \pm 25\sqrt{2}$$

Both solutions are positive, so it does not matter which you choose. If you let

$$x = 100 + 25\sqrt{2} \approx 135.4 \text{ meters}$$

the length of the other leg is

$$200 - x \approx 200 - 135.4 \approx 64.6 \text{ meters.}$$

In Example 5, notice that you obtain the same dimensions if you choose the other value of x. That is, if the length of one leg is

$$x = 100 - 25\sqrt{2} \approx 64.6 \text{ meters}$$

the length of the other leg is

$$200 - x \approx 200 - 64.6 \approx 135.4 \text{ meters.}$$

Example 6   Work-Rate Problem

An office contains two copy machines. machine B is known to take 12 minutes longer than machine A to copy the company's monthly report. Using both machines together, it takes 8 minutes to reproduce the report. How long would it take each machine alone to reproduce the report?

Solution

Verbal Model:

$$\dfrac{\text{Work done by}}{\text{machine A}} \;+\; \dfrac{\text{Work done by}}{\text{machine B}} \;=\; \dfrac{\text{1 complete}}{\text{job}}$$

$$\dfrac{\text{Rate}}{\text{for A}} \cdot \dfrac{\text{Time}}{\text{for both}} \;+\; \dfrac{\text{Rate}}{\text{for B}} \cdot \dfrac{\text{Time}}{\text{for both}} = 1$$

Labels:

Time for machine A $= t$	(minutes)
Rate for machine A $= \dfrac{1}{t}$	(job per minute)
Time for machine B $= t + 12$	(minutes)
Rate for machine B $= \dfrac{1}{t + 12}$	(job per minute)
Time for both machines $= 8$	(minutes)
Rate for both machines $= \dfrac{1}{8}$	(job per minute)

Equation:

$$\dfrac{1}{t}(8) + \dfrac{1}{t + 12}(8) = 1 \qquad \text{\color{red}Original equation}$$

$$8\left(\dfrac{1}{t} + \dfrac{1}{t + 12}\right) = 1 \qquad \text{\color{red}Distributive Property}$$

$$8\left[\dfrac{t + 12 + t}{t(t + 12)}\right] = 1 \qquad \text{\color{red}Rewrite with common denominator.}$$

$$8t(t + 12)\left[\dfrac{2t + 12}{t(t + 12)}\right] = t(t + 12) \qquad \text{\color{red}Multiply each side by } t(t + 12).$$

$$8(2t + 12) = t^2 + 12t \qquad \text{\color{red}Simplify.}$$

$$16t + 96 = t^2 + 12t \qquad \text{\color{red}Distributive Property}$$

$$0 = t^2 - 4t - 96 \qquad \text{\color{red}Subtract } 16t + 96 \text{ from each side.}$$

$$0 = (t - 12)(t + 8) \qquad \text{\color{red}Factor right side of equation.}$$

$$t - 12 = 0 \;\Longrightarrow\; t = 12 \qquad \text{\color{red}Set 1st factor equal to 0.}$$

$$t + 8 = 0 \;\Longrightarrow\; t = -8 \qquad \text{\color{red}Set 2nd factor equal to 0.}$$

By choosing the positive value for t, you can conclude that the times for the two machines are

Time for machine A $= t = 12$ minutes

Time for machine B $= t + 12 = 12 + 12 = 24$ minutes.

Check these solutions in the original statement of the problem.

Example 7   The Height of a Model Rocket

A model rocket is projected straight upward from ground level according to the height equation

$$h = -16t^2 + 192t, \, t \geq 0$$

where h is the height in feet and t is the time in seconds.

a. After how many seconds is the height 432 feet?

b. After how many seconds does the rocket hit the ground?

c. What is the maximum height of the rocket?

Solution

a.
$$h = -16t^2 + 192t \qquad \text{Write original equation.}$$
$$432 = -16t^2 + 192t \qquad \text{Substitute 432 for } h.$$
$$16t^2 - 192t + 432 = 0 \qquad \text{Write in general form.}$$
$$t^2 - 12t + 27 = 0 \qquad \text{Divide each side by 16.}$$
$$(t - 3)(t - 9) = 0 \qquad \text{Factor.}$$

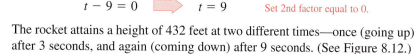

$$t - 3 = 0 \quad \Longrightarrow \quad t = 3 \qquad \text{Set 1st factor equal to 0.}$$
$$t - 9 = 0 \quad \Longrightarrow \quad t = 9 \qquad \text{Set 2nd factor equal to 0.}$$

The rocket attains a height of 432 feet at two different times—once (going up) after 3 seconds, and again (coming down) after 9 seconds. (See Figure 8.12.)

b. To find the time it takes for the rocket to hit the ground, let the height be 0.

$$0 = -16t^2 + 192t \qquad \text{Substitute 0 for } h \text{ in original equation.}$$
$$0 = t^2 - 12t \qquad \text{Divide each side by } -16.$$
$$0 = t(t - 12) \qquad \text{Factor.}$$
$$t = 0 \quad \text{or} \quad t = 12 \qquad \text{Solutions}$$

The rocket hits the ground after 12 seconds. (Note that the time of $t = 0$ seconds corresponds to the time of lift-off.)

c. The maximum value for h in the equation $h = -16t^2 + 192t$ occurs when $t = -\dfrac{b}{2a}$. So, the t-coordinate is

$$t = \frac{-b}{2a} = \frac{-192}{2(-16)} = 6$$

and the h-coordinate is

$$h = -16(6)^2 + 192(6) = 576.$$

So, the maximum height of the rocket is 576 feet.

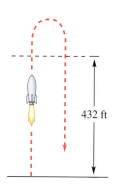

432 ft

Figure 8.12

8.5 Exercises

Review   Concepts, Skills, and Problem Solving

Keep mathematically in shape by doing these exercises *before* the problems of this section.

Properties and Definitions

1. *Writing* 🖉 Define the slope of the line through the points (x_1, y_1) and (x_2, y_2).

2. Give the following forms of an equation of a line.

   (a) Slope-intercept form

   (b) Point-slope form

   (c) General form

   (d) Horizontal line

Equations of Lines

In Exercises 3–10, find the general form of the equation of the line through the two points.

3. $(0, 0), (4, -2)$

4. $(0, 0), (100, 75)$

5. $(-1, -2), (3, 6)$

6. $(1, 5), (6, 0)$

7. $\left(\frac{3}{2}, 8\right), \left(\frac{11}{2}, \frac{5}{2}\right)$

8. $(0, 2), (7.3, 15.4)$

9. $(0, 8), (5, 8)$

10. $(-3, 2), (-3, 5)$

Problem Solving

11. *Endowment*   A group of people agree to share equally in the cost of a \$250,000 endowment to a college. If they could find two more people to join the group, each person's share of the cost would decrease by \$6250. How many people are presently in the group?

12. *Current Speed*   A boat travels at a speed of 18 miles per hour in still water. It travels 35 miles upstream and then returns to the starting point in a total of 4 hours. Find the speed of the current.

Solving Problems

1. *Selling Price*   A store owner bought a case of eggs for \$21.60. By the time all but 6 dozen of the eggs had been sold at a profit of \$0.30 per dozen, the original investment of \$21.60 had been regained. How many dozen eggs did the owner sell, and what was the selling price per dozen? See Example 1.

2. *Selling Price*   A manager of a computer store bought several computers of the same model for \$27,000. When all but three of the computers had been sold at a profit of \$750 per computer, the original investment of \$27,000 had been regained. How many computers were sold, and what was the selling price of each?

3. *Selling Price*   A store owner bought a case of video games for \$480. By the time he had sold all but eight of them at a profit of \$10 each, the original investment of \$480 had been regained. How many video games were sold, and what was the selling price of each game?

4. *Selling Price*   A math club bought a case of sweatshirts for \$850 to sell as a fundraiser. By the time all but 16 sweatshirts had been sold at a profit of \$8 per sweatshirt, the original investment of \$850 had been regained. How many sweatshirts were sold, and what was the selling price of each sweatshirt?

🔺 *Geometry*   In Exercises 5–14, complete the table of widths, lengths, perimeters, and areas of rectangles.

Width	Length	Perimeter	Area
5. $1.4l$	l	54 in.	
6. w	$3.5w$	60 m	
7. w	$2.5w$		250 ft^2
8. w	$1.5w$		216 cm^2
9. $\frac{1}{3}l$	l		192 in.^{2}
10. $\frac{3}{4}l$	l		2700 in.^{2}
11. w	$w + 3$	54 km	
12. $l - 6$	l	108 ft	

	Width	Length	Perimeter	Area
13.	$l - 20$	l		12,000 m²
14.	w	$w + 5$		500 ft²

15. ▲ *Geometry*   A picture frame is 4 inches taller than it is wide and has an area of 192 square inches. What are the dimensions of the picture frame? See Example 2.

16. ▲ *Geometry*   The height of a triangle is 8 inches less than its base. The area of the triangle is 192 square inches. Find the dimensions of the triangle.

17. *Storage Area*   A retail lumberyard plans to store lumber in a rectangular region adjoining the sales office (see figure). The region will be fenced on three sides, and the fourth side will be bounded by the wall of the office building. There is 350 feet of fencing available, and the area of the region is 12,500 square feet. Find the dimensions of the region.

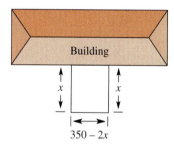

18. ▲ *Geometry*   Your home is on a square lot. To add more space to your yard, you purchase an additional 20 feet along the side of the property (see figure). The area of the lot is now 25,500 square feet. What are the dimensions of the new lot?

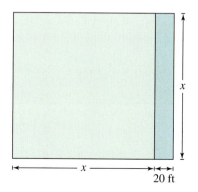

19. *Fenced Area*   A family built a fence around three sides of their property (see figure). In total, they used 550 feet of fencing. By their calculations, the lot is 1 acre (43,560 square feet). Is this correct? Explain your reasoning.

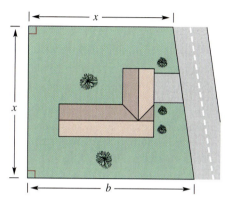

20. *Fenced Area*   You have 100 feet of fencing. Do you have enough to enclose a rectangular region whose area is 630 square feet? Is there enough to enclose a circular area of 630 square feet? Explain.

21. *Open Conduit*   An open-topped rectangular conduit for carrying water in a manufacturing process is made by folding up the edges of a sheet of aluminum 48 inches wide (see figure). A cross section of the conduit must have an area of 288 square inches. Find the width and height of the conduit.

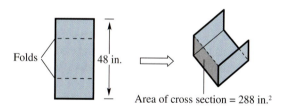

Area of cross section = 288 in.²

22. *Photography*   A photographer has a photograph that is 6 inches by 8 inches. He wishes to reduce the photo by the same amount on each side such that the resulting photo will have an area that is half the area of the original photo. By how much should each side be reduced?

Compound Interest   In Exercises 23–28, find the interest rate r. Use the formula $A = P(1 + r)^2$, where A is the amount after 2 years in an account earning r percent (in decimal form) compounded annually, and P is the original investment. See Example 3.

23. $P = \$10,000$
    $A = \$11,990.25$

24. $P = \$3000$
    $A = \$3499.20$

25. $P = \$500$
    $A = \$572.45$

26. $P = \$250$
    $A = \$280.90$

27. $P = \$6500$
    $A = \$7372.46$

28. $P = \$8000$
    $A = \$8421.41$

29. *Reduced Rates*   A service organization pays \$210 for a block of tickets to a ball game. The block contains three more tickets than the organization needs for its members. By inviting three more people to attend (and share in the cost), the organization lowers the price per ticket by \$3.50. How many people are going to the game? See Example 4.

30. *Reduced Rates*   A service organization buys a block of tickets to a ball game for \$240. After eight more people decide to go to the game, the price per ticket is decreased by \$1. How many people are going to the game?

31. *Reduced Fares*   A science club charters a bus to attend a science fair at a cost of \$480. To lower the bus fare per person, the club invites nonmembers to go along. When two nonmembers join the trip, the fare per person is decreased by \$1. How many people are going on the excursion?

32. *Venture Capital*   Eighty thousand dollars is needed to begin a small business. The cost will be divided equally among the investors. Some have made a commitment to invest. If three more investors are found, the amount required from each will decrease by \$6000. How many have made a commitment to invest in the business?

33. *Delivery Route*   You are asked to deliver pizza to an insurance office and an apartment complex (see figure), and you keep a log of all the mileages between stops. You forget to look at the odometer at the insurance office, but after getting to the apartment complex you record the total distance traveled from the pizza shop as 18 miles. The return distance from the apartment complex to the pizza shop is 16 miles. The route approximates a right triangle. Estimate the distance from the pizza shop to the insurance office. See Example 5.

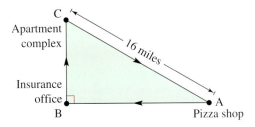

Figure for 33

34. ▲ *Geometry*   An L-shaped sidewalk from the library (point A) to the gym (point B) on a high school campus is 100 yards long, as shown in the figure. By cutting diagonally across the grass, students shorten the walking distance to 80 yards. What are the lengths of the two legs of the sidewalk?

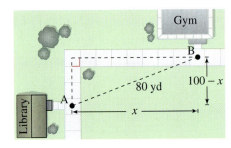

35. ▲ *Geometry*   An adjustable rectangular form has minimum dimensions of 3 meters by 4 meters. The length and width can be expanded by equal amounts x (see figure).

(a) Write an equation relating the length d of the diagonal to x.

(b) ▦ Use a graphing calculator to graph the equation.

(c) ▦ Use the graph to approximate the value of x when $d = 10$ meters.

(d) Find x algebraically when $d = 10$.

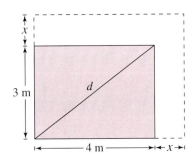

36. Solving Graphically and Numerically   A meteorologist is positioned 100 feet from the point where a weather balloon is launched (see figure).

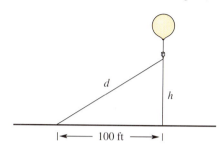

(a) Write an equation relating the distance d between the balloon and the meteorologist to the height h of the balloon.

(b) ▦ Use a graphing calculator to graph the equation.

(c) ▦ Use the graph to approximate the value of h when $d = 200$ feet.

(d) Complete the table.

h	0	100	200	300
d				

37. Work Rate   Working together, two people can complete a task in 5 hours. Working alone, one person takes 2 hours longer than the other. How long would it take each person to do the task alone? See Example 6.

38. Work Rate   Working together, two people can complete a task in 6 hours. Working alone, one person takes 2 hours longer than the other. How long would it take each person to do the task alone?

39. Work Rate   An office contains two printers. Machine B is known to take 3 minutes longer than Machine A to produce the company's monthly financial report. Using both machines together, it takes 6 minutes to produce the report. How long would it take each machine to produce the report?

40. Work Rate   A builder works with two plumbing companies. Company A is known to take 3 days longer than Company B to do the plumbing in a particular style of house. Using both companies, it takes 4 days. How long would it take to do the plumbing using each company individually?

Free-Falling Object   In Exercises 41–44, find the time necessary for an object to fall to ground level from an initial height of h_0 feet if its height h at any time t (in seconds) is given by $h = h_0 - 16t^2$.

41. $h_0 = 169$ **42.** $h_0 = 729$

43. $h_0 = 1454$ (height of the Sears Tower)

44. $h_0 = 984$ (height of the Eiffel Tower)

45. Height   The height h in feet of a baseball hit 3 feet above the ground is given by $h = 3 + 75t - 16t^2$, where t is time in seconds. Find the time when the ball hits the ground. See Example 7.

46. Height   You are hitting baseballs. When you toss the ball into the air, your hand is 5 feet above the ground (see figure). You hit the ball when it falls back to a height of 4 feet. You toss the ball with an initial velocity of 25 feet per second. The height h of the ball t seconds after leaving your hand is given by $h = 5 + 25t - 16t^2$. How much time will pass before you hit the ball?

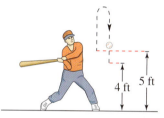

47. Height   A model rocket is projected straight upward from ground level according to the height equation $h = -16t^2 + 160t$, where h is the height of the rocket in feet and t is the time in seconds.

(a) After how many seconds is the height 336 feet?

(b) After how many seconds does the rocket hit the ground?

(c) What is the maximum height of the rocket?

48. Height   A tennis ball is tossed vertically upward from a height of 5 feet according to the height equation $h = -16t^2 + 21t + 5$, where h is the height of the tennis ball in feet and t is the time in seconds.

(a) After how many seconds is the height 11 feet?

(b) After how many seconds does the tennis ball hit the ground?

(c) What is the maximum height of the ball?

Number Problems In Exercises 49–54, find two positive integers that satisfy the requirement.

49. The product of two consecutive integers is 182.

50. The product of two consecutive integers is 1806.

51. The product of two consecutive even integers is 168.

52. The product of two consecutive even integers is 2808.

53. The product of two consecutive odd integers is 323.

54. The product of two consecutive odd integers is 1443.

55. *Air Speed* An airline runs a commuter flight between two cities that are 720 miles apart. If the average speed of the planes could be increased by 40 miles per hour, the travel time would be decreased by 12 minutes. What air speed is required to obtain this decrease in travel time?

56. *Average Speed* A truck traveled the first 100 miles of a trip at one speed and the last 135 miles at an average speed of 5 miles per hour less. The entire trip took 5 hours. What was the average speed for the first part of the trip?

57. *Speed* A small business uses a minivan to make deliveries. The cost per hour for fuel for the van is $C = v^2/600$, where v is the speed in miles per hour. The driver is paid $5 per hour. Find the speed if the cost for wages and fuel for a 110-mile trip is $20.39.

58. *Distance* Find any points on the line $y = 14$ that are 13 units from the point $(1, 2)$.

59. ▲ *Geometry* The area of an ellipse is given by $A = \pi ab$ (see figure). For a certain ellipse, it is required that $a + b = 20$.

(a) Show that $A = \pi a(20 - a)$.

(b) Complete the table.

a	4	7	10	13	16
A					

(c) Find two values of a such that $A = 300$.

(d) ⊞ Use a graphing calculator to graph the area equation. Then use the graph to verify the results in part (c).

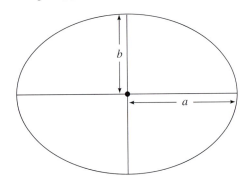

Figure for 59

60. ⊞ *Data Analysis* For the years 1993 through 2000, the sales s (in millions of dollars) of recreational vehicles in the United States can be approximated by $s = 156.45t^2 - 1035.5t + 6875$, $3 \le t \le 10$, where t is time in years, with $t = 3$ corresponding to 1993. (Source: National Sporting Goods Association)

(a) Use a graphing calculator to graph the model.

(b) Use the graph in part (a) to determine the year in which sales were approximately $6.3 billion.

Explaining Concepts

61. *Writing*✏ In your own words, describe strategies for solving word problems.

62. *Writing*✏ List the strategies that can be used to solve a quadratic equation.

63. *Unit Analysis* Describe the units of the product.

$$\frac{9 \text{ dollars}}{\text{hour}} \cdot (20 \text{ hours})$$

64. *Unit Analysis* Describe the units of the product.

$$\frac{20 \text{ feet}}{\text{minute}} \cdot \frac{1 \text{ minute}}{60 \text{ seconds}} \cdot (45 \text{ seconds})$$

8.6 Quadratic and Rational Inequalities

Will Hart/PhotoEdit, Inc.

What You Should Learn

1. Determine test intervals for polynomials.
2. Use test intervals to solve quadratic inequalities.
3. Use test intervals to solve rational inequalities.
4. Use inequalities to solve application problems.

Why You Should Learn It

Rational inequalities can be used to model and solve real-life problems. For instance, in Exercise 116 on page 566, a rational inequality is used to model the temperature of a metal in a laboratory experiment.

1. Determine test intervals for polynomials.

Finding Test Intervals

When working with polynomial inequalities, it is important to realize that the value of a polynomial can change signs only at its **zeros**. That is, a polynomial can change signs only at the x-values for which the value of the polynomial is zero. For instance, the first-degree polynomial $x + 2$ has a zero at $x = -2$, and it changes signs at that zero. You can picture this result on the real number line, as shown in Figure 8.13.

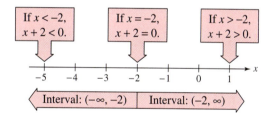

Figure 8.13

Note in Figure 8.13 that the zero of the polynomial partitions the real number line into two **test intervals.** The value of the polynomial is negative for every x-value in the first test interval $(-\infty, -2)$, and it is positive for every x-value in the second test interval $(-2, \infty)$. You can use the same basic approach to determine the test intervals for any polynomial.

Finding Test Intervals for a Polynomial

1. Find all real zeros of the polynomial, and arrange the zeros in increasing order. The zeros of a polynomial are called its **critical numbers.**

2. Use the critical numbers of the polynomial to determine its test intervals.

3. Choose a representative x-value in each test interval and evaluate the polynomial at that value. If the value of the polynomial is negative, the polynomial will have negative values for *every* x-value in the interval. If the value of the polynomial is positive, the polynomial will have positive values for *every* x-value in the interval.

2 Use test intervals to solve quadratic inequalities.

Quadratic Inequalities

The concepts of critical numbers and test intervals can be used to solve nonlinear inequalities, as demonstrated in Examples 1, 2, and 4.

Example 1   Solving a Quadratic Inequality

Solve the inequality $x^2 - 5x < 0$.

Solution

First find the *critical numbers* of $x^2 - 5x < 0$ by finding the solutions of the equation $x^2 - 5x = 0$.

$$x^2 - 5x = 0 \qquad \text{Write corresponding equation.}$$

$$x(x - 5) = 0 \qquad \text{Factor.}$$

$$x = 0, x = 5 \qquad \text{Critical numbers}$$

This implies that the test intervals are $(-\infty, 0)$, $(0, 5)$, and $(5, \infty)$. To test an interval, choose a convenient number in the interval and determine if the number satisfies the inequality.

Test interval	Representative x-value	Is inequality satisfied?
$(-\infty, 0)$	$x = -1$	$(-1)^2 - 5(-1) \overset{?}{<} 0$ $6 \not< 0$
$(0, 5)$	$x = 1$	$1^2 - 5(1) \overset{?}{<} 0$ $-4 < 0$
$(5, \infty)$	$x = 6$	$6^2 - 5(6) \overset{?}{<} 0$ $6 \not< 0$

Because the inequality $x^2 - 5x < 0$ is satisfied only by the middle test interval, you can conclude that the solution set of the inequality is the interval $(0, 5)$, as shown in Figure 8.14.

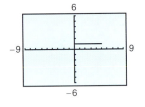
Study Tip

In Example 1, note that you would have used the same basic procedure if the inequality symbol had been $\le$, $>$, or $\ge$. For instance, in Figure 8.14, you can see that the solution set of the inequality $x^2 - 5x \ge 0$ consists of the union of the half-open intervals $(-\infty, 0]$ and $[5, \infty)$, which is written as $(-\infty, 0] \cup [5, \infty)$.

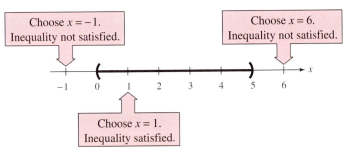

Figure 8.14

Just as in solving quadratic *equations*, the first step in solving a quadratic *inequality* is to write the inequality in **general form,** with the polynomial on the left and zero on the right, as demonstrated in Example 2.

Example 2   Solving a Quadratic Inequality

Solve the inequality $2x^2 + 5x \geq 12$.

Solution

Begin by writing the inequality in general form.

$$2x^2 + 5x - 12 \geq 0 \qquad \text{Write in general form.}$$

Next, find the critical numbers of $2x^2 + 5x - 12 \geq 0$ by finding the solutions to the equation $2x^2 + 5x - 12 = 0$.

$$2x^2 + 5x - 12 = 0 \qquad \text{Write corresponding equation.}$$

$$(x + 4)(2x - 3) = 0 \qquad \text{Factor.}$$

$$x = -4, \, x = \frac{3}{2} \qquad \text{Critical numbers}$$

This implies that the test intervals are $(-\infty, -4)$, $\left(-4, \frac{3}{2}\right)$, and $\left(\frac{3}{2}, \infty\right)$. To test an interval, choose a convenient number in the interval and determine if the number satisfies the inequality.

Test interval	Representative x-value	Is inequality satisfied?
$(-\infty, -4)$	$x = -5$	$2(-5)^2 + 5(-5) \overset{?}{\geq} 12$ $25 \geq 12$
$\left(-4, \frac{3}{2}\right)$	$x = 0$	$2(0)^2 + 5(0) \overset{?}{\geq} 12$ $0 \not\geq 12$
$\left(\frac{3}{2}, \infty\right)$	$x = 2$	$2(2)^2 + 5(2) \overset{?}{\geq} 12$ $18 \geq 12$

From this table you can see that the inequality $2x^2 + 5x \geq 12$ is satisfied for the intervals $(-\infty, -4]$ and $\left[\frac{3}{2}, \infty\right)$. So, the solution set of the inequality is $(-\infty, -4] \cup \left[\frac{3}{2}, \infty\right)$, as shown in Figure 8.15.

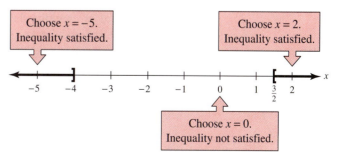

Figure 8.15

Study Tip

In Examples 1 and 2, the critical numbers are found by factoring. With quadratic polynomials that do not factor, you can use the Quadratic Formula to find the critical numbers. For instance, to solve the inequality

$$x^2 - 2x - 1 \leq 0$$

you can use the Quadratic Formula to determine that the critical numbers are

$$1 - \sqrt{2} \approx -0.414$$

and

$$1 + \sqrt{2} \approx 2.414.$$

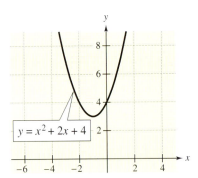

Figure 8.16

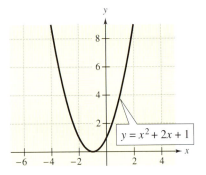

Figure 8.17

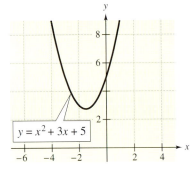

Figure 8.18

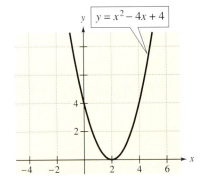

Figure 8.19

The solutions of the quadratic inequalities in Examples 1 and 2 consist, respectively, of a single interval and the union of two intervals. When solving the exercises for this section, you should watch for some unusual solution sets, as illustrated in Example 3.

Example 3   Unusual Solution Sets

Solve each inequality.

a. The solution set of the quadratic inequality

$$x^2 + 2x + 4 > 0$$

consists of the entire set of real numbers, $(-\infty, \infty)$. This is true because the value of the quadratic $x^2 + 2x + 4$ is positive for every real value of x. You can see in Figure 8.16 that the entire parabola lies above the x-axis.

b. The solution set of the quadratic inequality

$$x^2 + 2x + 1 \leq 0$$

consists of the single number $\{-1\}$. This is true because $x^2 + 2x + 1 = (x + 1)^2$ has just one critical number, $x = -1$, and it is the only value that satisfies the inequality. You can see in Figure 8.17 that the parabola meets the x-axis at $x = -1$.

c. The solution set of the quadratic inequality

$$x^2 + 3x + 5 < 0$$

is empty. This is true because the value of the quadratic $x^2 + 3x + 5$ is not less than zero for any value of x. No point on the parabola lies below the x-axis, as shown in Figure 8.18.

d. The solution set of the quadratic inequality

$$x^2 - 4x + 4 > 0$$

consists of all real numbers *except* the number 2. In interval notation, this solution set can be written as $(-\infty, 2) \cup (2, \infty)$. You can see in Figure 8.19 that the parabola lies above the x-axis *except* at $x = 2$, where it meets the x-axis.

Remember that checking the solution set of an inequality is not as straightforward as checking the solutions of an equation, because inequalities tend to have infinitely many solutions. Even so, you should check several x-values in your solution set to confirm that they satisfy the inequality. Also try checking x-values that are not in the solution set to verify that they do not satisfy the inequality.

For instance, the solution set of $x^2 - 5x < 0$ is the interval $(0, 5)$. Try checking some numbers in this interval to verify that they satisfy the inequality. Then check some numbers outside the interval to verify that they do not satisfy the inequality.

③ Use test intervals to solve rational inequalities.

Rational Inequalities

The concepts of critical numbers and test intervals can be extended to inequalities involving rational expressions. To do this, use the fact that the value of a rational expression can change sign only at its *zeros* (the x-values for which its numerator is zero) and its *undefined values* (the x-values for which its denominator is zero). These two types of numbers make up the **critical numbers** of a rational inequality. For instance, the critical numbers of the inequality

$$\frac{x - 2}{(x - 1)(x + 3)} < 0$$

are $x = 2$ (the numerator is zero), and $x = 1$ and $x = -3$ (the denominator is zero). From these three critical numbers you can see that the inequality has *four* test intervals: $(-\infty, -3)$, $(-3, 1)$, $(1, 2)$, and $(2, \infty)$.

Example 4   Solving a Rational Inequality

To solve the inequality $\dfrac{x}{x - 2} > 0$, first find the critical numbers. The numerator is zero when $x = 0$, and the denominator is zero when $x = 2$. So, the two critical numbers are 0 and 2, which implies that the test intervals are $(-\infty, 0)$, $(0, 2)$, and $(2, \infty)$. To test an interval, choose a convenient number in the interval and determine if the number satisfies the inequality, as shown in the table.

Test interval	Representative x-value	Is inequality satisfied?	
$(-\infty, 0)$	$x = -1$	$\dfrac{-1}{-1 - 2} \overset{?}{>} 0$	$\dfrac{1}{3} > 0$
$(0, 2)$	$x = 1$	$\dfrac{1}{1 - 2} \overset{?}{>} 0$	$-1 \not> 0$
$(2, \infty)$	$x = 3$	$\dfrac{3}{3 - 2} \overset{?}{>} 0$	$3 > 0$

You can see that the inequality is satisfied for the intervals $(-\infty, 0)$ and $(2, \infty)$. So, the solution set of the inequality is $(-\infty, 0) \cup (2, \infty)$. See Figure 8.20.

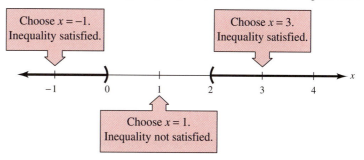

Choose $x = -1$.
Inequality satisfied.

Choose $x = 3$.
Inequality satisfied.

Choose $x = 1$.
Inequality not satisfied.

Figure 8.20

Study Tip

When solving a rational inequality, you should begin by writing the inequality in general form, with the rational expression (as a single fraction) on the left and zero on the right. For instance, the first step in solving

$$\frac{2x}{x + 3} < 4$$

is to write it as

$$\frac{2x}{x + 3} - 4 < 0$$

$$\frac{2x - 4(x + 3)}{x + 3} < 0$$

$$\frac{-2x - 12}{x + 3} < 0.$$

Try solving this inequality. You should find that the solution set is $(-\infty, -6) \cup (-3, \infty)$.

4 Use inequalities to solve application problems.

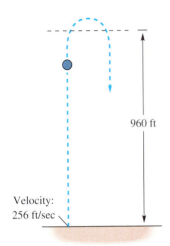

Velocity:
256 ft/sec

Figure 8.21

Application

Example 5   The Height of a Projectile

A projectile is fired straight upward from ground level with an initial velocity of 256 feet per second, as shown in Figure 8.21, so that its height h at any time t is given by

$$h = -16t^2 + 256t$$

where h is measured in feet and t is measured in seconds. During what interval of time will the height of the projectile exceed 960 feet?

Solution

To solve this problem, begin by writing the inequality in general form.

$-16t^2 + 256t > 960$	Write original inequality.
$-16t^2 + 256t - 960 > 0$	Write in general form.

Next, find the critical numbers for $-16t^2 + 256t - 960 > 0$ by finding the solution to the equation $-16t^2 + 256t - 960 = 0$.

$-16t^2 + 256t - 960 = 0$	Write corresponding equation.
$t^2 - 16t + 60 = 0$	Divide each side by -16.
$(t - 6)(t - 10) = 0$	Factor.
$t = 6, t = 10$	Critical numbers

This implies that the test intervals are

$(-\infty, 6), (6, 10),$ and $(10, \infty).$	Test intervals

To test an interval, choose a convenient number in the interval and determine if the number satisfies the inequality.

Test interval	Representative x-value	Is inequality satisfied?
$(-\infty, 6)$	$t = 0$	$-16(0)^2 + 256(0) \overset{?}{>} 960$ $0 \not> 960$
$(6, 10)$	$t = 7$	$-16(7)^2 + 256(7) \overset{?}{>} 960$ $1008 > 960$
$(10, \infty)$	$t = 11$	$-16(11)^2 + 256(11) \overset{?}{>} 960$ $880 \not> 960$

So, the height of the projectile will exceed 960 feet for values of t such that $6 < t < 10$.

8.6 Exercises

Review *Concepts, Skills, and Problem Solving*

Keep mathematically in shape by doing these exercises *before* the problems of this section.

Properties and Definitions

1. *Writing* Is 36.83×10^8 written in scientific notation? Explain.

2. *Writing* The numbers $n_1 \times 10^2$ and $n_2 \times 10^4$ are in scientific notation. The product of these two numbers must lie in what interval? Explain.

Simplifying Expressions

In Exercises 3–8, factor the expression.

3. $6u^2v - 192v^2$

4. $5x^{2/3} - 10x^{1/3}$

5. $x(x - 10) - 4(x - 10)$

6. $x^3 + 3x^2 - 4x - 12$

7. $16x^2 - 121$

8. $4x^3 - 12x^2 + 16x$

Mathematical Modeling

▲ *Geometry* In Exercises 9–12, write an expression for the area of the figure.

9.

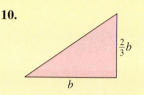

10.

11.

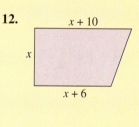

12.

Developing Skills

In Exercises 1–10, find the critical numbers.

1. $x(2x - 5)$

2. $5x(x - 3)$

3. $4x^2 - 81$

4. $9y^2 - 16$

5. $x(x + 3) - 5(x + 3)$

6. $y(y - 4) - 3(y - 4)$

7. $x^2 - 4x + 3$

8. $3x^2 - 2x - 8$

9. $4x^2 - 20x + 25$

10. $4x^2 - 4x - 3$

In Exercises 11–20, determine the intervals for which the polynomial is entirely negative and entirely positive.

11. $x - 4$

12. $3 - x$

13. $3 - \frac{1}{2}x$

14. $\frac{2}{3}x - 8$

15. $2x(x - 4)$

16. $7x(3 - x)$

17. $4 - x^2$

18. $x^2 - 9$

19. $x^2 - 4x - 5$

20. $2x^2 - 4x - 3$

In Exercises 21–60, solve the inequality and graph the solution on the real number line. (Some of the inequalities have no solution.) See Examples 1–3.

21. $3x(x - 2) < 0$

22. $2x(x - 6) > 0$

23. $3x(2 - x) \geq 0$

24. $2x(6 - x) > 0$

25. $x^2 > 4$

26. $z^2 \leq 9$

27. $x^2 - 3x - 10 \geq 0$

28. $x^2 + 8x + 7 < 0$

29. $x^2 + 4x > 0$

30. $x^2 - 5x \geq 0$

31. $x^2 + 3x \leq 10$

32. $t^2 - 4t > 12$

33. $u^2 + 2u - 2 > 1$

34. $t^2 - 15t + 50 < 0$

35. $x^2 + 4x + 5 < 0$

36. $x^2 + 6x + 10 > 0$

37. $x^2 + 2x + 1 \geq 0$

38. $y^2 - 5y + 6 > 0$

39. $x^2 - 4x + 2 > 0$

40. $-x^2 + 8x - 11 \leq 0$

41. $x^2 - 6x + 9 \geq 0$

42. $x^2 + 8x + 16 < 0$

43. $u^2 - 10u + 25 < 0$

44. $y^2 + 16y + 64 \leq 0$

45. $3x^2 + 2x - 8 \leq 0$

46. $2t^2 - 3t - 20 \geq 0$

47. $-6u^2 + 19u - 10 > 0$

48. $4x^2 - 4x - 63 < 0$

49. $-2u^2 + 7u + 4 < 0$

50. $-3x^2 - 4x + 4 \leq 0$

51. $4x^2 + 28x + 49 \leq 0$

52. $9x^2 - 24x + 16 \geq 0$

53. $(x - 5)^2 < 0$

54. $(y + 3)^2 \geq 0$

55. $6 - (x - 5)^2 < 0$

56. $(y + 3)^2 - 6 \geq 0$

57. $16 \leq (u + 5)^2$

58. $25 \geq (x - 3)^2$

59. $x(x - 2)(x + 2) > 0$

60. $x(x - 1)(x + 4) \leq 0$

⊞ In Exercises 61–68, use a graphing calculator to solve the inequality. Verify your result algebraically.

61. $x^2 - 6x < 0$

62. $2x^2 + 5x > 0$

63. $0.5x^2 + 1.25x - 3 > 0$

64. $\frac{1}{3}x^2 - 3x < 0$

65. $x^2 + 4x + 4 \geq 9$

66. $x^2 - 6x + 9 < 16$

67. $9 - 0.2(x - 2)^2 < 4$

68. $8x - x^2 > 12$

In Exercises 69–72, find the critical numbers.

69. $\dfrac{5}{x - 3}$

70. $\dfrac{-6}{x + 2}$

71. $\dfrac{2x}{x + 5}$

72. $\dfrac{x - 2}{x - 10}$

In Exercises 73–94, solve the inequality and graph the solution on the real number line. See Example 4.

73. $\dfrac{5}{x - 3} > 0$

74. $\dfrac{3}{4 - x} > 0$

75. $\dfrac{-5}{x - 3} > 0$

76. $\dfrac{-3}{4 - x} > 0$

77. $\dfrac{3}{y - 1} \leq -1$

78. $\dfrac{2}{x - 3} \geq -1$

79. $\dfrac{x + 3}{x - 1} > 0$

80. $\dfrac{x - 5}{x + 2} < 0$

81. $\dfrac{y - 4}{y - 1} \leq 0$

82. $\dfrac{y + 6}{y + 2} \geq 0$

83. $\dfrac{4x - 2}{2x - 4} > 0$

84. $\dfrac{3x + 4}{2x - 1} < 0$

85. $\dfrac{x + 2}{4x + 6} \leq 0$

86. $\dfrac{u - 6}{3u - 5} \le 0$

87. $\dfrac{3(u - 3)}{u + 1} < 0$

88. $\dfrac{2(4 - t)}{4 + t} > 0$

89. $\dfrac{6}{x - 4} > 2$

90. $\dfrac{1}{x + 2} > -3$

91. $\dfrac{4x}{x + 2} < -1$

92. $\dfrac{6x}{x - 4} < 5$

93. $\dfrac{x - 1}{x - 3} \le 2$

94. $\dfrac{x + 4}{x - 5} \ge 10$

In Exercises 95–102, use a graphing calculator to solve the rational inequality. Verify your result algebraically.

95. $\dfrac{1}{x} - x > 0$

96. $\dfrac{1}{x} - 4 < 0$

97. $\dfrac{x + 6}{x + 1} - 2 < 0$

98. $\dfrac{x + 12}{x + 2} - 3 \ge 0$

99. $\dfrac{6x - 3}{x + 5} < 2$

100. $\dfrac{3x - 4}{x - 4} < -5$

101. $x + \dfrac{1}{x} > 3$

102. $4 - \dfrac{1}{x^2} > 1$

Graphical Analysis   In Exercises 103–106, use a graphing calculator to graph the function. Use the graph to approximate the values of *x* that satisfy the specified inequalities.

Function	Inequalities

103. $f(x) = \dfrac{3x}{x - 2}$ (a) $f(x) \le 0$ (b) $f(x) \ge 6$

104. $f(x) = \dfrac{2(x - 2)}{x + 1}$ (a) $f(x) \le 0$ (b) $f(x) \ge 8$

105. $f(x) = \dfrac{2x^2}{x^2 + 4}$ (a) $f(x) \ge 1$ (b) $f(x) \le 2$

106. $f(x) = \dfrac{5x}{x^2 + 4}$ (a) $f(x) \ge 1$ (b) $f(x) \ge 0$

Solving Problems

107. *Height*   A projectile is fired straight upward from ground level with an initial velocity of 128 feet per second, so that its height *h* at any time *t* is given by $h = -16t^2 + 128t$, where *h* is measured in feet and *t* is measured in seconds. During what interval of time will the height of the projectile exceed 240 feet?

108. *Height*   A projectile is fired straight upward from ground level with an initial velocity of 88 feet per second, so that its height *h* at any time *t* is given by $h = -16t^2 + 88t$, where *h* is measured in feet and *t* is measured in seconds. During what interval of time will the height of the projectile exceed 50 feet?

109. *Compound Interest*   You are investing $1000 in a certificate of deposit for 2 years and you want the interest for that time period to exceed $150. The interest is compounded annually. What interest rate should you have? [*Hint:* Solve the inequality $1000(1 + r)^2 > 1150$.]

110. *Compound Interest*   You are investing $500 in a certificate of deposit for 2 years and you want the interest for that time to exceed $50. The interest is compounded annually. What interest rate should you have? [*Hint:* Solve the inequality $500(1 + r)^2 > 550$.]

111. ▲ *Geometry* You have 64 feet of fencing to enclose a rectangular region. Determine the interval for the length such that the area will exceed 240 square feet.

112. ▲ *Geometry* A rectangular playing field with a perimeter of 100 meters is to have an area of at least 500 square meters. Within what bounds must the length of the field lie?

113. *Cost, Revenue, and Profit* The revenue and cost equations for a computer desk are given by

$$R = x(50 - 0.0002x) \text{ and } C = 12x + 150,000$$

where R and C are measured in dollars and x represents the number of desks sold. How many desks must be sold to obtain a profit of at least $1,650,000?

114. *Cost, Revenue, and Profit* The revenue and cost equations for a digital camera are given by

$$R = x(125 - 0.0005x) \text{ and } C = 3.5x + 185,000$$

where R and C are measured in dollars and x represents the number of cameras sold. How many cameras must be sold to obtain a profit of at least $6,000,000?

115. *Average Cost* The cost C of producing x calendars is $C = 3000 + 0.75x$, $x > 0$.

(a) Write the average cost $\overline{C} = C/x$ as a function of x.

(b) ⊞ Use a graphing calculator to graph the average cost function in part (a).

(c) How many calendars must be produced if the average cost per unit is to be less than $2?

116. *Data Analysis* The temperature T (in degrees Fahrenheit) of a metal in a laboratory experiment was recorded every 2 minutes for a period of 16 minutes. The table shows the experimental data, where t is the time in minutes.

t	0	2	4	6	8
T	250	290	338	410	498

t	10	12	14	16
T	560	530	370	160

A model for this data is

$$T = \frac{248.5 - 13.72t}{1.0 - 0.13t + 0.005t^2}.$$

(a) ⊞ Use a graphing calculator to plot the data and graph the model in the same viewing window.

(b) Use the graph to approximate the times when the temperature was at least $400°$F.

Explaining Concepts

117. ◆ Answer part (e) of Motivating the Chapter on page 506.

118. *Writing* Explain the change in an inequality when each side is multiplied by a negative real number.

119. *Writing* Give a verbal description of the intervals $(-\infty, 5] \cup (10, \infty)$.

120. *Writing* Define the term *critical number* and explain how critical numbers are used in solving quadratic and rational inequalities.

121. *Writing* In your own words, describe the procedure for solving quadratic inequalities.

122. Give an example of a quadratic inequality that has no real solution.

What Did You Learn?

Key Terms

double or repeated solution, *p. 508*

quadratic form, *p. 511*

discriminant, *p. 525*

parabola, *p. 536*

standard form of a quadratic function, *p. 536*

vertex of a parabola, *p. 536*

axis of a parabola, *p. 536*

zeros of a polynomial, *p. 557*

test intervals, *p. 557*

critical numbers, *p. 561*

Key Concepts

8.1 ○ Square Root Property

The equation $u^2 = d$, where $d > 0$, has exactly two solutions:

$$u = \sqrt{d} \quad \text{and} \quad u = -\sqrt{d}.$$

8.1 ○ Square Root Property (complex square root)

The equation $u^2 = d$, where $d < 0$, has exactly two solutions:

$$u = \sqrt{|d|}\,i \quad \text{and} \quad u = -\sqrt{|d|}\,i.$$

8.2 ○ Completing the square

To complete the square for the expression $x^2 + bx$, add $(b/2)^2$, which is the square of half the coefficient of x. Consequently

$$x^2 + bx + \left(\frac{b}{2}\right)^2 = \left(x + \frac{b}{2}\right)^2.$$

8.3 ○ The Quadratic Formula

The solutions of $ax^2 + bx + c = 0$, $a \neq 0$, are given by the Quadratic Formula

$$x = \frac{-b \pm \sqrt{b^2 - 4ac}}{2a}.$$

The expression inside the radical, $b^2 - 4ac$, is called the discriminant.

1. If $b^2 - 4ac > 0$, the equation has two real solutions.

2. If $b^2 - 4ac = 0$, the equation has one (repeated) real solution.

3. If $b^2 - 4ac < 0$, the equation has no real solutions.

8.3 ○ Using the discriminant

The discriminant of the quadratic equation

$$ax^2 + bx + c = 0, a \neq 0$$

can be used to classify the solutions of the equation as follows.

Discriminant	Solution Type
1. Perfect square	Two distinct rational solutions
2. Positive nonperfect square	Two distinct irrational solutions
3. Zero	One repeated rational solution
4. Negative number	Two distinct complex solutions

8.4 ○ Sketching a parabola

1. Determine the vertex and axis of the parabola by completing the square or by using the formula $x = -b/(2a)$.

2. Plot the vertex, axis, x- and y-intercepts, and a few additional points on the parabola. (Using the symmetry about the axis can reduce the number of points you need to plot.)

3. Use the fact that the parabola opens upward if $a > 0$ and opens downward if $a < 0$ to complete the sketch.

8.6 ○ Finding test intervals for inequalities.

1. For a polynomial expression, find all the real zeros. For a rational expression, find all the real zeros and those x-values for which the function is undefined.

2. Arrange the numbers found in Step 1 in increasing order. These numbers are called critical numbers.

3. Use the critical numbers to determine the test intervals.

4. Choose a representative x-value in each test interval and evaluate the expression at that value. If the value of the expression is negative, the expression will have negative values for every x-value in the interval. If the value of the expression is positive, the expression will have positive values for every x-value in the interval.

Review Exercises

8.1 Solving Quadratic Equations: Factoring and Special Forms

① Solve quadratic equations by factoring.

In Exercises 1–10, solve the equation by factoring.

1. $x^2 + 12x = 0$ 2. $u^2 - 18u = 0$

3. $4y^2 - 1 = 0$ 4. $2z^2 - 72 = 0$

5. $4y^2 + 20y + 25 = 0$

6. $x^2 + \frac{8}{3}x + \frac{16}{9} = 0$

7. $2x^2 - 2x - 180 = 0$

8. $15x^2 - 30x - 45 = 0$

9. $6x^2 - 12x = 4x^2 - 3x + 18$

10. $10x - 8 = 3x^2 - 9x + 12$

② Solve quadratic equations by the Square Root Property.

In Exercises 11–16, solve the equation by using the Square Root Property.

11. $4x^2 = 10{,}000$ 12. $2x^2 = 98$

13. $y^2 - 12 = 0$ 14. $y^2 - 8 = 0$

15. $(x - 16)^2 = 400$ 16. $(x + 3)^2 = 900$

③ Solve quadratic equations with complex solutions by the Square Root Property.

In Exercises 17–22, solve the equation by using the Square Root Property.

17. $z^2 = -121$ 18. $u^2 = -25$

19. $y^2 + 50 = 0$ 20. $x^2 + 48 = 0$

21. $(y + 4)^2 + 18 = 0$

22. $(x - 2)^2 + 24 = 0$

④ Use substitution to solve equations of quadratic form.

In Exercises 23–30, solve the equation of quadratic form. (Find all real *and* complex solutions.)

23. $x^4 - 4x^2 - 5 = 0$

24. $x^4 - 10x^2 + 9 = 0$

25. $x - 4\sqrt{x} + 3 = 0$

26. $x + 2\sqrt{x} - 3 = 0$

27. $(x^2 - 2x)^2 - 4(x^2 - 2x) - 5 = 0$

28. $\left(\sqrt{x} - 2\right)^2 + 2\left(\sqrt{x} - 2\right) - 3 = 0$

29. $x^{2/3} + 3x^{1/3} - 28 = 0$

30. $x^{2/5} + 4x^{1/5} + 3 = 0$

8.2 Completing the Square

① Rewrite quadratic expressions in completed square form.

In Exercises 31–36, add a term to the expression so that it becomes a perfect square trinomial.

31. $x^2 + 24x + $ ▢   32. $y^2 - 80y + $ ▢

33. $x^2 - 15x + $ ▢   34. $x^2 + 21x + $ ▢

35. $y^2 + \frac{2}{5}y + $ ▢   36. $x^2 - \frac{3}{4}x + $ ▢

② Solve quadratic equations by completing the square.

In Exercises 37–42, solve the equation by completing the square. Give the solutions in exact form and in decimal form rounded to two decimal places. (The solutions may be complex numbers.)

37. $x^2 - 6x - 3 = 0$

38. $x^2 + 12x + 6 = 0$

39. $x^2 - 3x + 3 = 0$

40. $u^2 - 5u + 6 = 0$

41. $y^2 - \frac{2}{3}y + 2 = 0$

42. $t^2 + \frac{1}{2}t - 1 = 0$

8.3 The Quadratic Formula

② Use the Quadratic Formula to solve quadratic equations.

In Exercises 43–48, solve the equation by using the Quadratic Formula. (Find all real *and* complex solutions.)

43. $y^2 + y - 30 = 0$

44. $x^2 - x - 72 = 0$

45. $2y^2 + y - 21 = 0$

46. $2x^2 - 3x - 20 = 0$

47. $5x^2 - 16x + 2 = 0$

48. $3x^2 + 12x + 4 = 0$

③ Determine the types of solutions of quadratic equations using the discriminant.

In Exercises 49–56, use the discriminant to determine the type of solutions of the quadratic equation.

49. $x^2 + 4x + 4 = 0$

50. $y^2 - 26y + 169 = 0$

51. $s^2 - s - 20 = 0$

52. $r^2 - 5r - 45 = 0$

53. $3t^2 + 17t + 10 = 0$

54. $7x^2 + 3x - 18 = 0$

55. $v^2 - 6v + 21 = 0$

56. $9y^2 + 1 = 0$

④ Write quadratic equations from solutions of the equations.

In Exercises 57–62, write a quadratic equation having the given solutions.

57. $3, -7$

58. $-1, 10$

59. $5 + \sqrt{7}, 5 - \sqrt{7}$

60. $2 + \sqrt{2}, 2 - \sqrt{2}$

61. $6 + 2i, 6 - 2i$

62. $3 + 4i, 3 - 4i$

8.4 Graphs of Quadratic Functions

① Determine the vertices of parabolas by completing the square.

In Exercises 63–66, write the equation of the parabola in standard form and find the vertex of its graph.

63. $y = x^2 - 8x + 3$

64. $y = x^2 + 12x - 9$

65. $y = 2x^2 - x + 3$

66. $y = 3x^2 + 2x - 6$

② Sketch parabolas.

In Exercises 67–70, sketch the parabola. Identify the vertex and any x-intercepts. Use a graphing calculator to verify your results.

67. $y = x^2 + 8x$

68. $y = -x^2 + 3x$

69. $f(x) = x^2 - 6x + 5$

70. $f(x) = x^2 + 3x - 10$

③ Write the equation of a parabola given the vertex and a point on the graph.

In Exercises 71–74, write an equation of the parabola $y = a(x - h)^2 + k$ that satisfies the conditions.

71. Vertex: $(2, -5)$; Point on the graph: $(0, 3)$

72. Vertex: $(-4, 0)$; Point on the graph: $(0, -6)$

73. Vertex: $(5, 0)$; Point on the graph: $(1, 1)$

74. Vertex: $(-2, 5)$; Point on the graph: $(0, 1)$

④ Use parabolas to solve application problems.

75. *Path of a Ball*   The height y (in feet) of a ball thrown by a child is given by $y = -\frac{1}{10}x^2 + 3x + 6$, where x is the horizontal distance (in feet) from where the ball is thrown.

   (a) 🖩 Use a graphing calculator to graph the path of the ball.

   (b) How high is the ball when it leaves the child's hand?

   (c) How high is the ball when it reaches its maximum height?

   (d) How far from the child does the ball strike the ground?

76. 🖩 *Graphical Estimation*   The number N (in thousands) of bankruptcies filed in the United States for the years 1996 through 2000 is approximated by $N = -66.36t^2 + 1116.2t - 3259$, $6 \le t \le 10$, where t is the time in years, with $t = 6$ corresponding to 1996.   (Source: Administrative Office of the U.S. Courts)

   (a) Use a graphing calculator to graph the model.

   (b) Use the graph from part (a) to approximate the maximum number of bankruptcies filed during 1996 through 2000. During what year did this maximum occur?

8.5 Applications of Quadratic Equations

1 Use quadratic equations to solve application problems.

77. *Selling Price*   A car dealer bought a fleet of used cars for a total of $80,000. By the time all but four of the cars had been sold, at an average profit of $1000 each, the original investment of $80,000 had been regained. How many cars were sold, and what was the average price per car?

78. *Selling Price*   A manager of a computer store bought several computers of the same model for $27,000. When all but five of the computers had been sold at a profit of $900 per computer, the original investment of $27,000 had been regained. How many computers were sold, and what was the selling price of each computer?

79. ▲ *Geometry*   The length of a rectangle is 12 inches greater than its width. The area of the rectangle is 108 square inches. Find the dimensions of the rectangle.

80. *Compound Interest*   You want to invest $35,000 for 2 years at an annual interest rate of r (in decimal form). Interest on the account is compounded annually. Find the interest rate if a deposit of $35,000 increases to $38,955.88 over a two-year period.

81. *Reduced Rates*   A Little League baseball team obtains a block of tickets to a ball game for $96. After three more people decide to go to the game, the price per ticket is decreased by $1.60. How many people are going to the game?

82. ▲ *Geometry*   A corner lot has an L-shaped sidewalk along its sides. The total length of the sidewalk is 51 feet. By cutting diagonally across the lot, the walking distance is shortened to 39 feet. What are the lengths of the two legs of the sidewalk?

83. *Work-Rate Problem*   Working together, two people can complete a task in 10 hours. Working alone, one person takes 2 hours longer than the other. How long would it take each person to do the task alone?

84. *Height*   An object is projected vertically upward at an initial velocity of 64 feet per second from a height of 192 feet, so that the height h at any time is given by $h = -16t^2 + 64t + 192$, where t is the time in seconds.

(a) After how many seconds is the height 256 feet?

(b) After how many seconds does the object hit the ground?

8.6 Quadratic and Rational Inequalities

1 Determine test intervals for polynomials.

In Exercises 85–88, find the critical numbers.

85. $2x(x + 7)$ **86.** $5x(x - 13)$

87. $x^2 - 6x - 27$

88. $2x^2 + 11x + 5$

2 Use test intervals to solve quadratic inequalities.

In Exercises 89–94, solve the inequality and graph the solution on the real number line.

89. $5x(7 - x) > 0$

90. $-2x(x - 10) \le 0$

91. $16 - (x - 2)^2 \le 0$

92. $(x - 5)^2 - 36 > 0$

93. $2x^2 + 3x - 20 < 0$

94. $3x^2 - 2x - 8 > 0$

3 Use test intervals to solve rational inequalities.

In Exercises 95–98, solve the inequality and graph the solution on the real number line.

95. $\dfrac{x + 3}{2x - 7} \ge 0$

96. $\dfrac{3x + 2}{x - 3} > 0$

97. $\dfrac{x + 4}{x - 1} < 0$ **98.** $\dfrac{2x - 9}{x - 1} \le 0$

4 Use inequalities to solve application problems.

99. *Height*   A projectile is fired straight upward from ground level with an initial velocity of 312 feet per second, so that its height h at any time t is given by $h = -16t^2 + 312t$, where h is measured in feet and t is measured in seconds. During what interval of time will the height of the projectile exceed 1200 feet?

100. *Average Cost*   The cost C of producing x notebooks is $C = 100,000 + 0.9x, x > 0$. Write the average cost $\overline{C} = C/x$ as a function of x. Then determine how many notebooks must be produced if the average cost per unit is to be less than $2.

Chapter Test

Take this test as you would take a test in class. After you are done, check your work against the answers in the back of the book.

In Exercises 1–6, solve the equation by the specified method.

1. Factoring:
   $x(x - 3) - 10(x - 3) = 0$

2. Factoring:
   $8x^2 - 21x - 9 = 0$

3. Square Root Property:
   $(x - 2)^2 = 0.09$

4. Square Root Property:
   $(x + 4)^2 + 100 = 0$

5. Completing the square:
   $2x^2 - 6x + 3 = 0$

6. Quadratic Formula:
   $2y(y - 2) = 7$

In Exercises 7 and 8, solve the equation of quadratic form.

7. $\dfrac{1}{x^2} - \dfrac{6}{x} + 4 = 0$

8. $x^{2/3} - 9x^{1/3} + 8 = 0$

9. Find the discriminant and explain what it means in terms of the type of solutions of the quadratic equation $5x^2 - 12x + 10 = 0$.

10. Find a quadratic equation having the solutions -4 and 5.

In Exercises 11 and 12, sketch the parabola. Identify the vertex and any x-intercepts. Use a graphing calculator to verify your results.

11. $y = -x^2 + 7$

12. $y = x^2 - 2x - 15$

In Exercises 13–15, solve the inequality and sketch its solution.

13. $16 \le (x - 2)^2$

14. $2x(x - 3) < 0$

15. $\dfrac{x + 1}{x - 5} \le 0$

16. The width of a rectangle is 8 feet less than its length. The area of the rectangle is 240 square feet. Find the dimensions of the rectangle.

17. An English club chartered a bus trip to a Shakespearean festival. The cost of the bus was $1250. To lower the bus fare per person, the club invited nonmembers to go along. When 10 nonmembers joined the trip, the fare per person decreased by $6.25. How many club members are going on the trip?

18. An object is dropped from a height of 75 feet. Its height h (in feet) at any time t is given by $h = -16t^2 + 75$, where t is measured in seconds. Find the time required for the object to fall to a height of 35 feet.

19. Two buildings are connected by an L-shaped protected walkway. The total length of the walkway is 140 feet. By cutting diagonally across the grass, the walking distance is shortened to 100 feet. What are the lengths of the two legs of the walkway?

Motivating the Chapter

 ## *Choosing the Best Investment*

You receive an inheritance of $5000 and want to invest it.

See Section 9.1, Exercise 99.

a. Complete the table by finding the amount A of the $5000 investment after 3 years with an annual interest rate of $r = 6\%$. Which form of compounding gives you the greatest balance?

Compounding	Amount, A
Annual	
Quarterly	
Monthly	
Daily	
Hourly	
Continuous	

b. You are considering two different investment options. The first investment option has an interest rate of 7% compounded continuously. The second investment option has an interest rate of 8% compounded quarterly. Which investment should you choose? Explain.

See Section 9.5, Exercise 139.

c. What annual percentage rate is needed to obtain a balance of $6200 in 3 years when the interest is compounded monthly?

d. With an interest rate of 6%, compounded continuously, how long will it take for your inheritance to grow to $7500?

e. What is the *effective yield* on your investment when the interest rate is 8% compounded quarterly?

f. With an interest rate of 6%, compounded continuously, how long will it take your inheritance to double? How long will it take your inheritance to quadruple (reach four times the original amount)?

Monika Graff/The Image Works

Exponential and Logarithmic Functions

9.1 ● Exponential Functions

9.2 ● Composite and Inverse Functions

9.3 ● Logarithmic Functions

9.4 ● Properties of Logarithms

9.5 ● Solving Exponential and Logarithmic Equations

9.6 ● Applications

9.1 Exponential Functions

Ryan McVay/Photodisc/Getty Images

What You Should Learn

1 Evaluate exponential functions.

2 Graph exponential functions.

3 Evaluate the natural base e and graph natural exponential functions.

4 Use exponential functions to solve application problems.

Why You Should Learn It

Exponential functions can be used to model and solve real-life problems. For instance, in Exercise 96 on page 585, you will use an exponential function to model the median price of a home in the United States.

1 Evaluate exponential functions.

Exponential Functions

In this section, you will study a new type of function called an **exponential function.** Whereas polynomial and rational functions have terms with variable bases and constant exponents, exponential functions have terms with *constant bases* and *variable exponents.* Here are some examples.

Polynomial or Rational Function

Constant Exponents

$$f(x) = x^2, \quad f(x) = x^{-3}$$

Variable Bases

Exponential Function

Variable Exponents

$$f(x) = 2^x, \quad f(x) = 3^{-x}$$

Constant Bases

Definition of Exponential Function

The **exponential function** f **with base** a is denoted by

$$f(x) = a^x$$

where $a > 0$, $a \neq 1$, and x is any real number.

The base $a = 1$ is excluded because $f(x) = 1^x = 1$ is a constant function, *not* an exponential function.

In Chapters 5 and 7, you learned to evaluate a^x for integer and rational values of x. For example, you know that

$$a^3 = a \cdot a \cdot a, \quad a^{-4} = \frac{1}{a^4}, \quad \text{and} \quad a^{5/3} = \left(\sqrt[3]{a}\right)^5.$$

However, to evaluate a^x for any real number x, you need to interpret forms with *irrational* exponents, such as $a^{\sqrt{2}}$ or a^π. For the purposes of this text, it is sufficient to think of a number such as $a^{\sqrt{2}}$, where $\sqrt{2} \approx 1.414214$, as the number that has the successively closer approximations

$$a^{1.4}, a^{1.41}, a^{1.414}, a^{1.4142}, a^{1.41421}, a^{1.414214}, \ldots.$$

The rules of exponents that were discussed in Section 5.1 can be extended to cover exponential functions, as described on the following page.

Study Tip

Rule 4 of the rules of exponential functions indicates that 2^{-x} can be written as

$$2^{-x} = \frac{1}{2^x}.$$

Similarly, $\dfrac{1}{3^{-x}}$ can be written as

$$\frac{1}{3^{-x}} = 3^x.$$

In other words, you can move a *factor* from the numerator to the denominator (or from the denominator to the numerator) by changing the sign of its exponent.

Rules of Exponential Functions

Let a be a positive real number, and let x and y be real numbers, variables, or algebraic expressions.

1. $a^x \cdot a^y = a^{x+y}$ Product rule

2. $\dfrac{a^x}{a^y} = a^{x-y}$ Quotient rule

3. $(a^x)^y = a^{xy}$ Power rule

4. $a^{-x} = \dfrac{1}{a^x} = \left(\dfrac{1}{a}\right)^x$ Negative exponent rule

To evaluate exponential functions with a calculator, you can use the exponential key $\boxed{y^x}$ (where y is the base and x is the exponent) or $\boxed{\wedge}$. For example, to evaluate $3^{-1.3}$, you can use the following keystrokes.

Keystrokes	Display	
3 $\boxed{y^x}$ 1.3 $\boxed{+/-}$ $\boxed{=}$	0.239741	Scientific
3 $\boxed{\wedge}$ $\boxed{(}$ $\boxed{(-)}$ 1.3 $\boxed{)}$ $\boxed{\text{ENTER}}$	0.239741	Graphing

Example 1   Evaluating Exponential Functions

Evaluate each function. Use a calculator only if it is necessary or more efficient.

Function	Values
a. $f(x) = 2^x$	$x = 3, x = -4, x = \pi$
b. $g(x) = 12^x$	$x = 3, x = -0.1, x = \frac{5}{7}$
c. $h(x) = (1.04)^{2x}$	$x = 0, x = -2, x = \sqrt{2}$

Solution

Evaluation	Comment
a. $f(3) = 2^3 = 8$	Calculator is not necessary.
$f(-4) = 2^{-4} = \dfrac{1}{2^4} = \dfrac{1}{16}$	Calculator is not necessary.
$f(\pi) = 2^\pi \approx 8.825$	Calculator is necessary.
b. $g(3) = 12^3 = 1728$	Calculator is more efficient.
$g(-0.1) = 12^{-0.1} \approx 0.780$	Calculator is necessary.
$g\left(\dfrac{5}{7}\right) = 12^{5/7} \approx 5.900$	Calculator is necessary.
c. $h(0) = (1.04)^{2\cdot 0} = (1.04)^0 = 1$	Calculator is not necessary.
$h(-2) = (1.04)^{2(-2)} \approx 0.855$	Calculator is more efficient.
$h\left(\sqrt{2}\right) = (1.04)^{2\sqrt{2}} \approx 1.117$	Calculator is necessary.

② Graph exponential functions.

Graphs of Exponential Functions

The basic nature of the graph of an exponential function can be determined by the point-plotting method or by using a graphing calculator.

Example 2   The Graphs of Exponential Functions

In the same coordinate plane, sketch the graph of each function. Determine the domain and range.

a. $f(x) = 2^x$

b. $g(x) = 4^x$

Solution

The table lists some values of each function, and Figure 9.1 shows the graph of each function. From the graphs, you can see that the domain of each function is the set of all real numbers and that the range of each function is the set of all positive real numbers.

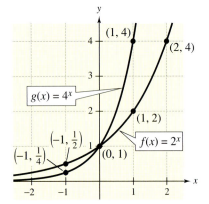

Figure 9.1

x	-2	-1	0	1	2	3
2^x	$\frac{1}{4}$	$\frac{1}{2}$	1	2	4	8
4^x	$\frac{1}{16}$	$\frac{1}{4}$	1	4	16	64

You know from your study of functions in Section 3.7 that the graph of $h(x) = f(-x) = 2^{-x}$ is a reflection of the graph of $f(x) = 2^x$ in the y-axis. This is reinforced in the next example.

Example 3   The Graphs of Exponential Functions

In the same coordinate plane, sketch the graph of each function.

a. $f(x) = 2^{-x}$

b. $g(x) = 4^{-x}$

Solution

The table lists some values of each function, and Figure 9.2 shows the graph of each function.

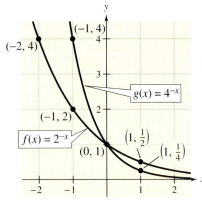

Figure 9.2

x	-3	-2	-1	0	1	2
2^{-x}	8	4	2	1	$\frac{1}{2}$	$\frac{1}{4}$
4^{-x}	64	16	4	1	$\frac{1}{4}$	$\frac{1}{16}$

Examples 2 and 3 suggest that for $a > 1$, the values of the function of $y = a^x$ increase as x increases and the values of the function $y = a^{-x} = (1/a)^x$ decrease as x increases. The graphs shown in Figure 9.3 are typical of the graphs of exponential functions. Note that each graph has a y-intercept at $(0, 1)$ and a **horizontal asymptote** of $y = 0$ (the x-axis).

Graph of $y = a^x$

- Domain: $(-\infty, \infty)$
- Range: $(0, \infty)$
- Intercept: $(0, 1)$
- Increasing
  (moves up to the right)
- Asymptote: x-axis

Graph of $y = a^{-x} = \left(\dfrac{1}{a}\right)^x$

- Domain: $(-\infty, \infty)$
- Range: $(0, \infty)$
- Intercept: $(0, 1)$
- Decreasing
  (moves down to the right)
- Asymptote: x-axis

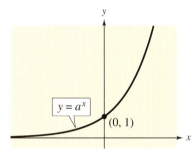

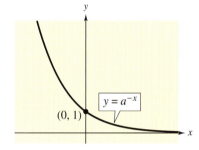

Figure 9.3   Characteristics of the exponential functions $y = a^x$ and $y = a^{-x}(a > 1)$

In the next two examples, notice how the graph of $y = a^x$ can be used to sketch the graphs of functions of the form $f(x) = b \pm a^{x+c}$. Also note that the transformation in Example 4(a) keeps the x-axis as a horizontal asymptote, but the transformation in Example 4(b) yields a new horizontal asymptote of $y = -2$. Also, be sure to note how the y-intercept is affected by each transformation.

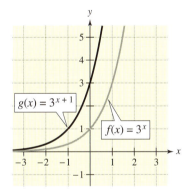

Figure 9.4

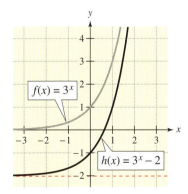

Figure 9.5

Example 4   Transformations of Graphs of Exponential Functions

Use transformations to analyze and sketch the graph of each function.

a. $g(x) = 3^{x+1}$ **b.** $h(x) = 3^x - 2$

Solution

Consider the function $f(x) = 3^x$.

a. The function g is related to f by $g(x) = f(x + 1)$. To sketch the graph of g, shift the graph of f one unit to the left, as shown in Figure 9.4. Note that the y-intercept of g is $(0, 3)$.

b. The function h is related to f by $h(x) = f(x) - 2$. To sketch the graph of g, shift the graph of f two units downward, as shown in Figure 9.5. Note that the y-intercept of h is $(0, -1)$ and the horizontal asymptote is $y = -2$.

Example 5   Reflections of Graphs of Exponential Functions

Use transformations to analyze and sketch the graph of each function.

a. $g(x) = -3^x$ **b.** $h(x) = 3^{-x}$

Solution

Consider the function $f(x) = 3^x$.

a. The function g is related to f by $g(x) = -f(x)$. To sketch the graph of g, reflect the graph of f about the x-axis, as shown in Figure 9.6. Note that the y-intercept of g is $(0, -1)$.

b. The function h is related to f by $h(x) = f(-x)$. To sketch the graph of h, reflect the graph of f about the y-axis, as shown in Figure 9.7.

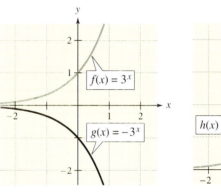

Figure 9.6   Figure 9.7

(3) Evaluate the natural base e and graph natural exponential functions.

The Natural Exponential Function

So far, integers or rational numbers have been used as bases of exponential functions. In many applications of exponential functions, the convenient choice for a base is the following irrational number, denoted by the letter "e."

$$e \approx 2.71828 \ldots \qquad \text{Natural base}$$

This number is called the **natural base.** The function

$$f(x) = e^x \qquad \text{Natural exponential function}$$

is called the **natural exponential function.** To evaluate the natural exponential function, you need a calculator, preferably one having a natural exponential key $\boxed{e^x}$. Here are some examples of how to use such a calculator to evaluate the natural exponential function.

Technology: Tip

Some calculators do not have a key labeled $\boxed{e^x}$. If your calculator does not have this key, but does have a key labeled $\boxed{\text{LN}}$, you will have to use the two-keystroke sequence $\boxed{\text{INV}}$ $\boxed{\text{LN}}$ in place of $\boxed{e^x}$.

Value	Keystrokes	Display	
e^2	2 $\boxed{e^x}$	7.3890561	Scientific
e^2	$\boxed{e^x}$ 2 $\boxed{\text{ENTER}}$	7.3890561	Graphing
e^{-3}	3 $\boxed{+/-}$ $\boxed{e^x}$	0.0497871	Scientific
e^{-3}	$\boxed{e^x}$ $\boxed{(-)}$ 3 $\boxed{)}$ $\boxed{\text{ENTER}}$	0.0497871	Graphing

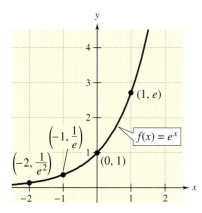

Figure 9.8

When evaluating the natural exponential function, remember that e is the constant number 2.71828 . . . ., and x is a variable. After evaluating this function at several values, as shown in the table, you can sketch its graph, as shown in Figure 9.8.

x	-2	-1.5	-1.0	-0.5	0.0	0.5	1.0	1.5
$f(x) = e^x$	0.135	0.223	0.368	0.607	1.000	1.649	2.718	4.482

From the graph, notice the following characteristics of the natural exponential function.

- Domain: $(-\infty, \infty)$
- Range: $(0, \infty)$
- Intercept: $(0, 1)$
- Increasing (moves up to the right)
- Asymptote: x-axis

Notice that these characteristics are consistent with those listed for the exponential function $y = a^x$ on page 577.

④ Use exponential functions to solve application problems.

Applications

A common scientific application of exponential functions is **radioactive decay.**

Example 6   Radioactive Decay

After t years, the remaining mass y (in grams) of 10 grams of a radioactive element whose half-life is 25 years is given by

$$y = 10\left(\frac{1}{2}\right)^{t/25}, \quad t \geq 0.$$

How much of the initial mass remains after 120 years?

Solution

When $t = 120$, the mass is given by

$$y = 10\left(\frac{1}{2}\right)^{120/25} \qquad \text{Substitute 120 for } t.$$

$$= 10\left(\frac{1}{2}\right)^{4.8} \qquad \text{Simplify.}$$

$$\approx 0.359. \qquad \text{Use a calculator.}$$

So, after 120 years, the mass has decayed from an initial amount of 10 grams to only 0.359 gram. Note in Figure 9.9 that the graph of the function shows the 25-year half-life. That is, after 25 years the mass is 5 grams (half of the original), after another 25 years the mass is 2.5 grams, and so on.

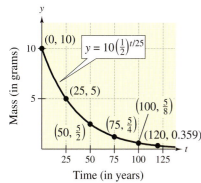

Figure 9.9

One of the most familiar uses of exponential functions involves **compound interest.** A principal P is invested at an annual interest rate r (in decimal form), compounded once a year. If the interest is added to the principal at the end of the year, the balance is

$$A = P + Pr$$
$$= P(1 + r).$$

This pattern of multiplying the previous principal by $(1 + r)$ is then repeated each successive year, as shown below.

Time in Years	*Balance at Given Time*
0	$A = P$
1	$A = P(1 + r)$
2	$A = P(1 + r)(1 + r) = P(1 + r)^2$
3	$A = P(1 + r)^2(1 + r) = P(1 + r)^3$
$\vdots$	$\vdots$
t	$A = P(1 + r)^t$

To account for more frequent compounding of interest (such as quarterly or monthly compounding), let n be the number of compoundings per year and let t be the number of years. Then the rate per compounding is r/n and the account balance after t years is

$$A = P\left(1 + \frac{r}{n}\right)^{nt}.$$

Example 7   Finding the Balance for Compound Interest

A sum of $10,000 is invested at an annual interest rate of 7.5%, compounded monthly. Find the balance in the account after 10 years.

Solution

Using the formula for compound interest, with $P = 10,000$, $r = 0.075$, $n = 12$ (for monthly compounding), and $t = 10$, you obtain the following balance.

$$A = 10{,}000\left(1 + \frac{0.075}{12}\right)^{12(10)}$$

$$\approx \$21{,}120.65$$

A second method that banks use to compute interest is called **continuous compounding.** The formula for the balance for this type of compounding is

$$A = Pe^{rt}.$$

The formulas for both types of compounding are summarized on the next page.

Formulas for Compound Interest

After t years, the balance A in an account with principal P and annual interest rate r (in decimal form) is given by the following formulas.

1. For n compoundings per year: $A = P\left(1 + \dfrac{r}{n}\right)^{nt}$

2. For continuous compounding: $A = Pe^{rt}$

Example 8   Comparing Three Types of Compounding

A total of $15,000 is invested at an annual interest rate of 8%. Find the balance after 6 years for each type of compounding.

a. Quarterly
b. Monthly
c. Continuous

Solution

a. Letting $P = 15,000$, $r = 0.08$, $n = 4$, and $t = 6$, the balance after 6 years at quarterly compounding is

$$A = 15,000\left(1 + \frac{0.08}{4}\right)^{4(6)}$$

$$\approx \$24,126.56.$$

b. Letting $P = 15,000$, $r = 0.08$, $n = 12$, and $t = 6$, the balance after 6 years at monthly compounding is

$$A = 15,000\left(1 + \frac{0.08}{12}\right)^{12(6)}$$

$$\approx \$24,202.53.$$

c. Letting $P = 15,000$, $r = 0.08$, and $t = 6$, the balance after 6 years at continuous compounding is

$$A = 15,000e^{0.08(6)}$$

$$\approx \$24,241.12.$$

Note that the balance is greater with continuous compounding than with quarterly or monthly compounding.

Example 8 illustrates the following general rule. For a given principal, interest rate, and time, the more often the interest is compounded per year, the greater the balance will be. Moreover, the balance obtained by continuous compounding is larger than the balance obtained by compounding n times per year.

9.1 Exercises

Developing Skills

In Exercises 1–8, simplify the expression.

1. $2^x \cdot 2^{x-1}$

2. $10e^{2x} \cdot e^{-x}$

3. $\dfrac{e^{x+2}}{e^x}$

4. $\dfrac{3^{2x+3}}{3^{x+1}}$

5. $(2e^x)^3$

6. $-4e^{-2x}$

7. $\sqrt[3]{-8e^{3x}}$

8. $\sqrt{4e^{6x}}$

In Exercises 9–16, evaluate the expression. (Round your answer to three decimal places.)

9. $4^{\sqrt{3}}$

10. $6^{-\pi}$

11. $e^{1/3}$

12. $e^{-1/3}$

13. $4(3e^4)^{1/2}$

14. $(9e^2)^{3/2}$

15. $\dfrac{4e^3}{12e^2}$

16. $\dfrac{6e^5}{10e^7}$

In Exercises 17–30, evaluate the function as indicated. Use a calculator only if it is necessary or more efficient. (Round your answer to three decimal places.) See Example 1.

17. $f(x) = 3^x$
   (a) $x = -2$
   (b) $x = 0$
   (c) $x = 1$

18. $F(x) = 3^{-x}$
   (a) $x = -2$
   (b) $x = 0$
   (c) $x = 1$

19. $g(x) = 3.8^x$
   (a) $x = -1$
   (b) $x = 3$
   (c) $x = \sqrt{5}$

20. $G(x) = 1.1^{-x}$
   (a) $x = -1$
   (b) $x = 1$
   (c) $x = \sqrt{3}$

21. $f(t) = 500\left(\frac{1}{2}\right)^t$
   (a) $t = 0$
   (b) $t = 1$
   (c) $t = \pi$

22. $g(s) = 1200\left(\frac{2}{3}\right)^s$
   (a) $s = 0$
   (b) $s = 2$
   (c) $s = \sqrt{2}$

23. $f(x) = 1000(1.05)^{2x}$
   (a) $x = 0$
   (b) $x = 5$
   (c) $x = 10$

24. $g(t) = 10{,}000(1.03)^{4t}$
   (a) $t = 1$
   (b) $t = 3$
   (c) $t = 5.5$

25. $h(x) = \dfrac{5000}{(1.06)^{8x}}$

(a) $x = 5$

(b) $x = 10$

(c) $x = 20$

26. $P(t) = \dfrac{10,000}{(1.01)^{12t}}$

(a) $t = 2$

(b) $t = 10$

(c) $t = 20$

27. $g(x) = 10e^{-0.5x}$

(a) $x = -4$

(b) $x = 4$

(c) $x = 8$

28. $A(t) = 200e^{0.1t}$

(a) $t = 10$

(b) $t = 20$

(c) $t = 40$

29. $g(x) = \dfrac{1000}{2 + e^{-0.12x}}$

(a) $x = 0$

(b) $x = 10$

(c) $x = 50$

30. $f(z) = \dfrac{100}{1 + e^{-0.05z}}$

(a) $z = 0$

(b) $z = 10$

(c) $z = 20$

In Exercises 31–36, match the function with its graph. [The graphs are labeled (a), (b), (c), (d), (e), and (f).]

(a)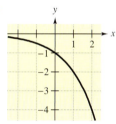

(b)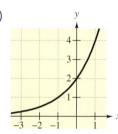

(c)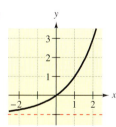

(d)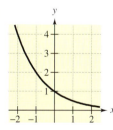

(e)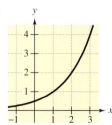

(f)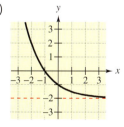

31. $f(x) = -2^x$

32. $f(x) = 2^{-x}$

33. $f(x) = 2^x - 1$

34. $f(x) = 2^{x-1}$

35. $f(x) = 2^{x+1}$

36. $f(x) = \left(\tfrac{1}{2}\right)^x - 2$

In Exercises 37–56, sketch the graph of the function. Identify the horizontal asymptote. See Examples 2 and 3.

37. $f(x) = 3^x$

38. $f(x) = 3^{-x} = \left(\tfrac{1}{3}\right)^x$

39. $h(x) = \tfrac{1}{2}(3^x)$

40. $h(x) = \tfrac{1}{2}(3^{-x})$

41. $g(x) = 3^x - 2$

42. $g(x) = 3^x + 1$

43. $f(x) = 5^x + 2$

44. $f(x) = 5^x - 4$

45. $g(x) = 5^{x-1}$

46. $g(x) = 5^{x+3}$

47. $f(t) = 2^{-t^2}$

48. $f(t) = 2^{t^2}$

49. $f(x) = -2^{0.5x}$

50. $h(t) = -2^{-0.5t}$

51. $h(x) = 2^{0.5x}$

52. $g(x) = 2^{-0.5x}$

53. $f(x) = -\left(\tfrac{1}{3}\right)^x$

54. $f(x) = \left(\tfrac{3}{4}\right)^x + 1$

55. $g(t) = 200\left(\tfrac{1}{2}\right)^t$

56. $h(x) = 27\left(\tfrac{2}{3}\right)^x$

 In Exercises 57–68, use a graphing calculator to graph the function.

57. $y = 7^{x/2}$

58. $y = 7^{-x/2}$

59. $y = 7^{-x/2} + 5$

60. $y = 7^{(x-3)/2}$

61. $y = 500(1.06)^t$

62. $y = 100(1.06)^{-t}$

63. $y = 3e^{0.2x}$

64. $y = 50e^{-0.05x}$

65. $P(t) = 100e^{-0.1t}$

66. $A(t) = 1000e^{0.08t}$

67. $y = 6e^{-x^2/3}$

68. $g(x) = 7e^{(x+1)/2}$

In Exercises 69–74, identify the transformation of the graph of $f(x) = 4^x$ and sketch the graph of h. See Examples 4 and 5.

69. $h(x) = 4^x - 1$

70. $h(x) = 4^x + 2$

71. $h(x) = 4^{x+2}$

72. $h(x) = 4^{x-4}$

73. $h(x) = -4^x$

74. $h(x) = -4^x + 2$

Solving Problems

75. *Radioactive Decay*   After t years, the remaining mass y (in grams) of 16 grams of a radioactive element whose half-life is 30 years is given by

$$y = 16\left(\frac{1}{2}\right)^{t/30}, \quad t \ge 0.$$

How much of the initial mass remains after 80 years?

76. *Radioactive Decay*   After t years, the remaining mass y (in grams) of 23 grams of a radioactive element whose half-life is 45 years is given by

$$y = 23\left(\frac{1}{2}\right)^{t/45}, \quad t \ge 0.$$

How much of the initial mass remains after 150 years?

Compound Interest   In Exercises 77–80, complete the table to determine the balance A for P dollars invested at rate r for t years, compounded n times per year.

n	1	4	12	365	Continuous compounding
A					

	Principal	Rate	Time
77.	$P = \$100$	$r = 7\%$	$t = 15$ years
78.	$P = \$400$	$r = 7\%$	$t = 20$ years
79.	$P = \$2000$	$r = 9.5\%$	$t = 10$ years
80.	$P = \$1500$	$r = 6.5\%$	$t = 20$ years

Compound Interest   In Exercises 81–84, complete the table to determine the principal P that will yield a balance of A dollars when invested at rate r for t years, compounded n times per year.

n	1	4	12	365	Continuous compounding
P					

	Balance	Rate	Time
81.	$A = \$5000$	$r = 7\%$	$t = 10$ years
82.	$A = \$100,000$	$r = 9\%$	$t = 20$ years
83.	$A = \$1,000,000$	$r = 10.5\%$	$t = 40$ years
84.	$A = \$2500$	$r = 7.5\%$	$t = 2$ years

85. *Demand*   The daily demand x and the price p for a collectible are related by $p = 25 - 0.4e^{0.02x}$. Find the prices for demands of (a) $x = 100$ units and (b) $x = 125$ units.

86. *Population Growth*   The population P (in millions) of the United States from 1970 to 2000 can be approximated by the exponential function $P(t) = 203.0(1.0107)^t$, where t is the time in years, with $t = 0$ corresponding to 1970. Use the model to estimate the population in the years (a) 2010 and (b) 2020.   (Source: U.S. Census Bureau)

87. *Property Value*   The value of a piece of property doubles every 15 years. You buy the property for $64,000. Its value t years after the date of purchase should be $V(t) = 64,000(2)^{t/15}$. Use the model to approximate the value of the property (a) 5 years and (b) 20 years after it is purchased.

88. *Inflation Rate*   The annual rate of inflation is predicted to average 5% over the next 10 years. With this rate of inflation, the approximate cost C of goods or services during any year in that decade will be given by $C(t) = P(1.05)^t$, $0 \le t \le 10$, where t is time in years and P is the present cost. The price of an oil change for your car is presently $24.95. Estimate the price 10 years from now.

89. *Depreciation*   After t years, the value of a car that originally cost $16,000 depreciates so that each year it is worth $\frac{3}{4}$ of its value for the previous year. Find a model for $V(t)$, the value of the car after t years. Sketch a graph of the model and determine the value of the car 2 years after it was purchased.

90. *Depreciation*   Straight-line depreciation is used to determine the value of the car in Exercise 89. Assume that the car depreciates $3000 per year.

(a) Write a linear equation for $V(t)$, the value of the car after t years.

(b) Sketch the graph of the model in part (a) on the same coordinate axes used for the graph in Exercise 89.

(c) If you were selling the car after 2 years, which depreciation model would you prefer?

(d) If you were selling the car after 4 years, which model would you prefer?

91. *Graphical Interpretation*   Investments of $500 in two different accounts with interest rates of 6% and 8% are compounded continuously.

(a) For each account, write an exponential function that represents the balance after t years.

(b) ▦ Use a graphing calculator to graph each of the models in the same viewing window.

(c) ▦ Use a graphing calculator to graph the function $A_2 - A_1$ in the same viewing window with the graphs in part (b).

(d) ▦ Use the graphs to discuss the rates of increase of the balances in the two accounts.

92. *Savings Plan*   You decide to start saving pennies according to the following pattern. You save 1 penny the first day, 2 pennies the second day, 4 the third day, 8 the fourth day, and so on. Each day you save twice the number of pennies you saved on the previous day. Write an exponential function that models this problem. How many pennies do you save on the thirtieth day? (In the next chapter you will learn how to find the total number saved.)

93. *Parachute Drop*   A parachutist jumps from a plane and opens the parachute at a height of 2000 feet (see figure). The distance h between the parachutist and the ground is $h = 1950 + 50e^{-0.4433t} - 22t$, where h is in feet and t is the time in seconds. (The time $t = 0$ corresponds to the time when the parachute is opened.)

(a) ▦ Use a graphing calculator to graph the function.

(b) Find the distances between the parachutist and the ground when $t = 0$, 25, 50, and 75.

2000 ft

Not drawn to scale

94. *Parachute Drop*   A parachutist jumps from a plane and opens the parachute at a height of 3000 feet. The distance h (in feet) between the parachutist and the ground is $h = 2940 + 60e^{-0.4021t} - 24t$, where t is the time in seconds. (The time $t = 0$ corresponds to the time when the parachute is opened.)

(a) ▦ Use a graphing calculator to graph the function.

(b) Find the distances between the parachutist and the ground when $t = 0$, 50, and 100.

95. ▦ *Data Analysis*   A meteorologist measures the atmospheric pressure P (in kilograms per square meter) at altitude h (in kilometers). The data are shown in the table.

h	0	5	10	15	20
P	10,332	5583	2376	1240	517

(a) Use a graphing calculator to plot the data points.

(b) A model for the data is given by $P = 10,958e^{-0.15h}$. Use a graphing calculator to graph the model in the same viewing window as in part (a). How well does the model fit the data?

(c) Use a graphing calculator to create a table comparing the model with the data points.

(d) Estimate the atmospheric pressure at an altitude of 8 kilometers.

(e) Use the graph to estimate the altitude at which the atmospheric pressure is 2000 kilograms per square meter.

96. *Data Analysis*   For the years 1994 through 2001, the median prices of a one-family home in the United States are shown in the table. (Source: U.S. Census Bureau and U.S. Department of Housing and Urban Development)

Year	1994	1995	1996	1997
Price	$130,000	$133,900	$140,000	$146,000

Year	1998	1999	2000	2001
Price	$152,500	$161,000	$169,000	$175,200

A model for this data is given by $y = 107{,}773e^{0.0442t}$, where t is time in years, with $t = 4$ representing 1994.

(a) Use the model to complete the table and compare the results with the actual data.

Year	1994	1995	1996	1997
Price				

Year	1998	1999	2000	2001
Price				

(b) 🖩 Use a graphing calculator to graph the model.

(c) If the model were used to predict home prices in the years ahead, would the predictions be increasing at a higher rate or a lower rate with increasing t? Do you think the model would be reliable for predicting the future prices of homes? Explain.

97. *Calculator Experiment*

(a) Use a calculator to complete the table.

x	1	10	100	1000	10,000
$\left(1 + \dfrac{1}{x}\right)^x$					

(b) Use the table to sketch the graph of the function

$$f(x) = \left(1 + \frac{1}{x}\right)^x.$$

Does this graph appear to be approaching a horizontal asymptote?

(c) From parts (a) and (b), what conclusions can you make about the value of

$$\left(1 + \frac{1}{x}\right)^x$$

as x gets larger and larger?

98. Identify the graphs of $y_1 = e^{0.2x}$, $y_2 = e^{0.5x}$, and $y_3 = e^x$ in the figure. Describe the effect on the graph of $y = e^{kx}$ when $k > 0$ is changed.

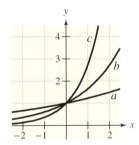

Explaining Concepts

99. �</> Answer parts (a) and (b) of Motivating the Chapter on page 572.

100. *Writing* ✎ Describe the differences between exponential functions and polynomial functions.

101. *Writing* ✎ Explain why 1^x is not an exponential function.

102. *Writing* ✎ Compare the graphs of $f(x) = 3^x$ and $g(x) = \left(\frac{1}{3}\right)^x$.

103. *Writing* ✎ Is $e = \dfrac{271{,}801}{99{,}990}$? Explain.

104. *Writing* ✎ Without using a calculator, explain why $2^{\sqrt{2}}$ is greater than 2 but less than 4?

105. 🖩 Use a graphing calculator to investigate the function $f(x) = k^x$ for $0 < k < 1$, $k = 1$, and $k > 1$. Discuss the effect that k has on the shape of the graph.

9.2 Composite and Inverse Functions

Superstock, Inc.

What You Should Learn

1. Form compositions of two functions and find the domains of composite functions.
2. Use the Horizontal Line Test to determine whether functions have inverse functions.
3. Find inverse functions algebraically.
4. Graphically verify that two functions are inverse functions of each other.

Why You Should Learn It

Inverse functions can be used to model and solve real-life problems. For instance, in Exercise 108 on page 600, you will use an inverse function to determine the number of units produced for a certain hourly wage.

1. Form compositions of two functions and find the domains of composite functions.

Composite Functions

Two functions can be combined to form another function called the **composition** of the two functions. For instance, if $f(x) = 2x^2$ and $g(x) = x - 1$, the composition of f with g is denoted by $f \circ g$ and is given by

$$f(g(x)) = f(x - 1) = 2(x - 1)^2.$$

Definition of Composition of Two Functions

The **composition** of the functions f and g is given by

$$(f \circ g)(x) = f(g(x)).$$

The domain of the **composite function** $(f \circ g)$ is the set of all x in the domain of g such that $g(x)$ is in the domain of f. See Figure 9.10.

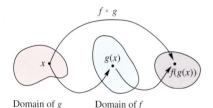

$f \circ g$

x $g(x)$ $f(g(x))$

Domain of g Domain of f

Figure 9.10

Study Tip

A composite function can be viewed as a function within a function, where the composition

$$(f \circ g)(x) = f(g(x))$$

has f as the "outer" function and g as the "inner" function. This is reversed in the composition

$$(g \circ f)(x) = g(f(x)).$$

Example 1   Forming the Composition of Two Functions

Given $f(x) = 2x + 4$ and $g(x) = 3x - 1$, find the composition of f with g. Then evaluate the composite function when $x = 1$ and when $x = -3$.

Solution

$$
\begin{aligned}
(f \circ g)(x) &= f(g(x)) &&\text{Definition of } f \circ g \\
&= f(3x - 1) &&g(x) = 3x - 1 \text{ is the inner function.} \\
&= 2(3x - 1) + 4 &&\text{Input } 3x - 1 \text{ into the outer function } f. \\
&= 6x - 2 + 4 &&\text{Distributive Property} \\
&= 6x + 2 &&\text{Simplify.}
\end{aligned}
$$

When $x = 1$, the value of this composite function is

$$(f \circ g)(1) = 6(1) + 2 = 8.$$

When $x = -3$, the value of this composite function is

$$(f \circ g)(-3) = 6(-3) + 2 = -16.$$

The composition of f with g is generally *not* the same as the composition of g with f. This is illustrated in Example 2.

Example 2   Comparing the Compositions of Functions

Given $f(x) = 2x - 3$ and $g(x) = x^2 + 1$, find each composition.

a. $(f \circ g)(x)$ **b.** $(g \circ f)(x)$

Solution

a. $(f \circ g)(x) = f(g(x))$ — Definition of $f \circ g$

$\qquad = f(x^2 + 1)$ — $g(x) = x^2 + 1$ is the inner function.

$\qquad = 2(x^2 + 1) - 3$ — Input $x^2 + 1$ into the outer function f.

$\qquad = 2x^2 + 2 - 3$ — Distributive Property

$\qquad = 2x^2 - 1$ — Simplify.

b. $(g \circ f)(x) = g(f(x))$ — Definition of $g \circ f$

$\qquad = g(2x - 3)$ — $f(x) = 2x - 3$ is the inner function.

$\qquad = (2x - 3)^2 + 1$ — Input $2x - 3$ into the outer function g.

$\qquad = 4x^2 - 12x + 9 + 1$ — Expand.

$\qquad = 4x^2 - 12x + 10$ — Simplify.

Note that $(f \circ g)(x) \neq (g \circ f)(x)$.

To determine the domain of a composite function, first write the composite function in simplest form. Then use the fact that its domain *either is equal to or is a restriction of the domain of the "inner" function*. This is demonstrated in Example 3.

Example 3   Finding the Domain of a Composite Function

Find the domain of the composition of f with g when $f(x) = x^2$ and $g(x) = \sqrt{x}$.

Solution

$(f \circ g)(x) = f(g(x))$ — Definition of $f \circ g$

$\qquad = f(\sqrt{x})$ — $g(x) = \sqrt{x}$ is the inner function.

$\qquad = (\sqrt{x})^2$ — Input $\sqrt{x}$ into the outer function f.

$\qquad = x, \ x \geq 0$ — Domain of $f \circ g$ is all $x \geq 0$.

The domain of the inner function $g(x) = \sqrt{x}$ is the set of all nonnegative real numbers. The simplified form of $f \circ g$ has no restriction on this set of numbers. So, the restriction $x \geq 0$ must be added to the composition of this function. The domain of $f \circ g$ is the set of all nonnegative real numbers.

2 Use the Horizontal Line Test to determine whether functions have inverse functions.

One-to-One and Inverse Functions

In Section 3.6, you learned that a function can be represented by a set of ordered pairs. For instance, the function $f(x) = x + 2$ from the set $A = \{1, 2, 3, 4\}$ to the set $B = \{3, 4, 5, 6\}$ can be written as follows.

$$f(x) = x + 2: \quad \{(1, 3), (2, 4), (3, 5), (4, 6)\}$$

By interchanging the first and second coordinates of each of these ordered pairs, you can form another function that is called the **inverse function** of f, denoted by f^{-1}. It is a function from the set B to the set A, and can be written as follows.

$$f^{-1}(x) = x - 2: \quad \{(3, 1), (4, 2), (5, 3), (6, 4)\}$$

Interchanging the ordered pairs for a function f will only produce another function when f is one-to-one. A function f is **one-to-one** if each value of the dependent variable corresponds to exactly one value of the independent variable. Figure 9.11 shows that the domain of f is the range of f^{-1} and the range of f is the domain of f^{-1}.

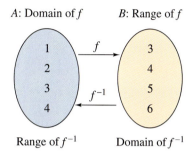

A: Domain of f B: Range of f

Range of f^{-1} Domain of f^{-1}

Figure 9.11 f is one-to-one and has inverse function f^{-1}.

Horizontal Line Test for Inverse Functions

A function f has an inverse function f^{-1} if and only if f is one-to-one. Graphically, a function f has an inverse function f^{-1} if and only if no *horizontal* line intersects the graph of f at more than one point.

Example 4   Applying the Horizontal Line Test

Use the Horizontal Line Test to determine if the function is one-to-one and so has an inverse function.

a. The graph of the function $f(x) = x^3 - 1$ is shown in Figure 9.12. Because no horizontal line intersects the graph of f at more than one point, you can conclude that f *is* a one-to-one function and *does* have an inverse function.

b. The graph of the function $f(x) = x^2 - 1$ is shown in Figure 9.13. Because it is possible to find a horizontal line that intersects the graph of f at more than one point, you can conclude that f *is not* a one-to-one function and *does not* have an inverse function.

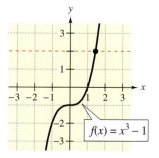

Figure 9.12

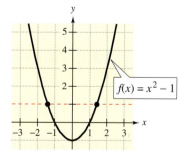

Figure 9.13

The formal definition of an inverse function is given as follows.

Definition of Inverse Function

Let f and g be two functions such that

$$f(g(x)) = x \quad \text{for every } x \text{ in the domain of } g$$

and

$$g(f(x)) = x \quad \text{for every } x \text{ in the domain of } f.$$

The function g is called the **inverse function** of the function f, and is denoted by f^{-1} (read "f-inverse"). So, $f(f^{-1}(x)) = x$ and $f^{-1}(f(x)) = x$. The domain of f must be equal to the range of f^{-1}, and vice versa.

Do not be confused by the use of -1 to denote the inverse function f^{-1}. Whenever f^{-1} is written, it *always* refers to the inverse function f and *not* to the reciprocal of $f(x)$.

If the function g is the inverse function of the function f, it must also be true that the function f is the inverse function of the function g. For this reason, you can refer to the functions f and g as being *inverse functions of each other.*

Example 5   Verifying Inverse Functions

Show that $f(x) = 2x - 4$ and $g(x) = \dfrac{x + 4}{2}$ are inverse functions of each other.

Solution

Begin by noting that the domain and range of both functions are the entire set of real numbers. To show that f and g are inverse functions of each other, you need to show that $f(g(x)) = x$ and $g(f(x)) = x$, as follows.

$$f(g(x)) = f\left(\frac{x + 4}{2}\right) \qquad \text{\color{red} $g(x) = (x + 4)/2$ is the inner function.}$$

$$= 2\left(\frac{x + 4}{2}\right) - 4 \qquad \text{\color{red} Input $(x + 4)/2$ into the outer function f.}$$

$$= x + 4 - 4 = x \qquad \text{\color{red} Simplify.}$$

$$g(f(x)) = g(2x - 4) \qquad \text{\color{red} $f(x) = 2x - 4$ is the inner function.}$$

$$= \frac{(2x - 4) + 4}{2} \qquad \text{\color{red} Input $2x - 4$ into the outer function g.}$$

$$= \frac{2x}{2} = x \qquad \text{\color{red} Simplify.}$$

Note that the two functions f and g "undo" each other in the following verbal sense. The function f first *multiplies* the input x by 2 and then *subtracts* 4, whereas the function g first *adds* 4 and then *divides* the result by 2.

Example 6 **Verifying Inverse Functions**

Show that the functions

$$f(x) = x^3 + 1 \text{ and } g(x) = \sqrt[3]{x - 1}$$

are inverse functions of each other.

Solution

Begin by noting that the domain and range of both functions are the entire set of real numbers. To show that f and g are inverse functions of each other, you need to show that $f(g(x)) = x$ and $g(f(x)) = x$, as follows.

$$f(g(x)) = f\left(\sqrt[3]{x - 1}\right) \qquad g(x) = \sqrt[3]{x - 1} \text{ is the inner function.}$$
$$= \left(\sqrt[3]{x - 1}\right)^3 + 1 \qquad \text{Input } \sqrt[3]{x - 1} \text{ into the outer function } f.$$
$$= (x - 1) + 1 = x \qquad \text{Simplify.}$$
$$g(f(x)) = g(x^3 + 1) \qquad f(x) = x^3 + 1 \text{ is the inner function.}$$
$$= \sqrt[3]{(x^3 + 1) - 1} \qquad \text{Input } x^3 + 1 \text{ into the outer function } g.$$
$$= \sqrt[3]{x^3} = x \qquad \text{Simplify.}$$

Note that the two functions f and g "undo" each other in the following verbal sense. The function f first *cubes* the input x and then *adds* 1, whereas the function g first *subtracts* 1 and then *takes the cube root* of the result.

---

③ Find inverse functions algebraically.

Finding an Inverse Function Algebraically

You can find the inverse function of a simple function by inspection. For instance, the inverse function of $f(x) = 10x$ is $f^{-1}(x) = x/10$. For more complicated functions, however, it is best to use the following steps for finding an inverse function. The key step in these guidelines is switching the roles of x and y. This step corresponds to the fact that inverse functions have ordered pairs with the coordinates reversed.

Study Tip

You can graph a function and use the Horizontal Line Test to see if the function is one-to-one before trying to find its inverse function.

Finding an Inverse Function Algebraically

1. In the equation for $f(x)$, replace $f(x)$ with y.

2. Interchange the roles of x and y.

3. If the new equation does not represent y as a function of x, the function f does not have an inverse function. If the new equation does represent y as a function of x, solve the new equation for y.

4. Replace y with $f^{-1}(x)$.

5. Verify that f and f^{-1} are inverse functions of each other by showing that $f(f^{-1}(x)) = x = f^{-1}(f(x))$.

Example 7 Finding an Inverse Function

Determine whether each function has an inverse function. If it does, find its inverse function.

a. $f(x) = 2x + 3$ **b.** $f(x) = x^3 + 3$

Solution

a.

$f(x) = 2x + 3$	Write original function.
$y = 2x + 3$	Replace $f(x)$ with y.
$x = 2y + 3$	Interchange x and y.
$y = \dfrac{x - 3}{2}$	Solve for y.
$f^{-1}(x) = \dfrac{x - 3}{2}$	Replace y with $f^{-1}(x)$.

You can verify that $f(f^{-1}(x)) = x = f^{-1}(f(x))$, as follows.

$$f(f^{-1}(x)) = f\left(\frac{x-3}{2}\right) = 2\left(\frac{x-3}{2}\right) + 3 = (x-3) + 3 = x$$

$$f^{-1}(f(x)) = f^{-1}(2x+3) = \frac{(2x+3)-3}{2} = \frac{2x}{2} = x$$

b.

$f(x) = x^3 + 3$	Write original function.
$y = x^3 + 3$	Replace $f(x)$ with y.
$x = y^3 + 3$	Interchange x and y.
$\sqrt[3]{x - 3} = y$	Solve for y.
$f^{-1}(x) = \sqrt[3]{x - 3}$	Replace y with $f^{-1}(x)$.

You can verify that $f(f^{-1}(x)) = x = f^{-1}(f(x))$, as follows.

$$f(f^{-1}(x)) = f(\sqrt[3]{x-3}) = (\sqrt[3]{x-3})^3 + 3 = (x-3) + 3 = x$$

$$f^{-1}(f(x)) = f^{-1}(x^3 + 3) = \sqrt[3]{(x^3 + 3) - 3} = \sqrt[3]{x^3} = x$$

Example 8 A Function That Has No Inverse Function

$f(x) = x^2$	Original equation
$y = x^2$	Replace $f(x)$ with y.
$x = y^2$	Interchange x and y.

Recall from Section 3.6 that the equation $x = y^2$ does not represent y as a function of x because you can find two different y-values that correspond to the same x-value. Because the equation does not represent y as a function of x, you can conclude that the original function f does not have an inverse function.

Technology: Discovery

Use a graphing calculator to graph $f(x) = x^3 + 1$, $f^{-1}(x) = \sqrt[3]{x - 1}$, and $y = x$ in the same viewing window.

a. Relative to the line $y = x$, how do the graphs of f and f^{-1} compare?

b. For the graph of f, complete the table.

x	-1	0	1
f			

For the graph of f^{-1}, complete the table.

x	0	1	2
f^{-1}			

What can you conclude about the coordinates of the points on the graph of f compared with those on the graph of f^{-1}?

④ Graphically verify that two functions are inverse functions of each other.

Graphs of Inverse Functions

The graphs of f and f^{-1} are related to each other in the following way. If the point (a, b) lies on the graph of f, the point (b, a) must lie on the graph of f^{-1}, and vice versa. This means that the graph of f^{-1} is a reflection of the graph of f in the line $y = x$, as shown in Figure 9.14. This "reflective property" of the graphs of f and f^{-1} is illustrated in Examples 9 and 10.

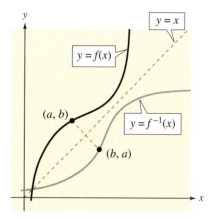

Figure 9.14   The graph of f^{-1} is a reflection of the graph of f in the line $y = x$.

Example 9   The Graphs of f and f^{-1}

Sketch the graphs of the inverse functions $f(x) = 2x - 3$ and $f^{-1}(x) = \frac{1}{2}(x + 3)$ on the same rectangular coordinate system, and show that the graphs are reflections of each other in the line $y = x$.

Solution

The graphs of f and f^{-1} are shown in Figure 9.15. Visually, it appears that the graphs are reflections of each other. You can further verify this reflective property by testing a few points on each graph. Note in the following list that if the point (a, b) is on the graph of f, the point (b, a) is on the graph of f^{-1}.

$f(x) = 2x - 3$	$f^{-1}(x) = \frac{1}{2}(x + 3)$
$(-1, -5)$	$(-5, -1)$
$(0, -3)$	$(-3, 0)$
$(1, -1)$	$(-1, 1)$
$(2, 1)$	$(1, 2)$
$(3, 3)$	$(3, 3)$

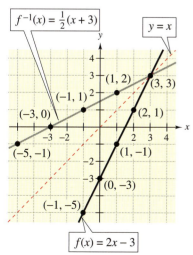

Figure 9.15

You can sketch the graph of an inverse function without knowing the equation of the inverse function. Simply find the coordinates of points that lie on the original function. By interchanging the x- and y-coordinates, you have points that lie on the graph of the inverse function. Plot these points and sketch the graph of the inverse function.

In Example 8, you saw that the function

$$f(x) = x^2$$

has no inverse function. A more complete way of saying this is "*assuming that the domain of f is the entire real line,* the function $f(x) = x^2$ has no inverse function." If, however, you *restrict* the domain of f to the nonnegative real numbers, then f does have an inverse function, as demonstrated in Example 10.

Example 10   Verifying Inverse Functions Graphically

Graphically verify that f and g are inverse functions of each other.

$$f(x) = x^2, \quad x \geq 0 \quad \text{and} \quad g(x) = \sqrt{x}$$

Solution

You can graphically verify that f and g are inverse functions of each other by graphing the functions on the same rectangular coordinate system, as shown in Figure 9.16. Visually, it appears that the graphs are reflections of each other in the line $y = x$. You can further verify this reflective property by testing a few points on each graph. Note in the following list that if the point (a, b) is on the graph of f, the point (b, a) is on the graph of g.

$f(x) = x^2, \quad x \geq 0$	$g(x) = f^{-1}(x) = \sqrt{x}$
$(0, 0)$	$(0, 0)$
$(1, 1)$	$(1, 1)$
$(2, 4)$	$(4, 2)$
$(3, 9)$	$(9, 3)$

Technology: Tip

A graphing calculator program for several models of graphing calculators that graphs the function f and its reflection in the line $y = x$ can be found at our website
math.college.hmco.com/students.

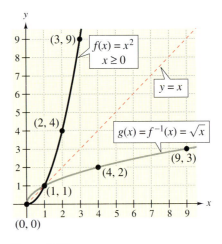

Figure 9.16

So, f and g are inverse functions of each other.

9.2 Exercises

Review   Concepts, Skills, and Problem Solving

Keep mathematically in shape by doing these exercises *before* the problems of this section.

Properties and Definitions

1. *Writing* ✎   Decide whether $x - y^2 = 0$ represents y as a function of x. Explain.

2. *Writing* ✎   Decide whether $|x| - 2y = 4$ represents y as a function of x. Explain.

3. *Writing* ✎   Explain why the domains of f and g are not the same.

$$f(x) = \sqrt{4 - x^2} \qquad g(x) = \frac{6}{\sqrt{4 - x^2}}$$

4. Determine the range of $h(x) = 8 - \sqrt{x}$ over the domain $\{0, 4, 9, 16\}$.

Simplifying Expressions

In Exercises 5–10, perform the indicated operations and simplify.

5. $-(5x^2 - 1) + (3x^2 - 5)$

6. $(-2x)(-5x)(3x + 4)$

7. $(u - 4v)(u + 4v)$ 8. $(3a - 2b)^2$

9. $(t - 2)^3$ 10. $\dfrac{6x^3 - 3x^2}{12x}$

Problem Solving

11. *Free-Falling Object*   The velocity of a free-falling object is given by $v = \sqrt{2gh}$, where v is the velocity measured in feet per second, $g = 32$ feet per second per second, and h is the distance (in feet) the object has fallen. Find the distance an object has fallen if its velocity is 80 feet per second.

12. *Consumer Awareness*   The cost of a long-distance telephone call is $0.95 for the first minute and $0.35 for each additional minute. The total cost of a call is $5.15. Find the length of the call.

Developing Skills

In Exercises 1–10, find the compositions. See Examples 1 and 2.

1. $f(x) = 2x + 3$, $g(x) = x - 6$
   (a) $(f \circ g)(x)$ (b) $(g \circ f)(x)$
   (c) $(f \circ g)(4)$ (d) $(g \circ f)(7)$

2. $f(x) = x - 5$, $g(x) = 6 + 2x$
   (a) $(f \circ g)(x)$ (b) $(g \circ f)(x)$
   (c) $(f \circ g)(3)$ (d) $(g \circ f)(3)$

3. $f(x) = x^2 + 3$, $g(x) = 2x + 1$
   (a) $(f \circ g)(x)$ (b) $(g \circ f)(x)$

   (c) $(f \circ g)(2)$ (d) $(g \circ f)(-3)$

4. $f(x) = 2x + 1$, $g(x) = x^2 - 5$
   (a) $(f \circ g)(x)$ (b) $(g \circ f)(x)$
   (c) $(f \circ g)(-1)$ (d) $(g \circ f)(3)$

5. $f(x) = |x - 3|$, $g(x) = 3x$
   (a) $(f \circ g)(x)$ (b) $(g \circ f)(x)$

   (c) $(f \circ g)(1)$ (d) $(g \circ f)(2)$

6. $f(x) = |x|$, $g(x) = 2x + 5$
   (a) $(f \circ g)(x)$ (b) $(g \circ f)(x)$
   (c) $(f \circ g)(-2)$ (d) $(g \circ f)(-4)$

7. $f(x) = \sqrt{x - 4}$, $g(x) = x + 5$
   (a) $(f \circ g)(x)$ (b) $(g \circ f)(x)$
   (c) $(f \circ g)(3)$ (d) $(g \circ f)(8)$

8. $f(x) = \sqrt{x + 6}$, $g(x) = 2x - 3$
   (a) $(f \circ g)(x)$ (b) $(g \circ f)(x)$

   (c) $(f \circ g)(3)$ (d) $(g \circ f)(-2)$

9. $f(x) = \dfrac{1}{x - 3}$, $g(x) = \dfrac{2}{x^2}$
   (a) $(f \circ g)(x)$ (b) $(g \circ f)(x)$

   (c) $(f \circ g)(-1)$ (d) $(g \circ f)(2)$

10. $f(x) = \dfrac{4}{x^2 - 4}, \quad g(x) = \dfrac{1}{x}$

    (a) $(f \circ g)(x)$ (b) $(g \circ f)(x)$

    (c) $(f \circ g)(-2)$ (d) $(g \circ f)(1)$

In Exercises 11–14, use the functions f and g to find the indicated values.

$f = \{(-2, 3), (-1, 1), (0, 0), (1, -1), (2, -3)\},$

$g = \{(-3, 1), (-1, -2), (0, 2), (2, 2), (3, 1)\}$

11. (a) $f(1)$ **12.** (a) $g(0)$

    (b) $g(-1)$ (b) $f(2)$

    (c) $(g \circ f)(1)$ (c) $(f \circ g)(0)$

13. (a) $(f \circ g)(-3)$ **14.** (a) $(f \circ g)(2)$

    (b) $(g \circ f)(-2)$ (b) $(g \circ f)(2)$

In Exercises 15–18, use the functions f and g to find the indicated values.

$f = \{(0, 1), (1, 2), (2, 5), (3, 10), (4, 17)\},$

$g = \{(5, 4), (10, 1), (2, 3), (17, 0), (1, 2)\}$

15. (a) $f(3)$ **16.** (a) $g(2)$

    (b) $g(10)$ (b) $f(0)$

    (c) $(g \circ f)(3)$ (c) $(f \circ g)(10)$

17. (a) $(g \circ f)(4)$ **18.** (a) $(f \circ g)(1)$

    (b) $(f \circ g)(2)$ (b) $(g \circ f)(0)$

In Exercises 19–26, find the compositions (a) $f \circ g$ and (b) $g \circ f$. Then find the domain of each composition. See Example 3.

19. $f(x) = 3x + 4$ **20.** $f(x) = x + 5$

    $g(x) = x - 7$ $g(x) = 4x - 1$

21. $f(x) = \sqrt{x}$ **22.** $f(x) = \sqrt{x - 5}$

    $g(x) = x - 2$ $g(x) = x + 3$

23. $f(x) = x^2 + 3$

    $g(x) = \sqrt{x - 1}$

24. $f(x) = \sqrt{3x + 1}$

    $g(x) = x^2 - 8$

25. $f(x) = \dfrac{x}{x + 5}$ **26.** $f(x) = \dfrac{x}{x - 4}$

    $g(x) = \sqrt{x - 1}$ $g(x) = \sqrt{x}$

 In Exercises 27–34, use a graphing calculator to graph the function and determine whether the function is one-to-one.

27. $f(x) = x^3 - 1$ **28.** $f(x) = (2 - x)^3$

29. $f(t) = \sqrt[3]{5 - t}$ **30.** $h(t) = 4 - \sqrt[3]{t}$

31. $g(x) = x^4 - 6$ **32.** $f(x) = (x + 2)^5$

33. $h(t) = \dfrac{5}{t}$ **34.** $g(t) = \dfrac{5}{t^2}$

In Exercises 35–40, use the Horizontal Line Test to determine if the function is one-to-one and so has an inverse function. See Example 4.

35. $f(x) = x^2 - 2$ **36.** $f(x) = \frac{1}{5}x$

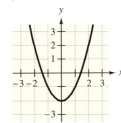   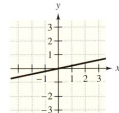

37. $f(x) = x^2, \quad x \geq 0$ **38.** $f(x) = \sqrt{-x}$

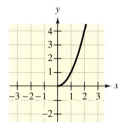   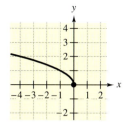

39. $g(x) = \sqrt{25 - x^2}$ **40.** $g(x) = |x - 4|$

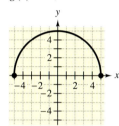   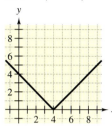

In Exercises 41–52, verify algebraically that the functions f and g are inverse functions of each other. See Examples 5 and 6.

41. $f(x) = -6x, \quad g(x) = -\frac{1}{6}x$

42. $f(x) = \frac{2}{3}x, \quad g(x) = \frac{3}{2}x$

43. $f(x) = x + 15, \quad g(x) = x - 15$

44. $f(x) = 3 - x, \quad g(x) = 3 - x$

45. $f(x) = 1 - 2x, \quad g(x) = \frac{1}{2}(1 - x)$

46. $f(x) = 2x - 1, \quad g(x) = \frac{1}{2}(x + 1)$

47. $f(x) = 2 - 3x, \quad g(x) = \frac{1}{3}(2 - x)$

48. $f(x) = -\frac{1}{4}x + 3, \quad g(x) = -4(x - 3)$

49. $f(x) = \sqrt[3]{x + 1}, \quad g(x) = x^3 - 1$

50. $f(x) = x^7, \quad g(x) = \sqrt[7]{x}$

51. $f(x) = \frac{1}{x}, \quad g(x) = \frac{1}{x}$

52. $f(x) = \frac{1}{x + 1}, \quad g(x) = \frac{1 - x}{x}$

In Exercises 53–64, find the inverse function of f. Verify that $f(f^{-1}(x))$ and $f^{-1}(f(x))$ are equal to the identity function. See Example 7.

53. $f(x) = 5x$ **54.** $f(x) = -3x$

55. $f(x) = -\frac{2}{5}x$ **56.** $f(x) = \frac{1}{3}x$

57. $f(x) = x + 10$ **58.** $f(x) = x - 5$

59. $f(x) = 3 - x$ **60.** $f(x) = 8 - x$

61. $f(x) = x^7$ **62.** $f(x) = x^5$

63. $f(x) = \sqrt[3]{x}$ **64.** $f(x) = x^{1/5}$

In Exercises 65–78, find the inverse function. See Example 7.

65. $f(x) = 8x$ **66.** $f(x) = \frac{1}{10}x$

67. $g(x) = x + 25$ **68.** $f(x) = 7 - x$

69. $g(x) = 3 - 4x$

70. $g(t) = 6t + 1$

71. $g(t) = \frac{1}{4}t + 2$

72. $h(s) = 5 - \frac{3}{2}s$

73. $h(x) = \sqrt{x}$

74. $h(x) = \sqrt{x + 5}$

75. $f(t) = t^3 - 1$

76. $h(t) = t^5 + 8$

77. $f(x) = \sqrt{x + 3}, \quad x \geq -3$

78. $f(x) = \sqrt{x^2 - 4}, \quad x \geq 2$

In Exercises 79–82, match the graph with the graph of its inverse function. [The graphs of the inverse functions are labeled (a), (b), (c), and (d).]

(a)

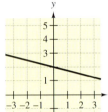

(b)

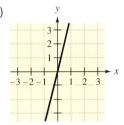

(c)

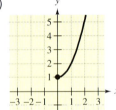

(d)

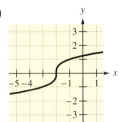

79.

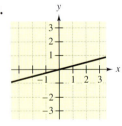

80.

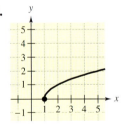

81.

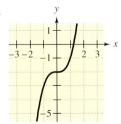

82.

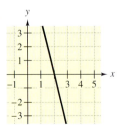

In Exercises 83–88, sketch the graphs of f and f^{-1} on the same rectangular coordinate system. Show that the graphs are reflections of each other in the line $y = x$. See Example 9.

83. $f(x) = x + 4, \ f^{-1}(x) = x - 4$

84. $f(x) = x - 7, \ f^{-1}(x) = x + 7$

85. $f(x) = 3x - 1, \ f^{-1}(x) = \frac{1}{3}(x + 1)$

86. $f(x) = 5 - 4x, \ f^{-1}(x) = -\frac{1}{4}(x - 5)$

87. $f(x) = x^2 - 1, \ x \geq 0,$

    $f^{-1}(x) = \sqrt{x + 1}$

88. $f(x) = (x + 2)^2, \ x \geq -2,$

    $f^{-1}(x) = \sqrt{x} - 2$

 In Exercises 89–96, use a graphing calculator to graph the functions in the same viewing window. Graphically verify that f and g are inverse functions of each other.

89. $f(x) = \frac{1}{3}x$

    $g(x) = 3x$

90. $f(x) = \frac{1}{5}x - 1$

    $g(x) = 5x + 5$

91. $f(x) = \sqrt{x + 1}$

    $g(x) = x^2 - 1, \ x \geq 0$

92. $f(x) = \sqrt{4 - x}$

    $g(x) = 4 - x^2, \ x \geq 0$

93. $f(x) = \frac{1}{8}x^3$

    $g(x) = 2\sqrt[3]{x}$

94. $f(x) = \sqrt[3]{x + 2}$

    $g(x) = x^3 - 2$

95. $f(x) = |3 - x|, \ x \geq 3$

    $g(x) = 3 + x, \ x \geq 0$

96. $f(x) = |x - 2|, \ x \geq 2$

    $g(x) = x + 2, \ x \geq 0$

In Exercises 97–100, delete part of the graph of the function so that the remaining part is one-to-one. Find the inverse function of the remaining part and find the domain of the inverse function. (*Note:* There is more than one correct answer.) See Example 10.

97. $f(x) = (x - 2)^2$

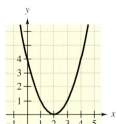

98. $f(x) = 9 - x^2$

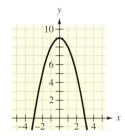

99. $f(x) = |x| + 1$

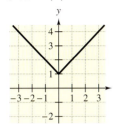

100. $f(x) = |x - 2|$

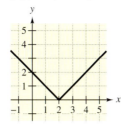

In Exercises 101 and 102, consider the function $f(x) = 3 - 2x$.

101. Find $f^{-1}(x)$.

102. Find $(f^{-1})^{-1}(x)$.

Solving Problems

103. ▲ *Geometry*   You are standing on a bridge over a calm pond and drop a pebble, causing ripples of concentric circles in the water. The radius (in feet) of the outer ripple is given by $r(t) = 0.6t$, where t is time in seconds after the pebble hits the water. The area of the circle is given by the function $A(r) = \pi r^2$. Find an equation for the composition $A(r(t))$. What are the input and output of this composite function?

104. *Sales Bonus*   You are a sales representative for a clothing manufacturer. You are paid an annual salary plus a bonus of 2% of your sales over \$200,000. Consider the two functions $f(x) = x - 200{,}000$ and $g(x) = 0.02x$. If x is greater than \$200,000, find each composition and determine which represents your bonus. Explain.

(a) $f(g(x))$

(b) $g(f(x))$

105. *Daily Production Cost*   The daily cost of producing x units in a manufacturing process is $C(x) = 8.5x + 300$. The number of units produced in t hours during a day is given by $x(t) = 12t$, $0 \le t \le 8$. Find, simplify, and interpret $(C \circ x)(t)$.

106. *Rebate and Discount*   The suggested retail price of a new car is p dollars. The dealership advertised a factory rebate of \$2000 and a 5% discount.

(a) Write a function R in terms of p, giving the cost of the car after receiving the factory rebate.

(b) Write a function S in terms of p, giving the cost of the car after receiving the dealership discount.

(c) Form the composite functions $(R \circ S)(p)$ and $(S \circ R)(p)$ and interpret each.

(d) Find $(R \circ S)(26{,}000)$ and $(S \circ R)(26{,}000)$. Which yields the smaller cost for the car? Explain.

107. *Rebate and Discount*   The suggested retail price of a plasma television is p dollars. The electronics store is offering a manufacturer's rebate of \$500 and a 10% discount.

(a) Write a function R in terms of p, giving the cost of the television after receiving the manufacturer's rebate.

(b) Write a function S in terms of p, giving the cost of the television after receiving the 10% discount.

(c) Form the composite functions $(R \circ S)(p)$ and $(S \circ R)(p)$ and interpret each.

(d) Find $(R \circ S)(6000)$ and $(S \circ R)(6000)$. Which yields the smaller cost for the plasma television? Explain.

108. *Hourly Wage*   Your wage is $9.00 per hour plus $0.65 for each unit produced per hour. So, your hourly wage y in terms of the number of units produced x is $y = 9 + 0.65x$.

(a) Find the inverse function.

(b) What does each variable represent in the inverse function?

(c) Determine the number of units produced when your hourly wage averages $14.20.

109. *Cost*   You need 100 pounds of two fruits: oranges that cost $0.75 per pound and apples that cost $0.95 per pound.

(a) Verify that your total cost is $y = 0.75x + 0.95(100 - x)$, where x is the number of pounds of oranges.

(b) Find the inverse function. What does each variable represent in the inverse function?

(c) Use the context of the problem to determine the domain of the inverse function.

(d) Determine the number of pounds of oranges purchased if the total cost is $84.

110. *Exploration*   Consider the functions $f(x) = 4x$ and $g(x) = x + 6$.

(a) Find $(f \circ g)(x)$.

(b) Find $(f \circ g)^{-1}(x)$.

(c) Find $f^{-1}(x)$ and $g^{-1}(x)$.

(d) Find $(g^{-1} \circ f^{-1})(x)$ and compare the result with that of part (b).

(e) Repeat parts (a) through (d) for $f(x) = x^3 + 1$ and $g(x) = 2x$.

(f) Make a conjecture about $(f \circ g)^{-1}(x)$ and $(g^{-1} \circ f^{-1})(x)$.

Explaining Concepts

True or False?   In Exercises 111–114, decide whether the statement is true or false. If true, explain your reasoning. If false, give an example.

111. If the inverse function of f exists, the y-intercept of f is an x-intercept of f^{-1}. Explain.

112. There exists no function f such that $f = f^{-1}$.

113. If the inverse function of f exists, the domains of f and f^{-1} are the same.

114. If the inverse function of f exists and its graph passes through the point $(2, 2)$, the graph of f^{-1} also passes through the point $(2, 2)$.

115. *Writing*   Describe how to find the inverse of a function given by a set of ordered pairs. Give an example.

116. *Writing*   Describe how to find the inverse of a function given by an equation in x and y. Give an example.

117. Give an example of a function that does not have an inverse function.

118. *Writing*   Explain the Horizontal Line Test. What is the relationship between this test and a function being one-to-one?

119. *Writing*   Describe the relationship between the graph of a function and its inverse function.

9.3 Logarithmic Functions

A.T. Willett/Alamy

What You Should Learn

① Evaluate logarithmic functions.

② Graph logarithmic functions.

③ Graph and evaluate natural logarithmic functions.

④ Use the change-of-base formula to evaluate logarithms.

Why You Should Learn It

Logarithmic functions can be used to model and solve real-life problems. For instance, in Exercise 128 on page 611, you will use a logarithmic function to determine the speed of the wind near the center of a tornado.

① Evaluate logarithmic functions.

Logarithmic Functions

In Section 9.2, you were introduced to the concept of an inverse function. Moreover, you saw that if a function has the property that no horizontal line intersects the graph of the function more than once, the function must have an inverse function. By looking back at the graphs of the exponential functions introduced in Section 9.1, you will see that every function of the form

$$f(x) = a^x$$

passes the Horizontal Line Test, and so must have an inverse function. To describe the inverse function of $f(x) = a^x$, follow the steps used in Section 9.2.

$y = a^x$ Replace $f(x)$ by y.

$x = a^y$ Interchange x and y.

At this point, there is no way to solve for y. A verbal description of y in the equation $x = a^y$ is "y equals the exponent needed on base a to get x." This inverse of $f(x) = a^x$ is denoted by the **logarithmic function with base a**

$$f^{-1}(x) = \log_a x.$$

Definition of Logarithmic Function

Let a and x be positive real numbers such that $a \neq 1$. The **logarithm of x with base a** is denoted by $\log_a x$ and is defined as follows.

$$y = \log_a x \quad \text{if and only if} \quad x = a^y$$

The function $f(x) = \log_a x$ is the **logarithmic function with base a.**

From the definition it is clear that

Logarithmic Equation   *Exponential Equation*

$y = \log_a x$ is equivalent to $x = a^y$.

So, to find the value of $\log_a x$, *think*

"$\log_a x$ = the exponent needed on base a to get x."

For instance,

$$y = \log_2 8 \qquad \text{Think: "The exponent needed on 2 to get 8."}$$

$$y = 3.$$

That is,

$$3 = \log_2 8. \qquad \text{This is equivalent to } 2^3 = 8.$$

By now it should be clear that *a logarithm is an exponent*.

Example 1   Evaluating Logarithms

Evaluate each logarithm.

a. $\log_2 16$ **b.** $\log_3 9$ **c.** $\log_4 2$

Solution

In each case you should answer the question, "To what power must the base be raised to obtain the given number?"

a. The power to which 2 must be raised to obtain 16 is 4. That is,

$$2^4 = 16 \quad \Longrightarrow \quad \log_2 16 = 4.$$

b. The power to which 3 must be raised to obtain 9 is 2. That is,

$$3^2 = 9 \quad \Longrightarrow \quad \log_3 9 = 2.$$

c. The power to which 4 must be raised to obtain 2 is $\frac{1}{2}$. That is,

$$4^{1/2} = 2 \quad \Longrightarrow \quad \log_4 2 = \frac{1}{2}.$$

Example 2   Evaluating Logarithms

Evaluate each logarithm.

a. $\log_5 1$ **b.** $\log_{10} \dfrac{1}{10}$ **c.** $\log_3(-1)$ **d.** $\log_4 0$

Solution

a. The power to which 5 must be raised to obtain 1 is 0. That is,

$$5^0 = 1 \quad \Longrightarrow \quad \log_5 1 = 0.$$

b. The power to which 10 must be raised to obtain $\frac{1}{10}$ is -1. That is,

$$10^{-1} = \frac{1}{10} \quad \Longrightarrow \quad \log_{10} \frac{1}{10} = -1.$$

c. There is no power to which 3 can be raised to obtain -1. The reason for this is that for any value of x, 3^x is a positive number. So, $\log_3(-1)$ is undefined.

d. There is no power to which 4 can be raised to obtain 0. So, $\log_4 0$ is undefined.

Study Tip

Study the results in Example 2 carefully. Each of the logarithms illustrates an important special property of logarithms that you should know.

The following properties of logarithms follow directly from the definition of the logarithmic function with base a.

Properties of Logarithms

Let a and x be positive real numbers such that $a \neq 1$. Then the following properties are true.

1. $\log_a 1 = 0$ because $a^0 = 1$.

2. $\log_a a = 1$ because $a^1 = a$.

3. $\log_a a^x = x$ because $a^x = a^x$.

The logarithmic function with base 10 is called the **common logarithmic function.** On most calculators, this function can be evaluated with the common logarithmic key [LOG], as illustrated in the next example.

Example 3 Evaluating Common Logarithms

Evaluate each logarithm. Use a calculator only if necessary.

a. $\log_{10} 100$ **b.** $\log_{10} 0.01$

c. $\log_{10} 5$ **d.** $\log_{10} 2.5$

Solution

a. The power to which 10 must be raised to obtain 100 is 2. That is,

$$10^2 = 100 \quad \Longrightarrow \quad \log_{10} 100 = 2.$$

b. The power to which 10 must be raised to obtain 0.01 or $\frac{1}{100}$ is -2. That is,

$$10^{-2} = \tfrac{1}{100} \quad \Longrightarrow \quad \log_{10} 0.01 = -2.$$

c. There is no simple power to which 10 can be raised to obtain 5, so you should use a calculator to evaluate $\log_{10} 5$.

Keystrokes	Display	
5 [LOG]	0.69897	Scientific
[LOG] 5 [ENTER]	0.69897	Graphing

So, rounded to three decimal places, $\log_{10} 5 \approx 0.699$.

d. There is no simple power to which 10 can be raised to obtain 2.5, so you should use a calculator to evaluate $\log_{10} 2.5$.

Keystrokes	Display	
2.5 [LOG]	0.39794	Scientific
[LOG] 2.5 [ENTER]	0.39794	Graphing

So, rounded to three decimal places, $\log_{10} 2.5 \approx 0.398$.

Study Tip

Be sure you see that the value of a logarithm can be zero or negative, as in Example 3(b), *but* you cannot take the logarithm of zero or a negative number. This means that the logarithms $\log_{10}(-10)$ and $\log_5 0$ are not valid.

② Graph logarithmic functions.

Graphs of Logarithmic Functions

To sketch the graph of

$$y = \log_a x$$

you can use the fact that the graphs of inverse functions are reflections of each other in the line $y = x$.

Example 4 Graphs of Exponential and Logarithmic Functions

On the same rectangular coordinate system, sketch the graph of each function.

a. $f(x) = 2^x$ **b.** $g(x) = \log_2 x$

Solution

a. Begin by making a table of values for $f(x) = 2^x$.

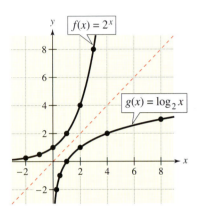

Figure 9.17 Inverse Functions

x	-2	-1	0	1	2	3
$f(x) = 2^x$	$\frac{1}{4}$	$\frac{1}{2}$	1	2	4	8

By plotting these points and connecting them with a smooth curve, you obtain the graph shown in Figure 9.17.

b. Because $g(x) = \log_2 x$ is the inverse function of $f(x) = 2^x$, the graph of g is obtained by reflecting the graph of f in the line $y = x$, as shown in Figure 9.17.

Notice from the graph of $g(x) = \log_2 x$, shown in Figure 9.17, that the domain of the function is the set of positive numbers and the range is the set of all real numbers. The basic characteristics of the graph of a logarithmic function are summarized in Figure 9.18. In this figure, note that the graph has one x-intercept at $(1, 0)$. Also note that $x = 0$ (y-axis) is a vertical asymptote of the graph.

Study Tip

In Example 4, the inverse property of logarithmic functions is used to sketch the graph of $g(x) = \log_2 x$. You could also use a standard point-plotting approach or a graphing calculator.

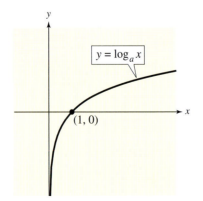

Graph of $y = \log_a x$, $a > 1$

- Domain: $(0, \infty)$
- Range: $(-\infty, \infty)$
- Intercept: $(1, 0)$
- Increasing (moves up to the right)
- Asymptote: y-axis

Figure 9.18 Characteristics of logarithmic function $y = \log_a x \, (a > 1)$

In the following example, the graph of $\log_a x$ is used to sketch the graphs of functions of the form $y = b \pm \log_a(x + c)$. Notice how each transformation affects the vertical asymptote.

Example 5   Sketching the Graphs of Logarithmic Functions

The graph of each function is similar to the graph of $f(x) = \log_{10} x$, as shown in Figure 9.19. From the graph you can determine the domain of the function.

a. Because $g(x) = \log_{10}(x - 1) = f(x - 1)$, the graph of g can be obtained by shifting the graph of f one unit to the right. The vertical asymptote of the graph of g is $x = 1$. The domain of g is $(1, \infty)$.

b. Because $h(x) = 2 + \log_{10} x = 2 + f(x)$, the graph of h can be obtained by shifting the graph of f two units upward. The vertical asymptote of the graph of h is $x = 0$. The domain of h is $(0, \infty)$.

c. Because $k(x) = -\log_{10} x = -f(x)$, the graph of k can be obtained by reflecting the graph of f in the x-axis. The vertical asymptote of the graph of k is $x = 0$. The domain of k is $(0, \infty)$.

d. Because $j(x) = \log_{10}(-x) = f(-x)$, the graph of j can be obtained by reflecting the graph of f in the y-axis. The vertical asymptote of the graph of j is $x = 0$. The domain of j is $(-\infty, 0)$.

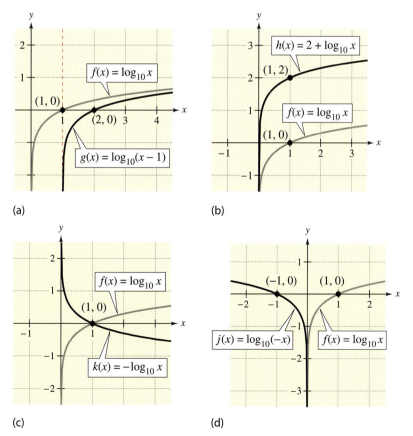

(a)   (b)

(c)   (d)

Figure 9.19

③ Graph and evaluate natural logarithmic functions.

The Natural Logarithmic Function

As with exponential functions, the most widely used base for logarithmic functions is the number e. The logarithmic function with base e is the **natural logarithmic function** and is denoted by the special symbol ln x, which is read as "el en of x."

The Natural Logarithmic Function

The function defined by

$$f(x) = \log_e x = \ln x$$

where $x > 0$, is called the **natural logarithmic function.**

Graph of $g(x) = \ln x$
- Domain: $(0, \infty)$
- Range: $(-\infty, \infty)$
- Intercept: $(1, 0)$
- Increasing (moves up to the right)
- Asymptote: y-axis

The definition above implies that the natural logarithmic function and the natural exponential function are inverse functions of each other. So, every logarithmic equation can be written in an equivalent exponential form and every exponential equation can be written in logarithmic form.

Because the functions $f(x) = e^x$ and $g(x) = \ln x$ are inverse functions of each other, their graphs are reflections of each other in the line $y = x$. This reflective property is illustrated in Figure 9.20. The figure also contains a summary of several characteristics of the graph of the natural logarithmic function.

Notice that the domain of the natural logarithmic function, as with every other logarithmic function, is the set of *positive real numbers*—be sure you see that ln x is not defined for zero or for negative numbers.

The three properties of logarithms listed earlier in this section are also valid for natural logarithms.

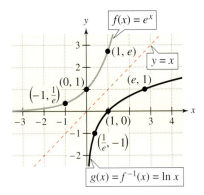

Figure 9.20 Characteristics of the natural logarithmic function $g(x) = \ln x$

Properties of Natural Logarithms

Let x be a positive real number. Then the following properties are true.

1. $\ln 1 = 0$ because $e^0 = 1$.

2. $\ln e = 1$ because $e^1 = e$.

3. $\ln e^x = x$ because $e^x = e^x$.

Technology: Tip

On most calculators, the natural logarithm key is denoted by ⌊LN⌋. For instance, on a scientific calculator, you can evaluate ln 2 as 2 ⌊LN⌋ and on a graphing calculator, you can evaluate it as ⌊LN⌋ 2 ⌊ENTER⌋. In either case, you should obtain a display of 0.6931472.

Example 6 **Evaluating Natural Logarithmic Functions**

Evaluate each expression.

a. $\ln e^2$ **b.** $\ln \dfrac{1}{e}$

Solution

Using the property that $\ln e^x = x$, you obtain the following.

a. $\ln e^2 = 2$ **b.** $\ln \dfrac{1}{e} = \ln e^{-1} = -1$

④ Use the change-of-base formula to evaluate logarithms.

Change of Base

Although 10 and e are the most frequently used bases, you occasionally need to evaluate logarithms with other bases. In such cases the following **change-of-base formula** is useful.

Change-of-Base Formula

Let a, b, and x be positive real numbers such that $a \neq 1$ and $b \neq 1$. Then $\log_a x$ is given as follows.

$$\log_a x = \frac{\log_b x}{\log_b a} \qquad \text{or} \qquad \log_a x = \frac{\ln x}{\ln a}$$

The usefulness of this change-of-base formula is that you can use a calculator that has only the common logarithm key [LOG] and the natural logarithm key [LN] to evaluate logarithms to any base.

Example 7   Changing Bases to Evaluate Logarithms

a. Use *common* logarithms to evaluate $\log_3 5$.

b. Use *natural* logarithms to evaluate $\log_6 2$.

Solution

Using the change-of-base formula, you can convert to common and natural logarithms by writing

$$\log_3 5 = \frac{\log_{10} 5}{\log_{10} 3} \qquad \text{and} \qquad \log_6 2 = \frac{\ln 2}{\ln 6}.$$

Now, use the following keystrokes.

a.

Keystrokes	Display	
5 [LOG] [÷] 3 [LOG] [=]	1.4649735	*Scientific*
[LOG] 5 [ )] [÷] [LOG] 3 [ )] [ENTER]	1.4649735	*Graphing*

So, $\log_3 5 \approx 1.465$.

b.

Keystrokes	Display	
2 [LN] [÷] 6 [LN] [=]	0.3868528	*Scientific*
[LN] 2 [ )] [÷] [LN] 6 [ )] [ENTER]	0.3868528	*Graphing*

So, $\log_6 2 \approx 0.387$.

At this point, you have been introduced to all the basic types of functions that are covered in this course: polynomial functions, radical functions, rational functions, exponential functions, and logarithmic functions. The only other common types of functions are *trigonometric functions*, which you will study if you go on to take a course in trigonometry or precalculus.

Technology: Tip

You can use a graphing calculator to graph logarithmic functions that do not have a base of 10 by using the change-of-base formula. Use the change-of-base formula to rewrite $g(x) = \log_2 x$ in Example 4 on page 604 (with $b = 10$) and graph the function. You should obtain a graph similar to the one below.

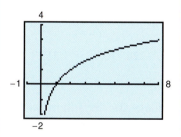

Study Tip

In Example 7(a), $\log_3 5$ could have been evaluated using natural logarithms in the change-of-base formula.

$$\log_3 5 = \frac{\ln 5}{\ln 3} \approx 1.465$$

Notice that you get the same answer whether you use natural logarithms or common logarithms in the change-of-base formula.

9.3 Exercises

Developing Skills

In Exercises 1–12, write the logarithmic equation in exponential form.

1. $\log_7 49 = 2$
2. $\log_{11} 121 = 2$

3. $\log_2 \frac{1}{32} = -5$
4. $\log_3 \frac{1}{27} = -3$

5. $\log_3 \frac{1}{243} = -5$
6. $\log_{10} 10{,}000 = 4$

7. $\log_{36} 6 = \frac{1}{2}$
8. $\log_{32} 4 = \frac{2}{5}$

9. $\log_8 4 = \frac{2}{3}$
10. $\log_{16} 8 = \frac{3}{4}$

11. $\log_2 2.462 \approx 1.3$
12. $\log_3 1.179 \approx 0.15$

In Exercises 13–24, write the exponential equation in logarithmic form.

13. $6^2 = 36$
14. $3^5 = 243$

15. $4^{-2} = \frac{1}{16}$
16. $6^{-4} = \frac{1}{1296}$

17. $8^{2/3} = 4$
18. $81^{3/4} = 27$

19. $25^{-1/2} = \frac{1}{5}$
20. $6^{-3} = \frac{1}{216}$

21. $4^0 = 1$
22. $6^1 = 6$

23. $5^{1.4} \approx 9.518$
24. $10^{0.12} \approx 1.318$

In Exercises 25–46, evaluate the logarithm without using a calculator. (If not possible, state the reason.) See Examples 1 and 2.

25. $\log_2 8$
26. $\log_3 27$
27. $\log_{10} 1000$
28. $\log_{10} 0.00001$
29. $\log_2 \frac{1}{4}$
30. $\log_3 \frac{1}{9}$
31. $\log_4 \frac{1}{64}$
32. $\log_5 \frac{1}{125}$
33. $\log_{10} \frac{1}{10{,}000}$
34. $\log_{10} \frac{1}{100}$
35. $\log_2(-3)$

36. $\log_4(-4)$

37. $\log_4 1$
38. $\log_3 1$
39. $\log_5(-6)$

40. $\log_2 0$

41. $\log_9 3$
42. $\log_{25} 125$
43. $\log_{16} 8$
44. $\log_{144} 12$
45. $\log_7 7^4$
46. $\log_5 5^3$

In Exercises 47–52, use a calculator to evaluate the common logarithm. (Round your answer to four decimal places.) See Example 3.

47. $\log_{10} 42$

48. $\log_{10} 6281$

49. $\log_{10} 0.023$

50. $\log_{10} 0.149$

51. $\log_{10}\left(\sqrt{2} + 4\right)$

52. $\log_{10} \dfrac{\sqrt{3}}{2}$

In Exercises 53–56, match the function with its graph. [The graphs are labeled (a), (b), (c), and (d).]

(a)
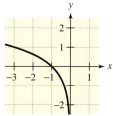

(b)
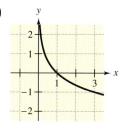

(c)
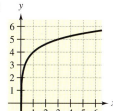

(d)
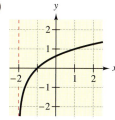

53. $f(x) = 4 + \log_3 x$

54. $f(x) = -\log_3 x$

55. $f(x) = \log_3(-x)$

56. $f(x) = \log_3(x + 2)$

In Exercises 57–60, sketch the graphs of f and g on the same set of coordinate axes. What can you conclude about the relationship between f and g? See Example 4.

57. $f(x) = \log_3 x$
$g(x) = 3^x$

58. $f(x) = \log_4 x$
$g(x) = 4^x$

59. $f(x) = \log_6 x$
$g(x) = 6^x$

60. $f(x) = \log_{1/2} x$
$g(x) = \left(\tfrac{1}{2}\right)^x$

In Exercises 61–66, identify the transformation of the graph of $f(x) = \log_2 x$ and sketch the graph of h. See Example 5.

61. $h(x) = 3 + \log_2 x$

62. $h(x) = -4 + \log_2 x$

63. $h(x) = \log_2(x - 2)$

64. $h(x) = \log_2(x + 4)$

65. $h(x) = \log_2(-x)$

66. $h(x) = -\log_2(x)$

In Exercises 67–76, sketch the graph of the function. Identify the vertical asymptote.

67. $f(x) = \log_5 x$

68. $g(x) = \log_8 x$

69. $g(t) = -\log_2 t$

70. $h(s) = -2 \log_3 s$

71. $f(x) = 3 + \log_2 x$

72. $f(x) = -2 + \log_3 x$

73. $g(x) = \log_2(x - 3)$

74. $h(x) = \log_3(x + 1)$

75. $f(x) = \log_{10}(10x)$

76. $g(x) = \log_4(4x)$

In Exercises 77–82, find the domain and vertical asymptote of the function. Sketch its graph.

77. $f(x) = \log_4 x$

78. $g(x) = \log_6 x$

79. $h(x) = \log_4(x - 3)$

80. $f(x) = -\log_6(x + 2)$

81. $y = -\log_3 x + 2$

82. $y = \log_5(x - 1) + 4$

 In Exercises 83–88, use a graphing calculator to graph the function. Determine the domain and the vertical asymptote.

83. $y = 5 \log_{10} x$

84. $y = 5 \log_{10}(x - 3)$

85. $y = -3 + 5 \log_{10} x$

86. $y = 5 \log_{10}(3x)$

87. $y = \log_{10}\left(\dfrac{x}{5}\right)$

88. $y = \log_{10}(-x)$

In Exercises 89–94, use a calculator to evaluate the natural logarithm. (Round your answer to four decimal places.) See Example 6.

89. $\ln 38$

90. $\ln 14.2$

91. $\ln 0.15$

92. $\ln 0.002$

93. $\ln\left(\dfrac{1 + \sqrt{5}}{3}\right)$

94. $\ln\left(1 + \dfrac{0.10}{12}\right)$

In Exercises 95–98, match the function with its graph. [The graphs are labeled (a), (b), (c), and (d).]

(a)
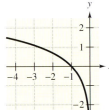

(b)
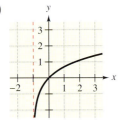

(c)
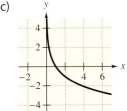

(d)
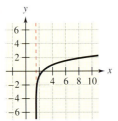

95. $f(x) = \ln(x + 1)$

96. $f(x) = \ln(-x)$

97. $f(x) = \ln\left(x - \dfrac{3}{2}\right)$

98. $f(x) = -\dfrac{3}{2} \ln x$

In Exercises 99–106, sketch the graph of the function. Identify the vertical asymptote.

99. $f(x) = -\ln x$

100. $f(x) = -2 \ln x$

101. $f(x) = 3 \ln x$

102. $h(t) = 4 \ln t$

103. $f(x) = 1 + \ln x$

104. $h(x) = 2 + \ln x$

105. $g(t) = 2 \ln(t - 4)$

106. $g(x) = -3 \ln(x + 3)$

 In Exercises 107–110, use a graphing calculator to graph the function. Determine the domain and the vertical asymptote.

107. $g(x) = -\ln(x + 1)$

108. $h(x) = \ln(x + 5)$

109. $f(t) = 7 + 3 \ln t$

110. $g(t) = \ln(5 - t)$

In Exercises 111–124, use a calculator to evaluate the logarithm by means of the change-of-base formula. Use (a) the common logarithm key and (b) the natural logarithm key. (Round your answer to four decimal places.) See Example 7.

111. $\log_9 36$

112. $\log_7 411$

113. $\log_4 6$

114. $\log_6 9$

115. $\log_2 0.72$

116. $\log_{12} 0.6$

117. $\log_{15} 1250$

118. $\log_{20} 125$

119. $\log_{1/2} 4$

120. $\log_{1/3} 18$

121. $\log_4 \sqrt{42}$

122. $\log_3 \sqrt{26}$

123. $\log_2(1 + e)$

124. $\log_4(2 + e^3)$

Solving Problems

125. *American Elk*   The antler spread a (in inches) and shoulder height h (in inches) of an adult male American elk are related by the model

$$h = 116 \log_{10}(a + 40) - 176.$$

Approximate the shoulder height of a male American elk with an antler spread of 55 inches.

126. *Sound Intensity*   The relationship between the number of decibels B and the intensity of a sound I in watts per centimeter squared is given by

$$B = 10 \log_{10}\left(\dfrac{I}{10^{-16}}\right).$$

Determine the number of decibels of a sound with an intensity of 10^{-4} watts per centimeter squared.

127. *Compound Interest*   The time t in years for an investment to double in value when compounded continuously at interest rate r is given by

$$t = \frac{\ln 2}{r}.$$

Complete the table, which shows the "doubling times" for several annual percent rates.

r	0.07	0.08	0.09	0.10	0.11	0.12
t						

128. *Meteorology*   Most tornadoes last less than 1 hour and travel about 20 miles. The speed of the wind S (in miles per hour) near the center of the tornado and the distance d (in miles) the tornado travels are related by the model $S = 93 \log_{10} d + 65$. On March 18, 1925, a large tornado struck portions of Missouri, Illinois, and Indiana, covering a distance of 220 miles. Approximate the speed of the wind near the center of this tornado.

129. *Tractrix*   A person walking along a dock (the y-axis) drags a boat by a 10-foot rope (see figure). The boat travels along a path known as a *tractrix*. The equation of the path is

$$y = 10 \ln\left(\frac{10 + \sqrt{100 - x^2}}{x}\right) - \sqrt{100 - x^2}.$$

(a) ▦ Use a graphing calculator to graph the function. What is the domain of the function?

(b) ▦ Identify any asymptotes.

(c) Determine the position of the person when the x-coordinate of the position of the boat is $x = 2$.

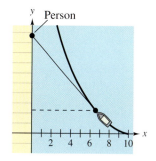

130. ▦ *Home Mortgage*   The model

$$t = 10.042 \ln\left(\frac{x}{x - 1250}\right), \quad x > 1250$$

approximates the length t (in years) of a home mortgage of \$150,000 at 10% interest in terms of the monthly payment x.

(a) Use a graphing calculator to graph the model. Describe the change in the length of the mortgage as the monthly payment increases.

(b) Use the graph in part (a) to approximate the length of the mortgage when the monthly payment is \$1316.35.

(c) Use the result of part (b) to find the total amount paid over the term of the mortgage. What amount of the total is interest costs?

Explaining Concepts

131. Write "logarithm of x with base 5" symbolically.

132. *Writing* ✎   Explain the relationship between the functions $f(x) = 2^x$ and $g(x) = \log_2 x$.

133. *Writing* ✎   Explain why $\log_a a = 1$.

134. *Writing* ✎   Explain why $\log_a a^x = x$.

135. *Writing* ✎   What are common logarithms and natural logarithms?

136. *Writing* ✎   Describe how to use a calculator to find the logarithm of a number if the base is not 10 or e.

Think About It   In Exercises 137–142, answer the question for the function $f(x) = \log_{10} x$. (Do not use a calculator.)

137. What is the domain of f?

138. Find the inverse function of f.

139. Describe the values of $f(x)$ for $1000 \le x \le 10{,}000$.

140. Describe the values of x, given that $f(x)$ is negative.

141. By what amount will x increase, given that $f(x)$ is increased by 1 unit?

142. Find the ratio of a to b when $f(a) = 3 + f(b)$.

Mid-Chapter Quiz

Take this quiz as you would take a quiz in class. After you are done, check your work against the answers in the back of the book.

1. Given $f(x) = \left(\frac{4}{3}\right)^x$, find (a) $f(2)$, (b) $f(0)$, (c) $f(-1)$, and (d) $f(1.5)$.

2. Find the domain and range of $g(x) = 2^{-0.5x}$.

In Exercises 3–6, sketch the graph of the function. Identify the horizontal asymptote. Use a graphing calculator for Exercises 5 and 6.

3. $y = \frac{1}{2}(4^x)$

4. $y = 5(2^{-x})$

5. ▦ $f(t) = 12e^{-0.4t}$

6. ▦ $g(x) = 100(1.08)^x$

7. Given $f(x) = 2x - 3$ and $g(x) = x^3$, find the indicated composition.
   (a) $(f \circ g)(x)$ (b) $(g \circ f)(x)$ (c) $(f \circ g)(-2)$ (d) $(g \circ f)(4)$

8. Verify algebraically and graphically that $f(x) = 3 - 5x$ and $g(x) = \frac{1}{5}(3 - x)$ are inverse functions of each other.

In Exercises 9 and 10, find the inverse function.

9. $h(x) = 10x + 3$

10. $g(t) = \frac{1}{2}t^3 + 2$

11. Write the logarithmic equation $\log_9 \frac{1}{81} = -2$ in exponential form.

12. Write the exponential equation $3^4 = 81$ in logarithmic form.

13. Evaluate $\log_5 125$ without a calculator.

▦ **In Exercises 14 and 15, use a graphing calculator to graph the function. Identify the vertical asymptote.**

14. $f(t) = -2 \ln(t + 3)$

15. $h(x) = 5 + \frac{1}{2} \ln x$

16. Use the graph of f shown at the left to determine h and k if $f(x) = \log_5(x - h) + k$.

17. Use a calculator and the change-of-base formula to evaluate $\log_3 782$.

18. You deposit \$750 in an account at an annual interest rate of $7\frac{1}{2}\%$. Complete the table showing the balance A in the account after 20 years for several types of compounding.

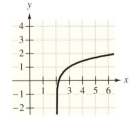

Figure for 16

n	1	4	12	365	Continuous compounding
A					

19. After t years, the remaining mass y (in grams) of 14 grams of a radioactive element whose half-life is 40 years is given by $y = 14\left(\frac{1}{2}\right)^{t/40}$, $t \geq 0$. How much of the initial mass remains after 125 years?

9.4 Properties of Logarithms

Charles Gupton/Corbis

What You Should Learn

1. Use the properties of logarithms to evaluate logarithms.
2. Use the properties of logarithms to rewrite, expand, or condense logarithmic expressions.
3. Use the properties of logarithms to solve application problems.

Why You Should Learn It

Logarithmic equations are often used to model scientific observations. For instance, in Example 8 on page 617, a logarithmic equation is used to model human memory.

1 Use the properties of logarithms to evaluate logarithms.

Properties of Logarithms

You know from the preceding section that the logarithmic function with base a is the *inverse function* of the exponential function with base a. So, it makes sense that each property of exponents should have a corresponding property of logarithms. For instance, the exponential property

$$a^0 = 1 \qquad \text{Exponential property}$$

has the corresponding logarithmic property

$$\log_a 1 = 0. \qquad \text{Corresponding logarithmic property}$$

In this section you will study the logarithmic properties that correspond to the following three exponential properties:

	Base a	*Natural Base*
1.	$a^m a^n = a^{m+n}$	$e^m e^n = e^{m+n}$
2.	$\dfrac{a^m}{a^n} = a^{m-n}$	$\dfrac{e^m}{e^n} = e^{m-n}$
3.	$(a^m)^n = a^{mn}$	$(e^m)^n = e^{mn}$

Properties of Logarithms

Let a be a positive real number such that $a \neq 1$, and let n be a real number. If u and v are real numbers, variables, or algebraic expressions such that $u > 0$ and $v > 0$, the following properties are true.

	Logarithm with Base a	*Natural Logarithm*
1. Product Property:	$\log_a(uv) = \log_a u + \log_a v$	$\ln(uv) = \ln u + \ln v$
2. Quotient Property:	$\log_a \dfrac{u}{v} = \log_a u - \log_a v$	$\ln \dfrac{u}{v} = \ln u - \ln v$
3. Power Property:	$\log_a u^n = n \log_a u$	$\ln u^n = n \ln u$

There is no general property of logarithms that can be used to simplify $\log_a(u + v)$. Specifically,

$$\log_a(u + v) \text{ does not equal } \log_a u + \log_a v.$$

Example 1 Using Properties of Logarithms

Use $\ln 2 \approx 0.693$, $\ln 3 \approx 1.099$, and $\ln 5 \approx 1.609$ to approximate each expression.

a. $\ln \dfrac{2}{3}$ **b.** $\ln 10$ **c.** $\ln 30$

Solution

a. $\ln \dfrac{2}{3} = \ln 2 - \ln 3$ Quotient Property

$\approx 0.693 - 1.099 = -0.406$ Substitute for ln 2 and ln 3.

b. $\ln 10 = \ln(2 \cdot 5)$ Factor.

$= \ln 2 + \ln 5$ Product Property

$\approx 0.693 + 1.609$ Substitute for ln 2 and ln 5.

$= 2.302$ Simplify.

c. $\ln 30 = \ln(2 \cdot 3 \cdot 5)$ Factor.

$= \ln 2 + \ln 3 + \ln 5$ Product Property

$\approx 0.693 + 1.099 + 1.609$ Substitute for ln 2, ln 3, and ln 5.

$= 3.401$ Simplify.

When using the properties of logarithms, it helps to state the properties *verbally*. For instance, the verbal form of the Product Property

$$\ln(uv) = \ln u + \ln v$$

is: *The log of a product is the sum of the logs of the factors.*
Similarly, the verbal form of the Quotient Property

$$\ln \frac{u}{v} = \ln u - \ln v$$

is: *The log of a quotient is the difference of the logs of the numerator and denominator.*

Example 2 Using Properties of Logarithms

Use the properties of logarithms to verify that $-\ln 2 = \ln \frac{1}{2}$.

Solution

Using the Power Property, you can write the following.

$-\ln 2 = (-1) \ln 2$ Rewrite coefficient as -1.

$= \ln 2^{-1}$ Power Property

$= \ln \dfrac{1}{2}$ Rewrite 2^{-1} as $\frac{1}{2}$.

② Use the properties of logarithms to rewrite, expand, or condense logarithmic expressions.

Rewriting Logarithmic Expressions

In Examples 1 and 2, the properties of logarithms were used to rewrite logarithmic expressions involving the log of a *constant*. A more common use of these properties is to rewrite the log of a *variable expression*.

Example 3   Rewriting Logarithmic Expressions

Use the properties of logarithms to rewrite each expression.

a. $\log_{10} 7x^3 = \log_{10} 7 + \log_{10} x^3$ Product Property

$\qquad\qquad = \log_{10} 7 + 3 \log_{10} x$ Power Property

b. $\ln \dfrac{8x^3}{y} = \ln 8x^3 - \ln y$ Quotient Property

$\qquad\qquad = \ln 8 + \ln x^3 - \ln y$ Product Property

$\qquad\qquad = \ln 8 + 3 \ln x - \ln y$ Power Property

When you rewrite a logarithmic expression as in Example 3, you are **expanding** the expression. The reverse procedure is demonstrated in Example 4, and is called **condensing** a logarithmic expression.

Example 4   Condensing Logarithmic Expressions

Use the properties of logarithms to condense each expression.

a. $\ln x - \ln 3$ **b.** $2 \log_3 x + \log_3 5$

Solution

a. $\ln x - \ln 3 = \ln \dfrac{x}{3}$ Quotient Property

b. $2 \log_3 x + \log_3 5 = \log_3 x^2 + \log_3 5$ Power Property

$\qquad\qquad\qquad = \log_3 5x^2$ Product Property

Technology: Tip

When you are rewriting a logarithmic expression, remember that you can use a graphing calculator to check your result graphically. For instance, in Example 4(a), try graphing the functions.

$$y_1 = \ln x - \ln 3 \text{ and } y_2 = \ln \dfrac{x}{3}$$

in the same viewing window. You should obtain the same graph for each function.

Example 5   **Expanding Logarithmic Expressions**

Use the properties of logarithms to expand each expression.

a. $\log_6 3xy^2, x > 0, y > 0$ **b.** $\ln \dfrac{\sqrt{3x - 5}}{7}$

Solution

a. $\log_6 3xy^2 = \log_6 3 + \log_6 x + \log_6 y^2$ Product Property

      $= \log_6 3 + \log_6 x + 2 \log_6 y$ Power Property

b. $\ln \dfrac{\sqrt{3x - 5}}{7} = \ln \left[\dfrac{(3x - 5)^{1/2}}{7} \right]$ Rewrite using rational exponent.

      $= \ln(3x - 5)^{1/2} - \ln 7$ Quotient Property

      $= \dfrac{1}{2} \ln(3x - 5) - \ln 7$ Power Property

Sometimes expanding or condensing logarithmic expressions involves several steps. In the next example, be sure that you can justify each step in the solution. Notice how different the expanded expression is from the original.

Example 6   **Expanding a Logarithmic Expression**

Use the properties of logarithms to expand $\ln \sqrt{x^2 - 1}, \quad x > 1$.

Solution

  $\ln \sqrt{x^2 - 1} = \ln(x^2 - 1)^{1/2}$ Rewrite using rational exponent.

     $= \dfrac{1}{2} \ln(x^2 - 1)$ Power Property

     $= \dfrac{1}{2} \ln[(x - 1)(x + 1)]$ Factor.

     $= \dfrac{1}{2}[\ln(x - 1) + \ln(x + 1)]$ Product Property

     $= \dfrac{1}{2} \ln(x - 1) + \dfrac{1}{2} \ln(x + 1)$ Distributive Property

Example 7   **Condensing Logarithmic Expressions**

Use the properties of logarithms to condense each expression.

a. $\ln 2 - 2 \ln x = \ln 2 - \ln x^2, \quad x > 0$ Power Property

      $= \ln \dfrac{2}{x^2}, \quad x > 0$ Quotient Property

b. $3(\ln 4 + \ln x) = 3(\ln 4x)$ Product Property

      $= \ln(4x)^3$ Power Property

      $= \ln 64x^3, \quad x \geq 0$ Simplify.

When you expand or condense a logarithmic expression, it is possible to change the domain of the expression. For instance, the domain of the function

$$f(x) = 2 \ln x \qquad \text{Domain is the set of positive real numbers.}$$

is the set of positive real numbers, whereas the domain of

$$g(x) = \ln x^2 \qquad \text{Domain is the set of nonzero real numbers.}$$

is the set of nonzero real numbers. So, when you expand or condense a logarithmic expression, you should check to see whether the rewriting has changed the domain of the expression. In such cases, you should restrict the domain appropriately. For instance, you can write

$$f(x) = 2 \ln x$$
$$= \ln x^2, \ x > 0.$$

3 Use the properties of logarithms to solve application problems.

Application

Example 8   Human Memory Model

In an experiment, students attended several lectures on a subject. Every month for a year after that, the students were tested to see how much of the material they remembered. The average scores for the group are given by the **human memory model**

$$f(t) = 80 - \ln(t + 1)^9, \quad 0 \le t \le 12$$

where t is the time in months. Find the average scores for the group after 2 months and 8 months.

Solution

To make the calculations easier, rewrite the model using the Power Property, as follows.

$$f(t) = 80 - 9 \ln(t + 1), \quad 0 \le t \le 12$$

After 2 months, the average score was

$$f(2) = 80 - 9 \ln(2 + 1) \qquad \text{Substitute 2 for } t.$$
$$\approx 80 - 9.9 \qquad \text{Simplify.}$$
$$= 70.1 \qquad \text{Average score after 2 months}$$

and after 8 months, the average score was

$$f(8) = 80 - 9 \ln(8 + 1) \qquad \text{Substitute 8 for } t.$$
$$\approx 80 - 19.8 \qquad \text{Simplify.}$$
$$= 60.2. \qquad \text{Average score after 8 months}$$

The graph of the function is shown in Figure 9.21.

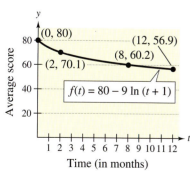

Human Memory Model
Figure 9.21

9.4 Exercises

Review Concepts, Skills, and Problem Solving

Keep mathematically in shape by doing these exercises *before* the problems of this section.

Properties and Definitions

In Exercises 1 and 2, use the rule for radicals to fill in the blank.

1. Product Rule: $\sqrt[n]{u}\,\sqrt[n]{v} = $ ☐

2. Quotient Rule: $\dfrac{\sqrt[n]{u}}{\sqrt[n]{v}} = $ ☐

3. *Writing* Explain why the radicals $\sqrt{2x}$ and $\sqrt[3]{2x}$ cannot be added.

4. *Writing* Is $1/\sqrt{2x}$ in simplest form? Explain.

Simplifying Expressions

In Exercises 5–10, perform the indicated operations and simplify. (Assume all variables are positive.)

5. $25\sqrt{3x} - 3\sqrt{12x}$

6. $\left(\sqrt{x} + 3\right)\left(\sqrt{x} - 3\right)$

7. $\sqrt{u}\left(\sqrt{20} - \sqrt{5}\right)$

8. $\left(2\sqrt{t} + 3\right)^2$

9. $\dfrac{50x}{\sqrt{2}}$

10. $\dfrac{12}{\sqrt{t+2} + \sqrt{t}}$

Problem Solving

11. *Demand* The demand equation for a product is given by $p = 30 - \sqrt{0.5(x - 1)}$, where x is the number of units demanded per day and p is the price per unit. Find the demand when the price is set at $26.76.

12. *List Price* The sale price of a computer is $1955. The discount is 15% of the list price. Find the list price.

Developing Skills

In Exercises 1–24, use properties of logarithms to evaluate the expression without a calculator. (If not possible, state the reason.)

1. $\log_{12} 12^3$

2. $\log_5 125$

3. $\log_4\left(\frac{1}{16}\right)^2$

4. $\log_7\left(\frac{1}{49}\right)^3$

5. $\log_5 \sqrt[3]{5}$

6. $\ln \sqrt{e}$

7. $\ln 14^0$

8. $\ln\left(\dfrac{7.14}{7.14}\right)$

9. $\ln e^{-6}$

10. $\ln e^7$

11. $\log_4 8 + \log_4 2$

12. $\log_6 2 + \log_6 3$

13. $\log_8 4 + \log_8 16$

14. $\log_{10} 5 + \log_{10} 20$

15. $\log_4 8 - \log_4 2$

16. $\log_5 50 - \log_5 2$

17. $\log_6 72 - \log_6 2$

18. $\log_3 324 - \log_3 4$

19. $\log_2 5 - \log_2 40$

20. $\log_3\left(\frac{2}{3}\right) + \log_3\left(\frac{1}{2}\right)$

21. $\ln e^8 + \ln e^4$

22. $\ln e^5 - \ln e^2$

23. $\ln \dfrac{e^3}{e^2}$

24. $\ln(e^2 \cdot e^4)$

In Exercises 25–36, use $\log_4 2 = 0.5000$, $\log_4 3 \approx 0.7925$, and the properties of logarithms to approximate the expression. Do not use a calculator. See Example 1.

25. $\log_4 4$

26. $\log_4 8$

27. $\log_4 6$

28. $\log_4 24$

29. $\log_4 \frac{3}{2}$

30. $\log_4 \frac{9}{2}$

31. $\log_4 \sqrt{2}$

32. $\log_4 \sqrt[3]{9}$

33. $\log_4(3 \cdot 2^4)$

34. $\log_4 \sqrt{3 \cdot 2^5}$

35. $\log_4 3^0$

36. $\log_4 4^3$

In Exercises 37–42, use $\ln 3 \approx 1.0986$, $\ln 12 \approx 2.4849$, and the properties of logarithms to approximate the expression. Use a calculator to verify your result.

37. $\ln 9$

38. $\ln \frac{1}{4}$

39. $\ln 36$

40. $\ln 144$

41. $\ln \sqrt{36}$

42. $\ln 5^0$

In Exercises 43–76, use the properties of logarithms to expand the expression. See Examples 3, 5, and 6.

43. $\log_3 11x$

44. $\log_2 3x$

45. $\log_7 x^2$

46. $\log_3 x^3$

47. $\log_5 x^{-2}$

48. $\log_2 s^{-4}$

49. $\log_4 \sqrt{3x}$

50. $\log_3 \sqrt[3]{5y}$

51. $\ln 3y$

52. $\ln 5x$

53. $\log_2 \frac{z}{17}$

54. $\log_{10} \frac{7}{y}$

55. $\log_9 \frac{\sqrt{x}}{12}$

56. $\ln \frac{\sqrt{x}}{x-9}$

57. $\ln x^2(y-2)$

58. $\ln y(y-1)^2$

59. $\log_4[x^6(x-7)^2]$

60. $\log_8[(x-y)^4 z^6]$

61. $\log_3 \sqrt[3]{x+1}$

62. $\log_5 \sqrt{xy}$

63. $\ln \sqrt{x(x+2)}$

64. $\ln \sqrt[3]{x(x+5)}$

65. $\ln\left(\frac{x+1}{x-1}\right)^2$

66. $\log_2\left(\frac{x^2}{x-3}\right)^3$

67. $\ln \sqrt[3]{\frac{x^2}{x+1}}$

68. $\ln \sqrt{\frac{3x}{x-5}}$

69. $\ln \frac{xy^2}{z^3}$

70. $\log_5 \frac{x^2 y^5}{z^7}$

71. $\log_3 \sqrt{\frac{x^7}{y^5 z^8}}$

72. $\ln \sqrt[3]{\frac{x^4}{y^3 z^2}}$

73. $\log_6\left[a\sqrt{b}(c-d)^3\right]$

74. $\ln[(xy)^2(x+3)^4]$

75. $\ln\left[(x+y)\frac{\sqrt[5]{w+2}}{3t}\right]$

76. $\ln\left[(u-v)\frac{\sqrt[3]{u-4}}{3v}\right]$

In Exercises 77–108, use the properties of logarithms to condense the expression. See Examples 4 and 7.

77. $\log_{12} x - \log_{12} 3$

78. $\log_6 12 - \log_6 y$

79. $\log_2 3 + \log_2 x$

80. $\log_5 2x + \log_5 3y$

81. $\log_{10} 4 - \log_{10} x$

82. $\ln 10x - \ln z$

83. $4 \ln b$

84. $10 \log_4 z$

85. $-2 \log_5 2x$

86. $-5 \ln(x + 3)$

87. $7 \log_2 x + 3 \log_2 z$

88. $2 \log_{10} x + \frac{1}{2} \log_{10} y$

89. $\log_3 2 + \frac{1}{2} \log_3 y$

90. $\ln 6 - 3 \ln z$

91. $2 \ln x + 3 \ln y - \ln z$

92. $4 \ln 3 - 2 \ln x - \ln y$

93. $5 \ln 2 - \ln x + 3 \ln y$

94. $4 \ln 2 + 2 \ln x - \frac{1}{2} \ln y$

95. $4(\ln x + \ln y)$

96. $\frac{1}{2}(\ln 8 + \ln 2x)$

97. $2[\ln x - \ln(x + 1)]$

98. $5\left[\ln x - \frac{1}{2} \ln(x + 4)\right]$

99. $\log_4(x + 8) - 3 \log_4 x$

100. $5 \log_3 x + \log_3(x - 6)$

101. $\frac{1}{2} \log_5(x + 2) - \log_5(x - 3)$

102. $\frac{1}{4} \log_6(x + 1) - 5 \log_6(x - 4)$

103. $5 \log_6(c + d) - \frac{1}{2} \log_6(m - n)$

104. $2 \log_5(x + y) + 3 \log_5 w$

105. $\frac{1}{5}(3 \log_2 x - 4 \log_2 y)$

106. $\frac{1}{3}[\ln(x - 6) - 4 \ln y - 2 \ln z]$

107. $\frac{1}{5} \log_6(x - 3) - 2 \log_6 x - 3 \log_6(x + 1)$

108. $3\left[\frac{1}{2} \log_9(a + 6) - 2 \log_9(a - 1)\right]$

In Exercises 109–114, simplify the expression.

109. $\ln 3e^2$

110. $\log_3(3^2 \cdot 4)$

111. $\log_5 \sqrt{50}$

112. $\log_2 \sqrt{22}$

113. $\log_4 \dfrac{4}{x^2}$

114. $\ln \dfrac{6}{e^5}$

⌨ In Exercises 115–118, use a graphing calculator to graph the two equations in the same viewing window. Use the graphs to verify that the expressions are equivalent. Assume $x > 0$.

115. $y_1 = \ln\left(\dfrac{10}{x^2 + 1}\right)^2$

  $y_2 = 2[\ln 10 - \ln(x^2 + 1)]$

116. $y_1 = \ln \sqrt{x(x + 1)}$

  $y_2 = \frac{1}{2}[\ln x + \ln(x + 1)]$

117. $y_1 = \ln[x^2(x + 2)]$

  $y_2 = 2 \ln x + \ln(x + 2)$

118. $y_1 = \ln\left(\dfrac{\sqrt{x}}{x - 3}\right)$

  $y_2 = \frac{1}{2} \ln x - \ln(x - 3)$

Solving Problems

119. *Sound Intensity*   The relationship between the number of decibels B and the intensity of a sound I in watts per centimeter squared is given by

$$B = 10 \log_{10}\left(\dfrac{I}{10^{-16}}\right).$$

Use properties of logarithms to write the formula in simpler form, and determine the number of decibels of a sound with an intensity of 10^{-10} watts per centimeter squared.

120. *Human Memory Model*   Students participating in an experiment attended several lectures on a subject. Every month for a year after that, the students were tested to see how much of the material they remembered. The average scores for the group are given by the human memory model

$$f(t) = 80 - \log_{10}(t + 1)^{12}, \quad 0 \le t \le 12$$

where t is the time in months.

(a) Find the average scores for the group after 2 months and 8 months.

(b) 🖩 Use a graphing calculator to graph the function.

Molecular Transport   In Exercises 121 and 122, use the following information. The energy E (in kilocalories per gram molecule) required to transport a substance from the outside to the inside of a living cell is given by

$$E = 1.4(\log_{10} C_2 - \log_{10} C_1)$$

where C_1 and C_2 are the concentrations of the substance outside and inside the cell, respectively.

121. Condense the expression.

122. The concentration of a substance inside a cell is twice the concentration outside the cell. How much energy is required to transport the substance from outside to inside the cell?

Explaining Concepts

True or False?   In Exercises 123–130, use properties of logarithms to determine whether the equation is true or false. If it is false, state why or give an example to show that it is false.

123. $\ln e^{2-x} = 2 - x$

124. $\log_2 8x = 3 + \log_2 x$

125. $\log_8 4 + \log_8 16 = 2$

126. $\log_3(u + v) = \log_3 u + \log_3 v$

127. $\log_3(u + v) = \log_3 u \cdot \log_3 v$

128. $\dfrac{\log_6 10}{\log_6 3} = \log_6 10 - \log_6 3$

129. If $f(x) = \log_a x$, then $f(ax) = 1 + f(x)$.

130. If $f(x) = \log_a x$, then $f(a^n) = n$.

True or False?   In Exercises 131–136, determine whether the statement is true or false given that $f(x) = \ln x$. If it is false, state why or give an example to show that the statement is false.

131. $f(0) = 0$

132. $f(2x) = \ln 2 + \ln x$

133. $f(x - 3) = \ln x - \ln 3, \quad x > 3$

134. $\sqrt{f(x)} = \frac{1}{2} \ln x$

135. If $f(u) = 2f(v)$, then $v = u^2$.

136. If $f(x) > 0$, then $x > 1$.

137. *Error Analysis*   Describe the error.

$$\log_b\left(\frac{1}{x}\right) = \log_b\left(\frac{x}{xx}\right)$$
$$= \log_b x - \log_b x + \log_b x$$
$$= \log_b x$$

138. *Think About It*   Explain how you can show that

$$\frac{\ln x}{\ln y} \ne \ln \frac{x}{y}.$$

139. *Think About It*   Without a calculator, approximate the natural logarithms of as many integers as possible between 1 and 20 using $\ln 2 \approx 0.6931$, $\ln 3 \approx 1.0986$, $\ln 5 \approx 1.6094$, and $\ln 7 \approx 1.9459$. Explain the method you used. Then verify your results with a calculator and explain any differences in the results.

9.5 Solving Exponential and Logarithmic Equations

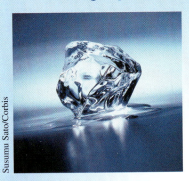

Susumu Sato/Corbis

What You Should Learn

1. Solve basic exponential and logarithmic equations.
2. Use inverse properties to solve exponential equations.
3. Use inverse properties to solve logarithmic equations.
4. Use exponential or logarithmic equations to solve application problems.

Why You Should Learn It

Exponential and logarithmic equations occur in many scientific applications. For instance, in Exercise 137 on page 631, you will use a logarithmic equation to determine how long it will take for ice cubes to form.

Exponential and Logarithmic Equations

In this section, you will study procedures for *solving equations* that involve exponential or logarithmic expressions. As a simple example, consider the exponential equation $2^x = 16$. By rewriting this equation in the form $2^x = 2^4$, you can see that the solution is $x = 4$. To solve this equation, you can use one of the following properties, which result from the fact that exponential and logarithmic functions are one-to-one functions.

1 Solve basic exponential and logarithmic equations.

One-to-One Properties of Exponential and Logarithmic Equations

Let a be a positive real number such that $a \neq 1$, and let x and y be real numbers. Then the following properties are true.

1. $a^x = a^y$ if and only if $x = y$.
2. $\log_a x = \log_a y$ if and only if $x = y$ $(x > 0, y > 0)$.

Example 1   Solving Exponential and Logarithmic Equations

Solve each equation.

a. $4^{x+2} = 64$ Original equation

     $4^{x+2} = 4^3$ Rewrite with like bases.

   $x + 2 = 3$ One-to-one property

       $x = 1$ Subtract 2 from each side.

The solution is $x = 1$. Check this in the original equation.

b. $\ln(2x - 3) = \ln 11$ Original equation

      $2x - 3 = 11$ One-to-one property

         $2x = 14$ Add 3 to each side.

          $x = 7$ Divide each side by 2.

The solution is $x = 7$. Check this in the original equation.

2 Use inverse properties to solve exponential equations.

Solving Exponential Equations

In Example 1(a), you were able to use a one-to-one property to solve the original equation because each side of the equation was written in exponential form with the same base. However, if only one side of the equation is written in exponential form or if both sides cannot be written with the same base, it is more difficult to solve the equation. For example, to solve the equation $2^x = 7$, you must find the power to which 2 can be raised to obtain 7. To do this, *rewrite the exponential equation in logarithmic form* by taking the logarithm of each side and use one of the following inverse properties of exponents and logarithms.

Solving Exponential Equations

To solve an exponential equation, first isolate the exponential expression, then **take the logarithm of each side of the equation** (or write the equation in logarithmic form) and solve for the variable.

Inverse Properties of Exponents and Logarithms

Base a	Natural Base e
1. $\log_a(a^x) = x$	$\ln(e^x) = x$
2. $a^{(\log_a x)} = x$	$e^{(\ln x)} = x$

Technology: Discovery

Use a graphing calculator to graph each side of each equation. What does this tell you about the inverse properties of exponents and logarithms?

1. (a) $\log_{10}(10^x) = x$
   (b) $10^{(\log_{10} x)} = x$
2. (a) $\ln(e^x) = x$
   (b) $e^{(\ln x)} = x$

Study Tip

Remember that to evaluate a logarithm such as $\log_2 7$ you need to use the change-of-base formula.

$$\log_2 7 = \frac{\ln 7}{\ln 2} \approx 2.807$$

Similarly,

$$\log_4 9 + 3 = \frac{\ln 9}{\ln 4} + 3$$
$$\approx 1.585 + 3$$
$$\approx 4.585$$

Example 2   Solving Exponential Equations

Solve each exponential equation.

a. $2^x = 7$ **b.** $4^{x-3} = 9$ **c.** $2e^x = 10$

Solution

a. To isolate the x, take the $\log_2$ of each side of the equation or write the equation in logarithmic form, as follows.

$2^x = 7$	Write original equation.
$x = \log_2 7$	Inverse property

The solution is $x = \log_2 7 \approx 2.807$. Check this in the original equation.

b.
$4^{x-3} = 9$	Write original equation.
$x - 3 = \log_4 9$	Inverse property
$x = \log_4 9 + 3$	Add 3 to each side.

The solution is $x = \log_4 9 + 3 \approx 4.585$. Check this in the original equation.

c.
$2e^x = 10$	Write original equation.
$e^x = 5$	Divide each side by 2.
$x = \ln 5$	Inverse property.

The solution is $x = \ln 5 \approx 1.609$. Check this in the original equation.

Technology: Tip

Remember that you can use a graphing calculator to solve equations graphically or check solutions that are obtained algebraically. For instance, to check the solutions in Examples 2(a) and 2(c), graph each side of the equations, as shown below.

Graph $y_1 = 2^x$ and $y_2 = 7$. Then use the *intersect* feature of the graphing calculator to approximate the intersection of the two graphs to be $x \approx 2.807$.

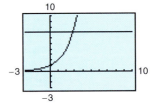

Graph $y_1 = 2e^x$ and $y_2 = 10$. Then use the *intersect* feature of the graphing calculator to approximate the intersection of the two graphs to be $x \approx 1.609$.

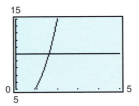

Example 3 **Solving an Exponential Equation**

Solve $5 + e^{x+1} = 20$.

Solution

$$5 + e^{x+1} = 20 \qquad \text{Write original equation.}$$
$$e^{x+1} = 15 \qquad \text{Subtract 5 from each side.}$$
$$\ln e^{x+1} = \ln 15 \qquad \text{Take the logarithm of each side.}$$
$$x + 1 = \ln 15 \qquad \text{Inverse property}$$
$$x = -1 + \ln 15 \qquad \text{Subtract 1 from each side.}$$

The solution is $x = -1 + \ln 15 \approx 1.708$. You can check this as follows.

Check

$$5 + e^{x+1} = 20 \qquad \text{Write original equation.}$$
$$5 + e^{-1 + \ln 15 + 1} \overset{?}{=} 20 \qquad \text{Substitute } -1 + \ln 15 \text{ for } x.$$
$$5 + e^{\ln 15} \overset{?}{=} 20 \qquad \text{Simplify.}$$
$$5 + 15 = 20 \qquad \text{Solution checks. } \checkmark$$

③ Use inverse properties to solve logarithmic equations.

Solving Logarithmic Equations

You know how to solve an exponential equation by *taking the logarithm of each side*. To solve a logarithmic equation, you need to **exponentiate** each side. For instance, to solve a logarithmic equation such as $\ln x = 2$ you can exponentiate each side of the equation as follows.

$$\ln x = 2 \qquad \text{Write original equation.}$$

$$e^{\ln x} = e^2 \qquad \text{Exponentiate each side.}$$

$$x = e^2 \qquad \text{Inverse property}$$

Notice that you obtain the same result by writing the equation in exponential form. This procedure is demonstrated in the next three examples. The following guideline can be used for solving logarithmic equations.

Solving Logarithmic Equations

To solve a logarithmic equation, first isolate the logarithmic expression, then **exponentiate each side of the equation** (or write the equation in exponential form) and solve for the variable.

Example 4   Solving Logarithmic Equations

a. $2 \log_4 x = 5$ Original equation

$$\log_4 x = \frac{5}{2} \qquad \text{Divide each side by 2.}$$

$$4^{\log_4 x} = 4^{5/2} \qquad \text{Exponentiate each side.}$$

$$x = 4^{5/2} \qquad \text{Inverse property}$$

$$x = 32 \qquad \text{Simplify.}$$

The solution is $x = 32$. Check this in the original equation, as follows.

$$2 \log_4 x = 5 \qquad \text{Original equation}$$

$$2 \log_4 (32) \overset{?}{=} 5 \qquad \text{Substitute 32 for } x.$$

$$2(2.5) \overset{?}{=} 5 \qquad \text{Use a calculator.}$$

$$5 = 5 \qquad \text{Solution checks. } ✓$$

b. $\dfrac{1}{4} \log_2 x = \dfrac{1}{2}$ Original equation

$$\log_2 x = 2 \qquad \text{Multiply each side by 4.}$$

$$2^{\log_2 x} = 2^2 \qquad \text{Exponentiate each side.}$$

$$x = 4 \qquad \text{Inverse property}$$

Example 5 Solving a Logarithmic Equation

Solve $3 \log_{10} x = 6$.

Solution

$3 \log_{10} x = 6$	Write original equation.
$\log_{10} x = 2$	Divide each side by 3.
$x = 10^2$	Exponential form
$x = 100$	Simplify.

The solution is $x = 100$. Check this in the original equation.

Example 6 Solving a Logarithmic Equation

> **Study Tip**
>
> When checking approximate solutions to exponential and logarithmic equations, be aware of the fact that because the solution is approximate, the check will not be exact.

Solve $20 \ln 0.2x = 30$.

Solution

$20 \ln 0.2x = 30$	Write original equation.
$\ln 0.2x = 1.5$	Divide each side by 20.
$e^{\ln 0.2x} = e^{1.5}$	Exponentiate each side.
$0.2x = e^{1.5}$	Inverse property
$x = 5e^{1.5}$	Divide each side by 0.2.

The solution is $x = 5e^{1.5} \approx 22.408$. Check this in the original equation.

The next two examples use logarithmic properties as part of the solutions.

Example 7 Solving a Logarithmic Equation

Solve $\log_3 2x - \log_3 (x - 3) = 1$.

Solution

$\log_3 2x - \log_3 (x - 3) = 1$	Write original equation.
$\log_3 \dfrac{2x}{x - 3} = 1$	Condense the left side.
$\dfrac{2x}{x - 3} = 3^1$	Exponential form
$2x = 3x - 9$	Multiply each side by $x - 3$.
$-x = -9$	Subtract $3x$ from each side.
$x = 9$	Divide each side by -1.

The solution is $x = 9$. Check this in the original equation.

Example 8   Checking For Extraneous Solutions

$$\log_6 x + \log_6(x - 5) = 2 \qquad \text{Original equation}$$

$$\log_6[x(x - 5)] = 2 \qquad \text{Condense the left side.}$$

$$x(x - 5) = 6^2 \qquad \text{Exponential form}$$

$$x^2 - 5x - 36 = 0 \qquad \text{Write in general form.}$$

$$(x - 9)(x + 4) = 0 \qquad \text{Factor.}$$

$$x - 9 = 0 \implies x = 9 \qquad \text{Set 1st factor equal to 0.}$$

$$x + 4 = 0 \implies x = -4 \qquad \text{Set 2nd factor equal to 0.}$$

From this, it appears that the solutions are $x = 9$ and $x = -4$. To be sure, you need to check each solution in the original equation, as follows.

First Solution

$$\log_6(9) + \log_6(9 - 5) \stackrel{?}{=} 2$$

$$\log_6(9 \cdot 4) \stackrel{?}{=} 2$$

$$\log_6 36 = 2 \checkmark$$

Second Solution

$$\log_6(-4) + \log_6(-4 - 5) \stackrel{?}{=} 2$$

$$\log_6(-4) + \log_6(-9) \neq 2 ✗$$

Of the two possible solutions, only $x = 9$ checks. So, $x = -4$ is extraneous.

④ Use exponential or logarithmic equations to solve application problems.

Application

Example 9   Compound Interest

A deposit of $5000 is placed in a savings account for 2 years. The interest on the account is compounded continuously. At the end of 2 years, the balance in the account is $5867.55. What is the annual interest rate for this account?

Solution

Using the formula for continuously compounded interest, $A = Pe^{rt}$, you have the following solution.

Formula: $A = Pe^{rt}$

Labels: Principal $= P = 5000$ \hfill (dollars)

Amount $= A = 5867.55$ \hfill (dollars)

Time $= t = 2$ \hfill (years)

Annual interest rate $= r$ \hfill (percent in decimal form)

Equation: $5867.55 = 5000e^{2r}$ \hfill Substitute for A, P, and t.

$$1.17351 = e^{2r} \qquad \text{Divide each side by 5000 and simplify.}$$

$$\ln 1.17351 = \ln(e^{2r}) \qquad \text{Take logarithm of each side.}$$

$$0.16 \approx 2r \implies 0.08 \approx r \qquad \text{Inverse property}$$

The annual interest rate is approximately 8%. Check this solution.

9.5 Exercises

Review *Concepts, Skills, and Problem Solving*

Keep mathematically in shape by doing these exercises *before* the problems of this section.

Properties and Definitions

1. *Writing* ✏ Is it possible for the system

   $$\begin{cases} 7x - 2y = 8 \\ x + y = 4 \end{cases}$$

   to have exactly two solutions? Explain.

2. *Writing* ✏ Explain why the following system has no solution.

   $$\begin{cases} 8x - 4y = 5 \\ -2x + y = 1 \end{cases}$$

Solving Equations

In Exercises 3–8, solve the equation.

3. $\frac{2}{3}x + \frac{2}{3} = 4x - 6$

4. $x^2 - 10x + 17 = 0$

5. $\frac{5}{2x} - \frac{4}{x} = 3$

6. $\frac{1}{x} + \frac{2}{x-5} = 0$

7. $|x - 4| = 3$

8. $\sqrt{x + 2} = 7$

Models and Graphing

9. *Distance* A train is traveling at 73 miles per hour. Write the distance d the train travels as a function of the time t. Graph the function.

10. ▲ *Geometry* The diameter of a right circular cylinder is 10 centimeters. Write the volume V of the cylinder as a function of its height h if the formula for its volume is $V = \pi r^2 h$. Graph the function.

11. ▲ *Geometry* The height of a right circular cylinder is 10 centimeters. Write the volume V of the cylinder as a function of its radius r if the formula for its volume is $V = \pi r^2 h$. Graph the function.

12. *Force* A force of 100 pounds stretches a spring 4 inches. Write the force F as a function of the distance x that the spring is stretched. Graph the function.

Developing Skills

In Exercises 1–6, determine whether the value of x is a solution of the equation.

1. $3^{2x-5} = 27$

   (a) $x = 1$
   (b) $x = 4$

2. $4^{x+3} = 16$

   (a) $x = -1$
   (b) $x = 0$

3. $e^{x+5} = 45$

   (a) $x = -5 + \ln 45$
   (b) $x = -5 + e^{45}$

4. $2^{3x-1} = 324$

   (a) $x \approx 3.1133$
   (b) $x \approx 2.4327$

5. $\log_9(6x) = \frac{3}{2}$

   (a) $x = 27$
   (b) $x = \frac{9}{2}$

6. $\ln(x + 3) = 2.5$

   (a) $x = -3 + e^{2.5}$
   (b) $x \approx 9.1825$

In Exercises 7–34, solve the equation. (Do not use a calculator.) See Example 1.

7. $7^x = 7^3$

8. $4^x = 4^6$

9. $e^{1-x} = e^4$

10. $9^{x+3} = 9^{10}$

11. $5^{x+6} = 25^5$

12. $2^{x-4} = 8^2$

13. $6^{2x} = 36$

14. $5^{3x} = 25$

15. $3^{2-x} = 81$

16. $4^{2x-1} = 64$

17. $5^x = \frac{1}{125}$

18. $3^x = \frac{1}{243}$

19. $2^{x+2} = \frac{1}{16}$

20. $3^{2-x} = 9$

21. $4^{x+3} = 32^x$

22. $9^{x-2} = 243^{x+1}$

23. $\ln 5x = \ln 22$

24. $\ln 3x = \ln 24$

25. $\log_6 3x = \log_6 18$

26. $\log_5 2x = \log_5 36$

27. $\ln(2x - 3) = \ln 15$

28. $\ln(2x - 3) = \ln 17$

29. $\log_2(x + 3) = \log_2 7$

30. $\log_4(x - 4) = \log_4 12$

31. $\log_5(2x - 3) = \log_5(4x - 5)$

32. $\log_3(4 - 3x) = \log_3(2x + 9)$

33. $\log_3(2 - x) = 2$ **34.** $\log_2(3x - 1) = 5$

In Exercises 35–38, simplify the expression.

35. $\ln e^{2x - 1}$ **36.** $\log_3 3^{x^2}$

37. $10^{\log_{10} 2x}$

38. $e^{\ln(x + 1)}$

In Exercises 39–82, solve the exponential equation. (Round your answer to two decimal places.) See Examples 2 and 3.

39. $3^x = 91$ **40.** $4^x = 40$

41. $5^x = 8.2$ **42.** $2^x = 3.6$

43. $6^{2x} = 205$ **44.** $4^{3x} = 168$

45. $7^{3y} = 126$ **46.** $5^{5y} = 305$

47. $3^{x+4} = 6$ **48.** $5^{3-x} = 15$

49. $10^{x+6} = 250$ **50.** $12^{x-1} = 324$

51. $3e^x = 42$ **52.** $6e^{-x} = 3$

53. $\frac{1}{4}e^x = 5$ **54.** $\frac{2}{3}e^x = 1$

55. $\frac{1}{2}e^{3x} = 20$ **56.** $4e^{-3x} = 6$

57. $250(1.04)^x = 1000$

58. $32(1.5)^x = 640$

59. $300e^{x/2} = 9000$

60. $6000e^{-2t} = 1200$

61. $1000^{0.12x} = 25,000$

62. $10,000e^{-0.1t} = 4000$

63. $\frac{1}{5}(4^{x+2}) = 300$

64. $3(2^{t+4}) = 350$

65. $6 + 2^{x-1} = 1$

66. $5^{x+6} - 4 = 12$

67. $7 + e^{2-x} = 28$

68. $9 + e^{5-x} = 32$

69. $8 - 12e^{-x} = 7$

70. $4 - 2e^x = -23$

71. $4 + e^{2x} = 10$ **72.** $10 + e^{4x} = 18$

73. $32 + e^{7x} = 46$

74. $50 - e^{x/2} = 35$

75. $23 - 5e^{x+1} = 3$

76. $2e^x + 5 = 115$

77. $4(1 + e^{x/3}) = 84$

78. $50(3 - e^{2x}) = 125$

79. $\dfrac{8000}{(1.03)^t} = 6000$

80. $\dfrac{5000}{(1.05)^x} = 250$

81. $\dfrac{300}{2 - e^{-0.15t}} = 200$

82. $\dfrac{500}{1 + e^{-0.1t}} = 400$

In Exercises 83–118, solve the logarithmic equation. (Round your answer to two decimal places.) See Examples 4–8.

83. $\log_{10} x = -1$ **84.** $\log_{10} x = 3$

85. $\log_3 x = 4.7$

86. $\log_5 x = 9.2$

87. $4 \log_3 x = 28$

88. $6 \log_2 x = 18$ **89.** $16 \ln x = 30$

90. $12 \ln x = 20$ **91.** $\log_{10} 4x = 2$

92. $\log_3 6x = 4$ **93.** $\ln 2x = 3$

94. $\ln(0.5t) = \frac{1}{4}$ **95.** $\ln x^2 = 6$

96. $\ln \sqrt{x} = 6.5$

97. $2 \log_4(x + 5) = 3$

98. $5 \log_{10}(x + 2) = 15$

99. $2 \log_8(x + 3) = 3$

100. $\frac{2}{3} \ln(x + 1) = -1$

101. $1 - 2 \ln x = -4$

102. $5 - 4 \log_2 x = 2$

103. $-1 + 3 \log_{10} \dfrac{x}{2} = 8$

104. $-5 + 2 \ln 3x = 5$

105. $\log_4 x + \log_4 5 = 2$

106. $\log_5 x - \log_5 4 = 2$

107. $\log_6(x + 8) + \log_6 3 = 2$

108. $\log_7(x - 1) - \log_7 4 = 1$

109. $\log_5(x + 3) - \log_5 x = 1$

110. $\log_3(x - 2) + \log_3 5 = 3$

111. $\log_{10} x + \log_{10}(x - 3) = 1$

112. $\log_{10} x + \log_{10}(x + 1) = 0$

113. $\log_2(x - 1) + \log_2(x + 3) = 3$

114. $\log_6(x - 5) + \log_6 x = 2$

115. $\log_4 3x + \log_4(x - 2) = \frac{1}{2}$

116. $\log_{10}(25x) - \log_{10}(x - 1) = 2$

117. $\log_2 x + \log_2(x + 2) - \log_2 3 = 4$

118. $\log_3 2x + \log_3(x - 1) - \log_3 4 = 1$

In Exercises 119–122, use a graphing calculator to approximate the x-intercept of the graph.

119. $y = 10^{x/2} - 5$

120. $y = 2e^x - 21$

121. $y = 6 \ln(0.4x) - 13$

122. $y = 5 \log_{10}(x + 1) - 3$

In Exercises 123–126, use a graphing calculator to solve the equation. (Round your answer to two decimal places.)

123. $e^x = 2$

124. $\ln x = 2$

125. $2 \ln(x + 3) = 3$

126. $1000e^{-x/2} = 200$

Solving Problems

127. *Compound Interest* A deposit of $10,000 is placed in a savings account for 2 years. The interest for the account is compounded continuously. At the end of 2 years, the balance in the account is $11,972.17. What is the annual interest rate for this account?

128. *Compound Interest* A deposit of $2500 is placed in a savings account for 2 years. The interest for the account is compounded continuously. At the end of 2 years, the balance in the account is $2847.07. What is the annual interest rate for this account?

129. *Doubling Time* Solve the exponential equation $5000 = 2500e^{0.09t}$ for t to determine the number of years for an investment of $2500 to double in value when compounded continuously at the rate of 9%.

130. *Doubling Rate* Solve the exponential equation $10,000 = 5000e^{10r}$ for r to determine the interest rate required for an investment of $5000 to double in value when compounded continuously for 10 years.

131. *Sound Intensity* The relationship between the number of decibels B and the intensity of a sound I in watts per centimeter squared is given by

$$B = 10 \log_{10}\left(\frac{I}{10^{-16}}\right).$$

Determine the intensity of a sound I if it registers 75 decibels on a decibel meter.

132. *Sound Intensity* The relationship between the number of decibels B and the intensity of a sound I in watts per centimeter squared is given by

$$B = 10 \log_{10}\left(\frac{I}{10^{-16}}\right).$$

Determine the intensity of a sound I if it registers 90 decibels on a decibel meter.

133. *Muon Decay* A muon is an elementary particle that is similar to an electron, but much heavier. Muons are unstable—they quickly decay to form electrons and other particles. In an experiment conducted in 1943, the number of muon decays m (of an original 5000 muons) was related to the time T by the model $T = 15.7 - 2.48 \ln m$, where T is in microseconds. How many decays were recorded when $T = 2.5$?

134. *Friction* In order to restrain an untrained horse, a person partially wraps a rope around a cylindrical post in a corral (see figure). The horse is pulling on the rope with a force of 200 pounds. The force F required by the person is $F = 200e^{-0.5\pi\theta/180}$, where F is in pounds and θ is the angle of wrap in degrees. Find the smallest value of θ if F cannot exceed 80 pounds.

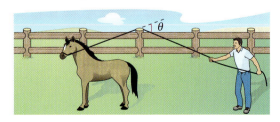

135. *Human Memory Model*   The average score A for a group of students who took a test t months after the completion of a course is given by the human memory model $A = 80 - \log_{10}(t + 1)^{12}$. How long after completing the course will the average score fall to $A = 72$?

(a) Answer the question algebraically by letting $A = 72$ and solving the resulting equation.

(b) ▦ Answer the question graphically by using a graphing calculator to graph the equations $y_1 = 80 - \log_{10}(t + 1)^{12}$ and $y_2 = 72$, and finding the point(s) of intersection.

(c) Which strategy works better for this problem? Explain.

136. ▦ *Car Sales*   The number N (in billions of dollars) of car sales at new car dealerships for the years 1994 through 2001 is modeled by the equation $N = 322.2e^{0.0689t}$, for $4 \leq t \leq 11$, where t is time in years, with $t = 4$ corresponding to 1994. (Source: National Automobile Dealers Association)

(a) Use a graphing calculator to graph the equation over the specified domain.

(b) Use the graph in part (a) to estimate the value of t when $N = 580$.

137. *Newton's Law of Cooling*   You place a tray of water at 60°F in a freezer that is set at 0°F. The water cools according to Newton's Law of Cooling

$$kt = \ln \frac{T - S}{T_0 - S}$$

where T is the temperature of the water (in °F), t is the number of hours the tray is in the freezer, S is the temperature of the surrounding air, and T_0 is the original temperature of the water.

(a) The water freezes in 4 hours. What is the constant k? (*Hint:* Water freezes at 32°F.)

(b) You lower the temperature in the freezer to -10°F. At this temperature, how long will it take for the ice cubes to form?

(c) The initial temperature of the water is 50°F. The freezer temperature is 0°F. How long will it take for the ice cubes to form?

138. *Oceanography*   Oceanographers use the density d (in grams per cubic centimeter) of seawater to obtain information about the circulation of water masses and the rates at which waters of different densities mix. For water with a salinity of 30%, the water temperature T (in °C) is related to the density by

$$T = 7.9 \ln(1.0245 - d) + 61.84.$$

Find the densities of the subantarctic water and the antarctic bottom water shown in the figure.

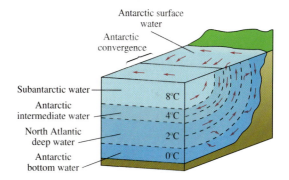

Figure for 138   This cross section shows complex currents at various depths in the South Atlantic Ocean off Antarctica.

Explaining Concepts

139. ⚡ Answer parts (c)–(f) of Motivating the Chapter on page 572.

140. *Writing* ✎   State the three basic properties of logarithms.

141. Which equation requires logarithms for its solution: $2^{x-1} = 32$ or $2^{x-1} = 30$?

142. *Writing* ✎   Explain how to solve $10^{2x-1} = 5316$.

143. *Writing* ✎   In your own words, state the guidelines for solving exponential and logarithmic equations.

9.6 Applications

Frank Siteman/PhotoEdit, Inc.

What You Should Learn

1. Use exponential equations to solve compound interest problems.
2. Use exponential equations to solve growth and decay problems.
3. Use logarithmic equations to solve intensity problems.

Why You Should Learn It

Exponential growth and decay models can be used in many real-life situations. For instance, in Exercise 55 on page 640, you will use an exponential growth model to represent the spread of a computer virus.

1 Use exponential equations to solve compound interest problems.

Compound Interest

In Section 9.1, you were introduced to two formulas for compound interest. Recall that in these formulas, A is the balance, P is the principal, r is the annual interest rate (in decimal form), and t is the time in years.

n Compoundings per Year	*Continuous Compounding*
$$A = P\left(1 + \frac{r}{n}\right)^{nt}$$	$$A = Pe^{rt}$$

Example 1   Finding the Annual Interest Rate

An investment of $50,000 is made in an account that compounds interest quarterly. After 4 years, the balance in the account is $71,381.07. What is the annual interest rate for this account?

Solution

Formula: $A = P\left(1 + \dfrac{r}{n}\right)^{nt}$

Labels:
Principal $= P = 50{,}000$ (dollars)
Amount $= A = 71{,}381.07$ (dollars)
Time $= t = 4$ (years)
Number of compoundings per year $= n = 4$
Annual interest rate $= r$ (percent in decimal form)

Equation:

$$71{,}381.07 = 50{,}000\left(1 + \frac{r}{4}\right)^{(4)(4)}$$ Substitute for A, P, n, and t.

$$1.42762 \approx \left(1 + \frac{r}{4}\right)^{16}$$ Divide each side by 50,000.

$$(1.42762)^{1/16} \approx 1 + \frac{r}{4}$$ Raise each side to $\frac{1}{16}$ power.

$$1.0225 \approx 1 + \frac{r}{4}$$ Simplify.

$$0.09 \approx r$$ Subtract 1 from each side and then multiply each side by 4.

The annual interest rate is approximately 9%. Check this in the original problem.

Example 2   Doubling Time for Continuous Compounding

An investment is made in a trust fund at an annual interest rate of 8.75%, compounded continuously. How long will it take for the investment to double?

Solution

$$A = Pe^{rt}$$ Formula for continuous compounding

$$2P = Pe^{0.0875t}$$ Substitute known values.

$$2 = e^{0.0875t}$$ Divide each side by P.

$$\ln 2 = 0.0875t$$ Inverse property

$$\frac{\ln 2}{0.0875} = t$$ Divide each side by 0.0875.

$$7.92 \approx t$$ Use a calculator.

It will take approximately 7.92 years for the investment to double.

Check

$$A = Pe^{rt}$$ Formula for continuous compounding

$$2P \stackrel{?}{=} Pe^{0.0875(7.92)}$$ Substitute $2P$ for A, 0.0875 for r, and 7.92 for t.

$$2P \stackrel{?}{=} Pe^{0.693}$$ Simplify.

$$2P \approx 1.9997P$$ Solution checks. ✔

Example 3   Finding the Type of Compounding

You deposit $1000 in an account. At the end of 1 year, your balance is $1077.63. The bank tells you that the annual interest rate for the account is 7.5%. How was the interest compounded?

Solution

If the interest had been compounded continuously at 7.5%, the balance would have been

$$A = 1000e^{(0.075)(1)} = \$1077.88.$$

Because the actual balance is slightly less than this, you should use the formula for interest that is compounded n times per year.

$$A = 1000\left(1 + \frac{0.075}{n}\right)^n = 1077.63$$

At this point, it is not clear what you should do to solve the equation for n. However, by completing a table like the one shown below, you can see that $n = 12$. So, the interest was compounded monthly.

n	1	4	12	365
$1000\left(1 + \dfrac{0.075}{n}\right)^n$	1075	1077.14	1077.63	1077.88

In Example 3, notice that an investment of $1000 compounded monthly produced a balance of $1077.63 at the end of 1 year. Because $77.63 of this amount is interest, the **effective yield** for the investment is

$$\text{Effective yield} = \frac{\text{Year's interest}}{\text{Amount invested}} = \frac{77.63}{1000} = 0.07763 = 7.763\%.$$

In other words, the effective yield for an investment collecting compound interest is the *simple interest rate* that would yield the same balance at the end of 1 year.

Example 4   Finding the Effective Yield

An investment is made in an account that pays 6.75% interest, compounded continuously. What is the effective yield for this investment?

Solution

Notice that you do not have to know the principal or the time that the money will be left in the account. Instead, you can choose an arbitrary principal, such as $1000. Then, because effective yield is based on the balance at the end of 1 year, you can use the following formula.

$$A = Pe^{rt}$$

$$= 1000e^{0.0675(1)}$$

$$= 1069.83$$

Now, because the account would earn $69.83 in interest after 1 year for a principal of $1000, you can conclude that the effective yield is

$$\text{Effective yield} = \frac{69.83}{1000} = 0.06983 = 6.983\%.$$

---

2 Use exponential equations to solve growth and decay problems.

Growth and Decay

The balance in an account earning *continuously* compounded interest is one example of a quantity that increases over time according to the **exponential growth model** $y = Ce^{kt}$.

Exponential Growth and Decay

The mathematical model for exponential growth or decay is given by

$$y = Ce^{kt}.$$

For this model, t is the time, C is the original amount of the quantity, and y is the amount after time t. The number k is a constant that is determined by the rate of growth. If $k > 0$, the model represents **exponential growth,** and if $k < 0$, it represents **exponential decay.**

One common application of exponential growth is in modeling the growth of a population. Example 5 illustrates the use of the growth model

$$y = Ce^{kt}, \quad k > 0.$$

Example 5   Population Growth

The population of Texas was 17 million in 1990 and 21 million in 2000. What would you predict the population of Texas to be in 2010?   (Source: U.S. Census Bureau)

Solution

If you assumed a *linear growth model*, you would simply predict the population in the year 2010 to be 25 million because the population would increase by 4 million every 10 years. However, social scientists and demographers have discovered that *exponential growth models* are better than linear growth models for representing population growth. So, you can use the exponential growth model

$$y = Ce^{kt}.$$

In this model, let $t = 0$ represent 1990. The given information about the population can be described by the following table.

t (year)	0	10	20
Ce^{kt} (million)	$Ce^{k(0)} = 17$	$Ce^{k(10)} = 21$	$Ce^{k(20)} = ?$

To find the population when $t = 20$, you must first find the values of C and k. From the table, you can use the fact that $Ce^{k(0)} = Ce^0 = 17$ to conclude that $C = 17$. Then, using this value of C, you can solve for k as follows.

$Ce^{k(10)} = 21$	From table
$17e^{10k} = 21$	Substitute value of C.
$e^{10k} = \dfrac{21}{17}$	Divide each side by 17.
$10k = \ln \dfrac{21}{17}$	Inverse property
$k = \dfrac{1}{10} \ln \dfrac{21}{17}$	Divide each side by 10.
$k \approx 0.0211$	Simplify.

Finally, you can use this value of k in the model from the table for 2010 (for $t = 20$) to predict the population in the year 2010 to be

$$17e^{0.0211(20)} \approx 17(1.53) = 26.01 \text{ million}.$$

Figure 9.22 graphically compares the exponential growth model with a linear growth model.

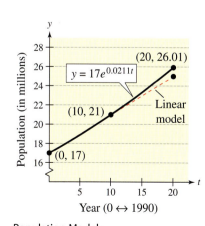

Population Models
Figure 9.22

Example 6   Radioactive Decay

Radioactive iodine is a by-product of some types of nuclear reactors. Its **half-life** is 60 days. That is, after 60 days, a given amount of radioactive iodine will have decayed to half the original amount. A nuclear accident occurs and releases 20 grams of radioactive iodine. How long will it take for the radioactive iodine to decay to a level of 1 gram?

Solution

To solve this problem, use the model for exponential decay.

$$y = Ce^{kt}$$

Next, use the information given in the problem to set up the following table.

t (days)	0	60	?
Ce^{kt} (grams)	$Ce^{k(0)} = 20$	$Ce^{k(60)} = 10$	$Ce^{k(t)} = 1$

Because $Ce^{k(0)} = Ce^0 = 20$, you can conclude that $C = 20$. Then, using this value of C, you can solve for k as follows.

$Ce^{k(60)} = 10$	From table
$20e^{60k} = 10$	Substitute value of C.
$e^{60k} = \dfrac{1}{2}$	Divide each side by 20.
$60k = \ln \dfrac{1}{2}$	Inverse property
$k = \dfrac{1}{60} \ln \dfrac{1}{2}$	Divide each side by 60.
$k \approx -0.01155$	Simplify.

Finally, you can use this value of k in the model from the table to find the time when the amount is 1 gram, as follows.

$Ce^{kt} = 1$	From table
$20e^{-0.01155t} = 1$	Substitute values of C and k.
$e^{-0.01155t} = \dfrac{1}{20}$	Divide each side by 20.
$-0.01155t = \ln \dfrac{1}{20}$	Inverse property
$t = \dfrac{1}{-0.01155} \ln \dfrac{1}{20}$	Divide each side by -0.01155.
$t \approx 259.4$ days	Simplify.

So, 20 grams of radioactive iodine will have decayed to 1 gram after about 259.4 days. This solution is shown graphically in Figure 9.23.

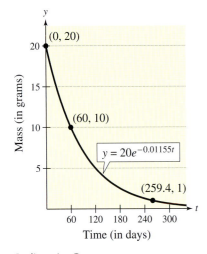

Radioactive Decay
Figure 9.23

Example 7   Website Growth

You created an algebra tutoring website in 2000. You have been keeping track of the number of hits (the number of visits to the site) for each year. In 2000, your website had 4080 hits, and in 2002, your website had 6120 hits. Use an exponential growth model to determine how many hits you can expect in 2008.

Solution

t (year)	Ce^{kt}
0	$Ce^{k(0)} = 4080$
2	$Ce^{k(2)} = 6120$
8	$Ce^{k(8)} = ?$

In the exponential growth model $y = Ce^{kt}$, let $t = 0$ represent 2000. Next, use the information given in the problem to set up the table shown at the left. Because $Ce^{k(0)} = Ce^0 = 4080$, you can conclude that $C = 4080$. Then, using this value of C, you can solve for k, as follows

$$Ce^{k(2)} = 6120 \qquad \textcolor{red}{\text{From table}}$$

$$4080e^{2k} = 6120 \qquad \textcolor{red}{\text{Substitute value of } C.}$$

$$e^{2k} = \tfrac{3}{2} \qquad \textcolor{red}{\text{Divide each side by 4080.}}$$

$$2k = \ln \tfrac{3}{2} \qquad \textcolor{red}{\text{Inverse property}}$$

$$k = \tfrac{1}{2} \ln \tfrac{3}{2} \approx 0.2027 \qquad \textcolor{red}{\text{Divide each side by 2 and simplify.}}$$

Finally, you can use this value of k in the model from the table to predict the number of hits in 2008 to be $4080e^{0.2027(8)} \approx 4080(5.06) \approx 20{,}645$ hits.

3 Use logarithmic equations to solve intensity problems.

Intensity Models

On the **Richter scale,** the magnitude R of an earthquake can be measured by the **intensity model** $R = \log_{10} I$, where I is the intensity of the shock wave.

Example 8   Earthquake Intensity

In 2001, Java, Indonesia experienced an earthquake that measured 5.0 on the Richter scale. In 2002, central Alaska experienced an earthquake that measured 7.9 on the Richter scale. Compare the intensities of these two earthquakes.

Solution

The intensity of the 2001 earthquake is given as follows.

$$5.0 = \log_{10} I \quad \Longrightarrow \quad 10^{5.0} = I \qquad \textcolor{red}{\text{Inverse property}}$$

The intensity of the 2002 earthquake can be found in a similar way.

$$7.9 = \log_{10} I \quad \Longrightarrow \quad 10^{7.9} = I \qquad \textcolor{red}{\text{Inverse property}}$$

The ratio of these two intensities is

$$\frac{I \text{ for 2002}}{I \text{ for 2001}} = \frac{10^{7.9}}{10^{5.0}} = 10^{7.9-5.0} = 10^{2.9} \approx 794.$$

So, the 2002 earthquake had an intensity that was about 794 times greater than the intensity of the 2001 earthquake.

9.6 Exercises

Review Concepts, Skills, and Problem Solving

Keep mathematically in shape by doing these exercises *before* the problems of this section.

Properties and Definitions

In Exercises 1–4, identify the type of variation given in the model.

1. $y = kx^2$

2. $y = \dfrac{k}{x}$

3. $z = kxy$

4. $z = \dfrac{kx}{y}$

Solving Systems

In Exercises 5–10, solve the system of equations.

5. $\begin{cases} x - y = 0 \\ x + 2y = 9 \end{cases}$

6. $\begin{cases} 2x + 5y = 15 \\ 3x + 6y = 20 \end{cases}$

7. $\begin{cases} y = x^2 \\ -3x + 2y = 2 \end{cases}$

8. $\begin{cases} x - y^3 = 0 \\ x - 2y^2 = 0 \end{cases}$

9. $\begin{cases} x - y = -1 \\ x + 2y - 2z = 3 \\ 3x - y + 2z = 1 \end{cases}$

10. $\begin{cases} 2x + y - 2z = 1 \\ x - z = 1 \\ 3x + 3y + z = 12 \end{cases}$

Graphs

In Exercises 11 and 12, use the function $y = -x^2 + 4x$.

11. (a) Does the graph open upward or downward? Explain.

(b) Find the *x*-intercepts algebraically.

(c) Find the coordinates of the vertex of the parabola.

12. Use a graphing calculator to graph the function and verify the results of Exercise 11.

Solving Problems

Compound Interest In Exercises 1–6, find the annual interest rate. See Example 1.

	Principal	Balance	Time	Compounding
1.	$500	$1004.83	10 years	Monthly
2.	$3000	$21,628.70	20 years	Quarterly
3.	$1000	$36,581.00	40 years	Daily
4.	$200	$314.85	5 years	Yearly
5.	$750	$8267.38	30 years	Continuous
6.	$2000	$4234.00	10 years	Continuous

Doubling Time In Exercises 7–12, find the time for the investment to double. Use a graphing calculator to verify the result graphically. See Example 2.

	Principal	Rate	Compounding
7.	$2500	7.5%	Monthly
8.	$900	$5\frac{3}{4}\%$	Quarterly
9.	$18,000	8%	Continuous
10.	$250	6.5%	Yearly

	Principal	Rate	Compounding
11.	$1500	$7\frac{1}{4}\%$	Monthly
12.	$600	9.75%	Continuous

Compound Interest In Exercises 13–16, determine the type of compounding. Solve the problem by trying the more common types of compounding. See Example 3.

	Principal	Balance	Time	Rate
13.	$750	$1587.75	10 years	7.5%
14.	$10,000	$73,890.56	20 years	10%
15.	$100	$141.48	5 years	7%
16.	$4000	$4788.76	2 years	9%

Effective Yield In Exercises 17–24, find the effective yield. See Example 4.

	Rate	Compounding
17.	8%	Continuous
18.	9.5%	Daily

	Rate	Compounding
19.	7%	Monthly
20.	8%	Yearly
21.	6%	Quarterly
22.	9%	Quarterly
23.	8%	Monthly
24.	$5\frac{1}{4}\%$	Daily

25. *Doubling Time* Is it necessary to know the principal P to find the doubling time in Exercises 7–12? Explain.

26. *Effective Yield*

(a) Is it necessary to know the principal P to find the effective yield in Exercises 17–24? Explain.

(b) When the interest is compounded more frequently, what inference can you make about the difference between the effective yield and the stated annual percentage rate?

Compound Interest In Exercises 27–34, find the principal that must be deposited in an account to obtain the given balance.

	Balance	Rate	Time	Compounding
27.	$10,000	9%	20 years	Continuous
28.	$5000	8%	5 years	Continuous
29.	$750	6%	3 years	Daily
30.	$3000	7%	10 years	Monthly
31.	$25,000	7%	30 years	Monthly
32.	$8000	6%	2 years	Monthly
33.	$1000	5%	1 year	Daily
34.	$100,000	9%	40 years	Daily

Monthly Deposits In Exercises 35–38, you make monthly deposits of P dollars in a savings account at an annual interest rate r, compounded continuously. Find the balance A after t years given that

$$A = \frac{P(e^{rt} - 1)}{e^{r/12} - 1}.$$

	Principal	Rate	Time
35.	$P = 30$	$r = 8\%$	$t = 10$ years
36.	$P = 100$	$r = 9\%$	$t = 30$ years
37.	$P = 50$	$r = 10\%$	$t = 40$ years
38.	$P = 20$	$r = 7\%$	$t = 20$ years

Monthly Deposits In Exercises 39 and 40, you make monthly deposits of $30 in a savings account at an annual interest rate of 8%, compounded continuously (see figure).

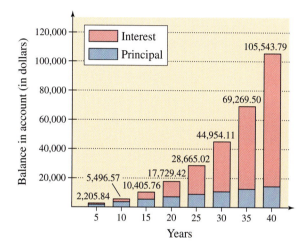

39. Find the total amount that has been deposited in the account in 20 years and the total interest earned.

40. Find the total amount that has been deposited in the account in 40 years and the total interest earned.

Exponential Growth and Decay In Exercises 41–44, find the constant k such that the graph of $y = Ce^{kt}$ passes through the points.

41.
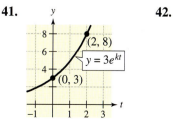

42.
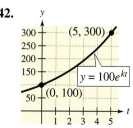

43.

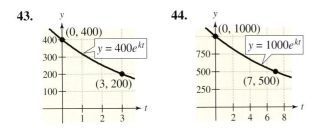

44.

54. World Population The figure shows the population P (in billions) of the world as projected by the U.S. Census Bureau. The bureau's projection can be modeled by

$$P = \frac{10.8}{1 + 1.03e^{-0.0283t}}$$

where $t = 0$ represents 1990. Use the model to estimate the population in 2020.

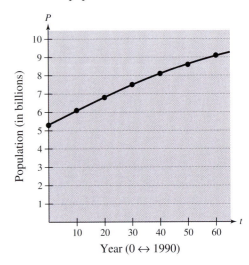

Population of a Country In Exercises 45–52, the population (in millions) of a country for 2000 and the predicted population (in millions) for the year 2025 are given. Find the constants C and k to obtain the exponential growth model $y = Ce^{kt}$ for the population. (Let $t = 0$ correspond to the year 2000.) Use the model to predict the population of the region in the year 2030. See Example 5. *(Source: United Nations)*

Country	2000	2025
45. Australia	19.1	23.5
46. Brazil	170.4	219.0
47. China	1275.1	1470.8
48. Japan	127.1	123.8
49. Ireland	3.8	4.7
50. Uruguay	3.3	3.9
51. United States of America	283.2	346.8
52. United Kingdom	59.4	61.2

53. Rate of Growth

(a) Compare the values of k in Exercises 47 and 49. Which is larger? Explain.

(b) What variable in the continuous compound interest formula is equivalent to k in the model for population growth? Use your answer to give an interpretation of k.

55. Computer Virus A computer virus tends to spread at an exponential rate. In 2000, the number of computers infected by the "Love Bug" virus spread from 100 to about 1,000,000 in 2 hours.

(a) Find the constants C and k to obtain the exponential growth model $y = Ce^{kt}$ for the "Love Bug."

(b) Use your model from part (a) to estimate how long it took the "Love Bug" virus to spread to 80,000 computers.

56. Painting Value In 1990, $82.5 million was paid for Vincent VanGogh's *Portrait of Dr. Gachet*. The same painting was sold for $58 million in 1987. Assume that the value of the painting increases at an exponential rate.

(a) Find the constants C and k to obtain the exponential growth model $y = Ce^{kt}$ for the value of Van Gogh's *Portrait of Dr. Gachet*.

(b) Use your model from part (a) to estimate the value of the painting in 2007.

(c) When will the value of the painting be $1 billion?

57. *Cellular Phones* The number of users of cellular phones in the United States in 1995 was 33,786,000. In 2000, the number of users was 109,478,000. If it is assumed that the number of users of cellular phones grows at an exponential rate, how many users will there be in 2010? (Source: Cellular Telecommunications & Internet Association)

58. *DVDs* The number of DVDs shipped by DVD manufacturers in the United States in 1998 was 500,000. In 2000, the number had increased to 3,300,000. If it is assumed that the number of DVDs shipped by manufacturers grows at an exponential rate, how many DVDs will be shipped in 2010? (Source: Recording Industry Association of America)

Radioactive Decay In Exercises 59–64, complete the table for the radioactive isotopes. See Example 6.

Isotope	Half-Life (Years)	Initial Quantity	Amount After 1000 Years
59. ^{226}Ra	1620	6 g	g
60. ^{226}Ra	1620	g	0.25 g
61. ^{14}C	5730	g	4.0 g
62. ^{14}C	5730	10 g	g
63. ^{230}Pu	24,360	4.2 g	g
64. ^{230}Pu	24,360	g	1.5 g

65. *Radioactive Decay* Radioactive radium (^{226}Ra) has a half-life of 1620 years. If you start with 5 grams of the isotope, how much will remain after 1000 years?

66. *Carbon 14 Dating* Carbon 14 dating assumes that the carbon dioxide on Earth today has the same radioactive content as it did centuries ago. If this is true, the amount of ^{14}C absorbed by a tree that grew several centuries ago should be the same as the amount of ^{14}C absorbed by a tree growing today. A piece of ancient charcoal contains only 15% as much of the radioactive carbon as a piece of modern charcoal. How long ago did the tree burn to make the ancient charcoal if the half-life of ^{14}C is 5730 years? (Round your answer to the nearest 100 years.)

67. *Radioactive Decay* The isotope ^{230}Pu has a half-life of 24,360 years. If you start with 10 grams of the isotope, how much will remain after 10,000 years?

68. *Radioactive Decay* Carbon 14 (^{14}C) has a half-life of 5730 years. If you start with 5 grams of this isotope, how much will remain after 1000 years?

69. *Depreciation* A sport utility vehicle that cost $34,000 new has a depreciated value of $26,000 after 1 year. Find the value of the sport utility vehicle when it is 3 years old by using the exponential model $y = Ce^{kt}$.

70. ▦ *Depreciation* After x years, the value y of a recreational vehicle that cost $8000 new is given by

$$y = 8000(0.8)^x.$$

(a) Use a graphing calculator to graph the model.

(b) Graphically approximate the value of the recreational vehicle after 1 year.

(c) Graphically approximate the time when the recreational vehicle's value will be $4000.

Earthquake Intensity In Exercises 71–74, compare the intensities of the two earthquakes. See Example 8.

Location	Date	Magnitude
71. Central Alaska	10/23/2002	6.7
Southern Italy	10/31/2002	5.9
72. Southern California	10/16/1999	7.2
Morocco	2/29/1960	5.8
73. Mexico City, Mexico	9/19/1985	8.1
New York	4/20/2002	5.0
74. Peru	6/23/2001	8.4
Armenia, USSR	12/7/1988	6.8

Acidity In Exercises 75–78, use the acidity model pH $= -\log_{10}[H^+]$, where acidity (pH) is a measure of the hydrogen ion concentration $[H^+]$ (measured in moles of hydrogen per liter) of a solution.

75. Find the pH of a solution that has a hydrogen ion concentration of 9.2×10^{-8}.

76. Compute the hydrogen ion concentration if the pH of a solution is 4.7.

77. A blueberry has a pH of 2.5 and an antacid tablet has a pH of 9.5. The hydrogen ion concentration of the fruit is how many times the concentration of the tablet?

78. If the pH of a solution is decreased by 1 unit, the hydrogen ion concentration is increased by what factor?

79. *Population Growth*   The population p of a species of wild rabbit t years after it is introduced into a new habitat is given by

$$p(t) = \frac{5000}{1 + 4e^{-t/6}}.$$

(a) ▦ Use a graphing calculator to graph the population function.

(b) Determine the size of the population of rabbits that was introduced into the habitat.

(c) Determine the size of the population of rabbits after 9 years.

(d) After how many years will the size of the population of rabbits be 2000?

80. ▦ *Sales Growth*   Annual sales y of a personal digital assistant x years after it is introduced are approximated by

$$y = \frac{2000}{1 + 4e^{-x/2}}.$$

(a) Use a graphing calculator to graph the model.

(b) Use the graph in part (a) to approximate annual sales of this personal digital assistant model when $x = 4$.

(c) Use the graph in part (a) to approximate the time when annual sales of this personal digital assistant model are $y = 1100$ units.

(d) Use the graph in part (a) to estimate the maximum level that annual sales of this model will approach.

81. *Advertising Effect*   The sales S (in thousands of units) of a brand of jeans after spending x hundred dollars in advertising are given by

$$S = 10(1 - e^{kx}).$$

(a) Write S as a function of x if 2500 jeans are sold when $500 is spent on advertising.

(b) How many jeans will be sold if advertising expenditures are raised to $700?

82. *Advertising Effect*   The sales S of a video game after spending x thousand dollars in advertising are given by

$$S = 4500(1 - e^{kx}).$$

(a) Write S as a function of x if 2030 copies of the video game are sold when $10,000 is spent on advertising.

(b) How many copies of the video game will be sold if advertising expenditures are raised to $25,000?

Explaining Concepts

83. If the equation $y = Ce^{kt}$ models exponential growth, what must be true about k?

84. If the equation $y = Ce^{kt}$ models exponential decay, what must be true about k?

85. *Writing* ✏   The formulas for periodic and continuous compounding have the four variables A, P, r, and t in common. Explain what each variable measures.

86. *Writing* ✏   What is meant by the effective yield of an investment? Explain how it is computed.

87. *Writing* ✏   In your own words, explain what is meant by the half-life of a radioactive isotope.

88. If the reading on the Richter scale is increased by 1, the intensity of the earthquake is increased by what factor?

What Did You Learn?

Key Terms

exponential function, *p. 574*
natural base, *p. 578*
natural exponential function,
  p. 578
composition, *p. 587*

inverse function, *p. 589*
one-to-one, *p. 589*
logarithmic function with base *a*,
  p. 601
common logarithmic function,
  p. 603

natural logarithmic function,
  p. 606
exponentiate, *p. 625*
exponential growth, *p. 634*
exponential decay, *p. 634*

Key Concepts

9.1 ● Rules of exponential functions

1. $a^x \cdot a^y = a^{x+y}$ 2. $\dfrac{a^x}{a^y} = a^{x-y}$

3. $(a^x)^y = a^{xy}$ 4. $a^{-x} = \dfrac{1}{a^x} = \left(\dfrac{1}{a}\right)^x$

9.2 ● Composition of two functions

The composition of two functions f and g is given by $(f \circ g)(x) = f(g(x))$. The domain of the composite function $(f \circ g)$ is the set of all x in the domain of g such that $g(x)$ is in the domain of f.

9.2 ● Horizontal Line Test for inverse functions

A function f has an inverse function f^{-1} if and only if f is one-to-one. Graphically, a function f has an inverse function f^{-1} if and only if no horizontal line intersects the graph of f at more than one point.

9.2 ● Finding an inverse function algebraically

1. In the equation for $f(x)$, replace $f(x)$ with y.

2. Interchange the roles of x and y.

3. If the new equation does not represent y as a function of x, the function f does not have an inverse function. If the new equation does represent y as a function of x, solve the new equation for y.

4. Replace y with $f^{-1}(x)$.

5. Verify that f and f^{-1} are inverse functions of each other by showing that $f(f^{-1}(x)) = x = f^{-1}(f(x))$.

9.3 ● Properties of logarithms and natural logarithms

Let a and x be positive real numbers such that $a \neq 1$. Then the following properties are true.

1. $\log_a 1 = 0$ because $a^0 = 1$.
   $\ln 1 = 0$ because $e^0 = 1$.

2. $\log_a a = 1$ because $a^1 = a$.
   $\ln e = 1$ because $e^1 = e$.

3. $\log_a a^x = x$ because $a^x = a^x$.
   $\ln e^x = x$ because $e^x = e^x$.

9.3 ● Change-of-base formula

Let a, b, and x be positive real numbers such that $a \neq 1$ and $b \neq 1$. Then $\log_a x = \dfrac{\log_b x}{\log_b a}$ or $\log_a x = \dfrac{\ln x}{\ln a}$.

9.4 ● Properties of logarithms

Let a be a positive real number such that $a \neq 1$, and let n be a real number. If u and v are real numbers, variables, or algebraic expressions such that $u > 0$ and $v > 0$, the following properties are true.

Logarithm with base a	*Natural logarithm*
1. $\log_a(uv) = \log_a u + \log_a v$	$\ln(uv) = \ln u + \ln v$
2. $\log_a \dfrac{u}{v} = \log_a u - \log_a v$	$\ln \dfrac{u}{v} = \ln u - \ln v$
3. $\log_a u^n = n \log_a u$	$\ln u^n = n \ln u$

9.5 ● One-to-one properties of exponential and logarithmic equations

Let a be a positive real number such that $a \neq 1$, and let x and y be real numbers. Then the following properties are true.

1. $a^x = a^y$ if and only if $x = y$.

2. $\log_a x = \log_a y$ if and only if $x = y$ $(x > 0, y > 0)$.

9.5 ● Inverse properties of exponents and logarithms

Base a	*Natural base e*
1. $\log_a(a^x) = x$	$\ln(e^x) = x$
2. $a^{(\log_a x)} = x$	$e^{(\ln x)} = x$

Review Exercises

9.1 Exponential Functions

1 Evaluate exponential functions.

In Exercises 1–4, evaluate the exponential function as indicated. (Round your answer to three decimal places.)

1. $f(x) = 2^x$
   (a) $x = -3$
   (b) $x = 1$
   (c) $x = 2$

2. $g(x) = 2^{-x}$
   (a) $x = -2$
   (b) $x = 0$
   (c) $x = 2$

3. $g(t) = 5^{-t/3}$
   (a) $t = -3$
   (b) $t = \pi$
   (c) $t = 6$

4. $h(s) = 1 - 3^{0.2s}$
   (a) $s = 0$
   (b) $s = 2$
   (c) $s = \sqrt{10}$

2 Graph exponential functions.

In Exercises 5–14, sketch the graph of the function. Identify the horizontal asymptote.

5. $f(x) = 3^x$

6. $f(x) = 3^{-x}$

7. $f(x) = 3^x - 1$

8. $f(x) = 3^x + 2$

9. $f(x) = 3^{(x+1)}$

10. $f(x) = 3^{(x-1)}$

11. $f(x) = 3^{x/2}$

12. $f(x) = 3^{-x/2}$

13. $f(x) = 3^{x/2} - 2$

14. $f(x) = 3^{x/2} + 3$

In Exercises 15–18, use a graphing calculator to graph the function.

15. $f(x) = 2^{-x^2}$

16. $g(x) = 2^{|x|}$

17. $y = 10(1.09)^t$

18. $y = 250(1.08)^t$

3 Evaluate the natural base e and graph natural exponential functions.

In Exercises 19 and 20, evaluate the exponential function as indicated. (Round your answer to three decimal places.)

19. $f(x) = 3e^{-2x}$
   (a) $x = 3$
   (b) $x = 0$
   (c) $x = -19$

20. $g(x) = e^{x/5} + 11$
   (a) $x = 12$
   (b) $x = -8$
   (c) $x = 18.4$

In Exercises 21–24, use a graphing calculator to graph the function.

21. $y = 5e^{-x/4}$

22. $y = 6 - e^{x/2}$

23. $f(x) = e^{x+2}$

24. $h(t) = \dfrac{8}{1 + e^{-t/5}}$

4 Use exponential functions to solve application problems.

Compound Interest In Exercises 25 and 26, complete the table to determine the balance A for P dollars invested at interest rate r for t years, compounded n times per year.

n	1	4	12	365	Continuous compounding
A					

	Principal	*Rate*	*Time*
25.	$P = \$5000$	$r = 10\%$	$t = 40$ years
26.	$P = \$10{,}000$	$r = 9.5\%$	$t = 30$ years

27. *Radioactive Decay*   After t years, the remaining mass y (in grams) of 21 grams of a radioactive element whose half-life is 25 years is given by $y = 21\left(\frac{1}{2}\right)^{t/25}$, $t \geq 0$. How much of the initial mass remains after 58 years?

28. *Depreciation*   After t years, the value of a truck that originally cost $29,000 depreciates so that each year it is worth $\frac{2}{3}$ of its value for the previous year. Find a model for $V(t)$, the value of the truck after t years. Sketch a graph of the model and determine the value of the truck 4 years after it was purchased.

9.2 Composite and Inverse Functions

① Form compositions of two functions and find the domains of composite functions.

In Exercises 29–32, find the compositions.

29. $f(x) = x + 2$, $g(x) = x^2$
   (a) $(f \circ g)(2)$ (b) $(g \circ f)(-1)$

30. $f(x) = \sqrt[3]{x}$, $g(x) = x + 2$
   (a) $(f \circ g)(6)$ (b) $(g \circ f)(64)$

31. $f(x) = \sqrt{x + 1}$, $g(x) = x^2 - 1$
   (a) $(f \circ g)(5)$ (b) $(g \circ f)(-1)$

32. $f(x) = \dfrac{1}{x - 5}$, $g(x) = \dfrac{5x + 1}{x}$

   (a) $(f \circ g)(1)$ (b) $(g \circ f)\left(\dfrac{1}{5}\right)$

In Exercises 33 and 34, find the compositions (a) $f \circ g$ and (b) $g \circ f$. Then find the domain of each composition.

33. $f(x) = \sqrt{x - 4}$, $g(x) = 2x$

34. $f(x) = \dfrac{2}{x - 4}$, $g(x) = x^2$

② Use the Horizontal Line Test to determine whether functions have inverse functions.

In Exercises 35–38, use the Horizontal Line Test to determine if the function is one-to-one and so has an inverse function.

35. $f(x) = x^2 - 25$

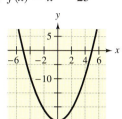

36. $f(x) = \frac{1}{4}x^3$

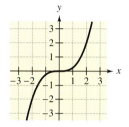

37. $h(x) = 4\sqrt[3]{x}$

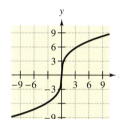

38. $g(x) = \sqrt{9 - x^2}$

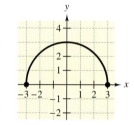

③ Find inverse functions algebraically.

In Exercises 39–44, find the inverse function.

39. $f(x) = 3x + 4$

40. $f(x) = 2x - 3$

41. $h(x) = \sqrt{x}$

42. $g(x) = x^2 + 2$, $x \geq 0$

43. $f(t) = t^3 + 4$

44. $h(t) = \sqrt[3]{t - 1}$

④ Graphically verify that two functions are inverse functions of each other.

In Exercises 45 and 46, use a graphing calculator to graph the functions in the same viewing window. Graphically verify that f and g are inverse functions of each other.

45. $f(x) = 3x + 4$
    $g(x) = \frac{1}{3}(x - 4)$

46. $f(x) = \frac{1}{3}\sqrt[3]{x}$
    $g(x) = 27x^3$

In Exercises 47–50, use the graph of f to sketch the graph of f^{-1}.

47.

48.

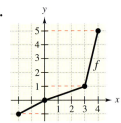

49.

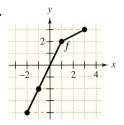

50.

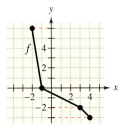

9.3 Logarithmic Functions

1 Evaluate logarithmic functions.

In Exercises 51–58, evaluate the logarithm.

51. $\log_{10} 1000$

52. $\log_9 3$

53. $\log_3 \frac{1}{9}$

54. $\log_4 \frac{1}{16}$

55. $\log_2 64$

56. $\log_{10} 0.01$

57. $\log_3 1$

58. $\log_2 \sqrt{4}$

2 Graph logarithmic functions.

In Exercises 59–64, sketch the graph of the function. Identify the vertical asymptote.

59. $f(x) = \log_3 x$

60. $f(x) = -\log_3 x$

61. $f(x) = -2 + \log_3 x$

62. $f(x) = 2 + \log_3 x$

63. $y = \log_2(x - 4)$

64. $y = \log_4(x + 1)$

3 Graph and evaluate natural logarithmic functions.

In Exercises 65 and 66, use your calculator to evaluate the natural logarithm.

65. $\ln e^7$

66. $\ln \dfrac{1}{e}$

In Exercises 67–70, sketch the graph of the function. Identify the vertical asymptote.

67. $y = \ln(x - 3)$

68. $y = -\ln(x + 2)$

69. $y = 5 - \ln x$

70. $y = 3 + \ln x$

4 Use the change-of-base formula to evaluate logarithms.

In Exercises 71–74, use a calculator to evaluate the logarithm by means of the change-of-base formula. Round your answer to four decimal places.

71. $\log_4 9$

72. $\log_{1/2} 5$

73. $\log_{12} 200$

74. $\log_3 0.28$

9.4 Properties of Logarithms

1 Use the properties of logarithms to evaluate logarithms.

In Exercises 75–80, use $\log_5 2 \approx 0.4307$ and $\log_5 3 \approx 0.6826$ to approximate the expression. Do not use a calculator.

75. $\log_5 18$

76. $\log_5 \sqrt{6}$

77. $\log_5 \frac{1}{2}$

78. $\log_5 \frac{2}{3}$

79. $\log_5(12)^{2/3}$

80. $\log_5(5^2 \cdot 6)$

2 Use the properties of logarithms to rewrite, expand, or condense logarithmic expressions.

In Exercises 81–88, use the properties of logarithms to expand the expression.

81. $\log_4 6x^4$

82. $\log_{10} 2x^{-3}$

83. $\log_5 \sqrt{x + 2}$

84. $\ln \sqrt[3]{\dfrac{x}{5}}$

85. $\ln \dfrac{x + 2}{x - 2}$

86. $\ln x(x - 3)^2$

87. $\ln\left[\sqrt{2x}(x + 3)^5\right]$

88. $\log_3 \dfrac{a^2 \sqrt{b}}{cd^5}$

In Exercises 89–98, use the properties of logarithms to condense the expression.

89. $-\dfrac{2}{3} \ln 3y$

90. $5 \log_2 y$

91. $\log_8 16x + \log_8 2x^2$

92. $\log_4 6x - \log_4 10$

93. $-2(\ln 2x - \ln 3)$

94. $4(1 + \ln x + \ln x)$

95. $4[\log_2 k - \log_2(k - t)]$

96. $\dfrac{1}{3}(\log_8 a + 2 \log_8 b)$

97. $3 \ln x + 4 \ln y + \ln z$

98. $\ln(x + 4) - 3 \ln x - \ln y$

True or False? **In Exercises 99–104, use properties of logarithms to determine whether the equation is true or false. If it is false, state why or give an example to show that it is false.**

99. $\log_2 4x = 2 \log_2 x$

100. $\dfrac{\ln 5x}{\ln 10x} = \ln \dfrac{1}{2}$

101. $\log_{10} 10^{2x} = 2x$

102. $e^{\ln t} = t$

103. $\log_4 \dfrac{16}{x} = 2 - \log_4 x$

104. $6 \ln x + 6 \ln y = \ln(xy)^6$

③ Use the properties of logarithms to solve application problems.

105. *Light Intensity* The intensity of light y as it passes through a medium is given by

$$y = \ln\left(\dfrac{I_0}{I}\right)^{0.83}.$$

Use properties of logarithms to write the formula in simpler form, and determine the intensity of light passing through this medium when $I_0 = 4.2$ and $I = 3.3$.

106. *Human Memory Model* A psychologist finds that the percent p of retention in a group of subjects can be modeled by

$$p = \dfrac{\log_{10}(10^{68})}{\log_{10}(t + 1)^{20}}$$

where t is the time in months from the subjects' initial testing. Use properties of logarithms to write the formula in simpler form, and determine the percent of retention after 5 months.

9.5 Solving Exponential and Logarithmic Equations

① Solve basic exponential and logarithmic equations.

In Exercises 107–112, solve the equation.

107. $2^x = 64$

108. $5^x = 25$

109. $4^{x-3} = \dfrac{1}{16}$

110. $3^{x-2} = 81$

111. $\log_7(x + 6) = \log_7 12$

112. $\ln(5 - x) = \ln(8)$

② Use inverse properties to solve exponential equations.

In Exercises 113–118, solve the exponential equation. (Round your answer to two decimal places.)

113. $3^x = 500$

114. $8^x = 1000$

115. $2e^{0.5x} = 45$

116. $100e^{-0.6x} = 20$

117. $12(1 - 4^x) = 18$

118. $25(1 - e^t) = 12$

③ Use inverse properties to solve logarithmic equations.

In Exercises 119–128, solve the logarithmic equation. (Round your answer to two decimal places.)

119. $\ln x = 7.25$

120. $\ln x = -0.5$

121. $\log_{10} 2x = 1.5$

122. $\log_2 2x = -0.65$

123. $\log_3(2x + 1) = 2$

124. $\log_5(x - 10) = 2$

125. $\frac{1}{3}\log_2 x + 5 = 7$

126. $4 \log_5(x + 1) = 4.8$

127. $\log_2 x + \log_2 3 = 3$

128. $2 \log_4 x - \log_4(x - 1) = 1$

④ Use exponential or logarithmic equations to solve application problems.

129. *Compound Interest*   A deposit of $5000 is placed in a savings account for 2 years. The interest for the account is compounded continuously. At the end of 2 years, the balance in the account is $5751.37. What is the annual interest rate for this account?

130. *Sound Intensity*   The relationship between the number of decibels B and the intensity of a sound I in watts per centimeter squared is given by

$$B = 10 \log_{10}\left(\frac{I}{10^{-16}}\right).$$

Determine the intensity of a sound I if it registers 125 decibels on a decibel meter.

9.6 Applications

① Use exponential equations to solve compound interest problems.

Annual Interest Rate   **In Exercises 131–136, find the annual interest rate.**

	Principal	Balance	Time	Compounding
131.	$250	$410.90	10 years	Quarterly
132.	$1000	$1348.85	5 years	Monthly
133.	$5000	$15,399.30	15 years	Daily
134.	$10,000	$35,236.45	20 years	Yearly
135.	$1500	$24,666.97	40 years	Continuous
136.	$7500	$15,877.50	15 years	Continuous

Effective Yield   **In Exercises 137–142, find the effective yield.**

	Rate	Compounding
137.	5.5%	Daily
138.	6%	Monthly
139.	7.5%	Quarterly
140.	8%	Yearly
141.	7.5%	Continuously
142.	4%	Continuously

② Use exponential equations to solve growth and decay problems.

Radioactive Decay   **In Exercises 143–148, complete the table for the radioactive isotopes.**

	Isotope	Half-Life (Years)	Initial Quantity	Amount After 1000 Years
143.	^{226}Ra	1620	3.5 g	▢ g
144.	^{226}Ra	1620	▢ g	0.5 g
145.	^{14}C	5730	▢ g	2.6 g
146.	^{14}C	5730	10 g	▢ g
147.	^{230}Pu	24,360	5 g	▢ g
148.	^{230}Pu	24,360	▢ g	2.5 g

③ Use logarithmic equations to solve intensity problems.

In Exercises 149 and 150, compare the intensities of the two earthquakes.

	Location	Date	Magnitude
149.	San Francisco, California	4/18/1906	8.3
	Napa, California	9/3/2000	4.9
150.	El Salvador	2/13/2001	6.5
	Colombia	1/25/1999	5.7

Chapter Test

Take this test as you would take a test in class. After you are done, check your work against the answers in the back of the book.

1. Evaluate $f(t) = 54\left(\frac{2}{3}\right)^t$ when $t = -1, 0, \frac{1}{2}$, and 2.

2. Sketch a graph of the function $f(x) = 2^{x/3}$ and identify the horizontal asymptote.

3. Find the compositions (a) $f \circ g$ and (b) $g \circ f$. Then find the domain of each composition.

$$f(x) = 2x^2 + x \qquad g(x) = 5 - 3x$$

4. Find the inverse function of $f(x) = 5x + 6$.

5. Verify algebraically that the functions f and g are inverse functions of each other.

$$f(x) = -\tfrac{1}{2}x + 3, \qquad g(x) = -2x + 6$$

6. Evaluate $\log_8 2$ without a calculator.

7. Describe the relationship between the graphs of $f(x) = \log_5 x$ and $g(x) = 5^x$.

8. Use the properties of logarithms to expand $\log_4\left(5x^2/\sqrt{y}\right)$.

9. Use the properties of logarithms to condense $\ln x - 4 \ln y$.

In Exercises 10–17, solve the equation. Round your answer to two decimal places, if necessary.

10. $\log_2 x = 5$

11. $9^{2x} = 182$

12. $400e^{0.08t} = 1200$

13. $3 \ln(2x - 3) = 10$

14. $8(2 - 3^x) = -56$

15. $\log_2 x + \log_2 4 = 5$

16. $\ln x - \ln 2 = 4$

17. $30(e^x + 9) = 300$

18. Determine the balance after 20 years if $2000 is invested at 7% compounded (a) quarterly and (b) continuously.

19. Determine the principal that will yield $100,000 when invested at 9% compounded quarterly for 25 years.

20. A principal of $500 yields a balance of $1006.88 in 10 years when the interest is compounded continuously. What is the annual interest rate?

21. A car that cost $18,000 new has a depreciated value of $14,000 after 1 year. Find the value of the car when it is 3 years old by using the exponential model $y = Ce^{kt}$.

In Exercises 22–24, the population p of a species of fox t years after it is introduced into a new habitat is given by

$$p(t) = \frac{2400}{1 + 3e^{-t/4}}.$$

22. Determine the population size that was introduced into the habitat.

23. Determine the population after 4 years.

24. After how many years will the population be 1200?

Motivating the Chapter

 Postal Delivery Route

You are a mail carrier for a post office that receives mail for everyone living within a five-mile radius. Your route covers the portions of Anderson Road and Murphy Road that pass through this region.

See Section 10.1, Exercise 101.

a. Assume that the post office is located at the point $(0, 0)$. Write an equation for the circle that bounds the region where the mail is delivered.

b. Sketch the graph of the circular region serviced by the post office.

See Section 10.3, Exercise 45.

c. Assume that Anderson Road follows one branch of a hyperbolic path given by $x^2 - y^2 - 4x - 23 = 0$. Find the center and vertices of this hyperbola.

d. On the same set of coordinate axes as the circular region, sketch the graph of the hyperbola that represents Anderson Road.

See Section 10.4, Exercise 85.

e. You begin your delivery on Anderson Road at the point $(-4, -3)$. Where on Anderson Road will you end your delivery? Explain.

f. You finish delivery on Anderson Road at the point where it intersects both the circular boundary of the post office and Murphy Road. At the intersection, you begin delivering on Murphy Road, which is a straight road that cuts through the center of the circular boundary and continues past the post office. Find the equation that represents Murphy Road.

g. Where on Murphy Road will you end your delivery? Explain.

Bill Aron/PhotoEdit, Inc.

Conics

10.1 ● **Circles and Parabolas**

10.2 ● **Ellipses**

10.3 ● **Hyperbolas**

10.4 ● **Solving Nonlinear Systems of Equations**

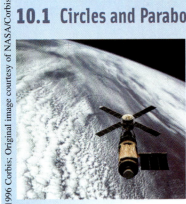

10.1 Circles and Parabolas

What You Should Learn

1 Recognize the four basic conics: circles, parabolas, ellipses, and hyperbolas.
2 Graph and write equations of circles centered at the origin.
3 Graph and write equations of circles centered at (h, k).
4 Graph and write equations of parabolas.

Why You Should Learn It

Circles can be used to model and solve scientific problems. For instance, in Exercise 93 on page 662, you will write an equation of the circular orbit of a satellite.

The Conics

In Section 8.4, you saw that the graph of a second-degree equation of the form $y = ax^2 + bx + c$ is a parabola. A parabola is one of four types of **conics** or **conic sections.** The other three types are circles, ellipses, and hyperbolas. All four types have equations of second degree. Each figure can be obtained by intersecting a plane with a double-napped cone, as shown in Figure 10.1.

1 Recognize the four basic conics: circles, parabolas, ellipses, and hyperbolas.

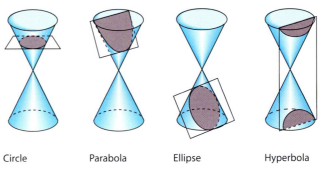

| Circle | Parabola | Ellipse | Hyperbola |

Figure 10.1

Conics occur in many practical applications. Reflective surfaces in satellite dishes, flashlights, and telescopes often have a parabolic shape. The orbits of planets are elliptical, and the orbits of comets are usually elliptical or hyperbolic. Ellipses and parabolas are also used in building archways and bridges.

2 Graph and write equations of circles centered at the origin.

Circles Centered at the Origin

Definition of a Circle

A **circle** in the rectangular coordinate system consists of all points (x, y) that are a given positive distance r from a fixed point, called the **center** of the circle. The distance r is called the **radius** of the circle.

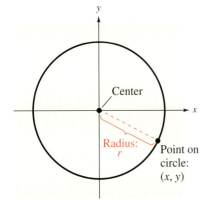

Figure 10.2

If the center of the circle is the origin, as shown in Figure 10.2, the relationship between the coordinates of any point (x, y) on the circle and the radius r is

$$r = \sqrt{(x - 0)^2 + (y - 0)^2} = \sqrt{x^2 + y^2}. \qquad \text{Distance Formula (See Section 3.1.)}$$

By squaring each side of this equation, you obtain the equation below, which is called the **standard form of the equation of a circle centered at the origin.**

Standard Equation of a Circle (Center at Origin)

The **standard form of the equation of a circle centered at the origin** is

$$x^2 + y^2 = r^2.$$ Circle with center at $(0, 0)$

The positive number r is called the **radius** of the circle.

Example 1   Writing an Equation of a Circle

Write an equation of the circle that is centered at the origin and has a radius of 2 (See Figure 10.3).

Solution

Using the standard form of the equation of a circle (with center at the origin) and $r = 2$, you obtain

$$x^2 + y^2 = r^2$$ Standard form with center at $(0, 0)$

$$x^2 + y^2 = 2^2$$ Substitute 2 for r.

$$x^2 + y^2 = 4.$$ Equation of circle

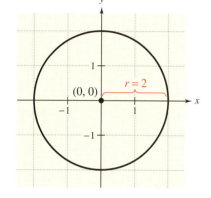

Figure 10.3

To sketch the circle for a given equation, first write the equation in standard form. Then, from the standard form, you can identify the center and radius and sketch the circle.

Example 2   Sketching a Circle

Identify the center and radius of the circle given by the equation $4x^2 + 4y^2 - 25 = 0$. Then sketch the circle.

Solution

Begin by writing the equation in standard form.

$$4x^2 + 4y^2 - 25 = 0$$ Write original equation.

$$4x^2 + 4y^2 = 25$$ Add 25 to each side.

$$x^2 + y^2 = \frac{25}{4}$$ Divide each side by 4.

$$x^2 + y^2 = \left(\frac{5}{2}\right)^2$$ Standard form

Now, from this standard form, you can see that the graph of the equation is a circle that is centered at the origin and has a radius of $\frac{5}{2}$. The graph of the equation of the circle is shown in Figure 10.4.

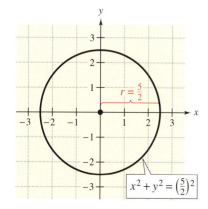

Figure 10.4

3 Graph and write equations of circles centered at (h, k).

Circles Centered at (h, k)

Consider a circle whose radius is r and whose center is the point (h, k), as shown in Figure 10.5. Let (x, y) be any point on the circle. To find an equation for this circle, you can use a variation of the Distance Formula and write

$$\text{Radius} = r = \sqrt{(x - h)^2 + (y - k)^2}. \qquad \textcolor{red}{\text{Distance Formula (See Section 3.1.)}}$$

By squaring each side of this equation, you obtain the equation shown below, which is called the **standard form of the equation of a circle centered at (h, k).**

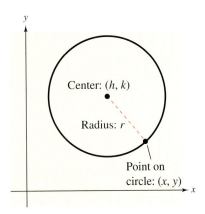

Figure 10.5

Standard Equation of a Circle [Center at (h, k)]

The **standard form of the equation of a circle centered at (h, k)** is

$$(x - h)^2 + (y - k)^2 = r^2.$$

When $h = 0$ and $k = 0$, the circle is centered at the origin. Otherwise, you can use the rules on horizontal and vertical shifts from Section 3.7 to shift the center of the circle h units horizontally and k units vertically from the origin.

Example 3   Writing an Equation of a Circle

The point $(2, 5)$ lies on a circle whose center is $(5, 1)$, as shown in Figure 10.6. Write the standard form of the equation of this circle.

Solution

The radius r of the circle is the distance between $(2, 5)$ and $(5, 1)$.

$$r = \sqrt{(2 - 5)^2 + (5 - 1)^2} \qquad \textcolor{red}{\text{Distance Formula}}$$

$$= \sqrt{(-3)^2 + 4^2} \qquad \textcolor{red}{\text{Simplify.}}$$

$$= \sqrt{9 + 16} \qquad \textcolor{red}{\text{Simplify.}}$$

$$= \sqrt{25} \qquad \textcolor{red}{\text{Simplify.}}$$

$$= 5 \qquad \textcolor{red}{\text{Radius}}$$

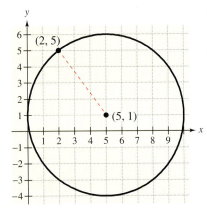

Figure 10.6

Using $(h, k) = (5, 1)$ and $r = 5$, the equation of the circle is

$$(x - h)^2 + (y - k)^2 = r^2 \qquad \textcolor{red}{\text{Standard form}}$$

$$(x - 5)^2 + (y - 1)^2 = 5^2 \qquad \textcolor{red}{\text{Substitute for } h, k, \text{ and } r.}$$

$$(x - 5)^2 + (y - 1)^2 = 25. \qquad \textcolor{red}{\text{Equation of circle}}$$

From the graph, you can see that the center of the circle is shifted five units to the right and one unit upward from the origin.

To write the equation of a circle in standard form, you may need to complete the square, as demonstrated in Example 4.

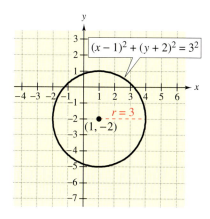

Figure 10.7

Example 4   Writing an Equation in Standard Form

Write the equation

$$x^2 + y^2 - 2x + 4y - 4 = 0$$

in standard form, and sketch the circle represented by the equation.

Solution

$$x^2 + y^2 - 2x + 4y - 4 = 0 \qquad \text{Write original equation.}$$

$$\left(x^2 - 2x + \boxed{}\right) + \left(y^2 + 4y + \boxed{}\right) = 4 \qquad \text{Group terms.}$$

$$[x^2 - 2x + (-1)^2] + (y^2 + 4y + 2^2) = 4 + 1 + 4 \qquad \text{Complete the square.}$$

$$\underbrace{}_{(\text{half})^2} \qquad \underbrace{}_{(\text{half})^2}$$

$$(x - 1)^2 + (y + 2)^2 = 3^2 \qquad \text{Standard form}$$

From this standard form, you can see that the circle has a radius of 3 and that the center of the circle is $(1, -2)$. The graph of the equation of the circle is shown in Figure 10.7. From the graph you can see that the center of the circle is shifted one unit to the right and two units downward from the origin.

Example 5   An Application: Mechanical Drawing

You are in a mechanical drawing class and are asked to help program a computer to model the metal piece shown in Figure 10.8. Part of your assignment is to find an equation for the semicircular upper portion of the hole in the metal piece. What is the equation?

Solution

From the drawing, you can see that the center of the circle is $(h, k) = (5, 2)$ and that the radius of the circle is $r = 1.5$. This implies that the equation of the entire circle is

$$(x - h)^2 + (y - k)^2 = r^2 \qquad \text{Standard form}$$

$$(x - 5)^2 + (y - 2)^2 = 1.5^2 \qquad \text{Substitute for } h, k, \text{ and } r.$$

$$(x - 5)^2 + (y - 2)^2 = 2.25. \qquad \text{Equation of circle}$$

To find the equation of the upper portion of the circle, solve this standard equation for y.

$$(x - 5)^2 + (y - 2)^2 = 2.25$$

$$(y - 2)^2 = 2.25 - (x - 5)^2$$

$$y - 2 = \pm\sqrt{2.25 - (x - 5)^2}$$

$$y = 2 \pm \sqrt{2.25 - (x - 5)^2}$$

Finally, take the positive square root to obtain the equation of the upper portion of the circle.

$$y = 2 + \sqrt{2.25 - (x - 5)^2}$$

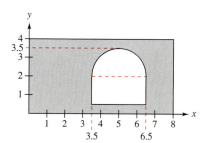

Figure 10.8

Study Tip

In Example 5, if you had wanted the equation of the lower portion of the circle, you would have taken the negative square root

$$y = 2 - \sqrt{2.25 - (x - 5)^2}.$$

④ Graph and write equations of parabolas.

Equations of Parabolas

The second basic type of conic is a **parabola.** In Section 8.4, you studied some of the properties of parabolas. There you saw that the graph of a quadratic function of the form $y = ax^2 + bx + c$ is a parabola that opens upward if a is positive and downward if a is negative. You also learned that each parabola has a vertex and that the vertex of the graph of $y = ax^2 + bx + c$ occurs when $x = -b/(2a)$.

In this section, you will study the technical definition of a parabola, and you will study the equations of parabolas that open to the right and to the left.

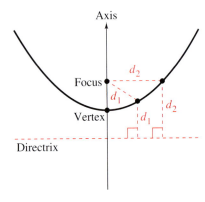

Figure 10.9

Definition of a Parabola

A **parabola** is the set of all points (x, y) that are equidistant from a fixed line (**directrix**) and a fixed point (**focus**) not on the line.

The midpoint between the focus and the directrix is called the **vertex,** and the line passing through the focus and the vertex is called the **axis** of the parabola. Note in Figure 10.9 that a parabola is symmetric with respect to its axis. Using the definition of a parabola, you can derive the **standard form of the equation of a parabola** whose directrix is parallel to the x-axis or to the y-axis.

Standard Equation of a Parabola

The **standard form of the equation of a parabola** with vertex at the origin $(0, 0)$ is

$x^2 = 4py, \quad p \neq 0$ Vertical axis

$y^2 = 4px, \quad p \neq 0.$ Horizontal axis

The focus lies on the axis p units (*directed distance*) from the vertex. If the vertex is at (h, k), then the standard form of the equation is

$(x - h)^2 = 4p(y - k), \quad p \neq 0$ Vertical axis; directrix: $y = k - p$
$(y - k)^2 = 4p(x - h), \quad p \neq 0.$ Horizontal axis; directrix: $x = h - p$

(See Figure 10.10.)

Study Tip

If the focus of a parabola is above or to the right of the vertex, p is positive. If the focus is below or to the left of the vertex, p is negative.

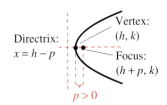

Focus: Vertex:
$(h, k + p)$ (h, k)

Directrix:
$y = k - p$

$p > 0$

Parabola with vertical axis

Directrix: Vertex:
$x = h - p$ (h, k)

Focus:
$(h + p, k)$

$p > 0$

Parabola with horizontal axis

Figure 10.10

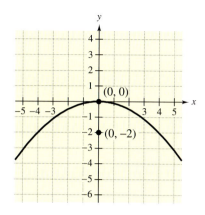

Figure 10.11

Example 6   Writing the Standard Equation of a Parabola

Write the standard form of the equation of the parabola with vertex $(0, 0)$ and focus $(0, -2)$, as shown in Figure 10.11.

Solution

Because the vertex is at the origin and the axis of the parabola is vertical, consider the equation

$$x^2 = 4py$$

where p is the directed distance from the vertex to the focus. Because the focus is two units *below* the vertex, you have $p = -2$. So, the equation of the parabola is

$x^2 = 4py$	Standard form
$x^2 = 4(-2)y$	Substitute for p.
$x^2 = -8y$.	Equation of parabola

Example 7   Writing the Standard Equation of a Parabola

Write the standard form of the equation of the parabola with vertex $(3, -2)$ and focus $(4, -2)$, as shown in Figure 10.12.

Solution

Because the vertex is at $(h, k) = (3, -2)$ and the axis of the parabola is horizontal, consider the equation

$$(y - k)^2 = 4p(x - h)$$

where $h = 3$, $k = -2$, and $p = 1$. So, the equation of the parabola is

$(y - k)^2 = 4p(x - h)$	Standard form
$[y - (-2)]^2 = 4(1)(x - 3)$	Substitute for h, k, and p.
$(y + 2)^2 = 4(x - 3)$.	

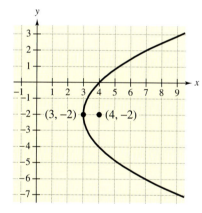

Figure 10.12

Technology: Tip

You cannot represent a circle or a parabola as a single function of x. You can, however, represent it by two functions of x. For instance, try using a graphing calculator to graph the equations below in the same viewing window. Use a viewing window in which $-1 \leq x \leq 10$ and $-8 \leq y \leq 4$. Do you obtain a parabola? Does the graphing calculator connect the two portions of the parabola?

$y_1 = -2 + 2\sqrt{x - 3}$	Upper portion of parabola
$y_2 = -2 - 2\sqrt{x - 3}$	Lower portion of parabola

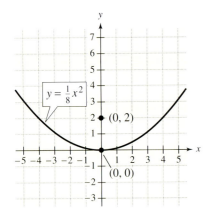

Figure 10.13

Example 8   Analyzing a Parabola

Sketch the graph of the parabola $y = \frac{1}{8}x^2$ and identify its vertex and focus.

Solution

Because the equation can be written in the standard form $x^2 = 4py$, it is a parabola whose vertex is at the origin. You can identify the focus of the parabola by writing its equation in standard form.

$$y = \frac{1}{8}x^2 \qquad \text{Write original equation.}$$

$$\frac{1}{8}x^2 = y \qquad \text{Interchange sides of the equation.}$$

$$x^2 = 8y \qquad \text{Multiply each side by 8.}$$

$$x^2 = 4(2)y \qquad \text{Rewrite 8 in the form } 4p.$$

From this standard form, you can see that $p = 2$. Because the parabola opens upward, as shown in Figure 10.13, you can conclude that the focus lies $p = 2$ units above the vertex. So, the focus is $(0, 2)$.

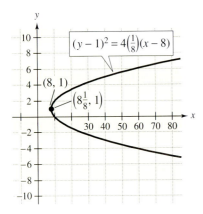

Figure 10.14

Example 9   Analyzing a Parabola

Sketch the parabola $x = 2y^2 - 4y + 10$ and identify its vertex and focus.

Solution

This equation can be written in the standard form $(y - k)^2 = 4p(x - h)$. To do this, you can complete the square, as follows.

$$x = 2y^2 - 4y + 10 \qquad \text{Write original equation.}$$

$$2y^2 - 4y + 10 = x \qquad \text{Interchange sides of the equation.}$$

$$y^2 - 2y + 5 = \frac{1}{2}x \qquad \text{Divide each side by 2.}$$

$$y^2 - 2y = \frac{1}{2}x - 5 \qquad \text{Subtract 5 from each side.}$$

$$y^2 - 2y + 1 = \frac{1}{2}x - 5 + 1 \qquad \text{Complete the square on left side.}$$

$$(y - 1)^2 = \frac{1}{2}x - 4 \qquad \text{Simplify.}$$

$$(y - 1)^2 = \frac{1}{2}(x - 8) \qquad \text{Factor.}$$

$$(y - 1)^2 = 4\left(\tfrac{1}{8}\right)(x - 8) \qquad \text{Rewrite } \tfrac{1}{2} \text{ in the form } 4p.$$

From this standard form, you can see that the vertex is $(h, k) = (8, 1)$ and $p = \frac{1}{8}$. Because the parabola opens to the right, as shown in Figure 10.14, the focus lies $p = \frac{1}{8}$ unit to the right of the vertex. So, the focus is $\left(8\frac{1}{8}, 1\right)$.

Parabolas occur in a wide variety of applications. For instance, a parabolic reflector can be formed by revolving a parabola around its axis. The resulting surface has the property that all incoming rays parallel to the axis are reflected through the focus of the parabola. This is the principle behind the construction of the parabolic mirrors used in reflecting telescopes. Conversely, the light rays emanating from the focus of a parabolic reflector used in a flashlight are all parallel to one another, as shown in Figure 10.15.

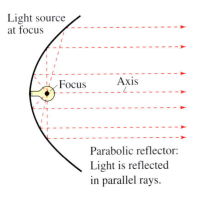

Parabolic reflector: Light is reflected in parallel rays.

Figure 10.15

10.1 Exercises

Review   Concepts, Skills, and Problem Solving

Keep mathematically in shape by doing these exercises *before* the problems of this section.

Expressions and Equations

In Exercises 1–4, expand and simplify the expression.

1. $(x + 6)^2 - 5$

2. $(x - 7)^2 - 2$

3. $12 - (x - 8)^2$

4. $16 - (x + 1)^2$

In Exercises 5–8, complete the square for the quadratic expression.

5. $x^2 - 4x + 1$

6. $x^2 + 12x - 3$

7. $-x^2 + 6x + 5$

8. $-2x^2 + 10x - 14$

In Exercises 9 and 10, find an equation of the line passing through the point with the specified slope.

9. $(-2, 5)$, $m = \frac{5}{8}$

10. $(1, 7)$, $m = -\frac{2}{5}$

Problem Solving

11. *Reduced Rates*   A service organization paid $288 for a block of tickets to a ball game. The block contained three more tickets than the organization needed for its members. By inviting three more people to attend and share the cost, the organization lowered the price per ticket by $8. How many people are going to the game?

12. *Investment*   To begin a small business, $135,000 is needed. The cost will be divided equally among the investors. Some people have made a commitment to invest. If three more investors could be found, the amount required from each would decrease by $1500. How many people have made a commitment to invest in the business?

Developing Skills

In Exercises 1–6, match the equation with its graph. [The graphs are labeled (a), (b), (c), (d), (e), and (f).]

(a)
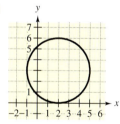

(b)
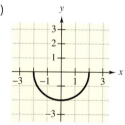

(c)
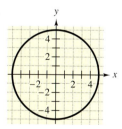

(d)
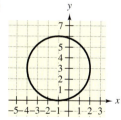

(e)
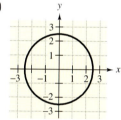

(f)
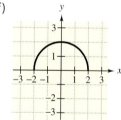

1. $x^2 + y^2 = 25$

2. $4x^2 + 4y^2 = 25$

3. $(x - 2)^2 + (y - 3)^2 = 9$

4. $(x + 1)^2 + (y - 3)^2 = 9$

5. $y = -\sqrt{4 - x^2}$

6. $y = \sqrt{4 - x^2}$

In Exercises 7–14, write the standard form of the equation of the circle with center at $(0, 0)$ that satisfies the criterion. See Example 1.

7. Radius: 5

8. Radius: 7

9. Radius: $\frac{2}{3}$

10. Radius: $\frac{5}{2}$

11. Passes through the point $(0, 8)$

12. Passes through the point $(-2, 0)$

13. Passes through the point $(5, 2)$

14. Passes through the point $(-1, -4)$

In Exercises 15–22, write the standard form of the equation of the circle with center at (h, k) that satisfies the criteria. See Example 3.

15. Center: $(4, 3)$
 Radius: 10

16. Center: $(-2, 5)$
 Radius: 6

17. Center: $(5, -3)$
 Radius: 9

18. Center: $(-5, -2)$
 Radius: $\frac{5}{2}$

19. Center: $(-2, 1)$
 Passes through the point $(0, 1)$

20. Center: $(8, 2)$
 Passes through the point $(8, 0)$

21. Center: $(3, 2)$
 Passes through the point $(4, 6)$

22. Center: $(-3, -5)$
 Passes through the point $(0, 0)$

In Exercises 23–42, identify the center and radius of the circle and sketch the circle. See Examples 2 and 4.

23. $x^2 + y^2 = 16$

24. $x^2 + y^2 = 1$

25. $x^2 + y^2 = 36$

26. $x^2 + y^2 = 10$

27. $4x^2 + 4y^2 = 1$

28. $9x^2 + 9y^2 = 64$

29. $25x^2 + 25y^2 - 144 = 0$

30. $\dfrac{x^2}{4} + \dfrac{y^2}{4} - 1 = 0$

31. $(x + 1)^2 + (y - 5)^2 = 64$

32. $(x + 10)^2 + (y + 1)^2 = 100$

33. $(x - 2)^2 + (y - 3)^2 = 4$

34. $(x + 4)^2 + (y - 3)^2 = 25$

35. $\left(x + \frac{5}{2}\right)^2 + (y + 3)^2 = 9$

36. $(x - 5)^2 + \left(y + \frac{3}{4}\right)^2 = 1$

37. $x^2 + y^2 - 4x - 2y + 1 = 0$

38. $x^2 + y^2 + 6x - 4y - 3 = 0$

39. $x^2 + y^2 + 2x + 6y + 6 = 0$

40. $x^2 + y^2 - 2x + 6y - 15 = 0$

41. $x^2 + y^2 + 8x + 4y - 5 = 0$

42. $x^2 + y^2 - 14x + 8y + 56 = 0$

 In Exercises 43–46, use a graphing calculator to graph the circle. (*Note:* Solve for *y*. Use the square setting so the circles appear correct.)

43. $x^2 + y^2 = 30$ **44.** $4x^2 + 4y^2 = 45$

45. $(x - 2)^2 + y^2 = 10$ **46.** $(x + 3)^2 + y^2 = 15$

In Exercises 47–52, match the equation with its graph. [The graphs are labeled (a), (b), (c), (d), (e), and (f).]

(a)

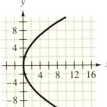

(b)
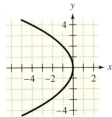

(c)

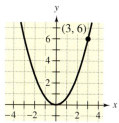

(d)

(e)

(f)

47. $y^2 = -4x$

48. $x^2 = 2y$

49. $x^2 = -8y$

50. $y^2 = 12x$

51. $(y - 1)^2 = 4(x - 3)$

52. $(x + 3)^2 = -2(y - 1)$

In Exercises 53–64, write the standard form of the equation of the parabola with its vertex at the origin. See Example 6.

53.

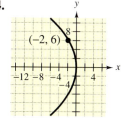

54.

55. Focus: $\left(0, -\frac{3}{2}\right)$

56. Focus: $(2, 0)$

57. Focus: $(-2, 0)$

58. Focus: $(0, -2)$

59. Focus: $(0, 1)$

60. Focus: $(-3, 0)$

61. Focus: $(4, 0)$

62. Focus: $(0, 2)$

63. Horizontal axis and passes through the point $(4, 6)$

64. Vertical axis and passes through the point $(-2, -2)$

In Exercises 65–74, write the standard form of the equation of the parabola with its vertex at (h, k). See Example 7.

65.

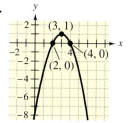

66.

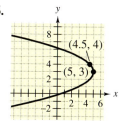

67.

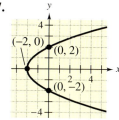

68.

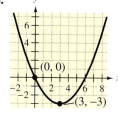

69. Vertex: $(3, 2)$;   Focus: $(1, 2)$

70. Vertex: $(-1, 2)$;   Focus: $(-1, 0)$

71. Vertex: $(0, 4)$;   Focus: $(0, 6)$

72. Vertex: $(-2, 1)$;   Focus: $(-5, 1)$

73. Vertex: $(0, 2)$;

Horizontal axis and passes through $(1, 3)$

74. Vertex: $(0, 2)$;

Vertical axis and passes through $(6, 0)$

In Exercises 75–88, identify the vertex and focus of the parabola and sketch the parabola. See Examples 8 and 9.

75. $y = \frac{1}{2}x^2$

76. $y = 2x^2$

77. $y^2 = -6x$

78. $y^2 = 3x$

79. $x^2 + 8y = 0$

80. $x + y^2 = 0$

81. $(x - 1)^2 + 8(y + 2) = 0$

82. $(x + 3) + (y - 2)^2 = 0$

83. $\left(y + \frac{1}{2}\right)^2 = 2(x - 5)$

84. $\left(x + \frac{1}{2}\right)^2 = 4(y - 3)$

85. $y = \frac{1}{4}(x^2 - 2x + 5)$

86. $4x - y^2 - 2y - 33 = 0$

87. $y^2 + 6y + 8x + 25 = 0$

88. $y^2 - 4y - 4x = 0$

In Exercises 89–92, use a graphing calculator to graph the parabola. Identify the vertex and focus.

89. $y = -\frac{1}{6}(x^2 + 4x - 2)$

90. $x^2 - 2x + 8y + 9 = 0$

91. $y^2 + x + y = 0$
92. $y^2 - 4x - 4 = 0$

Solving Problems

93. *Satellite Orbit* Write an equation of the circular orbit of a satellite 500 miles above the surface of Earth. Place the origin of the rectangular coordinate system at the center of Earth and assume the radius of Earth is 4000 miles.

94. *Architecture* The top portion of a stained glass window is in the form of a pointed Gothic arch (see figure). Each side of the arch is an arc of a circle of radius 12 feet and center at the base of the opposite arch. Write an equation of one of the circles and use it to determine the height of the point of the arch above the horizontal base of the window.

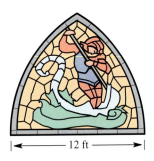

|← —— 12 ft —— →|

95. *Architecture* A semicircular arch for a tunnel under a river has a diameter of 100 feet (see figure). Write an equation of the semicircle. Determine the height of the arch 5 feet from the edge of the tunnel.

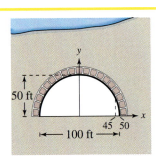

Figure for 95

96. *Graphical Estimation* A rectangle centered at the origin with sides parallel to the coordinate axes is placed in a circle of radius 25 inches centered at the origin (see figure). The length of the rectangle is $2x$ inches.

(a) Show that the width and area of the rectangle are given by $2\sqrt{625 - x^2}$ and $4x\sqrt{625 - x^2}$, respectively.

(b) Use a graphing calculator to graph the area function. Approximate the value of x for which the area is maximum.

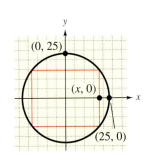

97. *Suspension Bridge*   Each cable of a suspension bridge is suspended (in the shape of a parabola) between two towers that are 120 meters apart, and the top of each tower is 20 meters above the road way. The cables touch the roadway at the midpoint between the two towers (see figure).

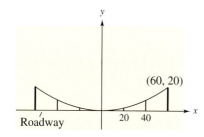

(a) Write an equation for the parabolic shape of each cable.

(b) Complete the table by finding the height of the suspension cables y over the roadway at a distance of x meters from the center of the bridge.

x	0	20	40	60
y				

98. *Beam Deflection*   A simply supported beam is 16 meters long and has a load at the center (see figure). The deflection of the beam at its center is 3 centimeters. Assume that the shape of the deflected beam is parabolic.

(a) Write an equation of the parabola. (Assume that the origin is at the center of the deflected beam.)

(b) How far from the center of the beam is the deflection equal to 1 centimeter?

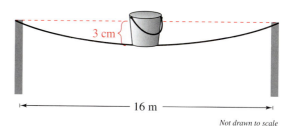

Figure for 98

99. 🖩 *Revenue*   The revenue R generated by the sale of x computer desks is given by $R = 375x - \frac{3}{2}x^2$.

(a) Use a graphing calculator to graph the function.

(b) Use the graph to approximate the number of sales that will maximize revenue.

100. 🖩 *Path of a Softball*   The path of a softball is given by $y = -0.08x^2 + x + 4$. The coordinates x and y are measured in feet, with $x = 0$ corresponding to the position from which the ball was thrown.

(a) Use a graphing calculator to graph the path of the softball.

(b) Move the cursor along the path to approximate the highest point and the range of the path.

Explaining Concepts

101. ⚡ Answer parts (a) and (b) of Motivating the Chapter on page 650.

102. *Writing*✏   Name the four types of conics.

103. *Writing*✏   Define a circle and write the standard form of the equation of a circle centered at the origin.

104. *Writing*✏   Explain how to use the method of completing the square to write an equation of a circle in standard form.

105. *Writing*✏   Explain the significance of a parabola's directrix and focus.

106. *Writing*✏   Is y a function of x in the equation $y^2 = 6x$? Explain.

107. *Writing*✏   Is it possible for a parabola to intersect its directrix? Explain.

108. *Writing*✏   If the vertex and focus of a parabola are on a horizontal line, is the directrix of the parabola vertical? Explain.

10.2 Ellipses

David Young-Wolff/PhotoEdit, Inc.

What You Should Learn

1 Graph and write equations of ellipses centered at the origin.

2 Graph and wrire equations of ellipses centered at (h, k).

Why You Should Learn It

Equations of ellipses can be used to model and solve real-life problems. For instance, in Exercise 58 on page 673, you will use an equation of an ellipse to model a chainwheel.

1 Graph and write equations of ellipses centered at the origin.

Ellipses Centered at the Origin

The third type of conic is called an *ellipse* and is defined as follows.

Definition of an Ellipse

An **ellipse** in the rectangular coordinate system consists of all points (x, y) such that the sum of the distances between (x, y) and two distinct fixed points is a constant, as shown in Figure 10.16. Each of the two fixed points is called a **focus** of the ellipse. (The plural of focus is *foci*.)

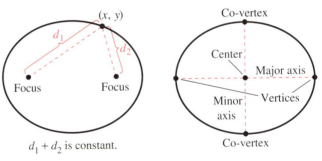

Figure 10.16   Figure 10.17

The line through the foci intersects the ellipse at two points, called the **vertices,** as shown in Figure 10.17. The line segment joining the vertices is called the **major axis,** and its midpoint is called the **center** of the ellipse. The line segment perpendicular to the major axis at the center is called the **minor axis** of the ellipse, and the points at which the minor axis intersects the ellipse are called **co-vertices.**

To trace an ellipse, place two thumbtacks at the foci, as shown in Figure 10.18. If the ends of a fixed length of string are fastened to the thumbtacks and the string is drawn taut with a pencil, the path traced by the pencil will be an ellipse.

Figure 10.18

The standard form of the equation of an ellipse takes one of two forms, depending on whether the major axis is horizontal or vertical.

Standard Equation of an Ellipse (Center at Origin)

The **standard form of the equation of an ellipse centered at the origin** with major and minor axes of lengths $2a$ and $2b$ is

$$\frac{x^2}{a^2} + \frac{y^2}{b^2} = 1 \quad \text{or} \quad \frac{x^2}{b^2} + \frac{y^2}{a^2} = 1, \qquad 0 < b < a.$$

The vertices lie on the major axis, a units from the center, and the co-vertices lie on the minor axis, b units from the center, as shown in Figure 10.19.

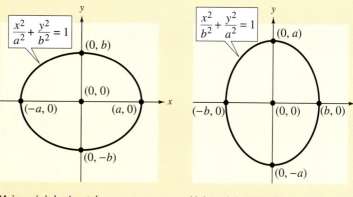

Major axis is horizontal.
Minor axis is vertical.

Major axis is vertical.
Minor axis is horizontal.

Figure 10.19

Example 1   Writing the Standard Equation of an Ellipse

Write an equation of the ellipse that is centered at the origin, with vertices $(-3, 0)$ and $(3, 0)$ and co-vertices $(0, -2)$ and $(0, 2)$.

Solution

Begin by plotting the vertices and co-vertices, as shown in Figure 10.20. The center of the ellipse is $(0, 0)$. So, the equation of the ellipse has the form

$$\frac{x^2}{a^2} + \frac{y^2}{b^2} = 1. \qquad \color{red}{\text{Major axis is horizontal.}}$$

For this ellipse, the major axis is horizontal. So, a is the distance between the center and either vertex, which implies that $a = 3$. Similarly, b is the distance between the center and either co-vertex, which implies that $b = 2$. So, the standard form of the equation of the ellipse is

$$\frac{x^2}{3^2} + \frac{y^2}{2^2} = 1. \qquad \color{red}{\text{Standard form}}$$

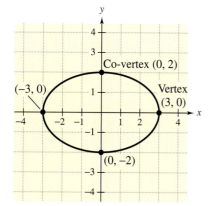

Figure 10.20

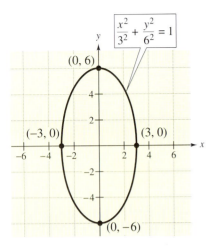

$$\frac{x^2}{3^2} + \frac{y^2}{6^2} = 1$$

Figure 10.21

Example 2   Sketching an Ellipse

Sketch the ellipse given by $4x^2 + y^2 = 36$. Identify the vertices and co-vertices.

Solution

To sketch an ellipse, it helps first to write its equation in standard form.

$4x^2 + y^2 = 36$	Write original equation.
$\dfrac{x^2}{9} + \dfrac{y^2}{36} = 1$	Divide each side by 36 and simplify.
$\dfrac{x^2}{3^2} + \dfrac{y^2}{6^2} = 1$	Standard form

Because the denominator of the y^2-term is larger than the denominator of the x^2-term, you can conclude that the major axis is vertical. Moreover, because $a = 6$, the vertices are $(0, -6)$ and $(0, 6)$. Finally, because $b = 3$, the co-vertices are $(-3, 0)$ and $(3, 0)$, as shown in Figure 10.21.

---

② Graph and write equations of ellipses centered at (h, k).

Ellipses Centered at (h, k)

Standard Equation of an Ellipse [Center at (h, k)]

The **standard form of the equation of an ellipse centered at (h, k)** with major and minor axes of lengths $2a$ and $2b$, where $0 < b < a$, is

$$\frac{(x - h)^2}{a^2} + \frac{(y - k)^2}{b^2} = 1 \qquad \text{Major axis is horizontal.}$$

or

$$\frac{(x - h)^2}{b^2} + \frac{(y - k)^2}{a^2} = 1. \qquad \text{Major axis is vertical.}$$

The foci lie on the major axis, c units from the center, with $c^2 = a^2 - b^2$.

Figure 10.22 shows the horizontal and vertical orientations for an ellipse.

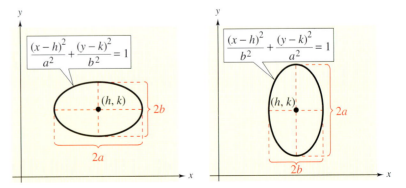

Figure 10.22

When $h = 0$ and $k = 0$, the ellipse is centered at the origin. Otherwise, you can use the rules for horizontal and vertical shifts from Section 3.7 to shift the center of the ellipse h units horizontally and k units vertically from the origin.

Example 3   Writing the Standard Equation of an Ellipse

Write the standard form of the equation of the ellipse with vertices $(-2, 2)$ and $(4, 2)$ and co-vertices $(1, 3)$ and $(1, 1)$, as shown in Figure 10.23.

Solution

Because the vertices are $(-2, 2)$ and $(4, 2)$, the center of the ellipse is $(h, k) = (1, 2)$. The distance from the center to either vertex is $a = 3$, and the distance to either co-vertex is $b = 1$. Because the major axis is horizontal, the standard form of the equation is

$$\frac{(x - h)^2}{a^2} + \frac{(y - k)^2}{b^2} = 1. \qquad \text{Major axis is horizontal.}$$

Substitute the values of h, k, a, and b to obtain

$$\frac{(x - 1)^2}{3^2} + \frac{(y - 2)^2}{1^2} = 1. \qquad \text{Standard form}$$

From the graph, you can see that the center of the ellipse is shifted one unit to the right and two units upward from the origin.

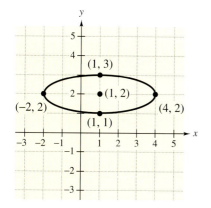

Figure 10.23

Technology: Tip

You can use a graphing calculator to graph an ellipse by graphing the upper and lower portions in the same viewing window. For instance, to graph the ellipse $x^2 + 4y^2 = 4$, first solve for y to obtain

$$y_1 = \frac{1}{2}\sqrt{4 - x^2}$$

and

$$y_2 = -\frac{1}{2}\sqrt{4 - x^2}.$$

Use a viewing window in which $-3 \le x \le 3$ and $-2 \le y \le 2$. You should obtain the graph shown below.

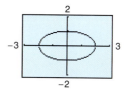

Use this information to graph the ellipse in Example 3 on your graphing calculator.

To write an equation of an ellipse in standard form, you must group the x-terms and the y-terms and then complete each square, as shown in Example 4.

Example 4   Sketching an Ellipse

Sketch the ellipse given by $4x^2 + y^2 - 8x + 6y + 9 = 0$.

Solution

Begin by writing the equation in standard form. In the fourth step, note that 9 and 4 are added to *each* side of the equation.

$$4x^2 + y^2 - 8x + 6y + 9 = 0 \qquad \text{Write original equation.}$$

$$(4x^2 - 8x +) + (y^2 + 6y +) = -9 \qquad \text{Group terms.}$$

$$4(x^2 - 2x +) + (y^2 + 6y +) = -9 \qquad \text{Factor 4 out of } x\text{-terms.}$$

$$4(x^2 - 2x + 1) + (y^2 + 6y + 9) = -9 + 4(1) + 9 \qquad \text{Complete each square.}$$

$$4(x - 1)^2 + (y + 3)^2 = 4 \qquad \text{Simplify.}$$

$$\frac{(x - 1)^2}{1} + \frac{(y + 3)^2}{4} = 1 \qquad \text{Divide each side by 4.}$$

$$\frac{(x - 1)^2}{1^2} + \frac{(y + 3)^2}{2^2} = 1 \qquad \text{Standard form}$$

Now you can see that the center of the ellipse is at $(h, k) = (1, -3)$. Because the denominator of the y^2-term is larger than the denominator of the x^2-term, you can conclude that the major axis is vertical. Because the denominator of the x^2-term is $b^2 = 1^2$, you can locate the endpoints of the minor axis one unit to the right and left of the center, and because the denominator of the y^2-term is $a^2 = 2^2$, you can locate the endpoints of the major axis two units upward and downward from the center, as shown in Figure 10.24. To complete the graph, sketch an oval shape that is determined by the vertices and co-vertices, as shown in Figure 10.25.

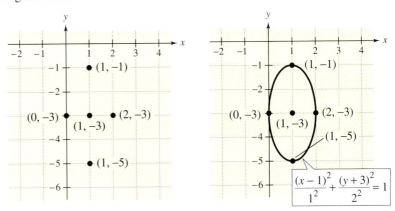

Figure 10.24   **Figure 10.25**

From Figure 10.25, you can see that the center of the ellipse is shifted one unit to the right and three units downward from the origin.

Figure 10.26

Example 5   An Application: Semielliptical Archway

You are responsible for designing a semielliptical archway, as shown in Figure 10.26. The height of the archway is 10 feet and its width is 30 feet. Write an equation of the ellipse and use the equation to sketch an accurate diagram of the archway.

Solution

To make the equation simple, place the origin at the center of the ellipse. This means that the standard form of the equation is

$$\frac{x^2}{a^2} + \frac{y^2}{b^2} = 1.$$ Major axis is horizontal.

Because the major axis is horizontal, it follows that $a = 15$ and $b = 10$, which implies that the equation is

$$\frac{x^2}{15^2} + \frac{y^2}{10^2} = 1.$$ Standard form

To make an accurate sketch of the ellipse, solve this equation for y as follows.

$$\frac{x^2}{225} + \frac{y^2}{100} = 1$$ Simplify denominators.

$$\frac{y^2}{100} = 1 - \frac{x^2}{225}$$ Subtract $\frac{x^2}{225}$ from each side.

$$y^2 = 100\left(1 - \frac{x^2}{225}\right)$$ Multiply each side by 100.

$$y = 10\sqrt{1 - \frac{x^2}{225}}$$ Take the positive square root of each side.

Next, calculate several y-values for the archway, as shown in the table. Then use the values in the table to sketch the archway, as shown in Figure 10.27.

x	± 15	± 12.5	± 10	± 7.5	± 5	± 2.5	0
y	0	5.53	7.45	8.66	9.43	9.86	10

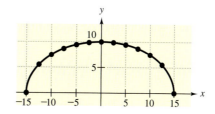

Figure 10.27

10.2 Exercises

Review   *Concepts, Skills, and Problem Solving*

Keep mathematically in shape by doing these exercises *before* the problems of this section.

Simplifying Expressions

In Exercises 1–4, simplify the expression.

1. $\dfrac{15y^{-3}}{10y^2}$

2. $\left(\dfrac{3x^2}{2y}\right)^{-2}$

3. $\dfrac{3x^2y^3}{18x^{-1}y^2}$

4. $(x^2 + 1)^0$

Solving Equations

In Exercises 5 and 6, solve the quadratic equation by completing the square.

5. $x^2 + 6x - 4 = 0$

6. $2x^2 - 16x + 5 = 0$

In Exercises 7–10, sketch the graph of the equation.

7. $y = \frac{2}{5}x + 3$

8. $y = -2x - 1$

9. $y = x^2 - 12x + 36$

10. $y = 25 - x^2$

Problem Solving

11. *Test Scores*   A student has test scores of 90, 74, 82, and 90. The next examination is the final examination, which counts as two tests. What score does the student need on the final examination to produce an average score of 85?

12. *Simple Interest*   An investment of $2500 is made at an annual simple interest rate of 5.5%. How much additional money must be invested at an annual simple interest rate of 8% so that the total interest earned is 7% of the total investment?

Developing Skills

In Exercises 1–6, match the equation with its graph. [The graphs are labeled (a), (b), (c), (d), (e), and (f).]

(a)
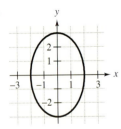

(b)
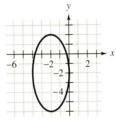

(c)
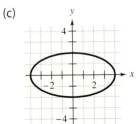

(d)
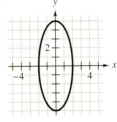

(e)
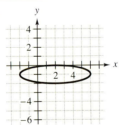

(f)
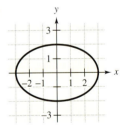

1. $\dfrac{x^2}{4} + \dfrac{y^2}{9} = 1$

2. $\dfrac{x^2}{9} + \dfrac{y^2}{4} = 1$

3. $\dfrac{x^2}{4} + \dfrac{y^2}{25} = 1$

4. $\dfrac{y^2}{4} + \dfrac{x^2}{16} = 1$

5. $\dfrac{(x - 2)^2}{16} + \dfrac{(y + 1)^2}{1} = 1$

6. $\dfrac{(x + 2)^2}{4} + \dfrac{(y + 2)^2}{16} = 1$

In Exercises 7–18, write the standard form of the equation of the ellipse centered at the origin. See Example 1.

Vertices	Co-vertices
7. $(-4, 0), (4, 0)$	$(0, -3), (0, 3)$
8. $(-4, 0), (4, 0)$	$(0, -1), (0, 1)$
9. $(-2, 0), (2, 0)$	$(0, -1), (0, 1)$
10. $(-10, 0), (10, 0)$	$(0, -4), (0, 4)$
11. $(0, -4), (0, 4)$	$(-3, 0), (3, 0)$
12. $(0, -5), (0, 5)$	$(-1, 0), (1, 0)$
13. $(0, -2), (0, 2)$	$(-1, 0), (1, 0)$
14. $(0, -8), (0, 8)$	$(-4, 0), (4, 0)$

15. Major axis (vertical) 10 units, minor axis 6 units

16. Major axis (horizontal) 24 units, minor axis 10 units

17. Major axis (horizontal) 20 units, minor axis 12 units

18. Major axis (horizontal) 50 units, minor axis 30 units

In Exercises 19–32, sketch the ellipse. Identify the vertices and co-vertices. See Example 2.

19. $\dfrac{x^2}{16} + \dfrac{y^2}{4} = 1$

20. $\dfrac{x^2}{25} + \dfrac{y^2}{9} = 1$

21. $\dfrac{x^2}{4} + \dfrac{y^2}{16} = 1$

22. $\dfrac{x^2}{9} + \dfrac{y^2}{25} = 1$

23. $\dfrac{x^2}{25/9} + \dfrac{y^2}{16/9} = 1$

24. $\dfrac{x^2}{1} + \dfrac{y^2}{1/4} = 1$

25. $\dfrac{9x^2}{4} + \dfrac{25y^2}{16} = 1$

26. $\dfrac{36x^2}{49} + \dfrac{16y^2}{9} = 1$

27. $16x^2 + 25y^2 - 9 = 0$

28. $64x^2 + 36y^2 - 49 = 0$

29. $4x^2 + y^2 - 4 = 0$

30. $4x^2 + 9y^2 - 36 = 0$

31. $10x^2 + 16y^2 - 160 = 0$

32. $16x^2 + 4y^2 - 64 = 0$

▦ In Exercises 33–36, use a graphing calculator to graph the ellipse. Identify the vertices. (Note: Solve for y.)

33. $x^2 + 2y^2 = 4$
34. $9x^2 + y^2 = 64$
35. $3x^2 + y^2 - 12 = 0$
36. $5x^2 + 2y^2 - 10 = 0$

In Exercises 37–40, write the standard form of the equation of the ellipse. See Example 3.

37.

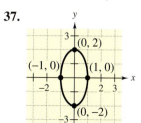

38.

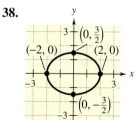

39.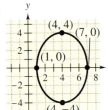

40.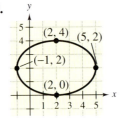

44. $\dfrac{(x + 2)^2}{1/4} + \dfrac{(y + 4)^2}{1} = 1$

45. $9x^2 + 4y^2 + 36x - 24y + 36 = 0$

46. $9x^2 + 4y^2 - 36x + 8y + 31 = 0$

47. $4(x - 2)^2 + 9(y + 2)^2 = 36$

48. $(x + 3)^2 + 9(y + 1)^2 = 81$

In Exercises 41–54, find the center and vertices of the ellipse and sketch the ellipse. See Example 4.

49. $12(x + 4)^2 + 3(y - 1)^2 = 48$

50. $16(x - 2)^2 + 4(y + 3)^2 = 16$

41. $\dfrac{(x + 5)^2}{16} + y^2 = 1$

51. $25x^2 + 9y^2 - 200x + 54y + 256 = 0$

42. $\dfrac{(x - 2)^2}{4} + \dfrac{(y - 3)^2}{9} = 1$

52. $25x^2 + 16y^2 - 150x - 128y + 81 = 0$

43. $\dfrac{(x - 1)^2}{9} + \dfrac{(y - 5)^2}{25} = 1$

53. $x^2 + 4y^2 - 4x - 8y - 92 = 0$

54. $x^2 + 4y^2 + 6x + 16y - 11 = 0$

Solving Problems

55. *Architecture* A semielliptical arch for a tunnel under a river has a width of 100 feet and a height of 40 feet (see figure). Determine the height of the arch 5 feet from the edge of the tunnel.

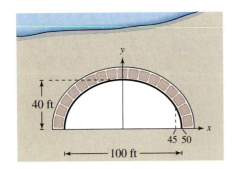

56. *Wading Pool* You are building a wading pool that is in the shape of an ellipse. Your plans give an equation for the elliptical shape of the pool measured in feet as

$$\frac{x^2}{324} + \frac{y^2}{196} = 1.$$

Find the longest distance and shortest distance across the pool.

57. *Sports* In Australia, football by *Australian Rules* (or rugby) is played on elliptical fields. The field can be a maximum of 170 yards wide and a maximum of 200 yards long. Let the center of a field of maximum size be represented by the point $(0, 85)$. Write an equation of the ellipse that represents this field. (Source: Oxford Companion to World Sports and Games)

58. *Bicycle Chainwheel*   The pedals of a bicycle drive a chainwheel, which drives a smaller sprocket wheel on the rear axle (see figure). Many chainwheels are circular. Some, however, are slightly elliptical, which tends to make pedaling easier. Write an equation of an elliptical chainwheel that is 8 inches in diameter at its widest point and $7\frac{1}{2}$ inches in diameter at its narrowest point.

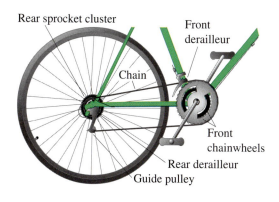

Rear sprocket cluster
Front derailleur
Chain
Front chainwheels
Rear derailleur
Guide pulley

59. *Area*   The area A of the ellipse

$$\frac{x^2}{a^2} + \frac{y^2}{b^2} = 1$$

is given by $A = \pi a b$. Write the equation of an ellipse with an area of 301.59 square units and $a + b = 20$.

60. Sketch a graph of the ellipse that consists of all points (x, y) such that the sum of the distances between (x, y) and two fixed points is 15 units and for which the foci are located at the centers of the two sets of concentric circles in the figure.

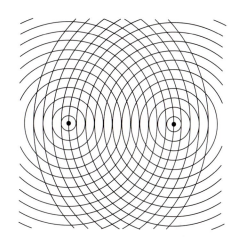

Explaining Concepts

61. *Writing*   Describe the relationship between circles and ellipses. How are they similar? How do they differ?

62. *Writing*   Define an ellipse and write the standard form of the equation of an ellipse centered at the origin.

63. *Writing*   Explain the significance of the foci in an ellipse.

64. *Writing*   Explain how to write an equation of an ellipse if you know the coordinates of the vertices and co-vertices.

65. *Writing*   From its equation, how can you determine the lengths of the axes of an ellipse?

Mid-Chapter Quiz

Take this quiz as you would take a quiz in class. After you are done, check your work against the answers in the back of the book.

1. Write the standard form of the equation of the circle shown in the figure.

2. Write the standard form of the equation of the parabola shown in the figure.

3. Write the standard form of the equation of the ellipse shown in the figure.

4. Write the standard form of the equation of the circle with center $(3, -5)$ and passing through the point $(0, -1)$.

5. Write the standard form of the equation of the parabola with vertex $(2, 3)$ and focus $(2, 1)$.

6. Write the standard form of the equation of the ellipse with vertices $(0, -10)$ and $(0, 10)$ and co-vertices $(-3, 0)$ and $(3, 0)$.

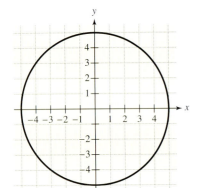

Figure for 1

In Exercises 7 and 8, write the equation of the circle in standard form, then find the center and the radius of the circle.

7. $x^2 + y^2 - 10x + 16 = 0$

8. $x^2 + y^2 + 2x - 4y + 4 = 0$

In Exercises 9 and 10, write the equation of the parabola in standard form, then find the vertex and the focus of the parabola.

9. $x = y^2 - 6y - 7$

10. $x^2 - 8x + y + 12 = 0$

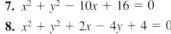

Figure for 2

In Exercises 11 and 12, write the equation of the ellipse in standard form, then find the center and the vertices of the ellipse.

11. $20x^2 + 9y^2 - 180 = 0$

12. $4x^2 + 9y^2 - 48x + 36y + 144 = 0$

In Exercises 13–18, sketch the graph of the equation.

13. $(x + 5)^2 + (y - 1)^2 = 9$

14. $\dfrac{x^2}{9} + \dfrac{y^2}{16} = 1$

15. $x = -y^2 - 4y$

16. $x^2 + (y + 4)^2 = 1$

17. $y = x^2 - 2x + 1$

18. $4(x + 3)^2 + (y - 2)^2 = 16$

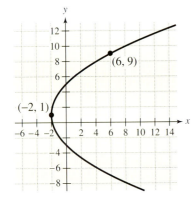

Figure for 3

10.3 Hyperbolas

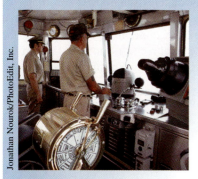

Jonathan Nourok/PhotoEdit, Inc.

What You Should Learn

① Graph and write equations of hyperbolas centered at the origin.

② Graph and write equations of hyperbolas centered at (h, k).

Why You Should Learn It

Equations of hyperbolas are often used in navigation. For instance, in Exercise 43 on page 682, a hyperbola is used to model long-distance radio navigation for a ship.

① Graph and write equations of hyperbolas centered at the origin.

Hyperbolas Centered at the Origin

The fourth basic type of conic is called a **hyperbola** and is defined as follows.

Definition of a Hyperbola

A **hyperbola** on the rectangular coordinate system consists of all points (x, y) such that the *difference* of the distances between (x, y) and two fixed points is a positive constant, as shown in Figure 10.28. The two fixed points are called the **foci** of the hyperbola. The line on which the foci lie is called the **transverse axis** of the hyperbola.

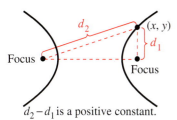

$d_2 - d_1$ is a positive constant.

Figure 10.28

Standard Equation of a Hyperbola (Center at Origin)

The **standard form of the equation of a hyperbola centered at the origin** is

$$\frac{x^2}{a^2} - \frac{y^2}{b^2} = 1 \qquad \text{or} \qquad \frac{y^2}{a^2} - \frac{x^2}{b^2} = 1$$

Transverse axis   Transverse axis
is horizontal.   is vertical.

where a and b are positive real numbers. The **vertices** of the hyperbola lie on the transverse axis, a units from the center, as shown in Figure 10.29.

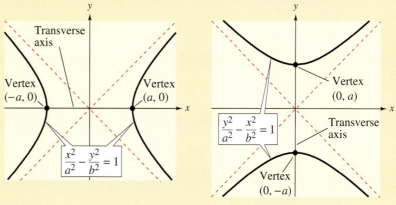

Figure 10.29

A hyperbola has two disconnected parts, each of which is called a **branch** of the hyperbola. The two branches approach a pair of intersecting lines called the **asymptotes** of the hyperbola. The two asymptotes intersect at the center of the hyperbola. To sketch a hyperbola, form a **central rectangle** that is centered at the origin and has side lengths of $2a$ and $2b$. Note in Figure 10.30 that the asymptotes pass through the corners of the central rectangle and that the vertices of the hyperbola lie at the centers of opposite sides of the central rectangle.

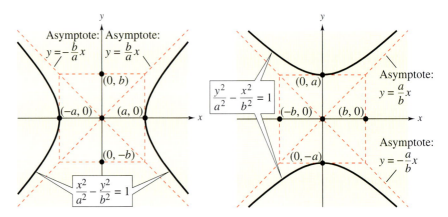

Tranverse axis is horizontal.   Tranverse axis is vertical.
Figure 10.30

Example 1   Sketching a Hyperbola

Identify the vertices of the hyperbola given by the equation, and sketch the hyperbola.

$$\frac{x^2}{36} - \frac{y^2}{16} = 1$$

Solution

From the standard form of the equation

$$\frac{x^2}{6^2} - \frac{y^2}{4^2} = 1$$

you can see that the center of the hyperbola is the origin and the transverse axis is horizontal. So, the vertices lie six units to the left and right of the center at the points

$$(-6, 0) \text{ and } (6, 0).$$

Because $a = 6$ and $b = 4$, you can sketch the hyperbola by first drawing a central rectangle with a width of $2a = 12$ and a height of $2b = 8$, as shown in Figure 10.31. Next, draw the asymptotes of the hyperbola through the corners of the central rectangle and plot the vertices. Finally, draw the hyperbola, as shown in Figure 10.32.

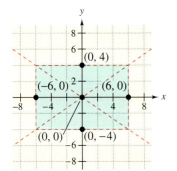

Figure 10.31

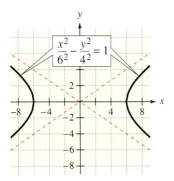

Figure 10.32

Writing the equation of a hyperbola is a little more difficult than writing equations of the other three types of conics. However, if you know the vertices and the asymptotes, you can find the values of a and b, which enable you to write the equation. Notice in Example 2 that the key to this procedure is knowing that the central rectangle has a width of $2b$ and a height of $2a$.

Example 2   Writing the Equation of a Hyperbola

Write the standard form of the equation of the hyperbola with a vertical transverse axis and vertices $(0, 3)$ and $(0, -3)$. The equations of the asymptotes of the hyperbola are $y = \frac{3}{5}x$ and $y = -\frac{3}{5}x$.

Solution

To begin, sketch the lines that represent the asymptotes, as shown in Figure 10.33. Note that these two lines intersect at the origin, which implies that the center of the hyperbola is $(0, 0)$. Next, plot the two vertices at the points $(0, 3)$ and $(0, -3)$. Because you know where the vertices are located, you can sketch the central rectangle of the hyperbola, as shown in Figure 10.33. Note that the corners of the central rectangle occur at the points

$$(-5, 3),\ (5, 3),\ (-5, -3),\ \text{and}\ (5, -3).$$

Because the width of the central rectangle is $2b = 10$, it follows that $b = 5$. Similarly, because the height of the central rectangle is $2a = 6$, it follows that $a = 3$. Now that you know the values of a and b, you can use the standard form of the equation of the hyperbola to write the equation.

$$\frac{y^2}{a^2} - \frac{x^2}{b^2} = 1 \qquad \text{Transverse axis is vertical.}$$

$$\frac{y^2}{3^2} - \frac{x^2}{5^2} = 1 \qquad \text{Substitute 3 for } a \text{ and 5 for } b.$$

$$\frac{y^2}{9} - \frac{x^2}{25} = 1 \qquad \text{Simplify.}$$

The graph is shown in Figure 10.34.

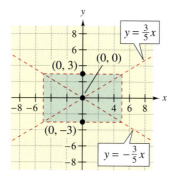

Figure 10.33

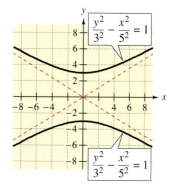

Figure 10.34

Hyperbolas Centered at (h, k)

Standard Equation of a Hyperbola [Center at (h, k)]

The **standard form of the equation of a hyperbola centered at (h, k)** is

$$\frac{(x - h)^2}{a^2} - \frac{(y - k)^2}{b^2} = 1 \qquad \text{Transverse axis is horizontal.}$$

or

$$\frac{(y - k)^2}{a^2} - \frac{(x - h)^2}{b^2} = 1 \qquad \text{Transverse axis is vertical.}$$

where a and b are positive real numbers. The vertices lie on the transverse axis, a units from the center, as shown in Figure 10.35.

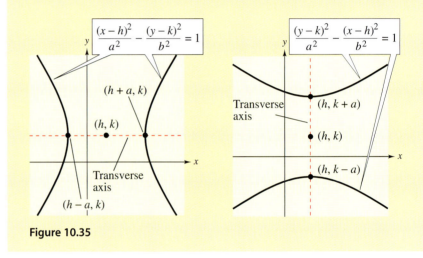

Figure 10.35

When $h = 0$ and $k = 0$, the hyperbola is centered at the origin. Otherwise, you can use the rules on horizontal and vertical shifts from Section 3.7 to shift the center of the hyperbola h units horizontally and k units vertically from the origin.

Example 3   Sketching a Hyperbola

Sketch the hyperbola given by $\dfrac{(y - 1)^2}{9} - \dfrac{(x + 2)^2}{4} = 1.$

Solution

From the form of the equation, you can see that the transverse axis is vertical. The center of the hyperbola is $(h, k) = (-2, 1)$. Because $a = 3$ and $b = 2$, you can begin by sketching a central rectangle that is six units high and four units wide, centered at $(-2, 1)$. Then, sketch the asymptotes by drawing lines through the corners of the central rectangle. Sketch the hyperbola, as shown in Figure 10.36. From the graph, you can see that the center of the hyperbola is shifted two units to the left and one unit upward from the origin.

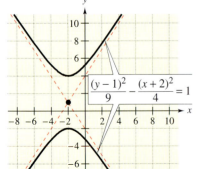

Figure 10.36

Example 4   Sketching a Hyperbola

Sketch the hyperbola given by $x^2 - 4y^2 + 8x + 16y - 4 = 0$.

Solution

Complete the square to write the equation in standard form.

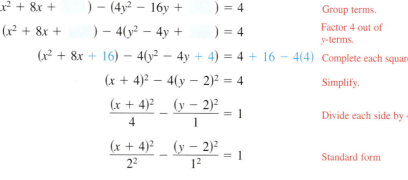

$$x^2 - 4y^2 + 8x + 16y - 4 = 0 \qquad \text{Write original equation.}$$

$$(x^2 + 8x + \quad) - (4y^2 - 16y + \quad) = 4 \qquad \text{Group terms.}$$

$$(x^2 + 8x + \quad) - 4(y^2 - 4y + \quad) = 4 \qquad \text{Factor 4 out of } y\text{-terms.}$$

$$(x^2 + 8x + 16) - 4(y^2 - 4y + 4) = 4 + 16 - 4(4) \qquad \text{Complete each square.}$$

$$(x + 4)^2 - 4(y - 2)^2 = 4 \qquad \text{Simplify.}$$

$$\frac{(x + 4)^2}{4} - \frac{(y - 2)^2}{1} = 1 \qquad \text{Divide each side by 4.}$$

$$\frac{(x + 4)^2}{2^2} - \frac{(y - 2)^2}{1^2} = 1 \qquad \text{Standard form}$$

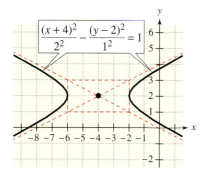

From this standard form, you can see that the transverse axis is horizontal and the center of the hyperbola is $(h, k) = (-4, 2)$. Because $a = 2$ and $b = 1$, you can begin by sketching a central rectangle that is four units wide and two units high, centered at $(-4, 2)$. Then, sketch the asymptotes by drawing lines through the corners of the central rectangle. Sketch the hyperbola, as shown in Figure 10.37. From the graph, you can see that the center of the hyperbola is shifted four units to the left and two units upward from the origin.

Figure 10.37

Technology: Tip

You can use a graphing calculator to graph a hyperbola. For instance, to graph the hyperbola $4y^2 - 9x^2 = 36$, first solve for y to obtain

$$y_1 = 3\sqrt{\frac{x^2}{4} + 1}$$

and

$$y_2 = -3\sqrt{\frac{x^2}{4} + 1}.$$

Use a viewing window in which $-6 \le x \le 6$ and $-8 \le y \le 8$. You should obtain the graph shown below.

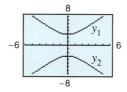

10.3 Exercises

Review　Concepts, Skills, and Problem Solving

Keep mathematically in shape by doing these exercises *before* the problems of this section.

Distance Formula

In Exercises 1 and 2, find the distance between the points.

1. $(5, 2), (-1, 4)$
2. $(-4, -3), (6, 10)$

Graphs

In Exercises 3–6, graph the lines on the same set of coordinate axes.

3. $y = \pm 4x$
4. $y = 6 \pm \frac{1}{3}x$
5. $y = 5 \pm \frac{1}{2}(x - 2)$
6. $y = \pm\frac{1}{3}(x - 6)$

Solving Equations

In Exercises 7–10, find the unknown in the equation $c^2 = a^2 - b^2$. (Assume that a, b, and c are positive.)

7. $a = 25, \ b = 7$
8. $a = \sqrt{41}, \ c = 4$
9. $b = 5, \ c = 12$
10. $a = 6, \ b = 3$

Problem Solving

11. *Average Speed* From a point on a straight road, two people ride bicycles in opposite directions. One person rides at 10 miles per hour and the other rides at 12 miles per hour. In how many hours will they be 55 miles apart?

12. *Mixture Problem* You have a collection of 30 gold coins. Some of the coins are worth $10 each, and the rest are worth $20 each. The value of the entire collection is $540. How many of each type of coin do you have?

Developing Skills

In Exercises 1–6, match the equation with its graph. [The graphs are labeled (a), (b), (c), (d), (e), and (f).]

(a)
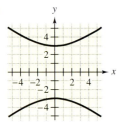

(b)
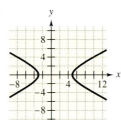

(c)
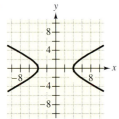

(d)
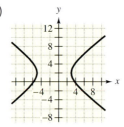

(e)
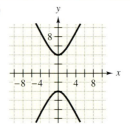

(f)
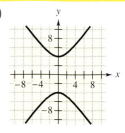

1. $\dfrac{x^2}{16} - \dfrac{y^2}{4} = 1$

2. $\dfrac{y^2}{16} - \dfrac{x^2}{4} = 1$

3. $\dfrac{y^2}{9} - \dfrac{x^2}{16} = 1$

4. $\dfrac{y^2}{16} - \dfrac{x^2}{9} = 1$

5. $\dfrac{(x - 1)^2}{16} - \dfrac{y^2}{4} = 1$

6. $\dfrac{(x + 1)^2}{16} - \dfrac{(y - 2)^2}{9} = 1$

In Exercises 7–18, sketch the hyperbola. Identify the vertices and asymptotes. See Example 1.

7. $x^2 - y^2 = 9$

8. $x^2 - y^2 = 1$

9. $y^2 - x^2 = 9$

10. $y^2 - x^2 = 1$

11. $\dfrac{x^2}{9} - \dfrac{y^2}{25} = 1$

12. $\dfrac{x^2}{4} - \dfrac{y^2}{9} = 1$

13. $\dfrac{y^2}{9} - \dfrac{x^2}{25} = 1$

14. $\dfrac{y^2}{4} - \dfrac{x^2}{9} = 1$

15. $\dfrac{x^2}{1} - \dfrac{y^2}{9/4} = 1$

16. $\dfrac{y^2}{1/4} - \dfrac{x^2}{25/4} = 1$

17. $4y^2 - x^2 + 16 = 0$

18. $4y^2 - 9x^2 - 36 = 0$

In Exercises 19–26, write the standard form of the equation of the hyperbola centered at the origin. See Example 2.

Vertices	Asymptotes	
19. $(-4, 0), (4, 0)$	$y = 2x$	$y = -2x$
20. $(-2, 0), (2, 0)$	$y = \frac{1}{3}x$	$y = -\frac{1}{3}x$
21. $(0, -4), (0, 4)$	$y = \frac{1}{2}x$	$y = -\frac{1}{2}x$
22. $(0, -2), (0, 2)$	$y = 3x$	$y = -3x$
23. $(-9, 0), (9, 0)$	$y = \frac{2}{3}x$	$y = -\frac{2}{3}x$
24. $(-1, 0), (1, 0)$	$y = \frac{1}{2}x$	$y = -\frac{1}{2}x$
25. $(0, -1), (0, 1)$	$y = 2x$	$y = -2x$
26. $(0, -5), (0, 5)$	$y = x$	$y = -x$

In Exercises 27–30, use a graphing calculator to graph the equation. (*Note:* Solve for *y*.)

27. $\dfrac{x^2}{16} - \dfrac{y^2}{4} = 1$

28. $\dfrac{y^2}{16} - \dfrac{x^2}{4} = 1$

29. $5x^2 - 2y^2 + 10 = 0$

30. $x^2 - 2y^2 - 4 = 0$

In Exercises 31–38, find the center and vertices of the hyperbola and sketch the hyperbola. See Examples 3 and 4.

31. $(y + 4)^2 - (x - 3)^2 = 25$

32. $(y + 6)^2 - (x - 2)^2 = 1$

33. $\dfrac{(x - 1)^2}{4} - \dfrac{(y + 2)^2}{1} = 1$

34. $\dfrac{(x - 2)^2}{4} - \dfrac{(y - 3)^2}{9} = 1$

35. $9x^2 - y^2 - 36x - 6y + 18 = 0$

36. $x^2 - 9y^2 + 36y - 72 = 0$

37. $4x^2 - y^2 + 24x + 4y + 28 = 0$

38. $25x^2 - 4y^2 + 100x + 8y + 196 = 0$

In Exercises 39–42, write the standard form of the equation of the hyperbola.

39.

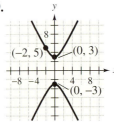

40.

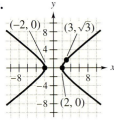

41.

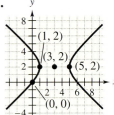

42.

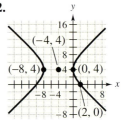

Solving Problems

43. *Navigation* Long-distance radio navigation for aircraft and ships uses synchronized pulses transmitted by widely separated transmitting stations. These pulses travel at the speed of light (186,000 miles per second). The difference in the times of arrival of these pulses at an aircraft or ship is constant on a hyperbola having the transmitting stations as foci. Assume that two stations 300 miles apart are positioned on a rectangular coordinate system at points with coordinates $(-150, 0)$ and $(150, 0)$ and that a ship is traveling on a path with coordinates $(x, 75)$, as shown in the figure. Find the x-coordinate of the position of the ship if the time difference between the pulses from the transmitting stations is 1000 microseconds (0.001 second).

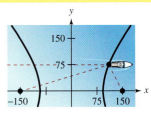

Figure for 43

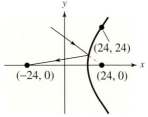

Figure for 44

44. *Optics* A hyperbolic mirror (used in some telescopes) has the property that a light ray directed at the focus will be reflected to the other focus. The focus of a hyperbolic mirror (see figure) has coordinates $(24, 0)$. Find the vertex of the mirror if its mount at the top edge of the mirror has coordinates $(24, 24)$.

Explaining Concepts

45. Answer parts (c) and (d) of Motivating the Chapter on page 650.

46. *Writing* Define a hyperbola and write the standard form of the equation of a hyperbola centered at the origin.

47. *Writing* Explain the significance of the foci in a hyperbola.

48. *Writing* Explain how the central rectangle of a hyperbola can be used to sketch its asymptotes.

49. *Think About It* Describe the part of the hyperbola

$$\frac{(x-3)^2}{4} - \frac{(y-1)^2}{9} = 1$$

given by each equation.
(a) $x = 3 - \frac{2}{3}\sqrt{9 + (y-1)^2}$
(b) $y = 1 + \frac{3}{2}\sqrt{(x-3)^2 - 4}$

50. Cut cone-shaped pieces of styrofoam to demonstrate how to obtain each type of conic section: circle, parabola, ellipse, and hyperbola. Discuss how you could write directions for someone else to form each conic section. Compile a list of real-life situations and/or everyday objects in which conic sections may be seen.

Chapter Test

Take this test as you would take a test in class. After you are done, check your work against the answers in the back of the book.

1. Write the standard form of the equation of the circle shown in the figure.

In Exercises 2 and 3, write the equation of the circle in standard form. Then sketch the circle.

2. $x^2 + y^2 - 2x - 6y + 1 = 0$

3. $x^2 + y^2 + 4x - 6y + 4 = 0$

4. Identify the vertex and the focus of the parabola $x = -3y^2 + 12y - 8$. Then sketch the parabola.

5. Write the standard form of the equation of the parabola with vertex $(7, -2)$ and focus $(7, 0)$.

6. Write the standard form of the equation of the ellipse shown in the figure.

In Exercises 7 and 8, find the center and vertices of the ellipse. Then sketch the ellipse.

7. $16x^2 + 4y^2 = 64$

8. $9x^2 + 4y^2 - 36x + 32y + 64 = 0$

In Exercises 9 and 10, write the standard form of the equation of the hyperbola.

9. Vertices: $(-3, 0), (3, 0)$; Asymptotes: $y = \pm\frac{2}{3}x$

10. Vertices: $(0, -2), (0, 2)$; Asymptotes: $y = \pm 2x$

In Exercises 11 and 12, find the center and vertices of the hyperbola. Then sketch the hyperbola.

11. $9x^2 - 4y^2 + 24y - 72 = 0$

12. $16y^2 - 25x^2 + 64y + 200x - 736 = 0$

In Exercises 13–15, solve the nonlinear system of equations.

13. $\begin{cases} x^2/16 + y^2/9 = 1 \\ 3x + 4y = 12 \end{cases}$ 14. $\begin{cases} x^2 + y^2 = 16 \\ x^2/16 - y^2/9 = 1 \end{cases}$ 15. $\begin{cases} x^2 + y^2 = 10 \\ x^2 = y^2 + 2 \end{cases}$

16. Write the equation of the circular orbit of a satellite 1000 miles above the surface of Earth. Place the origin of the rectangular coordinate system at the center of Earth and assume the radius of Earth to be 4000 miles.

17. A rectangle has a perimeter of 56 inches and a diagonal of length 20 inches. Find the dimensions of the rectangle.

Figure for 1

Figure for 6

Take this test as you would take a test in class. After you are done, check your work against the answers in the back of the book.

In Exercises 1–4, solve the equation by the specified method.

1. Factoring:

$4x^2 - 9x - 9 = 0$

2. Square Root Property:

$(x - 5)^2 - 64 = 0$

3. Completing the square:

$x^2 - 10x - 25 = 0$

4. Quadratic Formula:

$3x^2 + 6x + 2 = 0$

5. Solve the equation of quadratic form: $x - \sqrt{x} - 12 = 0$.

In Exercises 6 and 7, solve the inequality and graph the solution on the real number line.

6. $3x^2 + 5x \le 3$

7. $\dfrac{3x + 4}{2x - 1} < 0$

8. Find a quadratic equation having the solutions -2 and 6.

9. Find the compositions (a) $f \circ g$ and (b) $g \circ f$. Then find the domain of each composition. $f(x) = 2x^2 - 3$, $g(x) = 5x - 1$

10. Find the inverse function of $f(x) = \dfrac{2x + 3}{8}$.

11. Evaluate $f(x) = 7 + 2^{-x}$ when $x = 1, 0.5$, and 3.

12. Sketch the graph of $f(x) = 4^{x-1}$ and identify the horizontal asymptote.

13. Describe the relationship between the graphs of $f(x) = e^x$ and $g(x) = \ln x$.

14. Sketch the graph of $\log_3(x - 1)$ and identify the vertical asymptote.

15. Evaluate $\log_4 \frac{1}{16}$ without using a calculator.

16. Use the properties of logarithms to condense $3(\log_2 x + \log_2 y) - \log_2 z$.

17. Use the properties of logarithms to expand $\ln \dfrac{5x}{(x + 1)^2}$.

In Exercises 18–21, solve the equation.

18. $\log_x \left(\frac{1}{9} \right) = -2$

19. $4 \ln x = 10$

20. $500(1.08)^t = 2000$

21. $3(1 + e^{2x}) = 20$

22. If the inflation rate averages 3.5% over the next 5 years, the approximate cost C of goods and services t years from now is given by

$$C(t) = P(1.035)^t, \quad 0 \le t \le 5$$

where P is the present cost. The price of an oil change is presently $24.95. Estimate the price 5 years from now.

23. Determine the effective yield of an 8% interest rate compounded continuously.

24. Determine the length of time for an investment of $1000 to quadruple in value if the investment earns 9% compounded continuously.

25. Write the equation of the circle in standard form and sketch the circle:

$$x^2 + y^2 - 6x + 14y - 6 = 0$$

26. Identify the vertex and focus of the parabola and sketch the parabola:

$$y = 2x^2 - 20x + 5.$$

27. Write the standard form of the equation of the ellipse shown in the figure.

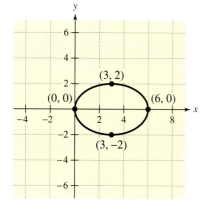

Figure for 27

28. Find the center and vertices of the ellipse and sketch the ellipse: $4x^2 + y^2 = 4$.

29. Write the standard form of the equation of the hyperbola with vertices $(-1, 0)$ and $(1, 0)$ and asymptotes $y = \pm 2x$.

30. Find the center and vertices of the hyperbola and sketch the hyperbola:

$$x^2 - 9y^2 + 18y = 153.$$

In Exercises 31 and 32, solve the nonlinear system of equations.

31. $\begin{cases} y = x^2 - x - 1 \\ 3x - y = 4 \end{cases}$

32. $\begin{cases} 2x^2 + 3y^2 = 6 \\ 5x^2 + 4y^2 = 15 \end{cases}$

33. A rectangle has an area of 32 square feet and a perimeter of 24 feet. Find the dimensions of the rectangle.

34. ▦ The path of a ball is given by $y = -0.1x^2 + 3x + 6$. The coordinates x and y are measured in feet, with $x = 0$ corresponding to the position from which the ball was thrown.

(a) Use a graphing calculator to graph the path of the ball.

(b) Move the cursor along the path to approximate the highest point and the range of the path.

Motivating the Chapter

⚡ *Ancestors and Descendants*

See Section 11.3, Exercise 123.

a. Your ancestors consist of your two parents (first generation), your four grandparents (second generation), your eight great-grandparents (third generation), and so on. Write a geometric sequence that describes the number of ancestors for each generation.

b. If your ancestry could be traced back 66 generations (approximately 2000 years), how many different ancestors would you have?

c. A common ancestor is one to whom you are related in more than one way. (See figure.) From the results of part (b), do you think that you have had no common ancestors in the last 2000 years?

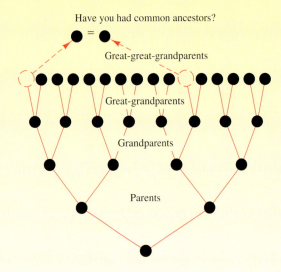

Have you had common ancestors?

Great-great-grandparents

Great-grandparents

Grandparents

Parents

Stewart Cohen/Index Stock

Sequences, Series, and the Binomial Theorem

11.1 ● Sequences and Series

11.2 ● Arithmetic Sequences

11.3 ● Geometric Sequences and Series

11.4 ● The Binomial Theorem

11

11.1 Sequences and Series

Macduff Everton/Corbis

What You Should Learn

1 Use sequence notation to write the terms of sequences.

2 Write the terms of sequences involving factorials.

3 Find the apparent nth term of a sequence.

4 Sum the terms of sequences to obtain series and use sigma notation to represent partial sums.

Why You Should Learn It

Sequences and series are useful in modeling sets of values in order to identify patterns. For instance, in Exercise 110 on page 713, you will use a sequence to model the depreciation of a sport utility vehicle.

1 Use sequence notation to write the terms of sequences.

Year	Contract A	Contract B
1	$20,000	$20,000
2	$22,200	$22,000
3	$24,400	$24,200
4	$26,600	$26,620
5	$28,800	$29,282
Total	$122,000	$122,102

Sequences

You are given the following choice of contract offers for the next 5 years of employment.

Contract A $20,000 the first year and a $2200 raise each year

Contract B $20,000 the first year and a 10% raise each year

Which contract offers the largest salary over the five-year period? The salaries for each contract are shown in the table at the left. Notice that after 5 years contract B represents a better contract offer than contract A. The salaries for each contract option represent a sequence.

A mathematical **sequence** is simply an ordered list of numbers. Each number in the list is a **term** of the sequence. A sequence can have a finite number of terms or an infinite number of terms. For instance, the sequence of positive odd integers that are less than 15 is a *finite* sequence

1, 3, 5, 7, 9, 11, 13   Finite sequence

whereas the sequence of positive odd integers is an *infinite* sequence.

1, 3, 5, 7, 9, 11, 13, . . .   Infinite sequence

Note that the three dots indicate that the sequence continues and has an infinite number of terms.

Because each term of a sequence is matched with its location, a sequence can be defined as a function whose domain is a subset of positive integers.

Sequences

An **infinite sequence** $a_1, a_2, a_3, \ldots, a_n, \ldots$ is a function whose domain is the set of positive integers.

A **finite sequence** $a_1, a_2, a_3, \ldots, a_n$ is a function whose domain is the finite set $\{1, 2, 3, \ldots, n\}$.

In some cases it is convenient to begin subscripting a sequence with 0 instead of 1. Then the domain of the infinite sequence is the set of nonnegative integers and the domain of the finite sequence is the set $\{0, 1, 2, \ldots, n\}$. The terms of the sequence are denoted by $a_0, a_1, a_2, a_3, a_4, \ldots, a_n, \ldots$.

$a_{(\,)} = 2(\,) + 1$

$a_{(1)} = 2(1) + 1 = 3$

$a_{(2)} = 2(2) + 1 = 5$

$\vdots$

$a_{(51)} = 2(51) + 1 = 103$

The subscripts of a sequence are used in place of function notation. For instance, if parentheses replaced the n in $a_n = 2n + 1$, the notation would be similar to function notation, as shown at the left.

Example 1   Writing the Terms of a Sequence

Write the first six terms of the sequence whose nth term is

$$a_n = n^2 - 1. \qquad \text{Begin sequence with } n = 1.$$

Solution

$$a_1 = (1)^2 - 1 = 0 \qquad a_2 = (2)^2 - 1 = 3 \qquad a_3 = (3)^2 - 1 = 8$$
$$a_4 = (4)^2 - 1 = 15 \qquad a_5 = (5)^2 - 1 = 24 \qquad a_6 = (6)^2 - 1 = 35$$

The entire sequence can be written as follows.

$$0, 3, 8, 15, 24, 35, \ldots, n^2 - 1, \ldots$$

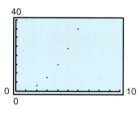
Example 2   Writing the Terms of a Sequence

Write the first six terms of the sequence whose nth term is

$$a_n = 3(2^n). \qquad \text{Begin sequence with } n = 0.$$

Solution

$$a_0 = 3(2^0) = 3 \cdot 1 = 3 \qquad a_1 = 3(2^1) = 3 \cdot 2 = 6$$
$$a_2 = 3(2^2) = 3 \cdot 4 = 12 \qquad a_3 = 3(2^3) = 3 \cdot 8 = 24$$
$$a_4 = 3(2^4) = 3 \cdot 16 = 48 \qquad a_5 = 3(2^5) = 3 \cdot 32 = 96$$

The entire sequence can be written as follows.

$$3, 6, 12, 24, 48, 96, \ldots, 3(2^n), \ldots$$

Example 3   A Sequence Whose Terms Alternate in Sign

Write the first six terms of the sequence whose nth term is

$$a_n = \frac{(-1)^n}{2n - 1}. \qquad \text{Begin sequence with } n = 1.$$

Solution

$$a_1 = \frac{(-1)^1}{2(1) - 1} = -\frac{1}{1} \qquad a_2 = \frac{(-1)^2}{2(2) - 1} = \frac{1}{3} \qquad a_3 = \frac{(-1)^3}{2(3) - 1} = -\frac{1}{5}$$

$$a_4 = \frac{(-1)^4}{2(4) - 1} = \frac{1}{7} \qquad a_5 = \frac{(-1)^5}{2(5) - 1} = -\frac{1}{9} \qquad a_6 = \frac{(-1)^6}{2(6) - 1} = \frac{1}{11}$$

The entire sequence can be written as follows.

$$-1, \frac{1}{3}, -\frac{1}{5}, \frac{1}{7}, -\frac{1}{9}, \frac{1}{11}, \ldots, \frac{(-1)^n}{2n - 1}, \ldots$$

Factorial Notation

Some very important sequences in mathematics involve terms that are defined with special types of products called **factorials.**

Definition of Factorial

If n is a positive integer, n **factorial** is defined as

$$n! = 1 \cdot 2 \cdot 3 \cdot 4 \cdot \cdots \cdot (n-1) \cdot n.$$

As a special case, zero factorial is defined as $0! = 1$.

The first several factorial values are as follows.

$$0! = 1 \qquad\qquad 1! = 1$$
$$2! = 1 \cdot 2 = 2 \qquad\qquad 3! = 1 \cdot 2 \cdot 3 = 6$$
$$4! = 1 \cdot 2 \cdot 3 \cdot 4 = 24 \qquad 5! = 1 \cdot 2 \cdot 3 \cdot 4 \cdot 5 = 120$$

Many calculators have a factorial key, denoted by $\boxed{n!}$. If your calculator has such a key, try using it to evaluate $n!$ for several values of n. You will see that the value of n does not have to be very large before the value of $n!$ becomes huge. For instance

$$10! = 3,628,800.$$

Example 4   A Sequence Involving Factorials

Write the first six terms of the sequence with the given nth term.

a. $a_n = \dfrac{1}{n!}$ Begin sequence with $n = 0$.

b. $a_n = \dfrac{2^n}{n!}$ Begin sequence with $n = 0$.

Solution

a. $a_0 = \dfrac{1}{0!} = \dfrac{1}{1} = 1$ $\qquad a_1 = \dfrac{1}{1!} = \dfrac{1}{1} = 1$

$a_2 = \dfrac{1}{2!} = \dfrac{1}{1 \cdot 2} = \dfrac{1}{2}$ $\qquad a_3 = \dfrac{1}{3!} = \dfrac{1}{1 \cdot 2 \cdot 3} = \dfrac{1}{6}$

$a_4 = \dfrac{1}{4!} = \dfrac{1}{1 \cdot 2 \cdot 3 \cdot 4} = \dfrac{1}{24}$ $\qquad a_5 = \dfrac{1}{5!} = \dfrac{1}{1 \cdot 2 \cdot 3 \cdot 4 \cdot 5} = \dfrac{1}{120}$

b. $a_0 = \dfrac{2^0}{0!} = \dfrac{1}{1} = 1$ $\qquad a_1 = \dfrac{2^1}{1!} = \dfrac{2}{1} = 2$

$a_2 = \dfrac{2^2}{2!} = \dfrac{2 \cdot 2}{1 \cdot 2} = \dfrac{4}{2} = 2$ $\qquad a_3 = \dfrac{2^3}{3!} = \dfrac{8}{1 \cdot 2 \cdot 3} = \dfrac{8}{6} = \dfrac{4}{3}$

$a_4 = \dfrac{2^4}{4!} = \dfrac{2 \cdot 2 \cdot 2 \cdot 2}{1 \cdot 2 \cdot 3 \cdot 4} = \dfrac{2}{3}$ $\qquad a_5 = \dfrac{2^5}{5!} = \dfrac{2 \cdot 2 \cdot 2 \cdot 2 \cdot 2}{1 \cdot 2 \cdot 3 \cdot 4 \cdot 5} = \dfrac{4}{15}$

③ Find the apparent *n*th term of a sequence.

Finding the *n*th Term of a Sequence

Sometimes you will have the first several terms of a sequence and need to find a formula (the *n*th term) that will generate those terms. Pattern recognition is crucial in finding a form for the *n*th term.

Study Tip

Simply listing the first few terms is not sufficient to define a unique sequence—the *n*th term must be given. Consider the sequence

$$\frac{1}{2}, \frac{1}{4}, \frac{1}{8}, \frac{1}{15}, \dots$$

The first three terms are identical to the first three terms of the sequence in Example 5(a). However, the *n*th term of this sequence is defined as

$$a_n = \frac{6}{(n + 1)(n^2 - n + 6)}.$$

Example 5   Finding the *n*th Term of a Sequence

Write an expression for the *n*th term of each sequence.

a. $\dfrac{1}{2}, \dfrac{1}{4}, \dfrac{1}{8}, \dfrac{1}{16}, \dfrac{1}{32}, \dots$ **b.** $1, -4, 9, -16, 25, \dots$

Solution

a.

n:	1	2	3	4	5	. . .	*n*
Terms:	$\dfrac{1}{2}$	$\dfrac{1}{4}$	$\dfrac{1}{8}$	$\dfrac{1}{16}$	$\dfrac{1}{32}$	. . .	a_n

Pattern: The numerator is 1 and the denominators are increasing powers of 2.

$$a_n = \frac{1}{2^n}$$

b.

n:	1	2	3	4	5	. . .	*n*
Terms:	1	-4	9	-16	25	. . .	a_n

Pattern: The terms have alternating signs, with those in the even positions being negative. The absolute value of each term is the square of *n*.

$$a_n = (-1)^{n+1} n^2$$

④ Sum the terms of sequences to obtain series and use sigma notation to represent partial sums.

Series

In the opening illustration of this section, the terms of the finite sequence were *added*. If you add all the terms of an infinite sequence, you obtain a **series.**

Definition of a Series

For an infinite sequence $a_1, a_2, a_3, \dots, a_n, \dots$

1. the sum of the first *n* terms

$$S_n = a_1 + a_2 + a_3 + \cdots + a_n$$

is called a **partial sum,** and

2. the sum of all the terms

$$a_1 + a_2 + a_3 + \cdots + a_n + \cdots$$

is called an **infinite series,** or simply a **series.**

Technology: Tip

Most graphing calculators have a built-in program that will calculate the partial sum of a sequence. Consult the user's guide for your graphing calculator for specific instructions.

Example 6   Finding Partial Sums

Find the indicated partial sums for each sequence.

a. Find S_1, S_2, and S_5 for $a_n = 3n - 1$.

b. Find S_2, S_3, and S_4 for $a_n = \dfrac{(-1)^n}{n + 1}$.

Solution

a. The first five terms of the sequence $a_n = 3n - 1$ are

$$a_1 = 2, a_2 = 5, a_3 = 8, a_4 = 11, \text{ and } a_5 = 14.$$

So, the partial sums are

$$S_1 = 2, S_2 = 2 + 5 = 7, \text{ and } S_5 = 2 + 5 + 8 + 11 + 14 = 40.$$

b. The first four terms of the sequence $a_n = \dfrac{(-1)^n}{n + 1}$ are

$$a_1 = -\frac{1}{2}, a_2 = \frac{1}{3}, a_3 = -\frac{1}{4}, \text{ and } a_4 = \frac{1}{5}.$$

So, the partial sums are

$$S_2 = -\frac{1}{2} + \frac{1}{3} = -\frac{1}{6},$$

$$S_3 = -\frac{1}{2} + \frac{1}{3} - \frac{1}{4} = -\frac{5}{12},$$

and

$$S_4 = -\frac{1}{2} + \frac{1}{3} - \frac{1}{4} + \frac{1}{5} = -\frac{13}{60}.$$

A convenient shorthand notation for denoting a partial sum is called **sigma notation.** This name comes from the use of the uppercase Greek letter sigma, written as Σ.

Definition of Sigma Notation

The sum of the first n terms of the sequence whose nth term is a_n is

$$\sum_{i=1}^{n} a_i = a_1 + a_2 + a_3 + a_4 + \cdots + a_n$$

where i is the **index of summation,** n is the **upper limit of summation,** and 1 is the **lower limit of summation.**

Summation notation is an instruction to add the terms of a sequence. From the definition above, the upper limit of summation tells you where to end the sum. Summation notation helps you generate the appropriate terms of the sequence prior to finding the actual sum.

Example 7   Sigma Notation for Sums

Find the sum $\displaystyle\sum_{i=1}^{6} 2i$.

Solution

$$\sum_{i=1}^{6} 2i = 2(1) + 2(2) + 2(3) + 2(4) + 2(5) + 2(6)$$

$$= 2 + 4 + 6 + 8 + 10 + 12$$

$$= 42$$

Study Tip

In Example 7, the index of summation is i and the summation begins with $i = 1$. Any letter can be used as the index of summation, and the summation can begin with any integer. For instance, in Example 8, the index of summation is k and the summation begins with $k = 0$.

Example 8   Sigma Notation for Sums

Find the sum $\displaystyle\sum_{k=0}^{8} \frac{1}{k!}$.

Solution

$$\sum_{k=0}^{8} \frac{1}{k!} = \frac{1}{0!} + \frac{1}{1!} + \frac{1}{2!} + \frac{1}{3!} + \frac{1}{4!} + \frac{1}{5!} + \frac{1}{6!} + \frac{1}{7!} + \frac{1}{8!}$$

$$= 1 + 1 + \frac{1}{2} + \frac{1}{6} + \frac{1}{24} + \frac{1}{120} + \frac{1}{720} + \frac{1}{5040} + \frac{1}{40,320}$$

$$\approx 2.71828$$

Note that this sum is approximately $e = 2.71828. \ldots$

Example 9   Writing a Sum in Sigma Notation

Write each sum in sigma notation.

a. $\dfrac{2}{2} + \dfrac{2}{3} + \dfrac{2}{4} + \dfrac{2}{5} + \dfrac{2}{6}$ **b.** $1 - \dfrac{1}{3} + \dfrac{1}{9} - \dfrac{1}{27} + \dfrac{1}{81}$

Solution

a. To write this sum in sigma notation, you must find a pattern for the terms. After examining the terms, you can see that they have numerators of 2 and denominators that range over the integers from 2 to 6. So, one possible sigma notation is

$$\sum_{i=1}^{5} \frac{2}{i+1} = \frac{2}{2} + \frac{2}{3} + \frac{2}{4} + \frac{2}{5} + \frac{2}{6}.$$

b. To write this sum in sigma notation, you must find a pattern for the terms. After examining the terms, you can see that the numerators alternate in sign and the denominators are integer powers of 3, starting with 3^0 and ending with 3^4. So, one possible sigma notation is

$$\sum_{i=0}^{4} \frac{(-1)^i}{3^i} = \frac{1}{3^0} + \frac{-1}{3^1} + \frac{1}{3^2} + \frac{-1}{3^3} + \frac{1}{3^4}.$$

11.1 Exercises

Review   *Concepts, Skills, and Problem Solving*

Keep mathematically in shape by doing these exercises *before* the problems of this section.

Properties and Definitions

1. Demonstrate the Multiplication Property of Equality for the equation $-7x = 35$.

2. Demonstrate the Addition Property of Equality for the equation $7x + 63 = 35$.

3. How do you determine whether $t = -3$ is a solution of the equation $t^2 + 4t + 3 = 0$?

4. What is the usual first step in solving an equation such as the one below?

$$\frac{3}{x} - \frac{1}{x + 1} = 10$$

Simplifying Expressions

In Exercises 5–10, simplify the expression.

5. $(x + 10)^{-2}$

6. $\dfrac{18(x - 3)^5}{(x - 3)^2}$

7. $(a^2)^{-4}$

8. $(8x^3)^{1/3}$

9. $\sqrt{128x^3}$

10. $\dfrac{5}{\sqrt{x} - 2}$

Graphs and Models

▦ ▲ *Geometry*   In Exercises 11 and 12, (a) write a function that represents the area of the region, (b) use a graphing calculator to graph the function, and (c) approximate the value of x if the area of the region is 200 square units.

11.
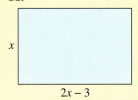

12.
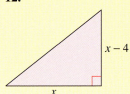

Developing Skills

In Exercises 1–22, write the first five terms of the sequence. (Assume that n begins with 1.) See Examples 1–4.

1. $a_n = 2n$

2. $a_n = 3n$

3. $a_n = (-1)^n 2n$

4. $a_n = (-1)^{n+1} 3n$

5. $a_n = \left(\frac{1}{2}\right)^n$

6. $a_n = \left(\frac{1}{3}\right)^n$

7. $a_n = \left(-\frac{1}{2}\right)^{n+1}$

8. $a_n = \left(\frac{2}{3}\right)^{n-1}$

9. $a_n = 5n - 2$

10. $a_n = 2n + 3$

11. $a_n = \dfrac{4}{n + 3}$

12. $a_n = \dfrac{5}{4 + 2n}$

13. $a_n = \dfrac{3n}{5n - 1}$

14. $a_n = \dfrac{2n}{6n - 3}$

15. $a_n = \dfrac{(-1)^n}{n^2}$

16. $a_n = \dfrac{1}{\sqrt{n}}$

17. $a_n = 5 - \dfrac{1}{2^n}$

18. $a_n = 7 + \dfrac{1}{3^n}$

19. $a_n = \dfrac{(n+1)!}{n!}$

20. $a_n = \dfrac{n!}{(n-1)!}$

21. $a_n = \dfrac{2 + (-2)^n}{n!}$

22. $a_n = \dfrac{1 + (-1)^n}{n^2}$

In Exercises 23–26, find the indicated term of the sequence.

23. $a_n = (-1)^n(5n - 3)$

$a_{15} = $

24. $a_n = (-1)^{n-1}(2n + 4)$

$a_{14} = $

25. $a_n = \dfrac{n^2 - 2}{(n-1)!}$

$a_8 = $

26. $a_n = \dfrac{n^2}{n!}$

$a_{12} = $

In Exercises 27–38, simplify the expression.

27. $\dfrac{5!}{4!}$

28. $\dfrac{18!}{17!}$

29. $\dfrac{10!}{12!}$

30. $\dfrac{5!}{8!}$

31. $\dfrac{25!}{20!5!}$

32. $\dfrac{20!}{15! \cdot 5!}$

33. $\dfrac{n!}{(n+1)!}$

34. $\dfrac{(n+2)!}{n!}$

35. $\dfrac{(n+1)!}{(n-1)!}$

36. $\dfrac{(3n)!}{(3n+2)!}$

37. $\dfrac{(2n)!}{(2n-1)!}$

38. $\dfrac{(2n+2)!}{(2n)!}$

In Exercises 39–42, match the sequence with the graph of its first 10 terms. [The graphs are labeled (a), (b), (c), and (d).]

(a)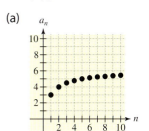

(b)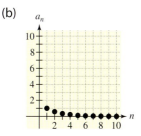

(c)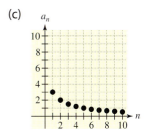

(d)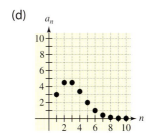

39. $a_n = \dfrac{6}{n+1}$

40. $a_n = \dfrac{6n}{n+1}$

41. $a_n = (0.6)^{n-1}$

42. $a_n = \dfrac{3^n}{n!}$

In Exercises 43–48, use a graphing calculator to graph the first 10 terms of the sequence.

43. $a_n = \dfrac{4n^2}{n^2 - 2}$

44. $a_n = \dfrac{2n^2}{n^2 + 1}$

45. $a_n = 3 - \dfrac{4}{n}$

46. $a_n = \dfrac{n+2}{n}$

47. $a_n = (-0.8)^{n-1}$

48. $a_n = 10\left(\dfrac{3}{4}\right)^{n-1}$

In Exercises 49–66, write an expression for the nth term of the sequence. (Assume that n begins with 1.) See Example 5.

49. $1, 3, 5, 7, 9, \ldots$ **50.** $2, 4, 6, 8, 10, \ldots$

51. $2, 6, 10, 14, 18, \ldots$ **52.** $5, 8, 11, 14, 17, \ldots$

53. $0, 3, 8, 15, 24, \ldots$ **54.** $1, 8, 27, 64, 125, \ldots$

55. $2, -4, 6, -8, 10, \ldots$ **56.** $1, -1, 1, -1, 1, \ldots$

57. $\frac{2}{3}, \frac{3}{4}, \frac{4}{5}, \frac{5}{6}, \frac{6}{7}, \ldots$ **58.** $\frac{2}{1}, \frac{3}{3}, \frac{4}{5}, \frac{5}{7}, \frac{6}{9}, \ldots$

59. $\frac{1}{2}, \frac{-1}{4}, \frac{1}{8}, \frac{-1}{16}, \ldots$ **60.** $1, \frac{1}{4}, \frac{1}{9}, \frac{1}{16}, \frac{1}{25}, \ldots$

61. $1, \frac{1}{2}, \frac{1}{4}, \frac{1}{8}, \ldots$ **62.** $\frac{1}{3}, \frac{2}{9}, \frac{4}{27}, \frac{8}{81}, \ldots$

63. $1 + \frac{1}{1}, 1 + \frac{1}{2}, 1 + \frac{1}{3}, 1 + \frac{1}{4}, 1 + \frac{1}{5}, \ldots$

64. $1 + \frac{1}{2}, 1 + \frac{3}{4}, 1 + \frac{7}{8}, 1 + \frac{15}{16}, 1 + \frac{31}{32}, \ldots$

65. $1, \frac{1}{2}, \frac{1}{6}, \frac{1}{24}, \frac{1}{120}, \ldots$

66. $1, 2, \frac{2^2}{2}, \frac{2^3}{6}, \frac{2^4}{24}, \frac{2^5}{120}, \ldots$

In Exercises 67–82, find the partial sum. See Examples 6–8.

67. $\displaystyle\sum_{k=1}^{6} 3k$ **68.** $\displaystyle\sum_{k=1}^{4} 5k$

69. $\displaystyle\sum_{i=0}^{6} (2i + 5)$ **70.** $\displaystyle\sum_{i=0}^{4} (2i + 3)$

71. $\displaystyle\sum_{j=3}^{7} (6j - 10)$ **72.** $\displaystyle\sum_{i=2}^{7} (4i - 1)$

73. $\displaystyle\sum_{j=1}^{5} \frac{(-1)^{j+1}}{j^2}$ **74.** $\displaystyle\sum_{j=0}^{3} \frac{1}{j^2 + 1}$

75. $\displaystyle\sum_{m=1}^{8} \frac{m}{m + 1}$ **76.** $\displaystyle\sum_{k=1}^{4} \frac{k - 2}{k + 3}$

77. $\displaystyle\sum_{k=1}^{6} (-8)$ **78.** $\displaystyle\sum_{n=3}^{12} 10$

79. $\displaystyle\sum_{i=1}^{8} \left(\frac{1}{i} - \frac{1}{i + 1} \right)$ **80.** $\displaystyle\sum_{k=1}^{5} \left(\frac{2}{k} - \frac{2}{k + 2} \right)$

81. $\displaystyle\sum_{n=0}^{5} \left(-\frac{1}{3} \right)^n$ **82.** $\displaystyle\sum_{n=0}^{6} \left(\frac{3}{2} \right)^n$

In Exercises 83–90, use a graphing calculator to find the partial sum.

83. $\displaystyle\sum_{n=1}^{6} 3n^2$ **84.** $\displaystyle\sum_{n=0}^{5} 2n^2$

85. $\displaystyle\sum_{j=2}^{5} (j! - j)$ **86.** $\displaystyle\sum_{i=0}^{4} (i! + 4)$

87. $\displaystyle\sum_{j=0}^{4} \frac{6}{j!}$ **88.** $\displaystyle\sum_{k=1}^{6} \left(\frac{1}{2k} - \frac{1}{2k - 1} \right)$

89. $\displaystyle\sum_{k=1}^{6} \ln k$ **90.** $\displaystyle\sum_{k=2}^{4} \frac{k}{\ln k}$

In Exercises 91–108, write the sum using sigma notation. (Begin with $k = 0$ or $k = 1$.) See Example 9.

91. $1 + 2 + 3 + 4 + 5$

92. $8 + 9 + 10 + 11 + 12 + 13 + 14$

93. $2 + 4 + 6 + 8 + 10$

94. $24 + 30 + 36 + 42$

95. $\dfrac{1}{2(1)} + \dfrac{1}{2(2)} + \dfrac{1}{2(3)} + \dfrac{1}{2(4)} + \cdots + \dfrac{1}{2(10)}$

96. $\dfrac{3}{1 + 1} + \dfrac{3}{1 + 2} + \dfrac{3}{1 + 3} + \cdots + \dfrac{3}{1 + 50}$

97. $\dfrac{1}{1^2} + \dfrac{1}{2^2} + \dfrac{1}{3^2} + \dfrac{1}{4^2} + \cdots + \dfrac{1}{20^2}$

98. $\dfrac{1}{2^0} + \dfrac{1}{2^1} + \dfrac{1}{2^2} + \dfrac{1}{2^3} + \cdots + \dfrac{1}{2^{12}}$

99. $\dfrac{1}{3^0} - \dfrac{1}{3^1} + \dfrac{1}{3^2} - \dfrac{1}{3^3} + \cdots - \dfrac{1}{3^9}$

100. $\left(-\dfrac{2}{3}\right)^0 + \left(-\dfrac{2}{3}\right)^1 + \left(-\dfrac{2}{3}\right)^2 + \cdots + \left(-\dfrac{2}{3}\right)^{20}$

101. $\dfrac{4}{1+3} + \dfrac{4}{2+3} + \dfrac{4}{3+3} + \cdots + \dfrac{4}{20+3}$

102. $\dfrac{1}{2^3} - \dfrac{1}{4^3} + \dfrac{1}{6^3} - \dfrac{1}{8^3} + \cdots + \dfrac{1}{14^3}$

103. $\dfrac{1}{2} + \dfrac{2}{3} + \dfrac{3}{4} + \dfrac{4}{5} + \dfrac{5}{6} + \cdots + \dfrac{11}{12}$

104. $\dfrac{2}{4} + \dfrac{4}{7} + \dfrac{6}{10} + \dfrac{8}{13} + \dfrac{10}{16} + \cdots + \dfrac{20}{31}$

105. $\dfrac{2}{4} + \dfrac{4}{5} + \dfrac{6}{6} + \dfrac{8}{7} + \cdots + \dfrac{40}{23}$

106. $\left(2 + \dfrac{1}{1}\right) + \left(2 + \dfrac{1}{2}\right) + \left(2 + \dfrac{1}{3}\right) + \cdots + \left(2 + \dfrac{1}{25}\right)$

107. $1 + 1 + 2 + 6 + 24 + 120 + 720$

108. $1 + 1 + \dfrac{1}{2} + \dfrac{1}{6} + \dfrac{1}{24} + \dfrac{1}{120} + \dfrac{1}{720}$

Solving Problems

109. *Compound Interest*   A deposit of \$500 is made in an account that earns 7% interest compounded yearly. The balance in the account after N years is given by

$$A_N = 500(1 + 0.07)^N, \quad N = 1, 2, 3, \ldots$$

(a) Compute the first eight terms of the sequence.

(b) Find the balance in this account after 40 years by computing A_{40}.

(c) ▦ Use a graphing calculator to graph the first 40 terms of the sequence.

(d) The terms are increasing. Is the rate of growth of the terms increasing? Explain.

110. *Depreciation*   At the end of each year, the value of a sport utility vehicle with an initial cost of \$32,000 is three-fourths what it was at the beginning of the year. After n years, its value is given by

$$a_n = 32{,}000\left(\dfrac{3}{4}\right)^n, \quad n = 1, 2, 3, \ldots$$

(a) Find the value of the sport utility vehicle 3 years after it was purchased by computing a_3.

(b) Find the value of the sport utility vehicle 6 years after it was purchased by computing a_6. Is this value half of what it was after 3 years? Explain.

111. *Sports*   The number of degrees a_n in each angle of a regular n-sided polygon is

$$a_n = \dfrac{180(n - 2)}{n}, \quad n \ge 3.$$

The surface of a soccer ball is made of regular hexagons and pentagons. When a soccer ball is taken apart and flattened, as shown in the figure, the sides don't meet each other. Use the terms a_5 and a_6 to explain why there are gaps between adjacent hexagons.

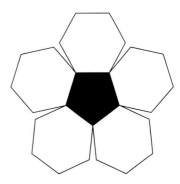

112. *Stars*   The number of degrees d_n in the angle at each point of each of the six n-pointed stars in the figure (on the next page) is given by

$$d_n = \dfrac{180(n - 4)}{n}, \quad n \ge 5.$$

Write the first six terms of this sequence.

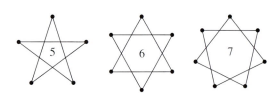

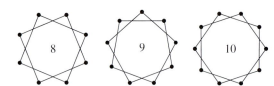

Figure for 112

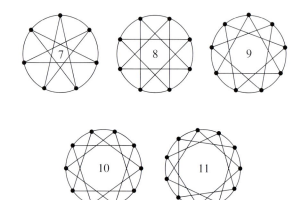

Figure for 113

113. *Stars*   The stars in Exercise 112 were formed by placing n equally spaced points on a circle and connecting each point with the second point from it on the circle. The stars in the figure for this exercise were formed in a similar way except that each point was connected with the third point from it. For these stars, the number of degrees d_n in the angle at each point is given by

$$d_n = \frac{180(n - 6)}{n}, \quad n \geq 7.$$

Write the first five terms of this sequence.

114. 🖩 *Number of Stores*   The number a_n of Home Depot stores for the years 1991 through 2001 is modeled by

$$a_n = 9.73n^2 - 3.0n + 180, \quad n = 1, 2, \ldots, 11$$

where n is the year, with $n = 1$ corresponding to 1991. Find the terms of this finite sequence and use a graphing calculator to construct a bar graph that represents the sequence. (Source: The Home Depot)

Explaining Concepts

115. Give an example of an infinite sequence.

116. *Writing* ✏️   State the definition of n factorial.

117. *Writing* ✏️   The nth term of a sequence is $a_n = (-1)^n n$. Which terms of the sequence are negative? Explain.

118. You learned in this section that a sequence is an ordered list of numbers. Study the following sequence and see if you can guess what its next term should be.

Z, O, T, T, F, F, S, S, E, N, T, E, T, . . .

In Exercises 119–121, decide whether the statement is true or false. Justify your answer.

119. $\displaystyle\sum_{i=1}^{4} (i^2 + 2i) = \sum_{i=1}^{4} i^2 + \sum_{i=1}^{4} 2i$

120. $\displaystyle\sum_{k=1}^{4} 3k = 3\sum_{k=1}^{4} k$

121. $\displaystyle\sum_{j=1}^{4} 2^j = \sum_{j=3}^{6} 2^{j-2}$

11.2 Arithmetic Sequences

Lynn Goldsmith/Corbis

What You Should Learn

1 Recognize, write, and find the nth terms of arithmetic sequences.

2 Find the nth partial sum of an arithmetic sequence.

3 Use arithmetic sequences to solve application problems.

Why You Should Learn It

An arithmetic sequence can reduce the amount of time it takes to find the sum of a sequence of numbers with a common difference. For instance, in Exercise 109 on page 722, you will use an arithmetic sequence to determine how much to charge for tickets for a concert at an outdoor arena.

1 Recognize, write, and find the nth terms of arithmetic sequences.

Arithmetic Sequences

A sequence whose consecutive terms have a common difference is called an **arithmetic sequence.**

Definition of an Arithmetic Sequence

A sequence is called **arithmetic** if the differences between consecutive terms are the same. So, the sequence

$$a_1, a_2, a_3, a_4, \ldots, a_n, \ldots$$

is arithmetic if there is a number d such that

$$a_2 - a_1 = d, \quad a_3 - a_2 = d, \quad a_4 - a_3 = d$$

and so on. The number d is the **common difference** of the sequence.

Example 1   Examples of Arithmetic Sequences

a. The sequence whose nth term is $3n + 2$ is arithmetic. For this sequence, the common difference between consecutive terms is 3.

$$5, 8, 11, 14, \ldots, 3n + 2, \ldots \qquad \text{Begin with } n = 1.$$

$8 - 5 = 3$

b. The sequence whose nth term is $7 - 5n$ is arithmetic. For this sequence, the common difference between consecutive terms is -5.

$$2, -3, -8, -13, \ldots, 7 - 5n, \ldots \qquad \text{Begin with } n = 1.$$

$-3 - 2 = -5$

c. The sequence whose nth term is $\frac{1}{4}(n + 3)$ is arithmetic. For this sequence, the common difference between consecutive terms is $\frac{1}{4}$.

$$1, \frac{5}{4}, \frac{3}{2}, \frac{7}{4}, \ldots, \frac{1}{4}(n + 3), \ldots \qquad \text{Begin with } n = 1.$$

$\frac{5}{4} - 1 = \frac{1}{4}$

Study Tip

The nth term of an arithmetic sequence can be derived from the following pattern.

$a_1 = a_1$ 1st term

$a_2 = a_1 + d$ 2nd term

$a_3 = a_1 + 2d$ 3rd term

$a_4 = a_1 + 3d$ 4th term

$a_5 = a_1 + 4d$ 5th term

1 less

⋮   ⋮

$a_n = a_1 + (n - 1)d$ nth term

1 less

The nth Term of an Arithmetic Sequence

The nth term of an arithmetic sequence has the form

$$a_n = a_1 + (n - 1)d$$

where d is the common difference between the terms of the sequence, and a_1 is the first term.

Example 2   Finding the nth Term of an Arithmetic Sequence

Find a formula for the nth term of the arithmetic sequence whose common difference is 2 and whose first term is 5.

Solution

You know that the formula for the nth term is of the form $a_n = a_1 + (n - 1)d$. Moreover, because the common difference is $d = 2$, and the first term is $a_1 = 5$, the formula must have the form

$$a_n = 5 + 2(n - 1).$$

So, the formula for the nth term is

$$a_n = 2n + 3.$$

The sequence therefore has the following form.

$$5, 7, 9, 11, 13, \ldots, 2n + 3, \ldots$$

If you know the nth term and the common difference of an arithmetic sequence, you can find the $(n + 1)$th term by using the **recursion formula**

$$a_{n+1} = a_n + d.$$

Example 3   Using a Recursion Formula

The 12th term of an arithmetic sequence is 52 and the common difference is 3.

a. What is the 13th term of the sequence?   **b.** What is the first term?

Solution

a. You know that $a_{12} = 52$ and $d = 3$. So, using the recursion formula $a_{13} = a_{12} + d$, you can determine that the 13th term of the sequence is

$$a_{13} = 52 + 3 = 55.$$

b. Using $n = 12, d = 3$, and $a_{12} = 52$ in the formula $a_n = a_1 + (n - 1)d$ yields

$$52 = a_1 + (12 - 1)(3)$$

$$19 = a_1.$$

② Find the *n*th partial sum of an arithmetic sequence.

The Partial Sum of an Arithmetic Sequence

The sum of the first n terms of an arithmetic sequence is called the **nth partial sum** of the sequence. For instance, the fifth partial sum of the arithmetic sequence whose nth term is $3n + 4$ is

$$\sum_{i=1}^{5} (3i + 4) = 7 + 10 + 13 + 16 + 19 = 65.$$

To find a formula for the nth partial sum S_n of an arithmetic sequence, write out S_n forwards and backwards and then add the two forms, as follows.

$$S_n = a_1 + (a_1 + d) + (a_1 + 2d) + \cdots + [a_1 + (n - 1)d] \quad \text{Forwards}$$
$$S_n = a_n + (a_n - d) + (a_n - 2d) + \cdots + [a_n - (n - 1)d] \quad \text{Backwards}$$
$$2S_n = (a_1 + a_n) + (a_1 + a_n) + (a_1 + a_n) + \cdots + [a_1 + a_n] \quad \text{Sum of two equations}$$
$$= n(a_1 + a_n) \quad \text{n groups of } (a_1 + a_n)$$

Dividing each side by 2 yields the following formula.

Study Tip

You can use the formula for the nth partial sum of an arithmetic sequence to find the sum of consecutive numbers. For instance, the sum of the integers from 1 to 100 is

$$\sum_{i=1}^{100} i = \frac{100}{2}(1 + 100)$$
$$= 50(101)$$
$$= 5050.$$

The nth Partial Sum of an Arithmetic Sequence

The nth partial sum of the arithmetic sequence whose nth term is a_n is

$$\sum_{i=1}^{n} a_i = a_1 + a_2 + a_3 + a_4 + \cdots + a_n$$
$$= \frac{n}{2}(a_1 + a_n).$$

Or equivalently, you can find the sum of the first n terms of an arithmetic sequence, by finding the average of the first and nth terms, and multiply by n.

Example 4   Finding the nth Partial Sum

Find the sum of the first 20 terms of the arithmetic sequence whose nth term is $4n + 1$.

Solution

The first term of this sequence is $a_1 = 4(1) + 1 = 5$ and the 20th term is $a_{20} = 4(20) + 1 = 81$. So, the sum of the first 20 terms is given by

$$\sum_{i=1}^{n} a_i = \frac{n}{2}(a_1 + a_n) \qquad \text{nth partial sum formula}$$
$$\sum_{i=1}^{20} (4i + 1) = \frac{20}{2}(a_1 + a_{20}) \qquad \text{Substitute 20 for n.}$$
$$= 10(5 + 81) \qquad \text{Substitute 5 for a_1 and 81 for a_{20}.}$$
$$= 10(86) \qquad \text{Simplify.}$$
$$= 860. \qquad \text{nth partial sum}$$

Example 5   Finding the nth Partial Sum

Find the sum of the even integers from 2 to 100.

Solution

Because the integers

$$2, 4, 6, 8, \ldots, 100$$

form an arithmetic sequence, you can find the sum as follows.

$$\sum_{i=1}^{n} a_i = \frac{n}{2}(a_1 + a_n) \qquad \text{nth partial sum formula}$$

$$\sum_{i=1}^{50} 2i = \frac{50}{2}(a_1 + a_{50}) \qquad \text{Substitute 50 for n.}$$

$$= 25(2 + 100) \qquad \text{Substitute 2 for a_1 and 100 for a_{50}.}$$

$$= 25(102) \qquad \text{Simplify.}$$

$$= 2550 \qquad \text{nth partial sum}$$

③ Use arithmetic sequences to solve application problems.

Application

Example 6   Total Sales

Your business sells $100,000 worth of handmade furniture during its first year. You have a goal of increasing annual sales by $25,000 each year for 9 years. If you meet this goal, how much will you sell during your first 10 years of business?

Solution

The annual sales during the first 10 years form the following arithmetic sequence.

$100,000, $125,000, $150,000, $175,000, $200,000,
$225,000, $250,000, $275,000, $300,000, $325,000

Using the formula for the nth partial sum of an arithmetic sequence, you find the total sales during the first 10 years as follows.

$$\text{Total sales} = \frac{n}{2}(a_1 + a_n) \qquad \text{nth partial sum formula}$$

$$= \frac{10}{2}(100{,}000 + 325{,}000) \qquad \text{Substitute for n, a_1, and a_n.}$$

$$= 5(425{,}000) \qquad \text{Simplify.}$$

$$= \$2{,}125{,}000 \qquad \text{Simplify.}$$

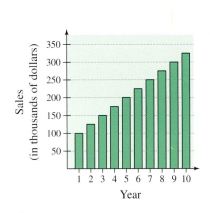

Figure 11.1

From the bar graph shown in Figure 11.1, notice that the annual sales for your company follows a *linear growth* pattern. In other words, saying that a quantity increases arithmetically is the same as saying that it increases linearly.

11.2 Exercises

Review Concepts, Skills, and Problem Solving

Keep mathematically in shape by doing these exercises *before* the problems of this section.

Properties and Definitions

1. *Writing* ✎   In your own words, state the definition of an algebraic expression.

2. *Writing* ✎   In your own words, state the definition of the terms of an algebraic expression.

3. Give an example of a trinomial of degree 3.

4. Give an example of a monomial of degree 4.

Domain

In Exercises 5–10, find the domain of the function.

5. $f(x) = x^3 - 2x$

6. $g(x) = \sqrt[3]{x}$

7. $h(x) = \sqrt{16 - x^2}$

8. $A(x) = \dfrac{3}{36 - x^2}$

9. $g(t) = \ln(t - 2)$

10. $f(s) = 630e^{-0.2s}$

Problem Solving

11. *Compound Interest*   Determine the balance in an account when $10,000 is invested at $7\frac{1}{2}\%$ compounded daily for 15 years.

12. *Compound Interest*   Determine the amount after 5 years if $4000 is invested in an account earning 6% compounded monthly.

Developing Skills

In Exercises 1–10, find the common difference of the arithmetic sequence. See Example 1.

1. $2, 5, 8, 11, \ldots$

2. $-8, 0, 8, 16, \ldots$

3. $100, 94, 88, 82, \ldots$

4. $3200, 2800, 2400, 2000, \ldots$

5. $10, -2, -14, -26, -38, \ldots$

6. $4, \frac{9}{2}, 5, \frac{11}{2}, 6, \ldots$

7. $1, \frac{5}{3}, \frac{7}{3}, 3, \ldots$

8. $\frac{1}{2}, \frac{5}{4}, 2, \frac{11}{4}, \ldots$

9. $\frac{7}{2}, \frac{9}{4}, 1, -\frac{1}{4}, -\frac{3}{2}, \ldots$

10. $\frac{5}{2}, \frac{11}{6}, \frac{7}{6}, \frac{1}{2}, -\frac{1}{6}, \ldots$

15. $32, 16, 0, -16, \ldots$

16. $32, 16, 8, 4, \ldots$

17. $3.2, 4, 4.8, 5.6, \ldots$

18. $8, 4, 2, 1, 0.5, 0.25, \ldots$

19. $2, \frac{7}{2}, 5, \frac{13}{2}, \ldots$

20. $3, \frac{5}{2}, 2, \frac{3}{2}, 1, \ldots$

21. $\frac{1}{3}, \frac{2}{3}, \frac{4}{3}, \frac{8}{3}, \frac{16}{3}, \ldots$

22. $\frac{9}{4}, 2, \frac{7}{4}, \frac{3}{2}, \frac{5}{4}, \ldots$

23. $1, \sqrt{2}, \sqrt{3}, 2, \sqrt{5}, \ldots$

24. $1, 4, 9, 16, 25, \ldots$

25. $\ln 4, \ln 8, \ln 12, \ln 16, \ldots$

26. $e, e^2, e^3, e^4, \ldots$

In Exercises 11–26, determine whether the sequence is arithmetic. If so, find the common difference.

11. $2, 4, 6, 8, \ldots$

12. $1, 2, 4, 8, 16, \ldots$

13. $10, 8, 6, 4, 2, \ldots$

14. $2, 6, 10, 14, \ldots$

In Exercises 27–36, write the first five terms of the arithmetic sequence. (Assume that n begins with 1.)

27. $a_n = 3n + 4$

28. $a_n = 5n - 4$

29. $a_n = -2n + 8$

30. $a_n = -10n + 100$

31. $a_n = \frac{5}{2}n - 1$

32. $a_n = \frac{2}{3}n + 2$

33. $a_n = \frac{3}{5}n + 1$

34. $a_n = \frac{3}{4}n - 2$

35. $a_n = -\frac{1}{4}(n - 1) + 4$

36. $a_n = 4(n + 2) + 24$

In Exercises 37–54, find a formula for the nth term of the arithmetic sequence. See Example 2.

37. $a_1 = 4, \quad d = 3$

38. $a_1 = 7, \quad d = 2$

39. $a_1 = \frac{1}{2}, \quad d = \frac{3}{2}$

40. $a_1 = \frac{5}{3}, \quad d = \frac{1}{3}$

41. $a_1 = 100, \quad d = -5$

42. $a_1 = -6, \quad d = -1$

43. $a_1 = 3, \quad d = \frac{3}{2}$

44. $a_6 = 5, \quad d = \frac{3}{2}$

45. $a_1 = 5, \quad a_5 = 15$

46. $a_2 = 93, \quad a_6 = 65$

47. $a_3 = 16, \quad a_4 = 20$

48. $a_5 = 30, \quad a_4 = 25$

49. $a_1 = 50, \quad a_3 = 30$

50. $a_{10} = 32, \quad a_{12} = 48$

51. $a_2 = 10, \quad a_6 = 8$

52. $a_7 = 8, \quad a_{13} = 6$

53. $a_1 = 0.35, \quad a_2 = 0.30$

54. $a_1 = 0.08, \quad a_2 = 0.082$

In Exercises 55–62, write the first five terms of the arithmetic sequence defined recursively. See Example 3.

55. $a_1 = 14$
$a_{k+1} = a_k + 6$

56. $a_1 = 3$
$a_{k+1} = a_k - 2$

57. $a_1 = 23$
$a_{k+1} = a_k - 5$

58. $a_1 = 12$
$a_{k+1} = a_k + 6$

59. $a_1 = -16$
$a_{k+1} = a_k + 5$

60. $a_1 = -22$
$a_{k+1} = a_k - 4$

61. $a_1 = 3.4$
$a_{k+1} = a_k - 1.1$

62. $a_1 = 10.9$
$a_{k+1} = a_k + 0.7$

In Exercises 63–72, find the partial sum. See Example 4.

63. $\displaystyle\sum_{k=1}^{20} k$

64. $\displaystyle\sum_{k=1}^{30} 4k$

65. $\displaystyle\sum_{k=1}^{50} (k + 3)$

66. $\displaystyle\sum_{n=1}^{30} (n + 2)$

67. $\displaystyle\sum_{k=1}^{10} (5k - 2)$

68. $\displaystyle\sum_{k=1}^{100} (4k - 1)$

69. $\displaystyle\sum_{n=1}^{500} \frac{n}{2}$

70. $\displaystyle\sum_{n=1}^{600} \frac{2n}{3}$

71. $\displaystyle\sum_{n=1}^{30} \left(\tfrac{1}{3}n - 4\right)$

72. $\displaystyle\sum_{n=1}^{75} (0.3n + 5)$

In Exercises 73–84, find the nth partial sum of the arithmetic sequence. See Example 5.

73. $5, 12, 19, 26, 33, \ldots, \quad n = 12$

74. $2, 12, 22, 32, 42, \ldots, \quad n = 20$

75. $2, 8, 14, 20, \ldots, \quad n = 25$

76. $500, 480, 460, 440, \ldots, \quad n = 20$

77. $200, 175, 150, 125, 100, \ldots, \quad n = 8$

78. $800, 785, 770, 755, 740, \ldots, \quad n = 25$

79. $-50, -38, -26, -14, -2, \ldots, \quad n = 50$

80. $-16, -8, 0, 8, 16, \ldots, \quad n = 30$

81. $1, 4.5, 8, 11.5, 15, \ldots, \quad n = 12$

82. $2.2, 2.8, 3.4, 4.0, 4.6, \ldots, \quad n = 12$

83. $a_1 = 0.5, \ a_4 = 1.7, \ldots, \quad n = 10$

84. $a_1 = 15, \ a_{100} = 307, \ldots, \quad n = 100$

In Exercises 85–90, match the arithmetic sequence with its graph. [The graphs are labeled (a), (b), (c), (d), (e), and (f).]

(a)

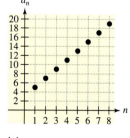

(b)

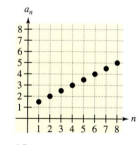

(c)

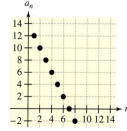

(d)

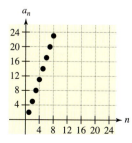

(e)

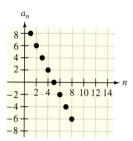

(f)

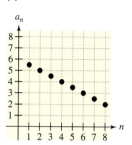

85. $a_n = \tfrac{1}{2}n + 1$

86. $a_n = -\tfrac{1}{2}n + 6$

87. $a_n = -2n + 10$

88. $a_n = 2n + 3$

89. $a_1 = 12$

$a_{n+1} = a_n - 2$

90. $a_1 = 2$

$a_{n+1} = a_n + 3$

In Exercises 91–96, use a graphing calculator to graph the first 10 terms of the sequence.

91. $a_n = -2n + 21$

92. $a_n = -25n + 500$

93. $a_n = \tfrac{3}{5}n + \tfrac{3}{2}$

94. $a_n = \tfrac{3}{2}n + 1$

95. $a_n = 2.5n - 8$

96. $a_n = 6.2n + 3$

In Exercises 97–102, use a graphing calculator to find the partial sum.

97. $\displaystyle\sum_{j=1}^{25} (750 - 30j)$

98. $\displaystyle\sum_{n=1}^{40} (1000 - 25n)$

99. $\displaystyle\sum_{i=1}^{60} \left(300 - \tfrac{8}{3}i\right)$

100. $\displaystyle\sum_{n=1}^{20} \left(500 - \tfrac{1}{10}n\right)$

101. $\displaystyle\sum_{n=1}^{50} (2.15n + 5.4)$

102. $\displaystyle\sum_{n=1}^{60} (200 - 3.4n)$

Solving Problems

103. *Number Problem*   Find the sum of the first 75 positive integers.

104. *Number Problem*   Find the sum of the integers from 35 to 100.

105. *Number Problem*   Find the sum of the first 50 positive odd integers.

106. *Number Problem*   Find the sum of the first 100 positive even integers.

107. *Salary*   In your new job as an actuary you are told that your starting salary will be $36,000 with an increase of $2000 at the end of each of the first 5 years. How much will you be paid through the end of your first six years of employment with the company?

108. *Wages*   You earn 25 cents on the first day of the month, 50 cents on the second day, 75 cents on the third day, and so on. Determine the total amount that you will earn during a 30-day month.

109. *Ticket Prices*   There are 20 rows of seats on the main floor of a an outdoor arena: 20 seats in the first row, 21 seats in the second row, 22 seats in the third row, and so on (see figure). How much should you charge per ticket in order to obtain $15,000 for the sale of all the seats on the main floor?

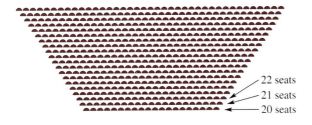

22 seats
21 seats
20 seats

110. *Pile of Logs*   Logs are stacked in a pile as shown in the figure. The top row has 15 logs and the bottom row has 21 logs. How many logs are in the pile?

— 15

— 21

111. *Baling Hay*   In the first two trips baling hay around a large field (see figure), a farmer obtains 93 bales and 89 bales, respectively. The farmer estimates that the same pattern will continue. Estimate the total number of bales made if there are another six trips around the field.

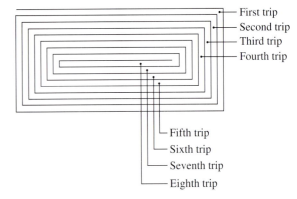

First trip
Second trip
Third trip
Fourth trip

Fifth trip
Sixth trip
Seventh trip
Eighth trip

112. *Baling Hay*   In the first two trips baling hay around a field (see figure), a farmer obtains 64 bales and 60 bales, respectively. The farmer estimates that the same pattern will continue. Estimate the total number of bales made if there are another four trips around the field.

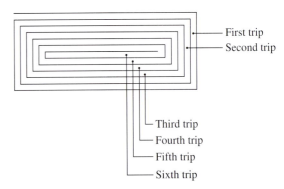

First trip
Second trip

Third trip
Fourth trip
Fifth trip
Sixth trip

113. *Clock Chimes*   A clock chimes once at 1:00, twice at 2:00, three times at 3:00, and so on. The clock also chimes once at 15-minute intervals that are not on the hour. How many times does the clock chime in a 12-hour period?

114. *Clock Chimes*   A clock chimes once at 1:00, twice at 2:00, three times at 3:00, and so on. The clock also chimes once on the half-hour. How many times does the clock chime in a 12-hour period?

115. *Free-Falling Object*   A free-falling object will fall 16 feet during the first second, 48 more feet during the second second, 80 more feet during the third second, and so on. What is the total distance the object will fall in 8 seconds if this pattern continues?

116. *Free-Falling Object*   A free-falling object will fall 4.9 meters during the first second, 14.7 more meters during the second second, 24.5 more meters during the third second, and so on. What is the total distance the object will fall in 5 seconds if this pattern continues?

Explaining Concepts

117. *Pattern Recognition*

(a) Complete the table.

Figure	Number of Sides	Sum of Interior Angles
Triangle	3	180°
Quadrilateral	4	
Pentagon	5	
Hexagon	6	

(b) Use the pattern in part (a) to determine the sum of the interior angles of a figure with n sides.

(c) Determine whether the sequence formed by the entries in the third column of the table in part (a) is an arithmetic sequence. If so, find the common difference.

118. *Writing*   In your own words, explain what makes a sequence arithmetic.

119. The second and third terms of an arithmetic sequence are 12 and 15, respectively. What is the first term?

120. *Writing*   Explain how the first two terms of an arithmetic sequence can be used to find the nth term.

121. *Writing*   Explain what is meant by a recursion formula.

122. *Writing*   Explain what is meant by the nth partial sum of a sequence.

123. *Writing*   Explain how to find the sum of the integers from 100 to 200.

124. *Pattern Recognition*

(a) Compute the sums of positive odd integers.

$1 + 3 = $

$1 + 3 + 5 = $

$1 + 3 + 5 + 7 = $

$1 + 3 + 5 + 7 + 9 = $

$1 + 3 + 5 + 7 + 9 + 11 = $

(b) Use the sums in part (a) to make a conjecture about the sums of positive odd integers. Check your conjecture for the sum

$1 + 3 + 5 + 7 + 9 + 11 + 13 = $.

(c) Verify your conjecture in part (b) analytically.

125. *Writing*   Each term of an arithmetic sequence is multiplied by a constant C. Is the resulting sequence arithmetic? If so, how does the common difference compare with the common difference of the original sequence?

Mid-Chapter Quiz

Take this quiz as you would take a quiz in class. After you are done, check your work against the answers in the back of the book.

In Exercises 1–4, write the first five terms of the sequence. (Assume that n begins with 1.)

1. $a_n = (-2)^{n+1}$

2. $a_n = n(n + 2)$

3. $a_n = 32\left(\dfrac{1}{4}\right)^{n-1}$

4. $a_n = \dfrac{(-3)^n n}{n + 4}$

In Exercises 5–10, find the sum.

5. $\displaystyle\sum_{k=1}^{4} 10k$

6. $\displaystyle\sum_{i=1}^{10} 4$

7. $\displaystyle\sum_{j=1}^{5} \dfrac{60}{j + 1}$

8. $\displaystyle\sum_{n=1}^{8} 8\left(-\dfrac{1}{2}\right)$

9. $\displaystyle\sum_{n=1}^{5} (3n - 1)$

10. $\displaystyle\sum_{k=1}^{4} (k^2 - 1)$

In Exercises 11–14, write the sum using sigma notation. (Begin with $k = 1$.)

11. $\dfrac{2}{3(1)} + \dfrac{2}{3(2)} + \dfrac{2}{3(3)} + \cdots + \dfrac{2}{3(20)}$

12. $\dfrac{1}{1^3} - \dfrac{1}{2^3} + \dfrac{1}{3^3} - \cdots + \dfrac{1}{25^3}$

13. $0 + \dfrac{1}{2} + \dfrac{2}{3} + \dfrac{3}{4} + \cdots + \dfrac{19}{20}$

14. $\dfrac{1}{2} + \dfrac{4}{2} + \dfrac{9}{2} + \cdots + \dfrac{100}{2}$

In Exercises 15 and 16, find the common difference of the arithmetic sequence.

15. $1, \frac{3}{2}, 2, \frac{5}{2}, 3, \ldots$

16. $100, 94, 88, 82, 76, \ldots$

In Exercises 17 and 18, find a formula for the nth term of the arithmetic sequence.

17. $a_1 = 20, \quad a_4 = 11$

18. $a_1 = 32, \quad d = -4$

19. Find the sum of the first 50 positive even numbers.

20. You save \$.50 on one day, \$1.00 the next day, \$1.50 the next day, and so on. How much will you have accumulated at the end of one year (365 days)?

11.3 Geometric Sequences and Series

Paul A. Souders/Corbis

Why You Should Learn It

A geometric sequence can reduce the amount of time it takes to find the sum of a sequence of numbers with a common ratio. For instance, in Exercise 121 on page 734, you will use a geometric sequence to find the total distance traveled by a bungee jumper.

What You Should Learn

1. Recognize, write, and find the nth terms of geometric sequences.
2. Find the nth partial sum of a geometric sequence.
3. Find the sum of an infinite geometric series.
4. Use geometric sequences to solve application problems.

Geometric Sequences

① Recognize, write, and find the nth terms of geometric sequences.

In Section 11.2, you studied sequences whose consecutive terms have a common *difference*. In this section, you will study sequences whose consecutive terms have a common *ratio*.

Definition of a Geometric Sequence

A sequence is called **geometric** if the ratios of consecutive terms are the same. So, the sequence $a_1, a_2, a_3, a_4, \ldots, a_n, \ldots$ is geometric if there is a number r, $r \neq 0$, such that

$$\frac{a_2}{a_1} = r, \quad \frac{a_3}{a_2} = r, \quad \frac{a_4}{a_3} = r$$

and so on. The number r is the **common ratio** of the sequence.

Example 1   Examples of Geometric Sequences

a. The sequence whose nth term is 2^n is geometric. For this sequence, the common ratio between consecutive terms is 2.

$$2, 4, 8, 16, \ldots, 2^n, \ldots \qquad \text{Begin with } n = 1.$$

$$\frac{4}{2} = 2$$

b. The sequence whose nth term is $4(3^n)$ is geometric. For this sequence, the common ratio between consecutive terms is 3.

$$12, 36, 108, 324, \ldots, 4(3^n), \ldots \qquad \text{Begin with } n = 1.$$

$$\frac{36}{12} = 3$$

c. The sequence whose nth term is $\left(-\frac{1}{3}\right)^n$ is geometric. For this sequence, the common ratio between consecutive terms is $-\frac{1}{3}$.

$$-\frac{1}{3}, \frac{1}{9}, -\frac{1}{27}, \frac{1}{81}, \ldots, \left(-\frac{1}{3}\right)^n, \ldots \qquad \text{Begin with } n = 1.$$

$$\frac{1/9}{-1/3} = -\frac{1}{3}$$

Study Tip

If you know the nth term of a geometric sequence, the $(n + 1)$th term can be found by multiplying by r. That is, $a_{n+1} = ra_n$.

The nth Term of a Geometric Sequence

The nth term of a geometric sequence has the form

$$a_n = a_1 r^{n-1}$$

where r is the common ratio of consecutive terms of the sequence. So, every geometric sequence can be written in the following form.

$$a_1, a_1 r, a_1 r^2, a_1 r^3, a_1 r^4, \ldots, a_1 r^{n-1}, \ldots$$

Example 2   Finding the nth Term of a Geometric Sequence

a. Find a formula for the nth term of the geometric sequence whose common ratio is 3 and whose first term is 1.

b. What is the eighth term of the sequence found in part (a)?

Solution

a. The formula for the nth term is of the form $a_n = a_1 r^{n-1}$. Moreover, because the common ratio is $r = 3$ and the first term is $a_1 = 1$, the formula must have the form

$$
\begin{aligned}
a_n &= a_1 r^{n-1} && \text{Formula for geometric sequence} \\
&= (1)(3)^{n-1} && \text{Substitute 1 for } a_1 \text{ and 3 for } r. \\
&= 3^{n-1}. && \text{Simplify.}
\end{aligned}
$$

The sequence therefore has the following form.

$$1, 3, 9, 27, 81, \ldots, 3^{n-1}, \ldots$$

b. The eighth term of the sequence is $a_8 = 3^{8-1} = 3^7 = 2187$.

Example 3   Finding the nth Term of a Geometric Sequence

Find a formula for the nth term of the geometric sequence whose first two terms are 4 and 2.

Solution

Because the common ratio is

$$r = \frac{a_2}{a_1} = \frac{2}{4} = \frac{1}{2}$$

the formula for the nth term must be

$$
\begin{aligned}
a_n &= a_1 r^{n-1} && \text{Formula for geometric sequence} \\
&= 4\left(\frac{1}{2}\right)^{n-1}. && \text{Substitute 4 for } a_1 \text{ and } \tfrac{1}{2} \text{ for } r.
\end{aligned}
$$

The sequence therefore has the form $4, 2, 1, \dfrac{1}{2}, \dfrac{1}{4}, \ldots, 4\left(\dfrac{1}{2}\right)^{n-1}, \ldots$

2 Find the nth partial sum of a geometric sequence.

The Partial Sum of a Geometric Sequence

The nth Partial Sum of a Geometric Sequence

The nth partial sum of the geometric sequence whose nth term is $a_n = a_1 r^{n-1}$ is given by

$$\sum_{i=1}^{n} a_1 r^{i-1} = a_1 + a_1 r + a_1 r^2 + a_1 r^3 + \cdots + a_1 r^{n-1} = a_1 \left(\frac{r^n - 1}{r - 1} \right).$$

Example 4   Finding the nth Partial Sum

Find the sum $1 + 2 + 4 + 8 + 16 + 32 + 64 + 128$.

Solution

This is a geometric sequence whose common ratio is $r = 2$. Because the first term of the sequence is $a_1 = 1$, it follows that the sum is

$$\sum_{i=1}^{8} 2^{i-1} = (1)\left(\frac{2^8 - 1}{2 - 1} \right) = \frac{256 - 1}{2 - 1} = 255. \qquad \text{Substitute 1 for } a_1 \text{ and 2 for } r.$$

Example 5   Finding the nth Partial Sum

Find the sum of the first five terms of the geometric sequence whose nth term is $a_n = \left(\frac{2}{3} \right)^n$.

Solution

$$\sum_{i=1}^{5} \left(\frac{2}{3} \right)^i = \frac{2}{3} \left[\frac{(2/3)^5 - 1}{(2/3) - 1} \right] \qquad \text{Substitute } \tfrac{2}{3} \text{ for } a_1 \text{ and } \tfrac{2}{3} \text{ for } r.$$

$$= \frac{2}{3} \left[\frac{(32/243) - 1}{-1/3} \right] \qquad \text{Simplify.}$$

$$= \frac{2}{3} \left(-\frac{211}{243} \right)(-3) \qquad \text{Simplify.}$$

$$= \frac{422}{243} \approx 1.737 \qquad \text{Use a calculator to simplify.}$$

3 Find the sum of an infinite geometric series.

Geometric Series

Suppose that in Example 5, you were to find the sum of all the terms of the infinite geometric sequence

$$\frac{2}{3}, \frac{4}{9}, \frac{8}{27}, \frac{16}{81}, \cdots, \left(\frac{2}{3} \right)^n, \cdots$$

A summation of all the terms of an infinite geometric sequence is called an **infinite geometric series,** or simply a **geometric series.**

In your mind, would this sum be infinitely large or would it be a finite number? Consider the formula for the nth partial sum of a geometric sequence.

$$S_n = a_1\left(\frac{r^n - 1}{r - 1}\right) = a_1\left(\frac{1 - r^n}{1 - r}\right)$$

Suppose that $|r| < 1$ and you let n become larger and larger. It follows that r^n gets closer and closer to 0, so that the term r^n drops out of the formula above. You then get the sum

$$S = a_1\left(\frac{1}{1 - r}\right) = \frac{a_1}{1 - r}.$$

Notice that this sum is not dependent on the nth term of the sequence. In the case of Example 5, $r = \left(\frac{2}{3}\right) < 1$, and so the sum of the infinite geometric sequence is

$$S = \sum_{i=1}^{\infty} \left(\frac{2}{3}\right)^i = \frac{a_1}{1 - r} = \frac{2/3}{1 - (2/3)} = \frac{2/3}{1/3} = 2.$$

> ### Technology: Tip
>
> Evaluate $\left(\frac{1}{2}\right)^n$ for $n = 1, 10,$ 100, and 1000. What happens to the value of $\left(\frac{1}{2}\right)^n$ as n increases? Make a conjecture about the value of $\left(\frac{1}{2}\right)^n$ as n approaches infinity.

Sum of an Infinite Geometric Series

If $a_1, a_1 r, a_1 r^2, \ldots, a_1 r^n, \ldots$ is an infinite geometric sequence, then for $|r| < 1$, the sum of the terms of the corresponding infinite geometric series is

$$S = \sum_{i=0}^{\infty} a_1 r^i = \frac{a_1}{1 - r}.$$

Example 6   Finding the Sum of an Infinite Geometric Series

Find each sum.

a. $\displaystyle\sum_{i=1}^{\infty} 5\left(\frac{3}{4}\right)^{i-1}$ **b.** $\displaystyle\sum_{n=0}^{\infty} 4\left(\frac{3}{10}\right)^n$ **c.** $\displaystyle\sum_{i=0}^{\infty} \left(-\frac{3}{5}\right)^i$

Solution

a. The series is geometric, with $a_1 = 5\left(\frac{3}{4}\right)^{1-1} = 5$ and $r = \frac{3}{4}$. So,

$$\sum_{i=1}^{\infty} 5\left(\frac{3}{4}\right)^{i-1} = \frac{5}{1 - (3/4)}$$

$$= \frac{5}{1/4} = 20.$$

b. The series is geometric, with $a_1 = 4\left(\frac{3}{10}\right)^0 = 4$ and $r = \frac{3}{10}$. So,

$$\sum_{n=0}^{\infty} 4\left(\frac{3}{10}\right)^n = \frac{4}{1 - (3/10)} = \frac{4}{7/10} = \frac{40}{7}.$$

c. The series is geometric, with $a_1 = \left(-\frac{3}{5}\right)^0 = 1$ and $r = -\frac{3}{5}$. So,

$$\sum_{i=0}^{\infty} \left(-\frac{3}{5}\right)^i = \frac{1}{1 - (-3/5)} = \frac{1}{1 + (3/5)} = \frac{5}{8}.$$

4 Use geometric sequences to solve
application problems.

Applications

Example 7   A Lifetime Salary

You have accepted a job as a meteorologist that pays a salary of $28,000 the first year. During the next 39 years, suppose you receive a 6% raise each year. What will your total salary be over the 40-year period?

Solution

Using a geometric sequence, your salary during the first year will be $a_1 = 28,000$. Then, with a 6% raise each year, your salary for the next 2 years will be as follows.

$$a_2 = 28,000 + 28,000(0.06) = 28,000(1.06)^1$$

$$a_3 = 28,000(1.06) + 28,000(1.06)(0.06) = 28,000(1.06)^2$$

From this pattern, you can see that the common ratio of the geometric sequence is $r = 1.06$. Using the formula for the nth partial sum of a geometric sequence, you will find that the total salary over the 40-year period is given by

$$\text{Total salary} = a_1\left(\frac{r^n - 1}{r - 1}\right)$$

$$= 28,000\left[\frac{(1.06)^{40} - 1}{1.06 - 1}\right]$$

$$= 28,000\left[\frac{(1.06)^{40} - 1}{0.06}\right] \approx \$4,333,335.$$

Example 8   Increasing Annuity

You deposit $100 in an account each month for 2 years. The account pays an annual interest rate of 9%, compounded monthly. What is your balance at the end of 2 years? (This type of savings plan is called an **increasing annuity.**)

Solution

The first deposit would earn interest for the full 24 months, the second deposit would earn interest for 23 months, the third deposit would earn interest for 22 months, and so on. Using the formula for compound interest, you can see that the total of the 24 deposits would be

$$\text{Total} = a_1 + a_2 + \cdots + a_{24}$$

$$= 100\left(1 + \frac{0.09}{12}\right)^1 + 100\left(1 + \frac{0.09}{12}\right)^2 + \cdots + 100\left(1 + \frac{0.09}{12}\right)^{24}$$

$$= 100(1.0075)^1 + 100(1.0075)^2 + \cdots + 100(1.0075)^{24}$$

$$= 100(1.0075)\left(\frac{1.0075^{24} - 1}{1.0075 - 1}\right) \qquad a_1\left(\frac{r^n - 1}{r - 1}\right)$$

$$= \$2638.49.$$

11.3 Exercises

Review   Concepts, Skills, and Problem Solving

Keep mathematically in shape by doing these exercises *before* the problems of this section.

Properties and Definitions

1. *Writing* Relative to the x- and y-axes, explain the meaning of each coordinate of the point $(-6, 4)$.

2. A point lies five units from the x-axis and ten units from the y-axis. Give the ordered pair for such a point in each quadrant.

3. *Writing* In your own words, define the graph of the function $y = f(x)$.

4. *Writing* Describe the procedure for finding the x- and y-intercepts of the graph of $f(x) = 2\sqrt{x} + 4$.

Solving Inequalities

In Exercises 5–10, solve the inequality.

5. $3x - 5 > 0$

6. $\frac{3}{2}y + 11 < 20$

7. $100 < 2x + 30 < 150$

8. $-5 < -\frac{x}{6} < 2$

9. $2x^2 - 7x + 5 > 0$

10. $2x - \frac{5}{x} > 3$

Problem Solving

11. ▲ *Geometry* A television set is advertised as having a 19-inch screen. Determine the dimensions of the square screen if its diagonal is 19 inches.

12. ▲ *Geometry* A construction worker is building the forms for the rectangular foundation of a home that is 25 feet wide and 40 feet long. To make sure that the corners are square, the worker measures the diagonal of the foundation. What should that measurement be?

Developing Skills

In Exercises 1–12, find the common ratio of the geometric sequence. See Example 1.

1. $7, 14, 28, 56, \ldots$

2. $2, 6, 18, 54, \ldots$

3. $5, -5, 5, -5, \ldots$

4. $-5, -0.5, -0.05, -0.005, \ldots$

5. $\frac{1}{2}, -\frac{1}{4}, \frac{1}{8}, -\frac{1}{16}, \ldots$

6. $\frac{2}{3}, -\frac{4}{3}, \frac{8}{3}, -\frac{16}{3}, \ldots$

7. $75, 15, 3, \frac{3}{5}, \ldots$

8. $12, -4, \frac{4}{3}, -\frac{4}{9}, \ldots$

9. $1, \pi, \pi^2, \pi^3, \ldots$

10. $e, e^2, e^3, e^4, \ldots$

11. $500(1.06), 500(1.06)^2, 500(1.06)^3, 500(1.06)^4, \ldots$

12. $1.1, (1.1)^2, (1.1)^3, (1.1)^4, \ldots$

In Exercises 13–24, determine whether the sequence is geometric. If so, find the common ratio.

13. $64, 32, 16, 8, \ldots$

14. $64, 32, 0, -32, \ldots$

15. $10, 15, 20, 25, \ldots$

16. $10, 20, 40, 80, \ldots$

17. $5, 10, 20, 40, \ldots$

18. $54, -18, 6, -2, \ldots$

19. $1, 8, 27, 64, 125, \ldots$

20. $12, 7, 2, -3, -8, \ldots$

21. $1, -\frac{2}{3}, \frac{4}{9}, -\frac{8}{27}, \ldots$

22. $\frac{1}{3}, -\frac{2}{3}, \frac{4}{3}, -\frac{8}{3}, \ldots$

23. $10(1 + 0.02), 10(1 + 0.02)^2, 10(1 + 0.02)^3, \ldots$

24. $1, 0.2, 0.04, 0.008, \ldots$

In Exercises 25–38, write the first five terms of the geometric sequence. If necessary, round your answers to two decimal places.

25. $a_1 = 4, \quad r = 2$

26. $a_1 = 3, \quad r = 4$

27. $a_1 = 6, \quad r = \frac{1}{2}$

28. $a_1 = 90, \quad r = \frac{1}{3}$

29. $a_1 = 5, \quad r = -2$

30. $a_1 = -12, \quad r = -1$

31. $a_1 = 1, \quad r = -\frac{1}{2}$

32. $a_1 = 3, \quad r = -\frac{3}{2}$

33. $a_1 = 1000, \quad r = 1.01$

34. $a_1 = 200, \quad r = 1.07$

35. $a_1 = 4000, \quad r = \frac{1}{1.01}$

36. $a_1 = 1000, \quad r = \frac{1}{1.05}$

37. $a_1 = 10, \quad r = \frac{3}{5}$

38. $a_1 = 36, \quad r = \frac{2}{3}$

In Exercises 39–52, find the specified term of the geometric sequence.

39. $a_1 = 6, \quad r = \frac{1}{2}, \quad a_{10} =$

40. $a_1 = 8, \quad r = \frac{3}{4}, \quad a_8 =$

41. $a_1 = 3, \quad r = \sqrt{2}, \quad a_{10} =$

42. $a_1 = 5, \quad r = \sqrt{3}, \quad a_9 =$

43. $a_1 = 200, \quad r = 1.2, \quad a_{12} =$

44. $a_1 = 500, \quad r = 1.06, \quad a_{40} =$

45. $a_1 = 120, \quad r = -\frac{1}{3}, \quad a_{10} =$

46. $a_1 = 240, \quad r = -\frac{1}{4}, \quad a_{13} =$

47. $a_1 = 4, \quad a_2 = 3, \quad a_5 =$

48. $a_1 = 1, \quad a_2 = 9, \quad a_7 =$

49. $a_1 = 1, \quad a_3 = \frac{9}{4}, \quad a_6 =$

50. $a_3 = 6, \quad a_5 = \frac{8}{3}, \quad a_6 =$

51. $a_2 = 12, \quad a_3 = 16, \quad a_4 =$

52. $a_4 = 100, \quad a_5 = -25, \quad a_7 =$

In Exercises 53–66, find a formula for the nth term of the geometric sequence. (Assume that n begins with 1.) See Examples 2 and 3.

53. $a_1 = 2, \quad r = 3$

54. $a_1 = 5, \quad r = 4$

55. $a_1 = 1, \quad r = 2$

56. $a_1 = 25, \quad r = 4$

57. $a_1 = 1, \quad r = -\frac{1}{5}$

58. $a_1 = 12, \quad r = -\frac{4}{3}$

59. $a_1 = 4, \quad r = -\frac{1}{2}$

60. $a_1 = 9, \quad r = \frac{2}{3}$

61. $a_1 = 8, \quad a_2 = 2$

62. $a_1 = 18, \quad a_2 = 8$

63. $a_1 = 14, \quad a_2 = \frac{21}{2}$

64. $a_1 = 36, \quad a_2 = \frac{27}{2}$

65. $4, -6, 9, -\frac{27}{2}, \ldots$

66. $1, \frac{3}{2}, \frac{9}{4}, \frac{27}{8}, \ldots$

In Exercises 67–70, match the geometric sequence with its graph. [The graphs are labeled (a), (b), (c), and (d).]

(a)
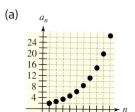

(b)
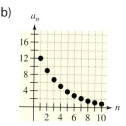

(c)
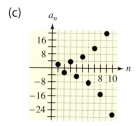

(d)
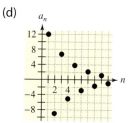

67. $a_n = 12\left(\frac{3}{4}\right)^{n-1}$

68. $a_n = 12\left(-\frac{3}{4}\right)^{n-1}$

69. $a_n = 2\left(\frac{4}{3}\right)^{n-1}$

70. $a_n = 2\left(-\frac{4}{3}\right)^{n-1}$

In Exercises 71–80, find the partial sum. See Examples 4 and 5.

71. $\displaystyle\sum_{i=1}^{10} 2^{i-1}$

72. $\displaystyle\sum_{i=1}^{6} 3^{i-1}$

73. $\displaystyle\sum_{i=1}^{12} 3\left(\frac{3}{2}\right)^{i-1}$

74. $\displaystyle\sum_{i=1}^{20} 12\left(\frac{2}{3}\right)^{i-1}$

75. $\displaystyle\sum_{i=1}^{15} 3\left(-\frac{1}{3}\right)^{i-1}$

76. $\displaystyle\sum_{i=1}^{8} 8\left(-\frac{1}{4}\right)^{i-1}$

77. $\displaystyle\sum_{i=1}^{12} 4(-2)^{i-1}$

78. $\displaystyle\sum_{i=1}^{20} 16\left(\frac{1}{2}\right)^{i-1}$

79. $\displaystyle\sum_{i=1}^{8} 6(0.1)^{i-1}$

80. $\displaystyle\sum_{i=1}^{24} 1000(1.06)^{i-1}$

In Exercises 81–92, find the *n*th partial sum of the geometric sequence.

81. $1, -3, 9, -27, 81, \ldots, n = 10$

82. $3, -6, 12, -24, 48, \ldots, n = 12$

83. $8, 4, 2, 1, \frac{1}{2}, \ldots, n = 15$

84. $9, 6, 4, \frac{8}{3}, \frac{16}{9}, \ldots, n = 10$

85. $4, 12, 36, 108, \ldots, n = 8$

86. $\frac{1}{36}, -\frac{1}{12}, \frac{1}{4}, -\frac{3}{4}, \ldots, n = 20$

87. $60, -15, \frac{15}{4}, -\frac{15}{16}, \ldots, n = 12$

88. $40, -10, \frac{5}{2}, -\frac{5}{8}, \frac{5}{32}, \ldots, n = 10$

89. $30, 30(1.06), 30(1.06)^2, 30(1.06)^3, \ldots, n = 20$

90. $100, 100(1.08), 100(1.08)^2, 100(1.08)^3, \ldots,$
$n = 40$

91. $500, 500(1.04), 500(1.04)^2, 500(1.04)^3, \ldots,$
$n = 18$

92. $1, \sqrt{2}, 2, 2\sqrt{2}, 4, \ldots, n = 12$

In Exercises 93–100, find the sum. See Example 6.

93. $\displaystyle\sum_{n=0}^{\infty} \left(\frac{1}{2}\right)^{n}$

94. $\displaystyle\sum_{n=0}^{\infty} 2\left(\frac{2}{3}\right)^{n}$

95. $\displaystyle\sum_{n=0}^{\infty} \left(-\frac{1}{2}\right)^{n}$

96. $\displaystyle\sum_{n=0}^{\infty} \left(\frac{1}{10}\right)^{n}$

97. $\displaystyle\sum_{n=0}^{\infty} 2\left(-\frac{2}{3}\right)^{n}$

98. $\displaystyle\sum_{n=0}^{\infty} 4\left(\frac{1}{4}\right)^{n}$

99. $8 + 6 + \frac{9}{2} + \frac{27}{8} + \cdots$

100. $3 - 1 + \frac{1}{3} - \frac{1}{9} + \cdots$

In Exercises 101–104, use a graphing calculator to graph the first 10 terms of the sequence.

101. $a_n = 20(-0.6)^{n-1}$

102. $a_n = 4(1.4)^{n-1}$

103. $a_n = 15(0.6)^{n-1}$

104. $a_n = 8(-0.6)^{n-1}$

Solving Problems

105. *Depreciation*   A company buys a machine for $250,000. During the next 5 years, the machine depreciates at the rate of 25% per year. (That is, at the end of each year, the depreciated value is 75% of what it was at the beginning of the year.)

  (a) Find a formula for the *n*th term of the geometric sequence that gives the value of the machine *n* full years after it was purchased.

  (b) Find the depreciated value of the machine at the end of 5 full years.

  (c) During which year did the machine depreciate the most?

106. *Population Increase*   A city of 500,000 people is growing at the rate of 1% per year. (That is, at the end of each year, the population is 1.01 times the population at the beginning of the year.)

  (a) Find a formula for the *n*th term of the geometric sequence that gives the population *n* years from now.

  (b) Estimate the population 20 years from now.

107. *Salary*   You accept a job as an archaeologist that pays a salary of $30,000 the first year. During the next 39 years, you receive a 5% raise each year. What would your total salary be over the 40-year period?

108. *Salary*   You accept a job as a biologist that pays a salary of $30,000 the first year. During the next 39 years, you receive a 5.5% raise each year.

(a) What would your total salary be over the 40-year period?

(b) How much more income did the extra 0.5% provide than the result in Exercise 107?

Increasing Annuity   In Exercises 109–114, find the balance *A* in an increasing annuity in which a principal of *P* dollars is invested each month for *t* years, compounded monthly at rate *r*.

109. $P = \$100$ $t = 10$ years $r = 9\%$

110. $P = \$50$ $t = 5$ years $r = 7\%$

111. $P = \$30$ $t = 40$ years $r = 8\%$

112. $P = \$200$ $t = 30$ years $r = 10\%$

113. $P = \$75$ $t = 30$ years $r = 6\%$

114. $P = \$100$ $t = 25$ years $r = 8\%$

115. *Wages*   You start work at a company that pays $.01 for the first day, $.02 for the second day, $.04 for the third day, and so on. The daily wage keeps doubling. What would your total income be for working (a) 29 days and (b) 30 days?

116. *Wages*   You start work at a company that pays $.01 for the first day, $.03 for the second day, $.09 for the third day, and so on. The daily wage keeps tripling. What would your total income be for working (a) 25 days and (b) 26 days?

117. *Power Supply*   The electrical power for an implanted medical device decreases by 0.1% each day.

(a) Find a formula for the *n*th term of the geometric sequence that gives the percent of the initial power *n* days after the device is implanted.

(b) What percent of the initial power is still available 1 year after the device is implanted?

(c) 🖩 The power supply needs to be changed when half the power is depleted. Use a graphing calculator to graph the first 750 terms of the sequence and estimate when the power source should be changed.

118. *Cooling*   The temperature of water in an ice cube tray is 70°F when it is placed in a freezer. Its temperature *n* hours after being placed in the freezer is 20% less than 1 hour earlier.

(a) Find a formula for the *n*th term of the geometric sequence that gives the temperature of the water *n* hours after being placed in the freezer.

(b) Find the temperature of the water 6 hours after it is placed in the freezer.

(c) 🖩 Use a graphing calculator to estimate the time when the water freezes. Explain your reasoning.

119. ▲ *Geometry*   A square has 12-inch sides. A new square is formed by connecting the midpoints of the sides of the square. Then two of the triangles are shaded (see figure). This process is repeated five more times. What is the total area of the shaded region?

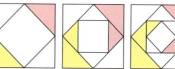 

120. ▲ *Geometry*   A square has 12-inch sides. The square is divided into nine smaller squares and the center square is shaded (see figure). Each of the eight unshaded squares is then divided into nine smaller squares and each center square is shaded. This process is repeated four more times. What is the total area of the shaded region?

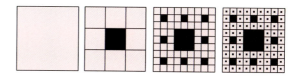

121. *Bungee Jumping*   A bungee jumper jumps from a bridge and stretches a cord 100 feet. Successive bounces stretch the cord 75% of each previous length (see figure). Find the total distance traveled by the bungee jumper during 10 bounces.

$$100 + 2(100)(0.75) + \cdots + 2(100)(0.75)^{10}$$

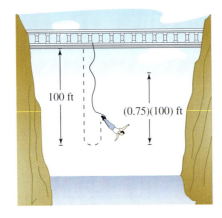

100 ft

(0.75)(100) ft

122. *Distance*   A ball is dropped from a height of 16 feet. Each time it drops h feet, it rebounds $0.81h$ feet.

(a) Find the total distance traveled by the ball.

(b) The ball takes the following time for each fall.

$s_1 = -16t^2 + 16,$ $s_1 = 0$ if $t = 1$

$s_2 = -16t^2 + 16(0.81),$ $s_2 = 0$ if $t = 0.9$

$s_3 = -16t^2 + 16(0.81)^2,$ $s_3 = 0$ if $t = (0.9)^2$

$s_4 = -16t^2 + 16(0.81)^3,$ $s_4 = 0$ if $t = (0.9)^3$

$$\vdots \qquad\qquad\qquad \vdots$$

$s_n = -16t^2 + 16(0.81)^{n-1},\ s_n = 0$ if $t = (0.9)^{n-1}$

Beginning with s_2, the ball takes the same amount of time to bounce up as it does to fall, and so the total time elapsed before it comes to rest is

$$t = 1 + 2\sum_{n=1}^{\infty} (0.9)^n.$$

Find this total.

Explaining Concepts

123. Answer parts (a)–(c) of Motivating the Chapter on page 702.

124. *Writing*   In your own words, explain what makes a sequence geometric.

125. What is the general formula for the nth term of a geometric sequence?

126. The second and third terms of a geometric sequence are 6 and 3, respectively. What is the first term?

127. Give an example of a geometric sequence whose terms alternate in sign.

128. *Writing*   Explain why the terms of a geometric sequence decrease when $a_1 > 0$ and $0 < r < 1$.

129. *Writing*   In your own words, describe an increasing annuity.

130. *Writing*   Explain what is meant by the nth partial sum of a sequence.

11.4 The Binomial Theorem

What You Should Learn

1. Use the Binomial Theorem to calculate binomial coefficients.
2. Use Pascal's Triangle to calculate binomial coefficients.
3. Expand binomial expressions.

Why You Should Learn It

You can use the Binomial Theorem to expand quantities used in probability. See Exercises 71–74 on page 742.

1. Use the Binomial Theorem to calculate binomial coefficients.

Binomial Coefficients

Recall that a **binomial** is a polynomial that has two terms. In this section, you will study a formula that provides a quick method of raising a binomial to a power. To begin, let's look at the expansion of $(x + y)^n$ for several values of n.

$$(x + y)^0 = 1$$

$$(x + y)^1 = x + y$$

$$(x + y)^2 = x^2 + 2xy + y^2$$

$$(x + y)^3 = x^3 + 3x^2y + 3xy^2 + y^3$$

$$(x + y)^4 = x^4 + 4x^3y + 6x^2y^2 + 4xy^3 + y^4$$

$$(x + y)^5 = x^5 + 5x^4y + 10x^3y^2 + 10x^2y^3 + 5xy^4 + y^5$$

There are several observations you can make about these expansions.

1. In each expansion, there are $n + 1$ terms.

2. In each expansion, x and y have symmetrical roles. The powers of x decrease by 1 in successive terms, whereas the powers of y increase by 1.

3. The sum of the powers of each term is n. For instance, in the expansion of $(x + y)^5$, the sum of the powers of each term is 5.

$$4 + 1 = 5 \quad 3 + 2 = 5$$

$$(x + y)^5 = x^5 + 5x^4y^1 + 10x^3y^2 + 10x^2y^3 + 5xy^4 + y^5$$

4. The coefficients increase and then decrease in a symmetrical pattern.

The coefficients of a binomial expansion are called **binomial coefficients.** To find them, you can use the **Binomial Theorem.**

Study Tip

Other notations that are commonly used for $_nC_r$ are

$\binom{n}{r}$ and $C(n, r)$.

The Binomial Theorem

In the expansion of $(x + y)^n$

$$(x + y)^n = x^n + nx^{n-1}y + \cdots + {_nC_r}x^{n-r}y^r + \cdots + nxy^{n-1} + y^n$$

the coefficient of $x^{n-r}y^r$ is given by

$$_nC_r = \frac{n!}{(n - r)!r!}.$$

Example 1 Finding Binomial Coefficients

Find each binomial coefficient.

a. $_8C_2$ **b.** $_{10}C_3$ **c.** $_7C_0$ **d.** $_8C_8$ **e.** $_9C_6$

Solution

a. $_8C_2 = \dfrac{8!}{6! \cdot 2!} = \dfrac{(8 \cdot 7) \cdot 6!}{6! \cdot 2!} = \dfrac{8 \cdot 7}{2 \cdot 1} = 28$

b. $_{10}C_3 = \dfrac{10!}{7! \cdot 3!} = \dfrac{(10 \cdot 9 \cdot 8) \cdot 7!}{7! \cdot 3!} = \dfrac{10 \cdot 9 \cdot 8}{3 \cdot 2 \cdot 1} = 120$

c. $_7C_0 = \dfrac{7!}{7! \cdot 0!} = 1$

d. $_8C_8 = \dfrac{8!}{0! \cdot 8!} = 1$

e. $_9C_6 = \dfrac{9!}{3! \cdot 6!} = \dfrac{(9 \cdot 8 \cdot 7) \cdot 6!}{3! \cdot 6!} = \dfrac{9 \cdot 8 \cdot 7}{3 \cdot 2 \cdot 1} = 84$

> **Technology: Tip**
>
> The formula for the binomial coefficient is the same as the formula for combinations in the study of probability. Most graphing calculators have the capability to evaluate a binomial coefficient. Consult the user's guide for your graphing calculator.

When $r \neq 0$ and $r \neq n$, as in parts (a) and (b) of Example 1, there is a simple pattern for evaluating binomial coefficients. Note how this is used in parts (a) and (b) of Example 2.

Example 2 Finding Binomial Coefficients

Find each binomial coefficient.

a. $_7C_3$ **b.** $_7C_4$ **c.** $_{12}C_1$ **d.** $_{12}C_{11}$

Solution

a. $_7C_3 = \dfrac{7 \cdot 6 \cdot 5}{3 \cdot 2 \cdot 1} = 35$

b. $_7C_4 = \dfrac{7 \cdot 6 \cdot 5 \cdot 4}{4 \cdot 3 \cdot 2 \cdot 1} = 35$ $_7C_4 = {_7C_3}$

c. $_{12}C_1 = \dfrac{12!}{11! \cdot 1!} = \dfrac{(12) \cdot 11!}{11! \cdot 1!} = \dfrac{12}{1} = 12$

d. $_{12}C_{11} = \dfrac{12!}{1! \cdot 11!} = \dfrac{(12) \cdot 11!}{1! \cdot 11!} = \dfrac{12}{1} = 12$ $_{12}C_{11} = {_{12}C_1}$

In Example 2, it is not a coincidence that the answers to parts (a) and (b) are the same and that the answers to parts (c) and (d) are the same. In general, it is true that

$$_nC_r = {_nC_{n-r}}.$$

This shows the symmetric property of binomial coefficients.

2 Use Pascal's Triangle to calculate binomial coefficients.

Pascal's Triangle

There is a convenient way to remember a pattern for binomial coefficients. By arranging the coefficients in a triangular pattern, you obtain the following array, which is called **Pascal's Triangle.** This triangle is named after the famous French mathematician Blaise Pascal (1623–1662).

```
                    1
                 1     1
              1     2     1            1 + 2 = 3
           1     3     3     1
        1     4     6     4     1
     1     5    10    10     5     1    10 + 5 = 15
  1     6    15    20    15     6     1
1     7    21    35    35    21     7     1
```

Study Tip

The top row in Pascal's Triangle is called the *zeroth row* because it corresponds to the binomial expansion

$(x + y)^0 = 1.$

Similarly, the next row is called the *first row* because it corresponds to the binomial expansion

$(x + y)^1 = 1(x) + 1(y).$

In general, the *nth row* in Pascal's Triangle gives the coefficients of $(x + y)^n$.

The first and last numbers in each row of Pascal's Triangle are 1. As shown above, every other number in each row is formed by adding the two numbers immediately above the number. Pascal noticed that numbers in this triangle are precisely the same numbers that are the coefficients of binomial expansions.

$$(x + y)^0 = 1$$ 0th row
$$(x + y)^1 = 1x + 1y$$ 1st row
$$(x + y)^2 = 1x^2 + 2xy + 1y^2$$ 2nd row
$$(x + y)^3 = 1x^3 + 3x^2y + 3xy^2 + 1y^3$$ 3rd row
$$(x + y)^4 = 1x^4 + 4x^3y + 6x^2y^2 + 4xy^3 + 1y^4$$
$$(x + y)^5 = 1x^5 + 5x^4y + 10x^3y^2 + 10x^2y^3 + 5xy^4 + 1y^5$$
$$(x + y)^6 = 1x^6 + 6x^5y + 15x^4y^2 + 20x^3y^3 + 15x^2y^4 + 6xy^5 + 1y^6$$
$$(x + y)^7 = 1x^7 + 7x^6y + 21x^5y^2 + 35x^4y^3 + 35x^3y^4 + 21x^2y^5 + 7xy^6 + 1y^7$$

You can use the seventh row of Pascal's Triangle to find the binomial coefficients of the eighth row.

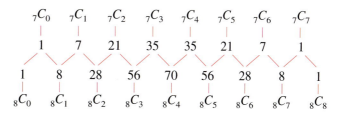

Example 3   Using Pascal's Triangle

Use the fifth row of Pascal's Triangle to evaluate $_5C_2$.

Solution

```
   1       5      10      10       5       1
   |       |       |       |       |       |
 5C0     5C1     5C2     5C3     5C4     5C5
```

So, $_5C_2 = 10$.

③ Expand binomial expressions.

Binomial Expansions

As mentioned at the beginning of this section, when you write out the coefficients for a binomial that is raised to a power, you are **expanding a binomial.** The formulas for binomial coefficients give you an easy way to expand binomials, as demonstrated in the next four examples.

Example 4   Expanding a Binomial

Write the expansion of the expression $(x + 1)^5$.

Solution

The binomial coefficients from the fifth row of Pascal's Triangle are

$1, 5, 10, 10, 5, 1.$

So, the expansion is as follows.

$$(x + 1)^5 = (1)x^5 + (5)x^4(1) + (10)x^3(1^2) + (10)x^2(1^3) + (5)x(1^4) + (1)(1^5)$$

$$= x^5 + 5x^4 + 10x^3 + 10x^2 + 5x + 1$$

To expand binomials representing *differences*, rather than sums, you alternate signs. Here are two examples.

$$(x - 1)^3 = x^3 - 3x^2 + 3x - 1$$
$$(x - 1)^4 = x^4 - 4x^3 + 6x^2 - 4x + 1$$

Example 5   Expanding a Binomial

Write the expansion of each expression.

a. $(x - 3)^4$ **b.** $(2x - 1)^3$

Solution

a. The binomial coefficients from the fourth row of Pascal's Triangle are

$1, 4, 6, 4, 1.$

So, the expansion is as follows.

$$(x - 3)^4 = (1)x^4 - (4)x^3(3) + (6)x^2(3^2) - (4)x(3^3) + (1)(3^4)$$

$$= x^4 - 12x^3 + 54x^2 - 108x + 81$$

b. The binomial coefficients from the third row of Pascal's Triangle are

$1, 3, 3, 1.$

So, the expansion is as follows.

$$(2x - 1)^3 = (1)(2x)^3 - (3)(2x)^2(1) + (3)(2x)(1^2) - (1)(1^3)$$

$$= 8x^3 - 12x^2 + 6x - 1$$

Example 6   Expanding a Binomial

Write the expansion of the expression.

$$(x - 2y)^4$$

Solution

Use the fourth row of Pascal's Triangle, as follows.

$$(x - 2y)^4 = (1)x^4 - (4)x^3(2y) + (6)x^2(2y)^2 - (4)x(2y)^3 + (1)(2y)^4$$
$$= x^4 - 8x^3y + 24x^2y^2 - 32xy^3 + 16y^4$$

Example 7   Expanding a Binomial

Write the expansion of the expression.

$$(x^2 + 4)^3$$

Solution

Use the third row of Pascal's Triangle, as follows.

$$(x^2 + 4)^3 = (1)(x^2)^3 + (3)(x^2)^2(4) + (3)x^2(4^2) + (1)(4^3)$$
$$= x^6 + 12x^4 + 48x^2 + 64$$

Sometimes you will need to find a specific term in a binomial expansion. Instead of writing out the entire expansion, you can use the fact that from the Binomial Theorem, the $(r + 1)$th term is

$$_nC_r\, x^{n-r}y^r.$$

Example 8   Finding a Term in the Binomial Expansion

a. Find the sixth term of $(a + 2b)^8$.

b. Find the coefficient of the term a^6b^5 in the expansion of $(3a - 2b)^{11}$.

Solution

a. In this case, $6 = r + 1$ means that $r = 5$. Because $n = 8$, $x = a$, and $y = 2b$, the sixth term in the binomial expansion is

$$_8C_5 a^{8-5}(2b)^5 = 56 \cdot a^3 \cdot (2b)^5$$
$$= 56(2^5)a^3b^5$$
$$= 1792\,a^3b^5.$$

b. In this case, $n = 11$, $r = 5$, $x = 3a$, and $y = -2b$. Substitute these values to obtain

$$_nC_r\, x^{n-r}y^r = {_{11}C_5}(3a)^6(-2b)^5$$
$$= 462(729a^6)(-32b^5)$$
$$= -10{,}777{,}536a^6b^5.$$

So, the coefficient is $-10{,}777{,}536$.

11.4 Exercises

Review   Concepts, Skills, and Problem Solving

Keep mathematically in shape by doing these exercises *before* the problems of this section.

Properties and Definitions

1. Is it possible to find the determinant of the following matrix? Explain.

$$\begin{bmatrix} 3 & 2 & 6 \\ 1 & -4 & 7 \end{bmatrix}$$

2. State the three elementary row operations that can be used to transform a matrix into a second, row-equivalent matrix.

3. Is the matrix in row-echelon form? Explain.

$$\begin{bmatrix} 1 & 2 & 6 \\ 0 & 1 & 7 \end{bmatrix}$$

4. Form the (a) coefficient matrix and (b) augmented matrix for the system of linear equations.

$$\begin{cases} x + 3y = -1 \\ 4x - y = 2 \end{cases}$$

Determinants

In Exercises 5–8, find the determinant of the matrix.

5. $\begin{bmatrix} 10 & 25 \\ 6 & -5 \end{bmatrix}$

6. $\begin{bmatrix} 3 & 7 \\ -2 & 6 \end{bmatrix}$

7. $\begin{bmatrix} 3 & -2 & 1 \\ 0 & 5 & 3 \\ 6 & 1 & 1 \end{bmatrix}$

8. $\begin{bmatrix} 4 & 3 & 5 \\ 3 & 2 & -2 \\ 5 & -2 & 0 \end{bmatrix}$

Problem Solving

9. Use determinants to find the equation of the line through $(2, -1)$ and $(4, 7)$.

10. Use a determinant to find the area of the triangle with vertices $(-5, 8)$, $(10, 0)$, and $(3, -4)$.

Developing Skills

In Exercises 1–12, evaluate the binomial coefficient $_nC_r$. See Examples 1 and 2.

1. $_6C_4$

2. $_7C_3$

3. $_{10}C_5$

4. $_{12}C_9$

5. $_{20}C_{20}$

6. $_{15}C_0$

7. $_{13}C_0$

8. $_{200}C_1$

9. $_{50}C_1$

10. $_{12}C_{12}$

11. $_{25}C_4$

12. $_{18}C_5$

In Exercises 13–22, use a graphing calculator to evaluate $_nC_r$.

13. $_{30}C_6$

14. $_{25}C_{10}$

15. $_{12}C_7$

16. $_{40}C_5$

17. $_{52}C_5$

18. $_{100}C_6$

19. $_{200}C_{195}$

20. $_{500}C_4$

21. $_{800}C_{797}$

22. $_{1000}C_2$

In Exercises 23–28, use Pascal's Triangle to evaluate $_nC_r$. See Example 3.

23. $_6C_2$ **24.** $_9C_3$

25. $_7C_3$ **26.** $_9C_5$

27. $_8C_4$ **28.** $_{10}C_6$

In Exercises 29–38, use Pascal's Triangle to expand the expression. See Examples 4–7.

29. $(a + 2)^3$

30. $(x + 3)^5$

31. $(m - n)^5$

32. $(r - s)^7$

33. $(2x - 1)^5$

34. $(4 - 3y)^3$

35. $(2y + z)^6$

36. $(3c + d)^6$

37. $(x^2 + 2)^4$

38. $(5 + y^2)^5$

In Exercises 39–50, use the Binomial Theorem to expand the expression.

39. $(x + 3)^6$

40. $(x - 5)^4$

41. $(x + y)^4$

42. $(u + v)^6$

43. $(u - 2v)^3$

44. $(2x + y)^5$

45. $(3a + 2b)^4$

46. $(4u - 3v)^3$

47. $\left(x + \dfrac{2}{y}\right)^4$

48. $\left(3s + \dfrac{1}{t}\right)^5$

49. $(2x^2 - y)^5$

50. $(x - 4y^3)^4$

In Exercises 51–58, find the specified term in the expansion of the binomial. See Example 8.

51. $(x + y)^{10}$, 4th term   **52.** $(x - y)^6$, 7th term

53. $(a + 4b)^9$, 6th term   **54.** $(a + 5b)^{12}$, 8th term

55. $(3a - b)^{12}$, 10th term   **56.** $(8x - y)^4$, 3rd term

57. $(3x + 2y)^{15}$, 7th term   **58.** $(4a - 3b)^9$, 8th term

In Exercises 59–66, find the coefficient of the term in the expansion of the binomial. See Example 8.

	Expression	*Term*
59.	$(x + 1)^{10}$	x^7
60.	$(x + 3)^{12}$	x^9
61.	$(x - y)^{15}$	x^4y^{11}
62.	$(x - 3y)^{14}$	x^3y^{11}
63.	$(2x + y)^{12}$	x^3y^9
64.	$(x + y)^{10}$	x^7y^3
65.	$(x^2 - 3)^4$	x^4
66.	$(3 - y^3)^5$	y^9

In Exercises 67–70, use the Binomial Theorem to approximate the quantity accurate to three decimal places. For example:
$(1.02)^{10} = (1 + 0.02)^{10} \approx 1 + 10(0.02) + 45(0.02)^2$.

67. $(1.02)^8$

68. $(2.005)^{10}$

69. $(2.99)^{12}$

70. $(1.98)^9$

Solving Problems

Probability In Exercises 71–74, use the Binomial Theorem to expand the expression. In the study of probability, it is sometimes necessary to use the expansion $(p + q)^n$, where $p + q = 1$.

71. $\left(\frac{1}{2} + \frac{1}{2}\right)^5$

72. $\left(\frac{2}{3} + \frac{1}{3}\right)^4$

73. $\left(\frac{1}{4} + \frac{3}{4}\right)^4$

74. $\left(\frac{2}{5} + \frac{3}{5}\right)^3$

75. *Pascal's Triangle* Describe the pattern.

76. *Pascal's Triangle* Use each encircled group of numbers to form a 2×2 matrix. Find the determinant of each matrix. Describe the pattern.

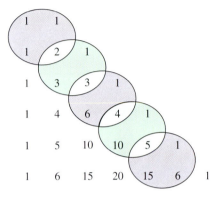

Explaining Concepts

77. How many terms are in the expansion of $(x + y)^n$?

78. How do the expansions of $(x + y)^n$ and $(x - y)^n$ differ?

79. Which of the following is equal to $_{11}C_5$? Explain.

(a) $\dfrac{11 \cdot 10 \cdot 9 \cdot 8 \cdot 7}{5 \cdot 4 \cdot 3 \cdot 2 \cdot 1}$ (b) $\dfrac{11 \cdot 10 \cdot 9 \cdot 8 \cdot 7}{6 \cdot 5 \cdot 4 \cdot 3 \cdot 2 \cdot 1}$

80. *Writing* What is the relationship between $_{n}C_r$ and $_{n}C_{n-r}$? Explain.

81. *Writing* In your own words, explain how to form the rows in Pascal's Triangle.

What Did You Learn?

Key Terms

sequence, *p. 704*
term (of a sequence), *p. 704*
infinite sequence, *p. 704*
finite sequence, *p. 704*
factorials, *p. 706*
series, *p. 707*
partial sum, *p. 707*
infinite series, *p. 707*

sigma notation, *p. 708*
index of summation, *p. 708*
upper limit of summation, *p. 708*
lower limit of summation, *p. 708*
arithmetic sequence, *p. 715*
common difference, *p. 715*
recursion formula, *p. 716*
*n*th partial sum, *pp. 717, 727*

geometric sequence, *p. 725*
common ratio, *p. 725*
infinite geometric series, *p. 727*
increasing annuity, *p. 729*
binomial coefficients, *p. 735*
Pascal's Triangle, *p. 737*
expanding a binomial, *p. 738*

Key Concepts

11.1 ◯ Definition of factorial

If *n* is a positive integer, *n* factorial is defined as

$$n! = 1 \cdot 2 \cdot 3 \cdot 4 \cdots (n-1) \cdot n.$$

As a special case, zero factorial is defined as $0! = 1$.

11.1 ◯ Definition of series

For an infinite sequence, $a_1, a_2, a_3, \ldots, a_n, \ldots$
1. the sum of the first *n* terms
$$S_n = a_1 + a_2 + a_3 + \cdots + a_n$$
is called a partial sum, and

2. the sum of all terms
$$a_1 + a_2 + a_3 + \cdots + a_n + \cdots$$
is called an infinite series, or simply a series.

11.1 ◯ Definition of sigma notation

The sum of the first *n* terms of the sequence whose *n*th term is a_n is

$$\sum_{i=1}^{n} a_i = a_1 + a_2 + a_3 + a_4 + \cdots + a_n$$

where *i* is the index of summation, *n* is the upper limit of summation, and 1 is the lower limit of summation.

11.2 ◯ The *n*th term of an arithmetic sequence

The *n*th term of an arithmetic sequence has the form $a_n = a_1 + (n-1)d$, where *d* is the common difference of the sequence, and a_1 is the first term.

11.2 ◯ The *n*th partial sum of an arithmetic sequence

The *n*th partial sum of the arithmetic sequence whose *n*th term is a_n is

$$\sum_{i=1}^{n} a_i = a_1 + a_2 + a_3 + a_4 + \cdots + a_n$$

$$= \frac{n}{2}(a_1 + a_n).$$

11.3 ◯ The *n*th term of a geometric sequence

The *n*th term of a geometric sequence has the form $a_n = a_1 r^{n-1}$, where *r* is the common ratio of consecutive terms of the sequence. So, every geometric sequence can be written in the following form.

$$a_1, a_1 r, a_1 r^2, a_1 r^3, a_1 r^4, \ldots, a_1 r^{n-1}, \ldots$$

11.3 ◯ The *n*th partial sum of a geometric sequence

The *n*th partial sum of the geometric sequence whose *n*th term is $a_n = a_1 r^{n-1}$ is given by

$$\sum_{i=1}^{n} a_1 r^{i-1} = a_1 + a_1 r + a_1 r^2 + a_1 r^3 + \cdots$$
$$+ a_1 r^{n-1} = a_1 \left(\frac{r^n - 1}{r - 1} \right).$$

11.3 ◯ Sum of an infinite geometric series

If $a_1, a_1 r, a_1 r^2, \ldots, a_1 r^n, \ldots$ is an infinite geometric sequence, then for $|r| < 1$, the sum of the terms of the corresponding infinite geometric series is

$$S = \sum_{i=0}^{\infty} a_1 r^i = \frac{a_1}{1 - r}.$$

11.4 ◯ The Binomial Theorem

In the expansion of $(x + y)^n$
$$(x + y)^n = x^n + nx^{n-1}y + \cdots +$$
$$_nC_r x^{n-r} y^r + \cdots + nxy^{n-1} + y^n$$
the coefficient of $x^{n-r} y^r$ is given by

$$_nC_r = \frac{n!}{(n-r)!r!}.$$

Review Exercises

11.1 Sequences and Series

1 Use sequence notation to write the terms of sequences.

In Exercises 1–4, write the first five terms of the sequence. (Assume that n begins with 1.)

1. $a_n = 3n + 5$

2. $a_n = \frac{1}{2}n - 4$

3. $a_n = \frac{1}{2^n} + \frac{1}{2}$

4. $a_n = 3^n + n$

2 Write the terms of sequences involving factorials.

In Exercises 5–8, write the first five terms of the sequence. (Assume that n begins with 1.)

5. $a_n = (n + 1)!$

6. $a_n = n! - 2$

7. $a_n = \dfrac{n!}{2n}$

8. $a_n = \dfrac{(n + 1)!}{(2n)!}$

3 Find the apparent nth term of a sequence.

In Exercises 9–18, write an expression for the nth term of the sequence. (Assume that n begins with 1.)

9. $1, 3, 5, 7, 9, \ldots$

10. $3, -6, 9, -12, 15, \ldots$

11. $\frac{1}{4}, \frac{2}{9}, \frac{3}{16}, \frac{4}{25}, \frac{5}{36}, \ldots$

12. $\frac{0}{2}, \frac{1}{3}, \frac{2}{4}, \frac{3}{5}, \frac{4}{6}, \ldots$

13. $4, \frac{9}{2}, \frac{14}{3}, \frac{19}{4}, \frac{24}{5}, \ldots$

14. $12, 4, \frac{4}{3}, \frac{4}{9}, \frac{4}{27}, \ldots$

15. $3, 1, -1, -3, -5, \ldots$

16. $3, 7, 11, 15, 19, \ldots$

17. $\frac{3}{2}, \frac{12}{5}, \frac{27}{10}, \frac{48}{17}, \frac{75}{26}, \ldots$

18. $-1, \frac{1}{2}, -\frac{1}{4}, \frac{1}{8}, -\frac{1}{16}, \ldots$

4 Sum the terms of sequences to obtain series and use sigma notation to represent partial sums.

In Exercises 19–22, find the partial sum.

19. $\displaystyle\sum_{k=1}^{4} 7$

20. $\displaystyle\sum_{k=1}^{4} \frac{(-1)^k}{k}$

21. $\displaystyle\sum_{n=1}^{4} \left(\frac{1}{n} - \frac{1}{n+1}\right)$

22. $\displaystyle\sum_{n=1}^{4} \left(\frac{1}{n} - \frac{1}{n+2}\right)$

In Exercises 23–26, write the sum using sigma notation. (Begin with $k = 0$ or $k = 1$.)

23. $[5(1) - 3] + [5(2) - 3] + [5(3) - 3] +$
   $[5(4) - 3]$

24. $[9 - 10(1)] + [9 - 10(2)] + [9 - 10(3)] +$
   $[9 - 10(4)]$

25. $\dfrac{1}{3(1)} + \dfrac{1}{3(2)} + \dfrac{1}{3(3)} + \dfrac{1}{3(4)} + \dfrac{1}{3(5)} + \dfrac{1}{3(6)}$

26. $\left(-\frac{1}{3}\right)^0 + \left(-\frac{1}{3}\right)^1 + \left(-\frac{1}{3}\right)^2 + \left(-\frac{1}{3}\right)^3 + \left(-\frac{1}{3}\right)^4$

11.2 Arithmetic Sequences

1 Recognize, write, and find the nth terms of arithmetic sequences.

In Exercises 27 and 28, find the common difference of the arithmetic sequence.

27. $30, 27.5, 25, 22.5, 20, \ldots$

28. $9, 12, 15, 18, 21, \ldots$

In Exercises 29–36, write the first five terms of the arithmetic sequence. (Assume that n begins with 1.)

29. $a_n = 132 - 5n$

30. $a_n = 2n + 3$

31. $a_n = \frac{3}{4}n + \frac{1}{2}$

32. $a_n = -\frac{3}{5}n + 1$

33. $a_1 = 5$
   $a_{k+1} = a_k + 3$

34. $a_1 = 12$
   $a_{k+1} = a_k + 1.5$

35. $a_1 = 80$
   $a_{k+1} = a_k - \frac{5}{2}$

36. $a_1 = 25$
   $a_{k+1} = a_k - 6$

In Exercises 37–40, find a formula for the nth term of the arithmetic sequence.

37. $a_1 = 10$, $d = 4$

38. $a_1 = 32$, $d = -2$

39. $a_1 = 1000$, $a_2 = 950$

40. $a_1 = 12$, $a_2 = 20$

2 Find the nth partial sum of an arithmetic sequence.

In Exercises 41–44, find the partial sum.

41. $\displaystyle\sum_{k=1}^{12} (7k - 5)$

42. $\displaystyle\sum_{k=1}^{10} (100 - 10k)$

43. $\displaystyle\sum_{j=1}^{100} \frac{j}{4}$

44. $\displaystyle\sum_{j=1}^{50} \frac{3j}{2}$

In Exercises 45 and 46, use a graphing calculator to find the partial sum.

45. $\displaystyle\sum_{i=1}^{60} (1.25i + 4)$

46. $\displaystyle\sum_{i=1}^{100} (5000 - 3.5i)$

3 Use arithmetic sequences to solve application problems.

47. *Number Problem* Find the sum of the first 50 positive integers that are multiples of 4.

48. *Number Problem* Find the sum of the integers from 225 to 300.

49. *Auditorium Seating* Each row in a small auditorium has three more seats than the preceding row. The front row seats 22 people and there are 12 rows of seats. Find the seating capacity of the auditorium.

50. *Pile of Logs* A pile of logs has 20 logs on the bottom layer and one log on the top layer. Each layer has one log less than the layer below it. How many logs are in the pile?

11.3 Geometric Sequences and Series

1 Recognize, write, and find the nth terms of geometric sequences.

In Exercises 51 and 52, find the common ratio of the geometric sequence.

51. $8, 12, 18, 27, \frac{81}{2}, \ldots$

52. $27, -18, 12, -8, \frac{16}{3}, \ldots$

In Exercises 53–58, write the first five terms of the geometric sequence.

53. $a_1 = 10$, $r = 3$

54. $a_1 = 2$, $r = -5$

55. $a_1 = 100$, $r = -\frac{1}{2}$

56. $a_1 = 12$, $r = \frac{1}{6}$

57. $a_1 = 4$, $r = \frac{3}{2}$

58. $a_1 = 32$, $r = -\frac{3}{4}$

In Exercises 59–64, find a formula for the nth term of the geometric sequence. (Assume that n begins with 1.)

59. $a_1 = 1$, $r = -\frac{2}{3}$

60. $a_1 = 100$, $r = 1.07$

61. $a_1 = 24$, $a_2 = 48$

62. $a_1 = 16$, $a_2 = -4$

63. $a_1 = 12$, $a_4 = -\frac{3}{2}$

64. $a_2 = 1$, $a_3 = \frac{1}{3}$

2 Find the nth partial sum of a geometric sequence.

In Exercises 65–72, find the partial sum.

65. $\displaystyle\sum_{n=1}^{12} 2^n$

66. $\displaystyle\sum_{n=1}^{12} (-2)^n$

67. $\displaystyle\sum_{k=1}^{8} 5\left(-\frac{3}{4}\right)^k$

68. $\displaystyle\sum_{k=1}^{10} 4\left(\frac{3}{2}\right)^k$

69. $\displaystyle\sum_{i=1}^{8} (1.25)^{i-1}$

70. $\displaystyle\sum_{i=1}^{8} (-1.25)^{i-1}$

71. $\displaystyle\sum_{n=1}^{120} 500(1.01)^n$

72. $\displaystyle\sum_{n=1}^{40} 1000(1.1)^n$

In Exercises 73 and 74, use a graphing calculator to find the partial sum.

73. $\displaystyle\sum_{k=1}^{50} 50(1.2)^{k-1}$

74. $\displaystyle\sum_{j=1}^{60} 25(0.9)^{j-1}$

3 Find the sum of an infinite geometric series.

In Exercises 75–78, find the sum.

75. $\displaystyle\sum_{i=1}^{\infty} \left(\frac{7}{8}\right)^{i-1}$

76. $\displaystyle\sum_{i=1}^{\infty} \left(\frac{1}{3}\right)^{i-1}$

77. $\sum_{k=1}^{\infty} 4\left(\frac{2}{3}\right)^{k-1}$ **78.** $\sum_{k=1}^{\infty} 1.3\left(\frac{1}{10}\right)^{k-1}$

④ Use geometric sequences to solve application problems.

79. *Depreciation* A company pays $120,000 for a machine. During the next 5 years, the machine depreciates at the rate of 30% per year. (That is, at the end of each year, the depreciated value is 70% of what it was at the beginning of the year.)

(a) Find a formula for the *n*th term of the geometric sequence that gives the value of the machine *n* full years after it was purchased.

(b) Find the depreciated value of the machine at the end of 5 full years.

80. *Population Increase* A city of 85,000 people is growing at the rate of 1.2% per year. (That is, at the end of each year, the population is 1.012 times what it was at the beginning of the year.)

(a) Find a formula for the *n*th term of the geometric sequence that gives the population *n* years from now.

(b) Estimate the population 50 years from now.

81. *Salary* You accept a job as an architect that pays a salary of $32,000 the first year. During the next 39 years, you receive a 5.5% raise each year. What would your total salary be over the 40-year period?

82. *Increasing Annuity* You deposit $200 in an account each month for 10 years. The account pays an annual interest rate of 8%, compounded monthly. What is your balance at the end of 10 years?

11.4 The Binomial Theorem

① Use the Binomial Theorem to calculate binomial coefficients.

In Exercises 83–86, evaluate the binomial coefficient $_nC_r$.

83. $_8C_3$ **84.** $_{12}C_2$
85. $_{12}C_0$ **86.** $_{100}C_1$

▦ In Exercises 87–90, use a graphing calculator to evaluate $_nC_r$.

87. $_{40}C_4$ **88.** $_{15}C_9$
89. $_{25}C_6$ **90.** $_{32}C_2$

② Use Pascal's Triangle to calculate binomial coefficients.

In Exercises 91–94, use Pascal's Triangle to evaluate $_nC_r$.

91. $_5C_3$ **92.** $_9C_9$
93. $_8C_4$ **94.** $_6C_5$

③ Expand binomial expressions.

In Exercises 95–98, use Pascal's Triangle to expand the expression.

95. $(x - 5)^4$
96. $(x + y)^7$

97. $(2x + 1)^3$
98. $(x - 3y)^4$

In Exercises 99–104, use the Binomial Theorem to expand the expression.

99. $(x + 1)^{10}$

100. $(y - 2)^6$

101. $(3x - 2y)^4$

102. $(2u + 5v)^4$

103. $(u^2 + v^3)^9$

104. $(x^4 - y^5)^8$

In Exercises 105 and 106, find the specified term in the expansion of the binomial.

105. $(x + 4)^9$, 4th term
106. $(2x - 3y)^5$, 4th term

In Exercises 107 and 108, find the coefficient of the term in the expansion of the binomial.

	Expression	*Term*
107.	$(x - 3)^{10}$	x^5
108.	$(x + 2y)^7$	x^4y^3

Take this test as you would take a test in class. After you are done, check your work against the answers in the back of the book.

1. Write the first five terms of the sequence $a_n = \left(-\frac{2}{3}\right)^{n-1}$. (Assume that n begins with 1.)

2. Write the first five terms of the sequence $a_n = 3n^2 - n$. (Assume that n begins with 1.)

In Exercises 3–5, find the partial sum.

3. $\displaystyle\sum_{n=1}^{12} 5$

4. $\displaystyle\sum_{j=0}^{4} (3j + 1)$

5. $\displaystyle\sum_{n=1}^{5} (3 - 4n)$

6. Use sigma notation to write $\dfrac{2}{3(1) + 1} + \dfrac{2}{3(2) + 1} + \cdots + \dfrac{2}{3(12) + 1}$.

7. Use sigma notation to write

$$\left(\frac{1}{2}\right)^0 + \left(\frac{1}{2}\right)^2 + \left(\frac{1}{2}\right)^4 + \left(\frac{1}{2}\right)^6 + \left(\frac{1}{2}\right)^8 + \left(\frac{1}{2}\right)^{10}.$$

8. Write the first five terms of the arithmetic sequence whose first term is $a_1 = 12$ and whose common difference is $d = 4$.

9. Find a formula for the nth term of the arithmetic sequence whose first term is $a_1 = 5000$ and whose common difference is $d = -100$.

10. Find the sum of the first 50 positive integers that are multiples of 3.

11. Find the common ratio of the geometric sequence: $2, -3, \frac{9}{2}, -\frac{27}{4}, \ldots$

12. Find a formula for the nth term of the geometric sequence whose first term is $a_1 = 4$ and whose common ratio is $r = \frac{1}{2}$.

In Exercises 13 and 14, find the partial sum.

13. $\displaystyle\sum_{n=1}^{8} 2(2^n)$

14. $\displaystyle\sum_{n=1}^{10} 3\left(\frac{1}{2}\right)^n$

In Exercises 15 and 16, find the sum of the infinite geometric series.

15. $\displaystyle\sum_{i=1}^{\infty} \left(\frac{1}{2}\right)^i$

16. $\displaystyle\sum_{i=1}^{\infty} 4\left(\frac{2}{3}\right)^{i-1}$

17. Evaluate: $_{20}C_3$

18. Use Pascal's Triangle to expand $(x - 2)^5$.

19. Find the coefficient of the term x^3y^5 in the expansion of $(x + y)^8$.

20. A free-falling object will fall 4.9 meters during the first second, 14.7 more meters during the second second, 24.5 more meters during the third second, and so on. What is the total distance the object will fall in 10 seconds if this pattern continues?

21. Fifty dollars is deposited each month in an increasing annuity that pays 8%, compounded monthly. What is the balance after 25 years?

Introduction to Graphing Calculators

Introduction   •   Using a Graphing Calculator   •   Using Special Features of a Graphing Calculator

Introduction

In Section 3.2, you studied the point-plotting method for sketching the graph of an equation. One of the disadvantages of the point-plotting method is that to get a good idea about the shape of a graph you need to plot *many* points. By plotting only a few points, you can badly misrepresent the graph.

For instance, consider the equation $y = x^3$. To graph this equation, suppose you calculated only the following three points.

x	-1	0	1
$y = x^3$	-1	0	1
Solution point	$(-1, -1)$	$(0, 0)$	$(1, 1)$

By plotting these three points, as shown in Figure A.1, you might assume that the graph of the equation is a line. This, however, is not correct. By plotting several more points, as shown in Figure A.2, you can see that the actual graph is not straight at all.

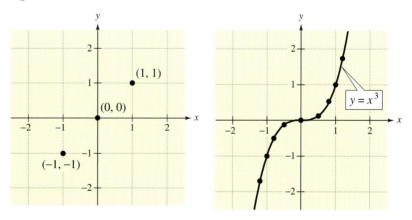

Figure A.1   Figure A.2

So, the point-plotting method leaves you with a dilemma. On the one hand, the method can be very inaccurate if only a few points are plotted. But, on the other hand, it is very time-consuming to plot a dozen (or more) points. Technology can help you solve this dilemma. Plotting several points (or even hundreds of points) on a rectangular coordinate system is something that a computer or graphing calculator can do easily.

Using a Graphing Calculator

There are many different graphing utilities: some are graphing packages for computers and some are hand-held graphing calculators. In this appendix, the steps used to graph an equation with a *TI-83* or *TI-83 Plus* graphing calculator are described. Keystroke sequences are often given for illustration; however, these may not agree precisely with the steps required by *your* calculator.*

Graphing an Equation with a *TI-83* or *TI-83 Plus* Graphing Calculator

Before performing the following steps, set your calculator so that all of the standard defaults are active. For instance, all of the options at the left of the MODE screen should be highlighted.

1. Set the viewing window for the graph. (See Example 3.) To set the standard viewing window, press ZOOM 6.

2. Rewrite the equation so that *y* is isolated on the left side of the equation.

3. Press the Y= key. Then enter the right side of the equation on the first line of the display. (The first line is labeled $Y_1 = .$)

4. Press the GRAPH key.

Example 1   Graphing a Linear Equation

Use a graphing calculator to graph $2y + x = 4$.

Solution

To begin, solve the equation for *y* in terms of *x*.

$$2y + x = 4 \qquad \textcolor{red}{\text{Write original equation.}}$$

$$2y = -x + 4 \qquad \textcolor{red}{\text{Subtract } x \text{ from each side.}}$$

$$y = -\frac{1}{2}x + 2 \qquad \textcolor{red}{\text{Divide each side by 2.}}$$

Press the Y= key, and enter the following keystrokes.

(−) X,T,θ,*n* ÷ 2 + 2

The top row of the display should now be as follows.

$$Y_1 = -X/2 + 2$$

Press the GRAPH key, and the screen should look like that shown in Figure A.3.

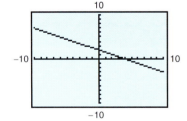

Figure A.3

---

*The graphing calculator keystrokes given in this section correspond to the *TI-83* and *TI-83 Plus* graphing calculators by Texas Instruments. For other graphing calculators, the keystrokes may differ. Consult your user's guide.

In Figure A.3, notice that the calculator screen does not label the tick marks on the *x*-axis or the *y*-axis. To see what the tick marks represent, you can press [WINDOW]. If you set your calculator to the standard graphing defaults before working Example 1, the screen should show the following values.

Xmin = -10	The minimum *x*-value is -10.
Xmax = 10	The maximum *x*-value is 10.
Xscl = 1	The *x*-scale is 1 unit per tick mark.
Ymin = -10	The minimum *y*-value is -10.
Ymax = 10	The maximum *y*-value is 10.
Yscl = 1	The *y*-scale is 1 unit per tick mark.
Xres = 1	Sets the pixel resolution

These settings are summarized visually in Figure A.4.

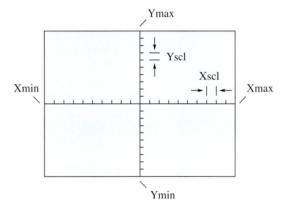

Figure A.4

Example 2   Graphing an Equation Involving Absolute Value

Use a graphing calculator to graph

$$y = |x - 3|.$$

Solution

This equation is already written so that *y* is isolated on the left side of the equation. Press the [Y=] key, and enter the following keystrokes.

[MATH] [▶] 1 [X,T,θ,*n*] [−] 3 [)]

The top row of the display should now be as follows.

$$Y_1 = \text{abs}(X - 3)$$

Press the [GRAPH] key, and the screen should look like that shown in Figure A.5.

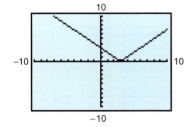

Figure A.5

Using Special Features of a Graphing Calculator

To use your graphing calculator to its best advantage, you must learn to set the viewing window, as illustrated in the next example.

Example 3   Setting the Viewing Window

Use a graphing calculator to graph

$$y = x^2 + 12.$$

Solution

Press $\boxed{Y=}$ and enter $x^2 + 12$ on the first line.

$\boxed{X,T,\theta,n}$ $\boxed{x^2}$ $\boxed{+}$ 12

Press the $\boxed{\text{GRAPH}}$ key. If your calculator is set to the standard viewing window, nothing will appear on the screen. The reason for this is that the lowest point on the graph of $y = x^2 + 12$ occurs at the point $(0, 12)$. Using the standard viewing window, you obtain a screen whose largest y-value is 10. In other words, none of the graph is visible on a screen whose y-values vary between -10 and 10, as shown in Figure A.6. To change these settings, press $\boxed{\text{WINDOW}}$ and enter the following values.

Xmin = -10	The minimum x-value is -10.
Xmax = 10	The maximum x-value is 10.
Xscl = 1	The x-scale is 1 unit per tick mark.
Ymin = -10	The minimum y-value is -10.
Ymax = 30	The maximum y-value is 30.
Yscl = 5	The y-scale is 5 units per tick mark.
Xres = 1	Sets the pixel resolution

Press $\boxed{\text{GRAPH}}$ and you will obtain the graph shown in Figure A.7. On this graph, note that each tick mark on the y-axis represents five units because you changed the y-scale to 5. Also note that the highest point on the y-axis is now 30 because you changed the maximum value of y to 30.

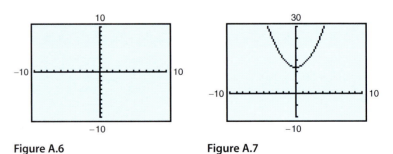

Figure A.6   Figure A.7

If you changed the y-maximum and y-scale on your calculator as indicated in Example 3, you should return to the standard setting before working Example 4. To do this, press $\boxed{\text{ZOOM}}$ 6.

Example 4   Using a Square Setting

Use a graphing calculator to graph $y = x$. The graph of this equation is a line that makes a 45° angle with the x-axis and with the y-axis. From the graph on your calculator, does the angle appear to be 45°?

Solution

Press $\boxed{Y=}$ and enter x on the first line.

$$Y_1 = X$$

Press the $\boxed{GRAPH}$ key and you will obtain the graph shown in Figure A.8. Notice that the angle the line makes with the x-axis doesn't appear to be 45°. The reason for this is that the screen is wider than it is tall. This makes the tick marks on the x-axis farther apart than the tick marks on the y-axis. To obtain the same distance between tick marks on both axes, you can change the graphing settings from "standard" to "square." To do this, press the following keys.

$\boxed{ZOOM}$ 5   Square setting

The screen should look like that shown in Figure A.9. Note in this figure that the square setting has changed the viewing window so that the x-values vary from -15 to 15.

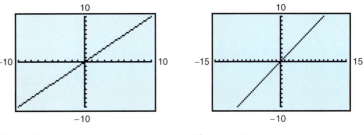

Figure A.8   Figure A.9

There are many possible square settings on a graphing calculator. To create a square setting, you need the following ratio to be $\frac{2}{3}$.

$$\frac{Ymax - Ymin}{Xmax - Xmin}$$

For instance, the setting in Example 4 is square because $(Ymax - Ymin) = 20$ and $(Xmax - Xmin) = 30$.

Example 5   Graphing More than One Equation in the Same Viewing Window

Use a graphing calculator to graph each equation in the same viewing window.

$$y = -x + 4, \quad y = -x, \quad \text{and} \quad y = -x - 4$$

Solution

To begin, press $\boxed{Y=}$ and enter all three equations on the first three lines. The display should now be as follows.

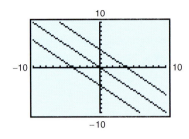

Figure A.10

$Y_1 = -X + 4$　　　$\boxed{(-)}\ \boxed{X,T,\theta,n}\ \boxed{+}\ 4$

$Y_2 = -X$　　　　　$\boxed{(-)}\ \boxed{X,T,\theta,n}$

$Y_3 = -X - 4$　　　$\boxed{(-)}\ \boxed{X,T,\theta,n}\ \boxed{-}\ 4$

Press the $\boxed{\text{GRAPH}}$ key and you will obtain the graph shown in Figure A.10. Note that the graph of each equation is a line, and that the lines are parallel to each other.

Another special feature of a graphing calculator is the *trace* feature. This feature is used to find solution points of an equation. For example, you can approximate the x- and y-intercepts of $y = 3x + 6$ by first graphing the equation, then pressing the $\boxed{\text{TRACE}}$ key, and finally pressing the $\boxed{\blacktriangleleft}\boxed{\blacktriangleright}$ keys. To get a better approximation of a solution point, you can use the following keystrokes repeatedly.

$$\boxed{\text{ZOOM}}\ 2\ \boxed{\text{ENTER}}$$

Check to see that you get an x-intercept of $(-2, 0)$ and a y-intercept of $(0, 6)$. Use the *trace* feature to find the x- and y-intercepts of $y = \frac{1}{2}x - 4$.

Appendix A　Exercises

⊞ In Exercises 1–12, use a graphing calculator to graph the equation. (Use a standard setting.) See Examples 1 and 2.

1. $y = -3x$　　　　　　　**2.** $y = x - 4$

3. $y = \frac{3}{4}x - 6$　　　　　**4.** $y = -3x + 2$

5. $y = \frac{1}{2}x^2$　　　　　　　**6.** $y = -\frac{2}{3}x^2$

7. $y = x^2 - 4x + 2$　　　**8.** $y = -0.5x^2 - 2x + 2$

9. $y = |x - 3|$　　　　　**10.** $y = |x + 4|$

11. $y = |x^2 - 4|$　　　　**12.** $y = |x - 2| - 5$

⊞ In Exercises 13–16, use a graphing calculator to graph the equation using the given window settings. See Example 3.

13. $y = 27x + 100$　　　**14.** $y = 50,000 - 6000x$

| Xmin = 0 |
| Xmax = 5 |
| Xscl = .5 |
| Ymin = 75 |
| Ymax = 250 |
| Yscl = 25 |
| Xres = 1 |

| Xmin = 0 |
| Xmax = 7 |
| Xscl = .5 |
| Ymin = 0 |
| Ymax = 50000 |
| Yscl = 5000 |
| Xres = 1 |

15. $y = 0.001x^2 + 0.5x$　　　**16.** $y = 100 - 0.5|x|$

| Xmin = -500 |
| Xmax = 200 |
| Xscl = 50 |
| Ymin = -100 |
| Ymax = 100 |
| Yscl = 20 |
| Xres = 1 |

| Xmin = -300 |
| Xmax = 300 |
| Xscl = 60 |
| Ymin = -100 |
| Ymax = 100 |
| Yscl = 20 |
| Xres = 1 |

⊞ In Exercises 17–20, find a viewing window that shows the important characteristics of the graph.

17. $y = 15 + |x - 12|$　　**18.** $y = 15 + (x - 12)^2$

19. $y = -15 + |x + 12|$　**20.** $y = -15 + (x + 12)^2$

⊞ In Exercises 21–24, use a graphing calculator to graph both equations in the same viewing window. Are the graphs identical? If so, what basic rule of algebra is being illustrated? See Example 5.

21. $y_1 = 2x + (x + 1)$　　　**22.** $y_1 = \frac{1}{2}(3 - 2x)$

$\quad\ \ y_2 = (2x + x) + 1$　　　　$\quad\ \ y_2 = \frac{3}{2} - x$

23. $y_1 = 2\left(\frac{1}{2}\right)$

$y_2 = 1$

24. $y_1 = x(0.5x)$

$y_2 = (0.5x)x$

In Exercises 25–32, use the *trace* feature of a graphing calculator to approximate the *x*- and *y*-intercepts of the graph.

25. $y = 9 - x^2$

26. $y = 3x^2 - 2x - 5$

27. $y = 6 - |x + 2|$

28. $y = |x - 2|^2 - 3$

29. $y = 2x - 5$

30. $y = 4 - |x|$

31. $y = x^2 + 1.5x - 1$

32. $y = x^3 - 4x$

Geometry   In Exercises 33–36, use a graphing calculator to graph the equations in the same viewing window. Using a "square setting," determine the geometrical shape bounded by the graphs.

33. $y = -4, \quad y = -|x|$

34. $y = |x|, \quad y = 5$

35. $y = |x| - 8, \quad y = -|x| + 8$

36. $y = -\frac{1}{2}x + 7, \quad y = \frac{8}{3}(x + 5), \quad y = \frac{2}{7}(3x - 4)$

Modeling Data   In Exercises 37 and 38, use the following models, which give the number of pieces of first-class mail and the number of pieces of Standard A (third-class) mail handled by the U.S. Postal Service.

First Class

$y = 0.008x^2 + 1.42x + 88.7 \quad 0 \le x \le 10$

Standard A (Third Class)

$y = 0.246x^2 + 0.36x + 62.5 \quad 0 \le x \le 10$

In these models, *y* is the number of pieces handled (in billions) and *x* is the year, with $x = 0$ corresponding to 1990.   (Source: U.S. Postal Service)

37. Use the following window setting to graph both models in the same viewing window of a graphing calculator.

Xmin = 0
Xmax = 10
Xscl = 1
Ymin = 0
Ymax = 120
Yscl = 10
Xres = 1

38. (a) Were the numbers of pieces of first-class mail and Standard A mail increasing or decreasing over time?

(b) Is the distance between the graphs increasing or decreasing over time? What does this mean to the U.S. Postal Service?

Answers to Reviews, Odd-Numbered Exercises, Quizzes, and Tests

Chapter 1

Section 1.1 *(page 9)*

1. (a) $1, 4, 6$ (b) $-10, 0, 1, 4, 6$
(c) $-10, -\frac{2}{3}, -\frac{1}{4}, 0, \frac{5}{8}, 1, 4, 6$
(d) $-\sqrt{5}, \sqrt{3}, 2\pi$

3. (a) 3 (b) $-\sqrt{4}, 0, 3$
(c) $-3.5, -\sqrt{4}, -\frac{1}{2}, -0.\overline{3}, 0, 3, 25.2$
(d) $\sqrt{5}, 3\pi$

5. $-5, -4, -3, -2, -1, 0, 1, 2, 3$

7. $1, 3, 5, 7, 9$

9. (a)   (b)

(c)   (d)

11. $-1 < 3$ **13.** $-\frac{9}{2} < -2$ **15.** $<$ **17.** $<$

19. $<$ **21.** $<$ **23.** $>$ **25.** $>$ **27.** $>$

29. $<$ **31.** 6 **33.** 19 **35.** 50 **37.** 8

39. 35 **41.** 3 **43.** 10 **45.** 225 **47.** -85

49. -16 **51.** $-\frac{3}{4}$ **53.** -3.5 **55.** π **57.** $>$

59. $>$ **61.** $=$ **63.** $>$ **65.** $-34, 34$

67. $160, 160$ **69.** $\frac{3}{11}, \frac{3}{11}$ **71.** $-\frac{5}{4}, \frac{5}{4}$ **73.** $-4.7, 4.7$

75.   **77.**

7 5

79.   **81.**

$\frac{3}{5}$ $\frac{5}{3}$

83.  ; 4.25

85. $x < 0$ **87.** $x \geq 0$ **89.** $2 < z \leq 10$

91. $p < 225$ **93.** True

95. False. $|3 + (-2)| = 1 \neq 5 = |3| + |-2|$

97. False. Zero is a whole number but not a natural number.

99. The set of integers includes the natural numbers, zero, and the negative integers.

101. Yes. The nonnegative real numbers include 0.

103. Plot them on the real number line. The number on the right is greater.

Section 1.2 *(page 19)*

1. 45 **3.** 4 **5.** -2.7 **7.** 7 **9.** -25.9

11. -22 **13.** -21 **15.** -20 **17.** 0.7 **19.** 22

21. $\frac{5}{4}$ **23.** $\frac{1}{2}$ **25.** $\frac{1}{10}$ **27.** $\frac{1}{24}$ **29.** $\frac{63}{8}$ **31.** $\frac{105}{8}$

33. 60 **35.** 45.95 **37.** -28 **39.** $5 + 5 + 5 + 5$

41. $(-4) + (-4) + (-4)$ **43.** $\frac{1}{2} + \frac{1}{2} + \frac{1}{2} + \frac{1}{2} + \frac{1}{2}$

45. $4 \cdot 9$ **47.** $6\left(\frac{1}{4}\right)$ **49.** $4\left(-\frac{1}{5}\right)$ **51.** -30

53. 48 **55.** -72 **57.** -40 **59.** 36 **61.** $\frac{1}{2}$

63. $-\frac{12}{5}$ **65.** $\frac{1}{12}$ **67.** $-\frac{1}{3}$ **69.** $-\frac{1}{2}$ **71.** 6

73. -3 **75.** -9 **77.** $-\frac{5}{2}$ **79.** $\frac{2}{5}$ **81.** $\frac{46}{17}$

83. $\frac{11}{12}$ **85.** $(4)(4)(4)$ **87.** $\left(-\frac{3}{4}\right)\left(-\frac{3}{4}\right)\left(-\frac{3}{4}\right)\left(-\frac{3}{4}\right)$

89. $(-0.8)(-0.8)(-0.8)(-0.8)(-0.8)(-0.8)$ **91.** $(-7)^3$

93. $(-5)^4$ **95.** -7^3 **97.** 16 **99.** -64

101. $\frac{64}{125}$ **103.** 0.027 **105.** -0.32 **107.** 0

109. 4 **111.** 22 **113.** 6 **115.** 57 **117.** 27

119. 135 **121.** -5 **123.** -6 **125.** 1

127. 161 **129.** $14,425$ **131.** 171.36 **133.** $\frac{17}{180}$

135. $\$2533.56$

137. (a)

Day	Daily gain or loss
Tuesday	$\$5$
Wednesday	$\$8$
Thursday	$-\$5$
Friday	$\$16$

(b) The stock gained $24 in value during the week. Find the difference between the first bar (Monday) and the last bar (Friday).

139. (a) $\$10,800$ (b) $\$13,008.64$ (c) $\$2208.64$

141. 15 square meters   **143.** 20 square inches

145. 6.125 cubic feet

147. (a) $20 \times 3 + 18 = 60 + 18 = 78$

    (b) Yes.

        $(0 + 6 + 3 + 2 + 0 + 2) \times 3$
        $+ (7 + 7 + 7 + 0 + 1) = 61$

        The next-highest multiple of 10 is 70. So, $70 - 61 = 9$, which is the check digit.

    (c) No. The check digit should be 4.

149. True. A nonzero rational number is of the form $\frac{a}{b}$, where a and b are integers and $a \neq 0$, $b \neq 0$. The reciprocal will be $\frac{b}{a}$, which is also rational.

151. False. A negative number raised to an odd power is negative.

153. Yes. $-3 + (-4) = -7$

155. If the numbers have like signs, the product or quotient is positive. If the numbers have unlike signs, the product or quotient is negative.

157. Evaluate additions and subtractions from left to right. For example, $6 - 5 - 2 = (6 - 5) - 2 = 1 - 2 = -1$. If this order were not understood, one might *incorrectly* write $6 - (5 - 2) = 6 - 3 = 3$.

159. To add fractions with unlike denominators, you first find the lowest common denominator.

161. Only common factors (not digits) of the numerator and denominator can be divided out.

Section 1.3   *(page 28)*

  1. Commutative Property of Addition

  3. Additive Inverse Property

  5. Multiplicative Inverse Property

  7. Commutative Property of Addition

  9. Associative Property of Addition

11. Distributive Property

13. Associative Property of Multiplication

15. Multiplicative Identity Property

17. Additive Identity Property

19. Associative Property of Addition

21. Distributive Property   **23.** Distributive Property

25. $(3 \cdot 6)y$ **27.** $-3(15)$ **29.** $5 \cdot 6 + 5 \cdot z$

31. $-x + 25$ **33.** $x + 8$ **35.** (a) -10 (b) $\frac{1}{10}$

37. (a) 16 (b) $-\frac{1}{16}$ **39.** (a) $-\frac{1}{2}$ (b) 2

41. (a) $\frac{5}{8}$ (b) $-\frac{8}{5}$ **43.** (a) $-6z$ (b) $1/(6z)$

45. (a) $7x$ (b) $-1/(7x)$

47. (a) $-x - 1$ or $-(x + 1)$ (b) $1/(x + 1)$

49. $x + (5 - 3)$ **51.** $(32 + 4) + y$ **53.** $(3 \cdot 4)5$

55. $(6 \cdot 2)y$ **57.** $20 \cdot 2 + 20 \cdot 5$

59. $5(3x) + 5(-4)$ or $15x - 20$

61. $x(-2) + 6(-2)$ or $-2x - 12$

63. $-6(2y) + (-6)(-5)$ or $-12y + 30$ **65.** $3x + 15$

67. $-2x - 16$ **69.** Answers will vary.

71. $x + 5 = 3$

Original equation

$(x + 5) + (-5) = 3 + (-5)$

Addition Property of Equality

$x + [5 + (-5)] = -2$

Associative Property of Addition

         $x + 0 = -2$

Additive Inverse Property

         $x = -2$

Additive Identity Property

73. $2x - 5 = 6$

Original equation

    $(2x - 5) + 5 = 6 + 5$

Addition Property of Equality

$2x + (-5 + 5) = 11$

Associative Property of Addition

       $2x + 0 = 11$

Additive Inverse Property

         $2x = 11$

Additive Identity Property

        $\frac{1}{2}(2x) = \frac{1}{2}(11)$

Multiplication Property of Equality

       $\left(\frac{1}{2} \cdot 2\right)x = \frac{11}{2}$

Associative Property of Multiplication

        $1 \cdot x = \frac{11}{2}$

Multiplicative Inverse Property

         $x = \frac{11}{2}$

Multiplicative Identity Property

75. 28 **77.** 434 **79.** 62.82

81. $a(b + c) = ab + ac$ **83.** $\$0.60$ **85.** $\$0.94$

87. Given two real numbers a and b, the sum a plus b is the same as the sum b plus a.

89. The multiplicative inverse of a real number a $(a \neq 0)$ is the number $1/a$. The product of a number and its multiplicative inverse is the multiplicative identity 1. For example, $8 \cdot \frac{1}{8} = 1$.

91. $0 \cdot a = 0$

93. $4 \odot 7 = 15 \neq 18 = 7 \odot 4$

    $3 \odot (4 \odot 7) = 21 \neq 27 = (3 \odot 4) \odot 7$

Mid-Chapter Quiz   (page 31)

1.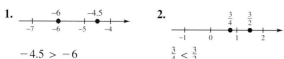

    $-4.5 > -6$

2. (number line with $\frac{3}{4}$ and $\frac{3}{2}$ marked)

    $\frac{3}{4} < \frac{3}{2}$

3. 22   4. 3.75   5. 3.2   6. -9.8 7. 14

8. -22 9. $\frac{5}{2}$ 10. $\frac{1}{2}$ 11. 60   12. $-\frac{3}{8}$

13. $\frac{7}{10}$ 14. $-\frac{27}{8}$ 15. 4   16. 2

17. (a) Distributive Property

    (b) Additive Inverse Property

18. (a) Associative Property of Addition

    (b) Multiplicative Identity Property

19. $1492.28   20. $3600   21. $\frac{7}{24}$

Section 1.4   (page 37)

1. $10x, 5$; $10, 5$ 3. $-6x^2, 12$; $-6, 12$

5. $-3y^2, 2y, -8$; $-3, 2, -8$ 7. $-4a^3, 1.2a$; $-4, 1.2$

9. $4x^2, -3y^2, -5x, 2y$; $4, -3, -5, 2$

11. $-5x^2y, 2y^2, xy$; $-5, 2, 1$ 13. $\frac{1}{4}x^2, -\frac{3}{8}x, 5$; $\frac{1}{4}, -\frac{3}{8}, 5$

15. $\frac{x^2}{7}, -\frac{2x}{5}, 3$; $\frac{1}{7}, -\frac{2}{5}, 3$

17. $x^2, -2.5x, -\frac{1}{x}$; $1, -2.5, -1$

19. Commutative Property of Addition

21. Associative Property of Multiplication

23. Distributive Property

25. $5(x + 6) = 5x + 30$ or $5x + 5 \cdot 6$

27. $6x + 6 = 6(x + 1)$ 29. $7x$ 31. $2x^2$ 33. $-4x$

35. $8y$ 37. $8x + 18y$ 39. $6x^2 - 2x$

41. $-2z^4 + 5z + 8$ 43. $-x^2 + 3xy + y$

45. $3x^2 + 5x^2y + xy^2$ 47. $8x^2 + 4x - 12$

49. $-18y^2 + 3y + 6$ 51. $-3x^2 + 2x - 4$

53. $5x^2 + 2x$ 55. $-12x^2 + 51x$ 57. $10t^2 - 35t$

59. $12x - 35$ 61. $-4x - 9$ 63. $a + 3$

65. $-7y - 7$ 67. $y^3 + 6$ 69. $x^3 - 12$

71. $2y^3 + y^2 + y$ 73. $44a - 22$ 75. $-6x + 96$

77. $12x^2 + 2x$ 79. $-2b^2 + 4b - 36$

81. $9x^2 + 15x - 3$ 83. (a) 3 (b) -10

85. (a) 6 (b) 9   87. (a) 7 (b) 11

89. (a) 0 (b) $\frac{3}{10}$ 91. (a) 13 (b) -36

93. (a) 7 (b) 7   95. (a) $\frac{12}{5}$ (b) 3   97. (a) 3 (b) 0

99. (a) 210 (b) 140   101. $\frac{1}{2}b^2 - \frac{3}{2}b$; 90

103. $2x^2 + 3x$ 105. $70,000 million, $70,721.40 million

107. $161 thousand, $161.4 thousand

109. 1440 square feet

111.

(d) $(0 + 3 + 1 + 2 + 6 + 7) \times 3 + (4 + 8 + 9 + 3 + a)$

(e) 5; No, because for any other value the check digit would be incorrect.

113. Add (or subtract) their respective coefficients and attach the common variable factor.

$5x^4 - 3x^4 = (5 - 3)x^4 = 2x^4$

115. $5x + 3x = (5 + 3)x = 8x$

117. No. When $y = 3$, the expression is undefined.

Section 1.5   (page 47)

1. $23 + n$ 3. $12 + 2n$ 5. $n - 6$ 7. $4n - 10$

9. $\frac{1}{2}n$ 11. $\frac{x}{6}$ 13. $8 \cdot \frac{N}{5}$ 15. $4c + 10$

17. $0.30L$ 19. $\frac{n + 5}{10}$ 21. $|n - 8|$ 23. $3x^2 - 4$

25. A number decreased by 2

27. A number increased by 50

29. Two decreased by three times a number

31. The ratio of a number to 2

33. Four-fifths of a number

35. Eight times the difference of a number and 5

37. The sum of a number and 10, divided by 3

39. The product of a number and the sum of the number and 7

41. $0.25n$ 43. $0.10m$ 45. $5m + 10n$ 47. $65t$

49. $\frac{230}{r}$ 51. $0.45y$ 53. $0.0125I$

55. $L - 0.20L = 0.80L$ 57. $8.25 + 0.60q$

59. $n + 3n = 4n$ 61. $(2n + 1) + (2n + 3) = 4n + 4$

63. $\frac{2n(2n + 2)}{4}$ 65. s^2 67. $\frac{1}{2}b(0.75b) = 0.375b^2$

69. Perimeter: $2(2w) + 2w = 6w$; Area: $2w(w) = 2w^2$

71. Perimeter: $6 + 2x + 3 + x + 3 + x = 4x + 12$

    Area: $3x + 6x = 9x$

73. $l(l - 6) = (l^2 - 6l)$ square feet

75.

n	0	1	2	3	4	5
$5n - 3$	-3	2	7	12	17	22
Differences		5	5	5	5	5

The differences are constant.

77. a **79.** Subtraction   **81.** a and c

83. Using a specific case may make it easier to see the form of the expression for the general case.

Review Exercises   *(page 52)*

1. (a) $\sqrt{9}, 52$ (b) $-4, 0, \sqrt{9}, 52$

    (c) $-4, -\frac{1}{8}, 0, \frac{3}{5}, \sqrt{9}, 52$ (d) $\sqrt{2}$

3. $\{1, 2, 3, 4\}$

5.   **7.**

$<$ $<$

9. 11   **11.** 7.3   **13.** 5   **15.** -7.2 **17.** 11

19. 230   **21.** -41.8 **23.** $\frac{11}{21}$ **25.** $\frac{1}{6}$ **27.** $\frac{17}{8}$

29. -28 **31.** -4200 **33.** $-\frac{1}{20}$ **35.** 14   **37.** 2

39. 6^7 **41.** 1296   **43.** -16 **45.** $\frac{1}{8}$ **47.** 20

49. 98   **51.** 1,165,469.01   **53.** $644

55. Additive Inverse Property   **57.** Distributive Property

59. Associative Property of Addition

61. Commutative Property of Multiplication

63. Distributive Property

65. $u - 3v$ **67.** $-3y^2 + 10y$

69. $4y^3, -y^2, \dfrac{17}{2}y; 4, -1, \dfrac{17}{2}$

71. $-1.2x^3, \dfrac{1}{x}, 52; -1.2, 1, 52$ **73.** $5x$ **75.** $5v$

77. $5x - 10$ **79.** $5x - y$ **81.** $18b - 15a$

83. (a) 0   (b) -3 **85.** $200 - 3n$ **87.** $y^2 + 49$

89. The sum of two times a number and 7

91. The difference of a number and 5, divided by 4

93. $0.18I$ **95.** $l(l - 5)$

Chapter Test   *(page 55)*

1. (a) $<$ (b) $>$ **2.** 11.9   **3.** -20 **4.** $-\frac{1}{2}$

5. -150 **6.** 60   **7.** $\frac{1}{6}$ **8.** $\frac{4}{27}$ **9.** $-\frac{27}{125}$ **10.** 15

11. (a) Associative Property of Multiplication

    (b) Multiplicative Inverse Property

12. $-12x + 6$ **13.** $-2x^2 + 5x - 1$ **14.** $x + 13$

15. a^2 **16.** $11t + 7$

17. Evaluating an expression is solving the expression when values are provided for its variables.
    (a) 4   (b) -12

18. 16 feet   **19.** 640 cubic feet   **20.** $5n - 8$

21. $2n + (2n + 2) = 4n + 2$

22. Perimeter: $3.2l$; Area: $0.6l^2$

Chapter 2

Section 2.1   *(page 65)*

Review   *(page 65)*

1. Commutative Property of Addition

2. Multiplicative Inverse Property

3. Distributive Property

4. Associative Property of Addition

5. 1   **6.** 4   **7.** $\frac{9}{2}$ **8.** 5   **9.** $\frac{4}{5}$

10. $-\frac{4}{9}$ **11.** $18,000   **12.** 9 feet

---

1. (a) Not a solution   (b) Solution

3. (a) Solution   (b) Not a solution

5. (a) Solution   (b) Not a solution

7. (a) Not a solution   (b) Solution

9. No solution   **11.** Identity   **13.** Linear

15. Not linear, because the exponent of the variable is -1.

17. Original equation

    Subtract 15 from each side.

    Combine like terms.

    Divide each side by 3.

    Simplify.

19. Original equation

    Subtract 5 from each side.

    Combine like terms.

    Divide each side by -2.

    Simplify.

21. 3   **23.** 4   **25.** -0.7 **27.** $-\frac{2}{3}$ **29.** -1

31. -1 **33.** 2   **35.** $-\frac{10}{3}$

37. No solution, because $-3 \neq 0$.

39. No solution, because $-4 \neq 0$.   **41.** $\frac{1}{3}$ **43.** -2

45. 2   **47.** 15   **49.** 11   **51.** -2 **53.** $-\frac{9}{2}$

55. $\frac{6}{5}$ **57.** -3 **59.** -3 **61.** $\frac{25}{3}$ **63.** 50

65. $\frac{19}{10}$ **67.** $-\frac{10}{3}$ **69.** $-\frac{20}{9}$ **71.** -20

73. $-\frac{8}{31}$ **75.** 23   **77.** 12   **79.** $\frac{1}{5}$ **81.** 125, 126

83. 82, 84   **85.** 1.5 hours   **87.** 1.5 seconds

89. 6 hours

91. (a)

t	1	1.5	2
Width	300	240	200
Length	300	360	400
Area	90,000	86,400	80,000

t	3	4	5
Width	150	120	100
Length	450	480	500
Area	67,500	57,600	50,000

(b) Because the length is t times the width and the perimeter is fixed, as t gets larger the length gets larger and the area gets smaller. The maximum area occurs when the length and width are equal.

93. 1995

Graphically: You can use the bar graph to estimate the expenditures per student to be about $6000 in 1995.

Numerically: Use the model $y = 233.6t + 4797$ to create a table showing values for $t = 3, 4, 5, 6, 7, 8, 9,$ and 10. The table should indicate that the expenditures reached $5965 in 1995.

Algebraically: Substitute 5965 for y and solve for t using the model $y = 233.6t + 4797$. The solution $t = 5$ indicates that the expenditures reached $5965 in 1995.

95. An equation whose solution set is not the entire set of real numbers is called a conditional equation. The solution set of an identity is all real numbers.

97. $ax + b = c$, $a \neq 0$. A linear equation is called a first-degree equation because the variable has an implied degree of 1.

99. False. This does not follow the Multiplication Property of Equality.

101. (a) No solution   (b) Conditional equation

(c) Identity

Examples will vary. Possibilities include: (b) trying to find the pre-tax price of an item from 5% sales tax and final price paid information, and (c) an example in algebra class illustrating the Distributive Property.

Section 2.2   *(page 75)*

Review   *(page 75)*

1. A collection of letters (called variables) and real numbers (called constants) combined using the operations of addition, subtraction, multiplication, and division is called an algebraic expression.

2. The terms of an algebraic expression are those parts separated by addition or subtraction.

3. -240 **4.** 34   **5.** 120   **6.** -56

7. $-\frac{1}{4}$ **8.** $\frac{6}{5}$ **9.** -27 **10.** $\frac{25}{64}$

11. $6x + 1$ **12.** $14x + 2$

1. (A number) $+ 30 = 82$

Equation: $x + 30 = 82$; Solution: 52

3. 26(biweekly pay) + (bonus) = 30,500

Equation: $26x + 2300 = 30,500$; Solution: $1084.62

Percent	Parts out of 100	Decimal	Fraction
5. 30%	30	0.30	$\frac{3}{10}$
7. 7.5%	7.5	0.075	$\frac{3}{40}$
9. $66\frac{2}{3}\%$	$66\frac{2}{3}$	0.66 . . .	$\frac{2}{3}$
11. 100%	100	1.00	1

13. 87.5   **15.** 69.36   **17.** 128   **19.** 600   **21.** 80

23. 350   **25.** 35   **27.** 12,000   **29.** 62%

31. 0.5%   **33.** 175%   **35.** $\frac{2}{3}$ **37.** $\frac{3}{4}$ **39.** $\frac{1}{25}$

41. $\frac{10}{3}$ **43.** 4   **45.** 6   **47.** $\frac{15}{2}$ **49.** 6   **51.** 4

53. 12,456 freshmen   **55.** 2   **57.** 15.625%

59. About 18.3%   **61.** About 21%   **63.** About 10%

65. About 7%   **67.** 200 parts   **69.** 177.77%, 56.25%

71. Boulder: 18.88%; Weld: 11.73%; Douglas: 11.39%; Mesa: 7.54%; Larimer: 16.30%; Jefferson: 34.16%

73. $\approx$ 2 pounds, 3.1%

75. $\approx$ 11,750 million pounds

77. $\frac{1}{50}$ **79.** $\frac{85}{4}$ **81.** $\frac{4}{9}$ or $\frac{9}{4}$ **83.** $0.0475

85. $0.0845   **87.** $14\frac{1}{2}$-ounce bag

89. six-ounce tube   **91.** $5\frac{1}{11}$ **93.** 3

95. $\frac{516}{11} \approx 46.9$ feet   **97.** 17.1 gallons   **99.** $2400

101. 2667 units   **103.** 2400 units   **105.** 83 units

107. Percent means parts out of 100.

109. No. $\frac{1}{2}\% = 0.5\% = 0.005$

111. 7.53%. Base number is smaller for percent increase.

Section 2.3 *(page 88)*

Review *(page 88)*

1. Negative. To add two real numbers with like signs, add their absolute values and attach the common sign to the result.

2. Negative. To add two real numbers with unlike signs, subtract the smaller absolute value from the greater absolute value and attach the sign of the number with the greater absolute value.

3. Positive. To multiply two real numbers with like signs, find the product of their absolute values. The product is positive.

4. Negative. To multiply two real numbers with unlike signs, find the product of their absolute values. The product is negative.

5. 14   **6.** 0   **7.** 0   **8.** $-10,000$

9. -200 **10.** 40   **11.** 0.7 mile

12. $78\frac{11}{30}$ tons

Cost	Selling Price	Markup	Markup Rate
1. $45.97	$64.33	$18.36	40%
3. 152.00	$250.80	$98.80	65%
5. $22,250.00	$26,922.50	$4672.50	21%
7. $225.00	$416.70	$191.70	85.2%

List Price	Sale Price	Discount	Discount Rate
9. $49.95	$25.74	$24.21	48.5%
11. $300.00	$111.00	$189.00	63%
13. $95.00	$33.25	$61.75	65%
15. $1145.00	$893.10	$251.90	22%

17. $26.19   **19.** 28%   **21.** $25   **23.** 20%

25. 9 minutes, $2.06   **27.** $54.15

29. (a) $267, $4717   (b) $3717   **31.** 2.5 hours

33. 3 hours   **35.** 50 gallons at 20%; 50 gallons at 60%

37. 8 quarts at 15%; 16 quarts at 60%

39. 75 pounds at $12 per pound; 25 pounds at $20 per pound

41. 100 children's tickets   **43.** $\frac{5}{6}$ gallon

	Distance, d	Rate, r	Time, t
45.	2275 mi	650 mi/hr	$3\frac{1}{2}$ hr
47.	1000 km	110 km/hr	$\frac{100}{11}$ hr
49.	385 mi	55 mi/hr	7 hr

51. $2\frac{1}{2}$ hours   **53.** $1\frac{1}{4}$ hours   **55.** $\frac{5}{6}$ hour = 50 minutes

57. 1440 miles   **59.** 17.14 minutes

61. 3 hours at 58 miles per hour; $2\frac{3}{4}$ hours at 52 miles per hour

63. (a) 8 pages per minute   (b) $\frac{15}{4}$ units per hour

65. (a) $\frac{1}{3}, \frac{1}{4}$ (b) $\frac{12}{7} = 1\frac{5}{7}$ hours

67. $R = \dfrac{E}{I}$ **69.** $L = \dfrac{S}{1-r}$ **71.** $a = \dfrac{2h - 96t}{t^2}$

73. $a = \dfrac{2h + 24t}{t^2}$ **75.** $a = \dfrac{2h + 30t - 19}{t^2}$

77. $147\pi \approx 461.8$ cubic centimeters   **79.** 0.926 foot

81. 43 centimeters   **83.** 15 degrees Celsius   **85.** $1950

87. $3571.43

89. (a) 1996, yes   (b) $0.264; Explanations will vary.

91. (a) (Monthly cost) = (standard service) + 11.91(number of channels)

(b) $C = 35.20 + 11.91x$

x	1	2	3	4	5
C	$47.11	$59.02	$70.93	$82.84	$94.75

(c) (Monthly cost) = (standard service) + (standard pay-per-view service) + 3.95(number of movies)

(d) $C = 38.19 + 3.95x$

x	1	2	3	4
C	$42.14	$46.09	$50.04	$53.99

x	5	6	7	8
C	$57.94	$61.89	$65.84	$69.79

(e) 40.4%

93. The sale price is the list price minus the discount rate times the list price.

95. Yes. The perimeter of a square of side s is $4s$. If the length of the sides is $2s$, the perimeter is $4(2s) = 8s$.

97. The area of the base times the height.

Mid-Chapter Quiz *(page 93)*

1. 2   **2.** 2   **3.** 2   **4.** Identity   **5.** $\frac{28}{5}$ **6.** $\frac{12}{7}$

7. $\frac{33}{2}$ **8.** $-\frac{29}{144}$ **9.** 6   **10.** 1.41   **11.** $\frac{9}{20}$, 45%

12. 200   **13.** $0.1958 per ounce

14. 2000 defective units   **15.** The computer from the store

16. 7 hours

17. 25% solution: 40 gallons; 50% solution: 10 gallons

18. 1.5 hours at 62 miles per hour; 4.5 hours at 46 miles per hour

19. $\frac{24}{7} \approx 3.43$ hours   **20.** 169 square inches

Section 2.4 *(page 103)*

Review *(page 103)*

1. Commutative Property of Multiplication

2. Additive Inverse Property

3. Distributive Property

4. Additive Identity Property

5. 7   **6.** 0   **7.** 0   **8.** 1   **9.** 4   **10.** 9

11. 19.8 square meters   **12.** 104 square feet

1. (a) Yes   (b) No   (c) Yes   (d) No

3. (a) No   (b) Yes   (c) Yes   (d) No

5. a   **6.** e   **7.** d   **8.** b   **9.** f   **10.** c

11.

13.

15.

17.

19.

21.

23.

25. $-15 + x < -24$

27. $x \geq 4$

29. $x \leq 2$

31. $x < 4$

33. $x \leq -4$

35. $x > 8$

37. $x \geq 7$

39. $x > 7.55$

41. $x > -\frac{2}{3}$

43. $x > \frac{9}{2}$

45. $x > \frac{20}{11}$

47. $x > \frac{8}{3}$

49. $x \leq -8$

51. $x > -15$

53. $\frac{5}{2} < x < 7$

55. $-3 \leq x < -1$

57. $-5 < x < 5$

59. $-\frac{3}{2} < x < \frac{9}{2}$

61. $1 < x < 10$

63. $-1 < x \leq 4$

65. $-5 \leq x < 1$

67. $x < \frac{52}{11}$ or $x > \frac{67}{12}$

69. $x < -\frac{8}{3}$ or $x \geq \frac{5}{2}$

71. $y \leq -10$

73. $-5 < x \leq 0$

75. $x < -3$ or $x \geq 2$, $\{x \mid x < -3\} \cup \{x \mid x \geq 2\}$

77. $-5 \leq x < 4$, $\{x \mid x \geq -5\} \cap \{x \mid x < 4\}$

79. $x \leq -2.5$ or $x \geq -0.5$, $\{x \mid x \leq -2.5\} \cup \{x \mid x \geq -0.5\}$

81. $\{x \mid x \geq -7\} \cap \{x \mid x < 0\}$

83. $\{x \mid x < -5\} \cup \{x \mid x > 3\}$

85. $\left\{x \mid x > -\frac{9}{2}\right\} \cap \left\{x \mid x \leq -\frac{3}{2}\right\}$ **87.** $x \geq 0$ **89.** $z \geq 8$

91. $10 \leq n \leq 16$ **93.** x is at least $\frac{5}{2}$.

95. y is at least 3 and less than 5.

97. z is more than 0 and no more than π.

99. $2600

101. The average temperature in Miami is greater than the average temperature in New York.

103. 26,000 miles   **105.** $x \geq 31$

107. The call must be less than or equal to 6.38 minutes. If a portion of a minute is billed as a full minute, the call must be less than or equal to 6 minutes.

109. $2 \leq x \leq 16$ **111.** $3 \leq n \leq \frac{15}{2}$

113. $12.50 < 8 + 0.75n$; $n > 6$ **115.** 1994, 1995

117. (f) 1 premium movie channel

(g) 2 pay-per-view movies

Answers will vary.

Sample answer:
Purchasing one premium channel each month is a better deal because it is likely to air more than two movies a month.

119. Yes. By definition, dividing by a number is the same as multiplying by its reciprocal.

121. $3x - 2 \leq 4$, $-(3x - 2) \geq -4$

Section 2.5 *(page 113)*

Review *(page 113)*

1. $2n$ is an even integer and $2n + 1$ is an odd integer.

2. No. $-2x^4 \neq 16x^4 = (-2x)^4$

3. $\dfrac{35}{14} = \dfrac{7 \cdot 5}{7 \cdot 2} = \dfrac{5}{2}$

4. $\dfrac{4}{5} \div \dfrac{z}{3} = \dfrac{4}{5} \cdot \dfrac{3}{z} = \dfrac{12}{5z}$

5. <   **6.** >   **7.** >   **8.** >   **9.** >

10. <   **11.** More than \$500   **12.** Less than \$500

1. Not a solution   **3.** Solution

5. $x - 10 = 17$; $x - 10 = -17$

7. $4x + 1 = \frac{1}{2}$; $4x + 1 = -\frac{1}{2}$ **9.** 4, -4

11. No solution   **13.** 0   **15.** 3, -3 **17.** 4, -6

19. 11, -14 **21.** $\frac{16}{3}$, 16   **23.** No solution   **25.** $\frac{4}{3}$

27. $\frac{15}{2}$, $-\frac{39}{2}$ **29.** 18.75, -6.25 **31.** $\frac{17}{5}$, $-\frac{11}{5}$

33. $-\frac{5}{3}$, $-\frac{13}{3}$ **35.** 3   **37.** $\frac{11}{5}$, -1 **39.** $\frac{15}{4}$, $-\frac{1}{4}$

41. 2, 3   **43.** 7, -3 **45.** $\frac{3}{2}$, $-\frac{1}{4}$ **47.** 11, 13

49. 28, $-\frac{12}{5}$ **51.** $\frac{1}{2}$ **53.** $|x - 5| = 3$

55. (a) Solution   (b) Not a solution

(c) Not a solution   (d) Solution

57. (a) Not a solution   (b) Solution

(c) Solution   (d) Not a solution

59. $-3 < y + 5 < 3$ **61.** $7 - 2h \geq 9$ or $7 - 2h \leq -9$

63. 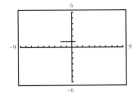   **65.**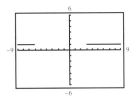

67. $-4 < y < 4$ **69.** $x \leq -6$ or $x \geq 6$

71. $-7 < x < 7$ **73.** $-9 \leq y \leq 9$

75. $-2 \leq y \leq 6$ **77.** $x < -16$ or $x > 4$

79. $-3 \leq x \leq 4$ **81.** $t \leq -\frac{15}{2}$ or $t \geq \frac{5}{2}$

83. $-\infty < x < \infty$ **85.** No solution

87. $-82 \leq x \leq 78$ **89.** $-104 < y < 136$

91. $z < -50$ or $z > 110$ **93.** $x < -\frac{11}{3}$ or $x > 1$

95. $-5 < x < 35$ **97.** $\frac{28}{3} \leq x \leq \frac{32}{3}$

99. $-\infty < x < \infty$ **101.** $-4 \leq x \leq 40$

103. $-\infty < x < \infty$

105. $-2 < x < \frac{2}{3}$ **107.** $x < -6$ or $x > 3$

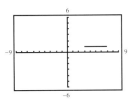

109. $3 \leq x \leq 7$

111. d   **112.** c   **113.** b   **114.** a   **115.** $|x| \leq 2$

117. $|x - 19| < 3$ **119.** $|x| < 3$ **121.** $|x - 5| > 6$

123.

Maximum: 82 degrees Fahrenheit; Minimum: 62 degrees Fahrenheit

125. $|x - 98.6| \leq 1$

127. The absolute value of a real number measures the distance of the real number from zero.

129. The solutions of $|x| = a$ are $x = a$ and $x = -a$. $|x - 3| = 5$ means $x - 3 = 5$ or $x - 3 = -5$. Thus, $x = 8$ or $x = -2$.

131. $|2x - 6| \leq 6$

133. Because $|3x - 4|$ is always nonnegative, the inequality is always true for *all* values of x. The student's solution eliminates the values $-\frac{1}{3} < x < 3$.

Review Exercises *(page 118)*

1. (a) Not a solution   (b) Solution

3. (a) Solution   (b) Not a solution

5. -7 **7.** 24   **9.** 3   **11.** -12 **13.** -24

15. 3   **17.** 2   **19.** 4   **21.** 4   **23.** 14

25. No solution   **27.** 2   **29.** -4.2 **31.** 73, 74

33. (Total earnings) = (weekly wage)(number of weeks) + (one-time bonus)

Equation: $2750 = 200x + 350$; Solution: 12 weeks

Percent	Parts out of 100	Decimal	Fraction
35. 87%	87	0.87	$\frac{87}{100}$
37. 60%	60	0.6	$\frac{3}{5}$

39. 65   **41.** 3000   **43.** 125%   **45.** 6%

47. $6\frac{3}{4}\%$ **49.** 32-ounce package   **51.** $\frac{1}{40}$ **53.** $\frac{7}{2}$

55. $\frac{10}{3}$ **57.** 3   **59.** 50 feet   **61.** $2000

	Cost	Selling Price	Markup	Markup Rate
63.	$99.95	$149.93	$49.98	50%
65.	$81.72	$125.85	$44.13	54%

	List Price	Sale Price	Discount	Discount Rate
67.	$71.95	$53.96	$17.99	25%
69.	$1995.50	$1396.85	$598.65	30%

71. $75, $1325, $1025

73. 30% solution: $3\frac{1}{3}$ liters; 60% solution: $6\frac{2}{3}$ liters

75. 3500 miles   **77.** 43.6 miles per hour

79. $\frac{18}{7} \approx 2.57$ hours   **81.** $340   **83.** $52,631.58

85. $30,000   **87.** 34.375 feet $\times$ 20.625 feet

89. 50.9 degrees Fahrenheit

91.
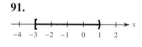

93.

95. $x \le 4$

97. $x > -6$

99. $x > 3$
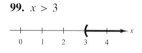

101. $x \le 6$

103. $y > -\frac{70}{3}$
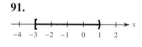

105. $x \le -3$

107. $-7 \le x < -2$

109. $-16 < x < -1$
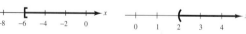

111. $-3 < x < 2$

113. At least $8333.33   **115.** ± 6 **117.** $4, -\frac{4}{3}$

119. $0, -\frac{8}{5}$ **121.** $\frac{1}{2}, 3$ **123.** $x < 1$ or $x > 7$

125. $x < -3$ or $x > 3$ **127.** $-4 < x < 11$

129. $b < -9$ or $b > 5$

131. $x \le 2$ or $x \ge 3$ **133.** $|x - 3| < 2$

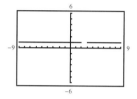

135.

Maximum: 116.6 degrees Fahrenheit

Minimum: 40 degrees Fahrenheit

Chapter Test *(page 123)*

1. 4   **2.** 3   **3.** 4   **4.** 24   **5.** 4000   **6.** 0.4%

7. $8000   **8.** 15-ounce package   **9.** $1466.67

10. $2\frac{1}{2}$ hours

11. $1.50 food: 15 pounds; $3.05 food: 10 pounds

12. 40 minutes   **13.** $2000

14. (a) $5, -11$ (b) $\frac{2}{3}, -\frac{4}{3}$ (c) No solution

15. (a) $x \ge -6$ (b) $x > 2$

(c) $-7 < x \le 1$ (d) $-1 \le x < \frac{5}{4}$

16. $t \ge 8$

17. (a) $1 \le x \le 5$ (b) $x < -\frac{9}{5}$ or $x > 3$

(c) $-\frac{44}{5} < x < -\frac{36}{5}$

18. 25,000 miles

Chapter 3

Section 3.1   *(page 135)*

Review   *(page 135)*

1. (a) $3x = 7$ is a linear equation because it can be written in the form $ax + b = 0$. (b) Because $x^2 + 3x = 2$ cannot be written in the form $ax + b = 0$, it is not a linear equation.

2. Substitute 3 for x in the equation. If the result is true, $x = 3$ is a solution.

3. $4x - 6$ 4. $2y^2 + 3y + 12$ 5. $t + 11$

6. $-m^2 + 8m - 10$ 7. $-6x + 1$

8. $-3x + 22$ 9. $12y$ 10. $0.02x + 100$

11. $\frac{1}{4}, \frac{1}{5}; \frac{20}{9} \approx 2.2$ hours   12. 45.65 miles per hour

1.

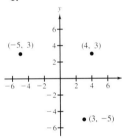

3.

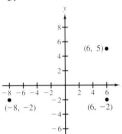

5.

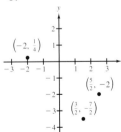

7.

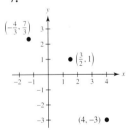

9. A: $(-2, 4)$, B: $(0, -2)$, C: $(4, -2)$

11. A: $(4, -2)$, B: $\left(-3, -\frac{5}{2}\right)$, C: $(3, 0)$

13.

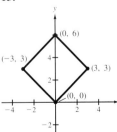

15.

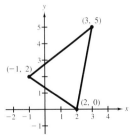

17.

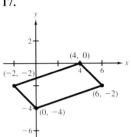

19.

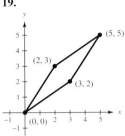

21. $(-5, 2)$ 23. $(3, -2)$ 25. $(-8, -8)$

27. $(10, 0)$ 29. Quadrant III   31. Quadrant II

33. Quadrant III   35. Quadrant IV

37. Quadrant I or II   39. Quadrant II or III

41. Quadrant I or III

43.

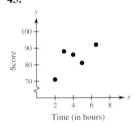

45.

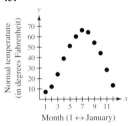

47. $(-3, -4) \implies (-1, 1)$
$(1, -3) \implies (3, 2)$
$(-2, -1) \implies (0, 4)$

49.

x	-2	0	2	4	6
$y = \frac{3}{4}x + 2$	$\frac{1}{2}$	2	$\frac{7}{2}$	5	$\frac{13}{2}$

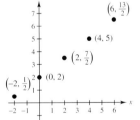

51.

x	-2	0	2	4	6
$y = \lvert 3x - 4 \rvert + 1$	11	5	3	9	15

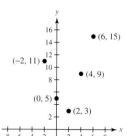

53.

x	-4	$\frac{2}{5}$	4	8	12
$y = -\frac{5}{2}x + 4$	14	3	-6	-16	-26

55. (a) Not a solution   (b) Solution
   (c) Not a solution   (d) Solution

57. (a) Not a solution   (b) Solution
   (c) Solution   (d) Not a solution

59. (a) Solution   (b) Solution
   (c) Not a solution   (d) Not a solution

61. 7; Vertical line   **63.** 7; Horizontal line

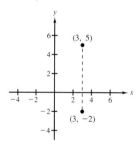   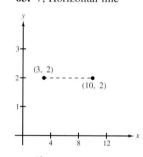

65. $\frac{3}{4}$; Vertical line   **67.** $\frac{13}{2}$; Horizontal line

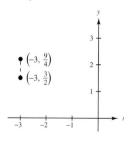   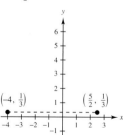

69. $\sqrt{5}$ **71.** 5   **73.** $\sqrt{58}$ **75.** $\sqrt{61}$ **77.** $\sqrt{29}$
79. $\left(\sqrt{13}\right)^2 + \left(\sqrt{13}\right)^2 = \left(\sqrt{26}\right)^2$
81. $\left(2\sqrt{5}\right)^2 + \left(2\sqrt{5}\right)^2 = \left(2\sqrt{10}\right)^2$

83. Not collinear   **85.** Collinear
87. $3 + \sqrt{26} + \sqrt{29} \approx 13.48$
89. $(1, 4)$ **91.** $\left(\frac{7}{2}, \frac{9}{2}\right)$

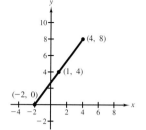   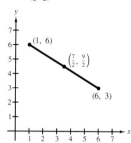

93.

x	1	2	3	4	5
$y = 200x + 450$	650	850	1050	1250	1450

The cost of installation is $450, plus $200 for every window installed.

95. $5\sqrt{61} \approx 39.05$ yards

97. The faster the car travels, up to 60 miles per hour, the less gas it uses. As the speed increases past 60 miles per hour, the car uses progressively more gas.

99. $2305.5 million

101. The word *ordered* is significant because each number in the pair has a particular interpretation. The first number measures horizontal distance, and the second number measures vertical distance.

103. The x-coordinate of any point on the y-axis is 0. The y-coordinate of any point on the x-axis is 0.

105. No. The scales on the x- and y-axes are determined by the magnitudes of the quantities being measured by x and y.

107.

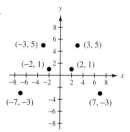

Reflection in the y-axis

Section 3.2 *(page 145)*

Review *(page 145)*

1. $t - 3 + c > 7 + c$ **2.** $(t - 3)c > 7c$

3. 1   **4.** Commutative Property of Addition

5. $x \geq 1$ **6.** $x < -3$ **7.** $-\frac{1}{2} < x < \frac{1}{2}$

8. $-\frac{1}{2} \leq x \leq \frac{3}{2}$ **9.** $-6 \leq x \leq 6$

10. $20 < x < 30$ **11.** \$29,018   **12.** \$108.50

1. e   **2.** b   **3.** f   **4.** a   **5.** d   **6.** c

7.
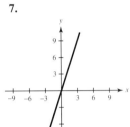

9.
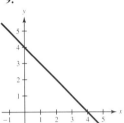

11.
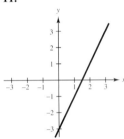

13.
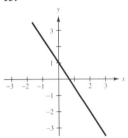

15.
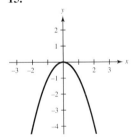

17.
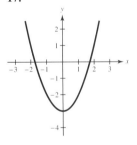

19.
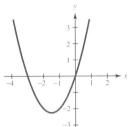

21.
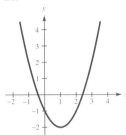

23.
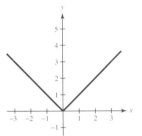

25.
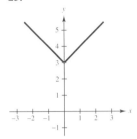

27.
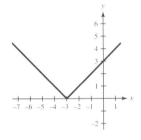

29.
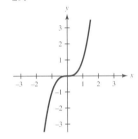

31. $\left(\frac{1}{2}, 0\right), (0, -3)$ **33.** $(30, 0), (0, 12)$

35. $(10, 0), (0, 5)$ **37.** $\left(-\frac{3}{4}, 0\right), (0, 3)$

39. $(\pm 1, 0), (0, -1)$ **41.** $(-5, 0), (0, -5)$

43. $(-2, 0), (4, 0), (0, -2)$ **45.** $(3, 0), (0, 2)$

47. $(0, 3)$ **49.** $(0, -2)$

51. $(2, 0), (0, 8)$ **53.** $(1, 0), (6, 0), (0, 6)$

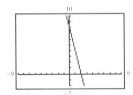
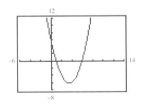

55. $(-2, 0), (-1, 0), (0, 4)$

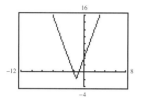

57.

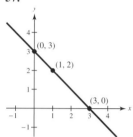

59.

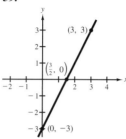

61.

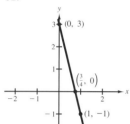

63.

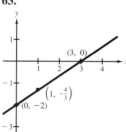

65.

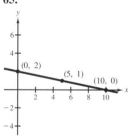

67.

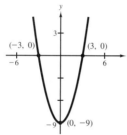

69.

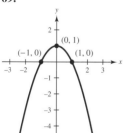

71.

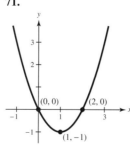

73.

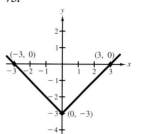

75.

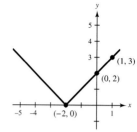

77.

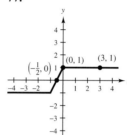

79.

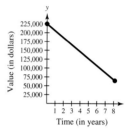

81. (a) $y = 40,000 - 5000t, \ 0 \le t \le 7$

(b)

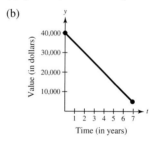

(c) $(0, 40,000)$; This represents the value of the delivery van when purchased.

83. (a)

x	0	3	6	9	12
F	0	4	8	12	16

(b)

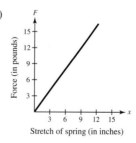

(c) F doubles because F is directly proportional to x.

85. The scales on the y-axes are different. From graph (a) it appears that sales have not increased. From graph (b) it appears that sales have increased dramatically.

87. The graph of an equation is the set of all solutions of the equation plotted on a rectangular coordinate system.

89. Yes

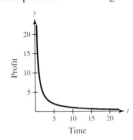

91. (a) 6 miles

(b) Stopped, because the graph remains constant on the interval $4 < t < 6$.

(c) $6 \le t \le 10$. The graph is steepest.

Section 3.3   *(page 158)*

Review   *(page 158)*

1. equivalent   **2.** 5   **3.** $\frac{8}{3}$ **4.** 27

5. 5   **6.** $\frac{10}{3}$ **7.** 18   **8.** -19

9. No solution   **10.** 0.1   **11.** $0 < t \le 23$

12. $m < 23{,}846$

1. $\frac{2}{3}$ **3.** -2 **5.** Undefined

7. (a) L_3 (b) L_2 (c) L_1

9. $m = \frac{5}{7}$; rises

11. $m = -\frac{4}{5}$; falls

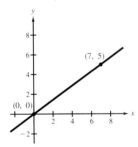

13. $m = 1$; rises

15. m is undefined; vertical

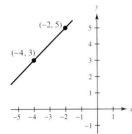

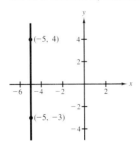

17. $m = 0$; horizontal

19. $m = -\frac{18}{17}$; falls

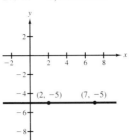

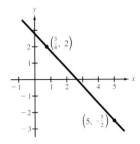

21. $m = \frac{1}{18}$; rises

23. $m = -\frac{5}{6}$; falls

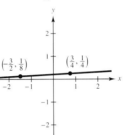

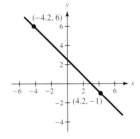

25. $m = 2$

27. $m = -\frac{1}{2}$

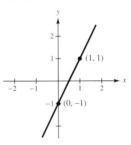

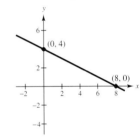

29. $m = -\frac{4}{5}$

31. $x = 1$ **33.** $y = -15$ **35.** $(6, 2), (10, 2)$

37. $(4, -1), (5, 2)$ **39.** $(1, 2), (2, 1)$

41. $(-2, 4), (1, 8)$ **43.** $y = 2x - 3$ **45.** $y = \frac{1}{4}x - 1$

47. $y = -\frac{2}{5}x + \frac{3}{5}$ **49.** $y = \frac{1}{2}x + 2$

51. $m = 3;\ (0, -2)$ **53.** $m = \frac{2}{3};\ (0, 1)$

55. $m = -\frac{5}{3};\ \left(0, \frac{2}{3}\right)$

57. $y = 3x - 2$

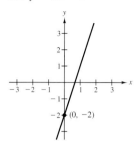

59. $y = -x$

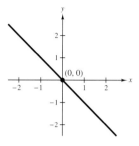

61. $y = -\frac{3}{2}x + 1$

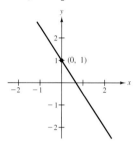

63. $y = \frac{1}{4}x + \frac{1}{2}$

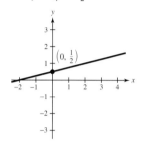

65. $y = 5x - 5$

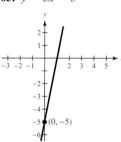

67.

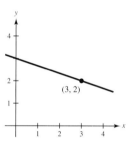

69.

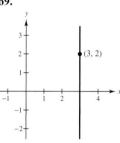

71.

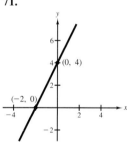

73.

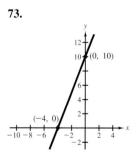

75. Parallel   **77.** Perpendicular   **79.** Parallel

81. Perpendicular   **83.** 16,667 feet

85. $\frac{45}{4} = 11.25$ feet

87. (a)

t	4	5	6	7
y	\$10,598.80	\$11,179.50	\$11,760.20	\$12,340.90

t	8	9	10
y	\$12,921.60	\$13,502.30	\$14,083

(b)

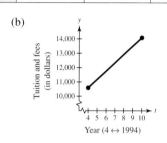

(c) \$580.70   (d) \$19,890

89. (a)

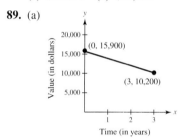

(b) -1900

(c) Annual depreciation

91. Negative slope: line falls to the right.

Zero slope: line is horizontal.

Positive slope: line rises to the right.

93. m is the slope; b is the y-intercept.

95. No. Their slopes must be negative reciprocals of each other.

Section 3.4 *(page 168)*

> ### Review *(page 168)*
>
> **1.** $\dfrac{a}{b}$　**2.** A proportion　**3.** 1.875　**4.** 9000
>
> **5.** 150%　**6.** 15.5%　**7.** $66\frac{2}{3}\%$　**8.** 128.6%
>
> **9.** 80,000　**10.** 275　**11.** 72 pounds
>
> **12.** 3 seconds

1. b　**2.** d　**3.** a　**4.** c　**5.** $y = 4x - 11$

7. $y = -\frac{1}{2}x - \frac{1}{2}$　**9.** $y = \frac{4}{5}x - \frac{8}{5}$　**11.** $y = -\frac{1}{2}x$

13. $y = 3x - 4$　**15.** $y = -\frac{3}{4}x + 6$　**17.** $y = -2x + 4$

19. $y = \frac{5}{4}x - 2$　**21.** $y = -4x - \frac{9}{2}$　**23.** $y = \frac{4}{3}x + \frac{3}{2}$

25. $y = -1$　**27.** $3x - 2y = 0$　**29.** $x + y - 4 = 0$

31. $x - 2y + 7 = 0$　**33.** $2x + 5y = 0$

35. $2x - 6y + 15 = 0$　**37.** $5x + 34y - 67 = 0$

39. $52x + 15y - 395 = 0$　**41.** $4x + 5y - 11 = 0$

43. $y = \frac{1}{2}x + 3$　**45.** $y = 3$　**47.** $x = -1$

49. $y = -5$　**51.** $x = -7$

53. (a) $y = 3x - 5$　(b) $y = -\frac{1}{3}x + \frac{5}{3}$

55. (a) $y = -\frac{5}{4}x - \frac{9}{4}$　(b) $y = \frac{4}{5}x + 8$

57. (a) $y = 4x - 5$　(b) $y = -\frac{1}{4}x + \frac{31}{4}$

59. (a) $x = \frac{2}{3}$　(b) $y = \frac{4}{3}$

61. (a) $y = 2$　(b) $x = -1$

63. $\dfrac{x}{3} + \dfrac{y}{2} = 1$　**65.** $-\dfrac{6x}{5} - \dfrac{3y}{7} = 1$

67. $C = 20x + 5000$; $13,000

69. $S = 100,000t$; $600,000

71. $S = 0.03M + 1500$; 3%

73. (a) $S = 0.70L$　(b) $94.50

75. (a) $V = 7400 - 1475t$　(b) $4450

77. (a) $N = 60t + 1200$　(b) 2400 students

(c) 1800 students

79. (a) and (b)

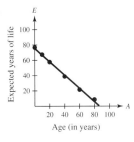

(c) Answers will vary.

Sample answer: $E = -0.87A + 73.7$

(d) 49.4 years

81. $x - 8y = 0$

Distance from the deep end	0	8	16	24	32	40
Depth of water	9	8	7	6	5	4

83. (d)

t	0	1	2	3	4
y	$15,900	$14,000	$12,100	$10,200	$8300

(e) $y = 15,900 - 1900x$

(f) No. $y = 6400$ when $x = 5$.

85. Point-slope form: $y - y_1 = m(x - x_1)$

Slope-intercept form: $y = mx + b$

General form: $ax + by + c = 0$

87. Any point on a vertical line is independent of y.

Mid-Chapter Quiz *(page 173)*

1. Quadrant I or II

2. (a) Not a solution　(b) Solution

(c) Solution　(d) Solution

3. Distance: 5　　　　　**4.** Distance: 13

Midpoint: $\left(1, \frac{7}{2}\right)$　　Midpoint: $\left(-\frac{1}{2}, 4\right)$

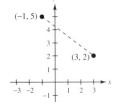

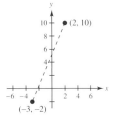

5.

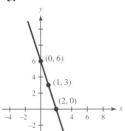

6.

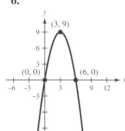

7.

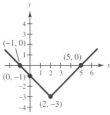

8. $m = 0$; Horizontal **9.** $m = \frac{5}{3}$; Rises

10. $m = -\frac{5}{3}$; Falls

11. $y = -\frac{1}{2}x + 1$ **12.** $y = \frac{1}{2}x - 2$

$\quad\ m = -\frac{1}{2}$ $\qquad\ m = \frac{1}{2}$

$\quad\ (0, 1)$ $\qquad\ (0, -2)$

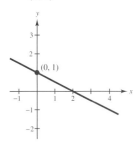

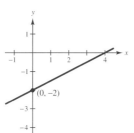

13. Perpendicular **14.** Neither **15.** $x - 2y - 8 = 0$

16. $V = 85{,}000 - 8100t,\ 0 \le t \le 10$

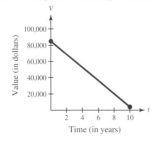

Section 3.5 *(page 179)*

Review *(page 179)*

1. The equation of a line in slope-intercept form is $y = mx + b$, where m is the slope and b is the y-intercept. To sketch the graph, determine another point on the line using the slope and draw a line through this point and the y-intercept.

2. The equation of a line in point-slope form passing through the point (x_1, y_1) with slope m is $y - y_1 = m(x - x_1)$. Isolating y determines the equation of the line.

3. $x < \frac{3}{2}$ **4.** $x < 5$ **5.** $x < \frac{12}{5}$ **6.** $x \le -11$

7. $1 < x < 5$ **8.** $x < 2$ or $x > 8$

9.

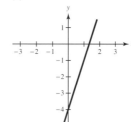

10.

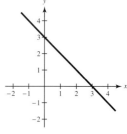

11.

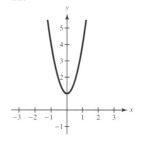

12.

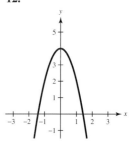

1. (a) Solution (b) Not a solution
   (c) Solution (d) Solution

3. (a) Not a solution (b) Not a solution
   (c) Solution (d) Solution

5. (a) Solution (b) Not a solution
   (c) Not a solution (d) Solution

7. (a) Not a solution (b) Solution
   (c) Not a solution (d) Solution

9. b **10.** a **11.** d **12.** e **13.** f **14.** c

15.

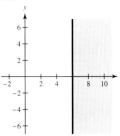

17.

35.

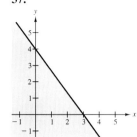

37.

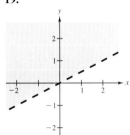

19.

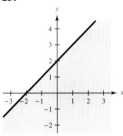

21.

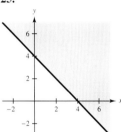

39.

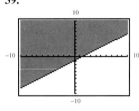

41.

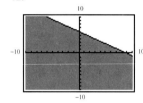

23.

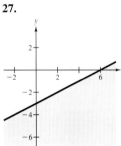

25.

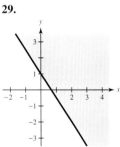

43.

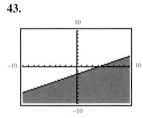

45.

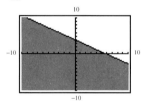

47. $3x + 4y > 17$ **49.** $y < 2$ **51.** $x - 2y < 0$

53. (a) $2x + 2y \leq 500$ or $y \leq -x + 250$
(Note: x and y cannot be negative.)

(b)

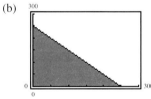

27.

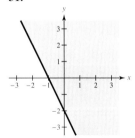

29.

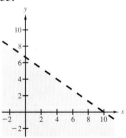

55. (a) $10x + 15y \leq 1000$ or $y \leq -\frac{2}{3}x + \frac{200}{3}$
(Note: x and y cannot be negative.)

(b)

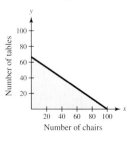

31.

33.

57. (a) $0.40x + 0.80y \le 10$
   (Note: x and y cannot be negative.)

(b)
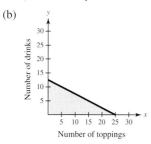
Number of toppings

(c) $(6, 6)$; yes

59. (a) $12x + 16y \ge 250$ (Note: x and y cannot be negative.)

(b)
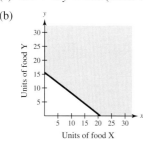
Units of food X

(x, y): $(1, 15)$, $(10, 9)$, $(15, 5)$

61. (a) $9x + 6y \ge 150$ (Note: x and y cannot be negative.)

(b)
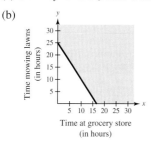
Time at grocery store
(in hours)

(x, y): $(2, 22)$, $(4, 19)$, $(10, 10)$

63. (a) $2w + t \ge 70$

(b)
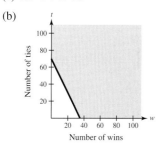
Number of wins

(w, t): $(10, 50)$, $(20, 30)$, $(30\ 10)$

65. $ax + by < c,\ ax + by \le c,\ ax + by > c,\ ax + by \ge c$

67. The graph of a line divides the plane into two half-planes. In graphing a linear inequality, graph the corresponding linear equation. The solution to the inequality will be one of the half-planes determined by the line. $x - 2y > 1$

69. Test a point in one of the half-planes.

Section 3.6   *(page 191)*

Review   *(page 191)*

1. $a < c$; Transitive Property

2. $9x = 36$
   $\frac{1}{9}(9x) = \frac{1}{9}(36)$
   $x = 4$

3. $y \le 45$ **4.** $x \ge 15$ **5.** $-4y$

6. $5(x - 2)$ **7.** $\frac{3}{2}t - \frac{5}{8}$ **8.** $\frac{7}{24}x + 8$

9. $-30x^2 + 23x + 3$

10. $4x^3 + 12x^2y + 4xy^2 + y^3$

11. $8\frac{3}{4}$ cups   **12.** 16 pints or 2 gallons

1. Domain: $\{-2, 0, 1\}$; Range: $\{-1, 0, 1, 4\}$

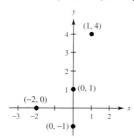

3. Domain: $\{0, 2, 4, 5, 6\}$; Range: $\{-3, 0, 5, 8\}$

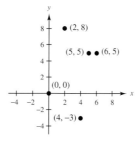

5. $(3, 150)$, $(2, 100)$, $(8, 400)$, $(6, 300)$, $\left(\frac{1}{2}, 25\right)$

7. $(1, 1)$, $(2, 8)$, $(3, 27)$, $(4, 64)$, $(5, 125)$, $(6, 216)$, $(7, 343)$

9. (1999, New York Yankees), (2000, New York Yankees), (2001, Arizona Diamondbacks), (2002, Anaheim Angels)

11. Not a function   **13.** Function   **15.** Not a function

17. Not a function   **19.** Function   **21.** Not a function

23. (a) Function from A to B

(b) Not a function from A to B

(c) Function from A to B

(d) Not a function from A to B

25. There are two values of y associated with one value of x.

27. There are two values of y associated with one value of x.

29. There is one value of y associated with each value of x.

31. There is one value of y associated with each value of x.

33. There is one value of y associated with each value of x.

35. (a) $2; 11$ (b) $-2; -1$

(c) $k; 3k + 5$ (d) $k + 1; 3k + 8$

37. (a) $0; 3$ (b) $-3; -6$

(c) $m; 3 - m^2$ (d) $2t; 3 - 4t^2$

39. (a) $3, 3; \dfrac{3}{5}$ (b) $-4, -4; 2$

(c) $s, s; \dfrac{s}{s + 2}$ (d) $s - 2, s - 2; \dfrac{s - 2}{s}$

41. (a) 29 (b) 11

(c) $12a - 2$ (d) $12a + 5$

43. (a) 2 (b) 2

(c) $4y^2 - 8y + 2$ (d) 16

45. (a) 2 (b) 3

(c) $\sqrt{z}$ (d) $\sqrt{5z + 5}$

47. (a) 4 (b) 4

(c) -7 (d) $8 - |x - 6|$

49. (a) 0 (b) $-\dfrac{3}{2}$

(c) $-\dfrac{5}{2}$ (d) $\dfrac{3x + 12}{x - 1}$

51. (a) 2 (b) -2 (c) 10 (d) -8

53. (a) 0 (b) $\dfrac{7}{4}$ (c) 3 (d) 0

55. (a) 2 (b) $\dfrac{2x - 12}{x}$

57. All real numbers x

59. All real numbers t such that $t \neq 0, -2$

61. All real numbers x such that $x \geq -4$

63. All real numbers x such that $x \geq \frac{1}{2}$

65. All real numbers t

67. Domain: $\{0, 2, 4, 6\}$; Range: $\{0, 1, 8, 27\}$

69. Domain: $\{-3, -1, 4, 10\}$; Range: $\left\{-\frac{17}{2}, -\frac{5}{2}, 2\right\}$

71. Domain: All real numbers r such that $r > 0$

Range: All real numbers C such that $C > 0$

73. Domain: All real numbers r such that $r > 0$

Range: All real numbers A such that $A > 0$

75. $P = 4x$ **77.** $V = x^3$

79. $C(x) = 1.95x + 8000, \ x > 0$ **81.** $d = 120t$

83. $d = 65t$; 260 miles

85. $V = x(24 - 2x)^2 = 4x(12 - x)^2$ **87.** $A = (32 - x)^2$

89. (a) 10,680 pounds   (b) 8010 pounds

91. Yes. Yes. For each year there is associated one public school enrollment and one private school enrollment.

93. (g) $y(x) = 15,900 - 1900x$; $2600

(h) Depreciates more slowly as the car ages

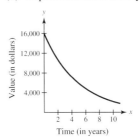

(i) Domain: All real numbers x such that $0 < x \leq 8.37$

Range: All real numbers y such that $0 < y \leq 15,900$

95. No. $\{(4, 3), (4, -2)\}$ is a relation, but not a function.

97. Sample answer:

Statement: The number of ceramic tiles required to floor a kitchen is a function of the floor's area.

Dependent variable: Number of ceramic tiles

Independent variable: Area of floor

This is a function.

99. (a) Correct   (b) Not correct

Section 3.7 *(page 203)*

Review *(page 203)*

1. Multiplicative Inverse Property

2. Additive Identity Property

3. Distributive Property

4. Associative Property of Addition

5. $-2x - 2$ **6.** $3y + 3$ **7.** $31x - 3$

8. $2y - 18$ **9.** $3t^2 - 8.7t$

10. $11.3 \ m^2 + 7.1 \ m$ **11.** Department store

12. $960.70

1.

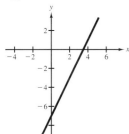

Domain: $-\infty < x < \infty$

Range: $-\infty < y < \infty$

3.

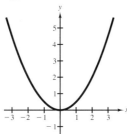

Domain: $-\infty < x < \infty$

Range: $0 \leq y < \infty$

17.

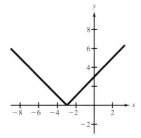

Domain: $-\infty < x < \infty$

Range: $0 \leq y < \infty$

19.

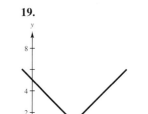

Domain: $-\infty < s < \infty$

Range: $1 \leq y < \infty$

5.

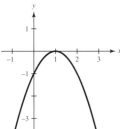

Domain: $-\infty < x < \infty$

Range: $-\infty < y \leq 0$

7.

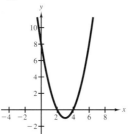

Domain: $-\infty < x < \infty$

Range: $-1 \leq y < \infty$

21.

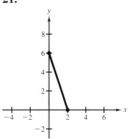

Domain: $0 \leq x \leq 2$

Range: $0 \leq y \leq 6$

23.

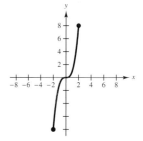

Domain: $-2 \leq x \leq 2$

Range: $-8 \leq y \leq 8$

9.

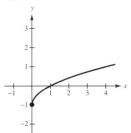

Domain: $0 \leq x < \infty$

Range: $-1 \leq y < \infty$

11.

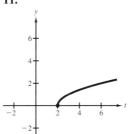

Domain: $2 \leq t < \infty$

Range: $0 \leq y < \infty$

25.

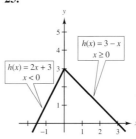

$h(x) = 3 - x$
$x \geq 0$

$h(x) = 2x + 3$
$x < 0$

Domain: $-\infty < x < \infty$

Range: $-\infty < y \leq 3$

27.

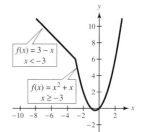

$f(x) = 3 - x$
$x < -3$

$f(x) = x^2 + x$
$x \geq -3$

Domain: $-\infty < x < \infty$

Range: $-\frac{1}{4} \leq y < \infty$

13.

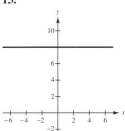

Domain: $-\infty < x < \infty$

Range: $y = 8$

15.

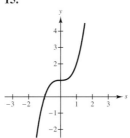

Domain: $-\infty < s < \infty$

Range: $-\infty < y < \infty$

29.

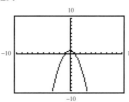

Domain: $-\infty < x < \infty$

Range: $-\infty < y \leq 1$

33. Function   **35.** Function

31.

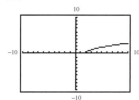

Domain: $2 \leq x < \infty$

Range: $0 \leq y < \infty$

37. Not a function

39. y is a function of x.

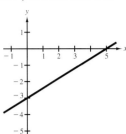

41. y is not a function of x.

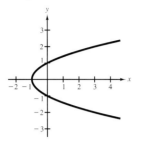

43. b   **44.** d   **45.** a   **46.** c

47. (a) Vertical shift
2 units upward

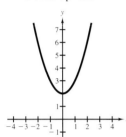

(b) Vertical shift
4 units downward

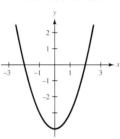

(c) Horizontal shift
2 units to the left

(d) Horizontal shift
4 units to the right

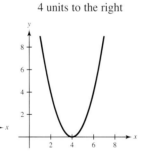

(e) Horizontal shift 3 units
to the right and a verti-
cal shift 1 unit upward

(f) Reflection in the x-axis
and a vertical shift
4 units upward

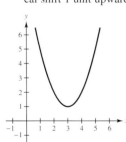

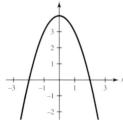

49. Horizontal shift 5 units
to the right

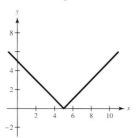

51. Vertical shift 5 units
downward

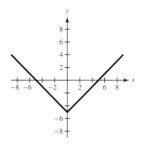

53. Reflection in the x-axis

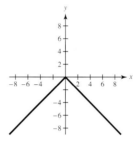

55. $h(x) = (x + 3)^2$ **57.** $h(x) = -x^2$

59. $h(x) = -(x + 3)^2$ **61.** $h(x) = -\sqrt{x}$

63. $h(x) = \sqrt{x + 2}$ **65.** $h(x) = \sqrt{-x}$

67. (a) $f(x) = (x + 3)^2 + 2$

(b) $f(x) = x^2 + 2 - 5 = x^2 - 3$

(c) $f(x) = x^2 + 2 + 1 = x^2 + 3$

(d) $f(x) = (x - 2)^2 + 2$

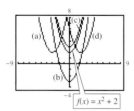

$f(x) = x^2 + 2$

69. Horizontal shift of $y = x^3$

$y = (x - 2)^3$

71. Reflection in the x-axis and horizontal and vertical
shifts of $y = x^2$

$y = -(x + 1)^2 + 1$

73. Vertical or horizontal shift of $y = x$

$y = x + 3$

75. (a)

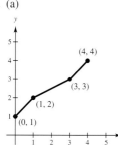

(b)

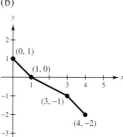

(c)

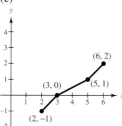

(d)

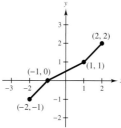

(e)

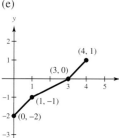

(f)

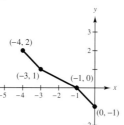

77. (a)

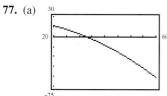

(b) 46%

79. (a) Perimeter:
$$P = 2l + 2w$$
$$200 = 2l + 2w$$
$$200 - 2l = 2w$$
$$100 - l = w$$
Area: $A = lw$
$$A = l(100 - l)$$

(b)

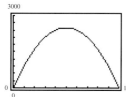

(c) $l = 50$. The figure is a square.

81. (a)

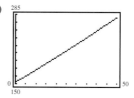

(b) 1980, because the transformation of the function shifts the original function 30 years to the left.

(c)

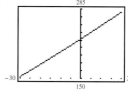

83. The range changes from $0 \leq y \leq 4$ to $0 \leq y \leq 8$

85. Vertical shift upward, vertical shift downward, horizontal shift to the left, horizontal shift to the right

87. Reflection in the y-axis

Review Exercises   *(page 209)*

1.

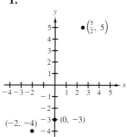

3.

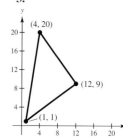

5. Quadrant IV   **7.** Quadrant I or IV

9. (a) Yes   (b) No   (c) No   (d) Yes

11. 5   **13.** $3\sqrt{5}$

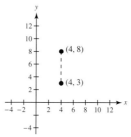

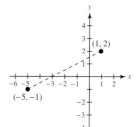

15. Collinear

17. $(4, 3)$

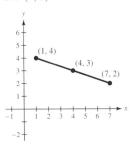

19. $(-3, -2)$

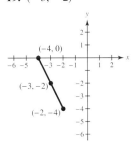

37. $(1.27, 0), (4.73, 0), (0, 6)$ **39.** $(0, -11)$

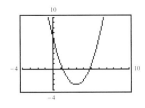   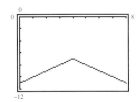

41. $(3, 0), (0, 1.73)$

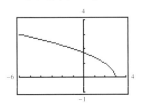

21.

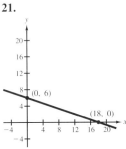

23.

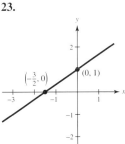

43. (a) $y = 28,000 - 3000t$ $0 \le t \le 5$

(b)

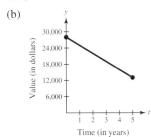

(c) $(0, 28,000)$; Initial purchase price

45. $\frac{2}{7}$ **47.** 0 **49.** $-\frac{3}{4}$ **51.** $(1, -1), (0, 2)$

53. $(7, 6), (11, 11)$ **55.** $(3, 0), (3, 5)$

57. $3\sqrt{145} \approx 36.12$ feet

25.

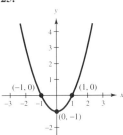

27.

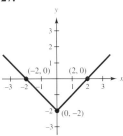

59. $y = \frac{5}{2}x - 2$ **61.** $y = -\frac{1}{2}x + 1$

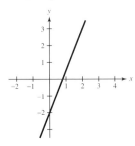   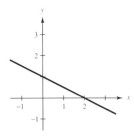

29. $\left(\frac{3}{2}, 0\right), (0, -6)$ **31.** $(-2, 0), (0, 7)$

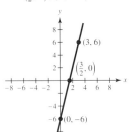

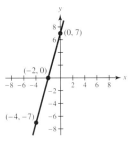

33. $(5, 0), (0, 5)$ **35.** $(-3, 0), (2, 0), (0, -4)$

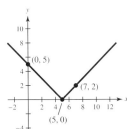

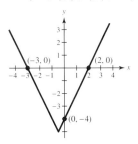

63. Neither   **65.** Perpendicular   **67.** Neither

69. $0.06   **71.** $2x - y - 6 = 0$ **73.** $4x + y = 0$

75. $2x + 3y - 17 = 0$ **77.** $4y - x - 26 = 0$

79. $y = -\frac{1}{2}x - 3$ **81.** $y = \frac{3}{2}x$ **83.** $y = \frac{3}{8}x - \frac{1}{3}$

85. $y = 7$ **87.** $x = -5$

89. (a) $3x + y - 1 = 0$ (b) $x - 3y - 3 = 0$

91. (a) $x - 12 = 0$ (b) $y - 1 = 0$

93. (a) $S = 28{,}500 + 1100(t - 8)$ (b) $37,300 (c) $32,900

95. (a) No (b) Yes (c) Yes (d) No

97.

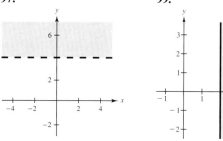

99.

101. **103.**

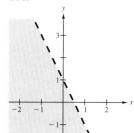 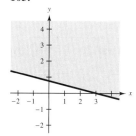

105. **107.**

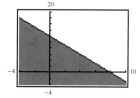 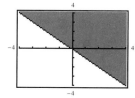

109. (a) $x + y \le 400$

(b)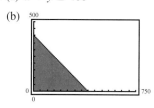

111. Domain: $\{-3, -1, 0, 1\}$; Range: $\{0, 1, 4, 5\}$

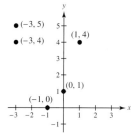

113. Not a function   **115.** Function

117. (a) 29   (b) 3

(c) $\dfrac{36 - 5t}{2}$ (d) $4 - \dfrac{5}{2}(x + h)$

119. (a) 3   (b) 0

(c) $\sqrt{2}$ (d) $\sqrt{5 - 5z}$

121. (a) -3 (b) 2

(c) 0   (d) -7

123. (a) -2 (b) $\dfrac{12 - 2x}{x}$

125. $-\infty < x < \infty$ **127.** $-\infty < x \le \frac{5}{2}$

129. $A = x(75 - x);\ 0 < x < \frac{75}{2}$

131. **133.**

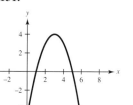 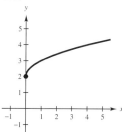

Domain: $-\infty < x < \infty$ Domain: $0 \le x < \infty$

Range: $-\infty < y \le 4$ Range: $2 \le y < \infty$

135. **137.**

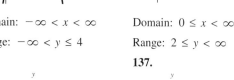

Domain: $-\infty < x < \infty$ Domain: $-2 \le x < 4$

Range: $-\infty < y \le 8$ Range: $-6 < y \le 12$

139.

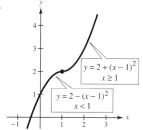

$y = 2 + (x - 1)^2$
$x \ge 1$

$y = 2 - (x - 1)^2$
$x < 1$

Domain: $-\infty < x < \infty$

Range: $-\infty < y < \infty$

141. e   **142.** c   **143.** a   **144.** f   **145.** d

146. b   **147.** Not a function   **149.** Function

151. Reflection in the x-axis

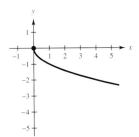

153. Horizontal shift 1 unit to the right

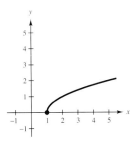

155. $y = x^2 - 2$ **157.** $y = -(x + 3)^2$

Chapter Test   (page 215)

1. Quadrant IV

2.

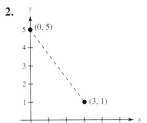

Distance: 5; Midpoint: $\left(\frac{3}{2}, 3\right)$

3. $(-1, 0), (0, -3)$

4.

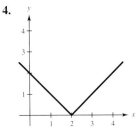

5. (a) $-\frac{2}{3}$ (b) Undefined

6.

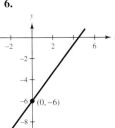

7.

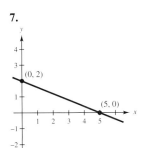

8. $x - 2y - 55 = 0$ **9.** $x + 2 = 0$

10. (a) $3x - 5y + 21 = 0$ (b) $5x + 3y + 1 = 0$

11.

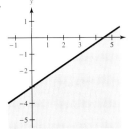

12. No. The graph does not pass the Vertical Line Test.

13. (a) Function   (b) Not a function

14. (a) -2 (b) 7   (c) $\dfrac{x + 2}{x - 1}$

15. (a) All real values of t such that $t \le 9$

   (b) All real values of x such that $x \ne 4$

16.

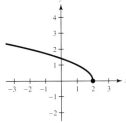

17. Reflection in the x-axis, horizontal shift 2 units to the right, vertical shift 1 unit upward

18. $V = 26{,}000 - 4000t$; 2.5 years since the car was purchased

19. (a) $y = |x - 2|$ (b) $y = |x| - 2$ (c) $y = 2 - |x|$

Chapter 4

Section 4.1   *(page 226)*

Review   *(page 226)*

1.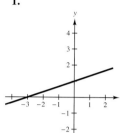

2.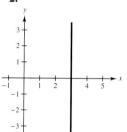

3. $\frac{3}{2}$

4. $m = -3$ because it has the greater absolute value.

5. $\frac{5}{11}$ **6.** $\frac{14}{11}$ **7.** 50   **8.** 64

9. $y = \frac{1}{4}(5 - 3x)$ **10.** $y = \frac{2}{3}(3 - x)$

11.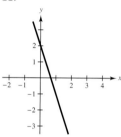

12.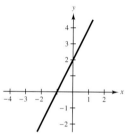

13.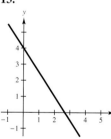

14.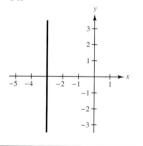

1. (a) Solution   (b) Not a solution

3. (a) Not a solution   (b) Solution

5. (a) Not a solution   (b) Solution

7. (a) Solution   (b) Not a solution

9. Inconsistent   **11.** Inconsistent   **13.** Consistent

15. Consistent

17.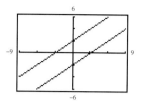

19.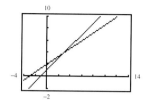

Inconsistent   One solution

21. No solution   **23.** $\left(1, \frac{1}{3}\right)$

25. Infinitely many solutions   **27.** $(1, 2)$ **29.** $(1, 2)$

31. $(2, 0)$ **33.** $(3, 1)$ **35.** Infinitely many solutions

37. $(10, 0)$ **39.** Infinitely many solutions

41. $(3, -1)$

43.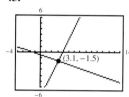

45.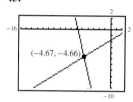

47. $(2, 1)$ **49.** $(4, 3)$ **51.** $(1, 2)$ **53.** $(4, -2)$

55. $(7, 2)$ **57.** $(10, 4)$ **59.** $(-2, -1)$ **61.** $\left(\frac{3}{2}, \frac{3}{2}\right)$

63. $\left(\frac{21}{5}, \frac{12}{5}\right)$ **65.** $\left(\frac{19}{8}, \frac{1}{8}\right)$ **67.** $\left(\frac{20}{3}, \frac{40}{3}\right)$ **69.** $(-4, 3)$

71.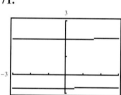

73.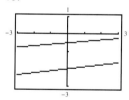

$\left(2992, \frac{798}{25}\right)$ $(50, 4)$

75. $\begin{cases} 2x - 3y = -7 \\ x + y = 9 \end{cases}$ **77.** $\begin{cases} 7x + y = -9 \\ -x + 3y = -5 \end{cases}$

79. $\begin{cases} 3x + 10y = 7 \\ 6x + 20y = 14 \end{cases}$ **81.** $\begin{cases} y + x = 4 \\ y + x = 3 \end{cases}$

83. \$125 per ton hay: 30 tons; \$75 per ton hay: 70 tons

85. 10,000 candy bars   **87.** 6250 bottles

89. \$4000 at 8.5%, \$8000 at 10%   **91.** 31, 49

93. 12, 15   **95.** 20, 32   **97.** 28, 43

99. 10 feet × 15 feet   **101.** 14 yards × 20 yards

103. 1994

105. A solution of a system of equations in two variables is an ordered pair (x, y) of real numbers that satisfies each equation in the system.

107. When Step (b) in the answer to Exercise 106 is performed, there will be a contradiction.

109. Any solutions to a system of equations correspond to the points of intersection of the graphs of the equations.

111. No. The lines represented by a consistent system of linear equations intersect at one point (one solution) or are identical lines (infinite number of solutions).

Section 4.2   *(page 236)*

Review   *(page 236)*

1. Distributive Property

2. Addition Property of Equality

3. $-2 < x < 2$ **4.** $4 \le x < 16$

5. $x < -2$ or $x > 2$ **6.** $-2 < x < 3$

7. $x < 3$ **8.** $x \ge \frac{5}{4}$ **9.** $m < 19{,}555.56$

10. $1500 + 0.04x > 2500;\ x > \$25{,}000$

1.

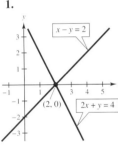

$(2, 0)$

3.

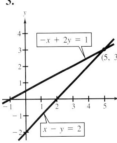

$(5, 3)$

5.

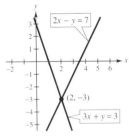

$(2, -3)$

7.

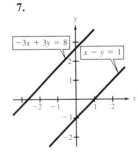

Inconsistent

9.

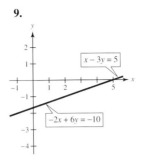

11.

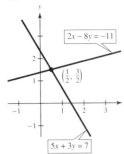

Infinitely many solutions $\left(\frac{1}{2}, \frac{3}{2}\right)$

13. $\left(\frac{47}{6}, \frac{11}{3}\right)$ **15.** $(2, -2)$ **17.** $(2, 1)$ **19.** $(3, -4)$

21. $(-1, -1)$ **23.** $(5, -1)$ **25.** $(7, -2)$

27. Inconsistent   **29.** $\left(\frac{3}{2}, 1\right)$ **31.** $(6, 3)$

33. $(-2, -1)$ **35.** Infinitely many solutions   **37.** $(4, 3)$

39. $\left(\frac{68}{5}, \frac{32}{5}\right)$ **41.** $(-3, 7)$ **43.** $(2, 7)$ **45.** $(15, 10)$

47. $(4, 3)$ **49.** Consistent   **51.** Consistent

53. Inconsistent   **55.** $k = 4$

57. $\begin{cases} x + 2y = 0 \\ 4x + 2y = 9 \end{cases}$ **59.** 122 weeks

61. 8% bond: $15,000; 9.5% bond: $5000   **63.** 4 hours

65. Speed of plane in still air: 550 miles per hour; Wind speed: 50 miles per hour

67. Adult: 375; Children: 125

69. Regular unleaded: $1.11; Premium unleaded: $1.22

71. 40% solution: 12 liters; 65% solution: 8 liters

73. 50% solution: 12 fluid ounces; 90% solution: 20 fluid ounces

75. 40% solution: 24 liters; 70% solution: 6 liters

77. Peanuts: 6.1 pounds; Cashews: 3.9 pounds

79. (a) $y = \frac{3}{2}x - \frac{1}{6}$

(b)

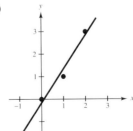

81. (a)

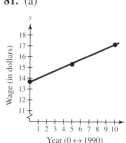

(b) $y = \frac{173}{500}x + \frac{4093}{300}$

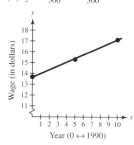

(c) The average annual increase in hourly wage

83. (a) $\begin{cases} x + y = 115 \\ 8x + 15y = 1445 \end{cases}$

(b) and (c) 40 student tickets, 75 nonstudent tickets

85. A system that has no solution

87. When you add the equations to eliminate one variable, both are eliminated, yielding a contradiction. For example,

$\begin{cases} x - y = 3 \\ x - y = 8 \end{cases} \Rightarrow \begin{array}{r} -x + y = -3 \\ x - y = 8 \\ \hline 0 = 5 \end{array}$

Section 4.3 *(page 249)*

Review *(page 249)*

1. One solution

2. Multiply each side of the equation by the lowest common denominator, 24.

3. $\frac{27}{4}$ **4.** 11   **5.** $-1, 5$ **6.** 2, 12

7. $-\infty < x < \infty$ **8.** $\frac{1}{2}$ **9.** $V = s^3$

10. $A = \dfrac{C^2}{4\pi}$

11. $d = 15t$ **12.** $P = 180 + 1.25n$

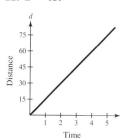

1. (a) Not a solution (b) Solution

   (c) Solution (d) Not a solution

3. $(22, -1, -5)$ **5.** $(14, 3, -1)$

7. No. When the first equation was multiplied by -2 and added to the second equation, the constant term should have been -11. Also, when the first equation was multiplied by -3 and added to the third equation, the coefficient of z should have been 2.

9. $\begin{cases} x - 2y = 8 \\ y = 14 \end{cases}$ Eliminated the x-term in Equation 2

11. $(1, 2, 3)$ **13.** $(1, 2, 3)$ **15.** $(2, -3, -2)$

17. No solution   **19.** $(-4, 8, 5)$ **21.** $(5, -2, 0)$

23. $\left(\frac{3}{10}, \frac{2}{5}, 0\right)$ **25.** $(-4, 2, 3)$ **27.** $\left(-\frac{1}{2}a + \frac{1}{2}, \frac{3}{5}a + \frac{2}{5}, a\right)$

29. $\left(-\frac{1}{2}a + \frac{1}{4}, \frac{1}{2}a + \frac{5}{4}, a\right)$ **31.** $(1, -1, 2)$

33. $\left(\frac{6}{13}a + \frac{10}{13}, \frac{5}{13}a + \frac{4}{13}, a\right)$ **35.** $\begin{cases} x + 2y - z = -4 \\ y + 2z = 1 \\ 3x + y + 3z = 15 \end{cases}$

37. $s = -16t^2 + 144$ **39.** $s = -16t^2 + 48t$

41. $88°, 32°, 60°$

43. \$17,404 at 6%, \$31,673 at 10%, \$30,923 at 15%

45. \$4200 at 6%, \$2100 at 8%, \$3200 at 9%

47. 20 gallons of spray X, 18 gallons of spray Y, 16 gallons of spray Z

49. \$20 arrangements: 400; \$30 arrangements: 100; \$40 arrangements: 350

51. (a) Not possible

   (b) 10% solution: 0 gallons; 15% solution: 6 gallons; 25% solution: 6 gallons

   (c) 10% solution: 4 gallons; 15% solution: 0 gallons; 25% solution: 8 gallons

53. Strings: 50; Winds: 20; Percussion: 8

55. (d) $\begin{cases} x + y + z = 200 \\ 8x + 15y + 100z = 4995 \\ x - 4z = 0 \end{cases}$

   (e) Students: 140; Nonstudents: 25; Major contributors: 35

   (f) No, it is not possible. To verify this, let $z = 18$ and solve the system of linear equations. The resulting values do not fulfill all the club's goals.

57. Substitute $y = 3$ in the first equation to obtain $x + 2(3) = 2$ or $x = 2 - 6 = -4$.

59. Answers will vary.

Mid-Chapter Quiz *(page 253)*

1. (a) Not a solution   (b) Solution

2.

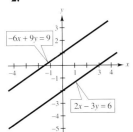

No solution

3.

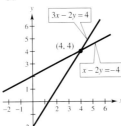

One solution

4.

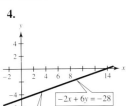

Infinitely many solutions

5.

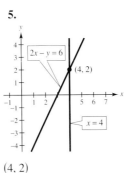

$(4, 2)$

6.

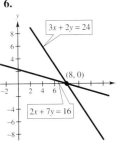

$(8, 0)$

7.

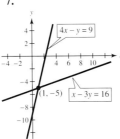

$(1, -5)$

8. $(5, 2)$ **9.** No solution   **10.** $\left(\frac{90}{13}, \frac{34}{13}\right)$ **11.** $(8, 1)$

12. $(-2, 4)$ **13.** $\left(\frac{1}{2}, -\frac{1}{2}, 1\right)$ **14.** $(5, -1, 3)$

15. $\begin{cases} x + y = -2 \\ 2x - y = 32 \end{cases}$ **16.** $\begin{cases} x + y - z = 11 \\ x + 2y - z = 14 \\ -2x + y + z = -6 \end{cases}$

17. 20% solution: $13\frac{1}{3}$ gallons, 50% solution: $6\frac{2}{3}$ gallons

18. $45°$; $80°$; $55°$

Section 4.4   *(page 262)*

Review   *(page 262)*

1. Additive Inverse Property
2. Multiplicative Identity Property
3. Commutative Property of Addition
4. Associative Property of Multiplication

5.

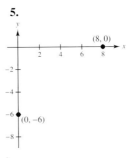

$\frac{3}{4}$

6.

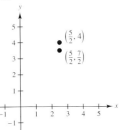

Undefined

7.

$-\frac{30}{13}$

8.

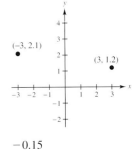

-0.15

9. 6; $(9, 8)$ **10.** $\dfrac{\sqrt{73}}{2}$; $\left(-\frac{9}{4}, 0\right)$

11. 7650 members   **12.** $940

---

1. 4×2 **3.** 2×2 **5.** 4×1 **7.** 1×1

9. 1×4 **11.** (a) $\begin{bmatrix} 4 & -5 \\ -1 & 8 \end{bmatrix}$ (b) $\begin{bmatrix} 4 & -5 & \vdots & -2 \\ -1 & 8 & \vdots & 10 \end{bmatrix}$

13. (a) $\begin{bmatrix} 1 & 10 & -3 \\ 5 & -3 & 4 \\ 2 & 4 & 0 \end{bmatrix}$ (b) $\begin{bmatrix} 1 & 10 & -3 & \vdots & 2 \\ 5 & -3 & 4 & \vdots & 0 \\ 2 & 4 & 0 & \vdots & 6 \end{bmatrix}$

15. (a) $\begin{bmatrix} 5 & 1 & -3 \\ 0 & 2 & 4 \end{bmatrix}$ (b) $\begin{bmatrix} 5 & 1 & -3 & \vdots & 7 \\ 0 & 2 & 4 & \vdots & 12 \end{bmatrix}$

17. $\begin{cases} 4x + 3y = 8 \\ x - 2y = 3 \end{cases}$ **19.** $\begin{cases} x \quad\quad + 2z = -10 \\ \quad 3y - z = 5 \\ 4x + 2y \quad = 3 \end{cases}$

21. $\begin{cases} 5x + 8y + 2z \quad\quad = -1 \\ -2x + 15y + 5z + w = 9 \\ x + 6y - 7z \quad\quad = -3 \end{cases}$

23. $\begin{cases} 13x + y + 4z - 2w = -4 \\ 5x + 4y \quad - w = 0 \\ x + 2y + 6z + 8w = 5 \\ -10x + 12y + 3z + w = -2 \end{cases}$

25. 2 **27.** $-2, \frac{2}{3}$ **29.** $-2, 6, 20, 4, 20, 4$

31. $\begin{bmatrix} 1 & 2 & 3 \\ 0 & 1 & 2 \end{bmatrix}$ **33.** $\begin{bmatrix} 1 & 0 & -\frac{7}{5} \\ 0 & 1 & \frac{11}{10} \end{bmatrix}$

35. $\begin{bmatrix} 1 & 1 & 0 & 5 \\ 0 & 1 & 2 & 0 \\ 0 & 0 & 1 & -1 \end{bmatrix}$ **37.** $\begin{bmatrix} 1 & -1 & -1 & 1 \\ 0 & 1 & 6 & 3 \\ 0 & 0 & 1 & \frac{4}{5} \end{bmatrix}$

39. $\begin{bmatrix} 1 & 1 & -1 & 3 \\ 0 & 1 & -4 & 1 \\ 0 & 0 & 0 & 0 \end{bmatrix}$

41. $\begin{cases} x - 2y = 4 \\ \quad\quad y = -3 \end{cases}$ **43.** $\begin{cases} x + 5y = 3 \\ \quad\quad y = -2 \end{cases}$

    $(-2, -3)$ $(13, -2)$

45. $\begin{cases} x - y + 2z = 4 \\ \quad\quad y - z = 2 \\ \quad\quad\quad\quad z = -2 \end{cases}$ **47.** $\left(\frac{9}{5}, \frac{13}{5}\right)$ **49.** $(1, 1)$

    $(8, 0, -2)$

51. $(-2, 1)$ **53.** No solution   **55.** $(2, -3, 2)$

57. $(2a + 1, 3a + 2, a)$ **59.** $(1, 2, -1)$ **61.** $(1, -1, 2)$

63. $(34, -4, -4)$ **65.** No solution

67. $(-12a - 1, 4a + 1, a)$ **69.** $\left(\frac{1}{2}, 2, 4\right)$ **71.** $\left(2, 5, \frac{5}{2}\right)$

73. 8%: \$800,000, 9%: \$500,000, 12%: \$200,000

75. Certificates of deposit: $250,000 - \frac{1}{2}s$

    Municipal bonds: $125,000 + \frac{1}{2}s$
    Blue-chip stocks: $125,000 - s$
    Growth stocks: s
    If $s = \$100,000$, then
        Certificates of deposit: \$200,000
        Municipal bonds: \$175,000
        Blue-chip stocks: \$25,000
        Growth stocks: \$100,000

77. Peanuts: 15 pounds; Almonds: 10 pounds;

    Pistachios: 25 pounds

79. 5, 8, 20

81. 15 computer chips, 10 resistors, 10 transistors

83. (a) Interchange two rows.

    (b) Multiply a row by a nonzero constant.
    (c) Add a multiple of a row to another row.

85. The one matrix can be obtained from the other by using the elementary row operations.

87. There will be a row in the matrix with all zero entries except in the last column.

Section 4.5 *(page 275)*

Review *(page 275)*

1. The set of inputs of the function.

2. The set of outputs of the function.

3. Relations may have ordered pairs with the same first component and different second components, while functions cannot.

4. The value of the function when $x = 4$.

5. -4 **6.** 5   **7.** 5   **8.** 4

9.

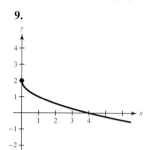

Domain: $0 \le x < \infty$; Range: $-\infty < y \le 2$

10.

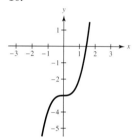

Domain: $-\infty < x < \infty$; Range: $-\infty < y < \infty$

11. $\dfrac{320}{r}$ **12.** $\dfrac{9}{2}$

1. 5   **3.** 27   **5.** 0   **7.** 6   **9.** -24

11. -0.16 **13.** -24 **15.** -2 **17.** -30 **19.** 3

21. 0   **23.** -58 **25.** 102   **27.** -0.22

29. $x - 5y + 2$ **31.** 248   **33.** 105.625   **35.** -6.37

37. $(1, 2)$ **39.** $(2, -2)$ **41.** $\left(\frac{3}{4}, -\frac{1}{2}\right)$

43. Not possible, $D = 0$ **45.** $\left(\frac{2}{3}, \frac{1}{2}\right)$ **47.** $(-1, 3, 2)$

49. $(1, -2, 1)$ **51.** Not possible, $D = 0$ **53.** $\left(\frac{22}{27}, \frac{22}{9}\right)$

55. $\left(\frac{51}{16}, -\frac{7}{16}, -\frac{13}{16}\right)$ **57.** 1, 6   **59.** 16   **61.** 7

63. $\frac{31}{2}$ **65.** $\frac{33}{8}$ **67.** 16   **69.** $\frac{53}{2}$

71. 250 square miles   **73.** Collinear   **75.** Collinear

77. Not collinear   **79.** $3x - 5y = 0$

81. $7x - 6y - 28 = 0$ **83.** $9x + 10y + 3 = 0$

85. $16x - 15y + 22 = 0$ **87.** $I_1 = 1, I_2 = 2, I_3 = 1$

89. $I_1 = 3, I_2 = 4, I_3 = 1$

91. (a) $\left(\dfrac{13}{2k + 6}, \dfrac{4 - 3k}{2k^2 + 6k}\right)$ (b) $0, -3$

93. (g) and (h) 140 students, 25 nonstudents, 35 major contributors

95. No. The matrix must be square.

97. The coefficient matrix of the system must be square and its determinant must be a nonzero real number.

Section 4.6 *(page 284)*

Review *(page 284)*

1. The graph of $h(x)$ is a vertical shift of $f(x)$ upward c units.

2. The graph of $h(x)$ is a vertical shift of $f(x)$ downward c units.

3. The graph of $h(x)$ is a horizontal shift of $f(x)$ to the right c units.

4. The graph of $h(x)$ is a horizontal shift of $f(x)$ to the left c units.

5. The graph of $h(x)$ is a reflection of $f(x)$ in the x-axis.

6. The graph of $h(x)$ is a reflection of $f(x)$ in the y-axis.

7. $h(x) = (x - 3)^2$ **8.** $h(x) = -(x + 1)^2$

9. 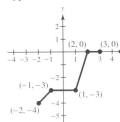   **10.**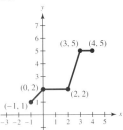

11. $230.52   **12.** $35.20

---

1. c   **2.** b   **3.** f   **4.** e   **5.** a   **6.** d

7. 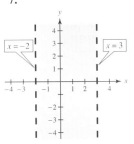   **9.**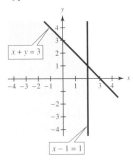

11. 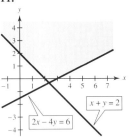   **13.**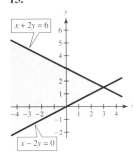

15. 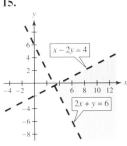   **17.**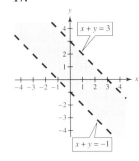

19. 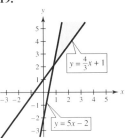   **21.**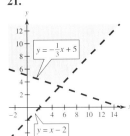

23. 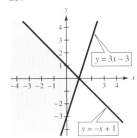   **25.**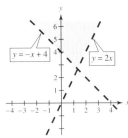

27.

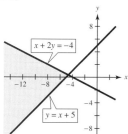

29.

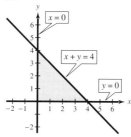

31.

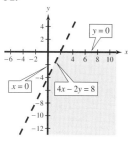

33.

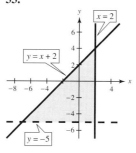

35.

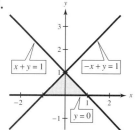

37. No solution

39.

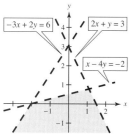

41.

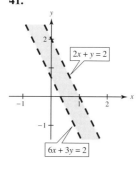

43.

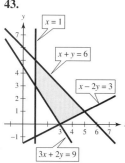

45.

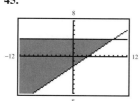

47.

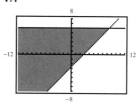

49.

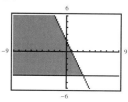

51. $\begin{cases} x \geq 1 \\ y \geq x - 3 \\ y \leq -2x + 6 \end{cases}$

53. $\begin{cases} x \geq 1 \\ x \leq 8 \\ y \geq -5 \\ y \leq 3 \end{cases}$

55. $\begin{cases} y \leq \frac{9}{10}x + \frac{42}{5} \\ y \geq 3x \\ y \geq \frac{2}{3}x + 7 \end{cases}$

57. $\begin{cases} x + \frac{3}{2}y \leq 12 \\ \frac{4}{3}x + \frac{3}{2}y \leq 15 \\ x \geq 0 \\ y \geq 0 \end{cases}$

59. $\begin{cases} x + y \leq 20{,}000 \\ x \geq 5000 \\ y \geq 2x \end{cases}$

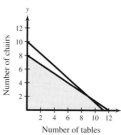

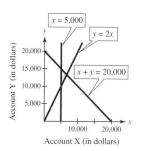

61. $\begin{cases} x + y \geq 15{,}000 \\ 15x + 25y \geq 275{,}000 \\ x \geq 8000 \\ y \geq 4000 \end{cases}$

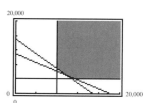

63. $\begin{cases} x \leq 90 \\ y \leq 0 \\ y \geq -10 \\ y \geq -\frac{1}{7}x \end{cases}$

65. $\begin{cases} 20x + 10y \geq 280 \\ 15x + 10y \geq 160 \\ 10x + 20y \geq 180 \\ x \geq 0 \\ y \geq 0 \end{cases}$

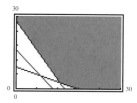

67. The graph of a linear equation splits the *xy*-plane into two parts, each of which is a half-plane. $y < 5$ is a half-plane.

69. Find all intersections between the lines corresponding to the inequalities.

Review Exercises   *(page 290)*

1. (a) Not a solution   (b) Solution

3. (a) Not a solution   (b) Not a solution

5.

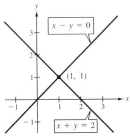

$(1, 1)$

7.

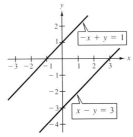

No solution

9.

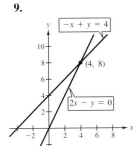

$(4, 8)$

11.

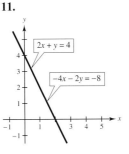

Infinitely many solutions

13.

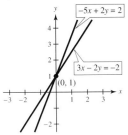

$(0, 1)$

15.

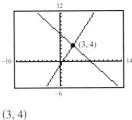

$(3, 4)$

17. $(2, -1)$ **19.** No solution   **21.** $(-10, -5)$

23. $(-1, 5)$ **25.** Infinitely many solutions

27. 16,667 units   **29.** $(0, 0)$ **31.** $\left(\frac{5}{2}, 3\right)$ **33.** $(-2, 5)$

35. $(-0.5, 0.8)$

37. 75% solution: 40 gallons; 50% solution: 60 gallons

39. 45 minutes   **41.** $(3, 2, 5)$ **43.** $(0, 3, 6)$

45. $(2, -3, 3)$ **47.** $(0, 1, -2)$

49. \$8000 at 7%, \$5000 at 9%, \$7000 at 11%

51. 1×2 **53.** 2×3

55. (a) $\begin{bmatrix} 3 & -2 \\ -1 & 1 \end{bmatrix}$ (b) $\begin{bmatrix} 3 & -2 & \vdots & 12 \\ -1 & 1 & \vdots & -2 \end{bmatrix}$

57. $\begin{cases} 4x - y = 2 \\ 6x + 3y + 2z = 1 \\ y + 4z = 0 \end{cases}$ **59.** $(10, -12)$ **61.** $(0.6, 0.5)$

63. $\left(\frac{24}{5}, \frac{22}{5}, -\frac{8}{5}\right)$ **65.** $\left(\frac{1}{2}, -\frac{1}{3}, 1\right)$ **67.** $(1, 0, -4)$

69. 5   **71.** -51 **73.** 1   **75.** $(-3, 7)$

77. $(2, -3, 3)$ **79.** 16   **81.** 7   **83.** Not collinear

85. $x - 2y + 4 = 0$

87.

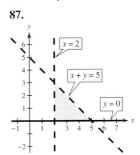

89.

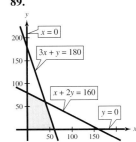

91. $\begin{cases} x + y \leq 1500 \\ x \geq 400 \\ y \geq 600 \end{cases}$

where *x* represents the number of bushels for Harrisburg and *y* for Philadelphia.

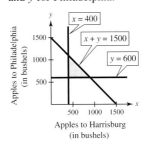

Chapter Test   *(page 294)*

1. (a) Not a solution   (b) Solution

2. $(3, 2)$ **3.** $(2, 4)$ **4.** $(2, 3)$ **5.** $(-2, 2)$

6. $(2, 2a - 1, a)$ **7.** $(-1, 3, 3)$ **8.** $(2, 1, -2)$

9. $\left(4, \frac{1}{7}\right)$ **10.** $(5, 1, -1)$ **11.** $\left(-\frac{11}{5}, \frac{56}{25}, \frac{32}{25}\right)$

12. -62 **13.** 12

14.

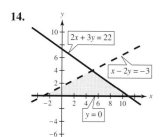

15. 16 feet $\times$ 18 feet

16. \$13,000 at 4.5%, \$9000 at 5%, \$3000 at 8%

17. $\begin{cases} 20x + 30y \geq 400,000 \\ x + y \geq 16,000 \\ y \geq 4000 \\ x \geq 9000 \end{cases}$

where x is the number of reserved tickets and y is the number of floor tickets.

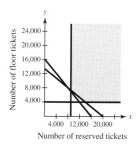

Cumulative Test: Chapters 1–4   *(page 295)*

1. (a) < (b) < (c) >   **2.** $3n - 8$ **3.** $3t^2 - 3t - 8$

4. $3x^3 - 5x^2 - 6x - 25$ **5.** $\frac{3}{2}$ **6.** $-\frac{3}{2}$

7. $x \leq -1$ or $x \geq 5$ **8.** $-\frac{3}{2} \leq x < 4$ **9.** \$1408.75

10. $\frac{13}{2}$ **11.** $x \geq 103$ **12.** Yes   **13.** $2 \leq x < \infty$

14. (a) 4   (b) $c^2 + 3c$

15. $m = \frac{3}{4}$; Distance: 10; Midpoint: $(0, 3)$

16. (a) $2x - y + 5 = 0$ (b) $2x - 3y + 7 = 0$

17.

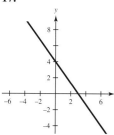

18.

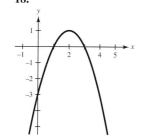

19. $(3, 3)$ **20.** $(4, -2)$ **21.** $(4, -1, 3)$

Chapter 5

Section 5.1   *(page 304)*

Review   *(page 304)*

1. The graph of an equation is the set of solution points of the equation on a rectangular coordinate system.

2. Create a table of solution points of the equation, plot those points on a rectangular coordinate system, and connect the points with a smooth curve or line.

3. $(2, 0), (6, 2)$

4. To find the x-intercept, let $y = 0$ and solve the equation for x. To find the y-intercept, let $x = 0$ and solve the equation for y.

5. (a) -15 (b) $-\frac{15}{2}$ **6.** (a) 20   (b) 2

7. (a) 0   (b) $-t^2 + 4t + 5$

8. (a) $\frac{2}{3}$ (b) $\frac{2 - z}{6 - z}$

9.

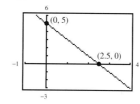

10.

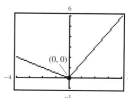

11.

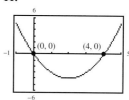

12.

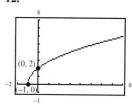

1. (a) $-3x^8$ (b) $9x^7$ **3.** (a) $-125z^6$ (b) $25z^8$

5. (a) $2u^3v^3$ (b) $-4u^9v$ **7.** (a) $-15u^8$ (b) $64u^5$

9. (a) $-m^{19}n^7$ (b) $-m^7n^3$ **11.** (a) $3m^4n^3$ (b) $3m^2n^3$

13. (a) $\frac{9x^2}{16y^2}$ (b) $\frac{125u^3}{27v^3}$ **15.** (a) $\frac{8x^4y}{9}$ (b) $-\frac{2x^2y^4}{3}$

17. (a) $\frac{25u^8v^2}{4}$ (b) $\frac{u^8v^2}{4}$

19. (a) $x^{2n-1}y^{2n-1}$ (b) $x^{2n-2}y^{n-12}$ **21.** $\frac{1}{25}$

23. $-\frac{1}{1000}$ **25.** 1   **27.** 64   **29.** -32 **31.** $\frac{3}{2}$

33. 1   **35.** 1   **37.** 729   **39.** 100,000   **41.** $\frac{1}{16}$

43. $\frac{1}{64}$ **45.** $\frac{3}{16}$ **47.** $\frac{64}{121}$ **49.** $\frac{16}{15}$ **51.** y^2 **53.** z^2

55. $\frac{7}{x^4}$ **57.** $\frac{1}{64x^3}$ **59.** x^6 **61.** $\frac{4}{3}a$ **63.** t^2

65. $\frac{1}{4x^4}$ **67.** $-\frac{12}{xy^3}$ **69.** $\frac{y^4}{9x^4}$ **71.** $\frac{10}{x}$ **73.** $\frac{x^5}{2y^4}$

75. $\frac{81v^8}{u^6}$ **77.** $\frac{b^5}{a^5}$ **79.** $\frac{1}{2x^8y^3}$ **81.** $6u$ **83.** x^8y^{12}

85. $\frac{2b^{11}}{25a^{12}}$ **87.** $\frac{v^2}{uv^2+1}$ **89.** $\frac{ab}{b-a}$ **91.** 3.6×10^6

93. 4.762×10^7 **95.** 3.1×10^{-4} **97.** 3.81×10^{-8}

99. 5.73×10^7 **101.** 9.4608×10^{12} **103.** 8.99×10^{-2}

105. $60{,}000{,}000$ **107.** 0.0000001359

109. $38{,}757{,}000{,}000$ **111.** $15{,}000{,}000$

113. 0.00000000048 **115.** 6.8×10^5 **117.** 2.5×10^9

119. 6×10^6 **121.** 9×10^{15} **123.** 1.6×10^{12}

125. 3.46×10^{10} **127.** 4.70×10^{11} **129.** 1.67×10^{14}

131. 2.74×10^{20} **133.** 9.3×10^7 miles

135. 1.59×10^{-5} year $\approx$ 8.4 minutes   **137.** 3.33×10^5

139. \$20,469   **141.** $3x$ is the base and 4 is the exponent.

143. Change the sign of the exponent of the factor.

145. When the numbers are very large or very small

Section 5.2   *(page 313)*

Review   *(page 313)*

1. Negative   **2.** 3 and 9   **3.** False. -36

4. True. 36   **5.** $x \geq 6$ **6.** $x > \frac{3}{2}$

7. $-3 < x < 3$ **8.** $-1 \leq x < 3$

9. $1 < x < 5$ **10.** $x < 2$ or $x > 8$

11. \$1489.66   **12.** $11\frac{3}{8}$ gallons

1. $-42x^3 - 10x^2 + 3x + 5; 3; -42$ **3.** $4y + 16; 1; 4$

5. $5x^2 + x - 7; 2; 5$ **7.** $-3x^3 - 2x^2 - 3; 3; -3$

9. $-4; 0; -4$ **11.** $-16t^2 + v_0t; 2; -16$

13. Binomial   **15.** Trinomial   **17.** Monomial

19. $3x^3$ **21.** $8x^2 + 5$

23. The first term is not of the form ax^k (k must be nonnegative).

25. The term is not of the form ax^k (k must be nonnegative).

27. $3x + 7$ **29.** $7x^2 + 3$ **31.** $4y^2 - y + 3$

33. $-2y^4 - 5y + 4$ **35.** 13   **37.** $6x^2 - 7x + 8$

39. $\frac{1}{6}x^3 + 3x + \frac{2}{5}$ **41.** $2.69t^2 + 7.35t - 4.2$

43. $2x^2 - 3x$ **45.** $4x^3 + 2x^2 + 9x - 6$

47. $2p^2 - 2p - 5$ **49.** $0.6b^2 - 0.6b + 7.1$

51. $-2y^3$ **53.** $x^2 - 3x + 2$ **55.** $7t^3 - t - 10$

57. $\frac{7}{4}y^2 - 9y - 12$

59. $9.37t^5 + 10.4t^4 - 5.4t^2 + 7.35t - 2.6$

61. $-2x^3 + x^2 + 2x$ **63.** $x^2 - 2x + 5$

65. $-4x^3 - 2x + 13$ **67.** $-11x^7 - 10x^5 + 8x^4 + 16$

69. $2x^3 - 2x + 3$ **71.** $-2x^3 - x^2 + 6x - 11$

73. $7y^2 - 9y + 2$ **75.** $7x^3 + 2x$ **77.** $3x^3 + 5x^2 + 2$

79. $3t^2 + 29$ **81.** $3v^2 + 78v + 27$ **83.** $29s + 8$

85. $5x^{2r} - 8x^r + 3$ **87.** $2x^{2m} + 6x^m - 11$

89. $3x^{4n} - x^{3n} + 3x^{2n} - 1$

91.

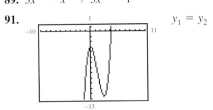

$y_1 = y_2$

93. $-x^3 - 4x^2 - x + 16$

95. (a) 64 feet   (b) 60 feet   (c) 48 feet   (d) 0 feet

At time $t = 0$, the object is dropped from a height of 64 feet and continues to fall, reaching the ground at time $t = 2$.

97. (a) 50 feet   (b) 146 feet   (c) 114 feet   (d) 50 feet

At time $t = 0$, the object projected upward from a height of 50 feet, reaches a maximum height, and returns downward. At time $t = 5$, the object is again at a height of 50 feet.

99. Dropped; 100 feet   **101.** Thrown downward; 50 feet

103. 1008 feet; 784 feet; 48 feet   **105.** \$15,000

107. $14x + 8$ **109.** $36x$ **111.** $5x + 72$

113. (a) $T = 0.0475x^2 + 1.099x + 0.23$

(b)

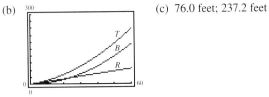

(c) 76.0 feet; 237.2 feet

115. The degree of the term ax^k is k. The degree of a polynomial is the degree of its highest-degree term.

117. $8x^2 - 3x^2 = (8 - 3)x^2 = 5x^2$

119. No. $x^3 + 2x + 3$

Section 5.3 *(page 323)*

Review *(page 323)*

1. The point is 2 units to the left of the y-axis and 3 units above the x-axis.

2. $(4, 3), (-4, 3), (-4, -3), (4, -3)$

3. 13   **4.** $-\frac{11}{3}$ **5.** 6.8   **6.** 7

7. (a) 5 (b) $x - 7$ **8.** (a) -10 (b) $-2x$

9. (a) $\dfrac{7}{12}$ (b) $\dfrac{c - 6}{c + 4}$ **10.** (a) -2 (b) $\sqrt{t - 1}$

11.
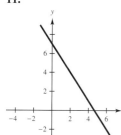
12.
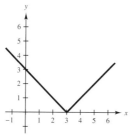

1. $16a^3$ **3.** $-2y^2 + 10y$ **5.** $8x^3 - 12x^2 + 20x$

7. $14x^5 - 6x^4 - 10x^2$ **9.** $-x^7 + 2x^6 - 5x^4 + 6x^3$

11. $75x^3 + 30x^2$ **13.** $3u^6v - 5u^4v^2 + 6u^3v^4$

15. $x^2 + 6x + 8$ **17.** $x^2 - x - 30$ **19.** $x^2 - 6x + 9$

21. $2x^2 + 7x - 15$ **23.** $10x^2 - 34x + 12$

25. $-12x^3 - 3x^2 + 32x + 8$ **27.** $48y^2 + 32y - 3$

29. $6x^2 + 7xy + 2y^2$ **31.** $4t^2 - 6t + 4$

33. $x^3 - 5x^2 + 10x - 6$ **35.** $3a^3 + 11a^2 + 9a + 2$

37. $8u^3 + 22u^2 - u - 20$ **39.** $x^4 - 2x^3 - 3x^2 + 8x - 4$

41. $5x^4 + 20x^3 - 3x^2 + 8x - 2$ **43.** $t^4 - t^2 + 4t - 4$

45. $28x^5 - 56x^4 + 36x^3 + 21x^2 - 42x + 27$

47. $2u^3 + u^2 - 7u - 6$ **49.** $-2x^3 + 3x^2 - 1$

51. $t^4 - t^2 + 4t - 4$ **53.** $x^2 - 4$ **55.** $x^2 - 64$

57. $4 - 49y^2$ **59.** $36 - 16x^2$ **61.** $4a^2 - 25b^2$

63. $36x^2 - 81y^2$ **65.** $4x^2 - \frac{1}{16}$ **67.** $0.04t^2 - 0.25$

69. $x^2 + 10x + 25$ **71.** $x^2 - 20x + 100$

73. $4x^2 + 20x + 25$ **75.** $36x^2 - 12x + 1$

77. $4x^2 - 28xy + 49y^2$

79. $x^2 + 2xy + y^2 + 4x + 4y + 4$

81. $u^2 - v^2 + 6v - 9$ **83.** $x^3 + 9x^2 + 27x + 27$

85. $u^3 + 3u^2v + 3uv^2 + v^3$ **87.** $15x^{3r} + 12x^{4r-1}$

89. $12x^{3m} - 10x^{2m} - 18x^m + 15$ **91.** $x^{m^2 - n^2}$

93.
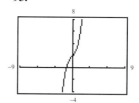
$y_1 = y_2$

95.
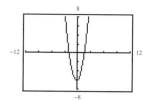
$y_1 = y_2$

97. (a) $t^2 - 8t + 15$ (b) $h^2 + 2h$

99. (a) $V(n) = n^3 + 6n^2 + 8n$ (b) 48 cubic inches

(c) $A(n) = n^2 + 2n$ (d) $A(n + 4) = n^2 + 10n + 24$

101. $8x^2 + 26x$ **103.** $1.2x^2$ **105.** (a) $5w$ (b) $\frac{3}{2}w^2$

107. $R = -0.02x^2 + 125x$; \$1050

109. $1000 + 2000r + 1000r^2$

111. $(x + a)(x + b) = x^2 + bx + ax + ab$; FOIL Method

113. (a) $x^2 - 1$ (b) $x^3 - 1$ (c) $x^4 - 1$

Yes, $x^5 - 1$

115. (a) $(x) \times (x + 5) \times (3x - 2)$

$V_B(x) = 3x^3 + 13x^2 - 10x$

(b) $(x - 3) \times (x - 1) \times 2(x - 3)$

$V_P(x) = \frac{2}{3}x^3 - \frac{14}{3}x^2 + 10x - 6$

(c) $V_S(x) = \frac{7}{3}x^3 + \frac{53}{3}x^2 - 20x + 6$

117. First, Outer, Inner, Last

119. (a) True. $(a^b)(c^d) = a^bc^d$, which is a monomial.

(b) False. $(x + 2)(x - 3) = x^2 - x - 6$

Mid-Chapter Quiz *(page 327)*

1. $4; -2$

2. The exponent in the term $-3x^{1/2}$ is not an integer.

3. $-10y^9$ **4.** $-54x^5$ **5.** $10n^5$ **6.** $-8x^{10}$

7. $-\dfrac{48}{x}$ **8.** $\dfrac{25x^2}{16y^4}$ **9.** $\dfrac{9}{a^4b^{10}}$ **10.** $\dfrac{8y^{33}}{125x^6}$

11. $3t^3 + 3t^2 + 7$ **12.** $7y^2 - 5y$ **13.** $9x^3 - 4x^2 + 1$

14. $2u^2 - u + 1$ **15.** $28y - 21y^2$ **16.** $x^2 - 4x - 21$

17. $24x^2 - 26xy + 5y^2$ **18.** $2z^2 + 3z - 35$

19. $36r^2 - 25$ **20.** $4x^2 - 12x + 9$ **21.** $x^3 + 1$

22. $x^4 + 2x^3 - 23x^2 + 40x - 20$

23. $\frac{1}{2}(x + 2)^2 - \frac{1}{2}x^2 = 2(x + 1)$

24. 116 feet; 100 feet   **25.** \$12,000

Section 5.4   *(page 334)*

Review   *(page 334)*

1. A function f from a set A to a set B is a rule of correspondence that assigns to each element x in the set A exactly one element y in the set B.

2. The set A (see Exercise 1) is called the domain (or set of inputs) of the function f, and the set B (see Exercise 1) contains the range (or set of outputs) of the function.

3.

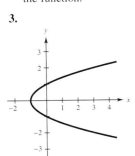

4.

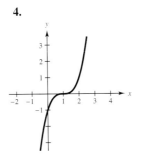

5.

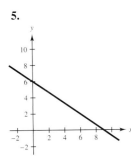

Function

6.

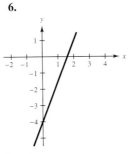

Function

7.

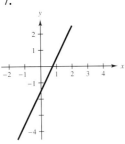

Function

8.

Function

9.

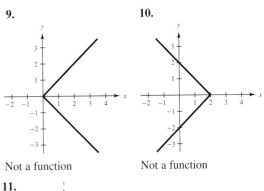

Not a function

10.

Not a function

11.

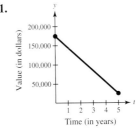

12. (a) $A = x(250 - x)$

(b)

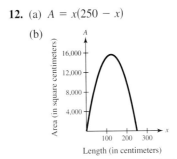

1. 6   **3.** $3x$ **5.** $6z^2$ **7.** $14b^2$ **9.** $21(x + 8)^2$

11. $4(x + 1)$ **13.** $2(3y - 10)$ **15.** $12(2t^2 - 3)$

17. $x(x + 9)$ **19.** $2t(t + 4)$

21. No common factor other than 1   **23.** $4(7x^2 + 4x - 2)$

25. $3y(x^2y - 5)$ **27.** $3xy(5y - x + 3)$

29. $x^2(14x^2y^3 + 21xy^2 + 9)$ **31.** $-(y^2 - 7)$

33. $-7(2x - 1)$ **35.** $-2(3x^2 - 2x - 8)$

37. $-y(3y^2 + 2y - 1)$ **39.** $10y - 3$ **41.** $6x + 5$

43. $(y - 4)(2y + 5)$ **45.** $(3x + 2)(5x - 3)$

47. $(7a + 6)(2 - 3a^2)$ **49.** $(4t - 1)^2(8t^3 + 3)$

51. $(4x + 9)(-2x - 9)$ **53.** $(x + 25)(x + 1)$

55. $(y - 6)(y + 2)$ **57.** $(x + 2)(x^2 + 1)$

59. $(a - 4)(3a^2 - 2)$ **61.** $(z + 3)(z^3 - 2)$

63. $(x - 2y)(5x^2 + 7y^2)$ **65.** $(x + 5)(x - 5)$

67. $(1 + a)(1 - a)$ **69.** $(4y + 3)(4y - 3)$

71. $(9 + 2x)(9 - 2x)$ **73.** $(2z + y)(2z - y)$

75. $(6x + 5y)(6x - 5y)$ **77.** $\left(u + \frac{1}{4}\right)\left(u - \frac{1}{4}\right)$

79. $\left(\frac{2}{3}x + \frac{4}{5}y\right)\left(\frac{2}{3}x - \frac{4}{5}y\right)$ **81.** $(x + 3)(x - 5)$

83. $(14 + z)(4 - z)$ **85.** $(x + 9)(3x + 1)$

87. $(x - 2)(x^2 + 2x + 4)$ **89.** $(y + 4)(y^2 - 4y + 16)$

91. $(2t - 3)(4t^2 + 6t + 9)$ **93.** $(3u + 1)(9u^2 - 3u + 1)$

95. $(4a + b)(16a^2 - 4ab + b^2)$

97. $(x + 3y)(x^2 - 3xy + 9y^2)$ **99.** $2(2 - 5x)(2 + 5x)$

101. $8(x + 2)(x^2 - 2x + 4)$

103. $(y - 3)(y + 3)(y^2 + 9)$ **105.** $3x^2(x + 10)(x - 10)$

107. $6(x^2 - 2y^2)(x^4 + 2x^2y^2 + 4y^4)$

109. $(2x^n + 5)(2x^n - 5)$ **111.** $2x^r(x^{2r} + 2x^r + 4)$

113. $y^{m+n}(7y^m - y^n + 4)$

115.

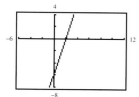

117.

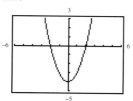

$y_1 = y_2$ $y_1 = y_2$

119. $x^2(3x + 4) - (3x + 4) = (3x + 4)(x + 1)(x - 1)$

$3x(x^2 - 1) + 4(x^2 - 1) = (x + 1)(x - 1)(3x + 4)$

121. $p = 800 - 0.25x$ **123.** $P(1 + rt)$

125. $w = 45 - l$ **127.** $S = 2x(x + 2h)$

129. $\pi(R - r)(R + r)$

131. The polynomial is written as a product of polynomials.

133. Determine the prime factorization of each integer. The greatest common factor is the product of each common prime factor raised to its lowest power in either one of the integers.

135. $x^2 + 2x = x(x + 2)$

Section 5.5   *(page 346)*

Review   *(page 346)*

1. A function can have only one value of y corresponding to $x = 0$.

2. 6   **3.** $|x| < 5$ **4.** $|x - 6| > 3$

5.

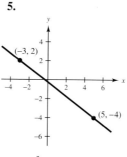

$(-3, 2)$ $(5, -4)$

$m = -\frac{3}{4}$

6.

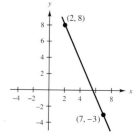

$(2, 8)$ $(7, -3)$

$m = -\frac{11}{5}$

7.

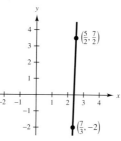

$\left(\frac{5}{2}, \frac{7}{2}\right)$ $\left(\frac{7}{3}, -2\right)$

$m = 33$

8.

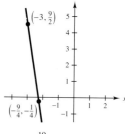

$\left(-3, \frac{9}{2}\right)$ $\left(-\frac{9}{4}, -\frac{1}{4}\right)$

$m = -\frac{19}{3}$

9.

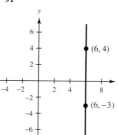

$(6, 4)$ $(6, -3)$

m is undefined.

10.

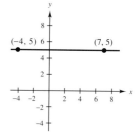

$(-4, 5)$ $(7, 5)$

$m = 0$

11. \$12,720   **12.** 49.1 miles per hour

1. $(x + 2)^2$ **3.** $(a - 6)^2$ **5.** $(5y - 1)^2$

7. $(3b + 2)^2$ **9.** $(u + 4v)^2$ **11.** $(6x - 5y)^2$

13. $5(x + 3)^2$ **15.** $2x(x + 6)^2$ **17.** $5v^2(2v - 3)^2$

19. $\frac{1}{36}(3x - 4)^2$ **21.** ± 18 **23.** ± 12 **25.** 16

27. 9   **29.** $x + 1$ **31.** $y - 5$ **33.** $x - 6$

35. $z - 2$ **37.** $(x + 3)(x + 1)$ **39.** $(x - 3)(x - 2)$

41. $(y + 10)(y - 3)$ **43.** $(t - 7)(t + 3)$

45. $(x - 8)(x - 12)$ **47.** $(x - 7y)(x + 5y)$

49. $(x + 12y)(x + 18y)$ **51.** $\pm 9, \pm 11, \pm 19$

53. $\pm 4, \pm 20$ **55.** $\pm 12, \pm 36$ **57.** $-16, 8$

59. $-18, 2$ **61.** $5x + 3$ **63.** $5a - 3$ **65.** $2y - 9$

67. $(3x + 5)(2x - 5)$ **69.** $(5y + 4)(2y - 3)$

71. $(4x - 1)(3x - 1)$ **73.** $(5z - 3)(z + 1)$

75. $(2t + 1)(t - 4)$ **77.** $(3b - 1)(2b + 7)$

79. $(2y + 3)(9y + 4)$ **81.** $(2 + x)(3 - 2x)$

83. $(1 + 4x)(1 - 15x)$ **85.** $3(x - 4)(2x + 7)$

87. $5y(3y - 2)(4y + 5)$ **89.** $(a + 2b)(10a + 3b)$

91. $(4x - 3y)(6x + y)$ **93.** $(3x + 4)(x + 2)$

95. $(2x - 1)(3x + 2)$ **97.** $(3x - 1)(5x - 2)$

99. $3x^3(x - 4)$ **101.** $2t(5t - 9)(t + 2)$

103. $2(3x - 1)(9x^2 + 3x + 1)$

105. $9ab^2(3ab + 2)(ab - 1)$ **107.** $(x + 2)(x + 4)(x - 4)$

109. $(3 - z)(9 + z)$ **111.** $(x - 5 + y)(x - 5 - y)$

113. $(x^4 + 1)(x^2 + 1)(x + 1)(x - 1)$

115. $(x^n - 8)(x^n + 3)$ **117.** $(x^n + 5)(x^n - 2)$

119. $(2y^n - 3)(2y^n + 1)$

121.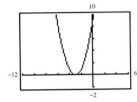

123.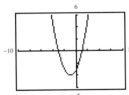

$y_1 = y_2$ $y_1 = y_2$

125. c **126.** a **127.** b **128.** d

129. $4(6 + x)(6 - x)$

131. (a) $(2n - 2)(2n)(2n + 2)$ (b) $18, 20, 22$

133. Begin by finding the factors of 6 whose sum is -5. They are -2 and -3. The factorization is $(x - 2)(x - 3)$.

135. Multiply the factors. The factors of $x^2 - 5x + 6$ are $x - 2$ and $x - 3$ because $(x - 2)(x - 3) = x^2 - 5x + 6$.

137. No. $x(x + 2) - 2(x + 2) = (x + 2)(x - 2)$

139. Problems will vary. It is possible to create factorable polynomials by working backward: first list several factors, and then multiply them to form a single polynomial.

Section 5.6 *(page 356)*

Review *(page 356)*

1. Additive Inverse Property

2. Multiplicative Identity Property

3. Distributive Property

4. Associative Property of Multiplication

5. -4

6. $353.\overline{33}$ **7.** No solution **8.** -19

9. 40 **10.** 24

11. (a) $P = -\frac{1}{4}x^2 + 8x - 12$

(b) 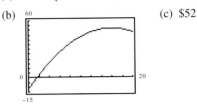 (c) $52

12. $1245.05 million

1. $0, 8$ **3.** $-10, 3$ **5.** $-4, 2$ **7.** $-\frac{5}{2}, -\frac{1}{3}$

9. $-\frac{25}{2}, 0, \frac{3}{2}$ **11.** $-4, -\frac{1}{2}, 3$ **13.** $0, 5$ **15.** $-\frac{5}{3}, 0$

17. $0, 16$ **19.** $0, 3$ **21.** ± 5 **23.** ± 4 **25.** $-2, 5$

27. $4, 6$ **29.** $-5, \frac{5}{4}$ **31.** $-\frac{1}{2}, 7$ **33.** $-1, \frac{2}{3}$

35. $4, 9$ **37.** 4 **39.** -8 **41.** $\frac{3}{2}$ **43.** $-2, 10$

45. ± 3 **47.** $-4, 9$ **49.** $-12, 6$ **51.** $-\frac{7}{2}, 5$

53. $-7, 0$ **55.** $-6, 5$ **57.** $-2, 6$ **59.** $-5, 1$

61. $-2, 8$ **63.** $-13, 5$ **65.** $0, 7, 12$ **67.** $-\frac{1}{3}, 0, \frac{1}{2}$

69. ± 2 **71.** $\pm 3, -2$ **73.** ± 3 **75.** $\pm 1, 0, 3$

77. $\pm 2, -\frac{3}{2}, 0$

79. $(-3, 0), (3, 0)$; the x-intercepts are solutions of the polynomial equation.

81. $(0, 0), (3, 0)$; the x-intercepts are solutions of the polynomial equation.

83.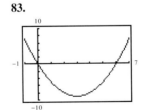

$(0, 0), (6, 0)$

85.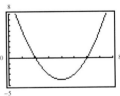

$(2, 0), (6, 0)$

87.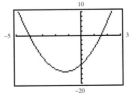

$(-4, 0), \left(\frac{3}{2}, 0\right)$

89.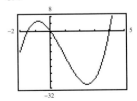

$\left(-\frac{3}{2}, 0\right), (0, 0), (4, 0)$

91. $-\frac{b}{a}, 0$ **93.** $x^2 - 2x - 15 = 0$ **95.** 15

97. 15 feet $\times$ 22 feet

99. Base: 8 inches; Height: 12 inches

101. (a) Length $= 5 - 2x$; Width $= 4 - 2x$; Height $= x$

Volume $=$ (Length)(Width)(Height)

$V = (5 - 2x)(4 - 2x)(x)$

(b) $0, 2, \frac{5}{2}$; $0 < x < 2$

(c)

x	0.25	0.50	0.75	1.00	1.25	1.50	1.75
V	3.94	6	6.56	6	4.69	3	1.31

(d) 1.50

(e) 0.74

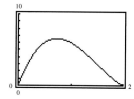

103. 9.75 seconds   **105.** 4 seconds   **107.** 3 seconds

109. 10 units, 20 units

111. (a) $-6, -\frac{1}{2}$ (b) $-6, -\frac{1}{2}$ (c) Answers will vary.

113. (d) 3 feet $\times$ 5 feet $\times$ 6 feet (e) 1026 cubic feet

(f) $x = 6$

115. False. This is not an application of the Zero-Factor Property, because there are an unlimited number of factors whose product is 1.

117. Maximum number: n. The third-degree equation $(x + 1)^3 = 0$ has only one real solution: $x = -1$.

Review Exercises   *(page 361)*

1. x^5 **3.** u^6 **5.** $-8z^3$ **7.** $4u^7v^3$ **9.** $2z^3$

11. $8u^2v^2$ **13.** $144x^4$ **15.** $\frac{1}{72}$ **17.** $\frac{125}{8}$ **19.** $12y$

21. $\frac{2}{x^3}$ **23.** 1   **25.** $\frac{b^9}{2a^8}$ **27.** $\frac{4x^6}{y^5}$ **29.** $\frac{405u^5}{v}$

31. 5.38×10^{-5} **33.** 483,300,000   **35.** 3.6×10^7

37. 500

39. $-x^4 + 6x^3 + 5x^2 - 4x$ **41.** $-7x^3 + 3x^2 - 6x + 14$

Leading coefficient: -1 Leading coefficient: -7

Degree: 4   Degree: 3

43. $-6x^5 + 2x - 4$ **45.** $x^2 + 13x + 8$

47. $-3x^3 - x^2 + 16$ **49.** $-2y^2 + 3y - 5$

51. $-9x^3 + 9x - 4$ **53.** $-2y - 15$

55. $x^5 - 4x^3 + 7x^2 - 9x + 3$ **57.** $-x^2 + 4x + 6$

59. $-t^2 + 4t$ **61.** $18x + 10$ **63.** \$6600

65. $-8x^4 - 32x^3$ **67.** $6x^3 - 15x^2 + 9x$

69. $x^2 + 5x - 14$ **71.** $15x^2 - 11x - 12$

73. $24x^4 + 22x^2 + 3$ **75.** $4x^3 - 5x + 6$

77. $u^2 - 8u + 7$ **79.** $16x^2 - 56x + 49$

81. $25u^2 - 64$ **83.** $u^2 - v^2 - 6u + 9$

85. $14x + 3$ **87.** (a) $P = 12w$ (b) $A = 5w^2$

89. $6(4x^2 - 3)$ **91.** $x(2x + 1)$

93. $3x(2x + 5x^2 - 1)$ **95.** $-42(x + 5)$

97. $(v + 1)(v - 1)(v - 2)$ **99.** $(t + 3)(t^2 + 3)$

101. $(x + 6)(x - 6)$ **103.** $(u - 3)(u + 15)$

105. $(u - 1)(u^2 + u + 1)$ **107.** $(2x + 3)(4x^2 - 6x + 9)$

109. $x(x + 1)(x - 1)$ **111.** $3(u + 2)(u^2 - 2u + 4)$

113. $(x - 9)^2$ **115.** $(2s + 10t)^2$ **117.** $(x + 7)(x - 5)$

119. $(2x - 3)(x - 2)$ **121.** $(3x + 2)(6x + 5)$

123. $(4x + 1)(x - 1)$ **125.** $(5x - 7)(x - 1)$

127. $(2s - 7)(s - 3)$ **129.** $4a(1 + 4a)(1 - 4a)$

131. $4(2x - 3)(2x - 1)$ **133.** $\left(\frac{1}{2}x + y\right)^2$

135. $(x + y - 5)(x - y - 5)$ **137.** $0, 2$ **139.** $-\frac{1}{2}, 3$

141. $-10, -\frac{9}{5}, \frac{1}{4}$ **143.** $-\frac{4}{3}, 2$ **145.** $0, 3$

147. $-4, 9$ **149.** ± 10 **151.** $-4, 0, 3$ **153.** $0, 2, 9$

155. $\pm 1, 6$ **157.** $\pm 3, 0, 5$ **159.** $13, 15$

161. 45 inches $\times$ 20 inches   **163.** 20 seconds

Chapter Test   *(page 365)*

1. Degree: 3; Leading coefficient: -5.2

2. The variable appears in the denominator.

3. (a) $\frac{20y^5}{x^5}$ (b) $\frac{1}{4x^4y^2z^6}$ **4.** (a) $-24u^9v^5$ (b) $\frac{27x^6}{2y^4}$

5. (a) $6a^2 - 3a$ (b) $-2y^2 - 2y$

6. (a) $8x^2 - 4x + 10$ (b) $11t + 7$

7. (a) $-3x^2 + 12x$ (b) $2x^2 + 7xy - 15y^2$

8. (a) $3x^2 - 6x + 3$ (b) $6s^3 - 17s^2 + 26s - 21$

9. (a) $16x^2 - 24x + 9$ (b) $16 - a^2 - 2ab - b^2$

10. $6y(3y - 2)$ **11.** $\left(v - \frac{4}{3}\right)\left(v + \frac{4}{3}\right)$

12. $(x + 2)(x - 2)(x - 3)$ **13.** $(3u - 1)^2$

14. $2(x - 5)(3x + 2)$ **15.** $(x + 3)(x^2 - 3x + 9)$

16. $-10, 7$ **17.** $1, -5$ **18.** $3, -\frac{4}{3}$

19. $-4, -1, 0$ **20.** $2x(x + 15) - x(x + 4) = x^2 + 26x$

21. 6 centimeters $\times$ 9 centimeters

22. Base: 5 feet; Height: 14 feet

23. 50 computer desks

Chapter 6

Section 6.1   *(page 375)*

Review   *(page 375)*

1. $m = \dfrac{y_2 - y_1}{x_2 - x_1}$

2. (a) $m > 0$ (b) $m < 0$

   (c) $m = 0$ (d) m is undefined.

3. 10 4. 12 5. $-8x - 10$ 6. $-2x^2 + 14x$

7. $\dfrac{25}{x^4}$ 8. $\dfrac{4u^3}{3}$

9. 30% solution: $13\frac{1}{3}$ gallons; 60% solution: $6\frac{2}{3}$ gallons

10. $500

---

1. $(-\infty, \infty)$ 3. $(-\infty, 5) \cup (5, \infty)$

5. $(-\infty, 4) \cup (4, \infty)$ 7. $(-\infty, -10) \cup (-10, \infty)$

9. $(-\infty, \infty)$ 11. $(-\infty, -3) \cup (-3, 0) \cup (0, \infty)$

13. $(-\infty, -4) \cup (-4, 4) \cup (4, \infty)$

15. $(-\infty, 0) \cup (0, 3) \cup (3, \infty)$

17. $(-\infty, 2) \cup (2, 3) \cup (3, \infty)$

19. $(-\infty, -1) \cup \left(-1, \frac{5}{3}\right) \cup \left(\frac{5}{3}, \infty\right)$

21. (a) 1 (b) -8

   (c) Undefined (division by 0)   (d) 0

23. (a) 0 (b) 0

   (c) Undefined (division by 0)

   (d) Undefined (division by 0)

25. (a) $\frac{25}{22}$ (b) 0

   (c) Undefined (division by 0)

   (d) Undefined (division by 0)

27. $(0, \infty)$ 29. $\{1, 2, 3, 4, \ldots\}$ 31. $[0, 100)$

33. $x + 3$ 35. $(3)(x + 16)^2$ 37. $(x)(x - 2)$

39. $x + 2$ 41. $\dfrac{x}{5}$ 43. $6y, \; y \neq 0$ 45. $\dfrac{6x}{5y^3}, \; x \neq 0$

47. $\dfrac{x - 3}{4x}$ 49. $x, \; x \neq 8, \; x \neq 0$ 51. $\dfrac{1}{2}, \; x \neq \dfrac{3}{2}$

53. $-\dfrac{1}{3}, \; x \neq 5$ 55. $\dfrac{1}{a + 3}$ 57. $\dfrac{x}{x - 7}$

59. $\dfrac{y(y + 2)}{y + 6}, \; y \neq 2$ 61. $\dfrac{x(x + 2)}{x - 3}, \; x \neq 2$

63. $-\dfrac{3x + 5}{x + 3}, \; x \neq 4$ 65. $\dfrac{x + 8}{x - 3}, \; x \neq -\dfrac{3}{2}$

67. $\dfrac{3x - 1}{5x - 4}, \; x \neq -\dfrac{4}{5}$ 69. $\dfrac{3y^2}{y^2 + 1}, \; x \neq 0$

71. $\dfrac{y - 8x}{15}, \; y \neq -8x$ 73. $\dfrac{5 + 3xy}{y^2}, \; x \neq 0$

75. $\dfrac{u - 2v}{u - v}, \; u \neq -2v$

77. $\dfrac{3(m - 2n)}{m + 2n}$

79.

x	-2	-1	0	1	2	3	4
$\dfrac{x^2 - x - 2}{x - 2}$	-1	0	1	2	Undef.	4	5
$x + 1$	-1	0	1	2	3	4	5

$\dfrac{x^2 - x - 2}{x - 2} = \dfrac{(x - 2)(x + 1)}{x - 2} = x + 1, \; x \neq 2$

81. $\dfrac{x}{x + 3}, \; x > 0$ 83. $\dfrac{1}{4}, \; x > 0$

85. (a) $C = 2500 + 9.25x$

   (b) $\overline{C} = \dfrac{2500 + 9.25x}{x}$

   (c) $\{1, 2, 3, 4, \ldots\}$

87. (a) Van: $45(t + 3)$; Car: $60t$ (b) $d = |15(9 - t)|$

   (c) $\dfrac{4t}{3(t + 3)}$ (d) $34.25

89. π 91. $\dfrac{1531.1t + 9358}{1.33t + 54.6}$

93. Let u and v be polynomials. The algebraic expression u/v is a rational expression.

95. The rational expression is in simplified form if the numerator and denominator have no factors in common (other than ± 1).

97. You can divide out only common factors.

99. (a) The student forgot to divide each term of the numerator by the denominator.

   Correct solution:

   $\dfrac{3x^2 + 5x - 4}{x} = \dfrac{3x^2}{x} + \dfrac{5x}{x} - \dfrac{4}{x} = 3x + 5 - \dfrac{4}{x}$

   (b) The student incorrectly divided out; the denominator may not be split up.

   Correct solution:

   $\dfrac{x^2 + 7x}{x + 7} = \dfrac{x(x + 7)}{x + 7} = x$

Section 6.2   *(page 385)*

Review   *(page 385)*

1. $u^2 - v^2 = (u + v)(u - v)$

$9t^2 - 4 = (3t + 2)(3t - 2)$

2. $u^2 - 2uv + v^2 = (u - v)^2$

$4x^2 - 12x + 9 = (2x - 3)^2$

3. $u^3 + v^3 = (u + v)(u^2 - uv + v^2)$

$8x^3 + 64 = (2x + 4)(4x^2 - 8x + 16)$

4. $(3x - 2)(x + 5)$. Multiply the binomial factors to see whether you obtain the original expression.

5. $5x(1 - 4x)$ **6.** $(2 + x)(14 - x)$

7. $(3x - 5)(5x + 3)$ **8.** $(4t + 1)^2$

9. $(y - 4)(y^2 + 4y + 16)$

10. $(2x + 1)(4x^2 - 2x + 1)$

11.

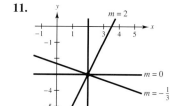

m is undefined.

12.

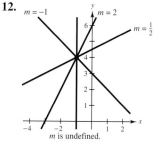

m is undefined.

1. x^2 **3.** $(x + 2)^2$ **5.** $u + 1$ **7.** $(-1)(2 + x)$

9. $\dfrac{9}{2}$ **11.** $\dfrac{s^3}{6}$, $s \neq 0$ **13.** $24u^2$, $u \neq 0$

15. 24, $x \neq -\dfrac{3}{4}$ **17.** $\dfrac{2uv(u + v)}{3(3u + v)}$, $u \neq 0$

19. -1, $r \neq 12$ **21.** $-\dfrac{x + 8}{x^2}$, $x \neq \dfrac{3}{2}$

23. $4(r + 2)$, $r \neq 3$, $r \neq 2$ **25.** $2t + 5$, $t \neq 3$, $t \neq -2$

27. $\dfrac{xy(x + 2y)}{x - 2y}$ **29.** $\dfrac{(x - y)^2}{x + y}$, $x \neq -3y$

31. $\dfrac{(x - 1)(2x + 1)}{(3x - 2)(x + 2)}$, $x \neq \pm 5$, $x \neq -1$

33. $\dfrac{x^2(x^2 - 9)(2x + 5)(3x - 1)}{2(2x + 1)(2x + 3)(3 - 2x)}$, $x \neq 0$, $x \neq \dfrac{1}{2}$

35. $\dfrac{(x + 3)^2}{x}$, $x \neq 3$, $x \neq 4$ **37.** $\dfrac{4x}{3}$, $x \neq 0$ **39.** $\dfrac{6}{x}$

41. $\dfrac{3y^2}{2ux^2}$, $v \neq 0$ **43.** $\dfrac{3}{2(a + b)}$

45. $x^4y(x + 2y)$, $x \neq 0$, $y \neq 0$, $x \neq -2y$

47. $-\dfrac{y - 5}{4}$, $y \neq \pm 3$ **49.** $\dfrac{x - 4}{x - 5}$, $x \neq -6$, $x \neq -5$, $x \neq 3$

51. $\dfrac{x + 4}{3}$, $x \neq -2$, $x \neq 0$ **53.** $\dfrac{1}{4}$, $x \neq -1$, $x \neq 0$, $y \neq 0$

55. $\dfrac{(x + 1)(2x - 5)}{x}$, $x \neq -1$, $x \neq -5$, $x \neq -\dfrac{2}{3}$

57. $\dfrac{x^4}{(x^n + 1)^2}$, $x^n \neq -3$, $x^n \neq 3$, $x \neq 0$

59.

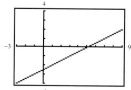

61. $\dfrac{2w^2 + 3w}{6}$ **63.** $\dfrac{x}{4(2x + 1)}$ **65.** $\dfrac{\pi x}{4(2x + 1)}$

67. (a) $\dfrac{1}{20}$ minute   (b) $\dfrac{x}{20}$ minutes   (c) $\dfrac{7}{4}$ minutes

69. Invert the divisor and multiply.

71. Invert the divisor, not the dividend.

Section 6.3   *(page 394)*

Review   *(page 394)*

1. (a) $y = \dfrac{3}{5}x + \dfrac{4}{5}$ (b) $y - 2 = \dfrac{3}{5}(x - 2)$

2. If the line rises from left to right, $m > 0$.
If the line falls from left to right, $m < 0$.

3. $42x^2 - 60x$ **4.** $6 + y - 2y^2$ **5.** $121 - x^2$

6. $16 - 25z^2$ **7.** $x^2 + 2x + 1$ **8.** $2t$

9. $x^3 - 8$ **10.** $2t^3 - 5t^2 - 12t$

11. $P = 12x + 6$; $A = 5x^2 + 9x$

12. $P = 12x$; $A = 6x^2$

1. $-\dfrac{x}{4}$ **3.** $-\dfrac{3}{a}$ **5.** $-\dfrac{2}{9}$ **7.** $\dfrac{2z^2 - 2}{3}$ **9.** $\dfrac{x + 6}{3x}$

11. $-\dfrac{4}{3}$ **13.** $1, \; y \neq 6$ **15.** $\dfrac{1}{x - 3}, \; x \neq 0$ **17.** $20x^3$

19. $36y^3$ **21.** $15x^2(x + 5)$ **23.** $126z^2(z + 1)^4$

25. $56t(t + 2)(t - 2)$ **27.** $6x(x + 2)(x - 2)$ **29.** x^2

31. $(u + 1)$ **33.** $-(x + 2)$ **35.** $\dfrac{2n^2(n + 8)}{6n^2(n - 4)}, \; \dfrac{10(n - 4)}{6n^2(n - 4)}$

37. $\dfrac{2(x + 3)}{x^2(x + 3)(x - 3)}, \; \dfrac{5x(x - 3)}{x^2(x + 3)(x - 3)}$

39. $\dfrac{3v^2}{6v^2(v + 1)}, \; \dfrac{8(v + 1)}{6v^2(v + 1)}$

41. $\dfrac{(x - 8)(x - 5)}{(x + 5)(x - 5)^2}, \; \dfrac{9x(x + 5)}{(x + 5)(x - 5)^2}$ **43.** $\dfrac{25 - 12x}{20x}$

45. $\dfrac{7(a + 2)}{a^2}$ **47.** $0, \; x \neq 4$ **49.** $\dfrac{3(x + 2)}{x - 8}$

51. $\dfrac{5(5x + 22)}{x + 4}$ **53.** $1, \; x \neq \dfrac{2}{3}$ **55.** $\dfrac{9x - 14}{2x(x - 2)}$

57. $\dfrac{x^2 - 7x - 15}{(x + 3)(x - 2)}$ **59.** $-\dfrac{2}{(x + 3)}, \; x \neq 3$

61. $\dfrac{5(x + 1)}{(x + 5)(x - 5)}$ **63.** $\dfrac{4}{x^2(x^2 + 1)}$

65. $\dfrac{x^2 + x + 9}{(x - 2)(x - 3)(x + 3)}$ **67.** $\dfrac{4x}{(x - 4)^2}$

69. $\dfrac{y - x}{xy}, \; x \neq -y$ **71.** $\dfrac{2(4x^2 + 5x - 3)}{x^2(x + 3)}$

73. $-\dfrac{u^2 - uv - 5u + 2v}{(u - v)^2}$ **75.** $\dfrac{x}{x - 1}, \; x \neq -6$

77.   **79.** $\dfrac{5t}{12}$

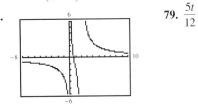

81. $A = -4, \; B = 2, \; C = 2$

83. $\dfrac{750.27t^2 + 5660.36t - 4827.2}{t(0.09t + 1.0)}$ (in thousands)

85. (a) Upstream: $\dfrac{10}{5 - x}$; Downstream: $\dfrac{10}{5 + x}$

(b) $f(x) = \dfrac{10}{5 - x} + \dfrac{10}{5 + x}$ (c) $f(x) = \dfrac{100}{(5 + x)(5 - x)}$

87. Rewrite each fraction in terms of the lowest common denominator, combine the numerators, and place the result over the lowest common denominator.

89. When the numerators are subtracted, the result should be $(x - 1) - (4x - 11) = x - 1 - 4x + 11.$

91. (a) $-\dfrac{7}{6}$ (b) $\dfrac{51}{7}$ (c) 8

Results are the same. Answers will vary.

Section 6.4   *(page 402)*

Review   *(page 402)*

1. Any expression with a zero exponent equals 1. Any expression with a negative exponent equals 1 divided by the expression.

2. The exponent is -6 since the decimal needs to be moved six positions to the right.

3. $\dfrac{14}{x}$ **4.** $\dfrac{1}{z^5}$ **5.** $\dfrac{a^5}{b^7}$ **6.** $x + 2$

7. Shifted 2 units down

8. Shifted 2 units to the right

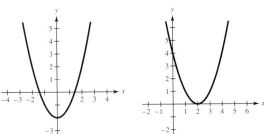

9. Reflected in x-axis

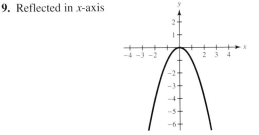

10. Reflected in the y-axis which produces an identical graph

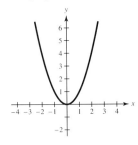

11. 5, 8, 20

12. Peanuts: 20 pounds; Almonds: 17.5 pounds; Pistachios: 12.5 pounds

1. $2x^2$, $x \neq 0$ **3.** $\dfrac{3x}{10}$, $x \neq 0$

5. $6xz^3$, $x \neq 0$, $y \neq 0$, $z \neq 0$ **7.** $\dfrac{2xy^2}{5}$, $x \neq 0$, $y \neq 0$

9. $-\dfrac{1}{y}$, $y \neq 3$ **11.** $-\dfrac{5x(x+1)}{2}$, $x \neq 0$, $x \neq 5$, $x \neq -1$

13. 2, $x \neq -1$, $x \neq 5$ **15.** $\dfrac{x+5}{3(x+4)}$, $x \neq 2$

17. $\dfrac{2(x+3)}{x-2}$, $x \neq 7$, $x \neq -3$

19. $\dfrac{(2x-5)(3x+1)}{3x(x+1)}$, $x \neq \pm\dfrac{1}{3}$

21. $\dfrac{(x+3)(4x+1)}{(3x-1)(x-1)}$, $x \neq -3$, $x \neq -\dfrac{1}{4}$

23. $x+2$, $x \neq \pm 2$, $x \neq -3$

25. $-\dfrac{(x-5)(x+2)^2}{(x+2)^2(x+3)}$, $x \neq -2$, $x \neq 7$ **27.** $\dfrac{y+3}{y^2}$

29. $\dfrac{x^2}{2(2x+3)}$, $x \neq 0$ **31.** $\dfrac{4+3x}{4-3x}$, $x \neq 0$

33. $\dfrac{3}{4}$, $x \neq 0$, $x \neq 3$ **35.** $\dfrac{5(x+3)}{2x(5x-2)}$

37. $y-x$, $x \neq 0$, $y \neq 0$, $x \neq -y$ **39.** $-\dfrac{(y-1)(y-3)}{y(4y-1)}$

41. $\dfrac{20}{7}$, $x \neq -1$ **43.** $\dfrac{1}{x}$, $x \neq -1$

45. $\dfrac{x(x+6)}{3x^3+10x-30}$, $x \neq 0$, $x \neq 3$ **47.** $\dfrac{y(2y^2-1)}{10y^2-1}$, $y \neq 0$

49. $\dfrac{x^2(7x^3+2)}{x^4+5}$, $x \neq 0$ **51.** $\dfrac{y+x}{y-x}$, $x \neq 0$, $y \neq 0$

53. $\dfrac{y-x}{(y+x)(x^2y^2)}$ **55.** $-\dfrac{1}{2(h+2)}$ **57.** $11x/60$

59. $11x/24$ **61.** $(b^2+5b+8)/(8b)$

63. $x/8$, $(5x)/36$, $(11x)/72$ **65.** $(R_1R_2)/(R_1+R_2)$

67. (a)

(b) $[250(1382.16t+5847.9)]/[3(4568.33t+1042.7)]$

69. A complex fraction is a fraction with a fraction in its numerator or denominator, or both.

$$\dfrac{\left(\dfrac{x+1}{2}\right)}{\left(\dfrac{x+1}{3}\right)}$$

Simplify by inverting the denominator and multiplying:

$$\left(\dfrac{x+1}{2}\right)\cdot\left(\dfrac{3}{x+1}\right)=\dfrac{3}{2}, \quad x \neq -1.$$

71. (a) Numerator: $\left(\dfrac{x-1}{5}\right)$;

Denominator: $\left(\dfrac{2}{x^2+2x-35}\right)$

(b) Numerator: $\left(\dfrac{1}{2y}+x\right)$;   Denominator: $\left(\dfrac{3}{y}+x\right)$

Mid-Chapter Quiz   (page 406)

1. $(-\infty, 0) \cup (0, 4) \cup (4, \infty)$

2. (a) 0 (b) $\dfrac{9}{2}$ (c) Undefined   (d) $\dfrac{8}{9}$

3. $\dfrac{3}{2}y$, $y \neq 0$ **4.** $\dfrac{2u^2}{9v}$, $u \neq 0$ **5.** $-\dfrac{2x+1}{x}$, $x \neq \dfrac{1}{2}$

6. $\dfrac{z+3}{2z-1}$, $z \neq -3$ **7.** $\dfrac{7+3ab}{a}$, $b \neq 0$

8. $\dfrac{n^2}{m+n}$, $2m-n \neq 0$ **9.** $\dfrac{t}{2}$, $t \neq 0$

10. $\dfrac{5x}{x-2}$, $x \neq -2$ **11.** $\dfrac{8x}{3(x-1)(x+3)(x-1)}$

12. $\dfrac{32x^7}{35zy^2}$, $x \neq 0$ **13.** $\dfrac{(a+1)^2}{9(a+b)^2}$, $a \neq b$

14. $\dfrac{4(u-v)^2}{5uv}$, $u \neq \pm v$ **15.** $\dfrac{7x-11}{x-2}$

16. $-\dfrac{4x^2-25x+36}{(x-3)(x+3)}$ **17.** 0, $x \neq 2$, $x \neq -1$

18. $-\dfrac{3t}{2}$, $t \neq 3$ **19.** $\dfrac{2(x+1)}{3x}$, $x \neq -2$, $x \neq -1$

20. $\dfrac{(3y-x)(x-y)}{xy}$, $x \neq y$

21. (a) $C = 25{,}000 + 144x$ (b) $\overline{C} = \dfrac{25{,}000+144x}{x}$

(c) $\$194$

22. $\dfrac{11x}{18}$

Section 6.5   *(page 413)*

Review   *(page 413)*

1. $\dfrac{120y}{90} = \dfrac{30 \cdot 4y}{30 \cdot 3} = \dfrac{4y}{3}$

2. $(2n + 1)(2n + 3) = 4n^2 + 8n + 3$

3. $(2n + 1) + (2n + 3) = 4n + 4$

4. $2n(2n + 2) = 4n^2 + 4n$

5. $\frac{3}{4}$ 6. $\frac{5}{2}$ 7. $\pm\frac{5}{2}$ 8. $0, 8$

9. $-7, 6$ 10. 5

11. $y = 1500 + 0.12x$ 12. $N = 3500 + 60t$

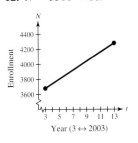

1. $7x^2 - 2x,\ x \neq 0$ 3. $-4x + 2,\ x \neq 0$

5. $m^3 + 2m - \dfrac{7}{m}$ 7. $-10z^2 - 6,\ z \neq 0$

9. $4z^2 + \frac{3}{2}z - 1,\ z \neq 0$ 11. $x^3 - \frac{3}{2}x^2 + 3x - 2,\ x \neq 0$

13. $\frac{5}{2}x - 4 + \frac{7}{2}y,\ x \neq 0,\ y \neq 0$ 15. $x - 5,\ x \neq 3$

17. $x + 10,\ x \neq -5$ 19. $x - 3 + \dfrac{2}{x - 2}$

21. $x + 7,\ x \neq 3$ 23. $5x - 8 + \dfrac{19}{x + 2}$

25. $4x + 3 - \dfrac{11}{3x + 2}$ 27. $6t - 5,\ t \neq \dfrac{5}{2}$

29. $y + 3,\ y \neq -\frac{1}{2}$ 31. $x^2 + 4,\ x \neq 2$

33. $3x^2 - 3x + 1 + \dfrac{2}{3x + 2}$ 35. $2 + \dfrac{5}{x + 2}$

37. $x - 4 + \dfrac{32}{x + 4}$ 39. $\dfrac{6}{5}z + \dfrac{41}{25} + \dfrac{41}{25(5z - 1)}$

41. $4x - 1,\ x \neq -\dfrac{1}{4}$ 43. $x^2 - 5x + 25,\ x \neq -5$

45. $x + 2$ 47. $4x^2 + 12x + 25 + \dfrac{52x - 55}{x^2 - 3x + 2}$

49. $x^5 + x^4 + x^3 + x^2 + x + 1,\ x \neq 1$

51. $x^3 - x + \dfrac{x}{x^2 + 1}$ 53. $2x,\ x \neq 0$

55. $7uv,\ u \neq 0,\ v \neq 0$ 57. $x + 3,\ x \neq 2$

59. $x^2 - x + 4 - \dfrac{17}{x + 4}$

61. $x^3 - 2x^2 - 4x - 7 - \dfrac{4}{x - 2}$

63. $5x^2 + 14x + 56 + \dfrac{232}{x - 4}$

65. $10x^3 + 10x^2 + 60x + 360 + \dfrac{1360}{x - 6}$

67. $0.1x + 0.82 + \dfrac{1.164}{x - 0.2}$ 69. $(x - 3)(x - 1)(x + 4)$

71. $(x - 1)(2x - 1)(3x - 2)$ 73. $(x + 3)^2(x - 3)(x + 4)$

75. $5\left(x - \frac{4}{5}\right)(3x + 2)$ 77. -8

79.

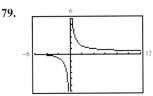

81. $x^{2n} + x^n + 4,\ x^n \neq -2$ 83. $x^3 - 5x^2 - 5x - 10$

85. $f(k)$ equals the remainder when dividing by $(x - k)$.

k	$f(k)$	Divisor $(x - k)$	Remainder
-2	-8	$x + 2$	-8
-1	0	$x + 1$	0
0	0	x	0
$\frac{1}{2}$	$-\frac{9}{8}$	$x - \frac{1}{2}$	$-\frac{9}{8}$
1	-2	$x - 1$	-2
2	0	$x - 2$	0

87. $x^2 - 3$ 89. $2x + 8$

91. x is not a factor of the numerator.

93. The remainder is 0 and the divisor is a factor of the dividend.

95. True. If $\dfrac{n(x)}{d(x)} = q(x)$, then $n(x) = d(x) \cdot q(x)$.

97.

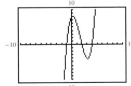

The polynomials in parts (a), (b), and (c) are all equivalent. The x-intercepts are $(-1, 0)$, $(2, 0)$, and $(4, 0)$.

Section 6.6   *(page 422)*

Review   *(page 422)*

1. Quadrant II or III   2. Quadrant I or II

3. x-axis   4. $(9, -6)$ 5. $x < \frac{3}{2}$ 6. $x < 5$

7. $1 < x < 5$ 8. $x < 2$ or $x > 8$

9. $x \leq -8$ or $x \geq 16$ 10. $-24 \leq x \leq 36$

11. 15 minutes, 2 miles

12. 7.5%: \$15,000; 9%: \$9000

1. (a) Not a solution   (b) Not a solution
   (c) Not a solution   (d) Solution

3. (a) Not a solution   (b) Solution
   (c) Solution   (d) Not a solution

5. 10   7. 1   9. 0   11. 8   13. $-\frac{9}{32}$

15. $-3, \frac{8}{3}$ 17. $-\frac{2}{9}$ 19. $\frac{7}{4}$ 21. $\frac{43}{8}$ 23. 61

25. $\frac{18}{5}$ 27. $-\frac{26}{5}$ 29. 3   31. 3   33. $-\frac{11}{5}$

35. $\frac{4}{3}$ 37. ± 6 39. ± 4 41. $-9, 8$ 43. 3, 13

45. No solution   47. -5 49. 8   51. 3   53. 5

55. $-\frac{11}{10}, 2$ 57. 20   59. $\frac{3}{2}$ 61. $3, -1$ 63. $\frac{17}{4}$

65. 2, 3   67. (a) and (b) $(-2, 0)$

69. (a) and (b) $(-1, 0), (1, 0)$

71. (a)   (b) $(4, 0)$

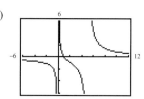

73. (a)   (b) $(1, 0)$

75. (a)   (b) $(-3, 0), (2, 0)$

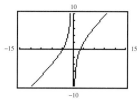

77. -12 79. $\dfrac{x^2 + 2x + 8}{(x + 4)(x - 4)}$ 81. $8, \dfrac{1}{8}$

83. 40 miles per hour

85. 8 miles per hour; 10 miles per hour

87. (d) 3 miles per hour, obtained by solving

$$\frac{10}{5 - x} + \frac{10}{5 + x} = 6.25$$

(e) Yes. When $x = 4$, the time for the entire trip is

$$f(4) = \frac{10}{5 - 4} + \frac{10}{5 + 4} \approx 11.1 \text{ hours.}$$

Because the result is less than 12 hours, you will be able to make the trip.

89. An extraneous solution is an extra solution found by multiplying both sides of the original equation by an expression containing the variable. It is identified by checking all solutions in the original equation.

91. When the equation involves only two fractions, one on each side of the equation, the equation can be solved by cross-multiplication.

Section 6.7   *(page 433)*

Review   *(page 433)*

1. $(-\infty, \infty)$ 2. $(-\infty, 0) \cup (0, \infty)$

3.   Yes, the graphs are the same.

4. Answers will vary.   5. Answers will vary.

6. Answers will vary.   7. $h + 4, \ h \neq 0$

8. $-\dfrac{3}{7(h + 7)}, \ h \neq 0$ 9. $C = 12{,}000 + 5.75x$

10. $P = 5w$

1. $I = kV$ 3. $V = kt$ 5. $u = kv^2$ 7. $p = k/d$

9. $A = k/t^4$ 11. $A = klw$ 13. $P = k/V$

15. Area varies jointly as the base and the height.

17. Volume varies jointly as the square of the radius and the height.

19. Average speed varies directly as the distance and inversely as the time.

21. $s = 5t$ 23. $F = \frac{5}{16}x^2$ 25. $n = 48/m$

27. $g = 4/\sqrt{z}$ **29.** $F = \frac{25}{6}xy$ **31.** $d = (120x^2)/r$

33. 4 miles per hour   **35.** 10 people   **37.** 9 hours

39. 15 hours; $22\frac{1}{2}$ hours

41. (a) $\{1, 2, 3, 4, \ldots\}$

(b)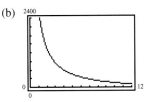

(c) 10d

(d) $153 = -139.1 + 2921/x$

$292.1 = 2921/x$

$x = 10$

43. \$4921.25; Price per unit

45. (a) 2 inches   (b) 15 pounds   **47.** 18 pounds

49. 32 feet per second per second   **51.** $208\frac{1}{3}$ feet

53. 3072 watts   **55.** 100

57. 0.36 pounds per square inch; 116 pounds

59. $T = \dfrac{4000}{d}, 0.91°C$

61.

x	2	4	6	8	10
$y = kx^2$	4	16	36	64	100

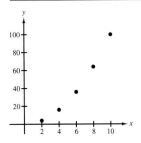

63.

x	2	4	6	8	10
$y = kx^2$	2	8	18	32	50

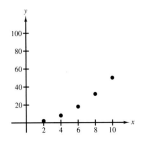

65.

x	2	4	6	8	10
$y = \dfrac{k}{x^2}$	5	$\dfrac{5}{4}$	$\dfrac{5}{9}$	$\dfrac{5}{16}$	$\dfrac{1}{5}$

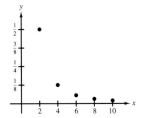

67.

x	2	4	6	8	10
$y = \dfrac{k}{x^2}$	$\dfrac{5}{4}$	$\dfrac{5}{16}$	$\dfrac{5}{36}$	$\dfrac{5}{64}$	$\dfrac{1}{20}$

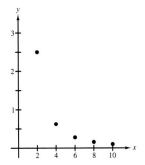

69. $y = k/x$ with $k = 4$

71. Increase. Because $y = kx$ and $k > 0$, the variables increase or decrease together.

73. The variable y will quadruple. If $y = kx^2$ and x is replaced with $2x$, you have $y = k(2x)^2 = 4kx^2$.

75. Answers will vary.

Review Exercises   *(page 439)*

1. $(-\infty, 8) \cup (8, \infty)$ **3.** $(-\infty, 1) \cup (1, 6) \cup (6, \infty)$

5. $(0, \infty)$ **7.** $\dfrac{2x^3}{5}$, $x \neq 0$, $y \neq 0$ **9.** $\dfrac{b - 3}{6(b - 4)}$

11. -9, $x \neq y$ **13.** $\dfrac{x}{2(x + 5)}$, $x \neq 5$ **15.** $3x^5y^2$

17. $\dfrac{y}{8x}$, $y \neq 0$ **19.** $12z(z - 6)$, $z \neq -6$

21. $-\frac{1}{4}$, $u \neq 0$, $u \neq 3$ **23.** $\frac{8}{5}x^3$, $x \neq 0$

25. $\dfrac{125y}{x}$, $y \neq 0$ **27.** $\dfrac{1}{3x - 2}$, $x \neq -2$, $x \neq -1$

29. $\dfrac{x(x - 1)}{x - 7}$, $x \neq -1$, $x \neq 1$ **31.** $3x$

33. $\dfrac{4}{x}$ **35.** $\dfrac{5y + 11}{2y + 1}$ **37.** $\dfrac{7x - 16}{x + 2}$

39. $\dfrac{4x + 3}{(x + 5)(x - 12)}$ **41.** $\dfrac{5x^3 - 5x^2 - 31x + 13}{(x - 3)(x + 2)}$

43. $\dfrac{2x + 17}{(x - 5)(x + 4)}$ **45.** $\dfrac{6(x - 9)}{(x + 3)^2(x - 3)}$

47.

49. $3x^2,\ x \neq 0$ **51.** $\dfrac{6(x + 5)}{x(x + 7)},\ x \neq \pm 5$

53. $\dfrac{3t^2}{5t - 2},\ t \neq 0,\ t \neq \dfrac{2}{5}$ **55.** $x - 1,\ x \neq 0,\ x \neq 2$

57. $\dfrac{-a^2 + a + 16}{(4a^2 + 16a + 1)(a - 4)},\ a \neq 0,\ a \neq -4$

59. $2x^2 - \dfrac{1}{2},\ x \neq 0$ **61.** $3xy - y + 1,\ x \neq 0,\ y \neq 0$

63. $2x^2 + \dfrac{4}{3}x - \dfrac{8}{9} + \dfrac{10}{9(3x - 1)}$ **65.** $x^2 - 2,\ x \neq \pm 1$

67. $x^2 - x - 3 - \dfrac{3x^2 - 2x - 3}{x^3 - 2x^2 + x - 1}$

69. $x^2 + 5x - 7,\ x \neq -2$

71. $x^3 + 3x^2 + 6x + 18 + \dfrac{29}{x - 3}$

73. $(x - 2)(x + 1)(x + 3)$ **75.** -120 **77.** $\dfrac{36}{23}$

79. 5 **81.** $-4, 6$ **83.** $-\dfrac{16}{3}, 3$ **85.** $-\dfrac{5}{2}, 1$

87. $-2, 2$ **89.** $-\dfrac{9}{5}, 3$ **91.** 56 miles per hour

93. 4 people   **95.** 8 years   **97.** 150 pounds

99. 2.44 hours   **101.** \$922.50

Chapter Test   *(page 443)*

1. $(-\infty, 2) \cup (2, 3) \cup (3, \infty)$ **2.** $-\dfrac{1}{3},\ x \neq 2$

3. $\dfrac{2a + 3}{5},\ a \neq 4$ **4.** $3x^3(x + 4)^2$ **5.** $\dfrac{5z}{3},\ z \neq 0$

6. $\dfrac{4}{y + 4},\ y \neq 2$ **7.** $(2x - 3)^2(x + 1),\ x \neq -\dfrac{3}{2},\ x \neq -1$

8. $\dfrac{14y^6}{15},\ x \neq 0$ **9.** $\dfrac{-2x^2 + 2x + 1}{x + 1}$

10. $\dfrac{5x^2 - 15x - 2}{(x - 3)(x + 2)}$ **11.** $\dfrac{5x^3 + x^2 - 7x - 5}{x^2(x + 1)^2}$

12. $4,\ x \neq -1$ **13.** $\dfrac{x^3}{4},\ x \neq 0,\ x \neq -2$

14. $-(3x + 1),\ x \neq 0,\ x \neq \dfrac{1}{3}$

15. $\dfrac{(3y + x^2)(x + y)}{x^2 y},\ x \neq -y$ **16.** $2x + 1 - \dfrac{7}{2x}$

17. $t^2 + 3 - \dfrac{6t - 6}{t^2 - 2}$

18. $2x^3 + 6x^2 + 3x + 9 + \dfrac{20}{x - 3}$ **19.** 22

20. $-1, -\dfrac{15}{2}$ **21.** No solution   **22.** $V = \dfrac{1}{4}\sqrt{u}$

23. 240 cubic meters

Chapter 7

Section 7.1   *(page 453)*

> ### Review   *(page 453)*
>
> **1.** a^{m+n} **2.** $a^m b^m$ **3.** a^{mn} **4.** a^{m-n}
>
> **5.** $y = 4 - 3x$ **6.** $y = \dfrac{2}{3}(1 - x)$
>
> **7.** $y = \dfrac{1}{7}(x - 7)$ **8.** $y = \dfrac{3}{2}(8x - 1)$ **9.** $-\dfrac{5}{3}, 0$
>
> **10.** $0, 10$ **11.** $-4, -2$ **12.** $-6, 7$

1. 8 **3.** -7 **5.** -3 **7.** Not a real number

9. Perfect square   **11.** Perfect cube   **13.** Neither

15. Irrational   **17.** Rational   **19.** 8 **21.** 10

23. Not a real number   **25.** $-\dfrac{2}{3}$ **27.** Not a real number

29. 5 **31.** -23 **33.** 5 **35.** 10 **37.** 6

39. $-\dfrac{1}{4}$ **41.** 11 **43.** -24 **45.** 3

47. Not a real number

Radical Form	*Rational Exponent Form*
49. $\sqrt{16} = 4$	$16^{1/2} = 4$
51. $\sqrt[3]{125} = 5$	$125^{1/3} = 5$

53. 5 **55.** -6 **57.** $\dfrac{1}{4}$ **59.** $\dfrac{1}{9}$ **61.** $\dfrac{4}{9}$ **63.** $\dfrac{3}{11}$

65. 9 **67.** -64 **69.** $t^{1/2}$ **71.** x^3 **73.** $u^{7/3}$

75. $x^{-1} = \dfrac{1}{x}$ **77.** $t^{-9/4} = \dfrac{1}{t^{9/4}}$ **79.** x^3 **81.** $y^{13/12}$

83. $x^{3/4} y^{1/4}$ **85.** $y^{5/2} z^4$ **87.** 3 **89.** $\sqrt[3]{2}$ **91.** $\dfrac{1}{2}$

93. $\sqrt{c}$ **95.** $\dfrac{3y^2}{4z^{4/3}}$ **97.** $\dfrac{9y^{3/2}}{x^{2/3}}$ **99.** $x^{1/4}$

101. $\sqrt[8]{y}$ **103.** $x^{3/8}$ **105.** $\sqrt{x + y}$

107. $\dfrac{1}{(3u - 2v)^{5/6}}$ **109.** 6.7082 **111.** 9.9845

113. 0.0038 **115.** 3.8158 **117.** 66.7213

119. 1.0420　　**121.** 0.7915

123. (a) 3　(b) 5　(c) Not a real number　(d) 9

125. (a) 2　(b) 3　(c) -2　(d) -4

127. (a) 2　(b) Not a real number　(c) 3　(d) 1

129. $[0, \infty)$　　**131.** $(0, \infty)$　　**133.** $(-\infty, \infty)$

135. $\left[-\frac{7}{3}, \infty\right)$　　**137.** $\left(-\infty, \frac{4}{9}\right]$

139. Domain: $(0, \infty)$　　**141.** Domain: $(-\infty, \infty)$

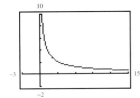　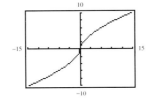

143. $2x^{3/2} - 3x^{1/2}$　　**145.** $1 + 5y$　　**147.** 0.128

149. 23 feet $\times$ 23 feet　　**151.** 10.49 centimeters

153. (a) 15 feet　(b) $h = \sqrt{15^2 - \left(\frac{a}{2}\right)^2}$

(c) 8.29 feet　(d) 25.38 feet

155. Given $\sqrt[n]{a}$, a is the radicand and n is the index.

157. No. $\sqrt{2}$ is an irrational number. Its decimal representation is a nonterminating, nonrepeating decimal.

159. 0, 1, 4, 5, 6, 9; Yes

Section 7.2　*(page 462)*

Review　*(page 462)*

1. Replace the inequality sign with an equal sign and sketch the graph of the resulting equation. (Use a dashed line for < or >, and a solid line for ≤ or ≥.) Test one point in each of the half-planes formed by the graph. If the point satisfies the inequality, shade the entire half-plane to denote that every point in the region satisfies the inequality.

2. The first includes the points on the line $3x + 4y = 4$, whereas the second does not.

3. $-(x - 3)(x^2 + 1)$　　**4.** $(2t + 13)(2t - 13)$

5. $(x - 1)(x - 2)$　　**6.** $(x - 1)(2x + 7)$

7. $(x + 1)(11x - 5)$　　**8.** $(2x - 7)^2$

9. 816 adults; 384 students　　**10.** 267 units

1. $2\sqrt{5}$　　**3.** $5\sqrt{2}$　　**5.** $4\sqrt{6}$　　**7.** $6\sqrt{6}$　　**9.** $13\sqrt{7}$

11. 0.2　　**13.** $0.06\sqrt{2}$　　**15.** $1.1\sqrt{2}$　　**17.** $\frac{\sqrt{13}}{5}$

19. $3x^2\sqrt{x}$　　**21.** $4y^2\sqrt{3}$　　**23.** $3\sqrt{13}|y^3|$

25. $2|x|y\sqrt{30y}$　　**27.** $8a^2b^3\sqrt{3ab}$　　**29.** $2\sqrt[3]{6}$

31. $2\sqrt[3]{14}$　　**33.** $2x\sqrt[3]{5x^2}$　　**35.** $3|y|\sqrt{2y}$　　**37.** $xy\sqrt[3]{x}$

39. $|x|\sqrt[4]{3y^2}$　　**41.** $2xy\sqrt[5]{y}$　　**43.** $\frac{\sqrt[3]{35}}{4}$　　**45.** $\frac{2\sqrt[5]{x^2}}{y}$

47. $\frac{3a\sqrt[3]{2a}}{b^3}$　　**49.** $\frac{4a^2\sqrt{2}}{|b|}$　　**51.** $3x^2$　　**53.** $\frac{\sqrt{3}}{3}$

55. $\frac{\sqrt{7}}{7}$　　**57.** $\frac{\sqrt[4]{20}}{2}$　　**59.** $\frac{3\sqrt[3]{2}}{2}$　　**61.** $\frac{\sqrt{y}}{y}$

63. $\frac{2\sqrt{x}}{x}$　　**65.** $\frac{\sqrt{2x}}{2x}$　　**67.** $\frac{2\sqrt{3b}}{b^2}$　　**69.** $\frac{\sqrt[3]{18xy^2}}{3y}$

71. $3\sqrt{5}$　　**73.** 89.44 cycles per second

75. $\sqrt{776} \approx 27.86$ feet　　**77.** 1

79. (a) All possible nth powered factors have been removed from each radical.

(b) No radical contains a fraction.

(c) No denominator of a fraction contains a radical.

81. When $x < 0$. $\sqrt{x^2}$ will always result in a positive value regardless of the sign of x. So, $\sqrt{x^2} \neq x$ when $x < 0$.

Example: $\sqrt{(-8)^2} = \sqrt{64} = 8$

Section 7.3　*(page 467)*

Review　*(page 467)*

1. An ordered pair (x, y) of real numbers that satisfies each equation in the system.

2. Yes, if the system is inconsistent.

3. Yes, if the system is dependent.

4. No, it must have one solution, no solution, or infinitely many solutions.

5.　　　　　　　　　**6.**

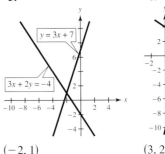

$(-2, 1)$　　　　　　　$(3, 2)$

7. $\left(-\frac{4}{5}, \frac{2}{5}\right)$　　**8.** No solution

9. Infinitely many solutions　　**10.** $(4, -8, 10)$

11. DVD: \$29; Videocassette tape: \$14

12. \$20 bills: 7; \$5 bills: 3; \$1 bills: 4

1. $2\sqrt{2}$ 3. $44\sqrt{2}$ 5. Cannot combine   7. $3\sqrt[3]{5}$
9. $13\sqrt[3]{y}$ 11. $14\sqrt[4]{s}$ 13. $9\sqrt{2}$
15. $-11\sqrt[4]{3} - 5\sqrt[4]{7}$ 17. $13\sqrt[3]{7} + \sqrt{3}$ 19. $21\sqrt{3}$
21. $23\sqrt{5}$ 23. $30\sqrt[3]{2}$ 25. $12\sqrt{x}$ 27. $13\sqrt{x+1}$
29. $13\sqrt{y}$ 31. $(10-z)\sqrt[3]{z}$ 33. $(6a+1)\sqrt{5a}$
35. $(x+2)\sqrt[3]{6x}$ 37. $4\sqrt{x-1}$ 39. $(x+2)\sqrt{x-1}$
41. $5a\sqrt[3]{ab^2}$ 43. $3r^3s^2\sqrt{rs}$ 45. 0   47. $\dfrac{2\sqrt{5}}{5}$
49. $\dfrac{9\sqrt{5}}{5}$ 51. $\dfrac{5\sqrt{3y}}{3}$ 53. $\dfrac{\sqrt{2x}(2x-3)}{2x}$
55. $\dfrac{\sqrt{7y}(7y^3-3)}{7y^2}$ 57. >   59. >   61. $12\sqrt{6x}$
63. $9x\sqrt{3} + 5\sqrt{3x}$
65. (a) $5\sqrt{10}$ feet (b) $400\sqrt{10} \approx 1264.9$ square feet
67. No; $\sqrt{2} + \sqrt{18} = \sqrt{2} + 3\sqrt{2} = 4\sqrt{2}$
69. No; $\sqrt{5} + \left(-\sqrt{5}\right) = 0$.
71. (a) The student combined terms with unlike radicands; can be simplified no further.

(b) The student combined terms with unlike indices; can be simplified no further.

Mid-Chapter Quiz (*page 470*)

1. 15   2. $\frac{3}{2}$ 3. 8   4. 9
5. (a) Not a real number   (b) 1   (c) 5
6. (a) 3   (b) 2   (c) Not a real number
7. $(-\infty, 0) \cup (0, \infty)$ 8. $\left[\frac{5}{4}, \infty\right)$ 9. $3|x|\sqrt{3}$
10. $3|x|\sqrt{x}$ 11. $\dfrac{2u\sqrt{u}}{3}$ 12. $\dfrac{2\sqrt[3]{2}}{u^2}$ 13. $5x|y|z^2\sqrt{5x}$
14. $4a^2b\sqrt[3]{2b^2}$ 15. $4\sqrt{3}$ 16. $\dfrac{2\sqrt{5x}}{x}$
17. $3\sqrt{3} - 4\sqrt{7}$ 18. $4\sqrt{2y}$ 19. $7\sqrt{3}$
20. $4\sqrt{x+2}$ 21. $6x\sqrt[3]{5x^2} + 4x\sqrt[3]{5x}$ 22. $4xy^2z^2\sqrt{xz}$
23. $23 + 8\sqrt{2}$ inches

Section 7.4 (*page 475*)

Review (*page 475*)

1. c   2. The signs are the same.
3. The signs are different.   4. b   5. $2x - y = 0$
6. $x + y - 6 = 0$ 7. $y - 3 = 0$
8. $x - 4 = 0$ 9. $6x + 11y - 96 = 0$
10. $x + y - 11 = 0$ 11. $\dfrac{360}{r}$ 12. $2L + 2\left(\dfrac{L}{3}\right)$

---

1. 4   3. $3\sqrt{2}$ 5. $2\sqrt[3]{9}$ 7. $2\sqrt[4]{3}$ 9. $3\sqrt{7} - 7$
11. $2\sqrt{10} + 8\sqrt{2}$ 13. $3\sqrt{2}$ 15. $4\sqrt{6} - 4\sqrt{10}$
17. $y + 4\sqrt{y}$ 19. $4\sqrt{a} - a$ 21. $2 - 7\sqrt[3]{4}$
23. -1 25. $\sqrt{15} - 5\sqrt{5} + 3\sqrt{3} - 15$
27. $8\sqrt{5} + 24$ 29. $2\sqrt[3]{3} + 3\sqrt[3]{6} - 3\sqrt[3]{4} - 9$
31. $2x + 20\sqrt{2x} + 100$ 33. $45x - 17\sqrt{x} - 6$
35. $9x - 25$ 37. $\sqrt[3]{4x^2} + 10\sqrt[3]{2x} + 25$
39. $y - 5\sqrt[3]{y} + 2\sqrt[3]{y^2} - 10$ 41. $t + 5\sqrt[3]{t^2} + \sqrt[3]{t} - 3$
43. $x^2y^2\left(2y - x\sqrt{y}\right)$ 45. $4xy^3\left(x^4\sqrt[3]{x} + y\sqrt[3]{2x^2y^2}\right)$
47. $x + 3$ 49. $4 - 3x$ 51. $2u + \sqrt{2u}$
53. $2 - \sqrt{5}, -1$ 55. $\sqrt{11} + \sqrt{3}, 8$
57. $\sqrt{15} - 3, 6$ 59. $\sqrt{x} + 3, x - 9$
61. $\sqrt{2u} + \sqrt{3}, 2u - 3$ 63. $2\sqrt{2} - \sqrt{4}, 4$
65. $\sqrt{x} - \sqrt{y}, x - y$ 67. (a) $2\sqrt{3} - 4$ (b) 0
69. (a) 0   (b) -1 71. $\dfrac{6\left(\sqrt{11} + 2\right)}{7}$
73. $\dfrac{7\left(5 - \sqrt{3}\right)}{22}$ 75. $\dfrac{5 + 2\sqrt{10}}{5}$ 77. $\dfrac{\sqrt{6} - \sqrt{2}}{2}$
79. $-\dfrac{9\left(\sqrt{3} + \sqrt{7}\right)}{4}$ 81. $\dfrac{4\sqrt{7} + 11}{3}$
83. $\dfrac{2x - 9\sqrt{x} - 5}{4x - 1}$ 85. $\dfrac{\left(\sqrt{15} + \sqrt{3}\right)x}{4}$
87. $\dfrac{2t^2\left(\sqrt{5} + \sqrt{t}\right)}{5 - t}$ 89. $4\left(\sqrt{3a} - \sqrt{a}\right), a \neq 0$
91. $\dfrac{3(x - 4)\left(x^2 + \sqrt{x}\right)}{x(x - 1)(x^2 + x + 1)}$
93. $-\dfrac{\sqrt{u + v}\left(\sqrt{u - v} + \sqrt{u}\right)}{v}$
95. 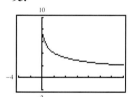   97.
99. $\dfrac{2}{7\sqrt{2}}$ 101. $\dfrac{5}{\sqrt{35x}}$ 103. $\dfrac{4}{5\left(\sqrt{7} - \sqrt{3}\right)}$
105. $\dfrac{y - 25}{\sqrt{3}\left(\sqrt{y} + 5\right)}$ 107. $192\sqrt{2}$ square inches
109. $\dfrac{500k\sqrt{k^2 + 1}}{k^2 + 1}$

111. $\sqrt{3}\left(1 - \sqrt{6}\right)$

$\quad = \sqrt{3} - \sqrt{3} \cdot \sqrt{6}$ Distributive Property

$\quad = \sqrt{3} - \sqrt{9 \cdot 2}$

$\quad = \sqrt{3} - 3\sqrt{2}$ Simplify radicals.

113. $\left(3 - \sqrt{2}\right)\left(3 + \sqrt{2}\right) = 9 - 2 = 7$

Multiplying the number by its conjugate yields the difference of two squares. Squaring a square root eliminates the radical.

Section 7.5   *(page 485)*

Review   *(page 485)*

1. The function is undefined when the denominator is zero. The domain is all real numbers x such that $x \neq -2$ and $x \neq 3$.

2. $\dfrac{2x^2 + 5x - 3}{x^2 - 9}$ is undefined if $x = -3$.

3. $36x^5y^8$ **4.** 1   **5.** $4rs^2$ **6.** $\dfrac{9x^2}{16y^6}$

7. $-\dfrac{x + 13}{5x^2}, \ x \neq 3$ **8.** $\dfrac{x^2 - 4}{25(x^2 - 9)}$

9. $\dfrac{2x + 5}{x - 5}$ **10.** $-\dfrac{5x - 8}{x - 1}$

11.   **12.**

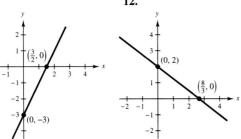

1. (a) Not a solution   (b) Not a solution

(c) Not a solution   (d) Solution

3. (a) Not a solution   (b) Solution

(c) Not a solution   (d) Not a solution

5. 144   **7.** 49   **9.** 27   **11.** 49   **13.** No solution

15. 64   **17.** 90   **19.** -27 **21.** $\frac{4}{5}$ **23.** 5

25. No solution   **27.** $\frac{44}{3}$ **29.** $\frac{14}{25}$ **31.** 4

33. No solution   **35.** 7   **37.** -15 **39.** $-\frac{9}{4}$

41. 8   **43.** 1, 3   **45.** 1   **47.** $\frac{1}{4}$ **49.** $\frac{1}{2}$ **51.** 4

53. 7   **55.** 4   **57.** 216   **59.** 4, -12 **61.** -16

63.   **65.**

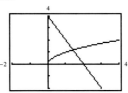   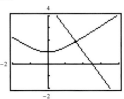

1.407   1.569

67.   **69.**

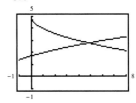   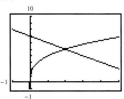

4.840   1.978

71.

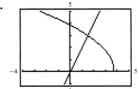

; 1.500

73. 25   **75.** 2, 6   **77.** 9.00   **79.** 12.00

81.

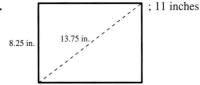

; 11 inches

83. $2\sqrt{10} \approx 6.32$ meters   **85.** 15 feet

87. 30 inches $\times$ 16 inches

89. $h = \dfrac{\sqrt{S^2 - \pi^2 r^4}}{\pi r}$; 34 centimeters

91. 64 feet   **93.** 56.57 feet per second

95. 56.25 feet   **97.** 1.82 feet   **99.** 500 units

101. (a)   (b) 1999

103. (e) \$12,708.73

105. No. It is not an operation that necessarily yields an equivalent equation. There may be extraneous solutions.

107. $\left(\sqrt{x} + \sqrt{6}\right)^2 \neq \left(\sqrt{x}\right)^2 + \left(\sqrt{6}\right)^2$

Section 7.6 *(page 495)*

Review *(page 495)*

1. Use the rule $\dfrac{u}{v} \cdot \dfrac{w}{z} = \dfrac{uw}{vz}$. That is, you multiply the numerators, multiply the denominators, and write the new fraction in simplified form.

2. Use the rule $\dfrac{u}{v} \div \dfrac{w}{z} = \dfrac{u}{v} \cdot \dfrac{z}{w}$. That is, you invert the divisor and multiply.

3. Rewrite the fractions so they have common denominators and then use the rule
$$\frac{u}{w} + \frac{v}{w} = \frac{u+v}{w}.$$

4. $\dfrac{t-5}{5-t} = \dfrac{-1(5-t)}{5-t} = -1$

5. $\dfrac{x}{5}$, $x \neq 0$, $x \neq -\dfrac{3}{2}$ 6. $\dfrac{x}{5(x+y)}$, $x \neq 0$, $x \neq y$

7. $\dfrac{9}{2(x+3)}$, $x \neq 0$ 8. $\dfrac{1}{x-2}$, $x \neq 0$, $x \neq -2$

9. $\dfrac{x^2 + 2x - 13}{x(x-2)}$, $x \neq \pm 3$

10. $\dfrac{(x+1)(x+3)}{3}$, $x \neq -1$ 11. $\dfrac{7x}{9}, \dfrac{19x}{18}$

12. $\dfrac{C_1 C_2}{C_1 + C_2}$

---

1. $2i$ 3. $-12i$ 5. $\dfrac{2}{5}i$ 7. $0.3i$ 9. $2\sqrt{2}\,i$

11. $\sqrt{7}\,i$ 13. 2 15. $\dfrac{3\sqrt{2}}{8}i$ 17. $10i$ 19. $3\sqrt{2}\,i$

21. $3\sqrt{3}\,i$ 23. -4 25. $-3\sqrt{6}$ 27. -0.44

29. $-2\sqrt{3} - 3$ 31. $5\sqrt{2} - 4\sqrt{5}$ 33. $4 + 3\sqrt{2}\,i$

35. -16 37. $-8i$ 39. $a = 3, b = -4$

41. $a = 2, b = -3$ 43. $a = -4, b = -2\sqrt{2}$

45. $a = 2, b = -2$ 47. $10 + 4i$ 49. $-14 - 40i$

51. $-14 + 20i$ 53. $9 - 7i$ 55. $3 + 6i$

57. $\dfrac{13}{6} + \dfrac{3}{2}i$ 59. $-3 + 49i$ 61. -36 63. 24

65. $-36i$ 67. $27i$ 69. -9 71. $-65 - 10i$

73. $20 - 12i$ 75. $4 + 18i$ 77. $-40 - 5i$

79. $-14 + 42i$ 81. 9 83. $-7 - 24i$

85. $-21 + 20i$ 87. $2 + 11i$ 89. $-i$ 91. 1

93. -1 95. i 97. -1 99. 5 101. 68

103. 31 105. 100 107. 4 109. 2.5

111. $-10i$ 113. $-\dfrac{1}{5} + \dfrac{2}{5}i$ 115. $2 + 2i$

117. $1 - 2i$ 119. $-\dfrac{24}{53} + \dfrac{84}{53}i$ 121. $\dfrac{6}{29} + \dfrac{15}{29}i$

123. $-\dfrac{23}{58} + \dfrac{43}{58}i$ 125. $\dfrac{9}{5} - \dfrac{2}{5}i$ 127. $\dfrac{47}{26} + \dfrac{27}{26}i$

129. $\dfrac{14}{29} - \dfrac{35}{29}i$ 131–133. (a) Solution and (b) Solution

135. (a) $\left(\dfrac{-5 + 5\sqrt{3}\,i}{2}\right)^3 = 125$

    (b) $\left(\dfrac{-5 - 5\sqrt{3}\,i}{2}\right)^3 = 125$

137. (a) $1, \dfrac{-1 + \sqrt{3}\,i}{2}, \dfrac{-1 - \sqrt{3}\,i}{2}$

    (b) $2, \dfrac{-2 + 2\sqrt{3}\,i}{2} = -1 + \sqrt{3}\,i,$

    $\dfrac{-2 - 2\sqrt{3}\,i}{2} = -1 - \sqrt{3}\,i$

    (c) $4, \dfrac{-4 + 4\sqrt{3}\,i}{2} = -2 + 2\sqrt{3}\,i,$

    $\dfrac{-4 - 4\sqrt{3}\,i}{2} = -2 - 2\sqrt{3}\,i$

139. $2a$ 141. $2bi$ 143. $i = \sqrt{-1}$

145. $\sqrt{-3}\sqrt{-3} = \left(\sqrt{3}\,i\right)\left(\sqrt{3}\,i\right) = 3i^2 = -3$

147. $x^2 + 1 = (x + i)(x - i)$

Review Exercises *(page 499)*

1. 7 3. -9 5. -2 7. -4 9. $\dfrac{5}{6}$ 11. $-\dfrac{1}{5}$

13. Not a real number

Radical Form	*Rational Exponent Form*
15. $\sqrt{49} = 7$	$49^{1/2} = 7$
17. $\sqrt[3]{216} = 6$	$216^{1/3} = 6$

19. 81 21. -125 23. $\dfrac{1}{16}$ 25. $x^{7/12}$ 27. $z^{5/3}$

29. $\dfrac{1}{x^{5/4}}$ 31. $ab^{2/3}$ 33. $x^{1/8}$ 35. $(3x + 2)^{1/3}$

37. 0.0392 39. 10.6301 41. (a) 3 (b) 9

43. (a) -1 (b) 3 45. $\left(-\infty, \dfrac{9}{2}\right]$ 47. $5u^2v^2\sqrt{3u}$

49. $0.5x^2\sqrt{y}$ 51. $2b\sqrt[4]{4a^2b}$ 53. $2ab\sqrt[3]{6b}$

55. $\dfrac{\sqrt{30}}{6}$ 57. $\dfrac{\sqrt{3x}}{2x}$ 59. $\dfrac{\sqrt[3]{4x^2}}{x}$ 61. $\sqrt{85}$

63. $\sqrt{7}$ 65. $-24\sqrt{10}$ 67. $14\sqrt{x} - 9\sqrt[3]{x}$

69. $7\sqrt[4]{y} + 3$ 71. $3x\sqrt[3]{3x^2y}$ 73. $x^3y^2\left(11\sqrt{y} - 4\sqrt{2y}\right)$

75. $21 + 12\sqrt{2}$ inches   77. $10\sqrt{3}$ 79. $5\sqrt{2} + 3\sqrt{5}$

81. $5\sqrt{2} + 2\sqrt{5}$ 83. $12\sqrt{5} + 41$ 85. $3 - x$

87. $3 + \sqrt{7}; 2$ 89. $\sqrt{x} - 20; x - 400$

91. $-3\left(1 + \sqrt{2}\right)$ 93. $\dfrac{6\left(4 - \sqrt{6}\right)}{5}$

95. $-\dfrac{\left(\sqrt{2} - 1\right)\left(\sqrt{3} + 4\right)}{13}$ **97.** $\dfrac{\left(\sqrt{x} + 10\right)^2}{x - 100}$

99. 225   **101.** No real solution   **103.** 105   **105.** 3

107. 5   **109.** $-5, -3$ **111.** 6   **113.** $\dfrac{3}{32}$

115. 12 inches $\times$ 5 inches   **117.** $9\sqrt{3} \approx 15.59$ feet

119. 576 feet   **121.** $4\sqrt{3}i$ **123.** $10 - 9\sqrt{3}i$

125. $\dfrac{3}{4} - \sqrt{3}i$ **127.** $15i$ **129.** $\left(11 - 2\sqrt{21}\right)i$

131. -5 **133.** $\sqrt{70} - 2\sqrt{10}$ **135.** $a = 10,\ b = -4$

137. $a = 4,\ b = 7$ **139.** $8 - 3i$ **141.** $8 + 4i$

143. 25   **145.** $11 - 60i$ **147.** $-\dfrac{7}{3}i$

149. $-\dfrac{8}{17} + \dfrac{2}{17}i$ **151.** $\dfrac{13}{37} - \dfrac{33}{37}i$

Chapter Test   *(page 503)*

1. (a) 64   (b) 10   **2.** (a) $\dfrac{1}{9}$ (b) 6

3. $f(-8) = 7, f(0) = 3$ **4.** $\left[\dfrac{3}{7}, \infty\right)$

5. (a) $x^{1/3}$ (b) 25   **6.** (a) $\dfrac{4}{3}\sqrt{2}$ (b) $2\sqrt[3]{3}$

7. (a) $2x\sqrt{6x}$ (b) $2xy^2\sqrt[4]{x}$ **8.** $\dfrac{\sqrt[3]{3y^2}}{3y}$

9. $\dfrac{5\left(\sqrt{6} + \sqrt{2}\right)}{2}$ **10.** $-10\sqrt{3x}$ **11.** $5\sqrt{3x} + 3\sqrt{5}$

12. $16 - 8\sqrt{2x} + 2x$ **13.** $3 + 4y$ **14.** 27

15. No solution   **16.** 9   **17.** $2 - 2i$ **18.** $-5 - 12i$

19. $-8 + 4i$ **20.** $13 + 13i$ **21.** $\dfrac{13}{10} - \dfrac{11}{10}i$

22. 100 feet

Cumulative Test: Chapters 5–7   *(page 504)*

1. $-\dfrac{y^2}{2x^5},\ z \neq 0$ **2.** $\dfrac{3s^7}{5t},\ s \neq 0$

3. $\dfrac{9x^{18}}{4y^{12}},\ x \neq 0,\ z \neq 0$ **4.** 1.6×10^7

5. $x^5 + 2x^2 - 11x + 4$

6. $-3x^3 + 15x^2 - 6x$ **7.** $2x^2 - 9x - 5$

8. $9x^3 + 3x^2 + x + 2$ **9.** $(3x + 7)(x - 5)$

10. $9(x + 4)(x - 4)$ **11.** $(y - 3)^2(y + 3)$

12. $2t(2t - 5)^2$ **13.** $-3, \dfrac{8}{3}$ **14.** $0, \pm 3$

15. $\dfrac{x(x + 2)(x + 4)}{9(x - 4)},\ x \neq -4,\ x \neq 0$

16. $\dfrac{x}{(2x - 1)(x - 4)},\ x \neq -4,\ x \neq 3$

17. $\dfrac{5x^2 - 15x - 2}{(x + 2)(x - 3)}$ **18.** $\dfrac{3x + 5}{x(x + 3)}$

19. $\dfrac{x^3}{4},\ x \neq -2,\ x \neq 0$ **20.** $x + y,\ x \neq 0,\ y \neq 0,\ x \neq y$

21. $2x^2 + x - 3 + \dfrac{4}{x + 3}$ **22.** $2x^3 - 2x^2 - x - \dfrac{4}{2x - 1}$

23. 2, 5   **24.** 2, 9   **25.** $2|x|y\sqrt{6y}$ **26.** $2a^5b^2\sqrt[3]{10b^2}$

27. $\dfrac{2|b^3|\sqrt{3}}{a^2}$ **28.** $\sqrt{t}$ **29.** $35\sqrt{5x}$

30. $2x - 6\sqrt{2x} + 9$ **31.** $\dfrac{3\left(\sqrt{10} + \sqrt{x}\right)}{10 - x}$ **32.** 41

33. 2   **34.** 11   **35.** 29   **36.** $-4 + 3\sqrt{2}i$

37. $-7 + 16i$ **38.** $-7 - 24i$ **39.** $\dfrac{2}{17} - \dfrac{9}{17}i$

40. $4(2x + 3)$ **41.** $\dfrac{1}{2}, \dfrac{2}{7}; \dfrac{14}{11}$ hours   **42.** 6.9 years

43. $60\sqrt{5} \approx 134.16$ feet   **44.** 256 feet

Chapter 8
Section 8.1   *(page 513)*

Review   *(page 513)*

1. -3. Coefficient of the term of highest degree

2. 5. $(y^2 - 2)(y^3 + 7) = y^5 - 2y^3 + 7y^2 - 14$

3.

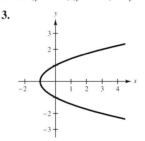

For some values of x there correspond two values of y.

4.

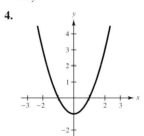

For each value of x there corresponds exactly one value of y.

5. $\dfrac{1}{x^3}$ **6.** $-\dfrac{15y^4}{x^2}$ **7.** $\dfrac{9y^2}{4x^2}$ **8.** $\dfrac{v^4}{u^5}$ **9.** $\dfrac{2u}{9v^6}$

10. $\dfrac{r^3}{7}$ **11.** 100   **12.** $\dfrac{29}{18} \approx 1.6$ hours; Distance

1. 5, 7   **3.** −5, 6   **5.** −9, 5   **7.** 6   **9.** −$\frac{4}{3}$

11. 0, 3   **13.** 9, 12   **15.** $\frac{5}{3}$, 6   **17.** 1, 6   **19.** −$\frac{5}{6}$, $\frac{1}{2}$

21. ±7   **23.** ±3   **25.** ±$\frac{4}{5}$ **27.** ±8   **29.** ±$\frac{5}{2}$

31. ±$\frac{15}{2}$ **33.** −12, 4   **35.** 2.5, 3.5   **37.** 2 ± $\sqrt{7}$

39. $\dfrac{-1 \pm 5\sqrt{2}}{2}$ **41.** $\dfrac{3 \pm 7\sqrt{2}}{4}$ **43.** ±6i   **45.** ±2i

47. ±$\dfrac{\sqrt{17}}{3}i$ **49.** 3 ± 5i   **51.** −$\dfrac{4}{3}$ ± 4i

53. −$\dfrac{3}{2}$ ± $\dfrac{3\sqrt{6}}{2}i$ **55.** −6 ± $\dfrac{11}{3}i$ **57.** 1 ± 3$\sqrt{3}i$

59. −1 ± 0.2i   **61.** $\dfrac{2}{3}$ ± $\dfrac{1}{3}i$ **63.** −$\dfrac{7}{3}$ ± $\dfrac{\sqrt{38}}{3}i$

65. 0, $\frac{5}{2}$ **67.** −4, $\frac{3}{2}$ **69.** ±30   **71.** ±30i   **73.** ±3

75. −5, 15   **77.** 5 ± 10i   **79.** −2 ± 3$\sqrt{2}i$

81.

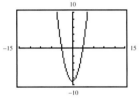

(−3, 0), (3, 0)

The result is the same.

83.

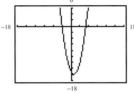

(−3, 0), (5, 0)

The result is the same.

85.

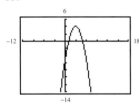

(1, 0), (5, 0)

The result is the same.

87.

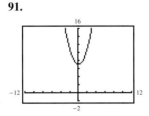

(2, 0), $\left(-\frac{3}{2}, 0\right)$

The result is the same.

89.

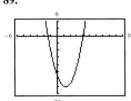

$\left(-\frac{4}{3}, 0\right)$, (4, 0)

The result is the same.

91.

±$\sqrt{7}i$, complex solution

93.

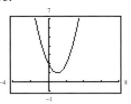

1 ± i, complex solution

95.

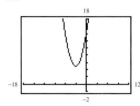

−3 ± $\sqrt{5}i$, complex solution

97. $f(x) = \sqrt{4 - x^2}$
    $g(x) = -\sqrt{4 - x^2}$

99. $f(x) = \frac{1}{2}\sqrt{4 - x^2}$
    $g(x) = -\frac{1}{2}\sqrt{4 - x^2}$

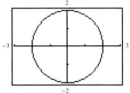

101. ±1, ±2   **103.** ±$\sqrt{2}$, ±$\sqrt{3}$ **105.** ±1, ±$\sqrt{5}$

107. 16   **109.** 4, 25   **111.** −8, 27   **113.** 1, $\frac{125}{8}$

115. 1, 32   **117.** $\frac{1}{32}$, 243   **119.** 729   **121.** 1, 16

123. $\frac{1}{2}$, 1   **125.** −1, $\frac{4}{5}$ **127.** $\frac{3}{2}$ ± $\dfrac{\sqrt{7}i}{2}$, $\frac{3}{2}$ ± $\dfrac{\sqrt{33}}{2}$

129. $\frac{12}{5}$ **131.** 17 feet   **133.** 4 seconds

135. 2$\sqrt{2}$ ≈ 2.83 seconds   **137.** 9 seconds   **139.** 6%

141. 1997

143. (a) 2$\sqrt{3}$ ≈ 3.46 seconds. Square root property, because the quadratic equation did not have a linear term.

(b) 0 seconds, 2 seconds. Factoring, because the quadratic equation did not have a constant term.

145. Factoring and the Zero-Factor Property allow you to solve a quadratic equation by converting it into two linear equations that you already know how to solve.

147. False. The solutions are x = 5 and x = −5.

149. To solve an equation of quadratic form, determine an algebraic expression u such that substitution yields the quadratic equation $au^2 + bu + c = 0$. Solve this quadratic equation for u and then, through back-substitution, find the solution of the original equation.

Section 8.2 (page 521)

Review (page 521)

1. a^4b^4 **2.** a^{rs} **3.** b^r/a^r **4.** $1/a^r$ **5.** 6

6. 3   **7.** −$\frac{2}{3}$, 5   **8.** −5, 8

9.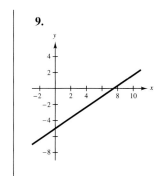

10.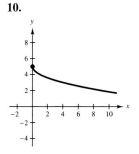

11.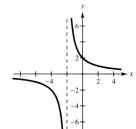

12.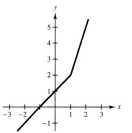

1. 16   **3.** 100   **5.** 64   **7.** $\frac{25}{4}$ **9.** $\frac{81}{4}$ **11.** $\frac{1}{36}$

13. $\frac{9}{100}$ **15.** 0.04   **17.** 0, 20   **19.** $-6, 0$

21. 0, 5   **23.** 1, 7   **25.** $-4, -3$ **27.** $-3, 6$

29. 1, 6   **31.** $-\frac{5}{2}, \frac{3}{2}$

33. $2 + \sqrt{7} \approx 4.65$ **35.** $-2 + \sqrt{7} \approx 0.65$
$2 - \sqrt{7} \approx -0.65$ $-2 - \sqrt{7} \approx -4.65$

37. $-7, 1$

39. $5 + \sqrt{47} \approx 11.86$ **41.** $-7, -1$ **43.** 3, 7
$5 - \sqrt{47} \approx -1.86$

45. $\dfrac{-5 + \sqrt{13}}{2} \approx -0.70$ **47.** $3 \pm i$ **49.** $-2 \pm 3i$

$\dfrac{-5 - \sqrt{13}}{2} \approx -4.30$

51. $\dfrac{1}{2} + \dfrac{\sqrt{3}}{2}i \approx 0.5 + 0.87i$ **53.** 3, 4

$\dfrac{1}{2} - \dfrac{\sqrt{3}}{2}i \approx 0.5 - 0.87i$

55. $\dfrac{1 + 2\sqrt{7}}{3} \approx 2.10$ **57.** $\dfrac{-3 + \sqrt{137}}{8} \approx 1.09$

$\dfrac{1 - 2\sqrt{7}}{3} \approx -1.43$ $\dfrac{-3 - \sqrt{137}}{8} \approx -1.84$

59. $\dfrac{-4 + \sqrt{10}}{2} \approx -0.42$ **61.** $\dfrac{-9 + \sqrt{21}}{6} \approx -0.74$

$\dfrac{-4 - \sqrt{10}}{2} \approx -3.58$ $\dfrac{-9 - \sqrt{21}}{6} \approx -2.26$

63. $\dfrac{-1 + \sqrt{10}}{2} \approx 1.08$

$\dfrac{-1 - \sqrt{10}}{2} \approx -2.08$

65. $\dfrac{3}{10} + \dfrac{\sqrt{191}}{10}i \approx 0.30 + 1.38i$

$\dfrac{3}{10} - \dfrac{\sqrt{191}}{10}i \approx 0.30 - 1.38i$

67. $\dfrac{7 + \sqrt{57}}{2} \approx 7.27$

$\dfrac{7 - \sqrt{57}}{2} \approx -0.27$

69. $-1 + \sqrt{3}i \approx -1 + 1.73i$ **71.** $-1 \pm 2i$
$-1 - \sqrt{3}i \approx -1 - 1.73i$

73. $1 \pm \sqrt{3}$ **75.** $1 \pm \sqrt{3}$ **77.** $4 + 2\sqrt{2}$

79. 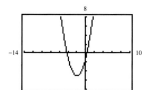   **81.**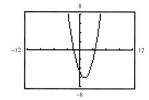

$\left(-2 \pm \sqrt{5}, 0\right)$ $\left(1 \pm \sqrt{6}, 0\right)$

The result is the same.   The result is the same.

83. 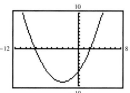   **85.**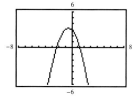

$\left(-3 \pm 3\sqrt{3}, 0\right)$ $\left(\dfrac{-1 \pm \sqrt{13}}{2}, 0\right)$

The result is the same.   The result is the same.

87. (a) $x^2 + 8x$ (b) $x^2 + 8x + 16$ (c) $(x + 4)^2$

89. 4 centimeters, 6 centimeters

91. 15 meters $\times$ $46\frac{2}{3}$ meters or 20 meters $\times$ 35 meters

93. 42 pairs, 58 pairs

95. A perfect square trinomial is one that can be written in the form $(x + k)^2$.

97. Use the method of completing the square to write the quadratic equation in the form $u^2 = d$. Then use the Square Root Property to simplify.

99. Divide each side of the equation by the leading coefficient. Dividing each side of an equation by a nonzero constant yields an equivalent equation.

101. (a) $d = 0$ (b) d is positive and a perfect square.

   (c) d is positive and is not a perfect square.   (d) $d < 0$

Section 8.3 *(page 530)*

Review *(page 530)*

1. $\sqrt{a}\sqrt{b}$ **2.** $\sqrt{a}/\sqrt{b}$

3. No. $\sqrt{72} = \sqrt{36 \cdot 2} = 6\sqrt{2}$

4. No. $\dfrac{10}{\sqrt{5}} = \dfrac{10\sqrt{5}}{(\sqrt{5})^2} = 2\sqrt{5}$ **5.** $23\sqrt{2}$

6. 150   **7.** 7   **8.** $11 + 6\sqrt{2}$ **9.** $\dfrac{4\sqrt{10}}{5}$

10. $\dfrac{5(1 + \sqrt{3})}{4}$ **11.** 10 inches × 15 inches

12. 200 units

---

1. $2x^2 + 2x - 7 = 0$ **3.** $-x^2 + 10x - 5 = 0$

5. $4, 7$ **7.** $-2, -4$ **9.** $-\dfrac{1}{2}$ **11.** $-\dfrac{3}{2}$ **13.** $-\dfrac{1}{2}, \dfrac{2}{3}$

15. $-15, 20$ **17.** $1 \pm \sqrt{5}$ **19.** $-2 \pm \sqrt{3}$

21. $-3 \pm 2\sqrt{3}$ **23.** $5 \pm \sqrt{2}$ **25.** $-\dfrac{3}{4} \pm \dfrac{\sqrt{15}}{4}i$

27. $-\dfrac{1}{3}, 1$ **29.** $\dfrac{-2 \pm \sqrt{10}}{2}$ **31.** $\dfrac{-1 \pm \sqrt{5}}{3}$

33. $\dfrac{-3 \pm \sqrt{21}}{4}$ **35.** $\dfrac{3}{8} \pm \dfrac{\sqrt{7}}{8}i$ **37.** $\dfrac{5 \pm \sqrt{73}}{4}$

39. $\dfrac{3 \pm \sqrt{13}}{6}$ **41.** $\dfrac{-3 \pm \sqrt{57}}{6}$ **43.** $\dfrac{1 \pm \sqrt{5}}{5}$

45. $\dfrac{-1 \pm \sqrt{10}}{5}$ **47.** Two distinct complex solutions

49. Two distinct irrational solutions

51. Two distinct complex solutions

53. One (repeated) rational solution

55. Two distinct complex solutions

57. ± 13 **59.** $-3, 0$ **61.** $\dfrac{9}{5}, \dfrac{21}{5}$ **63.** $-\dfrac{3}{2}, 18$

65. $-4 \pm 3i$ **67.** $8, 16$ **69.** $\dfrac{13}{6} \pm \dfrac{13\sqrt{11}}{6}i$

71. $\dfrac{-5 \pm 5\sqrt{17}}{12}$ **73.** $\dfrac{-11 \pm \sqrt{41}}{8}$

75. $x^2 - 3x - 10 = 0$ **77.** $x^2 - 8x + 7 = 0$

79. $x^2 - 2x - 1 = 0$ **81.** $x^2 + 25 = 0$

83. $x^2 - 24x + 144 = 0$

85.
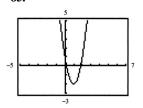

$(0.18, 0), (1.82, 0)$
The result is the same.

87.

$(2.50, 0)$
The result is the same.

89.
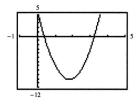

$(3.23, 0), (0.37, 0)$
The result is the same.

91.

$(99.80, 0), (0.20, 0)$
The result is the same.

93.
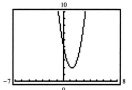

No real solutions

95.
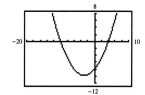

Two real solutions

97. $\dfrac{7 \pm \sqrt{17}}{4}$ **99.** No real values   **101.** $\dfrac{5 \pm \sqrt{185}}{8}$

103. $\dfrac{3 + \sqrt{17}}{2}$ **105.** (a) $c < 9$ (b) $c = 9$ (c) $c > 9$

107. (a) $c < 16$ (b) $c = 16$ (c) $c > 16$

109. 5.1 inches × 11.4 inches

111. (a) 2.5 seconds   (b) $\dfrac{5 + 5\sqrt{3}}{4} \approx 3.415$ seconds

113. (a) $\dfrac{5}{4}$ or 1.25 seconds   (b) $\dfrac{5 + 5\sqrt{5}}{8} \approx 2.023$ seconds

115. (a) $\dfrac{5}{16}$ or 0.3125 second

   (b) $\dfrac{5 + 5\sqrt{257}}{32} \approx 2.661$ seconds

117. (a)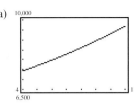

(b) 1999

(c) 10,080,000

119.

	x_1, x_2	$x_1 + x_2$	$x_1 x_2$
(a)	$-2, 3$	1	-6
(b)	$-3, \frac{1}{2}$	$-\frac{5}{2}$	$-\frac{3}{2}$
(c)	$-\frac{3}{2}, \frac{3}{2}$	0	$-\frac{9}{4}$
(d)	$5 + 3i, 5 - 3i$	10	34

121. (c) $\dfrac{2 + 3\sqrt{3}}{2} \approx 3.6$ seconds; Quadratic Formula, because the numbers were large and the equation would not factor.

(d) Yes

123. $b^2 - 4ac$. If the discriminant is positive, the quadratic equation has two real solutions; if it is zero, the equation has one (repeated) real solution; and if it is negative, the equation has no real solutions.

125. The four methods are factoring, the Square Root Property, completing the square, and the Quadratic Formula.

Mid-Chapter Quiz *(page 535)*

1. ± 6 **2.** $-4, \frac{5}{2}$ **3.** $\pm 2\sqrt{3}$ **4.** $-1, 7$

5. $-5 \pm 2\sqrt{6}$ **6.** $\dfrac{-3 \pm \sqrt{19}}{2}$ **7.** $-2 \pm \sqrt{10}$

8. $\dfrac{3 \pm \sqrt{105}}{12}$ **9.** $-\dfrac{5}{2} \pm \dfrac{\sqrt{3}}{2}i$ **10.** $-2, 10$

11. $-3, 10$ **12.** $-2, 5$ **13.** $\frac{3}{2}$ **14.** $\dfrac{-5 \pm \sqrt{10}}{3}$

15. 36 **16.** $\pm 2i, \pm\sqrt{3}i$ **17.** $9 + 4\sqrt{5}$

18. $\pm\sqrt{3}, \pm 3$

19.

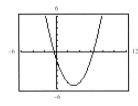

$(-0.32, 0), (6.32, 0)$

The result is the same.

20.

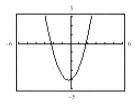

$(-2.24, 0), (1.79, 0)$

The result is the same.

21. 50 alarm clocks   **22.** 35 meters $\times$ 65 meters

Section 8.4 *(page 541)*

Review *(page 541)*

1. $2b, b^2$

2. $\frac{25}{4}$, The constant is found by squaring half the coefficient of x.

3. $-11x$ **4.** $-41v$ **5.** $6x^2 + 9$ **6.** -4

7. $2|x|y\sqrt{6y}$ **8.** $3\sqrt[3]{5}$ **9.** $\dfrac{2\sqrt{3}b^3}{a^2}$ **10.** 2

11. 4 liters

12. 30-second commercials: 6; 60-second commercials: 6

1. e   **2.** f   **3.** b   **4.** c   **5.** d   **6.** a

7. $y = (x - 0)^2 + 2, \ (0, 2)$ **9.** $y = (x - 2)^2 + 3, \ (2, 3)$

11. $y = (x + 3)^2 - 4, \ (-3, -4)$

13. $y = -(x - 3)^2 - 1, \ (3, -1)$

15. $y = -(x - 1)^2 - 6, \ (1, -6)$

17. $y = 2\left(x + \frac{3}{2}\right)^2 - \frac{5}{2}, \ \left(-\frac{3}{2}, -\frac{5}{2}\right)$

19. $(4, -1)$ **21.** $(-1, 2)$ **23.** $\left(-\frac{1}{2}, 3\right)$

25. Upward, $(0, 2)$ **27.** Downward, $(10, 4)$

29. Upward, $(0, -6)$ **31.** Downward, $(3, 0)$

33. Downward, $(3, 9)$ **35.** $(\pm 5, 0), (0, 25)$

37. $(0, 0), (9, 0)$ **39.** $(-3, 0), (1, 0), (0, -3)$

41. $\left(\frac{3}{2}, 0\right), (0, 9)$ **43.** $(0, 3)$

45. $\left(\dfrac{-3 \pm \sqrt{19}}{2}, 0\right), (0, 5)$

47.

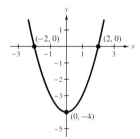

49.

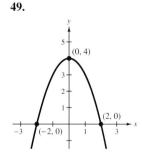

51.

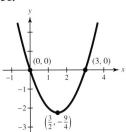

53.

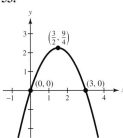

71. Vertical shift

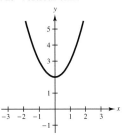

73. Horizontal shift

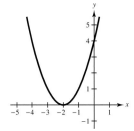

55.

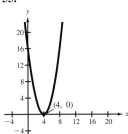

57.

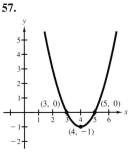

75. Horizontal shift and reflection in the x-axis

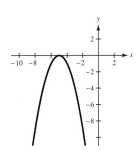

77. Horizontal and vertical shifts, reflection in the x-axis

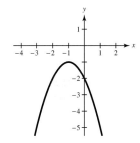

59.

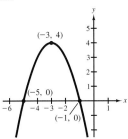

61.

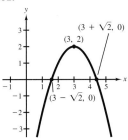

79.

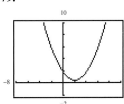

81.

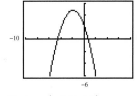

Vertex: $(2, 0.5)$ Vertex: $(-1.9, 4.9)$

83. $y = -x^2 + 4$ **85.** $y = x^2 + 4x + 2$

87. $y = x^2 - 4x + 5$ **89.** $y = x^2 - 4x$

91. $y = \frac{1}{2}x^2 - 3x + \frac{13}{2}$ **93.** $y = -4x^2 - 8x + 1$

95. (a) 4 feet (b) 16 feet (c) $12 + 8\sqrt{3} \approx 25.9$ feet

97. (a) 3 feet (b) 48 feet (c) $15 + 4\sqrt{15} \approx 30.5$ feet

99. (a) 6 feet (b) 56 feet (c) $100 + 40\sqrt{7} \approx 205.8$ feet

101. 14 feet

63.

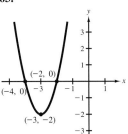

65.

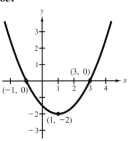

103.

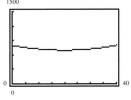

$x = 20$ when C is minimum

67.

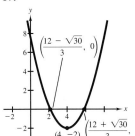

69.

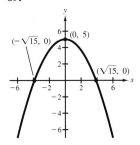

105. (a)

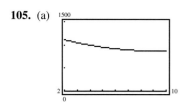

(b) 1992; 1,110,000 military reserves

107. (a) 1000 feet (b) 400 feet

(c)

x	± 100	± 200	± 300	± 400	± 500
y	16	64	144	256	400

109. Parabola

111. To find any x-intercepts, set $y = 0$ and solve the resulting equation for x. To find any y-intercepts, set $x = 0$ and solve the resulting equation for y.

113. If the discriminant is positive, the parabola has two x-intercepts; if it is zero, the parabola has one x-intercept; and if it is negative, the parabola has no x-intercepts.

115. Find the y-coordinate of the vertex of the graph of the function.

Section 8.5 *(page 552)*

Review *(page 552)*

1. $m = \dfrac{y_2 - y_1}{x_2 - x_1}$

2. (a) $y = mx + b$ (b) $y - y_1 = m(x - x_1)$

(c) $ax + by + c = 0$ (d) $y - b = 0$

3. $x + 2y = 0$ **4.** $3x - 4y = 0$

5. $2x - y = 0$ **6.** $x + y - 6 = 0$

7. $22x + 16y - 161 = 0$

8. $134x - 73y + 146 = 0$

9. $y - 8 = 0$ **10.** $x + 3 = 0$ **11.** 8 people

12. 3 miles per hour

---

1. 18 dozen, $1.20 per dozen **3.** 16 video games, $30

	Width	Length	Perimeter	Area
5.	1.4l	l	54 in.	$177\frac{3}{16}$ in.^{2}
7.	w	2.5w	70 ft	250 ft^2
9.	$\frac{1}{3}l$	l	64 in.	192 in.^{2}
11.	w	$w + 3$	54 km	180 km^2
13.	$l - 20$	l	440 m	12,000 m^2

15. 12 inches × 16 inches

17. 50 feet × 250 feet or 100 feet × 125 feet

19. No.

Area $= \frac{1}{2}(b_1 + b_2)h = \frac{1}{2}x[x + (550 - 2x)] = 43,560$

This equation has no real solution.

21. Height: 12 inches; Width: 24 inches **23.** 9.5%

25. 7% **27.** $\approx 6.5\%$ **29.** 15 people **31.** 32 people

33. 15.86 miles or 2.14 miles

35. (a) $d = \sqrt{(3 + x)^2 + (4 + x)^2}$

(b)

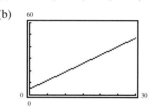

(c) $x \approx 3.55$ when $d = 10$

(d) $\dfrac{-7 + \sqrt{199}}{2} \approx 3.55$ meters

37. 9.1 hours, 11.1 hours **39.** 10.7 minutes, 13.7 minutes

41. $3\frac{1}{4}$ seconds **43.** 9.5 seconds **45.** 4.7 seconds

47. (a) 3 seconds, 7 seconds (b) 10 seconds

(c) 400 feet

49. 13, 14 **51.** 12, 14 **53.** 17, 19

55. 400 miles per hour

57. 46 miles per hour or 65 miles per hour

59. (a) $b = 20 - a$; $A = \pi ab$; $A = \pi a(20 - a)$

(b)

a	4	7	10	13	16
A	201.1	285.9	314.2	285.9	201.1

(c) 7.9, 12.1

(d)

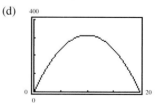

61. (a) Write a verbal model that describes what you need to know.

(b) Assign labels to each part of the verbal model—numbers to the known quantities and letters to the variable quantities.

(c) Use the labels to write an algebraic model based on the verbal model.

(d) Solve the resulting algebraic equation and check your solution.

63. Dollars

Section 8.6   *(page 563)*

Review   *(page 563)*

1. No. 3.683×10^9 **2.** $[10^6, 10^8)$

3. $6v(u^2 - 32v)$ **4.** $5x^{1/3}(x^{1/3} - 2)$

5. $(x - 10)(x - 4)$ **6.** $(x + 2)(x - 2)(x + 3)$

7. $(4x + 11)(4x - 11)$ **8.** $4x(x^2 - 3x + 4)$

9. $\frac{3}{2}h^2$ **10.** $\frac{1}{3}b^2$ **11.** $5x^2$ **12.** $x^2 + 8x$

1. $0, \frac{5}{2}$ **3.** $\pm \frac{9}{2}$ **5.** $-3, 5$ **7.** $1, 3$ **9.** $\frac{5}{2}$

11. Negative: $(-\infty, 4)$; Positive: $(4, \infty)$

13. Negative: $(6, \infty)$; Positive: $(-\infty, 6)$

15. Negative: $(0, 4)$; Positive: $(-\infty, 0) \cup (4, \infty)$

17. Negative: $(-\infty, -2) \cup (2, \infty)$; Positive: $(-2, 2)$

19. Negative: $(-1, 5)$; Positive: $(-\infty, -1) \cup (5, \infty)$

21. $(0, 2)$

23. $[0, 2]$

25. $(-\infty, -2) \cup (2, \infty)$

27. $(-\infty, -2] \cup [5, \infty)$

29. $(-\infty, -4) \cup (0, \infty)$

31. $[-5, 2]$

33. $(-\infty, -3) \cup (1, \infty)$

35. No solution

37. $(-\infty, \infty)$

39. $(-\infty, 2 - \sqrt{2}) \cup (2 + \sqrt{2}, \infty)$

41. $(-\infty, \infty)$ **43.** No solution

45. $\left[-2, \frac{4}{3}\right]$ **47.** $\left(\frac{2}{3}, \frac{5}{2}\right)$

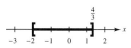

49. $\left(-\infty, -\frac{1}{2}\right) \cup (4, \infty)$ **51.** $-\frac{7}{2}$

53. No solution

55. $\left(-\infty, 5 - \sqrt{6}\right) \cup \left(5 + \sqrt{6}, \infty\right)$

57. $(-\infty, -9] \cup [-1, \infty)$ **59.** $(-2, 0) \cup (2, \infty)$

61.

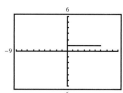

$(0, 6)$

63.

$(-\infty, -4) \cup \left(\frac{3}{2}, \infty\right)$

65.

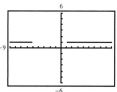

$(-\infty, -5] \cup [1, \infty)$

67.

$(-\infty, -3) \cup (7, \infty)$

69. 3   **71.** $0, -5$

73. $(3, \infty)$ **75.** $(-\infty, 3)$

77. $[-2, 1)$

79. $(-\infty, -3) \cup (1, \infty)$

81. $(1, 4]$

83. $\left(-\infty, \frac{1}{2}\right) \cup (2, \infty)$

85. $\left[-2, -\frac{3}{2}\right)$

87. $(-1, 3)$

89. $(4, 7)$

91. $\left(-2, -\frac{2}{5}\right)$

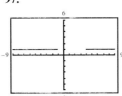

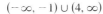

93. $(-\infty, 3) \cup [5, \infty)$

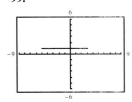

95.

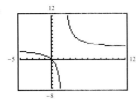

$(-\infty, -1) \cup (0, 1)$

97.

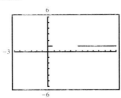

$(-\infty, -1) \cup (4, \infty)$

99.

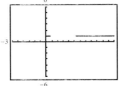

$\left(-5, \frac{13}{4}\right)$

101.

$(0, 0.382) \cup (2.618, \infty)$

(a) $[0, 2)$ (b) $(2, 4]$

103.

105.

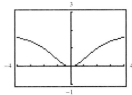

(a) $(-\infty, -2] \cup [2, \infty)$

(b) $(-\infty, \infty)$

107. $(3, 5)$ **109.** $r > 7.24\%$ **111.** $(12, 20)$

113. $90{,}000 \le x \le 100{,}000$

115. (a) $\overline{C} = \dfrac{3000}{x} + 0.75, \; x > 0$

(b)

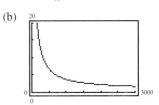

(c) $x > 2400$ calendars

117. (e) $[0, 3)$

119. All real numbers less than or equal to 5, and all real numbers greater than 10.

121. (a) Find the critical numbers of the quadratic polynomial.

(b) Use the critical numbers to determine the test intervals.

(c) Choose a representative x-value from each test interval and evaluate the quadratic polynomial.

Review Exercises   *(page 568)*

1. $-12, 0$ **3.** $\pm\frac{1}{2}$ **5.** $-\frac{5}{2}$ **7.** $-9, 10$ **9.** $-\frac{3}{2}, 6$

11. ± 50 **13.** $\pm 2\sqrt{3}$ **15.** $-4, 36$ **17.** $\pm 11i$

19. $\pm 5\sqrt{2}\,i$ **21.** $-4 + 3\sqrt{2}\,i$ **23.** $\pm\sqrt{5}, \pm i$

25. $1, 9$ **27.** $1, \; 1 \pm \sqrt{6}$ **29.** $-343, 64$ **31.** 144

33. $\frac{225}{4}$ **35.** $\frac{1}{25}$

37. $3 + 2\sqrt{3} \approx 6.46; \; 3 - 2\sqrt{3} \approx -0.46$

39. $\frac{3}{2} + \frac{\sqrt{3}}{2}i \approx 1.5 + 0.87i; \; \frac{3}{2} - \frac{\sqrt{3}}{2}i \approx 1.5 - 0.87i$

41. $\frac{1}{3} + \frac{\sqrt{17}}{3}i \approx 0.33 + 1.37i; \; \frac{1}{3} - \frac{\sqrt{17}}{3}i \approx 0.33 - 1.37i$

43. $-6, 5$ **45.** $-\frac{7}{2}, 3$ **47.** $\dfrac{8 \pm 3\sqrt{6}}{5}$

49. One repeated rational solution

51. Two distinct rational solutions

53. Two distinct rational solutions

55. Two distinct complex solutions

57. $x^2 + 4x - 21 = 0$

59. $x^2 - 10x + 18 = 0$ **61.** $x^2 - 12x + 40 = 0$

63. $y = (x - 4)^2 - 13$; Vertex: $(4, -13)$

65. $y = 2\left(x - \frac{1}{4}\right)^2 + \frac{23}{8}$; Vertex: $\left(\frac{1}{4}, \frac{23}{8}\right)$

67.

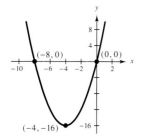

69.

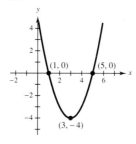

71. $y = 2(x - 2)^2 - 5$ **73.** $y = \frac{1}{16}(x - 5)^2$

75. (a)

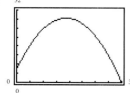

(b) 6 feet

(c) 28.5 feet

(d) 31.9 feet

77. 16 cars; \$5000   **79.** 6 inches × 18 inches

81. 15 people

83. $9 + \sqrt{101} \approx 19$ hours, $11 + \sqrt{101} \approx 21$ hours

85. $-7, 0$ **87.** $-3, 9$

89. $(0, 7)$

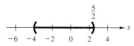

91. $(-\infty, -2] \cup [6, \infty)$

93. $\left(-4, \frac{5}{2}\right)$

95. $(-\infty, -3] \cup \left(\frac{7}{2}, \infty\right)$

97. $(-4, 1)$

99. $(-5.3, 14.2)$

Chapter Test   *(page 571)*

1. $3, 10$ **2.** $-\frac{3}{8}, 3$ **3.** $1.7, 2.3$ **4.** $-4 \pm 10i$

5. $\dfrac{3 \pm \sqrt{3}}{2}$ **6.** $\dfrac{2 \pm 3\sqrt{2}}{2}$ **7.** $\dfrac{3 \pm \sqrt{5}}{4}$ **8.** $1, 512$

9. -56; A negative discriminant tells us the equation has two imaginary solutions.

10. $x^2 - x - 20 = 0$

11.

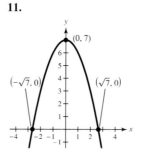

12.

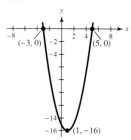

13. $(-\infty, -2] \cup [6, \infty)$

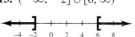

14. $(0, 3)$

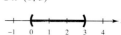

15. $[-1, 5)$

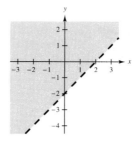

16. 12 feet × 20 feet

17. 40   **18.** $\dfrac{\sqrt{10}}{2} \approx 1.58$ seconds   **19.** 60 feet, 80 feet

Chapter 9

Section 9.1   *(page 582)*

Review   *(page 582)*

1. Test one point in each of the half-planes formed by the graph of $x + y = 5$. If the point satisfies the inequality, shade the entire half-plane to denote that every point in the region satisfies the inequality.

2. The first contains the boundary and the second does not.

3.

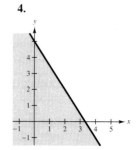

4.

5.

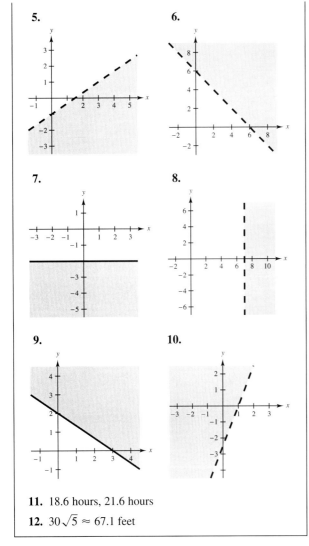

6.

7.

8.

9.

10.

11. 18.6 hours, 21.6 hours

12. $30\sqrt{5} \approx 67.1$ feet

1. 2^{2x-1} **3.** e^2 **5.** $8e^{3x}$ **7.** $-2e^x$ **9.** 11.036

11. 1.396   **13.** 51.193   **15.** 0.906

17. (a) $\frac{1}{9}$ (b) 1   (c) 3

19. (a) 0.263   (b) 54.872   (c) 19.790

21. (a) 500   (b) 250   (c) 56.657

23. (a) 1000   (b) 1628.895   (c) 2653.298

25. (a) 486.111   (b) 47.261   (c) 0.447

27. (a) 73.891   (b) 1.353   (c) 0.183

29. (a) 333.333   (b) 434.557   (c) 499.381

31. a   **32.** d   **33.** c   **34.** e   **35.** b   **36.** f

37.

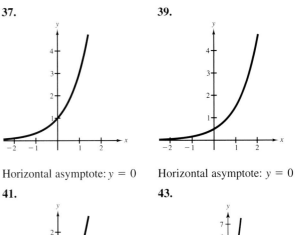

39.

Horizontal asymptote: $y = 0$ Horizontal asymptote: $y = 0$

41.

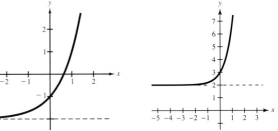

43.

Horizontal asymptote: $y = -2$ Horizontal asymptote: $y = 2$

45.

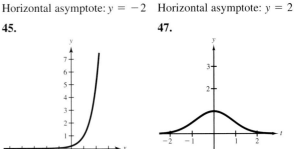

47.

Horizontal asymptote: $y = 0$ Horizontal asymptote: $y = 0$

49.

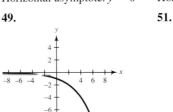

51.

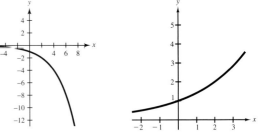

Horizontal asymptote: $y = 0$ Horizontal asymptote: $y = 0$

53.

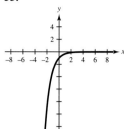

Horizontal asymptote: $y = 0$

55.

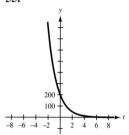

Horizontal asymptote: $y = 0$

73. Reflection in the x-axis   **75.** 2.520 grams

57.

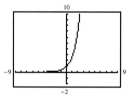

59.

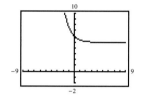

77.

n	1	4	12	365	Continuous
A	\$275.90	\$283.18	\$284.89	\$285.74	\$285.77

79.

n	1	4	12
A	\$4956.46	\$5114.30	\$5152.11

n	365	Continuous
A	\$5170.78	\$5171.42

61.

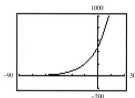

63.

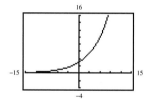

81.

n	1	4	12
P	\$2541.75	\$2498.00	\$2487.98

n	365	Continuous
P	\$2483.09	\$2482.93

65.

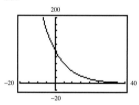

67.

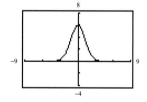

83.

n	1	4	12
P	\$18,429.30	\$15,830.43	\$15,272.04

n	365	Continuous
P	\$15,004.64	\$14,995.58

69. Vertical shift

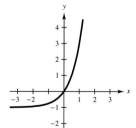

71. Horizontal shift

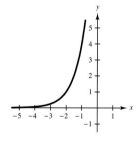

85. (a) \$22.04   (b) \$20.13

87. (a) \$80,634.95   (b) \$161,269.89

89. $V(t) = 16,000\left(\frac{3}{4}\right)^t$ \$9000

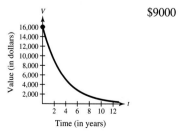

91. (a) $A_1 = 500e^{0.06t}$, $A_2 = 500e^{0.08t}$

(b) and (c)

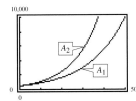 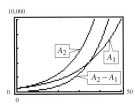

(d) The difference between the functions increases at an increasing rate.

93. (a)

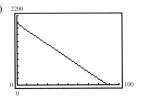

(b)

t	0	25	50	75
h	2000 ft	1400 ft	850 ft	300 ft

95. (a)

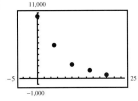

(b)

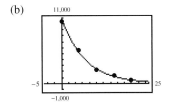

The model is a good fit for the data.

(c)

h	0	5	10	15	20
P	10,332	5583	2376	1240	517
Approx.	10,958	5176	2445	1155	546

(d) 3300 kilograms per square meter

(e) 11.3 kilometers

97. (a)

x	1	10	100	1000	10,000
$\left(1 + \dfrac{1}{x}\right)^x$	2	2.5937	2.7048	2.7169	2.7181

(b)

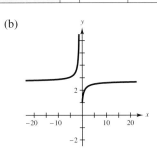

The graph appears to be approaching a horizontal asymptote.

(c) The value approaches e.

99. (a) Continuous

Compounding	Amount, A
Annual	\$5955.08
Quarterly	\$5978.09
Monthly	\$5983.40
Daily	\$5986.00
Hourly	\$5986.08
Continuous	\$5986.09

(b) Compounding quarterly, because the balance is greater at 8% than with continuous compounding at 7%. 7% continuous: \$6168.39; 8% quarterly: \$6341.21

101. By definition, the base of an exponential function must be positive and not equal to 1. If the base is 1, the function simplifies to the constant function $y = 1$.

103. No; e is an irrational number.

105. When $0 < k < 1$, the graph falls from left to right.

When $k = 1$, the graph is the straight line $y = 1$.

When $k > 1$, the graph rises from left to right.

Section 9.2 *(page 595)*

Review *(page 595)*

1. y is not a function of x because for some values of x there correspond two values of y. For example, $(4, 2)$ and $(4, -2)$ are solution points.

2. y is a function of x because for each value of x there corresponds exactly one value of y.

3. The domain of f is $-2 \le x \le 2$ and the domain of g is $-2 < x < 2$. g is undefined at $x = \pm 2$.

4. $\{4, 5, 6, 8\}$ 5. $-2x^2 - 4$ 6. $30x^3 + 40x^2$

7. $u^2 - 16v^2$ 8. $9a^2 - 12ab + 4b^2$

9. $t^3 - 6t^2 + 12t - 8$ 10. $\frac{1}{2}x^2 - \frac{1}{4}x$

11. 100 feet   12. 13 minutes

---

1. (a) $2x - 9$ (b) $2x - 3$ (c) -1 (d) 11

3. (a) $4x^2 + 4x + 4$ (b) $2x^2 + 7$ (c) 28 (d) 25

5. (a) $|3x - 3|$ (b) $3|x - 3|$ (c) 0 (d) 3

7. (a) $\sqrt{x + 1}$ (b) $\sqrt{x - 4} + 5$ (c) 2 (d) 7

9. (a) $\dfrac{x^2}{2 - 3x^2}$ (b) $2(x - 3)^2$ (c) -1 (d) 2

11. (a) -1 (b) -2 (c) -2

13. (a) -1 (b) 1

15. (a) 10 (b) 1 (c) 1

17. (a) 0 (b) 10

19. (a) $(f \circ g)(x) = 3x - 17$
   Domain: $(-\infty, \infty)$
   (b) $(g \circ f)(x) = 3x - 3$
   Domain: $(-\infty, \infty)$

21. (a) $(f \circ g)(x) = \sqrt{x - 2}$
   Domain: $[2, \infty)$
   (b) $(g \circ f)(x) = \sqrt{x} - 2$
   Domain: $[0, \infty)$

23. (a) $(f \circ g)(x) = x + 2$
   Domain: $[1, \infty)$
   (b) $(g \circ f)(x) = \sqrt{x^2 + 2}$
   Domain: $(-\infty, \infty)$

25. (a) $(f \circ g)(x) = \dfrac{\sqrt{x - 1}}{\sqrt{x - 1} + 5}$
   Domain: $[1, \infty)$
   (b) $(g \circ f)(x) = \sqrt{-\dfrac{5}{x + 5}}$
   Domain: $(-\infty, -5)$

27.
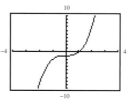
Yes

29.
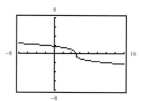
Yes

31.
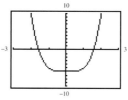
No

33.
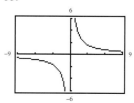
Yes

35. No   37. Yes   39. No

41. $f(g(x)) = f\left(-\frac{1}{6}x\right) = -6\left(-\frac{1}{6}x\right) = x$
   $g(f(x)) = g(-6x) = -\frac{1}{6}(-6x) = x$

43. $f(g(x)) = f(x - 15) = (x - 15) + 15 = x$
   $g(f(x)) = g(x + 15) = (x + 15) - 15 = x$

45. $f(g(x)) = f\left[\frac{1}{2}(1 - x)\right] = 1 - 2\left[\frac{1}{2}(1 - x)\right]$
   $= 1 - (1 - x) = x$
   $g(f(x)) = g(1 - 2x) = \frac{1}{2}[1 - (1 - 2x)] = \frac{1}{2}(2x) = x$

47. $f(g(x)) = f\left[\frac{1}{3}(2 - x)\right] = 2 - 3\left[\frac{1}{3}(2 - x)\right]$
   $= 2 - (2 - x) = x$
   $g(f(x)) = g(2 - 3x) = \frac{1}{3}[2 - (2 - 3x)] = \frac{1}{3}(3x) = x$

49. $f(g(x)) = f(x^3 - 1) = \sqrt[3]{(x^3 - 1) + 1} = \sqrt[3]{x^3} = x$
   $g(f(x)) = g(\sqrt[3]{x + 1}) = \left(\sqrt[3]{x + 1}\right)^3 - 1$
   $= x + 1 - 1 = x$

51. $f(g(x)) = f\left(\frac{1}{x}\right) = \dfrac{1}{(1/x)} = x$
   $g(f(x)) = g\left(\frac{1}{x}\right) = \dfrac{1}{(1/x)} = x$

53. $f^{-1}(x) = \frac{1}{5}x$ 55. $f^{-1}(x) = -\frac{5}{2}x$

57. $f^{-1}(x) = x - 10$ 59. $f^{-1}(x) = 3 - x$

61. $f^{-1}(x) = \sqrt[7]{x}$ 63. $f^{-1}(x) = x^3$ 65. $f^{-1}(x) = \dfrac{x}{8}$

67. $g^{-1}(x) = x - 25$ 69. $g^{-1}(x) = \dfrac{3 - x}{4}$

71. $g^{-1}(t) = 4t - 8$ 73. $h^{-1}(x) = x^2, \ x \ge 0$

75. $f^{-1}(t) = \sqrt[3]{t + 1}$ 77. $f^{-1}(x) = x^2 - 3, \ x \ge 0$

79. b   80. c   81. d   82. a

83.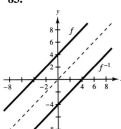

85.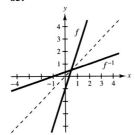

87.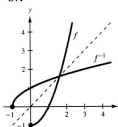

89.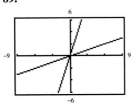

91.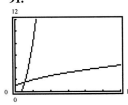

93.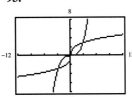

95.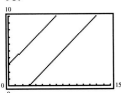

97. $x \geq 2$; $f^{-1}(x) = \sqrt{x} + 2$;

Domain of f^{-1}: $x \geq 0$

99. $x \geq 0$; $f^{-1}(x) = x - 1$; Domain of f^{-1}: $x \geq 1$

101. $f^{-1}(x) = \frac{1}{2}(3 - x)$

103. Input: time; Output: area

$A(r(t)) = 0.36\pi t^2$

105. $(C \circ x)(t) = 102t + 300$

Production cost after t hours of operation.

107. (a) $R = p - 500$

(b) $S = 0.9p$

(c) $(R \circ S)(p) = 0.9p - 500$; 10% discount followed by the \$500 rebate.

$(S \circ R)(p) = 0.9(p - 500)$; 10% discount after the price is reduced by the rebate.

(d) $(R \circ S)(6000) = 4900$; $(S \circ R)(6000) = 4950$

$R \circ S$ yields a lower cost because the dealer discount is calculated on a larger base.

109. (a) Total cost = Cost of oranges + Cost of apples

$y = 0.75x + 0.95(100 - x)$

(b) $y = 5(95 - x)$

x: total cost

y: number of pounds of oranges at \$0.75 per pound

(c) $75 \leq x \leq 95$ (d) 55 pounds

111. True. The x-coordinate of a point on the graph of f becomes the y-coordinate of a point on the graph of f^{-1}.

113. False. $f(x) = \sqrt{x - 1}$; Domain: $[1, \infty)$;

$f^{-1}(x) = x^2 + 1$, $x \geq 0$; Domain: $[0, \infty)$

115. Interchange the coordinates of each ordered pair. The inverse of the function defined by $\{(3, 6), (5, -2)\}$ is $\{(6, 3), (-2, 5)\}$.

117. $f(x) = x^4$

119. They are reflections in the line $y = x$.

Section 9.3   *(page 608)*

Review   *(page 608)*

1. Horizontal shift 4 units to the right

2. Reflection in the x-axis

3. Vertical shift 1 unit upward

4. Horizontal shift 3 units to the left and a vertical shift 5 units downward

5. $2x(x^2 - 3)$ **6.** $(2 - y)(6 + y)$

7. $(t + 5)^2$ **8.** $(5 - u)(1 + u^2)$

9.

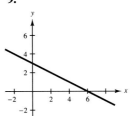

10.

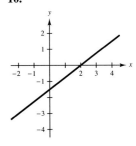

11. 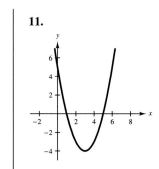   **12.**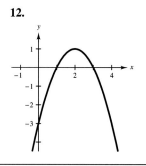

1. $7^2 = 49$ **3.** $2^{-5} = \frac{1}{32}$ **5.** $3^{-5} = \frac{1}{243}$

7. $36^{1/2} = 6$ **9.** $8^{2/3} = 4$ **11.** $2^{1.3} \approx 2.462$

13. $\log_6 36 = 2$ **15.** $\log_4 \frac{1}{16} = -2$ **17.** $\log_8 4 = \frac{2}{3}$

19. $\log_{25} \frac{1}{5} = -\frac{1}{2}$ **21.** $\log_4 1 = 0$

23. $\log_5 9.518 \approx 1.4$ **25.** 3   **27.** 3   **29.** -2

31. -3 **33.** -4

35. There is no power to which 2 can be raised to obtain -3.

37. 0

39. There is no power to which 5 can be raised to obtain -6.

41. $\frac{1}{2}$ **43.** $\frac{3}{4}$ **45.** 4   **47.** 1.6232   **49.** -1.6383

51. 0.7335   **53.** c   **54.** b   **55.** a   **56.** d

57. 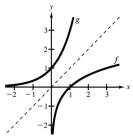   **59.**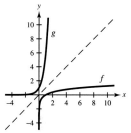

Inverse functions   Inverse functions

61. The graph is shifted 3 units upward.

63. The graph is shifted 2 units to the right.

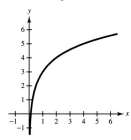   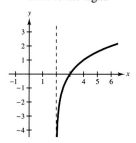

65. The graph is reflected in the y-axis.

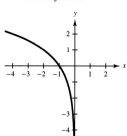

67.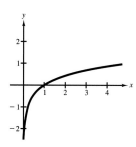

Vertical asymptote: $x = 0$

69.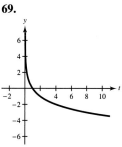

Vertical asymptote: $t = 0$

71.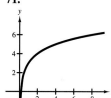

Vertical asymptote: $x = 0$

73.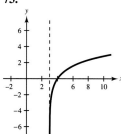

Vertical asymptote: $x = 3$

75.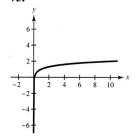

Vertical asymptote: $x = 0$

77. Domain: $(0, \infty)$
Vertical asymptote:
$x = 0$

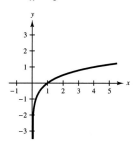

79. Domain: $(3, \infty)$
Vertical asymptote:
$x = 3$

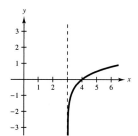

81. Domain: $(0, \infty)$

Vertical asymptote:
$x = 0$

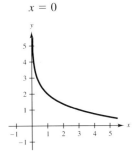

83. Domain: $(0, \infty)$

Vertical asymptote:
$x = 0$

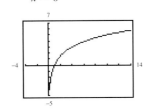

85. Domain: $(0, \infty)$

Vertical asymptote:
$x = 0$

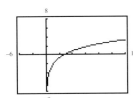

87. Domain: $(0, \infty)$

Vertical asymptote:
$x = 0$

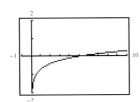

89. 3.6376   **91.** -1.8971 **93.** 0.0757   **95.** b

96. a   **97.** d   **98.** c

99. Vertical asymptote:
$x = 0$

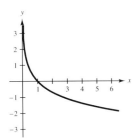

101. Vertical asymptote:
$x = 0$

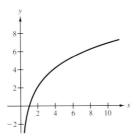

103. Vertical asymptote:
$x = 0$

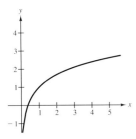

105. Vertical asymptote:
$t = 4$

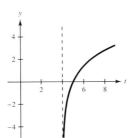

107.

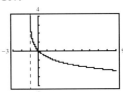

Domain: $(-1, \infty)$

Vertical asymptote: $x = -1$

109.

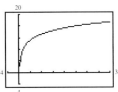

Domain: $(0, \infty)$

Vertical asymptote: $t = 0$

111. 1.6309   **113.** 1.2925   **115.** -0.4739

117. 2.6332   **119.** -2 **121.** 1.3481   **123.** 1.8946

125. 53.4 inches

127.

r	0.07	0.08	0.09	0.10	0.11	0.12
t	9.9	8.7	7.7	6.9	6.3	5.8

129. (a)

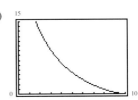

Domain: $(0, 10]$

(b) $x = 0$ (c) $(2, y) \approx (2, 13.1)$

131. $\log_5 x$ **133.** $a^1 = a$

135. Common logarithms are base 10 and natural logarithms are base e.

137. $(0, \infty)$ **139.** $3 \le f(x) \le 4$ **141.** A factor of 10

Mid-Chapter Quiz   *(page 612)*

1. (a) $\dfrac{16}{9}$ (b) 1   (c) $\dfrac{3}{4}$ (d) $\dfrac{8\sqrt{3}}{9}$

2. Domain: $(-\infty, \infty)$; Range: $(0, \infty)$

3.

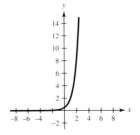

Horizontal asymptote: $y = 0$

4.

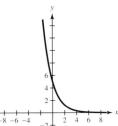

Horizontal asymptote: $y = 0$

5.

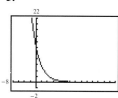

6.

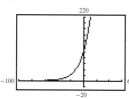

Horizontal asymptote: $y = 0$ Horizontal asymptote: $y = 0$

7. (a) $2x^3 - 3$ (b) $(2x - 3)^3$ (c) -19 (d) 125

8. $f(g(x)) = 3 - 5\left[\frac{1}{5}(3 - x)\right]$
$= 3 - 3 + x = x$
$g(f(x)) = \frac{1}{5}[3 - (3 - 5x)]$
$= \frac{1}{5}(5x) = x$

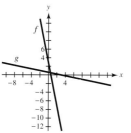

9. $h^{-1}(x) = \frac{1}{10}(x - 3)$ **10.** $g^{-1}(t) = \sqrt[3]{2(t - 2)}$

11. $9^{-2} = \frac{1}{81}$ **12.** $\log_3 81 = 4$ **13.** 3

14.

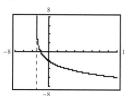

15.

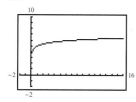

Vertical asymptote: $t = -3$ Vertical asymptote: $x = 0$

16. $h = 2, k = 1$ **17.** 6.0639

18.

n	1	4	12	365	Continuous compounding
A	\$3185.89	\$3314.90	\$3345.61	\$3360.75	\$3361.27

19. 1.60 grams

Section 9.4 *(page 618)*

Review *(page 618)*

1. $\sqrt[n]{uv}$ **2.** $\sqrt[n]{\dfrac{u}{v}}$ **3.** Different indices

4. No; $\dfrac{1}{\sqrt{2x}} = \dfrac{\sqrt{2x}}{2x}$ **5.** $19\sqrt{3x}$ **6.** $x - 9$

7. $\sqrt{5u}$ **8.** $4t + 12\sqrt{t} + 9$ **9.** $25\sqrt{2}x$

10. $6\left(\sqrt{t + 2} - \sqrt{t}\right)$ **11.** 22 units   **12.** \$2300

1. 3   **3.** -4 **5.** $\frac{1}{3}$ **7.** 0   **9.** -6 **11.** 2

13. 2   **15.** 1   **17.** 2   **19.** -3 **21.** 12   **23.** 1

25. 1   **27.** 1.2925   **29.** 0.2925   **31.** 0.2500

33. 2.7925   **35.** 0   **37.** 2.1972   **39.** 3.5835

41. 1.7918   **43.** $\log_3 11 + \log_3 x$ **45.** $2 \log_7 x$

47. $-2 \log_5 x$ **49.** $\frac{1}{2}(\log_4 3 + \log_4 x)$ **51.** $\ln 3 + \ln y$

53. $\log_2 z - \log_2 17$ **55.** $\frac{1}{2}\log_9 x - \log_9 12$

57. $2 \ln x + \ln(y - 2)$ **59.** $6 \log_4 x + 2 \log_4(x - 7)$

61. $\frac{1}{3}\log_3(x + 1)$ **63.** $\frac{1}{2}[\ln x + \ln(x + 2)]$

65. $2[\ln(x + 1) - \ln(x - 1)]$ **67.** $\frac{1}{3}[2 \ln x - \ln(x + 1)]$

69. $\ln x + 2 \ln y - 3 \ln z$

71. $\frac{1}{2}(7 \log_3 x - 5 \log_3 y - 8 \log_3 z)$

73. $\log_6 a + \frac{1}{2} \log_6 b + 3 \log_6(c - d)$

75. $\ln(x + y) + \frac{1}{5}\ln(w + 2) - (\ln 3 + \ln t)$

77. $\log_{12} \dfrac{x}{3}$ **79.** $\log_2 3x$ **81.** $\log_{10} \dfrac{4}{x}$

83. $\ln b^4, \ b > 0$ **85.** $\log_5(2x)^{-2}, \ x > 0$

87. $\log_2 x^7 z^3$ **89.** $\log_3 2\sqrt{y}$

91. $\ln \dfrac{x^2 y^3}{z}, \ x > 0, y > 0, z > 0$

93. $\ln \dfrac{2^5 y^3}{x}, \ x > 0, y > 0$ **95.** $\ln(xy)^4, \ x > 0, y > 0$

97. $\ln\left(\dfrac{x}{x + 1}\right)^2, \ x > 0$ **99.** $\log_4 \dfrac{x + 8}{x^3}, \ x > 0$

101. $\log_5 \dfrac{\sqrt{x + 2}}{x - 3}$ **103.** $\log_6 \dfrac{(c + d)^5}{\sqrt{m - n}}$

105. $\log_2 \sqrt[5]{\dfrac{x^3}{y^4}}, \ y > 0$ **107.** $\log_6 \dfrac{\sqrt[5]{x - 3}}{x^2(x + 1)^3}, \ x > 3$

109. $2 + \ln 3$ **111.** $1 + \frac{1}{2}\log_5 2$ **113.** $1 - 2 \log_4 x$

115.

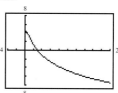

117.

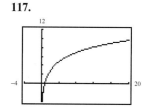

119. $B = 10(\log_{10} I + 16)$; 60 decibels

121. $E = \log_{10}\left(\dfrac{C_2}{C_1}\right)^{1.4}$ **123.** True   **125.** True

127. False; $\log_3(u + v)$ does not simplify.   **129.** True

131. False; 0 is not in the domain of f.

133. False; $f(x - 3) = \ln(x - 3)$.

135. False; If $f(u) = 2f(v)$, then $\ln u = 2 \ln v = \ln v^2 \implies$
$u = v^2$.

137. $\log_b\left(\dfrac{x}{xx}\right) = \log_b x - (\log_b x + \log_b x)$

139. $\ln 1 = 0$ $\ln 9 \approx 2.1972$

      $\ln 2 \approx 0.6931$ $\ln 10 \approx 2.3025$

      $\ln 3 \approx 1.0986$ $\ln 12 \approx 2.4848$

      $\ln 4 \approx 1.3862$ $\ln 14 \approx 2.6390$

      $\ln 5 \approx 1.6094$ $\ln 15 \approx 2.7080$

      $\ln 6 \approx 1.7917$ $\ln 16 \approx 2.7724$

      $\ln 7 \approx 1.9459$ $\ln 18 \approx 2.8903$

      $\ln 8 \approx 2.0793$ $\ln 20 \approx 2.9956$

Explanations will vary. Any differences are due to round-off errors.

Section 9.5   *(page 628)*

Review   *(page 628)*

1. No. A system of linear equations has no solutions, one solution, or an infinite number of solutions.

2. The equations represent parallel lines and therefore have no point of intersection.

3. 2   **4.** $5 \pm 2\sqrt{2}$ **5.** $-\dfrac{1}{2}$ **6.** $\dfrac{5}{3}$

7. 1, 7   **8.** 47

9. $d = 73t$ **10.** $V = 25\pi h$

11. $V = 10\pi r^2$ **12.** $F = 25x$

1. (a) Not a solution   (b) Solution

3. (a) Solution   (b) Not a solution

5. (a) Not a solution   (b) Solution

7. 3   **9.** -3 **11.** 4   **13.** 1   **15.** -2

17. -3 **19.** -6 **21.** 2   **23.** $\dfrac{22}{5}$ **25.** 6

27. 9   **29.** 4   **31.** No solution   **33.** -7

35. $2x - 1$ **37.** $2x, \; x > 0$ **39.** 4.11   **41.** 1.31

43. 1.49   **45.** 0.83   **47.** -2.37 **49.** -3.60

51. 2.64   **53.** 3.00   **55.** 1.23   **57.** 35.35

59. 0.80   **61.** 12.22   **63.** 3.28   **65.** No solution

67. -1.04 **69.** 2.48   **71.** 0.90   **73.** 0.38

75. 0.39   **77.** 8.99   **79.** 9.73   **81.** 4.62

83. 0.10   **85.** 174.77   **87.** 2187.00   **89.** 6.52

91. 25.00   **93.** 10.04   **95.** ± 20.09 **97.** 3.00

99. 19.63   **101.** 12.18   **103.** 2000.00   **105.** 3.20

107. 4.00   **109.** 0.75   **111.** 5.00   **113.** 2.46

115. 2.29   **117.** 6.00

119. $(1.40, 0)$ **121.** $(21.82, 0)$

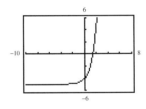   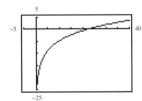

123. 0.69   **125.** 1.48

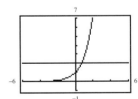   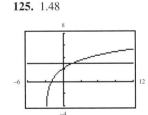

127. 9%   **129.** 7.70 years

131. $10^{-8.5}$ watts per square centimeter   **133.** 205

135. (a) 3.64 months

     (b) 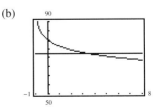   (c) Answers will vary.

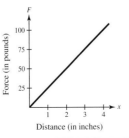

137. (a) $k = \frac{1}{4}\ln\frac{8}{15} \approx -0.1572$ (b) ≈ 3.25 hours

(c) ≈ 2.84 hours

139. (c) 7.2% (d) $\dfrac{\ln 1.5}{0.06} \approx 6.7578$ or $6\frac{3}{4}$ years (e) 8.24%

(f) Double: 11.6 years; Quadruple: 23.1 years

141. $2^{x-1} = 30$

143. To solve an exponential equation, first isolate the exponential expression, then take the logarithms of both sides of the equation, and solve for the variable.

To solve a logarithmic equation, first isolate the logarithmic expression, then exponentiate both sides of the equation, and solve for the variable.

Section 9.6 *(page 638)*

Review *(page 638)*

1. Direct variation as nth power

2. Inverse variation **3.** Joint variation

4. Combined variation **5.** $(3, 3)$ **6.** $\left(\frac{10}{3}, \frac{5}{3}\right)$

7. $(2, 4), \left(-\frac{1}{2}, \frac{1}{4}\right)$ **8.** $(0, 0), (8, 2)$ **9.** $\left(\frac{3}{5}, \frac{8}{5}, \frac{2}{5}\right)$

10. $(4, -1, 3)$

11. (a) Downward, because the equation is of quadratic type $y = ax^2 + bx + c$, and $a < 0$.

(b) $(0, 0), (4, 0)$ (c) $(2, 4)$

12.

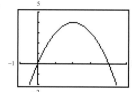

1. 7% **3.** 9% **5.** 8% **7.** 9.27 years

9. 8.66 years **11.** 9.59 years **13.** Continuous

15. Quarterly **17.** 8.33% **19.** 7.23% **21.** 6.14%

23. 8.30%

25. No. Each time the amount is divided by the principal, the result is always 2.

27. $\$1652.99$ **29.** $\$626.46$ **31.** $\$3080.15$

33. $\$951.23$ **35.** $\$5496.57$ **37.** $\$320,250.81$

39. Total deposits: $\$7200.00$; Total interest: $\$10,529.42$

41. $k = \frac{1}{2}\ln\frac{8}{3} \approx 0.4904$ **43.** $k = \frac{1}{3}\ln\frac{1}{2} \approx -0.2310$

45. $y = 19.1e^{0.0083t}$; 24.5 million

47. $y = 1275.1e^{0.0057t}$; 1512.9 million

49. $y = 3.8e^{0.0085t}$; 4.9 million

51. $y = 283.2e^{0.0081t}$; 361.1 million

53. (a) k is larger in Exercise 49, because the population of Ireland is increasing faster than the population of China.

(b) k corresponds to r; k gives the annual percent rate of growth.

55. (a) $y = 100e^{4.6052t}$ $(t = 0 \leftrightarrow 2000)$ (b) 1.45 hours

57. $1,149,000,000$ users

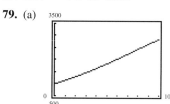

Isotope	Half-Life (Years)	Initial Quantity	Amount After 1000 Years
59. ^{226}Ra	1620	6 g	3.91 g
61. ^{14}C	5730	4.51 g	4.0 g
63. ^{230}Pu	24,360	4.2 g	4.08 g

65. 3.3 grams **67.** 7.5 grams **69.** $\$15,203$

71. The earthquake in Alaska was about 6.3 times greater.

73. The earthquake in Mexico was about 1259 times greater.

75. 7.04 **77.** 10^7 times

79. (a)

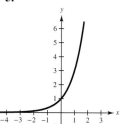

(b) 1000 rabbits

(c) 2642 rabbits

(d) 5.88 years

81. (a) $S = 10(1 - e^{-0.0575x})$ (b) 3314 jeans

83. $k > 0$

85. A is the balance, P is the principal, r is the annual interest rate, and t is the time in years.

87. The time required for the radioactive material to decay to half of its original amount

Review Exercises *(page 644)*

1. (a) $\frac{1}{8}$ (b) 2 (c) 4 **3.** (a) 5 (b) 0.185 (c) $\frac{1}{25}$

5.

7.

Horizontal asymptote:
$y = 0$

Horizontal asymptote:
$y = -1$

9.

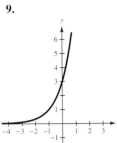

Horizontal asymptote:
$y = 0$

11.

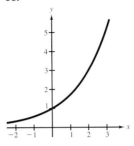

Horizontal asymptote:
$y = 0$

13.

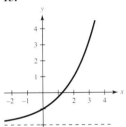

Horizontal asymptote: $y = -2$

15.

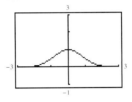

17.

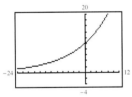

19. (a) 0.007   (b) 3   (c) 9.56×10^{16}

21.

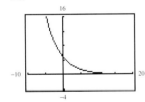

23.

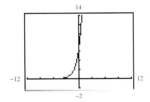

25.

n	1	4	12
A	\$226,296.28	\$259,889.34	\$268,503.32

n	365	Continuous
A	\$272,841.23	\$272,990.75

27. ≈ 4.21 grams   **29.** (a) 6   (b) 1

31. (a) 5   (b) -1

33. (a) $(f \circ g)(x) = \sqrt{2x - 4}$ (b) $(g \circ f)(x) = 2\sqrt{x - 4}$

   Domain: $[2, \infty)$ Domain: $[4, \infty)$

35. No   **37.** Yes   **39.** $f^{-1}(x) = \frac{1}{3}(x - 4)$

41. $h^{-1}(x) = x^2,\ x \geq 0$ **43.** $f^{-1}(t) = \sqrt[3]{t - 4}$

45.

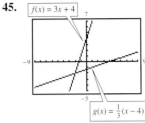

47.

x	0	1	3	4
f^{-1}	6	4	2	0

49.

x	-4	-2	2	3
f^{-1}	-2	-1	1	3

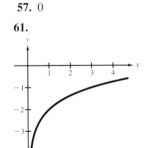

51. 3   **53.** -2 **55.** 6   **57.** 0

59.

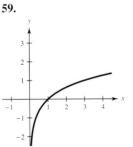

Vertical asymptote: $x = 0$

61.

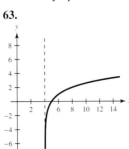

Vertical asymptote: $x = 0$

63.

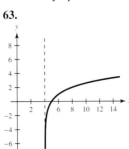

Vertical asymptote: $x = 4$

65. 7

67.

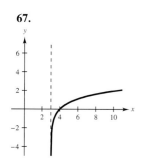

69.

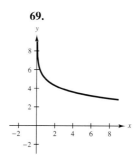

2.

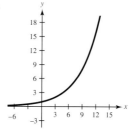

Horizontal asymptote: $y = 0$

Vertical asymptote: $x = 3$ Vertical asymptote: $x = 0$

71. 1.5850   **73.** 2.1322   **75.** 1.7959   **77.** -0.4307

79. 1.0293   **81.** $\log_4 6 + 4 \log_4 x$ **83.** $\frac{1}{2}\log_5(x + 2)$

85. $\ln(x + 2) - \ln(x - 2)$

87. $\frac{1}{2}(\ln 2 + \ln x) + 5 \ln(x + 3)$ **89.** $\ln\left(\dfrac{1}{3y}\right)^{2/3}$

91. $\log_8 32x^3$ **93.** $\ln\dfrac{9}{4x^2}$, $x > 0$

95. $\log_2\left(\dfrac{k}{k - t}\right)^4$, $k > t$

97. $\ln(x^3 y^4 z)$, $x > 0$, $y > 0$, $z > 0$

99. False; $\log_2 4x = 2 + \log_2 x$.   **101.** True

103. True   **105.** $y = 0.83(\ln I_0 - \ln I)$; 0.20   **107.** 6

109. 1   **111.** 6   **113.** 5.66   **115.** 6.23

117. No solution   **119.** 1408.10   **121.** 15.81

123. 4   **125.** 64   **127.** 2.67   **129.** 7%   **131.** 5%

133. 7.5%   **135.** 7%   **137.** 5.65%   **139.** 7.71%

141. 7.79%

Isotope	Half-Life (Years)	Initial Quantity	Amount After 1000 Years
143. ^{226}Ra	1620	3.5 g	2.282 g
145. ^{14}C	5730	2.934g	2.6 g
147. ^{230}Pu	24,360	5 g	4.860 g

149. The earthquake in San Francisco was about 2512 times greater.

Chapter Test   *(page 649)*

1. $f(-1) = 81$;

$f(0) = 54$;

$f\left(\frac{1}{2}\right) = 18\sqrt{6} \approx 44.09$;

$f(2) = 24$

3. (a) $(f \circ g)(x) = 18x^2 - 63x + 55$;

Domain: $(-\infty, \infty)$

(b) $(g \circ f)(x) = -6x^2 - 3x + 5$;

Domain: $(-\infty, \infty)$

4. $f^{-1}(x) = \frac{1}{5}(x - 6)$

5. $(f \circ g)(x) = -\frac{1}{2}(-2x + 6) + 3$

$= (x - 3) + 3$

$= x$

$(g \circ f)(x) = -2\left(-\frac{1}{2}x + 3\right) + 6$

$= (x - 6) + 6$

$= x$

6. $\frac{1}{3}$

7. $g = f^{-1}$

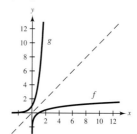

8. $\log_4 5 + 2 \log_4 x - \frac{1}{2} \log_4 y$ **9.** $\ln \dfrac{x}{y^4}$, $y > 0$

10. 32   **11.** 1.18   **12.** 13.73   **13.** 15.52

14. 2   **15.** 8   **16.** 109.20   **17.** 0

18. (a) $8012.78   (b) $8110.40   **19.** $10,806.08

20. 7%   **21.** $8469.14   **22.** 600   **23.** 1141

24. 4.4 years

Chapter 10

Section 10.1   (page 659)

Review   (page 659)

1. $x^2 + 12x + 31$ 2. $x^2 - 14x + 47$
3. $-x^2 + 16x - 52$ 4. $-x^2 - 2x + 15$
5. $(x - 2)^2 - 3$ 6. $(x + 6)^2 - 39$
7. $-(x - 3)^2 + 14$ 8. $-2\left(x - \frac{5}{2}\right)^2 - \frac{3}{2}$
9. $y = \frac{5}{8}x + \frac{25}{4}$ 10. $y = -\frac{2}{5}x + \frac{37}{5}$
11. 12 people   12. 15 people

1. c   2. e   3. a   4. d   5. b   6. f
7. $x^2 + y^2 = 25$ 9. $x^2 + y^2 = \frac{4}{9}$ 11. $x^2 + y^2 = 64$
13. $x^2 + y^2 = 29$ 15. $(x - 4)^2 + (y - 3)^2 = 100$
17. $(x - 5)^2 + (y + 3)^2 = 81$
19. $(x + 2)^2 + (y - 1)^2 = 4$
21. $(x - 3)^2 + (y - 2)^2 = 17$
23. Center: $(0, 0)$; $r = 4$ 25. Center: $(0, 0)$; $r = 6$

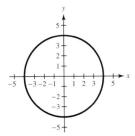 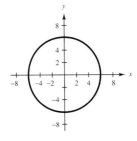

27. Center: $(0, 0)$; $r = \frac{1}{2}$ 29. Center: $(0, 0)$; $r = \frac{12}{5}$

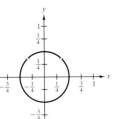 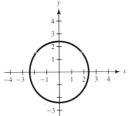

31. Center: $(-1, 5)$; $r = 8$ 33. Center: $(2, 3)$; $r = 2$

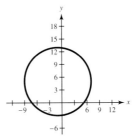 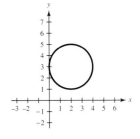

35. Center: $\left(-\frac{5}{2}, -3\right)$;   37. Center: $(2, 1)$;
    $r = 3$ $r = 2$

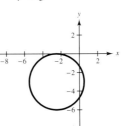 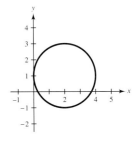

39. Center: $(-1, -3)$;   41. Center: $(-4, -2)$;
    $r = 2$ $r = 5$

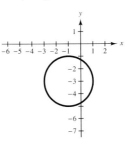 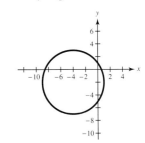

43.   45.

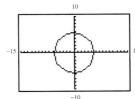 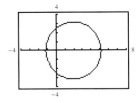

47. b   48. c   49. e   50. a   51. f   52. d
53. $x^2 = \frac{3}{2}y$ 55. $x^2 = -6y$ 57. $y^2 = -8x$
59. $x^2 = 4y$ 61. $y^2 = 16x$ 63. $y^2 = 9x$
65. $(x - 3)^2 = -(y - 1)$ 67. $y^2 = 2(x + 2)$
69. $(y - 2)^2 = -8(x - 3)$ 71. $x^2 = 8(y - 4)$
73. $(y - 2)^2 = x$

75. Vertex: $(0, 0)$

Focus: $\left(0, \frac{1}{2}\right)$

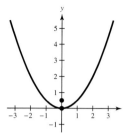

77. Vertex: $(0, 0)$

Focus: $\left(-\frac{3}{2}, 0\right)$

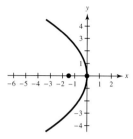

79. Vertex: $(0, 0)$

Focus: $(0, -2)$

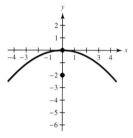

81. Vertex: $(1, -2)$

Focus: $(1, -4)$

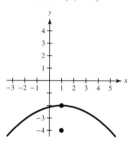

83. Vertex: $\left(5, -\frac{1}{2}\right)$

Focus: $\left(\frac{11}{2}, -\frac{1}{2}\right)$

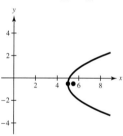

85. Vertex: $(1, 1)$

Focus: $(1, 2)$

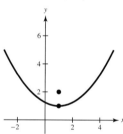

87. Vertex: $(-2, -3)$

Focus: $(-4, -3)$

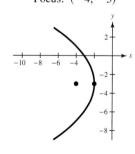

89. Vertex: $(-2, 1)$

Focus: $\left(-2, -\frac{1}{2}\right)$

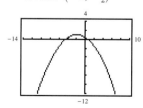

91. Vertex: $\left(\frac{1}{4}, -\frac{1}{2}\right)$

Focus: $\left(0, -\frac{1}{2}\right)$

93. $x^2 + y^2 = 4500^2$

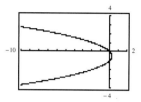

95. $y = \sqrt{2500 - x^2}$; $5\sqrt{19} \approx 21.8$ feet

97. (a) $y = \dfrac{x^2}{180}$　(b)

x	0	20	40	60
y	0	$2\frac{2}{9}$	$8\frac{8}{9}$	20

99. (a) 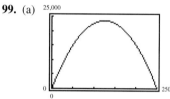　(b) $x = 125$

101. (a) $x^2 + y^2 = 25$　(b)

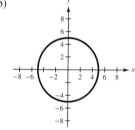

103. A circle is the set of all points (x, y) that are a given positive distance r from a fixed point (h, k) called the center.　$x^2 + y^2 = r^2$

105. All points on the parabola are equidistant from the directrix and the focus.

107. No. If the graph intersected the directrix, there would exist points nearer the directrix than the focus.

Section 10.2　(*page 670*)

Review　(*page 670*)

1. $\dfrac{3}{2y^5}$　**2.** $\dfrac{4y^2}{9x^4}$　**3.** $\dfrac{x^3 y}{6}$　**4.** 1

5. $-3 \pm \sqrt{13}$　**6.** $4 \pm \dfrac{3\sqrt{6}}{2}$

7.

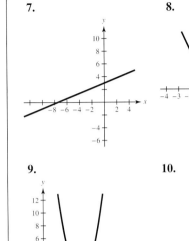

8.

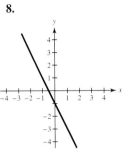

9.

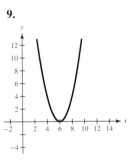

10.

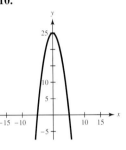

11. 87   **12.** $3750

23.

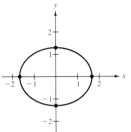

Vertices: $\left(\pm\frac{5}{3}, 0\right)$
Co-vertices: $\left(0, \pm\frac{4}{3}\right)$

25.

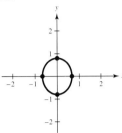

Vertices: $\left(0, \pm\frac{4}{5}\right)$
Co-vertices: $\left(\pm\frac{2}{3}, 0\right)$

27.

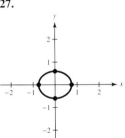

Vertices: $\left(\pm\frac{3}{4}, 0\right)$;
Co-vertices: $\left(0, \pm\frac{3}{5}\right)$

29.

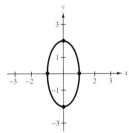

Vertices: $(0, \pm 2)$;
Co-vertices: $(\pm 1, 0)$

1. a   **2.** f   **3.** d   **4.** c   **5.** e   **6.** b

7. $\dfrac{x^2}{16} + \dfrac{y^2}{9} = 1$ **9.** $\dfrac{x^2}{4} + \dfrac{y^2}{1} = 1$ **11.** $\dfrac{x^2}{9} + \dfrac{y^2}{16} = 1$

13. $\dfrac{x^2}{1} + \dfrac{y^2}{4} = 1$ **15.** $\dfrac{x^2}{9} + \dfrac{y^2}{25} = 1$

17. $\dfrac{x^2}{100} + \dfrac{y^2}{36} = 1$

31.

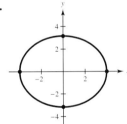

Vertices: $(\pm 4, 0)$
Co-vertices: $\left(0, \pm\sqrt{10}\right)$

19.

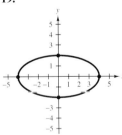

Vertices: $(\pm 4, 0)$
Co-vertices: $(0, \pm 2)$

21.

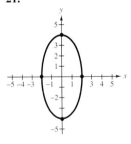

Vertices: $(0, \pm 4)$
Co-vertices: $(\pm 2, 0)$

33.

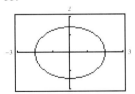

Vertices: $(\pm 2, 0)$

35.

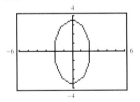

Vertices: $\left(0, \pm 2\sqrt{3}\right)$

37. $\dfrac{x^2}{1} + \dfrac{y^2}{4} = 1$ **39.** $\dfrac{(x-4)^2}{9} + \dfrac{y^2}{16} = 1$

41. Center: $(-5, 0)$

Vertices: $(-9, 0), (-1, 0)$

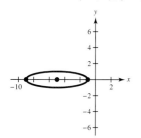

43. Center: $(1, 5)$

Vertices: $(1, 0), (1, 10)$

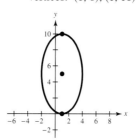

45. Center: $(-2, 3)$

Vertices: $(-2, 6), (-2, 0)$

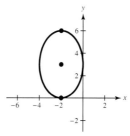

47. Center: $(2, -2)$

Vertices: $(-1, -2), (5, -2)$

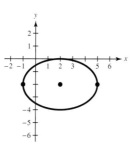

49. Center: $(-4, 1)$

Vertices: $(-4, -3), (-4, 5)$

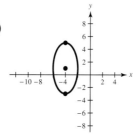

51. Center: $(4, -3)$

Vertices: $(4, -8), (4, 2)$

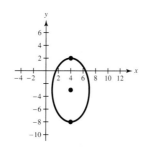

53. Center: $(2, 1)$

Vertices: $(-8, 1), (12, 1)$

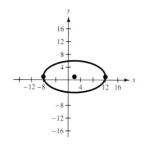

55. $\sqrt{304} \approx 17.4$ feet

57. $\dfrac{x^2}{7225} + \dfrac{(y - 85)^2}{10,000} = 1$ or $\dfrac{x^2}{10,000} + \dfrac{(y - 85)^2}{7225} = 1$

59. $\dfrac{x^2}{144} + \dfrac{y^2}{64} = 1$

61. A circle is an ellipse in which the major axis and the minor axis have the same length. Both circles and ellipses have foci; however, in a circle the foci are both at the same point, whereas in an ellipse they are not.

63. The sum of the distances between each point on the ellipse and the two foci is a constant.

65. Major axis: $2a$; Minor axis: $2b$

Mid-Chapter Quiz   *(page 674)*

1. $x^2 + y^2 = 25$ **2.** $(y - 1)^2 = 8(x + 2)$

3. $\dfrac{(x + 2)^2}{16} + \dfrac{(y + 1)^2}{4} = 1$

4. $(x - 3)^2 + (y + 5)^2 = 25$

5. $(x - 2)^2 = -8(y - 3)$ **6.** $\dfrac{x^2}{9} + \dfrac{y^2}{100} = 1$

7. $(x - 5)^2 + y^2 = 9$; Center: $(5, 0)$; $r = 3$

8. $(x + 1)^2 + (y - 2)^2 = 1$; Center: $(-1, 2)$; $r = 1$

9. $(y - 3)^2 = x + 16$; Vertex: $(-16, 3)$; Focus: $\left(-\frac{63}{4}, 3\right)$

10. $(x - 4)^2 = -(y - 4)$; Vertex: $(4, 4)$; Focus: $\left(4, \frac{15}{4}\right)$

11. $\dfrac{x^2}{9} + \dfrac{y^2}{20} = 1$

Center: $(0, 0)$

Vertices: $\left(0, -2\sqrt{5}\right), \left(0, 2\sqrt{5}\right)$

12. $\dfrac{(x - 6)^2}{9} + \dfrac{(y + 2)^2}{4} = 1$

Center: $(6, -2)$

Vertices: $(3, -2), (9, -2)$

13.

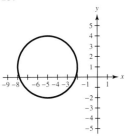

14.

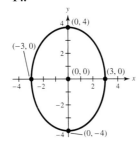

15.

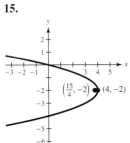

16.

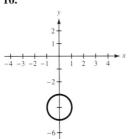

17.

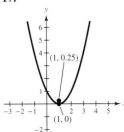

18.

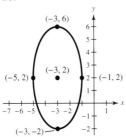

Section 10.3 *(page 680)*

Review *(page 680)*

1. $2\sqrt{10}$ **2.** $\sqrt{269}$

3.

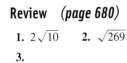

4.

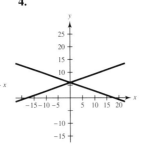

5.

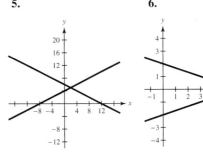

6.

7. 24   **8.** 5   **9.** 13   **10.** $3\sqrt{3}$

11. $2\frac{1}{2}$ hours   **12.** $10 coins: 6; $20 coins: 24

1. c   **2.** e   **3.** a   **4.** f   **5.** b   **6.** d

7.

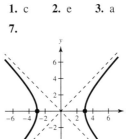

Vertices: $(\pm 3, 0)$
Asymptotes: $y = \pm x$

9.

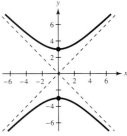

Vertices: $(0, \pm 3)$
Asymptotes: $y = \pm x$

11.

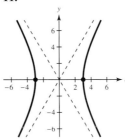

Vertices: $(\pm 3, 0)$
Asymptotes: $y = \pm \frac{5}{3}x$

13.

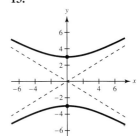

Vertices: $(0, \pm 3)$
Asymptotes: $y = \pm \frac{3}{5}x$

15.

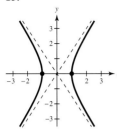

17.

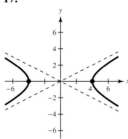

Vertices: $(\pm 1, 0)$

Vertices: $(\pm 4, 0)$

Asymptotes: $y = \pm \frac{3}{2}x$

Asymptotes: $y = \pm \frac{1}{2}x$

19. $\dfrac{x^2}{16} - \dfrac{y^2}{64} = 1$ **21.** $\dfrac{y^2}{16} - \dfrac{x^2}{64} = 1$

23. $\dfrac{x^2}{81} - \dfrac{y^2}{36} = 1$ **25.** $\dfrac{y^2}{1} - \dfrac{x^2}{1/4} = 1$

27.

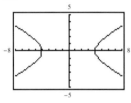

29.

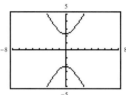

31. Center: $(3, -4)$

   Vertices: $(3, 1), (3, -9)$

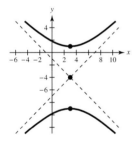

33. Center: $(1, -2)$

   Vertices: $(-1, -2), (3, -2)$

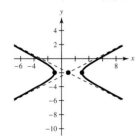

35. Center: $(2, -3)$

   Vertices: $(3, -3), (1, -3)$

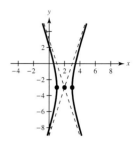

37. Center: $(-3, 2)$

   Vertices: $(-4, 2), (-2, 2)$

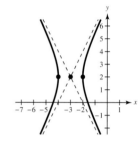

39. $\dfrac{y^2}{9} - \dfrac{x^2}{9/4} = 1$ **41.** $\dfrac{(x-3)^2}{4} - \dfrac{(y-2)^2}{16/5} = 1$

43. $x \approx 110.28$

45. (c) Center: $(2, 0)$

   Vertices: $\left(2 + 3\sqrt{3}, 0\right), \left(2 - 3\sqrt{3}, 0\right)$

   (d)

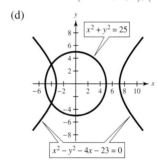

47. The difference of the distances between each point on the hyperbola and the two foci is a positive constant.

49. (a) Left half   (b) Top half

Section 10.4   *(page 690)*

Review   *(page 690)*

1. The second row in the new matrix was formed by multiplying the second row of the original matrix by -2.

2. Rows one and two were swapped.

3. The second row in the new matrix was formed by subtracting 3 times the first row from the second row of the original matrix.

4. Gaussian elimination is the process of using elementary row operations to rewrite a matrix representing the system of linear equations in row-echelon form.

5. $x = -2, y = 3$ **6.** $x = 9, y = -4$

7. $x = -\frac{22}{3}, y = \frac{37}{9}$ **8.** $x = \frac{41}{40}, y = -\frac{11}{40}$

9. $x = 2, y = -1, z = 3$

10. $x = -2, y = \frac{3}{5}, z = \frac{12}{5}$ **11.** $\$5520$

12. Cashews: 20 pounds; Brazil nuts: 30 pounds

1.

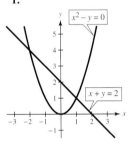

$(-2, 4), (1, 1)$

3.

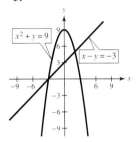

$(2, 5), (-3, 0)$

5.

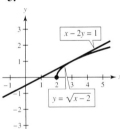

$(3, 1)$

7.

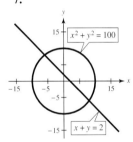

$(-6, 8), (8, -6)$

9.

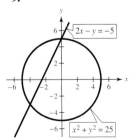

$(0, 5), (-4, -3)$

11.

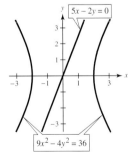

No real solution

13.

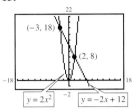

$(-3, 18), (2, 8)$

15.

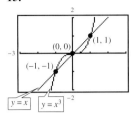

$(0, 0), (1, 1), (-1, -1)$

17.

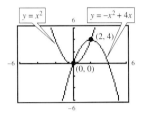

$(0, 0), (2, 4)$

19.

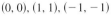

$(-4, 14), (1, -1)$

21.

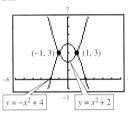

$(-1, 3), (1, 3)$

23.

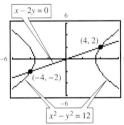

$(4, 2), (-4, -2)$

$(0, 0), (1, 1)$

25.

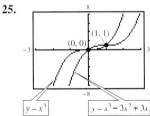

27. $(1, 2), (2, 8)$ **29.** $(0, 5), (2, 1)$ **31.** No real solution

33. $(0, 5)$ **35.** $(0, 8), \left(-\frac{24}{5}, -\frac{32}{5}\right)$

37. $\left(-\frac{17}{2}, -6\right), \left(-\frac{7}{2}, 4\right)$ **39.** $\left(-\frac{9}{5}, \frac{12}{5}\right), (3, 0)$

41. $(-14, -20), (2, -4)$ **43.** $(-4, 11), \left(\frac{5}{2}, \frac{5}{4}\right)$

45. $(0, 2), (3, 1)$ **47.** No real solution

49. $\left(\pm\sqrt{5}, 2\right), (0, -3)$ **51.** $(0, 4), (3, 0)$

53. No real solution   **55.** $\left(\pm\sqrt{3}, -1\right)$

57. $\left(2, \pm 2\sqrt{3}\right), (-1, \pm 3)$ **59.** $\left(\pm 2, \pm\sqrt{3}\right)$

61. $(\pm 2, 0)$ **63.** $(\pm 3, \pm 2)$ **65.** $(\pm 3, 0)$

67. $\left(\pm \sqrt{6}, \pm \sqrt{3}\right)$ **69.** No real solution

71. $\left(\pm \sqrt{5}, \pm 2\right)$ **73.** $\left(\pm \dfrac{2\sqrt{5}}{5}, \pm \dfrac{2\sqrt{5}}{5}\right)$

75. $\left(\pm \sqrt{3}, \pm \sqrt{13}\right)$ **77.** $(3.633, 2.733)$

79. ≈ 21 inches $\times 36$ inches   **81.** 15 inches $\times 8$ inches

83. Between points $\left(-\frac{3}{5}, -\frac{4}{5}\right)$ and $\left(\frac{4}{5}, -\frac{3}{5}\right)$

85. (e) $(-4, 3)$; This is a point of intersection between the hyperbola and the circle.

    (f) $y = -\frac{3}{4}x$

    (g) $(4, -3)$; This is the other point at which the line representing Murphy Road intersects the circle.

87. Multiply Equation 2 by a factor that makes the coefficients of one variable equal. Subtract Equation 2 from Equation 1. Write the resulting equation, and solve. Substitute the solution into either equation. Solve for the value of the other variable.

Review Exercises *(page 695)*

1. Hyperbola   **3.** Circle   **5.** Circle   **7.** Parabola

9. $x^2 + y^2 = 144$

11.

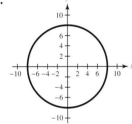

Center: $(0, 0)$; $r = 8$

13. $(x - 3)^2 + (y - 5)^2 = 25$

15.

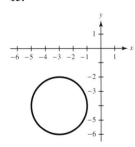

Center: $(-3, -4)$; $r = 2$

17. $y^2 = -8x$

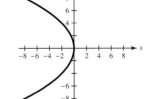

19. $(x + 6)^2 = -20(y - 4)$ **21.** $(x + 1)^2 = \frac{1}{2}(y - 3)$

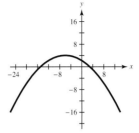   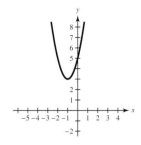

23.

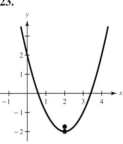

Vertex: $(2, -2)$; Focus: $\left(2, -\frac{7}{4}\right)$

25. $\dfrac{x^2}{4} + \dfrac{y^2}{25} = 1$ **27.** $\dfrac{x^2}{4} + \dfrac{y^2}{9} = 1$

29.   **31.**

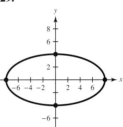   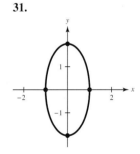

Vertices: $(\pm 8, 0)$ Vertices: $(0, \pm 2)$

Co-vertices: $(0, \pm 4)$ Co-vertices: $(\pm 1, 0)$

33. $\dfrac{(x - 3)^2}{25} + \dfrac{(y - 4)^2}{16} = 1$ **35.** $\dfrac{x^2}{9} + \dfrac{(y - 4)^2}{16} = 1$

37.   **39.**

Center: $(1, -2)$ Center: $(0, -3)$

Vertices: $(1, -11), (1, 7)$ Vertices: $(0, -7), (0, 1)$

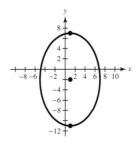   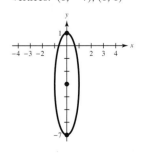

41.

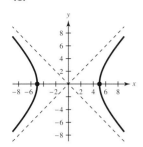

43.

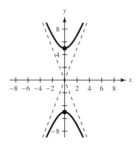

Vertices: $(\pm 5, 0)$

Asymptotes: $y = \pm x$

Vertices: $(0, \pm 5)$

Asymptotes: $y = \pm \frac{5}{2}x$

45. $\dfrac{x^2}{4} - \dfrac{y^2}{9} = 1$

47. $\dfrac{y^2}{25} - \dfrac{x^2}{4} = 1$

49. Center: $(3, -1)$

Vertices: $(0, -1), (6, -1)$

51. Center: $(4, -3)$

Vertices: $(4, -1), (4, -5)$

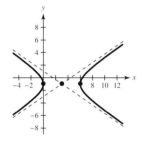

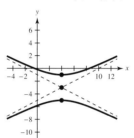

53. $\dfrac{(x + 4)^2}{4} - \dfrac{(y - 6)^2}{12} = 1$

55.

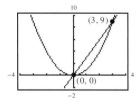

57.

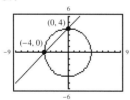

$(0, 0), (3, 9)$

$(-4, 0), (0, 4)$

59. $(-1, 5), (-2, 20)$ **61.** $(-1, 0), (0, -1)$

63. $(0, 2), \left(-\frac{16}{5}, -\frac{6}{5}\right)$ **65.** $(\pm 5, 0)$

67. 4 inches $\times$ 5 inches

69. 6 centimeters $\times$ 8 centimeters

71. Piece 1: 38.48 inches; Piece 2: 61.52 inches

Chapter Test   *(page 699)*

1. $(x + 1)^2 + (y - 3)^2 = 4$

2. $(x - 1)^2 + (y - 3)^2 = 9$ **3.** $(x + 2)^2 + (y - 3)^2 = 9$

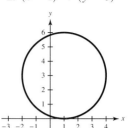

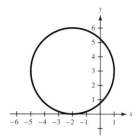

4. Vertex: $(4, 2)$; Focus: $\left(\frac{47}{12}, 2\right)$

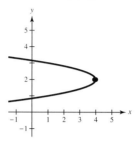

5. $(x - 7)^2 = 8(y + 2)$ **6.** $\dfrac{(x - 2)^2}{25} + \dfrac{y^2}{9} = 1$

7. Center: $(0, 0)$;

Vertices: $(0, \pm 4)$

8. Center: $(2, -4)$;

Vertices: $(2, -7), (2, -1)$

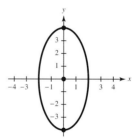

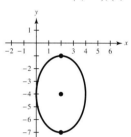

9. $\dfrac{x^2}{9} - \dfrac{y^2}{4} = 1$ **10.** $\dfrac{y^2}{4} - x^2 = 1$

11. Center: $(0, 3)$

Vertices: $(\pm 2, 3)$

12. Center: $(4, -2)$

Vertices: $(4, -7), (4, 3)$

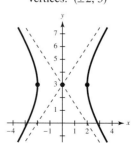

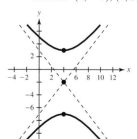

13. $(0, 3), (4, 0)$ **14.** $(\pm 4, 0)$

15. $\left(\sqrt{6}, 2\right), \left(\sqrt{6}, -2\right), \left(-\sqrt{6}, 2\right), \left(-\sqrt{6}, -2\right)$

16. $x^2 + y^2 = 25,000,000$ **17.** 16 inches $\times$ 12 inches

Cumulative Test: Chapters 8–10 *(page 700)*

1. $x = -\frac{3}{4}, 3$ **2.** $x = -3, 13$

3. $x = 5 \pm 5\sqrt{2}$ **4.** $x = -1 \pm \dfrac{\sqrt{3}}{3}$ **5.** $x = 16$

6. $\dfrac{-5 - \sqrt{61}}{6} \le x \le \dfrac{-5 + \sqrt{61}}{6}$

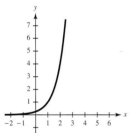

7. $-\dfrac{4}{3} < x < \dfrac{1}{2}$

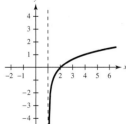

8. $x^2 - 4x - 12 = 0$

9. (a) $(f \circ g)(x) = 50x^2 - 20x - 1$; Domain: $(-\infty, \infty)$

 (b) $(g \circ f)(x) = 10x^2 - 16$; Domain: $(-\infty, \infty)$

10. $f^{-1}(x) = 4x - \dfrac{3}{2}$

11. $f(1) = \dfrac{15}{2}$, $f(0.5) = \dfrac{14 + \sqrt{2}}{2}$, $f(3) = \dfrac{57}{8}$

12.   Horizontal asymptote: $y = 0$

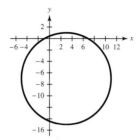

13. The graphs are reflections of each other in the line $y = x$.

14.   Vertical asymptote: $x = 1$

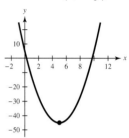

15. -2 **16.** $\log_2 \dfrac{x^3 y^3}{z}$ **17.** $\ln 5 + \ln x - 2\ln(x + 1)$

18. $x = 3$ **19.** $x = e^{5/2} \approx 12.182$ **20.** $t \approx 18.013$

21. $x \approx 0.867$ **22.** \$29.63   **23.** $\approx 8.33\%$

24. ≈ 15.403 years

25. $(x - 3)^2 + (y + 7)^2 = 64$ **26.** Vertex: $(5, -45)$

Focus: $\left(5, -\dfrac{359}{8}\right)$

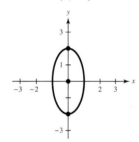

27. $\dfrac{(x - 3)^2}{9} + \dfrac{y^2}{4} = 1$

28. Center: $(0, 0)$ **29.** $x^2 - \dfrac{y^2}{4} = 1$

Vertices: $(0, \pm 2)$

30. Center: $(0, 1)$ **31.** $(1, -1), (3, 5)$

Vertices: $(\pm 12, 1)$

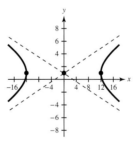

32. $\left(\pm\sqrt{3}, 0\right)$ **33.** 8 feet $\times$ 4 feet

34. (a)

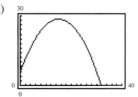

 (b) Highest point: 28.5 feet; Range: $[0, 28.5]$

Chapter 11

Section 11.1 *(page 710)*

Review *(page 710)*

1. $-7x = 35$

$$\frac{-7x}{-7} = \frac{35}{-7}$$

$$x = -5$$

2. $7x + 63 = 35$

$$7x + 63 - 63 = 35 - 63$$

$$7x = -28$$

3. It is a solution if the equation is true when -3 is substituted for t.

4. Multiply each side of the equation by the lowest common denominator; in this example, it is $x(x + 1)$.

5. $\dfrac{1}{(x + 10)^2}$ **6.** $18(x - 3)^3, x \neq 3$ **7.** $\dfrac{1}{a^8}$

8. $2x$ **9.** $8x\sqrt{2x}$ **10.** $\dfrac{5(\sqrt{x} + 2)}{x - 4}$

11. (a) $A = x(2x - 3)$

(b)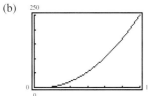

(c) $\dfrac{3 + \sqrt{1609}}{4} \approx 10.8$

12. (a) $A = \frac{1}{2}x(x - 4)$

(b)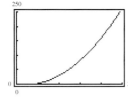

(c) $2\left(1 + \sqrt{101}\right) \approx 22.1$

1. $2, 4, 6, 8, 10$ **3.** $-2, 4, -6, 8, -10$

5. $\frac{1}{2}, \frac{1}{4}, \frac{1}{8}, \frac{1}{16}, \frac{1}{32}$ **7.** $\frac{1}{4}, -\frac{1}{8}, \frac{1}{16}, -\frac{1}{32}, \frac{1}{64}$

9. $3, 8, 13, 18, 23$ **11.** $1, \frac{4}{5}, \frac{2}{3}, \frac{4}{7}, \frac{1}{2}$ **13.** $\frac{3}{4}, \frac{2}{3}, \frac{9}{14}, \frac{12}{19}, \frac{5}{8}$

15. $-1, \frac{1}{4}, -\frac{1}{9}, \frac{1}{16}, -\frac{1}{25}$ **17.** $\frac{9}{2}, \frac{19}{4}, \frac{39}{8}, \frac{79}{16}, \frac{159}{32}$

19. $2, 3, 4, 5, 6$ **21.** $0, 3, -1, \frac{3}{4}, -\frac{1}{4}$ **23.** -72

25. $\frac{31}{2520}$ **27.** 5 **29.** $\frac{1}{132}$ **31.** $53{,}130$ **33.** $\dfrac{1}{n + 1}$

35. $n(n + 1)$ **37.** $2n$ **39.** c   **40.** a   **41.** b

42. d

43. 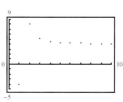   **45.**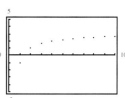

47. 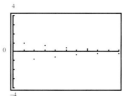   **49.** $a_n = 2n - 1$

51. $a_n = 4n - 2$ **53.** $a_n = n^2 - 1$

55. $a_n = (-1)^{n+1}2n$ **57.** $a_n = \dfrac{n + 1}{n + 2}$

59. $a_n = \dfrac{(-1)^{n+1}}{2^n}$ **61.** $a_n = \dfrac{1}{2^{n-1}}$ **63.** $a_n = 1 + \dfrac{1}{n}$

65. $a_n = \dfrac{1}{n!}$ **67.** 63 **69.** 77 **71.** 100 **73.** $\frac{3019}{3600}$

75. $\frac{15{,}551}{2520}$ **77.** -48 **79.** $\frac{8}{9}$ **81.** $\frac{182}{243}$ **83.** 273

85. 852 **87.** 16.25 **89.** 6.5793 **91.** $\displaystyle\sum_{k=1}^{5} k$

93. $\displaystyle\sum_{k=1}^{5} 2k$ **95.** $\displaystyle\sum_{k=1}^{10} \frac{1}{2k}$ **97.** $\displaystyle\sum_{k=1}^{20} \frac{1}{k^2}$ **99.** $\displaystyle\sum_{k=0}^{9} \frac{1}{(-3)^k}$

101. $\displaystyle\sum_{k=1}^{20} \frac{4}{k + 3}$ **103.** $\displaystyle\sum_{k=1}^{11} \frac{k}{k + 1}$ **105.** $\displaystyle\sum_{k=1}^{20} \frac{2k}{k + 3}$

107. $\displaystyle\sum_{k=0}^{6} k!$

109. (a) \$535, \$572.45, \$612.52, \$655.40, \$701.28, \$750.37, \$802.89, \$859.09

(b) \$7487.23

(c)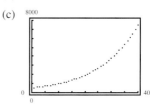

(d) Yes. Investment earning compound interest increases at an increasing rate.

111. $a_5 = 108°$, $a_6 = 120°$; At the point where any two hexagons and a pentagon meet, the sum of the three angles is $a_5 + 2a_6 = 348° < 360°$. Therefore, there is a gap of $12°$.

113. $25.7°, 45°, 60°, 72°, 81.8°$

115. $a_n = 3n$: $3, 6, 9, 12, \ldots$

117. Terms in which n is odd, because $(-1)^n = -1$ when n is odd and $(-1)^n = 1$ when n is even.

119. True.

$$\sum_{i=1}^{4}(i^2 + 2i) = (1 + 2) + (4 + 4) + (9 + 6) + (16 + 8)$$
$$= (1 + 4 + 9 + 16) + (2 + 4 + 6 + 8)$$
$$= \sum_{i=1}^{4}i^2 + \sum_{i=1}^{4}2i$$

121. True.

$$\sum_{j=1}^{4}2^j = 2^1 + 2^2 + 2^3 + 2^4$$
$$= 2^{3-2} + 2^{4-2} + 2^{5-2} + 2^{6-2}$$
$$= \sum_{j=3}^{6}2^{j-2}$$

Section 11.2 *(page 719)*

Review *(page 719)*

1. A collection of letters (called variables) and real numbers (called constants) combined with the operations of addition, subtraction, multiplication, and division is called an algebraic expression.

2. The terms of an algebraic expression are those parts separated by addition or subtraction.

3. $2x^3 - 3x^2 + 2$ **4.** $7x^4$ **5.** $(-\infty, \infty)$

6. $(-\infty, \infty)$ **7.** $[-4, 4]$

8. $(-\infty, -6) \cup (-6, 6) \cup (6, \infty)$ **9.** $(2, \infty)$

10. $(-\infty, \infty)$ **11.** \$30,798.61   **12.** \$5395.40

1. 3   **3.** -6 **5.** -12 **7.** $\frac{2}{3}$ **9.** $-\frac{5}{4}$

11. Arithmetic, 2   **13.** Arithmetic, -2

15. Arithmetic, -16 **17.** Arithmetic, 0.8

19. Arithmetic, $\frac{3}{2}$ **21.** Not arithmetic

23. Not arithmetic   **25.** Not arithmetic

27. $7, 10, 13, 16, 19$ **29.** $6, 4, 2, 0, -2$

31. $\frac{3}{2}, 4, \frac{13}{2}, 9, \frac{23}{2}$ **33.** $\frac{8}{5}, \frac{11}{5}, \frac{14}{5}, \frac{17}{5}, 4$ **35.** $4, \frac{15}{4}, \frac{7}{2}, \frac{13}{4}, 3$

37. $a_n = 3n + 1$ **39.** $a_n = \frac{3}{2}n - 1$

41. $a_n = -5n + 105$ **43.** $a_n = \frac{3}{2}n + \frac{3}{2}$

45. $a_n = \frac{5}{2}n + \frac{5}{2}$ **47.** $a_n = 4n + 4$

49. $a_n = -10n + 60$ **51.** $a_n = -\frac{1}{2}n + 11$

53. $a_n = -0.05n + 0.40$ **55.** $14, 20, 26, 32, 38$

57. $23, 18, 13, 8, 3$ **59.** $-16, -11, -6, -1, 4$

61. $3.4, 2.3, 1.2, 0.1, -1$ **63.** 210   **65.** 1425

67. 255   **69.** 62,625   **71.** 35   **73.** 522

75. 1850   **77.** 900   **79.** 12,200   **81.** 243   **83.** 23

85. b   **86.** f   **87.** e   **88.** a   **89.** c   **90.** d

91.
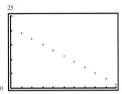

93.
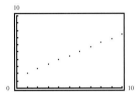

95.

97. 9000

99. 13,120   **101.** 3011.25   **103.** 2850   **105.** 2500

107. \$246,000   **109.** \$25.43   **111.** 632 bales

113. 114   **115.** 1024 feet

117. (a)

Figure	Number of Sides	Sum of Interior Angles
Triangle	3	180°
Quadrilateral	4	360°
Pentagon	5	540°
Hexagon	6	720°

(b) $180(n - 2)°$ (c) It is arithmetic; $d = 180°$

119. 9

121. A recursion formula gives the relationship between the terms a_{n+1} and a_n.

123. Use the formula for the nth partial sum of an arithmetic sequence to find the sum of the integers from 100 to 200. So,

$$\sum_{i=1}^{101}i + 99 = \frac{101}{2}(100 + 200).$$

125. Yes. C times the common difference of the original sequence.

Mid-Chapter Quiz   (page 724)

1. $4, -8, 16, -32, 64$ **2.** $3, 8, 15, 24, 35$

3. $32, 8, 2, \frac{1}{2}, \frac{1}{8}$ **4.** $-\frac{3}{5}, 3, -\frac{81}{7}, \frac{81}{2}, -135$

5. 100 **6.** 40 **7.** 87 **8.** -32 **9.** 40

10. 26 **11.** $\displaystyle\sum_{k=1}^{20} \frac{2}{3k}$ **12.** $\displaystyle\sum_{k=1}^{25} \frac{(-1)^{k-1}}{k^3}$

13. $\displaystyle\sum_{k=1}^{20} \frac{k-1}{k}$ **14.** $\displaystyle\sum_{k=1}^{10} \frac{k^2}{2}$ **15.** $\frac{1}{2}$ **16.** -6

17. $-3n + 23$ **18.** $-4n + 36$ **19.** 2550

20. $\$33,397.50$

Section 11.3   (page 730)

Review   (page 730)

1. The point is 6 units to the left of the y-axis and 4 units above the x-axis.

2. $(10, 5), (-10, 5), (-10, -5), (10, -5)$

3. The graph of f is the set of ordered pairs $(x, f(x))$, where x is in the domain of f.

4. To find the x-intercept(s), set $y = 0$ and solve the equation for x. To find the y-intercept(s), set $x = 0$ and solve the equation for y.

5. $x > \frac{5}{3}$ **6.** $y < 6$ **7.** $35 < x < 60$

8. $-12 < x < 30$ **9.** $x < 1$ or $x > \frac{5}{2}$

10. $-1 < x < 0$ or $x > \frac{5}{2}$

11. $\dfrac{19\sqrt{2}}{2} \approx 13.4$ inches   **12.** $5\sqrt{89} \approx 47.2$ feet

---

1. 2 **3.** -1 **5.** $-\frac{1}{2}$ **7.** $\frac{1}{5}$ **9.** π **11.** 1.06

13. Geometric, $\frac{1}{2}$ **15.** Not geometric   **17.** Geometric, 2

19. Not geometric   **21.** Geometric, $-\frac{2}{3}$

23. Geometric, 1.02 **25.** $4, 8, 16, 32, 64$

27. $6, 3, \frac{3}{2}, \frac{3}{4}, \frac{3}{8}$ **29.** $5, -10, 20, -40, 80$

31. $1, -\frac{1}{2}, \frac{1}{4}, -\frac{1}{8}, \frac{1}{16}$

33. $1000, 1010, 1020.1, 1030.30, 1040.60$

35. $4000, 3960.40, 3921.18, 3882.36, 3843.92$

37. $10, 6, \frac{18}{5}, \frac{54}{25}, \frac{162}{125}$ **39.** $\frac{3}{256}$ **41.** $48\sqrt{2}$

43. 1486.02 **45.** -0.00610 **47.** $\frac{81}{64}$ **49.** $\pm\frac{243}{32}$

51. $\frac{64}{3}$ **53.** $a_n = 2(3)^{n-1}$ **55.** $a_n = 2^{n-1}$

57. $a_n = \left(-\frac{1}{5}\right)^{n-1}$ **59.** $a_n = 4\left(-\frac{1}{2}\right)^{n-1}$

61. $a_n = 8\left(\frac{1}{4}\right)^{n-1}$ **63.** $a_n = 14\left(\frac{3}{4}\right)^{n-1}$

65. $a_n = 4\left(-\frac{3}{2}\right)^{n-1}$ **67.** b   **68.** d   **69.** a   **70.** c

71. 1023 **73.** 772.48 **75.** 2.25 **77.** -5460

79. 6.67 **81.** $-14,762$ **83.** 16 **85.** $13,120$

87. 48 **89.** 1103.57 **91.** $12,822.71$ **93.** 2

95. $\frac{2}{3}$ **97.** $\frac{6}{5}$ **99.** 32

101. 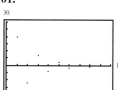   **103.**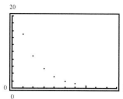

105. (a) There are many correct answers.

   $a_n = 187,500(0.75)^{n-1}$ or $a_n = 250,000(0.75)^n$

   (b) $\$59,326.17$ (c) The first year

107. $\$3,623,993$ **109.** $\$19,496.56$ **111.** $\$105,428.44$

113. $\$75,715.32$

115. (a) $\$5,368,709.11$ (b) $\$10,737,418.23$

117. (a) $P = (0.999)^n$ (b) 69.4%

   (c) 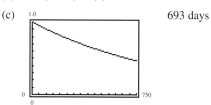   693 days

119. 70.875 square inches   **121.** 666.21 feet

123. (a) $a_n = 2^n$

   (b) $2 + 2^2 + 2^3 + 2^4 + \cdots + 2^{66} \approx 1.48 \times 10^{20}$ ancestors

   (c) It is likely that you have had common ancestors in the last 2000 years.

125. $a_n = a_1 r^{n-1}$ **127.** $a_n = \left(-\frac{2}{3}\right)^{n-1}$

129. An increasing annuity is an investment plan in which equal deposits are made in an account at equal time intervals.

Section 11.4   (page 740)

Review   (page 740)

1. No. The matrix must be square.

2. Interchange two rows.

   Multiply a row by a nonzero constant. Add a multiple of one row to another row.

3. Yes, because the matrix takes on a "stair-step" pattern with leading coefficients of 1.

4. (a) $\begin{bmatrix} 1 & 3 \\ 4 & -1 \end{bmatrix}$ (b) $\begin{bmatrix} 1 & 3 & \vdots & -1 \\ 4 & -1 & \vdots & 2 \end{bmatrix}$

5. -200 **6.** 32 **7.** -60 **8.** -126

9. $y = 4x - 9$ **10.** 58

---

1. 15 **3.** 252 **5.** 1 **7.** 1 **9.** 50

11. $12{,}650$ **13.** $593{,}775$ **15.** 792 **17.** $2{,}598{,}960$

19. $2{,}535{,}650{,}040$ **21.** $85{,}013{,}600$ **23.** 15 **25.** 35

27. 70 **29.** $a^3 + 6a^2 + 12a + 8$

31. $m^5 - 5m^4n + 10m^3n^2 - 10m^2n^3 + 5mn^4 - n^5$

33. $32x^5 - 80x^4 + 80x^3 - 40x^2 + 10x - 1$

35. $64y^6 + 192y^5z + 240y^4z^2 + 160y^3z^3 + 60y^2z^4$
$\quad + 12yz^5 + z^6$

37. $x^8 + 8x^6 + 24x^4 + 32x^2 + 16$

39. $x^6 + 18x^5 + 135x^4 + 540x^3 + 1215x^2 + 1458x + 729$

41. $x^4 + 4x^3y + 6x^2y^2 + 4xy^3 + y^4$

43. $u^3 - 6u^2v + 12uv^2 - 8v^3$

45. $81a^4 + 216a^3b + 216a^2b^2 + 96ab^3 + 16b^4$

47. $x^4 + \dfrac{8x^3}{y} + \dfrac{24x^2}{y^2} + \dfrac{32x}{y^3} + \dfrac{16}{y^4}$

49. $32x^{10} - 80x^8y + 80x^6y^2 - 40x^4y^3 + 10x^2y^4 - y^5$

51. $120x^7y^3$ **53.** $129{,}024a^4b^5$ **55.** $-5940a^3b^9$

57. $6{,}304{,}858{,}560x^9y^6$ **59.** 120 **61.** -1365

63. 1760 **65.** 54 **67.** 1.172 **69.** $510{,}568.785$

71. $\frac{1}{32} + \frac{5}{32} + \frac{10}{32} + \frac{10}{32} + \frac{5}{32} + \frac{1}{32}$

73. $\frac{1}{256} + \frac{12}{256} + \frac{54}{256} + \frac{108}{256} + \frac{81}{256}$

75. The difference between consecutive entries increases by 1. $2, 3, 4, 5$

77. $n + 1$ **79.** (a) $_{11}C_5 = \dfrac{11!}{6!5!} = \dfrac{11 \cdot 10 \cdot 9 \cdot 8 \cdot 7}{5 \cdot 4 \cdot 3 \cdot 2 \cdot 1}$

81. The first and last numbers in each row are 1. Every other number in the row is formed by adding the two numbers immediately above the number.

Review Exercises   *(page 744)*

1. $8, 11, 14, 17, 20$ **3.** $1, \frac{3}{4}, \frac{5}{8}, \frac{9}{16}, \frac{17}{32}$

5. $2, 6, 24, 120, 720$ **7.** $\frac{1}{2}, \frac{1}{2}, 1, 3, 12$ **9.** $a_n = 2n - 1$

11. $a_n = \dfrac{n}{(n+1)^2}$ **13.** $a_n = \dfrac{5n-1}{n}$

15. $a_n = -2n + 5$ **17.** $a_n = \dfrac{3n^2}{n^2 + 1}$ **19.** 28

---

21. $\frac{4}{5}$ **23.** $\displaystyle\sum_{k=1}^{4} (5k - 3)$ **25.** $\displaystyle\sum_{k=1}^{6} \frac{1}{3k}$ **27.** -2.5

29. $127, 122, 117, 112, 107$ **31.** $\frac{5}{4}, 2, \frac{11}{4}, \frac{7}{2}, \frac{17}{4}$

33. $5, 8, 11, 14, 17$ **35.** $80, \frac{155}{2}, 75, \frac{145}{2}, 70$

37. $4n + 6$ **39.** $-50n + 1050$ **41.** 486 **43.** $\frac{2525}{2}$

45. 2527.5 **47.** 5100 **49.** 462 seats   **51.** $\frac{3}{2}$

53. $10, 30, 90, 270, 810$ **55.** $100, -50, 25, -12.5, 6.25$

57. $4, 6, 9, \frac{27}{2}, \frac{81}{4}$ **59.** $a_n = \left(-\frac{2}{3}\right)^{n-1}$

61. $a_n = 24(2)^{n-1}$ **63.** $a_n = 12\left(-\frac{1}{2}\right)^{n-1}$ **65.** 8190

67. -1.928 **69.** 19.842 **71.** $116{,}169.54$

73. 2.275×10^6 **75.** 8 **77.** 12

79. (a) There are many correct answers. $a_n = 120{,}000(0.70)^n$
$\quad$ (b) \$20,168.40

81. \$4,371,379.65   **83.** 56 **85.** 1 **87.** $91{,}390$

89. $177{,}100$ **91.** 10 **93.** 70

95. $x^4 - 20x^3 + 150x^2 - 500x + 625$

97. $8x^3 + 12x^2 + 6x + 1$

99. $x^{10} + 10x^9 + 45x^8 + 120x^7 + 210x^6 + 252x^5$
$\quad + 210x^4 + 120x^3 + 45x^2 + 10x + 1$

101. $81x^4 - 216x^3y + 216x^2y^2 - 96xy^3 + 16y^4$

103. $u^{18} + 9u^{16}v^3 + 36u^{14}v^6 + 84u^{12}v^9 + 126u^{10}v^{12}$
$\quad + 126u^8v^{15} + 84u^6v^{18} + 36u^4v^{21} + 9u^2v^{24} + v^{27}$

105. $5376x^6$ **107.** $-61{,}236$

Chapter Test   *(page 747)*

1. $1, -\frac{2}{3}, \frac{4}{9}, -\frac{8}{27}, \frac{16}{81}$ **2.** $2, 10, 24, 44, 70$ **3.** 60

4. 35 **5.** -45 **6.** $\displaystyle\sum_{k=1}^{12} \frac{2}{3k + 1}$ **7.** $\displaystyle\sum_{k=1}^{6} \left(\frac{1}{2}\right)^{2n-2}$

8. $12, 16, 20, 24, 28$ **9.** $a_n = -100n + 5100$

10. 3825 **11.** $-\frac{3}{2}$ **12.** $a_n = 4\left(\frac{1}{2}\right)^{n-1}$ **13.** 1020

14. $\frac{3069}{1024}$ **15.** 1 **16.** 12 **17.** 1140

18. $x^5 - 10x^4 + 40x^3 - 80x^2 + 80x - 32$

19. 56 **20.** 490 meters   **21.** \$47,868.33

Appendix

Appendix A   *(page A6)*

1.   **3.**

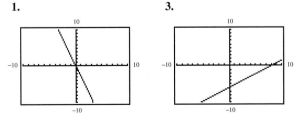

5. 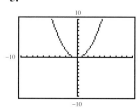   **7.**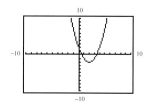

25. $(-3, 0), (3, 0), (0, 9)$ **27.** $(-8, 0), (4, 0), (0, 4)$

29. $\left(\frac{5}{2}, 0\right), (0, -5)$ **31.** $(-2, 0), \left(\frac{1}{2}, 0\right), (0, -1)$

33. Triangle   **35.** Square

9. 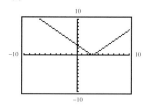   **11.**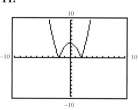

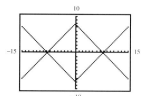

37.

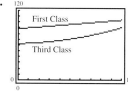

13. 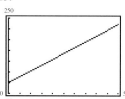   **15.**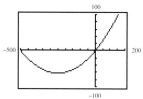

17. Sample answer:

Xmin = 4
Xmax = 20
Xscl = 1
Ymin = 14
Ymax = 22
Yscl = 1

19. Sample answer:

Xmin = -20
Xmax = -4
Xscl = 1
Ymin = -16
Ymax = -8
Yscl = 1

21. Yes, Associative Property of Addition

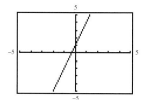

23. Yes, Multiplicative Inverse Property

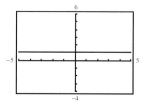

Index of Applications

Biology and Life Sciences

American elk, 610
Ancestors and descendants, 702
Body temperature, 116
Environment
    air pollutant emissions, 106
    oil spill, 436
Human memory model, 617, 621, 631, 647
Learning curve, 435
Number of air sacs in lungs, 306
Number of endangered and threatened plant species, 312
Nutrition, 181, 283, 288
Pollution removal, 376, 434
Population growth
    of deer, 505
    of fish, 442
    of fox, 649
    of insect culture, 434
    of wild rabbits, 642
Predator-prey relationship, 436, 513
Rattlesnake's pit-organ sensory system, 112
Width of human hair, 306

Business

Advertising effect, 642
Automobile depreciation, 124
Average cost, 370, 376, 378, 406, 439, 442, 566, 570
Break-even analysis, 224, 229, 238, 290, 359, 365
Budget, 113
Cost, 170, 195, 433, 442, 467, 544, 600
Cost, revenue, and profit, 105, 295, 316, 327, 356, 362, 566
Cost-benefit model, 426
Daily production cost, 599
Demand, 171, 430, 431, 436, 442, 488, 530, 584, 618
Depreciation, 455, 584, 641, 645, 649, 713, 732, 746
Discount, 89, 170
Discount rate, 81, 89
Dividends per share, Gillette Company, 30
Inventory cost, 293, 376
Investment, 546, 659
List price, 123, 618
Market research, 484
Markup, 88
Markup rate, 80, 89
Net profit, Calloway Golf Company, 21

Net sales
    PetsMart, Inc., 139
    Wal-Mart, 136
Number of stores, Home Depot, 714
Operating cost, 105, 123, 158, 236
Original price, 375
Partnership costs, 434, 442
Price of bananas, 89
Price-earnings ratio, 78, 119
Production, 112, 266, 286
Profit, 20, 195, 206, 544
Quality control, 77, 79, 93, 119, 462
Real estate commission, 71, 76, 77, 119
Reimbursed expenses, 170, 171
Retail price, 284
Revenue, 322, 326, 336, 427, 435, 516, 524, 535, 663, 689
    Best Buy Co., 139
    Merrill Lynch, 306
Sale price, 284
Sales, 170, 356, 718
    of Home Depot, Inc., 167
    of recreational vehicles, 556
    of sporting goods, 39
Sales bonus, 599
Sales commission, 170
Sales goal, 122
Sales growth, 642
Selling price, 552, 570
Stock values, 20
Straight-line depreciation, 144, 147, 171, 173, 210, 215, 334
Ticket prices, 722
Ticket sales, 90, 239, 287, 294, 462
Universal Product Code, 0
Venture capital, 554

Chemistry and Physics

Acid mixture, 239, 291
Acidity, 641
Alcohol mixture, 239, 291, 541
Antifreeze mixture, 90
Automobile engines, 206
Boyle's Law, 433, 443
Bungee jumping, 734
Capacitance, 495
Carbon 14 dating, 641
Charge of electron, 306
Chemical mixture, 251
Chemical reaction, 336
Compression ratio, 78
Cooling, 733
Earthquake intensity, 637, 641, 648
Electrical networks, 278

Electrical resistance, 405
Electricity, 483
Electrons, 307
Engineering, 437
Falling object, 488
Force, 278, 478, 628
Free-falling object, 311, 315, 316, 327, 355, 358, 359, 364, 435, 487, 502, 505, 506, 516, 532, 533, 555, 595, 723, 747
Frequency, 463
Friction, 630
Frictional force, 436
Fuel, 191
Fuel efficiency, 136, 138, 313
Fuel mixture, 90
Fuel usage, 79
Gasoline mixture, 239
Gear ratio, 78
Hooke's Law, 147, 428, 435, 442
Hyperbolic mirror, 692
Interior temperature of the sun, 306
Light intensity, 647
Metal expansion, 307
Meteorology
    atmospheric pressure and altitude, 585
    average daily high temperature for Boston, Massachusetts, 91
    average daily temperature for Boise, Idaho, 121
    average daily temperature for Jacksonville, Florida, 92
    average daily temperature in January for Atlanta, Georgia, 121
    average daily temperature in July for Baltimore, Maryland, 87
    average temperatures for Miami, Washington, DC, and New York City, 105
    monthly normal temperatures for Duluth, Minnesota, 136
    temperature of Pacific Ocean, 436
    tornadoes, 611
Mixture problem, 89, 93, 120, 252, 253, 375
Molecular transport, 621
Molecules in a drop of water, 302
Muon decay, 630
Newton's Law of Cooling, 631
Newton's Law of Universal Gravitation, 433
Oceanography, 631
Optics, 682
Period of a pendulum, 463, 487, 501

Power generation, 436
Power supply, 733
Pressure, 436
Radioactive decay, 579, 584, 612, 636, 641, 645, 648
Relative density of hydrogen, 306
Satellite orbit, 662, 699
Sound intensity, 610, 620, 630, 648
Spring length, 79
Temperature, 116, 122
    conversion, 170
    in a home, 207
    of metal, 566
Thickness of soap bubble, 306
Tractrix, 611
Velocity, 168, 456, 484, 503
Vertical motion, 247, 251, 292
Weight of an astronaut, 436
Width of air molecule, 306

Construction
Architecture
    semielliptical archway, 669, 672
    stained glass window, 662
    tunnel arch, 662
Auditorium seating, 745
Beam deflection, 663
Bridge design, 545
Building a greenhouse, 444
Carpentry, 172
Cement bags, 168
Depth markers, 172
Fenced area, 553
Floor space, 77
Golden Gate Bridge, 540
Highway design, 545
Mechanical drawing, 655
Open conduit, 553
Road grade, 160, 210
Roof pitch, 160
Safe load, 195
Shingle and roofing paper mixture, 223
Sidewalks, 86
Storage area, 553
Storage space, 181
Suspension bridge, 663
Wading pool, 672

Consumer
Account balance, 20, 31
Amount financed, 89, 120
Annual salary, 69, 75, 211
Appliance repair, 67
Budget, 105, 178
Cable television prices, 56
Car repair, 67
Company layoff, 76
Comparing costs, 102
Consumer awareness, 93, 203, 262, 376, 595
    better buy, 78, 119, 123

gasoline cost, 211
    T-bone steak cost, 211
Consumer spending, 488
Consumerism, 181
Contract offer, 704
Cost
    of a tire, 89
    of window installation, 138
Gasoline cost, 74
Gross income, 145
Heating a home, 139
Hourly wage, 75, 93, 106, 118, 600
Income tax, 77, 119
Inflation, 145
Inflation rate, 584, 701
Insurance premium, 89, 203, 295
Investment, 287
Labor
    for repairs, 81, 89, 120, 123
Lifetime salary, 729
Long-distance charges, 106, 122
Long-distance rate, 89
Monthly rent, 76
Monthly wage, 106, 236, 413
Pension fund, 76, 119
Price inflation, 77
Property tax, 79, 120, 123, 313
Property value, 584
Rebate and discount, 599
Reduced fares, 554
Reduced rates, 548, 554, 570, 571, 659
Rent, 212
Salary, 722, 732, 733, 746
Sales tax, 119
Savings, 724
Telephone charges, 158
Tip rate, 76
Total charge, 53
Unit price, 72, 78, 93
Wages, 196, 722, 733
Weekly pay, 89, 138, 182, 212, 249
Weekly wage, 118

Geometry
Accuracy of measurements, 116
Adjustable rectangular form, 554
Area of an advertising banner, 49
Area of a billiard table, 49
Area of a circle, 78, 194, 249
Area of an ellipse, 556, 673
Area of a figure, 21, 39, 48, 374, 378, 563
Area of a rectangle, 30, 67, 207, 213, 333, 334, 434, 544
Area of rectangular cross section of a beam, 477
Area of a region, 277, 316, 710
Area of a roof, 40, 469
Area of a rug, 77

Area of a shaded region, 316, 325, 327, 349, 363, 365, 387, 733
Area of a square, 93, 194, 195
Area of a trapezoid, 103
Area of a triangle, 78, 272, 277, 293, 326, 434
Area and volume of a box, 322, 325
Base of a triangle, 121
Base and height of a triangle, 358, 365
Chorus platform, 287
Circumference of a circle, 194
Collinear points, 293
Cross section of a swimming area, 287
Delivery route, 554
Diagonal of a foundation, 730
Dimensions of the base of a box, 358
Dimensions of a box, 524
Dimensions of a ceramic tile, 698
Dimensions of a cereal box, 520
Dimensions of a circuit board, 698
Dimensions of a container, 190
Dimensions of corrals, 523, 698
Dimensions of a floor, 358
Dimensions of an ice rink, 693
Dimensions of a lot, 553
Dimensions of a mirror, 456
Dimensions of a picture, 547, 553
Dimensions of a picture frame, 358
Dimensions of a piece of carpet, 456
Dimensions of a piece of wood, 693
Dimensions of a rectangle, 67, 91, 121, 229, 294, 364, 365, 487, 501, 523, 530, 532, 535, 570, 571, 698, 699, 701
Dimensions of a room, 355
Dimensions of a sail, 693
Dimensions of a television screen, 692, 730
Dimensions of a triangle, 523, 553
Distance between two points, 556
Floor space, 416
Geometric modeling, 326, 348, 523
Geometric probability, 387
Geometric shape bounded by graphs, A7
Golden section, 478
Height of a cylinder, 434
Height of a flagpole, 119
Height of a ladder, 487, 501
Height of a picture frame, 91
Height of a silo, 120
Height of a tree, 78
Land area of Earth, 306
Lateral surface area of a cone, 487
Length of a computer screen, 487
Length of the diagonal of a basketball court, 487
Length of the diagonal of a solid, 456
Length of the diagonal of a square, 195
Length of the diagonal of a volleyball court, 582

Length of a guy wire, 487, 505
Length of the hypotenuse of a triangle, 463, 500
Length of a ladder, 463, 487
Length of a ramp, 160, 210
Length of a rectangle, 106, 336
Length of a rectangular playing field, 566
Length of a rectangular region, 566
Length of a rope, 55, 65
Length of a shadow, 79
Length of a side of a right triangle, 486
Length of a sidewalk, 549, 554, 570
Length of a sign, 91
Length of a string, 463
Length of a walkway, 571
Length of wire, 698
Masses of Earth and sun, 307
Maximum width of a package, 102
Micron, 306
Nail sizes, 435
Perimeter and area of a figure, 394
Perimeter and area of a rectangle, 55, 326, 363, 552
Perimeter and area of a region, 48
Perimeter of a figure, 75, 316, 362, 468, 505
Perimeter of a piece of paper, 470, 500
Perimeter of a piece of plywood, 469
Perimeter of a rectangle, 181, 212, 376, 433, 439, 475
Perimeter of a square, 195
Perimeter of a trapezoid, 312
Perimeter of a triangle, 275, 466, 500
Perimeter of a walkway, 207
Pond ripples, 599
Radius of a circle, 478, 487
Rectangle inscribed in a circle, 662
Similar triangles, 73, 78, 119, 295
Slope of a plank, 487
Stars, 713, 714
Storage bin, 296
Sum of the angles of a triangle, 248, 251, 253
Surface area of a basketball, 515
Surface area of a cube, 195
Surface area of a rectangular solid, 336
Surface area of a right circular cylinder, 336
Surface area of a softball, 512
Surface area of a spherical float, 515
Vertices of a right triangle, 137
Volume of a bale of hay, 21
Volume of a box, 195, 358
Volume of a circular cylinder, 91
Volume of a cord of wood, 55
Volume of a cube, 195, 249, 337
Volume of a right circular cylinder, 434, 628
Volume of a solid, 416
Volume of a sphere, 91, 194, 434

Volume of two swimming pools, 378, 379
Water area of Earth, 306
Width of a rectangle, 106, 336, 416
Width of a stained glass window, 91

Interest Rate
Annual interest rate, 648
Choosing the best investment, 572
Compound interest, 326, 363, 516, 547, 554, 565, 570, 580, 581, 584, 585, 611, 612, 627, 630, 632, 633, 638, 639, 644, 648, 649, 701, 713, 719
Doubling rate, 630
Doubling time, 630, 633, 638, 639
Effective yield, 634, 638, 639, 648, 701
Home mortgage, 611
Increasing annuity, 729, 733, 746, 747
Investment, 225, 229, 238, 251, 265, 292, 294, 422
Investment portfolio, 261, 265, 266
Monthly deposits, 639
Monthly payment, 405
Savings plan, 21, 53, 585
Simple interest, 87, 92, 120, 121, 123, 161, 336, 346, 432, 442, 670, 690

Miscellaneous
Agriculture, 88
    total assets for farming sector, 147
Airline passengers, 488
Athletics, 88
Baling hay, 722
Bicycle chainwheel, 673
Busing boundary, 693
Canoe trip, 366
Clock chimes, 722
Coffee, 252
Coin mixture, 680
College enrollment, 76
Computer virus, 640
Cooking, 191
Cost to seize illegal drugs, 376
Delivery vans, 234
Dog food mixture, 123
Dog registrations, 393
Education
    college enrollment, 171, 413
Elections, votes cast, 76
Elevation, 105
Endowment, 552
Exam scores, 136
Exercise, 182
Fertilizer mixture, 251
Floral arrangements, 251
Flow rate, 435
Football pass, 134, 138
Fruit distribution, 293
Grades of paper, 248
Hay mixture, 229

Membership drive, 262
Money, 467
Nut mixture, 89, 239, 266, 402, 690
Painting value, 640
Passing grade, 76
Photocopy rate, 388
Photography, 553
Pile of logs, 722, 745
Population increase, 732, 746
Postal delivery route, 650
Poultry production, 307
Probability, 742
Product design, 337
Public opinion poll, 79
Pumping rate, 388
Quiz and exam scores, 171
Recipe, 79, 120
School orchestra, 252
Seed mixture, 82, 89, 229, 290
Soccer club fundraiser, 216
Soccer pass, 138
Sports, 252
    hockey, 182
    miniature golf, 692
    participants in high school athletic programs, 397
    rugby, 672
    soccer ball, 713
Super Bowl winners, 184
Television, 541
Test scores, 670
Time study, 116
Typing speed, 191
Website growth, 637
Work rate, 67, 84, 90, 91, 93, 120, 135, 396, 426, 435, 442, 505, 550, 555, 570, 582
World Series winners, 191

Time and Distance
Air speed, 238, 556
Average speed, 135, 238, 275, 291, 346, 424, 425, 434, 442, 556, 680
Current speed, 235, 434, 552
Distance, 83, 90, 120, 195
    a ball rolls down an inclined plane, 429
    a ball travels, 734
    between catcher and 2nd base, 461
    between two cars, 123
    between two houses, 483
    between two joggers, 422
    from Earth to sun, 307
    from star to Earth, 307
    a ship travels, 249
    a train travels, 628
Distance traveled, 378
Driving distance, 148
Height
    of a baseball, 555
    of a free-falling object, 487, 501

of a model rocket, 551, 555
of an object, 67, 570, 571
of a projectile, 562, 565, 570
of a stream of water, 67
of a tennis ball, 555
of a weather balloon, 555
Light year, 306, 307
Navigation, 682
Parachute drop, 585
Path of a ball, 543, 544, 569, 701
Path of a diver, 544
Path of an object, 543, 544
Path of a softball, 663
Speed, 120, 556
    of a commuter plane and a passenger
        jet, 424
    of two runners, 424
Speed of light, 90
Stopping distance, 317, 436, 442
Travel time, 90, 93, 120, 191, 442, 475,
    513
Wind speed, 424

U.S. Demographics
Amount spent on books and maps, 384
Average annual expenditures per
    student for primary and secondary
    public schools, 68
Average cost of a hotel room, 157
Average monthly rate for basic cable
    television service, 68
Average wage
    for cafeteria workers at public
        schools, 92
    for custodians at public schools, 92

Beverage consumption, 317
Cable TV revenue, 379
Car sales at new car dealerships, 631
Cellular phone subscribers, 533
Cellular phone users, 641
Cellular telephone subscribers and
    annual service revenue, 405
DVDs shipped by manufacturers, 641
Education
    annual amount of student tuition and
        fees, 161
Employment
    in the cellular telephone
        industry, 291
    in the construction industry, 533
Energy use, 77
Federal debt, 307
Hourly wages, employees in mining
    industry, 240
Life expectancy, 171
Married women in the labor force, 192
Median price of a one-family home,
    585
Median sale prices for homes, 39
National health expenditures, 516
Number of bankruptcies filed, 569
Number of daily morning and evening
    newspapers, 230
Number of military reserve personnel,
    544
Number of pieces of first-class mail
    and Standard A (third-class) mail
    handled by the U.S. Postal
    Service, A7
Number of trucks and buses sold, 64

Per capita food consumption of meat,
    77
Percent of public schools with Internet
    access, 36
Population
    of Alabama and Arizona, 230
    of California, 128
    of a country, 640
    of Georgia and North Carolina, 693
    of six counties in Colorado, 77
    of the United States, 207
Population growth
    of Texas, 635
    of the United States, 584
SAT and ACT participants, 466
Students enrolled at public and private
    schools, 196
U.S. governors, 185
U.S. presidents inaugurated, 191
Weekly earnings, employees in
    manufacturing industry, 240
World population, 640

Index

A

Absolute value, 7
  equation, 107
    solving, 107
    standard form of, 108
  function, graph of, 198
  inequality, solving, 110
Add, 11
Adding rational expressions
  with like denominators, 389
  with unlike denominators, 391
Addition
  Associative Property of, 23
  Commutative Property of, 23
  of fractions, 13
    with like denominators, 13
    with unlike denominators, 13
  Property of Equality, 25
  repeated, 14
  of two real numbers, 11
    with like signs, 11
    with unlike signs, 11
Addition and subtraction of fractions,
  13
Addition and Subtraction Properties of
  Inequalities, 96
Additional properties of real
  numbers, 25
Additive
  Identity Property, 23
  inverse, 7
  Inverse Property, 23
Algebraic approach to problem
  solving, 129
Algebraic expression(s), 32
  constructing, 41
  evaluating, 35
  simplifying an, 33
  terms of, 32
Algebraic inequalities, 94
Approximately equal to, 3
Area
  common formulas for, 85
  formulas, 85
  of a triangle, 272
Area, perimeter, and volume, common
  formulas for, 85
Arithmetic sequence, 715
  common difference of, 715
  nth partial sum of, 717
  nth term of, 716
Associative Property
  of Addition, 23
  of Multiplication, 23
Asymptote, 577
  horizontal, 577
  of a hyperbola, 676

Augmented matrix, 255
Average rate of change, 157
Axis of a parabola, 536, 656

B

Back-substitute, 221
Base, 16
  constant, 574
  natural, 578
  number, 70
Binomial, 308, 735
  coefficients, 735
  expanding, 738
  special products, 321
    square of a binomial, 321
    sum and difference of two terms,
    321
Binomial Theorem, 735
Bounded, 94
Bounded intervals, 94
  on the real number line, 94
Branch of a hyperbola, 676
Break-even point, 224

C

Calculator,
  graphing, 18
  scientific, 18
Cancellation Property
  of Addition, 25
  of Multiplication, 25
Cartesian plane, 126
Center,
  of a circle, 652
  of an ellipse, 664
Central rectangle of a hyperbola, 676
Change-of-base formula, 607
Characteristics of a function, 185
Check, 58
  a solution of an equation, 58
Circle, 652
  center of, 652
  radius of, 652, 653
  standard form of the equation of
    center at (h, k), 654
    center at origin, 653
Clear an equation of fractions, 63
Coefficient(s), 32, 308
  binomial, 735
  matrix, 255
  of variable term, 32
Collinear points, 133
  test for, 273
Combined variation, 431
Common difference, 715
Common formulas for area, perimeter,
  and volume, 85

Common logarithmic function, 603
Common ratio, 725
Commutative Property
  of Addition, 23
  of Multiplication, 23
Completely factored polynomial, 333
Completing the square, 517, 529
Complex conjugate, 493
Complex fraction, 398
Complex number(s), 2, 491
  imaginary part of, 491
  real part of, 491
  standard form of, 491
Composite function, 587
Composition of two functions, 587
Compound inequality, 99
Compound interest, 580
  continuous, 580
  formulas for, 581
Compression ratio, 78
Condensing a logarithmic expression,
  615
Conditional equation, 58
Conic, 652
  circle, 652
  ellipse, 664
  hyperbola, 675
  parabola, 656
Conic section, 652
Conjugate, 472
  complex, 493
Conjunctive, 99
Consecutive integers, 45
Consistent system of equations, 219
Constant, 32
  base, 574
  function, graph of, 198
  of proportionality, 427
  rate of change, 157
  term, 32
    of a polynomial, 308
Construct algebraic expressions, 41
Continuous compounding, 580
  formula for, 581
Coordinate, 126
  x-coordinate, 126
  y-coordinate, 126
Cost, 80
Co-vertices of an ellipse, 664
Cramer's Rule, 270
Critical numbers
  of a polynomial, 557
  of a rational inequality, 561
Cross-multiplying, 73, 421
Cube root, 446
Cubing a number, 16
Cubing function, graph of, 198

D

Decimal form, 44
Declining balances method, 455
Degree *n*, of a polynomial, 308
Denominator(s), 15
    least common, 13, 391
    like, 13
    rationalizing, 460
    unlike, 13
Dependent
    system of equations, 219
    variable, 186
Determinant, 267
    expanding by minors, 268
    of a 2×2 matrix, 267
Difference, 12
    common, 715
    of two cubes, 332
    of two squares, 331
Direct variation, 427
    as *n*th power, 429
Directly proportional, 427
    to the *n*th power, 429
Directrix of a parabola, 656
Discount, 81
    rate, 81
Discrete mathematics, 186
Discriminant, 525
    using, 528
Disjunctive, 99
Distance
    between *a* and *b*, 6
    between two real numbers, 6
Distance Formula, 132
Distance-rate-time formula, 85
Distributive Property, 23
Divide, 15
    evenly, 411
Dividend, 15, 408
Dividing
    a polynomial by a monomial, 407
    rational expressions, 383
Division
    long, of polynomials, 409
    Property of Zero, 25
    synthetic, 411
    of two real numbers, 15
    by zero is undefined, 25
Divisor, 15, 408
Domain
    of a function, 184, 187
    implied, 189
    of a radical function, 452
    of a rational expression, 368
    of a rational function, 368
    of a relation, 183
Double inequality, 99
Double negative, 7
Double solution, 508

E

Effective yield, 634
Elementary row operations, 256
Elimination
    Gaussian, 242
        with back-substitution, 258
    method of, 231, 232, 687
Ellipse, 664
    center of, 664
    co-vertices of, 664
    focus of, 664
    major axis of, 664
    minor axis of, 664
    standard form of the equation of
        center at (h, k), 666
        center at origin, 665
    vertex of, 664
Endpoints, 94
Entry of a matrix, 254
    minor of, 268
Equality, Properties of, 25, 59
Equation(s), 58
    absolute value, 107
        solving, 107
        standard form of, 108
    clearing of fractions, 63
    conditional, 58
    equivalent, 59
    first-degree, 60
    graph of, 140
    of a horizontal line, 166
    of a line
        general form of, 163, 166
        horizontal, 166
        intercept form of, 170
        point-slope form of, 163, 166
        slope-intercept form of, 154, 166
        summary of, 166
        two-point form of, 164, 274
        vertical, 166
    linear, 60, 140
    percent, 70
    position, 247
    quadratic, 351
        guidelines for solving, 352
    quadratic form of, 511
    raising each side to *n*th power, 479
    satisfy, 58
    solution of, 58, 129
    solution point of, 129
    solution set of, 58
    solving, 58
        absolute value, 107
    standard form
        absolute value, 108
        circle, 653, 654
        ellipse, 665, 666
        hyperbola, 675, 678
        parabola, 656

system of, 218
    consistent, 219
    dependent, 219
    equivalent, 242
    inconsistent, 219
    row-echelon form of, 241
    solution of, 218
    solving, 218
    of a vertical line, 166
Equivalent
    equations, 59
    form, 107
    inequalities, 96
    systems, 242
        operations that produce, 242
Evaluate, 35
    an algebraic expression, 35
Evaluating a function, 187
Even
    integers, 45
    number of negative factors, 14
    power, 17
Expanding
    a binomial, 738
    a logarithmic expression, 615
    by minors, 268
Exponent(s), 16
    inverse property of, 623
    negative, 299
    rational, 449
    rules of, 298, 575
        power, 300, 449
        power-to-power, 298
        product, 298
        product and quotient, 300, 449
        product-to-power, 298
        quotient, 298
        quotient-to-power, 298
        summary of, 300, 449
        zero and negative, 300, 449
    variable, 574
    zero, 299
Exponential decay, 634
Exponential equation
    one-to-one property of, 622
    solving, 623
Exponential form, 16, 298
Exponential function, 574
    natural, 578
    rules of, 575
    with base *a*, 574
Exponential growth, 634
    model, 634
Exponential notation, 16
    base, 16
    exponent, 16
Exponentiate each side of an equation, 625

Expression(s),
  algebraic, 32
  simplifying, 33
  condensing a logarithmic, 615
  expanding a logarithmic, 615
  rational, 368
    dividing, 383
    domain of, 368
    least common denominator of, 391
    multiplying, 380
    reciprocal of, 383
    reduced form of, 371
    simplified form of, 371
    simplifying, 371
Extracting square roots, 509
Extraneous solution, 420
Extremes, 73
  of a proportion, 73

F

Factor, 14, 32
  greatest common, 328
    monomial, 328, 329
  rationalizing, 460
Factorial, 706
Factoring, 529
  out the greatest common monomial
    factor, 328
  polynomials, 328
    difference of two squares, 331
    by grouping, 330
    guidelines for, 345
    perfect square trinomials, 338
Factors
  even number of negative, 14
  odd number of negative, 14
Finding an inverse function
  algebraically, 591
Finding test intervals for a polynomial,
  557
Finite sequence, 704
First-degree equation, 60
Focus
  of an ellipse, 664
  of a hyperbola, 675
  of a parabola, 656
FOIL Method, 319
Forming equivalent equations, 59
  properties of equality, 59
Formula(s), 85
  for area, 85
  change-of-base, 607
  common, 85
    miscellaneous, 85
  for compound interest, 581
  Distance, 132
  distance-rate-time, 85
  Midpoint, 134
  for perimeter, 85

Quadratic, 525
  recursion, 716
  simple interest, 85
  temperature, 85
  for volume, 85
Fraction(s), 3
  addition of, 13
    with like denominators, 13
    with unlike denominators, 13
  clear an equation of, 63
  complex, 398
  multiplication of, 15
  subtraction of, 13
    with like denominators, 13
    with unlike denominators, 13
Function(s), 184
  characteristics of, 185
  composite, 587
  composition of, 587
  domain of, 184, 187
    implied, 189
  evaluating, 187
  exponential, 574
    natural, 578
    rules of, 575
    with base a, 574
  graph of, 197
  inverse, 589, 590
    finding algebraically, 591
    Horizontal Line Test for, 589
  logarithmic, 601
    common, 603
    natural, 606
    with base a, 601
  name of, 187
  notation, 187
  one-to-one, 589
  piecewise-defined, 188
  position, 311
  quadratic
    graph of, 536
    standard form of, 536
  radical, 451
    domain of, 452
  range of, 184, 187
  rational, 368
    domain of, 368
  Vertical Line Test for, 199

G

Gaussian elimination, 242
  with back-substitution, 258
Gear ratio, 78
General form
  of the equation of a line, 163, 166
  of a polynomial equation, 352
  of a quadratic inequality, 559
Geometric sequence, 725
  common ratio of, 725

nth partial sum of, 727
  nth term of, 726
Geometric series, 727
  infinite, 727
  sum of an infinite, 728
Golden section, 478
Graph, 94
  of an absolute value function, 198
  of a constant function, 198
  of a cubing function, 198
  of an equation, 140
  of a function, 197
  of an identity function, 198
  of an inequality, 94
  of a linear inequality, 175
    half-plane, 175
  point-plotting method, 140, 141
  of a quadratic function, 536
  of a square root function, 198
  of a squaring function, 198
Graphical approach to problem
  solving, 129
Graphing
  an equation with a *TI-83* or *TI-83
    Plus* graphing calculator, A2
  a system of linear inequalities, 280
Graphing calculator, 18
Greater than, 5
  or equal to, 5
Greatest common factor (GCF), 328
Greatest common monomial factor,
  328, 329
  factoring out, 328
Guidelines
  for factoring $ax^2 + bx + c$, 342
    by grouping, 344
  for factoring polynomials, 345
  for factoring $x^2 + bx + c$, 340
  for solving quadratic equations, 352
  for verifying solutions, 130

H

Half-life, 636
Half-plane, 175
Hidden product, 80
Horizontal asymptote, 577
Horizontal line, 151, 166
Horizontal Line Test for inverse
  functions, 589
Horizontal shift, 200
Human memory model, 617
Hyperbola, 675
  asymptotes of, 676
  branch of, 676
  central rectangle of, 676
  focus of, 675
  standard form of the equation of
    center at (h, k), 678
    center at origin, 675

transverse axis of, 675
vertex of, 675

I

Identity, 58
  function, graph of, 198
  Property
    Additive, 23
    Multiplicative, 23
i-form, 489
Imaginary number(s), 2, 491
  pure, 491
Imaginary part of a complex number, 491
Imaginary unit *i*, 489
Implied domain, 189
Inconsistent system of equations, 219
Increasing annuity, 729
Independent variable, 186
Index
  of a radical, 446
  of summation, 708
Inequality (inequalities)
  absolute value, solving, 110
  algebraic, 94
  compound, 99
  conjunctive, 99
  disjunctive, 99
  double, 99
  equivalent, 96
  graph of, 94
  linear, 97, 174
    graph of, 175
    solution of, 174
  properties of, 96
  quadratic, general form of, 559
  rational, critical numbers of, 561
  satisfy an, 94
  solution set of, 94
  solutions of, 94
  solve an, 94
  symbols, 5
Infinite
  geometric series, 727
    sum of, 728
  interval, 94
  sequence, 704
  series, 707
Infinitely many solutions, 64
Infinity
  negative, 95
  positive, 95
Initial
  height, 311
  velocity, 311
Input-output ordered pairs, 184
Integers, 2
  consecutive, 45
  even, 45
  labels for, 45
    consecutive, 45

even, 45
odd, 45
negative, 2
odd, 45
positive, 2
Intensity model, 637
Intercept, 143
  x-intercept, 143
  y-intercept, 143
Intercept form of the equation of a line, 170
Intersection, 100, 369
Interval(s)
  bounded, 94
  infinite, 94
  length of, 94
  test, 557
  unbounded, 94, 95
Inverse, additive, 7
Inverse function, 589, 590
  finding algebraically, 591
  Horizontal Line Test for, 589
Inverse properties
  of exponents and logarithms, 623
  of *n*th powers and *n*th roots, 448
Inverse Property
  Additive, 23
  Multiplicative, 23
Inverse variation, 430
Inversely proportional, 430
Irrational numbers, 3
Isolate *x*, 60

J

Joint variation, 432
Jointly proportional, 432

K

Key words and phrases, translating, 41
Kirchhoff's Laws, 278

L

Labels for integers, 45
  consecutive, 45
  even, 45
  odd, 45
Law of Trichotomy, 8
Leading 1, 257
Leading coefficient of a polynomial, 308
Least common denominator (LCD), 13, 391
Least common multiple (LCM), 13, 390
Left-to-Right Rule, 17
Length of an interval, 94
Less than, 5
  or equal to, 5
Like
  denominators, 13
  radicals, 464

signs, 11, 14
terms, 33
Line(s)
  horizontal, 151, 166
  parallel, 156, 166
  perpendicular, 156, 166
  slope of, 151, 166
  vertical, 151, 166
Linear equation(s), 60, 140
  general form of, 163, 166
  intercept form of, 170
  point-slope form of, 163, 166
  slope-intercept form of, 154, 166
  summary of, 166
  two-point form of, 164, 274
Linear extrapolation, 167
Linear inequality, 97
  graph of, 175
    half-plane, 175
  solution of, 174
  system of, 279
    graphing, 280
    solution of, 279
    solution set of, 279
  in two variables, 174
Linear interpolation, 167
Linear system, number of solutions of, 245
Logarithm(s)
  inverse property of, 623
  natural, properties of, 606, 613
    power, 613
    product, 613
    quotient, 613
  properties of, 603, 613
    power, 613
    product, 613
    quotient, 613
  of *x* with base *a*, 601
Logarithmic equation
  one-to-one property of, 622
  solving, 625
Logarithmic expression
  condensing, 615
  expanding, 615
Logarithmic function, 601
  common, 603
  natural, 606
  with base *a*, 601
Long division of polynomials, 409
Lower limit of summation, 708

M

Major axis of an ellipse, 664
Markup, 80
  rate, 80
Mathematical
  modeling, 69
  system, 24
Matrix (matrices), 254
  augmented, 255
  coefficient, 255

determinant of, 267
    expanding by minors, 268
determinant of 2×2, 267
elementary row operations on, 256
entry of, 254
    minor of, 268
leading 1, 257
order of, 254
row-echelon form of, 257
row-equivalent, 256
square, 254
Means, 73
    of a proportion, 73
Method
    of elimination, 231, 232
    of substitution, 222, 686
Midpoint Formula, 134
Midpoint of a line segment, 134
Minor axis of an ellipse, 664
Minor of an entry, 268
Minus sign, 34
Miscellaneous common formulas, 85
Mixture problem, 82
Modeling, mathematical, 69
Monomial, 308
Multiple, least common, 13, 390
Multiplication, 14
    Associative Property of, 23
    by -1, 25
    Commutative Property of, 23
    Property of Equality, 25
    Property of Zero, 25
    repeated, 16, 298
    of two fractions, 15
    of two real numbers, 14
        with like signs, 14
        with unlike signs, 14
Multiplication and Division Properties
    of Inequalities, 96
    negative quantities, 96
    positive quantities, 96
Multiplicative
    Identity Property, 23
    Inverse Property, 23
Multiply, 14
Multiplying rational expressions, 380

N

Name of a function, 187
Natural base, 578
Natural exponential function, 578
Natural logarithm, properties of, 606,
    613
    inverse, 623
    power, 613
    product, 613
    quotient, 613
Natural logarithmic function, 606
Natural numbers, 2
Negation, Properties of, 25
Negative, 4, 14
    double, 7

exponent, 299
    rules for, 300, 449
factors
    even number of, 14
    odd number of, 14
infinity, 95
integers, 2
number, square root of, 489
No solution, 64
Nonlinear system of equations, 683
    solving by elimination, 687
    solving graphically, 683
    solving by method of substitution,
        686
Nonnegative real number, 4
Notation
    exponential, 16
    function, 187
    sigma, 708
nth partial sum
    of an arithmetic sequence, 717
    of a geometric sequence, 727
nth power(s)
    inverse properties of, 448
    raising each side of an equation to,
        479
nth root
    inverse properties of, 448
    of a number, 446
    principal, 446
    properties of, 447
nth term
    of an arithmetic sequence, 716
    of a geometric sequence, 726
Number(s)
    base, 70
    complex, 2, 491
        imaginary part of, 491
        real part of, 491
        standard form of, 491
    critical, 557, 561
    imaginary, 2, 491
        pure, 491
    natural, 2
    nonnegative real, 4
    nth root of a, 446
    rational, 3
    real, 2
        nonnegative, 4
    set of, 24
    whole, 2
Number line, real, 4
Number of solutions of a linear system,
    245
Numerator, 15
Numerical approach to problem
    solving, 129

O

Odd
    integers, 45
    number of negative factors, 14

power, 17
One-to-one
    function, 589
    properties of exponential and
        logarithmic equations, 622
Operations, 24
    order of, 17
    that produce equivalent systems, 242
Opposites, 7
    and additive inverses, 7
Order, 5
    of a matrix, 254
    of operations, 17
    on the real number line, 5
Ordered pair, 126
    input-output, 184
Ordered triple, 241
Origin, 4, 126

P

Parabola, 141, 536, 656
    axis of, 536, 656
    directrix of, 656
    focus of, 656
    sketching, 538
    standard form of the equation of,
        656
    vertex of, 536, 656
Parallel lines, 156, 166
Partial sum, 707
Pascal's Triangle, 737
Percent, 70
    equation, 70
Perfect
    cube, 447
    square, 447
    square trinomial, 338
Perimeter
    common formulas for, 85
    formulas, 85
Perpendicular lines, 156, 166
Piecewise-defined function, 188
Placement of negative signs, 25
Plotting, 4, 127
Plus sign, 34
Point-plotting method, of sketching a
    graph, 140, 141
Points of intersection, 219
Point-slope form of the equation of a
    line, 163, 166
Polynomial(s)
    completely factored, 333
    constant term of, 308
    critical numbers of, 557
    degree of, 308
    dividing by a monomial, 407
    equation, general form of, 352
    factoring, 328
        difference of two squares, 331
        by grouping, 330
        guidelines for, 345
        perfect square trinomials, 338

sum or difference of two cubes, 332
finding test intervals for, 557
greatest common monomial factor of, 328, 329
factoring out, 328
in *x*, 308
leading coefficient of, 308
long division of, 409
prime, 333
standard form of, 308
synthetic division of a third-degree, 411
zeros of, 557
Position equation, 247
Position function, 311
Positive, 4, 14
infinity, 95
integers, 2
Power property
of logarithms, 613
of natural logarithms, 613
Power rules of exponents, 300, 449
Power-to-Power Rule of Exponents, 298
Price, 80
Price-earnings ratio, 78
Prime polynomial, 333
Principal *n*th root, 446
Problem
mixture, 82
rate, 83
Problem solving
algebraic approach, 129
graphical approach, 129
numerical approach, 129
Product, 14
Property of logarithms, 613
Property of natural logarithms, 613
Rule of Exponents, 298
rule for radicals, 457
of two opposites, 25
Product and quotient rules of exponents, 300, 449
Product-to-Power Rule of Exponents, 298
Proofs, 25
Properties, 24
Additive
Identity, 23
Inverse, 23
Associative
of Addition, 23
of Multiplication, 23
Cancellation
of Addition, 25
of Multiplication, 25
Commutative
of Addition, 23
of Multiplication, 23

Distributive, 23
of Equality, 25, 59
Addition, 25
Cancellation of Addition, 25
Cancellation of Multiplication, 25
Multiplication, 25
of Inequalities, 96
Addition and Subtraction, 96
Multiplication and Division, 96
negative quantities, 96
positive quantities, 96
Transitive, 96
of logarithms, 603, 613
inverse, 623
power, 613
product, 613
quotient, 613
Multiplicative
Identity, 23
Inverse, 23
of natural logarithms, 606, 613
inverse, 623
power, 613
product, 613
quotient, 613
of negation, 25
multiplication by -1, 25
placement of negative signs, 25
product of two opposites, 25
of *n*th roots, 447
of real numbers, 23
additional, 25
Square Root, 509, 510, 529
complex square root, 510
of Zero, 25
Division, 25
division by zero is undefined, 25
Multiplication, 25
Zero-Factor, 350
reverse of, 529
Proportion, 73
extremes of, 73
means of, 73
Pure imaginary number, 491
Pythagorean Theorem, 131, 461

Q
Quadrant, 126
Quadratic equation, 351
guidelines for solving, 352
solving
completing the square, 517, 529
extracting square roots, 509, 529
factoring, 529
Quadratic Formula, 525, 529
Square Root Property, 509, 510, 529
summary of methods, 529
Quadratic form, 511
Quadratic Formula, 525, 529
discriminant, 525

Quadratic function
graph of, 536
standard form of, 536
Quadratic inequality, general form, 559
Quotient, 15, 408
Property
of logarithms, 613
of natural logarithms, 613
rule of exponents, 298
rule for radicals, 457
Quotient-to-Power Rule of Exponents, 298

R
Radical(s)
expression, simplifying, 460
function, 451
domain of, 452
index of, 446
like, 464
product rule for, 457
quotient rule for, 457
removing perfect square factors from, 457
symbol, 446
Radicand, 446
Radioactive decay, 579
Radius of a circle, 652, 653
Raising
a to the *n*th power, 16
each side of an equation to the *n*th power, 479
Range
of a function, 184, 187
of a relation, 183
Rate
discount, 81
markup, 80
problem, 83
of work, 84
Rate of change
average, 157
constant, 157
Ratio, 72
common, 725
compression, 78
gear, 78
price-earnings, 78
Rational, 3
Rational exponent, 449
Rational expression(s), 368
adding
with like denominators, 389
with unlike denominators, 391
complex fraction, 398
dividing, 383
domain of, 368
least common denominator of, 391
multiplying, 380
reciprocal of, 383
reduced form of, 371

simplified form of, 371
simplifying, 371
subtracting
    with like denominators, 389
    with unlike denominators, 391
Rational function, 368
    domain of, 368
Rational inequality, critical numbers, 561
Rational numbers, 3
Rationalizing
    the denominator, 460
    factor, 460
Real number line, 4
    bounded intervals on, 94
    order on, 5
    unbounded intervals on, 95
Real numbers, 2
    addition of, 11
    additional properties of, 25
    distance between two, 6
    division of, 15
    multiplication of, 14
    nonnegative, 4
    properties of, 23
    subtraction of, 12
Real part of a complex number, 491
Reciprocal, 15, 84, 383
Rectangular coordinate system, 126
Recursion formula, 716
Red herring, 84
Reduced form of a rational expression, 371
Reflection(s), 201
    in the coordinate axes, 201
    in the x-axis, 201
    in the y-axis, 201
Relation, 183
    domain of, 183
    range of, 183
Remainder, 408
Removing perfect square factors from the radical, 457
Repeated
    addition, 14
    multiplication, 16, 298
    solution, 508
Repeating decimal, 3
Reverse of Zero-Factor Property, 529
Rewriting an exponential equation in logarithmic form, 623
Rhombus, 135
Richter scale, 637
Root(s)
    cube, 446
    nth, 446
        inverse properties of, 448
        principal, 446
        properties of, 447
    square, 446
        extracting, 509

Round down, 3
Round up, 3
Rounded, 3
Row operations, 242
Row-echelon form
    of a matrix, 257
    of a system of equations, 241
Row-equivalent matrices, 256
Rules of exponents, 298, 575
    power, 300, 449
    power-to-power, 298
    product, 298
    product and quotient, 300, 449
    product-to-power, 298
    quotient, 298
    quotient-to-power, 298
    summary of, 300, 449
    zero and negative, 300, 449

S

Satisfy
    an equation, 58
    an inequality, 94
Scatter plot, 128
Scientific calculator, 18
Scientific notation, 302
Sequence, 704
    arithmetic, 715
    finite, 704
    geometric, 725
    infinite, 704
    term of, 704
Series, 707
    geometric, 727
        infinite, 727
    infinite, 707
Set, 2
    of numbers, 24
Sigma notation, 708
Sign(s)
    like, 11, 14
    minus, 34
    plus, 34
    unlike, 11, 14
Simple interest formula, 85
Simplified form of a rational expression, 371
Simplify, 33
    an algebraic expression, 33
Simplifying
    radical expressions, 460
    rational expressions, 371
Sketching
    the graph of a linear inequality in two variables, 175
    a parabola, 538
Slope, 149
    of a line, 151, 166
Slope-intercept form of the equation of a line, 154, 166

Solution(s), 58
    double, 508
    of an equation, 58, 129
    extraneous, 420
    guidelines for verifying, 130
    of an inequality, 94
    infinitely many, 64
    of a linear inequality, 174
    point, 129
    repeated, 508
    of a system of equations, 218
    of a system of linear inequalities, 279
Solution set, 58, 94
    of an equation, 58
    of an inequality, 94
    of a system of linear inequalities, 279
Solve
    an equation in x, 60
    an inequality, 94
Solving, 58
    an absolute value equation, 107
    an absolute value inequality, 110
    an equation, 58
    exponential equations, 623
    logarithmic equations, 625
    a nonlinear system graphically, 683
    quadratic equations
        by completing the square, 517, 529
        by extracting square roots, 509, 529
        by factoring, 529
        guidelines for, 352
        using the Quadratic Formula, 525, 529
        using the Square Root Property, 509, 510, 529
        summary of methods for, 529
    a system of equations, 218
        using Cramer's Rule, 270
        by Gaussian elimination, 242
        by Gaussian elimination with back-substitution, 258
        by the method of elimination, 231, 232
        by the method of substitution, 222
    word problems, strategy for, 74
Special products
    of binomials, 321
        square of a binomial, 321
        sum and difference of two terms, 321
Square
    of a binomial, 321
    completing the, 517, 529
    matrix, 254
Square root, 446
    extracting, 509

function, graph of, 198
    of a negative number, 489
Square Root Property, 509, 510, 529
    complex square root, 510
Squaring
    a number, 16
    function, graph of, 198
Standard form
    of an absolute value equation, 108
    of a complex number, 491
    of the equation of a circle
        center at (h, k), 654
        center at origin, 653
    of the equation of an ellipse
        center at (h, k), 666
        center at origin, 665
    of the equation of a hyperbola
        center at (h, k), 678
        center at origin, 675
    of the equation of a parabola, 656
    of a polynomial, 308
    of a quadratic function, 536
Straight-line depreciation, 144
Strategy for solving word problems, 74
Subset, 2
Substitution, method of, 222, 686
Subtract, 12
Subtracting rational expressions
    with like denominators, 389
    with unlike denominators, 391
Subtraction
    of fractions, 13
        with like denominators, 13
        with unlike denominators, 13
    of two real numbers, 12
Sum, 11
    of an infinite geometric series, 728
Sum and difference of two terms, 321
Sum or difference of two cubes, 332
Summary
    of equations of lines, 166
    of methods for solving quadratic
        equations, 529
    of rules of exponents, 300, 449
Summation
    index of, 708
    lower limit of, 708
    upper limit of, 708
Synthetic division, 411
    of a third-degree polynomial, 411
System, mathematical, 24
Systems of equations, 218
    consistent, 219
    dependent, 219
    equivalent, 242
        operations that produce, 242
    inconsistent, 219
    nonlinear, 683
    row-echelon form of, 241
    solution of, 218

solving, 218
    using Cramer's Rule, 270
    by Gaussian elimination, 242
    by Gaussian elimination with
        back-substitution, 258
    by the method of elimination, 231,
        232
    by the method of substitution, 222
System of linear inequalities, 279
    graphing, 280
    solution of, 279
    solution set of, 279

T

Table of values, 129
Taking the logarithm of each side of an
    equation, 625
Temperature formula, 85
Term(s), 11, 32
    of an algebraic expression, 32
    constant, 32
    like, 33
    of a sequence, 704
    variable, 32
Terminating decimal, 3
Test for collinear points, 273
Test intervals, 557
    finding, 557
Theorems, 25
Three approaches to problem solving,
    129
Transitive Property of Inequality, 96
Translating key words and phrases, 41
    addition, 41
    division, 41
    multiplication, 41
    subtraction, 41
Transverse axis of a hyperbola, 675
Trinomial, 308
    perfect square, 338
Two-point form of the equation of a
    line, 164, 274

U

Unbounded, 94
Unbounded intervals, 94, 95
    on the real number line, 95
Undefined, division by zero, 25
Union, 100, 369
Unit analysis, 43
Unlike
    denominators, 13
    signs, 11, 14
Upper limit of summation, 708
Using the discriminant, 528

V

Variable(s), 32
    dependent, 186
    exponent, 574

independent, 186
    term, 32
        coefficient of, 32
Variation
    combined, 431
    direct, 427
        as nth power, 429
    inverse, 430
    joint, 432
Varies directly, 427
    as the nth power, 429
Varies inversely, 430
Varies jointly, 432
Vertex
    of an ellipse, 664
    of a hyperbola, 675
    of a parabola, 536, 656
    of a region, 280
Vertical line, 151, 166
Vertical Line Test, 199
Vertical shift, 200
Volume
    common formulas for, 85
    formulas, 85

W

Whole numbers, 2
Word problems, strategy for solving,
    74
Work, rate of, 84

X

x-axis, 126
    reflection in, 201
x-coordinate, 126
x-intercept, 143

Y

y-axis, 126
    reflection in, 201
y-coordinate, 126
y-intercept, 143

Z

Zero, 2
    exponent, 299
        rule, 300, 449
    of a polynomial, 557
    Properties of, 25
Zero and negative exponent rules, 300,
    449
Zero-Factor Property, 350
    reverse of, 529